IFMBE Proceedings

Volume 85

The IFMBE Proceedings Book Series is an official publication of *the International Federation for Medical and Biological Engineering* (IFMBE). The series gathers the proceedings of various international conferences, which are either organized or endorsed by the Federation. Books published in this series report on cutting-edge findings and provide an informative survey on the most challenging topics and advances in the fields of medicine, biology, clinical engineering, and biophysics.

The series aims at disseminating high quality scientific information, encouraging both basic and applied research, and promoting world-wide collaboration between researchers and practitioners in the field of Medical and Biological Engineering.

Topics include, but are not limited to:

- Diagnostic Imaging, Image Processing, Biomedical Signal Processing
- Modeling and Simulation, Biomechanics
- Biomaterials, Cellular and Tissue Engineering
- Information and Communication in Medicine, Telemedicine and e-Health
- Instrumentation and Clinical Engineering
- Surgery, Minimal Invasive Interventions, Endoscopy and Image Guided Therapy
- Audiology, Ophthalmology, Emergency and Dental Medicine Applications
- Radiology, Radiation Oncology and Biological Effects of Radiation

IFMBE proceedings are indexed by by SCOPUS, EI Compendex, Japanese Science and Technology Agency (JST), SCImago.

Proposals can be submitted by contacting the Springer responsible editor shown on the series webpage (see “Contacts”), or by getting in touch with the series editor Ratko Magjarevic.

More information about this series at http://www.springer.com/series/7403

Vo Van Toi · Thi-Hiep Nguyen · Vong Binh Long ·
Ha Thi Thanh Huong
Editors

8th International Conference on the Development of Biomedical Engineering in Vietnam

Proceedings of BME 8, 2020, Vietnam: Healthcare Technology for Smart City in Low- and Middle-Income Countries

Springer

Editors
Vo Van Toi
School of Biomedical Engineering
International University—Vietnam
National Universities
Ho Chi Minh, Vietnam

Thi-Hiep Nguyen
School of Biomedical Engineering
International University—Vietnam
National Universities
Ho Chi Minh, Vietnam

Vong Binh Long
School of Biomedical Engineering
International University—Vietnam
National Universities
Ho Chi Minh, Vietnam

Ha Thi Thanh Huong
School of Biomedical Engineering
International University—Vietnam
National Universities
Ho Chi Minh, Vietnam

ISSN 1680-0737 ISSN 1433-9277 (electronic)
IFMBE Proceedings
ISBN 978-3-030-75505-8 ISBN 978-3-030-75506-5 (eBook)
https://doi.org/10.1007/978-3-030-75506-5

This Springer imprint is published by the registered company Springer Nature Switzerland AG
The registered company address is: Gewerbestrasse 11, 6330 Cham, Switzerland

Preface

Vietnam is a rapidly developing, socially dynamic country, where interest in Biomedical Engineering (BME) activities has grown steadily. Located in a low- and middle-income countries (LMIC) ecosystem, Vietnam yet has an intellectual workforce of an advanced country. Hence, these conferences organized in Vietnam offer unique forums for the BME international community to share experiences, develop support and collaboration networks to improve educational, research and entrepreneurship in LMIC and advanced countries.

In **January 2004**, under the sponsorship of the US National Science Foundation, Prof. Vo Van Toi of the Biomedical Engineering Department of Tufts University, Medford Massachusetts USA, led a US delegation that consisted of Biomedical Engineering professors from different universities in the USA and visited several universities and research institutions in Vietnam to assess the state of development of this field. This delegation proposed a five-year plan that was enthusiastically embraced by the international scientific communities to actively develop collaborations with Vietnam.

Within this framework, **in July 2005, the First International Conference on the Development of Biomedical Engineering in Vietnam** was organized by Prof. Vo Van Toi at the University of Technology in Ho Chi Minh City. From that conference, a Consortium of Vietnam-International Universities was created to advise and assist the development of Biomedical Engineering in Vietnamese universities.

In **July 2007**, **the Second International Conference** on the Development of Biomedical Engineering in Vietnam was held at the University of Technology in Hanoi with the participations of the Asia-Pacific International Molecular Biology Network (AIMBN), Biomedical Engineering Society Singapore (BESS), International Federation for Medical and Biological Engineering (IFMBE), Société Française de Génie Biologique et Médical (SFGBM) and IFMBE Asia-Pacific Working Group.

In **March 2009, International University (IU) established its Biomedical Engineering (BME) Department** and the **Engineer degree program in BME**. The BME Department at IU has since established the International Conference on the Development of Biomedical Engineering a biennial event.

In **January 2010, the Third International Conference** on the Development of Biomedical Engineering in Vietnam was organized by IU in Ho Chi Minh City. It reflected the steady growth of the activities in this field in Vietnam and featured the contributions of researchers of 21 countries, including: Australia, Belgium, Canada, Denmark, France, India, Japan, Korea, Malaysia, New Zealand, Philippines, Poland, Russia, Singapore, Spain, Switzerland, Taiwan, Thailand, UK, the USA and Vietnam. The conference was endorsed by the International Federation for Medical and Biological Engineering (IFMBE). It also hosted the Clinical Engineering Workshop of the IFMBE Asia-Pacific Working Group. The contributed papers were published in the IFMBE Proceedings Series by Springer (ISBN 978-3-642-12019-0).

In **January 2012, the Fourth International Conference** on the Development of Biomedical Engineering in Vietnam was organized in Ho Chi Minh City as a mega-conference. It was kicked off by the Regenerative Medicine Conference (January 8–10, 2012) with the theme "BUILDING A FACE: USING A REGENERATIVE MEDICINE APPROACH", endorsed mainly by the Tissue Engineering and Regenerative Medicine International Society (TERMIS) and co-organized by Prof. Stephen E. Feinberg, University of Michigan Health System, USA, Prof. Anh Le, University of Southern California, USA, and Prof. Vo Van Toi, International University-VNU-HCM, Vietnam. It was followed by the Computational Medicine Conference, endorsed mainly by the Computational Surgery International Network (COSINE) and the Computational Molecular Medicine of German National Funding Agency, and the General Biomedical Engineering Conference, endorsed mainly by the International Federation for Medical and Biological Engineering (IFMBE) (January 10–12) and co-organized by Prof. Paolo Carloni, German Research School for Simulation Sciences GmbH, Germany, Prof. Marc Garbey, University of Houston, USA, and Prof. Vo Van Toi, International University-VNU-HCM, Vietnam. It featured the contributions of 435 scientists from 30 countries, including: Australia, Austria, Belgium, Canada, China, Finland, France, Germany, Hungary, India, Iran, Italy, Japan, Jordan, Korea, Malaysia, Netherlands, Pakistan, Poland, Russian Federation, Singapore, Spain, Switzerland, Taiwan, Turkey, Ukraine, UK, USA, Uruguay and Vietnam. The contributed papers were published in the IFMBE Proceedings Series by Springer (ISBN 978-3-642-32182-5).

In **June 2014, the Fifth International Conference** on the Development of Biomedical Engineering in Vietnam occurred at IU of VNU-HCM. It officially opened the season for the celebration of the 20th Anniversary of Vietnam National Universities—Ho Chi Minh City. It also marked the 10th Anniversary of International University and the 5th Anniversary of the Biomedical Engineering Department. This conference features 231 papers of 532 authors and co-authors from 26 countries including Australia, Bangladesh, Belgium, Canada, China, Croatia, Czech Republic, Denmark, Finland, France, Germany, India, Israel, Italy, Japan, Korea, Malaysia, Norway, Singapore, Slovenia, Switzerland, Taiwan, Turkey, UK, USA and Vietnam. Almost all Vietnamese institutions have their delegations. Besides Vietnam, the two countries that have the most contributors were the USA and Australia. The plenary session featured the lectures on the progress in different fields of Biomedical Engineering by eight distinguished keynote speakers such as Prof.

Ratko Magjarevic, President of IFMBE; Dr. Robert A. Lieberman, Vice-President of International Society for Optics and Photonics SPIE; Prof. Vo-Dinh Tuan, Director of Fitzpatrick Institute for Photonics, Duke University, USA; Prof. Christian Griesinger, Director of Max Planck Institute for Biophysical Chemistry, Germany; Prof. Anja Boisen, Director of VKR Center Of Excellence "NAMEC," Denmark; Prof. Yin Xiao, Director of Australia-China Center for Tissue Engineering and Regenerative Medicine, Queensland University of Technology, Australia; Prof. Yukio Nagasaki, Department of Materials Science and Medical Sciences, University of Tsukuba, Japan; and Prof. Fong-Chin Su, President of Taiwanese Society of Biomedical Engineering, Director, Medical Device Innovation Center, National Cheng Kung University, Taiwan. The contributed papers were published in the IFMBE Proceedings Series by Springer (Vol. 46, Editors: Vo Van Toi and Tran Ha Lien Phuong, 2015, ISSN 1680-0737 ISSN 1433-9277 (electronic), ISBN 978-3-319-11775-1 ISBN 978-3-319-11776-8 (eBook), DOI 10.1007/978-3-319-11776-8).

The **Sixth International Conference** on the Development of Biomedical Engineering in Vietnam occurred from June 27 to 29, 2016, at IU of VNU-HCM. The theme of the conference was Healthcare Technologies for Developing Countries. The contributed papers were from 417 authors and co-authors, of whom 253 were corresponding authors from 23 countries including Australia, Canada, China, Finland, France, Germany, Hong Kong, India, Indonesia, Italy, Japan, Korea, Malaysia, Norway, Portugal, Singapore, Spain, Taiwan, Thailand, UK, USA and Vietnam. Besides Vietnam, three countries had the most contributors were Australia, USA and Korea. Ten distinguished keynote speakers gave their talks in plenary sessions and special topics sessions. They were: Prof. Jeff W. M. Bulte, Professor of Radiology, Oncology, Biomedical Engineering and Chemical and Biomolecular Engineering, Director, Cellular Imaging Section, Institute for Cell Engineering, Johns Hopkins University School of Medicine, USA; Prof. Paolo Carloni, German Research School for Simulation Sciences GmbH and Institute for Advanced Simulation (IAS), Germany; Prof. John Huguenard, Director, Neuroscience Graduate Program, Stanford University, Professional Advisory Board, Epilepsy Foundation, USA; Prof. Ryuji Kohno, Director, Medical Information and Communication Technology Center, Yokohama National University, Japan, Distinguished Professor, University of Oulu, Finland; Dr. Sajeda Meghji, Emeritus Reader in Oral Biology, University College London (UCL), UK; Prof. Beom-Jin Lee, Dean of College of Pharmacy, Ajou University, Head of the Pharmaceutical Research and Development Agency, Korean Pharmaceutical Manufacturing Associations, Korea; Prof. Yu-Lung Lo, Distinguished Professor, Head of Department of Mechanical Engineering, National Cheng Kung University, Taiwan. Prof. Yasuhiko Tabata, Chairman, Department of Biomaterials, Institute for Frontier Medical Sciences, Kyoto University, Board of Governors, Tissue Engineering Society International, Japanese Society of Biomaterials, Japanese Tissue Engineering Society, Society for Hard Tissue Regenerative Medicine, Japanese Society of Inflammation and Regeneration, Japan Society of Drug Delivery System, Japanese Regenerative Medicine Society, Japan; Prof. Christopher Wildrick Woods, Director of Master of Science in Global Health, Director of Graduate Studies, Duke Global Health Institute, Duke University, Chief,

Infectious Diseases Division, Durham VA Medical Center, USA; Prof. Hairong Zheng, Director of National Engineering Laboratory—Medical Imaging Technology and Equipment, Director of SIAT—Institute of Biomedical and Health Engineering, Director of Paul C. Lauterbur Research Center for Biomedical Imaging, Shenzhen Institutes of Advanced Technology (SIAT), Chinese Academy of Sciences, China.

Furthermore, 44 invited speakers gave their talks or tutorial lectures in the parallel sessions including materials for biomedical applications, biomechanics, lab-on-a-chip and point-of-care technologies, mathematical modeling in medicine, biophotonics, public health, pharmaceutical sciences and biomedicine, medical instrumentations, healthcare information technology and bioinformatics, biomedical signal and image processing, neuroscience and neuroengineering, advanced in stem cell and cell reprogramming, molecular and cellular techniques in medicine, ultrasonic characterization of bone tissues, and advanced molecular simulation and experimental biophysical approaches for drug design.

In addition, the Vietnam–UK workshop in the BME field, which was funded by the Newton Fund Researcher Links of British Council, conjointly organized by Prof. Le Hoai Quoc (SHTP), Dr. Le Chi Hieu (University of Greenwich) and Prof. Vo Van Toi (BME-IU) and coordinated by Dr. Le Quoc Trung (BME-IU). Researchers from ten UK universities: (1) University of Greenwich, (2) Cardiff University, (3) North Umbria University, (4) Sheffield University, (5) University of Hertfordshire, (6) Newcastle University, (7) Oxford University, (8) London South Bank University, (9) University of Derby, and (10) Aston University, and representatives of British Council participated. The last day of the workshop was devoted to a round-table discussion among 21 Vietnamese and 12 UK researchers.

Other activities included a field trip to visit SHTP and a new medical device manufacturer, and social events were also organized to introduce to the international guests Vietnamese culture.

The contributed articles were published in the "IFMBE Proceedings Series," Vol. 63 by Springer, Editors: Vo Van Toi, Nguyen Le Thanh An and Nguyen Duc Thang, ISBN: 978-981-10-4360-4, 2017. Sponsors included VNU-HCM, IU, Office of Naval Research Global, US Army International Technology Center—Pacific, Springer, IFMBE, SHTP, NAFOSTED, Newton Fund, Korea United Pharm. Inc., Global IMD Center, ESTC and others.

The **Seventh International Conference** (BME7) on the Development of Biomedical Engineering in Vietnam occurred from June 27 to 29, 2018, at IU of VNU-HCM. The theme of BME7 was Translational Health Science and Technology for Developing Countries. This conference featured 202 papers, including 145 oral talks and 57 posters, of about 500 authors and co-authors from 19 countries including Australia, Canada, China, Czech, France, Germany, Israel, Italy, Japan, Korea, the Netherlands, Portugal, Singapore, Switzerland, Taiwan, Thailand, UK, USA and Vietnam. Besides Vietnam, three countries that had the most contributors were the UK, Korea and the USA. Fifteen distinguished keynote speakers gave their talks either in plenary sessions for the general public or in parallel sessions on in-depth specific topics. They were: (1) Jeff Bulte, Professor of Radiology, Oncology, Biomedical Engineering and Chemical and Biomolecular Engineering,

Johns Hopkins University School of Medicine, USA; (2) Nigel Culkin, Professor of Enterprise and Entrepreneurial Development, Fellow and Past-President of the Institute of Small Business and Entrepreneurship (ISBE), University of Hertfordshire, UK; (3) Guillaume Haiat, Senior Research Director, French National Center for Scientific Research, Paris, France; (4) Sunderesh S. Heragu, Professor and Head of School of Industrial Engineering and Management, Donald and Cathey Humphreys Chair, Oklahoma State University, USA; (5) Beom-Jin Lee, Professor and Dean of College of Pharmacy, Ajou University, Head of the Pharmaceutical Research and Development Agency, Korean Pharmaceutical Manufacturing Associations, Korea; (6) Yung-cang Li, Professor of School of Engineering, RMIT University, Australia; (7) Paul Milgram, Professor of University of Toronto, Department of Mechanical and Industrial Engineering, Institute of Biomaterials and Biomedical Engineering, Toronto, Canada; (8) Nam-Trung Nguyen, Professor and Director of Queensland Micro- and Nanotechnology Center, Griffith University, Queensland, Australia; (9) Ruth Nussinov, Professor and Senior Principal Investigator of National Cancer Institute and Tel Aviv University, Israel; (10) Wellington Pham, Associate Professor of Radiology and Radiological Sciences, Biomedical Engineering, Vanderbilt University School of Medicine, USA; (11) Evan Y. Snyder, Professor of Stanford Burnham Prebys (SBP) Medical Discovery Institute, Director of Center for Stem Cells and Regenerative Medicine, Director of Stem Cell Research Center, Stanford Children's Health Research Center, USA; (12) Masahiro Takei, Professor and Vice Dean of Department of Mechanical Engineering, Medical System Engineering, Chiba University, Japan; (13) Alex Vitkin, Professor of Medical Biophysics and Radiation Oncology, University of Toronto, Canada; (14) Cui Wen, Distinguished Professor of School of Engineering, RMIT University, Australia; and (15) Ping Xue, Professor of the Department of Physics, Tsinghua University, China.

Furthermore, 15 invited speakers gave their talks or tutorial lectures in the parallel sessions including: Medical instrumentations and entrepreneurship, biomaterials and 3D printing, nanomedicine and drug delivery systems, biophotonics, biomechanics, translational health science and technology for developing countries, recent computational and experimental advances in molecular medicine, regenerative medicine and tissue engineering, lap-on-a-chip and point-of-care technologies, biomedical signal and image processing, public health: cancer therapy and reconstructive surgery, and advanced technologies in sleep diagnosis and sleep medicine.

In addition to these activities, a memorable event at the BME7 was the Award Ceremony of the Keylab to Prof. Ruth Nussinov, Professor and Senior Principal Investigator of National Cancer Institute (US) and Tel Aviv University, Israel; and the Signing Ceremony of the Memorandum of Understanding (MOU) between IU and Chung Hsing National University, Taiwan, as well as the MOU between the Biomedical Engineering Departments of IU and of Gwanju Institute of Science and Technology (GIST), Korea.

In addition, the Vietnam-United Kingdom (VN-UK) Newton Workshop in the BME field was organized under the auspice of the Researcher Links workshop grant delivered by the British Council. Researchers from the UK and Vietnam met together to determine the opportunities for long term collaboration. The mentors of the workshop included Prof. Le Hoai Quoc (SHTP), Prof. Vo Van Toi (BME-IU), Prof. Tony Cass (Imperial College London) and Prof. Nguyen T. K Thanh (University College London). The coordinators were Dr. Cecile Perrault-Lecturer, Department of Mechanical Engineering, University of Sheffield, UK, and Dr. Huynh Chan Khon, Lecturer at BME-IU. The workshop aimed to establish Microfluidics, Nanomaterials, and Point-of-Care networks and stimulate collaborations on the R&D, technology transfer, and entrepreneurship between the UK and Vietnam. It gathered 15 experts from 11 UK well-known institutions, more than 20 experts from different research, education and business institutions in Vietnam, and representatives of British Council. Eleven UK universities including: (1) University College London, (2) Imperial College London, (3) University of Sheffield, (4) Newcastle University, (5) Coventry University, (6) University of Brighton, (7) University of Oxford, (8) University of Cambridge, (9) Manchester Metropolitan University, (10) University of Portsmouth and (11) Oxford University Clinical Research Unit in Vietnam. Mr. Nguyen Minh Tuan, Head of Vietnam Ministry of Health's Medical Equipment and Construction Department, was the main invited speaker. The last day of the workshop was devoted to a round-table discussion among researchers. Several funding agencies in Vietnam and the UK were introduced. Many collaborative projects were identified to pave the way for follow-up concrete research proposals to be submitted for funding.

Furthermore, Prof. Vo Van Toi and BME Department faculty met with interested BME groups elsewhere and individual researchers below to discuss future mutual collaborations: (1) Dr. Tani Tohru's group from Japan (image-guided surgery, novel surgery methods); (2) Prof. Guillaume Haiat from France (biomechanics, quantitative ultrasound); (3) Prof. Sun Kim and three other professors from the BME Dept, Hanyang Uni., South Korea (biosignal acquisition and processing system for educational purpose, ubiquitous healthcare...); and (4) Prof. Jae Gwan Kim and four other professors from the BMSE Department, GIST, South Korea (biophotonics, genomic medicine, sleep medicine…).

Several BME companies exhibited their products during the conference. Other activities included a field trip to visit manufacturers in SHTP. A banquet and social events were also organized to introduce to the international guests Vietnamese culture. Sponsors: VNU-HCM, IU, Office of Naval Research Global, VNPT, IFMBE, NAFOSTED, SHTP, Newton Fund, British Council, Korea United Pharm. Inc., CUBE DCM Korea, Vitech Development Co. Ltd., T&N Trading and Investment Co. Ltd., Tan Mai Thanh Medical and Instrument Co. Ltd., BCE Vietnam Co. Ltd., Ho Chi Minh City Department of Science and Technology, Cao Thang International Eye Hospital, SISG Group, EST Co. JSC., PTK Co. Ltd. and many others. More information on the BME7 can be found at: http://csc.hcmiu.edu.vn/bmeconf/bme2018/.

The contributed articles were published in the "IFMBE Proceedings Series," Vol. 69 by Springer, Editors: Vo Van Toi, Le Quoc Trung, Ngo Thanh Hoan and Nguyen Thi-Hiep, ISBN: 978-981-13-5859-3. 2020. This is a book series, owned by the International Federation of Medical and Biological Engineering (IFMBE). The book series publishes the proceedings of all IFMBE conferences as printed books as well as electronic books on CD/DVD and Springer content platform link.springer.com. Every paper receives a DOI and is fully citable and recognized by several indexing services. IFMBE Proceedings are indexed by ISI Proceedings (Thomson Reuters), Scopus, Scimago, and Google scholar.

The **Eighth International Conference** (BME8) on the Development of Biomedical Engineering in Vietnam originally was planned for July 2020. However due to the COVID-19 pandemic, it was canceled. Yet, its contributed papers are published in this volume of "IFMBE Proceedings Series."

All submissions have been warranted for the novelty, soundness, and validity of the research by careful single-blinded peer-reviewing of the program committee and additional experts. Each submission was reviewed by the Program Committee (PC) before it was reviewed by one to two independent reviewers. Authors of all the accepted submissions were asked to submit a revised version based on the review comments. The revised submissions were finally checked by the PC members one more time before being accepted by publication.

The conference committee is grateful to EasyChair for providing access to the conference management platform.

We would like to thank all PC members and additional reviewers, who devoted their valuable time to review and provide constructive feedback to the authors.

We would like to thank all the authors and congratulate those whose submissions were accepted.

Finally, we would like to thank Springer Publisher for their best assistance.

Editors

Vo Van Toi	International University, VNU-HCM, Vietnam
Ha Thi Thanh Huong	International University, VNU-HCM, Vietnam
Vong Binh Long	International University, VNU-HCM, Vietnam
Thi-Hiep Nguyen	International University, VNU-HCM, Vietnam

PC Members

Vo Van Toi	International University, VNU-HCM, Vietnam
Ha Thi Thanh Huong	International University, VNU-HCM, Vietnam
Vong Binh Long	International University, VNU-HCM, Vietnam
Thi-Hiep Nguyen	International University, VNU-HCM, Vietnam
Pham Thi Thu Hien	International University, VNU-HCM, Vietnam
Ngo Thanh Hoan	International University, VNU-HCM, Vietnam
Huynh Chan Khon	International University, VNU-HCM, Vietnam
Phuoc Long Truong	International University, VNU-HCM, Vietnam
Ngo Thi Lua	International University, VNU-HCM, Vietnam
Vu Duy Hai	Ha Noi University of Science and Technology, Vietnam
Nhi Thao-Ngoc Dang	International University, VNU-HCM, Vietnam
Do Minh Thai	International University, VNU-HCM, Vietnam
Nguyen Thanh Tam	International University, VNU-HCM, Vietnam
Trinh Nhu Thuy	International University, VNU-HCM, Vietnam
Nguyen Hong Van	International University, VNU-HCM, Vietnam

Trinh Thi Ngoc Diep	International University, VNU-HCM, Vietnam
Phan Gia Hoang	International University, VNU-HCM, Vietnam
Nguyen Minh Nam	School of Medicine, VNU-HCM, Vietnam
Nguyen Phuong Nam	International University, VNU-HCM, Vietnam

And Editing Team

Nguyen Hoang Phuc	International University, VNU-HCM, Vietnam
Nguyen Thi Ngoc Hanh	Hong Bang International University, Vietnam
Le Minh Khoa	International University, VNU-HCM, Vietnam
Le Xuan Hieu	International University, VNU-HCM, Vietnam

Notes

Vietnam National Universities in Ho Chi Minh City (VNU-HCM) is one of the two leading university networks in Vietnam reporting directly to the prime minister of the country. It was founded in January 1995 and currently has seven university members and several research institutions for a total of more than 5600 staff (including 2600 teaching staff), and 55,000 full-time undergraduates, 9500 master and 680 Ph.D. students.

International University (IU) of VNU-HCM was established in 2003 to be a platform to promote the reform of higher education in Vietnam. It is the first public university that teaches all courses in English. It has extensive collaborations with many universities in Australia, New Zealand, Thailand, the UK and the USA. At IU, there are about 9000 students, 250 faculty and 100 staff.

The School of Biomedical Engineering Department (BME) at IU was established in 2009 as the Department of IU and became a school in 2019. It offers the accredited degrees of Engineer, Master and Ph.D. in BME. Its activities concentrate on the design and applications of medical devices to satisfy the urgent need of the country. Other activities include signal and image processing, pharmaceutical engineering and regenerative medicine. The school promotes the close relationship between education, research and entrepreneurship. Its motto is: high quality, sustainability and usefulness. It has more than 370 undergraduate and graduate students and 26 faculty and staff. In December 2015, the BME-IU undergraduate program was assessed by the ASEAN University Network Quality Assessment (AUN-QA) and ranked as the best in Vietnam and 2nd best in ASEAN of all programs assessed by AUN till that time. In 2019, it was accredited by the American-based ABET.

Professor Vo Van Toi, Vice-Provost for Life and Health Science, Engineering and Technology Development of IU and founder of the BME Department, obtained his Ph.D. in Micro-Engineering at the Swiss Federal Institute of Technology—Lausanne (EPFL), Switzerland, in 1983. From 1983 to 1984, he was a postdoctoral fellow at the Health Science and Technology Division (HST), a joint program of Harvard-MIT

(USA). From 1984 to 2009, he was a faculty of the School of Engineering at Tufts University. He was co-chair of the joint educational programs between the School of Engineering and School of Medicine and between the School of Engineering and School of Dental Medicine. From 1991 to 1992, he was on sabbatical from Tufts to be a research professor at the Scheie Eye Research Institute of University of Pennsylvania (USA). From 1992 to 1994, he helped create and was the vice-director of the Eye Research Institute in Sion (Switzerland). He was instrumental in establishing the BME Department at Tufts in 2003. From 2004 to 2007, he was nominated by President G. Bush to be a member of the Board of Directors of the Vietnam Education Foundation (VEF), a US federal agency established by the US Congress to bring the USA and Vietnam closer through educational exchanges related to science, engineering, mathematics, medicine and technology. From 2007 to 2009, he was on leave from Tufts to be the executive director of VEF. In 2009, he resigned from VEF and took an early retirement from Tufts to go back to Vietnam to establish and chair the BME Department at IU until 2018. He initiated the International Conferences on the Development of Biomedical Engineering. His research interests include: Design and applications of medical devices, mechanism of human visual system, ophthalmology and telemedicine.

For more information, please contact:

Prof. Vo Van Toi
School of Biomedical Engineering
International University—Vietnam National Universities
Ho Chi Minh City, Vietnam
Tel: (84-8)-37 24 42 70 Extn. 3237
Website: www.hcmiu.edu.vn/bme
Email: vvtoi@hcmiu.edu.vn

Ho Chi Minh, Vietnam

Vo Van Toi
Thi-Hiep Nguyen
Vong Binh Long
Ha Thi Thanh Huong

Contents

Neuroengineering

Medical Instrumentations

Device to Detect Acupuncture Points in the Feet Soles for Massage Treatment

Quyen Manh Nguyen, Tuan Hoang Nguyen, Lam Van Tran, Tin Thanh Dang, and Anh Dinh

Abstract This paper presents a massage-treatment prototype on acupuncture points whose method of detecting is based on differences of blood oxygen saturation ratio. Sensor MAX30100 is used to measure blood oxygen saturation (SpO2) of more than 20 people in bright room conditions with an intensity of 115 lx and an ideal dark room condition. Then based on Beer's law of absorbance and experiment, we determined that the points with SPO2 concentration over 99.2% are the locations of the acupuncture ones in order to perform biological massaging pulse. The main scientific principle is the diffuse remittances of diverse red led from meridian and acupuncture point that make blood oxygen saturation levels higher than other areas. In the experiment, we also consider prudently the results of tissues surrounding acupuncture points with the distance of 5, 10 and 20 mm and the influence of that to the final outcome. In general, the final result is reliable with moderate accuracy. However, there are some drawbacks in this system such as errors for people who have inappropriate skin surface for optical detection.

Keywords Oxygen saturation · Acupuncture points · Beer's law · SpO2

1 Introduction

Acupressure, especially foot massage, also known as "foot reflexology" is a popular oriental medical treatment and described as a magical therapy. This therapy has been known to indigenous people in Africa, also been practiced by ancient Chinese and Egyptians a long time ago. There are also numerous studies that verify foot

Q. M. Nguyen · T. H. Nguyen · L. V. Tran · T. T. Dang (✉)
Faculty of Electrical and Electronics Engineering, Ho Chi Minh City University of Technology, Ho Chi Minh City, Vietnam
e-mail: dttin@hcmut.edu.vn

A. Dinh
Department of Electrical and Computer Engineering, University of Saskatchewan, Saskatoon, Canada
e-mail: anh.dinh@usask.ca

V. Van Toi et al. (eds.), *8th International Conference on the Development of Biomedical Engineering in Vietnam*, IFMBE Proceedings 85,
https://doi.org/10.1007/978-3-030-75506-5_1

reflexology not only promotes circulation of local blood, improves the metabolism of nutrients, makes the muscles, bones, joints soft, supple, but also enhances resistance and immune system. Acupressure is considered as a proper remedy in hospitals performed by professional acupuncturists in order to impact a distinguish acupuncture point for desired health recovery. However, it is not easy to detect and influence precisely the appropriate points due to the complexity of the human body and the Chinese acupuncture map. Therefore, most of the acupuncturists follow their instinct and experience. The unclear proofs of acupuncture points and *qi* make this therapy seem suspicious to a group of people although having great beneficial health impacts. As a result, many scientists urge to find proof that people need to specify the unique characteristic of the acupuncture points. Throughout various experiments and hypotheses, there are two methods that are considered to be reliable: electrical properties and the optical absorption features of the acupuncture points.

In this paper, based on the optical absorption features of the acupuncture points, we measured the concentration of saturated oxygen in the feet soles and tried to find the different characteristics between the acupuncture points and the other nits. We also investigated the tissue properties around that location for the best overview. As a result, we have created a product that can both detect acupuncture points and massage with a biological pulse at that point. The structure of this paper consists of the following main parts: Sect. 2 lists several previous research works and experimental results. Section 3 details the hardware structure of our model system. Section 4 explains the algorithm to calculate oxygen saturation of a specific point in foot soles and determine a threshold for an acupuncture point via a descriptive statistics method. Section 5 provides empirical results in two experimental conditions with comparisons of results evaluation. Finally, the last part will summarize the assessment of the pros and cons and the prototype applicability in practice.

2 Related Works

In the past, there were many methods to study the optical absorption features of the acupuncture points in the soles of feet. Langevin et al. [1] carried out physiological studies and show the dissecting relationships of acupuncture points and meridians with connective tissue planes. By physical aroma John et al. [2] showed that meridians have lower resistance or impedance than others. Han et al. [3] presented a fiber-optic diagnostic method to identify acupuncture points under the thin cuticles. A fiber with a diameter of 600 μm is used in connecting with the cuticles to get information from the scattered light of the tissue layer. The light is emitted from halogen-halogen lamps (FOK-100W, Optical fiber, Korea) to single fiber with a diameter of 600 mm and then to the sample. Backlight scattering from samples is recovered by a spectrophotometer (HR4000G, Ocean Optics, USA). The analysis results show that the reflexion of the meridian point of Wai Guan is measured larger than the neighboring tissue in almost the entire wavelength. However, these methods have the disadvantage of limiting the ability to reproduce on the same subject because they have to contact the skin tissue.

Some of the following methods use a non-invasive sensor system that has the advantage of not causing any discomfort or irritation. Baik et al. [4] proposed a new procedure to observe the long-standing floating fiber structure inside biological photon engineering vessels. The main technique is to inject acridine-orange into the femoral vein and dye the nucleus of the intravascular fibers inside the blood vessel, then observe it under a fluorescence microscope. Lo [5] used infrared imaging to describe the effect of acupuncture on a patient's acupuncture points. He used a Meditherm2000 with the degree Celsius resolution of 0.01 °C, and a detector kept at 13 K measuring all infrared radiation from points on the body. The patient is acupuncture needles used by a doctor at a point located a certain distance from the area of pain, while the infrared camera constantly observes hot spots. Observation results showed that the temperature in the pain zone continuously decreases in the range of 0.5–2 °C.

Han et al. [6] have presented a new way to detect non-invasive human meridians. The main technique is to use three laser beams of He–Ne 633 nm (Melles Griot), 658 and 785 nm laser diodes (LQC, Newport) that have been modulated in 10 Hz by an optical cutting machine (SR540, Research System Stanford). The collected fibers are a double fiber investigation containing two strands with a main diameter of 600 mm (Ocean Optics) connected to PMT (Oriel, USA). The signals can be recorded and analyzed later. Research shows that light attenuation is less transmitted along pericardial meridians than along non-meridional directions.

3 Hardware Platform

See Fig. 1.

3.1 *Sensor Max30100*

I^2C address of the MAX30100 sensor is 0 × 57. Data is stored in a FIFO buffer up to 16 measurements, where each sample is a size of 4 bytes [7]. In the project, the sensor is set up to operate in Heart Rate and Oxygen Saturation mode with 100 Hz sampling rate, 1600 μs LED pulse width, 50 mA infrared led current and 27.1 mA red led current.

3.2 *Vibration Motor*

The speed of the motor is 6000 rpm at the voltage of 3 V, which is powerful enough to be a massage probe. This small-sized vibration motor is attached to a probe for massage purposes. PWM port 10 control on–off BJT C1815 in order to create a pulse

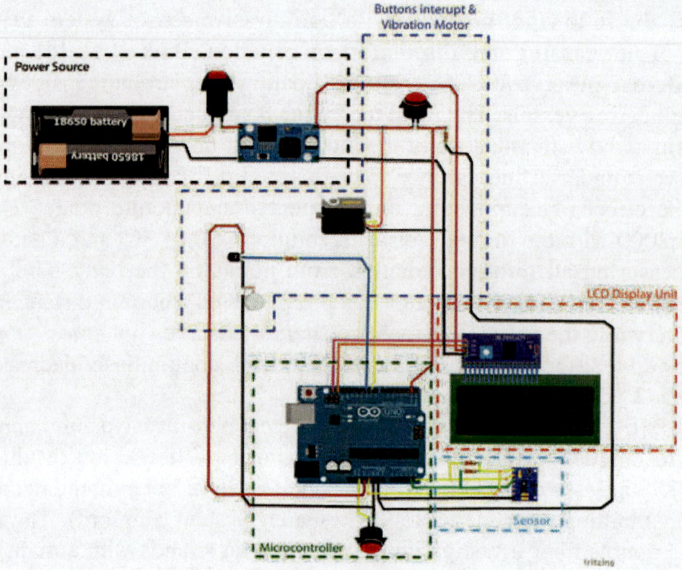

Fig. 1 The detailed wiring diagram for the system

Ultra Reiz (frequency of 143 Hz) with positive biological impact on the human body. The voltage source V_{CE} is 3.3 V from the Arduino Uno R3.

4 Acupressure Detection Method

4.1 Principle of Measuring SpO2

Oxygen saturation is the ratio of the hemoglobin component that contains oxygen to the total hemoglobin in the blood. Normal oxygen saturation ranges from 94 to 100%. Chromatographic measurements work based on the light absorption of Hemoglobin (Hb) and Oxyhemoglobin (HbO2) is different for 2 different wavelengths of Red light (660 nm) and Infrared light (905 nm). HbO2 absorbs more Infrared light than Red light [8]. According to BEER's law, the total amount of light absorbed will be proportional to the concentration of that light absorber. The oxygen saturation can be determined by Beer-Lambert Law:

$$SpO2 = \frac{R.\text{Ext}(Hb, \lambda_2) - \text{Ext}(Hb, \lambda_1)}{R.(\text{Ext}(Hb, \lambda_2) - \text{Ext}(HbO_2, \lambda_2)) + (\text{Ext}(HbO_2, \lambda_2) - \text{Ext}(Hb, \lambda_2))} \quad (1)$$

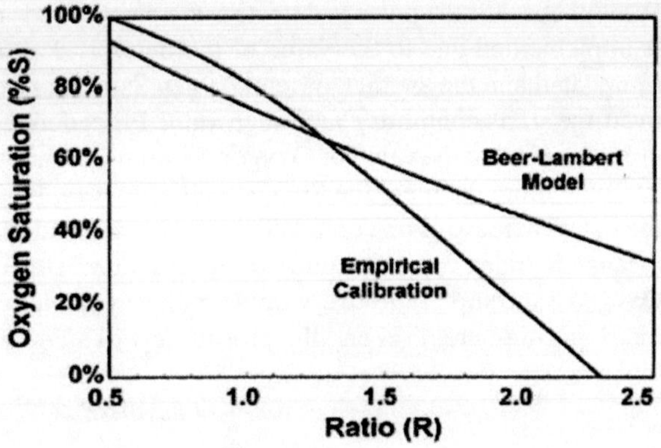

Fig. 2 Relationship between R and Oxygen Saturation [9]

where R is the ratio of Red light to Infrared light received from the sensor:

$$R = \frac{\log(AC_{red})}{\log(AC_{Infrared})} \tag{2}$$

In the prototype, the above Beer-Lambert Law can be simplified by this equation:

$$SpO2 = A + B.R \tag{3}$$

where A and B are two crucial constants required to calculate in order to find precisely the SpO2 ratio. Therefore, we use a medical device Accuro II (Charmcare) to calibrate. Following the Beer-Lambert model, we assume that there is a linear relationship between the SpO2 ratio that is larger than 80% and R. The calibrating procedure of the sensor Max30100 is experimented in an ideal environment (darkroom) to reduce the light noise that interferes final result of the sensor (Fig. 2).

4.2 *Algorithms for Acupuncture Points*

There are two operating modes in our device: mode 1 for detecting acupuncture point and mode 2 for triggering a massaging pulse to determinate one. The change of 2 operating modes can be easily implemented by the interrupt mechanism when the button is pressed.

Mode 2 is for massage purpose; the PWM will be created at port 10 to control the vibration motor at a desired frequency and duty cycle. The massage pulse is a modified pulse Ultra Reiz (frequency of 1.43 Hz). That means the condition of port 10 will be HIGH for 0.2 s and LOW for 0.5 s.

Mode 1 is for detecting acupuncture point, which is the starting mode of the loop. When activating mode 1 for the first time, all the results of the sensor are updated.

We will check if the pulse is detected, the procedure for stabilizing sensor (delay 5 s) is implemented due to the value of the index is 0. After that, SpO2 results will be re-updated and the average of ten Oxygen Saturation results from sensor Max30100 overtime starts computing (Average-value Procedure). The crucial condition for this procedure is that the value of Oxygen Saturation cannot be null: *Result.SaO2*! = '\0'.

To be more detailed, the SpO2 result after each loop is added to a sum following the index increase of one unit. When the sum is an addition of ten SpO2 results which means the index reaches value 10, the condition statement begins. In this, the final Oxygen Saturation ratio that would be displayed on LCD is the average of sum, the index and sum are reset and the procedure will start again.

$$sum+ = result.SaO2;$$

$$SpO2 = sum/10;$$

Furthermore, in order to ensure precisely the location of an acupuncture point, we propose a solution using the oxygen-saturation threshold and continuous data stability. The principle of continuous data stability is quite simple: the detecting point in the foot sole would be considered as an acupuncture point if after every average-value procedure, continuously numerous return values of SpO2 were always greater than a threshold. In our algorithm, by using a global variable: times, at least three consecutive values responded to the condition would be believed to determine whether the point is an acupuncture one or not. That means after every average-value procedure, SpO2 will be compared with the threshold, the times would increase if SpO2 satisfied the condition. Any value that is smaller than the threshold will change the determination by setting the times back to 0. LCD will display the notification of the founded acupuncture point when times reaches 3. Finally, times are reset for the beginning of the new loop.

We believe that our algorithm would build up a reliable method to detect and study the acupuncture point. However, the problem of this is the figure of the proposed threshold. It is required the various number of reference data to do the statistic and result in a trustworthy outcome. The process of finding the threshold will be presented in the experimental result section (Fig. 3).

5 Experimental Setup and Results

5.1 *Experimental Setup*

The experiment was performed in 20 people with the ages from 13 to 50 and the gender was random. Pulse Oximeter and Heart-Rate Sensor IC (Max30100) were used to find the difference between normal points and acupuncture points. An assumption is made that the acupuncture points have higher SpO2 than normal points and

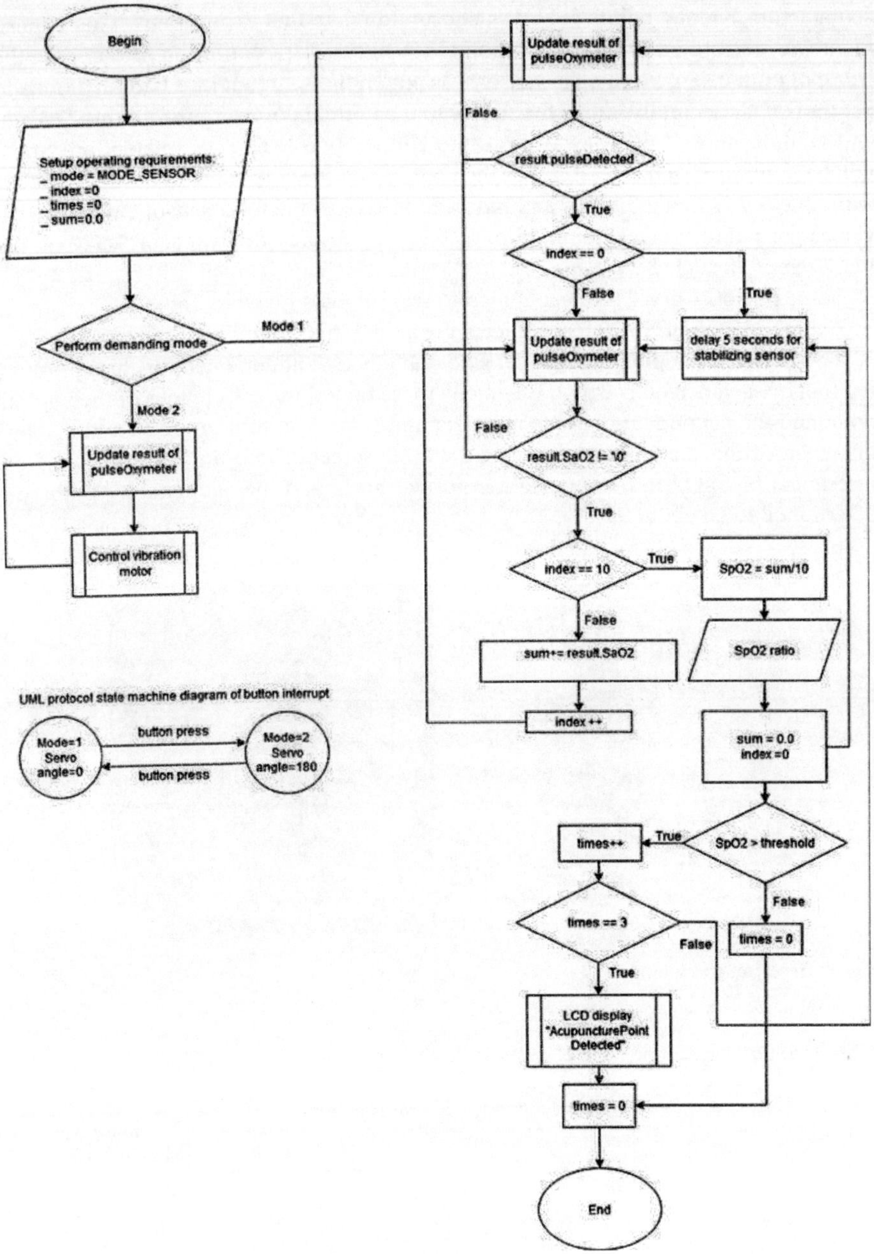

Fig. 3 Flowchart showing the operation sequence of the system

acupuncture Kidney points [11] was chosen to make this experiment. This experiment was performed in light condition (115 lx) and dark condition to compare the result of both cases. Firstly, the sensor was put on the Acupuncture Kidney Points to get the SpO2. Secondly, move the sensor to the normal points 5 mm, 10 mm, 20 mm and random normal point away from the Kidney point. The parameters are SpO2 will be monitored and save 10 times then output the average value in the sensor. This experiment was made 5 times and has been statistics; the total sensor calculated 50 values per point. It takes a minimum of 6.33 s to display the result of SpO2 on the LCD screen in such a way (Fig. 4).

Value of SpO2 in a 21-year-old man to test subjects given in Table 1.

Average Value of SpO2 test subjects given in Tables 2 and 3.

These are all people with normal and stable health and the acupuncture point in the foot of each person is equal. Basing on these tables, the differences of the normal point and the acupuncture points are clear and it takes minimum of 34 s to display the result of the acupuncture point on the LCD screen. From the above results, the thresholds of SpO2 to distinguish the normal point and the acupuncture point are determined to be about 99.1%.

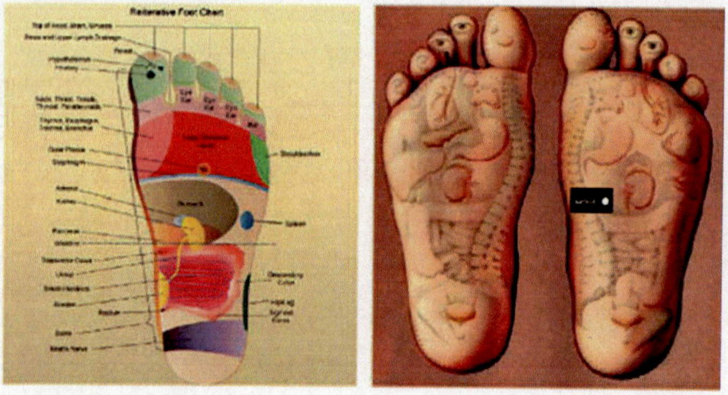

Fig. 4 Experimental setup [10]

Table 1 SpO2 concentration of a 21-year-old man in the standard room (115 lx)

Time	Acupuncture points	Normal points			
		5 mm away	10 mm away	20 mm away	Random normal point (heel)
1	99.44	98.73	98.61	98.25	98.17
2	99.47	98.64	97.73	98.37	98.74
3	99.48	98.76	98.42	98.5	98.69
4	99.39	98.67	98.16	98.59	98.79
5	99.31	98.89	98.52	98.45	99.01
Average	99.418	98.738	98.288	98.432	98.68

Table 2 SpO2 concentration in the standard room (115 lx)

Gender	Age	Acupuncture points	Normal points			
			5 mm away	10 mm away	20 mm away	Random normal point (heel)
Female	21	99.2	98.76	98.25	98.94	98.32
Male	21	99.42	98.74	98.29	98.43	98.68
Male	23	99.46	98.93	98.97	98.93	99.01
Female	13	99.29	98.60	99.18	98.72	98.35
Male	50	99.59	98.03	99.14	98.86	99.05

Table 3 SpO2 concentration in the darkroom

Gender	Age	Acupuncture points	Normal points			
			5 mm away	10 mm away	20 mm away	Random normal point (heel)
Female	21	99.29	98.65	98.76	98.91	98.59
Male	21	99.54	98.79	98.53	98.82	98.92
Male	23	99.36	98.8	98.97	98.71	99.04
Female	13	99.62	99.12	98.87	99.07	98.53
Male	50	99.56	98.76	98.93	98.45	98.73

5.2 *Accuracy Verification Methods*

In this part, the sensor was put on the different acupunctures to compare the variance between acupuncture points. Basing on the acupuncture map, we chose five acupuncture points to compare with the results of the sensor in the standard condition because this is the best condition for measurement. Noise parameters may occur when the external condition affects such as the humidity of skin, dust, malformed skin surface, etc. Besides, the sensor needs to be kept fixed to minimize the diffraction phenomenon. From that result, we can evaluate the accuracy of the threshold and algorithm. We conducted experiments on five different points including Kidney, Lung, Eye, Head, and Small Intestine. The results were shown in the table below.

Average Value of SpO2 in five acupuncture points test subjects given in Table 4.

The maximum variance of a person is 0.71% to give us an overview about the accuracy of the threshold. Basing on that results we can choose the threshold as the smallest SpO2 value at all points. However, this approach can be risky because depending on the host factor, the minimum of SpO2 in the acupuncture can vary. The accuracy of the algorithm can be improved when the number of samples is very big and the algorithm needs a filter of noise values. Otherwise, we can set up a system

Table 4 SpO2 concentration of five acupuncture points in the standard room

Gender	Age	Acupuncture points				
		Kidney	Lung	Eye	Head	Small intestine
Female	21	99.2	99.24	99.42	99.91	99.57
Male	21	99.41	99.84	99.36	99.82	99.93
Male	23	99.46	99.73	99.51	99.74	99.41
Female	13	99.29	99.66	99.43	99.59	99.75
Male	50	99.59	99.64	99.87	99.83	99.67

to consist of many sensors that work at the same time to measure SpO2 in the foot. By that way, the system can find the situation of acupuncture on each person.

Because the MAX30100 sensor is unprofessional in the medical equipment field, the accuracy of it has to be adjusted by more typical devices. For more details, we used a device which names "ACCURO II™". This device can enhance SPO2 measuring by the latest RT-SAT™ patented technology which belonged Charmcare. The SPO2 values of it are used as a standard model so that we adjusted the MAX30100 sensor and the algorithm. As a result, the accuracy of Max30100 sensor equals approximately to ACCURO II™.

6 Conclusion

To sum up, we presented a feasible non-contact approach for measuring the concentration of oxygen saturation in the soles of the feet with the MAX30100 sensor. The purpose of the method is to find and massage the acupressure points there. The system was tested on 20 objects with the device at distances of 5 mm, 10 mm and 20 mm from the point of pressure on the soles of their feet. The accuracy of the results is relatively high at about 80% and can be developed to detect blood circulation disorders. Moreover, it is feasible to bring this product to the market thanks to its affordable cost. However, this device only works correctly under mandatory conditions and cannot collect enough relevant data to give the most accurate SpO2 threshold for determining acupuncture points.

With further research and improvements in noise correction such as light, temperature and humidity to enhance detection accuracy in all cases. We hope that this method can be used in the early diagnosis of disorders of blood circulation. We are confident that this system will help the physician's better cure than pure reflexology. This device can add some new features like heating acupuncture points.

Conflict of Interest The authors have no conflict of interest to declare.

References

1. Langevin HM, Yandow JA (2002) Relationship of acupuncture points and meridians to connective tissue planes. J Anta Rec (New Anta) 257–265
2. John HM, Cho JH, Shin HS et al (2002) Frequency dependence of impedances at the acupuncture point use (PC3). IEEE Eng Med Biol Mag 21:33–36
3. Han Y-H et al (2006) Detection of the acupuncture points in Skin by differential path-length spectroscopy. In: IFMBE proceedings, vol 14/6
4. Baik KY, Lee JW, Lee BC, Johng HM, Nam TJ, Sung B, Cho SI, Soh KS (2005) Acupuncture meridian and intravascular Bingham duct. Key engineering materials, vol 277–279. Trans Tech Publications, Ltd., pp 125–129
5. Lo S-y (2002) Meridians in acupuncture and infrared imaging. J Med Hypotheses 58(1):72–76
6. Han Y-H et al (2007) Differences in optical transport properties between human meridian and non-meridian. Am J Chin Med 35(5):743–752
7. All datasheet. https://www.alldatasheet.com/datasheet-pdf/pdf/879178/MAXIM/MAX30100.html. Accessed 12 Nov 2019
8. Alharbi S et al (2018) An applicable approach for extracting human heart rate and oxygen saturation during physical movements using a multi-wavelength illumination optoelectronic sensor system. Proc SPIE 10486, Design and Quality for Biomedical Technologies XI, 104860S
9. Chan V, Underwood S (2012) A single-chip pulsoximeter design using the MSP430. SLAA274B, Texas Instruments, Dallas
10. Mary Juliet A (2005) Foot reflexology positions to control diabetic foot ulcers. J Chem Pharm Res 7:866–871
11. Thambirajah R (2018) Energetics in acupuncture: five element acupuncture made easy, 2nd edn. Christl Kiener, Munich

Wireless Heart Rate Monitoring Using Sound Sensor

Minh Cong Le, Luom Trung Le, Tin Thanh Dang, and Anh Dinh

Abstract This paper presents a wireless health care device to monitor the patients' heart rate. There are a variety of methods to measure cardiac cycle: optical sensor, one lead of ECG, sound sensor, etc. This proposed general algorithm is to measure the sounds of the heart rhythm transmitted from the heart to the patient's chest, and these sound signals are used to calculate the heart rate. This chosen method is simpler and lower cost compared to the others. The most important necessary thing is that the designed device has high accuracy shown in the results. This device is used to replace the workday routine of the doctors. Although they do not need to visit the patient's room, the doctors can still receive information about their patients' heart rate. These data can transfer from the measuring devices to the display devices. By following this method, doctors or nurses can always update the status of patients and immediately improvise them. This system can be worn easily on the patients' chest. The patients can wear it all the time while still feeling so comfortable. This device is expected to be used in the hospitals to monitor the heart rate of the patients for a long time. At the hospital settings where medical equipment is quite limited, this device is a good solution because it is compact, convenient and affordable.

Keywords Heart rate · BPM monitor · Sound sensor · Wireless RF · Health applications

M. C. Le · L. T. Le · T. T. Dang (✉)
Faculty of Electrical and Electronics Engineering, Ho Chi Minh City University of Technology, Ho Chi Minh City, Vietnam
e-mail: dttin@hcmut.edu.vn

A. Dinh
Department of Electrical and Computer Engineering, University of Saskatchewan, Saskatoon, Canada
e-mail: anh.dinh@usask.ca

V. Van Toi et al. (eds.), *8th International Conference on the Development of Biomedical Engineering in Vietnam*, IFMBE Proceedings 85,
https://doi.org/10.1007/978-3-030-75506-5_2

1 Introduction

The sound inside of a human body contains plenty of information which can be used to monitor the health and diagnose of the diseases. For example, pulmonary heart sounds have been widely used in clinical practice to diagnose heart and lung diseases. The fetal heart rate, calculated from the sound of the fetal heart, is an important indicator to recognize how healthy the fetus is. Bowel sounds may indicate whether abdominal surgery is successful without any tissue adhesion. The sound of the heart is no exception; doctors often use the stethoscope to listen to the heart sounds and diagnose certain diseases. This design follows that method by using the sound of the heart to determine the heart rate.

Currently, in the hospitals, monitoring the health of a patient is quite inconvenient as the doctors or nurses must visit the patients regularly to check the heartbeats. However, they may forget to check the patients' status sometimes since they are so busy, or they have some other urgent emergency cases. Moreover, when the doctor examines the heart rate, he has to wake up the patients if they are sleeping, so it will disturb them [1]. Therefore, it is useful to have a wireless monitor that can observe the heart rate remotely. So this device is expected to help doctors easily surveil the health of patients.

The device consists of two main parts; the first one is attached to a patient's chest for measuring heart rate and transmitting the rate to the other part of the device. The rest will display the heart rate on the screen for the doctor to monitor. The chest strap uses the audio sensor along with the chestpiece [2] of a medical stethoscope to capture audio signals from the heartbeat based on which we can accurately measure the heart rate. One of the advantages is the patient can wear it continuously without many impacts on the body when doing daily activities. And of course, it is convenient for the doctor to observe.

The paper is organized following: Sect. 2 presents the popular heart rate measurement methods and the works the team consults. Section 3 describes the main elements of the device, while Sect. 4 shows the prototype for monitoring the heart rate. Section 5 details the measuring method which our group use in the project. Section 6 is the empirical outcome and the final section concluding our paper.

2 Background and Related Work

Heart rate using one lead of ECG: ECG [3] is a method that uses the bioelectric signal generated by the heart to calculate and measure. The heart's contraction mechanism produces different potentials on the different points of the heart that spread to surrounding soft tissues, from which the doctor will use hardware such as electrodes, voltage amplifiers to take the result is an electrocardiogram [4]. ECG method can be used to accurately evaluate the heart rate and ECG signal waveform. It provides important information related to cardiac muscle function, suggesting different conditions when present. However, obtaining good ECG is a quite complicated approach as it requires to use electrodes attached to the body and a special amplifier circuit to record the signal. In addition to the inconvenience for the patient, the electrodes pasted to the bodily for a long time and continuously will cause physical signal loss.

Heart rate using optical sensor: When the heart contracts, blood is pushed throughout the body through arteries, veins, and capillaries, creating pressure changes on the walls of the vessels. Heart rate can be determined by measuring that change. When the heart is diastolic, the blood flow through the vessels is very small, so the incoming light is less absorbed and more light escapes and vice versa, when the ventricular contract, blood flows heavily through the vessels, so more incident light is absorbed, the results is less light going out. The light intensity is detected and heart rate is to be determined. Measuring heart rate by optical [5] is also very popular today, it is often integrated into wearable devices such as smart bracelets and smartwatches. This is a simple and less expensive method but the accuracy is not very high since the signal is prone to interference.

3 A Design for Wireless Heart Rate Monitoring

We separate the application into 3 parts of the heart rate monitoring system: the first part is the doctor's device for display the heart rate, a sensor measure heart rate on a patient and a RF module to connect them. Overview of design is shown in Fig. 1.

3.1 Doctor's Side for Monitoring the Heart Rate

As mentioned above, every time a doctor or a nurse wants to measure the heart rate, it is necessary to enter the patient room in order to measure it. With the development

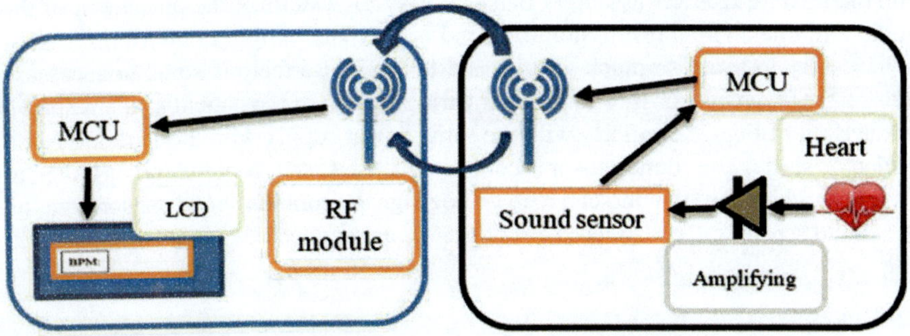

Fig. 1 Wireless heart rate monitoring system

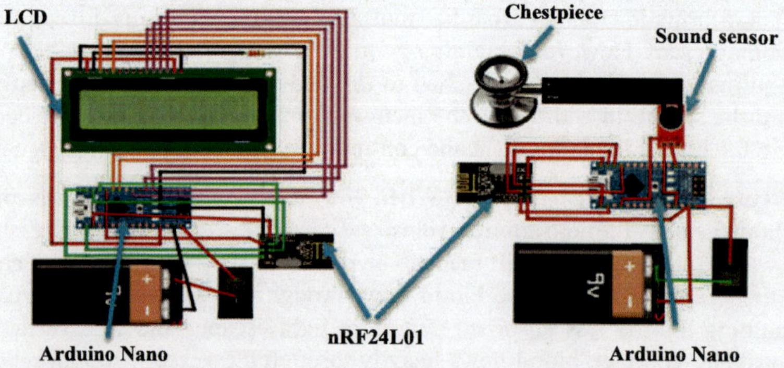

Fig. 2 The specific wiring design for heart rate monitoring

of wireless devices in recent years, the doctors do not need to visit patients directly as he can sit in the office or examine another patient and still be able to monitor the patient's heart rate. This compact device can easily move and display heart rate numbers via an LCD screen. Using the wireless transmitter, the heart rate value is continuously updated. The left of Fig. 1 is a device that displays the heart rate for the physician to monitor. The detailed connected wires is shown in the left of Fig. 2. The Arduino Nano uses power supply 9 V DC. The LCD is directly connected to the Arduino Nano through D2 to D7 pins and is supplied by 5 V DC from Arduino.

3.2 Patient's Side with the Sound Sensor

For ease of measurement, a body strap was designed. The patient wears a strap attached to the device on his chest, this device is presented in the right of Fig. 1. To orient sound and avoid external noise, a physical amplifier is used and will be introduced in Sect. 4. The audio amplifier head is put under the shirt and in direct contact with the skin of the chest. The tube attached to the audio amplifier head and the flexible receiver assembly makes it easy to customize the placement of the audio amplifier. The transmission of sound is very susceptible to interference so it is necessary to avoid so much activity and talk because there is some unnecessary noise from the vocal cords. The detailed wiring system of the patient's side is shown in the right of Fig. 2. Similarly with the same power supply with the Arduino Nano of doctor's side. The sound sensor is connected 5 V DC and its output pin is linked to A0 pin at which the Arduino can read analog signals from the sound sensor. Finally, the sound sensor and chestpiece of stethoscope are connected by a flexible tube.

3.3 Wireless RF Module for Communication

RF waves [6] are used for wireless transmission. The use of this wave is very simple and the cost of buying equipment is very minimal. The module consists of two parts: a transmitter and a receiver. The transmitter is attached to the head of the patient device. The receiver is attached to the doctor's side. The RF module nRF24L0 is connected with each Arduino Nano by Serial Peripheral Interface (SPI) [7]. The SPI will use CE, CSN, SCK, MOSI, MISO pin each component which is Arduino and RF module. Different from the rest, the RF module only must connect to 3.3 V DC of Arduino.

4 A Prototype for Heart Rate Monitoring

4.1 The Amplifier and the Sound Sensor

To amplify the heartbeat, a stethoscope's chestpiece to transfer one-way sound from the chest to the sound sensor is used. The stethoscope is made by Kenzmedico in Japan. This amplifier also undertakes the removing of noise from the outside sound. A diaphragm is placed against the skin on the chest. The good diaphragm oscillation ensures a constant pressure flow from the chest piece to the sound sensor. In addition to the diaphragm surrounding, there is a metal layer, so that the sound is guaranteed to be transmitted in the tube without external noise. The sound sensor is made up of an important part which has the microphone to receive sound. The microphone captures the sound and converts the physical signal into an electrical signal for ease of handling. The audio sensor is a micro integrated module and analog and digital pins for output to the MCU for processing (Fig. 3).

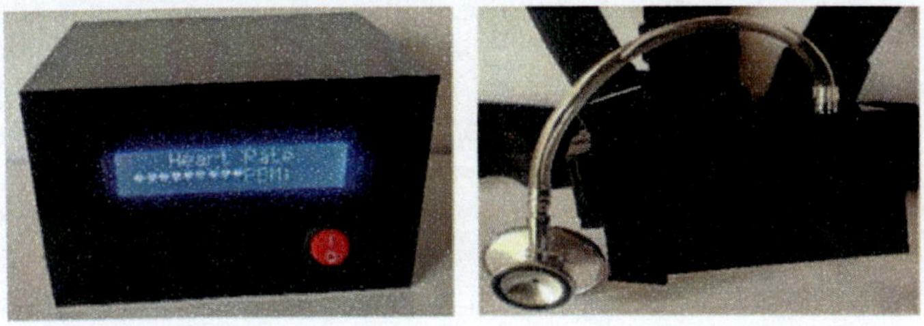

Fig. 3 Prototype with patient-side and doctor-side

4.2 The MCU

The Arduino Nano [8] is used as the main processor for both parts of the device. The Arduino Nano is a small, complete, and breadboard-friendly board based on the ATmega328P (Arduino Nano 3.x). Because the Arduino Nano is a compact processor, it is suitable for design and packages the product. Moreover, the Arduino is fully functional and easy to communicate with the Arduino IDE software.

4.3 The RF Module

The RF module nRF24 [9] is used for wireless data transmission. The nRF24L01 has a chip 2.4 GHz transceiver including an inserted baseband custom generator, devised for low-power wireless statements. This material is created for performance in the extensive ISM frequency band at 2.400–2.4835 GHz. An MCU (microcontroller) are necessitated to compose a communication mode with the nRF24L01. This module is also used on both parts of the device, can be transmitted in 2 directions and automatically connected. Our equipment only uses a 1-way transmission from the measurement to the display.

5 Measurement Method

5.1 Scheme for Measuring the Heart Rate

There are 4 steps to find and display the heartbeat. Figure 4 is a diagram to present these steps. Each step is arranged in a continuous sequence. Firstly, *step 1*—Amplify the sound: The doctor puts the amplifier which is a chestpiece of a stethoscope on the left chest. After the signal was physical amplified, it is transmitted to the sound sensor on the patient's side device. Second, *step 2*—Sample the signal: The Arduino Nano takes and saves the data from the sound sensor through analog signal pin A0. The signal is then sampled with sampling rate with frequency basic of the Arduino Nano. Third, *step 3*—Count the signal: The Arduino detects a peak value by maximum finding algorithm, this peak value is usually saved and updated to determine a threshold. Arduino based on this threshold to counting the heartbeat. Fourth, *step 4*—Output: After measuring the heart rate, the data is processed and

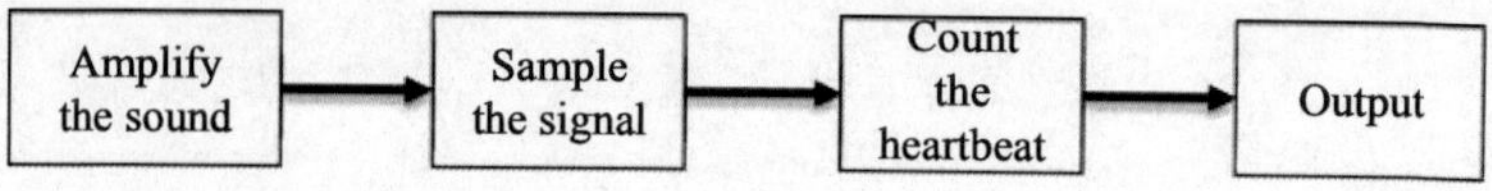

Fig. 4 Scheme for measuring heart rate

transmitted to the doctor's side device through RF waves. The doctor can view the heart rate and check the patient health status.

5.2 *Amplifying and Sampling the Sound of the Heart*

The chestpiece is placed at the bottom of the chest near the stomach so the sensor will receive the highest quality audio. Contemporary, the patient will bring the device which connects the amplifier. To avoid unexpected errors, the chestpiece must be fixed placed and do not oscillate so much. The sound signal will be transmitted through a flexible tube. The end of the tube is a sound sensor. Summary, the sound signal will be amplified and transfer into a micro of the sound sensor. The micro converts audio signals into electrical signals. This analog signal quantized into decimal values which are set between 0 and 1024. Because the sampling rate is about 9600 Hz, the heart rate, which is about several Hertz completely does not affect by the aliasing [10]. An audio signal is indicated in Fig. 5.

The periodic signal consists of two peaks with a high degree of slope and an upper vertex with a smaller slope in addition to the two lower vertices. The lower peak is created by the structure of the microphone that vibrates the membrane when the sound of the first beat of the heart causes the microphone membrane to oscillate on one side and then elastic against the other side to create the lower peak. To easily count the heart rate, it is necessary to find the lower peak value of the graph. This is quite simple, compare the successive values obtained and choose the smallest value.

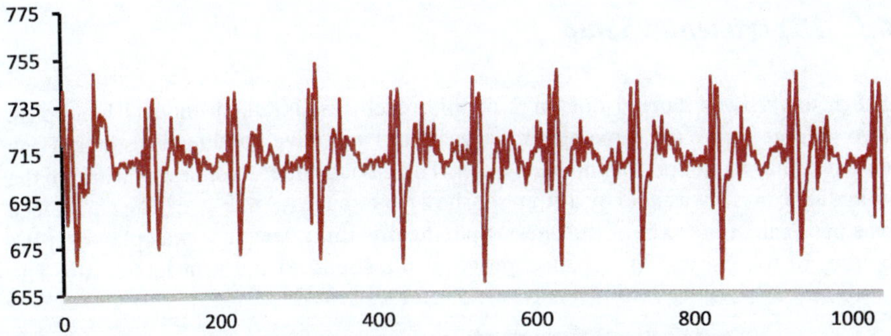

Fig. 5 The audio waveform of the heartbeat

5.3 Counting and Display the Heartbeat

Each cardiac cycle has one atrial contraction and one ventricular contraction. As shown in the picture, there will be five peaks generated by these 2 beats. However, taking these 3 beats to count is quite complicated, so we chose to count the lower peaks. Finding the lower peak value has been done so that it will provide a threshold value for comparison, and that threshold will be slightly larger than the peak value. Because the reason for this peak value will change randomly, the default threshold once difficult to make by comparison. Therefore, the constant threshold value must be changed according to the lower peak value, and this value will be updated continuously. When sampling, values that are just smaller than the previously determined threshold will be counted. For counting to the 5th vertex, the timestamp value at this vertex will be taken. Finally, to determine the BPM [11], a conversion formula must be performed as follows:

$$BPM = 5\frac{6000}{beatfinal - beatfirst} \tag{1}$$

The heart rate values are continuously updated. Due to the random nature, the bottom will be difficult to respond quickly, so there will be times when the error is not worth it.

6 Experimental Results

6.1 Experimental Setup

The research was carried out on 2 people; each test is taken about 10 samples. Two volunteers are the same age of 21 years old and have good health status. Each volunteer has to sit upright and keep quiet. The device patient's side was worn on the volunteers, and it was held by a strap set shown by image (a) in Fig. 6. The chestpiece was put against the skin of the chest. Specifically, this chestpiece was placed at the bottom of the chest. This location provides the sound of the ventricle easily. The

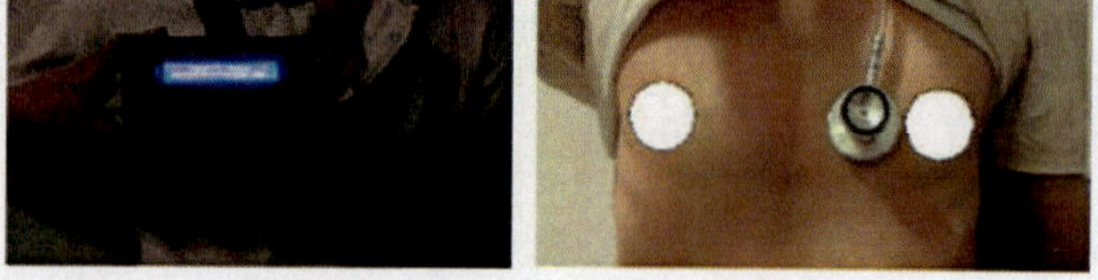

(a) The device was worn on the volunteer. (b) Put the chestpiece on volunteer's chest.

Fig. 6 Place the patient's side device on the volunteer

Table 1 Comparison of the results between the Accurate Heart Rate Monitor Apps and manual counting method

No.		1	2	3	4	5	6	7	8	9	10
Method	Phone	90	91	93	92	90	87	88	86	88	86
	Hand	90	90	93	94	90	90	88	86	86	88
Error (%)		0	1.11	0	2.13	0	3.33	0	0	2.23	2.23
Average error		1.123%									

doctor or nurse must adjust the chestpiece until the values of BPM is stable. The flexible tube, which is connected to the chestpiece, threads through the shirt slot of the volunteer.

Determining the test results of the device with manual counting by hand is difficult because when the hand is pressed against the chest, it will cause measurement errors. So we use an application of cell phones to verify our device. The application's name is Accurate Heart Rate Monitor which was run on an Android phone and based on a flash-light on the cell phone. The flashlight combined with a camera, which becomes the optical sensor. Verifying the result of the optical sensor on cell phones is very easy through manual counting methods by hand. Proofs of BPM was shown in Table 1. Test results of the optical sensor were given with quite low errors which are about lower 3%.

6.2 *Result*

We have examined the measurement of the heart rate of the device and compared to optical methods. The device is experimented on two volunteers to measure their heart rate, the first person is in an active normal state, and the other person has just finished the vigorous activity. We use the app measurement method on the cellphone as a standard and calculate the absolute error of the device. The data which we measured and displayed in Tables 2 and 3, the data showed that the average absolute error of volunteer 1 is less than 1% and volunteer 2 is less than 2%, the biggest error of the device is 5%, this error is acceptable, and that proves our device to be accurate. The above results indicate that the proposed system has high accuracy,

Table 2 Comparison of the results between cellphone's app and our device on volunteer 1

No.		1	2	3	4	5	6	7	8	9	10
Parameter	Device	88	76	64	68	80	68	72	80	76	80
	Phone	86	77	64	68	81	68	72	80	80	80
Error (%)		2.33	1.3	0	0	1.23	0	0	0	5	0
Average error		0.99%									

Table 3 Comparison of the results between cellphone's app and our device on volunteer 2

No.		1	2	3	4	5	6	7	8	9	10
Parameter	Device	99	100	96	100	100	100	96	96	96	96
	Phone	100	101	101	100	101	101	99	98	99	98
Error (%)		1	0.99	4.95	0	0.99	0.99	3.03	2.04	3.03	2.04
Average error		1.91%									

with the average accuracy of 98.55% on two volunteers and the accuracy of 95% for each measurement. Measurement results as above measured in the case of silent and inactive users. In reality, when the patient talk, cough or motion, the measurement will not be accurate and leads to much larger errors.

7 Conclusion and Future Work

The measurement results of the designed device are highly accurate. Moreover, the device can be applied to monitor the heart rate of the patients. This device, with many advantages of wireless data transmission, helps to observe the heart rate from a distance. This is not only absolutely convenient for the doctors' works but also will spare them plenty of time to spend to complete other important tasks. With all of the benefits above as well as a moderate price, this device is expected to be used in the future. However, this device still has some restrictions, such as patients must keep silent and avoid heavy activities to maintain accurate measurements. In the future, the filtering must be improved in order to eliminate the noises caused by the voice and other activities of the patients. Moreover, this device can be extended its function to process data obtained from the heart to detect and diagnose some cardiovascular diseases.

Conflicts of Interest This paper is not conflicted with any other contributions of the authors mentioned.

References

1. Wikipedia. Heart rate. https://en.wikipedia.org/wiki/Heart_rate. Accessed 03 Dec 2019
2. Wikipedia. ECG. https://en.wikipedia.org/wiki/Electrocardiography. Accessed 18 Dec 2019
3. Wikipedia. Anatomy and physiology of the stehoscope. https://pucgorbital.word-press.com/2017/05/13/anatomy-and-physiology-of-the-stehoscope/. Accessed 18 Dec 2019
4. Cristea C, Pasarica A, Andruseac G, Dionisie B, Ro-tariu C (2015) A wireless ECG acquisition device for remote monitoring of heart rate and arrhythmia detection. In: The 5th IEEE International Conference on E-Health and Bioengineering (EHB 2015), Iaci, Romania, 19–21 Nov 2015. https://doi.org/10.1109/EHB.2015.7391543

5. Shirzadfar H, Ghaziasgar MS, Piri Z et al (2018) Heart beat rate monitoring using optical sensors. Int J Biosen Bioelectron 4(2):45–51. https://doi.org/10.15406/ijbsbe.2018.04.00097
6. Wikipedia. Radio wave. https://en.wikipedia.org/wiki/Radio_wave. Accessed 11 Dec 2019
7. Aidan. Our Arduino SPI tutorial. https://core-electronics.com.au/tutorials/spi-arduino-tutorial.html. Accessed 21 Dec 2019
8. Arduino. Arduino nano. https://store.arduino.cc/usa/nano-every-pack. Accessed 18 Dec 2019
9. Roboindia. nRF24L01. https://roboindia.com/tutorials/arduino-nrf24l01/. Accessed 18 Dec 2019
10. Wikipedia. Aliasing. https://en.wikipedia.org/wiki/Aliasing. Accessed 18 Dec 2019
11. Medical news today. What should my heart rate be. https://www.medicalnewstoday.com/articles/235710.php. Accessed 18 Dec 2019

Equipment for Contact-Free Respiratory Measurement and Obstructive Sleep Apnea Detection

Tri Luong Phung, Minh Nghia Phan, Tin Thanh Dang, and Anh Dinh

Abstract Respiratory rate is always evaluated as an indispensable physiological parameter to know about the health status of a person. Also, the monitoring of human breathing activity for a long period has numerous foundational applications in medicine. This paper proposes a prototype whose functions are measuring respiration rate and detecting sleep apnea periods with a view to monitoring sleep quality. The main principle of this device is using an infrared sensor to detect respiratory activity based on the movement of the abdomen. The distance between the sensor and the patient is from 20 to 35 cm. The distance permits supervising the breathing activity without any awareness by the subject and any modification of the sleep quality, which is very important for sleep disorders diagnostic applications. The experiments were conducted on three people during about 10 minutes each, one of whom has no sleep apnea syndrome while two others are suffering from such disease. The average accuracy of the proposed system is 95.72%, the average sensitivity is 94.57%, and the average specificity is 96.92% in terms of respiratory measurement. The sleep apnea detection reaches high average parameters which are 100%. Generally speaking, the results obtained are moderately acceptable with high precision. However, the proposed system only performs well if the users sleep stably on their back, and there is no obstacle, such as a thick blanket between them and the sensor. These problems need improving in further research to create an optimal product which offers potential application in home and mobile healthcare.

Keywords Respiratory monitoring · Contactless · Infrared sensor · Obstructive sleep apnea

T. L. Phung · M. N. Phan · T. T. Dang (✉)
Faculty of Electrical and Electronics Engineering, Ho Chi Minh City University of Technology, Ho Chi Minh City, Vietnam
e-mail: dttin@hcmut.edu.vn

A. Dinh
Department of Electrical and Computer Engineering, University of Saskatchewan, Saskatoon, Canada
e-mail: anh.dinh@usask.ca

V. Van Toi et al. (eds.), *8th International Conference on the Development of Biomedical Engineering in Vietnam*, IFMBE Proceedings 85,
https://doi.org/10.1007/978-3-030-75506-5_3

1 Introduction

Respiration is an indispensable physiological task in living organisms. For humans, this process results in the inhalation of air containing oxygen into the lungs, where the diffusion of gas occurs through the alveolar-capillary membrane [1]. In fact, respiratory rate is one of the vital signs used to the monitor the sickness development, and an irregular breathing rate is a crucial marker of serious diseases. Considerably proven by many research papers, the changes in this parameter can be used to predict potentially alarming clinical incidents which are cardiac arrests or admission to the intensive care unit.

In particular, some sleep disorders may occur that have negative impacts on the sleep quality as well as seriously affect patient's health. According to Bucklin et al. [2], sleep apnea disorders are classified into three main types: central sleep apnea (CSA) occurs due to the absence of proper signal between the brain, and the muscle groups whose function is to regulate the breathing activity, obstructive sleep apnea (OSA) is caused by the complete or partial obstruction of the upper airway during sleep, and mixed sleep apnea (MSA) is the combination of both CSA and OSA. The obstructive sleep apnea is the most common sleep disorder among three types. As claimed by Javaheri et al. [3], OSA is confirmed when the inspiratory activity is missing for at least 10 seconds. The patients who suffer from OSA have numerous health issues such as cardiovascular diseases, depression, and dementia, as indicated by Surrel et al. [4]. Consequently, Gami et al. [5] found a greater risk of sudden deaths for the patients. Thus, it is essential to have a system that demonstrates sleep quality and diagnoses sleep apnea.

There are several available methods to monitor sleep quality, which are divided into two main types: contact monitoring (or intrusive) and non-contact (or non-intrusive). Among those, a typical type of intrusive monitoring is Polysomnography (PSG). It uses multiple electrodes and respirators to record biological parameters such as the electrocardiogram (ECG), electromyogram (EMG), electroencephalography (EEG), eye movement, nasal airway [1]. However, this method is very expensive and inconvenient due to the high requirements of equipment, operators, and operating space. In terms of contactless monitoring, because collecting information is more difficult, most studies focus on algorithms and signal processing.

This paper focuses on studying one of the methods for non-contact measuring breathing activity. It takes advantage of the movement of the abdominal wall or chest wall to infer a patient's breathing rate. This is considered a reliable method and can be applied in medical devices. Within the framework of this study, we conducted a prototype with the most important component is the infrared sensor which is used to measure the distance. The fundamental principle of this method is that when the subject inhales, the abdominal wall will stretch, and when he/she exhales, the abdominal wall will collapse. This makes a difference in the distance from the abdomen to the sensor to mark the appearance of respiration. In addition, this device is equipped with features to detect obstructive sleep apnea (OSA). If the interval between two inhalations is 10 seconds, apnea is determined, and a buzzer indicates the patient is in a state of illness. This simple and inexpensive prototype is powered by the national electric grid; therefore, it eliminates the utilization of toxic batteries.

The organization of this paper is as follows. Section 2 describes the related works. Section 3 presents the hardware platform. Section 4 shows the technique used for counting respiratory rate and detecting sleep apnea. Section 5 illustrates the experimental setup and shows the corresponding results. Finally, Sect. 6 draws some conclusions.

2 Related Work

Over the past decades, some methods have been studied as well as devices that have been developed to perform respiratory process monitoring. Other than contact-required methods, non-invasive sensing systems are beneficial since it does not confine the subject with cables, and there are no electrodes or straps which may cause discomfort or skin irritation. Moreover, patients tend to alter their respiration during the process when they are aware of the measurement, as commonly occurs during face-mask respiration technique [6]. Hence, the sensors become superior in long-term continuous breathing monitoring.

One of the first noncontact respiration rate monitoring systems called the Radar Vital Signs Monitor (RVSM) was reported by Greneker [7]. It was developed to monitor the performance of Olympic athletes from over 10 meters. Based on the Doppler phenomenon, the RVSM detected breathing-induced movements of the chest. A restriction of this method is the corruption of the breathing signals due to the motion artefact. Se Dong Min et al. [6] presented the ultrasonic proximity sensor to measure respiratory rate. This kind of sensor measures breathing signatures and rates in real-time for a long period. Basically, a short burst of ultrasonic sound from the device is transmitted toward a target; then it reflects the sound back to the sensor. After that, by measuring the time for the echo to come back to the sensor, the system calculates the distance from the device to the subject using the average speed of sound. They utilized a 240 kHz ultrasonic sensor to measure the time of a sound wave between the transmitted and received signals inferred from the abdominal wall motion during the breathing process. Respiratory rates of ten people given by the proposed device were compared with those measured with a thermocouple sensor. However, when the patients were clothed, they had difficulties measuring respiration with the ultrasonic sensor because the texture of the clothing scattered and absorbed sound wave. Consequently, this point led to incorrect results from the analyzed area.

A similar device was also introduced by Maki Uenoyama et al. [8]. Using a microwave antenna attached to the room ceiling, they carried out contactless monitoring of thoracic wall motions of patients covered by a thin blanket in bed, in order to measure their vital signs under nursing care in a welfare institution. Their system included a 1215 MHz-microwave radar (LDR1), an antenna box attached to the ceiling and a personal computer with analyzing software. After that, eight healthy volunteers were examined. In conclusion, this system could monitor subtle changes in respiration rate, and detecting increases in respiratory rate caused by disorders such as pneumonia would be possible.

3 Hardware Platform

We use Arduino Uno R3 which includes Atmega328 as the main processor and which supports 32 KB Flash memory as well as 2 KB SRAM. A 9 V DC adapter is used to power directly the Arduino and other components. Sharp GP2Y0A02YK0F Analog Distance Sensor measures the distance between our system and the user's abdomen with a range of 200–1500 mm. The results are displayed by an I2C 1602 LCD module. A 5 V DC buzzer is used for warning in case of sleep apnea detected. The detailed wiring diagram of the system and the processing blocks in the proposed system are illustrated in Figs. 1 and 2, respectively.

Arduino UNO is used to power other electronic components except for the buzzer by connecting them to 5 V pin and GND pin of the Arduino. The output of GP2Y0A02YK0F Distance Sensor and two ports which are SDA and SCL of the I2C LCD are respectively connected to the A0, A4 and A5 pin of the Arduino. The input power of the buzzer is connected to pin 13 of the Arduino. We designed a button with one pin connecting to the Reset port and another pin connecting to Ground in order to restart the system. During the operation, the distance sensor transmits the analog signal to the Arduino; the signal will be processed by applying breath counting algorithm and sleep apnea detecting algorithm to obtain required results. Then, the

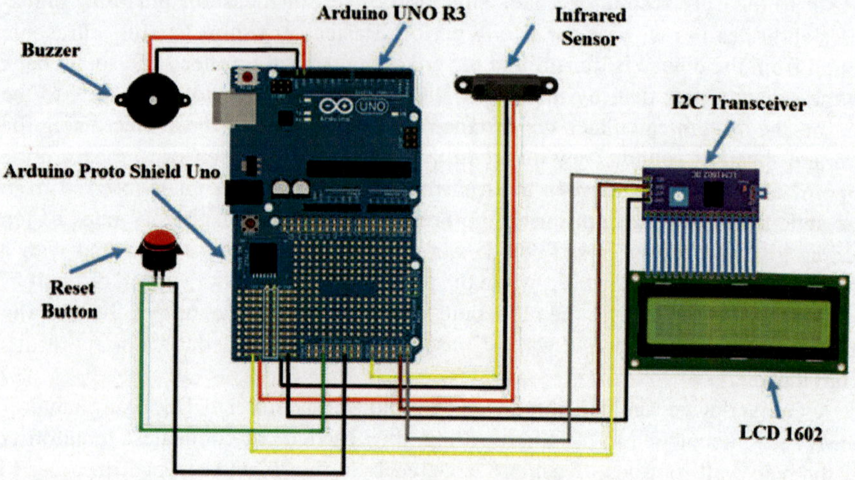

Fig. 1 The detailed wiring diagram for respiratory rate counting

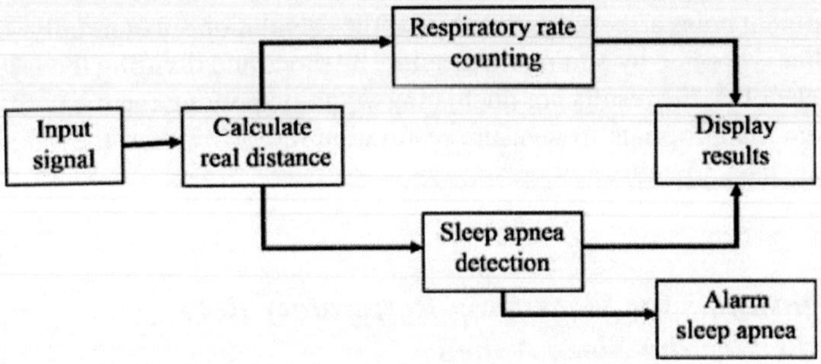

Fig. 2 Processing blocks in the proposed system

breathing rate and obstructive sleep apnea will be displayed on the LCD screen. The buzzer will go off whenever the sleep apnea is detected.

4 Respiratory Rate Counting and Sleep Apnea Detecting Method

4.1 Calculating Real Distance Values

The output of the GP2Y0A02YK0F Distance Sensor is the conventional integer values which depend on the distance between the sensor and the object. As a result, calibration is needed by finding a regression equation formula. This problem requires data samples; therefore, we measured the distance between the sensor and the object 16 times. We limit the distance between 200 and 350 mm, which will be suitable for respiration measuring experiment to obtain the high accuracy rate. Then, several non-linear regression methods are applied to find the regression equation formula. We use four proposed options which are Exponential, Logarithmic, Polynomial, and Power in Excel to obtain the required equation.

The R^2 value of Logarithmic regression method (0.9973) is the highest value among the four methods. As a result, the Logarithmic regression equation which is

$$y = 0.0004x^2 - 1.3971x + 783.69 \tag{1}$$

is used as a formula to determine real distance values and x is the output voltage from the sensor.

To ensure that the distance is in the required range of [200, 350], the two threshold values are set to 342 and 477. Values that are lower than 342 or higher than 477 will be set into 342 or 477. Equation (1) is then applied to calculate the real distance value between the sensor and the user's abdomen. Finally, the result is divided by 10 to convert to a centimeter in order to be easily read and displayed in the LCD.

The ambient noise affects the accuracy of the calculation substantially. Therefore, a mean filter is applied by adding the results 100 times and dividing the last result by 100. It is clear that the results are fluctuating without applying the mean filter, which makes them inappropriate to measure respiration rate. However, the values remain stable while using the mean filter.

4.2 Algorithms for Measuring Respiratory Rate and Detecting Sleep Apnea

Respiratory rate measurement. Totally 200 samples showing the distances between the range sensor and a user's abdomen are plotted. During each inhalation, the abdomen expands in the upper direction [9], which leads to the decrease in the distance value. Meanwhile, the distance values increase during exhalation because the abdomen goes down. The increase and decrease of the distance values are considered to be two signs to recognize the respiratory activity. We choose the inhalation as a sign for counting respiratory rate, which bases on the significant decrease in the obtained values in a short period of time. The change between the maximum value and its corresponding minimum value in a breathing rhythm is 1–1.5 cm in average, which is used to determine the threshold value of 0.3 to distinguish between the significant decrease and the slight decrease due to noise. If the value of the recent distance is less than that of the previous distance and the change between them is greater than the threshold value, which means.

$$dis < predis \text{ and } (dis\text{-}predis \geq threshold) \tag{2}$$

where *dis* is the value of recent distance, and *predis* is the value of previous distance, the counter will increase by 1. However, the condition in the inequality (2) may be satisfied multiple times during an inspiratory activity; the counter will continue to increase by 1, which means that the value of the counter is more than the actual respiratory rate. Therefore, a flag is required to mark the status of the breathing process. The flag's value is HIGH during the inhalation and LOW in other cases. If the condition (2) is satisfied and the value of the flag is LOW, the counter will increase by 1, and the flag is set to HIGH.

$$\begin{cases} dis < predis \\ |dis\text{-}predis| \geq threshold \\ flag = LOW \end{cases} \tag{3}$$

The condition in (3) is the inspiratory condition, and it is only satisfied one time during an inspiratory activity, which keeps the value of the counter the same as the actual respiratory rate. The breath counting algorithm flowchart is shown in Fig. 3.

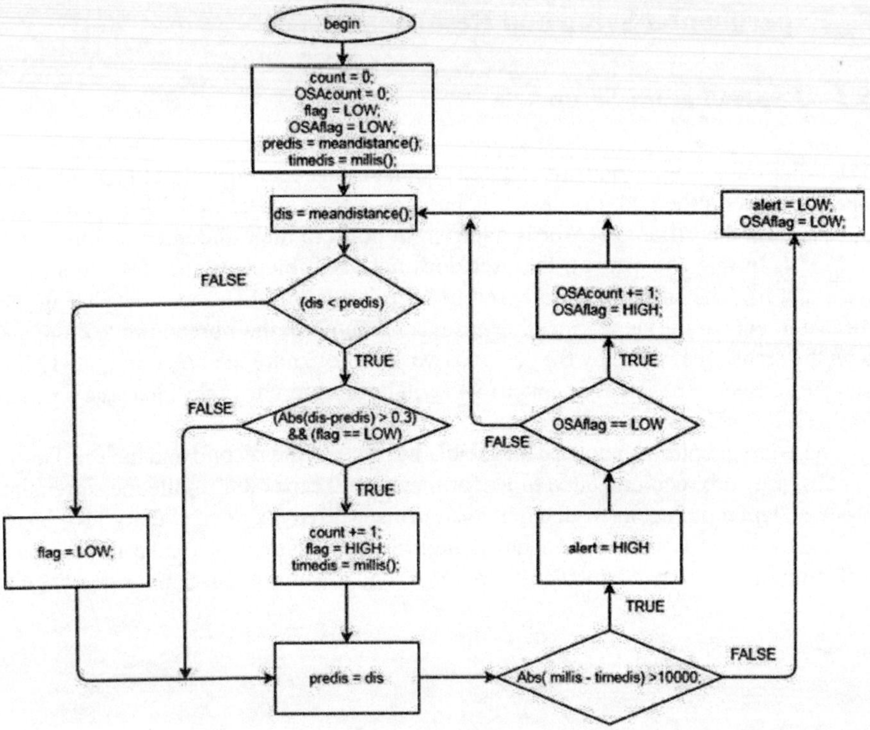

Fig. 3 Respiratory rate counting algorithm flowchart

Sleep apnea detection. The amount of time between the current time and the nearest previous inhalation is considered to detect sleep apnea disorder. The time when inhalation occurs is stored in a variable called *timedis*. The value of *timedis* is only updated when the inspiratory conditions are satisfied. If the change between *timedis* value and the actual time is greater than 10 seconds, which means that there is no inhalation occurring during the last 10 seconds, the value of a variable named alert will be set to HIGH to inform that the sleep apnea is happening. It will trigger the buzzer to go off. Otherwise, the value of alert is LOW, which means that the respiratory activity is normally happening.

When the sleep apnea occurs, a variable called *OSAcount* representing the number of occurrences of sleep apnea will increase by 1. Simultaneously, a flag is required to ensure that the increase is already taken into account by the system. It prevents the mismatch between the value of *OSAcount* and the actual number of occurrences of sleep apnea. At the time when the value of alert is set to HIGH, which means that the sleep apnea is happening, if the flag's value is LOW, the *OSAcount* value is then increased by 1, and the flag value is also set to HIGH. In other cases, the flag value is set to LOW.

5 Experimental Setup and Results

5.1 Experimental Setup

The experiment was carried out on three people; each experiment lasted ten minutes. The user lies on the bed, the device is hung on a rack and placed 20–35 cm above the user and the infrared sensor is adjusted to point to their abdomen as illustrated in Fig. 4. In this experiment, the user does not use a blanket, and the patient can wear a shirt. The subjects were asked to keep it comfortable and quiet during the measurement. At the same time, they also had to monitor their breathing to compare with the results measured by the device. Two parameters are the breathing rate (RR), and the number of episodes of apnea (SA) will be monitored. Table 1 lists the profiles for the 3 test subjects.

These are people with normal and stable health. For the second and the third individuals, they are recommended to perform apnea 10 times during the measurement process. Three indicators which are accuracy, sensitivity, and specificity [10] were defined for each RR and sleep apnea monitoring. The accuracy of a test is its ability to differentiate correctly the inhalation (or OSA) and the non-inhalation (or non OSA).

Fig. 4 Experimental setup

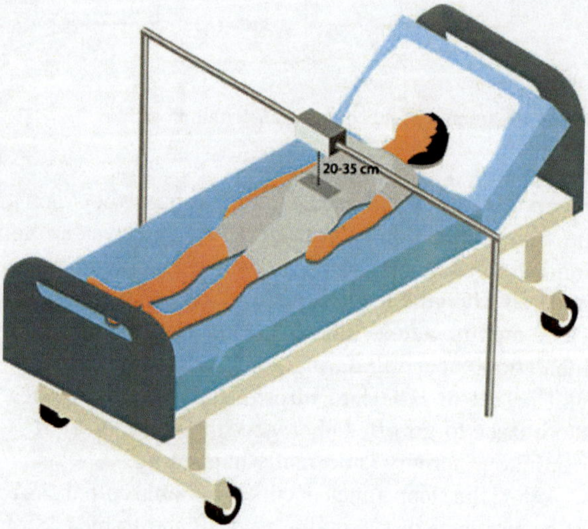

Table 1 Profiles of the three users

User	Gender	Age	Height (cm)	Weight (kg)	Health status
1	Male	10	135	40	Non OSA
2	Male	21	173	60	OSA
3	Female	50	167	65	OSA

Then, the sensitivity of this experiment is its ability to determine the case of inhalation (or OSA) precisely. Moreover, the specificity is its ability to verify exactly the non-inhalation (or non OSA). Three following formulas are given as follows:

$$Accuracy = \frac{TP + TN}{TP + TN + FP + FN} \tag{4}$$

$$Sensitivity = \frac{TP}{TP + FN} \tag{5}$$

$$Specificity = \frac{TN}{TN + FP} \tag{6}$$

where *TP* is the number of times correctly identified as inhalation (or sleep apnea), *FP* is the number of times incorrectly identified as inhalation (or sleep apnea), *TN* is the number of times correctly identified as non-inhalation (or non-sleep apnea), *FN* is the number of times incorrectly identified as non-inhalation (or non-sleep apnea).

5.2 Results

Each of the parameters *TP*, *FP*,*TN*, and *FN* of each user, corresponding to each case of breathing measurement and apnea detection (Table 2), is determined by the total value of each of these parameters in each experiment with each person. Table 3 shows three indicators of *Accuracy, Sensitivity* and *Specificity* for each case corresponding to each subject. These data are obtained by comparing the results of the device shown on the LCD and the results that the users self-verify.

The first user has no sleep apnea syndrome; therefore, there is no corresponding value for the sensitivity of sleep apnea detection. In terms of respiratory rate measurement, the proposed system has a high average accuracy of 95.72%, the sensitivity of 94.57% and the specificity of 96.92%. The sleep apnea detection also reaches 100% for all three parameters. In general, the apnea determination of the device is relatively accurate, 100% in all the tests. Besides, counting the breathing rate of each object has the difference in three parameters. As can be seen, the results for the first

Table 2 Parameters TP, FP, TN, FN of each user

Parameter	User 1		User 2		User 3	
	RR	SA	RR	SA	RR	SA
TP	207	0	142	10	102	10
FP	5	0	3	0	5	0
TN	208	10	138	12	100	15
FN	6	0	8	0	9	0

Table 3 Accuracy, sensitivity and specificity for each case

User	Parameter	Accuracy (%)	Sensitivity (%)	Specificity (%)
1	RR	97.42	97.18	97.65
	SA	100	–	100
2	RR	96.22	94.66	97.87
	SA	100	100	100
3	RR	93.52	91.89	95.24
	SA	100	100	100
Average	RR	95.72	94.57	96.02
	SA	100	100	100

test were the most accurate, sensitive and specific of the three. Meanwhile, the third person with weak breathing due to age leads to lower precision, as shown in Table 3.

6 Conclusion

This paper presents a viable non-contact method for measuring respiratory rate and detecting sleep apnea using the Sharp infrared sensor. The system was tested on three subjects with the device at 25 cm above their bellies. The result's accuracy is relatively high and can be deployed for patient's respiration detecting. Moreover, it is feasible for long-term sleep monitoring thanks to its affordable cost. However, this device only operates correctly under the required condition and could not collect enough related data to measure breathing rate for the subjects wearing a quilt or a thick blanket. The system is ideal for patients with little to none movement while sleeping like older people or patients who have difficulty in moving. With further improvements and investigations in the approach for the enhancement of detection accuracy in any circumstances of the subjects, it is expected that the method can be used in early diagnosis of cardio-respiratory sleep disorders. Furthermore, a combination of the system and available communication technologies can help to transmit important information to the doctors in real-time. Also, users can constantly understand their physical conditions. This equipment can also be applied to remote health monitoring applications.

Conflicts of Interest The authors declare that they have no conflict of interest.

References

1. Al-Khalidi F, Saatchi R, Burke D et al (2011) Respiration rate monitoring methods: a review. Pediatr Pulmonol 46:523–529. https://doi.org/10.1002/ppul.21416
2. Bucklin CL, Das M, Luo SL (2010) An inexpensive accelerometer-based sleep-apnea screening technique. In: Proceedings of the IEEE 2010 national aerospace & electronics conference. https://doi.org/10.1109/naecon.2010.5712984
3. Javaheri S, Barbe F, Campos-Rodriguez F et al (2017) Sleep apnea: types, mechanisms, and clinical cardiovascular consequences. J Am Coll Cardiol 69(7):841–858. https://doi.org/10.1016/j.jacc.2016.11.069
4. Surrel G, Aminifar A, Rincon F et al (2018) Online obstructive sleep apnea detection on medical wearable sensors. IEEE Trans Biomed Circuits Syst 12:762–773. https://doi.org/10.1109/tbcas.2018.2824659
5. Gami A, Howard D, Olson E, Somers V (2005) Day-night pattern of sudden death in obstructive sleep apnea. ACC Curr J Rev 14:30. https://doi.org/10.1016/j.accreview.2005.06.052
6. Min SD, Kim JK, Shin HS et al (2010) Noncontact respiration rate measurement system using an ultrasonic proximity sensor. IEEE Sens J 10:1732–1739. https://doi.org/10.1109/jsen.2010.2044239
7. Greneker E (1997) Radar sensing of heartbeat and respiration at a distance with applications of the technology. Radar Systems (RADAR 97). https://doi.org/10.1049/cp:19971650
8. Uenoyama M, Matsui T, Yamada K et al (2006) Non-contact respiratory monitoring system using a ceiling-attached microwave antenna. Med Biol Eng Comput 44:835–840. https://doi.org/10.1007/s11517-006-0091-8
9. Wheatley I (2018) Respiratory rate 4: breathing rhythm and chest movement | Nursing Times. Nursing Times. https://www.nursingtimes.net/clinical-archive/respiratory-clinical-archive/respiratory-rate-4-breathing-rhythm-and-chest-movement-20-08-2018/. Accessed 22 Oct 2019
10. Alireza Baratloo G (2019) Part 1: Simple definition and calculation of accuracy, sensitivity and specificity. In: PubMed Central (PMC). https://www.ncbi.nlm.nih.gov/pmc/arti-cles/PMC4614595/. Accessed 10 Dec 2019

Non-invasive Pressure Support Ventilator for Patients with Respiratory Failure in Under Resourced Regions

Jorge Otero, Ramon Farré, Onintza Garmendia, David Gozal, and Anh-Tuan Dinh-Xuan

Abstract Most hospitals and patients in low/middle-income countries (LMICs) cannot afford the current pricing schedules of commercial mechanical ventilators. This has resulted in a marked restriction of their availability and left patients with respiratory failure with restricted access to this life-saving treatment. Here, we present and discuss a novel, especially designed low-cost and easy-to-build non-invasive bilevel pressure ventilator that should reduce the serious shortage of ventilators in LMICs. The ventilator is built using off-the-shelf components available via e-commerce and is based on a high-pressure blower, two pressure transducers and an Arduino Nano controller with a digital display (total retail cost < 75 US$). The designed ventilator allows independent setting of inspiratory and expiratory pressures up to 20 and 10 cmH_2O, respectively, selectable cycling threshold and a back-up rescue frequency (assisted-controlled ventilation mode). All construction details are provided following an open-source hardware and code approach for free replication. This non-invasive ventilator, with equivalent performance to high-quality commercial devices, could facilitate the application of this life-saving therapy to LMIC patients who otherwise would not be treated.

J. Otero · R. Farré (✉) · O. Garmendia
Unitat de Biofísica i Bioenginyeria, Facultat de Medicina i Ciències de la Salut, Universitat de Barcelona, Casanova 143, 08036 Barcelona, Spain
e-mail: rfarre@ub.edu

J. Otero · R. Farré
CIBER de Enfermedades Respiratorias, Madrid, Spain

O. Garmendia
Sleep Lab, Hospital Clinic, Universitat de Barcelona, Barcelona, Spain

D. Gozal
Department of Child Health, The University of Missouri School of Medicine, Columbia, MO, USA

A.-T. Dinh-Xuan
Service de Physiologie-Explorations Fonctionnelles, Hôpital Cochin, Assistance Publique-Hôpitaux de Paris (AP-HP), Paris, France

V. Van Toi et al. (eds.), *8th International Conference on the Development of Biomedical Engineering in Vietnam*, IFMBE Proceedings 85,
https://doi.org/10.1007/978-3-030-75506-5_4

Keywords Mechanical ventilation · Open source hardware · Medical device · Low-medium income countries

1 Introduction

Patients with respiratory failure, a condition associated with a variety of diseases, are unable to achieve adequate blood concentrations of O_2 and CO_2 resulting from ineffective blood gas exchange in the lungs. The most common cause for such condition is that, in the context of underlying lung inflammation and infection, the pulmonary airways are narrower and the lung tissues are stiffer than in the healthy condition, resulting in increased resistance to air flows and in impeded lung inflation. The induced augmentation in the mechanical impedance of the lungs may reduce the ability of the respiratory muscles to provide sufficient ventilation for achieving adequate O_2 and CO_2 concentrations in blood. Additionally, the increased effort required from the inspiratory muscles for ventilating a mechanical load higher than normal may lead them to fatigue with the consequence that the patient is unable to breathe by himself/herself. Mechanical ventilation is a treatment aimed at helping these patients to achieve enough ventilation by supporting their inspiratory muscular pump with an artificial external pump [1]. The most severe cases require that the patient be tracheally intubated and connected to a mechanical ventilator, which takes full ventilatory control. In less severe cases, when the patient is still able to keep a certain level of respiratory activity, mechanical ventilation can be provided non-invasively through a nasal or face mask. In this case, the ventilator does not fully replace the patient's inspiratory pump but supports him/her to achieve adequate ventilation [1].

Non-invasive mechanical ventilation is commonly applied by controlling the pressure applied at the nasal mask [2]. Hence, the ventilator is set to generate high pressure during inspiration, and thus contribute to the action of the patient's inspiratory muscles, and also to apply a lower pressure during expiration to facilitate the lungs to recover their resting volume without collapsing. This modality of mechanical ventilation is usually known as pressure support ventilation. From an engineering viewpoint, the ventilator designed for such a treatment should be able to adequately carry out two main functions. On the one hand, by means of a well-controlled pressure generator, the ventilator must be able to provide the pressure level set by the clinician for each specific patient's need. To this end, current ventilators are usually based on a variable speed blower which is controlled to generate the desired pressure [3]. On the other hand, the ventilator must be capable of precisely detecting the time when the patient initiates the inspiration by activating his/her inspiratory muscles. Inspiratory detection is commonly based on measuring the deflection in pressure/flow caused at the mask when the patient starts inspiration. Moreover, the ventilator must be able to detect the time when the patient finishes inspiration so that the device starts its expiratory mode (a shift known as ventilator cycling). Although different procedures can be used for ventilator cycling, a usual way is to shift the ventilator to the expiratory mode when inspiratory flow decreases to a given percentage of the maximal inspiratory flow. Correct ventilator triggering and cycling is fundamental for applying the values of inspiratory and expiratory pressure set by the physician with a timing that is synchronous with the activity of the patient's respiratory muscles [4]. Otherwise, ventilator vs patient "fight", as this asynchrony is commonly referred to, results in patient discomfort and intolerance making the treatment inapplicable [3, 5].

The need for non-invasive pressure support ventilators is considerably high in low- and middle-income countries (LMIC) [6, 7]. The interest of this treatment to provide respiratory support in patients living in developing economies is being increasingly recognized because of its clinical benefits, relatively low cost, and ease of use. Remarkably, the need for mechanical ventilation, and consequently of pressure support devices, is expected to further increase in LMIC owing to growing urbanization and expanding access of patients to health care services. Nevertheless, the mechanical ventilators available in the market, which are mainly focused on customers in developed countries, are extremely expensive, particularly when taking into account budget capacity in LMIC. Hence, ventilators availability is reduced, with the consequence that a considerable number of patients with respiratory failure in these under resourced areas cannot receive the treatment they need. The chronic lack of mechanical ventilators in LMIC is further exacerbated during epidemic events, such as the current Covid-19 pandemic, which cause a tremendous increase in the number of patients suffering from respiratory failure [8].

Given the interest of providing non-invasive pressure support ventilators to patients in under resourced regions where conventional commercial devices are not available, we here describe a novel, especially designed ventilator which is both low-cost and easy to build. We also provide open source hardware and software access, therefore allowing for free and unrestricted replication and improvement of the ventilator [9–13].

2 General Description of the Device

Figure 1 shows a diagram of the designed pressure support ventilator which is based on a conventional rationale: use of a blower, single air tubing and an exhalation port (intended leak orifice). The blower takes room air and generates the airflow through a flexible tubing connected to the patient's mask. When the patient is not breathing, all the airflow generated by the blower reaches the atmosphere again through the exhalation port. Accordingly, the nasal pressure externally applied is the product of the blower-generated airflow times the resistance of the exhalation port.

Therefore, the pressure applied at the mask can be modulated by simply varying the flow generated by the blower. It is important to note that, in addition to being the nasal pressure source, the airflow generated by the blower plays the important role in avoiding rebreathing. When the patient breathes, the airflow through the

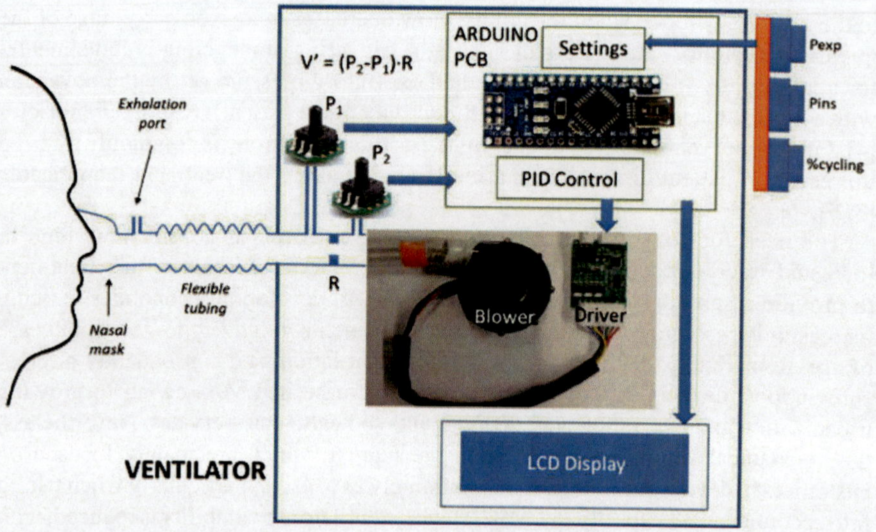

Fig. 1 Diagram of the mechanical ventilator setting. P_1 and P_2 are pressure transducers. R is an orifice-type resistor. Flow (V′) is computed as $V' = R \cdot (P_2 - P_1)$. See text for explanation

exhalation port changes because inspiratory/expiratory flow is added to the blower-generated flow. To keep a well-controlled nasal pressure according to the inspiratory and expiratory values set into the control panel of the ventilator, the blower should be able to continuously and automatically modify the blower-generated airflow. To this end, the pressure at the ventilator outlet, which is close to the mask pressure given the low resistance of the flexible tubing) is measured by a pressure transducer (P_1 in Fig. 1) and fed into a controller unit to provide an error signal to the blower driver, so that mask pressure is effectively adjusted. The ventilator air circuit includes a resistor, playing the role of a pneumotachograph, and an additional pressure transducer to allow for estimating flow (and hence triggering and cycling inspiration/expiration) from the pressure drop across the resistor (Fig. 1).

The front panel of the device is shown and described in Fig. 2. Although the device can also operate in a mode to provide continuous positive airway pressure (CPAP), in the following, it is assumed that the device is set to provide pressure support ventilation. After switching on, the ventilator takes a few seconds to auto-check and auto-correct any zero offset in their pressure sensors. The value of inspiratory pressure is set by keeping pushed the button ‘Setting’ and by simultaneously turning the knob ‘P-ins’ until the desired inspiratory pressure is shown in the front panel display (Fig. 2).

Setting the expiratory pressure value is carried out similarly: while pushing button ‘Setting’, the knob ‘P-esp’ should be turned until the desired expiratory pressure value is displayed. The range available for both inspiratory and expiratory pressures is 4–25 cmH_2O, but the device controller does not allow setting expiratory pressure higher than inspiratory pressure. Ventilator cycling (i.e. the time to finish inspiration

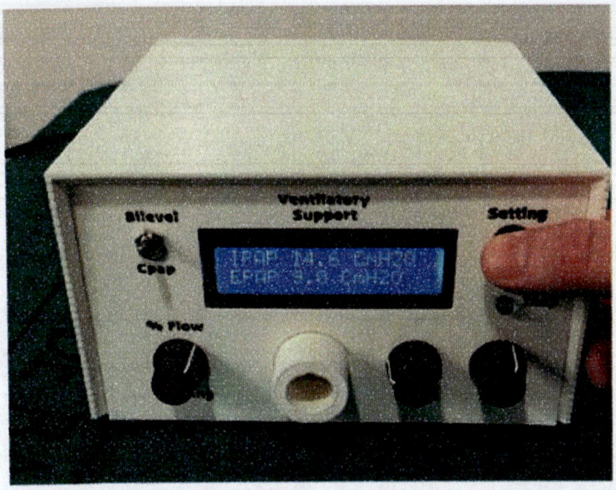

Fig. 2 Front panel of the ventilator. The Bilevel/CPAP switch allows selecting the type of pressure support: Bilevel pressure ventilation or continuous positive airway pressure, respectively. 'P-exp', 'P-ins': turning these knobs, while pushing button 'Setting', allows setting expiratory and inspiratory pressure, respectively. Green led on during inspiration. '%Flow' cycling: Turning the knob allows to set the % of maximum inspiratory flow threshold at which the device cycles to start expiration

and hence to start expiration) is set by simply turning knob '%Flow' (Fig. 2) to the threshold value of inspiratory flow indicated as a percentage of maximum inspiratory flow in the same cycle. The three settings (inspiratory and expiratory pressures and cycling flow threshold) can be independently modified during patient use, as required to optimize clinical treatment.

Once the ventilator settings are established, the device can be connected to a patient wearing a nasal mask (Fig. 3). Two obvious but important statements are required at this point. The first one is that non-invasive ventilation must be applied under the supervision of a health care professional who is legally qualified to provide this medical treatment. Second, the ventilator prototype must be connected to a conventional patient's interface consisting of a flexible tubing and a nasal mask incorporating the corresponding intended air leak orifice (exhalation port) to avoid rebreathing. Moreover, a suitable protecting air filter at the ventilator-patient interface should be used as required. Immediately after a patient is connected, the device detects pressure in the mask and ventilation is initiated.

Figure 4 illustrates representative nasal pressure and flow signals in the non-invasive pressure support ventilation. An inspiration (i.e. inspiratory pressure is applied to the nasal mask) is triggered when a positive deflection in flow ($\Delta V'$ in Fig. 4) which is greater than an internally established set threshold is observed. Activation of the inspiratory phase is indicated by the green led light labeled 'Insp' (Fig. 1). After inspiration is triggered, maximum inspiratory flow is detected (Fig. 4) and when inspiratory flow decreases to the set threshold value of '%Cycling' (Fig. 1) the ventilator shifts to expiration phase and the panel-set value of expiratory pressure

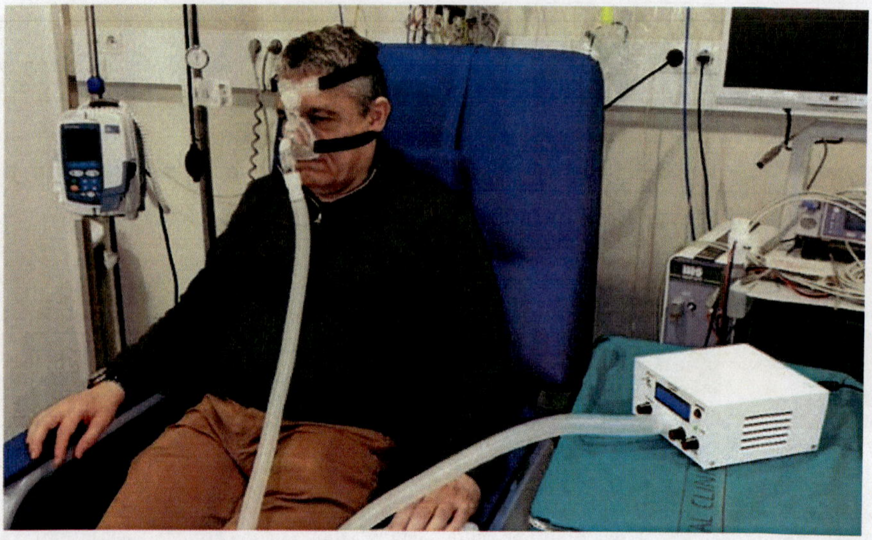

Fig. 3 Example of ventilator use

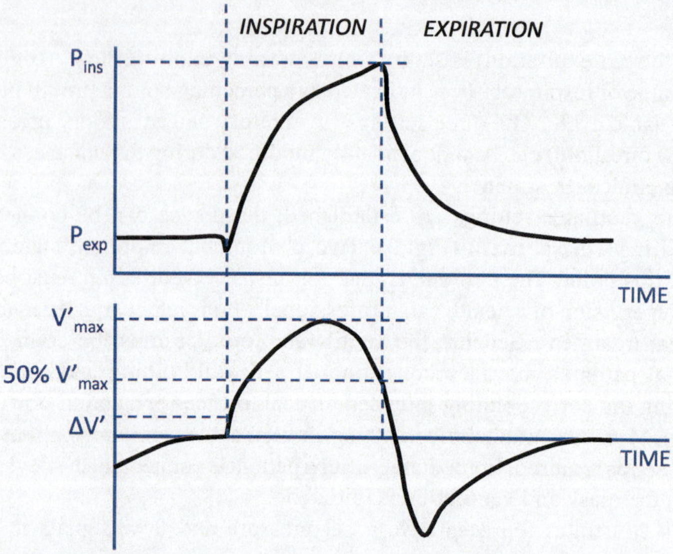

Fig. 4 Representative signals of flow and nasal pressure recorded along a breathing cycle during non-invasive pressure support. V'_{max}: maximum inspiratory flow; 40% V'_{max}: flow value threshold set to trigger ventilator cycling (change from inspiration to expiration). Pins: inspiratory pressure; P_{exp}: expiratory pressure; $\Delta V'$: flow deflection induced by patient's spontaneously initiating inspiration by respiratory muscles

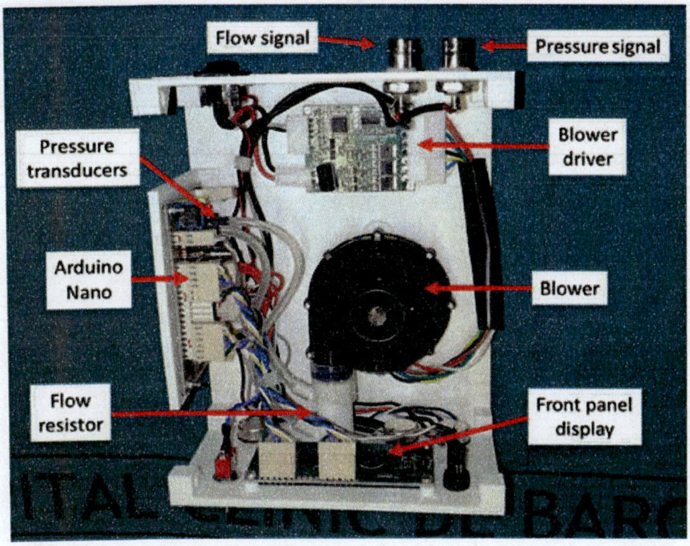

Fig. 5 Top view of the ventilator components

is applied to the nasal mask until a new inspiration is detected. In case that no inspiratory effort is detected after 5 s of the previously triggered inspiration, a back-up inspiration is automatically generated, corresponding to a rescue frequency of 12 breaths/min. Although not necessary for normal function, it is possible to monitor the pressure and flow signals in real-time from BNC connectors placed at the rear part of the ventilator (Fig. 5).

3 Ventilator Components, Data Processing and Control

3.1 Blower

The ventilator is driven by a low-cost, small (7 × 4 cm) high-pressure blower and its factory-provided driver (WM7040, Ning Bo Feng Hua Wei Cheng Motor Factory, Zhejiang, China) (Fig. 5) [14]. Given that the pressure that the blower can generate is maximal when it is occluded (zero flow) and that this pressure progressively decreases as the generated flow increases, the performance of the blower as pressure generator was measured when powered at 12, 15 and 24 V. Figure 6 shows the corresponding pressure-flow relationship. When the blower is powered at 15 V, the maximum pressure it can generate is 30 cmH_2O, which ensures a safe high-pressure limit for non-invasive ventilation. Interestingly, for flows up to 100 l/min (1.67 l/s)—which could account for peak inspiratory flow plus intended leak flow through exhalation port plus potential unintended leak caused by a poor fitting of the nasal mask on patient's

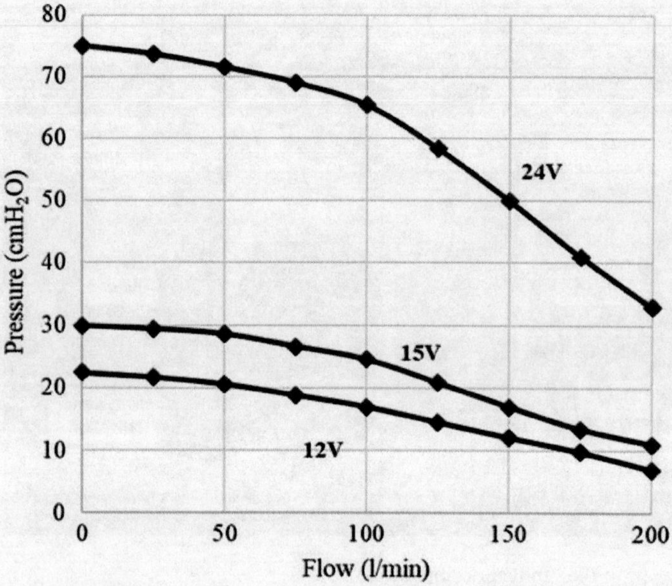

Fig. 6 Top view of the ventilator components. Performance of the blower as a pressure generator. Pressure-flow curves when the bower was powered by 12, 15, and 24 V

skin face, the generated pressure decreases to 25 cmH_2O, which is entirely sufficient for a non-invasive pressure support ventilator. For the same generated flow, pressures are 17 and 66 cmH_2O when the blower is powered at 12 and 24 V, respectively. These data suggest that the same blower (if powered at 24 V) could produce the level of pressure that would be required to implement an invasive mechanical ventilator for intubated ventilated patients (particularly taking into account that in this case there is no intended leak and unintended leaks are almost completely avoided). To control the output pressure, a pulse width modulation (PWM) control together with a proportional-integrative-derivative (PID) controller are used [15].

3.2 *Pressure Transducers*

Low-cost vacuum (absolute pressure measurement) pressure sensors can be either capacitive or piezoresistive. Although capacitive transducers present a typical sensitivity 50 times higher than piezoresistive ones [16], their acquisition circuit is more complex and expensive. The pressure sensors chosen were then piezoresistive (XGZP SOP6), P1 and P2 in Fig. 1, together with a specific ASIC for acquisition, sensitivity pre-calibration settings, and auto-offset correction (XGZP6847005KPG, CFSensor, Wuhu, Anhui, China) [14]. They present enough accuracy (± 1% full range, which is 50 cmH_2O) for the ventilator device.

3.3 Flow Measurement

Flow is sensed with a pneumotachograph consisting of a slight constriction in the tube section separating the two pressure transducers (Fig. 1). Given the well-known quadratic pressure-flow relationship in this type of resistors, the pressure drop signal measured across the resistor is linearized by computing its square root. In addition to low-pass filtering for noise reduction, the flow signal is high-pass filtered to eliminate the contribution of the quasi-static flow through the exhalation port. Given that cycling (shifting to expiration) is determined when the inspiratory flow reaches a value below a certain percentage (usually 30–50%, selectable by the user, Fig. 2) of the inspiratory peak, the flow signal does not require calibration of the resistor. This fact simplifies the construction process since the exact dimensions of the orifice-like resistor are not critical. However, it should be mentioned that if the resistor impedance is too low, it would provide a low-amplitude flow signal, which would be more noise-sensitive. By contrast, a too constricted resistor, while optimizing signal-to-noise ratio, could increase the mechanical load imposed to the blower and hence could reduce the maximum (inspiratory) pressure achievable at the nasal mask in case of maximum inspiratory flow and high intended and unintended air leak levels. The built-in resistor (a 10-mm diameter constriction within a 14-mm internal diameter tube; in the 3D printer construct) presents dimensions which provide a reasonable balance between signal-to-noise ratio and mechanical load. Interestingly, a potential way for improving the ventilator could be to replace the uncalibrated non-linear resistor by a linear pneumotachograph (for instance one based on a mesh wire/nylon screen). This change, together with proper calibration of the pressure-flow relationship in the exhalation port, would allow to actually measure flow, and hence inspiratory volume. However, this potential improvement would incur an increase in cost and would require establishing and following a routine for periodic pneumotach calibration.

3.4 Microcontroller

The program routine has been implemented to run in a simple 8-bit microcontroller: Arduino Nano based in a 16 MHz Atmega328 microprocessor (Fig. 5) [14]. It incorporates eight analog inputs (five are currently used in the implementation for the two pressure sensors and the three controls for 'P-ins', 'P-esp' and '%cycling'), 16 digital I/Os (three used) and 6 PWM outputs (just one used to control the blower). With 32 kB of memory, it is enough for the ventilator application and new functionalities can be added like wireless communication (for instance by using the HC-05 Bluetooth module) with a smartphone/computer for data monitoring or to substitute the LCD display.

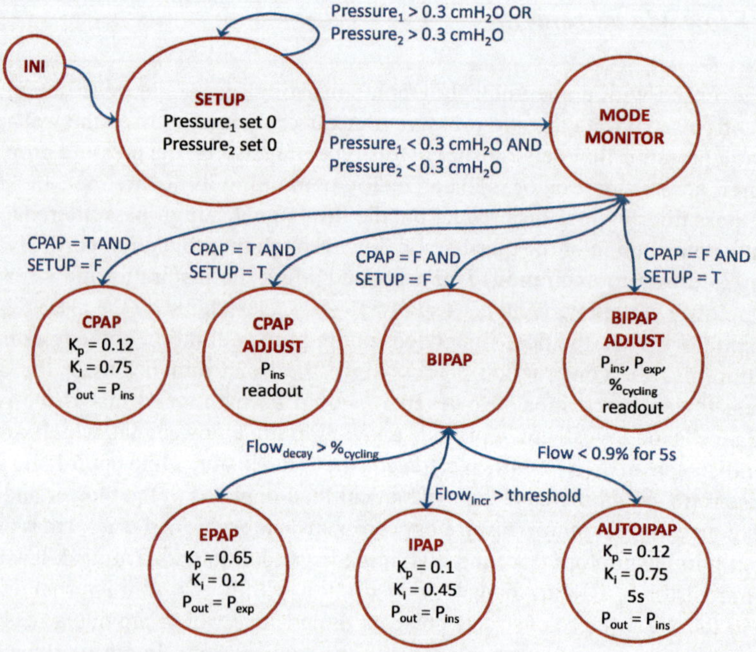

Fig. 7 State control diagram (see text for explanation)

3.5 Operator Interface

For simplicity, and to be able to work as a stand-alone equipment (without the need to use a computer, tablet or smartphone), the ventilator incorporates a low-cost LCD display (SODIAL(R) LCD display model 015085 with HD44780 controller) and potentiometers and switches to set up the different parameters of the ventilation (Fig. 5) [14]. However, they could be replaced by a USB connection if a PC is used to control the ventilator.

3.6 Ventilator State Control

The code for the Arduino microcontroller has been kept as simple as possible [14]. In this way, future improvements of the software (such as adding wireless communication or incorporating a linear pneumotachograph) would not need to upgrade the ventilator with a microcontroller with more memory. The program is based in a Moore state machine (for simplicity and upgradability of the code; to reduce the number of states and the memory occupied, a Mealy machine could be implemented)

[17]. The diagram of the implemented program can be seen in Fig. 7. After initialization, the pressures sensors are zero-corrected (until the pressure read from both is lower than 0.3 cmH_2O), and then the machine transitions to the mode monitor state. In addition to CPAP (constant positive airway pressure), there are two main modes, the BIPAP main working mode and the ADJUST mode, where the values for 'P-ins', 'P-esp' and '%cycling' are read from the inputs set by the operator. In the main BIPAP state, the parameters for the PID controller and the selection of the output pressure ('P-esp' or 'P-ins') are determined by the flow and pressure decays as shown in Fig. 4. If the flow is very low (0.9% of the measurement range) for more than 5 s, the AUTOIPAP state is activated.

From the technical point of view, the design of the device has been kept so simple that the ventilator is robust enough to work with minimal maintenance. Moreover, all the components are through-hole, and hence repair of the device can be easily done with the only aid of a standard multimeter and an iron solder. No internal adjustment of potentiometers or other components is needed (except for the contrast of the display), so it can be ready-to-use once built and tested by technical personnel. The sampling frequency has been kept as small as possible (100 Hz) while maintaining good performance of the ventilator. In this way, heating of the microcontroller is as low as possible to reduce the risk of malfunctioning and hence to length its lifetime. Similarly, the blower is always working well below its maximum power. Flow and pressure signals are low- and high-pass digitally filtered. Notch filters can be easily implemented in case the ventilator needs to work in environments with electromagnetically noisy equipment (in this case, we also recommend to build the device in a grounded metallic enclosure). The choice for the powering of the device is an external 15 V transformer for two main reasons: it reduces the electronic noise inside the device, and it allows the device to be used in countries with different AC voltages in the power line without the need to make any change in the device design.

4 Suitability of the Ventilator for Low Resource Regions

It goes without saying that patients with respiratory failure must be ventilated with conventional devices when this option is possible. However, when lack of resources generates a situation that precludes treating the patient, with the ensuing consequences, it is ethically mandatory to find a contingency solution. In this context, the ventilator design presented here seems particularly suitable when conventional commercialized devices are not available in under resourced regions or in emergencies because of pandemic respiratory infections. It is interesting to note that this ventilator was designed by a multidisciplinary team composed of experienced investigators covering all the involved scientific disciplines: biomedical engineering, respiratory physiology, physiotherapy and clinical respiratory medicine. Such a collaboration, which is important to design conventional ventilators, is fundamental when the aim is to produce a low-cost ventilator for under-resourced regions since in this case it is crucial to strike an optimal balance between acceptable performance and maximal simplicity and robustness. In this context, it is remarkable that the technical solutions taken in the design of the ventilator resulted in a device that performs adequately, as shown when a prototype was recently evaluated with the usual procedures employed in the literature to assess ventilators before their clinical application [18]. Indeed, a bench test on simulated patient models and a test in healthy volunteers mimicking patients with respiratory mechanics alterations showed that the prototype performs similarly to a high-quality commercially available device aimed at the same application.

The designed ventilator is suitable for being assembled by engineering hospital services or by local technical partners since it is made with off-the-shelf, low-cost components (75 US$). It is based on a modular structure (Fig. 5) with compact interconnecting electrical wiring thereby enabling simple replacement of each module independently. Interestingly, the device does not require calibration, and when it is switched on, an automatic routine tests and corrects any offset of the transducers. The maintenance required by the ventilator is minor since, contrary to linear pneumotachographs, the orifice-based resistor employed to assess breathing flow is minimally sensitive to dust deposition, which could be of interest in some setting where emergency devices are used. Noteworthy, the hardware design and the source code for the Arduino microcontroller of the ventilator is distributed in open source [14], thereby allowing for simple updating of the ventilator functions in LMICs, with locally or remotely updated code. For instance, the ventilator control can be adapted to incorporate cheap high or low-pressure acoustic alarms or to take into account the pressure drop induced by a humidifier in case this component is connected between the ventilator output and the tube connecting to the patient's mask.

The approach proposed here to help in providing LMIC patients with ventilators is different from the one consisting of philanthropic donation of commercially available medical devices. Although expensive, this humanitarian support is very effective to provide urgent solutions, but it has medium-long term limitations since it has been reported that up to 50% of donated devices become unusable due to lack of adequate maintenance and inability to obtain spare parts [19]. Moreover, to be sustainable such donations require long-term commitments such as to provide device servicing. The approach described herein tries to circumvent these disadvantages. Indeed, as the device is cheap, donated ventilators (and spare modules) assembled by external providers could be more effectively provided. But most importantly, the device proposed here is very simple, and thus can be assembled on-site by technical teams in LMIC, which is a rational approach to ensure proper servicing [20, 21] and, remarkably, to promote local industrial development [22, 23]. It should be mentioned, however, that the proposed prototype device has not been approved by official agencies in developed countries (e.g. FDA, CE) which requires a complex and expensive labelling process that would be generally unaffordable to LMIC. Alternatively, clinical authorization can be provided by local regulatory authorities or by hospital ethical boards. This is an open issue about the requirements for medical devices to be used in resource-poor areas, which requires to provide a solution to a difficult but fundamental ethical-legal balance. Whereas it is important that these devices have the same high standard levels as those approved for use in developed countries, it is also crucial that, until the market can provide commercial medical devices at affordable prices for LMICs, patients are not deprived of life-saving therapies.

5 Conclusions

A low-cost easy-assembly ventilator has been designed with open-source details available for enabling the provision of non-invasive respiratory support to patients in LMICs who otherwise would have no access to this potentially life-saving therapy under normal conditions and particularly during more strenuous circumstances such as those imposed by respiratory infection pandemics.

Acknowledgements The authors wish to thank Mr. Miguel A. Rodríguez-Lázaro (Unit of Biophysics and Bioengineering, University of Barcelona) for his excellent technical support.

Conflicts of Interest The authors have no conflict of interest to declare.

References

1. Rochwerg B, Brochard L, Elliott MW, Hess D, Hill NS, Nava S et al (2017) Official ERS/ATS clinical practice guidelines: noninvasive ventilation for acute respiratory failure. Eur Respir J 50(2):1602426
2. Schönhofer B, Sortor-Leger S (2002) Equipment needs for noninvasive mechanical ventilation. Eur Respir J 20(4):1029–1036
3. Dellaca RL, Veneroni C, Farre' R (2017) Trends in mechanical ventilation: are we ventilating our patients in the best possible way? Breathe 13(2):84–98
4. Gentile MA (2011) Cycling of the mechanical ventilator breath. Respir Care 56(1):52–60
5. Farré R, Navajas D, Montserrat JM (2016) Technology for noninvasive mechanical ventilation: looking into the black box. ERJ Open Res 2(1):00004
6. Mandelzweig K, Leligdowicz A, Murthy S, Lalitha R, Fowler RA, Adhikari NKJ (2018) Non-invasive ventilation in children and adults in low- and low-middle income countries: a systematic review and meta-analysis. J Crit Care 47:310–319
7. Murthy S, Leligdowicz A, Adhikari NKJ (2015) Intensive care unit capacity in low-income countries: a systematic review. In: Azevedo LCP (ed) PLoS One 10(1):e0116949
8. Truog RD, Mitchell C, Daley CQ The toughest triage—allocating ventilators in a pandemic. New Engl J Med (in press). https://doi.org/10.1056/NEJMp2005689
9. Pearce JM (2017) Maximizing returns for public funding of medical research with open-source hardware. Health Policy Technol 6(4):381–382
10. Eslambolchilar P, Thimbleby H (2016) Open-source hardware for medical devices. BMJ Innov 2:78–83
11. DePasse JW, Caldwell A, Santorino D et al (2016) Affordable medical technologies: bringing Value-Based Design into global health. BMJ Innov 2:4–7
12. Mackintosh M, Tibandebage P, Karimi Njeru M et al (2018) Rethinking health sector procurement as developmental linkages in East Africa. Soc Sci Med 200:182–189
13. De Maria C, Mazzei D, Ahluwalia A (2014) Open source biomedical engineering for sustainability in African healthcare: combining academic excellence with innovation. In: Proceedings of the ICDS 2014, the eighth international conference on digital society, 23–27 Mar 2014. Barcelona, Spain, pp 48–53

14. Garmendia O, Rodríguez-Lazaro MA, Otero J et al. Open source technical description of the ventilator. https://erj.ersjournals.com/highwire/filestream/163802/field_highwire_adjunct_files/0/ERJ-00846-2020.Technical_Description.zip
15. Kim S-H (2017) Chapter 2—Control of direct current motors. In: Kim S-H (ed) Electric motor control. Elsevier, pp 39–93
16. Sikarwar S, Satyendra SS et al (2017) Review on pressure sensors for structural health monitoring. Photonic Sens 7:294–304
17. Gomez M (2008) Chapter 6—Embedded state machine implementation. In: Ganssle J, Ball S, Berger AS, Curtis KE, Edwards LARW, Gentile R, Gomez M, Holland JM, Katz DJ, Keydel C, La Brosse J, Meding O, Oshana R, Wilson P (eds) Embedded systems. Newnes, Burlington, pp 247–257
18. Garmendia O, Rodríguez-Lazaro MA, Otero J et al (2020) Low-cost, easy-to-build non-invasive pressure support ventilator for under-resourced regions: open source hardware description, performance and feasibility testing [published online ahead of print, 2020 Apr 20]. Eur Respir J 2000846
19. Howie SRC, Hill SE, Peel D, Sanneh M, Njie M, Hill PC et al (2008) Beyond good intentions: lessons on equipment donation from an African hospital. Bull World Health Organ 86:52–56
20. Farré R, Trias G, Solana G, Ginovart G, Gozal D, Navajas D (2019) Novel approach for providing pediatric continuous positive airway pressure devices in low-income, under-resourced regions. Am J Respir Crit Care Med 199(1):118–120
21. Farré R, Montserrat JM, Solana G, Gozal D, Navajas D (2019) Easy-to-build and affordable continuous positive airway pressure CPAP device for adult patients in low-income countries. Eur Respir J 53(5):1802290
22. Clifford KL, Zaman MH (2016) Engineering, global health, and inclusive innovation: focus on partnership, system strengthening, and local impact for SDGs. Glob Health Action 9:30175. ds_2014_2_40_10173
23. Pearce JM (2020) A review of open source ventilators for COVID-19 and future pandemics [version 1; peer review: 2 approved, 1 approved with reservations]. F1000Research 9:218. https://doi.org/10.12688/f1000research.22942.1

Implementation of a Drum Collector for Electrospinning Machines Based on Embedded System

N. T. H. Phuc, Do Minh Thai, Thi-Hiep Nguyen, and Vo Van Toi

Abstract Nowadays, there has been an enormous rise in the request for polymeric nanofibers, which are encouraging candidates for numerous applications, namely blood vessels, drug delivery, tissue engineering, etc. Therefore, the development of electrospinning is also compulsory for the growing of tissue engineering area in Vietnam. The majority parts of electrospinning involve a collector, a syringe pump, a DC-high-voltage source. In this research, we aim to implement a low-cost grounded collector that is a crucial component in an electrospinning machine. The mechanical design includes a collector, ground facility and electrical control panel box. The hardware consists of motor driver boards. The firmware is developed based on stm32f103rbt6 microcontroller IC with the PID algorithm to automatically apply an accurate and responsive correction to a speed control function when the desired speed is the unequal set speed with the error about $\pm$ 3 rpm. The rotational speed can be adjusted from 0 to 2000 rpm for both random and aligned nanofiber applications. As expected, the prototype is a user-friendly and cost-effective platform; its cost is much less than the commercial laboratory devices one. The system has been undergoing experimental electrospinning processes with different polymer concentration at different collector speed. This system is also examined the quality and investigate in a set of electrospinning parameters that suit a variety of applications requiring different average diameter and the fiber's alignment degree.

Keywords Electrospinning · Grounded collector · High voltage power supply

---

N. T. H. Phuc · D. M. Thai (✉) · T.-H. Nguyen · V. Van Toi
School of Biomedical Engineering, International University, Ho Chi Minh 700000, Vietnam
e-mail: dmthai@hcmiu.edu.vn

Vietnam National University Ho Chi Minh Cityity (VNU-HCMC), Ho Chi Minh 700000, Vietnam

V. Van Toi et al. (eds.), *8th International Conference on the Development of Biomedical Engineering in Vietnam*, IFMBE Proceedings 85,
https://doi.org/10.1007/978-3-030-75506-5_5

1 Introduction

Because submicron-sized fibers have played an essential role in produce fiber from submicron to nanoscale regarding applications, including those in protective clothing, pharmaceutical, tissue engineering, and regenerative medicine, scientists are fascinated in investigating and progressing nanotechnology [1, 2]. Electrospinning is advanced nanotechnology, especially for the fabrication of various polymeric materials, including natural and synthetic polymers. The technique has been using a high voltage field to create nanofibers [3]. These ultrafine fibers have a crucial role due to the relationships of features of the extracellular matrix inside human tissue.

1.1 The Artificial Approach to Cure Wound Healing

With the aid of progress in tissue engineering, the production and study of artificial cellular via scaffolds are generated by biocompatible matters which assist would healing. This way is a right method of skin reconstruction [4], and the structure of scaffolds play a crucial role in the generation of tissues. These membranes are ordinarily inserted with high growth factors. The scaffold novel membrane is gradually restored by a venerable tissue. Thus, the structure of implant cells can be applied to integrate tissues, and then it is implanted in vitro to eliminate wound tissues inappropriately. The implantation is immediately into the lost tissue in human beings. This procedure is illustrated in Fig. 1.

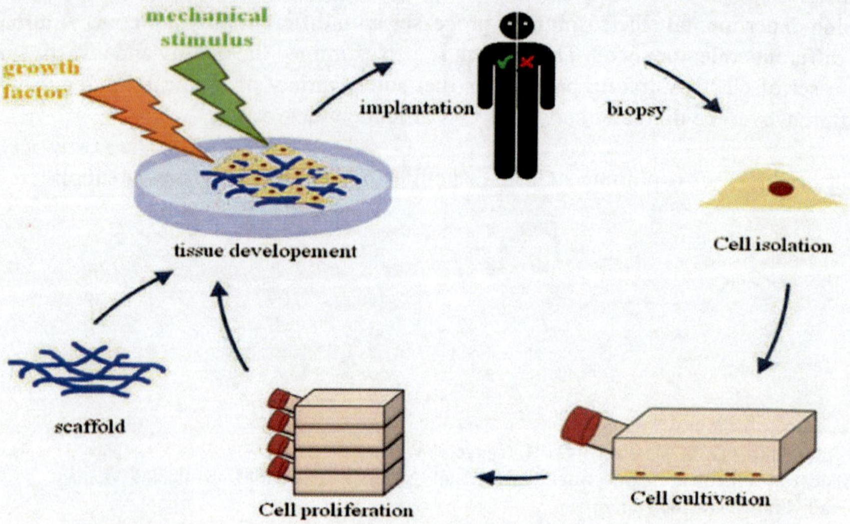

Fig. 1 Demonstrated the manner of synthetic skin development on the scaffold [4]

1.2 The Principal of Electrospinning Machines

Over the past two decades, it is witnessed the blossom of the conventional electrospinning method. A variety of derivative techniques have appeared, namely needleless electrospinning, multijet electrospinning, electro-blowing, bubble electrospinning, melt electrospinning, coaxial electrospinning, force spinning, etc. [5]. The essential electrospinning allows researchers to investigate a low-cost, adaptable and portable electrospinning setup consisting of three central elements: an emitter attached to a positive-high-voltage power supply, a collector connected to a negative high-voltage power supply, and a solution feeding system such as a syringe pump. The principal of electrospinning is demonstrated in Fig. 2.

By utilizing this system, numerous polymers and composite matters have a promising to be used in electrospun systems. This method is really simple and cost-effective. Nevertheless, in Vietnam, there is just several types of research associated with designing electrospinning machines. So, this machine which we aim to develop plays a vital role and remarkable influences on Vietnam's tissue engineering [6]. In this paper, we used a high voltage power supply which is designed in our previous project [6] to investigate the condition of nano- membrane fabrication.

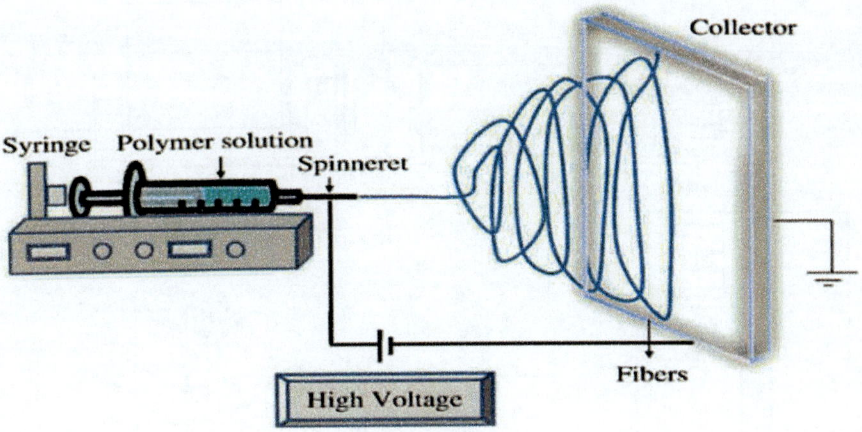

Fig. 2 Electrospinning system for nanofiber production [2]

2 Designing of a Drum Collector for the Electrospinning Machine Mechanical Design

2.1 Mechanical Design

The mechanical design includes a collector, ground facility and electrical control panel box.

The rotating drum collector parameters:

- The external diameter of the drum: 100 mm
- The internal diameter of the drum: 15 mm
- The length of the drum: 200 mm
- The length of the shaft: 45 mm.

The ground facility is made of iron material and is very good at negative charge conductivity via a crocodile clip attached to the designable metal holder. The electrical control box panel displays the value of set speed and real speed. This design is showed in Figs. 3 and 4.

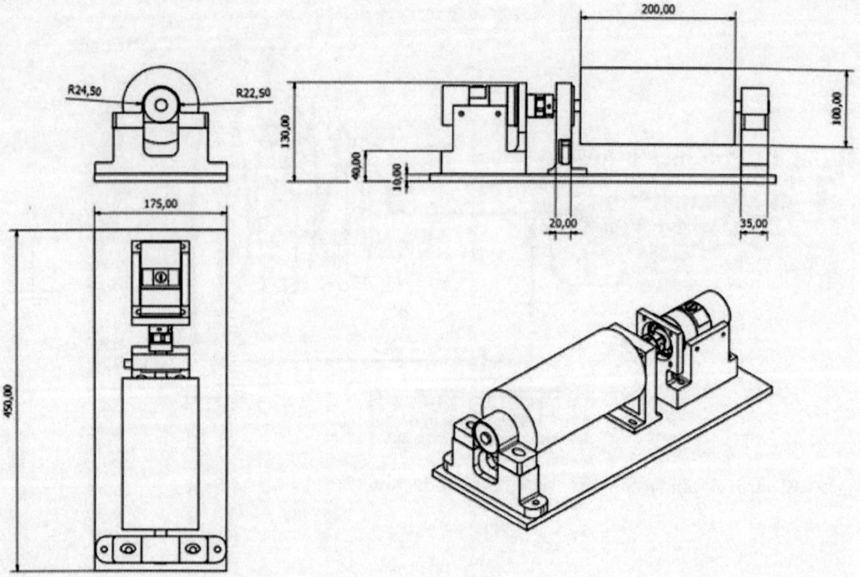

Fig. 3 Mechanical design by SOLIDWORKS

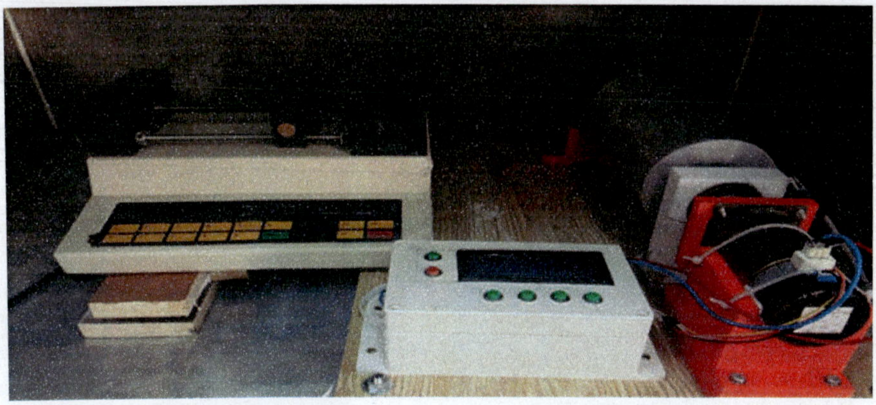

Fig. 4 The drum collector system

2.2 *Hardware Development*

Power Supply Block: The system requires + 3.3 V, + 5 V, and + 24 V DC power, the main power supply is taken from DC adapter + 24 V need to convert to lower voltage to perform the microprocessors and the rest components on board. To this end, the isolated DC-DC B2412S-1W is taken into consideration to convert from + 24 V to + 12 V and overcome the electromagnetic noise of the high voltage power lines. They then turn to + 5 and + 3.3 V by LM2576-5v and AMS1117-3.3v regulator executing respectively. This circuit is illustrated in Fig. 5.

Microcontroller Block: The design of control systems of DC servo motor based on the embedded microprocessor STM32F10x showed in Fig. 6, and the embedded integrated develop environment MDK. By executing powerful features, ARM Cortex-M3 chip achieves throughputs approaching a higher speed which guarantees us a more effective response in a motor control system allowing the system designer to optimize power consumption versus processing speed. The two timers TIM1-CH4 and TIM2-CH3 are used for PWM generating. Encoder mode is active by timer encoder TIM3 and TIM8 respectively.

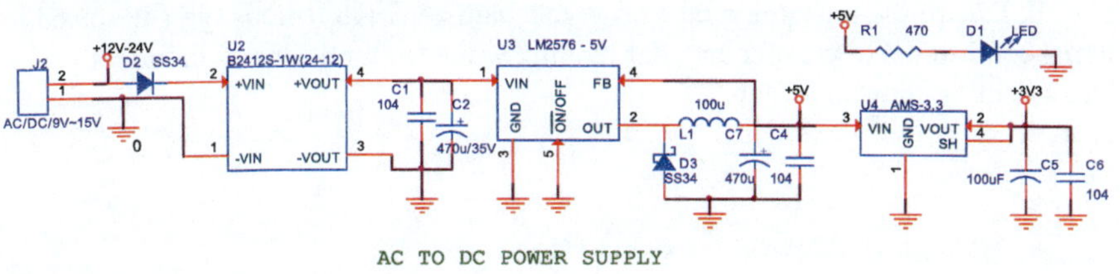

Fig. 5 Power supply circuit

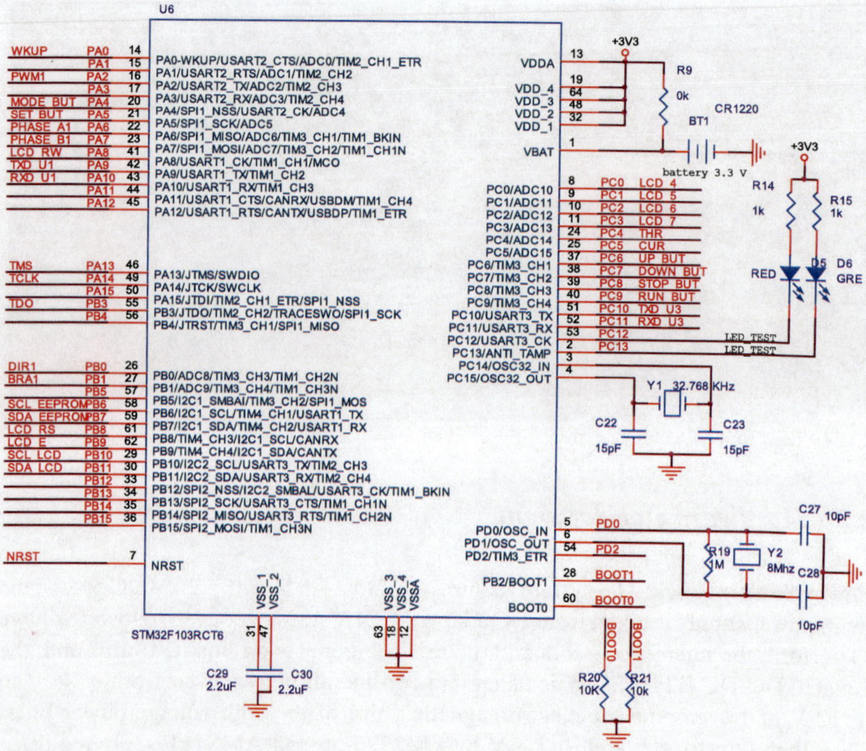

Fig. 6 The microcontroller circuit

Servo Motor Drivers Block: The servo motor has a rotation detector (encoder) mounted on the back-side shaft of the engine to detect the position and the speed of the rotor. The LMD18200T chip provides the intuitive visual in pin configuration. A most basic way would be to utilize only three pins: DIR, PWM, and BREAK; they then become more accessible to program the firmware for operating. Hence, the codes for controlling the motor driver will be simplified and easy to debug.

Optical isolator TLP280-4 is being to isolate two separated low voltage (+5 V and + 3V3) and high voltage in order to remove electrical unwanted noise and collision as well. Due to the unstable motor drive mechanism, 74HC14D is typically used in signal conditional to prevent encoder reading value from missing or incorrect pulse. This circuit is showed in Fig. 7.

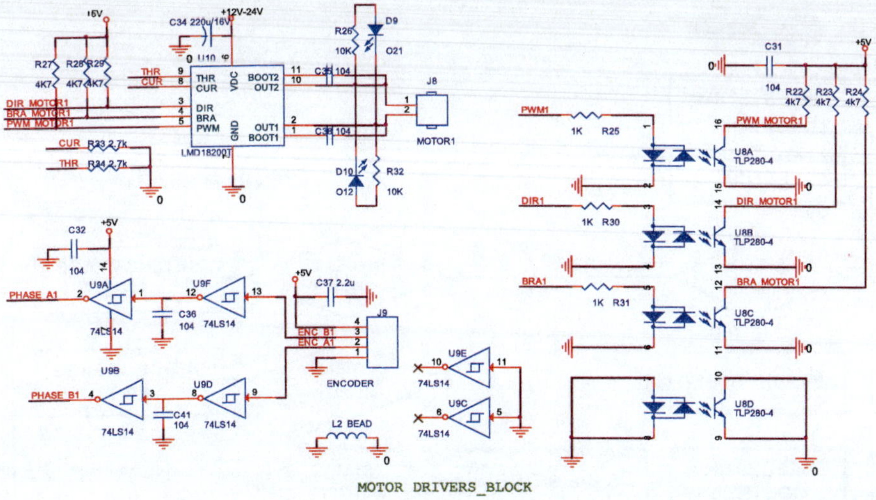

Fig. 7 Servo motor driver circuit

3 Methodology

After the set speed (unit: rpm) being entered from buttons, an optocoupler continues to be treating these driving signals, intending to eliminate undesired electrical collision. The LMD18200T controls the DC servo motor by following accessible programming features: PWM, DIR, BRAKE. The Schmitt trigger 74HC14D processes unwanted noises due to the mechanical system vibration.

Take the encoder of 200 pulses per revolution, so the formula has been implementing to estimate values of the real speed illustrated as:

$$\text{actualSpeed} = (\text{encoderPulse} * 6000)/200; \tag{1}$$

In which "encoder pulse"—the value is reading from timer every ten milliseconds. Most notably, the PID and Moving Average algorithm applied to enhance the accuracy and optimized automatic control. What is more, aiming at the increase of the accessibility of the following manipulation, the EEPROM was enhanced. While accessing this mode, the motor speed update automatically and drive the system. This flowchart is demonstrated in Fig. 8.

4 Result of Experiment

Many types of research have tried to interpret electrospinning parameters that affect fiber morphology. In working conditions of electrospinning, the morphology and

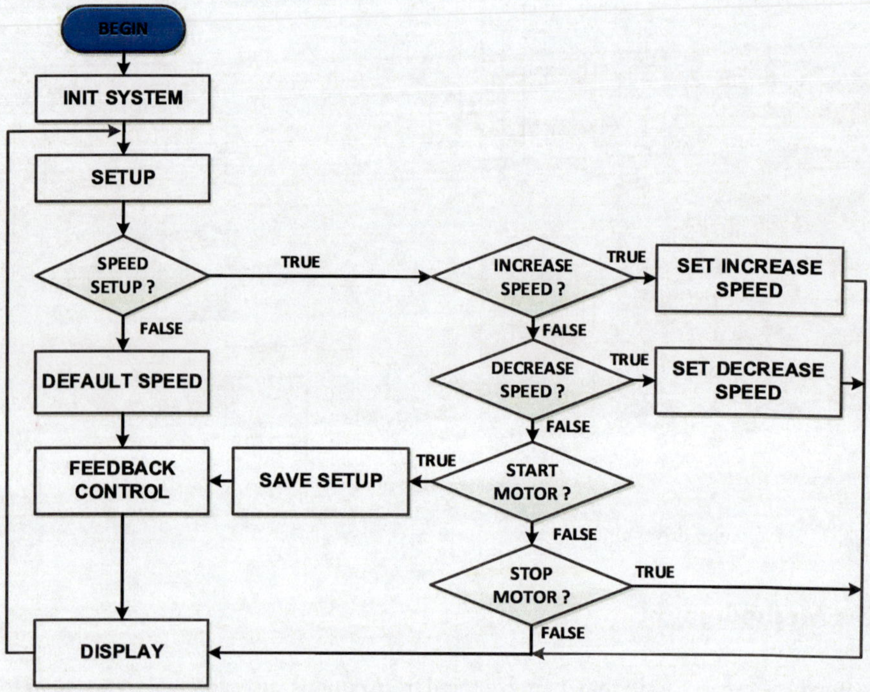

Fig. 8 The state machine diagram of drum control system

uniformity of nanofiber membranes are determined by investigating several listed factors: polymer solution parameters, setup parameters, and ambient conditions; Polymer solution parameters comprise solution viscosity, molecular weight, covering tension, dielectric constant, and solution conductivity while setup parameters contain the applied voltage, temperature, the flow rate, the needle diameter, and tip-to-collector distance and speed of collector. Besides, the ambient conditions involve humidity, pressure, air velocity in the chamber, and type of atmosphere. Diverse nanofibers with many morphologies, structures, and arrangements can be produced by means of judgment and varying these parameters.

4.1 Optimized Parameters

Optimization parameters are consolidated by investing the morphology of nanofibers under different electrospinning parameters which were listed below. This parameter is illustrated in Table 1.

Table 1 Electrospinning parameters setup

Process parameter	No. 1	No. 2	No. 3	No. 4	No. 5
Type of solvent used	CH_3COOH: $(CH_3)_2CO$ (7:3 ratio)	CH_3COOH: $(CH_3)_2CO$ (7:3 ratio)	CH_3COOH: $(CH_3)_2CO$ (7:3 ratio)	CH_3COOH: $(CH_3)_2CO$ (7:3 ratio)	$(CH_3)_2CO$
Polymer solution concentration	(20%) PCL	(22%) PCL	(25%) PCL	(30%) PCL	(22%) PCL
Flow rate (ml/h)	1	0.8	0.8	1	1
Applied voltage (kV)	15.4	15.4	15.4	15.4	15
Tip to collector distance (cm)	10	10	10	10	10
Needle type	20G	20G	20G	20G	20G
Number of samples	6	1	1	3	3
Speed of collector plate (RPM)	0, 300, 600, 900, 1200, 1200	0	0	0, 600, 1200	0, 600, 1200

4.2 *Morphology Analysis*

With applying the above parameters, the electrospinning was done, and the fibers were obtained on the rotating drum collector. In the spinning process of detailed above, a rotation rate between 300 and 900 rpm of the collector shows the fabrication of random nanofibers. From the aspect of aligned nanofibers production, the optimal rotation speed for the collector was found to be over 1000 rpm. The adjustment could easily be performed to any electrospinning machine setup with just partial change on the collector feature.

4.3 *Membranes Under SEM Image Observation*

Nanofiber Orientation

SEM images of random PCL nanofibers at a speed (a) 0 rpm, (b) 300 rpm, (c) 600 rpm and (d) 900 rpm respectively on 24th July 2019. (e) SEM images of aligned nanofibers at speed 1200 rpm on 24th July 2019. (f) SEM images of aligned nanofibers at a rate of 1500 rpm on 26th July 2019.

From Figs. 9 and 10, the fiber alignment in membranes can be effortlessly manipulated by altering the speed of the rotating drum collector. When the drum speed

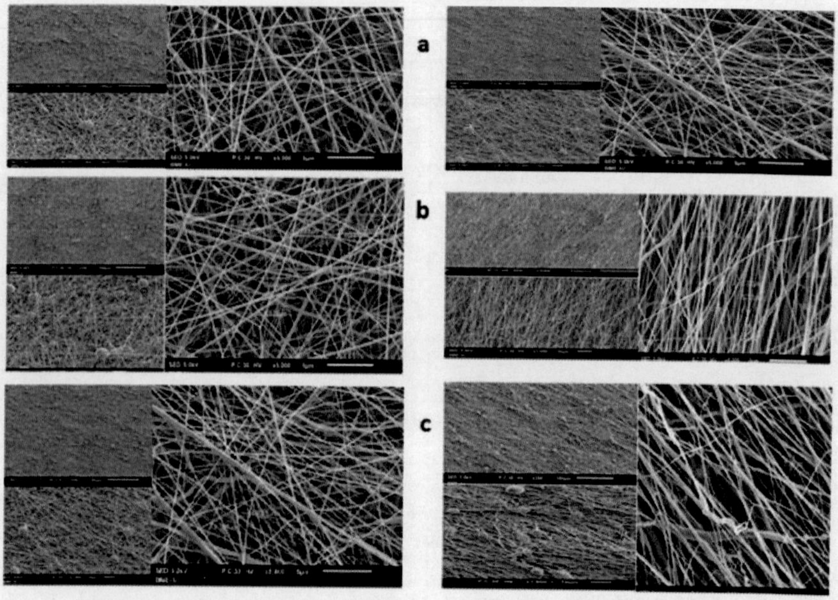

Fig. 9 Morphological analysis of PCL 22%

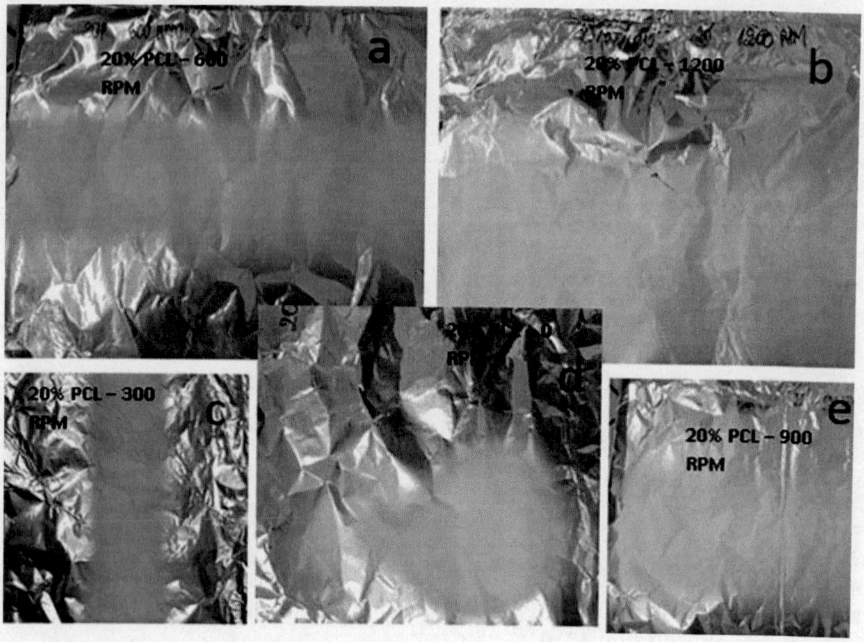

Fig. 10 Surface SEM of electrospun fiber mats at different speeds

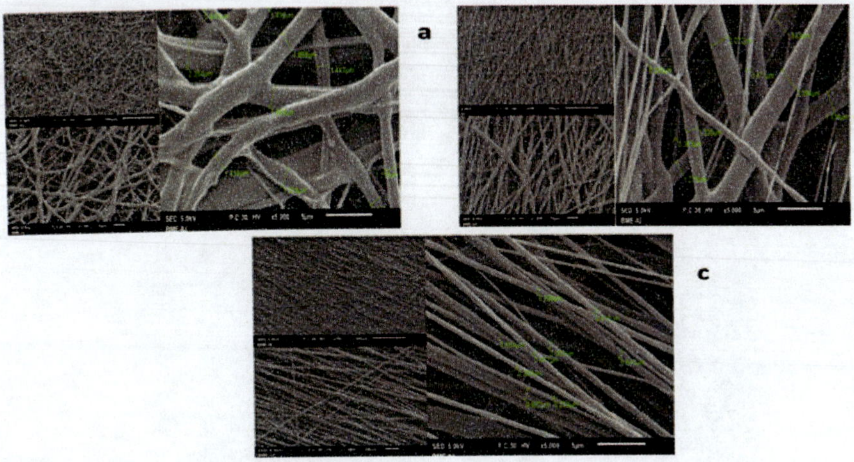

Fig. 11 Morphological analysis of PCL 30%

is slow, randomly oriented fibers are obtained. It generates a larger sample of the material than the static collector (Fig. 10d). However, by increasing the speed of the drum, the degree alignment (as Fig. 11) of fiber membranes improves. If the collector has controlled at higher rates over 1000 rpm, fibers are collected on the surface in the preferred route. This enables larger-dimensioned materials to be obtained, the internal structure of which is, on average, well-aligned.

SEM images of random PCL nanofibers at a speed (a) 0 rpm, (b) 600 rpm respectively on 07th Aug 2019, aligned nanofibers at a speed (c) 1200 rpm on 07th Aug 2019.

The occurrence of defects, such as beads, is a significant problem. The beaded fiber structure can be observed in comparison with the previous static drum on the same condition of solution properties. Hence, we tried to vary electrospinning parameters to prevent the formation of the beads. For the ground facility test, we set up the collector speed at 0 rpm and observed that there are no redundant nanofibers in the chamber well showed in Fig. 12.

Although collector under good negative charged conductivity and stationary, it is clear to observe the formation of beads. As per our discussion, the flow rate of the polymer in the syringe and solution concentration is taken into consideration as components affecting bead production. For detail, the low density and high flow rate resulted in the beaded fiber structure. At low solution concentrations, a mixture of fibers and beads are obtained. As the concentration increases, the shape of the bead changes from spherical to spindle-like (Fig. 14); in other words, the higher polymer concentration increases, the fewer beads form. Consequently, we increase the concentration of polymer solution concentration from 20 to 22%, 25% and 30% respectively and decrease flow rate from 1 to 0.8 ml/h; it is optimistic that there are decreasing in bead-on-string fibers even though beads still occurring (Fig. 13).

Fig. 12 Ground facility

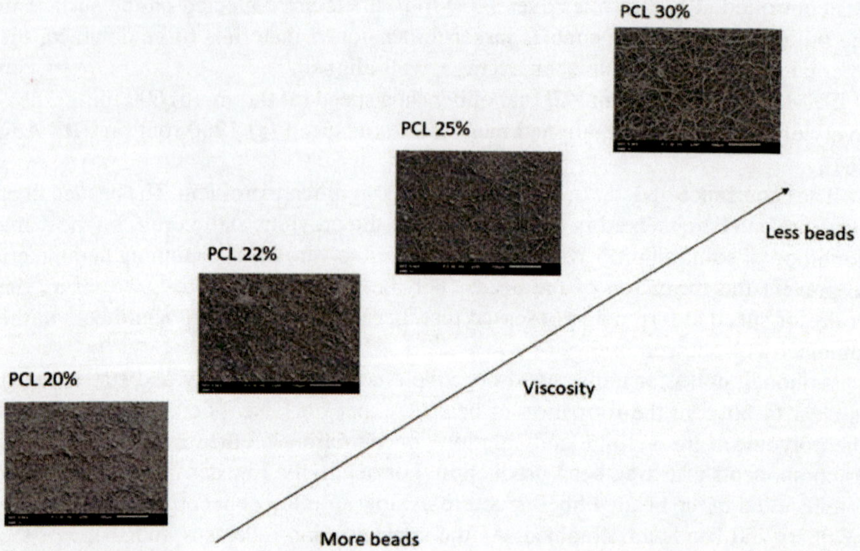

Fig. 13 SEM pictures of PCL nanofiber mat from different polymer concentration solutions at 250x

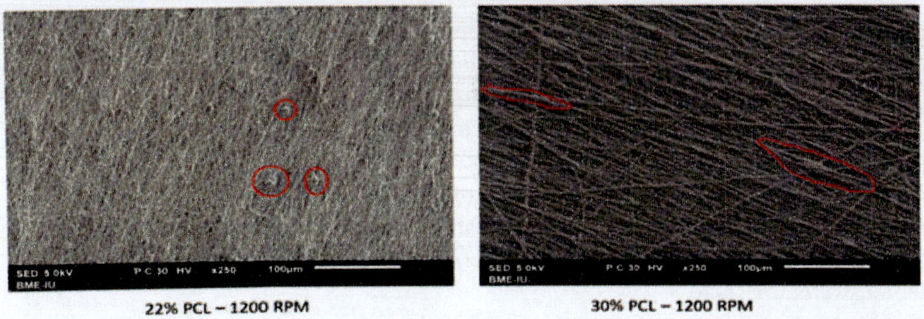

Fig. 14 SEM pictures of PCL nanofiber demonstrate the shape of the bead changes from spherical to spindle-like from increasing concentration solution at 250x

Then, we increased the polymer concentration from 20 to 22% but different solvent (acetone solvent) to test the effect of solution concentration and solvent's characteristic on beads formation and found that no beads were observed (Fig. 15). It can be explained by the effect of surface tensions. Different solvents may contribute to different surface tension. It obtained smoother nanofibers by reducing the surface tension and increasing the concentration, rather than beaded structure at higher surface tensions.

Nanofiber Diameter

Being observed by SEM, we gained real experience that the nanofiber diameter is proportional to the polymer concentration as higher polymer concentration leads

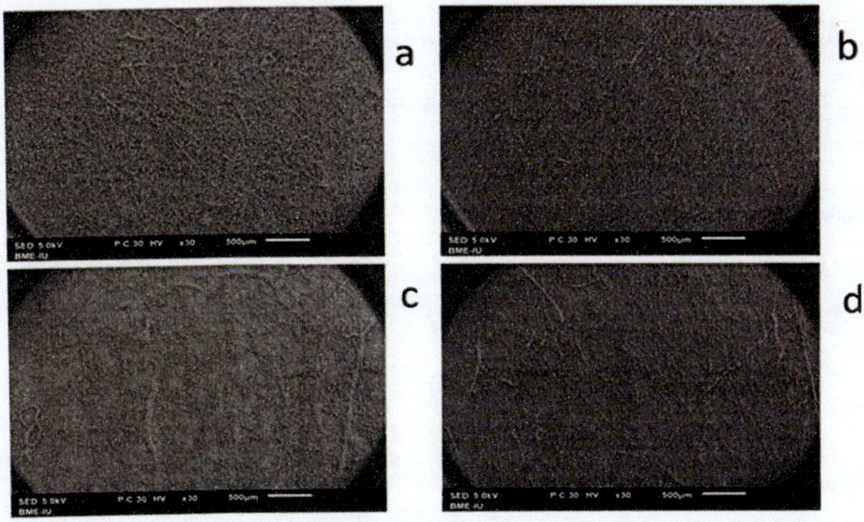

Fig. 15 PCL nanofiber from acetone solvent at 30x

to larger fiber diameter. What is more, keeping the concentration fixed, increasing collector plate speed will give a smaller diameter. The results were listed in Table 2.

Using acetone as dissolving solvent gained no beaded nanofiber mat but a much larger nanofiber diameter. However, due to lack of needle-tip cleaning equipment throughout PCL nanofiber electrospinning, acetone was used as a dissolving solvent. It is an inconvenience to use acetone as a dissolving solvent. As a final observation, it has been successfully created an adjustment of the electrospinning processing parameters to obtain desired fiber diameters and morphologies. This effect could be demonstrated in Table 3.

Table 2 Effect of motor speed on fiber diameter

Type of solvent	Motor speed (RPM)		Average diameter of the resulted fiber (nm)
$CH_3COOH:(CH_3)_2CO$ (7:3 ratio)	20% PCL + flow rate (1 ml/h)	0	234.4
		300	193.9
		600	190.6
		900	185.3
		1200	178.8
	22% PCL + flow rate (0.8 ml/h)	0	278.7
	25% PCL + flow rate (0.8 ml/h)	0	293.4
	30% PCL + flow rate (1 ml/h)	0	1706
		600	1028
		1200	530
$(CH_3)_2CO$ (100%)	20% PCL + flow rate (1 ml/h)	600	2413
	22% PCL + flow rate (1 ml/h)	1200	1930

Table 3 Effects of electrospinning parameters on fiber morphology

Parameter	Effect on fiber morphology
Motor speed	Fiber diameter , alignment
Polymer concentration (viscosity)	Fiber diameter (within optimal range), Bead formation
Solvent volatility (CH_3COOH)	Bead formation
Flow rate	Fiber diameter, bead formation

5 Conclusion

In conclusion, the experiment has brought a promising approach to support one option of the collection module of the electrospinning machine—Rotating Drum Collector. This particular design of Rotating Drum Collector is generally efficient to revolve in 2000 rpm in practical with a closed-loop system in aspect to obtain aligned polymeric fibers with a user-friendly interface and affordable cost.

Acknowledgements This research is funded by Vietnam National University HoChiMinh City (VNU-HCM) under grant number B2019-28-04.

Conflicts of Interest The authors have no conflict of interest to declare.

References

1. Call C (2009) The study of electrospun nanofibers and the application of electrospinning in engineering education
2. Zhu N, Chen X (2013) Biofabrication of tissue scaffolds
3. Boudriot U et al (2006) Electrospinning approaches toward scaffold engineering—a brief overview. 0160-564X (Print)
4. Killian M et al (2012) Recent advances in shoulder research, vol 14, p 214
5. Barhoum A, Bechelany M, Makhlouf ASH (2019) Handbook of nanofibers. Springer, Cham
6. Do TM et al (2020) A low cost high voltage power supply to use in electrospinning machines. In: 7th international conference on the development of biomedical engineering in Vietnam (BME7), Springer, Singapore

Development of a Low-Cost and Portable Real-Time PCR Machine for Developing Countries

Huy L. Ngo, Huy D. Nguyen, Viet N. Tran, and Hoan T. Ngo

Abstract Real-time Polymerase Chain Reaction (realtime PCR) is an important medical diagnostic technique in which target DNA is amplified and the amplification process is monitored in realtime. However, most realtime PCR systems are expensive, especially for laboratories in developing countries. To increase accessibility to the realtime PCR technique, we developed a low-cost and portable real-time PCR system. The system was composed of two main parts: a thermocycler and an optical system. The thermocycler's core was a Peltier and a heatsink fan, in tandem with an electronic circuit to control the heating process's temperature. A custom-made fluorescence optical setup was built using a blue LED chip (450 nm), an excitation filter, a dichroic mirror, an emission filter, and a CMOS camera. The results were displayed on a screen connected to a Raspberry Pi module. The thermocycler worked stably and accurately under the control of a PID controller. The system's performance was tested using samples of rice phospholipase D (PLD) gene and maize Bt11 gene. The results showed that, at the end of the amplification process, fluorescence signals from positive samples were significantly higher than from negative samples. Nucleic acid amplification products were verified using gel electrophoresis. With the cost of ~$320 USD, the system could be a suitable candidate for realtime PCR tests in laboratories in rural and remote areas of developing countries.

Keywords Polymerase chain reaction (PCR) · Real-time PCR · Molecular diagnosis · Disease diagnosis

1 Introduction

Polymerase Chain Reaction (PCR) was developed in the 1980s by Kary Mullis, whose pioneering work had awarded him the Nobel Prize in Chemistry in 1994. *Science* magazine said in 1989 that the PCR assay was "revolutionizing the approaches

H. L. Ngo · H. D. Nguyen · V. N. Tran · H. T. Ngo (✉)
School of Biomedical Engineering, International University, Vietnam National University, Ho Chi Minh City, Vietnam
e-mail: nthoan@hcmiu.edu.vn

V. Van Toi et al. (eds.), *8th International Conference on the Development of Biomedical Engineering in Vietnam*, IFMBE Proceedings 85,
https://doi.org/10.1007/978-3-030-75506-5_6

researchers are taking to many problems in biology." PCR is an in vitro method that amplifies specific DNA sequences using DNA polymerase's ability to synthesize a new strand of DNA complementary to a given template strand. Due to its feature to add a nucleotide onto a preexisting 3′-OH group, DNA polymerase demands oligonucleotide primers to which it can add the first nucleotide. This condition makes it possible to flank the region of interest in the target DNA. PCR is one of the most popular and robust assays in diagnostic labs to amplify DNA fragments and detect viruses like Ebola [1] and Zika [2]. However, conventional PCR assay requires post-PCR analysis [3], such as gel electrophoresis and image analysis to detect target amplicons' presence and quantity. The post-PCR analysis is labour-intensive and not suitable for high-throughput screening.

Realtime PCR, which is a modification of the PCR technique with the introduction of fluorescence reading during the amplification process, has overcome the aforementioned limits. The ability to monitor the amplification process in real-time makes real-time PCR superior to its predecessor. During the real-time PCR process, the amplification of target sequences generates an increase of fluorescent light directly proportional to the number of PCR amplicons yielded after each PCR cycle. This technique is the gold standard in detecting many different diseases and is very important, especially when the COVID-19 pandemic is threatening lives all around the world.

Nevertheless, in developing countries, real-time PCR is usually only available in laboratories in big cities due to its high cost. Currently, as the COVID-19 is spreading quickly with over 1.7 million infected cases and over 100,000 deaths (as of April 12 2020) [4], there is an urgent need for a low-cost real-time PCR system for laboratories in rural and remote areas of developing countries [5].

In an effort to address this need, we designed an affordable realtime PCR system made of a 3D-printed case and off-the-shelf electronics that provides real-time detection of PCR products. Our realtime PCR system consists of three main parts: a thermal cycler system to regulate temperature, an optical setup to both excite and capture fluorescence signal, and a graphical user interface for system control and result display. The whole size of the device is approximately 350 mm × 170 mm × 265 mm. We characterized our system by carrying out real-time PCR reactions. The results showed that the system successfully amplified target DNA and detected fluorescence signal, as verified by gel-electrophoresis results and the acquired fluorescence images.

2 Related Works

Regarding the development of low-cost and portable real-time PCR instrument, several research groups have achieved great advances in making fully functional low-cost real-time PCR systems [6–8]. Many of the groups depended on Peltier as the core of their thermocyclers for heating and cooling and used thermistor to measure temperature [6–11]. This proved to be an effective and economical way to perform the thermocycling process. Regarding optical set-up for fluorescence detection, different fluorescence excitation methods have been deployed, such as putting

LEDs on the sides of the test tubes for direct excitation [6–8, 11–13] or using LEDs array in combination with dichroic mirror for excitation from the top of the test tubes. Each method has its advantages and disadvantages. Despite the differences, the results of these methods are quite encouraging [6–13].

3 Materials and Methods

Our real-time PCR system consists of three main parts, including: (1) a thermocycler system; (2) an optical system; (3) an electronic system. Figure 1 shows the block diagram of the electronic system. At the heart of the electronic system is an MCU board that control the temperature of the Peltier-based thermocycler using PID algorithm. The temperature was monitored by a resistance thermometer PT100 (temperature range from −243 to 600 °C) connected to the MCU board. Sample solutions were excited by a LED light source, and the generated fluorescence could be detected in realtime by a camera. The system was powered by a 12 V power supply.

3.1 Thermocycler

Figure 2 shows 3D design of the thermocycler. The basis of the thermocycler is a 12 V, 150 W TEC1-12715 Peltier module and a computer heat sink cooler. One side of the Peltier was placed directly on the heatsink with fan blower to disperse

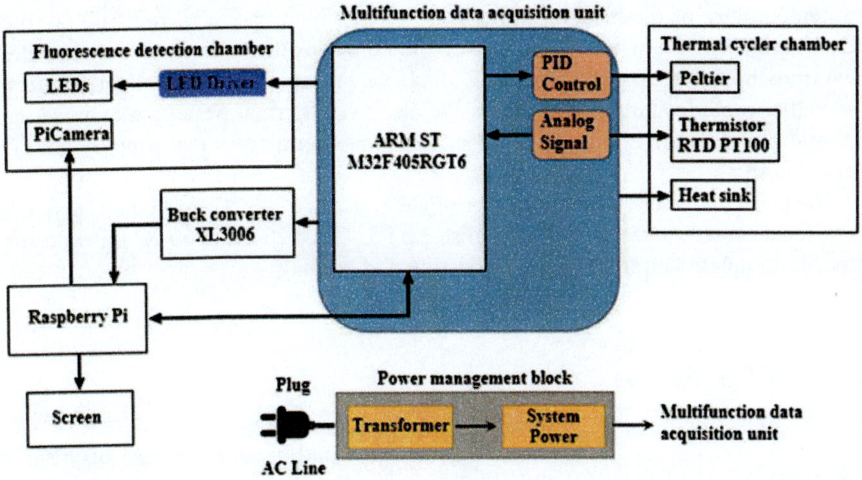

Fig. 1 Block diagram of the electronic system

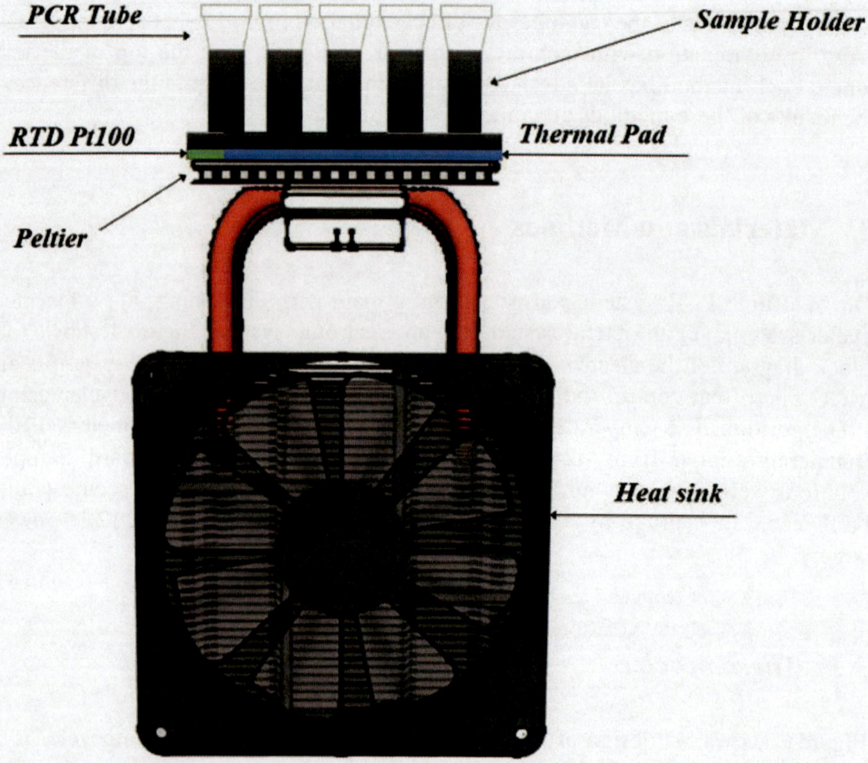

Fig. 2 The thermocycler of the system

the temperature of the heatsink. The heat block, which was salvaged from an old benchtop PCR instrument, was placed on the Peltier's other side. To evenly transfer heat from the Peltier module to the heat block, we put a piece of thermal pad between the Peltier module and the heat block. We also used thermal paste to maximize the thermal transmission effect. The temperature was recorded in real-time by a RTD Pt100 sensor.

The thermal cycler started with a pre-PCR process of 40 °C for 900 s. Then, the PCR reaction was performed with two steps: 95 °C for 20 s and 60 °C for 60 s; the process completed after 40 cycles of heating and cooling.

3.2 *Optical System*

Aiming to provide a uniform and intensive illumination over the field of view, a 7B10C COB blue LED with 470 nm center wavelength and power capacity up to 30 W was used for fluorescence excitation. Despite the fact that Xeon light source is chosen by many commercial PCR systems [14–16], we believe that LED is more compact, energy efficient, and more cost-effective for our purpose of building a low-cost device for developing countries. Moreover, the heat generated from Xeon lamp would require an additional cooling system that would increase the whole system's cost.

However, COB LED generates a divergent light beam. Therefore, to focus the light, a cone-shaped beam-restricting tube (Figs. 3 and 4) was designed as a simple light collimator to focus the light and enhance the uniformity of light illumination. The inner surface of the tube was painted with silver paint to increase reflectivity of the surface.

The use of the collimator significantly improved the uniformity of the illumination area created by the COB LED. Intensity maps of the illuminated area with and without the collimator were captured by a Samsung Galaxy S8 smartphone camera and plotted using Matlab. The light source was placed 120 mm away from the projected surface.

Figure 5 shows that, with the collimator, uniformity of intensity and magnitude was significantly improved.

Figure 6 shows the design of the optical system. Light from the blue COB LED went through an 480 nm excitation filter (bandwidth 30 nm), a collimator as described above, and reflected off a 500 nm high-pass dichroic mirror before exciting the samples. Emitted fluorescence with a wavelength over 500 nm will pass through the dichroic mirror, an 525 nm emission filter (bandwidth 15 nm) and captured by a CMOS camera. The emission filter's role is to suppress any residual blue excitation

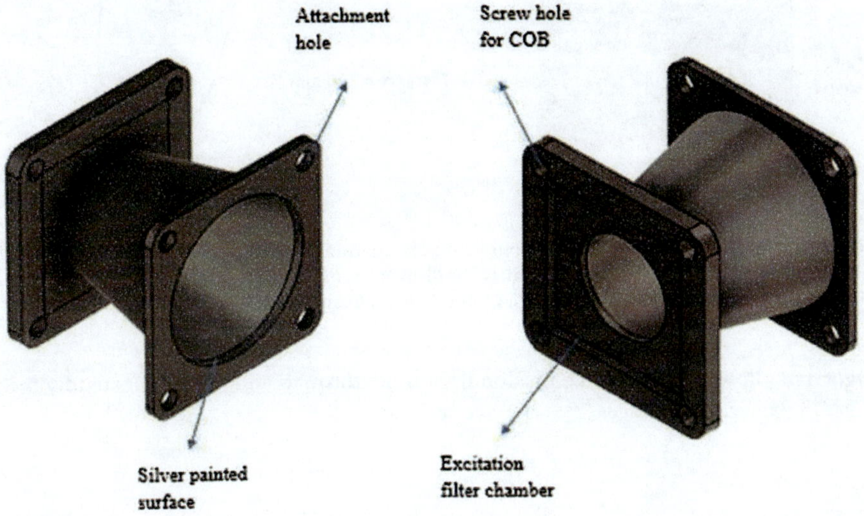

Fig. 3 The 3D model of the collimator viewed from two different angles. The design includes a cavity for excitation filter placement and holes for COB LED attachment. The inner surface of the cone is painted silver, enhancing light reflectivity

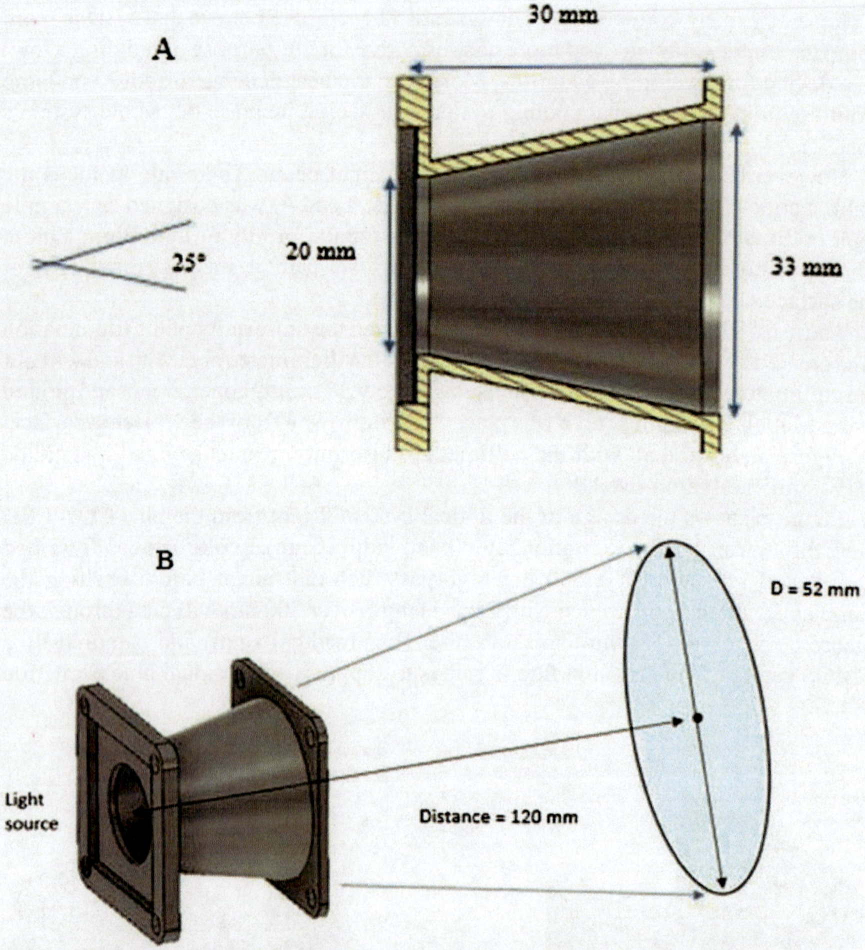

Fig. 4 **a** Cross section of cone-shaped collimator. The cone angle is 25°, the light aperture diameter is 20 mm and the output diameter is 33 mm. **b** The distance from the light source to sample is 120 mm. From the calculation, the circular illumination area has a diameter of 52 mm

light and allow fluorescence emission light to go through and to be collected by the CMOS camera.

3.3 Heat Lid

A heat lid is an important feature of a PCR instrument that prevents samples from condensing inside PCR test tubes. During the denaturation phase of a PCR cycle, the

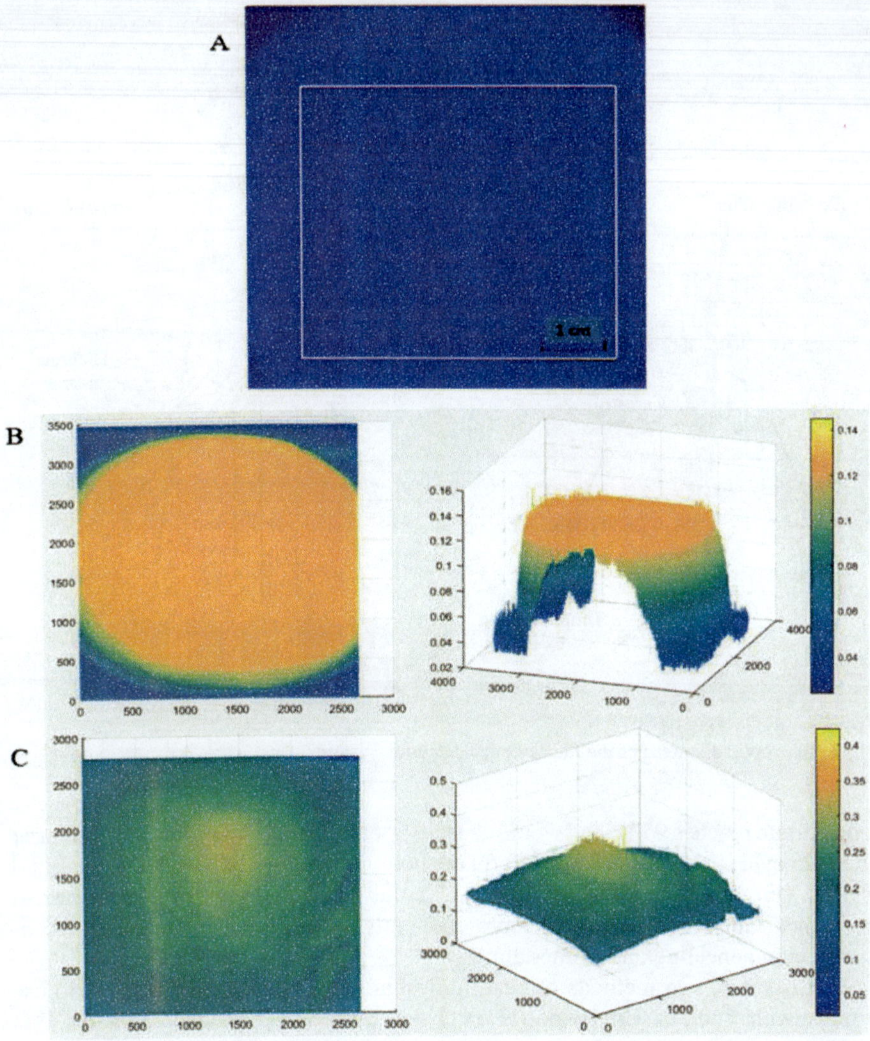

Fig. 5 **a** Area of illumination while attaching the collimator captured by Samsung Galaxy S8 smartphone camera. The white box represents the size of PCR heating block. **b** Total intensity 2D and 3D maps of the illumination area with collimator installed. **c** Total intensity 2D and 3D maps of the illumination area without attaching the collimator. The image is showed in the *xy* plane, the pixel intensity is demonstrated within the *xyz* 3D grid coordinates. The color bars demonstrate the pixel intensity of images. The intensity of the pixels represents the luminous intensity level of the lighting area

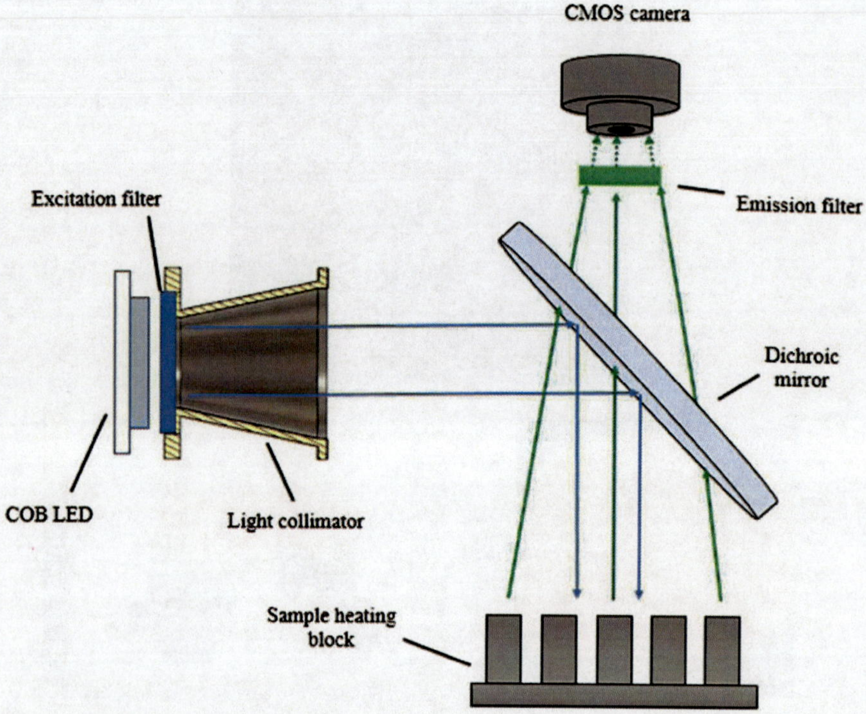

Fig. 6 The optical system of the fluorescence detection system

temperature reaches 95 °C and would cause evaporation and loss of sample, thus after about 25 cycles, there is nothing left on the bottom of the test tubes if no mineral oil is used. Our system has a heated lid to solve this problem. Furthermore, the heated lid allows samples to be heated more efficiently, which reduces the formation of nonspecific annealing and primer dimers.

In this study, two methods of heat lid design were considered. First was to use a transparent Indium Tin Oxide (ITO) coated glass as a heat generator. An ITO heater glass often operates between 80 and 130 °C, it allows 80% of light to transmit through. As a result, ITO heater provides a transparent heat lid that allows light to pass through. This method simplifies the design of the heat lid. However, during our experiments, ITO glass has shown its lack of stability: the temperature generated was not always stable and the soldering process of electric cable to ITO glass was arduous.

Furthermore, it was fragile when being pressed onto test tubes' caps.

Second method was to use an aluminium heater with holes that allow light to transmit through. Figure 7 shows the design of our heat lid. The main component is a Positive Temperature Coefficient (PTC) heater. PTC heater has high heating efficiency, low electricity consumption, cost-effectiveness, and self-regulating ability.

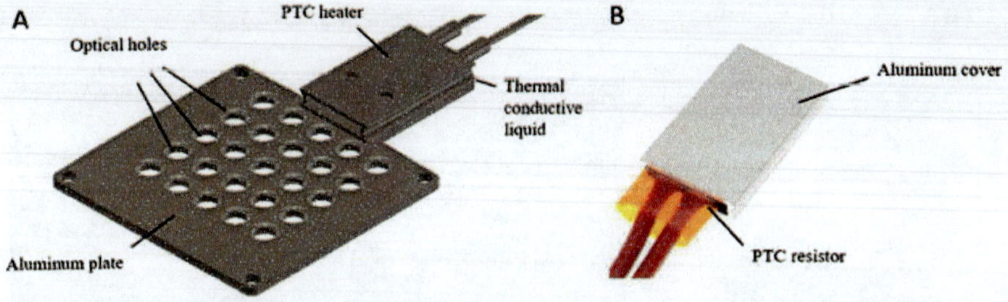

Fig. 7 **a** Heat lid design. An aluminum plate was screwed up with a PTC heater, with a thermal conductive layer in between. **b** PTC heater. PTC resistor is placed inside an aluminum cover, generating heat as electric current passes through

Temperature stability is a crucial reason that we chose PTC heater as the heat generator for our heat lid design. PTC heater was attached to an aluminum plate. Thermal conductive paste was applied between the two components in order to increase heat conductivity.

Heat distribution was not uniform (distant locations have a lower temperature) due to heat dissipation, but the average temperature was stable around 110 °C, which was enough to prevent condensation in test tubes.

4 Results and Discussion

4.1 Fluorescence Acquisition System

Our goal in this project was to develop a low cost and portable realtime PCR instrument that could be easily deployed in low-resource settings to assist the fight against infectious pandemics like Ebola [1], Zika [2] and COVID19 [17]. Our device is powered by an external power supply; however, a portable battery bank can be used as an alternative. Several research groups have reported portable real-time PCR instruments; yet to reduce the size, they relied on costly micro fabrication techniques to produce the heat blocks [7, 11] of the thermocycler. In contrast, our thermocycler leveraged two simple and low-cost components, a peltier module (US$4.3) and a computer heat sink ($30) to heat and cool down the heat block. Moreover, the heat block and the optical system was covered with a 3D-printed enclosure to prevent interference from the outside environment.

For detection of fluorescence signal, we utilized a COB LED ($2) and a CMOS camera ($30) along with an emission filter ($45), an excitation filter ($45) and a dichroic mirror ($60) to filter lights. In terms of controlling the device, an MCU board and a single-board computer, the Raspberry Pi, were used. Our instrument's weight is 3 kg, which can be easily transported and held by an adult. The complete

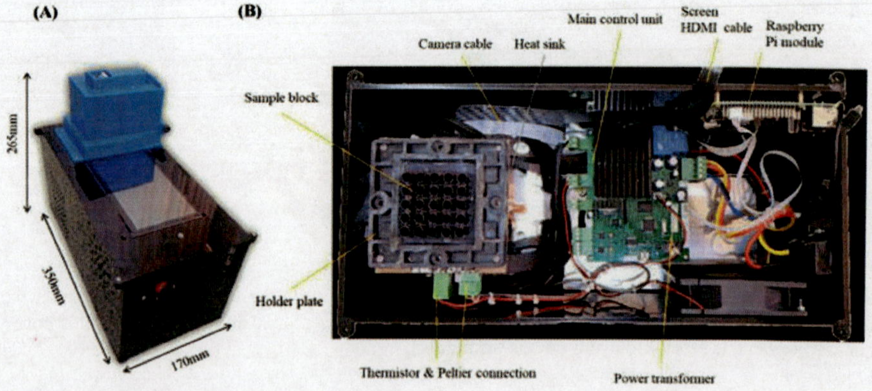

Fig. 8 **a** Our realtime PCR system looks from outside. **b** The internal structure of the device

system is shown in Fig. 8. In its current form, the total cost of the device is $320, which is much cheaper than commercial benchtop real-time PCR (>$4500) [18].

4.2 Temperature Control

To assess performance of the thermocycler and the PID controller, we implemented a PCR protocol that begins with a 15-min hot start at 95 °C, followed by 40 cycles of denaturing and annealing/extension stage at 95 °C and 60 °C, respectively.

Heating and cooling rate of the system could reach 1.4 °C/s (Fig. 9). A 40-cycle PCR reaction is finished in 100 min with 130 s for each cycle to be executed. Besides, the thermal system witnessed a few seconds of temperature overshoot in the first two minutes of the hot start stage. Then the control algorithm was able to cancel the temperature overshoot. To improve temperature ramp rate, Peltier may be replaced with a better quality one because the one that we were using has stated power of 150 W according to its datasheet. However, when it ran at full power, the Peltier was overheated and broken in several trials. So we ran the Peltier at only 120 W power (12 V and 10 A).

To analyze the accuracy of temperature control of the device, three PCR test tubes filled with 25 μL of distilled water were placed at random positions in the heat block during thermal cycling. The temperature of water inside the PCR tubes were measured using a thermocouple connected to a digital multimeter (True RMS Pro's kit MT-1707). Results showed that the system achieved good accuracy in controlling temperature of the solutions in the PCR test tubes (Fig. 10). As for the denaturation step in which set temperature was 95 °C, the actual temperature was 94 °C on average. As for the the annealing/extension step in which set temperature was 60 °C, the actual temperature was 59.8 °C on average.

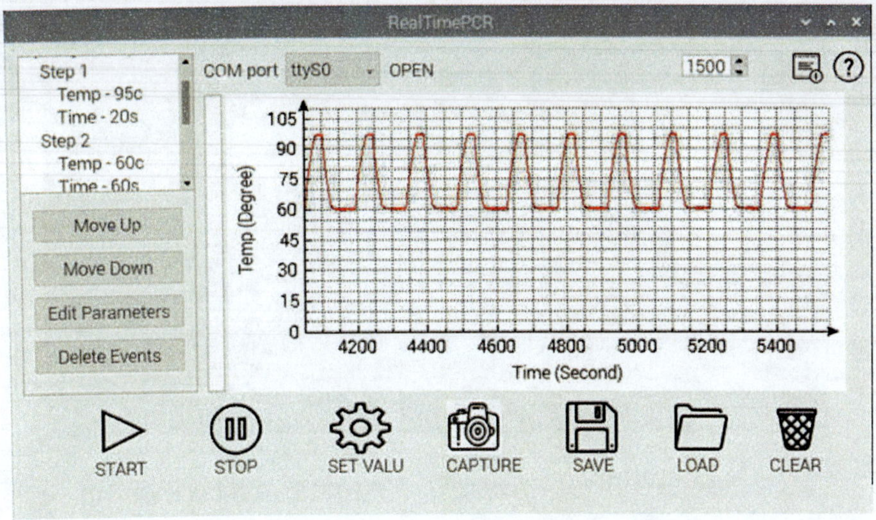

Fig. 9 Shows the thermal graph of the heat block as measured by an RTD Pt100

In terms of temperature ramp rate, our custom-built thermocycler has lower temperature ramp rate in comparison to commercial devices as shown in Table 1. However, our system has a much lower cost. The system could be easily adapted to perform isothermal assays which are gaining popularity recently [2, 10]. For example, Abbott has just released a 5-min diagnostic device for COVID-19 testing, called the Abbott ID Now™. The device uses an isothermal assay operating at around 56 °C for 5 min to give a positive result and 8 more minutes for negative one [17].

4.3 Optical Setup

The optical setup successfully detected fluorescent signal (Fig. 11). Figure 11a shows the captured image of 8 PCR samples with 4 positive samples on the left and 4 negative samples on the right. The tubes were put horizontally, and DNA samples stayed at the bottom of each tube. Figure 11b shows another test with 1 negative sample on the left and 1 positive sample on the right. This time the test tubes were put into the heat block. The images show clear difference between fluorescence of positive sample and fluorescence of negative sample. Gel electrophoresis results confirmed the successful amplification of target DNA in positive samples. It can be concluded that the optical setup functioned effectively.

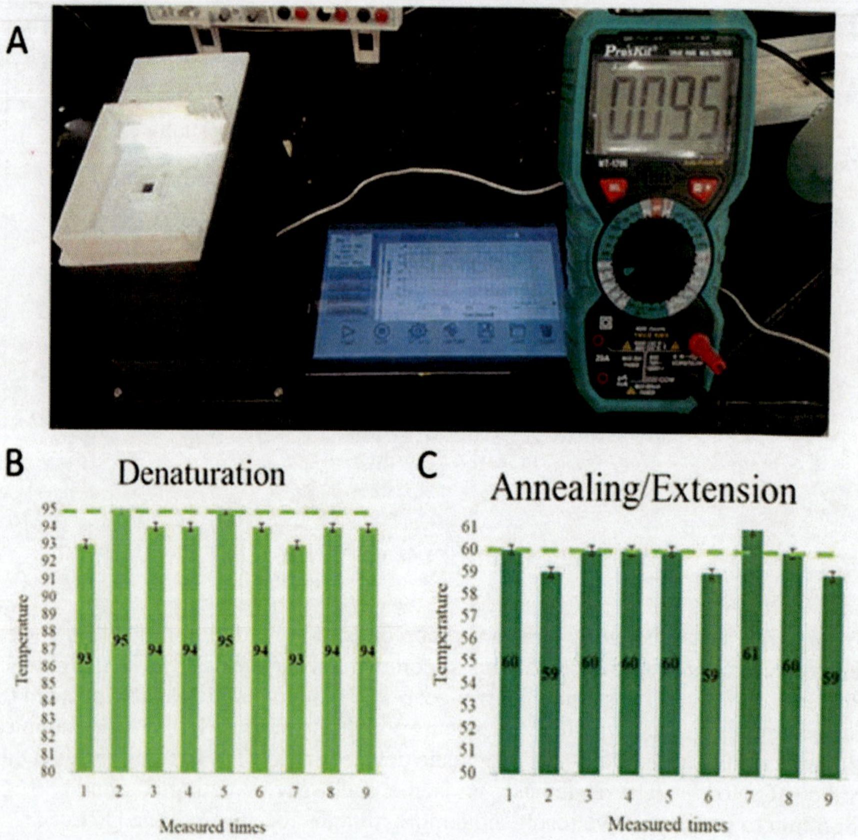

Fig. 10 **a** A Pro'sKit digital multimeter connected to a thermocouple was used for verifying temperature of solutions inside PCR test tubes; temperature of the pre-denaturation step was verified to be at 95 °C. Bar chart illustrates the set values and the actual temperatures of the denaturation stage **b**, and the annealing/extension stage **c**

4.4 Real-Time PCR

To determine the system's practical value, AccuPid Rice PLD Detection Kit [19] and AccuPid Bt11 Maize Detection Kit [20] were used. Positive and negative control samples were taken from the kits and run under two devices: our realtime PCR system and a commercial PCR Eppendorf MasterCycler Nexus Gradient with the same protocol: 15 min at 95 °C, followed by 40 cycles of 95 °C for denaturation and 60 °C for annealing/extension. In our system, after every 5 cycles, fluorescence images were captured. Each image was analyzed with ImageJ to find the average fluorescence intensity of the negative samples and the positive samples. Figure 12 shows that the fluorescence intensity of the positive samples increased over 40 PCR

Table 1 List of thermal cyclers in the study "Eppendorf Mastercyclers Save Time and Cost in PCR"

Thermal cycler	Max block heating rate (°C/s)
Eppendorf MasterCycler nexus GSX1	5
Eppendorf MasterCycler nexus gradient	5
Applied Biosystem Veriti™ Fast	3
Life Technologies SimpliAmp™	4
Bio-Rad T100	4
"An Affordable and Portable Thermocycler for RealTime PCR Made of 3D-printed Parts and Off-the-Shelf Electronics" [6]	1.4
'Smartphone-based mobile digital PCR device for DNA quantitative analysis with high accuracy" [10]	4
Our system	1.4

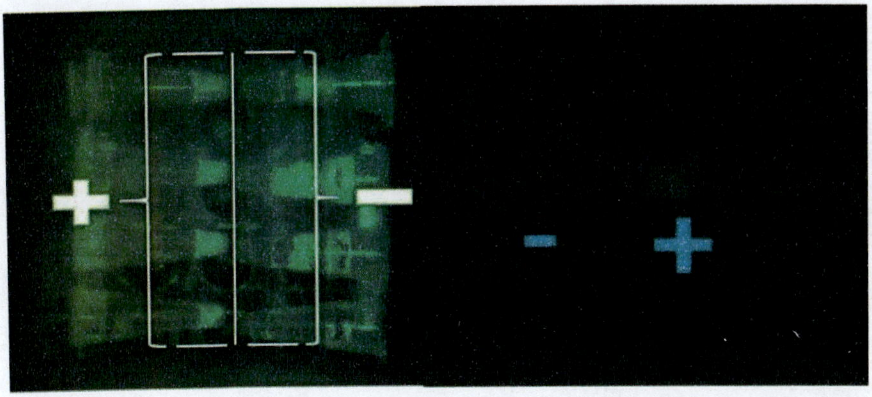

Fig. 11 Results images of tests, **a** 4 positive (+) and 4 negative (−) samples were put horizontally to test the ability to capture fluorescence differences, **b** 1 positive and 1 negative sample were put in the heat block with heated lid installed

cycles. Positive samples were also successfully amplified in the commercial PCR system (data not shown).

5 Conclusions

In this study, a highly integrated, low-cost, and portable realtime PCR system for genetic testing has been proposed and successfully built. The system achieved heating and cooling rate of 1.4 °C/s. The optical setup successfully detected fluorescence signal in realtime during the amplification process. We aim to increase the temperature ramp rate in future work by using a higher power Peltier module in combination

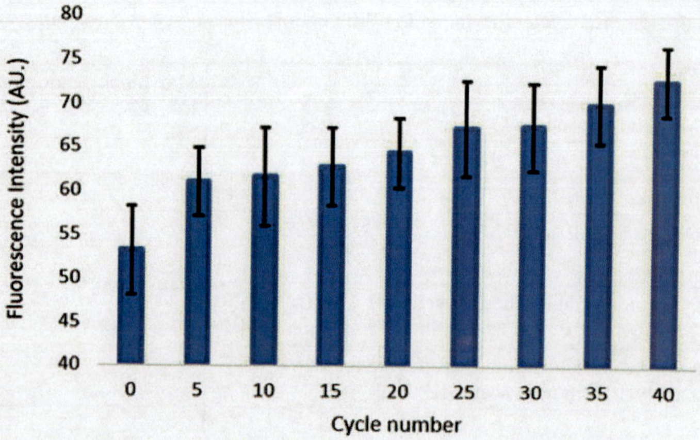

Fig. 12 Average fluorescence intensity of positive samples over 40 PCR cycles

with a better heat insulation design for the heat block. The optical setup design could also be improved to provide better contrast between negative samples and positive samples as well as lower variation in fluorescence intensity between replicates in a same run and in different runs.

Acknowledgements We would like to express our special thanks to the Khoa Thuong Biotech Ltd. for their precious support of PCR reagents and samples during the development of this realtime PCR system.

Conflict of Interest The authors declare that they have no conflict of interest.

References

1. Ahrberg CD, Manz A, Neuzil P (2016) Palm-sized device for point-of-care Ebola detection. Anal Chem 88(9):4803–4807
2. Shan C, Xie X, Barrett AD, Garcia-Blanco MA, TeshRB, Vasconcelos PF et al (2016) Zika virus: diagnosis, therapeutics, and vaccine. ACS Infect Dis 2(3):170–172
3. Garibyan L, Avashia N (2013) Polymerase chain reaction. J Invest Dermatol 133(3):1–4
4. Bản đồ lây nhiễm vi-rút corona (COVID-19). Retrieved 14 Apr 2020. https://www.google.com/covid19-map/
5. Wong G, Wong I, Chan K, Hsieh Y, Wong S (2015) A rapid and low-cost PCR thermal cycler for low resource settings. PLoS One 10(7):e0131701
6. Mendoza-Gallegos RA, Rios A, Garcia-Cordero JL (2018) An affordable and portable thermocycler for real-time PCR made of 3D-printed parts and off-the-shelf electronics. Anal Chem 90(9):5563–5568
7. Ahrberg CD, Manz A, Chung BG (2016) Polymerase chain reaction in microfluidic devices. Lab Chip 16(20):3866–3884
8. Zhou S, Gou T, Hu J, Wu W, Ding X, Fang W et al (2019) A highly integrated realtime digital PCR device for accurate DNA quantitative analysis. Biosens Bioelectron 128:151–158

9. Chung KH, Park SH, Choi YH (2010) A palmtop PCR system with a disposable polymer chip operated by the thermosiphon effect. Lab Chip 10(2):202–210
10. Song J, Pandian V, Mauk MG, Bau HH, Cherry S, Tisi LC, Liu C (2018) Smartphone based mobile detection platform for molecular diagnostics and spatiotemporal disease mapping. Anal Chem 90(7):4823–4831
11. Zhang Z, Zhao S, Hu F, Yang G, Li J, Tian H, Peng N (2020) An LED-driven AuNPsPDMS microfluidic chip and integrated device for the detection of digital loop-mediated isothermal DNA amplification. Micromachines (Basel) 11(2)
12. Walker FM, Ahmad KM, Eisenstein M, Soh HT (2014) Transformation of personal computers and mobile phones into genetic diagnostic systems. Anal Chem 86(18):9236–9241
13. Novak L, Neuzil P, Pipper J, Zhang Y, Lee S (2007) An integrated fluorescence detection system for lab-on-a-chip applications. Lab Chip 7(1):27–29
14. Illumina (2018) Eco real-time PCR system. Retrieved 25 Apr 2018, from https://support.illumina.com/real_time_pcr/eco_qpcr.html
15. Applied Biosystems (2018) Applied biosystems 7300/7500/7500 fast realtime PCR systems. Retrieved 25 Apr 2018, from http://tools.thermofisher.com/content/sfs/manuals/cms_042676.pdf
16. Fluidigm (2018). BioMark HD system. Retrieved 25 Apr 2018, from https://www.fluidigm.com/binaries/content/assets/fluidigm/biomark-hdsystem.pdf
17. Abbott Ltd (2020) Abbott ID NOW™. Retrieved 27 Mar 2020 from https://www.abbott.com/corpnewsroom/product-and-innovation/detect-covid-19-in-as-little-as-5minutes.html
18. Chai Ltd (2018) Chai open qPCR from https://www.chaibio.com/openqpcr
19. "AccuPid Bt11 maize detection kit—kt-biotech," kt. [Online]. Available: https://kt-biotech.com/sanpham/accupid-bt11-maize-detection-kit/. Accessed: 20 Aug 2020
20. "AccuPid rice PLD detection kit—kt-biotech," kt. [Online]. Available: https://kt-biotech.com/sanpham/accupid-rice-pld-detection-kit/. Accessed: 20 Aug 2020

Design and Fabrication of a Complete Electrospinning System

T. N. Viet, Do Minh Thai, P. T. Phuc, and Vo Van Toi

Abstract During the electrospinning process, the most crucial and impactful factors affecting the quality of the PCL membrane on collectors include humidity, temperature, syringe pump speed. Additionally, some buffers may generate toxic gas that must be immediately evacuated from the environment, and the high voltage part of the device is dangerous that must be safely isolated from the users. Therefore, designing a complete electrospinning system that generates high-quality membranes and safety to use becomes vital. We designed a complete electrospinning system that includes: a protecting glass chamber equipped, an electronic controller to supervise and display environmental parameters. The chamber was built from profiled aluminum and glass, and a fume hood ventilator was used to suck gas out of the chamber. A controller system consisting of hardware design and firmware programming was implemented to measure temperature and humidity in electrospinning chamber; to control factors affecting the nanofiber membrane such as drum collector speed, syringe pump, temperature, and humidity. The zero-crossing technique was used to control temperature and speed of fume hood fan. Furthermore, the PID controller was applied to regulate the rotational movement of drum collector. We successfully built the complete electrospinning system. The temperature and humidity ranges varied from −40 to +125 °C and 0–100% RH, respectively. Rotational speed of the aluminum drum was from 0 to 2000 rpm. And the volume error of the electrical syringe pump on the average is 0.60% and the speed is from 0.1 to 20 ml/h.

Keywords Electrospinning · High-voltage power · Fume hood

T. N. Viet · D. M. Thai · P. T. Phuc · V. Van Toi (✉)
School of Biomedical Engineering, International University, Ho Chi Minh City 700000, Vietnam
e-mail: vvtoi@hcmiu.edu.vn

Vietnam National University Ho Chi Minh City (VNU-HCMC), Ho Chi Minh City 700000, Vietnam

V. Van Toi et al. (eds.), *8th International Conference on the Development of Biomedical Engineering in Vietnam*, IFMBE Proceedings 85,
https://doi.org/10.1007/978-3-030-75506-5_7

1 Introduction

Nanosized fibers have been playing a vital role in manufacturing practical applications related to both microfiber and nanofiber, including protective clothing, pharmaceutical products, tissue engineering, and regenerative medicine, and have granted great interest from scientists, researchers, and nanotechnologists [1]. Electrospinning is an advanced and special nanotechnology, which produces different polymer material, including natural and synthetic polymer. This technology bases on a high-voltage environment to produce nanofibers, and these super smooth fibers are crucial due to their resemblance to the extracellular matrix in human tissue. Apart from the high-voltage part, the electrospinning machine required a controller for factors that affected the generated fibers such as the pumping speed of electrical syringe pump, the position between the pump (positive electrode of the high-voltage source) and the receiver (negative electrode of the high-voltage source), the air exchange to vacuum the poisonous solvent, spinning speed of the Drum Collector receiver (−), the internal light of the fume hood, and the measurement of environmental parameters, including internal temperature and humidity.

1.1 Some Methods to Treat Wound Healing

Wound healing has become much more effective for skin reproduction thanks to the advancement in tissue technology and the production and research of saving cells by biological scaffold, which is created by biologically compatible material [2]. The biological scaffold is actually a support for the formation of tissues and usually implanted with high growth factors. The newly produced tissues of the scaffold are rapidly replaced by a great number of new tissues. Therefore, the scaffold of the implanted tissues has the ability to be taken advantage of for synthesizing tissues; then, it will be injected into a test-tube to eliminate the lost tissues in particular. The meaning of this direct implant into the wound using the patient's own tissues is to reproduce the lost tissues. The process is illustrated in Fig. 1. Furthermore, antibacterial substances can be included on the surface of the biological scaffold in order to prevent bacteria from penetrating into the wound surface.

1.2 Basic Component of Electrospinning Machines

Traditional electrospinning and its modifications have received massive interest from the science community, especially in the biomedical field [4]. Methods such as multijet electrospinning, needleless electrospinning, electro-blowing, coaxial electrospinning, force spinning, etc. award researchers with a low-cost, flexible, and convenient set-up. A fundamental electrospinning system consists of three main

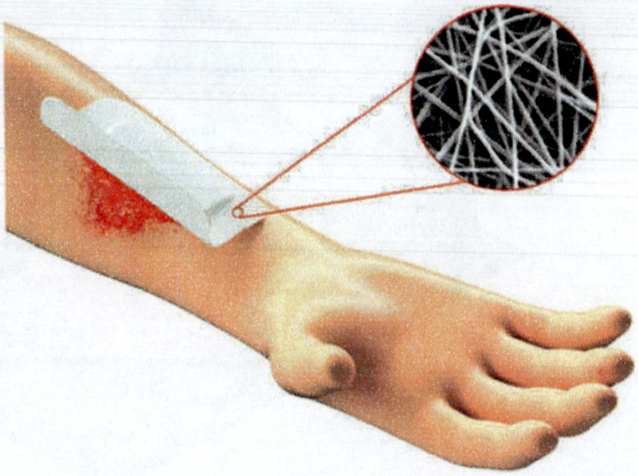

Fig. 1 This image illustrates the usage of the bandage applied electrospinning technology in treating open wound [3]

parts: an emitter, a collector, and a solution controlling pumping system [5]. This system has yielded a simple and economical assay to do electrospinning with various and composite materials, bringing enormous potential to the tissue engineering field in Vietnam (Fig. 2).

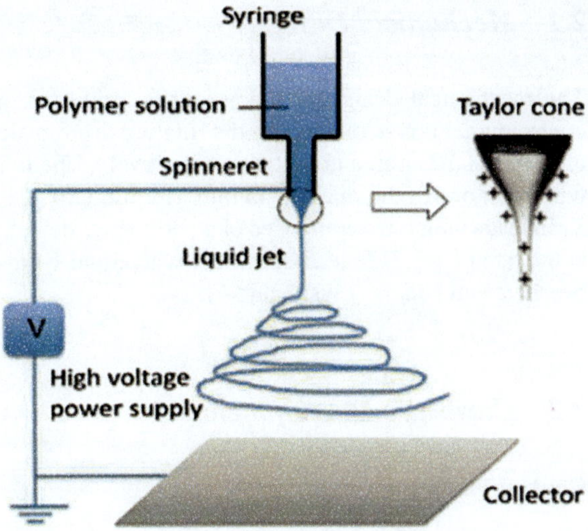

Fig. 2 Electrospinning system for nanofiber production [6, 7]

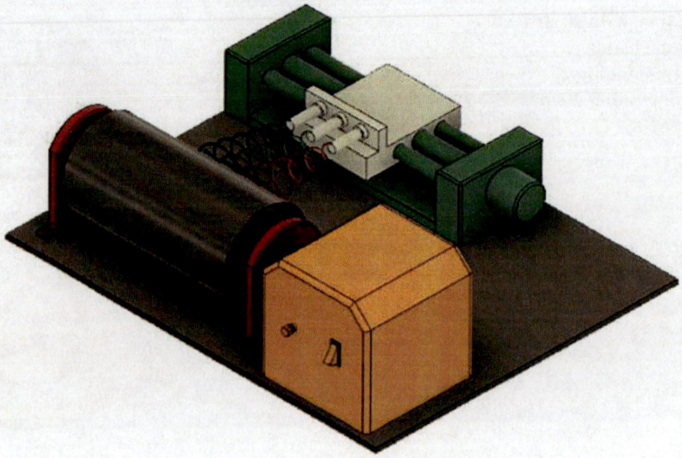

Fig. 3 Drum collector design

2 Design an Electrospinning System Consists of a Fume Hood Controller

2.1 Mechanical Design

The mechanical design includes a drum collector, ground facility, and electrical control panel box. Considering the rotating drum collector, the external and internal diameter is 100 mm and 15 mm, respectively. The length of the drum is at 200 mm while that one of the shafts is 45 mm. Additionally, the electrical syringe pump owns 3 pinheads which are composed of a 1.8-degree step motor with decelerating gearbox at the rate of 1:3.71, a lead-screw bar with equal 1-cm distance among the teeth, and bearings and sliders (Figs. 3 and 4).

2.2 Hardware Development

Control Panel Board

The control panel board consisted of a 3 × 4 button matrix and a Rotate Encoder rotating button, which were used to input values such as pumping speed of the electrical syringe pump, driving speed of the slider to maneuver the relative position between the pinhead and the drum collector, and manage the rotating speed of the drum collector, the lamp, and vacuum fan. An MCU controller collected the input values from users and displayed them on Text LCD 20 × 4, and simultaneously, sent orders to command the MCU. A display screen also took charge in keeping track of

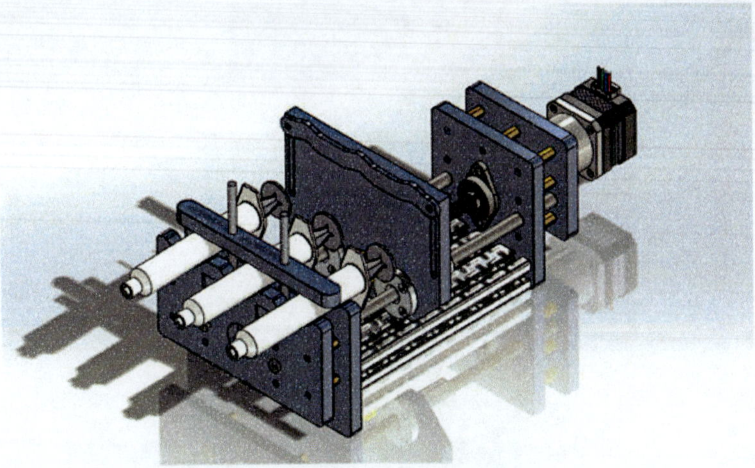

Fig. 4 Syringe pump design

the temperature, humidity, practical speed of Drum Collector, and remaining solution releasing time of the electrical syringe bump (Figs. 5 and 6).

Main Board Control

Central controlling board was in charge of regulating the speed of the Drum Collector from 10 to 2000 rounds/min based on Control Feedback (PID) algorithm. To control the temperature inside the chamber, and fan speed to vacuum the internal air out to external environment, we adjusted voltage AC220V by applying Zero-crossing method with the help of Dimer and One Pulse Mode function of ARM Cortex M4. Default setup and users' setup values were stored on an EEPROM for the purpose of further usage. There was also an additional function inside the central controlling board to monitor on-screen time of users and exhibited the time in real life through the RTC system inside the ARM STM32F407VET6.

Another function of the central controller was to measure the temperature and humidity inside the chamber using SHT30 sensor, and actual speed of the drum collector by encoder, which was for users to follow aspects that may influence the quality of electrospinning fibers.

The Main Control Unit supported users to communicate with the computer in order to trace the calibrated parameters by the device. Furthermore, the improvement of the central controller is on the way for the purpose of remote electrospinning production in the future.

Microcontroller STM32F407VET6 also interacted with a Slaver MCU through serial to conduct the electrical syringe pump and the slider (Fig. 7).

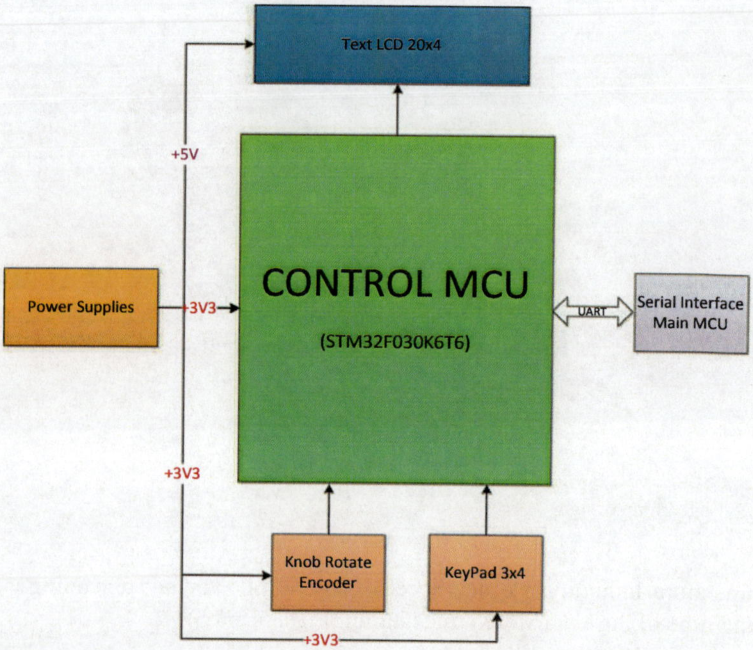

Fig. 5 Control panel board and block diagram

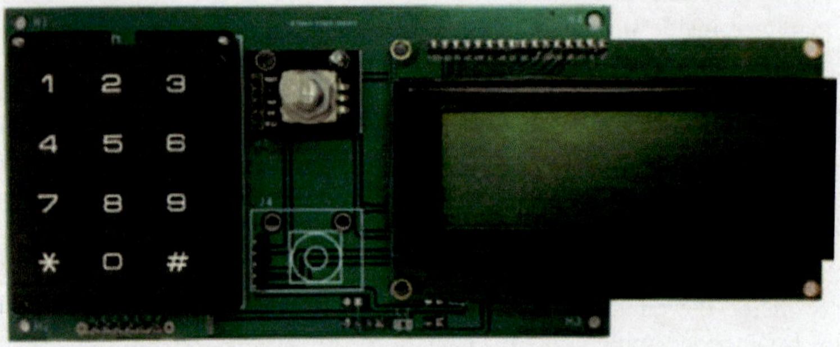

Fig. 6 The result product of the control panel board

Control Syringe Pump and Slider

The electrical syringe pump and slider controller exploited the Slaver MCU (STM32F030F4P6–ARM Cortex M0) to regulate pumping speed of the electrical syringe pump from 0.1 to 20 ml/h with all the available sizes on the market in Vietnam. Meanwhile, the slider lead-screw transmission system played as a manipulator to adjust the relative distance between electrical syringe pump and Drum Collector (Figs. 8 and 9).

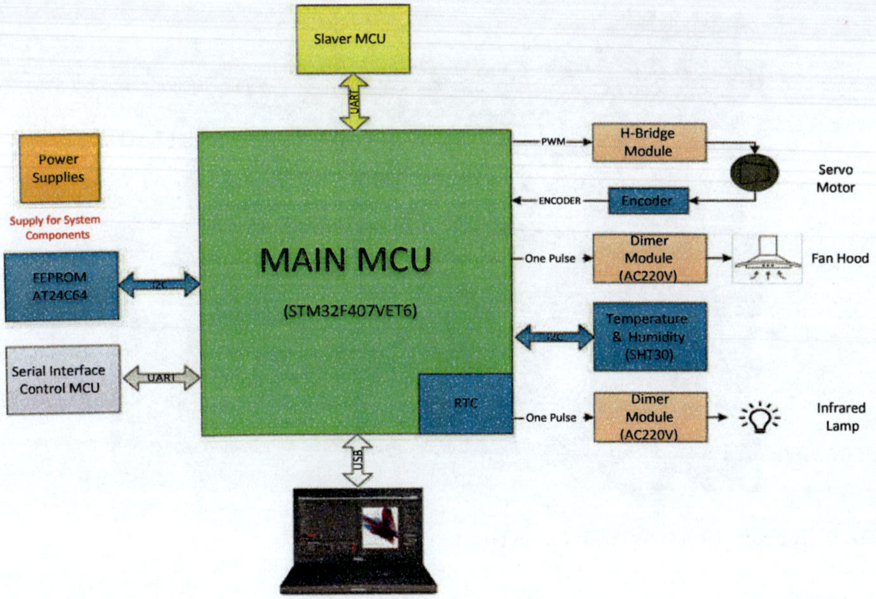

Fig. 7 Main board control block diagram

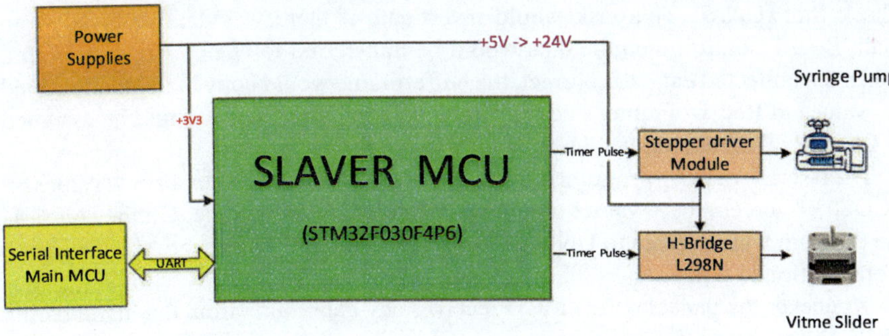

Fig. 8 Slaver control block diagram and main board

3 Firmware Development

3.1 The Operation of the Control Panel Board

Once begun, the program started building peripheries such as GPIO, UART, KeyPad and LCD20×4. The loop program analyzed whether the flags of Update Display LCD blocks were set; if they were, the program began processing them and updated new values on the LCD.

Fig. 9 The complete slaver control main board

Knob Rotate Encoder was in charge of confirming if users had input the values into the buttons or not; if they did, the buttons would execute corresponding orders. Vice versa, the buttons would skip the action and check ongoing tasks, known as Check RxTxUART. This task would investigate if there was data received from Main Board Control, then the data would be transferred to UART Break program. When the collected data was correct, the bufferFrame would copy them, then deleted the values in ReceiveFrame. The remaining data in bufferFrame would be assigned to CheckRxTxUART to handle.

The last task in Loop program is to verify KeyPad4×3. If users input new values on LCD20×4 screen, these values would be updated onto the storage. On the contrary, the program would return to Update Display LCD to do the verification and operate aforementioned tasks.

Whenever the panel board's MCU receives any data frame from the mainboard's MCU, the program of the panel board's MCU will go to the UART Interrupt Subroutine in order to process the received data frame. The program will evaluate to find out the right frame, then it will be copied to the BufferFrame and delete the Receive-Frame. Otherwise, the wrong receive frame will be deleted and waiting for the new coming data frame (Fig. 10).

Fig. 10 The flow chart of the control panel board

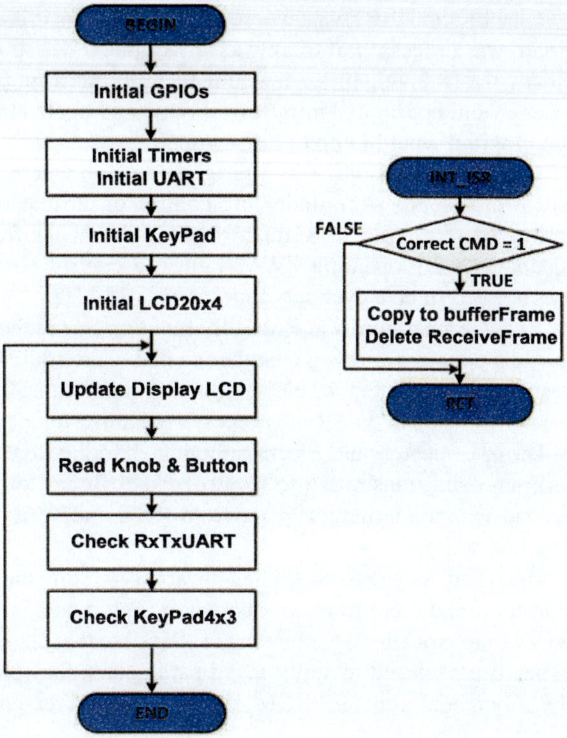

3.2 The Description of How the Main Board Control Operated

At the beginning, the program constructed Initial GPIOs for the system and internal Real Time Clock of STM32F407VET6 to keep track of the electrospinning shooting time. In the meantime, EEPROM was in use to save the setup of the machine and Kp, Ki, and Kd of PID were also activated to control the Motor Servo of Drum Collector. Then the Timer counter, System tick and UART were all initialized to do timing job for the controller and Initial Moving Average to measure the speed of Servo motor. Moreover, the Servo and Stepper motor were under the control of Initial H-Bridge.

In the first main loop, infrared light and fume hood fan availability would be examined; and if there were an order from the Control Panel Board, the program proceeded according to parameters input by users. On the other hand, the program skipped the previous step and perform the next algorithm, Control Slaver MCU.

The function of the Slaver MCU was to regulate the Syringe Pump Controller and Slider for the Electrospinning device. The orders went through the path from Control Panel Board to Main MCU and continued going to Slaver MCU. If there were no orders, the program advanced to CheckRxTxUART.

CheckRxTxUART was where all of the information received from Control Panel Board was processed; if there was any Frame orders transmitted from Control Panel Board, the receiving process would occur in Interrupt UART program, and Receive-Frame would be copied into BufferFrame and treated in Main loop program; or else, the program would jump to the next part.

Control Servo motor was a program section where the speed of Drum Collector and Motor servo were under the control of closed loop based on PID algorithm through determining the number of pulses read back from the encoder and checking feedbacks from updating PWM values to control H-Bridge; the frequency, which was applied to take over servo motor, was 15 kHz.

Read Temperature and Humidity function traced them temperature and humidity inside the electrospinning chamber so that users could adopt the most suitable environmental parameters for electrospinning membrane. The activation of this function happened while in the "Run" process, remaining temperature, humidity, actual speed of Drum Collector, and Electrospinning shooting time would be grafted with TranferFrame and transmitted to Control Panel Board to display on LCD screen. The system would automatically move to Read and Write EEPROM when the former function stopped.

Read and Write EEPROM had the ability to store the used controlling parameters; if users found them to be productive of high quality of electrospinning membrane, they would input these parameters in "Save" so that the setup be stored into EEPROM. When users wished to apply saved parameters, they chose "Read" function to read the parameters from EEPROM; the MCU would read these values and send them to Control Panel Board.

If Main MCU receives any transmit frame from Control Panel Board, Program will come to subroutine for process Data.

Whenever the mainboard's MCU receives any data frame from the panel board's MCU, the program of the mainboard's MCU will go to the UART Interrupt Subroutine in order to process the received data frame. The program will evaluate to find out the right frame, then it will be copied to the BufferFrame and delete the Receive-Frame. Otherwise, the wrong receive frame will be deleted and waiting for the new coming data frame (Fig. 11).

3.3 *The Control Syringe Pump and Slider Procedures*

When the program started, Initial GPIOs had charge of Syringe Pump and Slider, Initial Timer, and System Tick to do timing for the shooting speed of Syringe Pump and transferring speed of Slider. Initial UART adopted the orders from Main MCU to manage Initial Syringe Pump and Slider.

When we accessed the main loop, the initial task was CheckRxTxUART for receiving orders from main board to run the electrical syringe pump and slider; if there was no signals from main MCU, the controller would proceed to Slider Execute.

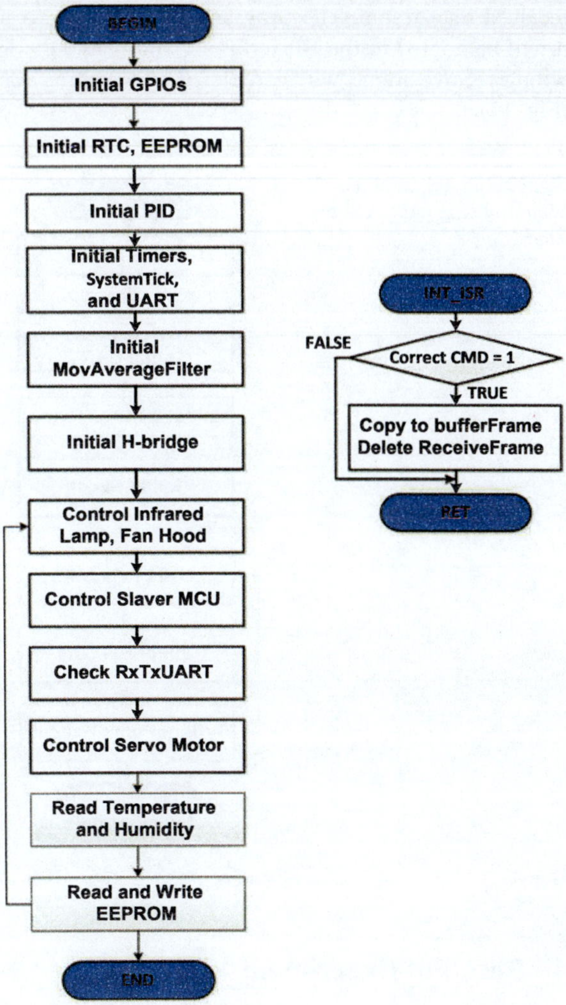

Fig. 11 The flow chart of the main board control

Slider Execute implemented the duty to adjust the relative positions between Syringe Pump and Drum Collector if it obtained the order from main MCU; if not, SyringePump Execute would be called.

SyringePump Execute was a function to administer the speed of the syringe to shoot out an appropriate amount of solution in the electrospinning application. The main MCU delivered order to the slaver MCU to regulate the syringe pump; however, when there was no orders, the program would move on to the next task: CheckTxRxUART.

Whenever the slaver's MCU, which using for controlling the Syringe Pump and Slider, receives any data frame from the mainboard's MCU, the program of the slaver's MCU will go to the UART Interrupt Subroutine in order to process the received data frame. The program will evaluate to find out the right frame, then it will be copied to the BufferFrame and delete the ReceiveFrame. Otherwise, the wrong receive frame will be deleted and waiting for the new coming data frame (Fig. 12).

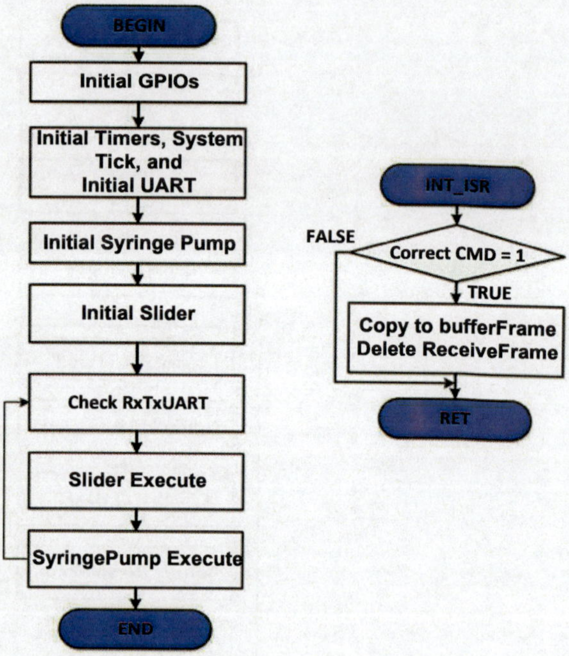

Fig. 12 Flow chart of the control syringe pump and the slider

4 Result of Experiment

The result is shown in Fig. 13 which is able to control the speed of Drum Collector from 10 to 2000 rounds/min and fume hood fan speed. The temperature and humidity has an active range from −40 to 125 °C with the precision up to ±0.3 °C and 0–100% with ±2% error (Following the SHT31's manufacturer datasheet).

4.1 Testing Methodology

We chose the 12 cc syringe with 15.8 mm diameter and pumping time is an hour for each experiment; the measurement is operated twice per experiment based on the following formula. The experiment was set up by using the caliper of Mitutoyo

Fig. 13 Final fume hood control product

Manufacturer (Model: CD-6″BS, with the accuracy, is ±0.002 mm). Firstly, the caliper was fixed position to the syringe clamp of Syringe Pump, then the machine ran 2 times with 1 time per 1 h. Secondly, the data was recorded after each time. Finally, we using the below formula to calculate the volume result (Table 1).

Table 1 Description of errors in the syringe pump

Test	Speed (ml/h)	Time (min)	Expected volume (ml/h)	Result volume (ml/h)	Volume error (%)
1	0.1	60	0.1	0.1025	2.50
2	0.5	60	0.5	0.4986	0.68
3	1	60	1	1.021	0.70
4	1.5	60	1.5	1.5072	0.48
5	2	60	2	2.0580	2.90
6	2.5	60	2.5	2.5103	0.41
7	3	60	3	2.9886	0.38
8	3.5	60	3.5	3.5128	0.37
9	4	60	4	4.0032	0.08
10	4.5	60	4.5	4.4825	0.39
11	5	60	5	5.0194	0.39
12	5.5	60	5.5	5.5022	0.04
13	6	60	6	6.0162	0.27
14	6.5	60	6.5	6.4985	0.02
15	7	60	7	7.0388	0.55
16	7.5	60	7.5	7.5086	0.11
17	8	60	8	8.0108	0.14
18	8.5	60	8.5	8.5022	0.03
19	9	60	9	9.0812	0.90
20	10	60	10	10.0607	0.61

$$R_{tb} = \frac{R1 + R2}{2} (ml/h) \tag{1}$$

R1 the first measurement.
R2 the second measurement.
R_{tb} Result Volume.

The Volume is evaluated in accordance with the following formula:

$$V = \pi * R^2 * D \quad So, D = \frac{V}{\pi * R^2} \tag{2}$$

V is the volume of the solution inside the cylinder.
R is the radius of the cylinder.
D is the distance of the piston shifting in an hour.

From (1), (2), we can conclude the substantial Volume by quantifying distance D of the electrical syringe pump as in Table 1.

Error calculation formula:

$$Error(\%) = \frac{abs(result - expected)}{expected} * 100 \tag{3}$$

The maximum and average error value are 2.9% and 0.60%, respectively based on the formula:

$$\text{Average Error} = \frac{Sum\ of\ Error}{Number\ Sample} (\%) \tag{4}$$

5 Conclusion

Overall, the system has proven its functionality in terms of controlling the speed of the drum collector and fume hood fan. The relative distance between the electrical syringe pump and drum collector has also been adjusted and managed by our algorithm. Furthermore, the system also has the ability to isolate the high voltage for the purpose of user's safety. Finally, our device is able to take fully charge of humidity and temperature, which is under the control of infrared light, inside the electrospinning chamber.

Acknowledgements This research is funded by Vietnam National University Ho Chi Minh City (VNU-HCM) under grant number C2020-28-08.

Conflicts of Interest The authors have no conflict of interest to declare.

References

1. Chou S-F, Carson D, Woodrow KA (2015) Current strategies for sustaining drug release from electrospun nanofibers. J Control Release 220:584–591
2. Mano J et al (2007) Natural origin biodegradable systems in tissue engineering and regenerative medicine: present status and some moving trends. J Roy Soc Interface 4(17):999–1030
3. Auger FA, Lacroix D, Germain L (2009) Skin substitutes and wound healing. Skin Pharmacol Physiol 22(2):94–102
4. Keirouz A, Chung M, Kwon J, Fortunato G, Radacsi N (2020) 2D and 3D electrospinning technologies for the fabrication of nanofibrous scaffolds for skin tissue engineering: a review. Wiley Interdisc Rev Nanomed Nanobiotechnol 12(4):e01626
5. Chimene D, Alge DL, Gaharwar AK (2015) Two-dimensional nanomaterials for biomedical applications: emerging trends and future prospects. Adv Mater 27(45):7261–7284
6. Call CC (2010) The study of electrospun nanofibers and the application of electrospinning in engineering education. Texas A & M University
7. Do TM, Ho MH, Do TB, Nguyen NP, Van Toi V (2018) A low cost high voltage power supply to use in electrospinning machines. In: International conference on the development of biomedical engineering in Vietnam. Springer, Berlin, pp 95–100

Design of a Soft Robotic Glove for Hand Rehabilitation Based on Pneumatic Network Method and Low Cost Electro-pneumatic Device

Boi Mai Quach, Vo Van Toi, and Hien Thi Thu Pham

Abstract The state-of-the-art field of bio-inspired soft robotics promises emergent and inventive methods to utilize robots made from soft, flexible materials in a wide range of applications. This research is on-trend, describes an advancement in the design of a wearable robotic glove to assist people suffering from hand impairments regain their ability to control their environment. This study built a streamlined, inexpensive electro-pneumatic system that can work with small and simple actuators to increase for finger movement flexibility. The use of pneumatic pressure to stretch and bend the actuators through inflation or deformation of elastic chambers, making this glove work effectively. The designed pneumatic device consists of STM32F101C8T6 microcontroller combined with a series of pumps and a matrix of solenoid valves. Along with the electro-pneumatic system, we also applied it to develop a wearable robotic glove (made from Ecoflex™ rubbers) which is suitable for the pneumatic system by the series of experiments that mechanically characterize the actuators. The results gathered in this study validate the feasibility of the first prototype of our soft robotic glove as an effective device to assist hand function in individuals. However, performance and ease of use of the system should be improved further in future development phases.

Keywords Soft robotics · Electro-pneumatic · Fluidic · Hydraulic

1 Introduction

The majority of hand disability patients undergo a stroke, Duchene/Becker muscular dystrophy or incomplete spinal cord injury [1, 2], experience an absence of hand motor ability, extremely restrict activities on a daily basis and markedly reducing living standards [3]. Physical therapies are currently involved in practicing required tasks repetitively for patients to improve hand strength, precisely, and movement

B. M. Quach (✉) · V. Van Toi · H. T. T. Pham
School of Biomedical Engineering, International University, Ho Chi Minh City, Vietnam

Vietnam National University, Ho Chi Minh City, Vietnam

V. Van Toi et al. (eds.), *8th International Conference on the Development of Biomedical Engineering in Vietnam*, IFMBE Proceedings 85,
https://doi.org/10.1007/978-3-030-75506-5_8

ability [3, 4]. However, these therapies need more support from technicians and more time to observe and analyze the patient's performances, which raises a high demand for the physical therapists [5]. From the limitations of traditional robotic instruments, medical specialists hope to find out alternative methods, and soft robotics seems to be an effective adoption. Soft robotics is made of materials which are similar in characteristics to human tissues. This method is an emerging field that has drawn increasing interest since their ability to interact with highly advanced eco-friendly environments without destructive impacts or self-damage compared with their heavy and redundant equivalents.

In this research, to operate the soft robotic hand glove, an inexpensive and portable electro-pneumatic control system was designed and included pneumatic (air) pressure sensors in line with a closed-loop controller to regulate the pressure. The problem to be solved here was reducing cost but keep relative its functions. Thus, all components had to be found easily with an economical cost. Another step to this project was to create a soft robotic glove which could lead to greater advances in physical therapy at home by providing: (a) more flexible in controlling hand movement, (b) an easily used robot for normal people, (c) safe interaction since the softness and compliance with their fabrication, (d) portability. It was also a perfect demo for the electro-pneumatic 's applications since it is reality and medical meaning.

The rest of paper is organized as follows: First, we briefly review the works on soft robotic glove and systems are used to control the robotic devices. The next section gives the methodology about designing a pneumatic network and making a soft robotic glove. We then exhibit the results of the whole procedure (hardware, calibration, software). Finally, we conclude the discussion and discuss the research plans for our future work.

2 Related Work

Soft robots can better adapt to their surroundings [6], perform differently and even multi-task, and can imitate the motion and function of biological systems, such as octopus's tentacles [7, 8] or movement of worms [9]. In biomedical engineering, and medical in general, there has been rising interest in numerous possible applications of soft robotics systems into various orientations and clinical settings. Actuator material and actuation method make it lightweight and easy to use for education and training of medical professionals, robotic organs, mechanotherapy, autonomous minimally invasive surgery, etc. In the literature, the soft robotic gloves for hand physical therapies, especially in stroke, are referred to as Table 1.

In these references, there are many different designs and actuation methods such as Shape memory alloys (SMA) driven actuators, Tendon driven actuators, Fluid driven actuator and Pneumatic actuators have been developed for robotic rehabilitation gloves. Most of the tendon driven cables support only daily living activities and have limited output force and hyperextension. SMA actuators have high operating temperatures ranging from 100 to 500 °C [12]. The complex design of SMA actuators

Table 1 Soft robotic gloves found in the literature

Reference, Developer	System	Type of usage	Supported movement
Exo-Glove [10]	Tendon routing system	Grasping of objects	Wrist, elbow, shoulder joints, full hand
AirExGlove [11]	Pneumatic actuators and Tendon routing system	Rehabilitation	Full hand
SMA's glove [12]	SMA (Shape memory alloys) actuators	Rehabilitation	Full hand
PneuGlove [13]	Pneumatic actuators	Rehabilitation	Full hand
Pneumatically actuated robotic glove controlled with EMG [14]	Pneumatic actuators and Surface Electromyography (sEMG)	Rehabilitation	Full hand
GRIPIT [15]	Tendon routing mechanism and Tension maintenance mechanism	Tripod grasp	Thumb, index finger, and middle finger
Soft Anthropomorphic Robotic Hand [16]	Kirigami-inspired system and Origami structure	Rehabilitation	One finger at one time
Gait rehabilitation soft robot [17]	Soft pneumatic actuators (SPAs)	Gait rehabilitation	Specialized animal
Fluidic actuated soft robotic glove [18]	Hydraulically actuated multi-segment soft actuators	Rehabilitation	Full hand
Wearable haptic device [19]	Haptic systems	Rehabilitation	One finger at one time

made it difficult to use in rehabilitation purposes and daily living activities. The fluid-driven actuator is no nifty portable and high user maintenance throughout the life of the system. Pneumatic actuators were selected due to higher stiffness, low weight and simpler design than the actuators mentioned above.

This paper presents soft robotic gloves for rehabilitation which used the inexpensive silicon for fabrication of these actuators. Air pressure used as actuation method in these actuators. Air pressure will help the robotic fingers for flexing and extending the hand. The glove will be attached on the hand's dorsal side which helps the patient feel the objects more naturally. Actuation energy source and electromechanical components are mounted separately to put them as low as the human finger's burden.

3 Methods

3.1 Design Elastomeric Materials

In this work, we proposed an easy and quick fabrication method of soft inflatable plastic actuators. Ecoflex™ rubbers, which are platinum-catalyzed silicones produced by Smooth-On, Inc, are commonly used in the making of prosthetic appliances, cushioning for orthotics and special effects applications. Its notable characteristics include the un- cured silicone (easy to mix, disperse and degas). The cured material is soft, flexible and remarkably strong, being able to stretch several times exceed the original size and then rebound back without deforming. Other beneficial traits include colorlessness, quick cure time, minimal shrinkage, and safety.

In order to cast silicon rubber, casting 3D molds are designed and include two main parts: an upper part with hollow chambers (air bladder) and a base made of pliable but non-elastic material such as fabrics or paper, or with the same material as the upper part but are completely dense. After the materials have been prepared, the single actuator's fabrication process is mainly developed in the laboratory and is described briefly in Fig. 1.

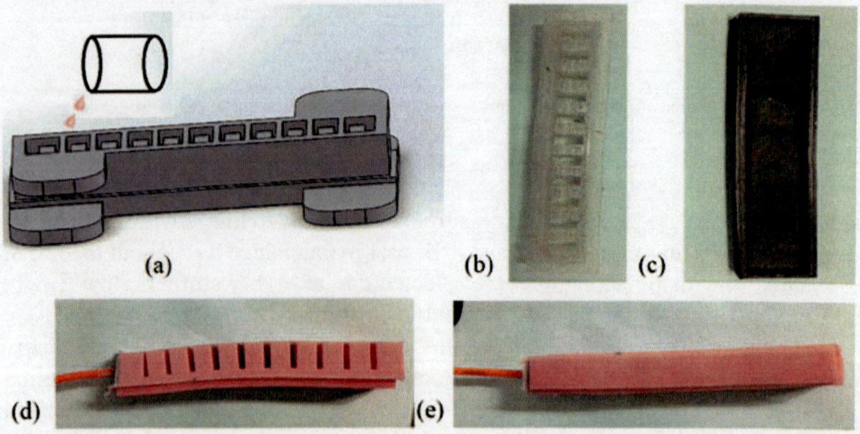

Fig. 1 Fabrication process. **a, c** Stirred liquid rubber was poured directly into the upper and bottom part and left to solidify. Then a paper or fabric layer was inserted and poured some more silicon which glued the base part. **b** Fabricated air chamber (top layer) after demolding. **d** Side view of fabricated actuator after inserting the pipe for air supply. **e** Bottom view of the actuator

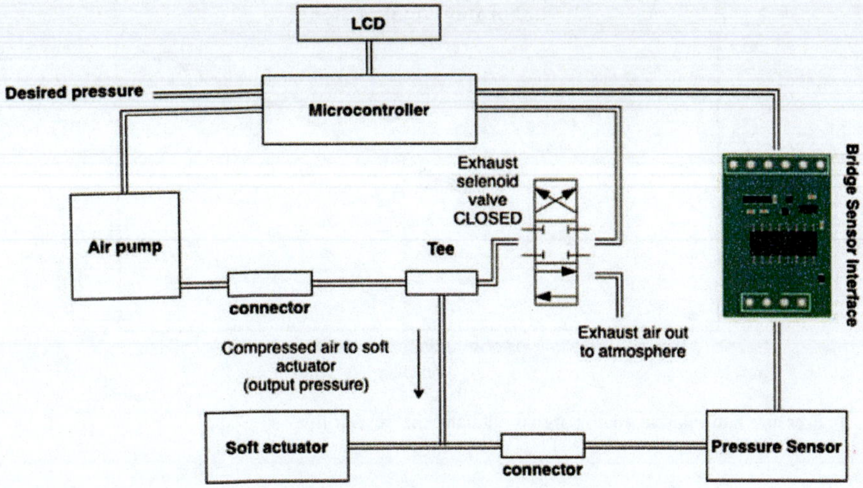

Fig. 2 Block diagram of the electro-pneumatic board

3.2 *Electro-pneumatic Design*

In order to control the hand prosthesis, it was assembled an electro-pneumatic circuit, presented in Fig. 2, consisting of a series of pumps which navigate the external airflow to the system and a matrix of solenoid valves which are used to open and close to regulate the flow of fluid in the system. The system's pressure is controlled based on PWM (Pulse-Width Modulation) principle, which involves the controlled timing of the activation and deactivation of the electrically controlled valves. Pressure sensor (XGZP101) provides feedback on system's behaviour after the analog signals are converted by HX711 (amplifier interface). May manually, change the state of knobs to control the device because each knob represents potential changes in the STM32F101C8T6 microcontroller. Additionally, the microcontroller will analyze the data collected from the sensor and convert them into digital signals before displaying on LCD.

Based on the control parts such as the matrix of buttons, pumps, and the valves system, the logic system will be designed in term of SPDT switches to control the actuators' inflation and deflation phases. These digital pins were connected to inputs. The chambers were able to inflate or deflate depending on the state of the outputs.

3.3 *Pressure Sensor Calibration*

The XGZP101 pressure sensor is an analogue pressure sensor. If the differential pressure is applied on the sensor, the sensor will deliver an analog output voltage proportional to the corresponding pressure. It is a gauge pressure sensor meaning

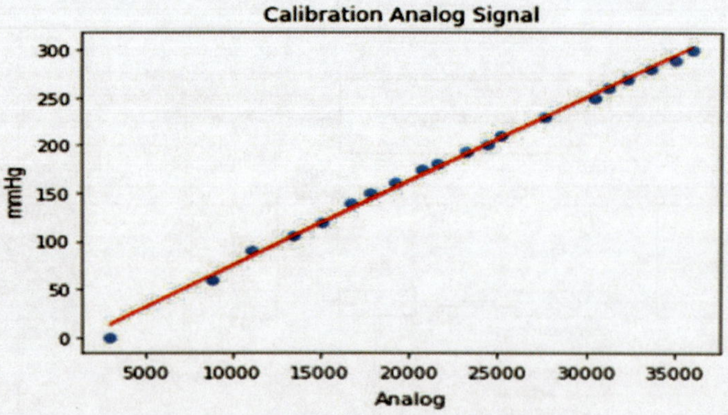

Fig. 3 Pressure and relative analog signal data and the best-fit line

that it measures the pressure in excess of atmospheric pressure from −15 to 15 psi. Using HX711 to convert analog signal up to 24 bit. After the collection of data, we indicated them on the graph and find the best-fit line as well as calculated the final Eq. (1):

$$y = -30.39 + 0.01 * x \tag{1}$$

where y is the pressure value (unit: mmHg), x is the analog signal. The collected data used for calibration and the best-fit line are shown in Fig. 3.

4 Experimental Test and Results

4.1 Electro-pneumatic System

Using SolidWorks drawings, the hardware was constructed by either mechanical work or 3D printing. An electro-pneumatic system (total weight is approximately 1.70 kg) allowed to navigate individual actuators, was assembled as Fig. 4.

The system has a two-dimensional array with white buttons, includes in two parts. Ten buttons on the left side will regulate solenoid valves and pumps connected with each finger, respectively. When we press the buttons on the top, the pumps active and the valves will be closed to release the air flows into the actuators. By contrast, the pumps will be stopped, and the solenoid valves are opened to release the airflow out of the environment if we press the buttons on the bottom. We designed two buttons on the right side and the same side with the power switch to change the state exercises. The one below will be active to control all fingers at the same time, and the other one will be the state which each finger can be activated independently. The blue rotate button in the top left corner of the box is a variable resistor to speed

Fig. 4 The portable control box

up the airflow movement from the pumps releasing. The LCD always displays all instructions to guide users, and the air pressure and the force value are also shown during the process.

Figure 5 illustrates the result of pressure after calibration. The green line describes the measured air pressure (P) and the pink line as the regression model's output. The trend of both two values follows a pattern.

Fig. 5 The result of calibration

4.2 Force Measurement

The blocking force was designed to measure the force generated by soft actuators as Fig. 6. One end of the actuator was fixed to act as a cantilever beam, and the other was free to measure the force. When the air is inhaled, the figure bent and touched the *FSG15N1A* sensor and then the analog data will be collected.

The experimental result of blocked force and input pressure relationship for the finger is shown in Fig. 7. The maximum pressure from the air which the finger, made from `Ecoflex`™ rubbers, can withstand is approximately 120 kPa.

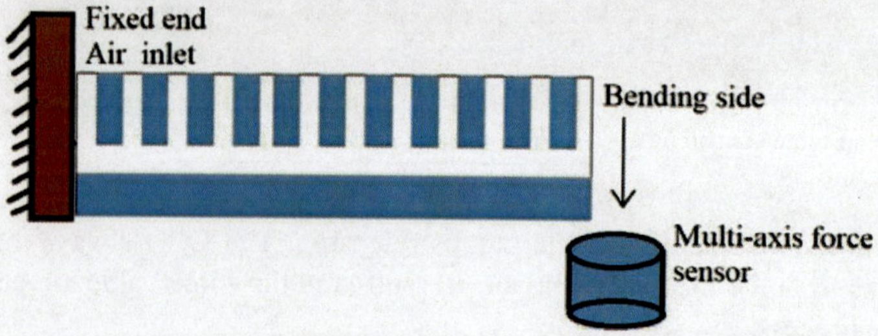

Fig. 6 Experimental setup for blocked force test

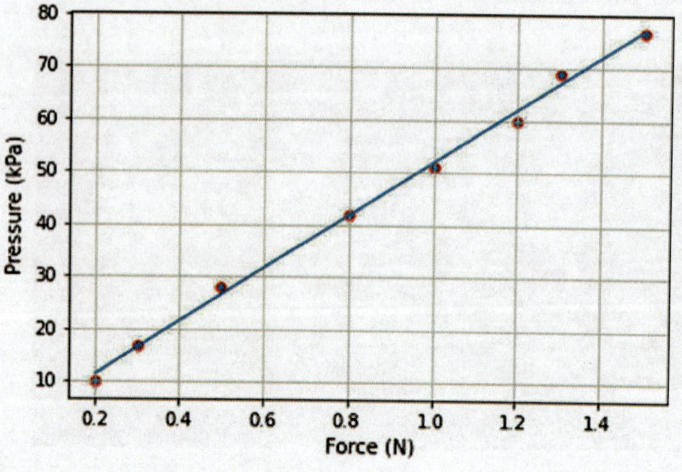

Fig. 7 Pressure and blocked force relationship

4.3 *Grasping of Objects of Various Shapes*

In the experiments, the participants are both a healthy subject and the disabled subject. In order for users to implement a series of grasp-transport-release practices, we selected four objects which are different size and weight: (1) a plastic bottle of water, (2) a baseball, (3) a mouse, and (4) a smartphone. During the experiment, individuals must navigate each finger's direction and synchronize fingers with whole hand operation. The soft robotic glove is more streamlined than the cumbersome robot hand, which is used in the hospital currently. Figure 8 shows the grasp performance of the disabled subject while using the glove. After the experiment, most participants were very satisfied with the aid hand glove since they could grasp various objects by themselves.

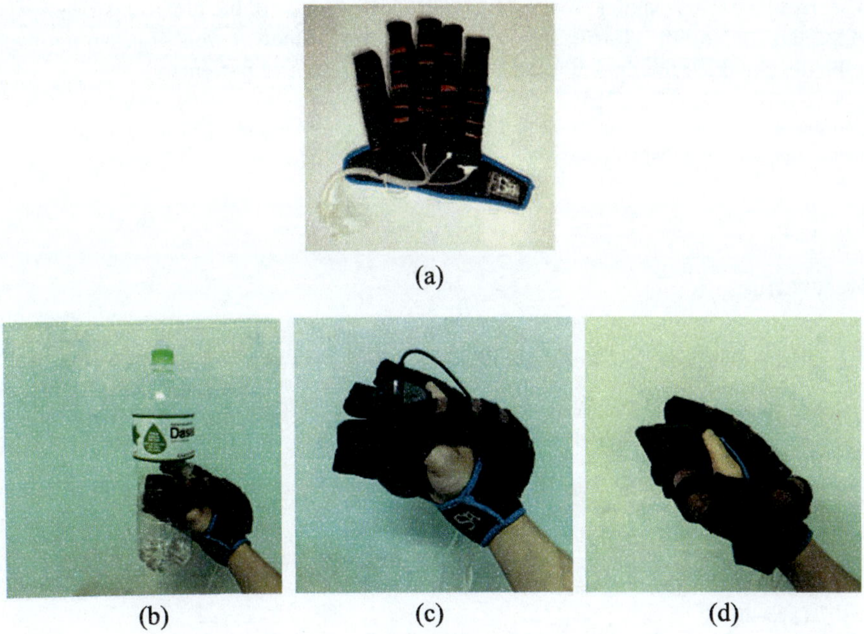

Fig. 8 **a** Soft robotic glove product. **b** Grasping ability of a plastic bottle of water with full hand. **c** Grasping ability of robotic glove for a mouse. **d** Picking up a mobile phone

5 Conclusion and Future Works

In the research, we presented a soft robotic hand glove designed to aid people who undergo a decline in hand function or have congenital hand deformities through rehabilitation exercises. A low-cost device is constructed by applying the novel electro-pneumatic actuator, which made of the silicone rubber. Consequently, experimental results had achieved successful performances of a wide range of target objects. Another advantage of this mechanism is easy and cheap to manufacture because the cost of elements is reasonable (less than $250). They are really popular in the market, or they may be recycled from used devices in the hospital. In addition, another key point is the use of hand glove allows us to exploit contacts with many difficult environments containing dirt, dust, or even liquids.

For further study, we will implement a larger study with many patients who suffer from hand muscle weakness to investigate the performance of the glove and the effects of human adaptation. More experimental exercises for many levels of grasping need to be designed for the device to be suitable with a range of patients.

Conflict of Interest The authors declare that they have no relevant financial interests in the manuscript, nor any other potential conflicts of interest.

References

1. W. G. Members, Roger VL, Go AS, Lloyd-Jones DM, Benjamin EJ, Berry JD, Borden WB, Bravata DM, Dai S, Ford ES et al (2012) Heart disease and stroke statistics—2012 update: a report from the American Heart Association. Circulation 125(1):e2
2. S. C. I. C. National (2014) Spinal cord injury facts and figures at a glance. J Spinal Cord Med 37(3):355
3. Takahashi CD, Der-Yeghiaian L, Le V, Motiwala RR, Cramer SC (2008) Robot-based hand motor therapy after stroke. Brain 131(2):425–437
4. Wolf SL, Winstein CJ, Miller JP, Thompson PA, Taub E, Uswatte G, Morris D, Blanton S, Nichols-Larsen D, Clark PC (2008) Retention of upper limb function in stroke survivors who have received constraint-induced movement therapy: the excite randomised trial. Lancet Neurol 7(1):33–40
5. Liepert J, Bauder H, Miltner WH, Taub E, Weiller C (2000) Treatment-induced cortical reorganization after stroke in humans. Stroke 31(6):1210–1216
6. Laschi C, Mazzolai B, Cianchetti M (2016) Soft robotics: technologies and systems pushing the boundaries of robot abilities. Sci Robot 1(1):eaah3690
7. Laschi C, Cianchetti M, Mazzolai B, Margheri L, Follador M, Dario P (2012) Soft robot arm inspired by the octopus. Adv Robot 26(7):709–727
8. Calisti M, Giorelli M, Levy G, Mazzolai B, Hochner B, Laschi C, Dario P (2011) An octopus-bioinspired solution to movement and manipulation for soft robots. Bioinspiration Biomimetics 6(3):036002
9. Jung K, Koo JC, Lee YK, Choi HR et al (2007) Artificial annelid robot driven by soft actuators. Bioinspiration Biomimetics 2(2):S42
10. In H, Kang BB, Sin M, Cho K-J (2015) Exo-glove: a wearable robot for the hand with a soft tendon routing system. IEEE Robot Autom Mag 22(1):97–105

11. Stilli A, Cremoni A, Bianchi M, Ridolfi A, Gerii F, Vannetti F, Wurdemann HA, Al-lotta B, Althoefer K (2018) Airexglove—a novel pneumatic exoskeleton glove for adaptive hand rehabilitation in post-stroke patients. In: 2018 IEEE international conference on soft robotics (RoboSoft). IEEE, pp 579–584
12. Shintake J, Cacucciolo V, Floreano D, Shea H (2018) Soft robotic grippers. Adv Mater 30(29):1707035
13. Connelly L, Jia Y, Toro ML, Stoykov ME, Kenyon RV, Kamper DG (2010) A pneumatic glove and immersive virtual reality environment for hand rehabilitative training after stroke. IEEE Trans Neural Syst Rehabil Eng 18(5):551–559
14. Polygerinos P, Galloway KC, Sanan S, Herman M, Walsh CJ (2015) EMG controlled soft robotic glove for assistance during activities of daily living. In: 2015 IEEE international conference on rehabilitation robotics (ICORR). IEEE, pp 55–60
15. Kim B, In H, Lee D-Y, Cho K-J (2017) Development and assessment of a hand assist device: Gripit. J Neuroeng Rehabil 14(1):15
16. Chan YH, Tse Z, Ren H (2017) Design evolution and pilot study for a Kirigami-inspired flexible and soft anthropomorphic robotic hand. In: 2017 18th international conference on advanced robotics (ICAR). IEEE, pp 432–437
17. Song YS, Sun YS, Van Den Brand R, Von Zitzewitz J, Micera S, Courtine G, Paik J (2013) Soft robot for gait rehabilitation of spinalized rodents. In: 2013 IEEE/RSJ international conference on intelligent robots and systems. IEEE, pp 971–976
18. Polygerinos P, Wang Z, Galloway KC, Wood RJ, Walsh CJ (2015) Soft robotic glove for combined assistance and at-home rehabilitation. Robot Auton Syst 73:135–143
19. Pacchierotti C, Sinclair S, Solazzi M, Frisoli A, Hayward V, Prattichizzo D (2017) Wearable haptic systems for the fingertip and the hand: taxonomy, review, and perspectives. IEEE Trans Haptics 10(4):580–600

Portable Hand Vein Finder System Based on Near-Infrared Illumination and Morphological Image Processing

Pham Van Quan, Phan Nguyen Nhue, Le Duy Tuan, Le Hoang Hai, Le Anh Tu, Dao Nguyen Thuan, and Ta Van Duong

Abstract Vein detection technology has significant applications in many areas such as healthcare, security and aesthetics. In healthcare, vein visualization is highly necessary for making an accurate venipuncture. This work demonstrates a portable hand vein finder system using noninvasive infrared technology that can display the real-time venous image on a monitor. The skin surface is illuminated by a near-infrared (NIR) light source that contains 5 LEDs (850 nm) and 5 LEDs (940 nm). These LEDs are arranged alternately in a circular form. Under NIR light exposition, vein and surrounding tissues are distinguished due to their difference in the brightness and contrast. A 3 MP camera is used to capture the illuminated skin area, and the obtained signal is then transferred to a Raspberry Pi 4 for image processing. The input image is processed with median filtering and contrast-limited adaptive histogram equalization (CLAHE) for displaying on a monitor. After processing, venous images with much better contrast are observed. To project the venous patterns on the skin surface, local adaptive threshold and subsequent morphological image processing are used. After the first algorithm, the venous patterns are segmented out of the image. Then, morphological processing, which has rarely been widely used for vein visualization, further reduces the noise signal, such as hair on the hand thus clearer and sharper venous patterns are obtained. Our work provides an effective and low-cost solution for nursing staff in low and middle-income countries to perform a safe and accurate venipuncture.

Keywords Vein finder · Near-infrared · Median filtering · Local adaptive threshold · Morphological processing

---

P. V. Quan · P. N. Nhue · L. D. Tuan · L. H. Hai · T. V. Duong (✉)
Department of Optical Devices, Le Quy Don Technical University, 236 Hoang Quoc Viet, Hanoi, Vietnam
e-mail: duong.ta@mta.edu.vn

L. A. Tu · D. N. Thuan
Institute of Materials Science, Vietnam Academy of Science and Technology (VAST), 18 Hoang Quoc Viet, Hanoi, Vietnam

V. Van Toi et al. (eds.), *8th International Conference on the Development of Biomedical Engineering in Vietnam*, IFMBE Proceedings 85,
https://doi.org/10.1007/978-3-030-75506-5_9

1 Introduction

Imaging of subcutaneous veins is important in many areas, including security and healthcare. In security, subcutaneous veins provide unique personal recognition schemes to either determine or confirm a person's identity [1]. Finger biometric using venous patterns has been demonstrated as a high-security method with an error rate of 0.13% [2]. In healthcare, vein visualization is highly necessary for vascular surgery and making an accurate venipuncture. Venipuncture is an everyday procedure that is performed for drug injection and the collection of blood for laboratory testing.

Near-infrared (NIR) technology has recently attracted a great deal of research interest as a non-invasive and effective method of imaging subcutaneous veins. It has been demonstrated that NIR imaging can reduce the error involved in puncturing the right vein by highlighting the superficially located veins [3]. Furthermore, with wireless communication, the digital vein images can be transferred to the clinic for diagnostics which opens the possibility of telehealthcare.

Nowadays, there are many vein finder products available on the market, for example, AV400 Vein Viewing System (AccuVein), VeinViewer Flex (Christie Medical Holdings), Vein Navigation Device (Novarix Ltd.). These products are very effective for finding the optimal venipuncture site, but they are highly expensive. As a result, an investigation of a cost-effective solution for healthcare in low and middle-income countries is necessary.

Herein, we demonstrate a simple, cost-effective design for a vein imaging system and vein finder prototype that can display the real-time venous image on a monitor. Our device costs only around $300, which is much cheaper than existing ones on the market.

2 Materials and Methods

2.1 Principle of Near-Infrared Imaging in Vein Finder

NIR is a region of the electromagnetic spectrum ranging from 780 to 2500 nm [4]. However, wavelengths between 780 and 950 nm are mainly used for biometrics and medical applications. The use of NIR light for clinical applications has been introduced since 1988 by the work of Cope and Delpy [5]. In their research, a long-term measurement of cerebral blood oxygenation in newborn infants' brains was carried out, and the result had made a great impact.

A NIR based vein finder system is composed of a light source (generally high power NIR-LEDs), a camera with NIR-sensitive sensor, a processing device and a monitor. In some cases, a digital light processing projector is necessary to project the venous images on the skin's surface. The LED light source is used to illuminate the vein. There are two types of illumination: transillumination and reflected light. The

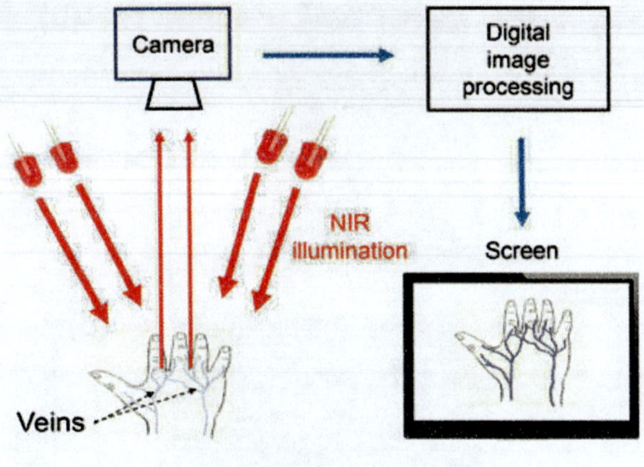

Fig. 1 Principle of a vein finder system using NIR illumination and working on reflected light

first method is employed for imaging deep veins. In this work, we apply the reflected light type for imaging vein near the skin's surface (Fig. 1).

Under the NIR illumination, vein and surrounding tissues are distinguished due to their difference in the brightness and contrast. In principle, NIR radiation penetrates the skin where veins are located. When the radiation goes inside a vein, it is strongly absorbed by deoxyhemoglobin in venous blood. As a result, the image formed by reflected and scattered light from the skin exhibits the veins as dark lines on the skin surface. This image signal is collected by a camera and subsequently delivered to a mini-computer for image processing. A series of image filtering techniques are employed to enhance the contrast between veins and skin background. Filtering algorithms are processed by using a standardized programming language Python. After processing, clearer and sharper venous images are obtained and displayed on a monitor to assist phlebotomists.

2.2 *Main Components of a Portable Vein Finder Prototype*

Figure 2 shows the main components used for building up a portable vein finder prototype. The light source contains 5 LEDs (850 nm) and 5 LEDs (940 nm) arranged alternately in a circular form (Fig. 2a). The circular arrangement of LEDs provides illumination with uniform intensity, thus allowing to obtain high contrast venous images. The image capturing device is a 3 MP camera with a focal length of 12 mm (Fig. 2b). The image signal is processed by a Raspberry Pi 4 (Fig. 2c) before displaying on a Raspberry Pi touch screen (Fig. 2d).

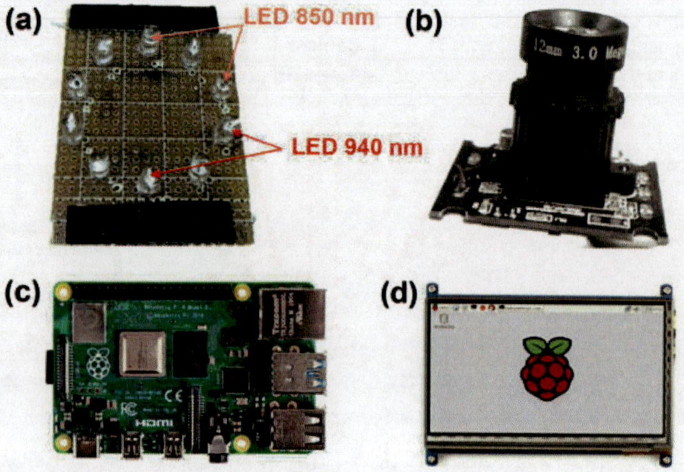

Fig. 2 Main components of a portable vein finder prototype. **a** Illuminated LEDs. **b** Camera. **c** Raspberry Pi 4. **d** Touch screen

3 Results and Discussion

3.1 Image Processing for Displaying on a Monitor

Venous images obtained from the camera generally have low contrast. Therefore, image processing is required to enhance the contrast and sharpness. It has been demonstrated that median filtering and histogram equalization are crucial for this purpose [6]. As shown in Fig. 3, the input image (obtained from the camera) with

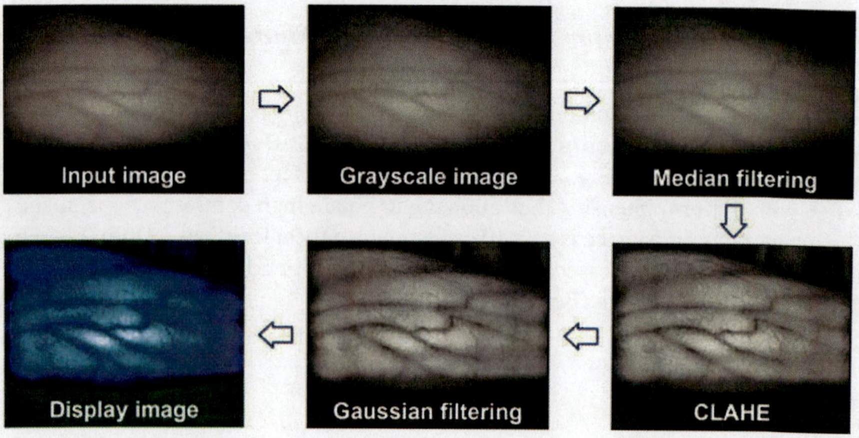

Fig. 3 Enhancement of venous images by using a series of different image processing methods

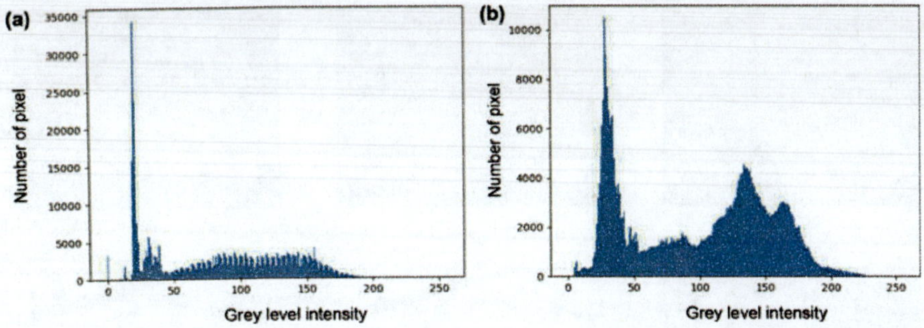

Fig. 4 Histogram of the input image and image after processing. **a** Input image. **b** Output image after Gaussian filtering

venous patterns is firstly converted to grayscale and then processed with median filtering. This process removes salt-and-pepper noise from the image while keeps the edges and venous lines. Next, contrast limited adaptive histogram equalization (CLAHE) [7] improves the contrast of venous images. The veins are now highlighted much better on the skin's surface as their colour is significantly darkened. The noise is further reduced by Gaussian filtering. Finally, a colourmap is added for better visualization, and the output image is displaying on a monitor. Compared with the initial image, the veins are now more easily to locate which can help medical practitioners make an accurate venipuncture.

To examine the effect of image processing, we have analyzed the initial image and the image after processing with Gaussian filtering. The result indicates that the contrast between the vein and the surrounding skin surface is only around 0.1. After processing, the contrast is about 0.3, which is 3 times higher. That means image processing is important for improving the image quality. In addition, the histogram of the image after processing also demonstrates a clear improvement. At the initial image, the number of pixels with a grey level intensity below 50 is quite large (Fig. 4a). After processing, those pixels decreases significantly, and grey level intensity (from 0 to 250) are well distributed in all pixels (Fig. 4b).

3.2 *Image Processing for Projecting the Venous Patterns on the Skin*

Projecting the venous patterns on the skin's surface would be very helpful for phlebotomists as they can perform venipuncture directly on the skin without the need to look at a display. To achieve this purpose, venous patterns are needed to extract from the initial image. Similarly to the previous image processing, salt-and-pepper noise is reduced in a typical pre-processing step to improve later processing results (Fig. 5). Subsequently, the local adaptive threshold [8] is applied to venous segment patterns

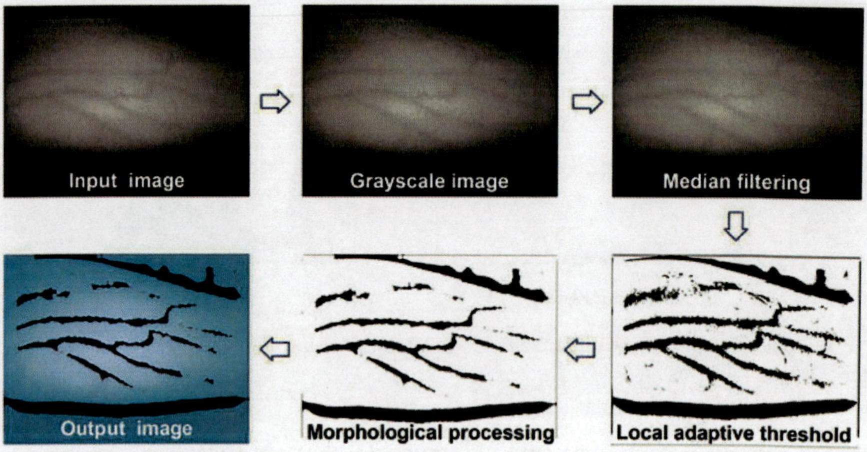

Fig. 5 Segmenting venous patterns for projecting it on the skin's surface using a series of different image processing algorithms

out of the image. This method's advantage is that it can remove the background effectively while keeping the vein images and its boundary. Next, morphological processing further reduces the noise signal, such as hair on the hand thus clearer and sharper venous images are obtained [9]. Finally, the colourmap is added for better visualization, and the output image is ready to project on the skin's surface. Although projecting venous patterns on the skin has not been done in this work, it could be realized using a pico-projector combined with an appropriate optical system [10]. It is an interesting direction for further study.

3.3 Design of a Portable Vein Finder Prototype

To create a portable device that can help doctors and medical practitioners detect blood vessels, we have designed a vein finder prototype (Fig. 6). This system includes all the main components that have been listed in Sect. 2.2.

3.4 Portable Vein Finder Prototype

Based on the above design, we have built up a portable vein finder prototype (Fig. 7a). The cover of this prototype was manufactured by a three-dimensional (3D) printing system. The advantage of this system is low-cost (all components cost only around 300 USD) but it can display reliably the real-time (around 25 frames per second), high-quality venous images/patterns on a monitor (Fig. 7b). Thanks to the image

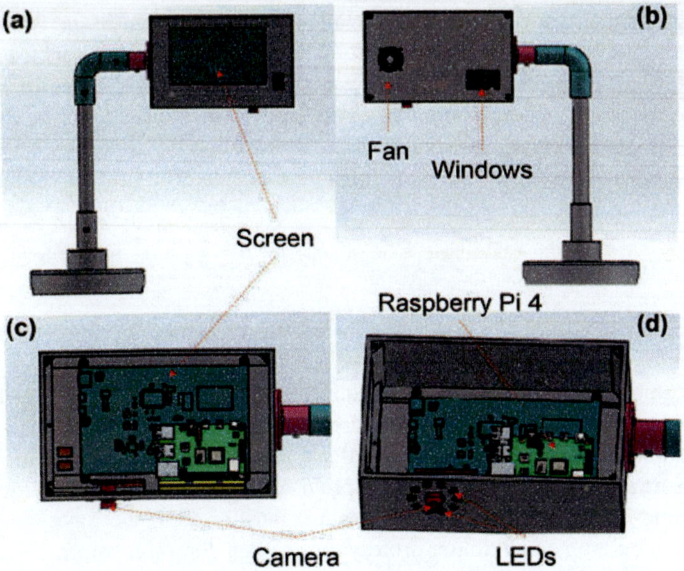

Fig. 6 Design of the portable vein finder prototype. **a** Front and **b** back view. **c** Position of the screen and Rasberry Pi 4 inside the vein finder prototype. **d** Illuminated LEDs and the camera are highlighted

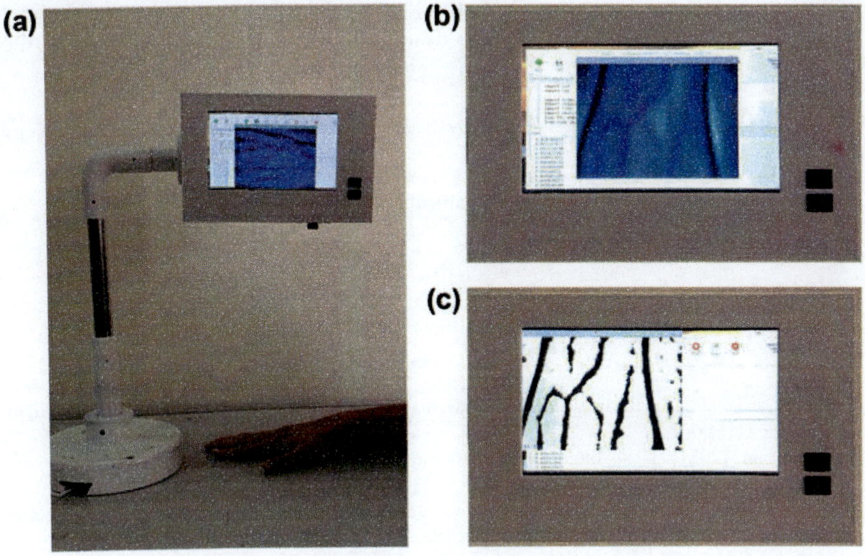

Fig. 7 **a** Real vein finder prototype. **b** Clear and sharp venous image with enhanced contrast is obtained by real-time image processing. **c** Binary image showing venous patterns that can be projected on the surface of the skin

processing, the contrast between the vein and surrounding skin tissue is around 0.3. The image can be turned to binary for projecting back on the skin surface (Fig. 7c). We have tested this prototype with about 20 volunteers, and the results show that blood vessel can be observed clearly at different skin areas including finger, the front and back of the hand and forearm. The system can be improved by adding an appropriate optical system connected with a pico-projector to project venous images on the skin's surface.

4 Conclusion

We have demonstrated that vein visualization by using non-invasive NIR illumination is useful for making safe venipuncture. The contrast between venous and skin background can be significantly enhanced via imaging processing techniques such as median filtering, contrast-limited adaptive histogram equalization, local adaptive threshold and morphological processing. Furthermore, we have successfully designed and built up a vein finder prototype based on common commercial components. The system can display the real-time venous image on a monitor, and venous patterns can be segmented from the image to be ready for projecting on the skin surface. Our work provides a cost-effective solution for the improvement of high-technology based healthcare in developing countries. We will develop this device further for commercial purpose. We are expected to provide quick and excellent service for customers as we master the design, the fabrication technology and the image processing techniques.

Acknowledgements This research is funded by the Vietnam's Ministry of Science and Technology under Project ĐTĐL.CN-40/19.

Conflicts of Interest The authors have no conflict of interest to declare.

References

1. Jain et al (2004) An introduction to biometric recognition. IEEE Trans Circuits Syst Video Technol 14(1):4–20
2. Lee et al (2011) New finger biometric method using near infrared imaging. Sensors 11(3):23192333
3. Marathe et al (2014) A novel wireless vein finder. In: International conference on circuits, communication, control and computing. IEEE, Bangalore, India, pp 277–280
4. Pan et al (2019) Vein pattern locating technology for cannulation: a review of the low-cost vein finder prototypes utilizing near infrared (NIR) light to improve peripheral subcutaneous vein selection for phlebotomy. Sensors 19(16):3573
5. Cope M, Delpy DT (1988) System for long-term measurement of cerebral blood and tissue oxygenation on newborn infants by near infra-red transillumination. Med Biol Eng Comput 26(3):289–294

6. Tran LT, Pham HTT (2018) Designing and building the vein finder system utilizing nearinfrared technique. In: 7th international conference in Vietnam on the development of biomedical engineering. Springer, Singapore, Vietnam, pp 383–387
7. Pizer et al (1990) Contrast-limited adaptive histogram equalization: speed and effectiveness. In: Proceedings of the first conference on visualization in biomedical computing. IEEE, Atlanta, USA, pp 337–345
8. Singh R et al (2011) A new local adaptive thresholding technique in binarization. Int J Comput Sci Issues 8:271–277
9. Oh J, Hwang H (2010) Feature enhancement of medical images using morphology-based homomorphic filter and differential evolution algorithm. Int J Control Autom 8(4):857–861
10. Tran VT et al (2017) Design and enhance the vein recognition using near infrared light and projector. Sci Technol Dev J 20(K2):91–95

Applying Soft Actuator Technology for Hand Rehabilitation

Quyen N. T. Vo, Tri T. M. Huynh, and Son V. T. Dao

Abstract In modern society, the ever-growing prevalence of stroke or hand disability, putting an extreme burden on the limited financial resources and capacities of health care providers in some of the Low and Middle-Income Countries (LMIC). Due to the large scale of flexibility and adaptability, soft robots turn into a smart solution for hand rehabilitation, especially soft pneumatic actuators. In this paper, we proposed a general design of soft robotic glove for hand rehabilitation with functional grasp pathologies by applying soft pneumatic actuator technology. Soft actuator included integrated channel connect each chamber to lead the air to go through the entire actuator and produce bending motion conforms with human finger shape and motion. Base on this aspect, before fabrication, a Finite Element Model (FEM) was built to analyze the bending curvature of these actuators. The soft actuator is cast in 3D printing molds using low-cost elastomer Ecoflex™ 0050. We carried out empirical tests to validate the findings of experimental actuators and compared them to FEM data displaying strong agreement. Finally, the design of a wearable for patients to practice rehabilitative activities such as grasping patterns. The outcome of this paper shows that the design of rehabilitation glove using soft actuator technology gave a successful solution for the health care system with low-cost material.

Keywords Soft pneumatic actuator · Soft robotic glove · Hand rehabilitation · Ecoflex™ 00-50 · FEM simulation

Q. N. T. Vo · S. V. T. Dao (✉)
School of Industrial Engineering and Management, International University HCMC, Ho Chi Minh City, Viet Nam
e-mail: dvtson@hcmiu.edu.vn

T. T. M. Huynh
School of Electrical Engineering, International University HCMC, Ho Chi Minh City, Viet Nam

V. Van Toi et al. (eds.), *8th International Conference on the Development of Biomedical Engineering in Vietnam*, IFMBE Proceedings 85,
https://doi.org/10.1007/978-3-030-75506-5_10

1 Introduction

According to the statistics of the Association of Stroke Prevention, in Vietnam, there are approximately two thousand new cases of stroke each year. Among chronic stroke survivors, up to 92% have hemiplegia who live in the community need to be rehabilitated. For most of these cases, patients suffer loss functionality of partial or total absence of hand motor, so they require long-term care from others in living. This hampers immensely daily activities that scale down the quality of life and simultaneously incur the highest costs to the healthcare system [1]. Traditional rehabilitation supports patients in improving hand functions by requiring them to perform repetitive task practice (RTP) [1–6]. Patients might practice many exercises, including individual tasks integrated from complex motion under the help of physical therapists. However, this method is expensive due to the long training time and the use of a variety of support equipment. Therefore, there should be an accessible system that assists physical therapy and helps the patients conduct exercises on their own at home or in the clinic.

Findings of some clinical studies have demonstrated that stroke patients who perform intense, repetitive movements under robot-assisted devices could gain a significant improvement in hand rehabilitation [2]. However, conventional rehabilitation robots are usually applied rigid-body components such as iPAM robots [3], which are inflexible, heavy, and bulky. For this reason, a new type of robot adopting hyperelastic materials for the main body creates a large scale of flexibility, adaptability, and deformability for hand rehabilitation robots [4, 5].

There are various types of soft robotic hands for rehabilitation have been proposed so far. Polygerinos et al. designed an open-palm glove configuration attached to soft actuators on its top [6]. A related approach was also investigated by the Polygerinos group consisting of molded elastomeric chambers with fiber reinforcements which provided the potential to increase user a wide range of motion of individual fingers [7]. Another design methodology is composed of 3D printable soft material based on the topology optimization method showed a powerful potential in designing and fabrication [8].

In this paper, we present soft hand rehabilitation applied pneumatic networks.

(PneuNets) technique fabricating by low cost and available material, EcoflexTM 0050. The design structure is modified from Polygerinos et al. [6] by making a soft robotic glove by integrating five soft pneumatic actuators to simulate real human hands and demonstrate its feasibility for rehabilitation task with grasping test.

2 Materials and Design

2.1 *Materials*

EcoflexTM 00-50 is being used to test the material properties affecting different fabrication presenting on the bending of the SPA, encompassing grasping object test, bending angle, and actuation speed characteristics in the real situation. Finite element modelling (FEM) method is used to analyze the mechanical properties and behavior of soft actuators.

2.2 *Soft Pneumatic Actuator*

In task-specific training or training of activities of daily living (ADL), grasping exercise is found as movement on activity level, which related to the performance of patients' abilities in actual life [9].

The design for a device assisting hand rehabilitation exercises is that a glove would provide tight-fitting soft fingers that match the shape of the human fingers. The intended purpose of this design is to achieve the grasping motion of the entire hand and fingers base on the bending posture of the soft actuator when it inflates. Therefore, actuators would be mounted on a glove or wearable device that conform to the fingers in a closed fist configuration.

The conceptual function of the proposed soft glove is operated by pressure-flow (Fig. 1b). When gas is pump into a chamber made by highly deformable materials (elastomers), the pressure exerted is obtained to bend the whole actuators. Additionally, the stiffness of the elastomer and the pressure-flow rate characteristic of the pump/compressor control the reaction time of actuators. Therefore, the fingers of patients are assisted with additional force to close their fist. In this study, the glove size was determined to a medium-sized hand with all five fingers have the length of 112 mm and use the same basic actuator design.

The mathematical model was inspired by the modelling method in Hao et al. [10] and Onal et al. [11]. With two assumptions to simplify are the total length of the actuator is assumed to be constant, and there is no radial expansion in modelling.

Because most of the chambers have similar geometrical structure, the bending angle is calculated by using parameters of one chamber and then integrated. Based on the geometries, the axial stress (σ_x) in the material for the single chamber is as follows

$$\sigma_x = P_A = \frac{\mathrm{Pwh}}{(2\mathrm{h} + \mathrm{w} + 2\mathrm{t})\mathrm{t}} \tag{1}$$

where P is the pressure exerted on the inside walls of the chamber; w, h, and t are described in Fig. 2.

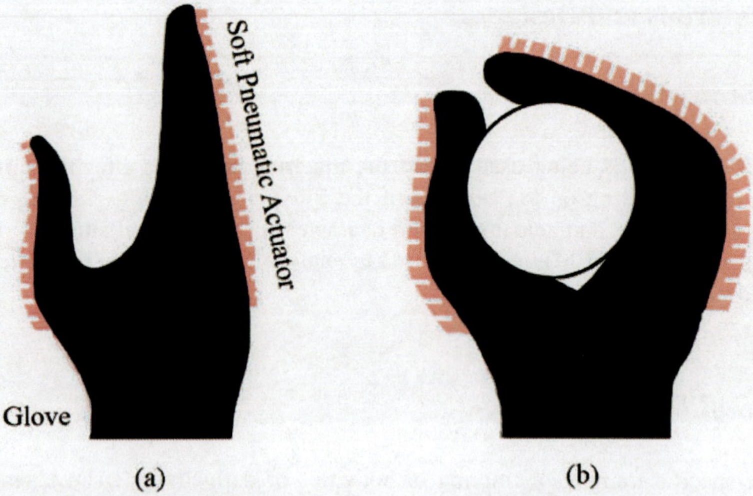

Fig. 1 Glove design for hand rehabilitation where the hand is in the open state (**a**) and close state (**b**) when grasping object

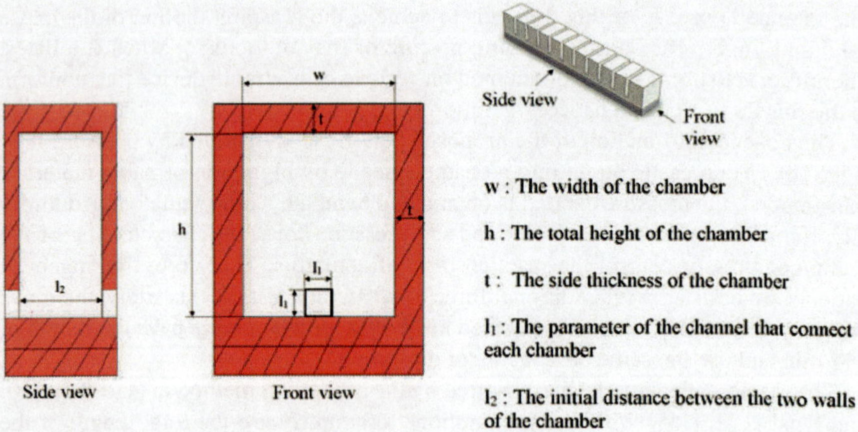

Fig. 2 The cross-sectional geometries of one chamber with the listed parameters (The left and right wall thickness of the side view are not considered in the model)

$\varepsilon(\sigma_x)$ is assumed to be the strain function of stress obtained by data extracted from the inside walls of the actuator modelling by FEM. The stress–strain results are modified by using Excel to get the function. Because the resulting strain$_x$ is a nonlinear function of the induced stresses, we used a fourth polynomial fitting function to express the relationship between stress (σ_x) and strain (ε_x) with data extracted from Abaqus by applying 9 kPa pressure to the actuator. Due to system configuration, stress–strain data of the entire element could not be obtained. Therefore, the stress–strain result was computed by taking the average number of three chambers, including the first, middle, and the last.

$$\varepsilon(\sigma_x) = -332,663\sigma_x^4 + 25,960\sigma_x^3 - 616.03\sigma_x^2 + 8.4002\sigma_x - 0.002 \quad (2)$$

According to the axial direction, the axial deformation of the individual chamber, resulting in the change of shape when inflating:

$$D_x = l_2\varepsilon(\sigma_x) \quad (3)$$

where l_2 is the distance between two walls before inflating.

Due to the restriction of the inextensible bottom layer (a paper which embedded in two-layer of elastomer), the soft actuator goes through bending deformation which creates the bending axis. Thus, the bending angle (θ) of the actuator could be calculated by integrating all the bending angles of the chambers:

$$\theta = 2n\tan^{-1}\frac{l_2\varepsilon x(\sigma x)}{2h} \quad (4)$$

where n is the number of channels. And the bending radius of the actuator:

$$\mathrm{R} = \mathrm{L} \quad (5)$$

With L is the total length of the actuator.

The design structures and parameters are modified from Soft Robotic Toolkit [12], which applied PneuNet principle of operation. This design is typical pneumatic networks that can only bend in one direction under pressurization. The fabrication of soft pneumatic actuator consists of an extensible top layer (main body) and inextensible bottom layer (Fig. 3a) which are cast separately with different mold design and then combined, including main body, which is the extensible layer of pneumatic actuator, expands when inflated. The main body is cast in the two-part mold shown in Fig. 3c. It has the gap between the inside walls, and the thickness of each chamber walls are thinner than the gap between the top wall and the inside wall; bottom layer, which is an inextensible layer, contains a paper layer which embedded in the elastomer.

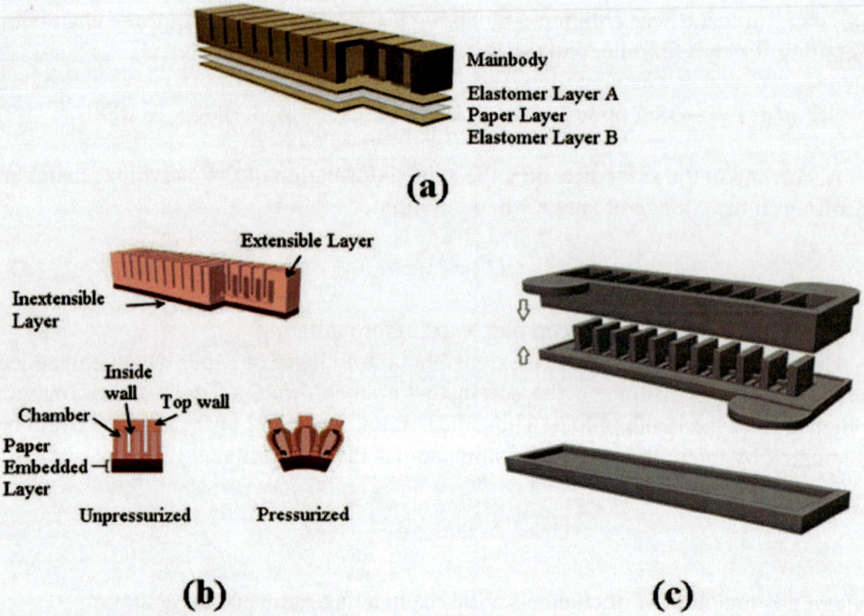

Fig. 3 Design structures of the soft pneumatic actuator [12]. An overview of design (**a**); Crosssectional view (**b**); Mold design to cast the actuator (**c**)

3 Results and Discussion

3.1 Mold Design

The solid model of mold design was carried out by using CAD through SolidWorks (SolidWorks.2018.SP2.0.Premium, Dassault Systèmes, French). The mold was made by 3D printing (from IN3DPLUS Co. Ltd) (see Fig. 4).

3.2 Simulation and Actual Results

The simulation software that we used is Abaqus (Abaqus 2017; SIMULIA Inc., France).

In the simulation, Ecoflex™ 00-50 (Smooth-On Inc., Macungie, Pennsylvania, USA) was set as the material properties. Due to the lack of equipment for fabricating the tensile test samples to find the best-fit material model, the mechanical properties using for simulation are implemented from Soft Robotic Toolkit [12] and Kulkarni [13]. Therefore, A reduced polynomial model (Yeoh) with coefficients of the fitted material models for Ecoflex series ($C10 = 1.27 \times 10^{-2}$; $C20 = 4.23 \times 10^{-4}$) and a

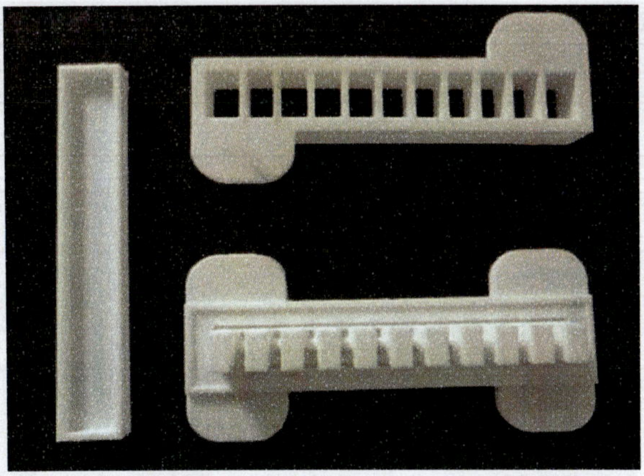

Fig. 4 Mold design by 3D printing

Young's Modulus with parameters for inextensible paper layer was used to build the properties of the material. Ecoflex™ 00-50 was assigned to two bottom layers and the main body; Paper was assigned to the inextensible layer, as shown in Fig. 5.

We carried out some empirical tests to validate the results of modelling. In the real experiment, one end of the actuator was kept fix by hand as the boundary condition "ENCASTRE" used in FEM. When applying pressure, the inflation of the actuator was observed and photographed by a smartphone camera (see Fig. 6).

According to the literature review, we measured the trajectory of the tip by recording its actuation to estimate the bending curve of the actuator. The measurement method was proposed by Polygerinos et al. [6].

We used a smartphone camera (Huawei) and a tripod which ensure the camera perpendicular to the plane. The actuator was filmed from the side to observe its tip when inflation. We use the size of the A4 paper to address lens distortion, and a ruler was fixed beside the actuator. Then, MATLAB program was used for analyzing the bending curvature between FEM and proposed (see Fig. 7).

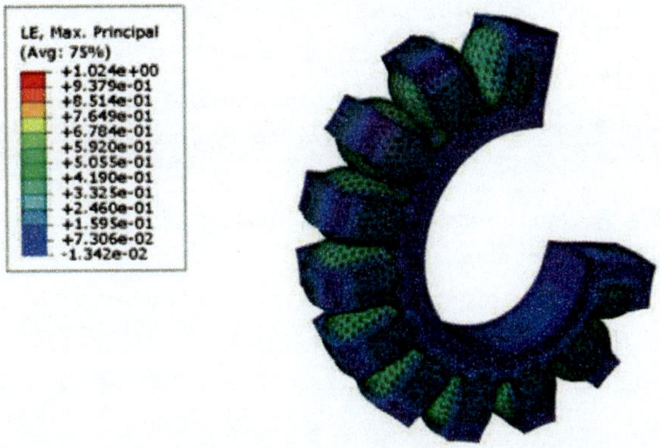

Fig. 5 The bending behavior of the soft pneumatic actuators

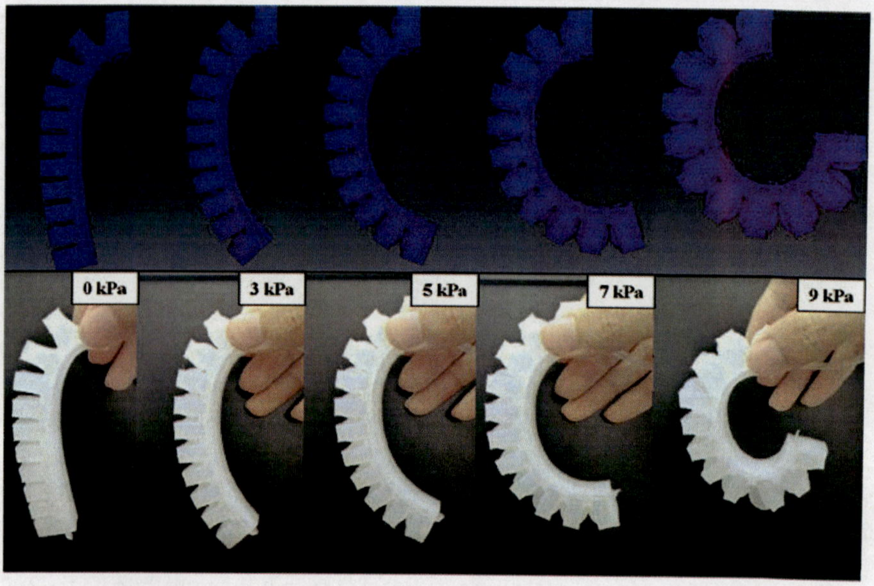

Fig. 6 Comparison between fabricated actuator and simulation model with the applied pressures are 0, 3, 5, 7, and 9 kPa, respectively

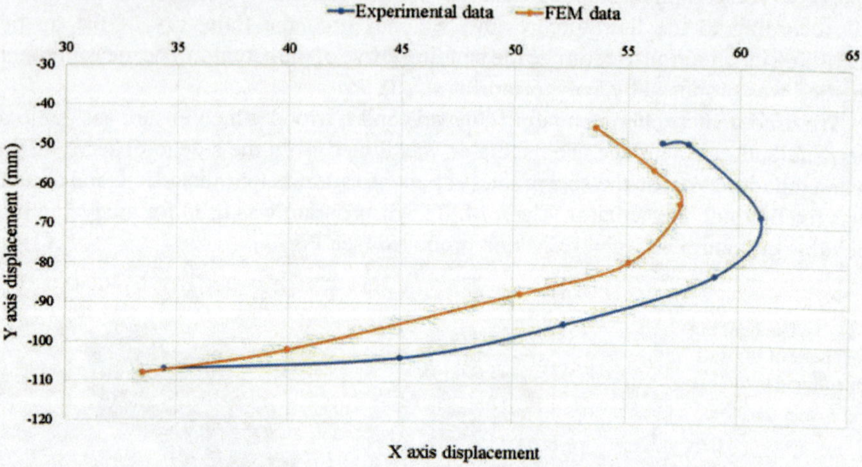

Fig. 7 Comparison of experimental bending curvature and FEM data when recording data of tip displacement

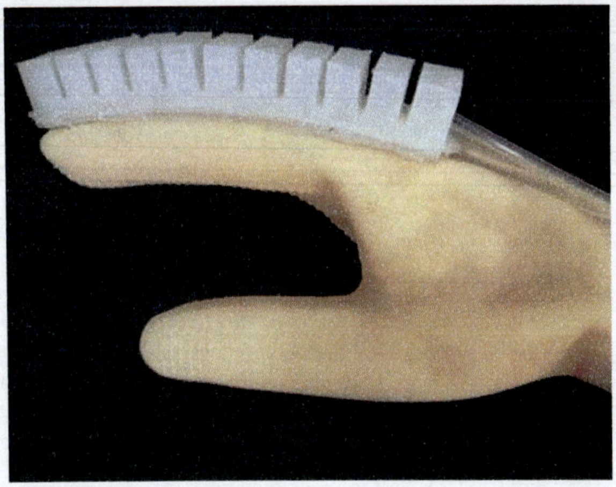

Fig. 8 A prototype of soft hand rehabilitation

3.3 Glove Design for Hand Rehabilitation

With the actuators attached on the top of each finger of the glove, its configuration is completely fitting to the human hand. Elastic fabric (vinyl chloride) was the main material for manufacturing glove. In this section, we demonstrated one of the applications of the soft pneumatic actuator, which is used in rehabilitation glove. As shown in Fig. 8, by placing the soft actuator on the top of the glove's finger, it allows greater concordance at all bending angles of the joints.

Due to the design requirement of rehabilitation glove, the output force of the actuator must be substantial to bend and actuate the finger joint effectively. Besides, the ability of glove should be tested for grasping with various objects in size and stiffness to control internal pressure and exert forces in different patients due to the complications of their diseases.

4 Conclusion

The objective of this work was to make a prototype robotic glove for hand rehabilitation therapies. The concept of inflating a soft pneumatic actuator to generate a guideline and help to develop this field in Vietnam. We presented the design, fabrication, and empirical tests of the soft pneumatic actuator. The finger design is referenced on the principle of the soft pneumatic actuator of but applied low-cost elastomer (Ecoflex™ 00-50). The FEM simulation results are used to compare with reality actuator to demonstrate that the mathematical model could be used to optimize the design and predict the behavior of soft actuators. In conclusion, the analysis of soft pneumatic actuator in this paper can be useful for future design and application of a soft biomedical glove for hand rehabilitation in the future.

Acknowledgements This work is self-funded. The authors did not receive any external funding for this work.

Conflicts of Interest The authors have no conflict of interest to declare.

References

1. Teasell R, Pereira S, Cotoi A (2018) Evidence-based review of stroke rehabilitation. In: The rehabilitation of severe stroke
2. Takahashi CD, Der-Yeghiaian L, Le V, Motiwala RR, Cramer SC (2008) Robot-based hand motor therapy after stroke. Brain 131:425–437
3. Kemna S, Culmer PR, Jackson AE et al (2009) Developing a user interface for the iPAM stroke rehabilitation system. In: IEEE international conference on rehabilitation robotics, pp 879–884
4. Majidi C (2014) Soft robotics: a perspective—current trends and prospects for the future. Soft Rob 1(1):5–11
5. Bao G, Fang H, Chen L, Wan Y, Xu F, Yang Q, Zhang L (2018) Soft robotics: academic insights and perspectives through bibliometric analysis. Soft Rob 5(3):229–241
6. Polygerinos P, Lyne S, Wang Z, Nicolini LF, Mosadegh B, Whitesides GM, Walsh CJ (2013) Towards a soft pneumatic glove for hand rehabilitation. In: 2013 IEEE/RSJ international conference on intelligent robots and systems
7. Polygerinos P, Wang Z, Galloway KC, Wood RJ, Walsh CJ (2015) Soft robotic glove for combined assistance and at-home rehabilitation. Rob Auton Syst (RAS) Spec Issue Wearable Rob 73:135–143
8. Zhang H, Wang Y, Wang MY, Fuh JYH, Kumar AS (2017) Design and analysis of soft grippers for hand rehabilitation. In: Bio and sustainable manufacturing, vol 4
9. Yue Z, Zhang X, Wang J (2017) Hand rehabilitation robotics on poststroke motor recovery. Behav Neurol 1–20
10. Hao Y, Wang T, Ren Z, Gong Z, Wang H, Yang X et al (2017) Modeling and experiments of a soft robotic gripper in amphibious environments. Int J Adv Robot Syst 14(3):172988141770714
11. Onal CD, Chen X, Whitesides GM, Rus D (2016) Soft mobile robots with on-board chemical pressure generation. Robot Res 525–540
12. Holland DP, Park EJ, Polygerinos P, Bennett GJ, Walsh CJ (2014) The soft robotics toolkit: shared resources for research and design. Soft Rob 1(3):224–230
13. Kulkarni P (2015) Centrifugal forming and mechanical properties of silicone-based elastomers for soft robotic actuators

Design of an Automatic System for Bioelectrical Impedance Phase Angle Measurement Using an Analog Multiplier

Vu Duy Hai, Vu Hoang Chuong, Hoang Trong Nam, Nguyen Ngoc Tram, Chu Quang Dan, and Dinh Thi Nhung

Abstract Phase angle (PA), related to bioelectrical impedance, detects changes in tissue electrical properties. Because of being dependent on body cell mass and cell membrane functions, PA has been considered a prognostic indicator in several clinical diseases such as HIV, chronic pulmonary disease, colorectal cancer, and breast cancer. PA has attracted considerable interest and has been frequently referred to as a global health marker within the clinical nutrition community research on cellular health. Therefore, this study is aimed to design a system for automatically measuring the phase angle with high accuracy of an individual body segment after excitation by a current source. The phase angle measuring system comprised a sine-wave generator, a low-intensity, high-frequency (100 kHz) current pump, a phase difference-to-voltage converter, and a high-resolution analog-to-digital converter (ADC). The design of the phase difference-to-voltage converter was based on a high-precision analog multiplier and a peak detector. The phase difference-to-voltage converter's performance assessment was conducted by applying two programmable phase-difference signals generated by the pulse generator to ensure the accuracy of the phase angle difference. The converted phase angle was then read by a 16-bit ADC and displayed. The two identical-frequency sine-wave signals were programmatically generated with phase angle difference step of 10° ranged from 10° to 90° for phase difference to voltage converter inputs. The measurement showed the error of $0.5° \pm 0.2°$ (mean $\pm$ standard deviation) for the overall nine phase angle difference steps.

Keywords Phase angle (PA) · Bioelectrical impedance analysis (BIA) · Analog multiplier · Peak detector

---

V. D. Hai (✉) · V. H. Chuong · H. T. Nam · N. N. Tram · C. Q. Dan · D. T. Nhung
Biomedical Electronics Center, Hanoi University of Science and Technology, Hanoi, Vietnam
e-mail: hai.vuduy@hust.edu.vn

V. Van Toi et al. (eds.), *8th International Conference on the Development of Biomedical Engineering in Vietnam*, IFMBE Proceedings 85,
https://doi.org/10.1007/978-3-030-75506-5_11

1 Introduction

Disease-related malnutrition is an extensive problem in most healthcare settings. In cancer patients, malnutrition is a common demonstration and a major contributor to morbidity and mortality [1]. Malnutrition is characterized by changes in cellular membrane integrity and alterations in fluid balance [2]. Hence, body composition analysis is an important part of the overall nutritional evaluation in cancer patients [3, 4]. Factually, nutrition status has been assessed by several methods, including using objective parameters such as anthropometric (weight change, height change, body mass index (BMI), body circumference and skinfold thickness), subjective global assessment (SGA), and laboratory measures (measuring serum proteins such as albumin, transferrin, retinol-binding protein) [5]. However, most of them have shown limitations in terms of application. Anthropometric measures are impracticable in clinical settings because they are time-consuming and require well-trained staff. Some of the objective measurements, such as albumin serum one, are probably affected by many non-nutritional factors [5–8].

Additionally, the assessment of nutritional status changes in a short period becomes difficult because some objective parameters, such as albumin, have long half-lives. Bioelectrical Impedance Analysis (BIA), a little-used tool to evaluate nutritional status, can overcome some of these challenges. This technique is a portable method developed in recent decades that combines morphological and functional evaluation [9]. BIA is a safe, non-invasive, and consistent method for evaluating body composition in clinical practice.

BIA has been considered a method to estimate the body composition and evaluate the nutritional status of many patients with various diseases, including cancer patients [1, 10]. Nevertheless, instead of measuring body composition directly, BIA measures two bioelectrical indicators: resistance (R) and reactance (X_C) [11]. When an alternating current with low amplitude and high frequency flows through tissues, R is the opposition of total body water to the flow of that current [12, 13], related to extracellular and intracellular fluid [14]. X_C is the opposition proffered by the current flowing through capacitance produced by tissue interface and cell membranes [14, 15]. Namely, it reflects cell membranes' ability to function as capacitors providing reactance when the alternating current flows through tissues [13]. X_C is related to the cell membrane structure and function [16], i.e., it is related to extracellular and intracellular fluid balance, which are independent of cell membrane integrity [17]. The relationship between capacitance and resistance reflects different electrical properties of tissues affected in different ways by disease, nutritional status, and hydration status [18]. There is a low-amplitude high-frequency electric current (500–800 μA at 100 kHz) flowing through some electrodes placed on the hand and foot to evaluate body composition. Capacitance is the inducement that causes the current phase to lag behind the voltage phase, thus creates a phase shift between them. Geometrically, this shift is calculated based on the angular transformation of the capacitance to resistance ratio called the Bioelectrical Impedance Phase Angle (BIPA) [11]. BIPA is the angle created by the impedance vector and the resistance vector; it is calculated using

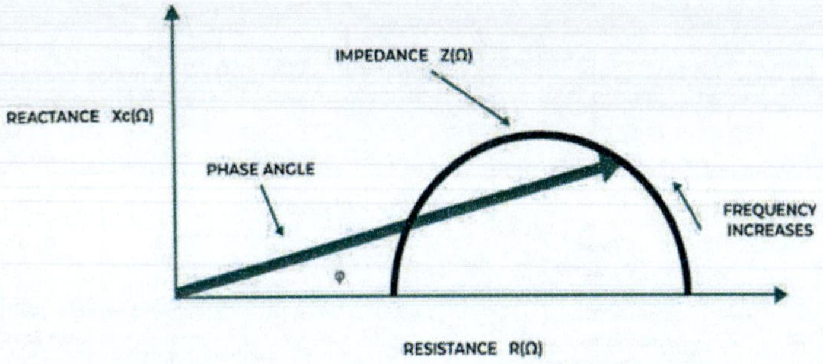

Fig. 1 BIA vector and phase angle

the arctangent of the ratio X_C in degrees [9]. The geometrical relationships among R impedance, reactance (X_C), resistance (R), phase angle, and frequency of the current are illustrated in Fig. 1.

When an electric current is applied to the human body, a portion of the current is stored in cell membranes, creating a BIPA, which is also the phase shift between the impedance and resistance vectors [11]. Consequently, the measured BIPA is governed by many objective factors such as the number of cells with their respective cell membranes, cell membrane integrity, related permeability, the amounts of extra and intracellular fluids [19], and body cell mass (BCM). BIPA might be changed if any changes occur in BCM, either functional defects of cell membranes [2]. By assessing BIPA, it can be concluded that it reflects the relative contributions of fluids (resistance) and cellular membranes (capacitance) of the human body. Low phase angle values suggest cell death or decreased cell integrity, while higher values suggest large quantities of intact cell membranes [9]. BIPA has been considered a prognostic indicator in various clinical situations like HIV, chronic pulmonary disease, colorectal cancer because it may indicate functional changes in the cell membrane and fluid balance [15, 18, 20, 21]. The present study's primary purpose is to design a system for automatically measuring the bioelectrical impedance phase angle with high accuracy of an individual body segment after excitation by a current source.

2 Methods and Materials

The phase angle measuring system is comprised of a sine-wave generator, a low intensity, high-frequency (100 kHz) current pump, a phase difference-to-voltage converter, and a high-resolution analog-to-digital converter (ADC). The design of the phase difference-to-voltage converter is based on a high-precision analog multiplier and a peak detector. The phase difference-to-voltage converter's performance assessment was conducted by applying two programmable phase-difference signals

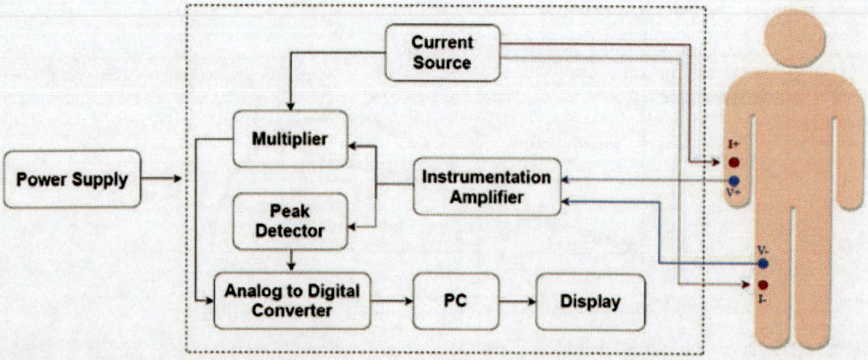

Fig. 2 Block diagram of hardware design

generated by the pulse generator to ensure the accuracy of the phase angle difference. The converted phase angle was then read by a 16-bit ADC and displayed. The proposed research aims to develop a system for measuring electrical impedance phase angle of body segments with help from a combination of some low cost, compact, reliable, and easy design circuits. The system block diagram and circuit design are illustrated in Fig. 2 [22–24].

2.1 Electrode Method

Four-electrode BIA is an informative, accurate, easy to perform, and affordable method of determining joints' impedance. The skin contact electrodes are constituted of Ag/AgCl material used mostly in bio-impedance applications [25].

2.2 Power Supply

The power unit is responsible for supplying power to the entire system. In which the current source block uses chips with a source of ±12V DC and ±15V DC. The analog signal receiving unit uses a source of ±15V DC. The digital processing unit and LCD use 5VDC power. The measurement circuit is powered by the DC-DC converter (JHM1524D12, XP Power, China) with voltage isolation of 4 kV, specifically designed for medical application, to ensure the electrical safety for measuring objects.

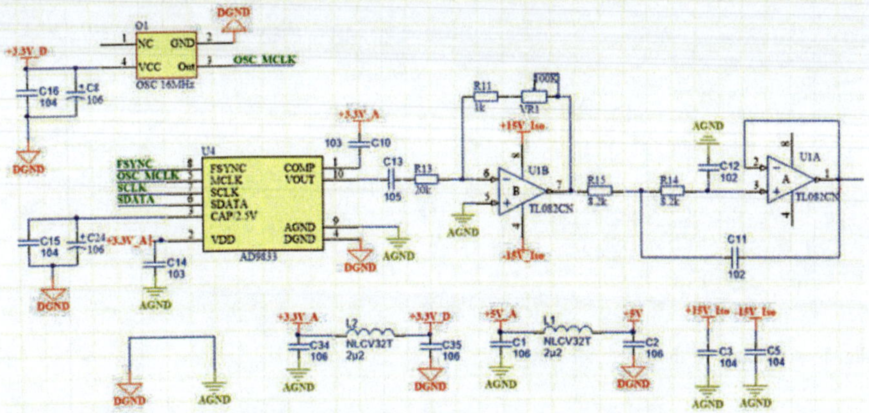

Fig. 3 Schematic of the current source unit

2.3 Current Source

In this design, the improved Howland Current Pump Topology (Fig. 3) is selected for the voltage-current converter due to its ability to provide bi-polar output current, wide frequency range, high output impedance and temperature stability. However, this topology has one drawback: it requires very precise resistors to match the ratio and an op-amp with high open loop gain and high CMRR to achieve a high output impedance current source, especially with high-frequency waveform. Because in bioimpedance measurements, the current source directly affects the quality of the recorded signal. The authors take careful consideration of PCB layout, selection of precise resistors (0.1% tolerance), and the op-amp to ensure the current source's quality. OPA2211 is used due to prominent features such as its low noise density, low voltage, low current noise and high speed.

In this study, a sine-wave high-frequency current source with an amplitude of 4 mA and a frequency of 100 kHz is injected into the body through two electrodes; the induced voltage is picked up by using another pair of electrodes. To ensure patient safety, any residual DC offset component is blocked by a high-pass filter [26–28].

2.4 Instrumentation Amplifier

The body segment's differential signals are fed to the instrumentation amplifier (INA129, Texas Instruments), which amplifies the signals from the body. A band-pass filter suppresses any unwanted noises outside of the signal band. This design's filter type is the 2nd-order Butterworth active filter, which ensures the maximal

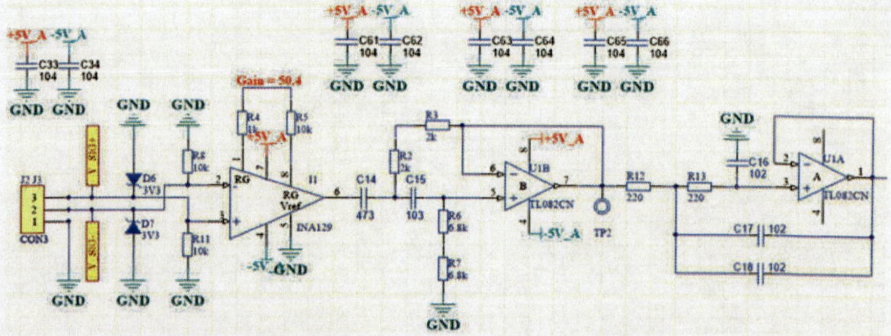

Fig. 4 Schematic of the amplifier unit

flatness and phase linearity, according to Sallen-Key topology, with a passband of 80–120 kHz. The schematic of the amplifier unit is shown in Fig. 4.

2.5 Bioelectrical Impedance Phase Angle Measurement Unit

The current is injected simultaneously into the body segment and a well-known reference resistor (470 Ω) in series. The voltage across the reference resistor and the instrumentation amplifier's signal is then fed to the multiplier IC (Analog Devices). The output voltage of AD633 is calculated as the following formula.

$$V = \frac{A.B}{2 \times SF} \cos \varphi \tag{1}$$

A is the amplitude of the reference signal, B is the amplitude of the signal attained from the body, SF is an adjusted scale factor of IC AD633, φ is the phase difference between the reference signal and the signal attained from the body. B is determined by a high-precision, high-speed peak detector unit. The schematic of the peak detector unit is illustrated in Fig. 5.

The phase difference-to-voltage converter includes an analog multiplier using a high-precision voltage reference generator and an active low-pass filter. The detailed schematic of the phase difference-to-voltage converter is illustrated in Fig. 6.

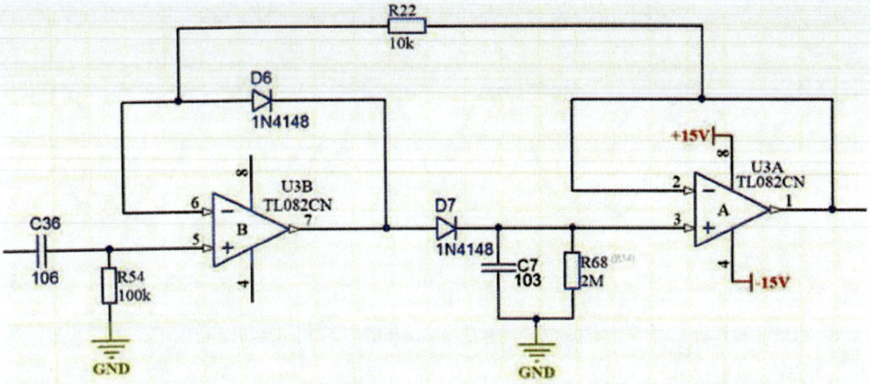

Fig. 5 Schematic of the peak detector unit

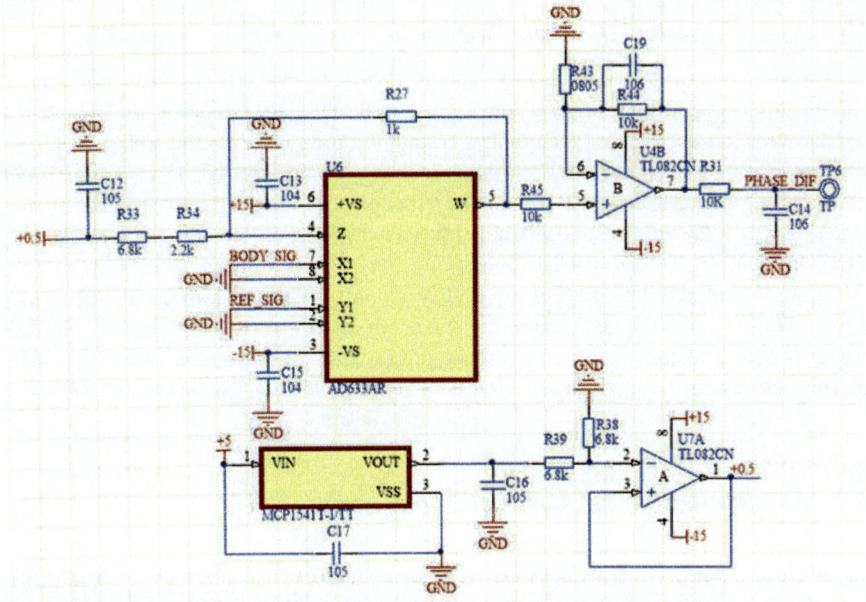

Fig. 6 Schematic of the multiplier unit

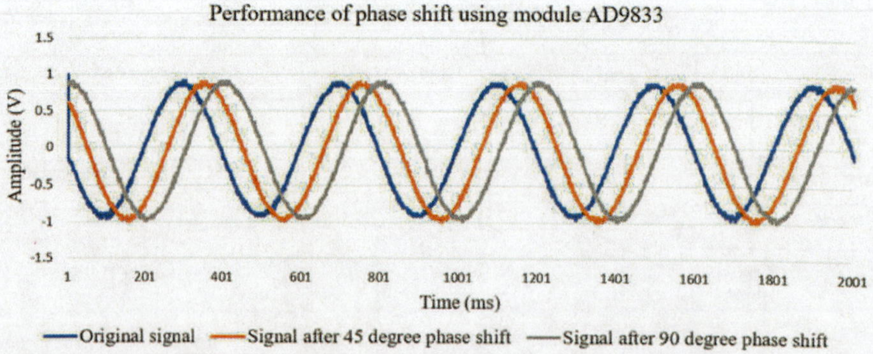

Fig. 7 The two generated sine-wave signals at a phase difference of 90° and 45°

3 Results and Discussion

3.1 Assessment of the Sine-Wave Generator Using AD9833

The designed generator using AD9833 can simultaneously generate two sine-wave signals with the precise programmable frequency range of 50–200 kHz and phase difference range of 0° to 180°. Figure 7 shows examples of the two generated sinewave signals at phase differences of 90° and 45°, respectively. The data used in the figures was sampled and extracted from a digital oscilloscope (MSOX2024A, KEYSIGHT).

Figure 7 demonstrates that setting up two signals with a custom phase difference using AD9833 gives the desired result. The accuracy of the two signals helps reduce the error of the results obtained from the phase shift measurement unit; thus, they are more accurate.

3.2 Assessment of the Current Source

The ability to supply a constant current independent of load resistance is the most important technical specification of the current source. To assess the current source, the current amplitude was fixed at the value of 1 mA; then the load resistors were changed within resistor values of 100 Ω, 220 Ω, 560 Ω, 680 Ω, 1 kΩ, 1.2 kΩ, 2 kΩ, 5.1 kΩ, and 10 kΩ. The voltage dropping in the load resistors was then measured to plot a dependency graph (Fig. 8) of the load resistance and the voltage to check the linearity in a certain range of resistors that the current source is designed to meet. From the figure, we can conclude that the curve is linear, with the load's resistance smaller than 4 kΩ. This means the maximum load which the current source can supply without distortion. It is approximately 4 kΩ.

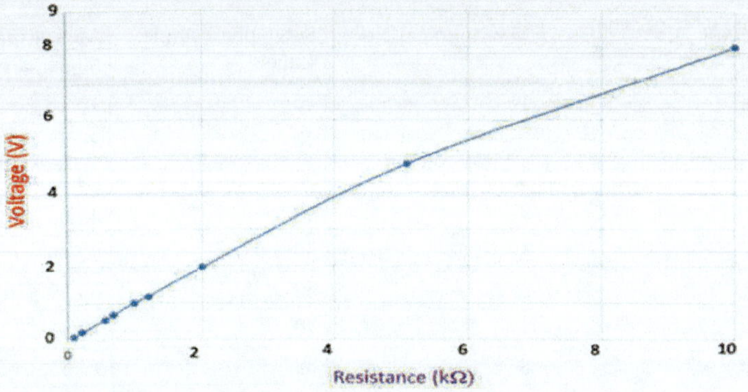

Fig. 8 Voltage–load dependency graph

While the body's internal impedance is less than 1181 Ω [29], the resistance of the load used in this assessment ranges from 0 to 4 kΩ in which the supplied constant current is independent of the load resistance. This means the current source will work effectively when the proposed system is used for body measurements without harming the patients.

3.3 *Assessment of the Phase Angle Measuring Unit*

This sub-section evaluates the accuracy of the phase angle measurement circuit. The evaluation is based on comparing the phase difference between the setup values and the measuring results. The two identical-frequency sine-wave signals were programmatically generated with phase angle difference step of 10° and ranged from 10° to 90° for phase difference-to-voltage converter inputs. A 12-bit ADC then sampled the signal with the sampling rate of 100 samples/s. The measuring phase angle values are shown in Table 1. The measuring results show the error of 0.5° ± 0.2° (mean ± standard deviation).

Figure 9 illustrates the correlation between measuring phase values and setup phase values. The proposed system's measuring results have a high correlation with the setup values (coefficient of determination $R^2 = 0.997$).

The results obtained from the measuring circuit give an acceptable error of 0.5° ± 0.2°over the range of 10° to 90°. Adjusting methods and the deviation of component values are the main factors causing the error above. These results show that the proposed system built on the given method is feasible and performs the desired function. When performing bioelectric impedance phase angle measurements, the results obtained in colorectal cancer patients range from 2° to 8° [18]; the thresholds used for classification at advanced cancer patients or liver cirrhosis ones are also below 10° [1, 9]. Hence the authors want to continue to improve the system, narrow the measuring range to 0 to 10° and increase the accuracy of the measurement results.

Table 1 The table of setup and measuring phase angle results

Measurement No.	Setup phase angle value	Measuring phase angle value 1
1	10	10.26
2	20	20.75
3	30	30.8
4	40	40.6
5	50	50.49
6	60	60.29
7	70	70.11
8	80	79.85
9	90	89.5

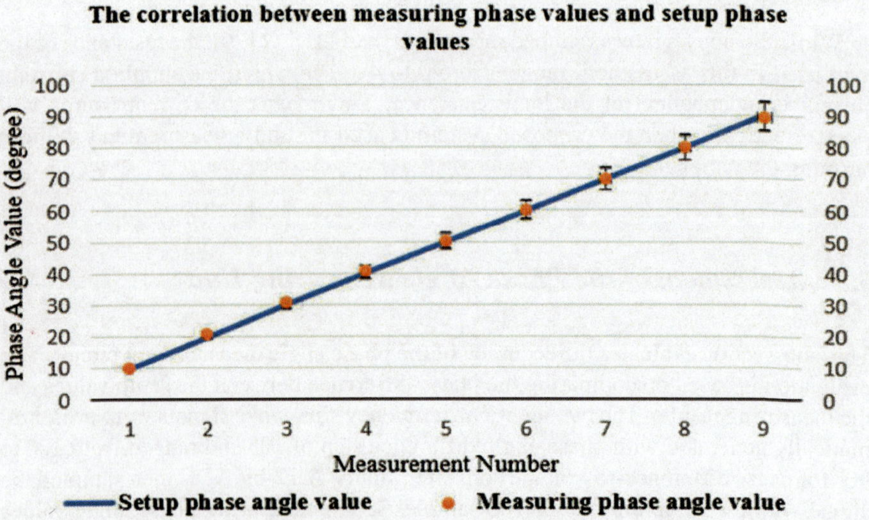

Fig. 9 The correlation between measuring phase values and setup phase values

Then, the direct measurement of the BIPA index on the body applied to monitor patients' nutritional status will bring great value.

4 Conclusion

Based on the results and analysis described earlier, the proposed measuring system works effectively in the measuring range and performs the desired function. This system offers an ability to automatically measure the bioelectrical impedance phase angle with low error. The initial results of this study are promising. They can be potentially developed to become instrumentation devices capable of detecting the bioelectrical impedance phase angle to support the clinical diagnosis and research purposes. The overall evaluation of the system could be temporarily performed on a well-known R-C model. In further studies, the authors would assess the system on individual body segments after meeting the safeguards and being approved by the ethics committee for clinical and research applications. The results would be published in further studies.

Acknowledgements The authors would like to acknowledge the support of the Biomedical Electronics Center (BMEC), School of Electronics and Telecommunications (SET), the university clinic, Hanoi University of Science and Technology (HUST).

Conflict of Interest No potential conflict of interest was reported by the authors.

References

1. Sarhill N, Mahmoud FA, Christie R, Tahir A (2003) Assessment of nutritional status and fluid deficits in advanced cancer. Am J Hosp Palliat Care 20:465–473
2. Barbosa-Silva MC, Barros AJ, Post CL, Waitzberg DL, Heymsfield SB (2003) Can bioelectrical impedance analysis identify malnutrition in preoperative nutrition assessment? Nutrition 19:422–426
3. Cox-Reijven PL, van Kreel B, Soeters PB (2003) Bioelectrical impedance measurements in patients with gastrointestinal disease: validation of the spectrum approach and a comparison of different methods for screening for nutritional depletion. Am J Clin Nutr 78:1111–1119
4. Fredrix EW, Saris WH, Soeters PB et al (1990) Estimation of body composition by bioelectrical impedance in cancer patients. Eur J Clin Nutr 44:749–752
5. Bauer J, Capra S, Ferguson M (2002) Use of the scored Patient Generated Subjective Global Assessment (PG-SGA) as a nutrition assessment tool in patients with cancer. Eur J Clin Nutr 56:779–785
6. Carney DE, Meguid MM (2002) Current concepts in nutritional assessment. Arch Surg 137:42–45
7. Waitzberg DL, Correia MI (2003) Nutritional assessment in the hospitalized patient. Curr Opin Clin Nutr Metab Care 6:531–538
8. Detsky AS, Baker JP, Mendelson RA, Wolman SL, Wesson DE, Jeejee-bhoy KN (1984) Evaluating the accuracy of nutritional assessment techniques applied to hospitalized patients: methodology and comparisons. JPEN J Parenter Enteral Nutr 8:153–159
9. Selberg O, Selberg D (2002) Norms and correlates of bioimpedance phase angle in healthy human subjects, hospitalized patients, and patients with liver cirrhosis. Eur J Appl Physiol 86(6):509–516
10. Simons JP, Schols AM, Westerterp KR, ten Velde GP, Wouters EF (1995) The use of bioelectrical impedance analysis to predict total body water in patients with cancer cachexia. Am J Clin Nutr 61:741–745
11. Baumgartner RN, Chumlea WC, Roche AF (1988) Bioelectric impedance phase angle and body composition. Am J Clin Nutr 48:16–23
12. Toso SPA, Gusella M, Menon D, Bononi A, Crepaldi G et al (2000) Altered tissue electric properties in lung cancer patients as detected by bioelectric impedance vector analysis. Nutrition 16(2):120–124
13. Mattar JÁ (1996) Application of total body bioimpedance to the critically ill patient. Brazilian Group for bioimpedance study. New Horiz 4(4):493–503

14. Piccoli A, Nescolarde LD, Rosell J (2002) Análisis convencional y vectorial de bioimpedância en la práctica clínica. Nefrologia 22(3):228–238
15. Scwenk A, Beisenherz A, Romer K, Kremer G, Salzberger B, Elia M (2002) Phase angle from bioelectrical impedance analysis remains an independent predictive marker in HIV infected patients in the era of highly active antiretroviral treatment. Am J Clin Nutr 72(2):496–501
16. Kyle UG, Bosaeus I, De Lorenzo AD, Deurenberg P, Elia M, Gómez JM et al (2004) Bioelectrical impedance analysis-part I: review of principles and methods. Clin Nutr 23(5):1226–1243
17. De Lorenzo A, Barra FPA, Sasso GF, Battistini NC, Deureberg P (1991) Body impedance measurements during dialysis. Eur J Clin Nutr 45:321–325
18. Gupta D, Lammersfeld CA, Burrows JL, Dahlk SL, Vashi PG, Grutsch JF et al (2004) Bioelectrical impedance phase angle in clinical practice: implications for prognosis in advanced colorectal cancer. Am J Clin Nutr 80(6):1634–1638
19. Barbosa-Silva MC, Barros AJ (2005) Bioelectrical impedance analysis in clinical practice: a new perspective on its use beyond body composition equations. Curr Opin Clin Nutr Metab Care 8:311–317
20. Faisy C, Rabbat A, Kouchakji B, Laaban JP (2000) Bioelectrical impedance analysis in estimating nutritional status and outcome of patients with chronic obstructive pulmonary disease and acute respiratory failure. Intensive Care Med 26:518–525
21. Mushnick R, Fein PA, Mittman N, Goel N, Chattopadhyay J, Avram MM (2003) Relationship of bioelectrical impedance parameters to nutrition and survival in peritoneal dialysis patients. Kidney Int Suppl 87:S53–S56
22. Ngoc PP, Hai VD, Bach NC et al (2015) EEG signal analysis and artifact removal by wavelet transform. In: The 5th international conference on biomedical engineering in Vietnam. IFMBE proceedings, vol 46. Springer, Cham
23. Hai VD, Thuan ND (2010) Design of laboratory information system for healthcare in Vietnam BK-LIS. In: The 3rd international conference on communications and electronics, art. no. 5670692, pp 110–114
24. Vu HD, Nguyen TD, Pham NP et al (2010) A design of renal dataflow control and patient record management system for renal department environment in Vietnam. In: The third international conference on the development of biomedical engineering in Vietnam. IFMBE proceedings, vol 27. Springer, Berlin
25. Hai VD, Hung PD, Hung PM et al (2019) Design of noninvasive hemodynamic monitoring equipment using impedance cardiography. In: The 7th international conference on the development of biomedical engineering in Vietnam. IFMBE proceedings, vol 69. Springer, Singapore
26. Hai VD, Hung PM, Trung LHP et al (2017) Design of software for wireless central patient monitoring system. In: Proceedings of KICS-IEEE international conference on information and communication with Samsung LTE&5G special workshop, Hanoi, Vietnam, pp 214–217
27. Hai VD (2021) Technique of receiving data from medical devices to create electronic medical records database. In: Soft computing applications and techniques in healthcare. CRC Press, Boca Raton. http://doi.org/10.1201/9781003003496
28. Hai VD, Hung PD, Dan CQ (2020) Modified electrode placements for measurement of hemodynamic parameters using impedance cardiography. J Med Eng Technol. http://doi.org/10.1080/03091902.2020.1799089
29. Agaa K, Taraoa H, Urushiharaa S (2016) Calculation of human body resistance at power frequency using anatomic numerical human model. Energy Procedia 89:401–407

Smart Blind Stick for Visually Impaired People

Hung Quoc Nguyen, Anh Hoang Lan Duong, Man Dinh Vu, Thinh Quoc Dinh, and Hoan Thanh Ngo

Abstract Detecting obstacles is always a difficult task for visually impaired people when they move. External guidance such as human, trained dog, or white cane, a.k.a. blind stick, plays an important role in the decision making of blind people. Due to its low cost, white cane is often used by visually impaired people. However, traditional white canes cannot accurately detect obstacles above knee level or at distance beyond the white cane's length. Our goal is to create an affordable, smart blind stick that can help blind people to navigate. The device consists of an ultrasonic sensor and infrared sensors for detection of obstacles in front of blind user and vibration motor + buzzer for alarming. One of the biggest challenges for blind people when they move indoor is to go up and downstair. We aim to address the challenge by integrating into our blind stick a function that alarms user in the presence of staircase. Moreover, this device also has a built-in GPS module and a GSM module that allows the device's and its user's location to be tracked and displayed on a smartphone app, a desirable feature for many family members of blind people. Ultrasonic and infrared sensors allowed our smart blind stick to detect obstacles at a distance from 5 cm to 150 cm from the user. Our design has several advantages including low-cost, capability to detect obstacles above knee level, staircase detection, location tracking via smartphone app, etc. In future work, more tests need to be conducted to determine its accuracy and reliability in real-world settings.

Keywords Smart blind stick · Smart white cane · Ultrasonic sensor · GPS module · Location tracking

---

H. Q. Nguyen · A. H. L. Duong · M. D. Vu · T. Q. Dinh · H. T. Ngo (✉)
School of Biomedical Engineering, International University, Ho Chi Minh City, Vietnam
e-mail: nthoan@hcmiu.edu.vn

Vietnam National University-Ho Chi Minh City (VNU-HCMC), Ho Chi Minh City, Vietnam

V. Van Toi et al. (eds.), *8th International Conference on the Development of Biomedical Engineering in Vietnam*, IFMBE Proceedings 85,
https://doi.org/10.1007/978-3-030-75506-5_12

1 Introduction

According to the global prevalence estimates for blindness and vision impairment of the World Health Organization (WHO), visual impairment and blindness accounted for at least 2.2 billion people all over the world in 2019 and this number is continuing to increase as time passes. Approximately half of visually impaired people suffer from unaddressed refractive error, cataract, diabetic retinopathy, and other severe optical problem [1, 2]. Since one of the leading causes of blindness is cataract [3] which is easily found in developing countries, vision impairment is found more often in low- and middle-income regions than in high-income regions. Because of the vision loss, obstacles become a huge problem for blind people. Hence, external guidance such as humans, trained dogs, or supporting devices like white cane plays an important role in the decision making of blind people [4]. Although there are several psychological and social benefits for having a guided dog [5], training cost for a guided dog is relatively high. A previous study showed that majority of visually impaired people who participated in the study were willing to use personal navigation devices and would make more trip with such devices [6]. Due to those reasons, white cane, or blink stick is likely to be used by people with visual impairment [7, 8].

Many personal navigation devices for blind people, especially devices based on GPS, have been developed [9]. In addition, efforts have been made to create smart blind stick with ultrasonic sensor for obstacle detection for blind people [10]. Another smart blind stick, generally known as electronic travel aid (ETA), was reported in [11]. GuideCane [12], an innovation of Borenstein and Ulrich, has an array of the ultrasonic sensor at the end of a stick to detect obstacles. Another ETA was presented by Mahmud et al. [13]. This device used a PIC microcontroller (PIC16F90) to control an ultrasonic sensor, proximity sensor, and other components. In [14], Nada et al. reported a low cost and fast response smart stick.

Learning from the previous works, especially [11], in this work, we propose a Smart Blind Stick that can (1) detect and alarm obstacles as well as upward and downward stairs, (2) help blind people navigate better, (3) send via GSM temperature, humidity, speed, and location of blind user to a mobile app on smartphone of family members of the blind user, and (4) be low cost, low power consumption.

2 Materials and Methods

This proposed smart blind stick contains two main characteristics, which includes obstacle detection and location tracking. In terms of detecting hindrance, the ultrasonic sensor was used thanks to its wide range of detection based on ultrasound wave transmission and detection [15]. For keeping track on user, GSM module and temperature sensor provide the location and information of the user to another person, like a family member, by transfering these information to the family member's smartphone via SMS messages.

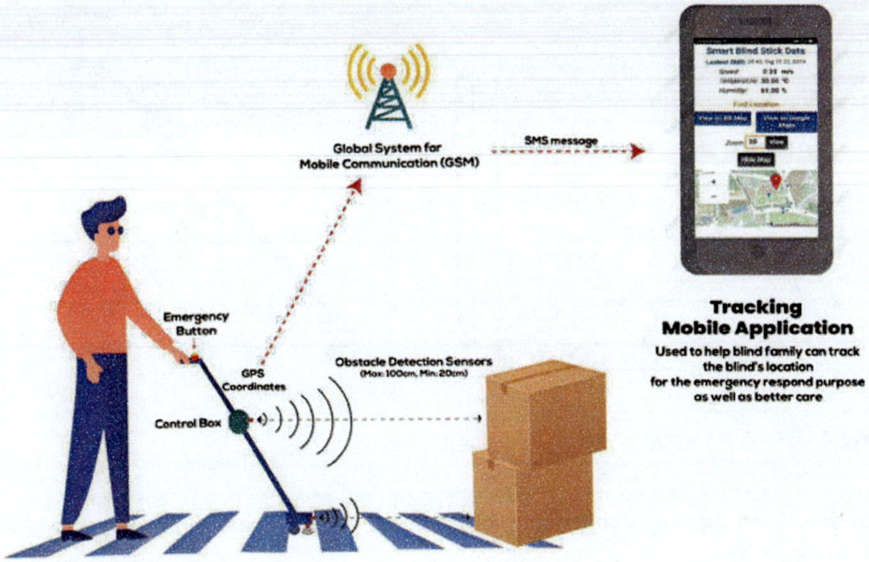

Fig. 1 A sketch of the proposed system

Figure 1 illustrates the working concept of our smart white cane. The structure includes three ultrasonic sensors, two of which are placed at the bottom of the cane for detecting obstacles in the nearest range of 2–10 cm. One ultrasonic sensor is mounted on the control box has a range from 30 to 150 cm (set to different ranges), which can detect things placed in the high position. A push-button (GSM switch) enables the user to send current status to a mobile application via SMS text messages. The control box contains a microcontroller in charge of controlling the system.

2.1 Stick Construction

According to a statistics of the Ministry of Health, Vietnam, the average height of Vietnamese people is 163.7 cm for men and 153 cm for women [16]. Thus, to make it easier to handle for both genders, height of this stick was designed to be approximately one meter. In term of material, a plastic rod made from polyvinyl chloride was used. This material is lighter and more durable than metal, and is able to withstand harsh weather conditions. The hollow rod allows reducing the weight and make room for wiring system.

At the end of the stick, a 3D plastic box was designed to fix the stick at an angle of 60° to the ground and protect internal electronic components. In the middle of the stick, there is a control box that was printed to cover a ultrasonic sensor, a GSM module, and a Arduino board. On the top of the white cane, there is a handle to make it possible for user to firmly hold the stick. In addition, there is also a push-button

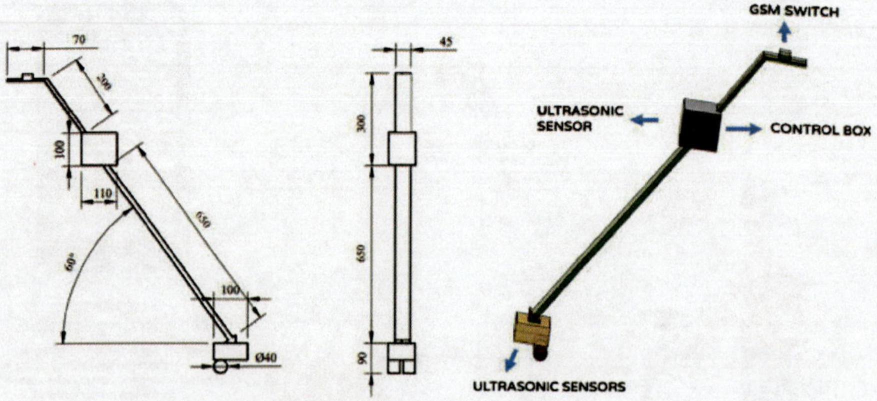

Fig. 2 Technical drawing of smart white cane prototype

to send SMS messages containing location, coordinate, speed (m/s) of the person holding the stick.

Figure 2 illustrates technical drawing of our prototype. The first and second sketch shows the front and side view of the prototype respectively while the final one is an illustration of the whole system.

2.2 *Obstacle Detection and Alarming System*

Ultrasonic Sensor

Ultrasonic sensors have many benefits including: high accuracy in range (2 cm–2 m), cost effective (about 1 $), good reliability in different environmental conditions. Compared to ultrasonic sensor, infrared one has some disadvantages, for example interference from other infrared sources (e.g., sunlight).

At the beginning, we used infrared sensor for detecting obstacles at close distances, because at close distance infrared sensor works well. However, when we tested the stick outdoor, infrared sensor alarmed whenever it was exposed to sunlight. Therefore, we chose ultrasonic sensors for all 3 positions (Table 1).

Ultrasonic sensor has working frequency at 40 kHz and requires 5 V and 15 mA current. Detection range is about 2–200 cm. Figure 3 illustrates working scheme of an ultrasonic sensor.

Ultrasonic sensors have 2 pins: Trigger and Echo. Specifically, the trigger pin initiates HIGH state in 10 μs, followed by a transmission sequence of 8 ultrasound burst at 40 kHz wave. Echo pin is activated immediately at the end of the eighth or the last burst, then acoustic wave bounces back from obstacle, generating a raise of voltage at the echo pin that can help to determine the time of wave traveled. Distance the wave traveled can be calculated from the relationship between distance,

Table 1 A comparison of different obstacle detection sensors

	Ultrasonic	Infrared	Laser	Radar
Principle	Transmission and reception of ultrasonic wave	Transmission and reception of infrared light	Transmission and reception of light	Transmission and reception of microwave
Range	2–200 cm	1–15 cm	SLR: 15–120 cm LLR: about 10–50 m	150–200 m
Beamwidth	Wide	Cramped	Cramped	Depend on the size of sensor
Noise	By the shape of obstacle and angle interaction	By other infrared sources	Yes	Yes
Cost	Low (1$)	Low (0.5$)	High (5$)	High

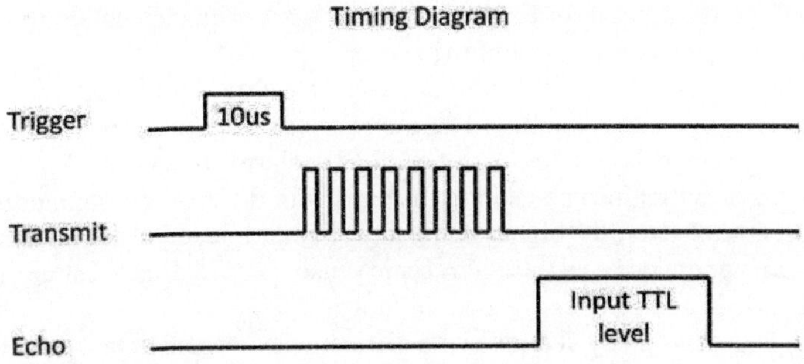

Fig. 3 Timing diagram of ultrasonic sensor to measure the distance

time, and velocity of ultrasound wave.

$$D = c \times t \quad (1)$$

Accordingly, D denotes the distance from the detector to the obstacle, c denotes the velocity of sound at 20 Celsius, which is 343 m/s, and t denotes the time that wave traveled.

Converting unit of wave velocity to centimeter per micro second, we got 0.0343 (cm/μs). About time, since the wave traveled the path twice (one for transmission and one for reflection), time to travel from the sensor to the obstacle is half the total time. In summary, the formulation for calculating distance is:

$$D = t/2 * 0.00343 \text{ (cm)} \quad (2)$$

In reality, ultrasonic sensor performance is affected by factors such as shape/material of obstacles, angle between wave's direction and obstacle's surface.

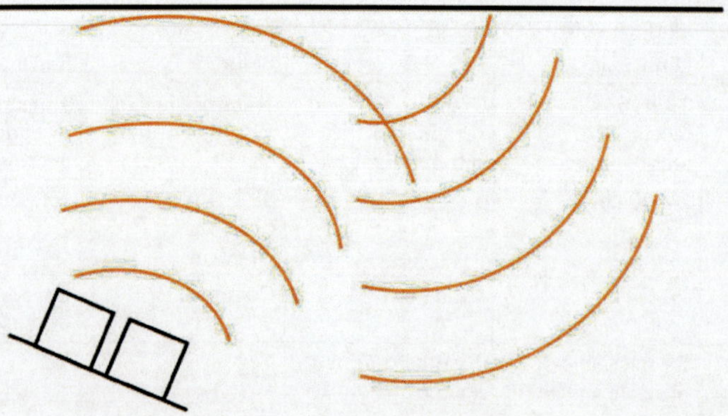

Fig. 4 Scenario in which echo wave does not return to the sensor

For example in the case in Fig. 4, ultrasonic sensor cannot detect the obstacle although there is obstacle in front of the sensor.

For detecting obstacles in front of blind user, two ultrasonic sensors are used and cross-checked to increase reliability. There are various ways to alarm blind user when an obstacle is detected. In our design, we decided to alarm blind user using two senses of human: touch (vibration) and sound (buzzer) with different combinations (Table 2). The closer the obstacle the higher frequency of vibration/sound alarm. Besides, for detecting stairs going down, a third ultrasonic sensor is used. Electrical wiring of the obstacle detection and alarming system is shown in Fig. 5.

Arduino board has many versions with different performances and purposes such as Arduino Mega, Arduino LilyPad … Among them, Arduino Uno R3 is the most widely used one due to its low cost and flexibility.

Table 2 Type of alarm signals corresponding to different distances of obstacle

Type of alarming	Distance (cm) from the 3rd ultrasonic sensor	Distance (cm) from the 2nd ultrasonic sensor	Delay (s)	Signal
Frontline	150	125	1.5	Vibration (Slow)
	120	95	1.0	Vibration (Quite slow)
	90	65	0.7	Vibration (Quite fast)
	60	35	0.5	Vibration (Fast)
	40	15	0.8	Buzzer (Quite fast)
	30	5	0.5	Buzzer (Fast)
Downstairs			0.2	Vibration + Buzzer (Fast)

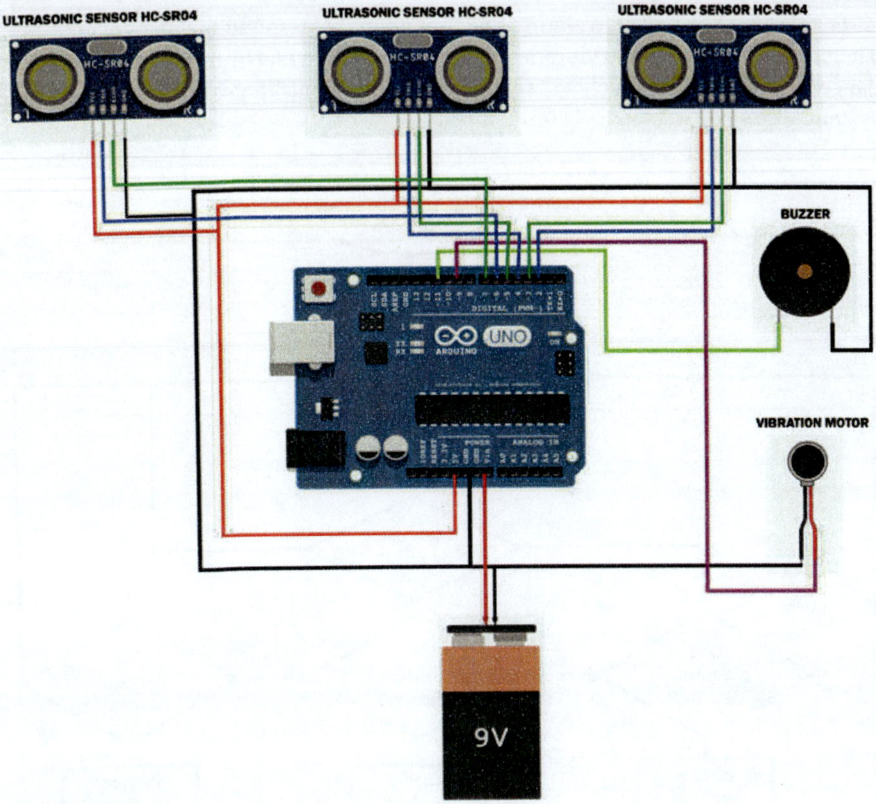

Fig. 5 Wiring diagram of the obstacle detection and alarming system

The Arduino Uno R3 Board

This board is built with the ATmega328P microcontroller as its core using quartz whose oscillation cycle is 16 MHz. With this microcontroller, we have a total of 14 IO pins numbered from 0 to 13 (including 6 pin PWM, marked ~ before the pin code). In parallel, we have 6 pins to receive analog signals marked with A0–A5. These 6 pins can also be used as normal input/output pins (like pin 0–13). Of the pins mentioned, pin 13 is special because it connects directly to a status LED on the board.

The way to connect all components:

First, Arduino is powered by a 9 V-battery via 2 pins Vin and ground (GND).

5 V pin (***Arduino***)—5 V pin (sensors).
GND pin (***Arduino***)—GND pin (sensors, buzzer, vibration motor).
11 (***Arduino***)—V+ (buzzer).
9 (***Arduino***)—V+ (vibration motor).
2, 4, 6 (***Arduino***)—Trigger pin (sensors).
3, 5, 7 (***Arduino***)—Echo pin (sensors).

For alarming, we want to alarm blind user not only by sound but also by vibration. Therefore, both buzzer and vibration motor were used so that blind user can recognize danger via two senses: touch and sound. Figure 6 presents the workflow of the whole system.

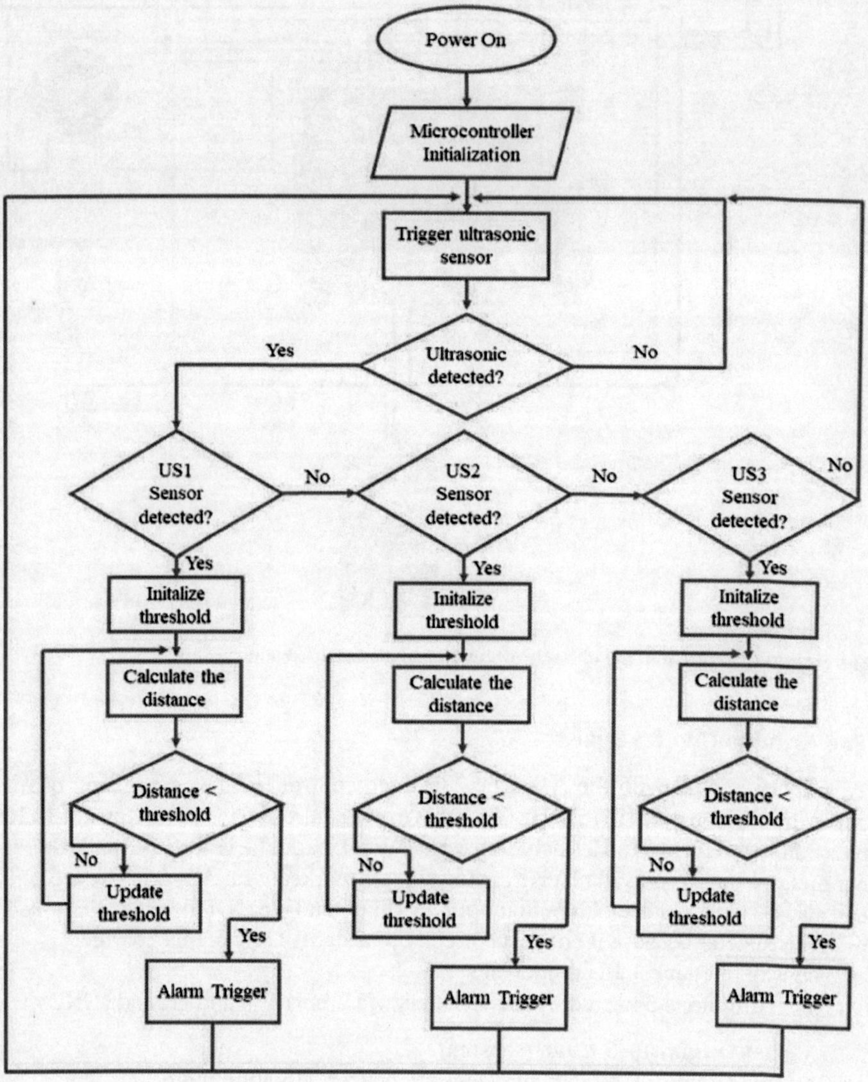

Fig. 6 Flow chart of the obstacle detection system

2.3 Data Transmission

For communication between the visually impaired people and their caretakers as well as location tracking, a GSM-based data transmission system was proposed in this work. When the blind user push the GSM button on the handle, temperature sensor (DHT11) and GPS module collect information about the speed of the user, environmental temperature, humidity, and location. Then, those data were coded in an SMS text message and send to the caretaker phone number. A specific Android-based application was used to read this code as well as display those data. The workflow of data transmission is summarized in Fig. 7.

GSM Module

The Global System for Mobile Communications (GSM) is a popular protocol developed by the European Telecommunications Standards Institute (ETSI) to define protocols for second-generation wireless mobile networks (2G). Used globally by mobile devices such as mobile phones and tablets. When a GSM module is attached to a microcontroller (ATmega328), this allows the microcontroller to use the GSM module to communicate over a mobile network. To control the GSM Module, AT command was used.

SIM800L GSM/GPRS module is a miniature GSM module, which could be integrated into a wide range of IoT devices. You'll be able to use this module to accomplish virtually anything a traditional mobile phone can: SMS text messages, create or receive phone calls, connecting to the web through GPRS. The module supports quad-band GSM/GPRS network, that means it works just about anywhere in the world. At the center of the module is a SIM800L GSM cellular chip from SimCom. The operational voltage of the chip is 3.4–4.4 V, which can be provided by a LiPo battery.

GPS NEO 6M Module

In reality, GPS receivers operate by finding out exactly how far they are from a range of satellites. At every given time, they are pre-programmed to learn where the GPS satellites are. The satellites relay information in the form of radio signals to the Earth about their location and the present time. Such signals mark the satellites and say where they are located to the receiver. The receiver then determines how far each satellite is by finding out how long it took for the signals to hit. It will identify your position on Earth after it has details about how far away at least three satellites are and where they are in the orbit. The process is called Trilateration [17] (Fig. 8).

At the GPS module's center is a U-blox NEO-6M GPS chip. The chip is less than the size of a postage stamp but can track up to 22 satellites on 50 channels and achieve the maximum degree of sensitivity in the industry by using just 45 mA supply current. Unlike other GPS modules, up to 5 position changes per second can be made with 2.5 m horizontal location accuracy. The U-blox 6 positioning engine also features a Time-To-First-Fix (TTFF) of less than 1 s. Power Saving Mode (PSM) is one of the key features that the chip has. It allows system power consumption to be minimized

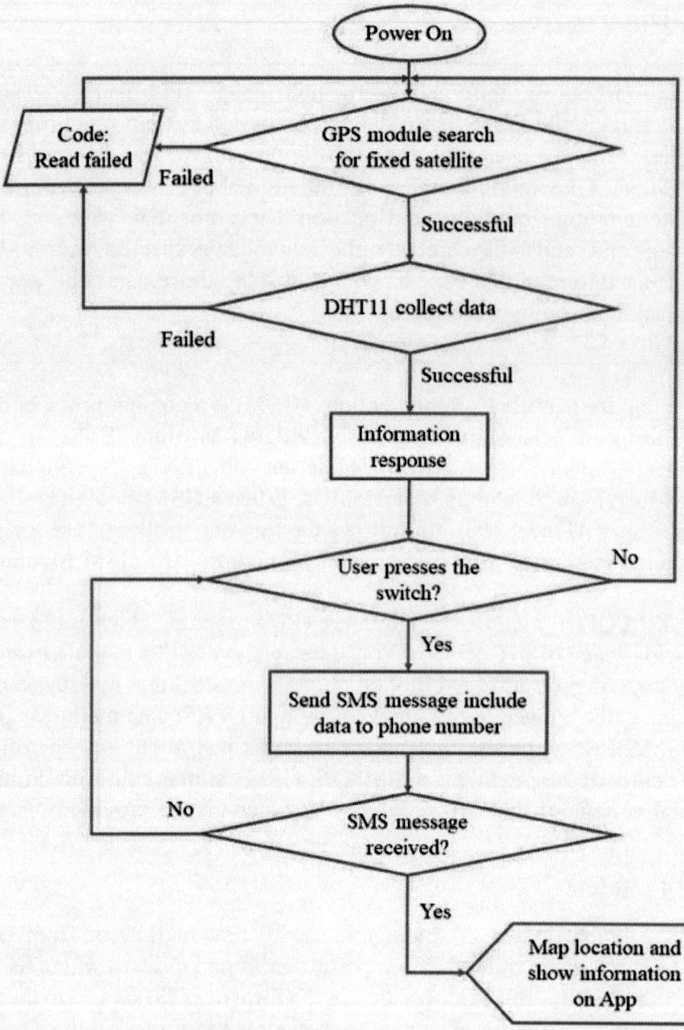

Fig. 7 Flow chart of data transmission system

by swapping the ON and OFF receiver sections selectively. This significantly lowers the module's power consumption to just 11 mA, making it ideal for applications that require high energy efficiency such as GPS wristwatch. The required NEO-6M GPS chip data pins are broken to a 0.1″ pitch header. This includes pins needed for communication with microcontroller via UART. The module supports baud rate from 4800 to 230,400 bps with a default baud rate of 9600 bps.

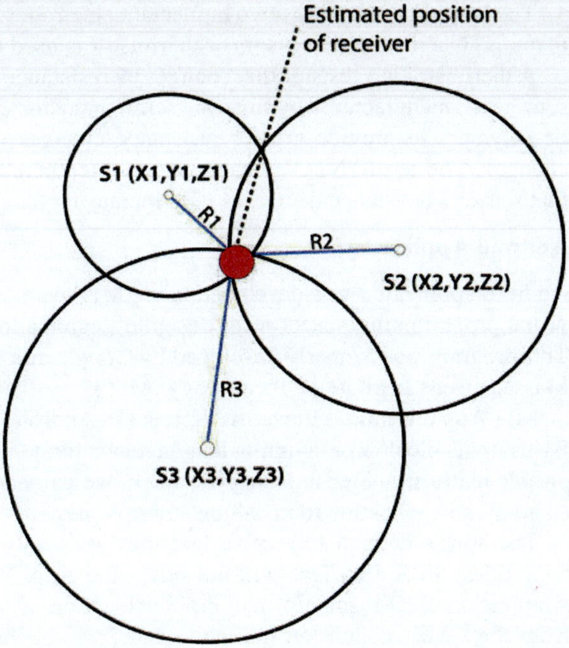

Fig. 8 The GPS trilateration technique, which applied signal received from at least three different satellites S1, S2, S3

Temperature Sensor

DHT11 is a common temperature and humidity sensor because it is affordable and easy to get data via 1-wire communication (digital 1-wire digital communication only). The sensor has a built-in signal preprocessor, making it convenient to use.

This sensor consists of a humidity sensor, an NTC temperature sensor (or thermistor), and an IC at the back of the sensor (Fig. 9).

Fig. 9 Internal structure of humidity sensor

The humidity sensor has two humectant electrodes to measure humidity between them. NTC temperature sensors or thermistor is used to measure temperature.

A thermistor is a resistor that changes its resistance to temperature change. These sensors are manufactured by sintering semiconductor with materials such as ceramics or polymers to provide greater resistance changes with only minor temperature changes. The term "NTC" means negative temperature coefficient, which means that sensor's resistance decreases with increasing temperature.

Android Application

Android application was developed using MIT App Inventor 2 [18, 19], which is a visual programming platform for designing applications for Android smartphones. This program was formerly developed by Google and is currently maintained by the Massachusetts Institute of Technology (MIT).

MIT App Inventor 2 allowed us to create an Android application via a web browser. By using its block programming language, we found it efficient to perform a project on this platform. After finishing the work, we can package our app and produce a stand-alone application to install on smartphones.

The app is defined to receive text messages only from the smart white cane; thus, other SMS messages will not affect the App. Regarding the function of the Application, since it got information about location, speed, temperature, and humidity from the GSM module on the white cane, we let the App show three variables: location, speed, and temperature. In terms of location tracking, there are two ways to show blind user's location. For people who prefer to use Google Maps app for searching location, we link the our Blind Stick App with Google Maps app so that app user can see blind user's location in Google Maps app. The other way is to have a map within the app by using the map function of MIT App Inventor 2 (BS Map) in which people can search for location together with other information.

2.4 *Power Source*

Our goal is to create a smart blind stick that can assist blind people continously for 4–6 h before battery recharging. To achieve this goal, we used a Lithium Polymer Battery (LiPo) in tandem with a charging circuit. Users can easily charge the battery through Micro-USB Type B charging port with relatively short charging time (in average 2–3 h). In our proposed system, there are two main parts which include (1) the data transmission system (Fig. 10) and (2) the obstacle detection and alarming system (Fig. 5). Each system is powered by a separate LiPo battery.

Power Supply for the Data Transmission System

At the beginning, we found it is challenging to power up the GSM module SIM800L. One of the most important parts of making the SIM800L module work is to give it enough power.

Depending on its status, the SIM800L is considered as a relatively current consuming device. Its current could be up to about 2 A during transmission burst. Normally, it is not required that amount of current but might ask for around 216 mA when calling or 80 mA during network transmission. The current consumption in different state is shown in the bellow datasheet summary (Table 3).

An external power supply between 3.4 and 4.4 V (Ideal 4.1 V) is required because the SIM800L module does not come with onboard voltage regulator. External power supply should also be able to provide a 2 A source of surge current, otherwise, the module would shut down. Therefore, we decide to use a LiPo battery which is in the range of 3.7–4.2 V. This battery is perfect for the GSM SIM800L Module because it can provide the right voltage range even during 2 A spikes.

The other module is the GPS NEO-6M Module which has the ultra-low dropout 3V3 regulator (MIC5205) from MICREL and consumes only 45 mA supply current. Therefore, by using just one LiPo battery 3.7 V–3000 mAh, we can power the whole data transmission system. This design not only ensures sufficient power supply for the two modules but also creates convenience for users during long periods of use and easy to integrate into the stick compactly and aesthetically.

Power Supply for the Obstacle Detection and Alarming System

In the obstacle detection and alarming system, we used a sensor system (consisting of three ultrasonic sensors) to detect obstacles at different distances. In terms of alarm, we used two methods to alarm blind user including sound by using a buzzer and vibration by using a vibration motor. The obstacle detection and alarming system is powered by a second lipo battery, more specifically a LiPo Battery 3.7 V (2000 mAh).

The Battery Charging System

With the aim of developing a portable device, we used LiPo batteries to power the smart blind stick. To recharge these batteries, we used commercially available charging circuits (TP4056) that is highly efficient, durable, stable, and easy to replace. The TP4056 is a complete constant-current/constant-voltage linear charger for single-cell lithium-ion batteries. The charge voltage is fixed at 4.2 V, and the charge current can be programmed externally with a single resistor. The TP4056 automatically terminates the charge cycle when the charge current drops to 1/10th the programmed value after the final float voltage is reached. Other features include current monitor,

Table 3 The current consumption according to mode (according to the GSM SIM800L datasheet)

Modes	Current consumption
Power down	60 μA
Sleep mode	1 mA
Standby	18 mA
Call	199 mA
GRPS	453 mA
Transmission burst	2 A

under-voltage lockout, automatic recharge, and two status pins to indicate charge termination and the presence of an input voltage. In our system, we are using two battery chargers for the two LiPo batteries that power two different parts of the smart blind stick. With that, we measured battery run-time of around three-to-four hours at 60% operation and three hours at 90% operation. Battery charging time is around two hours for full battery charging. To charge batteries, we used a 5 VDC power adapter (5 VDC–2 A) which will charge the two LiPo batteries through two battery chargers TP4056 (Fig. 11).

The circuit of the Power Supply is represented in Fig. 11.

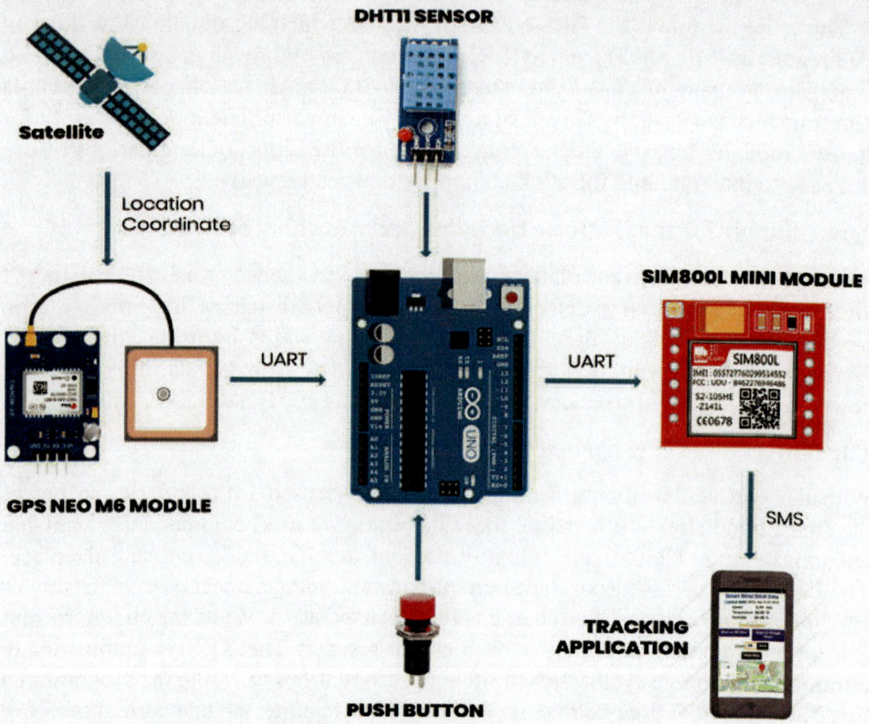

Fig. 10 Block diagram of the data transmission system

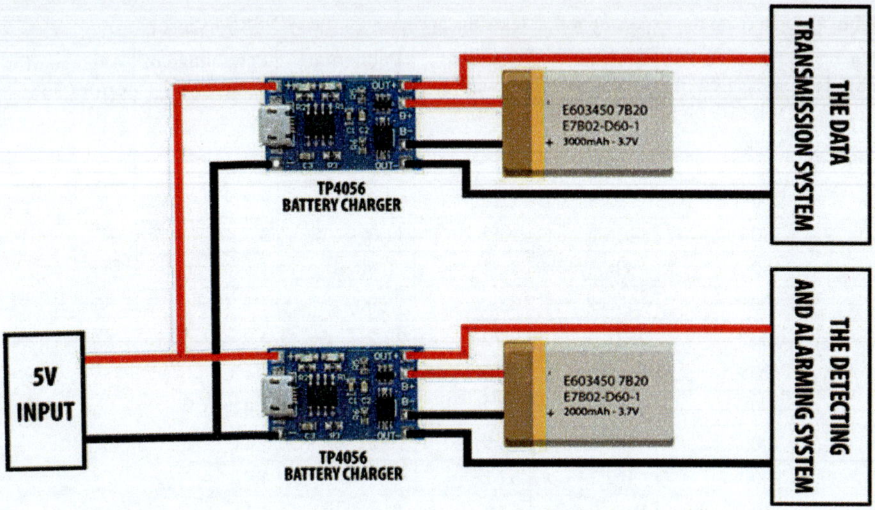

Fig. 11 The battery charger wiring diagram

3 Results and Discussions

3.1 *Testing Procedure*

One of the key components of our smart blind stick is the ultrasonic sensor (HC-SR04) which is used to detect obstacles. Based on its working principle, we designed experiments to test and calibrate distance measured by the sensor. After making measurements with the sensor, we measured and tested the ability to read the GPS value from the GPS-NEO 6M module, determined the error about the location value, accuracy, and stability of the system. All experiments were performed indoor in conjunction with measurements in outdoor environment, where the stick would be primarily used.

3.2 *Obstacle Detection Experiment*

Results of the Proposed Method

First, we experimented to determine the accuracy of the HC-SR04 sensor's ability to measure distances. For reference, a steel ruler was used to measure the actual distance in centimeters. Room walls were used as obstacle from which the sensor will measure distance. The sensor was moved away from the wall and then actual distances was measured using the steel ruler. Distances measured by the sensor will be shown on serial monitor screen of Arduino IDE. We designed experiments based

Table 4 Data from the experiment that tests the accuracy of the HC-SR04 sensor

No. of objects	Distance measured by ruler (Dm) (mm)	Distance measured by HC-SR04 (Ds) (mm)	Standard deviation (mm)	Delta (mm)	Percentage of error (%)	Average of error (%)
1	200	195.3	1.25	4.70	2.3	3.8
2	300	288.8	1.69	11.2	3.7	
3	400	393.3	3.53	6.70	1.7	
4	500	506.9	6.61	6.90	1.4	
5	600	574.8	2.70	25.2	4.2	
6	700	669.0	3.16	31.0	4.4	
7	800	761.2	4.52	38.8	4.8	
8	900	860.2	1.69	39.8	4.4	
9	1000	952.1	6.61	47.9	4.8	
10	1100	1052.2	7.19	47.8	4.3	
11	1200	1148.1	7.09	51.9	4.3	
12	1300	1241.5	4.48	58.5	4.5	
13	1400	1339.5	6.22	60.5	4.3	
14	1500	1438.1	6.49	61.9	4.1	

on experiments that [20] performed. Totally, we took 14 sample distances within the desired working range. For each distance, we measured three times to compute the mean value and standard deviation, as shown in Table 4. From the results, we analyzed percentage of error (ε) for each distance and average percentage of error (εa) as shown in Table 4. Percentage of error was computed by using the Eq. (3).

$$\varepsilon = \frac{D_m - D_s}{D_m} \times 100\% \tag{3}$$

Table 4 shows a very good agreement between steel ruler and HC-SR04 sensor. In all measurements, error percentage is less than 5% and so is the average. Based on Table 4, we can see that the further the distance between the object and the sensor, the greater the percentage of errors. Moreover, the percentage of errors can still be affected by external environmental conditions (i.e. Temperature, Humidity, Materials, etc.). Therefore, we proposed the measurement distance for our system should be in the range of (0–100 cm) in order to provide the most accurate results as well as minimize the percentage of errors. That will help to improve user experience as well as ensure user safety.

We also plotted the relation between fixed distance (Dm) measured by ruler and distance measured by HC-SR04 (Ds). From the plot, we obtained a regression equation where regression coefficient was 0.9996 using Microsoft Excel software, as shown in Fig. 12.

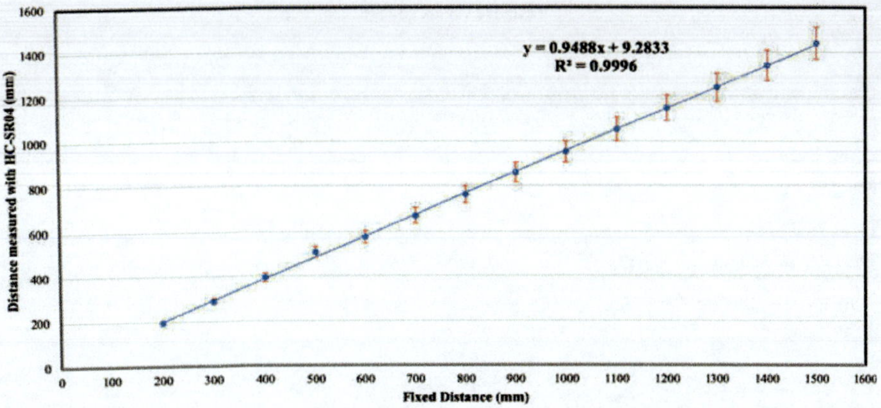

Fig. 12 Plot of fixed distances measured by ruler versus measured distances measured by the HC-SR04 sensor

Table 5 Voltage measurement at different distances

Distance (in.)	Voltage measured (mV)
1	40
2	85
3	124
4	155
5	190
6	230
7	264
8	300
9	337
10	372
11	408
20	747
30	1080
40	1350

The value of coefficient of determination (R2) is 0.9996 as extracted from Microsoft Excel software, and the slope of the regression line (i.e. regression coefficient, a_0) is 0.9488, which signifies that the sensor was performing very well.

In addition, we measured voltage of the ECHO pin of the ultrasonic sensor (HC-SR04) at different distances between the sensor and the obstacle. As shown in Table 5 and Fig. 13, we see that the larger the distance, the higher the measured voltage.

Next, we tested our smart blind stick's ability to detect different obstacles made of different materials (10 times for each one). The results are shown in Table 6.

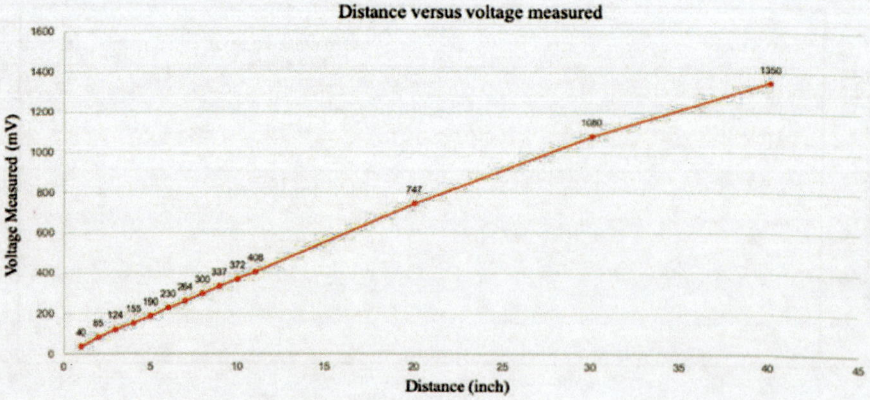

Fig. 13 Distance versus measured voltage

Table 6 Ability to detect different obstacles made of different materials

Number of material	Result (%)	Obstacle materials
1	100	Wall
2	100	Fridge
3	100	Closet
4	100	Curtains
5	50	Rough surface
6	100	Wooden table
7	100	Inox pot
8	100	Glass
9	100	Mica

Results in Table 6 reveals that the ultrasonic sensors are good for various types of material as long as the surface is flat. Whereas rough surface is difficult for sensor to detect (as explained Fig. 4). The sensor could detect only five times in the total of ten times tested. In general, the stick works well in most of the cases.

Comparison with previous works

The range of obstacle detection in our proposed method has shown promising results compared to the previous method. This can be analyzed from the system of Mahmud et al. [21] which also contains three ultrasonic sensors for detecting obstacles and uses vibration to inform the users. The threshold of our method is higher than the one from Mahmud et al. (20–100 cm and 30–70 cm respectively). Moreover, although there are three levels of vibration to help the user notice the different in distance, the previous method can lead to misapprehension, which has been improved in our prototype by using both buzzer and vibration. Nada et al. [14] represented an effective way to create a smart white cane by using infrared sensors and voice record for alert.

Table 7 A comparison with other devices

Method	Minimum measured distance (cm)	Maximum measured distance (cm)	Alert signal
Mahmud et al	30	70	Vibration (Highest, moderate and minimal speed)
Nada et al	Less than 20	80	Speech warning message
Proposed method	**20**	**100**	**Vibration and buzzer**

The bold values illustrates results of our method

This method provide the cane to have the priority in detecting obstacles in less than 20 cm while our method can identify at the minimum distance of 20 cm; however, Nada et al. show limited ability to work with object that is more than 80 cm from the stick. The comparison is illustrated in Table 7.

3.3 Data Transmission Experiment

For our proposed system, we also experimented to validate the accuracy of our GPS module (GPS NEO 6M). In this experiment, we took 100 location measurements both in good conditions and difficult conditions. Table 8 shows standard deviation of latitude, longitude, altitude, and the localization error in meter.

The localization error was computed by Eq. (4):

$$d = R\sqrt{a^2 + b} \tag{4}$$

where R is the radius of the Earth (6371 km) and $a = \Delta\lambda \cos\varphi_m$; $b = \Delta\varphi$. Accordingly, φ *denotes latitude,* λ *is longitude and* φ_m *denotes the mean value of latitude.*

From the table, we can observe that the precision of GPS data depends on both the GPS receiver and the observing conditions. The radio interference also might have a very pronounced impact on the accuracy of the results. Another key thing is

Table 8 Standard deviation of latitude, longitude, altitude, and the localization error of GPS NEO 6M in both good conditions and difficult conditions

GPS receiver	STD of latitude (s)	STD of longitude (s)	STD of altitude (m)	Localization error (m)
GPS NEO 6M (in good conditions)	0.17	0.17	10.45	6.785
GPS NEO 6M (in difficult conditions)	0.678	1.545	8.25	39.567

that the GPS module is unable to work under indoor environment because it needs to communicate with at least 2–3 satellites to receive the signal. So it is impossible to get GPS signal under indoor environment. This problem prevented the stick from transmitting location data to smartphone app when user moves inside buildings and future work is needed to address this issue.

4 Conclusion

This paper present a design and implementation of a smart blind stick that can detect and alarm the presence of obstacles for visual impaired people. Furthermore, the stick is able to send information including location of the blind user and temperature and humidity of the surrounding environment to a remote smartphone app so that family member of the blind user can track the blind user's location when needed. Multiple tests were conducted, and the results showed high accuracy and high reliability of the device.

Our design offers a low-cost and robust system for navigation and location tracking for blind people. In future work, weight of the stick could be reduced for better user experience.

Conflicts of Interest The authors have no conflict of interest to declare.

References

1. Bourne RRA, Flaxman SR, Braithwaite T, Cicinelli MV, Das A, Jonas JB, Keeffe J, Kempen JH, Leasher J, Limburg H, Naidoo K, Pesudovs K, Resnikoff S, Silvester A, Stevens GA, Tahhan N, Wong TY, Taylor HR (2017) Vision Loss Expert Group: magnitude, temporal trends, and projections of the global prevalence of blindness and distance and near vision impairment: a systematic review and meta-analysis. Lancet Glob Health 5:e888–e897. https://doi.org/10.1016/S2214-109X(17)30293-0
2. Fricke TR, Tahhan N, Resnikoff S, Papas E, Burnett A, Ho SM, Naduvilath T, Naidoo KS (2018) Global prevalence of presbyopia and vision impairment from uncorrected presbyopia: systematic review, meta-analysis, and modelling. Ophthalmology 125:1492–1499. https://doi.org/10.1016/j.ophtha.2018.04.013
3. Congdon NG, Friedman DS, Lietman T (2003) Important causes of visual impairment in the world today. JAMA 290:2057–2060. https://doi.org/10.1001/jama.290.15.2057
4. Sheth R, Rajandekar S, Laddha S, Chaudhari R (2014) Smart white cane—an elegant and economic walking aid. Am J Eng Res 3:84–89
5. Whitmarsh L (2005) The benefits of guide dog ownership. Vis Impair Res 7:27–42
6. Golledge RG, Marston JR, Loomis JM, Klatzky RL (2004) Stated preferences for components of a personal guidance system for nonvisual navigation. J Visual Impairment Blindness 98:135–147. https://doi.org/10.1177/0145482X0409800304
7. Kang S-J, Ho Y, Moon IH (2001) Development of an intelligent guide-stick for the blind. In: Proceedings 2001 ICRA. IEEE international conference on robotics and automation (Cat. No. 01CH37164). IEEE, pp 3208–3213

8. Dey N, Paul A, Ghosh P, Mukherjee C, De R, Dey S (2018) Ultrasonic sensor based smart blind stick. In: 2018 international conference on current trends towards converging technologies (ICCTCT), pp 1–4
9. Otaegui O, Loyo E, Carrasco E, Fösleitner C, Spiller J, Patti D, Adela M, Olmedo R, Dubielzig M (2012) ARGUS: Assisting peRsonal GUidance System for people with visual impairment
10. Nowshin N, Shadman S, Joy S, Aninda S, Minhajul IM (2017) An intelligent walking stick for the visually-impaired people. Int J Online Biomed Eng (iJOE) 13:94–101
11. Chaurasia S, Kavitha KVN (2014) An electronic walking stick for blinds. In: International conference on information communication and embedded systems (ICICES2014). IEEE, pp 1–5
12. Borenstein J, Ulrich I (1997) The guide cane—a computerized travel aid for the active guidance of blind pedestrians. In: Proceedings of international conference on robotics and automation. IEEE, Albuquerque, NM, USA, pp 1283–1288
13. Mahmud MH, Saha R, Islam S (2013) Smart walking stick—an electronic approach to assist visually disabled persons. Int J Sci Eng Res 4:111–114
14. Nada A, Mashelly S, Fakhr MA, Seddik AF (2015) Effective fast response smart stick for blind people. In: Proceedings of the second international conference on advances in bio-informatics and environmental engineering–ICABEE
15. Kim SY, Cho K (2013) Usability and design guidelines of smart canes for users with visual impairments. Int J Des 7:99–110
16. VietnamPlus: National health programme to improve Vietnamese stature, well-being | Health | Vietnam+ (VietnamPlus). https://en.vietnamplus.vn/national-health-programme-to-improve-vietnamese-stature-wellbeing/137797.vnp
17. Hoque MI, Al Amin A, Rahaman MJ, Hossam-E-Haider M (2017) Designing and performance evaluation of GSM/GPS based helicopter tracking device. In: 2017 IEEE international conference on telecommunications and photonics (ICTP), pp 77–81
18. Wolber D (2011) App inventor and real-world motivation. In: Proceedings of the 42nd ACM technical symposium on computer science education, pp 601–606
19. Wolber D, Abelson H, Spertus E, Looney L (2011) App inventor. O'Reilly Media, Inc., USA
20. Titlee R, Bhuyan M (2016) Design, implementation and testing of ultrasonic high precision contactless distance measurement system using microcontroller. Southeast Univ J Sci Eng (SEUJSE) 10:6–11
21. Mahmud N, Saha RK, Zafar RB, Bhuian MBH, Sarwar SS (2014) Vibration and voice operated navigation system for visually impaired person. In: 2014 international conference on informatics, electronics & vision (ICIEV). IEEE, Dhaka, Bangladesh, pp 1–5

Biomechanics

A Study of Polishing Parameters to Surface Roughness of Magnetorheological Polishing Methods

Duc-Nam Nguyen, Truong-Giang Nguyen, and Cong-Truyen Duong

Abstract The study aims to determine the effect of polishing parameters as well as polishing speed, abrasive sizes and electric current on surface roughness of titan alloy in the magnetorheological polishing (MRP) methods. When the electric field is applied, the rheological properties of MRP fluid will be changed in the machining process. As a result, the viscosity and shear yield stress of MRP fluid was greatly improved under applied electric current. The workpiece surface can be polished by the abrasive particles with adhering to MR fluid. The influence of polishing parameters on the surface roughness was investigated by experiments. The experimental results indicated that the polishing speed has a great influence on the surface roughness of workpiece. With the increase of the polishing speed, the best surface roughness can be obtained by using the MRP method. The abrasive size had very little effect on surface quality improvement when it was changed in the experimental process. In addition, the electric current also has a strong effect on the surface quality of workpiece. As a result, the surface roughness of ball titan alloy workpiece (Ø32 mm) was reduced rapidly from Ra = 120 nm to Ra = 18 nm under the appropriate machining conditions.

Keywords Magnetorheological polishing · Curve surfaces · Artificial joints · Surface roughness · Titan alloy · MR fluid

1 Introduction

The curved surface plays an important role in a wide range of applications, such as aerospace, astronomy, mould, automobile, biomedical implant and plain spherical bearings [1–5]. In particular, the number of implanted artificial hip joints is increasing significantly in the biomedical field. There are two parts of the hip joint, such as acetabulum or cup and a metalhead attached to the human body. The surfaces of

D.-N. Nguyen (✉) · T.-G. Nguyen · C.-T. Duong
Faculty of Mechanical Engineering, Industrial University of Ho Chi Minh City, Ho Chi Minh City, Vietnam
e-mail: nguyenducnam@iuh.edu.vn

V. Van Toi et al. (eds.), *8th International Conference on the Development of Biomedical Engineering in Vietnam*, IFMBE Proceedings 85,
https://doi.org/10.1007/978-3-030-75506-5_13

artificial joints are mostly spherical shapes. In order to increase the working life of the components, these surfaces need to be machined to achieve better surface roughness.

In addition, materials are also an important factor to improve the biocompatibility of the implant products. Titanium alloys are known as materials with excellent mechanical properties, high corrosion resistance, good hardness, and high applicability. Moreover, these alloys are more biocompatible and support cell attachment as well. Therefore, it is widely used in biomedical fields [6, 7]. However, the machining process of this material faces many difficulties due to unfavorable machining conditions when the chip was generated at elevated temperatures [8–10]. The surface quality of workpieces will not be satisfactory, and especially the surface roughness does not meet the product specifications. In that reason, the polishing process should be performed to improve the surface quality of the product.

There are numerous numbers of conventional machining methods such as grinding, boring shaping, milling and turning etc., used for manufacturing the complex surfaces. These conventional methods have limited application due to generated high stresses and large machining times. As a result, the surface damage and poor topography were generated on the product surfaces. In order to improve the machining efficiency, a computer numerical control (CNC) techniques have been applied to produce workpieces with high dimensional tolerances and perform machining with optimization machining trajectories. Further, many technologies have been carried out to significantly improve the surface roughness of the curved surface, such as chemical mechanical polishing (CMP) [11], magnetic abrasive finishing (MAF) [12], elastic deformation machining [13–16] and so on. However, they have various limitation in processing time and machining cost. In addition, the shear thickening polishing (STP) was also used for increasing the surface quality of the complex surfaces [17–21]. The best surface roughness can be obtained by using the appropriate abrasive sizes, polish-ing speed, and inclination angle of the workpiece, but the viscosity and properties of the polishing fluid were difficult to control during the machining process.

Therefore, the magnetorheological polishing (MRP) methods are considered and applied for processing the curved surfaces [22–24]. MRP is a flexible machining method by using magnetorheological fluid while the magnetic field is applied during the processing. When the electric field is provided, the microstructure and rheological properties of MRP fluid will be changed in the machining process. As a result, the viscosity and shear stress of MRP fluid were greatly improved. The workpiece surface can be polished by the abrasive particles with adhering to MR fluid. In this method, using external field-assisted polishing is a good method to polish curved surfaces.

2 Principle of MRP

The principle of the MRP method for processing the curve surface is presented in Fig. 1. In the MRP process, the machining fluid contains magnetorheological fluid

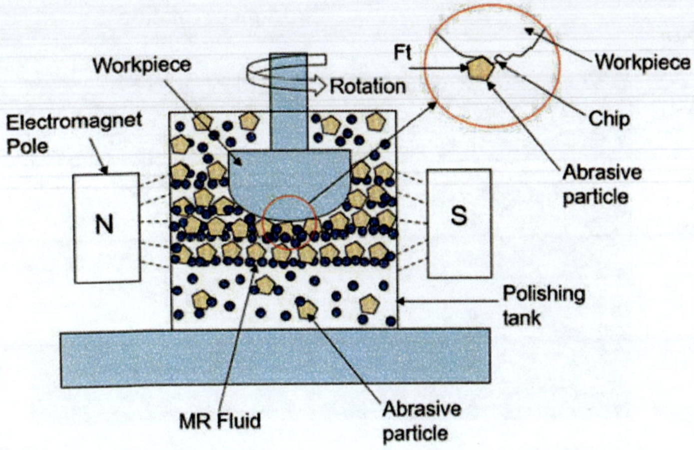

Fig. 1 Mechanism of the MRP process

(MR fluid) combine with abrasive particles. The cutting process will occur when the workpiece was relatively rotated in the MRP fluid. Under the influence of an external magnetic field, the viscosity of MRP fluid was generated and increased with increasing the intensity of the electric current. There, MRP fluid stiffens and forms a flexible magnetic abrasive brush as a polishing cutter. When the workpieces is rotated in the polishing tank, the workpiece surfaces will interact with the polishing fluid. Based on the rotation speed of workpiece and viscosity of the polishing fluid, the shearing stress is generated during the MRP machining.

The benefits of the method are that the viscosity of polishing fluid was easily controlled by the external magnetic field. In the machining process, the viscosity of the polishing fluid and the abrasive particles will be selected in accordance with the shape and material of the workpiece. As a result, the best surface roughness and high efficiency can be obtained during the MRP machining process.

3 Experimental Setup

In the experimental process, the MRP device was set up and developed, which can provide a relative motion between the workpiece and MRP fluid. As shown in Fig. 2, the workpiece was clamped by the spindle, which was driven by a stepping motor. The step motor and spindle were attached on the Z-axis, which can be moved in the vertical direction by a lead screw stage. In addition, two motors were attached on the X and Y axes, which used to control the position of the workpiece during the machining process.

A half-spherical workpiece which made from titan alloy was used in the study. The workpiece diameter of 32 mm was selected. The maximum rotation of the workpiece

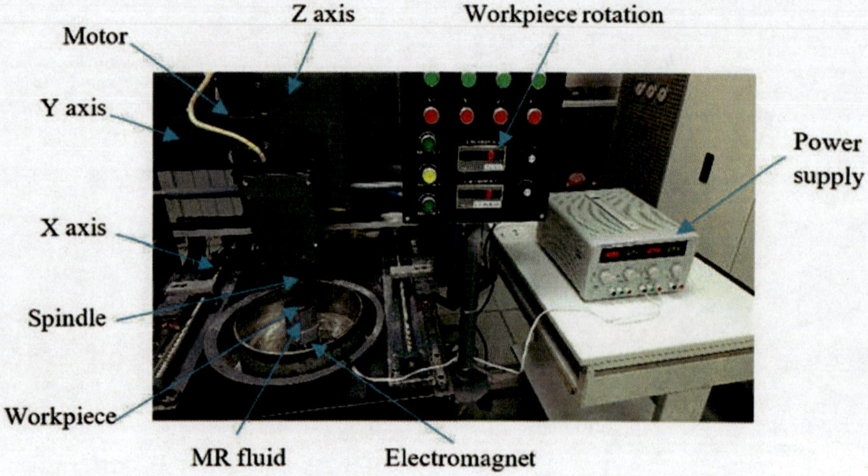

Fig. 2 The magnetorheological polishing (MRP) devices

is 350 rpm. The working gap between the tip of the workpiece and the bottom surface of the polishing tank was chosen of 8 mm. The original surface roughness (Ra) of the workpiece was 120 ± 10 nm. After the polishing process, the MarSurf-M400C measuring device was again used to test the change of surface roughness of the workpiece. After each experiment, five positions on the workpiece surface will be selected to measure the surface roughness. The final value of Ra is calculated by the average of these five positions. The physical properties of titan alloy are presented in Table 1.

In the MRP process, the machining parameters that affect the surface roughness of the workpiece are workpiece rotation, working gap and electric current [25, 26]. Therefore, three polishing parameters such as polishing speed, abrasive sizes and electric current values were chosen and investigated in this study. The single variable method was applied in the experiments. The experimental parameters are listed in Table 2.

Table 1 Physical properties of titan alloy

Items	Parameters
Modulus of elasticity (GPa)	120
Poisson ratio	0.34
Density (kg/m^3)	4500
Thermal conductivity (W/(m × K))	15
Thermal expansion (m/(m × K))	9.2 (300 °C)
Rockwell Hardened (HRC)	33

Table 2 Experimental parameters

Experimental conditions	Parameters
Workpiece	Titan alloy
Workpiece diameter (mm)	32
Abrasive slurry	Al_2O_3
Abrasive concentration (wt%)	25
Abrasive sizes	1000#; 2000#; 4000#
Rotational speed of the workpiece (rpm)	40, 50, 60, 70, 80, 90
Electric current (A)	0.6, 0.8, 1.0, 1.2, 1.4
MR fluid	MRF-132DG
Processing time per trial (min)	30

4 Results and Discussion

4.1 *Effect of Polishing Speed*

The relative movement between the workpiece and MRP fluid is an important factor that affects the surface roughness of the workpieces. The experimental studies were carried out on polishing the titan alloy ball with different polishing speed. The Al_2O_3 abrasive with the size of 4000# and abrasive concentration of 25%, which suspended in MR fluid was chosen in this experimental process. The electric current of 1.2 A was chosen in this section. The rotational speeds of the workpiece were set as 40, 50, 60, 70, 80 and 90 rpm, respectively. The relationship between polishing speed and surface roughness of workpiece is shown in Fig. 3.

As shown in Fig. 3, the surface roughness of workpiece tends to decrease with increasing the polishing speed. The surface roughness decreases rapidly during the first one hours. However, the surface roughness changes slightly during the last two hours of polishing. The best surface roughness with Ra of 18 nm can be obtained under the polishing speed of 70 rpm. When the polishing speed was more than 70 rpm, the surface roughness not only deteriorates but also increases. As a result, the polishing efficiency will not be satisfactory.

In the polishing process, the shear rate is upgraded by the increasing of polishing speed. This will lead to an increase in the contact force between the abrasive slurry and the workpiece. As a result, the cutting ability of the abrasives will be increased. The surface roughness is, therefore, significantly improved.

4.2 *Effect of Abrasive Sizes*

The abrasive sizes are an important specification of the MRP fluid. It determines the surface quality of the workpiece and the performance of the machining process.

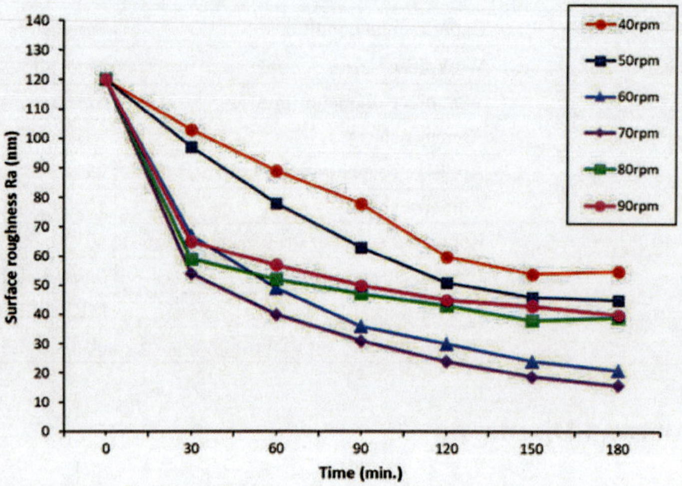

Fig. 3 Relationship between the polishing speed and the surface roughness

In this experimental process, the 1000#, 2000# and 4000# Al_2O_3 abrasives were applied for the experiment with the rotational speed of the workpiece of 70 rpm, the abrasive concentration of 25% and electric current of 1.2 A. The relationship between abrasive sizes and surface roughness of workpiece is shown in Fig. 4.

Base on the experimental results, the curves of surface roughness were almost the same when the workpiece was polished with different abrasive sizes. After three hours of polishing, the surface roughness of workpiece has little change under the

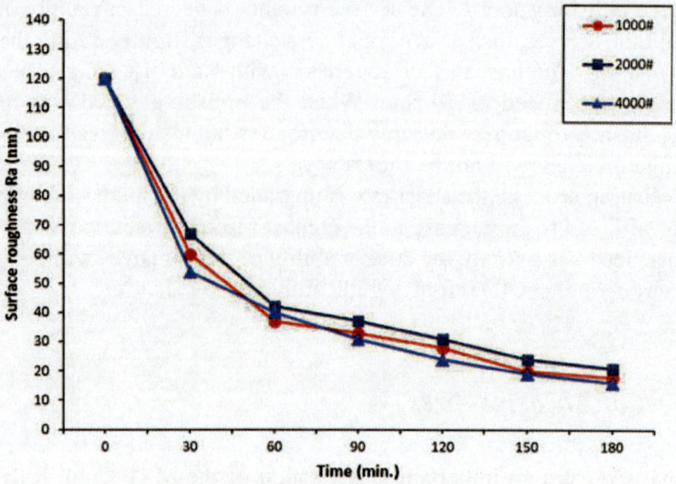

Fig. 4 Relationship between the abrasive sizes and the surface roughness

condition of machining parameters. The surface roughness Ra of 20 nm, 23 nm and 18 nm can be achieved when the abrasive size was 1000#, 2000# and 4000#, respectively.

Basically, the polishing force is the same with different abrasive sizes. The abrasive slurry is applied in the polishing process, so this process is considered as soft. Hence, there is very little difference of micro-scratches on the workpiece surface, which are produced under different abrasive sizes. It means that the abrasive size does not significantly affect the surface roughness of the workpiece.

4.3 *Effect of Electric Current*

The viscosity of MR fluid will be changed under applied electric current through the coil of the electromagnet. As a result, the viscosity of MR fluid is increased and combined with the abrasive particles to arrange into a flexible polishing tool. The rotational speed of the workpiece of 70 rpm, the abrasive size of 4000#, and the abrasive concentration of 25% were applied during the experiments. The electric current of the coil was set as 0.6 A, 0.8 A, 1.0 A, 1.2 A, and 1.4 A, respectively. The relationship between electric current and surface roughness of workpiece is shown in Fig. 5.

From Fig. 5, the surface roughness of workpiece was decreased when the electric current was increased. When the value of electric current is small, the viscosity of the MR fluid is insufficient to create the shear stress for the machining process. As a result, the surface roughness does not reach the desired surface quality. When the value of the electric current is increased, the MR fluid viscosity will be improved.

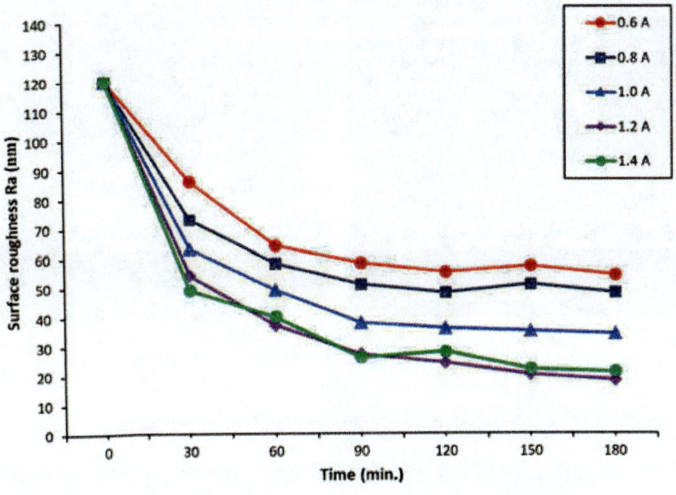

Fig. 5 Relationship between the electric current and the surface roughness

Therefore, the larger cutting force is generated to overcome viscosity resistance when the workpiece is moved in the polishing fluid. The best surface roughness with Ra of 18 nm that can be achieved when the electric current exceeds 1.2 A.

According to the results above, polishing speed and electric current were very important factors in MRP process. In addition, the abrasive size does not seem to affect on the surface roughness of workpiece. The surface topography of workpiece and the pictures of workpiece's surface before and after MR polishing were given in Figs. 6 and 7, respectively.

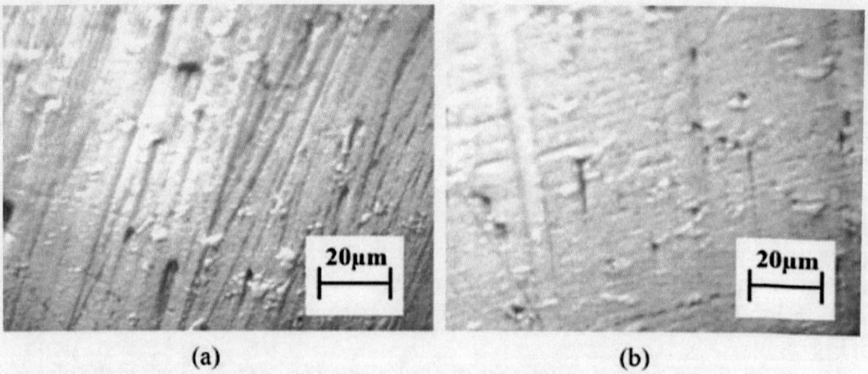

Fig. 6 Surface topography of workpiece: **a** before processing, **b** after processing

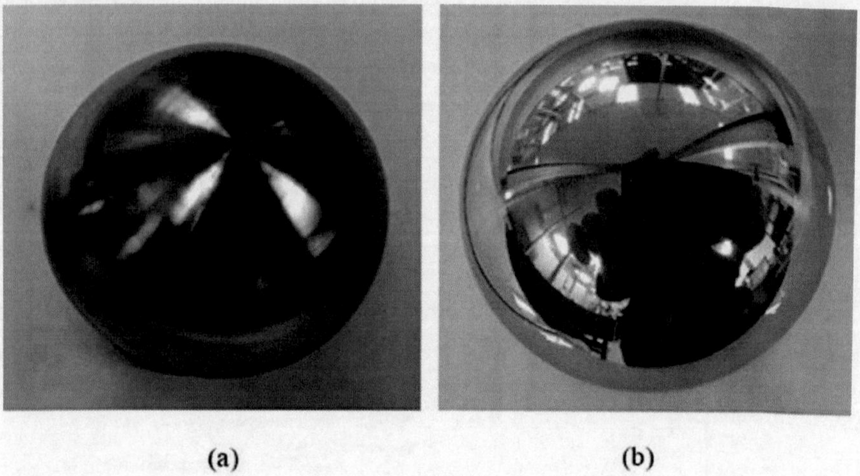

Fig. 7 Picture of workpiece surface: **a** before processing, **b** after processing

5 Conclusions

Experimental studies were carried out to determine the influence of polishing speeds, abrasive sizes and electric current on the surface roughness of workpieces. The conclusions were summarized as follows:

- The polishing speed is an important parameter that greatly influences the surface roughness of the workpiece. The best surface roughness can be achieved under the polishing speed of 70 rpm. When the polishing speed exceeds 70 rpm, the surface quality of workpiece is reduced. As a result, the polishing efficiency will not be satisfactory.
- The abrasive size has very little effect on surface quality improvement when it was changed in the experimental process.
- The electric current also has a strong effect on the surface roughness of workpiece. The best surface roughness Ra can be achieved of 18 nm with the abrasive size of 4000# Al_2O_3 and electric current over 1.2 A.

This proves that the MRP technique is the appropriate method for polishing the curved surfaces.

Acknowledgements This research is funded by the Industrial University of Ho Chi Minh City, Vietnam, through the University Research Foundation (No. 20/1.1CK03).

Conflicts of Interest The authors have no conflict of interest to declare.

References

1. Lin XH, Wang ZZ, Guo YB, Peng YF, Hu CL (2014) Research on the error analysis and compensation for the precision grinding of large aspheric mirror surface. Int J Adv Manuf Technol 71(1):233–239
2. Chen WK, Kuriyagawa T, Huang H, Yosihara N (2005) Machining of micro aspherical mould inserts. Precis Eng 29(3):315–323
3. Zeng SY, Blunt L (2014) Experimental investigation and analytical modelling of the effects of process parameters on material removal rate for bonnet polishing of cobalt chrome alloy. Precis Eng 38(1):348–355
4. Akay M, Aslan N (1996) Numerical and experimental stress analysis of a polymeric composite hip joint prosthesis. J Biomed Res 31:167–182
5. Ozturk O, Turkan U, Eroglu AE (2006) Metal ion release from nitrogen ion implanted CoCrMo orthopedic implant material. Surf Coat Technol 200(20–21):5687–5697
6. Hao YL, Yang R, Niinomi M, Kuroda D, Zhou YL, Fukunaga K, Suzuki A (2003) Aging response of the young's modulus and mechanical properties of Ti-29Nb-13Ta-4.6Zr for biomedical applications. Metall Mat Trans A 34:1007–1012
7. Rack HJ, Qazi JI (2006) Titanium alloys for biomedical applications. Mater Sci Eng C 26:1269–1277
8. Sun S, Brandt M, Dargusch MS (2010) Thermally enhanced machining of hard-to-machine materials—a review. Int J Mach Tools Manuf 50:663–680

9. Sun S, Brandt M, Dargusch MS (2010) Machining Ti–6Al–4V alloy with cryogenic compressed air cooling. Int J Tool Manuf 50:933–942
10. Wang ZG, Wong YS, Rahman M (2005) High-speed milling of titanium alloys using binderless CBN tools. Int J Mach Tools Manuf 45:105–114
11. Hu Z, Qin C, Chen ZC, Yang Z, Fang T, Mao M (2018) Experimental study of chemical mechanical polishing of the final surfaces of cemented carbide inserts for effective cutting austenitic stainless steel. Int J Adv Manuf Technol 95(9–12):4129–4140
12. Chang GW, Biing-Hwa Y, Rong-Tzong H (2002) Study on cylindrical magnetic abrasive finishing using unbonded magnetic abrasives. Int J Mach Tools Manuf 42:575–583
13. Wu Z, Yuan JL, Zhu YF, Nguyen DN (2016) Fracture analysis of brittle workpiece during elastic deformation molding process. Int J Adv Manuf Technol 86(9–12):3193–3202
14. Nguyen DN (2017) Study on improving the precision of form surface produced in elastic deformation molding process. Int J Adv Manuf Technol 93(9–12):3473–3484
15. Nguyen DN, Yuan JL, Wu Z (2016) Experimental study on elastic deformation molding process for generating aspheric surface glass. Int J Adv Manuf Technol 82(5–8):859–866
16. Nguyen DN, Lv BH, Yuan JL, Wu Z, Lu HZ (2013) Experimental study on elastic deformation machining process for aspheric surface glass. Int J Adv Manuf Technol 65(1–4):525–531
17. Lyu B, Dong C, Yuan J (2017) Experimental study on shear thickening polishing method for curved surface. Int J Nanomanuf 13(1):81–95
18. Li M, Lyu B, Yuan J, Dong C, Dai W (2015) Shear-thickening polishing method. Int J Mach Tools Manuf 94:88–99
19. Lyu BH, He QK, Chen SH, Shao Q, Chen Y, Geng ZY (2019) Experimental study on shear thickening polishing of cemented carbide insert with complex shape. Int J Adv Manuf Technol 103(1–4):585–595
20. Nguyen D (2020) Simulation and experimental study on polishing of spherical steel by non-Newtonian fluids. Int J Adv Manuf Technol 107:763–773
21. Nguyen DN, Dao TP, Chander P, Sunpreet S, Pramanik A, Grzegorz K, Catalin P (2020) Machining parameter optimization in shear thickening polishing of gear surfaces. J Mater Res Technol 1–15
22. Yi CJ, Peng XH, Zhao CW (2010) A magnetic-dipoles-based micro-macro constitutive model for MRFs subjected to shear deformation. Rheol Acta 49:815–825
23. Laun HM, Gabriel C, Schmidt G (2008) Primary and secondary normal stress differences of a magnetorheological fluid (MRF) up to magnetic flux density of 1 T. J Non-Newtonian Fluid Mech 148:47–56
24. Guo CY, Gong XL, Xuan SH, Qin LJ, Yan QF (2013) Compression behaviors of magnetorheological fluids under nonuniform magnetic field. Rheol Acta 52(2):165–176
25. Wang Y, Yin S, Hu T (2018) Ultra-precision finishing of optical mold by magnetorheological polishing using a cylindrical permanent magnet. Int J Adv Manuf Technol 97:3583–3594
26. Chanal A, Singh AK (2019) Magnetorheological nano-finishing of tube extrusion punch for improving its functional applications in press machine. Int J Adv Manuf Technol 103:2037–2052

Effect of Applied Load and Surface Texture on Friction of UHMWPE in Total Joint Replacement

Cong-Truyen Duong and Duc-Nam Nguyen

Abstract The objective of this study is to experimentally examine the effect of applied load and surface texture on friction of UHMWPE in total joint replacement by measuring the coefficient of friction (COF) between CoCrMo pin and UHMWPE discs. A CoCrMo pin with the dimensions of 11 mm in diameter and 18 mm in length was machined to obtain the surface roughness of 0.25 μm. For preparing different surface texture, ten UHMWPE discs with the dimensions of 88 mm in diameter and 5 mm in thickness were machined by two different machining processes of milling and turning to obtain two roughness values of 2.5 and 0.25 μm. 25 mg/ml BSA was made by mixing BSA powder in 10% PBS. A frictional testing apparatus was developed and carried out for the calculation of COF. Results show that the COF increases with the increase of the applied load in both cases of the UHMWPE roughness of 0.25 and 2.5 μm. With the UHMWPE roughness of 0.25 μm, the COF decreases when the surface texture machined by milling compared to that by turning in both cases of the applied loads of 4 and 16 N; however, with the UHMWPE roughness of 2.5 μm, the COF increases when the surface textures machined by milling compared to that by turning. These results suggest that both the applied loads and the surface textures have important effects on the friction of UHMWPE.

Keywords Applied load · Surface texture · Friction · UHMWPE · Total joint replacement

1 Introduction

Total joint replacements (TJR) have been accepted to be one of the most popular surgical procedures performed for replacements of damaged joints [1–3]. There were more than 790,000 TJR performed from 2003 to 2015 in England, and more than 60,000 TJR were performed from 2010 to 2017 in Korea [1]. In the United States of America, more than 300,000 TJR operations were performed annually [4].

C.-T. Duong (✉) · D.-N. Nguyen
Industrial University of Ho Chi Minh City, Ho Chi Minh City, Vietnam
e-mail: duongcongtruyen@iuh.edu.vn

V. Van Toi et al. (eds.), *8th International Conference on the Development of Biomedical Engineering in Vietnam*, IFMBE Proceedings 85,
https://doi.org/10.1007/978-3-030-75506-5_14

The popularly used materials for TJR include ultra-high molecular weight polyethylene (UHMWPE), metal alloys (such as CoCr, CoCrMo, titanium and stainless steel), ceramics (such as alumina, zirconia and silicon nitride) and ultra-hard coatings on metal materials (such as diamond-like carbon and titanium nitride) [1, 2]. The material combinations between metal and UHMWPE, metal and metal, ceramic and plastic, ceramic and ceramic have been widely used for the bearing surfaces in TJR [4, 5]. Mechanical and tribological properties, as well as biocompatibility and corrosion resistance, are the properties that should be considered for the selection of the material combinations of TJR [2].

Many previous studies have reported the tribological properties of material combinations under different testing conditions, and they generally accepted that tribological characteristics of material combinations depend on the type of materials, lubricant, roughness as well as their surface properties including surface texture [5–14]. However, the effect of the surface texture on the frictional properties of UHMWPE has not been clearly identified. Therefore, the objective of this study is to experimentally examine the effect of applied load and surface texture on friction of UHMWPE in total joint replacement by measuring the coefficient of friction between CoCrMo pin and UHMWPE discs in 25 mg/ml BSA. Our hypothesis is that increasing applied load will lead to an increase in the friction of UHMWPE and frictional properties of UHMWPE depend on UHMWPE surface texture.

2 Materials and Methods

2.1 *Preparation of Specimens*

CoCrMo material (diameter = 11 mm), which used as a pin, was purchased and the flat ends were machined to obtain the length of 18 mm and roughness (R_a) of 0.25 µm. For preparing different surface texture, UHMWPE material was purchased, and ten discs (diameter = 88 mm, thickness = 5 mm) were machined by two different machining processes of milling (five samples) and turning (five samples) to obtain two roughness (R_a) values of 2.5 and 0.25 µm. A SJ-310 Mitutoyo device was used in order to measure the roughness of pin and discs.

2.2 *Preparation of 25 mg/ml BSA*

Figure 1 shows the PBS (phosphate buffered saline) and BSA (bovine serum albumin) powders which were used for the preparation of 25 mg/ml BSA using as a lubricant. Firstly, 10% PBS solution was made from PBS powder. Then, by mixing BSA powder in 10% PBS, 25 mg/ml BSA was made. The lubricant of 25 mg/ml BSA for experiments in the current study was selected according to the range of protein in human

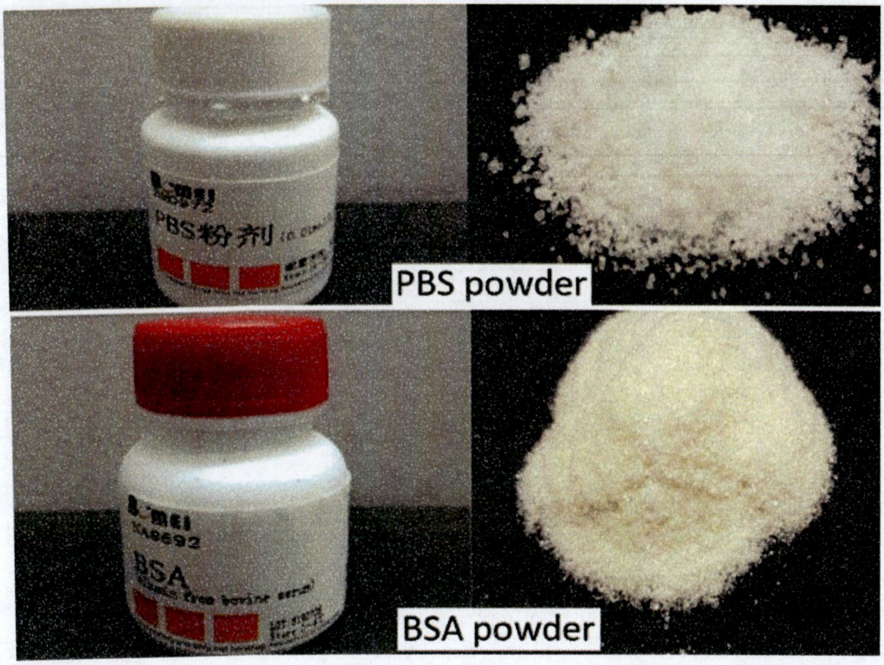

Fig. 1 PBS and BSA powders were used for preparing the lubricant of 25 mg/ml BSA

synovial fluid [15]. The lubricant of 25 mg/ml BSA was kept at room temperature (25 °C) and used for the experiments within 24 h.

2.3 *Frictional Testing Apparatus and Procedure*

Figure 2 shows a frictional testing apparatus, developed and carried out by our group, which we use to calculate the coefficient of friction (COF). The 25 mg/ml BSA was used as the lubricant for all experiments. The same methodology, which was described in detail in our recent paper, were used to calculate COF between CoCrMo pin and UHMWPE discs [16]. Briefly, in the current study, the rotation of UHMWPE discs was controlled at 0.3 rev/s, the applied loads were controlled at 4 N and 16 N and the frictional force was measured. Finally, the COF was finally calculated and averaged.

Fig. 2 A frictional testing apparatus was carried out for the measurement of COF

2.4 *Data Analyses*

Student's *t*-test was used to evaluate the statistical difference in the average values of COF between the applied load of 4 N and 16 N as well as between the UHMWPE surface texture machined by milling and by turning with $p < 0.05$ considered statistically significant.

3 Results and Discussion

3.1 *Effect of Applied Load on Friction of UHMWPE*

The variations of COF between CoCrMo pin and UHMWPE discs were shown in Fig. 3. Average COF $\pm$ SD between CoCrMo pin and UHMWPE discs with different applied loads and UHMWPE surface textures were described in Table 1. The results show that the COF increases with the increase of the applied loads in all cases of UHMWPE roughness and machining processes; the higher the applied loads, the higher the values of COF in all cases. The Student's *t*-test showed the statistical differences in COF between the applied load of 4 and 16 N ($p < 0.0001$, $n = 18$) in all cases (milling with the roughness of 0.25 μm, milling with the roughness of 2.5 μm, turning with the roughness of 0.25 μm and turning with the roughness of 2.5 μm) (Fig. 4). These results in this study indicate the important effect of the applied load on the friction of UHMWPE.

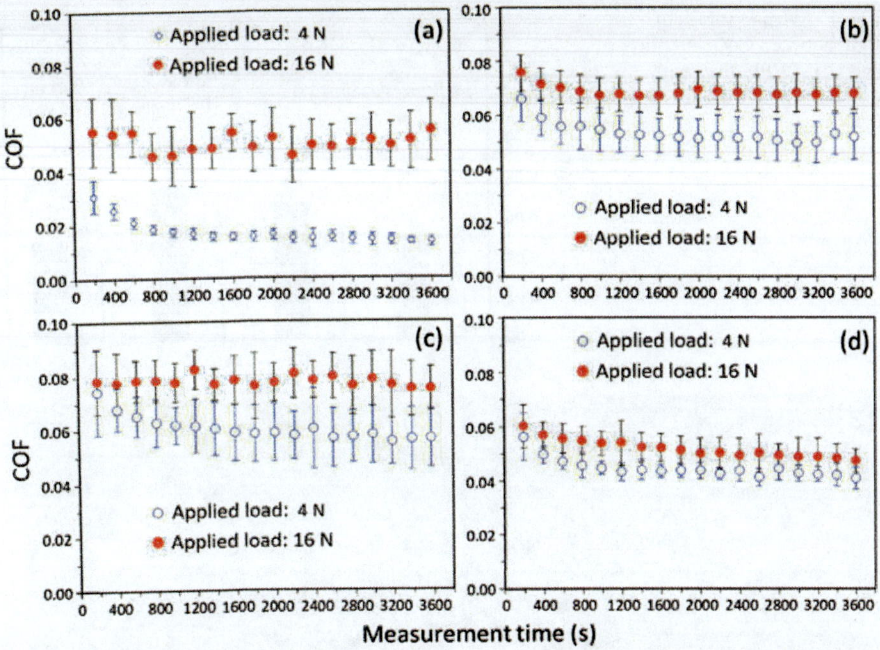

Fig. 3 Variations of COF between CoCrMo pin and UHMWPE discs indicating that COF increases with the increase of the applied loads in all cases: **a** $R_a = 0.25$ µm and surface texture machined by milling, **b** $R_a = 2.5$ µm and surface texture machined by milling, **c** $R_a = 0.25$ µm and surface texture machined by turning, and **d** $R_a = 2.5$ µm and surface texture machined by turning

Table 1 Average COF ± standard deviation (COF ± SD) between CoCrMo pin and UHMWPE discs with different applied loads and UHMWPE surface textures

Machining processes of UHMWPE discs	COF ± SD		p value
	Applied load: 4 N	Applied load: 16 N	
Milling, $R_a = 0.25$ µm	0.017 ± 0.002	0.051 ± 0.010	<0.0001
Milling, $R_a = 2.5$ µm	0.053 ± 0.008	0.069 ± 0.006	<0.0001
Turning, $R_a = 0.25$ µm	0.059 ± 0.010	0.076 ± 0.009	<0.0001
Turning, $R_a = 2.5$ µm	0.044 ± 0.004	0.052 ± 0.006	<0.0001

3.2 *Effect of Surface Texture on Friction of UHMWPE*

SEM images of UHMWPE disc surfaces from Fig. 5 showed the different surface textures when the UHMWPE surface machined by milling and turning operations. The variations of COF between CoCrMo pin and UHMWPE discs were described in Fig. 6.

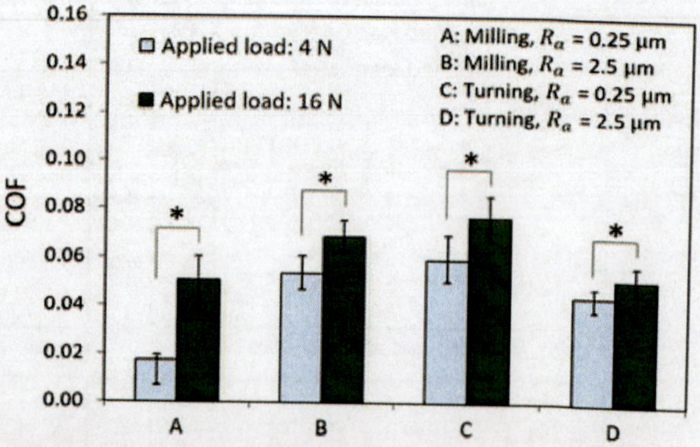

Fig. 4 Average values of COF between CoCrMo pin and UHMWPE discs with different applied loads (*$p <$ 0.0001)

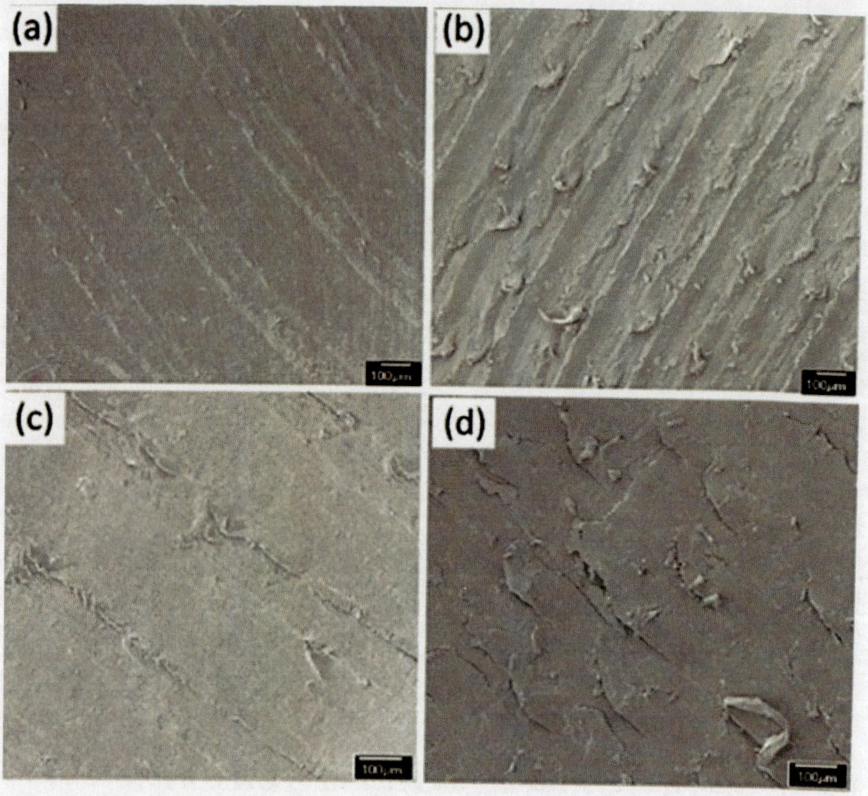

Fig. 5 SEM images of UHMWPE disc surfaces machined by two different machining processes indicating different surface textures: **a** milling before the experiment, **b** turning before the experiment, **c** milling after the experiment, **d** turning after the experiment

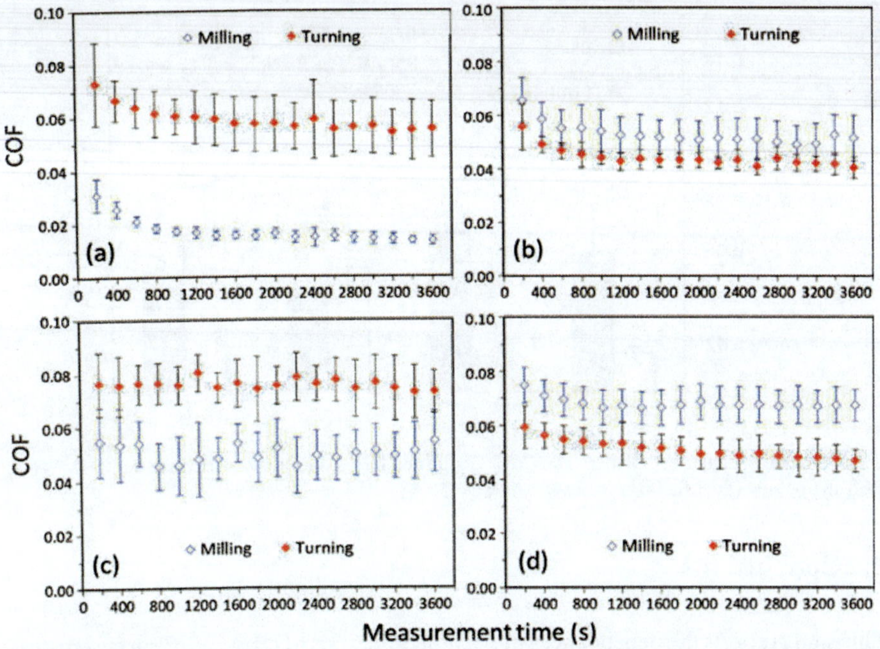

Fig. 6 Variations of COF between CoCrMo pin and UHMWPE discs indicating that COF depends on UHMWPE surface textures in all cases: **a** $R_a = 0.25$ μm and applied load of 4 N, **b** $R_a = 2.5$ μm and applied load of 4 N, **c** $R_a = 0.25$ μm and applied load of 16 N, and **d** $R_a = 2.5$ μm and applied load of 16 N

The average COF ± SD were also described in Table 1. The results show that the COF depends on the changes of the UHMWPE surface textures. With roughness of 0.25 μm, the surface textures machined by milling decreased the COF values compared to that by the surface textures machined by turning in both cases of applied load values (Fig. 6a, c). However, the opposite results occurred with the roughness of 2.5 μm (Fig. 6b, d). The Student's *t*-test showed the statistical differences in COF between surface textures machined by milling and turning ($p < 0.0001$, $n = 18$) in all cases (applied load 4 N and roughness of 0.25 μm, applied load 16 N and roughness of 0.25 μm, applied load 4 N and roughness of 2.5 μm and applied load 16 N and roughness of 2.5 μm) (Fig. 7). These results in the current study indicate the important effect of surface textures on the friction of UHMWPE that is also suitable to published frictional researches [2, 6, 13, 14].

The potential limitation of this study is the use of the pin-on-disc frictional testing apparatus, which does not accurately simulate the joint structure and its activity in the body. However, it is typically used as a comparative test by many previous studies [5, 9, 11, 14].

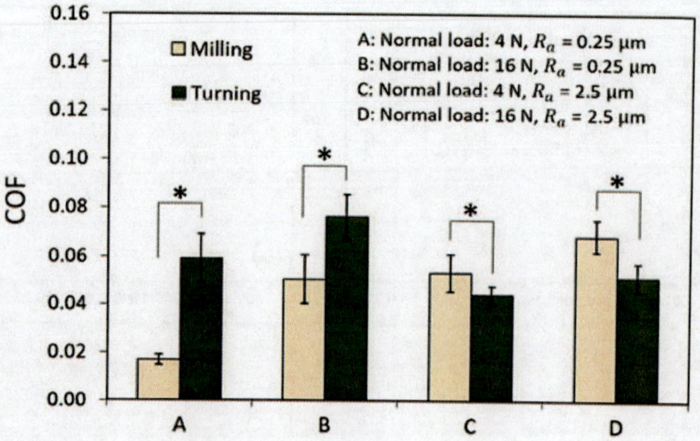

Fig. 7 Average values of COF between CoCrMo pin and UHMWPE discs with different UHMWPE surface textures (*$p < 0.0001$)

4 Conclusions

This study reports the dependences of friction properties of UHMWPE surface on the applied loads as well as on the surface textures by measuring the coefficient of friction between CoCrMo pin and UHMWPE discs in 25 mg/ml BSA. The main conclusions are that the coefficient of friction statistically significant increases with the increase of the applied load in both cases of the UHMWPE roughness of 0.25 and 2.5 μm. With the UHMWPE roughness of 0.25 μm, the coefficient of friction statistically significant decreases when the surface texture machined by milling compared to that by turning in both cases of the applied loads of 4 and 16 N; however, with the UHMWPE roughness of 2.5 μm, the coefficient of friction statistically significant increases when the surface textures machined by milling compared to that by turning in both cases of the applied loads of 4 and 16 N. The results suggest that both the applied loads and the surface textures have the important effects on the friction of UHMWPE.

Conflict of Interest The authors have no conflict of interest to declare.

References

1. Hu CT, Yoon TR (2018) Recent updates for biomaterials used in total hip arthroplasty. Biomater Res 22(33):1–12
2. Shen G, Fang F, Kang C (2018) Tribological performance of bioimplants: a comprehensive review. Nanotechnol Precis Eng 1(2):107–122

3. Shen G, Zhang JF, Fang FZ (2019) In vitro evaluation of artificial joints: a comprehensive review. Adv Manuf 7(1):1–14
4. Puccio FD, Mattei L (2015) Biotribology of artificial hip joints. World J Orthop 6(1):77–94
5. Borjali A, Langhorn J, Monson K, Raeymaekers B (2017) Using a patterned microtexture to reduce polyethylene wear in metal-on-polyethylene prosthetic bearing couples. Wear 292–293(1):77–83
6. Jin ZM, Stone M, Ingham E, Fisher J (2006) Biotribology. Curr Orthop 20(1):32–40
7. González-Mora VA, Hoffmann M, Stroosnijder R, Gil FJ (2011) The role of hardness and roughness on the wear of different CoCrMo counterfaces on UHMWPE for artificial joints. Biomed Sci Eng 4(1):651–656
8. Linn F (1968) Lubrication of animal joints II. The mechanism. Biomechanics 1(3):193–205
9. Gispert MP, Serro AP, Colaco R, Saramago B (2006) Friction and wear mechanisms in hip prosthesis: comparison of joint materials behaviour in several lubricants. Wear 260(1–2):149–158
10. Saikko V (2003) Effect of lubricant protein concentration on the wear of ultra-high molecular weight polyethylene sliding against a CoCr counterface. ASME J Tribol 125(3):638–642
11. Mishina H, Kojima M (2008) Changes in human serum albumin on arthroplasty frictional surfaces. Wear 265(5–6):655–663
12. Wang A, Essner A, Schmidig G (2004) The effects of lubricant composition on in vitro wear testing of polymeric acetabular components. Biomed Mater Res B 68(1):45–52
13. Choudhury D, Vrbk M, Mamat AB, Stavness I, Roye CK, Mootanah R, Krupk I (2017) The impact of surface and geometry on coefficient of friction of artificial hip joints. Mech Behav Biomed Mater 72(1):192–199
14. Borjali A, Monson K, Raeymaekers B (2018) Friction between a polyethylene pin and a microtextured CoCrMo disc. Tribol Int 127(1):568–574
15. McPherson RA, Pincus MR (2017) Henry's clinical diagnosis and management by laboratory methods, 23rd edn. Elsevier, Philadelphia
16. Duong CT, Le CD, Nguyen DN (2020) Effect of surface roughness on friction of CoCrMo-on-UHMWPE bearing in total hip arthroplasty under lubrication of bovine serum albumin. In: IFMBE proceedings, vol 69, no 1, pp 247–251

An Automatic Control System for Measuring Stokes Polarization and Utilizing UV Light Source

Y. Nguyen Le, Yen-Nhi Nguyen, and Thi-Thu-Hien Pham

Abstract Polarization is an elementary property of light waves and has many applications in medicine and health sciences. In this research, an automatic control UV light source Stokes polarimeter system is built using simple and inexpensive optical devices. The optical system includes one UV laser source, three polarizers, two quarter-waveplate, and one optical sensor applying the Mueller-Stokes decomposition for extracting linear birefringence (LB), linear dichroism (LD), circular birefringence (CB), circular dichroism (CD), linear depolarization (L-Dep), and circular depolarization (C-Dep) of a biological sample. The system can measure the intensity of 180 data points then calculate the output Stokes vector of measured samples from 4 inputs polarized lights (i.e., 0°, 45°, 90° of linear polarized lights and right-hand circular). The experimental results showed that the system could automatically measure Stokes parameters of a biological sample with the accuracy at ±5% to compare with the commercial device, Stokes polarimeter. Therefore, the designed system has the benefits not only of extracting the optical parameters of the biological samples but also of improving the accuracy of results by reducing the error effect on the measurement.

Keywords Stokes-Mueller · Stokes polarimeter · Polarized light measurement system

Y. Nguyen Le—Equally contributed.

Y. Nguyen Le · Y.-N. Nguyen · T.-T.-H. Pham (✉)
School of Biomedical Engineering, International University, Ho Chi Minh City 700000, Vietnam
e-mail: ptthien@hcmiu.edu.vn

Vietnam National University Ho Chi Minh City (VNU-HCMC), Ho Chi Minh City 700000, Vietnam

V. Van Toi et al. (eds.), *8th International Conference on the Development of Biomedical Engineering in Vietnam*, IFMBE Proceedings 85,
https://doi.org/10.1007/978-3-030-75506-5_15

1 Introduction

Polarized light has significant potential for applications in the cosmetic industry, chemistry, biology, engineering science, especially medicine [1–5]. Classic applications of the effective parameters include linear birefringence (LB) of the peripapillary retinal nerve fibre layer provides useful information for early glaucoma diagnosis and the complete understanding of scanning laser polarimetry [6, 7]. Moreover, linear birefringence measurement on the myocardium of rat hearts demonstrates the potential application for monitoring regenerative therapy of myocardial infarction [3]. Circular birefringence (CB) measurement of human blood can distinguish the diabetic severity status in patients [2]. Linear dichroism (LD) measurements are useful for characterizing significant properties of molecules that are hidden or support tumour diagnosis [8–10]. In contrast, the circular dichroism (CD) spectrum can be used to analyse and give valuable information about the secondary structure and function of proteins [11–13].

Furthermore, linear depolarization (L-Dep) and circular depolarization (C-Dep) measurements can provide useful knowledge for studying and characterizing turbid media [14]. Many research papers have shown that any polarization state of light can be completely described by four measurable quantities known as Stokes polarization parameters (S0, S1, S2, and S3) [15]. Applying the Mueller matrix method, the proposed system was designed and built for measuring the Stokes polarization parameters from 4 inputs polarized lights (right-hand circular and linear polarized lights in 0°, 45°, 90°) to determine the fundamental optical properties of biological samples including LB, LD, CB, CD, L-Dep and C-Dep [16]. In this research, an automatic control UV laser- Stokes polarimeter system was built to develop a medical device capable of detecting specific diseases, especially tumour or cancer at their early stages.

2 Methods

2.1 Methodology

Based on the analytical technique of Pham and Lo [17, 18], we design an automatic control system for measuring the Stokes parameters of biomedical samples. This control system can determine the optical properties of samples. A rotating half-wave plate will produce various orientations of linearly polarized light. The polarizing element was placed in a rotatable mount during the measurements and then inserted between the light source and the analyser. By applying Stokes polarimetry and Mueller matrix decomposition method [19], the optical parameters were extracted namely linear birefringence (LB), circular birefringence (CB), linear dichroism (LD), circular dichroism (CD), linear depolarization (L-Dep), and circular depolarization (C-Dep). The automatic control system of Stokes polarimeter is shown in Fig. 1.

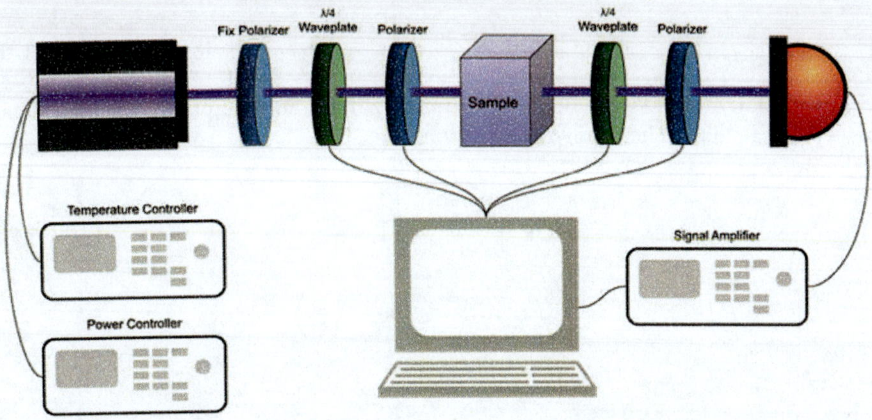

Fig. 1 Schematic illustration of the model of the automatic control UV light—Stokes polarimeter

The system includes a Laser diode source (LDM 785, Co. Thorlabs) with the wavelength of 375 nm firstly generate the light source to a fixed linearly vertical polarizer (LPUV050-MP2, Thorlabs Co.) to normalize and filter the light source, the power controller (LDC200C, Thorlabs Co.) and temperature controller (TEC200C, Thorlabs Co.) were utilized to control the temperature of the laser. Then, the polarization state generator (PSG) with a rotating quarter-wave plate (AQWP05M, Co. Thorlabs) and a rotating linear polarizer (LPUV050-MP2, Thorlabs Co.) generates two circular polarized lights (LH and RH) and four linear polarized lights at 0°, 45°, 90°, 135°. The generated lights interact with the sample and reach the polarization state analyser (PSA) part including another rotating quarter-wave plate (AQWP05M, Co. Thorlabs), a fixed vertical polarizer (LPUV050-MP2, Thorlabs Co.) and a biased Gallium Phosphide photodetector (DET25K/M-GaP Detector, 150–550 nm, Co. Thorlabs) for measuring the light intensity and output the photocurrent. This current was amplified by a photodiode amplifier (PDA200C, Thorlabs Co.) before reading by a computer. Based on the rotating quarter-wave plate polarimeter Stokes method, the rotating quarter-wave plate in the PSA rotates from 0° to 180° and 180 data sample points were obtained by the detector. It will rotate with the steps of 1° ($\theta n + 1 - \theta n = 180/180 = 1°$). Therefore, the angles of rotation are: $\theta 1 = 0°$, $\theta 2 = 1°$, $\theta 3 = 2°$ … $\theta 179 = 178°$, $\theta 180 = 179°$; with every angle of rotation the system reads the analogue signal. Lastly, for each polarized light, 180 measurement data points will be converted to digital signals and computed to Stokes vectors by the equations:

$$\mathrm{A} = \frac{2}{\mathrm{N}} \sum_{\mathrm{n}=1}^{\mathrm{N}} \mathrm{In} | \mathrm{B} = \frac{4}{\mathrm{N}} \sum_{\mathrm{n}=1}^{\mathrm{N}} \mathrm{I_n} \sin 2\theta_\mathrm{n} \tag{1}$$

$$C = \frac{4}{N} \sum_{n=1}^{N} I_n \cos 4\theta_\mathrm{n} | \mathrm{D} = \frac{4}{\mathrm{N}} \sum_{\mathrm{n}=1}^{\mathrm{N}} \mathrm{I_n} \sin 2\theta_\mathrm{n} \tag{2}$$

$$S_0 = A - C, \quad S_1 = 2C, \quad S_2 = 2D, \quad S_3 = B \tag{3}$$

Additionally, the Degree of Polarization, Azimuth and Ellipticity of each data point was calculated the following equations:

$$DOP = \frac{\sqrt{S_1^2 + S_2^2 + S_3^2}}{S_0} \tag{4}$$

$$Azimuth = 90^\circ - \frac{\tan^{-1}\left(\frac{S_1}{S_2}\right)}{2} \tag{5}$$

$$Ellipticity = \frac{\tan^{-1}\left(\frac{S_3}{\sqrt{S_1^2+S_2^2}}\right)}{2} \tag{6}$$

A graphical user interface application was designed to control the system and display the output conveniently algorithm written in MATLAB by Thi-Thu-Hien Pham was used to extract the output results, i.e., Stokes vectors, the Degree of Polarization, Azimuth and Ellipticity of the sample. Brief processes of MATLAB program were shown in Fig. 2.

The Stokes parameters were processed in MATLAB program by applying the Mueller matrix decomposition for extracting nine effective optical parameters including orientation angle of the fast axis of LB, linear birefringence of LB, optical rotation of CB, orientation angle of LD, linear dichroism of LD, circular dichroism of CD, linear depolarization, and circular depolarization.

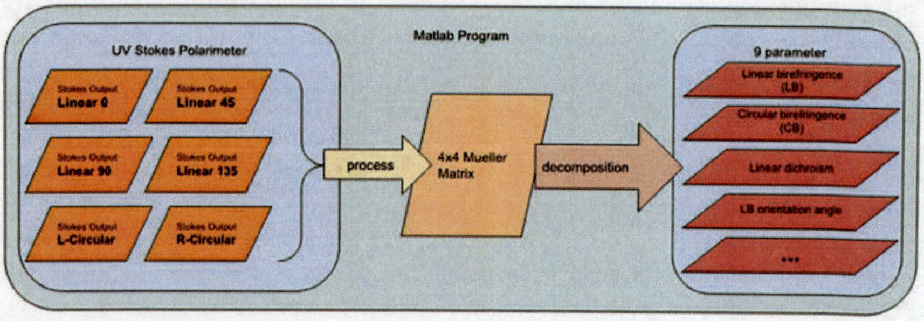

Fig. 2 The processing pipeline of the algorithm written in MATLAB

2.2 *Hardware Design*

The hardware of the system is designed in two main parts. The first part is an optic mounting, including the gear part controlled by a stepper motor. The holder was linking with the gear part using a magnet and the optic holder to secure the positions of the plates without damaging. With the magnet, the optic mounting can easily disassemble its components or change the tilt angle of the optic. The second part is the rotating optic module, which was created to control the optic angle, including two blocks fixed two stepper motors, which were placed in symmetry and connected with the optic mount by a belt drive.

All optic mounts and the rotating optic module were designed using Autodesk Fusion 360 and printed by 3D printers.

2.3 *Firmware and Software Development*

Targeting to an automatic control system, the firmware and software were developed to have the most ability to do measurement with less control from the user. The position of the fast axis of each optic was saved in Electrically Erasable Programmable Read-Only Memory of microcontroller to reduce the calibration of the system. The microcontroller was used to control hardware components, read and convert the analogue signal from the sensor. The software is designed by using C Sharp, giving it the ability to fast process and controlling the system. The flowchart of the software algorithm to control the system for generating six polarized light states and analysing the output light after going through the sample, as illustrated in Fig. 3.

3 Results and Discussion

3.1 *Designed Hardware*

Optic Mounting

In Fig. 4, the optic mount was designed to secure optical components without damaging the optic and controlled by a stepper motor. Thanks to the design of optic mounting with magnets, the optical elements are easily removed or replaced. The tilt angle of optical can be changed from −20 to 20°.

The gear part operated by a stepper motor was designed to link with the holder by magnets. This part has a home point for marking the optical angle. The optic holder was designed for the insertion of the plates, linear polarizer mount LPUV050-MP2 and quarter-wave plate AQWP05M from Thorlabs Co., as shown in Fig. 5.

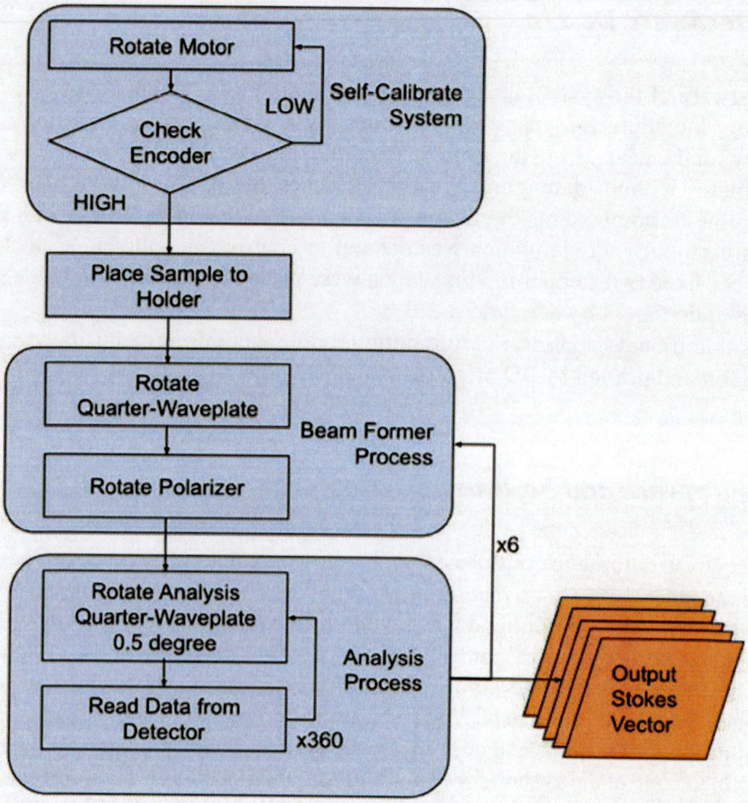

Fig. 3 The flow chart of the automatic Stokes measurement system

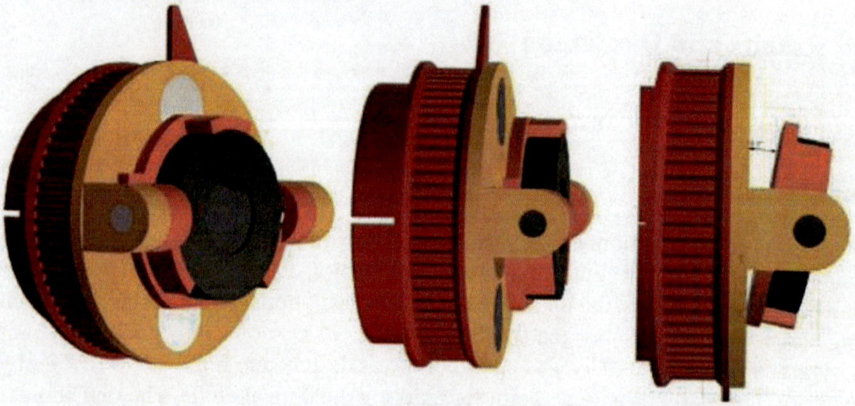

Fig. 4 The design of the optic mounting which was designed allowing the user to change the tilt angle of retarder from −20 to 20° with removable characteristic due to magnet linking

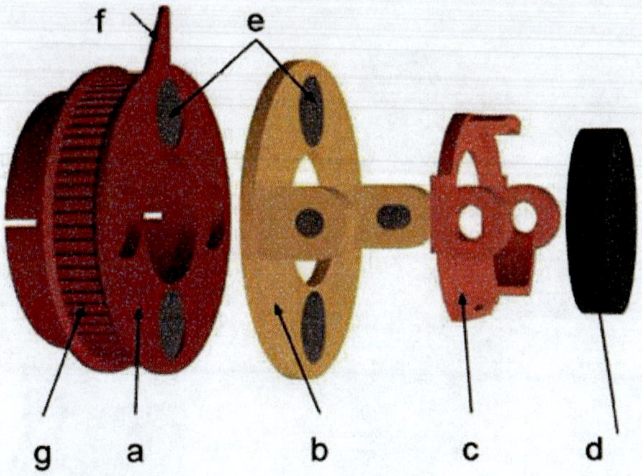

Fig. 5 **a** The gear part with magnets; **b** The holder with magnets to link with (**a**); **c** The optic holder part, designed to tilt the optic from −20 to 20°; **d** The optic; **e** Magnets; **f** The home point to mark the angle of optic; **g** Gear teeth

According to Meadowlark Optics Incorporated [20], the source errors, namely wavelength, angle of incidence, thermal, and the presence of multiple reflections can change the retardation. To avoid inaccurate retardance error, the optic mounting is designed to be able to alter the tilt angle of the optic. The equation calculates the retardance:

$$\delta = (n_e - n_o)\frac{t}{\lambda} = \frac{\beta t}{\lambda} \quad (7)$$

where t is the thickness of optic and λ is the wavelength of light, n_e is the extraordinary index, n_o is the ordinary index, ß is a birefringence. This equation proves that wavelength has a significant impact on retardation. The angle of incidence is another source error because it can change the extraordinary index. Retardation can change by temperature since it can change the thickness of the retarder and the birefringence of the material. The retardation is also affected by the presence of multiple reflections caused by the similarity of the coherence length of the light source and the thickness of the retarder.

Therefore, the optic mounting is designed to have the capability to change the tilt angle of the retarder from −20 to 20° for optimizing quarter-wave of retardance for each wavelength.

Optic Rotation Module

The block was designed to fit the stepper motor to control the polarizer lens angle. The stepper motor-operated the optic mounting through belt drives. The system has two optic rotation modules; each module has two motors and two optic holders placed in symmetry. This symmetry design was optimized to reduce the length of the light path or distance between two optics, the unbalancing of the mass balance of the module

Fig. 6 Design of the optic rotation module using Autodesk Fusion CAD. The module was designed to rotate the optic holder, consisting of two motors and two optic holders placed in symmetry

was also reduced. The optic rotation module was designed as following shown in Fig. 6.

The design of the automatic measurement of the UV Laser system was shown in Fig. 7. This design consists of 2 main parts. The first part is PSG including a Laser diode source and a first module attaching a rotating quarter-wave plate with a rotating linear polarizer. This part served as the generator of 6 inputs polarized lights. To calibrate the light source, a fixed linearly vertical polarizer is inserted before the first module. The second part is PSA consisting of a second module, which has a rotating quarter-wave plate, a fixed vertical polarizer, and a photodetector for measuring the analogue signals and calculating the Stokes parameters.

The light with the wavelength of 375 nm is firstly created by the light source, passes through a rotating quarter-wave plate and a rotating linear polarizer to produce sequentially six polarized light (two circular polarized lights and four linear polarized light). Subsequently, these lights interact with the sample, followed by reaching another rotating quarter-wave plate, a fixed vertical polarizer, and a photodetector for obtaining the analogue signal and calculating the intensity.

Fig. 7 The design of the system in autodesk fusion CAD. The system is designed into two main parts. The first part is the polarization state generator (PSG) including a laser diode source and a first optic rotation module for generating six inputs polarized lights. The second part is the polarization state analyser (PSA) including a second module and a photodetector for receiving the analogue signal and calculating the Stokes vectors

3.2 *Validation for Accuracy of the System*

Accuracy of Optic Rotation and Photodetector

A first experiment that is used to validate the accuracy of the motor and photodetector was set up to measure the light intensity of UV laser through a fixed polarizer and rotate another polarizer. Hypothetically, the output intensity recorded will be calculated by $I = I_o*(\cos(a))^2$, where (a) is the angle formed by the polarizer and the light axis; I_o is the intensity when the polarizer is parallel with the light axis. The error of the system was determined by subtracting the measured values to expected values, as shown in Eq. (8):

$$\%error = \left|\frac{Measured\ values - Expected\ values}{Expected\ values}\right| * 100\% \quad (8)$$

In our experiment, the measured data were compared with the expected values. The results of the system are well fitted with the mathematical assumption, as shown in Fig. 8. The accuracy of the motor and the light detector were perfectly validated in this experiment with the error is 0.25% of reading.

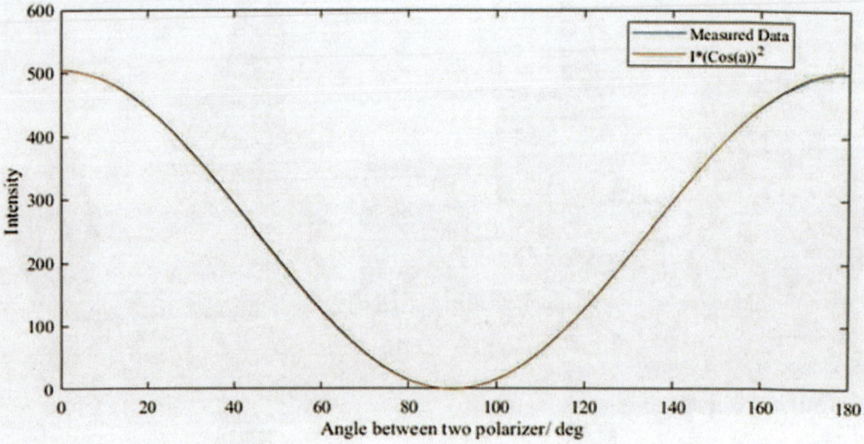

Fig. 8 The comparison of the results of the designed system and the expected values

The Output of Polarization State Generator (PSG)

The outcomes of the generated light from PSG, including Stokes vector, the degree of polarization, azimuth and ellipticity were measured and shown in Table 1.

The generated lights had an error from −0.2% to 1.1% about DOP satisfied the project goal to have the error below 5%. An experiment to extract the properties of the half-wave plate with generated lights was done and reported in the below part of the paper. The result was very good but can improve in the future.

Table 1 The outcomes of the generated polarized lights from PSG

	S_0	S_1	S_2	S_3	Azimuth	Ellipticity	Degree of Polarization
Linearly polarized 0°	1	1.007	0.009	−0.026	45.26	−0.73	100.8
Linearly polarized 45°	1	−0.027	1.011	−0.024	90.76	−0.66	101.1
Linearly polarized 90°	1	−0.998	−0.021	−0.017	45.59	−0.47	99.8
Linearly polarized 135°	1	0.003	−0.988	−0.011	90.76	−0.67	101.1
Left-hand circularly polarized	1	−0.059	0.007	−1	131.71	−44.91	100.2
Right-hand circularly polarized	1	0.039	0.029	0.997	63.23	43.60	99.8

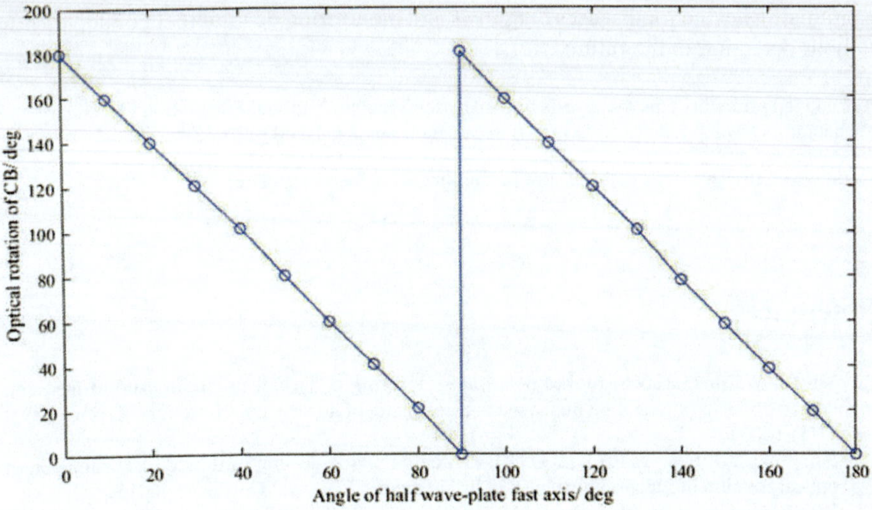

Fig. 9 The extracted optical rotation of CB property of a rotating half-wave plate when it rotated from 0° to 180°

Decomposition of Rotating Half Waveplate

To validate the accuracy of the motor and photodetector, a standard Achromatic Half-Wave Plate (AHWP05M-340, Thorlabs Co.) was positioned at the sample location (see Fig. 1) and rotated from 0 to 180°. After that, through extracting the LB, LD and CB properties of the half-wave plate, the accuracy was assessed. The results are illustrated in Fig. 9.

For each rotation of the half-wave plate, the optical rotation of CB decreases 10°, which is equal to the increasing angle of the half-wave plate, followed by starting a new cycle at the next 90° of the optic.

4 Conclusion

In conclusion, the system was designed, built and tested for automatically measuring the Mueller matrix of the samples with error ±0.25% for reading data. Thanks to the Stokes-Mueller matrix decomposition, nine effective parameters in LB, LD, CB, CD, L-Dep or C-Dep were extracted for investigating the optical properties of the sample. For future works, more rigorous experiments with biological samples and chemical compounds need setting up to characterize their optical properties that could be obscured by using other methods or compare the differences between the healthy and abnormal tissues which is contributed to the cancer diagnosis. This system not only minimizes the time consumption and reduces the errors by users in experiments. It is a promising device for diseases diagnosis and monitoring development or supporting tumour detection in the future.

Acknowledgements This research is funded by the Vietnam National Foundation for Science and Technology Development (NAFOSTED) under grant number 103.03-2019.381.

Conflicts of Interest The authors have no conflict of interest to declare.

References

1. Wu PJ, Walsh JT (2006) Stokes polarimetry imaging of rat tail tissue in a turbid medium: degree of linear polarization image maps using incident linearly polarized light. J Biomed Opt 11(1):14031
2. Sun P, Ma Y, Liu W, Yang Q, Jia Q (2014) Mueller matrix decomposition for determination of optical rotation of glucose molecules in turbid media. J Biomed Opt 19(4):46015
3. Wood MF, Ghosh N, Wallenburg MA, Li SH, Weisel RD (2010) Polarization birefringence measurements for characterizing the myocardium, including healthy, infarcted, and stem-cell-regenerated tissues. J Biomed Opt 15:047009
4. Bargo PR, Prahl SA, Goodell TT, Sleven RA, Koval G, Blair G, Jacques SL (2005) In vivo determination of optical properties of normal and tumor tissue with white light reflectance and an empirical light transport model during endoscopy. J Biomed Opt 10:034018-1–034018-15
5. Moffitt T, Chen YC, Prahl SA (2006) Preparation and characterization of polyurethane optical phantoms. J Biomed Opt 11:041103-1–041103-10
6. Huang X-R, Knighton RW (2002) Linear birefringence of the retinal nerve fiber layer measured in vitro with a multispectral imaging micropolarimeter. J Biomed Opt 7(2):199–204
7. Huang X-R, Bagga H, Greenfield DS, Knighton RW (2004) Variation of peripapillary retinal nerve fiber layer birefringence in normal human subjects. Invest Ophthalmol Vis Sci 45(9):30733080
8. Liu GL, Li Y, Cameron BD (2002) Polarization-based optical imaging and processing techniques with application to the cancer diagnostics. In: Proceedings SPIE, San Jose, CA, USA, pp 208–220
9. Wemyss AM, Razmkhah K, Chmel NP, Rodger A (2018) Fluorescence detected linear dichroism of small molecules oriented on polyethylene film
10. Nordén B (1978) Applications of linear dichroism spectroscopy. Appl Spectrosc Rev 14(2):157–248
11. Berova N, Nakanishi K, Woody RW (2000) Circular dichroism: principles and applications. Wiley-VCH, Weinheim
12. Kelly SM, Jess TJ, Price NC (2005) How to study proteins by circular dichroism. Biochim Biophys Acta (BBA) Proteins Proteomics 1751(2):119–139
13. Kelly SM, Price NC (2000) The use of circular dichroism in the investigation of protein structure and function. Curr Protein Pept Sci 1(4):349–384
14. Ghosh N, Wood MF, Vitkin IA (2008) Mueller matrix decomposition for extraction of individual polarization parameters from complex turbid media exhibiting multiple scattering, optical activity, and linear birefringence. J Biomed Opt 13:044036
15. Bickel WS, Bailey WM (1985) Stokes vectors, Mueller matrices, and polarized scattered light. Am J Phys 53(5):468–478
16. Firdous S, Ikram M (2012) Stokes polarimetry for the characterization of biomaterials using liquid crystal variable retarders. In: Proceedings SPIE 2012, vol 6632, p 66320F
17. Pham T-T-H, Lo Y-L (2012) Extraction of effective parameters of anisotropic optical materials using decoupled analytical method. J Biomed Opt 17(2):25006-1–25006-17

18. Pham T-T-H, Lo Y-L (2012) Extraction of effective parameters of turbid media utilizing the Mueller matrix approach: study of glucose sensing. J Biomed Opt 17(9):0970021–09700215
19. Lu SY, Chipman RA (1996) Interpretation of Mueller matrices based on polar decomposition. J Opt Soc Am A 13(5):1106–1113
20. Sources of error in retarders and waveplates. Meadowlark.com

A Statistical Approach to Evaluate Beta Response in Motor Imagery-Based Brain-Computer Interface

Minh Tran Duc Nguyen, Cuong Quoc Pham, Hai Ngoc Nguyen, Khai Quoc Le, and Linh Quang Huynh

Abstract Beta has been scarcely defined as a potential feature for motor imagery-based Brain-Computer Interface use due to its unstable nature in the temporal and spectral domain. Specifying or narrowing down beta sub-bands cannot detect beta occurring non-uniformly in time and frequency for different tasks. This study aims to evaluate beta response quantitatively in terms of event-related synchronization during 3 continuous stages (pre-task, on-task and post-task) of each single imagery trial. Electroencephalogram of thirteen healthy college-aged students were measured on 3 channels (C3, Cz and C4) while the subject performs 3 imagery movements (left hand, right hand and feet). The study proposes an alternative statistical approach utilizing standard event-related synchronization analysis and Fast Fourier Transform based on summarizing all measurement data sets. Broad beta power spectrum, varying from 13 to 35 Hz, is analyzed in order to get trial-channel-specific beta frequency response, and is visualized by using time–frequency color map. The new approach shows the mean percentage of beta response during on-task stage when combining all tasks are 21.8%, 23.1% and 21.1%, for 3 channels C3, Cz and C4, respectively. The results from ANOVA analysis shows that there is a significant difference ($p < 0.0001$) in the percentage of beta response during 3 continuous stages for all 3 tasks. Besides, feet imagery task has a difference in the percentage of beta response in 3 channels, while left/right hand task has no difference. Lastly, the most common beta frequency response is found to be different ($p < 0.05$) during 3 continuous stages for left hand task, while the right hand and feet task has no difference.

Keywords Event-related synchronization · Brain-computer interface · Motor imagery · Beta response · Beta rebound

M. T. D. Nguyen (✉) · C. Q. Pham · H. N. Nguyen · K. Q. Le · L. Q. Huynh
Department of Biomedical Engineering, Faculty of Applied Science, Ho Chi Minh City University of Technology (HCMUT), 268 Ly Thuong Kiet Street, District 10, Ho Chi Minh City, Vietnam

V. Van Toi et al. (eds.), *8th International Conference on the Development of Biomedical Engineering in Vietnam*, IFMBE Proceedings 85,
https://doi.org/10.1007/978-3-030-75506-5_16

1 Introduction

The role of brain-computer interface (BCI) is to establish a non-muscular means of communication for patients with severe limb paralysis or rehabilitation in patients with locked-in syndrome [1, 2]. People who are unable to control voluntary movements, but are cognitively functional, could have an ability to use computer [1], control artificial limbs [3], or operate a motorized wheelchair [4]. In the past few decades, a wide range of electrophysiological signal components have been utilized in BCI systems as inputs [1, 5], visual evoked potentials [6], P300 evoked potentials [7, 8], mu and beta rhythms [9–11].

Motor imagery (MI) has been claimed to be a potential method of BCI system to help people with severe motor disabilities, such as amyotrophic lateral sclerosis (ALS) or spinal cord injury (SCI). MI tasks can create amplitude suppression (ERD) and enhancement (ERS) in brain waves over primary motor cortex [12]. ERD/ERS, also known as event-related desynchronization/synchronization, is EEG power pattern when there is a decrease/increase in rhythm amplitude [13]. For instance, left or right hand MI of a healthy subject [14] elicited mu suppression (8–13 Hz) in contralateral hemisphere together with beta enhancement (18–24 Hz) in ipsilateral hemisphere; or mu ERD (8–12 Hz) and beta ERD (13–28 Hz) in patients with severe motor disabilities [15].

However, there are still obstacles along the way, first of all is the instability of beta oscillation in the spectral domain. Regardless of the difference in beta ERD/ERS pattern and frequency among subjects, it was also found to non-uniformly distribute in both temporal and spectral domain within a measurement of one subject in this study. Therefore, the quantification of ERD/ERS in time and space suggested in [9] is not applicable to the case. Averaging across trials over time would pose a levelized pattern as beta ERD/ERS elicits at different time for different trials. Furthermore, the frequency is not localized for one specific task at a specific channel, and it varies across trials, thus, narrowing into beta sub-band for a subject is not practical either.

Secondly, the expected imagery-related EEG changes have not been consistent in all studies. Followed by an ERD occurring shortly before and during the movement is beta rebound (beta ERS), the phenomena in which bursts of beta oscillations appear within a 1 s interval after movement offset [16]. One study found that only foot imagery (in three different types of motor imagery, namely hand, foot, and tongue) produced a clear beta rebound for the majority of subjects. In contrast, a significant short-lasting beta power burst after one-handed imagery can only be seen in two subjects [17]. Several studies reported on motor event-related desynchronization and synchronization (ERD/ERS) of central beta rhythms in the range from 13 to 35 Hz [18, 19].

The literature review above shows that more studies should be carried out on the changes in beta amplitudes in order to (1) examine beta response in time and frequency domain in a single trial and (2) verify beta oscillations pattern in the sensorimotor during motor imagery tasks. This study aims to: (1) propose a new approach of quantifying the beta ERS pattern dealing with single-trial variation, and

(2) evaluate beta response quantitatively in terms of event-related synchronization during 3 continuous stages (pre-task, on-task and post-task) of each single imagery trial.

2 Methodology

This research applies the Declaration of Helsinki principles in human studies.

2.1 *Subjects and Environment*

Subjects were 13 healthy undergraduate students (5 women, 8 men; age range 20–22 years, mean 20.79 $\pm$ 0.77). All subjects were surveyed and reported to have none neurological medical records which could potentially affect EEG recording. None had participated in the experiments relating to EEG or BCI before. All subjects were trained 2–3 days before acquiring data, and they were paid $2 per measurements for their participation. They were surveyed about mental condition before and after each session to ensure the alertness state of their brains throughout the recording. Furthermore, participants are suggested not to consume caffeine or any stimulants or to have any negative feelings and sleep deprivation.

Subjects sat on a chair, with their hands placed on chair's arms or their knees, in Faraday room to minimize electromagnetic wave and observed computer screen. The room was a quiet environment equipped with a dimmed light to make subjects focus on the screen. They were instructed to keep all muscles relaxed and avoid any eye movements, blinks, body adjustments or any other movements during the visual cue onset. Subjects were allowed to give signs to skipped trials by noise generated by unexpected movements or conditions (cough, scratch, sleep, tired, etc.).

2.2 *Data Acquisition*

Data were recorded from monopolar electrodes placed on or near sites C3, Cz and C4 (NicoletOne V32, Natus) using the International 10–20 system of electrode placement [20]. A reference electrode was attached to the left earlobe and a ground electrode was placed on the right earlobe. All impedances were kept below 5 kΩ and the sampling frequency was 500 Hz. Studies [21, 22] also used the same electrodes and montage as these brain regions elicit the most noticeable event-related (de-)synchronization in mu and beta band when performing imaginary tasks. Before electrode attachment, Nuprep was used to clean the skin, and conductive gel Ten20 was used to reduce electrode resistance.

2.3 Experiment Paradigm

After initial preparation, subjects were asked to seat in Faraday room, on a chair which was about 1 m away from the laptop screen. There were 5 measurements in total for each subject, though some could not complete the full procedure. Each measurement consisted of 5 sessions and a 10-min break between each session. In each session, subjects observed a stimulation program, which was run by MATLAB for 8 min, generating 30 segments (see Fig. 1). Each segment lasted for 16 s with 10 s rest and 6 s mental task. Within the former 10 s, subjects were asked not to perform any movements or focus on anything (faded white cross appeared). For the next 6 s, subjects had to focus on imagining the feelings of executing the given task (kinesthetic motor imagery) which they trained before (the bold white cross and red arrow appeared). All of the tasks were generated randomly and marked into a trigger file. The arrow could be either left or right or downward indicating left hand (LH), or right hand (RH) or both feet (F) motor imagery, respectively. There was a beep cue onset and offset.

3 Data Processing

Figure 2a showed the principle of temporal relative power processing. The signal underwent artifacts removal (spike peak, muscular movement, skipped trials) based on exceeding amplitude and annotation marked during recording. 13–35 Hz Butterworth band-pass filter was applied to get desired beta band. All trials were then narrowed down to 12 s (3 s pre-task—6 s on-task—3 s post-task) instead of 16 s to eliminate transitional effect between trials. Short-time Fourier Transform (STFT)

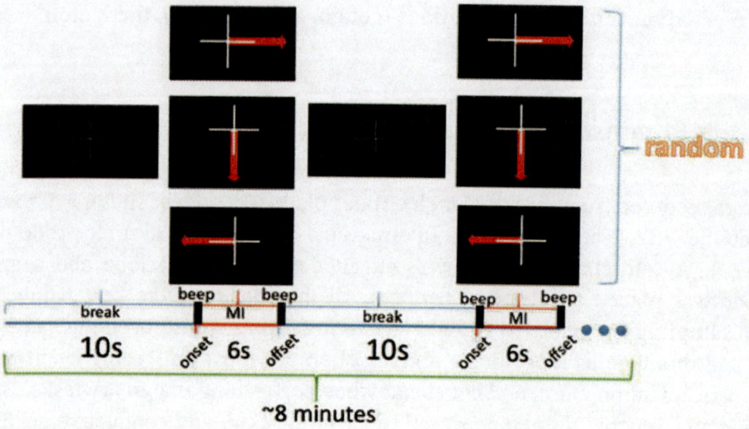

Fig. 1 Experiment paradigm

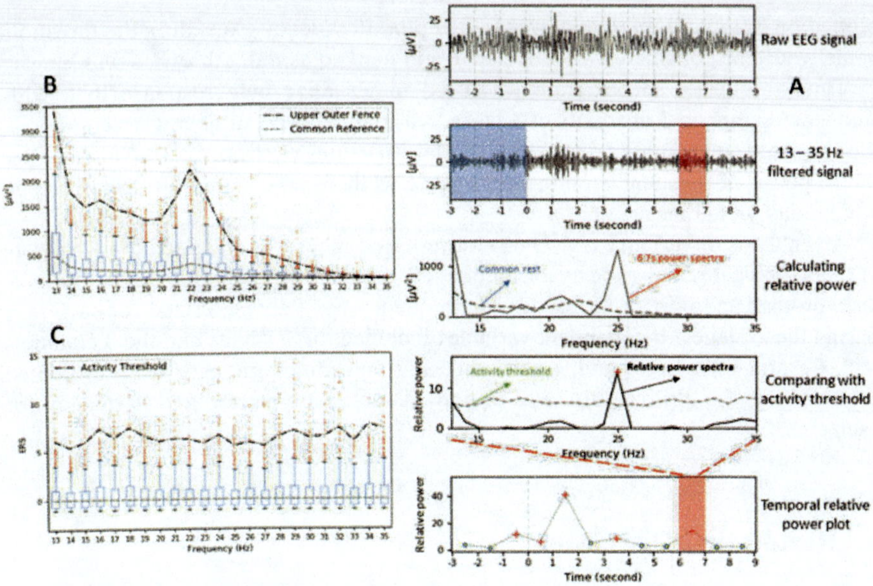

Fig. 2 **a** Principle of temporal relative power processing; **b** Statistic method to find the common rest; **c** Statistic method to find the activity threshold

was applied for 1-s time window in each trial to acquire power spectra for every second. Hence it generated 12 spectrums in total per trial.

In 3 s pre-task, according to the paradigm, the subject was supposed to be in the rest state; however, it was not the case for every trial. A few responses were showing dramatic changes in beta power some seconds before the cue appear, which was not good to be a reference interval. It could be rooted from the subject's anxiety, preparatory thought of movement, etc. To solve the problem, this study defines a term "common rest", where all trials use 1 common reference interval, and "activity threshold", whose frequency exceeding the value would be considered as the active state. Common rest and activity threshold within 1 measurement (1 day of recording) was calculated using boxplot for 3 s pre-task of all trials for each channel on frequency domain (see Fig. 2b, c). Therefore, there were 3 common rest spectrums and 3 thresholds corresponding to 3 channels C3, Cz and C4 for each subject within a day. Next, relative power of each power spectra was computed by the equation:

$$E_i = \frac{A_i - R_i}{R_i}$$

where A_i is the power of ith frequency in activity period, R_i is the power of ith frequency in common rest. Relative power in each power spectra E_i was then compared with the activity threshold with corresponding frequency. The E_i value of the frequency that had the most positive distance from activity threshold was plotted on Temporal relative power (TRP) plot. Frequency exceeding the threshold value would be considered as active state and marked as red, otherwise was blue.

Time–frequency spectrogram was used to visualize beta response in a wide frequency range on 3 channels of 3 tasks. All trials of each subject were computed, though some were omitted due to the criteria. All values were normalized to visualize the response of 3 channels on the same scale as their values are quite inconsistent, which may yield bias result.

An analysis of variance (ANOVA) for repeated-measures was used to investigate (1) beta ERS pattern elicited by motor imagery tasks during 3 continuous stages (pre-task, on-task and post-task) and (2) the frequency band characterized for each task during these stages. Independent variables included the 3 stages and the 3 channels (C3, Cz and C4). Left hand, right hand, and feet motor imagery were evaluated separately as the ultimate aim was to find characteristic pattern and frequency for each task.

4 Results and Discussion

4.1 Temporal Relative Power Plot

TRP plot provided a simpler look but still kept the general characteristic of beta band power for a single trial. The red star represented the increase in beta power in 1-s time window, regardless of which frequency. The dotted horizontal line marked the cue onset and offset respectively. In Fig. 3, the beta synchronization occurred non-uniformly and inconsistently in almost 3 stages among the same task or the same channel. Due to this reason, the quantification of ERD/ERS in time and space suggested in study [9] was not applicable for the case. Averaging across trials over time would pose a levelized pattern as beta ERD/ERS elicits at different time for different trials. Furthermore, the frequency was not localized for one specific task at a specific channel, and it varied across trials, thus, narrowing into beta sub-band for a subject was not practical either. Therefore, it required a statistical approach to evaluate beta response based on TRP plot of every single trial. However, there were some trials, which did not yield any responses, had to be omitted before further analysis according to the criteria: trial having at least 1 red star on any channels from the cue onset onward (on-task and post-task) was considered.

Based on TRP plot, ~ 96% of total trials were spotted with the increase in beta power according to the criteria and progressed to further analysis (see Table 1). This shows that the TRP plot was highly sensitive with the changes in beta power, regardless of which frequency.

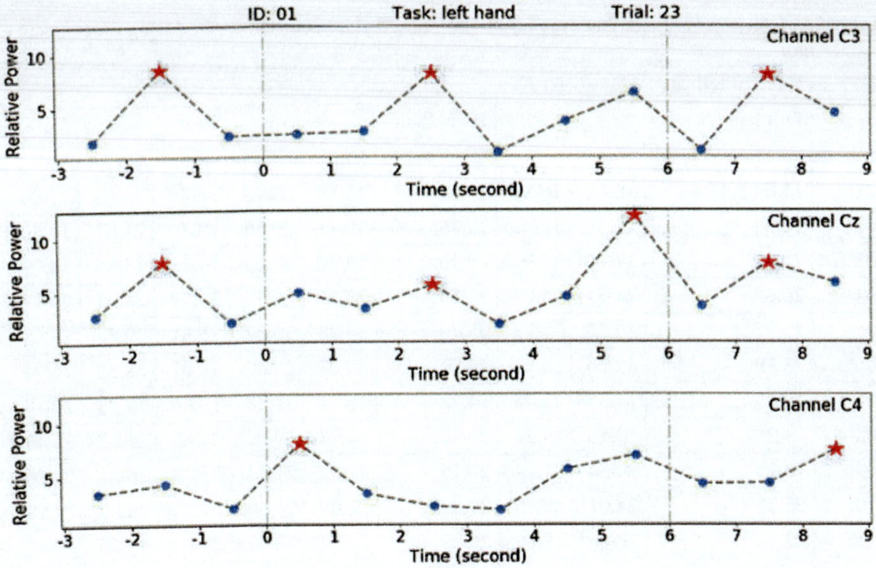

Fig. 3 Temporal relative power (TRP) plot of a single trial, x axis is the time within 1 trial, and y axis is the percentage of relative power, of 1 task for 3 chosen channels: C3, Cz and C4 respectively top-down

Table 1 Selected trials for further analysis by TRP plot of all 13 subjects

	LH	RH	F
Total available trials	2818	2818	2815
Selected trials criteria-based	2705	2698	2731
Ratio of selected trials (%)	95.99	95.74	97.02

4.2 *Beta Response During Motor Imagery*

When trials of all measurements were considered, the ratio of red star occurrence in TPR plot of on-task stage could be seen as a way to quantify beta response in that stage. Table 2 showed the ratio (in percentage) of 13 subjects of 3 different tasks on 3 channels. 13 subjects did not show any clear or characteristic pattern of mean beta response during on-task between channels or tasks. Subject 05, however, had significant ratios on channel Cz for 3 tasks.

Among the selected trials, the new approach showed the mean percentage of beta ERS during on-task stage when combining all tasks were 21.8%, 23.1% and 21.1%, for 3 channels C3, Cz and C4, respectively. This implies beta power were only synchronized in the first two seconds after cue onset and mostly suppressed during motor imagery. Beta ERS mostly occurs in the post-task, where imagery tasks were terminated, called beta rebound.

Table 2 Mean beta response in terms of the ratio of red star occurrence (%) during on-task stage in TRP plot

ID	Ratio of red star occurrence (%)								
	Left hand			Right hand			Feet		
	C3	Cz	C4	C3	Cz	C4	C3	Cz	C4
01	17.64	18.68	19.81	19.59	18.54	22.90	16.89	18.95	21.99
02	20.24	21.57	18.91	18.25	19.11	20.26	18.50	21.61	18.50
03	21.83	17.20	11.94	18.10	14.90	12.03	19.13	15.98	14.99
04	26.65	25.77	24.67	31.55	28.21	28.35	31.79	29.61	29.18
05	25.97	36.55	27.78	25.70	35.88	26.37	26.51	39.56	28.45
06	28.19	29.26	24.47	28.57	30.07	28.40	28.07	28.58	26.94
07	17.69	14.85	15.48	14.42	16.74	19.38	19.66	18.93	20.15
08	21.26	19.56	19.28	20.22	22.32	20.22	27.85	21.67	21.47
09	25.25	24.54	21.24	20.16	21.62	21.62	23.48	26.01	23.26
10	35.76	28.01	29.86	29.89	22.07	24.60	27.13	26.09	22.99
11	23.12	23.66	19.89	26.84	21.89	28.85	25.00	41.60	27.42
12	21.35	21.07	21.76	20.00	22.22	19.86	22.45	28.24	21.49
13	14.22	17.17	16.24	15.24	19.24	17.02	13.40	27.87	15.32
Mean	23.01	22.91	20.87	22.19	22.52	22.30	23.07	26.52	22.47

The general pattern of beta ERS according to the ratio of beta ERS was an increase in post-task after a decrease in on-task stage. In left hand task, the decrease in on-task stage was more dramatic on channel C4 than C3 and Cz. In feet MI and right-hand MI, the beta rebound was more discriminative in channel Cz. The pattern could be observed more specific in Fig. 4.

In each 1-s window, the number of red stars was calculated for each subject. Figure 4 shows the average ratio of 13 subjects in each continuous second of 3 channels corresponding to 3 tasks. The ratio of beta ERS occurrence of all tasks decreased in on-task stage and increased in post-task stage. In other words, the beta power declined before and while subjects were performing an MI task and went up again right after the task terminates. Left hand showed that beta desynchronized more on C4 when the cue onset, but it was not different from other two channels when the cue offset. Right hand and feet MI show the synchronization of beta wave was significant on Cz than C3 and C4 when the imagination diminished. Furthermore, beta rebound of RH task was more dominant on C3 than C4. This result is consistent with study [17], who found most subjects elicited beta rebound at the vertex during both feet MI while a few cases were reported that beta rebound appeared at the contralateral hand representation area during hand MI. The result is also consistent with the study [23], who found a significant pre-movement suppression 14–30 Hz beta band (beta ERD) in both brisk and continuous movements.

Table 3 shows whether there were significant differences in beta response of 3 stages and 3 channels in each task. The results from ANOVA analysis shows that there

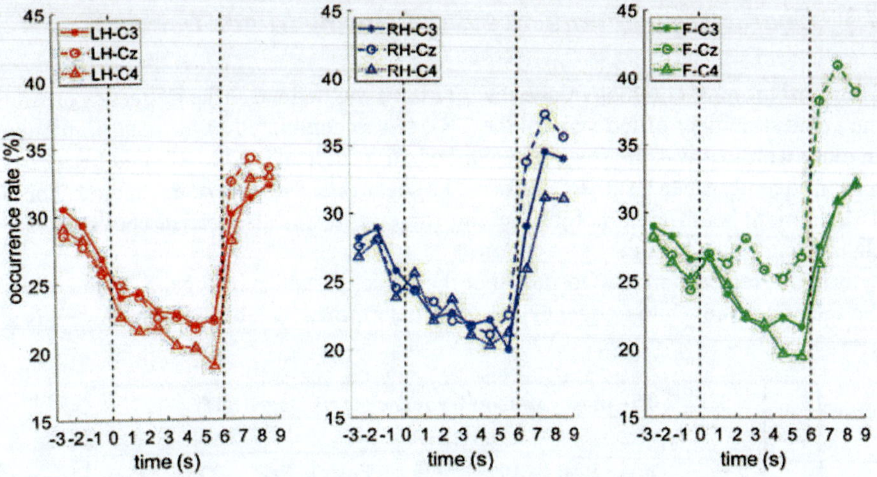

Fig. 4 Ratio of beta ERS of all subjects in 3 continuous stages breaking down into 1-s window. Left hand (left), right hand (middle) and feet (right)

Table 3 Summary of ANOVA results (mean beta response)

Tasks	Effect	F value	*P*-value
LH	Stages (PRE, ON, POST)	F(2,108) = 29.15	< 0.0001
	Channels (C3, Cz, C4)	F(2,108) = 0.61	0.543
RH	Stages (PRE, ON, POST)	F(2,108) = 31.28	< 0.0001
	Channels (C3, Cz, C4)	F(2,108) = 1.68	0.337
F	Stages (PRE, ON, POST)	F(2,108) = 19.59	< 0.0001
	Channels (C3, Cz, C4)	F(2,108) = 5.37	< 0.01

was a significant difference ($p < 0.0001$) in the percentage of beta response during 3 continuous stages for all 3 types of tasks. This could be explained in Fig. 4, where all 3 tasks experienced suppression in pre-task and on-task and rapid synchronization in post-task. Besides, feet imagery task also had a significant difference ($p < 0.01$) in the percentage of beta response in 3 channels, while left/right hand task had no difference. Interestingly, the beta response of F task on Cz was not suppressed during pre-task and on-task, but increased far greater in post-task than C3 and C4 (see Fig. 4). Only a few cases were reported that beta rebound appeared at the contralateral hand representation area during hand MI, but it was not significant.

4.3 Characteristic Frequency Band During Motor Imagery

The term "the most common frequency" (TMCF) was defined as the frequency having the greatest number of red stars in the TRP plot accumulated of all measurements. In other words, it was the frequency which had most activity and elicited beta ERS most frequently when a subject performed a specific task on a certain channel. Thus, TMCF might be different not just among subjects. It was also varied between task, channel and stage (or time).

It is of great importance to note that the values in tables in Figs. 5, 6 and 7 are the specific value while frequency has to be determined by a bandwidth. TMCF is a

ID	The most common frequency for 3 stages (Hz)								
	LH-C3			LH-Cz			LH-C4		
	Pre	On	Post	Pre	On	Post	Pre	On	Post
01	26	29	13	13	33	13	30	29	27
02	15	35	31	15	16	16	35	35	16
03	14	13	17	14	13	14	32	15	13
04	29	29	18	29	27	29	30	27	20
05	27	29	17	29	29	28	29	29	25
06	16	15	13	15	15	22	15	15	13
07	13	13	21	35	13	35	20	14	22
08	19	17	28	35	18	23	35	32	22
09	35	14	22	21	14	19	22	15	21
10	25	16	13	13	14	13	28	22	13
11	13	31	13	34	28	13	18	17	13
12	27	31	14	27	18	21	24	14	21
13	16	15	21	32	14	34	20	15	22

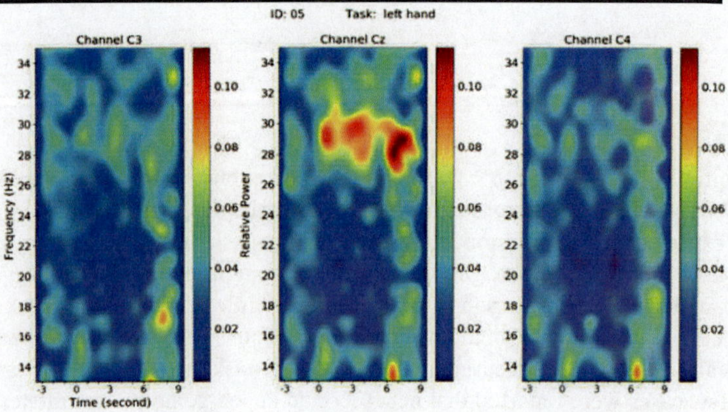

Fig. 5 Table of the most common frequency (above) and time–frequency spectrogram of a representative of task LH (below)

ID	The most common frequency for 3 stages (Hz)								
	RH-C3			RH-Cz			RH-C4		
	Pre	On	Post	Pre	On	Post	Pre	On	Post
01	29	26	13	18	31	13	15	30	29
02	15	31	16	35	29	20	35	29	15
03	16	13	13	14	35	13	27	14	31
04	14	29	13	29	29	23	28	19	21
05	18	29	13	17	29	29	29	29	28
06	16	15	15	14	17	15	13	17	15
07	13	14	13	20	17	13	21	14	22
08	25	19	27	24	17	26	13	18	16
09	13	15	22	13	15	20	13	15	29
10	32	25	30	29	14	15	24	14	15
11	21	31	13	23	28	14	21	20	13
12	25	32	20	28	31	24	28	22	20
13	33	16	23	26	16	23	15	16	22

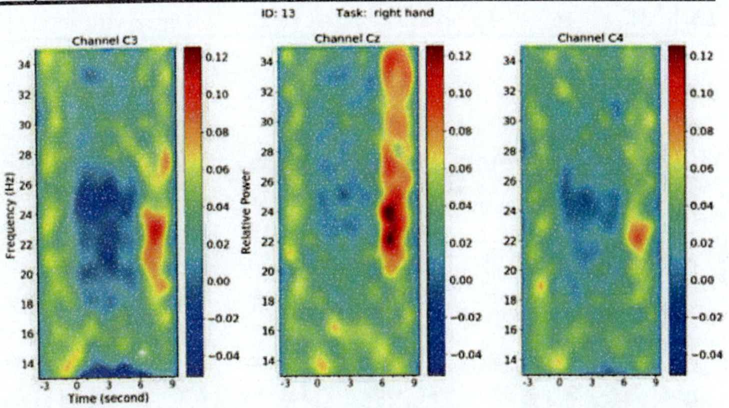

Fig. 6 Table of the most common frequency (above) and time–frequency spectrogram of a representative of task RH (below)

specific value, and its bandwidth is the upper and lower value of around 2 or 3 Hz, it depends. For that reason, time–frequency spectrograms were used to visualize and to determine a subject's specific frequency band for a specific task on a certain channel. The TMCF tables showed frequency of 13 subjects throughout 3 stages of 3 channels when performing 3 tasks LH, RH and F respectively.

The time–frequency spectrogram in Fig. 5 shows subject 05 had beta response at around 29 Hz on Cz throughout the 3 stages whilst a slight beta rebound occurred on C3 around 17 Hz when performing LH task. C4 did not show any signs on the spectrogram since the response was not as significant as the other two. It could be said that the characteristic frequency band of this subject was localized.

ID	The most common frequency for 3 stages (Hz)								
	F-C3			F-Cz			F-C4		
	Pre	**On**	**Post**	**Pre**	**On**	**Post**	**Pre**	**On**	**Post**
01	13	28	31	13	32	31	35	31	29
02	16	35	16	35	26	35	15	26	35
03	13	13	13	16	35	14	29	35	35
04	17	29	13	27	29	29	30	29	13
05	29	27	18	30	29	28	30	14	26
06	14	16	15	15	16	25	14	14	15
07	13	13	22	13	13	13	20	18	21
08	18	18	22	13	18	20	20	18	20
09	13	22	29	20	21	20	21	30	20
10	22	13	25	18	13	13	13	14	13
11	34	31	14	15	28	24	15	20	13
12	24	17	26	27	17	20	32	20	24
13	31	16	33	25	21	23	20	15	15

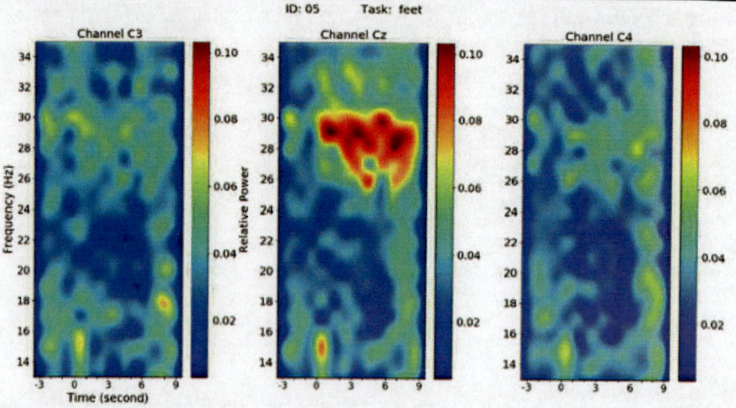

Fig. 7 Table of the most common frequency (above) and time–frequency spectrogram of a representative of task F (below)

The beta response of subject 13 was more scattered when performing a RH task (see Fig. 6). Mostly beta synchronized at around 23 Hz when the imagination of limb movement terminated. In addition, a wide beta synchronization occurred on Cz, varied from 20 to 35 Hz but most significant was at 20–25 Hz.

Subject 05 showed a significant and localized beta ERS at around 29 Hz when imagining feet movement. Interestingly, the response was most significant on Cz while there was little or no response on C3 and C4 (see Fig. 7). Hence, it is clear that for different task, the beta response of each person yielded different patterns at different frequency and on different channels.

Table 4 Summary of ANOVA results (ratio frequency)

Tasks	Effect	F value	*P*-value
LH	Stages (PRE, ON, POST)	F(2,108) = 5.25	< 0.001
	Channels (C3, Cz, C4)	F(2,108) = 3.3	0.04
RH	Stages (PRE, ON, POST)	F(2,108) = 3.3	0.04
	Channels (C3, Cz, C4)	F(2,108) = 0.17	0.83
F	Stages (PRE, ON, POST)	F(2,108) = 2.28	0.1
	Channels (C3, Cz, C4)	F(2,108) = 3.6	0.02

As mentioned above, TMCF might be different not just among subjects, it was also varied between task, channel and stage (or time). Table 4 shows whether there were significant differences in ratios of TMCF of 3 stages and of 3 channels in each task. The most common beta frequency response was found to be different ($p <$ 0.001) during 3 continuous stages for left hand task, while right hand and feet task had little or no significant differences.

5 Conclusion

In the needs of helping disabled patients getting back to their lives, BCI, which utilizes EEG, was born and is becoming a potential in providing means of communication such as: speller, computer cursor, wheelchair, artificial limbs, etc. However, EEG and extracted features can be influenced user's performance in terms of cognitive ability and methodology, which consequently would affect classification accuracy [11, 14]. Therefore, the relationship between individual cognitive tasks and accompanying changes in BCI signal was highly recommended to look into [23].

This study was carried out to investigate beta response with two aims: (1) propose a new approach of quantifying the beta ERS pattern dealing with single-trial variation, and (2) evaluate beta ERS response quantitatively during 3 continuous stages (pre-task, on-task and post-task) of each single imagery trial. Data were recorded while subjects were performing three mental tasks: imagination of left hand, right hand and both feet movement. The proposed method was to use TRP plot, first of all, to evaluate beta ERD/ERS pattern by the occurrence ratios in 1-s time window. Secondly, TRP plot was used to find the most common frequency of each subject doing a specific task over a certain channel. ANOVA for repeated measures was also used. There are two main outcomes in this study:

First, the beta response of 3 continuous stages for all 3 types of tasks was significantly different. Three tasks experienced suppression in pre-task and on-task and rapid synchronization in post-task. A beta rebound was reported during hand MI at the contralateral hand representation area, but it was not significant. Left hand showed that beta desynchronized more on C4 when the cue onset, but it was not different from the other two channels when the cue offset. The beta rebound of RH task was more dominant on C3 than C4. Right hand and feet MI show the synchronization of beta wave was significant on Cz than C3 and C4 when the imagination diminished. Feet imagery task also had a significant difference in the percentage of beta response in 3 channels, while left/right hand task had no difference. Interestingly, the beta response of feet imagery task on Cz was not suppressed during pre-task and on-task, but increased far greater in post-task than C3 and C4. This phenomenon was hypothetically explained in study [17] that supplementary motor area (SMA) and foot representation areas may be involved in both feet MI. It is because these cortical areas are close to each other and the response of the corresponding networks in both areas may be synchronized, beta rebound occurring after foot imagery movement results in large amplitude [17].

Secondly, TMCF and the time–frequency spectrogram helped to find subject-specific frequency band for certain tasks corresponding to stages and channels. Furthermore, the most common beta frequency response was found to be different ($p < 0.001$) during 3 continuous stages for left hand task, while right hand and feet task had little or no significant differences.

The proposed method and results accomplished the study's goal and proved themselves to be reliable, although further studies are still needed to ensure. To develop effective BCI applications, future studies should take these factors into consideration: ERD/ERS pattern, training and feedback of users, comparison subject's performance with and without receiving feedback, different reference period and optimal value [23].

Acknowledgements We acknowledge the support of time and facilities from Ho Chi Minh City University of Technology (HCMUT), VNU—HCM for this study.

Conflict of Interests The authors declare no conflict of interest.

References

1. Wolpaw JR, Birbaumer N, McFarland DJ, Pfurtscheller G, Vaughana TM (2002) Brain–computer interfaces for communication and control. Clin Neurophysiol 113(6):767–791
2. Scherer R, Lee F, Schlogl A, Leeb R, Bischof H, Pfurtscheller G (2008) Toward self-paced brain-computer communication: navigation through virtual worlds. IEEE Trans Biomed Eng 55(2):675–682
3. Obermaier B, Guger C, Neuper C, Pfurtscheller G (2001) Hidden Markov models for online classification of single trial EEG data. Pattern Recogn Lett 22(12):1299–1309
4. Galán F, Nuttin M, Lew E, Ferrez PW, Vanacker G, Philips J, del Millán, JR (2008) A brain-actuated wheelchair: Asynchronous and non-invasive brain–computer interfaces for continuous control of robots. Clin Neurophysiol 119(9):2159–2169
5. Graimann B, Allison BZ, Pfurtscheller G (2010) Brain-computer interfaces: revolutionizing human-computer interaction. Springer, New York
6. Middendorf M, McMillan G, Calhoun G, Jones KS (2000) Brain-computer interfaces based on the steady-state visual-evoked response. IEEE Trans Rehabil Eng 8(2):211–214
7. Donchin E, Spencer KM, Wijesinghe R (2000) The mental prosthesis: assessing the speed of a P300-based brain-computer interface. IEEE Trans Rehabil Eng 8(2):174–179

8. Li Y, Nam CS, Shadden BB, Johnson SL (2010) A P300-Based brain-computer interface: effects of interface type and screen size. Int J Human-Comput Inter 27(1):52–68
9. Pfurtscheller G, Lopes da Silva FH (1999) Event-related EEG/MEG synchronization and desynchronization: basic principles. Clin Neurophysiol 110(11):1842–1857
10. Neuper C, Müller-Putz GR, Scherer R, Pfurtscheller G (2006) Motor imagery and EEG-based control of spelling devices and neuroprostheses. Prog Brain Res 159:393–409
11. Neuper C, Scherer R, Wriessnegger S, Pfurtscheller G (2009) Motor imagery and action observation: Modulation of sensorimotor brain rhythms during mental control of a brain–computer interface. Clin Neurophysiol 120(2):239–247
12. Neuper C, Scherer R, Reiner M, Pfurtscheller G (2005) Imagery of motor actions: differential effects of kinesthetic and visual–motor mode of imagery in single-trial EEG. Cogn Brain Res 25(3):668–677
13. Neuper C, Pfurtscheller G (2001) Evidence for distinct beta resonance frequencies in human EEG related to specific sensorimotor cortical areas. Clin Neurophysiol 112(11):2084–2097
14. Pfurtscheller G, Neuper Ch, Flotzinger D, Pregenzer M (1997) EEG-based discrimination between imagination of right and left hand movement. Electroencephalogr Clin Neurophysiol 103(6):642–651
15. Neuper C, Pfurtscheller G (1998) ERD/ERS based brain computer interface (BCI): effects of motor imagery on sensorimotor rhythms. Int J Psychophysiol 30:53–54
16. Pfurtscheller G, Neuper C (1997) Motor imagery activates primary sensorimotor area in humans. Neurosci Lett 239(2–3):65–68
17. Pfurtscheller G, Neuper C, Brunner C, Lopes da Silva F (2005) Beta rebound after different types of motor imagery in man. Neurosci Lett 378:156–159
18. Alegre M, Labarga A, Gurtubay IG, Iriarte J, Malanda A, Artieda J (2002) Beta electroencephalograph changes during passive movements: sensory afferences contribute to beta event-related desynchronization in humans. Neurosci Lett 331(1):29–32
19. Pfurtscheller G, Neuper C, Pichler-Zalaudek K, Edlinger G, Lopes da Silva FH (2000) Do brain oscillations of different frequencies indicate interaction between cortical areas in humans? Neurosci Lett 286(1):66–68
20. Jasper HH (1958) The ten-twenty electrode system of the International Federation. Electroencephalogr Clin Neurophysiol 10:370–375
21. Cinar E, Sahin F (2013) New classification techniques for electroencephalogram (EEG) signals and a real-time EEG control of a robot. Neural Comput Appl 22:29–39
22. Kaiser V, Kreilinger A, Müller-Putz GR, Neuper C (2011) First steps toward a motor imagery based stroke BCI: new strategy to set up a classifier. Front Neurosci 5:86
23. Jeon Y, Namb CS, Kim Y-J, Whang MC (2011) Event-related (De)synchronization (ERD/ERS) during motor imagery tasks: implications for brain-computer interfaces. Int J Ind Ergon 41:428–436

Internet of Things and Telemedicine

A New Design of an Automated System to Collect and Monitor Temperature and Humidity

Khoa Che, Doan Dat Tran, Huynh Nguyen, Tin Thanh Dang, and Anh Dinh

Abstract Temperature and humidity are crucial to food manufacture and preservation processes. The traditional way to acquire this information is to enter the food storage and perform measurements manually. This approach could be relatively time-consuming compromising the desired isolation of food storage with the outside environment, yet it cannot provide real-time data. This paper presents a sufficient, concrete design of an automated system whose objective is to monitor temperature and humidity in real-time to reduce such shortcomings. The system consists of a handset and multiple sensor nodes. It operates on an established wireless local area network in the user's facility and is highly scalable in the number of sensor nodes. To prove the design, a prototype was built and evaluated of data accuracy, data latency and power consumption: the MAE of temperature is 0.5 °C, of relative humidity is 1.0375%, of data latency is less than 1 s, and of power consumption is 1020 mW. This result is sufficient for our intended application. After the experiments, we realized it is possible to generalize the design to monitor and record any environment property for further analysis. To uncover that possibility, the various aspects of such a system have been discussed: data integrity, ease of deployment, security concern, robustness, scalability, and extensibility. These discussions should help the creation of formal technical specifications to guide industry-level implementation.

Keywords Remote monitoring · WLAN · Temperature · Humidity · IoT

K. Che · D. D. Tran · H. Nguyen · T. T. Dang (✉)
Faculty of Electrical and Electronics Engineering, Ho Chi Minh City University of Technology, Ho Chi Minh City, Vietnam
e-mail: dttin@hcmut.edu.vn

A. Dinh
Department of Electrical and Computer Engineering, University of Saskatchewan, Saskatoon, Canada
e-mail: anh.dinh@usask.ca

V. Van Toi et al. (eds.), *8th International Conference on the Development of Biomedical Engineering in Vietnam*, IFMBE Proceedings 85,
https://doi.org/10.1007/978-3-030-75506-5_17

1 Introduction

The rapid development of IoT in the field of healthcare can be reasoned by the vast importance of health care itself and the natural ease of design of healthcare-related IoT systems, which is often small scale around a human body. Considering large scale, some design models and concrete systems are well suited for professional applications, typically for environmental scientists with high accuracy [1, 2], or high efficiency such as Wireless Sensor Network [3]. However, these large-scale designs have two major disadvantages: high cost and complicated deployment process. Facilities that want to keep track of their environment need a middle-scale solution, where multiple small-scale deployments are not automated enough, and large-scale solutions are unattainable or exaggerated. For instance, temperature and humidity are important environmental properties that highly affect food manufacture and preservation processes. Traditional instruments require measurement conducted manually by humans, which is time-consuming and requires to open the isolated environment to the outside, yet they cannot deliver accurate data in real-time. This measurement should be better performed automatically, and results are sent to the outside for monitoring and control purposes.

In this paper, the above issue was addressed, and an automated system was designed to help monitor temperature and humidity remotely, which is integrated into facilities' workflow. The system works well in a stable environment such as wine cellars, features an easy deployment process and high scalability that enables incremental additions of sensor nodes. A proof-of-concept prototype was also implemented, and certain experiments were conducted on data accuracy, data latency and power consumption. Afterward, we realized it is possible to develop the design into a generalized version that can track any property of the environment. In addition, the possibility in various aspects was considered: security and robustness for mission-critical applications, fostering the relationship of both collaboration and competition among technology service providers. The design will take into consideration aspects that realize a large-scale system for mission-critical applications: how to ease deployment process, scaling scenario and its solutions, the importance of enabling third-party extension and how to achieve that.

This paper is organized into five sections. After the introduction, Sect. 2 presents a concrete system design: hardware design, software design, and communication protocol. In Sect. 3, various experiments and their results are shown, then the generalized design and its realizations are discussed. Finally, Sect. 4 contains the research conclusion and scope for future development.

2 Experimental Design

2.1 System Design

Our design target is a distributed system to monitor temperature and humidity wirelessly. To perform experiments in the early stage of the design, a minimal yet complete system was desired, we favored choices that were easy to implement. Here the system comprises distributed nodes: several sensor nodes—each of which acquires data from a sensor module and a handset to collect such data and display to the user. Consider our target application; the handset must collect data from sensor nodes inside an isolated environment, wireless LAN, which was already established in the facility, was chosen instead of point-to-point wireless connections such as raw RF and Bluetooth. This approach will pass the concern of distance to local network infrastructure. To ensure system integrity, TCP was chosen as the transport layer protocol for the guarantee of reliable data transport.

With the presence of reliable data transport, advanced data interchange could be performed. HTTP was chosen as the application layer protocol for a stateless request/response communication mechanism. The next aspect to consider was the data serialization format: the text format JSON was chosen for ease of implementation. To represent temperature and humidity data, the JSON number has been consistent enough, although it does not ensure fixed-point precision [4] as the aforementioned sensor module had its smallest division of just 0.1 °C and 0.1%RH. Besides, in order not to have to configure IP addresses for nodes, multicast DNS protocol was made use of, so that each node could own a hostname without the presence of a central DNS server [5]. Finally, for faster prototyping, some WiFi credentials without a way to modify them on-the-fly were decided to be hardcoded.

Each sensor node would constantly send to the handset pieces of information with three fields: temperature, humidity, and timestamp. The timestamp was very useful for our latency experiment and analysis of data collected in the time domain. Leverage the many-to-one relationship of sensor nodes and handset; the handset was modeled as an HTTP server and sensor nodes as clients. That way, the communication protocol between the handset and sensor nodes could be conducted by just one URL endpoint (`/update`). Furthermore, to optimize the network bandwidth usage, the concept of interest level was introduced that the handset would respond to each update request with the interest of it with the data (interested or not), so the sensor node could adjust its update frequency. This result can be sent as plain text "true" or "false" to ease the implementation.

In conclusion, our system design can be summarized into a small communication protocol between the handset and the sensors. To send a data update to the handset, the sensor node makes an HTTP request to path `/update`, with a JSON document as the request body, which contains these fields: timestamp, temperature, and humidity. The handset will respond with a plain text string "true" or "false" to indicate the interest level.

2.2 Hardware Design

Handset. In order to collect data from the sensor nodes, a WiFi hardware module was needed. To ease development, the NodeMCU (Fig. 1(1a)) was chosen. This is a compact microcontroller circuit with integrated WiFi modules and TCP/IP network stack available. It also had a sufficient number of GPIO pins for our usage [6]. Data collected from the sensor nodes are displayed on an LCD screen. A 20 × 4LCD module (Fig. 1(5a)) was chosen because it can display data from 2 sensors node at the same time. However, there was a small problem; the module occupies at least 12 GPIO pins. To save pins for other purposes, an LCD I2C-PCF8574 module was used (Fig. 1(4a)) to proxy the original parallel communication as I^2C communication, which occupied only two pins.

To allow users to navigate the list of sensor data on the screen and configure which sensor nodes to monitor, a button pad (Fig. 1(6a)) was included. The button pad has 4 buttons: UP and DOWN for navigating, MODE to switch between display view and configuration view, and ENTER to toggle between displaying or not displaying data from a specific sensor node. Each button outputs a voltage level of high or low. While a button is down, it outputs a low voltage; otherwise, it outputs a high voltage.

The handset is a mobile device, so it requires a mobile power source. LCD screen and NodeMCU operate optimally at 5 V DC. The usage of 2 Li-ion batteries (Fig. 1(2a)) in combination with a buck converter (Fig. 1(3a)) is to provide a stable source to power above components.

Sensor Node. The sensor node was composed of a NodeMCU (Fig. 1(1b)) and a sensor module. Along with a NodeMCU, a DHT22 module (Fig. 1(1a)) was used

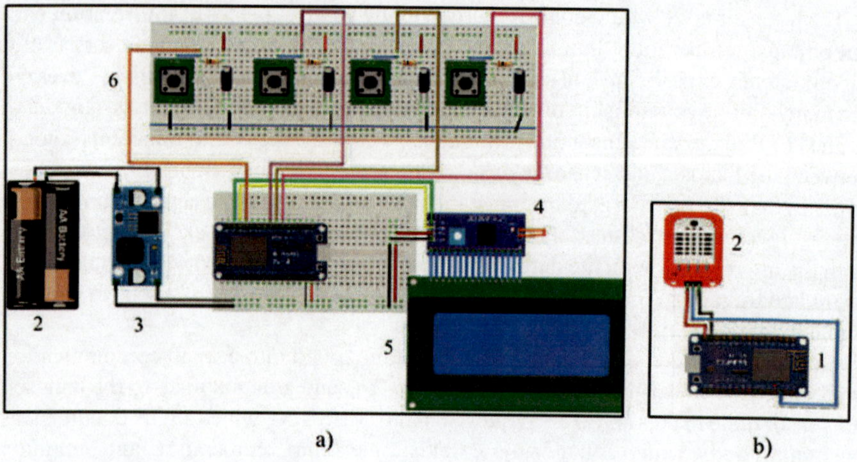

Fig. 1 Wiring diagrams of the physical devices: **a** Handset (1a: NodeMCU; 2a: Batteries; 3a: Voltage converter; 4a: I2C module; 5a: LCD screen; 6a: buttons), **b** sensor node (1b: NodeMCU; 2b: sensor DHT 22)

to collect temperature and humidity data. This is a sensor that provided sufficient accuracy for our application (humidity ± 2%Rh; temperature < ± 0.5 °C) and a large operating range (humidity 0–100%Rh; temperature 40–80 °C) [7]. The ideal temperature for wine storage is 12–14 °C, with an average of 60% relative humidity [8]. The DHT22 operates optimally when powered by a 3.3 V source. The sensor node uses 5 V DC Adapter to power directly through a MicroUSB port.

2.3 *Software Implementation*

The software layer for the above hardware was implemented with Arduino IDE as the development platform. Our work depended on some existing software packages: ESP8266 Arduino Core to make use of NodeMCU hardware features, especially the integrated WiFi module with the ability to establish network transport, perform HTTP request, create HTTP servers and mDNS responder; Adafruit's DHT Sensor Library to acquire data from DHT sensor modules; LiquidCrystal_I2C to control the LCD screen display; ArduinoJson to parse and serialize JSON documents.

Figure 2 depicts the high-level communication between software components of the system (sensor node, handset's API, and handset's LCD view), in which a sensor node sends an update request to the handset API, the handset API, in turn, update the LCD view and respond the sensor with its interest level; finally, the sensor node interprets the interest level received and enforce its update frequency.

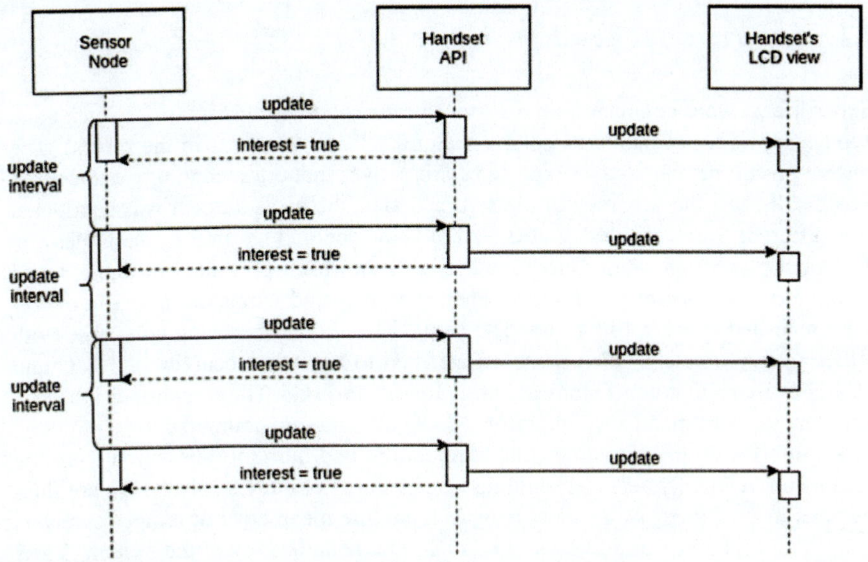

Fig. 2 Implemented system operation sequence diagram

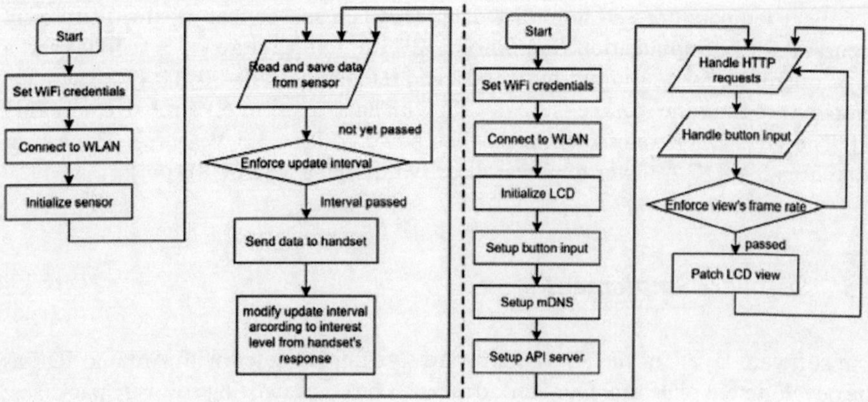

Fig. 3 Sensor node flowchart (left) and handset flowchart (right)

Figure 3 presents the program flowcharts of sensor node and handset, in which there are a setup phase and the main loop—to read and send data from the sensor in the case of sensor node; or to listen for updates from sensor nodes and buttons, then flush updates to LCD view in the case of the handset.

3 Result and Discussion

3.1 Experimental Result

Experiments were conducted on a system prototype that consists of a handset and two sensor nodes (denoted as columns: Sensor 1 and Sensor 2 in the tables). The system prototype operated as expected. This proves that our technology choices are feasible. To test the accuracy of the sensor nodes, the measurement was conducted at 4 different times of the day at 0, 9, 12, 17 h (denoted as rows in the tables), in 2 distinct places: an indoor environment with the result presented in Table 1 and an outdoor environment in Table 2. Then the measured temperature and humidity were recorded every 2 s in 4 min. The result was verified by being compared with a thermo-hygrometer, which has been certified to have the accuracy of 10 °C and 4%RH by Petro Vietnam Technical Safety Registration Co. The comparison is shown in the tables with mean absolute error (MAE) of 2 sensors computed.

As inferred from both tables, the temperature and humidity measured from the sensors are relatively accurate, but these figures are not quite stable. There are fluctuations in the range. When measuring indoors, the mean error of temperature was 0.5 °C, and relative humidity was 1.0375%; the mean latency of the system is less than 1 s. However, with DHT22—a low-cost sensor, the result is acceptable for our intended application related to closed spaces.

Table 1 Temperature and humidity measured by sensors in indoor (indoor: student accommodation at 10th floor, Bach Khoa dormitory) environment

Time (h)	Thermo-hygrometer		Sensor 1		Sensor 2		Mean error	
	T (°C)	Rh (%)	T (°C)	Rh (%)	T (°C)	Rh (%)	T (°C)	Rh (%)
0	28	70	28.1	68.2	28.4	67.8	0.25	2.00
9	31	60	30.1	60.7	31.0	59.8	0.45	0.45
12	31	61	31.3	60.4	30.9	62.1	0.20	0.85
17	30	62	29.1	62.2	28.7	63.5	1.10	0.85

Table 2 Temperature and humidity measured by sensors in outdoor (outdoor: in Bach Khoa dormitory campus) environment

Time (h)	Thermo-hygrometer		Sensor 1		Sensor 2		Mean error	
	T (°C)	Rh (%)	T (°C)	Rh (%)	T (°C)	Rh (%)	T (°C)	Rh (%)
0	27	75	26.7	74.7	26.4	74.1	0.45	0.60
9	29	58	29.4	59.2	28.6	58.7	0.40	0.95
12	32	60	32.5	58.8	33.6	61.0	1.05	1.10
17	30	60	30.5	58.8	29.9	60.3	0.30	0.75

Data latency of our system consists of sensor latency and network latency, in which we have confirmed that sensor latency is not significant compared to network latency, so network latency was chosen to represent total data latency. Network latency was measured by calculating the difference of two timestamps fetched from an NTP server [9], at a handset and a sensor node—two terminals of the network transmission. This measurement was performed ten times on the public WLAN of the HCMUT campus, results in the average data latency of 0.93 s and a standard deviation of 0.5 s, which is acceptable for real-time operation.

To determine power consumption, the handset was powered with two 3.4 V batteries connected in series (total voltage is 6.8 V). Then output current of batteries was measured ten times, which yielded the average value of 150 mA. Therefore, the power consumption was 1020 mW (6.8 V multiplied with 150 mA)—quite small because the electronic components used in this device were low-power. Similarly, to determine the power consumption of the sensor node, each node was powered by a 5 V adapter. Then the output current of the adapter was measured ten times, which yielded the average value of 80 mA so that the power consumption was 400 mW.

To determine the operating duration of the handset via batteries, the handset was powered with two popular Li-ion batteries with total capacity 6800 mAh. The system was fully-charged then is letting operated continuously until batteries run out. The measurement was performed ten times, yielding the average operating duration of 34.94 h and a standard deviation of 2.17 h, which is stable.

3.2 Discussion

In the previous subsection, a remote monitoring system for temperature and humidity in isolated environments was successfully designed and implemented. After the works, we realized there is a possibility to generalize and further extend the design into a system to collect, monitor, record and analyze any properties of the physical environment. As mentioned in Sect. 1, there are some existing designs for the collecting and usage of environment information, but they either in small scale or too-large scale and professional deployments that they do not fit the application of monitoring properties of the environment within a facility. In the future, this design could be enhanced to be compatible with the storage of various medical and scientific materials, including blood plasma, reagent and more.

Figure 4 demonstrates a larger system that can be used in medical facilities to track medicine stock and medical equipment stock: a system composed of many sensor nodes to track not only temperature, humidity, but also the brightness. The properties are monitored in a hierarchical manner. The monitored data are aggregated and then uploaded to ThingSpeak™ analysis platform, and also archived in a local database. As inferred from the figure, the analysis layer is a part of the existing or desired workflow of the facility—customer; hence the system is integrated into the customer's workflow.

The sensor nodes from our design, which implemented a tiny communication protocol, can be easily integrated into any other system as well. To further improve the quality of this general design. Other system aspects will be taken into consideration. Regarding data integrity, it can be observed that the most measurable properties of the environment can be represented as a real number. That number should be conceived

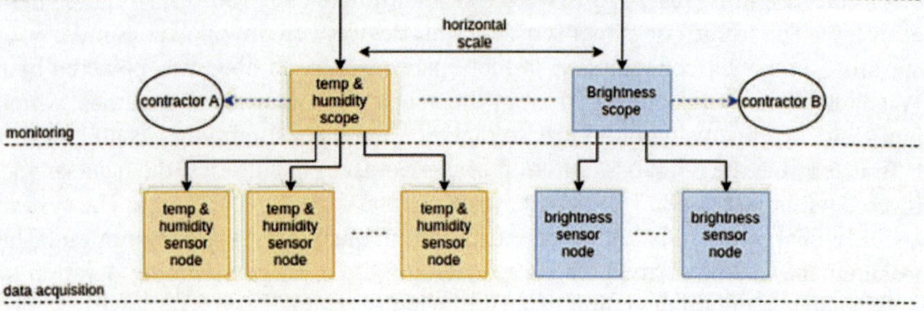

Fig. 4 Design realization with ThinkSpeak™

and stored and transmitted as a fixed-point number rounded by a consistent rounding policy (e.g., all round-up or all round-down). That means any representation of the number in the system must ensure fixed-point integrity.

Regarding extensibility, the system should use open and popular standards to implement communication between its nodes, even if their performance is worse than proprietary and unpopular choices. There is often no low-power or strict performance constraints in our intended application. Another benefit is to eliminate the risk of vendor locked-in, which resulted in a more durable deployment. High extensibility also enables indirect collaboration between service providers to provide a single system.

Regarding scalability, optimizing the system's implementation for performance can lead to better scalability; let our prototype above as an example; after optimization, the handset can track hundreds of sensor nodes at the same time. Moreover, there is a more systematic scaling model in the number of monitoring layer devices.

Regarding security, to ensure the confidentiality of data collected and secret keys through network transport, Transport Layer Security should be forced to create safe channels. A common practice in public-facing network application is to use TLS certificates signed by publicly-trusted Certificate Authority (CA); however, it comes with the maintenance burden of keeping these CA lists in sync all over the system. The use of self-signed certificates is recommended in this case, in which each vendor should explicitly express their trust to another vendor to integrate two systems. More nodes added to the existing system should be authenticated manually by the system administrator. Token bearer authorization is enough to authorize access to the system. The above methods come along with sufficient physical and network access restriction should result in a secured system.

4 Conclusion

The paper describes the design process of a temperature and humidity remote monitoring system. Experimental measures afterward ensure it is feasible for the tasks of monitoring temperature and humidity, both in data accuracy and data latency. Furthermore, the design has been being discussed and generalized into a broader application: a system that consolidates the monitoring and recording of any property of the environment in a specific facility. An example realization of such a system in medical facilities is presented to envision potential users and implementers. To create the foundation of general design, many aspects were taken into consideration: data integrity, extensibility, scalability, and security. The performance of the system was not thoroughly addressed because it requires much more research to be done. The design could be further improved to sufficient quality for mission-critical applications and then be developed into a formal specification to guide implementation.

Conflicts of Interest The authors have no conflict of interest to declare.

References

1. Athanasiadis IN, Mitkas PA (2004) An agent-based intelligent environmental monitoring system. Manage Environ Qual Int J 15:238–249
2. Fang S, Xu LD, Zhu Y, Ahati J, Pei H, Yan J, Liu Z (2014) An integrated system for regional environmental monitoring and management based on internet of things. IEEE Trans Industr Inf 10:1596–1605
3. Dahane A, Berrached N-E (2019) Wireless sensor networks: a survey. Mobile Wireless Sens Networks 1–24
4. Bray T. The JavaScript object notation (JSON) data interchange format. Available: https://tools.ietf.org/html/rfc7159. Last accessed 22 Dec 2019
5. Cheshire S, Krochmal M. Multicast DNS. Available: https://tools.ietf.org/html/rfc6762. Last accessed 22 Dec 2019
6. Handson Technology: ESP8266 NodeMCU WiFi Devkit User Manual V1.2. Available: https://www.handsontec.com/pdf_learn/esp8266-V10.pdf. Last accessed 20 Dec 2019
7. Aosong Electronics Co., Ltd: Digital-output relative humidity & temperature sensor/module DHT22 (AM2302) Datasheet. Available: https://www.sparkfun.com/datasheets/Sensors/Temperature/DHT22.pdf. Last accessed 20 Dec 2019
8. Wine Guardian: Proper wine storage temperature and humidity. Available: https://wineguardian.com/proper-wine-storage-temperature-and-humidity. Last accessed 27 Dec 2019
9. Mills D, Martin J, Burbank J, Kasch W (eds). Network time protocol Version 4: protocol and algorithms specification. Available: https://tools.ietf.org/html/rfc5905. Last accessed 22 Dec 2019

Design of NIBP Holter System Based on NIBP Device and Smartphone

Vu Duy Hai, Nguyen Anh Duc, and Vuong Ngoc Tan

Abstract The measurement of non-invasive blood pressure (NIBP) parameters is currently being used in many medical facilities for examination and treatment. In addition, the need to use these devices at home for patients to check their own blood pressure is also increasing. However, the measurement of blood pressure values at home is only for the purpose of patient's reference. Meanwhile, if this data is sent back to the doctor regularly, it will be a very useful source of information for doctors to monitor, analyze and recommend a good treatment regimen for patients with hypertension. This article presents the design of a system that continuously measures and sends blood pressure values to the doctor by combining a personal blood pressure device with a smartphone. The result of system design includes an application on the smartphone for the patient, able to connect to receive data from iChoice blood pressure monitor; a database used to store measurement results on the server and a smartphone application for doctors to analyze and process data. The system has been deployed in trial measurement on five volunteers with a total of 100 measurements. Test results show that the system works very stable, accurate and reliable, has practical applications to support the diagnosis and treatment of blood pressure diseases.

Keywords NIBP · Holter NIBP · Systolic pressure · Diastolic pressure · Heart rate

1 Introduction

Blood pressure is the pressure exerted by the circulation of blood in the blood vessels and is one of the main signs that the body is alive or dead. When the heart beats, blood pressure changes from the maximum (systolic pressure) to the minimum (diastolic pressure). Mean blood pressure, which is caused by the pumping of the heart and the resistance in the blood vessels, is reduced when blood in the arteries goes away from the heart. Blood pressure drops the fastest when blood runs in the small arteries and

V. D. Hai (✉) · N. A. Duc · V. N. Tan
Biomedical Electronics Center, Hanoi University of Science and Technology, Hanoi, Vietnam
e-mail: hai.vuduy@hust.edu.vn

V. Van Toi et al. (eds.), *8th International Conference on the Development of Biomedical Engineering in Vietnam*, IFMBE Proceedings 85,
https://doi.org/10.1007/978-3-030-75506-5_18

arterioles and continues to decrease as blood passes through the capillaries, and blood pressure reaches the smallest level in the vein returning to the heart. Factors affecting blood pressure include gravity, valves in the vein, respiration, muscle contraction [1].

There were 1.13 billion people living with high blood pressure worldwide in 2015. The majority of them come from low and middle income countries. Especially according to the research results, the rate of men with hypertension is higher than the rate of women. Studies emphasize the clear relationship between high-income and low-income people. High-income countries have shown a sharp decrease in blood pressure in their population in recent decades. While low and middle-income countries are on the rise, especially in Southern Asia and Africa. Countries like the United States, Canada and South Korea have the lowest rates in the world. The UK has the highest incidence of anaemia in Europe. The authors added that this is most likely due to the difference between making healthy food choices or also due to the diagnosis and treatment services in the health system [2, 3].

In the human body, in order to maintain a healthy body, there are many factors, but blood pressure is an important factor among them. Changes in blood pressure in the body reflect the condition of the patient. In addition, blood pressure at different times throughout the day was significant in terms of pathological diagnosis. Therefore, saving blood pressure results at all time of day is necessary to evaluate human health as well as diagnosing some diseases such as high blood pressure, low blood pressure, cardiovascular disease, etc. Blood pressure imbalance and abnormal changes in blood pressure are factors that determine the health of the human body. The blood pressure monitors and measurement system that we mentioned can solve the problem. The system allows patients to measure and record blood pressure results. From there, the doctor can review the patient's results to determine the most appropriate treatment regimen. In addition, this technique is non-invasive, low cost and easy to conduct. It also does not need the practice to conduct measuring. In this research, we present the design of a system that continuously measures and sends blood pressure values to the doctor by combining a personal blood pressure device with a smartphone [4, 5].

2 Design System

2.1 Principle Diagram Design

The overall objectives of designing the Holter NIBP system are comprised of ensures the electric safety for patients during measuring process; portable and can be used at the bedside; measurement blood pressure and pulse rate; store blood pressure values on database and doctor can give advices to patient. From the above requirements, the principle diagram of the system is designed, as shown in Fig. 1. The system has four main blocks, including Blood pressure monitor, App for the patient, App for the doctor, and Database [6].

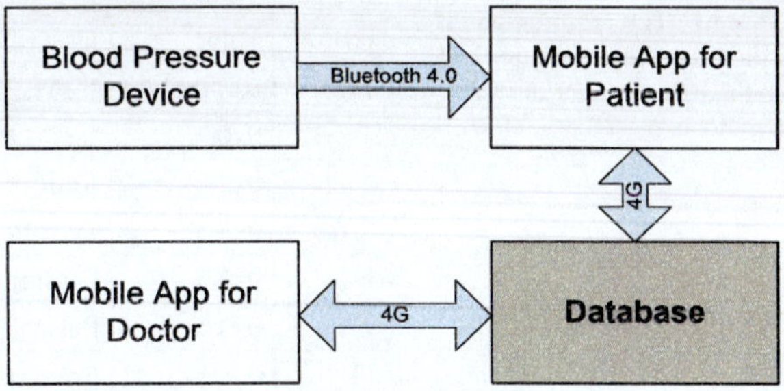

Fig. 1 Principle diagram of the holter NIBP system

Blood pressure monitor block: This block has a function to measure the blood pressure of patients. This monitor shows the result of measurement include systolic, diastolic and pulse. This monitor also sends the result to the mobile app through Bluetooth 4.0 protocol. In this research, we choose the monitor is iChoice BP1 device. Mobile app for patient block: This block synchronizes measurement data from blood pressure monitor iChoice. It stores measurement results, both offline and online. It displays results in graph and list. It also communicates and receives feedback from doctors. In addition, it reminds for measurement. Mobile app for doctor block: This block can access patient's data from anywhere with internet. It also displays the result in the graph, list and gives feedback to patients. Database block: This block stores and secure patient, doctor profiles and health record.

2.2 *iChoice BP1 Device*

With the iChoice BP1 monitor, we can take accurate blood pressure and heart rate reading in our own home or on the go. Bluetooth technology allows us to connect and sync our iChoice BP1 monitor easily to our iOS and Android device. This device communicates over Bluetooth 4.0 smart (Bluetooth low energy) profile using proprietary commands and data formats. In all communications with the device, bytes are packed in a little-endian mode; low byte comes first followed by high byte. For example, this sequence 0x04030201 should be transmitted from the gateway to the device as 0x01, 0x02, 0x03, 0x04 (from low byte to high byte). Similarly, when data is received from the device, the sequence 0x05, 0x06, 0x07, 0x08 should be interpreted by the gateway as 0x08070605. If not mentioned otherwise, all integers in the communication protocol should be interpreted as unsigned. The single arbitrary byte is denoted as 0xXX. The sequence of arbitrary bytes is denoted as 0xYY. There is no instruction for receiving measurement as the device sends blood pressure as soon as it is measured, so it is the only case when the communication is initiated by the device—gateway just waits for receiving blood pressure via cd04 characteristic. Here is a structure of the blood pressure data package received from the device. Table 1 describes the requirements of blood pressure measurement [7].

Table 1 The requirement of blood pressure measurement of the iChoice BP1 device

Byte #	Value	Description	Notes
1–2	0x5 0xAA	Instructions	
3	0x09	Instruction length	
4	0x34	Code	
5–6	0xYY	Systolic blood pressure	Little-endian
7–8	0xYY	Diastolic blood pressure	Little-endian
9–10	0xYY	Pulse	Little-endian
11	0x00	Reserved	
12	0xXX	Checksum	

2.3 Design the App for Patient

In the app for patients, we analyzed the flowchart into two categories are UI flowchart and module flowchart. UI flowchart includes open app screen, login screen, result display screen, feedback screen and profile screen. Module flowchart includes AS module, Internet module and BLE module. When the user opens the application, the app will display the auto-loading screen. The app will check the archive to see if the user has logged in from the previous session. If the user has logged in before, the app switched to the home screen. If this is the first-time user login in or user logout at the last session, the app will display the login screen. At the login screen, firstly, the app asks the user to log in with Facebook or Google. Secondly, the app checks the internet connection. Without an internet connection, the app will notice that there is no internet connection. If there is an internet connection, the app will wait for the callback from Facebook or Google. Then, the app will send all the measured data to the server. Next, the app will log in with the local account. In the three steps above, any error messages will give you an error message. Then, the app will store the token code. Next, the app will check user information on the server. If there is user information, the app will switch to the home screen. In contrast, the app will move through the fill information screen. After filling in the information, the app sends the data to the server (if this step fails, it also sends the error message).

Finally, the app stores information and displays the home screen. In the result display screen, the app loads the previous measurement from the storage. If there are previous archived results, the app will display them as graphs and lists. In contrast, the program displays no data. When the user presses the device connection button. The app will inform the user that the device is in progress. If it does not find the device, the app will give the message not found. In contrast, the app will connect to the device. Then, the app notifies characteristic. Next, the app waits for measurement results from the device. In the three steps above, any error messages will give the user an error message. Next, the app displays the measured results. Users can select cancel to cancel the measurement or choose to save the measurement to the storage. Finally, the app updates the measurement results to the home screen. Figure 2 shows the connecting device flowchart of the mobile app for the patient [8].

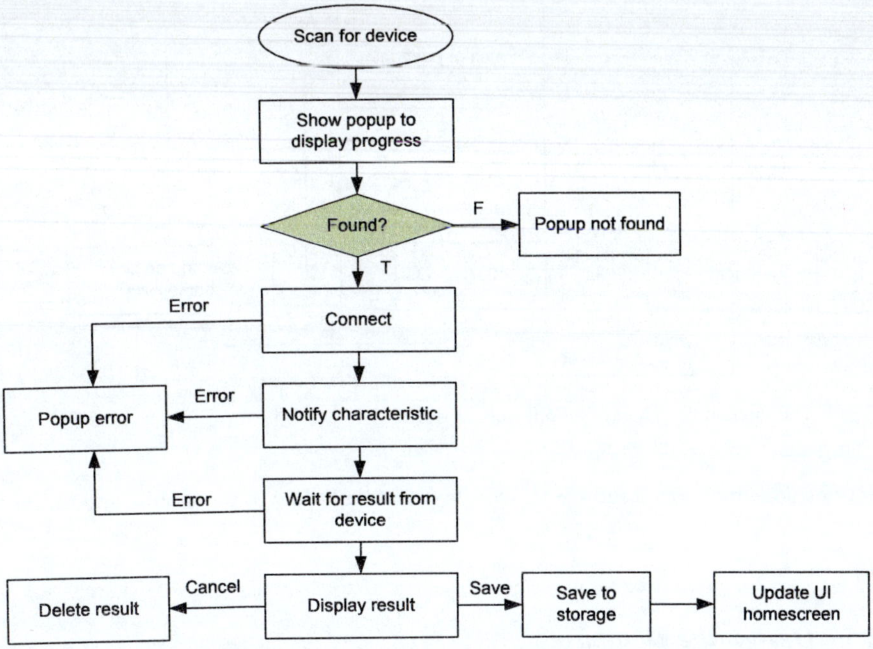

Fig. 2 Connecting device flowchart of mobile app for patient

2.4 Design the App for Doctor

In the app for the doctor, we analyzed the flowchart into two categories, UI flowchart and module flowchart. UI flowchart includes open app screen, login screen, choose patient screen, and patient profile screen. Module flowchart includes an AS module and Internet module. When the user opens the application, the app will display the auto-loading screen. The app will check the archive to see if the user has logged in from the previous session. If the user has logged in before, the app switched to the home screen. If this is the first-time user login in or user logout at the last session, the app will display the login screen [9].

In the patient selection screen, the app displays the patient list. Doctors can search for patients. If no patient needs to search, the app will display empty. If you find the patient, the app will display the list. At the patient profile screen, the app loads patient information by ID. If there is a patient, the program will display. With the AS module, the app reads the data by searching for the key of data. With the AS module, the app writes data and key of data. With the internet module, the app will fetch or post data. Then, the app transmits or receives the URL of data and data to the server. If the server informs the user not to log in, ask the user to login. If the server returns an error, the warning message is displayed. Finally, if the server return is true, the app

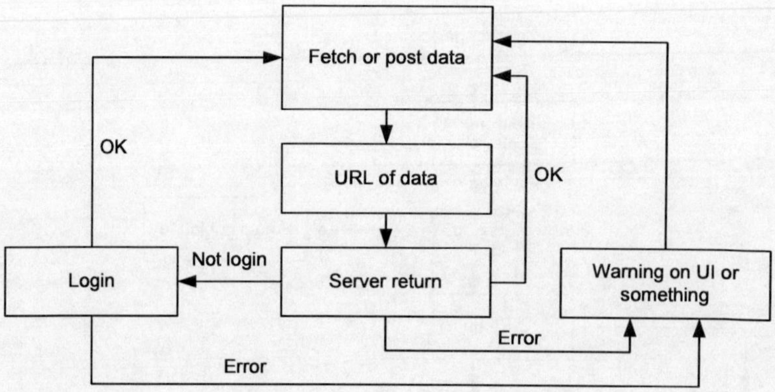

Fig. 3 The Internet flowchart of mobile app for doctor

notifies done message. Figure 3 shows the Internet module of the mobile app for the doctor.

2.5 Design the Database

We manage the data in four tables include a record table, feedback table, doctor table, and user table. Record table stores date, user ID and result list. Feedback table store date, patient ID, doctor ID, doctor name and content. Doctor table store username, password, name, phone, email, date of birth and gender. User table store account ID, token, pair ID, name, phone, email, date of birth, gender, Facebook ID and Google ID. With the database, every time the server receives incoming fetch. The server will check if the user has logged in or not. If the user has not logged in, the app will return the error. In contrast, the server will search by user ID to return the expected data. Figure 4 shows the Database flowchart of the system [10–12].

3 Designing and Testing Results

3.1 Designing Result

By using Visual studio code of Microsoft, we completed two mobile application for patients and doctors using JavaScript programming language. The holter NIBP system has been completely designed according to the above requirements. The figures show the designing result of the system.

Fig. 4 Thc Database flowchart of the system

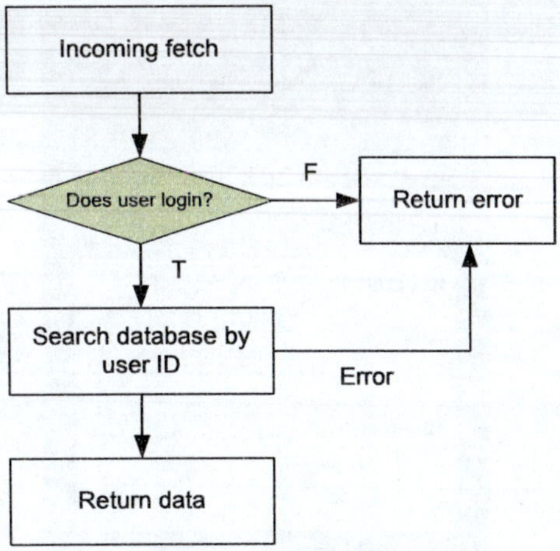

3.2 *Testing Result*

The experiments were conducted using the commercial device, the iChoice BP1 device of Medaval, China (https://medaval.ie/device/choicemmed-ichoice-bp1) to acquire data as Fig. 5. The data acquisition included systolic value, diastolic value and heart rate value (Figs. 6 and 7).

We measured and collected data on five volunteers, including three men and two women. Volunteers are all healthy people, without any medical treatment, aged

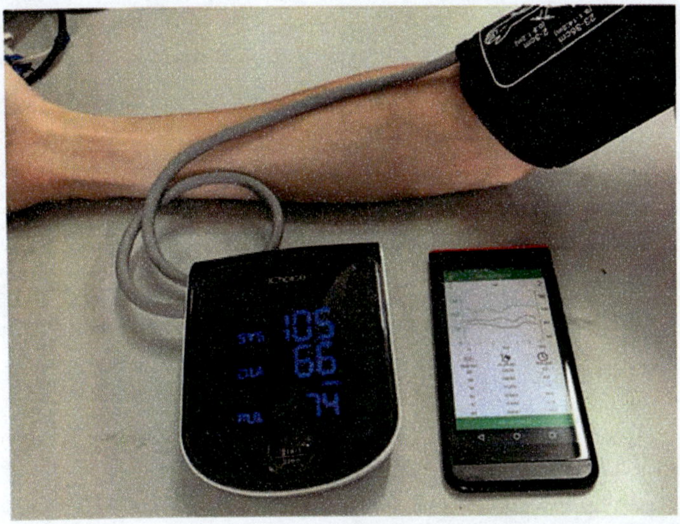

Fig. 5 The holter NIBP system with iChoice device and smartphone

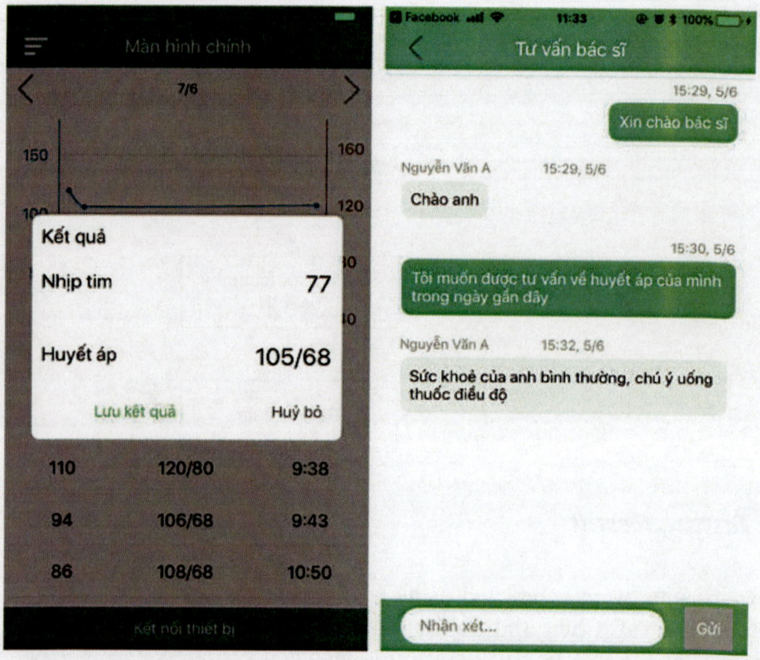

Fig. 6 Receive data from the divice interface and message with doctor interface of app for patient

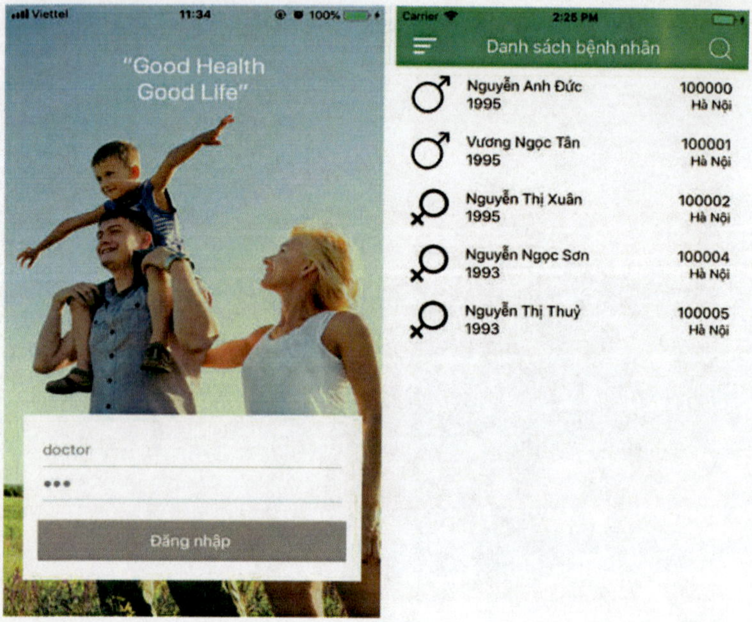

Fig. 7 Login interface and review patient data interface of app for doctor

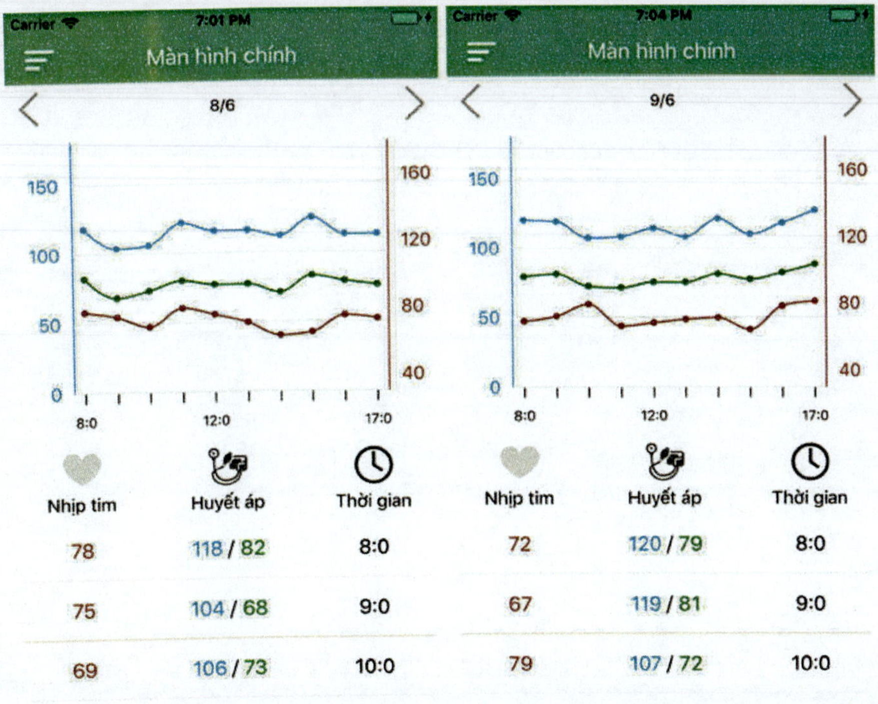

Fig. 8 Display the data of volunteer#1 on the app for doctor including Systolic Pressure (blue), Diastolic Pressure (green) and Heart Rate (red), they were received and kept continuously for two days from 8:00 to 17:00. The data is displayed in two graphs and numerical values over time for the doctor to follow. These measured values all coincide with those measured from the iChoice BP1 device

between 23 and 25 years and all the experiments were approved by the Ethics committee at our University Clinic and done by the doctors at our campus. For each volunteer, we measured for two days, ten times per day and the time between two measurements is one hour, from 8 am to 5 pm. The measurement results of volunteers are presented in the Figs. 8, 9, 10, 11 and 12.

4 Conclusions

For five candidates, we have 100 measurements. From the results show that connection between App for patient and iChocie is reliable and established up to three seconds and 100% of connections are successful. The results display on iChoice's screen, and App for the patient are always the same. Record from iChocie is transferred to App for the patient immediately in no time. Transmission between App for the patient and sever is stable. Every result on App for the patient is uploaded

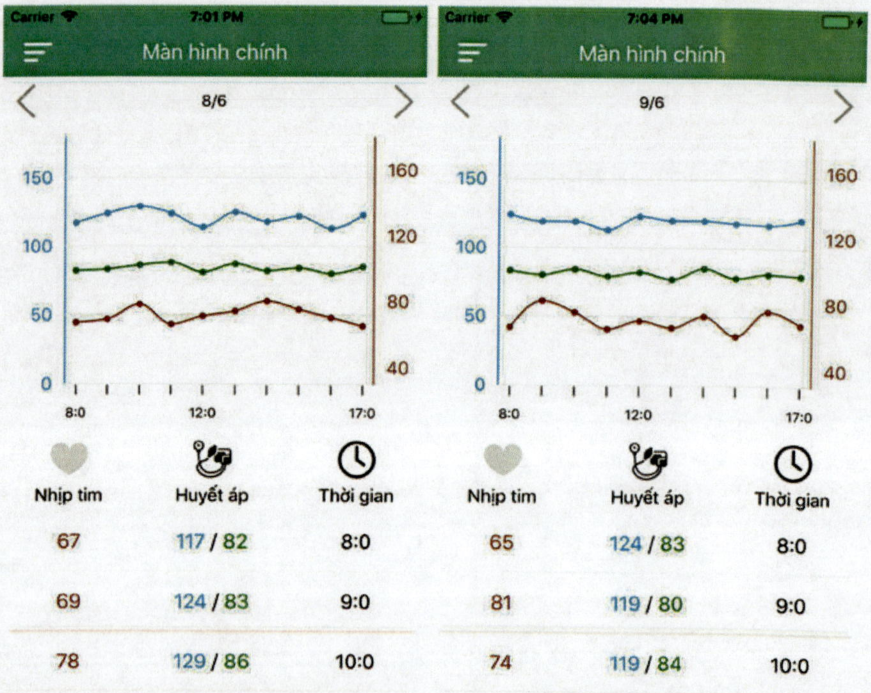

Fig. 9 Display the data of volunteer#2 on the app for doctor including Systolic Pressure (blue), Diastolic Pressure (green) and Heart Rate (red), they were received and kept continuously for 2 days from 8:00 to 17:00. The data is displayed in two graphs and numerical values over time for the doctor to follow. These measured values all coincide with those measured from the iChoice BP1 device

immediately and stored on the server side when the App for the patient has access to the internet. Transmission between App for doctor and server is stable. App for the doctor can access to patient record anytime without no error when it has access to the internet. App for doctor and App for the patient has no crash during testing. The system has practical applications to support the diagnosis and treatment of blood pressure diseases as a holter NIBP system.

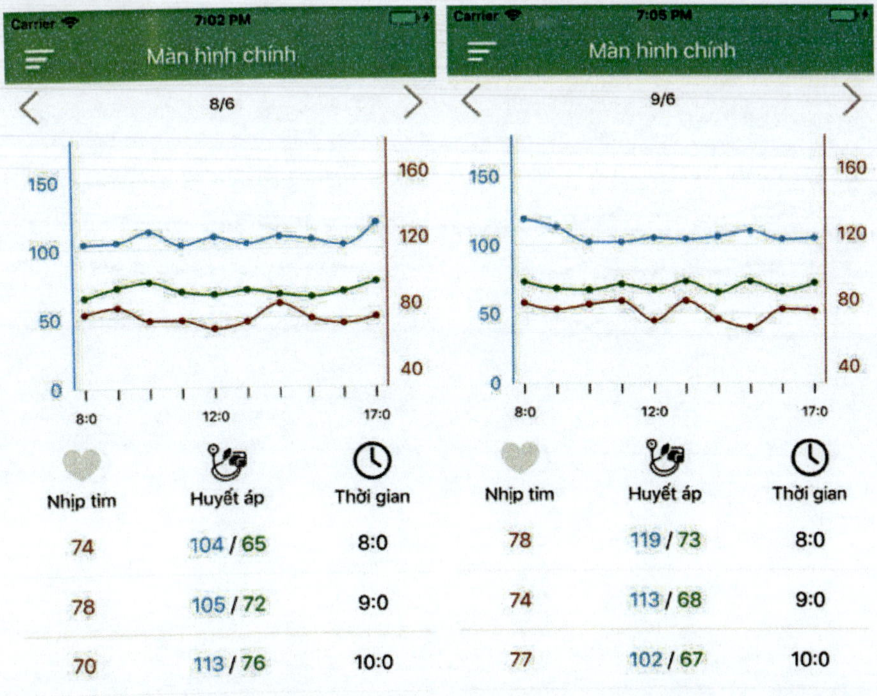

Fig. 10 Display the data of volunteer#3 on the app for doctor including Systolic Pressure (blue), Diastolic Pressure (green) and Heart Rate (red), they were received and kept continuously for two days from 8:00 to 17:00. The data is displayed in two graphs and numerical values over time for the doctor to follow. These measured values all coincide with those measured from the iChoice BP1 device

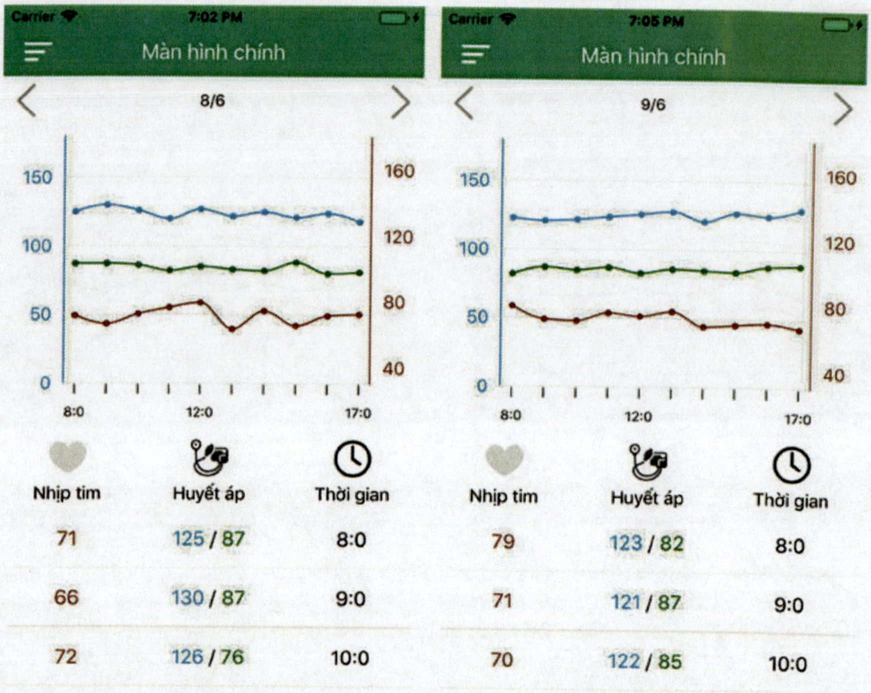

Fig. 11 Display the data of volunteer#4 on the app for doctor including Systolic Pressure (blue), Diastolic Pressure (green) and Heart Rate (red), they were received and kept continuously for two days from 8:00 to 17:00. The data is displayed in two graphs and numerical values over time for the doctor to follow. These measured values all coincide with those measured from the iChoice BP1 device

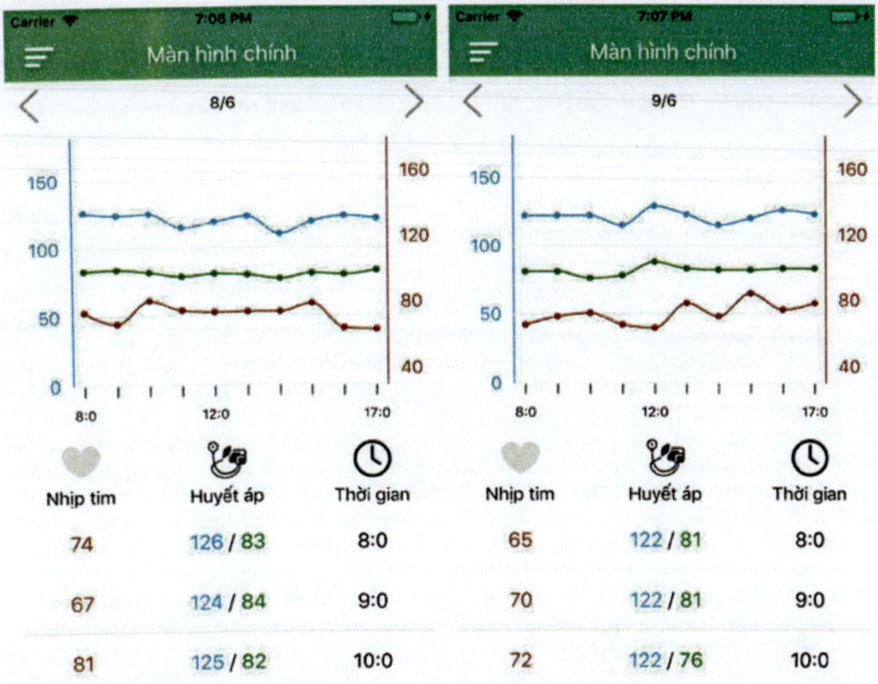

Fig. 12 Display the data of volunteer#5 on the app for doctor including Systolic Pressure (blue), Diastolic Pressure (green) and Heart Rate (red), they were received and kept continuously for two days from 8:00 to 17:00. The data is displayed in two graphs and numerical values over time for the doctor to follow. These measured values all coincide with those measured from the iChoice BP1 device

Acknowledgements The authors would like to acknowledge the support of Biomedical Electronics Center (BMEC), School of Electronics and Telecommunications (SET), the university clinic, Hanoi University of Science and Technology (HUST).

Conflicts of Interest No potential conflict of interest was reported by the authors.

References

1. Roger VL, Go AS, Lloyd Jones DM et al (2012) Heart disease and stroke statistics–2012 update: a report from the American Heart Association. Circulation 125:2–220
2. Ringrose JS, Polley G, Mc-Lean D et al (2017) An assessment of the accuracy of home blood pressure monitors when used in device owners, Am J Hypertension
3. Forouzanfar M, Dajani HR, Groza VZ et al (2015) Oscillometric blood pressure estimation: past, present, and future. IEEE Rev Biomed Eng 8:44–63
4. Van Monfrans GA (2001) Oscillometric blood pressure measurement: progress and problems. Blood Press Monit 6:287–290

5. Thompson AM, Eguchi K, Reznik ME et al (2007) Validation of an oscillometric home blood pressure monitor in an end-stage renal disease population and the effect of arterial stiffness on its accuracy. Blood Press Monit 12:227–232
6. Vu HD, Nguyen TD, Pham NP et al (2010) A design of renal dataflow control and patient record management system for renal department environment in Vietnam. In: The third international conference on the development of biomedical engineering in Vietnam. IFMBE Proceedings, vol 27. Springer, Berlin
7. Ngoc PP, Hai VD, Bach NC et al (2015) EEG signal analysis and artifact removal by wavelet transform. In: 5th international conference on biomedical engineering in Vietnam. IFMBE Proceedings, vol 46. Springer, Cham
8. Hai VD, Hung PD, Hung PM et al (2019) Design of non-invasive hemodynamic monitoring equipment using impedance cardiography. In: 7th international conference on the development of biomedical engineering in Vietnam. IFMBE Proceedings, vol 69. Springer, Singapore
9. Hai VD, Thuan ND (2010) Design of laboratory information system for healthcare in Vietnam BK-LIS. In: The 3rd international conference on communications and electronics, art. no. 5670692, pp 110–114
10. Hai VD, Hung PM, Trung LHP et al (2017) Design of software for wireless central patient monitoring system. In: Proceedings of KICS-IEEE international conference on information and communication with Samsung LTE&5G special workshop, Hanoi, Vietnam, pp 214–217
11. Hai VD (2020) Technique of receiving data from medical devices to create electronic medical records database. Soft Computing applications and techniques in healthcare. CRC Press, Boca Raton. https://doi.org/10.1201/9781003003496
12. Hai VD, Hung PD, Dan CQ (2020) Modified electrode placements for measurement of hemodynamic parameters using impedance cardiography. J Med Eng Technol. https://doi.org/10.1080/03091902.2020.1799089

Research and Design of Braille Learning Aids for Visually Impaired People Applying IoT Technology

Thi Bich Diep Nguyen, Quy Long Dinh, and The Dung Nguyen

Abstract Vietnam has about 2 million visual impairments people. In which only 8% go to school, 15% are trained in vocational centers and 20% have job opportunities. The blinds have many difficulties in living as well as learning and socializing. For the blind, learning Braille is the best way to minimize difficulties in all aspects of life. Each Braille is made up of 6 points; these points form a rectangular frame of two columns and three rows. The combination of six floatingsink points creates encoding for all characters. The scientific and technological solution that we have proposed with the goal of supporting Braille children to learn Braille in IoT applications. The product system includes many Braille coding devices for the blind; each module is used by a blind person. Each device is connected to the system's server which is designed on the internet.

Keywords IoT technology · Vietnamese braille · Braille for the visual impairments people · Braille code table · Braille learning support system

1 Introduction

At present, the application of electronic products to life is very common especially in developing smart devices that help people with disabilities in general and for the blind in particular. Products that help the visually impaired have been paid attention to by many research groups, organizations and countries around the world [1]. Some current research in this field is assistive technology for hearing impaired people learning math [2] or Braille education support systems with tablet devices [3]... Vietnam is also starting to focus more on the development of projects and equipment to support the disabled. In Vietnam, the rate of blind people is still very high and is a concern for agencies and centers for the visually impaired people [4].

Braille assistive devices are the solution to help improve the learning effect for the blind when they are young. With this device, people with visual impairments,

T. B. D. Nguyen (✉) · Q. L. Dinh · T. D. Nguyen
Thai Nguyen University of Information and Communication Technology, Thai Nguyen, Vietnam
e-mail: ntbdiep@ictu.edu.vn

V. Van Toi et al. (eds.), *8th International Conference on the Development of Biomedical Engineering in Vietnam*, IFMBE Proceedings 85,
https://doi.org/10.1007/978-3-030-75506-5_19

especially visually impaired children, will easily be exposed to Braille easily for later learning.

Braille encodings are difficult to learn for children and need the support of relatives or teachers at the blind center. So we offer a multifunctional device with a learning function to help the blind learn to learn Braille. This device includes entertainment and calculation functions right on the device to help visually impaired people, especially when the children study they will not be bored. Therefore, learning Braille with this device will be more effective than the traditional method.

2 Design System

Each device is connected to the system's server which is designed by us over the internet (Fig. 1).

In classroom mode, each student has access to a separate device in the production system. When operating in this mode, each device will be muted. The system's website will be managed by instructors directly in the classroom. Teachers can observe, evaluate, test directly or through the learning history and grade for each student. In this mode, teachers minimize the effort of teaching and can cover the entire learning process of students in the classroom. At home, every student can use the device to self-study without the help or support of a family member or instructor (Fig. 2).

Braille uses different arrangements of three rows and two dots, numbered from 1 to 6, to replace characters used in languages such as letters, tones, characters. Braille is specifically for the visually impaired, tactile sensation with the fingertips. So it is written on special paper, thicker than plain paper, written on the surface of the paper. To write Braille, a specialized writing board or typewriter is required [5] (Fig. 3).

In Vietnamese Braille, in addition to the usual arrangement of letters in syllables, it is important to note that tones (if any) are placed after the first consonant and before the vowel.

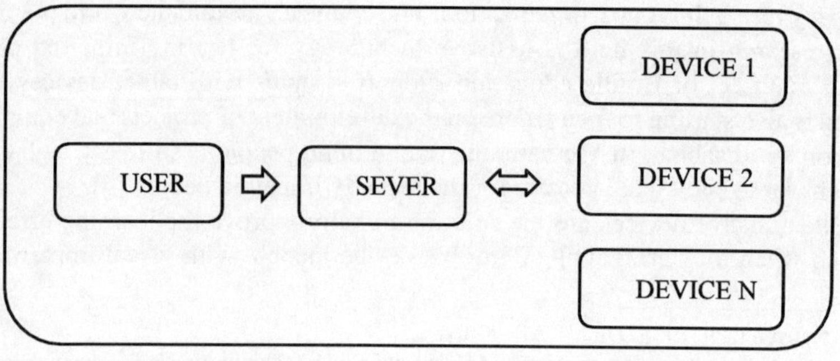

Fig. 1 System structure diagram

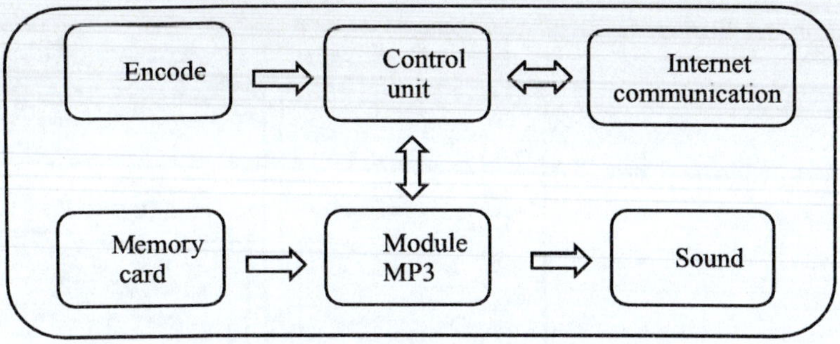

Fig. 2 Block diagram of Braille learning device

Fig. 3 The Braille Alphabet

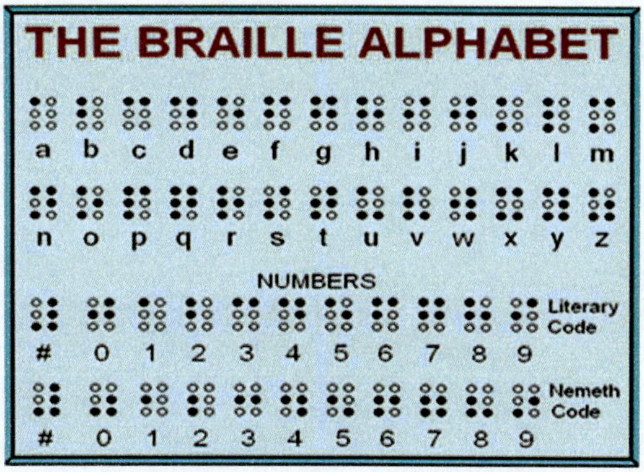

The coding block consists of a set of buttons designed on the basis of the Braille code table with six buttons arranged in three rows and two columns (Fig. 4).

Its control unit uses the Atmega328 chip—a common line of Atmel chips. The control unit has the function of receiving coded information from the encoder processing information and output compatible audio packets and sending data to the server [6] (Fig. 5).

DFPlayer is a module integrated hard decoding block, supporting popular audio formats such as MP3, WAV, WMA. It also supports TF card with FAT16. Maximum sampling frequency up to 48 kHz. 24-bit output supports reading resolutions up to 90 dB (Fig. 6).

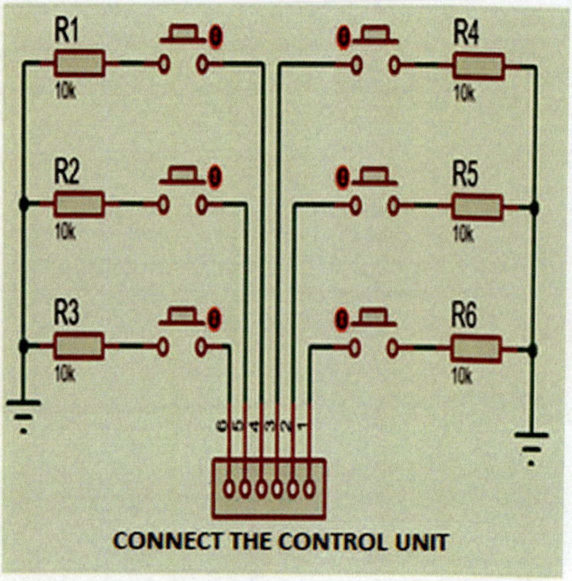

Fig. 4 Principle diagram of the encoder

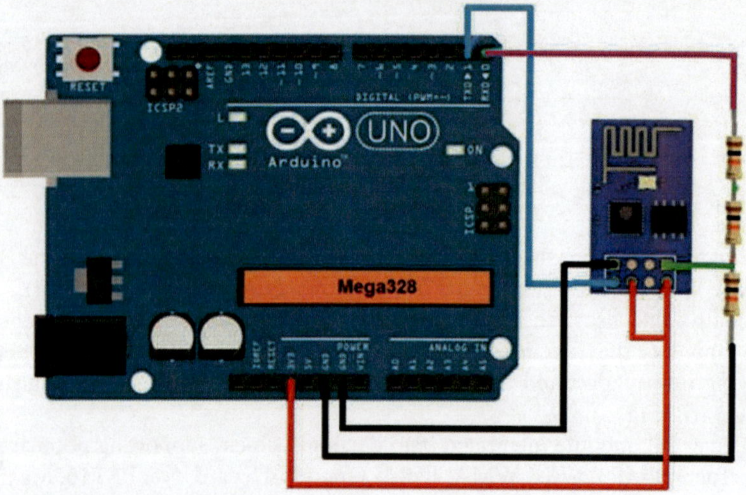

Fig. 5 Principle diagram block wifi connection control module

3 System Modes and IoT Technology Applications

The system offers two modes: self-study mode and classroom mode. With self-learning mode, the product is connected to external speakers to help children study. After system initialization, the user will enter a code of Braille characters. The central

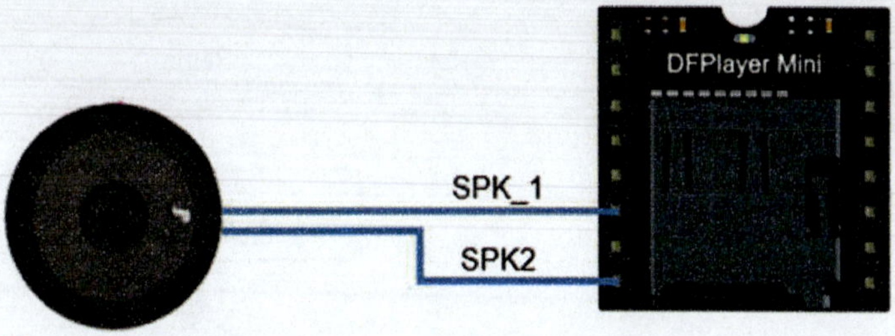

Fig. 6 Principle diagram MP3 module connection sound module

processor checks the code. If the code is within 53 basic characters, the central control unit will find the corresponding file.mp3. This file is saved in the DFPlayer Mini MP3 module. Then the speaker emits a sound corresponding to the character. If the code is wrong, the microcontroller will alert and perform the input process again (Fig. 7).

In classroom mode (mute) after initializing the system, the device will perform wifi connection checks. When there is a wifi connection, the central processing units on the device will constantly scan and read the encrypted signal (both true and false compared to Braille). Then it sent the information to the server and displayed on the website. Through this, teachers can supervise and check the learning level of all members of the class at the same time (Fig. 8).

Fig. 7 Flowchart of the algorithm in self-learning mode

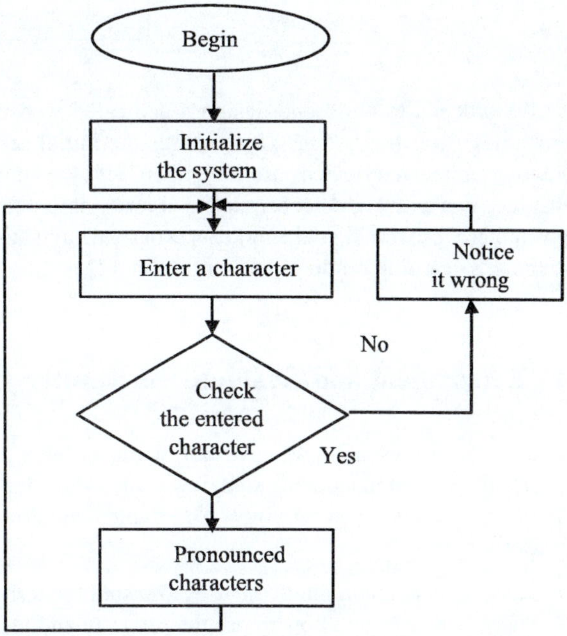

Fig. 8 Algorithm flowchart in classroom mode

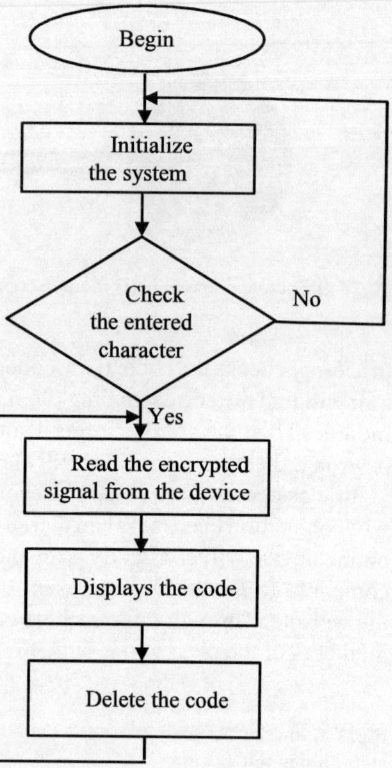

We proceed to build and design a dedicated website for the application of IoT technology for the system. This ensures the initial set of functions and goals. Our Internet-connected devices are integrated with the sensor chips described above so that they can convert, detect signals and turn it into data in the Internet environment. From there, processing data and executing the navigation in that internet, according to the self-learning modes that users desire [7].

4 Experiment and Evaluate the Results

To evaluate our research, we have carried out a trial of the system with a target group of visually impaired students studying at the Education for Disadvantaged Children School in Thai Nguyen province. The implementation process includes three main phases:

- Phase 1: install and guide the user to install the test program.
- Phase 2: training on how to use the program and testing equipment.

- Phase 3: distribute the survcy to evaluate the results and synthesize the evaluation results.

Through survey data on 85 learners, the research team received some general assessments as follows:

- The testing equipment and the system's curriculum are relatively easy to use.
- Surveys show that 100% of users agreed with the accuracy of the Braille encoding.
- The majority of test users (82.4%) rated the system as useful in helping the visually impaired learn.

In its solution, the team achieved the following results:

- Learn and apply the Braille coding table.
- Designing and constructing a product system.
- Successfully fabricating devices that help assist visually impaired children to learn Braille applications using IoT.

The future development direction of this product is that besides supporting children to learn to read, they will integrate more entertainment functions (music, storytelling, etc.). Especially, we will develop products that integrate with multiple languages (Japanese, Korean, Chinese, etc.).

Conflicts of Interest The authors have no conflict of interest to declare.

References

1. Wolffe KE (2003) An analysis of technology training for people with visual impairments. J Vis Impairment Blindness
2. Asebriy Z, Raghay S, Bencharef O (2018) An assistive technology for Braille users to support mathematical learning: a semantic retrieval system. Symmetry MDPI
3. Shintaro W, Paul H, Tetsuak B, Kumiko K (2016) Braille pad project: proposal of a Braille education support system using a tablet device. Int J Asia Digital Art Design
4. Bac NT (2005) Services for the blind in the public libraries of Vietnam: marking Vietnamese public libraries more accessible to visually impaired people. In: IFLA71
5. Martiniello N, Wittich W (2018) The perception and use of technology within braille instruction: A preliminary study of braille teaching professionals. Brit J Vis Impair 195–206
6. Khodzhaev Z (2016) Monitoring different sensors with ATmega328 microprocessor. Istanbul Technical University, Faculty of Science and Letters, Advanced Physics Project Report
7. Karl H, Willig A (2005) Protocols and architectures for wireless sensor networks. Wiley, London, pp 17–56

Internet of Things Application for Online Surveillance and Control of Attacks by Sea Lice

Nhut Tran-Minh, Bard Haug, Bendik Fyhn Terjesen, and Frank Karlsen

Abstract The production potential of Atlantic salmon in Norway is influenced by a large number of variables that may best be controlled if it is possible to follow representative samples, collected randomly from fish that are still alive inside the production plant. The most optimal or objective way to follow stress or well-being of a fish is to be able to analyze in-vitro, different molecular changes within representative samples automatically. The first generation of automated fish sample collection system was developed to combine with refining technology from the partner in the CtrlAQUA center (www.ctrlaqua.no) in order to give a high analytical sensitivity, a low enough number of false negatives and positives. Result of the experiments illustrates that non-clogging separation and high concentration ratio can be achieved.

Keywords Internet of things · Bluetooth low energy · Refining · Concentration

1 Introduction

About 71% of the Earth's surface is covered by water, but there is only 2% of foods come from the ocean. Growing fish in either a fish farm net cage or a closed containment system is necessary to increase fisheries production and reduce mortality of fishes. According to climate change, such as global warming, the number of fish parasites is rising and causing negative and unwanted stress and diseases for fish [1–5]. Harsh chemical treatments are becoming less effective since only a limited number of microparasites can be controlled. In Norway, about 15% of salmon fish die of stress caused by chemical treatments while only 5% of salmon fish are killed by parasites. Fish farmers are highly recommended to improve their management

N. Tran-Minh (✉) · B. F. Terjesen
Norwegian Institute of Food, Fisheries and Aquaculture Research, NO-1431 As, Norway

F. Karlsen
University of South-Eastern Norway, NO-3184 Borre, Norway

N. Tran-Minh · B. Haug
Oslofjord Ressurspark AS, NO-3184 Borre, Norway

V. Van Toi et al. (eds.), *8th International Conference on the Development of Biomedical Engineering in Vietnam*, IFMBE Proceedings 85,
https://doi.org/10.1007/978-3-030-75506-5_20

techniques and only turn to chemical treatments to kill off the parasites. Preventive measures are clearly needed for monitoring stress and disease to leave salmon fish in peace [6, 7]. These measurements should be lower cost and more environmentally sustainable.

In this paper, we focus on the design and evaluation of the automated Water Sample Collection system (WASACO). WASACO is a secure end-to-end IoT-based system for collecting particles from the surrounding seawater in a representative way. WASACO is able to concentrate identified particles from a large and representative volume of sample down to a very small volume. The most advantage of WASACO compared with other refining system is a novel design with arrays of micropillars that help to prevent clogging phenomenon. Furthermore, WASACO also reduces the salinity of the sample to a neglectable level by adding freshwater during the operation process.

2 Materials and Methods

WASACO has two main parts, including the electronic and microfluidic module. Electronic module plays as an important role in WASACO since it is considered as a central processor that controls the amount of seawater flow in and out of WASACO. Meanwhile, the microfluidic module implements both two processes, such as fluid concentration and particle separation.

2.1 *Electronic Module*

WASACO can be defined as an internet of things' device with four functions in one unit. Depend on user scenarios, WASACO can be run with four different operation modes: Manual mode, Bluetooth mode, WiFi mode and Mobile Telecommunication mode (Fig. 1).

In land-based aquaculture fish farming, WASACO's user can select either Manual mode or Bluetooth mode when the distance between WASACO and user is quite short. While Manual mode does not need any additional devices, Bluetooth mode requires Bluetooth enable devices such as smart phone, tablet, etc. Bluetooth Low Energy (BLE) technology is integrated into WASACO in order to provide lower cost and less power consumption while maintaining a similar communication range. Another reason of using BLE rather than classic Bluetooth technology is for compatibility with mobile operating systems including iOS (Apple Inc., USA), Android (Google and Open Handset Alliance, USA) which natively support Bluetooth Low Energy. Classic Bluetooth technology may reduce the cost of WASACO and increase the number of Bluetooth enable control devices, but WASACO must be approved by Apple's MFi Program before it is permitted for connecting to Apple's devices [8].

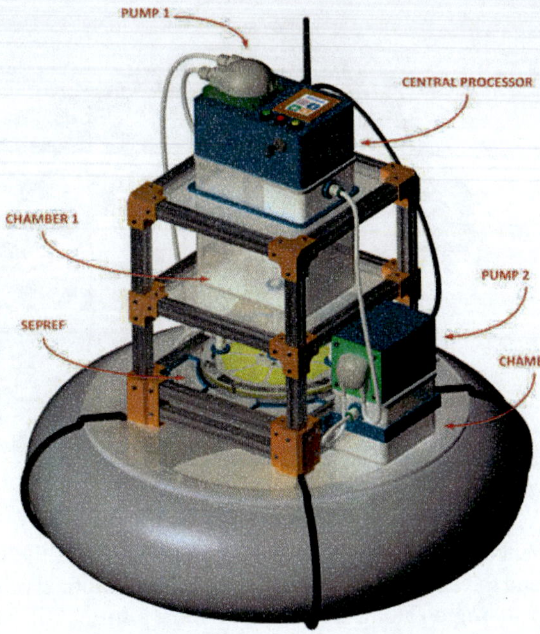

Fig. 1 3D model of automated Water Sample Collection system

When Recirculating Aquaculture Systems (RAS) and the user are located on different floors in the same building, WASACO can be operated with WiFi mode. WASACO supports wireless networking standard of 802.11 b/g/n that can operate on the 2.4 GHz band. With WiFi mode, the number of WASACO's control devices extends to personal computers or any devices have the ability to run Hypertext Markup Language (HTML) (World Wide Web Consortium, USA).

In sea-based aquaculture fish farming, WASACO will take advantage of telecommunication networks during its operations. WASACO can receive instruction from users or transmit measurement results to users even though users are located in different cities or countries. It may also reduce working hours for users when they can start WASACO before leaving home and receive measurement results after arriving at working place.

2.2 *Microfluidic Module*

The most advantage of WASACO compared with other similar device is not only a non-clogging sorting system but also a high concentration system. WASACO includes the module which is named as SepRef (Oslofjord Ressurspark AS, Norway). SepRef contains hundreds of refining units (see Fig. 2) that has 15 micropillars arranged on the contour of ellipse with the left half major axis shorter than the right half major axis [9].

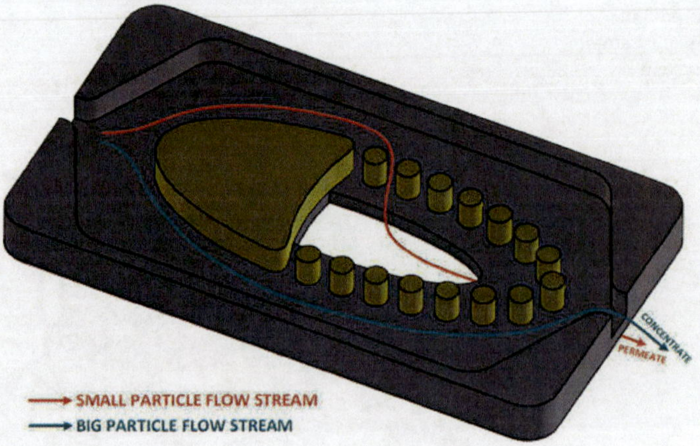

Fig. 2 Refining unit in WASACO

When the fluid flows through the unit, it faces the blocking head and is divided around the elliptical profile. The optimized elliptical profile provides the largest high velocity region along the two sides of the ellipse [10]. This high velocity region aims to prevent the particle clogging at the penetrated channels. The main flow of fluid is able to push any large particles which are tending to block the entrance of the penetration channels. In the other hand, the particles which are smaller than the gap between two adjacent micropillars will enter the penetration channels to the hole at the center.

2.3 *Process Flow*

When WASACO starts, both ultrasonic sensor 1 and 2 are activated. If the distance between sensor 1 and the water level in chamber 1 is greater than 7 cm, seawater from the ocean or indoor fish tank is pumped into chamber 1. This process is only paused until above distance is less than 7 cm. In the other side, a process for sensor 2 is same as process for sensor 1, but the comparison condition is opposite (Fig. 3).

While seawater from chamber 1 flows to SepRef via a plastic tube, seawater from chamber 2 is transferred to chamber 1 via pump 2 and then flows through SepRef one more time. This looping process aims to concentrate seawater sample. It helps to improve the analytical sensitivity of further Nucleic Acid Sequence Based Amplification (NASBA) process. SepRef has two outlet channels, named as Permeate and Concentrate. In Concentrate channel, water sample with larger particle will go to chamber 2. On the other hand, a large amount of water sample, including smaller particles, flows back to the ocean/tank via Permeate channel.

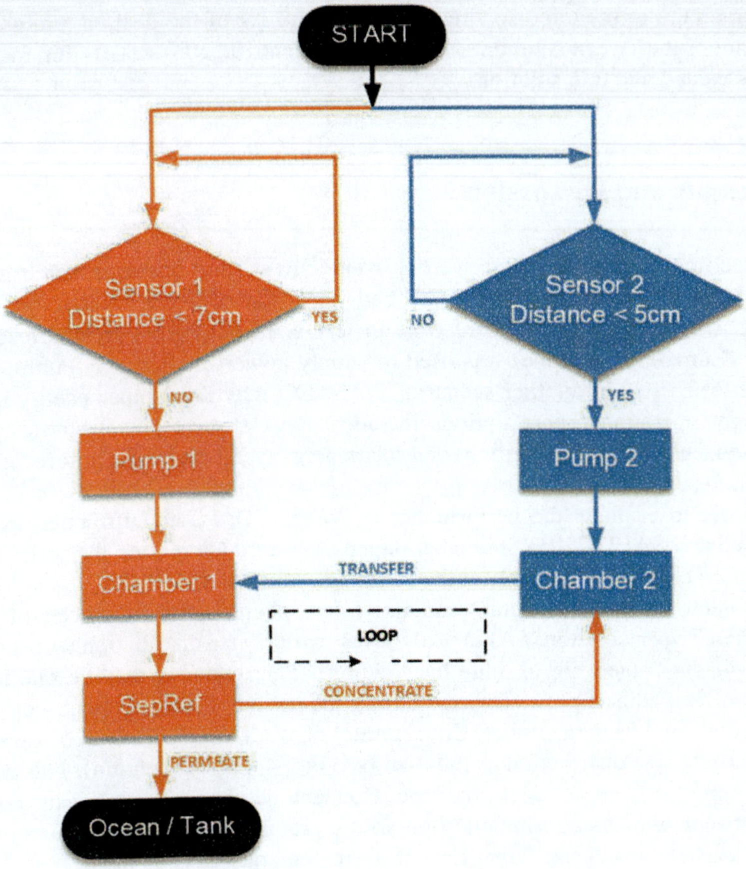

Fig. 3 Process flowchart of WASACO's operation

2.4 *Water Sample Preparation*

Since this paper mainly focuses on the design and pre-evaluation of the WASACO, artificial particles were prepared as a water sample for testing. As future work, more complex water sample such as water containing sea lice from land-based or sea-based aquaculture fish farming can be used to validate the effectiveness and reliability of WASACO. The following is the procedure to create sample including particles behave similarly to sea lice. De-ionized water is heated on a hot plate until it reaches a rolling boil. Tween Biocompatible Surfactants (Merck, Germany) is slowly mixed with hot de-ionized water with the ratio of 0.1 g of Tween per each 500 ml of water, and wait until the water cools and any bubbles have settled before using Tween solution. This Tween solution aims to suspend Violet Polyethylene Microspheres 1.00 g/cc 212–250 μm (Cospheric, USA). A pipette or syringe is used to dispense prepared Tween solution on top of the microspheres with ratio of 0.1 g of the desired microspheres with 2 ml of the Tween solution. The mixture is centrifuged for 5–10 min to get the spheres wetted and into solution.

3 Results and Discussion

All experiments in this paper were carried out at land-based research station at Sunndalsora (Norway). WASACO had been started with Manual mode for 60 min. Due to a short period of time to test WASACO, 12-V sealed lead acid battery (EuroGlobe, Europe) was used to supply power for Pump 1, Pump 2, and Central Processor. In another scenario, WASACO may need other energy option for longer operation. These options include energy storage power station (land-based aquaculture research station) and solar energy (sea-based aquaculture research station).

In order to evaluate the performance of WASACO, Laser Diffraction Particle Size Analyzer SALD-2201 (Shimadzu, Japan) was used to measure the particle size distribution of the water sample at the inlet and outlets of WASACO.

For water sample at the inlet contains microspheres with a diameter of 212–250 μm in Tween solution, SALD-2201 gives particle size distribution with a range of 189–482 μm (see Fig. 4, blue bar column). This variance may be due to the overlap of two adjacent microspheres during the period of measurement. For water sample collected from the outlet (Concentrate) of WASACO, SALD-2201 gives the same particle size distribution as the inlet (see Fig. 4, red bar column). This means almost particles from the inlet flows into Concentrate channel. It is because of the gaps between two neighboring micropillars are around 180 μm. These micropillars prevent particle which size larger than 180 μm penetrate through them.

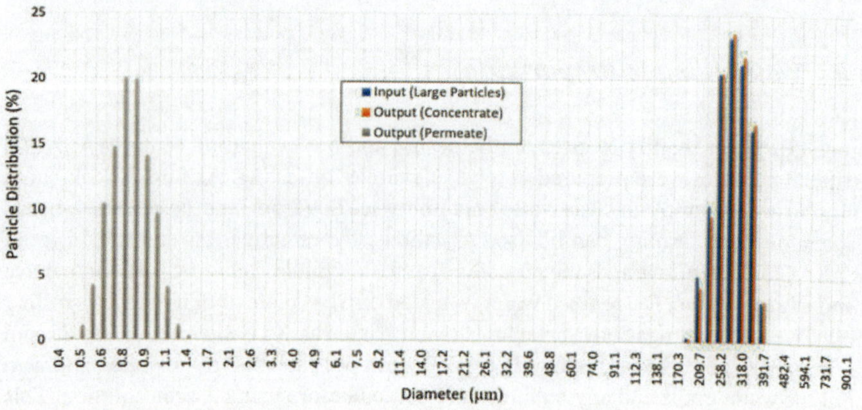

Fig. 4 Particle distribution at the inlet and outlets of WASACO

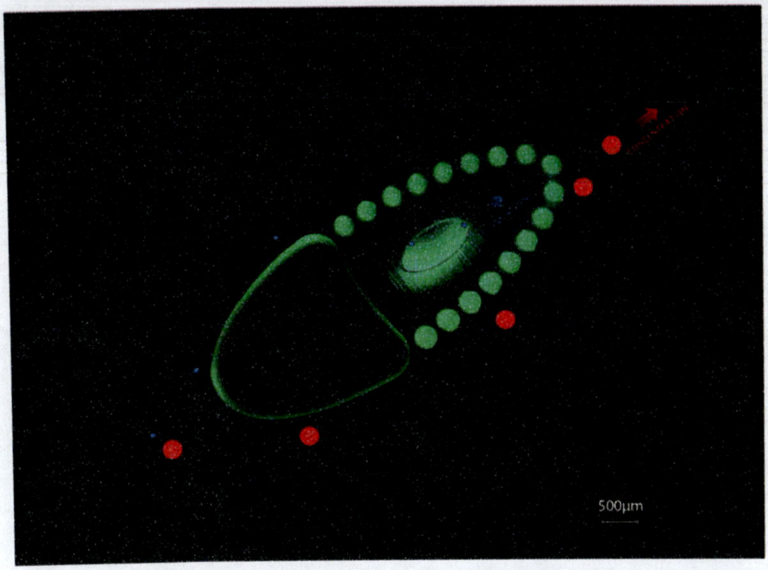

Fig. 5 Microscopic image of refining unit with visual illustration pathways of particles

Furthermore, particle size distribution at the outlet (Permeate) appears at a lower range of diameter. Water sample flow into Permeate channel is purely liquid without any particles, but SALD-2201 also give a result with a range of 0.45–1.74 μm (see Fig. 4, gray bar column). These small particles may be dust or micro animals. They come from WASACO itself since it is not 100% clean.

In the second experiment, water sample at the inlet contains microspheres with a diameter of 63–75 μm and 212–250 μm in Tween solution. Since the water sample including microspheres with a diameter of 63–75 μm which are smaller than a gap (180 μm) between two adjacent micropillars, these particles penetrate through these pillars and reach to Permeate channel, see illustration on Fig. 5.

Two water samples at the outlet of Permeate and Concentrate channel were collected and shown in Fig. 6b. Compared with the first experiment (Fig. 6a), microparticles appear both at Permeate and Concentrate channel in this second experiment. Especially, smaller particles with a diameter of 63-75 μm are not expected to present at Concentrate outlet. This issue may occur when the pathways of smaller particles are far away from the refining unit. In order to overcome this issue, the water sample at the outlet of Concentrate channel should be transferred back to the inlet tank and start a new refining process. This work can be assisted with WASACO.

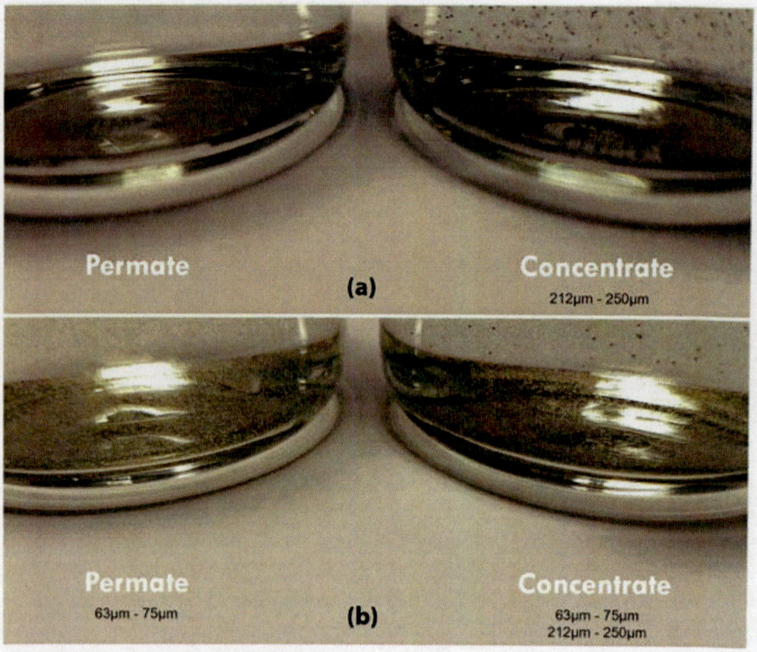

Fig. 6 Sample at the outlet of Permeate and Concentrate channel. **a** Inlet sample contains only microspheres with a diameter of 212–250 μm (violet particles). **b** Inlet sample contains both microspheres with a diameter of 63–75 μm (yellow particles) and 212–250 μm (violet particles)

4 Conclusion

This paper presented an automated water sample collection system named as WASACO, which is for continuous sorting and concentration of particles or cells in the large volume of sea, waste, or freshwater. This system is a standalone machine with its own central processor. It helps to reduce the labor force who need to collect representative water samples frequently. With the practical experiments, WASACO shows that non-clogging separation and high concentration ratio can be achieved. For future work, more complex water containing sea lice is planning to use for evaluating other aspects of WASACO's performance. Furthermore, WASACO is also possible to be used for studying about harmful algal blooms (HABs), especially in the southeastern Vietnamese coastal waters. Although these HAB event has a few oceanographic studies, but it caused large economic losses for the Vietnamese fisheries and aquaculture industry. Combining of cutting-edge of IoT with non-clogging refining system, WASACO may contribute promising results for HAB studies.

Acknowledgements The Research Council of Norway is acknowledged for the support through the Personal Post-doctoral Research Fellowships (HAVBRUK 2) [grant number 254862].

Conflicts of Interest The authors have no conflict of interest to declare.

References

1. Watson SB, Whitton BA, Higgins SN et al (2015) Harmful algal blooms. In: Freshwater algae of North America, 2nd edn. Academic, London, pp 873–920. https://doi.org/10.1016/B978-0-12-385876-4.00020-7
2. Getchell RG, Smolowitz RM, McGladdery SE et al (2016) Diseases and parasites of scallops. In: Developments in aquaculture and fisheries science. Elsevier, Amsterdam, pp 425–467. https://doi.org/10.1016/B978-0-444-62710-0.00010-9
3. Subramani PA, Michael RD (2017) Prophylactic and prevention methods against diseases in aquaculture. In: Fish diseases: prevention and control strategies. Academic, London, pp 81–117. https://doi.org/10.1016/B978-0-12-804564-0.00004-1
4. Sitja-Bobadilla A, Oidtmann B (2017) Integrated pathogen management strategies in fish farming. In: Fish diseases—prevention and control strategies. Academic, London, pp 119–144. https://doi.org/10.1016/B978-0-12-804564-0.00005-3
5. Joseph A (2017) Marine creatures with a difference. In: Investigating seafloors and oceans: from mud volcanoes to giant squid. Elsevier, Amsterdam, pp 443–491. https://doi.org/10.1016/B978-0-12-809357-3.00008-4
6. Marshall Adams S (2002) Biological Indicators of aquatic ecosystem stress: introduction and overview. In: Biological indicators of aquatic ecosystem stress. American Fisheries Society, pp 1–11
7. Helfman G, Collette BB, Facey DE et al (2009) Homeostasis. In: The diversity of fishes: biology, evolution, and ecology. Wiley, London, pp 91–109
8. Apple, MFi program. [online] Available: https://developer.apple.com/programs/mfi/. Accessed: Dec 2017
9. Tran-Minh N, Dong T, Karlsen F (2014) An efficient passive planar micromixer with ellipse-like micropillars for continuous mixing of human blood. Comput Methods Programs Biomed 117(1):20–29. https://doi.org/10.1016/j.cmpb.2014.05.007
10. Tran-Minh N, Dong T, Su Q et al (2011) Design and optimization of non-clogging counter-flow microconcentrator for enriching epidermoid cervical carcinoma cells. Biomed Microdevice 13(1):179–190. https://doi.org/10.1007/s10544-010-9483-5

Efficiency Evaluation of a Pilot Telemedicine System to Monitor High Blood Pressures in Binh Duong Province (Vietnam)

Vien Vinh Phu, Do Minh Thai, Nguyen Le Thanh An, Tran Ngoc Viet, Nguyen Phuong Nam, and Vo Van Toi

Abstract Cardiovascular disease and stroke due to high blood pressure continue to be one of the leading causes of death in Vietnam and elsewhere. Continuous control of patients' blood pressures plays, hence, a vital role in reducing the risks of these diseases. We implemented a telemedicine system to monitor the blood pressure of hypertensive patients in Binh Duong province. This cyber medical system consisted of a Tele-Blood Pressure device, a software program for servers, and an application for smartphones. Besides, it provided considerable utilities for doctors to better take care of patients. In the project described here, we evaluated the efficiency. First, two groups (control and treatment) of 100 patients each who had histories of high blood pressure were selected using randomized controlled trials (RCT) method. Second, pretest–posttest experiments were conducted. During 18 months both groups were monitored by the same 3 physicians and 4 nurses from the beginning to the end. The obtained results were compared using paired t-test or chi-square test. At the end of the study, a survey of the satisfaction of all aspect of the system was conducted to all patients of the treatment group, physicians and nurses. The percentages of satisfactions were well above 82%. This study will be extended at a larger scale, especially in the remote areas where the healthcare needs to be improved.

Keywords Telemedicine · Internet of things · Cyber medical system · High blood pressure · Tele blood pressure device · Heart rate

---

V. V. Phu · V. Van Toi
School of Medicine, Vietnam National Universities HCMC (VNU-HCM), Ho Chi Minh City, Vietnam

D. M. Thai (✉) · N. Le Thanh An · T. N. Viet · N. P. Nam · V. Van Toi
School of Biomedical Engineering, International University—VNU-HCM, Ho Chi Minh City, Vietnam
e-mail: dmthai@hcmiu.edu.vn

V. Van Toi et al. (eds.), *8th International Conference on the Development of Biomedical Engineering in Vietnam*, IFMBE Proceedings 85,
https://doi.org/10.1007/978-3-030-75506-5_21

1 Introduction

Telemedicine is an effective treatment method and prevention, particularly in third-world countries where the healthcare systems are usually inadequate [1]. Among many aspects of the telemedicine we developed an Internet of Things (IoT) system which consisted of a Tele-Blood Pressure device and a Cloud Telemedicine Information System (CTIS) to link patients and healthcare providers together from distance. This IoT system was implemented at the "Ban Chăm sóc và Bảo vệ Sức khỏe Cán bộ Trung Cao tỉnh Bình Dương" (Retired High and Middle Ranking Public Officials Clinics of Binh Duong Province or RPOC for short) to monitor its outpatients suffered from high blood pressure [2]. In the project described here, during 18 months, we evaluated the efficiency of this system with two groups of high blood pressure patients: control and treatment groups. Each group had 100 patients selected using randomized controlled trials (RCT) method. At the end of the study, a survey of the satisfaction of all aspects of the system was conducted to all patients, physicians and nurses. The percentage of satisfaction was well above 82%. This study was approved by the Human Subject Committee of VNU-HCM.

2 Materials and Method

2.1 *The Tele-blood Pressure Device*

Figure 1 shows our tele-blood pressure device and its mass production. It allows the patient, at home or on the move, to measure him/herself blood pressures and heart rate at the arm. Its use is simple and automatic with one button to do all. The obtained data will then be automatically sent via ADSL or 3–4G to the central server to be analyzed and stored. The patient can also share the data to other people e.g., parents and family doctor. These people, from a distance, can follow in real time the data displayed on a smartphone, tablet or laptop, and can interact with the patient immediately as if they were next to the patient. The device can also function as a holter to regularly check during 24 h the above values (Fig. 2). It can be used by other people. In this case, each user has his/her own account and website. The qualities of the tele-blood pressure, including its accuracy, safety and compliance, and of the whole system were controlled and certified by an official agency of the Vietnamese governmental Ministry of Information and Communications.

2.2 *The Cloud Telemedicine Information System (CTIS)*

The cyber medical system CTIS structure implemented at the RPOC is shown in Fig. 3. Its software and firmware allow the administrator to monitor, process

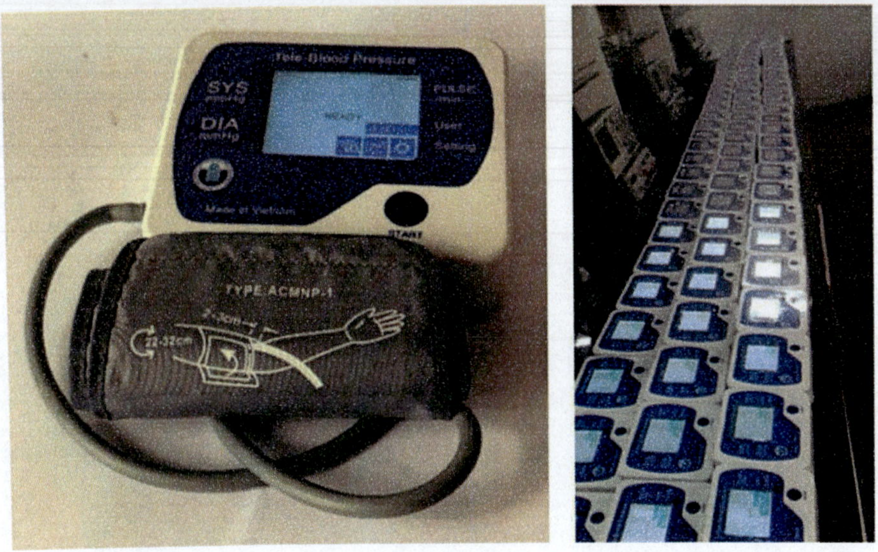

Fig. 1 The tele-blood pressure device used in this project and its mass production

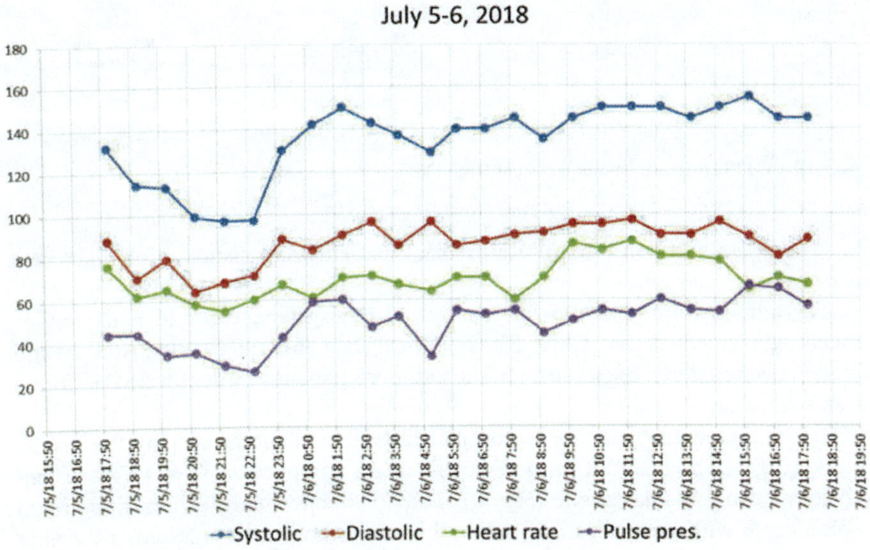

Fig. 2 The results when the tele-blood pressure device used as a pressure holter

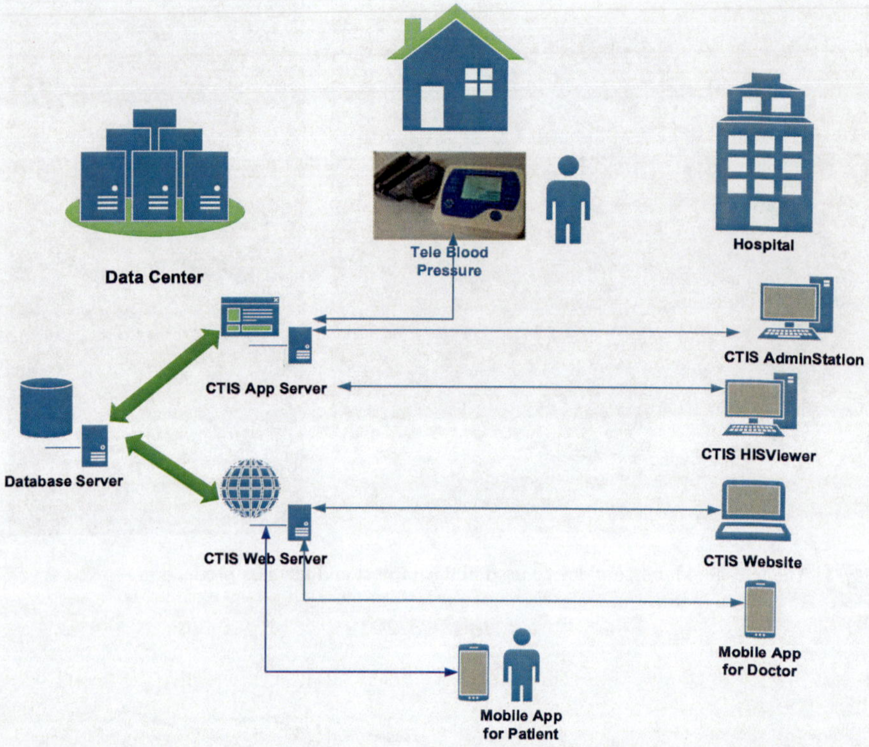

Fig. 3 The CTIS structure implemented at Binh Duong province RPOC

and stored data and permitted users to check patients data using a computer or a smartphone.

CTIS administrator has the full control of the system, can add and monitor patients data which content: Code, Name, Phone, Date of Birth, Gender, Address and Blood Pressure Alert Thresholds. He/she assigns a patient to a doctor and performs decentralization.

Doctors review and monitor decentralized patients and can display the results in time (Fig. 4), in statistics form (Fig. 5) using a computer to view CTIS system website or using a smartphone (Fig. 6). They can enter medical prescription and communicate with patients. If the patient has abnormal blood pressure, a warning sign will alarm the doctors to take appropriate treatment decisions immediately.

Patients, of course, can also access their information the same way as the doctors do.

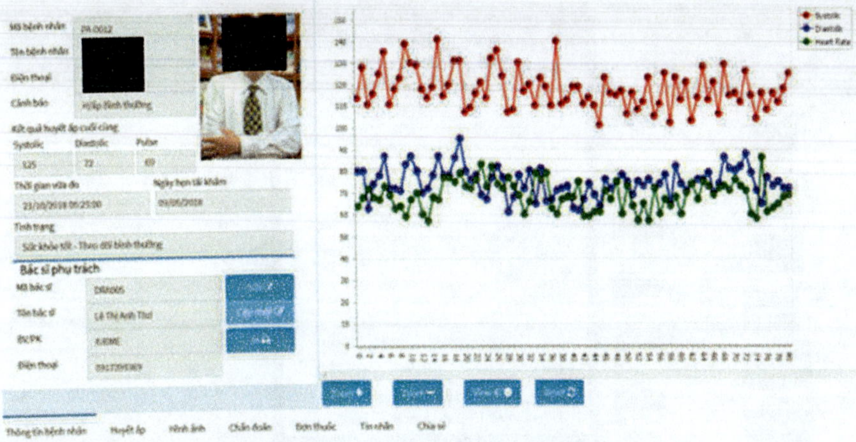

Fig. 4 Website of a patient with graphs of systolic, diastolic pressure and heart rate

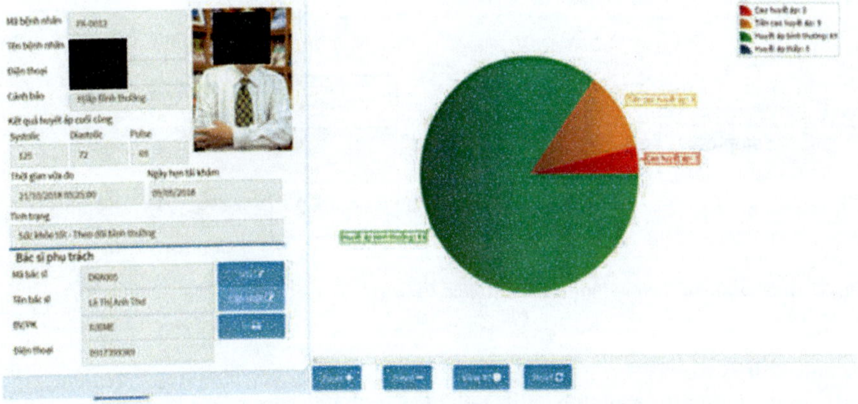

Fig. 5 Website of a patient with graphs of statistics of blood pressures fluctuations

2.3 The Patients Section Process for the Study

Randomized controlled trials (RCT) method was used to select patients of control and treatment groups. Figure 7 illustrates this method. The pretest–posttest experiments results were compared using paired t-test or chi-square test.

With the collaboration of RPOC staff, hypertensive outpatients with systolic pressure higher than 140 mmHg or diastolic pressure higher than 90 mmHg at the time of selection were invited to participate to the study. When the patients agreed, they signed a consent form. They were randomly selected into control or treatment group until each group reached 100 patients. Exclusion criteria were patients: with secondary hypertension; have just had a stroke; had a stroke 6 months ago; with

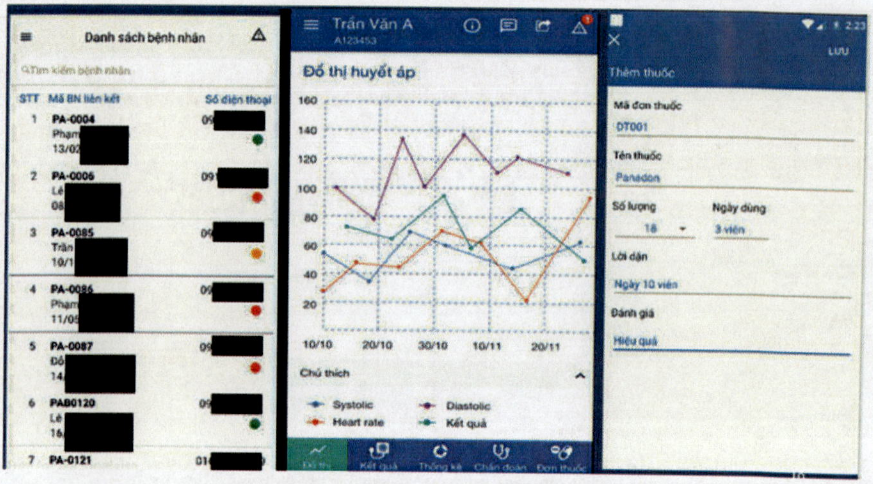

Fig. 6 Displays on the doctor smartphone of the information of their patients

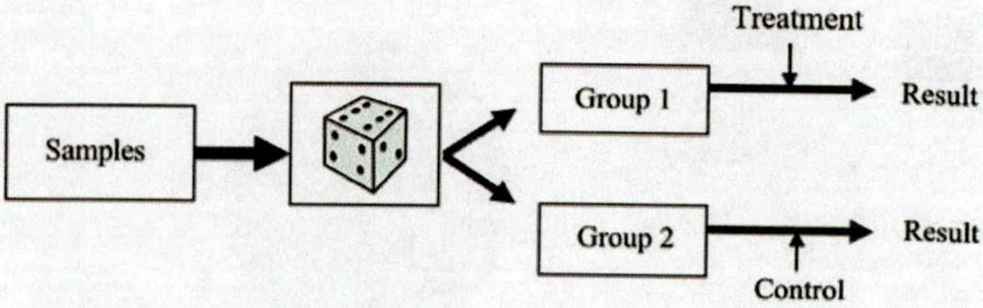

Fig. 7 Illustration of the randomized controlled trials (RCT) method [3]

chronic kidney disease on stages 4, 5; with dementia and need nursing care; and with arrhythmia. Other information collected, once as the baseline time, from the selected patients included:

- General characteristics:
  - Age
  - Gender
  - Weight
  - Height
  - Education (High school/University/Postgraduate)
  - Average income/month (<5 million, 5–10 million, > 10 million VND)
  - Family (Single / Living with spouse / Living with spouse and descendants)
  - Comorbidities
- Clinical characteristics:
  - Systolic pressure, diastolic blood pressure, heart rate

- Number of visits (excluding visits only to receive drugs under health insurance)
- Number of referrals
- Number of emergency transfers
- Number of hospitalizations
- Medications for hypertension.

2.4 *Clinical Monitoring Process*

The process consisted of 2 phases:

Phase 1: Pre-intervention period, lasting 6 months.

Participants in both groups were monitored and treated in the usual way at Binh Duong Province RPOC. This period is called O1 thereafter and used as a baseline of the study.

Phase 2: Intervention period, lasting 12 months.

At the end of O1, participants in the treatment group received our tele-blood pressure device and were in contact with their healthcare providers at will, at the regular schedule or based on the results obtained by the device. By the contrary, participants in the control group did not receive our devices and continued to use their owns, and communicated with the healthcare provider at a regular schedule or on-demand. The benefits of the two-phase study were that we could make a comparison between groups and within the treatment group. During this intervention period, we designated as O2: when the treatment group first received our Tele-Blood Pressure devices i.e. 6th month after O1; O3 and O4—12th month and 18th month respectively after O1 or 6th month and 12th month respectively after O2.

2.5 *Users Survey*

At the end of the study, a questionnaire was distributed to patients of the treatment group and healthcare providers. It aims to survey users opinions of the quality of the website, app on cell phone, blood pressure device and the cyber medical system as well as the patients' compliance.

3 Results

3.1 *Characteristics of Participants in 2 Groups*

The characteristics of participants in 2 groups, collected at O2 period, are indicated in Table 1. The last column indicates the *p*-value of the test (t-test or chi-square test) comparing the difference between control and treatment groups. It shows that

Table 1 The characteristics of participants in 2 groups

Feature	Control group		Treatment group		*p*-value
	Mean	SE	Mean	SE	
Age	64.29	0.80	63.22	0.71	0.315
% Male	78.0	–	73.7	–	0.482
% Female	22.0	–	26.3	–	
Weight (kg)	64.07	0.90	65.74	1.05	0.227
Height (cm)	161.78	0.61	161.26	0.62	0.552
BMI	24.45	0.28	25.18	0.33	0.091
Academic Level					
% High School	7.3	–	18.4	–	0.047
% College	6.2		2.0		
% Undergraduate	67.7		66.3		
% Graduate	17.7		10.2		
% Other	1.0		3.1		
Personal income (VNĐ)					
% < 5 million/month	3.1		2.1		0.166
% 5–10 million/month	73.5		62.2		
% > 10 million/month	23.4		35.7		
Number of people living in the patient house	3.94	0.18	4.01	0.19	0.800
Systolic blood pressure	125.68	1.36	128.48	1.40	0.154
Diastolic blood pressure	76.77	0.67	76.07	0.86	0.516
% Having hypertension	93.0	–	89.9	–	0.434

there is no statistically significant difference between the two groups ($p > 0.05$). This indicates that the randomization in the selection was successful. Note that most patients have both hypertension and type 2 diabetes.

3.2 *Comparison of the Results of the Two Research Groups at Each Study Time*

The patients in the two research groups were interviewed and collected their data at four times: O1—baseline was at the beginning of the study, O2—the 6th month when the treatment team started using Tele-Blood Pressure Devices, O3 and O4—months 12th and 18th from the beginning of the study.

The results in Table 2 show that the average patient's blood pressure gradually decreased during the study time. In the control group, the blood pressures at the times of O1, O2, O3, O4 were 126/77 mmHg, 122/75 mmHg, 122/75 mmHg, and

Table 2 Blood pressure of patients at the study time

		O1	O2	O3	O4	*p*-value O1 and O2	*p*-value O1 and O3	*p*-value O1 and O4
Systolic blood pressure								
Control	Mean	125.68	122.00	121.94	119.80	0.028*	0.009*	0.001*
	N	97	75	72	66			
	SE	1.36	1.21	1.35	1.08			
Treatment	Mean	128.48	125.86	123.54	122.59	0.063	0.002*	0.000*
	N	89	90	89	75			
	SE	1.40	1.35	1.08	1.15			
p-value treatment and control		0.154	0.035*	0.353	0.083			
Diastolic blood pressure								
Control	Mean	76.77	74.60	75.00	73.42	0.015*	0.069	0.003*
	N	96	75	72	66			
	SE	0.67	0.58	0.73	0.69			
Treatment	Mean	76.07	76.19	75.30	74.26	0.758	0.212	0.007*
	N	89	90	89	74			
	SE	0.86	0.74	0.68	0.57			
p-value treatment and control		0.516	0.091	0.762	0.353			

120/74 mmHg respectively. In the treatment group, the values were 128/76 mmHg, 126/76 mmHg, 124/75 mmHg, and 123/74 mmHg, respectively. This reduction of blood pressure has a statistically significant meaning (*). It showed that reasonable control of blood pressure in both groups had remained.

3.3 *Subclinical Results*

Subclinical results show that there was no difference between control group and treatment group at the beginning of the study. During the monitor time, there was a slight decrease in blood lipid indices (Cholesterol and LDL) in both groups. Other subclinical indicators did not change after the monitor time. Table 3 shows that, with some exceptions, there was no difference in subclinical results in the two study groups.

Table 3 Subclinical results of patients at the study time

		O1	O2	O3	O4	*p*-value O1 and 2	*p*-value O1 and 3	*p*-value O1 and 4
Blood sugar								
Control	Mean	6.48	6.12	6.67	6.53	0.158	0.187	0.447
	N	93	81	77	65			
	SE	0.21	0.16	0.20	0.21			
Treatment	Mean	6.38	6.40	6.38	6.40	0.854	0.433	0.680
	N	91	91	90	78			
	SE	0.18	0.14	0.15	0.18			
p-value treatment and control		0.723	0.214	0.233	0.644			
Cholesterol								
Control	Mean	4.68	4.49	4.65	4.49	0.144	0.644	0.009*
	N	93	81	77	65			
	SE	0.14	0.16	0.15	0.15			
Case	Mean	4.74	4.65	4.44	4.47	0.682	0.014*	0.017*
	N	91	91	90	78			
	SE	0.14	0.15	0.12	0.11			
p-value treatment and control		0.770	0.472	0.226	0.903			
LDL								
Control	Mean	2.63	2.59	2.66	2.31	0.767	0.717	0.035*
	N	90	81	76	65			
	SE	0.13	0.14	0.14	0.14			
Treatment	Mean	2.60	2.66	2.34	2.36	0.692	0.048*	0.316
	N	90	86	87	75			
	SE	0.13	0.14	0.12	0.14			
p-value treatment and control		0.846	0.724	0.087	0.802			
HDL								
Control	Mean	1.50	1.50	1.63	1.66	0.979	0.246	0.233
	N	91	79	76	64			
	SE	0.09	0.08	0.11	0.14			
Treatment	Mean	1.54	1.54	1.62	1.52	0.652	0.653	0.557
	N	88	89	90	76			
	SE	0.09	0.09	0.10	0.08			
p-value treatment and control		0.768	0.729	0.923	0.415			

(continued)

Table 3 (continued)

		O1	O2	O3	O4	*p*-value O1 and 2	*p*-value O1 and 3	*p*-value O1 and 4
Triglyceride								
Control	Mean	2.13	1.93	2.20	1.88	0.787	0.066	0.444
	N	92	81	77	64			
	SE	0.12	0.13	0.13	0.11			
Treatment	Mean	2.39	3.32	2.21	2.15	0.312	0.167	0.057
	N	90	91	90	75			
	SE	0.16	0.94	0.12	0.16			
p-value treatment and control		0.181	0.165	0.953	0.177			
Acid uric								
Control	Mean	393.24	394.16	405.28	394.70	0.637	0.942	0.907
	N	85	70	62	34			
	SE	10.37	13.26	12.06	14.68			
Treatment	Mean	397.08	410.81	375.06	385.86	0.250	0.231	0.468
	N	78	71	77	50			
	SE	11.27	13.71	10.75	16.82			
p-value treatment and control		0.802	0.384	0.063	0.711			
SGOT								
Control	Mean	28.44	27.28	27.54	27.17	0.946	0.690	0.396
	N	92	81	77	66			
	SE	1.18	1.51	1.22	1.39			
Treatment	Mean	29.75	29.36	30.63	29.72	0.771	0.742	0.867
	N	90	91	90	77			
	SE	1.36	1.21	1.49	2.05			
p-value treatment and control		0.468	0.278	0.119	0.321			
SGPT								
Control	Mean	31.47	27.38	29.39	28.70	0.249	0.541	0.257
	N	92	81	77	66			
	SE	2.07	1.69	1.73	1.98			
Treatment	Mean	30.47	30.65	32.67	29.98	0.869	0.657	0.986
	N	89	91	90	77			
	SE	1.72	1.96	1.80	1.70			
p-value treatement and control		0.712	0.214	0.195	0.621			

3.4 Evaluate the Effectiveness of Health Care in the Performance of the Telemedicine System

A. Disease management status (compliance of treatment, monitor of blood pressure)

Regarding treatment compliance in using of hypertension medication, as shown in Fig. 8, most patients self-assessed as taking enough medicine. Besides, they have never forgotten or rarely forgot (more 80% in both groups). The proportion of patients who sometimes forgot to take the drug in the control group was higher than the treatment group at the study time (18.8% compared to 9.4%). However, this difference is not statistically significant ($p = 0.088$).

After 12 months, when the patient had been using Tele-Blood pressure devices for six months, the control group had one patient with improvement and one patient with worse conditions. While the treatment group had four patients with improvement and two patients with no improvement, the difference of treatment compliance between O1 and O3 period was also not statistically significant in both groups ($p = 0.251$ and 0.096, respectively). The blood pressure monitored at the home of patients was poorer, however. 81% of patients in both groups failed to measure their blood pressure at least once a day, recommended at O1. In the control group, 16.8% of patients never measured blood pressure, and 64.2% of patients only measured when they felt ill. In the treatment group, 14.6% of patients never measured blood pressure, and 67.7% of patients only measured when they were exhausted. This difference between the two groups was not statistically significant at the time of O1 ($p = 0.519$). However, at the time of O3 and O4 (as shown in Fig. 9), there was a statistically significant difference between the control group and treatment group ($p = 0.000$).

The utilization of tele-blood pressure devices is expected to help patients improve their habits of measuring blood pressure at home. The results showed that the proportion of patients who did not measure their blood pressure at least once a day in the treatment group decreased from 82.3 to 49.5% at 12th month of O3 (meaning 6 months after using the devices). Meanwhile, the percentage in the control group increased slightly from 81 to 84.9%. The treatment group had a minimum rate of blood pressure measurement at home once daily had risen from 17.7 to 50.5%. This improvement was statistically significant (p-value of 0.000). At 18th month O4 (i.e. 12 months after using the devices), the rate of not measuring blood pressure at least once a day in the control group was still high at 81.5%, while in the treatment group was only 29.9%. The rate of blood pressure measurement at home once a day in the treatment group was 70.1% in comparison with 19.5% in the control group. The improvement in the treatment group was statistically significant between O3 and O4 ($p = 0.004$).

B. The situation of using Tele-Blood Pressure Devices in the treatment group

To objectively assess the effectiveness of Tele-blood pressure devices in improving the patient's blood pressure measurement habits, and using the iHIS website which

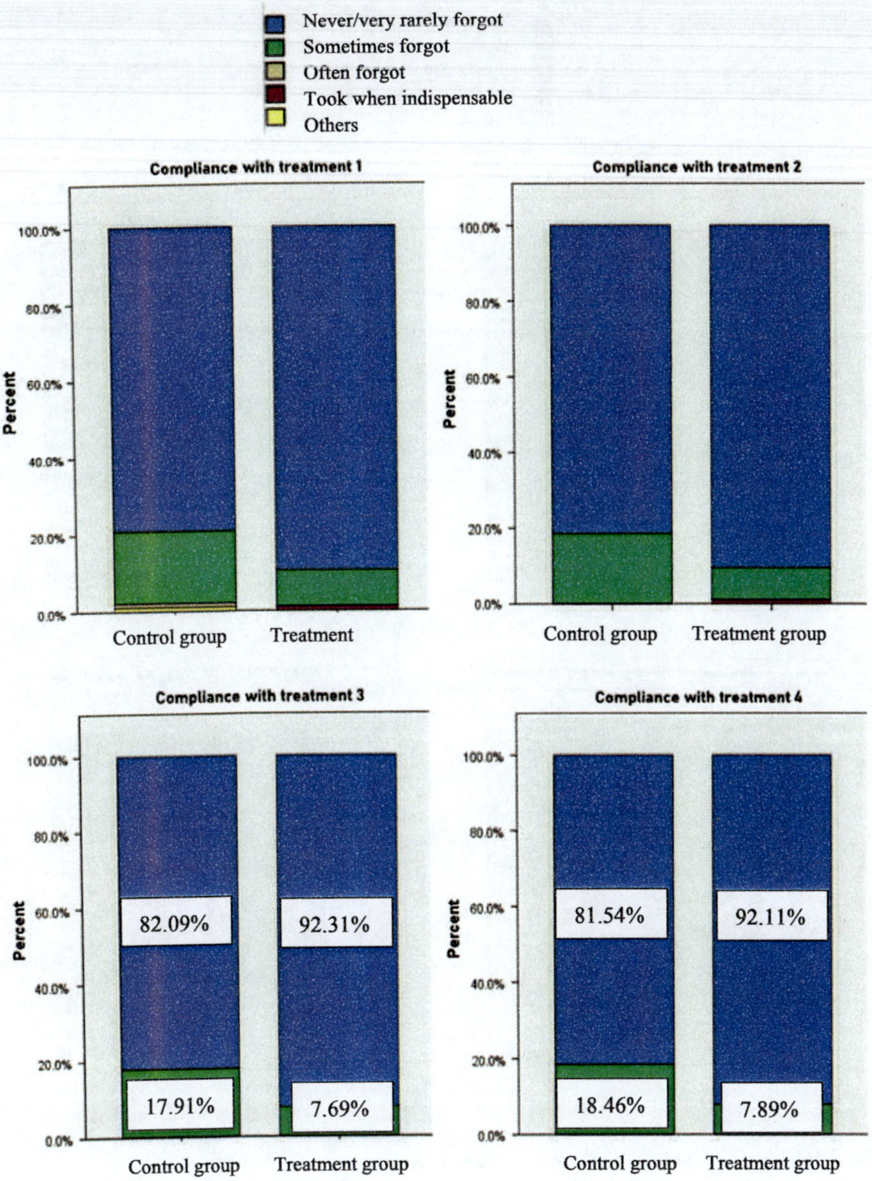

Fig. 8 Compliance to the treatment in the data record

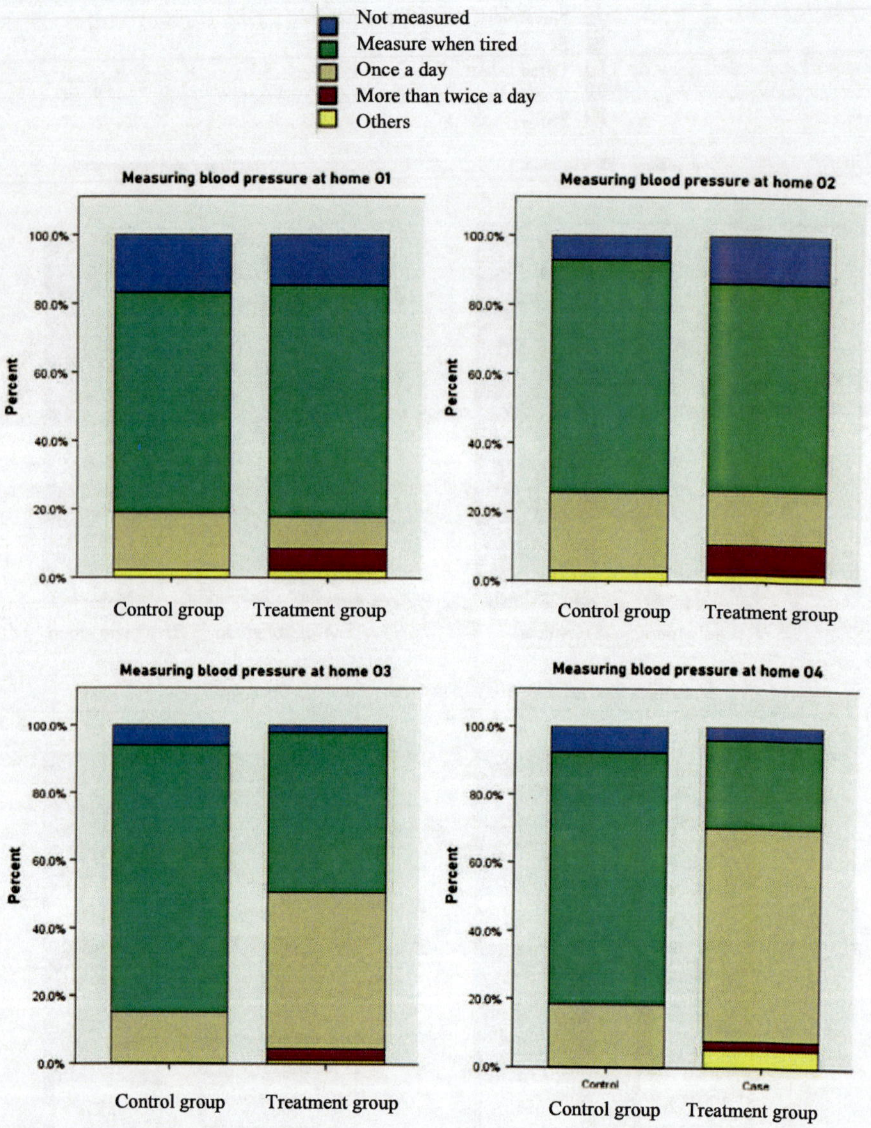

Fig. 9 Measuring the blood pressure of patients at home

stores all measurement results from the patient, patients' use monitoring of the treatment group is necessary. The results from Table 4 show that the frequency of using Tele-Blood Pressure Devices to measure at home is not too much. On average, about 4–5 times per month (average of 4.69, SE 0.68), Half of the patients only use the

Table 4 Frequency of using Tele-blood pressure devices

Number of measurements using Tele Blood Pressure device	Monthly	Before months	After 6 months	*p*-value of comparison before and after 6 months
Mean	5.65	5.87	3.85	0.036
SD	6.43	6.62	5.88	
SE	0.65	0.66	0.72	
Median	3.11	3.17	0.33	
Min	0.00	0.00	0.00	
Max	23.50	23.50	22.67	

device about three times a month (median 3.11). 2 patients received the device but not using it at all. Only 5 patients almost regularly measured daily (over 20 times/month).

To evaluate the maintenance of using Tele-Blood Pressure Devices, we compared the usage situation between the first 6 months and the next 6 months as in the last 3 columns in Table 4. The results show a gradual decrease in the use of the devices: monthly average measurements decreased from 5.87 to 3.85. This difference is statistically significant ($p = 0.036$).

C. **Effectiveness of the telemedicine system**

According to the survey results presented in Table 5, 97.1% of patients said that the telemedicine system could help to prevent hypertension complications, and 80.3% stated that the new treatment did not increase the cost. Most (87.5%) patients felt that their blood pressure was better controlled, and 81.7% confirmed that they would recommend this device for others to use.

4 Conclusions

We developed an IoT system that consisted of a Tele-Blood Pressure device, a software program for servers and an application for smartphones to nurture patients from distance and immediately. To evaluate its efficiency, in this project, we conducted a pilot study at Binh Duong province RPOC to monitor its outpatients suffered from high blood pressure. Using randomized controlled trials (RCT), we selected 100 patients for the control group and another 100 patients for the treatment group. Both groups were taken care by 3 medical doctors and 4 nurses of RPOC. Each patient of the treatment group received our Tele-Blood Pressure device and involved in our cyber medical system deployment. The study was performed during 18 months, and at the end of the study a survey was conducted to all users: participants of the treatment groups, medical doctors and nurses. The technical results showed that the Tele-blood pressure, the CTIS and the cyber medical system CTIS performed well, accurately and reliably. The survey results showed that all users gave a high mark to

Table 5 Patients' assessment of efficiency of the telemedicine system

		Frequency	Percent	Valid Percent	Cumulative Percent
Preventing hypertension complications					
Valid	Useless	2	2.7	2.9	2.9
	Effective	67	90.5	97.1	100.0
	Total	69	93.2	100.0	
Missing	System	5	6.8		
Total		74	100.0		
Expenses for monitoring and curing of hypertension					
Valid	Less	12	16.2	16.9	16.9
	Unchanged	57	77.0	80.3	97.2
	More	2	2.7	2.8	100.0
	Total	71	95.9	100.0	
Missing	System	3	4.1		
Total		74	100.0		
A feeling of blood pressure under controlled					
Valid	No	9	12.2	12.5	12.5
	Yes	63	85.1	87.5	100.0
	Total	72	97.3	100.0	
Missing	System	2	2.7		
Total		74	100.0		
Recommending for other people to use					
Valid	No	13	17.6	18.3	18.3
	Yes	58	78.4	81.7	100.0
	Total	71	95.9	100.0	
Missing	System	3	4.1		
Total		74	100.0		

the performance of the device as well as the telemedicine method. Evidence of the effectiveness of the telemedicine indicates that the new technique is user-friendly. This study will be extended at a larger scale, especially in the remote areas where the healthcare needs to be improved.

Acknowledgements This research is funded by the grant 02/HĐ –SKHCN from the Department of Science and Technology of Bình Dương province and by the grant NCM2020-28-01 from the Vietnam National Universities-HCM City. We would like to express our gratitude to the collaborations of Dr. Đoàn Thị Hồng Thơm, the healthcare staff and patients of the Ban Chăm sóc và Bảo vệ Sức khỏe Cán bộ Trung Cao tỉnh Bình Dương, and all staff and students of the School of Biomedical Engineering of International University of Vietnam National Universities-HCM. Conflicts of Interest: The authors have no conflict of interest to declare.

References

1. Combi C, Pozzani G, Pozzi G (2016) Telemedicine for developing countries. a survey and some design issues. Appl Clin Inform 7(4):1025–1050
2. Viet TN et al (2020) Implementation of a pilot cyber medical system with blood pressure devices for telemedicine in Binh Duong Province (Vietnam). In: 7th international conference on the development of biomedical engineering in Vietnam (BME7). Springer, Singapore
3. Kendall JM (2003) Designing a research project: randomised controlled trials and their principles. Emerg Med J 20(2):164

Pharmaceutical Engineering and Nanomaterials in Medical Applications

Improvement of Curcumin Solubility and Bioactivities by the Designed Redox Nanoparticles

Khoa Minh Le, Vinh Dinh-Xuan Nguyen, Anh Duc-The Doan, Nguyen Van Tien Dat, Nguyen Thi Thu Ha, Nhu-Thuy Trinh, Vo Van Toi, and Vong Binh Long

Abstract Chronic inflammation plays a central role in challenging ailments, from cardiovascular diseases to cancer. The connection has been made in progressed inflammation and the overproduction of reactive oxygen species (ROS), or oxidative stress, through various inflammatory pathways. Curcumin is well-recognized as a natural supplement with diverse pharmacological effects, especially antioxidant and anti-inflammatory activities. However, curcumin's extremely poor bioavailability and rapid oxidation by ROS has hindered promising clinical applications. Previously, silica-containing redox nanoparticles (siRNP) was developed and prepared from amphiphilic block copolymer possessing drug absorptive silica moiety and ROS-scavenging nitroxide radical moiety in the hydrophobic segment. In this research, curcumin loaded siRNP (CUR@siRNP) with the size of about 120 nm in diameter were employed as an efficient nanocarrier for curcumin via dialysis method. After encapsulation, CUR@siRNP showed a great improvement in solubility while drug release assay in varying pH demonstrated a slightly pH-sensitive property of siRNP in the gastric medium. The CUR@siRNP also showed an enhancement in antioxidant activity, which was confirmed through 2,2-diphenyl-1-picrylhydrazyl assay. The anti-inflammatory effect was also evaluated in vitro using albumin denaturation inhibition assay. The CUR@siRNP combination showed the potential to be further investigated and developed for a novel antioxidant and anti-inflammatory nano-formulation in treatment of chronic inflammation.

Keywords Inflammation · Antioxidant · Redox nanoparticles · Curcumin · Reactive oxygen species

K. M. Le · V. D.-X. Nguyen · A. D.-T. Doan · N. Van Tien Dat · N. T. T. Ha · N.-T. Trinh · V. Van Toi · V. B. Long (✉)
School of Biomedical Engineering, International University, 700000 Ho Chi Minh, Vietnam
e-mail: vblong@hcmiu.edu.vn

Vietnam National University Ho Chi Minh City (VNU-HCM), 700000 Ho Chi Minh, Vietnam

V. Van Toi et al. (eds.), *8th International Conference on the Development of Biomedical Engineering in Vietnam*, IFMBE Proceedings 85,
https://doi.org/10.1007/978-3-030-75506-5_22

1 Introduction

According to the World Health Organization (WHO), chronic diseases, including chronic inflammation, is well-recognized as one of the greatest threats to patients. Specifically, chronic inflammation plays a central role in some of the most challenging diseases, including cancer, heart disease, diabetes and even Alzheimer's. In developed countries, such as the United States, 60% of American had at least one chronic condition, while 42% had more than one (2000, Rand Corporation) [1]. As for developing countries, Vietnam is undergoing a rapid epidemiological transition of chronic diseases. In 2002, 66% of all deaths were recorded due to chronic diseases, which was the leading cause [2]. Reactive oxygen species (ROS) has shown active roles in several pathways among different inflammatory molecules. However, overproduction of ROS as a result of environmental stress (e.g., UV, toxin or heat exposure), leads to an oxidative stress imbalance that originates toxic effects through the development of peroxides and free radicals, which directly and adversely affect cell structures, including DNA, proteins and lipids [3]. It has been reported that oxidative stress accumulatively induces chronic inflammation as both signaling molecule and inflammatory mediator. Consequently, these processes not only lead to endothelial dysfunction and tissue injury but also internal damages and even apoptosis [4].

On the other hand, curcumin ($C_{21}H_{20}O_6$) is a polyphenolic compound derived from the rhizome of turmeric—a traditional herb and spice common around Southeast Asia. In constituents' use, curcumin accounts for 77% of the product, besides other curcuminoids being demethoxycurcumin and bis-demethoxycurcumin. Curcumin has diverse pharmacological effects, including anti-inflammatory, antioxidant and antiangiogenic activities [5]. Previously shown to be a highly pleiotropic molecule, curcumin displays interactions with myriad inflammatory responses targets. Most notably, curcumin has demonstrated inhibition of the pro-inflammatory pathways associated with chronic diseases by blocking the development of tumor necrosis factor (TNF) and TNF-mediated cell signaling [6]. In addition, curcumin can reduce the secretion of several interleukins, induce apoptosis and inhibit the activity of protein kinase C, epidermal growth factor receptor tyrosine kinase in the inflammation site via the nuclear factor kappa B (NF-κB) pathway [7]. Despite the admirable therapeutic effects of curcumin, many of its drawbacks have inhibited the use of curcumin in clinical applications. Curcumin's poor water solubility, slow absorption, rapid metabolism, and low biological half-life were connected to its extremely low bioavailability [6], which lead to poor bioactivity. Previously, adjuvants such as piperine, quercetin, or silibinin were used to enhance cell absorption; however, they exhibited significant toxicity in vivo research [8, 9]. As opposed to the mentioned approaches, the application of nanoparticle-based drug delivery systems in the situation displayed distinguished advancement in targeting effect and better solubility of the hydrophobic drugs. Some studies such as Thercurmin and formulated curcumin loaded solid lipid nanoparticles had been employed as nanocarriers for curcumin. Yet, they have been found to shared similar drawbacks: low stability in the gastrointestinal environment, low loading capacity, and ROS-induced degradation [10, 11].

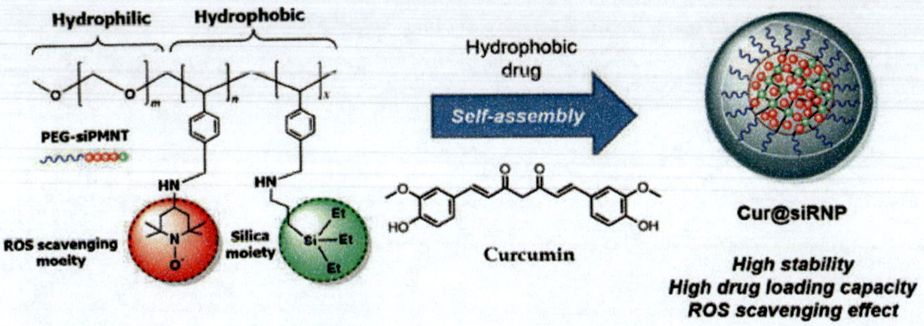

Fig. 1 Design of silica-containing redox nanoparticle (siRNP) [12]. Curcumin is encapsulated into the nanocarrier by self-assembly of hydrophilic PEG segment and hydrophobic segment with ROS scavenging feature

Previously, silica-containing redox nanoparticle (siRNP) (Fig. 1) was developed, featuring a ROS scavenging fragment of nitroxide radical and an absorptive drug fragment of silica that can optimize drug loading capability [12, 13]. siRNP should be potential nanocarrier of hydrophobic drugs such as curcumin for improvement of therapeutic application. In this research, the synthesized curcumin-loaded siRNP (CUR@siRNP) characterization was investigated, along with bioactivities such as antioxidant and anti-inflammatory capacity.

2 Methods

2.1 *Synthesis of PEG-siPMNT*

The synthesis of amphiphilic redox polymer (PEG-siPMNT) was described previously [13] as shown in Fig. 2. Firstly, PEG-*b*-chain transfer agent (PEG-CTA) was synthesized by using $C_6H_5CS_2MgBr$ reactant. Then, PEG-CTA was polymerized with chloromethyl styrene (CMS) with azobisisobutyronitrile (AIBN), a free radical initiator, at 60 °C for 24 h exposed to an atmosphere of N_2. Finally, the chloromethyl groups of PEG-PCMS were converted to a nitroxide radical moiety and silica moiety by an amination reaction with the NH_2-TEMPO radical and 3-aminopropyl)triethoxysilane (APTES), respectively.

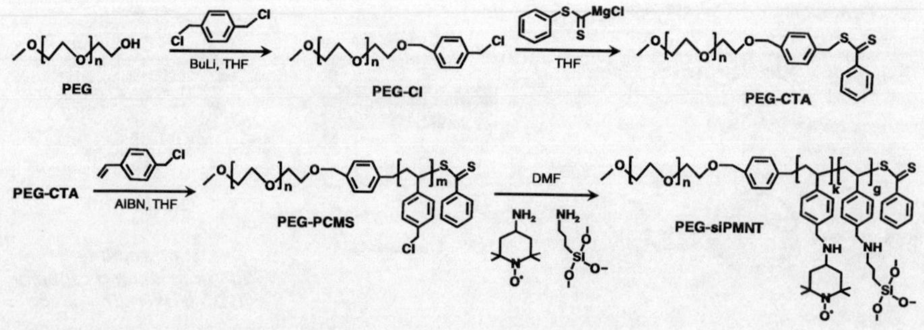

Fig. 2 The synthesis of the amphiphilic redox polymer PEG-siPMNT from polyethylene(glycol) (PEG) [12]

2.2 *Synthesis of siRNP and Curcumin Encapsulated siRNP (CUR@siRNP)*

siRNP was synthesized through dialysis of PEG-siPMNT in dimethylformamide (DMF) against water. Hydrophilic PEG segment facing outward while covering hydrophobic PMNT segment with nitroxide group and silica group within its core. Curcumin was encapsulated via the hydrophobic interaction and absorption in the silica core in of siRNP [12].

2.3 *Characterization of siRNP*

Encapsulation Efficiency (EE) and Loading Capacity (LC)
Polymers (PEG-siPMNT) was mixed with curcumin in dimethylformamide solution, followed by 24-h dialysis (3500 molecular weight cutoff; Spectra/Por dialysis membrane) to yield into CUR@siRNP. Pure curcumin was diluted in dimethyl sulfoxide into a range of concentrations of 0, 0.2, 0.4, 0.6, 0.8, 1 (μg/mL) for the standard curve. Encapsulation efficacy (EE) and loading capacity (LC) were calculated based on curcumin concentration measured within siRNP at the absorbance of 450 nm by the formula:

$$\text{EE}\ \% = \frac{\text{Total mass of drug encapsulated}}{\text{Initial drug mass}} \times 100\%$$

$$\text{LC}\ \% = \frac{\text{Total mass of drug encapsulated}}{\text{Initial polymer mass}} \times 100\%$$

The Size of siRNP and CUR@siRNP

Dynamic light scattering (DLS) was used to determine the size of siRNP and CUR@siRNP with an angle of 173° using Zetasizer ZS (Malvern, UK).

Drug Release Capacity

To investigate the ability of nanoparticles to release in different pH environments (pH 2.5 corresponds to the gastric environment and pH 7.4 corresponds to the physiological environment) using dialysis method was widely used for simplicity and to avoid drug separation from nanoparticles. Amount of curcumin released was determined by measuring the absorbance of 450 nm.

2.4 *Investigation of Nanoparticle Bioactivities*

The Antioxidant Assay

The antioxidant effect of curcumin and CUR@siRNP was investigated using 1,1diphenyl-2-picrylhydrazyl (DPPH, Fujifilm Wako Chemicals) assay. A reaction mixture consisting of DPPH (40 ug/ml) and various concentrations of curcumin in dimethyl sulfoxide, CUR@siRNP, siRNP (20, 50, and 80 μg/mL) was incubated at room temperature in dark condition for 30 min. A similar volume of distilled water served as control. At the same time, vitamin C at concentrations of 20, 50, and 80 μg/mL was used as a reference substance and treated similarly for determination of absorbance. The antioxidant effect was calculated using the following equation:

$$\%\ \text{Antioxidant effect} = \left(\frac{\text{OD}_{\text{test}} - \text{OD}_{\text{control}}}{\text{OD}_{\text{control}}}\right) \times 100$$

whereas the OD is the optical density of tested samples measured at 517 nm. Measurements are repeated in triplicate.

The Anti-inflammatory Activity In Vitro

Anti-inflammatory effect of curcumin and CUR@siRNP was studied using albumin denaturation inhibition assay [14]. Briefly, the mixture containing albumin (from fresh hen's egg) in distilled water, and samples (curcumin and CUR@siRNP) in varying concentration (20, 50, 80, and 100 μg/mL) was incubated at 37 °C for 15 min and then heated at 70 °C for 5 min in a thermal shaker. Ibuprofen (2-(4-isobutylphenyl) propanoic acid) was employed as a positive control of the anti-inflammatory drug. After cooling, the absorbances were measured at 660 nm for turbidity. The percentage of anti-inflammatory effect was calculated using the following equation:

$$\%\ \text{Antiinflammatory effect} = \left(\frac{\text{OD}_{\text{test}} - \text{OD}_{\text{control}}}{\text{OD}_{\text{control}}}\right) \times 100$$

whereas the OD is the optical density of tested samples measured at 660 nm. Measurements are repeated in triplicate.

2.5 *Statistical Analysis*

All experiments were conducted at least thrice, and the data represent the mean ± SEM. The results were analyzed and considered statistically significant differences when $p < 0.05$ using a two-tailed *t*-test.

3 Results and Discussion

4 Improvement in the Solubility of Curcumin

It is known that curcumin is water-insoluble, resulting in its low bioavailability. Therefore, improvement of the water solubility of curcumin can increase its bioavailability and bioactivities. As shown in Fig. 3a, free curcumin remained insoluble in distilled water, evidenced by small particles accumulated at the bottom of the vial. While CUR@siRNP resulted in a relatively dark brown solution and the solution was transparent (Fig. 3b), indicating the improvement of water solubility of curcumin. This data suggests that siRNP has the potential to increase the bioavailability of curcumin by overcoming the drug's low solubility.

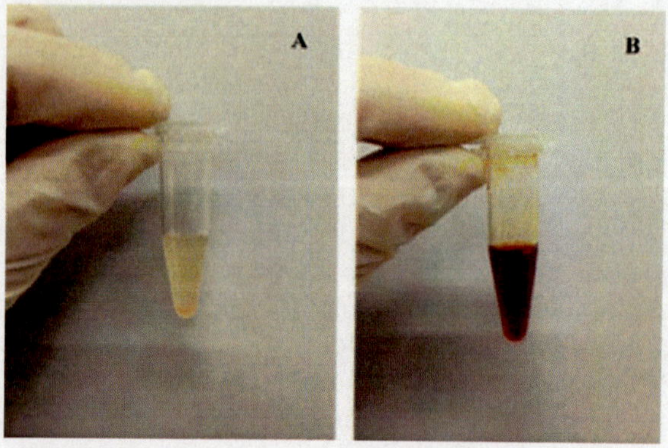

Fig. 3 The solubility of 1 mg/mL curcumin (**a**) and CUR@siRNP (**b**) in distilled water. The red circle indicates the insoluble curcumin

4.1 Characterization of siRNP

The DLS measured the results of the nanoparticle size with the assessment of siRNP before and after encapsulating curcumin. As shown in Fig. 4a and Table 1, it can be seen that the size of siRNP slightly increased from 121.2 ± 3.247 nm (siRNP) to 156.1 ± 2.563 nm (CUR@siRNP), which suggested that curcumin was encapsulated into the core of siRNP. In addition, monodispersity was observed with low PDI value (<0.3), suggesting the narrow size distribution of siRNP and CUR@siRNP. This data is reported as suitable nanosize for application in the treatment of inflammation and cancer via the improvement of enhanced permeability and retention effect. We next evaluated the encapsulation efficacy (EE) and drug loading capacity (LC) of CUR@siRNP. As shown in Table 1, EE and LC were 78% and 7.8%, respectively. As compared to the conventional polymeric nanoparticle, siRNP exhibited higher EE and LC due to the drug absorption silica core. Next, the release of CUR@siRNP was tested in the phosphate buffer with pH 2.5 and pH 7.4 by dialysis method. As shown in Fig. 4b, the release of CUR@siRNP was more rapid at pH 2.5 (≈50% released after 4 h) compared to pH 7.4 (≈20% released after 4 h). This data suggested the pH-sensitive character of siRNP despite the cross-liking of silica moiety in the core. The protonation of the amino group occurs partially under acidic environment, resulting

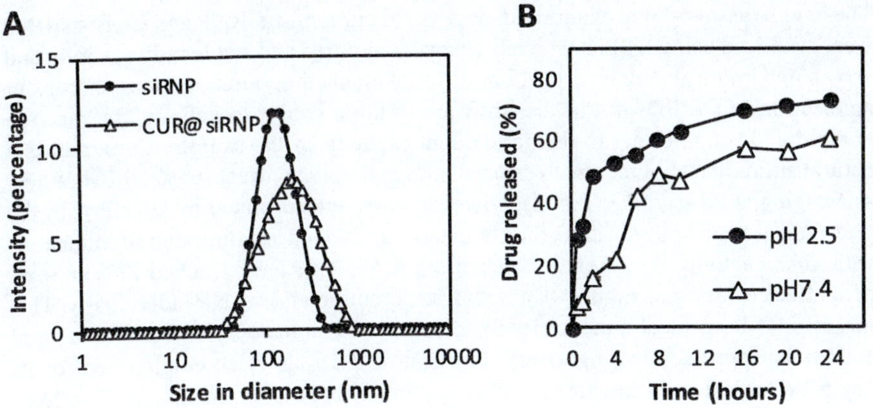

Fig. 4 Characterization of CUR@siRNP. **a** The size of siRNP and CUR@siRNP. **b** The release profile of CUR@siRNP at varying pH 2.5 and 7.4

Table 1 Summary of the size of siRNP and CUR@siRNP, and encapsulation efficacy (EE) and drug loading capacity (LC) of CUR@siRNP

	Size (d.nm)	PDI	EE (%)	LC (%)
CUR@siRNP	156.1 ± 2.563	0.283	78%	7.8%
siRNP	121.2 ± 3.247	0.253	–	–

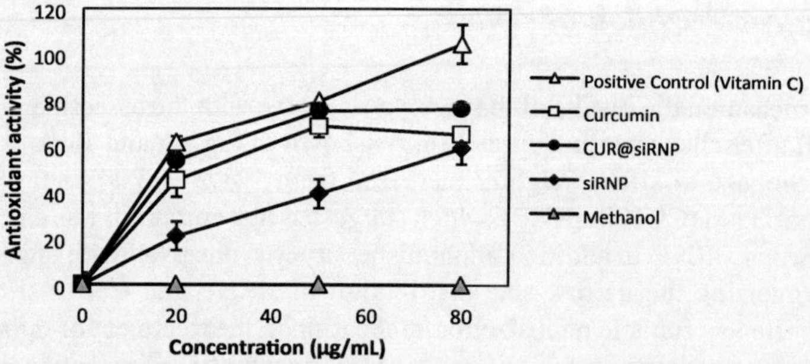

Fig. 5 The antioxidant activity of curcumin and CUR@siRNP using DPPH assay (*$p < 0.05$ and ***$p < 0.01$ by Student's t-test)

in higher drug release rate. The drug release also depends on nature character of drug that affects the absorption and detachment of drug and silica surface.

4.2 The Bioactivities of Curcumin and CUR@siRNP

The in vitro assay results of antioxidant effect of curcumin, siRNP and CUR@siRNP were evaluated using DPPH, which is a purple-coloured and stable radical cation and shown maximum absorbance at 517 nm. Antioxidant compounds, which can transfer an electron to DMPD• +, hence, shift the solution colour to yellow. The reaction is rapid and proportional to the antioxidant capacity of the sample. Curcumin is a natural antioxidant agent, which exerted high antioxidant effect (65% DPPH radical scavenging at 80 µg/mL) (Fig. 5); however, when encapsulated by siRNP with the free-radical scavenging feature, they both have developed an admirable simultaneous antioxidant activity. At the same concentration, CUR@siRNP reached 77% antioxidant effect, which was higher than both free curcumin and siRNP individually. This makes siRNP an ideal nanocarrier for curcumin by enhancing curcumin's natural antioxidant profile. Also, the overproduction of ROS has been emphasized for its key pathological roles in chronic inflammation. Thus, the improvement of antioxidant efficiency via the combination of curcumin and siRNP held a great promise for a potential anti-inflammatory therapy.

Curcumin and CUR@siRNP anti-inflammatory capacity were next assessed based on the inhibition of egg's albumin denaturation in vitro illustrated in Fig. 6. Protein denaturation has been well documented as a direct correlation to the occurrence of inflammation and tissue damage [15]. Hence, the ability to protect the egg's albumin from thermal-induced coagulation signifies the apparent potential for anti-inflammatory effect. With the increase in sample concentrations, the anti-inflammatory effect raised correspondently; however, the anti-inflammatory effect of curcumin remains approximately constant at higher concentrations (50, 80 and 100 µg/mL). Interestingly, CUR@siRNP demonstrated significantly higher

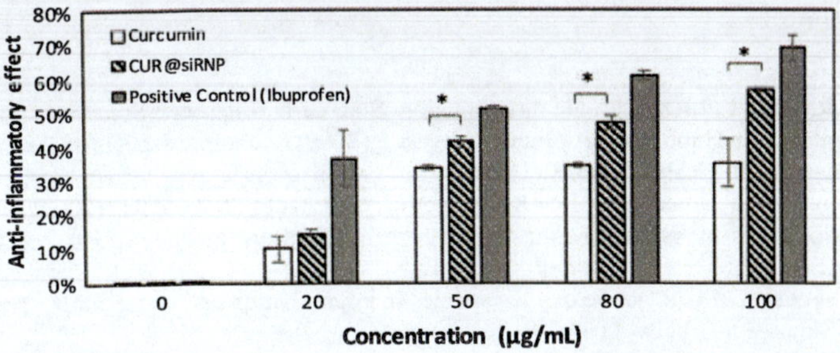

Fig. 6 The anti-inflammatory effect of curcumin and CUR@siRNP using albumin denaturation inhibition assay [14]. Ibuprofen (2-(4-Isobutylphenyl)propanoic acid) was used as reference (*$p <$ 0.05)

efficiency (57%) than curcumin sample (35%) at high concentrations (50, 80 and 100 µg/mL). This should be understood that the encapsulation of curcumin into siRNP has greatly improved curcumin anti-inflammatory profile, enhancing the valuable therapeutic effect.

Clinically, the majority of anti-inflammatory and pain-relief agents are non-steroid inflammatory drugs (NSAIDs) since they inhibit protein denaturation [16]. However, they also exhibit side effects on the gastrointestinal tract, and prolongation leads to gastric ulcers and cardiovascular complications [17–19]. Interestingly, turmeric (containing curcumin) had been well recognized as a medical herb for its anti-inflammatory properties. With siRNP as a novel carrier, CUR@siRNP can be a potential candidate for a safe and low-risk anti-inflammatory drug, with enhanced anti-inflammatory effect.

5 Conclusion

In this research, siRNP nanocarrier improved the stability of the entrapped curcumin for oral drug delivery to treat chronic inflammation. The results have evaluated CUR@siRNP basic yet fundamental characteristics and observed the improvement in its solubility, bioactivity (antioxidant and anti-inflammatory), hence improvement in bioavailability. Further investigations, including pharmacokinetic study, antiinflammatory effect in vitro cell and in vivo model will be performed to demonstrate the application potential of CUR@siRNP in treatment of chronic inflammation.

Acknowledgements This work was supported by the Vietnam National University Ho Chi Minh city (VNU-HCM) under grant number C2019-18-24. This was also partly supported by International University under grant number SV2019-BME08.

Conflicts of Interest The authors have no conflict of interest to declare.

References

1. Pahwa R, Jialal I (2018) Chronic inflammation. StatPearls [Internet]: StatPearls Publishing
2. Van Minh H, Lan Huong D, Bao Giang K, Byass P (2009) Economic aspects of chronic diseases in Vietnam. Glob Health Action 2(1):1965
3. Devasagayam TP, Tilak JC, Boloor KK, Sane KS, Ghaskadbi SS, Lele RD (2004) Free radicals and antioxidants in human health: current status and future prospects. J Assoc Phys India 52:794–804
4. Mittal M et al (2014) Reactive oxygen species in inflammation and tissue injury. Antioxidants Redox Signal 20(7):1126–1167
5. Anand P, Kunnumakkara AB, Newman RA, Aggarwal BB (2007) Bioavailability of curcumin: problems and promises. Mol Pharm 4(6):807–818
6. He Y, Yue Y, Zheng X, Zhang K, Chen S, Du Z (2015) Curcumin, inflammation, and chronic diseases: how are they linked? Molecules 20(5):9183–9213
7. Giordano S, Darley-Usmar V, Zhang J (2014) Autophagy as an essential cellular antioxidant pathway in neurodegenerative disease. Redox Biol 2:82–90
8. Kakkar V, Singh S, Singla D, Sahwney S, Chauhan AS, Singh G et al (2010) Pharmacokinetic applicability of a validated liquid chromatography tandem mass spectroscopy method for orally administered curcumin loaded solid lipid nanoparticles to rats. J Chromatogr B 878(32):3427–3431
9. Kanai M, Yoshimura K, Asada M, Imaizumi A, Suzuki C, Matsumoto S et al (2011) A phase I/II study of gemcitabine-based chemotherapy plus curcumin for patients with gemcitabineresistant pancreatic cancer. Cancer Chemother Pharmacol 68(1):157–164
10. Moorthi C, Kathiresan K (2013) Curcumin–Piperine/Curcumin–Quercetin/Curcumin–Silibinin dual drug-loaded nanoparticulate combination therapy: a novel approach to target and treat multidrug-resistant cancers. J Med Hypotheses Ideas 7(1):15–20
11. Mulik RS, Mönkkönen J, Juvonen RO, Mahadik KR, Paradkar AR (2010) Transferrin mediated solid lipid nanoparticles containing curcumin: enhanced in vitro anticancer activity by induction of apoptosis. Int J Pharm 398(1–2):190–203
12. Vong LB, Kimura S, Nagasaki Y (2017) Newly Designed silica-containing redox nanoparticles for oral delivery of novel TOP2 catalytic inhibitor for treating colon cancer. Adv Healthcare Mater 6(20):1700428
13. Vong LB, Tomita T, Yoshitomi T, Matsui H, Nagasaki Y (2012) An orally administered redox nanoparticle that accumulates in the colonic mucosa and reduces colitis in mice. Gastroenterology 143(4):1027–1036
14. Mizushima Y, Kobayashi M (1968) Interaction of anti-inflammatory drugs with serum proteins, especially with some biologically active proteins. J Pharm Pharmacol 20(3):169–173
15. Opie EL (1962) On the relation of necrosis and inflammation to denaturation of proteins. J Experim Med 115(3):597
16. Saso L, Valentini G, Casini ML, Grippa E, Gatto MT, Leone MG, Silvestrini B (2001) Inhibition of heat-induced denaturation of albumin by nonsteroidal antiinflammatory drugs (NSAIDs): pharmacological implications. Arch Pharmacal Res 24(2):150–158
17. Arroyo M, Lanas A (2006) NSAIDs-induced gastrointestinal damage. Review. Minerva Gastroenterol Dietol 52:249–259 [PubMed] [Google Scholar]
18. Lanas A, Ferrandez A (2006) NSAID-induced gastrointestinal damage: current clinical management and recommendations for prevention. Chin J Dig Dis 7:127–133 [PubMed] [Google Scholar]
19. Sostres C, Gargallo CJ, Arroyo MT, Lanas A (2010) Adverse effects of non-steroidal antiinflammatory drugs (NSAIDs, aspirin and coxibs) on upper gastrointestinal tract. Best Pract Res Clin Gastroenterol 24:121–132 [PubMed] [Google Scholar]

Preparation of Trimethyl Chitosan-Based Polyion Complex Micelle as Drug Delivery System for Cancer Therapy

Quynh-Nhu Nguyen-Trinh, Nhu-Thuy Trinh, Hanh Thi Ngoc Nguyen, and Vong Binh Long

Abstract Chitosan (CS) has been widely studied as polycation to form the polyion complex (PIC) micelles, which have emerged as promising delivery systems for sustained release of hydrophilic drugs. However, poor solubility in aqueous solutions limits the application of CS. In this study, trimethyl chitosan (TMC), the partially quaternized derivative of CS, was synthesized and investigated the PIC formation coupled with poly(acrylic acid) (PAA) at different cationic: anionic ratios (named as TMC/PAA micelles). The encapsulation of hydrophilic anticancer drug, doxorubicin (DOX), release profile, and cytotoxicity against murine colon cancer cells were also evaluated. The results showed that the sustained release of DOX from TMC/PAA micelles was confirmed. We found that the release of DOX from TMC/PAA micelles in acidic pH buffer (pH 5.5) was significantly higher than in physiological pH (7.4). Finally, the anticancer activity of DOX-loaded TMC/PAA micelles was remarkably higher than that of DOX. The results in this study suggest the potential application of TMC/PAA micelles as controlled drug delivery for cancer therapy.

Keywords Chitosan · TMC · Polyion complex micelles · Drug delivery · Self-assembly

---

Q.-N. Nguyen-Trinh
Department of Biochemistry, Faculty of Biology and Biotechnology, University of Science, Ho Chi Minh 700000, Vietnam

Q.-N. Nguyen-Trinh · N.-T. Trinh · V. B. Long (✉)
Vietnam National University-Ho Chi Minh City (VNU-HCM), Ho Chi Minh 700000, Vietnam
e-mail: vblong@hcmiu.edu.vn

N.-T. Trinh · V. B. Long
Department of Tissue Engineering and Regenerative Medicine, School of Biomedical Engineering, International University, Ho Chi Minh 700000, Vietnam

H. T. N. Nguyen
Faculty of Biotechnology and Environment, Hong Bang International University, Ho Chi Minh 717000, Vietnam

V. Van Toi et al. (eds.), *8th International Conference on the Development of Biomedical Engineering in Vietnam*, IFMBE Proceedings 85,
https://doi.org/10.1007/978-3-030-75506-5_23

1 Introduction

Chitosan (CS), a deacetylated derivative of chitin, is a copolymer composed of β-(1,4)-2-acetamido-*D*-glucose monomer and β-(1,4)-2-amino-*D*-glucose monomer. CS has attracted great attention in biomedical applications, including the development of a safe drug delivery tool since it exhibits high biocompatibility and biodegradability [1, 2]. However, the well-known limitation of CS is the poor water solubility at neutral and alkaline conditions, and it basically is soluble in only organic acids such as acetic acid [3]. Numerous studies have reported that the modification of chitosan to form new derivatives alters the physical, chemical, and biological properties of original CS. For example, *N,N,N*-trimethyl chitosan chloride (TMC), which is the partially quaternized derivative of CS, is highly soluble at neutral pH and improves biological activity and physicochemical stability of CS due to the high charge density [4, 5]. Since TMC possesses a highly positive recharge of quaternary amine groups, TMC achieves considerable attention as a drug and gene carrier. For example, TMC is used as polycation coupled with another polyanion to prepare the polyion complex (PIC) micelles improving stability and delivering the therapeutic agents including chemotherapy, protein, or nucleic acids [6, 7].

Doxorubicin (DOX) is well-known as an anticancer chemotherapy used in the treatment of a large variety of cancers. The anticancer mechanism of DOX is to interact with DNA to inhibit the DNA replication via the disruption of the topoisomerase II binding to DNA molecules during DNA synthesis. Besides, DOX releases free radicals that lead to damage to cellular membranes, DNA, and induces intracellular oxidative stress. However, since DOX is a low-molecular-weight agent, DOX distributes non-specially in the target tissues. Therefore, frequent administration of DOX causes multiple side effects consisting of myelotoxicity and haematological toxicity [8, 9], which limits its use in clinical practice.

In this study, TMC was synthesized, and a water-soluble PIC (named TMC/PAAc micelle) was prepared via the electrostatic interactions between TMC and poly(acrylic acid) (PAAc) to deliver DOX for the anticancer application. The DOX release profile from DOX-loaded TMC/PAAc micelle with different ratio of cation: anion polymer at varying physiological pH environments (pH 5.5 and pH 7.4) was investigated. In addition, the antitumor activity of DOX-loaded TMC/PAAc micelle was also evaluated.

2 Materials and Methods

2.1 *Synthesis of TMC*

Chitosan (80% deacetylation), iodomethane (CH_3I), sodium iodide (NaI), *N*-methyl-2-pyrrolidone (NMP), isopropanol (IPA), and acetic acid were purchased from Wako Pure Chemical Industries, Ltd., Japan. The TMC was synthesized via the methylation

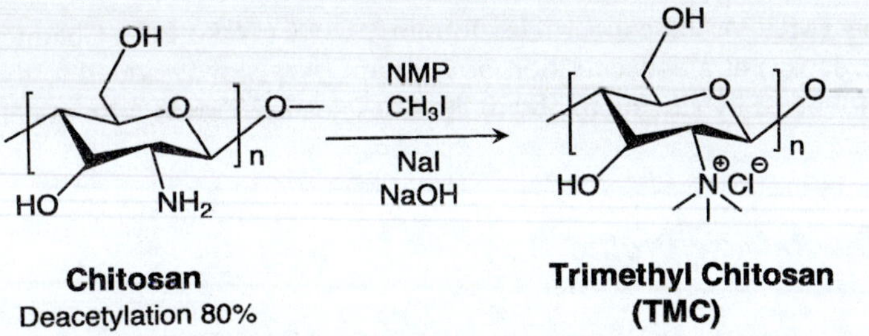

Fig. 1 Schemes of *N*-trimethyl chitosan (TMC)

of CS by using CH_3I (Fig. 1), as previously described with slight modification [10, 11]. Briefly, 5 g of was dissolved in 120 mL of NMP in an oil bath at 40 °C overnight with stirring. After, 4.8 g of NaI and 11 mL NaOH 15% was added to the reaction for 24 h at 60 °C. The TMC was collected by cold IPA reprecipitation following by dissolving in water and lyophilizing. The product was dissolved in D_2O, and the chemical structure was confirmed by ^1H NMR (JEOL ECS-400, JEOL Ltd., Tokyo, Japan).

2.2 *Preparation of TMC/PAAc Micelle and DOX-Loaded TMC/PAAc Micelle*

TMC/PAAc micelles were formed by ionic interactions between the positively charged TMC backbone and negatively charged PAAc (Fig. 2). The TMC/PAAc micelle formation were investigated by dissolving PAAc and TMC solution (10 mg/mL) in different ratio of cation and anion polymer (0.25, 0.5, 1, 2, 4 and 10). The average particle size, the size distribution of TMC/PAAc micelle was determined by dynamic lighting scattering (DLS, Malvern Zetasizer, UK) measurements.

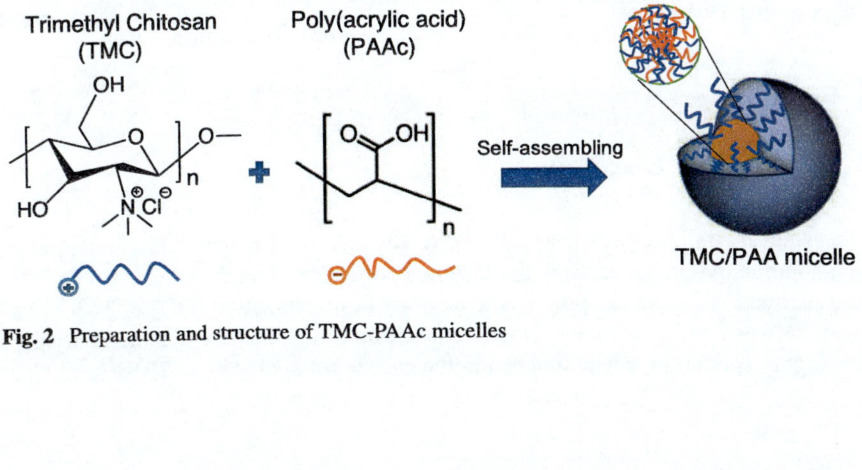

Fig. 2 Preparation and structure of TMC-PAAc micelles

During the TMC/PAAc nanoparticles forming, DOX (Wako Pure Chemical Industries, Ltd., Japan) at a concentration of 1 mg/mL was slowly added to the micelle solution, followed by stirring to obtain the DOX-loaded TMC/PAAc micelles.

2.3 *DOX Release Profile*

The in vitro release of DOX from the DOX-loaded TMC/PAAc nanoparticles was investigated by a dialysis method using the membrane tube (Spectrum, molecular weight cut-off 3.5 kDa). In brief, the DOX-loaded TMC/PAAc micelles were added in the membrane tube and dialyzed against a PBS solution at pH 5.5 and 7.4. At predetermined time intervals, the amount of released DOX was measured with a UV–vis spectrophotometer at a wavelength of 490 nm.

2.4 *Anticancer Activity* In Vitro

The anticancer effect of nanoparticles on the murine colon cancer cell C-26 (RIKEN, Japan) was investigated using 3-(4,5-dimethylthiazol-2-yl)-2,5-diphehyltetrazolium bromide (MTT) assay (Roche Diagnostics, Tokyo, Japan), as previously reported [12]. Briefly, C-26 cells were added in 96-well plates with a density of 1×10^4 cells/well. After 24 h incubation for cell attachment, C-26 cells were treated with DOX-loaded TMC/PAAc micelle or free DOX with different concentrations (0.5, 1, 5, 10 μg/mL) and blank TMC/PAAc micelle (5, 10, 50, 100 μg/mL) for 24 h. Then, treatment solutions were removed, and MTT (0.2 mg/mL) was added to each well for 4 h. Subsequently, MTT crystals were solubilized with solubilization buffer or dimethyl sulfoxide in each well. Thereafter, cell viability was measured using a microplate reader at a wavelength of 540 nm.

Percent cell viability curves using the following formula:

$$\text{Cell viability (\%)} = \frac{\text{mean treatment absorbance} - \text{mean blank absorbance}}{\text{mean control absorbance} - \text{mean blank absorbance}} \times 100\%$$

3 Results and Discussion

The synthesis of TMC was based on the methylation of CS with CH_3I in a strong base solution. After the reaction, the obtained polymer was analyzed using ^{1}H NMR measurement, as shown in Fig. 3. The degree of quaternization (DQ) of TMC was about 24% as compared to the ^{1}H NMR spectrum of CS, and the obtained polymer was highly soluble in water, suggesting the successful synthesis of TMC from CS.

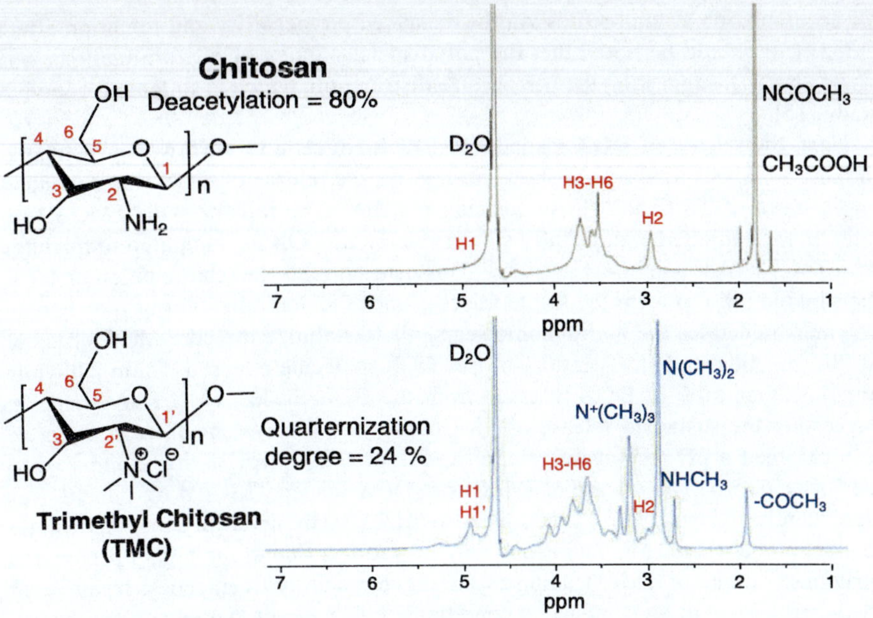

Fig. 3 The ^{1}H NMR spectrum of chitosan (CS) and trimethyl chitosan (TMC)

The obtained TMC was further used for the preparation of PIC by coupling with polyanion PAAc at different ratios of TMC and PAAc. As shown in Fig. 4, the size of micelles was increased from about 68 nm to 170 nm with the increase in TMC:PAAc ratio. The data suggests that changing different amount of polymer, different sizes of the micelle can be obtained, those sizes of TMC/PAAc micelle provide the potential

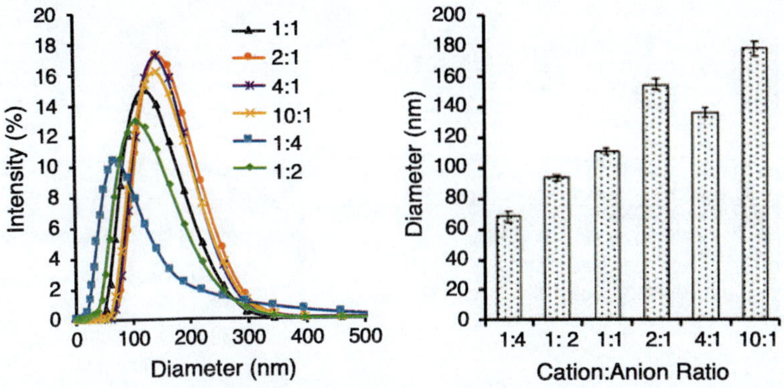

Fig. 4 The size distribution of TMC/PAAc micelle with a different ratio of TMC and PAAc. Data were expressed as mean ± SD with n = 3

for accumulation at tumor sites via the enhanced permeability and retention effect [13, 14]. It should be noted that the polydispersity index of all formulations was lower than 0.2, suggesting the narrow size distribution, which is suitable for in vivo study [15].

Next, the release of DOX-loaded TMC/PAAc micelle in vitro was investigated through a dialysis method. As shown in Fig. 5a, the release of DOX was depended on the ratio of TMC and PAAc forming micelles. The micelle with TMC:PAAc ratio of 1:1 exhibited significantly slower release of DOX as compared to micelles with TMC:PAAc ratio of 4:1 and 10:1. This data suggests that cation:anion of 1:1 is the suitable ratio to form the stable micelle since the neutralization of the charged polymers generates the hydrophobic segments to stabilize the core–shell structure of PIC micelle [16, 17]. Over 80% of free DOX molecule released within 2 h while only less than 50% of DOX released from the DOX-loaded TMC/PAAc micelle, suggesting the sustained release of TMC/PAAc micelle system. Interestingly, we also obtained a pH-responsive release profile of DOX at pH 5.5 as compared to pH 7.4 (Fig. 5b). The release of DOX from DOX-loaded TMC/PAAc micelle was significantly higher at pH 7.4 as compared to pH 5.5. At the acidic pH, the electrostatic interaction between TMC and PAAc may be stronger due to the highly protonated quaternary amine of TMC leading to the higher stability of micelles, resulting in the lower release of DOX at acidic conditions. This nano-carrier can be considered to apply for oral drug delivery to suppress the release of drug in the gastric acid environment.

Finally, we investigated the anticancer potential of DOX-loaded TMC/PAAc micelle using a murine colon cancer cell line (C-26). In cytotoxicity assay, treating C-26 cells with blank TMC/PAAc micelle was over 95% cell viability for 24 h at all tested concentrations (Fig. 6), suggesting low toxicity of this nano-carrier. Compared with free DOX, the DOX-loaded TMC/PAAc micelle exhibited significantly higher cytotoxicity on C-26 colon carcinoma cells at a range of concentration from 0.5 to

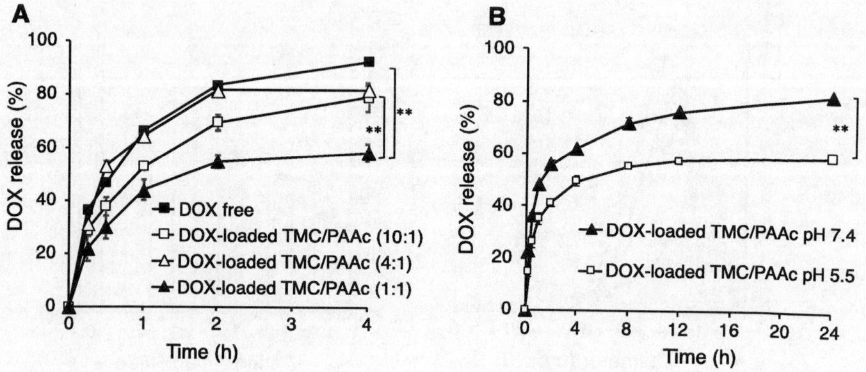

Fig. 5 The release profile of DOX from DOX-loaded TMC/PAAc micelles in difference rations of polycation and polyanion (**a**) and difference pH values (**b**). Data were expressed as mean ± SD with n = 3 and $^{**}p < 0.01$

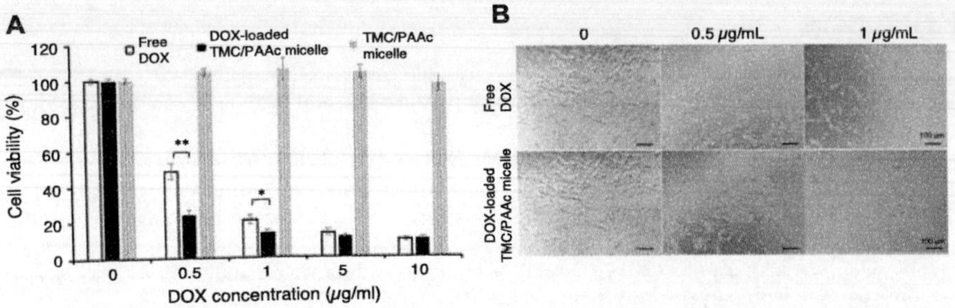

Fig. 6 The anticancer activity of DOX-loaded TMC-PAA micelle and free DOX against colon cancer C-26 cell by MTT assay. **a** Quantitative cell viability (%) was evaluated by comparing cells treated with samples with control. Data was expressed as mean ± SD with n = 3 and $*p < 0.05$, $**p < 0.01$. **b** Phase contrast images of C-26 cells at 24 h after treated with samples

1 μg/mL, indicating a high potential of DOX@TMC/PAAc micelle in the anticancer treatment. When higher DOX concentrations were used (5 and 10 μg/mL), the cell viability in DOX and DOX-loaded TMC/PAAc was not different due to high anticancer activity of DOX. This result suggested that a suitable dose of DOX should be considered to apply in vitro and in vivo to minimize the toxicity of DOX.

4 Conclusion

In this study, the water-soluble TMC was successfully synthesized by methylation of the amino groups in the water-insoluble CS. The PIC micelle with different sizes could be prepared by changing the ratio of positively recharged TMC and negatively recharge PAAc. The release of DOX was dependent on the charged ratio of TMC/PAAc micelles, and the pH-responsive release DOX from DOX-loaded TMC/PAAc micelle was obtained. The anticancer activity of DOX-loaded TMC/PAAc was significantly higher than that of free DOX treatment in a colon cancer cell. Further investigations, including cellular uptake, toxicity, or side effects, and the anticancer efficacy in the in vivo animal model should be investigated to clearly evaluate the potential of DOX-loaded TMC-PAAc micelle in the cancer therapy.

Acknowledgements This work was supported by the Vietnam National Foundation of Science and Technology Development (NAFOSTED) under grant number 108.05-2017.327.

Conflicts of Interest The authors have no conflict of interest to declare.

References

1. Bowman K, Leong KW (2006) Chitosan nanoparticles for oral drug and gene delivery. Int J Nanomedicine 1(2):117–128
2. Mohammed MA, Syeda JTM, Wasan KM, Wasan EK (2017) An overview of chitosan nanoparticles and its application in non-parenteral drug delivery. Pharmaceutics 9(4):53
3. Sogias IA, Khutoryanskiy VV, Williams AC (2010) Exploring the factors affecting the solubility of chitosan in water. Macromol Chem Phys 211(4):426–433
4. Wu QX, Lin DQ, Yao SJ (2014) Design of chitosan and its water soluble derivatives-based drug carriers with polyelectrolyte complexes. Mar Drugs 12(12):6236–6253
5. Mourya VK, Inamdar NN (2009) Trimethyl chitosan and its applications in drug delivery. J Mater Sci Mater Med 20(5):1057–1079
6. Germershaus O, Mao S, Sitterberg J, Bakowsky U, Kissel T (2008) Gene delivery using chitosan, trimethyl chitosan or polyethylenglycol-graft-trimethyl chitosan block copolymers: establishment of structure-activity relationships in vitro. J Control Release 125(2):145–154
7. Zhou Z, Zhong C, Dongyuan S, Ting X, Xiang L, Fengbo W (2018) Synthesis, antitumor activity and molecular mechanism of doxorubicin conjugated trimethyl-chitosan polymeric micelle loading Beclin1 siRNA for drug-resisted bladder cancer therapy. RSC Adv 8(62):35395–35402
8. Thorn CF, Oshiro C, Marsh S, Hernandez-Boussard T, McLeod H, Klein TE, Altman RB (2011) Doxorubicin pathways: pharmacodynamics and adverse effects. Pharmacogenet Genomics 21(7):440–446
9. Patel AG, Kaufmann SH (2012) How does doxorubicin work? eLife 1:e00387
10. Sieval AB, Thanou MM, Kotzé AF, Verhoef JC, Brussee J, Junginger HE (1998) Preparation and NMR characterization of highly substituted N-trimethyl chitosan chloride. Carbohyd Polym 36:157–165
11. Ignacio I, Andrew W, Francisco FT (2016) Polyion complex (PIC) particles: preparation and biomedical applications. Eur Polymer J 81:198–215
12. Vong LB, Nagasaki Y (2016) Combination treatment of murine colon cancer with doxorubicin and redox nanoparticles. Mol Pharm 13(2):449–455
13. Matsumura Y, Maeda HA (1986) A new concept for macromolecular therapeutics in cancer chemotherapy: mechanism of tumoritropic accumulation of proteins and the antitumor agent Smancs. Cancer Res 46(12):6387–6392
14. Fang J, Nakamura H, Maeda H (2011) The EPR effect: Unique features of tumor blood vessels for drug delivery, factors involved, and limitations and augmentation of the effect. Adv Drug Deliv Rev 63(3):136–151
15. Vong LB, Tomita T, Yoshitomi T, Matsui H, Nagasaki Y (2012) An orally administered redox nanoparticle that accumulates in the colonic mucosa and reduces colitis in mice. Gastroenterology 143(4):1027–1036
16. Vong LB, Ibayashi Y, Lee Y, Ngo DN, Nishikawa Y, Nagasaki Y (2019) Poly(ornithine)-based self-assembling drug for recovery of hyperammonemia and damage in acute liver injury. J Control Release 310:74–81
17. Kudo S, Nagasaki Y (2015) Facile and quantitative synthesis of a poly(ethylene glycol)-*b*-poly(L-arginine) block copolymer and its use for the preparation of polyion complex micelles with polyanions for biomedical applications. Macromol Rapid Commun 36(21):1916–1922

Enhancing In Vitro Bioavailability of Berberine by Incorporation of Beta–Cyclodextrin Complex into Solid Dispersion System

Hang Thu Than, Thuy Phan-Quoc Nguyen, Phat Dong Le, Phi Hong Tran, and Van Hong Nguyen

Abstract Berberine is widely used in Eastern countries thanks to low cost and various biological activities. However, low solubility has limited its bioavailability. The purpose of this study is to prepare and evaluate solid dispersion loading berberine—β-cyclodextrin complex to improve the effective treatment of berberine. Two techniques, inclusion complex and solid dispersion, are applied to increase the bioavailability of berberine for further applications. The inclusion complex is prepared by grinding berberine and β-cyclodextrin with various ratios. Solid dispersion loading berberine—β-cyclodextrin complex is fabricated from polyethylene glycol (PEG 6000) by melting completely and then spray-drying with lactose for powder formation. The complex and solid dispersion is evaluated about solubility and drug release in distilled water. Moreover, the final product is also analyzed with powder X-ray diffraction (PXRD). As a result, the complex which has ratio 1:1 between berberine and β-cyclodextrin expresses the highest solubility and optimal dissolution profiles in distilled water in comparison with pure berberine. Solid dispersion samples show many advantages in solubility and dissolution rate compared to pure berberine; the solubility reaches nearly three times over pure berberine. The results of the sample of incorporation of the β-cyclodextrin complex into solid dispersion have significant improvement compared to pure berberine. In general, solid dispersion containing berberine β-cyclodextrin complex expresses the considerable enhancement in vitro bioavailability of berberine. Hence, it could be applied as a promising system for improving the therapeutic treatment of berberine.

Keywords Berberine · β-cyclodextrin · Inclusion complex · Solid dispersion

H. T. Than · T. P.-Q. Nguyen · P. Dong Le · P. H. Tran · V. H. Nguyen (✉)
Pharmaceutical Engineering Lab, School of Biomedical Engineering, International University—Vietnam National University, Quarter 6, Linh Trung Ward, Thu Duc District, Ho Chi Minh City, Vietnam
e-mail: nhvan@hcmiu.edu.vn

V. Van Toi et al. (eds.), *8th International Conference on the Development of Biomedical Engineering in Vietnam*, IFMBE Proceedings 85,
https://doi.org/10.1007/978-3-030-75506-5_24

1 Introduction

Solubility, a chemical property referring to the amount of a solute being dissolved in a given volume of solvent at a specified temperature to form a uniform system, is believed to play an important role in achieving desired concentration of drug absorption into the body for optimal pharmacological response. Low aqueous solubility is a noticeable problem for oral administration route because in this condition, the amount of necessary drug which is delivered is larger, and this increases toxicity in the body. One important objective is to enhance the solubility of poorly soluble drugs, which leads to the discovery of many techniques including physical modifications of drugs that are particle size reduction (micronization, nanosuspension, sonocrystalisation, supercritical fluid process), drug dispersion in carries (eutectic mixtures, solid dispersion, solid solution), complex (use of complexing agents) and chemical ones that are pH adjustment, salt formation, use of a buffer, prodrug formation as well as other methods such as co-crystallization, precipitation, nanotechnology approaches [1] and so forth.

Berberine, an alkaloid naturally extracted from coptidis rhizome and amoorcorn tree bark [2], is the main active ingredient in many Eastern countries' drugs, such as coptis and phellodendron [3, 4]. Due to its low cost and abundant supply, berberine is widely used in clinical trials and demonstrates various biological activities like antibacterial [5, 6], antifungal [7], antimalarial [8], antileishmanial [9], anticancer [10] and antidiarrheal [11] properties, cholesterol-lowering effect [12] and hypoglycemic effect [13]. Table 1 introduces several products that are on the market. This indicates the widely uses and various dosage forms of berberine in many kinds of health condition treatment. Nevertheless, berberine suffers from several drawbacks such as bitterness, slight solubility (1 mg/mL in water) [14], gastrointestinal stimulation and low bioavailability [15]. Because of the disadvantages above, it is necessary that a big dose of berberine is delivered to ensure the therapeutic range of drug to the body, and it may cause discomfort for the gastrointestinal. The structure of berberine is represented in Fig. 1.

Although the widely uses of berberine, the low solubility is an adverse feature. Therefore, to take advantages the most of it, the berberine solubility, dissolution profile, and bioavailability strengthening are investigated in this study. The chosen technique is an inclusion complex method since inclusion-complex formation is a more facile technique to improve the solubility, dissolution rate, and bioavailability of low water-soluble drugs compared to other ones. Inclusion complexes are formed by the integration of non-polar molecule (active agent) (guest) into the cavity of the complexing agent (host) and the most common "host" is cyclodextrins (CDs). [17] Cyclodextrins are supramolecular oligosaccharides, divided into three types: α-cyclodextrin, β-cyclodextrin, or γ-cyclodextrin are composed of 6, 7, or 8 glucopyranoside units, respectively, which are linked by α-1,4-glycosidic bonds to form the ring shape. Toroid cone structures of cyclodextrins result in the hydrophobic core and hydrophilic exterior. The solubility and bioavailability can be increased by getting hydrophobic drug molecules interconnected in the cavity of cyclodextrins due to

Table 1 Drugs that contain berberine were on the market

Drug	Dosage form	Dose (mg)	Usage
Siro Berberis 100 mL	Solution (syrup)	15	Treating diarrhea for children
VK Berrine (berberine chloride)	Sugar-coated tablet	50	Treatment for bacterial dysentery, diarrhea, enteritis, cholangitis
Dr Whitaker: clinical grade Berberine 500 mg	Capsule	500	Support blood sugar, cholesterol, and insulin sensitivity
Havasu nutrition: berberine + cinnamon for blood sugar maintenance	Capsule	500	Helps aids glucose metabolism and immune function
Nature pure: support blood sugar, cholesterol, and insulin sensitivity	Capsule	500	Cardiovascular, gastrointestinal, immune weight loss support
Double wood supplements: Berberine HCL 500 mg	Capsule	500	Powerful AMPK activator, blood sugar support (Vegetarian, Gluten-Free, Non-GMO) complex extract supplement

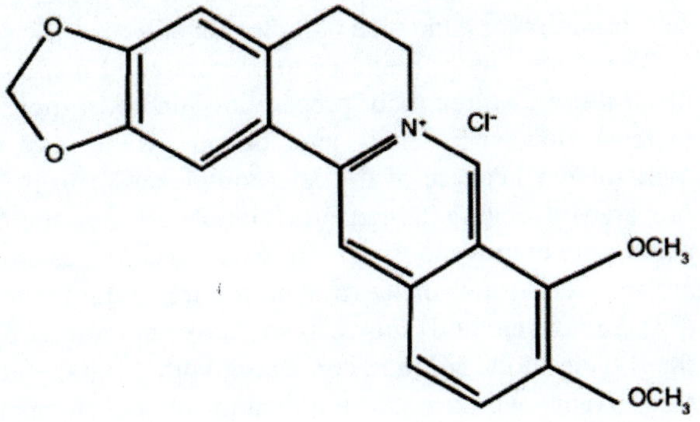

Fig. 1 Structure of berberine [16]

noncovalent bonds, which results in the change of physical and chemical properties of the drug. [18] Among these, thanks to perfect cavity size, efficient drug complex and loading, availability, and relatively low cost, β-cyclodextrin (β-CD) is the most appropriate chemical for the complex. [19, 20] β-CD is used to form the complex with berberine to enhance its solubility, reduce the dose and lower the bitterness [21–24]. The interaction between β-CD and berberine is executed due to the hydrophobic effect and van der Waals forces [25, 26]. Figure 2 shows the chemical structure and toroidal shape of β-CD as well as the inclusion complex of berberine.

On the other hand, solid dispersion has been considered as one of the most useful and effective strategies to improve the dissolution rate of poorly soluble drugs. [29]

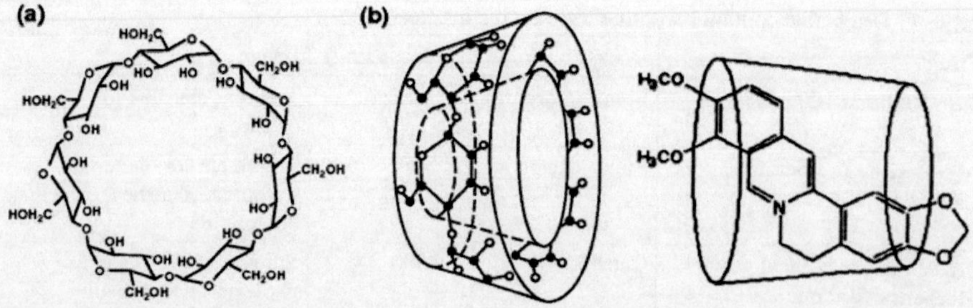

Fig. 2 Schematic depiction of **a** chemical structure and **b** the toroidal shape of β-CD [27] and **c** berberine-β-CD complex [28]

Solid dispersion—this term has been researched and developed through various timelines. Advantages of this method compared to other methods like salt formulation, milling for reducing the particle size, using solubilization (co-solvent, emulsions agents, micelles),... is solid dispersion can reduce the size of drugs into molecular level while methods of reducing particle size only downscale to 2–5 μm, which easily agglutinate in the formulation, dissolution process or during storage [30, 31]. Moreover, the interaction between drugs and carrier excipients prevents the agglutination of drug particles; therefore drugs can release and absorb rapidly. Last but not least, methods for the preparation of solid dispersion is simple and applicable than salt formation or nano-size particle, but still have the same efficiency [32].

There are a lot of methods to prepare solid dispersion systems, but can include of all those as three main preparation methods: melting, solvent and melt-solvent method. However, melting and solvent are the most pickable methods other than melt-solvent because of the less complexation in preparation. Melting method is prepared by melting the carrier(s) completely then adding the API in and mixing until getting the homogenous mixture. Some techniques are mainly used for this method are hot-melting, hot-melt extrusion, meltrex, liquid nitrogen, melt agglomeration, etc. [33]. Solvent method (known as solvent evaporation), this is obtained by evaporating the solvent of the solution containing both drug(s) and carrier(s). The advantage of the solvent method is the elimination of high temperature providing to melt the ingredients, this solve the different melting point between API and carrier(s) or ingredients that have a very high melting point. Some techniques mainly use for this method are: drying using the oven, vacuum, spraying, freezing, etc. [29]. The melt-solvent method is a combination between melting and solvent methods, basically, carrier(s) that heat-able are melting completely and mixing with others in a suitable solvent. This solution is evaporated for the final product.

By loading berberine complex into solid dispersion system via spray drying method, this study is expected to enhance the bioavailability by increasing the solubility and dissolution rate of berberine due to the advantage of two different main methods. Figure 3 indicates the depiction of incorporation of inclusion complex and solid dispersion system, in which the inclusion complex is loaded into the polymer matrix.

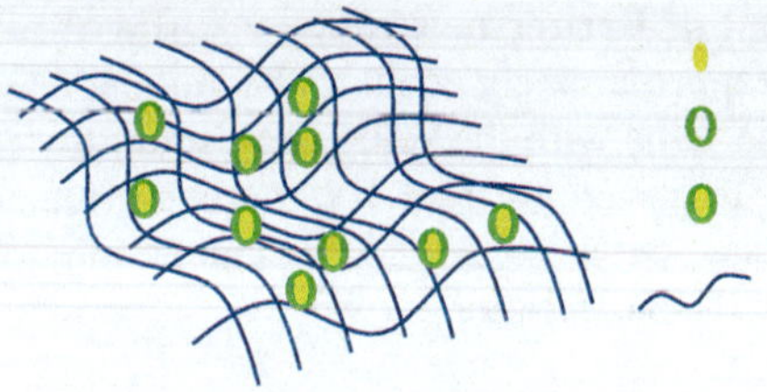

Fig. 3 Structure of berberine complex within the polymer by using solid dispersion

2 Materials

Berberine has a pharmaceutical ingredient grade. Other excipients were β-CD obtained from Roquette (France), PEG 6000 (Polyethylene Glycol) was purchased from Sigma Aldrich (USA). Lactose was obtained from Agropur Inc. (USA).

3 Preparation and Evaluation

3.1 Preparation and Evaluation of Berberine Complex

Preparation of berberine complex. Preparing berberine and β-CD with the molecular ratio as 1:1, 1:2, and 2:1. Firstly, β-CD is ground with distilled water in the mortar until getting homogeneous paste (about 3 min). Then berberine is added into the mixture and ground in 10 min with water as the assisted solvent. During this process, water is adjusted to the mixture to maintain the suitable consistency. Generally, the ratio between the weight of berberine plus β-CD and the total amount of water added is 1:1 (w/v). After that, the samples take about 24 h in −80 °C refrigerator (Panasonic MDF-U33V, Japan) and 24 h freeze-drying (Labconco, USA) to reach the required state. The three samples are evaluated, and the best one will be chosen as sample F1 for further works.

Quantification of Berberine Complex Products after Freeze Drying. Dissolve sample of 10 mg berberine content after freeze-drying into 10 mL volumetric flask containing ethanol 70% (sonicate if needed). Then, dilute the solution to 100 times and using UV–Vis spectrophotometer (Jasco V-730, Japan) at 345 nm to quantify the samples.

Solubility Test of Berberine Complex. After freeze-drying, the solubility test of samples is conducted, which berberine content is 5 mg/mL in each testing sample. Next, each sample is poured into distilled bottles that have been warmed up at 37 °C in the water shaking bath. After 72 h, measure the absorbance of samples to check whether the solubility increases. Finally, the solution is filtrated, diluted and measured with UV–VIS at 345 nm.

Dissolution Test of Berberine Complex. Using dissolution tester, apparatus with 900 mL water (approximately equal to the amount of solution in the intestinal tract) are prepared the same with the solution in solubility test and warmed up to 37 °C. The samples with dosage form of 100 mg berberine content are tested, and 10 mL of solution is taken out at 10, 20, 30, 45, and 60 min to measure the absorbance and then calculate the concentration.

3.2 *Solid Dispersion Preparation*

Preparation of berberine solid dispersion. The hot-melting method was used for the preparation of SDs. For evaluation of drug release, different ratio between berberine and PEG 6000 (1:3, 1:5 and 1:10) were prepared. In this method, the polymer was melted completely at 70 °C. After that berberine (with the required amount after being dried for 24 h) was added in and stirred continuously in 30 min until getting the homogenous dispersion mixture (the hole process is done at 70 °C). Lastly, samples are immediately stored in −20 °C condition in 24 h for stabilizing the molecular mobility. The samples are then collected and stored by 15 mL falcons at 25 °C and humidity 60%. The sample has the most optical results after evaluation will be F2.

Berberine—β-CD Complex Solid Dispersion Preparation. Solid dispersion loading berberine—β-CD complex was prepared by spray drying method using berberine—β-CD complex above, PEG 6000, and lactose. The materials are stirred until homogenous. Lastly, lactose solution is generally added in and stirred continuously until the mixture completely turns to solution before being sprayed (Eyela SD-1000, Japan) for powder. The temperature was set at 120 °C for inlet (nearly 90 °C for outlet), atomizing at 22 and pump speed at 30 mL/s. The sample is labelled as F3 and stored at −20 °C for further study.

Solubility Test of Solid Dispersion System. The sample is prepared. The sample has the content of berberine equal to 5 mg/ mL in distilled water and tests in condition at 37 °C (±0.5) in 72 h in the water shaking bath. Finally, filtrate, dilute and measure this solution with the UV–VIS at 345 nm.

Dissolution Test of Solid Dispersion System. The sample has the amount of berberine equal to 100 mg, then add to each container 900 mL of distilled water in the dissolution tester (Pharma Test DT70, Germany). Collect and filter this solution at five periods of time: 10, 20, 30, 45, 60 min. Then, those samples are measured with UV–VIS device at 345 nm.

Powder X-ray Diffraction (PXRD). To confirm the existing form of berberine with the incorporation of β–CD complexation and solid dispersion system, PXRD is applied to the analysis. The X-ray diffractometer (Brucker D8 Advance, USA) generates the peaks at diffraction angles (2θ) ranges from 5° to 50°, the voltage of

40 kV and a current of 40 mA is applied to generate X-ray source, the scanning rate is set up at 4.8°/ min. The samples analyzed in this test are berberine, berberine inclusion complex (F1), berberine solid dispersion (F2) and berberine with the incorporation of β–CD complexation and solid dispersion system (F3).

4 Results

4.1 *Solubility of Berberine Delivery System*

A glance at Table 2, it is easy to see that all the samples with the incorporation of β-CD have the improvement in solubility compared to pure berberine. The sample of ratio 1:1 of berberine: β-CD is the ideal ratio for the inclusion complex method. This is because the solubility of this ratio is higher than the other two (about 1.5 times higher than F12 and double F21). Therefore, ratio 1:1 is chosen for further study in solubility tests and labelled as F1.

According to Table 3, the solubility of the combination of berberine and PEG 6000 increases approximately three times compared to pure berberine. It is easy to see that sample of ratio 1:10 of berberine and PEG 6000 has the highest result on the solubility test. After 72 h, the result reaches to 3.68 mg/mL. However, the ratio of 1:3 and 1:5 both reach to 3.5 mg/mL, and the difference between the three results are not significant. Moreover, the solubility of ratio 1:3 and 1:5 is nearly the same. Therefore, ratio 1:3 is chosen for further study in the solubility test and labelled as F2.

By conducting the inclusion complex method of berberine with β-CD, the result of berberine solubility is evidently enhanced, which is presented in Table 4. The solubility of berberine complex is evaluated bigger than pure berberine in distilled water after 72 h. Additionally, the fluctuation of solubility between pure berberine and inclusion complex sample is approximately 1 mg/mL; in other words, the inclusion complex method increases the solubility of berberine by 26.69%. Therefore, based on this result, it determines the effective aspect of the inclusion complex

Table 2 The solubility of berberine and inclusion complex with various ratio (berberine: β-CD) (unit: mg/mL)

Berberine	F11 (ratio 1:1)	F12 (ratio 1:2)	F21 (ratio 2:1)
1.20 ± 0.03	4.05 ± 0.03	2.74 ± 0.09	2.03 ± 0.10

Table 3 The solubility of berberine and hot-melting samples with various ratio (berberine: PEG 6000) (unit: mg/mL)

Berberine	F13 (ratio 1:3)	F15 (ratio 1:5)	F110 (ratio 1:10)
1.20 ± 0.03	3.51 ± 0.05	3.5 ± 0.1	3.68 ± 0.11

Table 4 The solubility of berberine and samples F1, F2, and F3 (unit: mg/ mL)

Berberine	F1	F2	F3
1.20 ± 0.03	4.05 ± 0.01	3.51 ± 0.05	3.85 ± 0.15

F1 berberine inclusion complex, *F2* berberine—PEG 6000 hot-melting, *F3* berberine—β-CD solid dispersion

in the improvement of berberine solubility. The sample F2 is prepared to see the effect of hydrophilic polymer on the enhancement of solubility of berberine. As the table shows, PEG 6000 has the ability to improve the solubility of berberine but not significant. Therefore, the solid dispersion method is applied with more excipients for better improvement. The solid dispersion method is investigated with the purpose of improving the berberine dissolution rate and permeability. Basing on the result, the solubility of berberine solid dispersion is quite higher than pure berberine in distilled water after 72 h. Specifically, the solid dispersion method has ability to enhance the berberine solubility equal 1.59 mg/mL or increase the soluble property in distilled water by 34.27%. It can be concluded that the solid dispersion method is also the beneficial improvement for berberine solubility. The resulting combination of inclusion complex and solid dispersion of berberine is investigated by the solubility test that compares with pure berberine. Basically, the table presents the result of three solubility tests with three methods which also shows the solubility of combination method higher than the pure berberine in distilled water after 72 h. Obviously, the solubility of the combination method increases by 20.78% compared to berberine. Consequently, the combination method between inclusion complex and solid dispersion has a positive result for the promising development.

4.2 Dissolution Profile of Berberine Delivery System

According to Fig. 4, the data shows that there is a difference in dissolution profile between different ratios. To be more specific, ratio 1:2 holds the first position in the first 20 min, followed by ratio 2:1, and the final one is ratio 1:1. However, the difference is not significant. Moreover, based on the results of solubility, and for the convenient preparation and logical linkage between molecules, ratio 1:1 is chosen for onward study and called F1.

Based on the results on the Fig. 5, the data shows that there is not a significant difference in dissolution profile between different ratios. The graph of drug release between the ratios of 1:3, 1:5 and 1:10 is quite similar. However, for further study, ratio 1:3 is chosen according to the amount of PEG 6000 used is less than ratio 1:10, leading to further fusion of other materials. Therefore, to ensure the economy of the product, the ratio 1:3 sample is labelled as F2 for later study.

The graph represents the percentage of drug release and the time presents the enhancement of berberine dissolution rate in distilled water from 10 to 60 min (Fig. 6). Evidently, the berberine dissolution rate is more quickly than berberine through

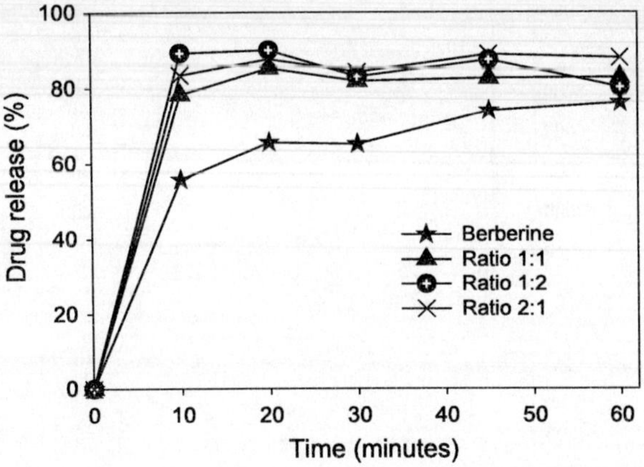

Fig. 4 The dissolution profile of pure berberine and inclusion complex with various ratio (berberine: β-CD)

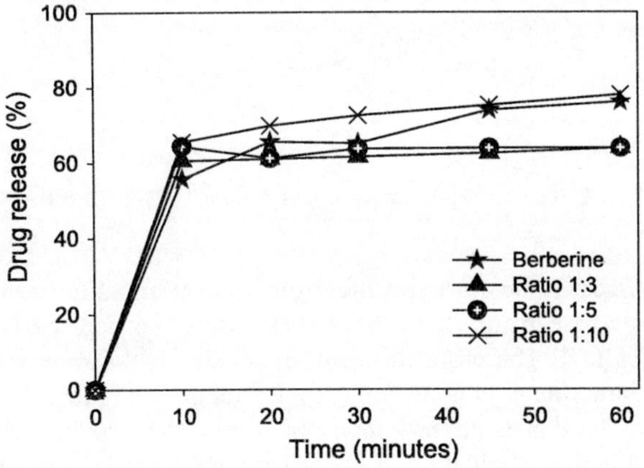

Fig. 5 The dissolution profile of pure berberine and hot-melting samples with various ratio (berberine: PEG 6000)

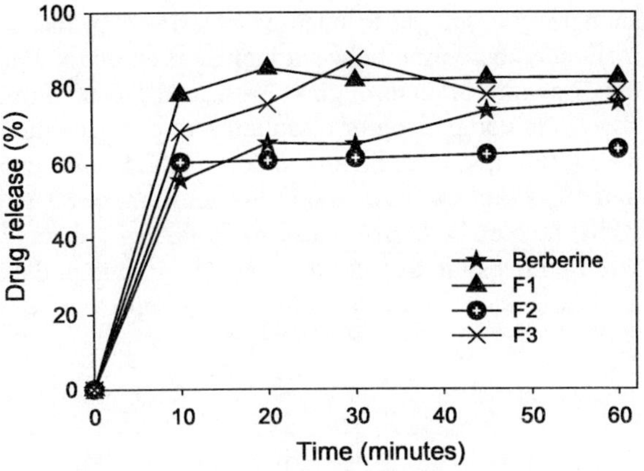

Fig. 6 The dissolution profile of pure berberine and samples F1, F2, and F3

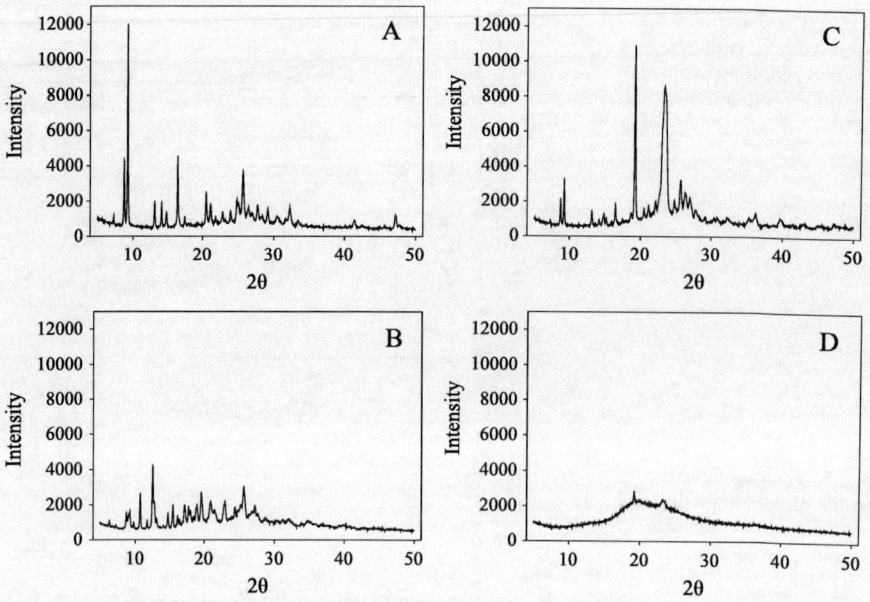

Fig. 7 X-ray diffractograms of **a** berberine, **b** F1, **c** F2, **d** F3

times. To be more specific, from 10 to 60 min, the berberine with inclusion complex increase from 78.25 to 82.76%, while the pure berberine increase from 55.73 to 76.11%. Therefore, the result shows clearly the improvement in the dissolution rate of berberine with inclusion complex method. Figure 6 indicates the dissolution profile of F2 that is slightly lower than that of berberine while it is the improvement in solubility evaluation. It can be explained that PEG increases berberine's hydrophilic, but due to polymer's characteristics, it swells in water first, and the time test is short for the sample to release berberine into the environment. The combination method of berberine between inclusion complex and solid dispersion is evaluated the dissolution rate through times from 10 to 60 min. Specifically, the below graph shows the comparison of dissolution rate among three methods, which present the dissolution rate of the combination method more rapid than pure berberine. To be more specific, from 10 to 60 min, the sample F3 increases from 68.57 to 78.9%, while the pure berberine enhances from 55.73 to 76.11%. This combination method also improves the dissolution rate of berberine in distilled water, consequently, has a positive narrow to develop promising improved berberine formulation.

4.3 *PXRD of Berberine Delivery System*

As can be seen from Fig. 7, the sharp characteristic peaks of berberine at diffraction angles (2θ) appear at ranges at 9°, 13°, 16°, 20°–30°, 32°, 41° and 48°. As the results above F2 (present for the evaluated of hot-melting) is still crystalline with many high variation peaks at 9°, 19°, 23° and 25°, the peak at 9° does reduce from 12,000 to less than 400; this shows PEG 6000 does impact to the elimination of crystalline of berberine. Based on the results of F1, F1 (presents for the evaluated of inclusion complex) is still crystalline; however, the peaks at diffraction angles (2θ) of 41° and 48° are disappeared. The highest peak of berberine at 9° is nearly eliminated in F1, in F1 there are still many other peaks that range from 10° to 28°, this due to the correspondence of the diffraction pattern of β-CD. F3 (the combination of F1 and F2) shows the best results when most of the peaks is eliminated, peak at 19° and 22° just at nearly 400 intensity, the peaks that correspond to berberine crystals are significantly weakened indicating that majority of berberine in the system of the combination of inclusion complex and solid dispersion is in an amorphous form. The reason for this phenomenon is based on the preparation technique of each sample. The preparation process of F3 immediately turns the amorphous solution into powder. As the results of PXRD analysis shows that F1 and F2 samples exist two very strong diffraction peaks at 19.3° and 23.3°, these peaks are nearly eliminated in the diffraction analysis of F3. F3 ingredients are the combination of F1 and F2; however, the preparation processes are different. The spray drying process of F3 is spray drying, which immediately turns the homogenous solution into powder. The process is significantly faster than the process of F2; the particle size of the sample is also smaller (depended to droplet speed management) [34]. Furthermore, PEG 6000 dispersed F1—which are the complexation of berberine and β-CD, this is based on its length of molecular structure. These advantages contribute to the amorphous stage of the sample compared to other samples which are enhanced by other methods.

5 Discussion

The limitation of berberine in bioavailability is due to poor solubility and poor permeability, which determines the designation of the biopharmaceutical classification system (BCS) class IV drug [35]. Particularly, the result of this research evaluates the low soluble property of berberine; therefore, the improving methods are applied to enhance the solubility of berberine.

The inclusion complex with β-CD is the effective method for improving solubility and dissolution rate of berberine. Structurally, the berberine molecule is loaded in the hydrophobic cavity of β-CD, which has the ability to interlock berberine molecule and increase the solubility of pure berberine. Clearly, it presents the higher solubility and dissolution rate significantly when comparing the results between pure berberine and inclusion complex sample. F2 (berberine and PEG 6000) showed lower results on solubility and dissolution according to F1; this is based on the sustained release on stabilized of PEG 6000 to solid dispersion method. Moreover, solid dispersion using spray drying method has shown many advantages in the improvement of both solubility and dissolution rate compared to the pure berberine. In addition, the results observed that berberine slightly soluble with aqueous solution approximated to 1 mg/mL. Therefore, by applying hydrophilic polymers such as PEG 6000, as carriers improving the dissolution rate, and drugs could dissolve better in an aqueous environment. Based on the positive results from the two methods, the sample with the combination between solid dispersion and inclusion complex is conducted to evaluate the improvement of berberine bioavailability. Similarly, the result of combining method shows the higher solubility and dissolution rate of the combined berberine than pure berberine. Consequently, the three successful methods for enhancing in vitro bioavailability of berberine achieve the promising result to develop the dosage forms of berberine for different treatments of human diseases.

6 Conclusion

Overall, the three methods for improving the bioavailability of berberine indicates the higher solubility and dissolution profile in comparison with pure berberine. The inclusion complex sample performs the best results on these aspects (4.05 mg/mL for solubility of F11 and 89.93% for dissolution rate of F12), while the combined methods between inclusion complex and solid dispersion F3 shows to advantage in cancelling peaks of pure berberine to contribute amorphous state formation in PXRD evaluation. This result concludes that these methods have the ability to progress pure berberine to overcome bioavailability limitation.

Acknowledgements This research is funded by International University, Vietnam National University, Ho Chi Minh City (VNU-HCMC) under grant number SV2019-BME-10 /HĐ-KHCN.

This research is funded by Vietnam National University HoChiMinh City (VNU-HCM) under grant number NCM2020-28-01.

Conflicts of Interest The authors have no conflict of interest to declare.

References

1. Jain P, Goel A, Sharma S, Parmar M (2010) Solubility enhancement techniques with special emphasis on hydrotrophy. Int J Pharma Profess Res 1(1):34–45
2. Ma S, Wang Y, Shang X, Yan F (2012) Formulation of berberine hydrochloride and hydroxypropyl-β-cyclodextrin inclusion complex with enhanced dissolution and reduced bitterness. Trop J Pharm Res 11(6):871–877
3. Tang J, Feng Y, Tsao S, Wang N, Curtain R, Wang Y (2009) Berberine and Coptidis rhizoma as novel antineoplastic agents: a review of traditional use and biomedical investigations. J Ethnopharmacol 126(1):5–17

4. Zhang JN, Zhao QL, You SJ, Zhang GD (2010) Power generation in bio-cathode microbial fuel cell with different cathode materials. Chem J Chin Univ 31:162–166
5. Iwasa K, Kamigauchi M, Ueki M, Taniguchi M (1996) Antibacterial activity and structure-activity relationships of berberine analogs. Eur J Med Chem 31(6):469–478
6. Samosorn S, Tanwirat B, Muhamad N, Casadei G, Tomkiewicz D, Lewis K et al (2009) Antibacterial activity of berberine-NorA pump inhibitor hybrids with a methylene ether linking group. Bioorganic Med Chem 17(11):3866–3872
7. Park KD, Lee JH, Kim SH, Kang TH, Moon JS, Kim SU (2006) Synthesis of 13-(substituted benzyl) berberine and berberrubine derivatives as antifungal agents. Bioorg Med Chem Lett 16(15):3913–3916
8. Iwasa K, Kim HS, Wataya Y, Lee DU (1998) Antimalarial activity and structure-activity relationships of protoberberine alkaloids. Eur J Med Chem 33(1):65–69
9. Vennerstrom JL, Lovelace JK, Waits VB, Hanson WL, Klayman DL (1990) Berberine derivatives as antileishmanial drugs. Antimicrob Agents Chemother 34(5):918–921
10. Letašiová S, Jantová S, Čipák LU, Múčková M (2006) Berberine—antiproliferative activity in vitro and induction of apoptosis/necrosis of the U937 and B16 cells. Cancer Lett 239(2):254–262
11. Yamamoto K, Takase H, Abe K, Saito Y, Suzuki A (1993) Pharmacological studies on antidiarrheal effects of a preparation containing berberine and geranii herba. Nihon Yakurigaku zasshi. Folia Pharmacologica Japonica 101(3):169–175
12. Yang P, Song DQ, Li YH, Kong WJ, Wang YX, Gao LM et al (2008) Synthesis and structure–activity relationships of berberine analogues as a novel class of low-density-lipoprotein receptor up-regulators. Bioorganic Med Chem Lett 18(16):4675–4677
13. Tang LQ, Wei W, Chen LM, Liu S (2006) Effects of berberine on diabetes induced by alloxan and a high-fat/high-cholesterol diet in rats. J Ethnopharmacol 108(1):109–115
14. Sahibzada MUK, Sadiq A, Faidah HS, Khurram M, Amin MU, Haseeb A, Kakar M (2018) Berberine nanoparticles with enhanced in vitro bioavailability: characterization and antimicrobial activity. Drug Des Dev Ther 12:303
15. Zhang DM, Liu HY, Xie L, Liu XD (2007) Effect of baicalin and berberine on transport of nimodipine on primary-cultured, rat brain microvascular endothelial cells 1. Acta Pharmacol Sin 28(4):573–578
16. Xue M, Yang MX, Zhang W, Li XM, Gao DH, Ou ZM et al (2013) Characterization, pharmacokinetics, and hypoglycemic effect of berberine loaded solid lipid nanoparticles. Int J Nanomed 8:4677
17. Savjani KT, Gajjar AK, Savjani JK (2012) Drug solubility: importance and enhancement techniques. ISRN Pharm
18. Pund S, Joshi A (2017) Nanoarchitectures for neglected tropical protozoal diseases: challenges and state of the art. In: Nano-and microscale drug delivery systems. Elsevier, Amsterdam, pp 439–480
19. Gidwani B, Vyas A (2015) A comprehensive review on cyclodextrin-based carriers for delivery of chemotherapeutic cytotoxic anticancer drugs. BioMed Res Int
20. Mangolim CS, Moriwaki C, Nogueira AC, Sato F, Baesso ML, Neto AM, Matioli G (2014) Curcumin–β-cyclodextrin inclusion complex: stability, solubility, characterization by FT-IR, FT-Raman, X-ray diffraction and photoacoustic spectroscopy, and food application. Food Chem 153:361–370
21. Chen W, Miao YQ, Fan DJ, Yang SS, Lin X, Meng LK, Tang X (2011) Bioavailability study of berberine and the enhancing effects of TPGS on intestinal absorption in rats. AAPS PharmSciTech 12(2):705–711
22. Li N, Xu L (2010) Thermal analysis of β-cyclodextrin/Berberine chloride inclusion compounds. Thermochim Acta 499(1–2):166–170
23. Dong-hui X, Xu-dong R, Long J, Hua L (2012) Inclusion Modes of berberine with beta-cyclodextrin in aqueous solution. Chem Res Chin Univ 28(2):282–286
24. Jia, B., Li, Y., Wang, D., & Duan, R.: Study on the interaction of β-cyclodextrin and berberine hydrochloride and its analytical application. PloS One 9(5):e95498

25. Cromwell WC, Bystrom K, Eftink MR (1985) Cyclodextrin-adamantanecarboxylate inclusion complexes: studies of the variation in cavity size. J Phys Chem 89(2):326–332
26. Sohi H, Sultana Y, Khar RK (2004) Taste masking technologies in oral pharmaceuticals: recent developments and approaches. Drug Dev Ind Pharm 30(5):429–448
27. Jambhekar SS, Breen P (2016) Cyclodextrins in pharmaceutical formulations I: structure and physicochemical properties, formation of complexes, and types of complex. Drug Discovery Today 21(2):356–362
28. Yu JS, Wei FD, Gao W, Zhao CC (2002) Thermodynamic study on the effects of β-cyclodextrin inclusion with berberine. Spectrochim Acta Part A Mol Biomol Spectrosc 58(2):249–256
29. Chiou WL, Riegelman S (1971) Pharmaceutical applications of solid dispersion systems. J Pharm Sci 60(9):1281–1302
30. Muhrer G, Meier U, Fusaro F, Albano S, Mazzotti M (2006) Use of compressed gas precipitation to enhance the dissolution behavior of a poorly water-soluble drug: generation of drug microparticles and drug–polymer solid dispersions. Int J Pharm 308(1–2):69–83
31. Karavas E, Ktistis G, Xenakis A, Georgarakis E (2006) Effect of hydrogen bonding interactions on the release mechanism of felodipine from nanodispersions with polyvinylpyrrolidone. Eur J Pharm Biopharm 63(2):103–114
32. Baird JA, Taylor LS (2012) Evaluation of amorphous solid dispersion properties using thermal analysis techniques. Adv Drug Deliv Rev 64(5):396–421
33. Chiou WL, Riegelman S (1969) Preparation and dissolution characteristics of several fast-release solid dispersions of griseofulvin. J Pharm Sci 58(12):1505–1510
34. Sollohub K, Cal K (2010) Spray drying technique: II. Current applications in pharmaceutical technology. J Pharm Sci 99(2):587–597
35. Mirhadi E, Rezaee M, Malaekeh-Nikouei B (2018) Nano strategies for berberine delivery, a natural alkaloid of Berberis. Biomed Pharmacother 104:465–473

Solid Lipid Nanoparticles Containing Berberine by Spray-Drying Method

Mai Chau-Ngoc Nguyen, Khoa Nguyen-Manh Le, Ngan Thi-Thu Huynh, Tham Thi-Thanh Nguyen, Hoa Thi-Xuan Bui, and Van Hong Nguyen

Abstract Berberine is commonly used in Eastern countries due to its various biological activities and low cost. However, low permeability has limited its bioavailability. This study was conducted to fabricate solid lipid nanoparticles (SLNs) containing berberine by spray-drying method to enhance its absorption in the human body. SLNs containing berberine was prepared by stearic acid as a solid lipid base and various surfactants. Then, the aqueous dispersion of SLNs was converted into powders to be stored over a long time by the spray-drying method with carbohydrate carriers. The spray-drying parameters such as the inlet and outlet temperatures, the blower speed, the atomizing pressure, the feeding rate were optimized. The nanoparticles physicochemical properties were evaluated before and after the spraying process. The lipid nanoparticles containing berberine were successfully prepared by the spray-drying method with a small particle size (around 230 nm), zeta potential (approximately −30 mV), and homogenous dispersion (polydispersity index just above 0.12). The optimized spray drying parameters were $T_{inlet/outlet}$ of 110/65 °C, the blower of 0.58 m^3/min, atomizing of 20 kPa, and feeding rate of 0.1 L/h. The resulted SLNs exhibited good physical stability and redispersion. The collected SLNs by spray-drying could be considered a potential drug delivery system to enhance berberine absorption into our human body.

Keywords Solid lipid nanoparticles · Spray-drying · Berberine

1 Introduction

The development of SLNs has shown the ability to combine the advantages of combination solid particles, liposomes, and emulsions at the beginning of the nineties [1,

M. C.-N. Nguyen · K. N.-M. Le · N. T.-T. Huynh · T. T.-T. Nguyen · H. T.-X. Bui · V. H. Nguyen (✉)
Pharmaceutical Engineering Lab, School of Biomedical Engineering, International University—Vietnam National University Ho Chi Minh City, Quarter 6, Linh Trung Ward, Thu Duc District, Ho Chi Minh City, Vietnam
e-mail: nhvan@hcmiu.edu.vn

V. Van Toi et al. (eds.), *8th International Conference on the Development of Biomedical Engineering in Vietnam*, IFMBE Proceedings 85,
https://doi.org/10.1007/978-3-030-75506-5_25

2]. Comparing with the traditional drug carriers, the advantages of lipid nanoparticles are lower cytotoxicity, high drug payload, large scale production, and the possibility of controlled drug release [3]. SLNs have been used in the form of pellets to achieve a retarded drug release for several years [4, 5] because drug mobility in a liquid oil should be considerably higher than with a solid lipid.

The structure of nanoparticles is organized by a solid lipid matrix, so the name "solid lipid nanoparticles" was derived from this reason [6]. The main ingredients of SLNs include solid lipid(s), emulsifier(s), and water. A mixture of lipids such as triglycerides (e.g. tristearin), fatty acids (e.g. stearic acid), partial glycerides (e.g. Imwitor), and waxes (e.g. cetyl palmitate) can be used to create the matrix lipid. To reduce the danger of chronic and acute toxicity, the lipid matrix should be made from physiological lipids. Besides, emulsifiers should be added to stabilize the lipid dispersion. Depend on the administration route, which is more limited for parenteral administration; emulsifiers should be chosen carefully. In addition, the size of these submicron colloidal carriers should range from 40 to 1000 nm technically. There are several methods to prepare SLNs such as microemulsion, high-pressure homogenization process, solvent displacement, etc. [7, 8] (Fig. 1)

High pressure homogenization (HPH) which is divided into hot and cold techniques, is known as the conventional method for the preparation of SLNs [10, 11]. To control the size of lipid nanoparticles, some crucial techniques use large mechanical forces such as high shear homogenization, and ultrasonication is used and combined together [12] (Fig. 2).

It is necessary to convert the liquid dispersion into a dry product when applying SLNs for some administration routes including i.v., oral administered system. This approach may be useful for SLNs loaded drugs that are straightforward to hydrolysis [13]. Spray- drying technique which has a lower cost than lyophilization was investigated for SLNs as a substitute method. The process usually includes 4 steps, which is the atomization of the feed into a spray, spray-air contact, then drying of the spray, and finally separation of the dried product from the drying gas [14]. Therefore, the liquid can be converted into fine, dust-free powders through this system.

Fig. 1 Structure of SLNs [9]

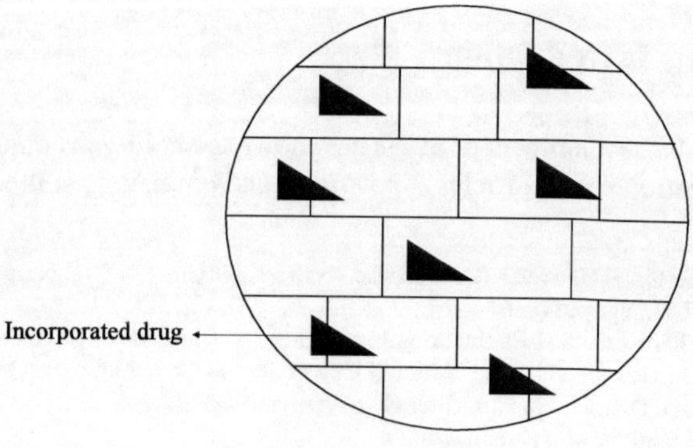

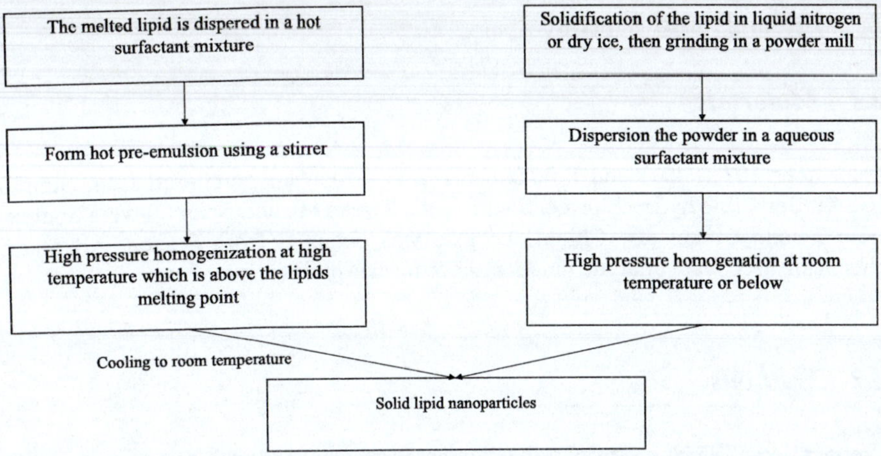

Fig. 2 Schematic procedure of hot homogenization (left) and cold homogenization technique (right) [1]

Herbal medicine has been proved the potentiality in treating and preventing many diseases [15]. Berberine which is extracted from the root and the stem bark of many medicinal plants such as Hydrastis Canadensis (goldenseal), Berberis Aquifolium (Oregon grape) [16], is a basic quaternary isoquinoline alkaloid with a pKa value of 2.47 [17]. It has been reported that berberine has a diversity of biological effects [18–20] including inhibiting acetylcholinesterase, lowering glucose and cholesterol, reducing mortality in patients with chronic congestive heart failure, immunomodulatory [21], antimicrobial [22] and anti-inflammatory [23].

However, berberine has low bioavailability, poor permeability, and slight absorption which limit its clinical application very much. Nanotechnology has been considered as the key approach for solving these problems. Berberine loaded SLNs were prepared to improve bioavailability. There are several parameters in the characterization of SLNs which affect stability directly and release kinetics. Particle size and zeta potential were determined by using Dynamic Light Scattering (Zetasizer Nano). In a previous study, spray drying of food-grade SLNs containing ω-3 fish oil, the size of particles was 180 nm [24].

In this research, the development of an SLNs encapsulating berberine by spray-drying technique was carried out in order to enhance the absorption of berberine in the human body. The SLNs powders should maintain the identical particle size distribution of the original dispersion.

2 Materials and Methods

2.1 Materials

Poloxamer 407 (P407) and Poloxamer 188 (P188) were purchased from Sigma (USA). Berberine hydrochloride, Stearic acid, Tween 80, and Span 80 were bought from Xilong Chemicals (China). Lactose was obtained from France. All other chemicals used were analytic grade without further purification.

2.2 Methods

Preparation of SLNs containing berberine. Berberine-loaded SLNs were prepared by using sonication. Table 1 showed the eight formulations investigated. For all formulation, water with an appropriate surfactant was added to the hot lipid phase at 70 °C. Then using ultrasonicator for 5 min and finally, the SLNs were cooled immediately at −20 °C.

Measurement of particle size and polydispersity index. Particle size (PS) and polydispersity index (PDI) were measured at 25 °C using Dynamic Light Scattering (Zetasizer Nano) at 173° after diluted samples with distilled water. Each sample was measured three times.

Measurement of zeta potential. Zeta potential indicates the electric charge on the particle surface and thus maybe reflects the physical stability of the colloidal system. The zeta potential (ZP) was conducted under an electrical field, at 25 °C, using the Zetasizer Nanomachine. Each sample was measured three times.

Table 1 Formulation of compositions of SLN

Code	Berberine (mg)	Stearic acid (mg)	P407 (mg)	P188 (mg)	Span 80 (mg)	Tween 80 (mg)	Lactose (mg)
SLN1	–	1000	200	–	–	–	–
SLN2	–	1000	–	200	–	–	–
SLN3	–	1000	–	–	200	–	–
SLN4	–	1000	–	–	–	200	–
BN1	40	1000	–	200	–	–	4000
BN2	80	1000	–	200	–	–	4000
BN3	100	1000	–	200	–	–	4000
BN4	125	1000	–	200	–	–	4000

Spray-drying the berberine-loaded SLNs. The SLNs that contained berberine mixtures were then sprayed with a laboratory-scale machine (Mini spray-dryer B-290, Büchi, Switzerland) equipped with nozzle atomizer (0.5 mm diameter). The spray drying parameters were the Tinlet/outlet of 110/65 °C, the blower is 0.58 m^3/min, the atomizing is 20 kPa, the feeding rate is 0.1 L/h.

Statistical analysis. All of the values were measured three times and data were expressed as mean ± standard deviation (SD). Statistical analyses were performed by the Student's *t*-test. A value of $p < 0.05$ was considered statistically significant.

3 Results and Discussion

3.1 *Effect of the Preparation Process on SLNs Physicochemical Properties*

The influence of time, rate of the ultrasonication process on particle size was considered. The ultrasonic electronic generator converts the electrical signal to mechanical vibration. Increasing the amplitude setting will increase the intensity within the sample [25]. Under ultrasonic irradiation, the formation and collapse of bubbles increase the pressure and temperature in the solution [26]. With the presence of particles, the cavitation phenomenon causes a change in the uniformity of bubble decomposition [27]. It was found that the particle size decreases proportionally to the increase in time on of sonication. Similarly, the size also decreases slightly when the sample was treated with higher ultrasound power (Table 2).

Table 2 The effect of different parameters process on the size of SLNs

Amplitude	Time on (seconds)	Time off (seconds)	Size (nm)	PDI	Zeta potential (mV)	Duration (minutes)
30 A	60	30	293.57 ± 1.61	0.11 ± 0.01	−28.57 ± 0.71	5
	15	30	484.07 ± 17.86	0.45 ± 0.03	−31.73 ± 0.06	
50 A	60	30	257.13 ± 2.14	0.10 ± 0.02	−36.40 ± 0.98	
	15	30	338.36 ± 4.18	0.20 ± 0.02	−32.53 ± 0.50	
70 A	60	30	496.80 ± 4.64	0.36 ± 0.02	−30.67 ± 0.25	
	15	30	294.30 ± 0.75	0.10 ± 0.01	−32.53 ± 50	

Note Data represented as mean ± standard deviation (SD)

3.2 Effect of Surfactant on SLNs Physicochemical Properties

The formation of SLNs is due to the crystalline lipid core of the nanoemulsion through the cooling process. Crystallization may be led to the aggregation and form larger particles. Size is one of the critical factors in determining the uptake pathway of nanoparticles into the cell. Nanoparticles with the size range from 250 nm to 3 μm can internalize to the cell through phagocytosis, whereas clathrin- or caveolin-mediated endocytosis uptake pathway requires smaller particles (120–150 nm) [28, 29]. Therefore, the SLNs should have a size of less than 300 nm.

It has been reported that the quality of SLNs was affected strongly by the type of surfactants and their concentration [30]. Specific surfactants were added to the lipid phase regulating stearic acid, such as P407, P188, Tween 80, Span 80. Then, we use the same ultrasonication parameters for all the formulation (from SLN1 to SLN4).

From Table 3, the size of SLN2 was smaller than the others, this can be explained by the different velocity of the coverage of the new lipid surface [1]. Moreover, each surfactant differences in their viscosity and the higher viscous samples require higher power than the other [25]. Therefore, different emulsifiers might need different sonication parameters.

Based on the result of the above tests, the formulation of stearic acid 1000 mg, P188 200 mg in 20 mL of distilled water gave a good influence on size. The formulation of berberine loaded SLNs was prepared by the method described before. The particle size and zeta potential of it were measured (Table 4).

The size was around 221–270 nm, and PDI was small (less than 0.2) which suggests that all particles are sized in a narrow range, and thus the SLNs may be considered homogeneous in size. This is because the PDI indicates the uniformity of a sample based on size [31].

Table 3 The effect of different surfactant on the size of SLNs

Surfactant	Size (nm)	PDI
Poloxamer 407	312.90 ± 8.19	0.13 ± 0.04
Poloxamer 188	257.13 ± 2.14	0.10 ± 0.02
Tween 80	314.90 ± 3.85	0.16 ± 0.03
Span 80	661.70 ± 5.48	0.40 ± 0.04

Note Data represented as mean ± standard deviation (SD)

Table 4 Particle size, PDI, and Zeta potential of berberine-loaded SLNs

Code	Size (nm)	PDI	Zeta potential (mV)
BN1	221.50 ± 2.02	0.10 ± 0.02	−28.03 ± 0.21
BN2	231.57 ± 0.25	0.12 ± 0.01	−23.07 ± 0.25
BN3	232.80 ± 1.55	0.12 ± 0.02	−19.93 ± 0.59
BN4	270.57 ± 0.85	0.18 ± 0.01	−15.23 ± 0.42

Note Data represented as mean ± standard deviation (SD)

Table 5 Particle size and PDI of berberine-loaded SLNs after spray drying

Code	Size (nm)	PDI
BN1	556.90 ± 39.64	0.56 ± 0.02
BN2	315.30 ± 13.24	0.39 ± 0.04
BN3	367.90 ± 12.20	0.46 ± 0.01
BN4	315.10 ± 7.68	0.41 ± 0.03

Note Data represented as mean ± standard deviation (SD)

The zeta potential calculation provides details on the dispersion storage stability. When the electrical repulsion is high, the aggregation of charged particles is less observed and thus enhance the stability of nanoparticles [32].

3.3 Effect of Spray-Drying Process on Berberine-Loaded SLNs Physicochemical Properties

The berberine loaded SLNs were subjected to spray-drying with the operating conditions mentioned above. From Table 5 and Fig. 3, the size and PDI of all berberine-loaded SLNs formulation were increased, so SLNs exhibited more agglomerated particles after the spray-drying process. The temperature during the spray-drying process is 110 °C, which is higher than the melting point of stearic acid, so the re-crystallization of the lipid core will occur when the temperature decreases in the

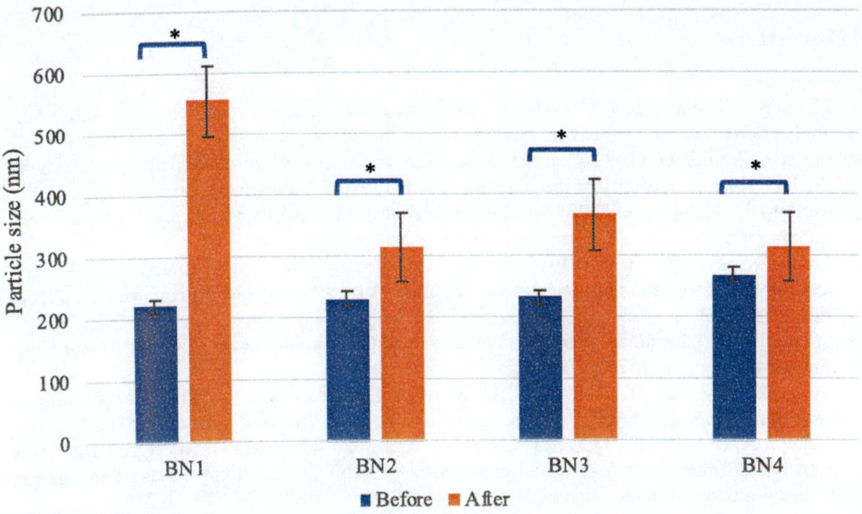

Fig. 3 Mean diameter of berberine-loaded SLNs before and after spraying with lactose. *Note* Data represented as mean ± standard deviation (SD); $*p < 0.05$

drying chamber. In addition, a strong shear force that is required in the spray-drying process can elevate the kinetic energy leading to frequent particle collision [33, 34]. Therefore, adding lactose can reduce the influence of temperature and shear forces, because they will form a layer around particles. This layer can prevent the agglomeration of melted lipid droplets and against shearing off from the particle surface [35].

4 Conclusion

In this study, four surfactants were applied to prepare SLNs, and each surfactant indicates different effects on the characteristics of solid lipid nanoparticles. SLNs dispersions can be converted to fine powders by spray drying technology. Increasing temperature and kinetic energy during the spray drying process result in particle growth. To reduce these effects, and prepare stable SLNs with minimal toxicity, carbohydrate was added as a natural and safe coating material.

Acknowledgements This research is funded by International University, Vietnam National University Ho Chi Minh City under grant number: SV2019-BME-09/HĐ-KHCN.

This research is funded by Vietnam National University HoChiMinh City (VNU-HCM) under grant number NCM2020-28-01".

Conflicts of Interest The authors have no conflicts of interest to declare.

References

1. Mehnert W, Mader K (2012) Solid lipid nanoparticles: production, characterization and applications. Adv Drug Deliv Rev 64:83–101
2. Freitas C, Muller RH (1999) Correlation between long-term stability of solid lipid nanoparticles (SLN^{TM}) and crystallinity of the lipid phase. Eur J Pharm Biopharm 47(2):125–132
3. Muller RH, Runge SA (1998) Solid lipid nanoparticles (SLN) for controlled drug delivery. Submicron Emulsions Drug targeting Delivery 22:219–234
4. Eldem T, Speiser P, Hincal A (1991) Optimization of spray-dried and congealed lipid micropellets and characterization of their surface morphology by scanning electron microscopy. Pharm Res 8(1):47–51
5. Speiser P (1990) Lipidnanopellets als Trägersystem für Arzneimittel zur peroralen Anwendung. European Patent EP 167825, 0167825
6. Muller RH, Shegokar R, Keck MC (2011) 20 years of lipid nanoparticles (SLN & NLC): present state of development & industrial applications. Curr Drug Discovery Technol 8(3):207–227
7. Beloqui A, Solinís MA, Delgado A, Evora C, del Pozo-Rodríguez A, Rodríguez Gascón A (2013) Biodistribution of nanostructured lipid carriers (NLCs) after intravenous administration to rats: influence of technological factors. Eur J Pharm Biopharm 84(2):309–314
8. Beloqui A, Solinís MÁ, Gascón AR, del Pozo-Rodríguez A, des Rieux A, Préat V (2013) Mechanism of transport of saquinavir-loaded nanostructured lipid carriers across the intestinal barrier. J Controlled Release 166(2):115–123

9. Keck CM, Hommoss A, Muller RH (2008) Lipid nanoparticles (SLN, NLC, LDC) for the enhancement of oral absorption. In: Walker RB, Hadgraft J, Roberts MS, Lane ME (eds) Modified release drug delivery systems, vol 1, 2nd ed, pp 269–286
10. Müller RH, Mäder K, Gohla S (2000) Solid lipid nanoparticles (SLN) for controlled drug delivery – a review of the state of the art. Eur J Pharm Biopharm 50(1):161–177
11. Corrias F, Lai F (2011) New methods for lipid nanoparticles preparation. Recent Pat Drug Deliv Formul 5(3):201–213
12. Ricci M, Puglia C, Bonina F, Di Giovanni C, Giovagnoli S, Rossi C (2005) Evaluation of indomethacin percutaneous absorption from nanostructured lipid carriers (NLC): in vitro and in vivo studies. J Pharm Sci 94(5):1149–1159
13. Freitas C, Lucks JS, Müller RH (1994) P238 effect of storage conditions on long-term stability of "solid lipid nanoparticles" (SLN) in aqueous dispersion. Eur J Pharm Sci 2(1–2):178
14. Broadhead J, Edmond Rouan SK, Rhodes CT (1992) The spray drying of pharmaceuticals. Drug Dev Ind Pharm 18(11–12):1169–1206
15. Freag MS, Saleh WM, Abdallah OY (2018) Laminated chitosan-based composite sponges for transmucosal delivery of novel protamine-decorated tripterine phytosomes: ex-vivo mucopenetration and in-vivo pharmacokinetic assessments. Carbohyd Polym 188:108–120
16. Yu F, Ao M, Zheng X, Li N, Xia J, Li Y et al (2017) PEG–lipid–PLGA hybrid nanoparticles loaded with berberine–phospholipid complex to facilitate the oral delivery efficiency. Drug Deliv 24(1):825–833
17. Spinozzi S, Colliva C, Camborata C, Roberti M, Ianni C, Neri F et al (2014) Berberine and its metabolites: relationship between physicochemical properties and plasma levels after administration to human subjects. J Nat Prod 77(4):766–772
18. Battu SK, Repka MA, Maddineni S, Chittiboyina AG, Avery MA, Majumdar S (2010) Physicochemical characterization of berberine chloride: a perspective in the development of a solution dosage form for oral delivery. AAPS PharmSciTech 11(3):1466–1475
19. Imanshahidi M, Hosseinzadeh H (2008) Pharmacological and therapeutic effects of *Berberis vulgaris* and its active constituent, berberine. Phytother Res 22(8):999–1012
20. Singh T, Vaid M, Katiyar N, Sharma S, Katiyar SK (2011) RETRACTED: Berberine, an isoquinoline alkaloid, inhibits melanoma cancer cell migration by reducing the expressions of cyclooxygenase-2, prostaglandin E2and prostaglandin E2receptors. Carcinogenesis 32(1):86–92
21. Birdsall TC (1997) Berberine: therapeutic potential of alkaloid found in several medicinal plants. Altern Med Rev 2:94–103
22. Kulkarni SK, Dandiya PC, Varandani NL (1972) Pharmacological investigations of berberine sulphate. Jpn J Pharmacol 22(1):11–16
23. Huang CG, Chu ZL, Yang ZM (1991) Effects of berberine on synthesis of platelet TXA2 and plasma PGI2 in rabbits. Zhongguo yao li xue bao Acta pharmacologica Sinica 12(6):526
24. Salminen H, Ankenbrand J, Zeeb B, Bönisch GB, Schäfer C, Kohlus R, Weiss J (2019) Influence of spray drying on the stability of food-grade solid lipid nanoparticles. Food Res Int 119:741–750
25. Qsonica. Sonicator ultrasonic processor operation manual. https://cdn.shopify.com/s/files/1/1726/3473/files/q700-manual.pdf?
26. Eskandarloo H, Badiei A, Behnajady MA, Ziarani GM (2016) Ultrasonic-assisted degradation of phenazopyridine with a combination of Sm-doped ZnO nanoparticles and inorganic oxidants. Ultrason Sonochem 28:169–177
27. Hao H, Wu M, Chen Y, Tang J, Wu Q (2004) Cavitation mechanism in cyanobacterial growth inhibition by ultrasonic irradiation. Colloids Surf, B 33(3–4):151–156
28. Panariti A, Miserocchi G, Rivolta I (2012) The effect of nanoparticle uptake on cellular behavior: disrupting or enabling functions? Nanotechnol Sci Appl 5:87
29. Rejman J, Oberle V, Zuhorn IS, Hoekstra D (2004) Size-dependent internalization of particles via the pathways of clathrin-and caveolae-mediated endocytosis. Biochem J 377(1):159–169
30. Müller RH, Mehnert W, Lucks JS, Schwarz C, Zur Mühlen A (1995) Solid lipid nanoparticles (SLN): an alternative colloidal carrier system for controlled drug delivery. Eur J Pharm Biopharm 41(1):62–69

31. Mudalige T, Qu H, Van Haute D, Ansar SM, Paredes A, Ingle T (2019) Characterization of nanomaterials: tools and challenges. In: Nanomaterials for food applications. Elsevier, Amsterdam, pp 313–353
32. Jain R, Savla H, Naik I, Maniar J, Punjabi K, Vaidya S, Menon M (2018) Novel nanotechnology based delivery systems for chemotherapy and prophylaxis of tuberculosis. In: Handbook of nanomaterials for industrial applications. Elsevier, Amsterdam, pp 587–620
33. Fang JY, Fang CL, Liu CH, Su YH (2008) Lipid nanoparticles as vehicles for topical psoralen delivery: solid lipid nanoparticles (SLN) versus nanostructured lipid carriers (NLC). Eur J Pharm Biopharm 70(2):633–640
34. Müller RH, Radtke M, Wissing SA (2002) Nanostructured lipid matrices for improved microencapsulation of drugs. Int J Pharm 242(1–2):121–128
35. Freitas C, Müller RH (1998) Spray-drying of solid lipid nanoparticles (SLNTM). Eur J Pharm Biopharm 46(2):145–151

Lipid-Based Vesicles Containing Rutin: Phytosome and Niosome

An Thi Tran, Dat Van-Tien Nguyen, Uyen Thi Nguyen, Long Minh Tran, Phuoc Long Truong, and Van Hong Nguyen

Abstract Rutin is a flavonoid which has a high effect on strengthening blood vessels, antioxidant, anti-inflammation, and even preventing cancer. However, the critical physicochemical properties of rutin limit its therapeutic effectiveness. This project aims to prepare two lipid-based nanoparticles: phytosome and niosome to increase the bioavailability of rutin. Lipid-based vesicles containing rutin were prepared by the thin-film hydration method. Rutin phytosome and niosome were formed by phospholipid, cholesterol, and surfactants. The effect of excipient combination and fabrication process parameters on the size and zeta potential were evaluated. Drug loading capacity and encapsulation efficiency were also investigated. Both phytosome and niosome formed nanosize vesicles which approximate 200–300 nm with an acceptable polydispersity index (PDI) (0.22–0.23) and zeta potential (absolute value nearly 30 mV), that indicates the good homogeneity and electronic surface interaction. Furthermore, these nanocarriers have succeeded in drug-loading and encapsulation, specifically, the percentages of loading drugs are from 26 to 31%, and the efficiency of encapsulation is 64 and 66%. Lipid-based vesicles were successfully prepared with lecithin in phytosome and Span 60 in niosome. Phytosome and niosome loading rutin are potential nano-drug delivery systems which can be practically applied in cosmetics and medicine.

Keywords Rutin · Phytosome · Niosome · Nano-carrier · Nanomaterials

A. T. Tran · D. V.-T. Nguyen · U. T. Nguyen · P. L. Truong · V. H. Nguyen (✉)
Pharmaceutical Engineering, School of Biomedical Engineering, International University, Vietnam National University Ho Chi Minh City (VNU-HCMC), Ho Chi Minh City 700000, Vietnam
e-mail: nhvan@hcmiu.edu.vn

L. M. Tran
School of Biotechnology, International University, Vietnam National University Ho Chi Minh City (VNU-HCMC), Ho Chi Minh City 700000, Vietnam

V. Van Toi et al. (eds.), *8th International Conference on the Development of Biomedical Engineering in Vietnam*, IFMBE Proceedings 85,
https://doi.org/10.1007/978-3-030-75506-5_26

1 Introduction

1.1 Rutin

Rutin is usually extracted from *Sophora japonica* flower buds, a kind of herb that has grown popular in Chinese and Japan [1]. Americans and Europeans usually use *Sophora japonica* for rutin extraction because of its wide medicinal potentials. In Vietnam, *Sophora japonica* is grown many in the North, especially, Thai Binh province for the use of anti-inflammatory, decreasing cholesterol in the blood, etc. mainly under oral administration [1]. Traditionally, rutin is applied widely in strengthening capillary wall endurance, supporting and treating bleeding syndrome, vascular sclerosis, purpura, etc.

Rutin expresses antioxidant, anti-inflammatory [2], anti-thrombotic, anti-platelet aggregation, anti-cancer, liver protection, vascular stability, lowering blood pressure, reducing blood fat bioactivities [3]. In the field of eye treatment, rutin can be used for cases of retinitis with hemorrhage, bleeding in the fundu [4]. For effects on the cardiovascular system [5], rutin inhibits platelet aggregation, reduces capillary permeability, increases vascular wall durability, improves blood flow, and improves blood circulation due to their ability to constrict blood vessels. In addition, rutin restores the elasticity of damaged capillaries, increases resistance, maintains the normal condition of capillaries, and ensures the capillaries retain their metabolic function [3, 4]. In the treatment of diabetes, rutin is effective in antidiabetic activity [6], by inhibiting cytokinematory inflammation and improving antioxidant and plasma lipid profiles in a high-fat diet in combination with streptozotocin causing type 2 diabetes. Rutin, therefore, may be useful to use in combination with standard diabetes medications [1, 4].

Rutin is presented molar mass 610.52 g/mol, melting point index 610.52 g/mol, 242 °C (468 °F, 515 °K). Its appearance is light yellow crystal. According to ChemAxon, Log P of rutin is equal to − 0.87. The solubility of rutin is shown by the following table, and its solubility in water is lower than solubility in an organic solvent (Table 1).

Because of low solubility and low permeability limiting rutin absorption in the human body, many studies have been concentrated on improving these drawbacks.

Rutin, or quercetin-O-glycoside, is a flavonoid consisting of aglycon that—quercetin with two sugar groups, glucose and rhamnose (Fig. 1). It is firstly found in *Ruta graveolens*, so the name of rutin appears from here. Chemically, rutin is a glycoside comprising of flavonoid aglycone quercetin along with disaccharide rutinose. It is demonstrated by several pharmacological activities such as antioxidant, cytoprotective, vasoprotective, etc. [7].

Table 1 Solubility of rutin in the different solvent [5]

Solvent	Solution at room temperature mg/100 mL	Solution at boiling temperature mg/100 mL
Water	0.013	0.50–0.55
Methanol	5.5	48.0
Ethanol	0.55	36.0
Isopropanol	1.45	7.0
n-propanol	0.6	4.0
n-butanol	…	70.0
Acetone	0.56	2.0
1,4-dioxane	…	12.6
Pyridine	37.3	…

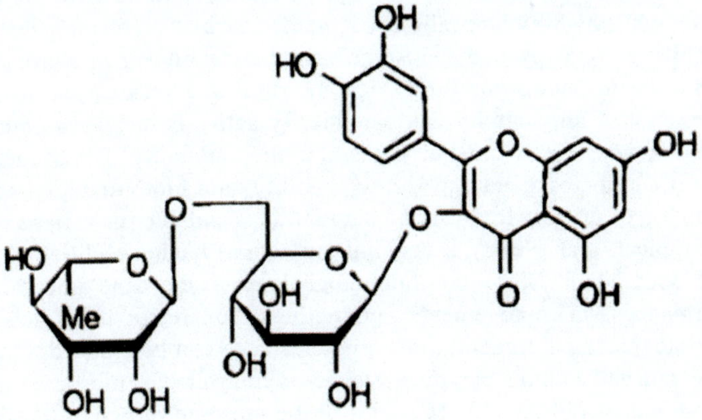

Fig. 1 Molecular structure of rutin [3]

1.2 Phytosome and Niosome

Advantages and disadvantages of phytosome and noisome: Phytosomes are novel drug carriers fabricated from phospholipids mainly containing herbal hydrophilic bioactive compounds. Phytosomes are potentially wide applications in functional food, cosmetics, and pharmaceutical products. Not only phytosomes improve drug absorption, but they also are safe [2] biodegradable and could be able for loading in various dosage forms [8, 9]. Phytosome rutin has some advantages that enhance the quality of drugs. First of all, it increases stability due to the chemical bond created between the polar head of the phospholipids and the herb. Secondly, the absorption of pharmaceutical substances is increased due to high solubility in oil, then increasing bioavailability, reducing dosage but increasing therapeutic effect. The next benefit is that the phytosome is good for the liver due to phospholipids carriers. Furthermore, the active ingredient is directed to the target tissue more effectively, and intracellular transport increased. It is suitable for various usages of parenteral, topical, oral, etc. In addition, it can increase permeability absorption through the skin due to the lipid properties of the shell of the phytosome; the active ingredient easily goes through lipid-rich biological membranes. Finally, phytosome increases the half-life of active ingredients that leads to the increase of active time of the active ingredient in the body, the biological effect of the compound [9–12]. Although phytosome rutin is a good vesicle, it has two disadvantages. Firstly, most phytosome preparation methods use toxic organic solvents such as dichloromethane, methanol, etc. to dissolve the lipid, affecting human health and the environment. Secondly, it makes phytosome by the supercritical method can overcome this disadvantage but requires a complicated technique and requires high pressure of 3000–4500 psi [9–12].

Niosome is formed by non-ionic surfactants as the main component and cholesterol in the case of both in contact with the aqueous medium. The surfactant molecules tend to orient and arrange themselves the way hydrophilic heads face outward creating an outer layer while hydrophobic tails face inward to each other, forming the structure of a closed bilayer. As a result, the niosome has spherical bilayer structure formation [13, 14]. Similar to phytosome rutin, niosome rutin is a vesicle, and it has some advantages. First of all, stability and osmotically active is increased compared to normal rutin powder due to smaller size and higher solubility. The second advantage is that it can increase entrapped drug stability, oral bioavailability, as well as skin penetration of drugs. Moreover, niosome rutin is suitable for various usages of parenteral, topical, oral. Nextly, it contains surfactants having abilities of biocompatibility, biodegradability, and non-immunogenicity. At the same time, therapeutic drug performance is an improvement by protecting it from restricting effects to target cells and biological environment; thus, drug clearance can be reduced. Finally, the drug release rate and administer typical vesicles is controlled by niosome rutin in the external non-aqueous phase [13, 15–17]. But the niosome rutin has several disadvantages. Firstly, some physical instability problems may occur like aggregation and fusion of vesicles. The other weakness of niosome is that the shelf life of the product may be affected due to niosomal vesicles hydrolysis and entrapped drug leakage. Finally, this is a time-consuming process, high production cost, and low practical production yield. In other words, it is costly and time-consuming to use sophisticated instruments such as extrusion, probe sonication for various size-reduction procedures [13–17].

Preparation of niosomes and phytosomes: The solvent evaporation method is most prevalent in preparing phytosomes and niosomes because of its convenience, easy to experiment, does not require high technology, easy to supplement, and improvements [9–11]. Phospholipids (in phytosomes) or surfactants (in niosomes), active ingredients, and other ingredients are dissolved in suitable organic solvents and mix them together. The requirement of the solvent mixture is for a homogeneous solution. The solution is then put into a round-bottom flask of the rotary evaporator; over a period and the appropriate temperature and velocity to form bonds in the complex. At the end of this process, the organic solvent is removed from the solution by evaporation with proper pressure at 100 Pa; then, a thin film is obtained. From this film, it is possible to hydrate directly in a rotating distillation using distilled water similar to the Bangham method to create a crude phytosome suspension. After that, the sample was reduced and uniformly sized particles by ultrasonic probe [9–11, 18].

Both phytosome and niosome are quite similar in structure (Fig. 2) and preparation, the key make a difference between niosome and phytosome is that the nonion surfactants are used to prepare niosome nanocarrier while phospholipid for fabricating phytosome. But they have a special structure. Niosome is built by the self-combination method, nonionic surfactants and cholesterol are combined in an aqueous phase, which results in closed bilayer structures. Phytosome is built by the reaction of a stoichiometric amount of the phospholipid with the standardized extractor polyphenolic constituents (like simple flavonoids) in a nonpolar solvent [7, 13–17, 19–21].

Evaluation: Evaluation is a significant step in any project because it appreciates the quality of the result of research. Phytosome and niosome are tested on several parameters: Size distribution and polydispersity index (PDI) indicate the size and the homogeneity of the molecules or particles in the liquid sample. Scanning electron microscopy (SEM) shows the view of surface morphology of nano-carrier, by

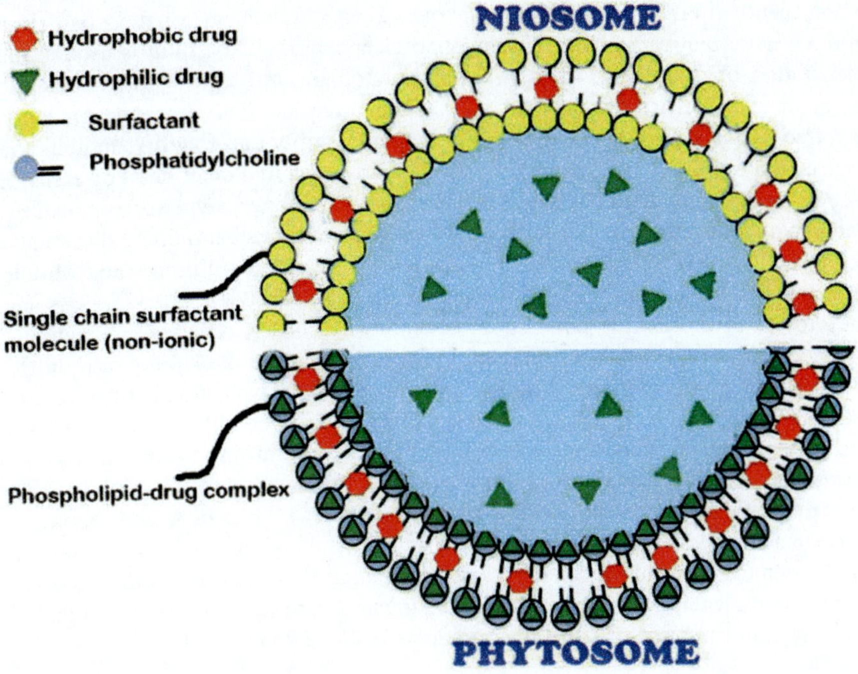

Fig. 2 Structure of niosome and phytosome

scanning the surface of the particles and creating an image through focusing electron beam: the sample interacts with the electron beam. It obtains information about morphology then various signals will be produced. Another morphology analysis is the transmission electron microscope (TEM) which helps researchers to have a view of thin specimens (tissue sections, molecules, etc.) through which electrons can pass generating a projection image. Zeta potential is one important characteristic that appreciates the electronic energy between the slipping plane among nano-carriers. Zeta potential is considered well if the number is lower than − 20 mV. Furthermore, several efficient parameters decide the quality of nano-carrier as encapsulate efficiency, which shows how many percent of phospholipid forming nanocarriers and loading capacity, which indicates the percentage of drug successfully loaded into the vesicle. Finally, stability is tested by letting the sample in a determined condition in a duration of time (10 days, 20 days, 1 month, etc.), then do the above evaluation method to consider whether the samples are still stable as the first day [13, 18, 22].

1.3 Market Products

U.S. Food and Drug Administration (FDA) had approved rutin due to its advanced pharmaceutical effects as mentioned above. Then the demand for rutin had risen in the pharmaceutical industry. In the pharmaceutical industry, rutin is used in the preparation of drugs. It is used as an antimicrobial, antifungal, and anti-allergic agent. It can increase the level of intercellular ascorbic acid, scavenge oxidants, and free radicals. Rutin decreases capillary permeability and fragility; inhibits the destruction of bones; and lowers the risk of heart attack and other heart diseases. It also helps in the treatment of major diseases such as diabetes, hypertension, cancer, and hypercholesterolemia [3]. In terms of geography, the global rutin market can be segregated into North America, Latin America, Asia Pacific, Europe, and Middle East and Africa. With an increase in the prevalence of lifestyle-related diseases and the growth of the elderly population, consumers in North America are inclining toward the use of antioxidant products. This is driving the market for rutin in the U.S. and other parts of North America, as rutin products have antioxidant properties and anti-aging effects. In Latin America, Chile has emerged to be a leading market for rutin, as the consumption pattern in the country is shifting from junk food to nutritious food. The demand for rutin is expected to increase with this change in the food trend. With the presence of developing economies such as India and China and a rise in awareness among consumers, the Asia Pacific region is likely to witness an increase in the demand for rutin products in the next few years. The rutin market in Europe is projected to witness significant growth during the forecast period due to the rising consumption of nutritional products in the region [1, 2, 14].

1.4 Research Objective

Rutin brings many benefits to human health and beauty. However, rutin has low solubility and bioavailability. The cause of this problem is that this pharmaceutical material has a large molecular size, poorly solute in water, so it is quite hard to absorb through both the skin and injection. Moreover, it is easily degradable before reaching gastrointestinal [2–5, 8, 9], so it is difficult to be absorbed through the oral. To overcome this drawback, recent studies around the world have been conducted in many different directions [2, 3]. The introduction of poorly soluble bioactive compounds into finely dispersed systems to increase transport has been studied in many formulations. This study aims to create phytosome and niosome drug-carrier systems that carry rutin to increase the solubility, permeability, bioavailability, and absorption of rutin in the human body.

2 Methodology

2.1 Materials

Rutin was obtained from Sigma Aldrich Company, Phosphatidylcholine (PC) (Egg lecithin), Cholesterol, Chloroform, Methanol, and Span 60 were purchased from Xilong Scientific, and distilled water.

2.2 Preparation of Phytosome and Niosome Loading Rutin

Rutin phytosome (RP) was prepared by using a thin layer hydration method. The experiment was based on molar ratios of Rutin: Lecithin: Cholesterol. Specifically, 50 mg rutin was dissolved in 10 mL ethanol with stirring 100 rpm at 50 °C in 1 h by a magnetic stirrer, while 125 mg lecithin and 25 mg cholesterol were dissolved in 40 mL chloroform. After that, the mixture was taken in a round bottom flask and evaporated in a rotary evaporator at 50 °C and stirring with 50 rpm in 2 h until evaporation of all solvents by rotary-evaporator machine (STRIKE 300 Vertical glassware) and producing a thin, dry film in the round bottom flask. Furthermore, prepared lipid films were placed at room temperature overnight to ensure that the solvent was completely removed. The film was hydrated with 50 mL distilled water in a rotary at 50 °C and stirred with 50 rpm in 3 h. The RP could be prepared by probe sonication (Qsonica Sonicator Q700) with 10,000 Hz in 15 min, off 30 s each minute. Cooling the sample during an ultrasound to limit overheating at 0 °C.

Rutin niosome (RN) has applied the same method to the phytosome but replaced lecithin by surfactants, (Tween and Span) Span 60, a non-ionic surfactant material.

2.3 Evaluation Method

In this research, the following parameters were collected:

Size distribution: indicated the average size and distribution of size peaks of nanoparticles.

Polydispersity index: The heterogeneity of sizes of molecules or particles in a mixture was measuring and called dispersity. The same size, shape, or mass samples is considered uniform. In opposites, non-uniform samples are samples that have inconsistent size, shape, and mass distribution.

Zeta Potential: Zeta potential was the electrical potential at the surface of the particle. This shows the electrokinetic potential in colloidal dispersions.

Entrapment Efficiency: percentages of rutin that are successfully entrapped into nanoparticles.

It is calculated as follows:

$$EE(\%) = \frac{total\ drugadded - unentrapped\ drug}{drug\ added} \times 100\% \quad (1)$$

The total drug added is the amount of drug added at the beginning.

The unentrapped drug can be calculated as follows: centrifuge the sample in 15 min, 25 °C, 10,000 rpm, then collect the liquid which is entrapped rutin in the phytosome; dry the precipitation and weigh them, it is entrapped drug.

Loading Capacity: Loading capacity shows the percent of the core rutin-loaded over the weight of a whole nanoparticle. It can be calculated by the formula:

$$LC(\%) = \frac{total\ entrapped\ drug\ weight}{total\ nanoparticles\ weight} \times 100\% \quad (2)$$

The total entrapped drug is calculated by total drug added minus unentrapped drug.

Total nanoparticles' weight is the total amount of rutin, phospholipid, and cholesterol added at the beginning.

All experiment was carried out triplicate. Collected data will be analyzed with ANOVA test.

3 Result and Discussion

Rutin phytosome and niosome were transparent solutions, in which the niosome sample was more slightly yellowish than the phytosome sample were transparent solutions.

Table 2 Physicochemical properties of RN at different ratio of cholesterol, ratio (rutin: lecithin: cholesterol)

	Z-ave (nm)	PDI	Zeta potential (mV)
RP 1 (1:2:0)	324.3 ± 3.700	0.162 ± 0.009	– 25.1 ± 0.265
RP 2 (1:2:0.1)	224.9 ± 0.808	0.211 ± 0.006	– 34.0 ± 0.346
RP 3 (1:2:0.2)	224.5 ± 1.358	0.237 ± 0.004	– 34.2 ± 0.611
RP 4 (1:2:0.4)	212.6 ± 3.617	0.244 ± 0.009	– 31.8 ± 0.551
RP 5 (1:2:0.8)	225.8 ± 2.597	0.275 ± 0.019	– 35.5 ± 1.020

3.1 Materials

Phytosomes formed from lecithin, cholesterol, and rutin with different ratios express various sizes, PDI and zeta potential. The effect of the presence of cholesterol on size, PDI, and zeta potential of phytosome was evaluated. As shown in Table 2, sample 1 without the presence with cholesterol got the best PDI (0.162 ± 0.009) which means it has the best homogeneity. The higher amount of cholesterol in phytosome formulation, the bigger PDI they got, as with ratio 0.1 in sample 2, PDI (0.2177 ± 0.006), and 0.275 ± 0.019 with ratio 0.8 in sample 5. Cholesterol increased surface charge of phytosomes which was indicated by the raising of the absolute value of zeta potential in phytosome containing this substance (more than 30 mV). Hence, incorporation of cholesterol improves the solidity of the phytosome membrane and stability of phytosomes neity. The higher amount of cholesterol in phytosome formulation, the bigger PDI they got, as others indicated lower electrical ability.

3.2 Evaluate the Stability of Prepared Phytosomes in 30 days (Table 3)

The stability of prepared phytosomes was evaluated in 30 days (Table 3). After one month of storage in the refrigerator, sample RP 1 and sample RP 4 continued to be highly stable with negligible change (the particle size increased about 5–7 nm). Because sample RP 1 was too large in size, further experiments will be carried out

Table 3 Stability of RP after 30 days stored in 4 °C

	Z-ave (nm)	PDI	Zeta potential (mV)
RP 1	343.38 ± 0.461	0.21 ± 0.005	– 27.24 ± 0.127
RP 2	265.23 ± 1.464	0.30 ± 0.002	– 32.23 ± 0.346
RP 3	295.95 ± 4.236	0.32 ± 0.035	– 37.85 ± 0.551
RP 4	236.17 ± 1.358	0.24 ± 0.002	– 39.41 ± 1.375
RP 5	252.92 ± 0.683	0.39 ± 0.009	– 41.76 ± 1.316

Table 4 Physicochemical properties of RN at different ratio of cholesterol, ratio (rutin: span 60: cholesterol)

	Z-average size (nm)	PDI	Zeta potential
RN 1 (1:2:0)	313.87 ± 0.321	0.32 ± 0.015	– 25.14 ± 0.372
RN 2 (1:2:0.2)	266.74 ± 3.351	0.24 ± 0.007	– 29.64 ± 1.671
RN 3 (1:2:0.4)	262.68 ± 2.902	0.31 ± 0.030	– 25.47 ± 0.984

with a sample RP 4 formulation. The particle size slightly increases 7 nm (from 229.2 ± 1.212 to 236.1 ± 1.358), while PDI and zeta potential can be considered unchangeable in sample RP 4 with rutin: lecithin: cholesterol = 1:2:0.4. Comparing to previous research about rutin phytosome (Hooresfand, Nov 2015) [22] that the size is smaller compare to this study (from 99 to 123 nm), however, the stability was worse. Specifically, the nano-size in that research increased up to 14 μm in just 7 days while in this research, the size does not change significantly in 30 days as below.

3.3 *Effect of Cholesterol on Size, PDI, and Zeta Potential of Niosome*

This result in Table 4 proves that cholesterol affects significantly on PDI and size. To be more detail, without cholesterol, both RN's size and PDI are big (size: 313.87 ± 0.321 nm and PDI: 0.32 ± 0.015), in the opposite, with cholesterol, both size and PDI are decreased (that the sizes are around 265 nm and PDIs are from 0.24 to 0.31) However, the zeta potentials are quite similar between the formula contained cholesterol and not contained cholesterol (around – 25 to – 30 mV).

3.4 *Effect of Sonication Time on Size, PDI, and Zeta Potential of RP and NP*

Sonication method in this study is applying ultrasound directly on the nanoparticles through a probe to arrange the molecules in the solution. In the illustration, after removing the sample out from the rotating flask, phospholipid film particles are dispersed unorganized in hydration media. Sonication energy enhances the mobility of phospholipid molecules, hence leading to the arrangement of phospholipid bilayer particles into forming phytosome and niosome structures. However, too much ultrasound energy can break the structure of particles. Therefore, the size and homogeneity of phytosomes are controlled by sonication time. This research surveyed the effect of time of sonication on the size and homogeneity of niosome and phytosome sample; the results are shown in Table 5.

Table 5 Physicochemical properties of RP, RN at different sonication time

	Sonication time (min)	Z-ave (nm)	PDI
Phytosome	5	277.63 ± 0.874	0.28 ± 0.035
	10	262.97 ± 3.460	0.25 ± 0.017
	15	253.42 ± 0.495	0.25 ± 0.001
	20	312.71 ± 6.686	0.33 ± 0.033
Niosome	5	298.90 ± 4.071	0.41 ± 0.009
	10	227.13 ± 4.120	0.27 ± 0.010
	15	266.73 ± 3.350	0.24 ± 0.007
	20	247.17 ± 3.014	0.25 ± 0.006

In both niosome and phytosome, the time of sonication affected the size and PDI significantly, which show detail in Table 5, and Fig. 3. For 5 min sonication, the sizes of niosome and phytosome were large; these were reduced when increasing the time of sonication. To be more specific, 10 min was the best time of sonication for niosome that the size is smaller (227.13 ± 4.120 nm) and PDI was acceptable (0.27 ± 0.010); after 10 min, both size and PDI were not changed as well (around 250 to 260 nm in size and PDI around 0.24 to 0.25 when sonicating in 15 and 20 min). In the phytosome, 15 min was the best time because it had the smallest size and PDI (253.42 ± 0.495 nm in size and PDI is 0.25 ± 0.001), however, when the sonicated process was too long, the size and PDI were getting worse, this case, both size and PDI increased quickly when sonicating in 20 min (312.71 ± 6.686 nm in size and PDI is 0.33 ± 0.033).

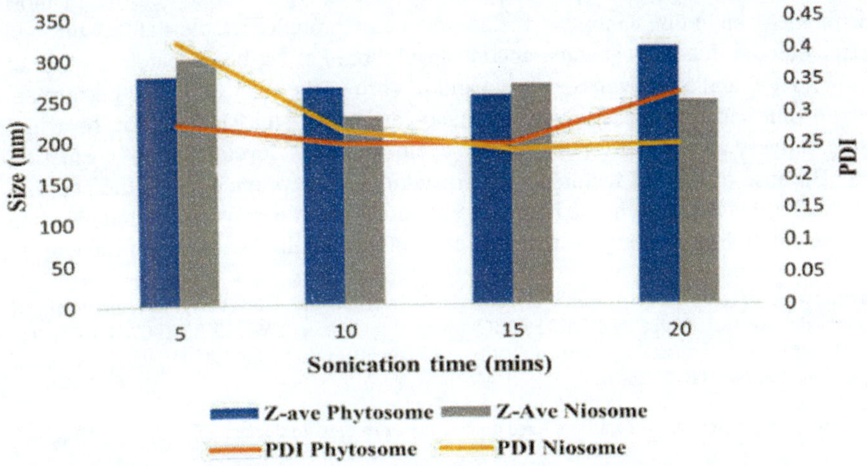

Fig. 3 Physicochemical properties of RP, RN at different sonication time

Table 6 Entrapment efficiency and loading capacity of samples

	Sample	EE%	LC%
Phytosome	RP 4	66.85 ± 2.44	26.70 ± 2.98
Niosome	RN 2	64.32 ± 2.93	31.09 ± 1.58

3.5 *Entrapment Efficiency and Loading Capacity*

This result in Table 6 indicated the efficiency of entrapment and the capacity of drug-loaded nanocarriers quite good; however, it needs to be optimized in further research.

Compared to Hooresfand's research [22], the entrapment efficiencies were from 70 to 80%, EE% in this study was approximate to that result. Besides, Das M. K. and Dalita B. had done a research "Design and evaluation of phyto-phospholipid complexes (phytosomes) of rutin for transdermal application" that the drug entrapment is almost absolutely (approximate 100%) however, the sizes were big, which varied from 600 to 1600 nm [23]. Hence, this research had acceptable drug loading capacity and entrapment efficiency. However, loading capacity in this research is quite low (approximately 30%); therefore, improving the loading capacity will be in further research of this study.

4 Conclusion

This research had loaded rutin successfully in two types of nano-carrier in most parameters, including nano-size, the homogeneity through PDI, the stability through zeta potential, loading capacity, and encapsulation. For further research, the test of drug release and effectiveness in vivo and in vitro of the drug should be performed. Phytosome and niosome are promising nano-carriers with massive potential for wider drug delivery applications, peptide and vaccine delivery, development of cosmetics, etc. Because of various factors like cost, stability, absorption, etc. and the pharmaceutical effects of rutin in the human body, rutin loaded in phytosome and niosome are potential drug delivery systems which would be applied widely over the world.

Acknowledgements This research is funded by International University, Vietnam National University, Ho Chi Minh City (VNU-HCMC) under grant number SV2019-BME-07/HĐ-KHCN.

This research is funded by Vietnam National University HoChiMinh City (VNU-HCM) under grant number NCM2020-28-01.

Conflicts and Interest The authors have no conflict of interest to declare.

References

1. Woerdenbag HJ, Nguyen TM, Vu DV, Tran H, Nguyen DT, Tran TV et al (2012) Vietnamese traditional medicine from a pharmacist's perspective. Exp Rev Clin Pharmacol 5(4):459–477
2. Yang J, Guo J, Yuan J (2008) In vitro antioxidant properties of rutin. LWT-Food Sci Technol 41(6):1060–1066
3. Huang Y, Feng F, Jiang J, Qiao Y, Wu T, Voglmeir J, Chen ZG (2017) Green and efficient extraction of rutin from tartary buckwheat hull by using natural deep eutectic solvents. Food Chem 221:1400–1405
4. Manach C, Morand C, Demigné C, Texier O, Régérat F, Rémésy C (1997) Bioavailability of rutin and quercetin in rats. FEBS Lett 409(1):12–16
5. Krewson CF, Naghski J (1952) Some physical properties of rutin. J Am Pharm Assoc 41(11):582–587
6. Dimov N, Dimova N (1993) Rutin purity determination by high-performance liquid chromatography. Chromatographia 35(9–12):506–508
7. Ganeshpurkar A, Saluja AK (2017) The pharmacological potential of rutin. Saudi Pharm J 25(2):149–164
8. Khanh NV, Phương ĐT, Huyền NT, Hải NT, Thị V, Hằng T (2018) Research on producing phytosome rutin 2:51–59
9. Kumar A, Kumar B, Singh SK, Kaur B, Singh S (2017) A review on phytosomes: novel approach for herbal phytochemicals. Asian J Pharm Clin Res 10(10):41–47
10. Sharma S (2010) Phytosome, an emergency of technique. Microscopy 5:1–7
11. Tripathy S, Patel DK, Barob L, Naira SK (2013) A review on phytosomes, their characterization, advancement and potential for transdermal application. J Drug Deliv Therap 3(3):147–152
12. Un RN, Barlas FB, Yavuz M, Ag Seleci D, Seleci M, Gumus ZP, Guler E, Demir B, Can M, Coskunol H, Timur S (2015) Phyto-niosomes: in vitro assessment of the novel nanovesicles containing marigold extract. Int J Polym Mater Polym Biomater 64(17):927–937
13. Moghassemi S, Hadjizadeh A (2014) Nano-niosomes as nanoscale drug delivery systems: an illustrated review. J Control Release 185:22–36
14. Amoabediny G, Haghiralsadat F, Naderinezhad S, Helder MN, Akhoundi Kharanaghi E, Mohammadnejad Arough J, Zandieh-Doulabi B (2018) Overview of preparation methods of polymeric and lipid-based (niosome, solid lipid, liposome) nanoparticles: a comprehensive review. Int J Polym Mater Polym Biomater 67(6):383–400
15. Ge X, Wei M, He S, Yuan WE (2019) Advances of non-ionic surfactant vesicles (niosomes) and their application in drug delivery. Pharmaceutics 11(2):55
16. Gharbavi M, Amani J, Kheiri-Manjili H, Danafar H, Sharafi A (2018) Niosome: a promising nanocarrier for natural drug delivery through blood-brain barrier. Adv Pharmacol Sci
17. Auda SH, Fathalla D, Fetih G, El-Badry M, Shakeel F (2016) Niosomes as transdermal drug delivery system for celecoxib: in vitro and in vivo studies. Polym Bull 73(5):1229–1245
18. Vankudri R, Habbu P, Hiremath M, Patil BS, Savant C (2016) Preparation and therapeutic evaluation of rutin-phospholipid complex for antidiabetic activity. J Appl Pharm Sci 6(01):090–101
19. Ag Seleci D, Seleci M, Walter JG, Stahl F, Scheper T (2016) Niosomes as nanoparticular drug carriers: fundamentals and recent applications. J Nanomater
20. Chandu VP, Arunachalam A, Jeganath S, Yamini K, Tharangini K, Chaitanya G (2012) Niosomes: a novel drug delivery system. Int J Novel Trend Pharm Sci 2(1):25–31
21. Yeo PL et al (2018) Niosomes: a review of their structure, properties, methods of preparation, and medical applications. Asian Biomed 11(4):301–314
22. Hooresfand Z, Ghanbarzadeh S, Hamishehkar H (2015) Preparation and characterization of rutin-loaded nano phytosomes. Pharm Sci 21(3):145–151
23. Das MK, Kalita B (2014) Design and evaluation of phyto-phospholipid complexes (phytosomes) of rutin for transdermal application. J Appl Pharm Sci 4(10):051–057

Synthesis and Evaluation of Thermoresponsive PLA-PEG-PLA and pH/Temperature-Sensitive OS-PLA-PEG-PLA-OS Hydrogels as Injectable Drug Delivery Application

Hien Thi-Thanh Nguyen, Van-Tien Bui, Anh Cam Ha, and Phu Dai Huynh

Abstract A series of pH/temperature-sensitive OS-PLA-PEG-PLA-OS copolymers that could create in situ hydrogels are synthesized by combining thermoresponsive PLA-PEG-PLA (poly lactide-poly ethylene glycol-poly lactide) with OS (oligoester serine—a pH-sensitive component). The characteristic features of the copolymers were systematically evaluated by analyzing the chemical structure, morphology, sol–gel transition, and in vitro degradation. Thanking to sulfonamide groups of OS blocks with excellent pH-sensitive property, the pentablock copolymers (OS-PLA-PEG-PLA-OS) possessed better stability related to pH and nature in buffer solution than those of the triblock copolymers (PLA-PEG-PLA). It is also found that the OS-PLA_{1960}-PEG_{1750}-PLA_{1960}-OS hydrogel only exhibited the thermoresponsive behavior due to too long hydrophobic block in its structure, whereas, the OS-PLA_{1750}-PEG_{1750}-PLA_{1750}-OS hydrogel (30 wt%) demonstrated dual pH/temperature-sensitive properties with superior characteristics including a

V.-T. Bui · P. D. Huynh (✉)
Department of Polymer Materials, Polymer Research Center, Faculty of Materials Technology, Ho Chi Minh City University of Technology, 268 Ly Thuong Kiet Street, District 10, Ho Chi Minh City, Vietnam
e-mail: hdphu@hcmut.edu.vn

V.-T. Bui
e-mail: tienbv@hcmut.edu.vn

H. T.-T. Nguyen (✉) · A. C. Ha
Faculty of Chemical Engineering, Ho Chi Minh City University of Technology, 268 Ly Thuong Kiet Street, District 10, Ho Chi Minh City, Vietnam
e-mail: hienntthanh@hufi.edu.vn; 1880314@hcmut.edu.vn

A. C. Ha
e-mail: hcanh@hcmut.edu.vn

H. T.-T. Nguyen · V.-T. Bui · A. C. Ha · P. D. Huynh
Vietnam National University Ho Chi Minh City, Linh Trung Ward, Thu Duc District, Ho Chi Minh City, Vietnam

H. T.-T. Nguyen
Faculty of Chemical Engineering, Ho Chi Minh City University of Food Industry, 140 Le Trong Tan Street, Tan Phu District, Ho Chi Minh City, Vietnam

V. Van Toi et al. (eds.), *8th International Conference on the Development of Biomedical Engineering in Vietnam*, IFMBE Proceedings 85,
https://doi.org/10.1007/978-3-030-75506-5_27

safe gel range under pH ranging from 6.0 to 7.8 at the temperature of 37 °C and slow degradation rate. We believe that the present pH/temperature-sensitive hydrogel has a potential application as an injectable carrier for prolonged release of drugs.

Keywords Hydrogels · Drugs delivery · pH/temperature sensitive

1 Introduction

PLA-PEG-PLA (poly lactide-poly ethylene glycol-poly lactide) copolymer has been known as a thermal sensitive hydrogel which can be applied in drug delivery owing to its special properties. The thermoresponsive hydrogel can change its microstructural features corresponding to changing ambient temperature. Accordingly, the hydrogel presents sol (solution) state at or below room temperature, while it possesses gel (non-flowing) state at the physiological temperature [1, 2]. Besides, PLA-PEG-PLA is a BAB type of the amphiphilic copolymer, which contains PEG blocks as hydrophilic components and PLA blocks as hydrophobic counterparts. Both PEG and PLA polymers are approved as synthetic degradable polymers by FDA (Food and Drug Administration) [3]. Moreover, the BAB hydrogel could form a core–shell micelle in aqueous solution with the hydrophobic part in core and the hydrophilic part on the shell [4]. Desirable amphiphilic block copolymers were achieved by varying their block ratio, molecular weight, or chemical structure [5]. Playing a role as a drug agent carrier, PLA-PEG-PLA hydrogel has been exploited with some kind of drugs such as paclitaxel, doxorubicin [6, 7]. It exhibited some critical advantages such as high loading capacity and prolonged release of drug. However, the copolymers also revealed fast degradation rate, forming lactic acid through hydrolysis process. The rapid degradation causes initial burst release of drug [7, 8].

There are several efforts have been made to control degradation rate in order to suppress the initial burst release. Among them, modifying chemical structure of the hydrogel is one of the most effective routes. According to Shim et al. [8], the pentablock copolymer containing sulfamethazine oligomer as pH-sensitive groups exhibited not only both temperature and pH-responsive properties but also demonstrate no pH change of the block copolymer solution even after 1 month of the hydrogel degradation. Therefore, it is expected that incorporating pH adjusting functional groups onto the PLA-PEG-PLA copolymer backbone would be beneficial for effective control of the physicochemical properties of the hydrogel.

There are several amide groups such as acrylamide, carboxamide or sulfonamide, which have been used as pH-sensitive components. These amide groups demonstrated a feature of ionizable characteristic. Among them, sulfonamide group (SO_2NH) can be ionized at high pH value (pH > pKa) and also deionized in low pH condition [9]. As a result, grafting sulfonamide group onto PLA-PEG-PLA triblock copolymer is expected to improve the stability of the hydrogel in low pH environment.

In this study, a series of thermosensitive PLA-PEG-PLA triblock copolymers and pH/temperature-sensitive OS-PLA-PEG-PLA-OS pentablock copolymers were synthesized. ^{1}H NMR (proton nuclear magnetic resonance) spectra are used to analyze the chemical structure of copolymers. Sol–gel transition, morphology, in vitro degradation of copolymers were examined and compared.

2 Materials and Methods

2.1 Materials

Poly (ethylene glycol) (PEG) (Mw = 1750) and Stannous octoate [$Sn(Oct)_2$] were purchased from Sigma–Aldrich. D,L-Lactide was bought from Tokyo Chemical Industry. Dicyclohexyl carboimide (99%), 4-(dimethyl amino) pyridine (99%), D,L-Serine and Benzenesulfonyl chloride (> 98%) were obtained from Aladdin (China). Phosphate-buffered saline (PBS) and dimethylformamide (DMF) solvent were supplied from Prolabo.

2.2 Methods

Synthesis of temperature sensitive PLA-PEG-PLA triblock copolymers. PLA-PEG-PLA triblock copolymers were synthesized following a procedure described previously [3, 6, 7]. The polymerization was introduced by ring opening of D,L-lactide in the presence of stannous octoate catalyst and PEG as initiator.

Synthesis of pH/temperature-sensitive OS-PLA-PEG-PLA-OS pentablock copolymers. Firstly, D,L-Serine was modified by benzenesulfonyl chloride to form mSerine following the procedure reported by Nguyen et al. [9, 10]. After that, PLA-PEG-PLA copolymer was reacted with mSerine with 0.1 molar ratio in DMF medium, using DCC and DMAP as catalysts. The synthesis process was conducted in nitrogen gas environment at room temperature for 48 h. Subsequently, the mixture was filtered to remove dicyclohexylurea (DCU) (a byproduct), and then DMF was taken off by rotary evaporation. Finally, the pentanblock copolymer was purified by using excess diethyl ether and dried under vacuum at 60 °C for 48 h.

Characterization. *^{1}H NMR (proton nuclear magnetic resonance) analysis.* The ^{1}H-NMR data was recorded from Bruker Advance machine at 500 MHz with DMSO, $CDCl_3$ containing 0.03% (v/v) Tetramethylsilane (TMS) as solvent signal. The ^{1}H-NMR spectra were used to determine the molecular structure of mSerine, copolymers. The composition of triblock copolymers was calculated by comparing the proton peak integration of PEG and PLA.

HR TEM (high resolution transmission electron microscope) analysis. Morphology and size of hydrogels were observed by HR TEM (TEM, JEM 2100, Jeol, Japan).

Sol–gel phase transition measurement. Sol–gel transition studies were performed via test tube inverting method. The sol phase was characterized by stage of flow, whereas gel phase was non flow stage. The copolymers were solubilized in PBS buffer at the given concentration and stored overnight at 10 °C. Then, tubes were placed in water bath, and temperature was elevated gradually from 15 to 60 °C with step of 1 °C/min. The sample vials were equilibrated at each temperature for 5 min. The sol–gel transition behavior was determined by tilting the vial horizontally after keeping it at a constant temperature for 10 s [2, 9].

In vitro degradation of copolymers. The polymeric aqueous solution (1 mL) was injected into a test tube and incubated (37 °C) in 1 h for formed gel. After that, 3 mL of PBS solution (pH 7.4) was added on the top of the gel. Each week, samples were observed on the surface of water-gel and measured pH [2].

3 Results and Discussion

3.1 Synthesis and Characterization of PLA-PEG-PLA and OS-PLA-PEG-PLA-OS Copolymers

A series of PLA-PEG-PLA copolymers (constant PEG molecular weight) with different LA/EO ratio were synthesized, which are denoted as T-1.89, T-2.00, and T-2.24. The chemical shift δ (ppm) of ^{1}H-NMR spectrum presented at 5.2 (1H, CO–CH (CH_3)–O of the lactic acid), 3.6 (4H, –CH_2–CH_2–O– of PEG), 1.5 (3H, CO–CH(CH_3)–O of the lactic acid) in Fig. 1b. Polymerization degree (DP) of each PLA block and molar weight of the synthesized triblock copolymers were calculated by comparing the intensity of the PLA characteristic resonance at 5.2 ppm and that of PEG at 3.6 ppm. The results were shown in Table 1.

As shown in Fig. 1c, ^{1}H NMR data of pentablock copolymers appeared chemical shift δ signals of triblock copolymers (Fig. 1b) and characteristic peaks of aromatic protons (7.6–7.9 ppm) of mSerine (Fig. 1a). In addition, typical methine proton signals of mSerine at 3.78 ppm (Fig. 1a) were replaced by the signals at 3.0 ppm (Fig. 1c) by the linking ester formed between OS and PLA. These results indicated that the reaction between triblock copolymers and mSerine to form pentablock copolymers was successfully performed.

3.2 Sol–gel Transition Behavior of Block Copolymers

All of synthesized PLA-PEG-PLA triblock copolymers were able to dissolve in water to form sol–gel phase at pH of 7.4. However, with different LA/EO ratio (ranging from 1.89 ÷ 2.24), the phase transition diagram of triblock copolymers is diverse,

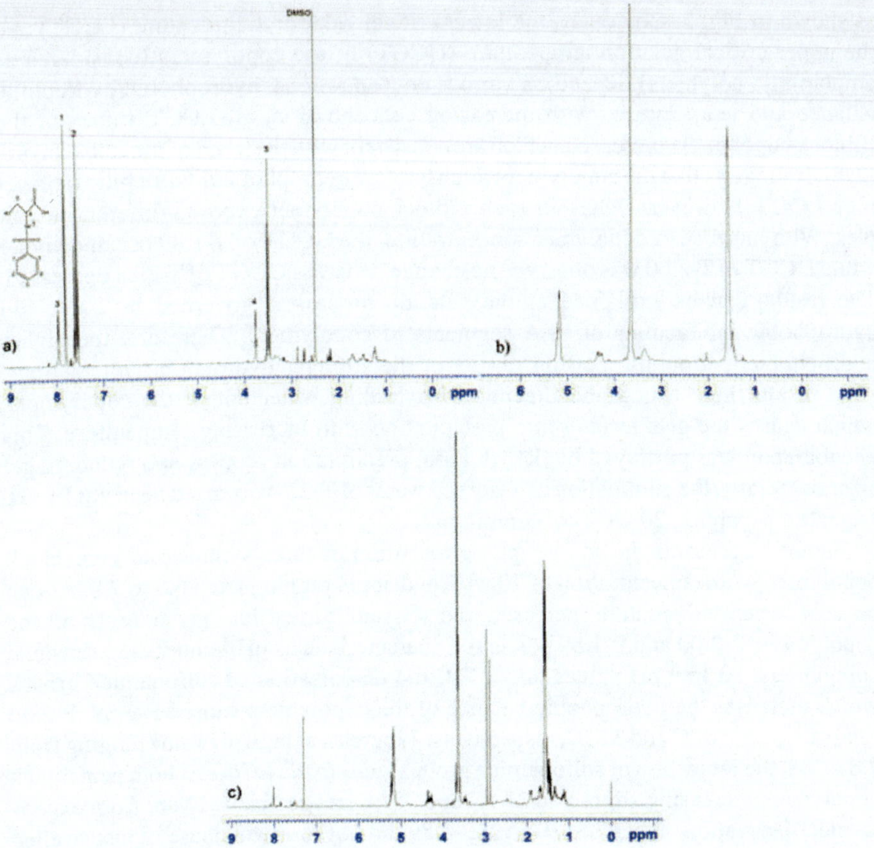

Fig. 1 ^{1}H NMR spectra of: **a** mSerine, **b** PLA-PEG-PLA, **c** OS-PLA-PEG-PLA-OS

Table 1 Molecular mass results of synthesized triblock copolymers

Samples[a]	PLA-PEG-PLA	LA/EO[b]	OS-PLA-PEG-PLA-OS
PLA_{1650}-PEG_{1750}-PLA_{1650}	T-1.89	1.89	P-1.89
PLA_{1750}-PEG_{1750}-PLA_{1750}	T-2.00	2.00	P-2.00
PLA_{1960}-PEG_{1750}-PLA_{1960}	T-2.24	2.24	P-2.24

[a]Derived from ^{1}H NMR results which number average molecular weight of PLA blocks $M_{nPLA} = 2 * DP_{PLA} * 72$ and $DP_{PLA} = DP_{PEG}/(EO/LA)$, EO/LA the ratio of ethylene oxide and lactyl units calculated from ^{1}H NMR spectra

[b]Mass fraction of PLA and PEG blocks

as shown in Fig. 2. Obviously, the lower critical gelation temperature (LCGT) and the upper critical gelation temperature (UCGT) of the copolymers (considered as amphiphilic polymers) depend on various concentrations, hydrophobic/hydrophilic balance and temperature. With increasing concentration, all LCGT values of the triblock copolymers increased, which are consistent with the trend of previous reports [2, 8]. It suggests that the copolymer which has low hydrophobic/hydrophilic ratio get high LCGT. However, UCGT of each triblock copolymer exposed different principles. With increase of copolymer concentration, the UCGT of T-1.89 became higher while UCGT of T-2.00 was observed no change, whereas UCGT of T-2.24 decreased. The results can be explained as the gelation process is governed by interchain hydrophobic interactions of PLA segments of copolymers. Therefore, the higher hydrophobic/hydrophilic portion results in the stronger hydrophobic interactions. As a result, there was a phenomenon of extruding water out of the copolymers, which causes the gels turning into precipitation with increasing temperature. This phenomenon was portrayed by dehydration, precipitation process and reducing gel size. Especially, the elimination of absorbed water of T-2.24 occurred nearly at LCGT (signal*), exception 20 wt% concentration.

Figure 3a presents the sol–gel phase transition of three synthesized pentablock copolymers with concentration of 20 wt% and at pH ranging from 6.8 to 7.8. As can be seen, when temperature increased, the sol state turned into gel state for all the copolymers. P-2.00 and P-1.89 possessed characteristic as pH/temperature sensitive copolymers. At low pH values under 7.4, the deionization of sulfonamide groups which increases the hydrophobic balance of the copolymers caused P-1.89, P-2.00 easily to gel at 30 °C and 31 °C, respectively. Whereas, at high pH value ranging from 7.6 to 7.8, the ionization of sulfonamide groups leads to the LCGT of both pentablock copolymers increasing slightly to 32 °C and 33 °C, respectively. Even, from experimental observation, at pH 7.8, 20 wt% P-1.89 did not form gel phase; it just swelled (still flowed) and excluded the buffer solution with increasing temperature.

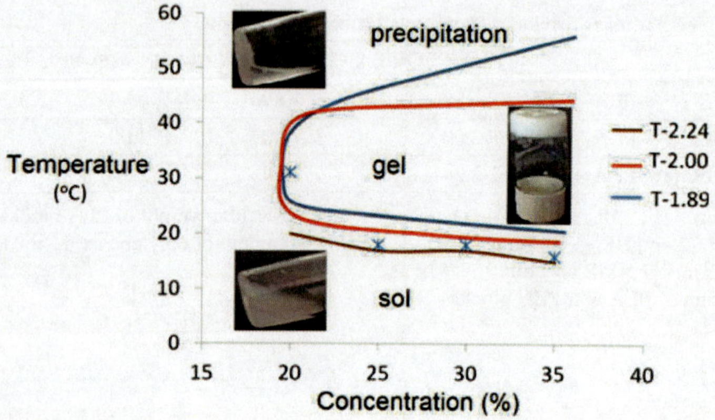

Fig. 2 Sol–gel phase diagram of PLA-PEG-PLA triblock copolymers

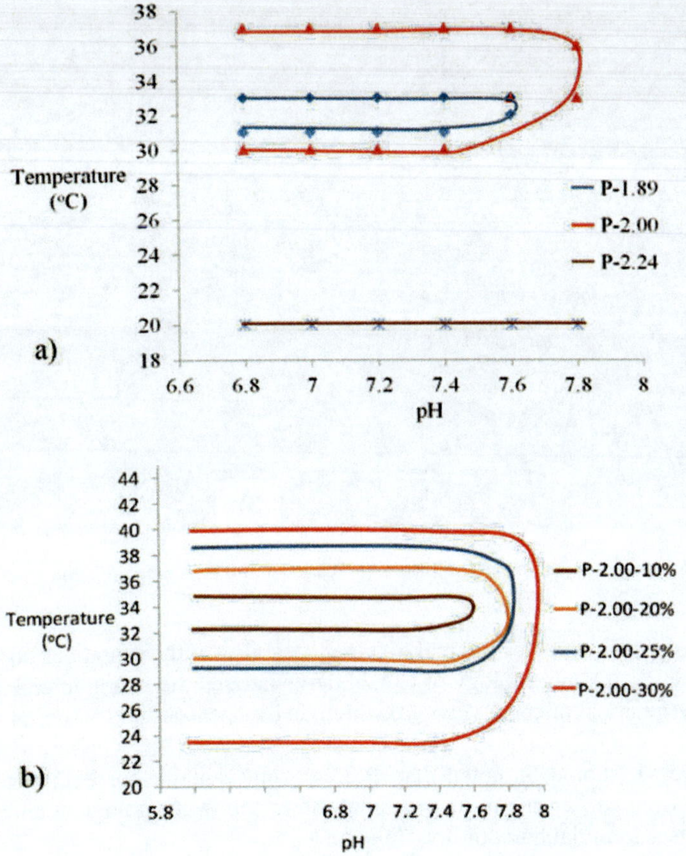

Fig. 3 Sol–gel phase diagram of: **a** 20 wt% of pentablock copolymers, **b** different concentration of P-2.00

In contrast to P-1.89 and P-2.00, P-2.24 pentablock copolymer just exhibited a thermosensitive hydrogel. With pH value ranging from 6.8 to 7.8, the LCGT was the same 20 °C in 20 wt% concentration and the gel immediately crashed out of solution. With high ratio of hydrophobic/hydrophilic, the ester linking strongly of –COOH groups of OS and –OH groups of P-2.24 induced increase of hydrophobic length and molar mass of the block copolymer. It should be noted that the gel formed from P-2.24 was more compact packing. According to Lu et al. [4], the major driving force behind self-assembling of amphiphilic copolymers in aqueous environment is hydrophobic interactions that lower the free energy of the system to form polymeric micelles. Therefore, it is said that P-2.24 formed pentablock micelles. It suggests that hydrophobic OS-PLA component played a role as core, while hydrophilic PEG block served as the shell of the micelle. Comparison to PLA-PEG-PLA block copolymers backbone, hydrophobic interaction induced by multiple

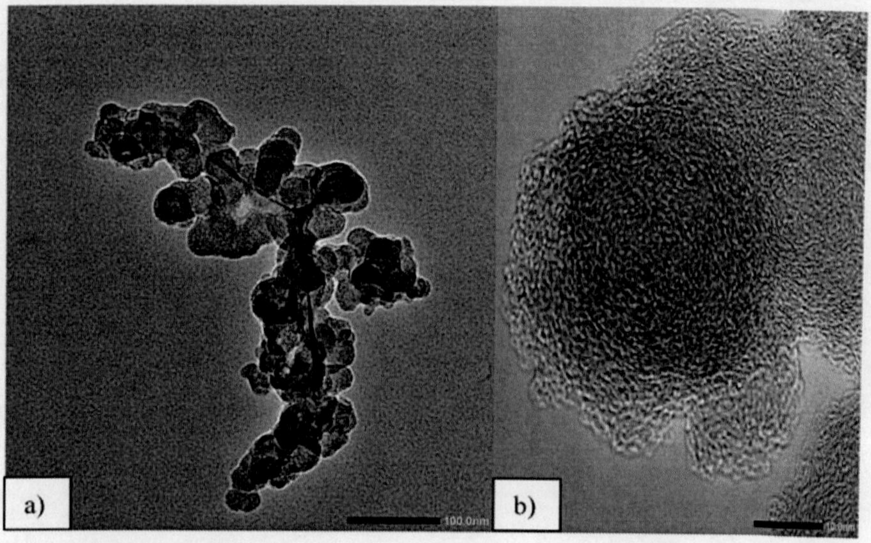

Fig. 4 TEM images of P-2.24 with **a** low magnification and **b** high magnification

sites on polymer molecules can make a kinetic stability of the pentabock copolymers [5]. Obviously, the morphology of P-2.24 demonstrated the strong interchain interactions by spherical micelles (Fig. 4a) which theirs size is around 50 nm in diameter (Fig. 4b).

However, it should be noted that, P-2.24 is not suitable for being used as an injectable drug carrier due to its precipitate state below physiological temperature, which causes to initial burst of drug release.

In general, pentablock copolymers had higher LCGT and lower UCGT with the same triblock copolymers backbone. It is explained by increase in of hydrophobic/hydrophilic ratio of hydrogels. Note that, pentablock copolymers also exhibited the same trend of concentration effect to sol–gel phase behavior, as mentioned previously in Fig. 3b. Higher concentration results in the broader gel range. More importantly, all P-2.00 hydrogels with different concentrations exhibited sharp response at high pH (pH > 7.4). Therefore, the P-2.00 hydrogel with concentration of 30 wt% demonstrated safe gel area in pH ranging from 6.0 to 7.8.

3.3 *In Vitro Degradation of Block Copolymers*

Figure 5 presented the in vitro degradation test of the block copolymers in PBS buffer (initial pH = 7.4) at 37 °C. Although P-2.24 (20 wt%) and T-2.24 (20 wt%) were firm gels (precipitate state), demonstrating good thermodynamic stability at 37 °C, pH 7.4, they were experimented in vitro degradation for comparison with

Fig. 5 Optical images representing in vitro degradation test **a** at beginning and **b** after 4 weeks of copolymers in PBS buffer (initial pH of 7.4) at 37 °C. (1) P-2.24 with 20 wt%, (2) P-2.00 with 30 wt%, (3) T-2.24 with 20 wt% and (4) T-2.00 with 20 wt%

gel state of P-2.00 (30 wt%) and T-2.00 (20 wt%). Obviously, the volumes of gels in tube 1 and tube 3 were smaller than those of tube 2 and tube 4. After 4 weeks, both of the triblock copolymers were disintegrated, and their surface was eroded. In contrarily, gel shapes of the pentablock copolymers were retained. Their colour just turned from initial orange into pale yellow. Considerably, T-2.00 was deemed of more deformation than T-2.24.

Figure 6 presented the change of pH value of the buffer solution corresponding to time of degradation test. As can be seen in Fig. 6, pH of all copolymer solutions decreased with increasing aging time, in which the sharp decrease was observed during the first week while slight change happened from the second week. Among them, the solutions containing triblock copolymers showed much higher decrease rate than pentablock copolymers because pH solutions turned from 7.4 into 3.0 and 5.7 for triblock and pentablock, respectively. These results can be explained by the generation of lactic acid as a degradation product of copolymers, resulting in a decrease of the pH solution. Therefore, these results indicate that triblock copolymer

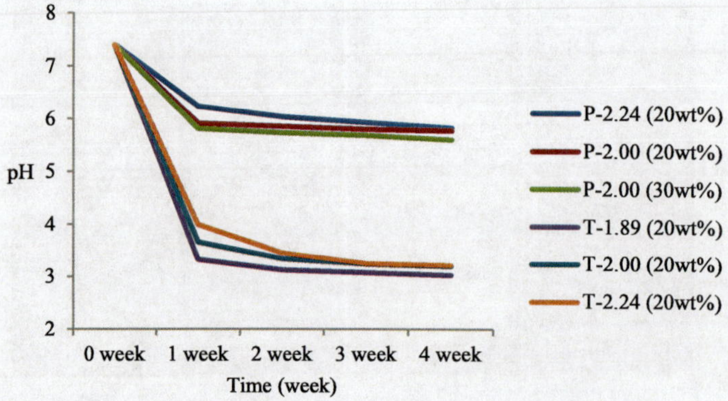

Fig. 6 Change of pH of copolymer solutions corresponding to degradation time

is less stable than pentablock copolymer in the in vitro testing conditions. The superior stability of pentablock copolymers may be attributed to the strong interchain interaction between the OS-PLA hydrophobic blocks. Besides, the degradation rate also depended on the chemical compositions of the copolymers. Correspondingly, the decrease of pH value of the copolymer solution (20 wt%) followed the order of T-1.89 > T-2.00 > T-2.24 > P-2.00 > P-2.24.

Here, P-2.00 (30 wt%) obtains all requirement as injectable drugs carrier.

4 Conclusion

In this work, both temperature sensitive PLA-PEG-PLA and pH/temperature-sensitive OS-PLA-PEG-PLA-OS copolymers were synthesized successfully, which was confirmed by ^{1}H-NMR analysis. Sol–gel transition analysis revealed that P-2.00 and P-1.89 are pH/temperature-sensitive copolymers where P-2.24 just exhibited as a thermosensitive copolymer. In addition, the broader gel range can be achieved with higher copolymer concentration. The TEM images revealed that P-2.24 formed spherical micelles with diameter of 50 nm. The degradation test indicated the higher stability of pentablock copolymers than triblock copolymer and P-2.00 hydrogel with concentration of 30 wt% was found to be suitable for application as injectable drugs carrier.

Acknowledgements We acknowledge the support of time and facilities from Ho Chi Minh city University of Technology (HCMUT), VNU-HCM and Ho Chi Minh City University of Food Industry. This research is funded by the Ministry of Science and Technology of Vietnam under grant number: NĐT.27.KR/17.

Conflicts of Interest The authors declare that they have no conflict of interest.

References

1. Almeida H, Amaral MH, Lobão P (2012) Temperature and pH stimuli-responsive polymers and their applications in controlled and self-regulated drug delivery. J Appl Pharm Sci 2(6):1–10
2. Shi K, Wang Y, Qu Y, Liao J, Chu B, Zhang H, Luo F, Qian Z (2016) Synthesis, characterization, and application of reversible PDLLA-PEG-PDLLA copolymer thermogels in vitro and in vivo. Sci Rep 6:1–15. https://doi.org/10.1038/srep19077
3. Shen X, Liu X, Li R, Yum P, Li C, Su F, Li S (2017) Biocompatibility of filomicelles prepared from poly (ethylene glycol)-polylactide diblock copolymers as potential drug carrier. J Biomater Sci Polym edn. https://doi.org/10.1080/09205063.2017.1344383
4. Lu Y, Park K (2014) Polymeric micelles and alternative nanonized delivery vehicles for poorly soluble drug. Int J Pharm 453(1):198–214
5. Kulthe SS, Choudhari YM, Inamdar NN, Mourya V (2012) Polymeric micelles: authoritative aspects for drug delivery. Des Monom Polym 1–57. https://doi.org/10.1080/1385772X.2012.688328
6. Jelonek K, Li S, Kasperczyk J, Wu X, Orchel A (2017) Effect of polymer degradation on prolonged release of paclitaxel from filomicelles of polylactide/poly (ethylene glycol) block copolymers. Mater Sci Eng, C 75:918–925
7. Mao H, Shan G, Bao Y, Wub ZL, Pan P (2016) Thermoresponsive physical hydrogels of poly(lactic acid)/poly(ethylene glycol) stereoblock copolymers tuned by stereostructure and hydrophobic block sequence. Soft Matter. https://doi.org/10.1039/c6sm00517a
8. Shim WS, Kim SW, Lee DS (2006) Sulfonamide-based pH- and temperature-sensitive biodegradable block copolymer hydrogels. Biomacromolecules 7(6):1935–1941
9. Nguyen DT, Huynh DP (2016) D, l-serine based pH-sensitive oligoester. J Sci Technol 54(5A):45–55
10. Nguyen DT, Phan VHG, Lee DS, Thambi T, Huynh DP (2019) Bioresorbable pH- and temperature-responsive injectable hydrogels incorporating electrosprayed particles for the sustained release of insulin. Polym Degrad Stab 162:36–46

A Colorimetric Aptasensor for Detection of Tetracycline Based on Aggregation of Silver Nanoparticles

Phuoc Long Truong

Abstract Tetracycline is a broad-spectrum antibiotic that can help control microbial infections in human and veterinary. Excessive use of tetracycline in veterinary medicine and animal feeds as antibiotics and growth promotors could lead to antibiotic residues in foodstuff. Tetracycline residues in the food chain can be responsible for undesirable side effects on human health and emergence of antibiotic resistance strains of bacteria in both veterinary and human. Therefore, there is an urgent need to develop a simple, rapid method for determination and quantification of antibiotic residues in animal-derived foods. In this study, a colorimetric aptasensor was designed for sensitive and fast detection of tetracycline based on aggregation of silver nanoparticles with a low cost. Our detection method exploits tetracycline-specific DNA aptamer as the recognition element of an optical sensor. In the absence of tetracycline, silver nanoparticles are still well—dispersed, stable against salt-induced aggregation. In the presence of tetracycline, DNA aptamers detach from the nanoparticle's surface and bind to its target, leading to the aggregation of silver nanoparticles by salt and a noticeable color change from yellow to colorless. The presented aptasensor showed high sensitivity and selectivity toward tetracycline with a limit of detection as low as 100 nM with an analysis time of ~ 20 min. With regard to demonstrating sensitivity and specificity of the sensor, our sensing platform shows a great potential application in rapid detection of antibiotics in the field of biomedicine and food safety.

Keywords Aptasensor · Silver nanoparticle · Tetracycline

P. L. Truong (✉)
School of Biomedical Engineering, International University, Vietnam National University-Ho Chi Minh City (VNU-HCMC), Ho Chi Minh City 700000, Vietnam
e-mail: tplong@hcmiu.edu.vn

V. Van Toi et al. (eds.), *8th International Conference on the Development of Biomedical Engineering in Vietnam*, IFMBE Proceedings 85,
https://doi.org/10.1007/978-3-030-75506-5_28

1 Introduction

Tetracycline is a family of broad-spectrum antibiotics that was produced by the Streptomyces genus of Actinobacteria. It was used for the control of microbial infections in human and veterinary [1]. The use of tetracycline in farming as a feed additive to prove growth rate of cattle and poultry led to its accumulation in the human food chain and foodstuff, which can cause serious threats and side effects to human health including liver damage, tooth discoloration and allergic reactions in persons with hypersensitivity. But most importantly, the misuse of tetracycline can lead to the emergence of harmful strains of bacteria that are resistant to antibiotics in both veterinary and human [2, 3]. To ensure the customer health and prevent the adverse health effects of tetracycline, several organizations including the Food and Agricultural Organization and World Health Organization (FAO/WHO) and the European Union (EU) have established a maximum residue limit (MRL) of tetracycline in various food products. Hence, to control tetracycline in animal-derived foodstuff effectively, it is essential to develop a simple, sensitive assay for rapid detection of tetracycline with a low cost.

Up to now, there are numerous techniques used for detection of antibiotic residues in foods including immunoassays [4], microbiological methods [5], high performance liquid chromatography (HPLC) [6], chemiluminescence [7], electrochemical and capillary electrophoresis [8, 9]. Despite their sensitivity and reliability, these methods still have their shortcomings for food safety control including prerequisite sample pre-treatment, time-consuming, complexity, high cost, sophisticated instrumentation, trained technical personnel and an inability for on-site testing [3, 10]. For these reasons, there is a great need to develop a simple, rapid and accurate detection methodology with a low cost.

Recently, noble nanoparticles have shown a great potential for the development of colorimetric assays owing to their unique optical properties originated from the interaction of free conductive electrons on the surface of nanoparticles with the electromagnetic field of the coming light. This interaction is known as the so-called localized surface plasmon resonance (LSPR) or the particle surface plasmon, which is in charge of the bright color of the nanoparticle colloids that attract people attention for over 1500 years [11–13]. This phenomenon only occurs in a few metals such as silver and gold, and this characteristic has attracted a great interest from the scientific community for biosensing applications [14]. In spite of good chemical stability and biocompatibility of gold nanoparticles for biosensing applications, silver nanoparticles offer better sensitivity for optical biosensing compared to gold nanoparticles [15]. Moreover, the development of colorimetric aptasensors using noble nanoparticles is of great interest in the field of optical biosensing due to its simple preparation for colorimetric detection with the unaided eye [16].

In this study, a simple colorimetric aptasensor was designed for specific and sensitive detection of tetracycline based on the aggregation-induced color change of aptamer modified silver nanoparticles. This strategy is based on the competitive binding between silver nanoparticle and tetracycline with DNA aptamer. The proposed approach showed high sensitivity and selectivity toward tetracycline with a detection limit as low as 100 nM with an analysis time of ~ 20 min. This result demonstrated that this aptasensor could be used as a robust assay for rapid determination and quantification of antibiotics in food.

2 Materials and Methods

2.1 *Materials*

DNA aptamer of tetracycline (5′-CCC CCG GCA GGC CAC GGC TTG GGT TGG TCC CAC TGC GC-3′) was obtained from PHUSA Biochem Co. (Vietnam). Tetracycline hydrochloride was purchased from Sigma-Aldrich, Inc. (USA). Amoxicillin, ampicillin, chloramphenicol and doxycycline were purchased from Bio Basic (Canada). Sodium borohydride ($NaBH_4$, 98%), sodium citrate dihydrate ($Na_3C_6H_5O_7 \cdot 2H_2O$), sodium chloride (NaCl), and silver nitrate ($AgNO_3$, 99%) were purchased from Sigma-Aldrich, Inc. (USA). All essential chemicals were available in AR grade. All glassware was intensively cleaned with a freshly prepared solution of aqua regia and rinsed carefully with ultrapure water.

2.2 *Methods*

Synthesis of silver nanoparticles. Silver nanoparticles (AgNPs) were synthesized by reducing $AgNO_3$ with a mixture of two chemicals, including sodium citrate and sodium borohydride [17]. In brief, under vigorous stirring, $AgNO_3$ (0.5 mL, 100 mM) is added dropwise to 49.5 mL of freshly prepared ice-cold solution that contains 0.30 mM trisodium citrate and 1 mM $NaBH_4$. The resulting solution was centrifuged at 14,000 rpm for 15 min at 4^0C to remove the excess of $NaBH_4$, and the silver nanoparticle was re-suspended in 2.2 mM sodium citrate. After that, the solution was filtered to remove aggregated AgNPs using a filter (0.22 μm). The shape and size of AgNPs were characterized by UV–vis spectrophotometer (Jasco V-730, Japan), and transmission electron microscopy (TEM; JEM-1400, USA).

Preparation of tetracycline aptamer functionalized AgNPs. In our research, DNA aptamer absorbs on the nanoparticle surface and protect AgNPs against the aggregation that was induced by adding a high level of salt. Hence, it is needed to determine the optimal amount of aptamers for biosensing. In brief, 6 μL of DNA aptamer with various concentration (0, 5, 10, 15, 20, 30, 40, 50, 60 pmol/μL) was mixed with AgNPs (350 μL, 2 nM). The solution was incubated at room temperature for 30 min following the addition of 350 μL PBS pH 7.4. Then, NaCl (35 μL, 1.7 M) was added and mixed with the above solution and incubating for 10 min. The stability of aptamer modified AgNPs was observed by the unaided eye and UV–vis spectrophotometer.

The lowest amount of DNA aptamer that can stabilize AgNPs was chosen for the biosensing [18].

Detection of tetracycline using aptasensor. 200 μL of different concentration of tetracycline ranging from 0 to 60 μM was mixed with 300 μL of tetracycline aptamer functionalized AgNPs. The solution was incubated at room temperature for 10 min. Then, NaCl (30 μL, 1.7 M) was added into the mixture, followed by 10 min incubation at room temperature. The result of this reaction was inspected by unaided eyes and UV–vis spectral measurement. To ensure the specificity of the proposed approach, the response of aptasensor to other antibiotics, including amoxicillin, ampicillin, chloramphenicol and doxycycline was investigated under the same procedure.

3 Results and Discussion

An experimental scheme for the tetracycline detection is illustrated in Fig. 1. Here, the DNA aptamer was utilized as a specific bioreceptor, and the silver nanoparticle was used as a signal transducer element, which senses the binding between DNA aptamer and tetracycline. Our sensing platform was designed based on the principle that target antibiotic and AgNPs competitively bind to DNA aptamer. In brief, in the absence of tetracycline, the DNA aptamer helps stabilize the nanoparticle against salt-induced aggregation via electrostatic repulsion [19]. The electrostatic repulsion assists in the prevention of the strong van der Waals attraction resulted in the enhancement of the stability of metallic nanoparticles. In this case, the silver nanoparticles are still well—dispersed in the presence of a high concentration of salt. In the presence of tetracycline, the interaction between DNA aptamer and its target leads to the detachment of tetracycline from the nanoparticle's surface and bind to its target, resulted in the aggregation of AgNPs by salt and an obvious color change of silver colloid from yellow to colorless.

In this research, the silver nanoparticle was used as a signal transducer, prepared by classical chemical synthesis method. Briefly, a mixture of two reducing agents, including trisodium citrate and sodium borohydride was used to reduce the silver

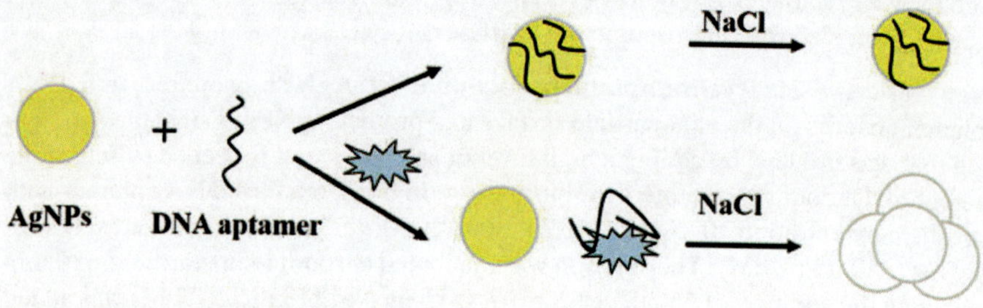

Fig. 1 A schematic procedure describing the fabricated aptasensor

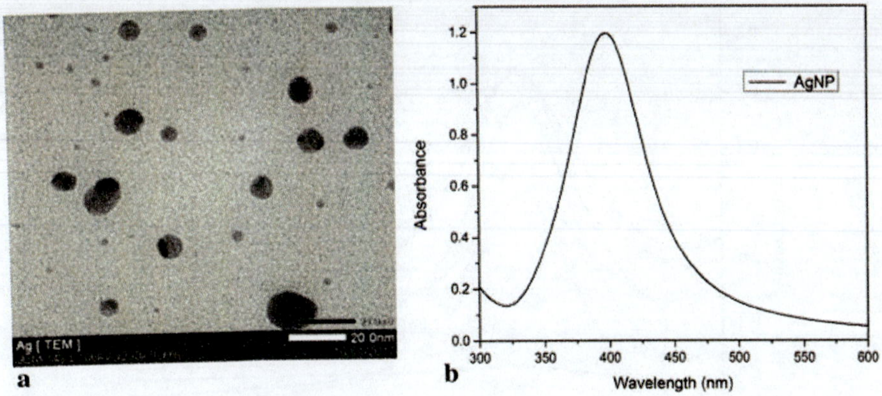

Fig. 2 **a** Morphology of AgNPs (TEM). **b** UV–Vis spectrum of citrate capped AgNPs

cation, and the remaining citrate anions act as a stabilizing agent. The nanoparticle morphology of synthesized nanoparticle was revealed by UV–Vis spectrophotometer and transmission electron microscopy (TEM). As illustrated in Fig. 2, the silver nanoparticle synthesized in this research was nearly spherical with a diameter of ~ 12 nm, the size dispersion of only a few nanometers, and the wavelength of surface plasmon resonance (SPR) occurred at ~ 393 nm. The concentration of colloidal silver was ~ 2 nM which was calculated by the Beer-Lambert law [20].

To get high performance of aptasensor, it is needed to optimize the ratio between the DNA aptamer and AgNPs at which the nanoparticles are still stable when adding a high amount of salt. As illustrated in Fig. 3, the DNA aptamer can prevent the aggregation of nanoparticles when adding salt with high concentration. This is because the hydrophobic interaction between the bases of nucleotides and the AgNPs induces the DNA aptamers to stick on the nanoparticle surface [19]. The experiments showed that in the presence of DNA aptamer of tetracycline (240 pmol DNA aptamer/350 μL of AgNPs), the colloidal silver was still stable in a solution containing high concentration of salt. The significant aggregation of AgNPs was observed when adding the lower amount of DNA aptamer in the colloidal silver. In case of the higher amount of aptamer, the colloidal silver keeps the original yellow colour and its absorbance at 393 nm that indicated no aggregation of the silver nanoparticles. From the above experiments, the optimal amount of aptamer to the nanoparticles applied for detection of tetracycline was 240 pmol DNA aptamer/350 μL of AgNPs with concentration of 2 nM.

For tetracycline analysis, different concentrations of tetracycline ranging from 0 to 100 μM were incubated with aptamer modified AgNPs, and the colour alteration as well as spectra response were recorded. As demonstrated in Fig. 4, in the presence of tetracycline, DNA aptamer of tetracycline interacts with its target, leading to the separation of aptamer from the nanoparticle surface. Therefore, the aptamer cannot shield the silver nanoparticle against salt-induced aggregation, resulted in the obvious color alteration of silver colloid from yellow to colorless and the decrease of the light

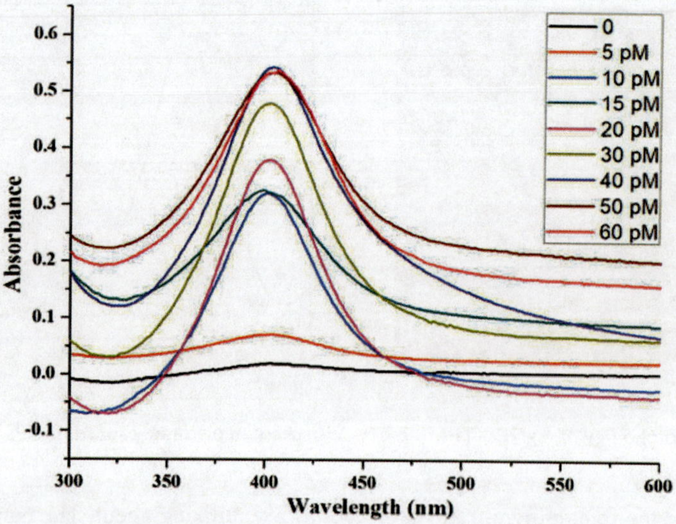

Fig. 3 UV–Vis spectra of silver nanoparticles (350 μL, 2 nM) stabilized by various amounts of DNA aptamer of tetracycline (0, 30, 60, 90, 120, 180, 240, 300 and 360 pmol) after salt addition (35 μL, 1.7 M)

absorbance at 393 nm. Hence, the sensitivity of proposed aptasensor was determined by governing the absorbance alteration at 393 nm after adding various amounts of tetracycline and salt. The absorbance of silver nanoparticles at various concentrations of tetracycline was shown in Fig. 4b. A linear range of proposed aptasensor for detection of tetracycline was in a concentration ranging from 0.1 μM to 6 μM, and the linear regression equation was $y = 0.04271x + 0.05152$ ($R^2 = 0.9946$), where y and x are the absorbance change at 393 nm and tetracycline concentration (μM), respectively (Fig. 4c). Moreover, the limit of detection (LOD) and quantification (LOQ) of aptasensor were estimated by the formula 3α/s and 10α/s, respectively, where α is the standard deviation of plank sample (n = 3) and s is the slope of the linear regression equation. Therefore, the obtained values were LOD ~ 10 nM and LOQ ~ 100 nM. This LOD corresponds to ~ 4.8 μg/L or ~ 4.8 μg/kg. The lowest concentration of tetracycline that can be detected by naked eyes was ~ 1 μM. It shows that the detection limit of this method is much lower than the detection limit for tetracycline obtained from the capillary electrophoresis method (0.5–1 mg/kg). Compared to the detection limit of other assays for detection of tetracycline that require costly instrumentation, expensive reagents, sample pre-treatment with the toxic organic reagents such as immunoassay (0.19 μg/L), carrier mediated hollow fiber liquid phase micro-extraction combined with HPLC–UV (0.5–1 μg/L), electrochemical aptasensor (2.4 μg/L) [1], our sensing platform shows its advantages in terms of simplicity, experimental cost, time for analysis and ability for on-site testing.

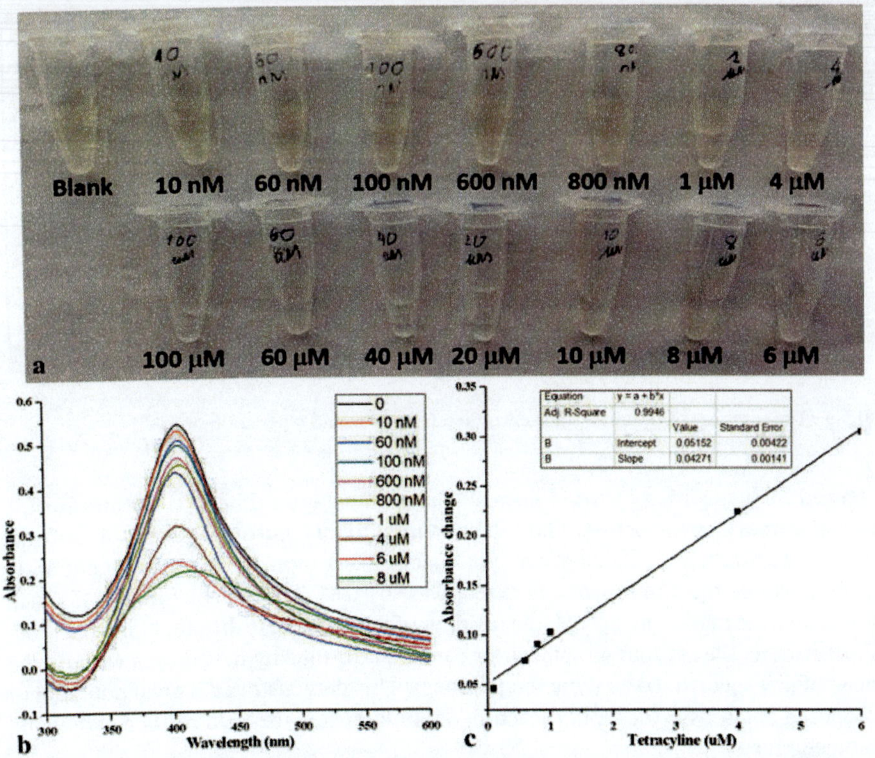

Fig. 4 **a** Colorimetric response of DNA aptamer modified AgNPs as a function of tetracycline concentration; **b** UV–vis spectral response of aptasensor in the presence of various amount of tetracycline; **c** The linear scale of calibration curve describing measurement of tetracycline at concentrations ranging from 0.1 to 6 µM

Beside a suitable sensitivity for practical applications, the selectivity of a biosensor is recognized as one of the most critical characteristics in analytical methods. To evaluate the selectivity, several antibiotics which are similar structures with tetracycline including amoxicillin, ampicillin, chloramphenicol and doxycycline were tested. As illustrated in Fig. 5, the decrease of absorbance at 393 nm for tetracycline was much larger than those of other antibiotics. This result indicated that the proposed sensing platform is able to determine tetracycline effectively.

4 Conclusion

We have designed and proposed a simple, cost-effective, colorimetric aptasensor for sensitive and rapid detection of tetracycline based on the principle that target antibiotic and AgNPs bind competitively to DNA aptamer. The proposed aptasensor

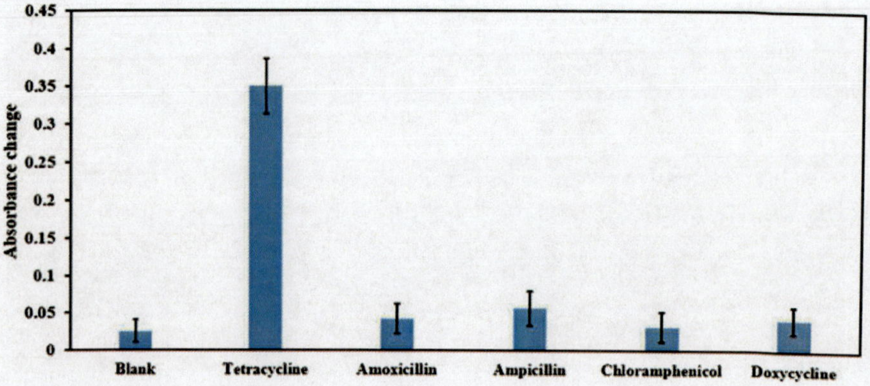

Fig. 5 The response of designed aptasensor to tetracycline and other antibiotics

showed high specificity toward tetracycline with analysis time of ~ 20 min for the step of tetracycline detection. The detection limit of this sensor was as low as 100 nM in case of absorbance measurement. For visual observation, the lowest concentration of tetracycline that the sensor can detect was ~ 1 μM. Those detection limits much lower than the maximum acceptable residues of tetracycline in different kinds of food. Furthermore, the readout of aptasensor can be performed by naked eyes without the need of any specialized analytical equipment. This demonstrated a great potential of proposed aptasensor for rapid detection of antibiotics in the field of food safety and biomedicine.

Acknowledgements This research is funded by Vietnam National University Ho Chi Minh City (VNU-HCM) under grant number NCM2020-28-01.

Conflicts of Interest The authors have no conflict of interest to declare.

References

1. Jalalian SH, Taghdisi SM, Danesh NM, Bakhtiari H, Lavaee P, Ramezani M, Abnous K (2015) Sensitive and fast detection of tetracycline using an aptasensor. Anal Methods 7(6):2523–2528
2. Tan H, Ma C, Song Y, Xu F, Chen S, Wang L (2013) Determination of tetracycline in milk by using nucleotide/lanthanide coordination polymer-based ternary complex. Biosens Bioelectron 50:447–452
3. Ramezani M, Danesh NM, Lavaee P, Abnous K, Taghdisi SM (2015) A novel colorimetric triple-helix molecular switch aptasensor for ultrasensitive detection of tetracycline. Biosens Bioelectron 70:181–187
4. Aga DS, Goldfish R, Kulshrestha P (2003) Application of ELISA in determining the fate of tetracyclines in land-applied livestock wastes. Analyst 128(6):658–662
5. Idowu F, Junaid K, Paul A, Gabriel O, Paul A, Sati N, Maryam M, Jarlath U (2010) Antimicrobial screening of commercial eggs and determination of tetracycline residue using two microbiological methods. Int J Poult Sci 9(10):959–962

6. Shariati S, Yamini Y, Esrafili A (2009) Carrier mediated hollow fiber liquid phase microextraction combined with HPLC–UV for preconcentration and determination of some tetracycline antibiotics. J Chromatogr B 877(4):393–400
7. Han S, Liu E, Li H (2006) Determination of tetracycline, chlortetracycline and oxytetracycline by flow injection with inhibitory chemiluminescence detection using copper (II) as a probe ion. Luminescence: J Biol Chem Lumines 21(2):106–111
8. Tong J, Rao Q, Zhu K, Jiang Z, Ding S (2009) Simultaneous determination of five tetracycline and macrolide antibiotics in feeds using HPCE. J Sep Sci 32(23–24):4254–4260
9. Gholivand MB, Khani H (2013) Determination of tetracycline at a UV-irradiated DNA film modified glassy carbon electrode. Electroanalysis 25(2):461–467
10. Mehlhorn A, Rahimi P, Joseph Y (2018) Aptamer-based biosensors for antibiotic detection: a review. Biosensors 8(2):54
11. Truong PL, Ma X, Sim SJ (2014) Resonant Rayleigh light scattering of single Au nanoparticles with different sizes and shapes. Nanoscale 6(4):2307–2315
12. Truong PL, Sim SJ (2015) Development of individual plasmonic nanosensors for clinical diagnosis. In: 5th international conference on biomedical engineering in vietnam. Springer, Cham, pp 1–6
13. Loiseau A, Asila V, Boitel-Aullen G, Lam M, Salmain M, Boujday S (2019) Silver-based plasmonic nanoparticles for and their use in biosensing. Biosensors 9(2):78
14. Pérez-Juste J, Pastoriza-Santos I, Liz-Marzán LM, Mulvaney P (2005) Gold nanorods: synthesis, characterization and applications. Coord Chem Rev 249(17–18):1870–1901
15. Rycenga M, Cobley CM, Zeng J, Li W, Moran CH, Zhang Q, Qin D, Xia Y (2011) Controlling the synthesis and assembly of silver nanostructures for plasmonic applications. Chem Rev 111(6):3669–3712
16. Kim CH, Lee LP, Min JR, Lim MW, Jeong SH (2014) An indirect competitive assay-based aptasensor for detection of oxytetracycline in milk. Biosens Bioelectron 51:426–430
17. Almaquer FEP, Ricacho JSY, Ronquillo RLG (2019) Simple and rapid colorimetric sensing of Ni (II) ions in tap water based on aggregation of citrate-stabilized silver nanoparticles. Sustain Environ Res 29(1):1–10
18. Wu WH, Li M, Wang Y, Ouyang HX, Wang L, Li CX, Cao YC, Meng QH, Lu JX (2012) Aptasensors for rapid detection of *Escherichia coli* O157: H7 and *Salmonella typhimurium*. Nanoscale Res Lett 7(1):658
19. Wang Y, Wang J, Yang F, Yang X (2010) Spectrophotometric determination of cysteine with gold nanoparticles stabilized with single-stranded oligonucleotides. Anal Sci 26:545–549
20. Paramelle D, Sadovoy A, Gorelik S, Free P, Hobley J, Fernig DG (2014) A rapid method to estimate the concentration of citrate capped silver nanoparticles from UV-visible light spectra. Analyst 139(19):4855–4861

Fast Detection of *Staphylococcus aureus* Using DNA Aptamer and Silver Nanoparticles

Phuoc Long Truong

Abstract *Staphylococcus aureus* is one of the ubiquitous bacterial pathogens in human that are able to cause various diseases from skin infections to serious diseases. This pathogen can make seven types of bacterial toxins that are accountable for food poisoning. Moreover, the pandemic of methicillin-resistant *Staphylococcus aureus* has been globally dangerous for chronically infectious diseases, and nosocomial infections with life-threatening problems, because of their resistance to most antibiotics. Therefore, rapid and accurate detection of *S. aureus* has emerged as an urgent need in fields of clinical diagnostics, biodefense and food safety. However, the current rapid tests are challenging with being limited their capacities in on-site detection for the increase of prevention results. In this study, we report an aptamer-based sensor for fast detection of *S. aureus*. Instead of trying to capture bacterial cells on the substrate or nanoparticle surface, our detection method allows the DNA aptamer to detach from the silver nanoparticle when interacting with bacteria and bind to its target, leading to the aggregation of silver nanoparticle by salt. The strategy dramatically enhances the selectivity and sensitivity of whole-cell based biosensor. The results demonstrated that aptasensor could detect *S. aureus* with a detection limit of ~ 10^5 CFU/mL within 20 min. The detection method showed many advantages compared with conventional methods with respect to cost, simplicity and analysis time. This biosensor can be used as a powerful bio-analytical tool for whole-cell detection.

Keywords DNA aptamer · Silver nanoparticle · *S. aureus* · Aptasensor

1 Introduction

Staphylococcus aureus (*S. aureus*), which is a ubiquitous bacterial pathogen, is able to cause various kinds of illnesses from skin infections to serious diseases. It is well-known that *S. aureus* is one of the most critical foodborne diseases worldwide

P. L. Truong (✉)
School of Biomedical Engineering, International University, Vietnam National University-Ho Chi Minh City (VNU-HCMC), Ho Chi Minh City 700000, Vietnam
e-mail: tplong@hcmiu.edu.com.vn

V. Van Toi et al. (eds.), *8th International Conference on the Development of Biomedical Engineering in Vietnam*, IFMBE Proceedings 85,
https://doi.org/10.1007/978-3-030-75506-5_29

[1]. *S. aureus* can make seven types of enterotoxins that are frequently in charge of food poisoning [2, 3]. This bacteria commonly contributes to infections after injury or surgery. Moreover, the pandemic of antibiotic-resistant bacteria such as Methicillin-resistant *Staphylococcus aureus* (MRSA) that leads to the difficulties in the treatment of diseases [4]. Hence, the development of appropriate detection methods for simple, fast and specific detection of *S. aureus* is a decisive factor to reduce the risks associated with food safety and public health.

The standard assays for determination of *S. aureus*, which involve bacterial cell culture followed by characterization of either genotype or phenotype, are time-consuming and labor-intensive [5, 6]. Currently, there are several approaches for the elimination of the time-consuming step of bacterial cell culture and shortening of the analysis time, including enzyme-linked immunosorbent assay (ELISA) and gene amplification assays. Among these, noble metal nanoparticles have demonstrated to be powerful tools for clinical applications due to outstanding optical properties of the metallic nanoparticles that attract great interest from the scientific community. These emerging properties of noble nanoparticles relate to the interaction between conductive free electrons on the nanoparticles surface and the electromagnetic field of the coming light [7–9]. The application of nanoparticles in biosensors is very promising because they significantly increase the performance of the biosensor. Furthermore, metallic nanoparticles functionalized with bio-receptors can specifically bind to the target cells or target molecules with high efficiency [10].

High-affinity antibodies with target analytes are often chosen for immunoassays. However, the use of antibodies in bio-sensing has some disadvantages, such as high-cost and poor stability [9]. Recently, the so-called aptamers, a new class of targeting compounds have emerged as an alternative to antibodies for bio-assays. The aptamers are short, single-stranded functional nucleic acids including DNAs or RNAs that possess high affinity and can fold into intricate 3D conformation when binding with a wide range of target analytes. In comparison with immunosensors, aptamer-based sensors bear several benefits such as high selectivity, high affinity and high stability against biodegradation and denaturation [11].

Herein, we reported an aptamer-based sensor for fast detection of *S. aureus* with a low cost. The detection method exploits a bacteria-specific DNA aptamer as the recognition element of optical aptasensor. The detection method allows the DNA aptamers, which are adsorbed on the silver nanoparticle's surface, to detach from the surface of the silver nanoparticle when interacting with target bacteria. This leads to bacteria-induced aggregation of silver nanoparticle upon high-salt concentrations. Using *S. aureus* as a model for the detection of bacteria, the DNA aptasensor could detect ~ 10^5 CFU/mL within 20 min. The results showed that this aptasensor could be used as a powerful bio-analytical tool for whole-cell detection. The method showed many advantages compared with conventional methods with respect to cost, simplicity, analysis time.

2 Materials and Methods

2.1 Materials

Sodium citrate dihydrate ($Na_3C_6H_5O_7 \cdot 2H_2O$), sodium borohydride ($NaBH_4$, 98%), sodium chloride (NaCl), phosphate buffer saline pH 7.4 (PBS buffer) and silver nitrate ($AgNO_3$, 99%) were purchased from Sigma-Aldrich, Inc. (USA). The DNA aptamer of *S. aureus* (5′-TCC CTA CGG CGC TAA CCT CCC AAC CGC TCC ACC CTG CCT CCG CCT CGC CAC CGT GCT ACA AC-3′) [12] was obtained from PHUSA Biochem Co. (Vietnam). The reagents and essential chemicals used in this study were available in AR grade. All glassware was intensively cleaned with a freshly prepared solution of aqua regia and rinsed carefully with ultrapure water.

2.2 Methods

Synthesis and Characterization of Silver Nanoparticles

Citrate-capped silver nanoparticles (AgNPs) were prepared by reducing silver nitrate with a mixture of two reducing agents, trisodium citrate and sodium borohydride [13]. Briefly, 0.5 mL of 100 mM $AgNO_3$ is added drop-wise to 49.5 mL of freshly prepared ice-cold solution that contains 0.30 mM trisodium citrate and 1 mM $NaBH_4$. The resulting solution was centrifuged at 14,000 rpm for 15 min to eliminate the excess of sodium borohydride, and the precipitation was re-dispersed in 2.2 mM trisodium citrate. Then, the solution was filtered by a 0.22 μm filter to remove any aggregated nanoparticles. The morphology of AgNPs was analyzed by UV–vis spectrophotometer (Jasco V-730, Japan), and transmission electron microscopy (TEM; JEM-1400, USA).

Preparation of Aptamer Wrapped Silver Nanoparticles

DNA aptamer-wrapped silver nanoparticles (Apta-AgNPs) were prepared based on previous studies with little modification [14]. Briefly, different amounts of DNA aptamer (0, 35, 70, 140, 210, 280, 350, 420 and 490 pmol) were thoroughly mixed with 350 μL AgNPs and incubated at room temperature for 30 min following addition of 350 μl PBS pH 7.4. Then, 35 μL of 1.7M NaCl was added and mixed with the above solution and incubating for 10 min. The stability of Apta-AgNPs was observed by the naked eye and UV–vis absorption spectrophotometer. The quality of Apta-AgNPs was checked by zeta potential analyzer (Malvern Zetasizer). The minimum concentration of DNA aptamer that can stabilize AgNPs was selected for the assay.

Detection of *S. aureus* Using Aptasensor

350 μL of Apta-AgNPs was mixed with 350 μL of various concentrations of *S. aureus* ATCC 25,923 (10^3, 10^4, 10^5, 10^6, 10^7 and 10^8 CFU/mL) and incubated at

room temperature (~ 25 °C) for 10 min. Then, 35 μL of 1.7M NaCl was added into the mixture, followed by either visual inspection or UV–vis spectral measurement. The sensitivity and selectivity of the aptamer sensor for determination of *S. aureus* was evaluated by exposing the aptamer sensor to *S. epidermidis*, which is the close relative with *S. aureus*.

3 Results and Discussion

The mechanism of the proposed sensing platform is shown in Fig. 1. In a previous study, Rothberg et al. demonstrated that the single-stranded DNA (ssDNA) has the propensity to adsorb on the metallic nanoparticles in colloidal solution [15]. ssDNA can uncoil its bases to make them exposed to the metallic nanoparticle's surface. The electrostatic interaction between the positively charged bases of ssDNA and the negatively charged nanoparticles keeps the ssDNA sticking to the nanoparticle surface. Hence, ssDNA can stabilize the nanoparticle against aggregation via electrostatic repulsion when adding salt with high concentration [14, 16]. In our approach, we used ssDNA aptamer as a specific recognition element and citrate capped silver nanoparticle (AgNPs) as a signal transducer element. As shown in Fig. 1, our detection method based on the principle that target bacteria and AgNPs bind competitively to ssDNA aptamer. The electrostatic repulsion helps prevent the strong van der Waals attraction and enhance the stability of metallic nanoparticles. The DNA aptamer turns its conformation from a randomly coiled structure to a folded 3D structure when adding target bacteria that made DNA aptamers detach from the AgNPs

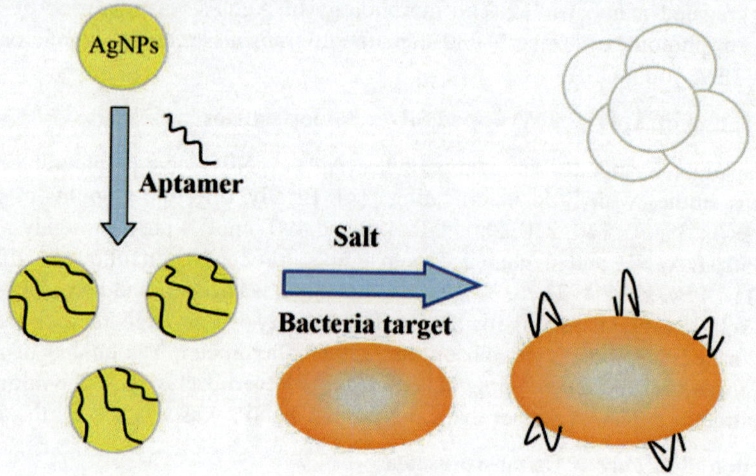

Fig. 1 A schematic procedure depicting the experimental processes

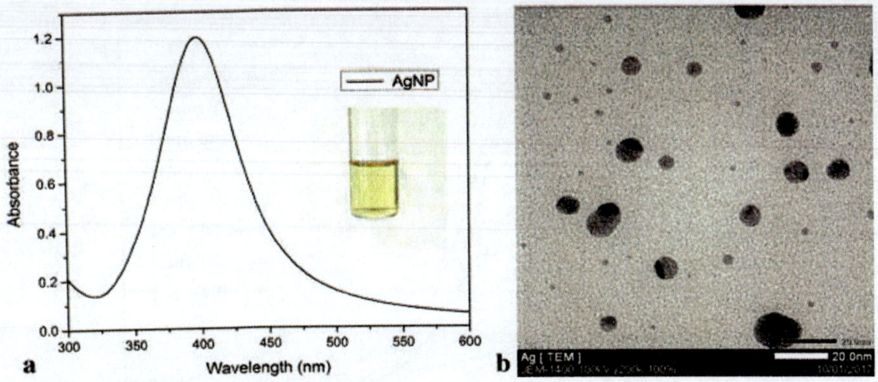

Fig. 2 **a** UV–Vis spectrum of citrate capped silver nanoparticles. **b** Morphology and size of Ag nanoparticles (TEM)

surface. This transition resulted in the subsequent aggregation of AgNPs that leads to the changes of the UV–Vis spectrum and the characteristic color change of AgNPs.

In this study, the colloidal silver was prepared by the chemical synthesis. Shortly, the silver nitrate was reduced by a mixture of two reducing agents, trisodium citrate and sodium borohydride, and remaining citrate anions act as a stabilizing agent. The reaction occurs at room temperature with a high-speed rate [16]. This method yielded spherical, pale yellow nanoparticles with a diameter of ~ 12 nm, as revealed by transmission electron microscopy (TEM) (Fig. 2). The silver nanoparticles were nearly spherical with the size dispersion of only a few nanometers, and the surface plasmon resonance (SPR) occurred at 393 nm. The concentration of AgNPs was ~ 2.0 nM estimated based on Beer-Lambert law and its average diameter [17].

To identify a suitable ratio between the DNA aptamer and AgNPs at which the AgNP is still stable in the absence of *S. aureus* after adding of a high amount of NaCl, various amounts of DNA aptamer in AgNPs were investigated. As shown in Fig. 3, DNA could protect the AgNPs from the aggregation with high salinity. In the presence of 280 pmol of aptamer in 350 μL AgNPs, the silver colloid could keep stable in a solution containing a high concentration of NaCl and the Apta-AgNPs suspension kept its original yellow color. The results also indicate that there is no significant difference regarding the UV–Vis spectrum of ssDNA wrapped AgNPs in the presence and absence of salt. With the lower amount of DNA aptamer, the absorbance of AgNPs at 393 nm gradually decreased, and the colour of silver solutions was changed from yellow to yellowish indicating aggregation of the nanoparticles. With the higher amount of DNA aptamer, the colloidal solutions of nanoparticles maintain the original yellow colour and the absorbance of AgNPs at 393 nm that indicated no aggregation of the nanoparticles. The successful assembly of the Apta-AgNPs was confirmed by zeta potential measurements. The zeta potential of citrate capped AgNPs was − 42.3 ± 3.1 mV (n = 3) due to the presence of citrate on the nanoparticle surface. After modification, the zeta potential of Apta-AgNPs was − 38.2 ± 1.7 mV (n =

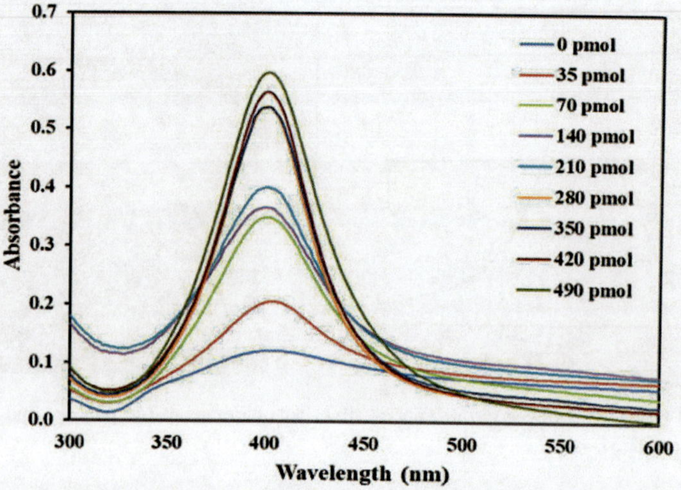

Fig. 3 UV–Vis absorption spectra of AgNPs stabilized by various amounts of DNA aptamer after addition of sodium chloride (NaCl)

3). This indicates that the AgNPs were stabilized by ssDNA aptamer. Therefore, the ratio (280 pmol DNA aptamer/350 μL AgNPs) was chosen in the subsequent study.

After determining the optimal ratio of the aptamer to silver colloid, the performance of aptasensor was checked by performing experiments with the positive control, negative control and a reagent blank. Positive control was investigated by mixing *S. aureus* ATCC 25,923 (10^8 CFU/mL) with apta-AgNPs. *S. epidermidis* ATCC 12,228 (10^8 CFU/mL) with apta-AgNPs was used as the negative control and the PBS buffer pH 7.4 was served as a reagent blank. As shown in Fig. 4a, the significant change in absorbance 393 nm was only observed for *S. aureus* with apta-AgNPs.

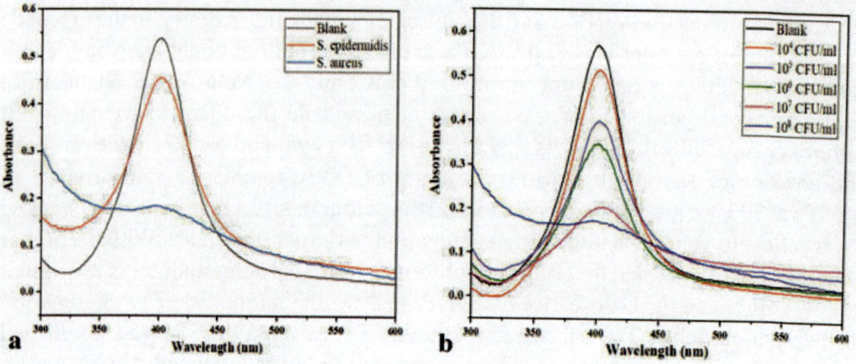

Fig. 4 **a** Specificity of DNA aptasensor; **b** UV–vis absorption spectra for detection of *S. aureus* based on apta-AgNPs

In the presence of *S. aureus*, DNA aptamers detach from the Ag nanoparticles, form a folded complex structure and bind to *S. aureus*. This leads to the aggregation of Ag nanoparticles when adding salt with a high concentration that results in the decrease of the absorbance at 393 nm, the changes of UV–Vis spectrum and the characteristic color change of apta-AgNPs. (data not shown). In case of the absence of *S. aureus*, apta-AgNPs remain stable and keep its original color that showed no nanoparticle aggregation. To investigate the detection sensitivity of aptasensor, various concentrations of *S. aureus* ATCC 25,923 ranging from 10^3 to 10^8 CFU/mL was mixed with apta-AgNPs solutions. The results showed that the characteristic absorbance at 393 nm decreased when increasing the concentration of *S. aureus*, indicating an increase in the aggregation of AgNPs (Fig. 4b). Therefore, the sensitivity of this assay was determined by governing the change of absorbance after adding various amounts of target bacteria and high amount of NaCl. Figure 4 also showed that the non-specific binding of *S. epidermidis* as compared to the specific binding between *S. aureus* and DNA aptamer, which was shown in a concentration ranging from10^3 to 10^8 CFU/mL. As obviously demonstrated in the figure, these binding were negligible. Hence, it can be concluded that the limit of the detection of the assay was 10^5 CFU/mL, corresponding to 3.5×10^4 CFU per test. This detection limit can be comparable to the SPR based biosensor with the same bacteria (10^4 CFU/mL) [18]. Compared to the detection limit of nucleic acid amplification assays for detection of *S. aureus* such as PCR (<10 CFU/reaction) [19], real-time PCR (6.8×10^1 CFU/mL) [20], our designed aptasensor has poorer detection limit. However, the proposed sensing platform takes ~ 20 min for the direct detection of the whole bacteria without specialized instrumentation and pretreatment steps such as cell lysis and DNA extraction. This dramatically reduces the analysis time in comparison with the conventional cultivation method that needs to take 1 to 4 days used for culture and identification of bacteria and take ~ 2 to 4 h in case of conventional PCR and real-time PCR [19, 20]. In comparison with conventional methods, this assay can detect *S. aureus* in the real sample with a total analytical time of ~ 24 h or less. Our sensing platform shows its advantages concerning simplicity, cost, time for analysis and ability for on-site testing.

4 Conclusion

We have illustrated a simple, low-cost, and straightforward method for detection of *S. aureus* based on the principle that target bacteria and AgNPs bind competitively to DNA aptamer. This aptasensor takes ~ 20 min for the step of bacteria detection in pure culture and 18–24 h for culture in selective enrichment media, which significantly reduces the analysis time compared with traditional culture methods and other rapid methods. In order to reduce analysis time in case of urgent need, the microfiltration and immunomagnetic separation techniques should be applied for the sample pretreatment step. The detection method showed many advantages compared with conventional methods with regard to cost, simplicity and analysis time. The aptasensor can be used as a powerful bio-analytical tool for whole-cell detection without any specialized instrumentation and complicated pretreatment steps.

Acknowledgements This research is funded by Vietnam National University Ho Chi Minh City (VNU-HCM) under grant number NCM2020-28-01.

Conflicts of Interest he authors have no conflict of interest to declare.

References

1. Lowy FD (1998) *Staphylococcus aureus* infections. N Engl J Med 339(8):520–532
2. Argudín MÁ, Mendoza MC, Rodicio MR (2010) Food poisoning and *Staphylococcus aureus* enterotoxins. Toxins 2(7):1751–1773
3. Balaban N, Rasooly A (2000) Staphylococcal enterotoxins. Int J Food Microbiol 61(1):1–10
4. Centers for Disease Control and Prevention (CDC) (2003) Outbreaks of community-associated methicillin-resistant *Staphylococcus aureus* skin infections. Los Angeles County, California, 2002–2003. MMWR Morb Mortal Wkly Rep 52(5):88
5. Shahdordizadeh M, Taghdisi SM, Ansari N, Langroodi FA, Abnous K, Ramezani M (2017) Aptamer based biosensors for detection of *Staphylococcus aureus*. Sens Actuators, B Chem 241:619–635
6. Štěpán J, Pantůček R, Doškař J (2004) Molecular diagnostics of clinically important staphylococci. Folia Microbiol 49(4):353–386
7. Truong PL, Sim SJ (2015) Development of individual plasmonic nanosensors for clinical diagnosis. In: 5th international conference on biomedical engineering in Vietnam. Springer, Cham, pp 1–6
8. Sun Y, Long TP, Wolff A, Bang DD (2016) A novel gold nanoparticle-DNA aptamer-based plasmonic chip for rapid and sensitive detection of bacterial pathogens. In: 20th international conference on miniaturized systems for chemistry and life sciences: MicroTAS 2016. Chemical and Biological Microsystems Society, pp 1222–1223
9. Truong PL, Choi SP, Sim SJ (2013) Amplification of resonant Rayleigh light scattering response using immunogold colloids for detection of lysozyme. Small 9(20):3485–3492
10. Lai HZ, Wang SG, Wu CY, Chen YC (2015) Detection of *Staphylococcus aureus* by functional gold nanoparticle-based affinity surface-assisted laser desorption/ionization mass spectrometry. Anal Chem 87(4):2114–2120
11. Abbaspour A, Norouz-Sarvestani F, Noori A, Soltani N (2015) Aptamer-conjugated silver nanoparticles for electrochemical dual-aptamer-based sandwich detection of *Staphylococcus aureus*. Biosens Bioelectron 68:149–155
12. Chang YC, Yang CY, Sun RL, Cheng YF, Kao WC, Yang PC (2013) Rapid single cell detection of *Staphylococcus aureus* by aptamer-conjugated gold nanoparticles. Sci Rep 3(1):1–7
13. Almaquer FEP, Ricacho JSY, Ronquillo RLG (2019) Simple and rapid colorimetric sensing of Ni (II) ions in tap water based on aggregation of citrate-stabilized silver nanoparticles. Sustain Environ Res 29(1):1–10
14. Wu WH, Li M, Wang Y, Ouyang HX, Wang L, Li CX, Cao YC, Meng QH, Lu JX (2012) Aptasensors for rapid detection of *Escherichia coli* O157: H7 and *Salmonella typhimurium*. Nanoscale Res Lett 7(1):658
15. Li H, Rothberg L (2004) Colorimetric detection of DNA sequences based on electrostatic interactions with unmodified gold nanoparticles. Proc Natl Acad Sci 101(39):14036–14039
16. Ramezani M, Danesh NM, Lavaee P, Abnous K, Taghdisi SM (2015) A novel colorimetric triple-helix molecular switch aptasensor for ultrasensitive detection of tetracycline. Biosens Bioelectron 70:181–187

17. Paramelle D, Sadovoy A, Gorelik S, Free P, Hobley J, Fernig DG (2014) A rapid method to estimate the concentration of citrate capped silver nanoparticles from UV-visible light spectra. Analyst 139(19):4855–4861
18. Balasubramanian S, Sorokulova IB, Vodyanoy VJ, Simonian AL (2007) Lytic phage as a specific and selective probe for detection of *Staphylococcus aureus*—a surface plasmon resonance spectroscopic study. Biosens Bioelectron 22(6):948–955
19. Brakstad OG, Aasbakk K, Maeland JA (1992) Detection of *Staphylococcus aureus* by polymerase chain reaction amplification of the nuc gene. J Clin Microbiol 30(7):1654–1660
20. Trnčíková T, Hrušková V, Oravcová K, Pangallo D, Kaclíková E (2009) Rapid and sensitive detection of *Staphylococcus aureus* in food using selective enrichment and real-time PCR targeting a new gene marker. Food Anal Methods 2(4):241

Application of Vancomycin Therapeutic Drug Monitoring in a Vietnamese Private Hospital

Nguyen Thi Thu Phuong, Tran Hai Long, Tran Van Anh, and Nguyen Thanh Hoi

Abstract Vancomycin is a primary regimen for treating MRSA and other Gram-positive β-lactam–resistant organisms. But the problem is the resistance to vancomycin, the clinical and microbiological characteristics in which it is used is always changing. Therefore, we conducted this retrospective study to analysis benefits of Vancomycin TDM in clinical outcomes and discovered the factors affecting on the vancomycin maintenance dose. We conducted a follow-up on 137 patients using vancomycin divided into two groups, including the TDM and non-TDM users. The evaluation criteria included panoramic vancomycin use, infection status, renal function and treatment results after ten days. The total number of samples taken from 79 patients applied TDM were 116 and 73.3% of the samples reached the target trough concentration. Compared to 58 patients of the non-TDM group, applying TDM will reduce c-reactive protein (CRP) and improve treatment outcome. We also found that the vancomycin trough concentration, creatinine clearance, and weight contributed significantly (24.7%, 16.8% and 26.6%, respectively) to the inter-individual variability in vancomycin dose. The multiple linear regression model for vancomycin maintenance dose was built as the formula: Vancomycin maintenance dose (mg/kg) = 21.4 + 0.24 * vancomycin trough concentration (μg/mL) + 0.17 * Clcr(mL/min) − 0.26 * weight(kg) + 0.17 * age(years) − 0.03 * ΔCRP. Our study demonstrated the benefits of vancomycin TDM technology on clinical outcomes and improvement of infection situation. Furthermore, we have successfully indicated a strong relationship between vancomycin trough concentration, creatinine clearance, age, weight, ΔCRP and maintenance dose of vancomycin. Clinicians in the process of designing a regimen dose of vancomycin should comprehensively consider these factors in order to optimize the dose and reduce the adverse effect of vancomycin.

Keywords Vancomycin · TDM · Multiple linear regression model

N. T. T. Phuong (✉) · T. H. Long · T. V. Anh · N. T. Hoi
Hai Phong University of Medicine and Pharmacy, Haiphong, Vietnam
e-mail: nttphuong@hpmu.edu.vn

N. T. T. Phuong · T. V. Anh · N. T. Hoi
Hai Phong International Hospital, Haiphong, Vietnam

V. Van Toi et al. (eds.), *8th International Conference on the Development of Biomedical Engineering in Vietnam*, IFMBE Proceedings 85,
https://doi.org/10.1007/978-3-030-75506-5_30

1 Introduction

Methicillin-resistant *Staphylococcus aureus* (MRSA) has been recognized as one of the most important risks bacterial related to the development of antimicrobial resistance (AMR) [1]. In addition, the side effects caused by vancomycin, such as acute kidney injury are also one of the issues that should be noted. Therapeutic drug monitoring (TDM) is an important solution in managing the effectiveness of treatment as well as limiting adverse drug reactions caused by vancomycin. In 2009, specific guidelines about TDM of Vancomycin in adult patients was published by American Society of Health-System Pharmacists (ASHP), the Infectious Diseases Society of America (IDSA) and the Society of Infectious Disease Pharmacists (SIDP) [2]. From that time until now, the guideline has been updated to change and supplement new recommendations and clinical evidence for special patients, antibiotic resistance and pharmacokinetic index selection. Some clinical studies, animal models, in vitro data proved AUC/MIC ratio's the target consensus of $\geq$ 400 for MRSA infections and having related an AUC/MIC ratio of 350–400 to successful outcome [3–8]. 15–20 μg/mL was recommended as target trough concentration of TDM by several of clinical practice guidelines for severe patients in which 10–15 μg/mL was suggested for patients with nonsevere infections [9]. Although TDM of Vancomycin has been applied routinely in most countries around the world. In Vietnam, the application of TDM is still very limited. Hai Phong International Hospital started implementing TDM of Vancomycin in 2019 to ensure successful treatment effectiveness and minimize the risk of toxicity in patients. Recognizing the importance of applying Therapeutic drug and monitoring of vancomycin in treatment, we implement the topic "Application of Vancomycin Therapeutic Drug Monitoring in a Vietnamese Private Hospital" aimed to analysis the ability to reach the target concentration during treatment and the risk of adverse renal events when applying Vancomycin TDM.

2 Materials and Methods

2.1 Materials

The study protocol obeyed to the principles of the Declaration of Helsinki and was approved by the Ethics Committee of Haiphong International Hospital (12/QĐ-BVĐKQT). Individuals indicated with vancomycin more than 24 h were eligible to participate in this study. This prospective study was designed to assess the efficacy of our vancomycin dosing recommendations to achieve target therapeutic concentration intervention. During the research process, there were 137 patients who satisfied the criteria of our study (Fig. 1), in which 79 subjects underwent Vancomycin TDM process (Fig. 2).

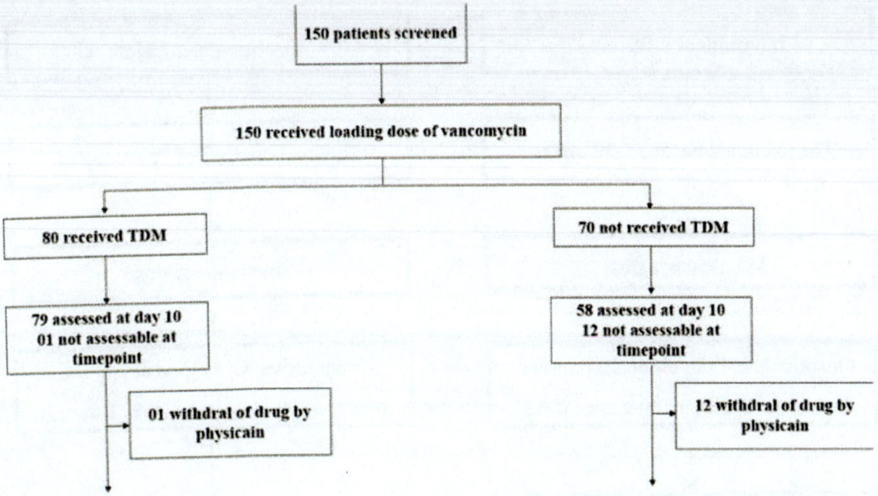

Fig. 1 Study design

2.2 *Methods*

To evaluate the ability to reach the target concentration during treatment and the risk of renal adverse effects when applying TDM of Vancomycin, we identified the vancomycin trough concentration, number of patients taking samples, number of sampling time per one patient, number of samples reached the target concentration and number of dose adjustment times in patients. The risk of nephrotoxicity experience: number of patients experiencing nephrotoxicity; the number of nephrotoxicity—induced drugs used during treatment (not including antibiotics); assessment of nephrotoxicity level (based on RIFLE). The RIFLE classification is calculated based on Serum Creatinin or Urine output determinants, and think carefully three severity classes of Acute Kidney Injury (AKI) (Risk, Injury and Failure), according to the variations in SCr and/or UO, and two outcome classes (end-stage kidney disease and loss of kidney function) [10].

Comparison of clinical outcomes and frequency of nephrotoxicity between patients applied Vancomycin TDM and patients have not applied Vancomycin TDM process: Hospital duration (day); $\Delta WBC = WBC_{Highest} - WBC_{Lowest}$; $\Delta CRP = CRP_{Hightest} - CRP_{Lowest}$; Nephrotoxicity experience; Outcomes.

Biochemistry Analysis. Vancomycin trough concentration was measured using the UniCel® DxC 600 (Beckman Coulter, USA) analyzer.

Statistical Analysis. Data was collected and analyzed using Excel 2019 software and SPSS version 20. Continuous variables with a normal distribution are described by mean ± standard deviations. Continuous variables with non-standard distributions are described by median and quartiles. The qualitative variables are described

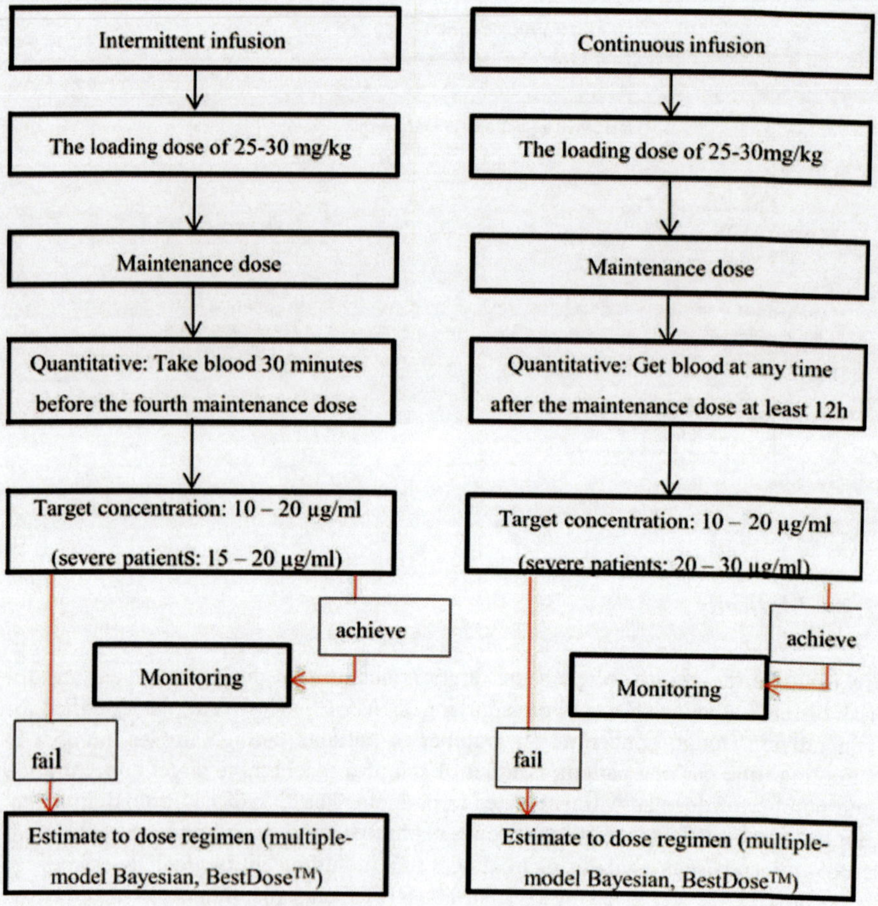

Fig. 2 Vancomycin TDM process

by quantity and percentage. Quantitative variables are evaluated for linear correlation and set up linear regression equations by Simple Linear Regression and Multiple Linear Regression. Testing the relationship between qualitative and quantitative variables is carried out with the Independent-Samples T-Test. Continuous variables with standard distributions were compared by t-test statistical tests for two independent samples. Continuous variables with non-standard distributions were compared by the non-parametric test Mann–Whitney U. The qualitative variables were scaled by statistical χ^2 for two independent samples.

3 Results and Discussion

Demographic and clinical data of the patients were presented in Table 1.

The age of patients in the study had median as 49 years old, and there was a big difference between the age of the patients with the highest age of 97 years old and the minimum age of 1 year old. Serum creatinine at the time of starting the use of vancomycin had median with 65.38 mmol/L. The average creatinine clearance was 91.42 mL/min, with large differences between patients ranging from 23.21 to 333.55 mL/min. Average CRP was also noted to be significantly different among patients at the time of initiation of vancomycin, with the highest index being 388.3 mg/L and the lowest index being 0.1 mg/L. The average WBC of patients had median with 9.8×10^9/L, ranging from 4.9×10^9/L to 25.5×10^9/L. There was a big difference between the number of days hospitalized, with the longest course of treatment recorded as 99 days and the shortest as two days. The patients were cured with 14.6%, remission was 70.8%, the unchanged condition was 13.1%, and two cases had serious condition transferring to higher level hospital.

Table 1 Demographic and clinical data of the patients (n = 137)

Characteristics	Result
Age (years) (Median, range)	49 (29–64)
Gender (N,%)	
Male	78 (56.9)
Female	59 (43.1)
Weight (kg) (Median, range)	58 (51–65)
Initial temperature (°C) (Mean ± SD)	37.5 ± 0.7
Initial SCr (μmol/L) (Median, range)	65.38 (54.02–78.42)
Initial ClCr (mL/min) (Mean ± SD)	91.42 ± 38.61
CRP (mg/L) (Median, range)	40.6 (14.8–66.1)
WBC (10^9/L) (Median, range)	9.8 (7.1–13.4)
Duration of hospitalization (days) (Median, range)	13 (9–19)
Vancomycin duration of treatment (Median, range)	8 (4.5–11)
Loading dose (mg/kg) (Mean ± SD)	20.82 ± 6.75
Maintenance dose (mg/day) (Mean ± SD)	1944 ± 653
Treatment outcome (N,%)	
Cured	25(18.25)
Remission	88(64.23)
Unchanged	19(13.87)
Severity	5(3.65)

There were 79 patients of 137 patients in the study applied Vancomycin TDM. The number of days using vancomycin ranged from 1 to 66 days; the median was 8. About the loading dose, there are five dose levels used in patients, of which the dose level 1000 mg is most applied at the rate of 65%. Loading dose based on mg/kg had an average intake dose recorded as 20.82 mg/kg. The maintenance dose used in patients before vancomycin trough concentration was measured with an average of 1944 mg/day, ranging from 165 mg/day to 4000 mg/day. After quantification of vancomycin trough concentration in patients, the average maintenance dose level increased to 1990 mg/day. However, when comparing the average maintenance dose with the study of Liang et al. [11] in 2018, the results were different when the maintenance dose was 1710 mg day, lower than ours. The difference due to the number of patients participating in the study of Xiaoyu Liang et al. was 370.

Characteristics of Vancomycin trough concentration in our study were described in Table 2.

A total of 79 patients were sampled to measure vancomycin trough quantitative measurements, of which 55 patients were sampled once, 16 patients were sampled twice, five patients were sampled three times, two patients were sampled four times, and one patient was sampled five times. The total number of samples taken from patients was 116, and 73.3% of the samples reached a trough concentration of 10 μg/mL or more. Vancomycin trough concentration fluctuated greatly, recorded

Table 2 Characteristics of Vancomycin trough concentration

Criteria	Value
Number of patients taking samples (N,%)	79 (100)
Taking one sample	55 (69.6)
Number of sample time (N)	116
Number of samples reached the target (N, %)	85/116 (73.3)
Number of patients having sample reached target concentration (N,%)	
At the first time	52/79 (65.8)
At the second time	8/79 (10.2)
At the third time	2/79 (2.5)
Vancomycin concentration (Median; range)	14.49 (9.66–21.93)
< 10 (μg/mL) (N, %)	31/116 (26.7)
≥ 10–20 (μg/mL) (N, %)	49/116 (42.2)
≥ 20–30 (μg/mL) (N, %)	26/116 (22.4)
> 30 (μg/mL) (N,%)	10/116 (8.7)
Dose adjustment in patients (N, %)	9 (11.4)

Number of patient taking samples: one patient can take one or more sample to do TDM depend on the results of vancomycin concentration. If patient's vancomycin concentration reach the target we don't request the second time of sample

Number of sample time: the total of sample time of our patients

Dose adjustment in patient: Vancomycin dose of 9 were adjusted based on the vancomycin concentration from 0.47 to 43.38 μg/mL; the median was 14.49 μg/mL. To achieve the target blood concentration, nine patients were carried out dose adjustments by our clinical pharmacists. In the study of Liang et al. [11], the trough concentration had a quartile range of 4.65–14.17 μg/mL, and the median of 8.64 μg/mL. The initial regimen of vancomycin was 1 g per 24–96 h for patients having CrCl rates between 10 and 50 ml/L and 1 g per 12 h for patients having CrCl rates > 50 ml/L; The loading dose calculated based on ClCr rather than mg/kg as in our study should lead to differences in concentration.

The comparison between the two patient groups were established: group 1 were the patients applied vancomycin Therapeutic drug monitoring; group 2 were the patients not applied. Factors to be compared include the number of days in the hospital, decreased WBC, decreased CRP, nephrotoxicity and treatment outcome (Table 3).

There was a statistically significant difference between the two opposing groups in CRP reduction with *p*-value < 0.05. The CRP reduced in the TDM applied group was higher than the non-applied group. Applying TDM will reduce CRP more than not applying. In the group applied TDM, 15(18.99%) cases were cured, 57(72.15%) subjects were in remission, but only 1 (1.26%) became severely. In the group not applied TDM, 10(17.24%) cases were cured, 31(53.45) subjects were in remission, but 4(6.9%) patients had a severe situation. The treatment outcome of the two

Table 3 Comparison of clinical outcomes and the frequency of nephrotoxicity

Criteria	Group 1 patients applied TDM (N = 79)	Group 2 patients not applied TDM (N = 58)	*P*-value
Hospital duration (Median; range)	14 (10–21)	12 (9–16)	0.07
ΔWBC (Median; range)	2.4 (0–5.9)	0.3 (0–4.1)	0.152
ΔCRP (Median; range)	25.9 (0–87.2)	0.5 (0–36.9)	**0.012**
Nephrotoxicity experience (N, %)	5 (11.3)	1 (12.5)	0.780
Outcome (N, %)			
Cured	15(18.99)	10(17.24)	**0.018**
Remission	57(72.15)	31(53.45)	
Unchanged	6(7.59)	13(22.41)	
Severity	1(1.26)	4(6.9)	

Hospital duration: are calculated by subtracting day of admission from day of discharge; ΔWBC: are calculated by subtracting WBC of admission day from WBC of discharge day; ΔCRP: are calculated by subtracting CRP of admission day from CRP of discharge day; Nephrotoxicity experience: number of patients got nephrotoxicity during hospitalization. Outcome: results of treatment groups differs significantly with *P*-value < 0.05 (table 2). Unfortunately, no significant differences were found between the groups on hospital duration, ΔWBC, and nephrotoxicity experience.

To build the multiple regression model for vancomycin maintenance dose, we initially discovered the association between all factors and the vancomycin maintenance dose. The factors with a linear regression *P*-value < 0.05 included age, weight, vancomycin level, CRP, Clcr. Finally, five factors combined in the regression model were listed in Table 4 and Fig. 3

In this model, the *vancomycin trough concentration, Clcr, and weight* contributed most (24.7%, 16.8% and 26.6%, respectively) to the inter-individual variability in vancomycin dose. To determine a patient's dose using our model, a clinician would fill out the following algorithm using the patient's clinical and genetic characteristics.

Table 4 Final multiple models of maintenance dose of vancomycin produced by stepwise regression analysis

	Estimate	Std	Error	*P*_value
Intercept	21.41785	8.16451	2.623	0.011457
Clcr	0.16875	0.04313	3.913	0.000271
Vancomycin concentration	0.24776	0.08171	3.032	0.003813
Weight	– 0.26698	0.09615	– 2.777	0.007659
Age	0.16828	0.08254	2.039	0.046679
ΔCRP	– 0.02947	0.01349	– 2.184	0.033617

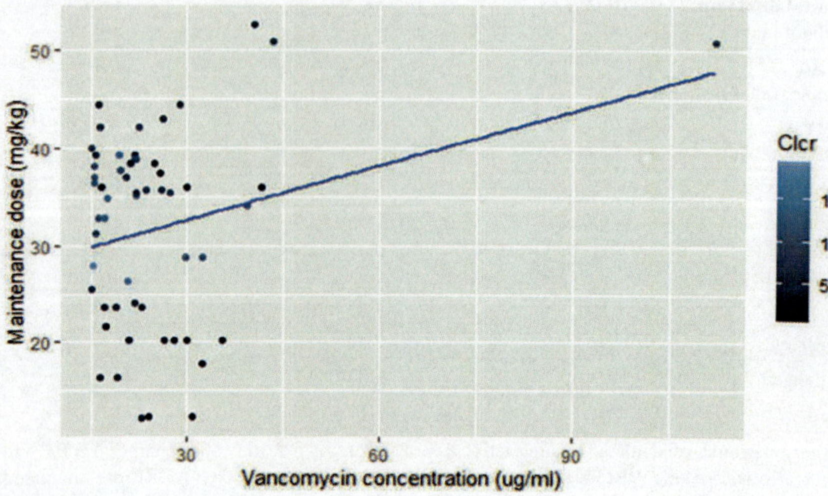

Fig. 3 The correlation of maintenance dose and vancomycin concentration and creatinine clearance

Vancomycin maintenance dose (mg/kg) = 21.4 + 0.24 × vancomycin trough concentration (μg/mL) + 0.17 × Clcr(mL/min) − 0.26 × weight(kg) + 0.17 × age(years) − 0.03 × ΔCRP.

In total, the algorithm could explain 43.9% (*P*-value < 0.001) of the variability in vancomycin maintenance dose. Agree with our research results, Lin and colleagues [12] confirmed creatinine, vancomycin dosage, body mass index, and age affected significantly on vancomycin dose. However, we firstly reported ΔCRP could slightly be influenced by vancomycin maintenance dose.

4 Conclusion

Our research conducted on a total of 137 patients with 79 patients who were administered TDM of vancomycin has demonstrated the benefits of vancomycin TDM technology on clinical outcomes and improvement of infection situation. Furthermore, we have successfully indicated a strong relationship between vancomycin trough concentration, creatinine clearance, age, weight, ΔCRP and maintenance dose of vancomycin. Clinicians in the process of designing a regimen dose of vancomycin should comprehensively consider these factors in order to optimize the dose and reduce the adverse effect of vancomycin.

Acknowledgements We thank all physicians and nurses of Haiphong International Hospital for their contributions to this study.

Conflicts of Interest The authors have no conflict of interest to declare.

References

1. Harkins CP et al (2017) Methicillin-resistant *Staphylococcus aureus* emerged long before the introduction of methicillin into clinical practice. Genome Biol 18(1):130–130
2. Rybak M et al (2009) Therapeutic monitoring of vancomycin in adult patients: a consensus review of the American Society of Health-System Pharmacists, the Infectious Diseases Society of America, and the Society of Infectious Diseases Pharmacists. Am J Health Syst Pharm 66(1):82–98
3. Brown J, Brown K, Forrest A (2012) Vancomycin AUC24/MIC ratio in patients with complicated bacteremia and infective endocarditis due to methicillin-resistant *Staphylococcus aureus* and its association with attributable mortality during hospitalization. Antimicrob Agents Chemother 56(2):634–638
4. Harigaya Y et al (2009) Pharmacodynamics of vancomycin at simulated epithelial lining fluid concentrations against methicillin-resistant Staphylococcus aureus (MRSA): implications for dosing in MRSA pneumonia. Antimicrob Agents Chemother 53(9):3894–3901
5. Holmes NE et al (2013) Vancomycin AUC/MIC ratio and 30-day mortality in patients with *Staphylococcus aureus* bacteremia. Antimicrob Agents Chemother 57(4):1654–1663
6. Kullar R et al (2011) Impact of vancomycin exposure on outcomes in patients with methicillin-resistant Staphylococcus aureus bacteremia: support for consensus guidelines suggested targets. Clin Infect Dis 52(8):975–981

7. LaPlante KL et al (2008) Activities of clindamycin, daptomycin, doxycycline, linezolid, trimethoprim-sulfamethoxazole, and vancomycin against community-associated methicillin-resistant *Staphylococcus aureus* with inducible clindamycin resistance in murine thigh infection and in vitro pharmacodynamic models. Antimicrob Agents Chemother 52(6):2156–2162
8. Rybak MJ (2006) Pharmacodynamics: relation to antimicrobial resistance. Am J Infect Control 34(5):S38–S45
9. Ye ZK, Li C, Zhai SD (2014) Guidelines for therapeutic drug monitoring of vancomycin: a systematic review. PLoS One 9(6):e99044
10. Bellomo R et al (2004) Acute renal failure–definition, outcome measures, animal models, fluid therapy and information technology needs: the second international consensus conference of the acute dialysis quality initiative (ADQI) group. Crit Care 8(4):R204
11. Liang X et al (2018) A prospective multicenter clinical observational study on vancomycin efficiency and safety with therapeutic drug monitoring. Clin Infect Dis 67(suppl_2):S249–S255
12. Lin Z et al (2014) Clinical research for trough value of serum vancomycin in critical patients. Zhonghua Wei Zhong Bing Ji Jiu Yi Xue 26(7):473–477

Tissue Engineering and Regenerative Medicine

Fabrication and Characterization of Porous Flow-Assembled Chitosan Membranes in Microfluidics

Khanh L. Ly and Xiaolong Luo

Abstract The aim of this research is to customize the porosity of biofabricated chitosan membrane (CM) employing co-assembled polystyrene nanoparticles (np) as a sacrificial template. CM with np (CM-np) was manufactured inside the microfluidic chip using the flow-assembly technique. Glutaraldehyde was used to crosslink the fabricated CM-np were then dissolved with dimethyl sulfoxide, leaving the porous chitosan membrane (pCM). The growth rate of CM and CM-np was investigated to determine the effects of np incorporation on the growth of the fabricated membrane. The morphology of the biofabricated CM and pCM were evaluated using scanning electron microscopy. The mass transport tests were also conducted to confirm the increase in pores size of pCM in comparison with pure CM. Thus, in this study, we have demonstrated the capability to manipulate the porosity of the biofabricared CM manufactured by flows inside microfluidic chips and characterized the properties of the fabricated membranes. This tuning process is promising and can enhance the applicability of biopolymer CM in biochemistry and biology researches.

Keywords Porous chitosan membranes · Flow-assembly · Co-assembled nanoparticles as a template

1 Introduction

The incorporation of membranes into microfluidic devices offers precise mass transport for many applications such as filtration, extraction, and separation [1, 2]. The driving force of the mass transport through the integrated membrane is the chemical potentials differences across the membrane, allowing selective subjects to diffuse through its pores [3]. Since the membrane porosity determines the mass transport and its selectivity, the ability to customize the membrane porosity is highly desired and can broaden the implementations of membranes-incorporated microfluidic platforms.

K. L. Ly · X. Luo (✉)
The Catholic University of America, Washington, DC 20064, USA
e-mail: luox@cua.edu

V. Van Toi et al. (eds.), *8th International Conference on the Development of Biomedical Engineering in Vietnam*, IFMBE Proceedings 85,
https://doi.org/10.1007/978-3-030-75506-5_31

There are several approaches to integrate membranes into microfluidic networks. Whilst the direct incorporation of the commercially available membrane is challenging to avoid leakage and chemical compatibility issues due to the insufficient layers sealing, the preparation of membrane as a part of the microfluidic chip fabrication process is complex and exorbitant [3, 4]. Therefore, the in-situ preparation of membrane on-chip has emerged as a promising alternative. Among the available materials for such an approach, the in-situ preparation of chitosan membranes (CM) based on interfacial polymerization reactions can be a facile approach to integrate biopolymer membranes to microfluidic devices [5]. Chitosan, a derivative of chitin, has been extensively used in agriculture and biomedicine thanks to its unique film forming and physiochemical properties [6–9]. Specifically, chitosan is water-soluble at acidic pH and becomes insoluble when the pH of the surrounding environment is higher than its pKa (~ 6.3) [10]. Based on that, an insoluble CM was in-situ biofabricated on a polyelectrolyte complex membrane (PECM), instantaneously formed through electrostatic interactions between alginate and chitosan macromolecules, in microfluidic devices by creating a localized pH gradient using a flow-assembly technique as previously reported [10–12].

Many methods have been employed to tune membrane porosity such as post-treatment of a porous structure, phase inversion, and interfacial polymerization [13]. Here, polystyrene nanoparticles (np) were utilized as sacrificial materials during the fabrication process to modulate the porosity of the biofabricated CM. Polystyrene np was chosen thanks to their highly controllable physicochemical properties and commercial availability in a diverse range of sizes and shapes [14–16], enabling the active manipulation the porosity of the biofabricated CM based on application needs. One such need is to enhance the mass transport of membranes towards biomacromolecules such as antibodies.

The aims of this study are to develop a process to manipulate the porosity of the CM assembled by flows in microfluidic networks and characterize the properties of the fabricated membranes. Figure 1b illustrates the fabrication process involving the direct co-assembly of CM with polystyrene np (CM-np) inside microchannels, and then the fabricated CM-np were treated with glutaraldehyde (GA), followed by the removal of the co-assembled np with dimethyl sulfoxide (DMSO), leaving the porous CM (pCM).

2 Materials and Methods

2.1 *Materials*

Sylgard 184 and its curing agent for polydimethylsiloxane (PDMS) chip fabrication was supplied by Ellsworth Adhesives (Germantown, WI, USA). PTFE tubing (interior diameter of 0.022″ and outside diameter of 0.042″) were supplied by Cole-Parmer (Vernon Hills, IL, USA). Hollow metal couplers (22-gauge size) were

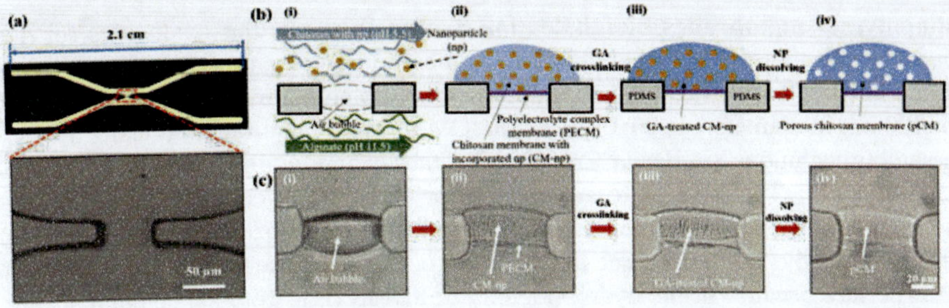

Fig. 1 The process to customize the porosity of chitosan membrane. **a** Design of the single-aperture PDMS device and the zoom-in image of the aperture region under microscopy; **b** illustrative schematics of the biofabrication; and **c** the representative images showing the fabrication process of porous chitosan membrane (pCM): (i) air bubble was trapped at the aperture; (ii) fabricated CM with np (CM-np); (iii) GA-treated CM-np; and (iv) porous chitosan membrane (pCM) after dissolving np with DMSO. Scale bars as indicated

supplied by Instech Laboratories Inc. (Plymouth Meeting, PA, USA). Disposable syringes were supplied by Becton, Dickinson and Company (Franklin Lakes, NJ, USA). Phosphate-buffered saline (PBS) 10X solution (pH = 7.4) was supplied by TissuePro Technology, LLC (Gainesville, FL, USA) and diluted into PBS 1X solution with deionized water. Sodium alginate powder, chitosan flakes (medium molecular weight, 85% deacetylated), and GA solution (25% in H_2O) were supplied by Sigma Aldrich (St. Louis, MO, USA) and diluted into 10% GA solution with PBS. DMSO solution was supplied by Amresco® Company (Solon, OH, USA). Polystyrene and fluorescein-labelled polystyrene np (F-polystyrene) solutions were supplied by Degradex® Phosphorex, Inc. (Hopkinton, MA, USA).

2.2 *Methods*

Microfluidic devices fabrication. The single-aperture silicon mold used to make PDMS devices were fabricated using photolithography technique while the PDMS microfluidic devices were fabricated using the soft-lithography technique as previously reported [17]. The single-aperture PDMS device consists of two microchannels with two inputs and two outputs with each channel's width, and height is 500 and 40 μm, correspondingly (Fig. 1a). To make PDMS devices, Sylgard 184 and its curing agent were thoroughly mixed at the ratio of 10:1, degassed, poured on the single-aperture silicon mold, and cured at 65 °C for 2–4 h. The solidified PDMS was delaminated from the silicon mold and sectioned into PDMS stabs containing the single-aperture channels that were punched for input and output ports. Next, Plasma Cleaner PDC-32G (Harrick Plasma) was employed to bind the PDMS stabs with either glass slides, or plain PDMS layers to make the complete PDMS devices that were then placed in an oven at 120 °C overnight prior to the membrane biofabrication.

Biopolymer membranes biofabrication. Figure 1b shows the schematics for the customizing the biofabricated chitosan membrane porosity process. The chitosan solution (0.5% w/v) was mixed with polystyrene np solution (molecular size of 50 or 100 nm in diameter) at 20:1 (v/v) ratio. CM-np were manufactured using a flow-assembly technique similar to CM. For CM fabrication, alginate (pH = 11.5) and chitosan (pH = 5.5) (or chitosan with np for CM-np fabrication) solutions were separately introduced into the two channels at the flow rate of 1 μL/min. In most of the cases, as the solutions came to the aperture, an air bubble was trapped between the aperture because of the hydrophobicity of PDMS (Fig. 1c(i)). The pumps were stopped to establish the pressure balance. The trapped air bubble was removed with the add-on vacuum layer, as previously reported [18]. Once the trapped air bubble was completely dissipated, a thin PECM was spontaneously established in the aperture. After that, the flows were restarted, which allowed the hydroxyl ions to diffuse from the alginate side to the chitosan side via the PECM, creating a local pH gradient for CM or CM-np formation (Fig. 1c(ii)). Within 1–2 min, the CM or CM-np reached the desired thickness, the flows were stopped, the add-on vacuum layer was removed, the tubing was disconnected, the channels were rinsed with PBS, and the biofabricated CM was stored at 4 °C. This is the end for CM fabrication process. The process continued for CM-np as follows. Next, the microchannels were filled with 10% GA solutions to crosslink CM-np in one hour (Fig. 1c(iii)) and rinsed with PBS. Then, DMSO solutions were supplied to the channels for two hours to dissolve np (Fig. 1c(iv)). Finally, all microchannels were rinsed with PBS, the biofabricated pCM from 50 or 100 nm np (pCM50 or pCM100, respectively) were then stored at 4 °C for further uses.

Membrane growth rate and morphology observation. The growth rate of CM and CM-np was determined by plotting the membrane thickness versus time. The membrane thickness was measured every 10 s using ImageJ (NIH, USA). Data were expressed as mean ± standard deviation (SD). The fabricated CM, CM-np, and pCM were fixed in 4% GA, rinsed with PBS, and dehydrated with the series of alcohol solutions from 50 to 100%. The two-layer PDMS devices containing the samples were left to dry at room temperature, then the samples were taken out using a simple punching technique with a biopsy punch. The punched PDMS pieces containing the samples were gently delaminated, mounted to carbon tape, and coated with a thin carbon layer prior to observation with scanning electron microscopy (SEM, Tescan XEIA3) at 1 kV.

Membrane permeability. The mass transport of CM and pCM was studied using fluorescent-labelled polystyrene (F-polystyrene) of different size of 50 and 100 nm as representative for biomacromolecules. The F-polystyrene solutions (0.1% w/v, the molecular size of 50 and 100 nm) were continuously supplied in one microchannel at 1 μL/min flow rate whilst the other was filled with PBS at static conditions (Fig. 3a(i)) for 30 min. The permeability of the pCMs fabricated from different chitosan versus np ratios (v/v) was tested with The average fluorescence of a 250 × 50 μm^2 region across CM or pCM was measure using ImageJ as shown in Fig. 4a(ii) and the fluorescent intensity versus distance profile across the membrane was plotted in Fig. 3b. All the images are taken with Ludesco EXI-310 inverted fluorescent microscopy with green-fluorescent filter.

3 Results and Discussion

The fabrication process and the transformation in the microscale of pCM through each step are shown in Fig. 1c. The trapped air bubble (Fig. 1c(i)) was dissipated with an add-on vacuum layer as reported [18]. The in situ biofabricated CM-np contained numerous of tiny polystyrene np, leaving its appearance more textured and less transparent under brightfield microscopy (Fig. 1c(ii)) than CM [5]. Next, the CM-np was treated with GA to enhance its mechanical strengths and prevent the collapse of the porous structure as np being removed [19–22]. The morphology of GA-treated CM-np200 is shown in Fig. 1c(iii) where no significant morphological difference was observed. After that, DMSO, a low toxic solvent, which possesses great solubility against polystyrene, while induces minimal swelling effects on PDMS [23, 24] was employed for np removal process. This allowed DMSO solution to stay in PDMS device long enough (2 h) to dissolve the incorporated np while minimizing the effects on PDMS device that might lead to unexpected leakage. Lastly, the microscale morphology of pCM200 is shown in Fig. 1c(iv), which experienced significant transformation. In particular, the membrane after np removal became less textured and more transparent, indicating that the np in CM-np blocking the transmitted light had been eliminated, resulting in the pCM.

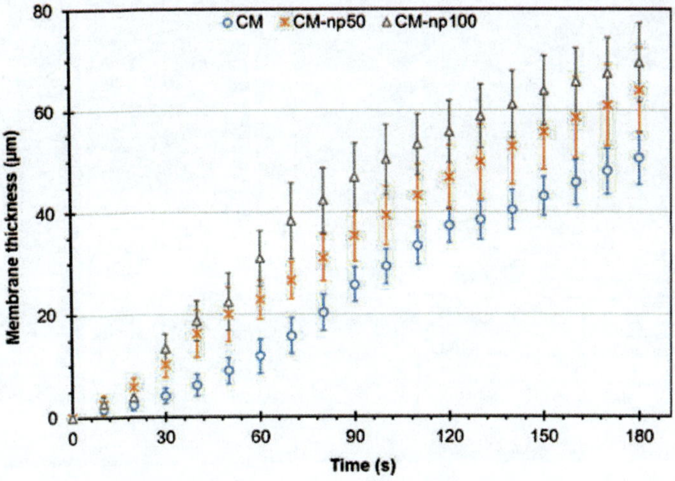

Fig. 2 Typical membrane growth curves of: CM (o, blue), CM-np50 (x, orange), and CM-np100 (Δ, gray) (data = mean ± SD, n = 14)

Figure 2 shows the typical membrane growth curves of CM and CM-np over the same period. Overall, the growth curves were similar among samples that increased steeply during the first 90 s, followed by a gradual rise afterwards. The growth of CM and CM-np depends on the deprotonation rate of chitosan molecules induced by the rate of hydroxyl ions diffusing from the high pH alginate solution. Since the diffusion of hydroxyl ions across PECM and formed membrane is one-dimensional and independent on the width and the height of the apertures, the effect of aperture dimensions on the membrane growth rate is, thus, minimized [18]. It is worth noticing that the membrane thickness of CM-np was significantly greater than that without np, and the membrane was thicker for np of bigger size. This was expected because the membrane growth was the result of the deprotonation of chitosan amine groups, which was consistent under the same experimental conditions such as the solution pH and flow rate. The additional np in chitosan solution would have occupied extra volume in the fabricated membrane, resulting in thicker membranes as the channel depth was fixed.

The morphological changes of CM, CM-np, and pCM in the nanoscale are examined using SEM. Figure 3(i) reveal the entire CM, CM-np50, CM-np100, pCM50, and pCM100 obtained from PDMS devices with the length x width x depth of around 60 × 30 × 40 (μm), respectively. In general, CM had relatively smooth surfaces (Fig. 3a(ii, iii)), while pCM (Fig. 3c, e) possessed significantly rougher morphologies. It is depicted in Fig. 3b, d(ii, iii) that as CM was co-assembled with np, an evenly distributed layer of np was observed on the top-view and the side-view of the CM-np. Then, as can be seen in Fig. 3c, e(ii, iii), the amount of np was considerably

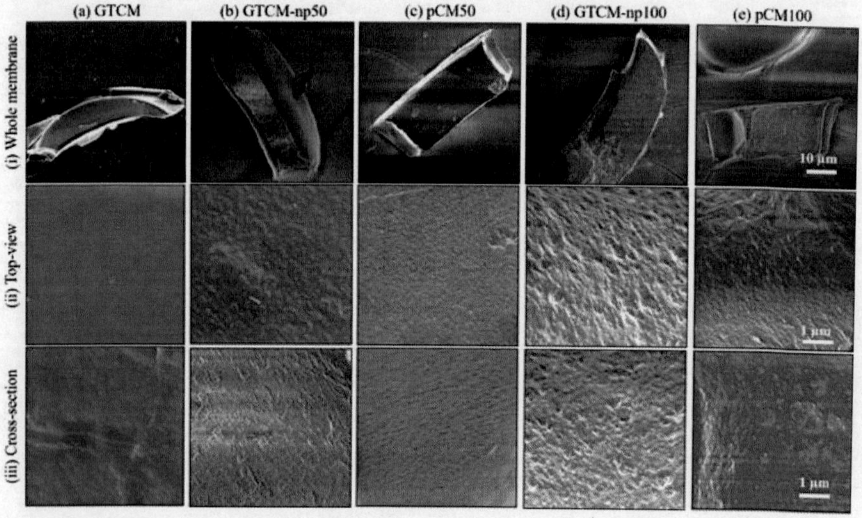

Fig. 3 SEM images of **a** CM, **b** CM-np50, **c** pCM50, **d** CM-np100, and **e** pCM100: (i) entire membranes obtained from PDMS devices; the morphology of membranes from (ii) the top-view, and (iii) the side-view. Scale bars as indicated

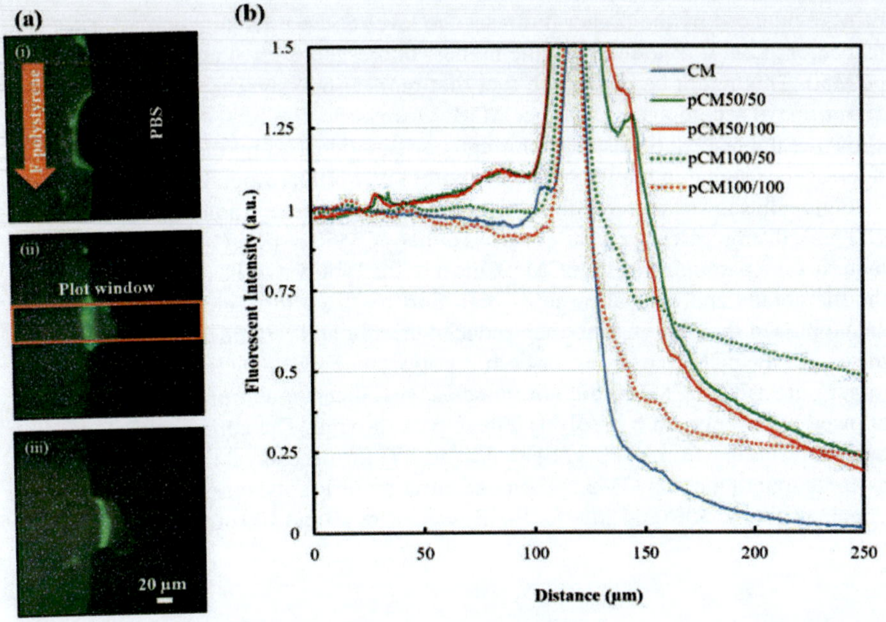

Fig. 4 **a** Representative fluorescence images of (i) CM, (ii) pCM50, and (iii) pCM100 under the transport test with fluorescent-labelled polystyrene (F-polystyrene). Scale bars as indicated. **b** The fluorescent signal passed through CM and pCM versus distance. The spectra are named as follows: membrane type/size of F-polystyrene in diameter (unit: nm)

reduced, leaving the tiny and shallow pores on either pCM surfaces after the removal process. Furthermore, it is highly desired if we could observe the internal structure of pCM, which could partially demonstrate the enhanced mass transport ability. This can be done by fabricating elongated membranes, involving the blocking of alginate side output to force the alginate solution to directly react with the chitosan solution, resulting in the elongated membranes [17].

According to SEM images, the removal of np significantly changed the nanoscale structure of pCM that might as well affect their mass transport. Therefore, the permeability of CM and pCM is an essential characteristic and needs to be investigated.

Figure 4a shows the selected fluorescence images of the fabricated CM, pCM50, and pCM100 underwent the mass transport test with F-polystyrene (molecular size of 50 nm), while Fig. 4b shows the corresponding fluorescent profiles of F-polystyrene (molecular size of 50 and 100 nm) passed across CM, pCM50, and pCM100. It can be seen in Fig. 4a that almost no fluorescence was observed in the PBS side across CM while significantly high fluorescence levels at the right channels across pCM were seen. As a barrier membrane, either CM or pCM tend to disable the penetration of F-polystyrene at the first layer of the membrane. Later on, under the continuous flows, F-polystyrene beads were able to pass through the pores and internal pores of pCM, evidenced by the higher fluorescence level in the PBS side across pCM than that across CM. It is noticed that the highest fluorescence level was observed within pCM50. This might be due to the fact that more F-polystyrene (molecular size of 50 nm) have accumulated inside pCM50. Meanwhile, the larger pores of pCM100 allow the fluorescent beads to go through easily, evidenced by the higher and lower fluorescence signal in the PBS side and within pCM100, respectively.

Figure 4b shows that regardless of the size F-polystyrene (molecular size of 50 or 100 nm), the fluorescence profiles across either pCM50 or pCM100 shared similarly patterned. In particular, as for pCM50, most of the F-polystyrene accumulated inside the membrane and only allowed a small portion to go through. The fluorescence signal passed through pCM50 then reduced steeply and remained less than 25% at the end of the pCM50 profiles for both F-polystyrene sizes. On the other hand, with larger pores size, pCM100 did not induce as much accumulation as in pCM50 and allowed noticeably more F-polystyrene to pass through. The concentration gradient of F-polystyrene from the right-sided pCM100 to the end of the plot was flatter as compared to that of pCM50. The concentration of F-polystyrene across pCM100 remained around 50% and 25% for F-polystyrene of 50 and 100 nm, correspondingly.

4 Conclusion

To conclude, the porosity of CM has been successfully customized in microfluidic networks using np as sacrificial materials. The reported biofabrication process is facile and reliable, resulting in the freestanding and robust biofabricated CM and pCM with the spatial and temporal controllability over their thickness. The co-assembly of np significantly increased the membranes' growth rate as compared to pure CM. Furthermore, the SEM and permeability results show that the pCM were more porous and allowed enhanced mass transport than CM. The pCM-incorporated microfluidic platforms can be used to study macromolecules gradients generation [17] and should be further exploited for biomolecular and cellular studies.

Acknowledgements This effort was supported in part by the National Science Foundation (NSF) under grant number CAREER 1553330 and the National Institute of Health (NIH) under grant number 1R15GM129766-01. We would like to acknowledge the support of the Maryland NanoCenter and its AIMLab.

Conflicts of Interest The authors report no conflicts of interest in this work.

References

1. Luo X, Berlin DL, Betz J, Payne GF, Bentley WE, Rubloff GW (2010) In situ generation of pH gradients in microfluidic devices for biofabrication of freestanding, semi-permeable chitosan membranes. Lab Chip 10(1):59–65. https://doi.org/10.1039/b916548g

2. Hsieh YC, Zahn JD (2007) On-chip microdialysis system with flow-through glucose sensing capabilities. J Diabetes Sci Technol 1(3):375–383. https://doi.org/10.1177/193229680700100310
3. de Jong J, Lammertink RG, Wessling M (2006) Membranes and microfluidics: a review. Lab Chip 6(9):1125–1139. https://doi.org/10.1039/b603275c
4. Ding W, Liang C, Sun S, He L, Gao D (2015) On-chip fabrication of carbon nanoparticle-chitosan composite membrane. J Mater Sci Technol 31(11):1087–1093. https://doi.org/10.1016/j.jmst.2015.09.004
5. Li K, Correa SO, Pham P, Raub CB, Luo X (2017) Birefringence of flow-assembled chitosan membranes in microfluidics. Biofabrication 9(3):034101. https://doi.org/10.1088/1758-5090/aa786e
6. Dragostin OM, Samal SK, Dash M, Lupascu F, Pânzariu A, Tuchilus C, Ghetu N, Danciu M, Dubruel P, Pieptu D, Vasile C, Tatia R, Profire L (2016) New antimicrobial chitosan derivatives for wound dressing applications. Carbohyd Polym 141:28–40. https://doi.org/10.1016/j.carbpol.2015.12.078
7. Nguyen TD, Nguyen TT, Ly KL, Tran AH, Nguyen TTN, Vo MT, Ho HM, Dang NTN, Vo VT, Nguyen DH, Nguyen TTH, Nguyen TH (2019) In vivo study of the antibacterial chitosan/polyvinyl alcohol loaded with silver nanoparticle hydrogel for wound healing applications. Int J Polym Sci 7382717. https://doi.org/10.1155/2019/7382717
8. Nguyen-My Le A, Nguyen TT, Ly KL, Luong TD, Ho MH, Minh-Phuong Tran N, Ngoc-Thao Dang N, Van Vo T, Tran QN, Nguyen TH (2020) Modulating biodegradation and biocompatibility of in situ crosslinked hydrogel by the integration of alginate into N,O-carboxylmethyl chitosan—aldehyde hyaluronic acid network. Polym Degrad Stab 180:109270. https://doi.org/10.1016/j.polymdegradstab.2020.109270
9. Hu P, Rooholghodos SA, Pham LH, Ly KL, Luo X (2020) Interfacial electrofabrication of freestanding biopolymer membranes with distal electrodes. Langmuir 36(37):11034–11043. https://doi.org/10.1021/acs.langmuir.0c01894
10. Cheng Y, Luo X, Betz J, Buckhout-White S, Bekdash O, Payne GF, Bentley WE, Rubloff GW (2010) In situ quantitative visualization and characterization of chitosan electrodeposition with paired sidewall electrodes. Soft Matter 6(14):3177–3183. https://doi.org/10.1039/C0SM00124D
11. Luo X, Wu H-C, Betz J, Rubloff GW, Bentley WE (2014) Air bubble-initiated biofabrication of freestanding, semi-permeable biopolymer membranes in PDMS microfluidics. Biochem Eng J 89:2–9. https://doi.org/10.1016/j.bej.2013.12.013
12. Ly KL, Hu P, Pham LHP, Luo X (2021) Flow-assembled chitosan membranes in microfluidics: recent advances and applications. J Mater Chem B 9(15):3258–3283. https://doi.org/10.1039/D1TB00045D
13. Paul M, Jons SD (2016) Chemistry and fabrication of polymeric nanofiltration membranes: a review. Polymer 103:417–456. https://doi.org/10.1016/j.polymer.2016.07.085
14. Nandiyanto ABD, Hagura N, Iskandar F, Okuyama K (2010) Design of a highly ordered and uniform porous structure with multisized pores in film and particle forms using a template-driven self-assembly technique. Acta Mater 58(1):282–289. https://doi.org/10.1016/j.actamat.2009.09.004
15. Nandiyanto A, Ogi T, Okuyama K (2014) Polystyrene spheres for template in the production of nanostructured materials, pp 241–267
16. Sandberg LIC, Gao T, Jelle BP, Gustavsen A (2013) Synthesis of hollow silica nanospheres by sacrificial polystyrene templates for thermal insulation applications. Adv Mater Sci Eng 2013:6. https://doi.org/10.1155/2013/483651
17. Ly KL, Raub CB, Luo X (2020) Tuning the porosity of biofabricated chitosan membranes in microfluidics with co-assembled nanoparticles as templates. Mater Adv 1(1):34–44. https://doi.org/10.1039/D0MA00073F
18. Pham P, Vo T, Luo X (2017) Steering air bubbles with an add-on vacuum layer for biopolymer membrane biofabrication in PDMS microfluidics. Lab Chip 17(2):248–255. https://doi.org/10.1039/C6LC01362G

19. Fu J, Yang F, Guo Z (2018) The chitosan hydrogels: from structure to function. New J Chem 42(21):17162–17180. https://doi.org/10.1039/C8NJ03482F
20. Hoare TR, Kohane DS (2008) Hydrogels in drug delivery: progress and challenges. Polymer 49(8):1993–2007. https://doi.org/10.1016/j.polymer.2008.01.027
21. Webster A, Halling MD, Grant DM (2007) Metal complexation of chitosan and its glutaraldehyde cross-linked derivative. Carbohyd Res 342(9):1189–1201. https://doi.org/10.1016/j.carres.2007.03.008
22. Hu P, Raub CB, Choy JS, Luo X (2020) Modulating the properties of flow-assembled chitosan membranes in microfluidics with glutaraldehyde crosslinking. J Mater Chem B 8(12):2519–2529. https://doi.org/10.1039/C9TB02527H
23. de Abreu Costa L, Henrique Fernandes Ottoni M, Dos Santos MG, Meireles AB, Gomes de Almeida V, de Fatima Pereira W, Alves de Avelar-Freitas B, Eustaquio Alvim Brito-Melo G (2017) Dimethyl sulfoxide (DMSO) decreases cell proliferation and TNF-alpha, IFN-gamma, and IL-2 cytokines production in cultures of peripheral blood lymphocytes. Molecules 22(11):1789. https://doi.org/10.3390/molecules22111789
24. Lee JN, Park C, Whitesides GM (2003) Solvent compatibility of poly(dimethylsiloxane)-based microfluidic devices. Anal Chem 75(23):6544–6554. https://doi.org/10.1021/ac0346712

Fabrication of Virgin Coconut Oil-Loaded Electrospun Polycaprolactone/Polyurethane Membrane for Application in Vascular Engineering

Thien Hai Nguyen, Thuan Ba Nguyen, Nam Minh-phuong Tran, and Thi-Hiep Nguyen

Abstract The mortality and morbidity caused by vascular diseases has resulted in an ever-persistent demand for artificial vascular graft replacing injured vessels. The electrospun combination of polyurethane (PU) and polycaprolactone (PCL) has proved its potential for the application as vascular graft but it lacks sufficient antithrombotic properties. Based on its biological properties and sustainable costs, virgin coconut oil (VCO) could be a promising additional anticoagulant agent for the PU/PCL composite.This research fabricates PU/PCL electrospun scaffold loaded with various concentrations of VCO. The parameters of electrospinning were varied to obtain suitable membranes. The effects of these parameters and the VCO on the morphological and mechanical properties of PU/PCL were investigated by scanning electron microscope and tensile strength assay.It was revealed that the condition set of 12 cm tip-to-collector, 20 kV applied and 0.5 mL/h flow rate yielded suitable electrospun membranes with various VCO percentages. The loaded VCO decreased the mechanical strength of the membranes. The ultimate stress and strain of the membranes decreased accordingly with higher content of VCO but did not drop below the acceptable limit. This study has obtained the electrospun PU/PCL membrane loaded with VCO processing appropriate morphology and mechanical properties. The loading of VCO altered the fabricated membranes but they still satisfied vascular graft requirements. Therefore, the PU/PCL loaded with VCO membranes are promising for further research as artificial blood vessels.

Keywords Polyurethane · Polycaprolactone · Virgin coconut oil · Electrospinning · Vascular graft

---

Thien Hai Nguyen and Thuan Ba Nguyen—contributed equally to the chapter.

T. H. Nguyen · T. B. Nguyen · N. M. Tran · T.-H. Nguyen (✉)
Department of Tissue Engineering and Regenerative Medicine, School of Biomedical Engineering, International University, Ho Chi Minh City 70000, Vietnam
e-mail: nthiep@hcmiu.edu.vn

Vietnam National University, Ho Chi Minh City, Vietnam

V. Van Toi et al. (eds.), *8th International Conference on the Development of Biomedical Engineering in Vietnam*, IFMBE Proceedings 85,
https://doi.org/10.1007/978-3-030-75506-5_32

1 Introduction

Cardiovascular disease (CVD) is a major factor causing immense global health and economic burdens at the moment and in the foreseeable future. On average, it is projected by the American Heart Association that there are 2303 deaths from CVD each day in the world [1]. Treatments for CVD range from dietary and lifestyle alterations to pharmaceutical intervention and clinical surgery. For specific cases, the vascular surgery may involve minimally invasive surgery including angioplasty, stenting, and vein ablation [2]. Despite the advancements of endovascular surgery in recent decades, vascular grafting still remains commonplace and is considered the optimal therapy for patients requiring long-term revascularization solutions in severe cases of damaged or occluded vessels.

Due to the drawbacks of autologous grafts, synthetic materials arose as potential candidates for fabrication of vascular graft in recent decades, especially for the ease and flexibility of tailoring their mechanical properties [3]. The existing grafts have shown satisfactory long-term results for replacement of large- and medium-diameter arteries, but still had poor patency rates when applied to small-diameter vessels (diameter of less than 6 mm). Thus, a sustainable small-diameter vascular graft is still a great challenge of tissue engineering and has thus attracted global research effort.

In previous studies, the electrospun combination of polyurethane (PU) and polycaprolactone (PCL) (PU/PCL) scaffold was shown to satisfy the requirements for vascular prosthesis due to the excellent mechanical strength and biocompatibility of both materials [4]. However, the PU/PCL required additional antithrombotic approach due to the deficit in their hemocompatibility. Thus, different strategies to incorporate conjugated linoleic acid (CLA)—an antithrombotic fatty acid—into the scaffold have been realized successfully to certain extent [5, 6]. The grafted CLA successfully increased the antithrombotic capability of the electrospun membrane without notably affecting their morphology or mechanical properties [6]. In spite of the researched advantages, the cost of CLA might overwhelm the sustainability for clinical application and manufacturing.

Thus, this study aims to employ virgin coconut oil (VCO)—a natural source of CLA and other beneficial fatty acids—for its potential biological properties and sustainable cost. VCO is renowned for its beneficial health effects associated with its phenolic acids and flavonoid contents such as Caffeic acid, p-coumaric acid, ferulic acid and (±) catechin [7]. Previous studies also revealed outstanding biological properties of VCO including its antioxidant and antithrombotic activity [8]. Additionally, VCO was proven to be applicable for vascular engineering [9].

As such, in this research, the solution containing VCO and PU/PCL was electrospun at various conditions to fabricate the membranes. The morphology of created fibers was observed by Scanning electron microscopy (SEM) to select the most suitable electrospinning conditions for further investigation. The project then investigated the mechanical strength of electrospun VCO-loaded membranes to determine the effect of the loading. It is expected to fabricate VCO-loaded membranes satisfying the physical requirements of vascular grafts.

2 Materials and Methods

2.1 *Material*

Polycaprolactone, poly[4,4′-methylenebis(phenyl isocyanate)-alt-1,4 butanediol/di(propylene glycol)/polycaprolactone] (PU), *N*,*N*-Dimethylformamide (DMF), and tetrahydrofuran (THF)—used to fabricate the membrane—were purchased from Sigma-Aldrich (USA). Virgin coconut oil was obtained from Nutiva (USA). All other chemicals were analytical graded.

2.2 *Methods*

The preparation of electrospinning PU/PCL loaded VCO solution

Firstly, the polymer solution was prepared by mixing the DMF: THF (1:1 v/v) solvent with PU/PCL pellets (1:1 w/w). Next, VCO is added to the mixture so that its final concentrations varied from 6% v/v to 10% v/v for investigation purposes. Then, the mixtures were stirred for 12 h with the temperature set at 50 °C for PU/PCL dissolution and then cooled to room temperature (Table 1).

The electrospinning process

The PU/PCL loading VCO solution was placed in a glass syringe that was attached to a peristaltic pump (Harvard Apparatus, Infusion Syringe Pump 980,638). A collector—a rotating steel drum wrapped with aluminum foil was used to collect the fibers. A high voltage of 20 kV between the tip of the needle and the collector was supplied by a DC voltage power supply. For investigation purposes, the parameters of this process was varied. The tip-to-collector distance was set at 10 cm or 12 cm; the flow rates of the pump were varied by 0.5, 1 and 1.5 mL/h.

Table 1 Content of each sample and their labels

Sample	PU/PCL concentration (w/v) (%)	VCO concentration (%v/v) (%)
PU/PCL	12	0
VCO 6	12	6
VCO 8	12	8
VCO 10	12	10

Scanning electron microscopy

The electrospun membranes (1 × 1 cm) were sputter coated with gold for 60 s prior to scanning electron microscopy (SEM). The SEM (JSM-IT100, JEOL, Japan) with an accelerating voltage of 10 kV was used to acquire the SEM images of the membranes.

Tensile strength

The mechanical strength of electrospun membranes were evaluated by the Exponent Stable Micro System, UK. 10 × 50 mm specimens were prepared and their thickness was measured by using the electronic microcaliper prior to testing. The specimen was clamped at both ends so that the stress length is 30 mm.

2.3 Determination of Clotting Times of Blood Plasma

Alteration in clotting times of blood plasma—including activated partial thromboplastin time (aPTT), prothrombin time (PT), and fibrinogen time (FT)—after its exposure to the samples of this study were investigated as described elsewhere [10, 11]. Briefly, platelet-poor plasma (PPP) was prepared by centrifuging whole human blood at 3500 rpm for 10 min. Then, a 1 cm × 1 cm membrane was immersed and incubated in 0.5 mL of PPP for 1 h at 37 °C. After the incubation time, the membrane was removed, and the coagulation parameters of the exposed PPP were evaluated with an automated coagulation analyzer (CA-540, Sysmex, Japan). Unexposed PPP was used as control. The experiment was repeated three times. Data were reported as mean ± standard deviation.

3 Results and Discussion

3.1 Electrospinning Diameter for PU/PCL Loaded VCO

XRD spectra of the nid PU/PCL and its treated samples were analyzed to identify the undesirable effect of the process on their structure and crystallinity. As shown in Fig. 1, the spectra were highly similar, with no spectrum showed an outstanding peak. The spectra all exhibited two diffraction peaks at the 2θ angle of 21.71° and 23.91°, and a broad diffraction peak is ranging from 12° to 17°. Although there was no new signal, the intensities ratio between the broad peak and the other two peaks varied notably among the samples. The VCO-loaded solutions of 12% w/w PU/PCL in DMF/THF were electrospun at different parameters to investigate their effects on the morphology of the membranes and to achieve a non-woven, uniform mesh. Figure 1 shows the SEM image of loaded 6% VCO PU/PCL membrane electrospun at 20 kV. At the 10 cm tip-to-collector distance, the fibers at all flow rates were flattened and not distinctly separated. Most of the fibers fused at their intersections, creating a

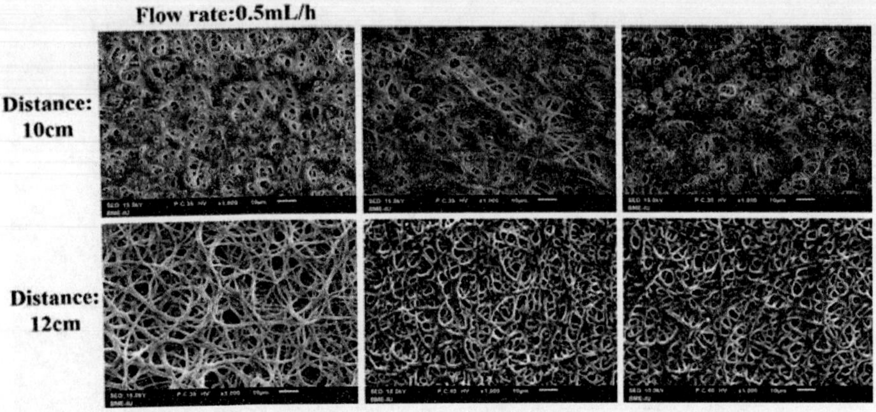

Fig. 1 SEM images of loaded 6% VCO PU/PCL membranes electrospun at 20 kV with different conditions

dense woven matrix with small pores. This phenomenon was also observed in 12 cm tip-to-collector condition at 1 and 1.5 mL/h flow rates but the surfaces were less obstructed. The fiber electrospun at 0.5 mL/h flow rate and tip-to-collector distance of 12 cm yielded the best result, where the membrane was non-woven with uniform, separated fibers.

The PU/PCL membranes loaded with 8% VCO were shown in Fig. 2. Similar to the 6% VCO membranes, at the 12 cm-distance condition, membranes electrospun at high flow rates (namely 1 and 1.5 mL/h) were dense and with fused, large fibers. This phenomenon also happened at all flow rates with the shorter distance of 10 cm. Deductively, with 8% of VCO, the membrane electrospun at the flow rate of 0.5 mL/h and 12 cm away from needle tip possessed the most uniform and porous morphology.

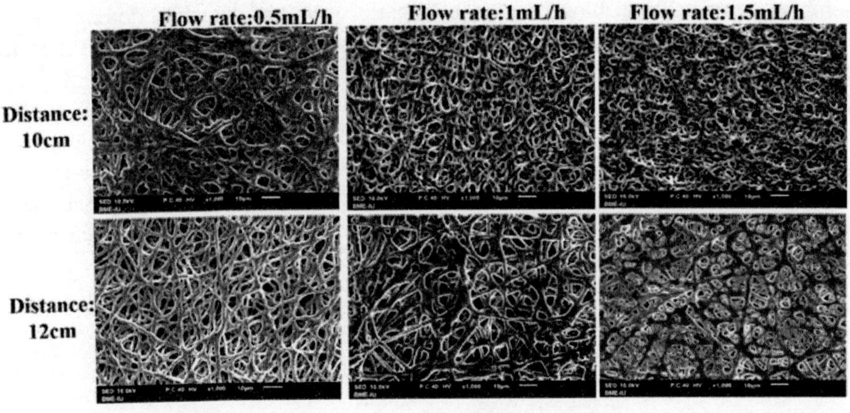

Fig. 2 SEM images of loaded 8% VCO PU/PCL membranes electrospun at 20 kV with different conditions

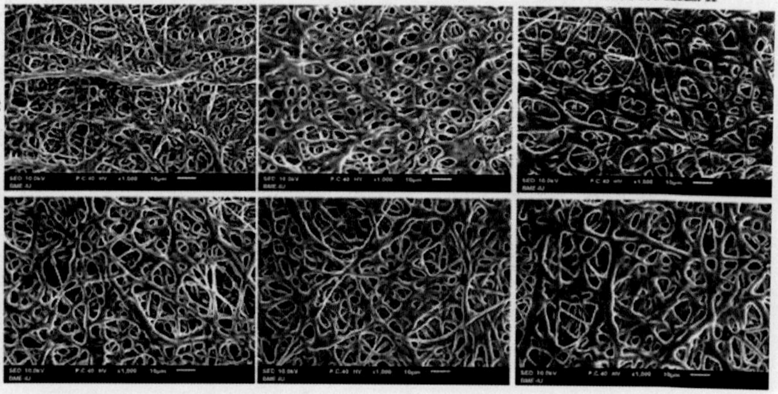

Fig. 3 SEM images of loaded 10% VCO PU/PCL membranes electrospun at 20 kV with different conditions

Figure 3 illustrates SEM images of loaded 10% VCO PU/PCL membranes electrospun at 20 kV with various flow rate and tip-to-collector distance. Overall, there is no remarkable differences between conditions. The electrospun fibers were all dense, fused and blended at many positions, creating a woven matrix. Despite of the substandards, the loaded 10% VCO PU/PCL membrane fabricated at flow rate of 0.5 mL/h and 12 cm of distance was still selected for further investigations in regards to the unification due to previous selected conditions (6 and 8% of VCO at 0.5 mL/h of flow rate and 12 cm tip-to-collector distance).

Electrospinning provides an efficient, controllable and scalable approach to fabricate scaffold in form of membrane. However, the method is sophisticated as the fabricated scaffold is highly sensitive to slight changes in operational conditions. Thus, it is essential to firstly optimize these parameters to yield the most appropriate membrane for artificial blood vessel application. As the data had demonstrated, slower flow rate, longer tip-to-collector distance and higher voltage all resulted in smaller, less sticky and more uniform fibers. As stated by Akduman et al. [10], sticky and blended fibers generally occurred because the solvent did not have sufficient time to evaporate completely before hitting the collector. Thus, reduce the flow rate and increase the tip-to-collector distance both prolong this duration, allowing the fibers to become completely dried. All these characteristics (uniform, small, non-fused fibers; high porosity) are desirable for electrospun membranes that are used as artificial blood vessels [11].

3.2 *Effect of VCO and Electrospinning Parameters on Mechanical Properties of the Membranes*

The mechanical properties of the membranes regarding strain and tensile strength were tested and compared to detect alterations resulted from the loading of VCO. As illustrated in Fig. 4, the VCO treatment dramatically decreased the strength behavior of polymeric membranes, compared to the PU/PCL sample. Specifically, PU/PCL preserved its elasticity against stress up to 6 MPa while VCO-loaded samples deformed plastically at 1 MPa. The PU/PCL membrane was torn at the stress of 14.96 MPa while the increase in concentration of VCO (VCO 6, VCO 8, VCO 10) was directly proportional to the decrease of break points (13.86 MPa, 10.45 MPa and 6.61 MPa, respectively).

Correspondingly, the alteration of ductility seemed to be influenced by VCO. The PU/PCL sample was torn completely at the strain of 256.16% while this phenomenon happened with 6% VCO sample at the strain of 382.91%. However, the ductility of loaded VCO membranes reduced when the concentration of VCO was over 6%. At 8% of VCO-loaded, the polymeric membrane could be torn completely at the strain of 297.63%. The 10% VCO membrane had the smallest ductility when the sample was completely torn at the strain of 235.52%. As a reference, the stress and strain of the native internal mammary artery—the main autologous vessels used for vascular graft—were reported as 4.3 ± 1.8 MPa and 59 ± 15% [12, 13]. Thus, the VCO-loaded membranes should satisfy the mechanical requirements for this purpose.

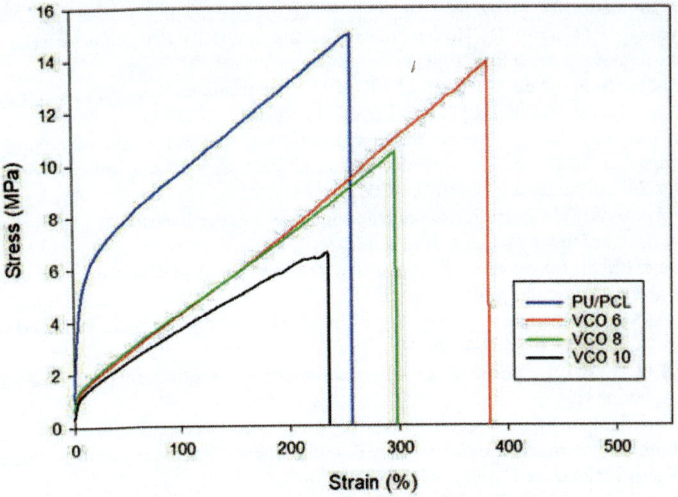

Fig. 4 Stress–strain graph of electrospun membranes

4 Conclusion

The study reported the successful fabrication of VCO-loaded electrospun PU/PCL membrane by varying several parameters of electrospinning and the VCO concentration. The electrospinning condition with the slowest flow rate (0.5 mL/h) and largest tip-to-collector distance (12 cm) yielded the suitable membranes with non-woven matrix, no bead formation and uniform fibers. On the other hand, VCO had notable effects—according to its concentration—on the morphology and mechanical strength of PU/PCL membranes. Higher concentration of VCO enlarged the size of fibers and fused the fibers. The loading also decreased the strain and tensile stress of the samples. However, these mechanical properties still satisfied the requirements of artificial vascular graft. Thus, the loading of VCO into electrospun polymers was feasible and promising for further investigation in vascular engineering.

Acknowledgements This research was supported by the National Foundation for Science and Technology Development (NAFOSTED, Vietnam) under grant number 108.06-2018.18.

Conflicts of Interest The authors declare that they have no conflict of interest.

References

1. Heart disease and stroke statistics—2017 update: a report from the American Heart Association (2017). Circulation 135(10):e146
2. Abdulhannan P, Russell D, Homer-Vanniasinkam S (2012) Peripheral arterial disease: a literature review. Br Med Bull 104(1)
3. Pashneh-Tala S, MacNeil S, Claeyssens F (2016) The tissue-engineered vascular graft—past, present, and future. Tissue Eng Part B Rev 22(1):68–100
4. Nguyen T-H et al (2013) A hybrid electrospun PU/PCL scaffold satisfied the requirements of blood vessel prosthesis in terms of mechanical properties, pore size, and biocompatibility. J Biomater Sci Polym Ed 24(14):1692–1706
5. Minh HH et al (2017) Fabrication of polycaprolactone/polyurethane loading conjugated linoleic acid and its antiplatelet adhesion. Int J Biomater
6. Tran N et al (2020) Polyurethane/polycaprolactone membrane grafted with conjugated linoleic acid for artificial vascular graft application. Sci Technol Adv Mater (just-accepted)
7. Marina A, Man YC, Amin I (2009) Virgin coconut oil: emerging functional food oil. Trends Food Sci Technol 20(10):481–487
8. Chua LS et al (2012) Hydrolysis of virgin coconut oil using immobilized lipase in a batch reactor. Enzyme Res
9. Jaganathan SK, Fauzi Ismail A (2018) Production and hemocompatibility assessment of novel electrospun polyurethane nanofibers loaded with dietary virgin coconut oil for vascular graft applications. J Bioactive Comp Polym 33(2):210–223
10. Akduman C, Kumbasar EPA (2017) Electrospun polyurethane nanofibers. In: Aspects of polyurethanes. InTech
11. Brown T et al (2012) 16 electrospinning for regenerative medicine

12. Stekelenburg M et al (2008) Dynamic straining combined with fibrin gel cell seeding improves strength of tissue-engineered small-diameter vascular grafts. Tissue Eng Part A 15(5):1081–1108
13. L'Heureux N et al (2006) Human tissue-engineered blood vessels for adult arterial revascularization. Nat Med 12(3):361

Parameters' Investigation of Tubular Electrospinning System for the Fabrication of Polyurethane (PU)/Polycaprolactone (PCL) Small Hollow Tube for Vascular Engineering's Applications

An Nguyen-My Le, Phu Dai Huynh, Phuoc Long Truong, Ha Thi Thanh Huong, and Thi-Hiep Nguyen

Abstract Electrospinning is a preferable technique to fabricate scaffold, which have the ability to mimic the structure of extracellular matrix. Small-diameter vascular grafts have been developed by co-electrospinning PCL and elastic PU. Among the suitable materials used to engineer small diameter vascular prosthesis, the polycaprolactone (PCL) and polyurethane (PU) are two appropriate candidates for artificial blood vessels application approved by Food and Drug Administration for biomedical applications with good mechanical properties and high biocompatibility. This electospun graft was adopted with long degradation time and high mechanical strength. PU/PCL nanofibrous membrane were successfully fabricated by the electrospinning system design by Biomedical Engineering School, International University, Ho Chi Minh National University (BME-IU). However, the fabrication of electrospun hollow tube at small diameter was affected by many parameters including applied voltage, distance from needle to collector, pumping speed, collector's rotation speed and the type of solvent. This research studies the parameters' effect of tubular electrospun system designed by BME-IU on the morphological properties of these vascular scaffolds using SEM and diameter analysis.

Keywords Electrospinning · Hollow tube · PU · PCL · Small-diameter vascular grafts

The authors declare that they have no conflict of interest.

A. N.-M. Le · P. L. Truong · H. T. T. Huong · T.-H. Nguyen (✉)
School of Biomedical Engineering, International University, Vietnam National University-Ho Chi Minh City (VNU-HCMC), Ho Chi Minh City 700000, Vietnam
e-mail: nthiep@hcmiu.edu.vn

P. D. Huynh
Department of Polymer Materials, Ho Chi Minh City University of Technology, Vietnam National University-Ho Chi Minh City (VNU-HCMC), Ho Chi Minh City 700000, Vietnam

V. Van Toi et al. (eds.), *8th International Conference on the Development of Biomedical Engineering in Vietnam*, IFMBE Proceedings 85,
https://doi.org/10.1007/978-3-030-75506-5_33

1 Introduction

Cardiovascular diseases is one of the leading causes of death globally [1]. It is estimated that by 2030, the cases of mortalities related to these diseases will reach 23.3 million cases per year [2] with the common root causes including: reduced blood flow and insufficient nutrient supply due to vessel narrowing or blockage. Vascular transplantation and bypass surgery are highly recommended in the circumstances with the small vessels obstruction [3]. Calling attention, small diameter vascular transplantation is not often sufficient to meet patient needs due to the lack of donor resources, the thrombosis occlusion, infection, and rejection of the vascular transplant [4, 5]. In line with these conditions, the need for small diameter artificial vessel is considered more than ever. Therefore, novel tissue engineering technology is necessary for vascular engineering.

Researchers were, are and will be investigated on a variety of material and method to fabricate novel artificial prosthesis. Nowadays, the use of electrospun technology for the development of tissue-engineered vascular grafts has raise attention tremendously since this technique can create scaffold with structure similar to extracellular matrix (ECM) of vascular tissue, which is composed of 40–400 nm diameter structural protein and polysaccharide fibers [6]. Moreover, the use of electrospun scaffold have been highlighted in recent researches due to their appropriate mechanical and biological behaviors [7, 8]. The method applies an electric field on a polymer solution to create a charged jet. As the jet is pulled from the dispenser through air to the metal collector, the solvent evaporates leaving behind charged fibers. By manipulating different parameters such as distance from dispenser to collector, voltage, solvent used, types of collector, researchers can collect fibers with different cross-sectional shapes, sizes or orientation [9, 10].

Among the suitable materials used to engineer small diameter vascular prosthesis, the polycaprolactone (PCL) and polyurethane (PU) are two appropriate candidates for artificial blood vessels application approved by Food and Drug Administration for biomedical applications with good mechanical properties and high biocompatibility [11]. Excellent strength and elasticity allow PU to provide an environment comparable to native vessels [12]. On the other hand, biodegradable PCL is known to possess decent mechanical properties and high biocompatibility. In previous study, PU was co-electrospun with PCL to enhanced the mechanical properties of the composite as well as controlled other features like hydrophobicity and pore size distribution [13]. PU/PCL nanofibrous membrane were successfully fabricated using the electrospinning system design by Biomedical Engineering School, International University, Ho Chi Minh National University (BME-IU).

Therefore, this study investigated electrospinning parameters on the composite PU/PCL to obtain the small diameter hollow tube ($\leq$ 6 mm) with appropriate physical/mechanical characteristics including morphology consistency, fiber diameter, etc. using electrospun designed by BME-IU. Applied voltage, distance from tip to collector, pumping speed, collector's rotation speed that affect to the fabrication of electrospun tube at small diameter were studied for the application of small-diameter vascular engineering.

2 Materials and Methods

2.1 *Materials*

Polycaprolactone (PCL) (440,752), Polyurethane (PU) (81,367), *N,N*-Dimethylformamide (DMF) (NSC 5356), and tetrahydrofuran (THF) (401,757) were purchased from Sigma-Aldrich (USA). Absolute ethanol was bought from Xilong manufacturer.

2.2 *Methods*

Investigation of parameters of electrospinning

1.2 g of PU/PCL polymer was dissolved mixed 10 mL DMF: THF (1:1 v/v) solvent until the homogeneous mixed pellets and solvents were obtained. The mixture was stirred overnight at room temperature until the solution became transparent and homogeneous. A glass syringe (Exmire, lure-lock type, 12 mL) with metal syringe needle was used to contain the prepared solution. After the solution was placed in, the syringe was attached to the pump and face the collector. The used collector was a rotating steel stick with small diameter ($\leq$ 6 mm) covered with aluminum foil. A high voltage between the syringe and the collector was supplied by a high DC voltage power supply. For investigation purposes, the following parameters have been varied: distance from needle to collector, pumping rate and voltage. The voltage was changed between 15 and 20 kV; the flow rate was set at 0.2, 0.4 and 0.8 mL/h; and the distance was varied from 7 to 13 cm. The best condition was chosen to prepare the membrane for further studies.

Scanning electron microscopy

The morphology of inner and outer surface of the electropsun tube were evaluated using scanning electron microscopy (JSM-IT100, JEOL, Japan). The electrospun inner and outer surface were coated with gold at first. Then, the image of the membrane was observed using SEM.

Fiber diameter analysis

The fiber diameter of the membranes was measured and analyzed using ImageJ software. The data were presented as average $\pm$ S.D.

All experiments were done in triplicate and SPSS Statistics software (IBM) was used to analyze data.

3 Results and Discussions

In recent years, the use of electrospinning, among these techniques, for the development of tissue-engineered vascular grafts has increased tremendously since this technique has a great potential to mimic the structure of extracellular matrix (ECM) of vascular tissue, which is composed of 50–500 nm diameter structural protein and polysaccharide fibers. Most of nanofiber scaffolds are prepared by electrospinning since electrospinning enables the fabrication of fibrous scaffolds with fiber diameter ranging from nanometers to micrometers, which have physical properties close to that of natural ECM [14, 15], including high porosity, pore-interconnectivity, and large surface areas, providing suitable surface sites to the cells to adhere, proliferate, and grow [16]. Porosity, interconnected pores to facilitate cell migration and the transport of nutrients during tissue regeneration. From a structural viewpoint, structures that can mimic these features are believed to be necessary for precisely guiding cell growth and tissue regeneration, which lead to the recovery of their natural tissue function.

3.1 Tubular Electrospun System

The three main blocks of an electrospinning machine include the high-voltage power supply, the pump syringe system, and the fiber collector system (Fig. 1). In electrospinning, the machine's high-voltage power supply is the core for its operation. This is a DC source of about 5–30 kV. These power supplies on the market are very expensive and are not suitable for our application. Therefore, we designed this high-voltage power supply [17]. Meanwhile, the syringe pump system on the market is stable over time, the pump speed is accurate, there are many options and affordable prices so the team will buy the machine available in the market. The design of the fiber collecting system is necessary to enable the team to actively create scaffolds

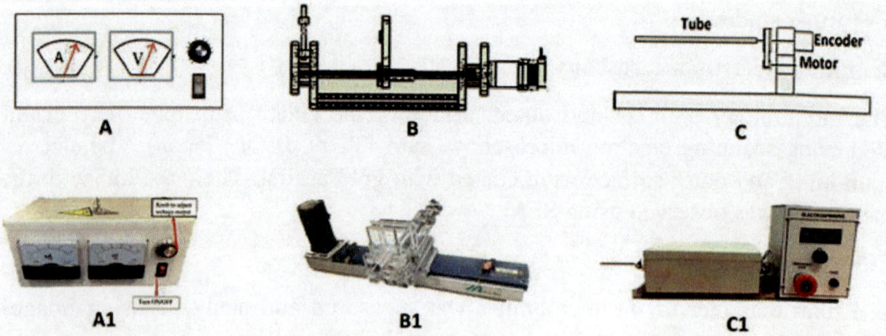

Fig. 1 Schematic (**a**, **b**, **c**) and image (**a1**, **b1**, **c1**) of power supply, syringe pump and rotating collector respectively of electrospinning machine

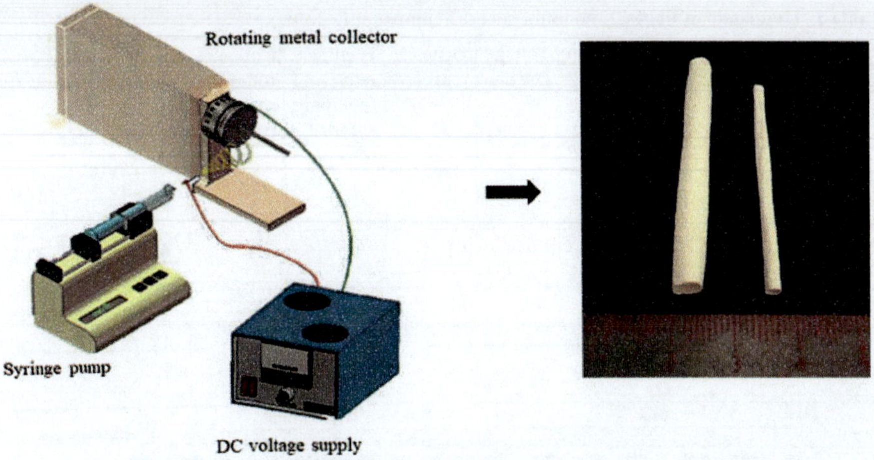

Fig. 2 Custom-made electrospinning setup including power supply, syringe pump, metal needle and tubular rotating collector for the fabrication of small diameter hollow tube

of different shapes, for tissue engineering applications. The fiber collecting system will be designed as centrifugal form for tubular hollow tube (Fig. 2).

The mechanical design includes:

- High-voltage power supply: output voltage range: 5–30 kV DC, input voltage: 220 V AC-50 Hz, output voltage display (Fig. 1a, a1).
- Pump syringe system: pumping speed: 1–20 mL/h, error: 2% (Fig. 1b, b1).
- Collector: tubes with sizes of 6, 2 mm in diameter, material: monolithic aluminum milling, speed of rotation 0-2000 rpm (Fig. 1c, c1).

3.2 *Effect of Applied Voltage*

Electrospinning is based on the fundamental principle of a high voltage generated into the polymer solution. According to Table 1, the first variable was supply voltage (kV), while other parameters were constant. In case of rotating tubular collector, flat topography of the electrospun tube's the inner surface was resulted from the direct contact of the fibers to the collector. In this process, the initial jet was formed when the electrostatic force overcomes the surface tension [18]. Table 1 and Fig. 3 showed the effect of different applied voltage (15, 17 and 20 kV) on the morphology of the electrospun mat and diameter of the fibers. Significant differences were observed between samples at different voltage rate, respectively. At higher voltage, fibers are elongated and the nano fibers were formed separately without bead formation. On the other hand, fibers with a large variation in size were formed at lower voltage since the polymer jet was unstable, which can be seen in diameter distribution analysis

Table 1 Diameters of PU/PCL nanofibers at different supply voltage

Sample	Solvent	Concentration (wt%)	Voltage (kV)	Distance (cm)	Feeding rate (pumping speed) (mL/h)	Collector rotation speed (rpm)	Mean nanofibers diameter (nm)
S1	DMF:THF (1:1)	12	**15**	10	0.2	800	1236.42 ± 2112.69
S2	DMF:THF (1:1)	12	**17**	10	0.2	800	439.42 ± 278.81
S3	DMF:THF (1:1)	12	**20**	10	0.2	800	770.97 ± 266.64

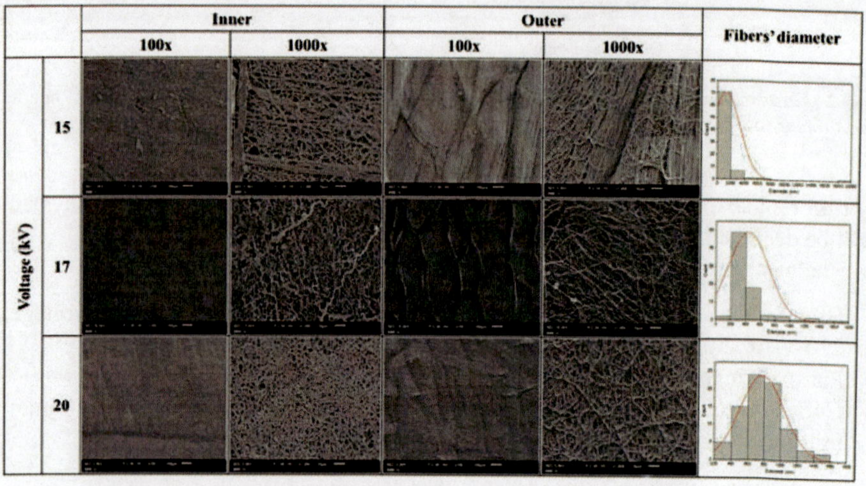

Fig. 3 SEM image of electrospun PU/PCL tube at different supply voltage

(Fig. 3). Therefore, the applied voltage that is suitable to electrospinning PU/PCL polymer solution is 20 kV.

3.3 Effect of Feeding Rate (Pumping Speed)

The pumping speed identifies the amount of solution feeding for electrospun process. As a result, fibers with smaller diameter were produced at slower feeding rate [19]. The feeding rate generate differences on the mean diameter and morphology of fibers, which is shown in Table 2 and Fig. 4. In this study, 0.2, 0.4 and 0.6 mL/h were chosen to be the feeding rates for the evaluation and its influence on the fibers' morphology and diameter is shown in Table 2 and Fig. 4. It can be easily noticed that mean fiber

Table 2 Diameters of PU/PCL nanofibers at different feeding rate (pumping speed)

Sample	Solvent	Concentration (wt%)	Voltage (kV)	Distance (cm)	Feeding rate (pumping speed) (mL/h)	Collector rotation speed (rpm)	Mean nanofibers diameter (nm)
S1	DMF:THF (1:1)	12	20	10	**0.2**	800	770.97 ± 266.64
S2	DMF:THF (1:1)	12	20	10	**0.4**	800	1357.51 ± 595.93
S3	DMF:THF (1:1)	12	20	10	**0.6**	800	152.78 ± 157.32

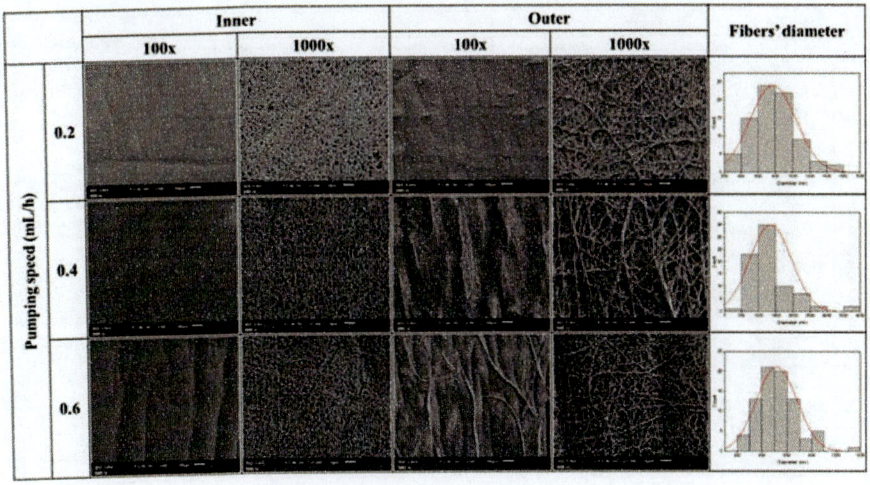

Fig. 4 SEM image of electrospun PU/PCL tube at different pumping speed (feeding rate)

diameters and the variation in diameters increase in corresponding to the increase in feeding rate. The SEM images (Fig. 4) also indicated that at high feeding rate, fibers tend to stick to each other, form uneven outer layer of electrospun mat and the fibers' distribution express an abnormal trend. Therefore, 0,2 ml/h is considered to be the most suitable feeding rate.

3.4 *Effect of Collector-Needle Distance*

Another parameter that has influence the electrospinning process is needle-to-collector distance since the solvent can evaporate at appropriate distance. The distances from needle to collector were 7, 10 and 13 cm (Table 3). SEM images of

Table 3 Diameters of PU/PCL nanofibers at different collector-to-needle distance

Sample	Solvent	Concentration (wt%)	Voltage (kV)	Distance (cm)	Feeding rate (pumping speed) (mL/h)	Collector rotation speed (rpm)	Mean nanofibers diameter (nm)
S1	DMF:THF (1:1)	12	20	**7**	0.2	800	837.27 ± 427.70
S2	DMF:THF (1:1)	12	20	**10**	0.2	800	770.97 ± 266.64
S3	DMF:THF (1:1)	12	20	**13**	0.2	800	429.36 ± 228.61

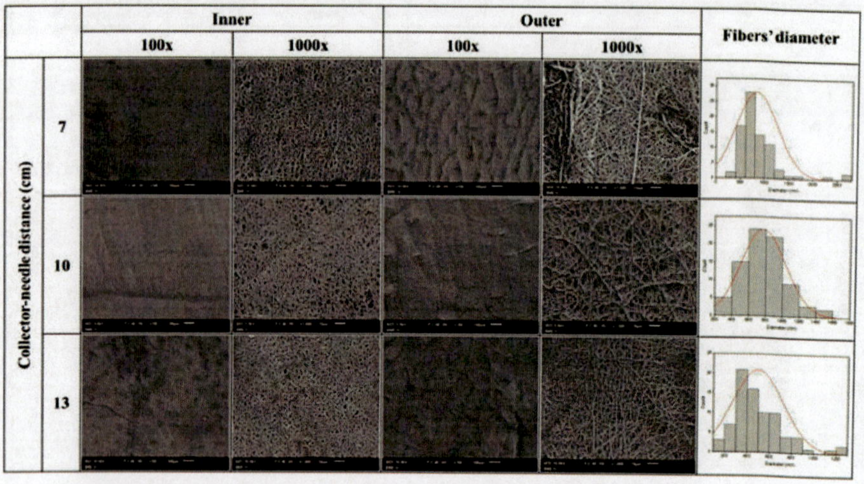

Fig. 5 SEM images of electrospun PU/PCL tube at different collector-needle distance

Fig. 5 indicated no significant difference in fiber morphology at difference distances. In addition, at the distance greater than 13 cm, the polymer solution can't reach the collector because of the poor conductivity. Nevertheless, the fiber diameter exhibited a descending trend when the distance from the needle tips to the collectors increased. For this parameter, 10 and 13 cm can suit the process.

3.5 *Effect of Rotation Speed*

In order to fabricate the artificial electrospun tube, this study use the rotating collector, hence the rotating speed of the collector is a essential variable to be investigated. In this study, the rotating collector was set with 500, 800 and rpm as rotation speeds

Table 4 Diameters of PU/PCL nanofibers at different collector's rotation speeed

Sample	Solvent	Concentration (wt%)	Voltage (kV)	Distance (cm)	Feeding rate (pumping speed) (mL/h)	Collector rotation speed (rpm)	Mean nanofibers diameter (nm)
S1	DMF:THF (1:1)	12	20	10	0.2	**500**	424.91 ± 156.26
S2	DMF:THF (1:1)	12	20	10	0.2	**800**	770.97 ± 266.64
S3	DMF:THF (1:1)	12	20	10	0.2	**1000**	5444.99 ± 233.22

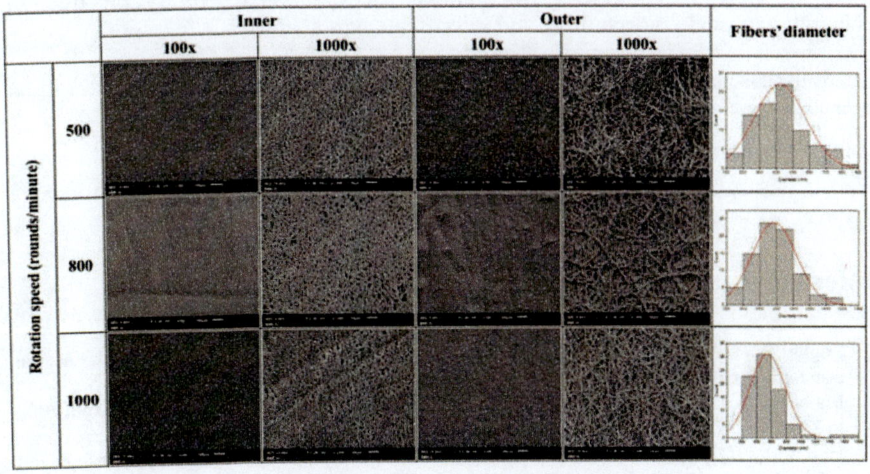

Fig. 6 SEM images of electrospun PU/PCL tube at different rotation speed

(Table 4). The rotating tubullar collector allows the fiber to be collected along the rotaion's direction, because the collector exerts a tangential force to stretch the polymer stretch and align the fibers onto collector. Under high rotating speed, the solvent did not evaporate completely before the fiber reaching the collector (Fig. 6). Therefore, 800 rpm can be set as rotation speed for electrospinning process.

4 Conclusion

This study initially investigated electrospinning parameters on the composite PU/PCL to obtain the small diameter hollow tube (≤ 6 mm) using BME-IU tubular electrospinning system. In general, electrospinning tubular scaffolds have presented an effective tendency toward vascular engineering since this technique exhibited a great potential to freely design scaffolds according to purposes. Furthemore, electrospinning combined with novel biocompatible materials, like biopolymers, opens a new approach for vascular engineering. In this way, by applying electrospinning technique, this research was focused on electrospinning parameters and primarily stabilized the process parameters in order to fabricate appropriate PU/PCL nanofibrous graft. It is found that several process parameters generate influence on morphology and fiber structure including applied voltage, collector-needle distance, pumping speed and collector's rotation speed.

In particular, the applied voltage and collector's rotation speed exhibited the most influence to the fiber size and morphology. By optimizing all the parameters of the tubular electrospinning system fabricated by BME-IU (**polymer concentration: 12 wt%, applied voltage: 20 kV, pumping speed: 0.2 mL/h, collector's rotation speed: 800 rpm and collector-to-tip distance: 10 or 13 cm**), a favorable electrospun tube with nanoscale morphology and microscale was achieved.

Acknowledgements This research is funded by Vietnam National Foundation for Science and Technology Development (NAFOSTED) under grant number 108.06-2018.18.

Conflict of Interest No conflict of interest regarding the publication of this paper.

References

1. Zoghbi W et al (2014) Sustainable development goals and the future of cardiovascular health: a statement from the global cardiovascular disease taskforce. Global Heart 35
2. Mathers CD, Loncar D (2006) Projections of global mortality and burden of disease from 2002 to 2030. PLoS Med 3(11):e442
3. Humphrey JD, McCulloch AD (2003) The cardiovascular system—anatomy, physiology and cell biology. In: Holzapfel GA, Ogden RW (eds) Biomechanics of soft tissue in cardiovascular systems. Springer, Vienna, pp 1–14
4. Müller-Nordhorn J, Willich SN (2017) Coronary heart disease. In Quah SR (ed) International encyclopedia of public health, 2nd edn. Academic Press, Oxford, pp 159–167
5. Verma S et al (2004) Should radial arteries be used routinely for coronary artery bypass grafting? 110(5):e40–e46
6. Bhardwaj N, Kundu SC (2010) Electrospinning: a fascinating fiber fabrication technique. Biotechnol Adv 28(3):325–347
7. Hasan A et al (2014) Electrospun scaffolds for tissue engineering of vascular grafts. Acta Biomater 10(1):11–25
8. In vivo applications of electrospun tissue-engineered vascular grafts: a review 20(6):628–640
9. Caracciolo PC et al (2009) Electrospinning of novel biodegradable poly(ester urethane)s and poly(ester urethane urea)s for soft tissue-engineering applications. J Mater Sci Mater Med 20(10):2129–2137
10. Doshi J, Reneker DH (1995) Electrospinning process and applications of electrospun fibers. J Electrostat 35(2):151–160
11. Williamson MR, Black R, Kielty C (2006) PCL–PU composite vascular scaffold production for vascular tissue engineering: attachment, proliferation and bioactivity of human vascular endothelial cells. Biomaterials 27(19):3608–3616
12. Punnakitikashem P et al (2014) Electrospun biodegradable elastic polyurethane scaffolds with dipyridamole release for small diameter vascular grafts. Acta Biomater 10(11):4618–4628

13. Thottappillil N, Nair PD (2015) Scaffolds in vascular regeneration: current status. Vas Health Risk Manag 11:79–91
14. Klinkert P et al (2004) Saphenous vein versus PTFE for above-knee femoropopliteal bypass. Rev Lit Eur J Vas Endovasc Surg 27(4):357–362
15. Ndreu A et al (2008) Electrospun biodegradable nanofibrous mats for tissue engineering. Nanomedicine (Lond) 3(1):45–60
16. Stupp SI et al (1997) Supramolecular materials: self-organized nanostructures. Science 276(5311):384–389
17. Do TM et al (2020) A low cost high voltage power supply to use in electrospinning machines. In: 7th international conference on the development of biomedical engineering in Vietnam (BME7). Springer, Singapore
18. Lee YS, Collins G, Arinzeh TL (2011) Neurite extension of primary neurons on electrospun piezoelectric scaffolds. Acta Biomater 7(11):3877–3886
19. Zong XS et al (2002) Structure and process relationship of electrospun bioabsorbable nanofiber membranes. Polymer 43:4403–4412

Chitosan Oligomer Mono-coated and Multi-coated Nanofibrous Polycaprolactone Toward the Characterization of Mechanical Strength for Wound Dressing Application

Nhi Thao-Ngoc Dang, Trinh Thi-Phuong Ho, Linh Kim-Khanh Nguyen, Vinh Khanh Doan, An Nguyen-My Le, and Thi-Hiep Nguyen

Abstract Chitosan oligomer (OCS) captured much attention for diverse applications in the biomedical and pharmaceutical fields. In this study, OCS, as an antimicrobial agent, was coated onto dressing based electrospun polycaprolactone (PCL) membrane to enhance the bio-function and physical properties of materials. The various content of the OCS coating layer in terms of concentration (2, 4, 8, 16%) or coating times (1, 3, and 6 times) was concerned in this study. A Fourier-transform infrared was used to analyze the functional structure OCS derived from chitosan. Agar diffusion test was performed to pretest the antibacterial ability of OCS coating solution with gram-positive (*S. aureus*) and gram-negative (*P. aeruginosa*). The morphology and physical strength of PCL membranes and OCS coating layer were observed by Scanning Electron Microscope (SEM) and the uniaxial tension, respectively. From the agar diffusion test, OCS with a low concentration (2%) inhibited both mentioned pathogens. SEM evaluation displayed the irregular surface by the OCS multi-coating layer, while smoother morphology by one-layer coating with high OCS concentration. In the physical test, the membrane became more inflexible as the higher OCS content in both the coating method as the different mechanical strength between PCL substrate and OCS coating film.

Keywords Chitosan oligomer · Polycaprolactone · Antimicrobial property

1 Introduction

The wound dressing is designed to protect the wound as a barrier against bacterial infection and further physical harm. The wound dressing now takes many other

N. T.-N. Dang (✉) · T. T.-P. Ho · T. T.-P. Ho · L. K.-K. Nguyen · V. K. Doan · A. N.-M. Le · T.-H. Nguyen
School of Biomedical Engineering International University, Vietnam National University-Ho Chi Minh City (VNU-HCMC), Ho Chi Minh City, Vietnam
e-mail: dntnhi@hcmiu.edu.vn

V. Van Toi et al. (eds.), *8th International Conference on the Development of Biomedical Engineering in Vietnam*, IFMBE Proceedings 85,
https://doi.org/10.1007/978-3-030-75506-5_34

advantages to facilitate healing, absorbent wound fluid, promoting gas exchange, eliminating pain, and even reasonable costs [1]. The antimicrobial property is the essential function of the wound dressing to reduce bacterial infection that prolongs the wound healing [2]. Therefore, a lot of agents with the potential antibacterial ability have been incorporated in the fabrication of modern wound dressings such as chitosan [3], curcumin [4], or silver nanoparticles [5].

Chitosan is one of the natural polymers with a structure of polysaccharide consisting of *N*-glucosamine and *N*-acetylglucosamine units linked by β-1,4 glycoside. Chitosan possesses many unique properties such as biocompatible, biodegradability, bio-function such as antifungal, antibacterial, antitumor activity, and promoting the wound healing process. Besides, chitosan shows the potential to prepare hydrogels, film-forming capability, and fibers in medical and pharmaceuticals fields [6]. However, the low water solubility and poor biodegradation performance restrict the application of chitosan. As the degraded products of chitosan, chitosan oligomer (OCS) is readily soluble in aqueous solutions due to smaller molecular weight than chitosan, making OCS perform valuable biological activities as well as chitosan [7, 8].

Although chitosan or OCS has generated enormous interest due to its various advantages, the rigid and brittleness in regulating the mechanical properties limit OCS usage alone [9]. Therefore, this study explored the incorporation of OCS with the electrospun elastomeric membrane PCL to combine the advantages of individual materials. PCL is a biomaterial polymer approved by FDA with high mechanical strength and long-term degradation [10]. PCL can be easily dissolved in common organic solvents such as acetone and acetic. In wound care application, electrospun PCL membranes have been broadly applied due to its fibrous structure, contributing to absorbing the wound fluid and exchange oxygen for cellular respiration [11]. Indeed, high mechanical properties of electrospun PCL fibers allow the dressing easily elongation to cover the wounds against the environment factor [12].

Moreover, a wide range of OCS content as the increase in OCS concentration and the number of OCS coating times was considered in this study. The coating technique is preferred as the uncomplicated process, low cost, and, especially, different solvents for dissolving OCS and PCL materials. The hypothesis is the OCS coating layer and PCL will give a novel electrospun PCL membranes coating OCS (PCL.OCS) where the physical properties of both materials are complemented.

2 Material and Method

2.1 Materials

The commercial low viscosity chitosan and Poly (ε-caprolactone) (PCL, Mn 80.000) were purchased from Sigma-Aldric. Mueller Hinton Broth (M391-500G) was supplied by Hi-Media (India). All other chemicals were obtained from Xilong Chemical Co., Ltd. (China).

2.2 Method

Preparation and characterization of chitosan oligomer

Chitosan oligomer (OCS) was obtained via the degradation of chitosan under microwave assistant. Chitosan powder was immersed into H_2O_2 15% solution in 10 min before the conduction of microwave radiation operated at 400 W for 3 min. Next, three amounts of ethanol were added into the OCS solution to collect and then filter the precipitation. The obtained filtrate was lyophilized for 24 h to get the OCS powder. Besides, OCS solutions with the concentration of 2% (OCS.2%), 4% (OCS.4%), 8% (OCS.8%) and 16% (OCS.16%) (w/v) were obtained by dissolving OCS powder in distilled water.

A Fourier-transform infrared (FT-IR) spectrometer (FT-IR, Spectrum GX, PerkinElmer Inc., USA) was used to analyze the functional structure changes between OCS and the commercial chitosan. Also, the antimicrobial pretest for the OCS solution at low concentration (OCS.2%) before coating on the PCL fibrous mat was performed by the agar diffusion. Briefly, a bacterial solution of gram-positive (*S. aureus*) and gram-negative (*P. aeruginosa*) (OD620 value of 0.080–0.1 equal to a 0.5 McFarland standards) was spread out onto the Mueller–Hinton agar plate. Then, the filter paper with 0.8 cm of diameter after 24 h immersion in the OCS.2% solution was placed on the surface of the plate. The inhibition zone around the membrane was recorded by photograph after 24 h of incubation at 37 °C.

Fabrication of electrospun PCL membranes coating OCS (PCL.OCS)

The method for the fabrication of PCL membrane was followed in a previous study [13]. Briefly, a 15% w/v PCL solution was prepared by dissolving PCL pellets in a mixture of acetic acid and acetone at a ratio of 7:3 (v/v). Next, the PCL solution was electrospun onto a stationary drum collector at a flow rate of 1 mL/h and an operating voltage of 15 kV. The obtained PCL membrane was then received air plasma treatment for 3 min under the following conditions: 30 W RF power and 13.56 MHz. To fabricate PCL.OCS, the first coat onto the PCL fibrous mat (PCL.1.OCS) by OCS solution, was performed with a ratio of 1 mL: 10 cm^2 per coating time and then dried at room temperature for 24 h. In terms of the multi-coating membrane, the previous steps were continuously repeated for three times (PCL.3.OCS) and six times (PCL.6.OCS).

Morphological and physicochemical characterization of PCL.OCS membrane

The morphology of PCL membranes and OCS coating layer were observed by Scanning Electron Microscope (SEM, JSM-IT100, JEOL Ltd., Japan). For physical strength, the uniaxial tension of the membranes was measured by a micro stable system (TA.XTplus, Co., UK).

3 Result and Dicussion

3.1 Characterization of OCS

In Fig. 1, the FT-IR result shows the differences in the chemical structure of commercial low viscosity chitosan and its derived OCS. For the FT-IR spectra of chitosan, the broadband of O–H and N–H bond was stretching, located at around 3332 cm^{-1}. The C–H stretching was assigned at 2881 cm^{-1}. The peaks at 1660 and 1562 cm^{-1} corresponded to the stretching vibration of amide I and amide II groups, respectively. The peaks at 1420 cm^{-1} were associated with the CH_2 bending as well as CH_3 deformation. The absorbance of the amide III band was indicated at 1322 $cm^{-1.}$ The absorption bands at around 1155 and 890 cm^{-1} were assigned to the C-O stretching. Compare with chitosan, the absorption peaks of chitosan were still present in the FT-IR result of OCS after the oxidative degradation. However, in OCS, the peaks assigned to amide I and amide II are significantly enhanced, which caused by the degradation reaction mostly occurred at the amino group of chitosan. Also, Du et al.

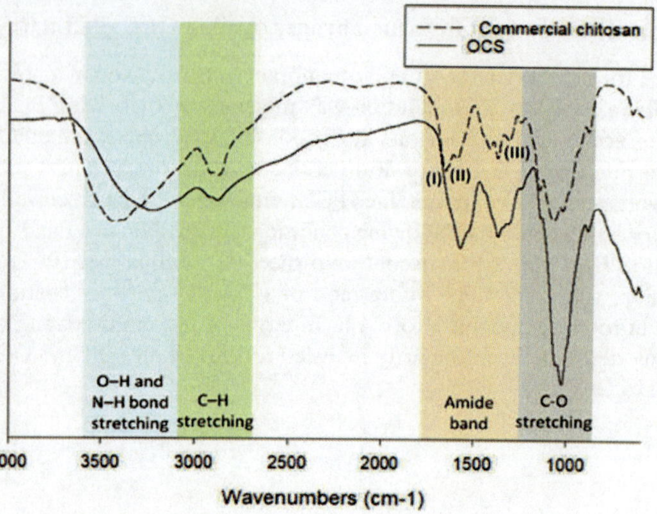

Fig. 1 FTIR of the commercial low viscosity chitosan and obtained OCS

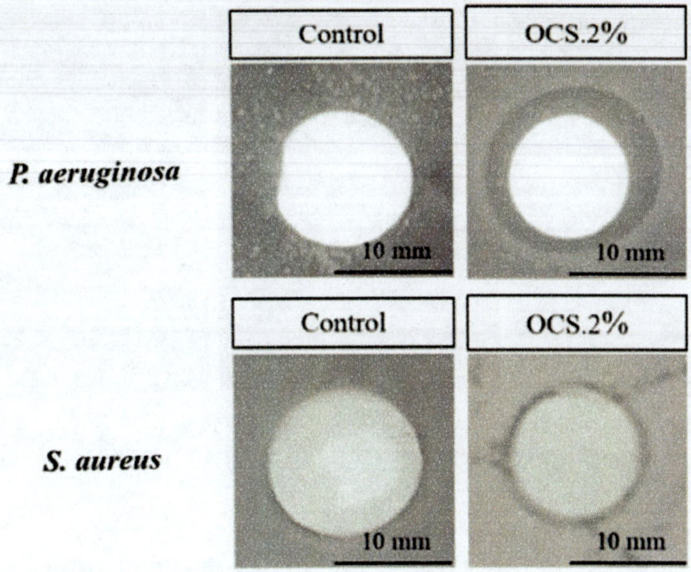

Fig. 2 The zone of inhibition test for OCS.2% coating solution against *S. aureus* and *P. aeruginosa*

[14] prepared OCS by combining H_2O_2 and gamma irradiation of the chitosan solution. As a result, they claimed that the hydroxyl radicals formed through the radiolysis of water and H_2O_2 were responsible for the degradation of chitosan.

Photographs in Fig. 2 present the test for antibacterial properties of the OCS against *S. aureus* and *P. aeruginosa* via agar diffusion. The results indicated that the obtained OCS could release into the media and had a preliminary antibacterial effect on both *P. aeruginosa* and *S. aureus* despite the low concentration of OCS (OCS.2%). Several antimicrobial mechanisms of chitosan have been proposed to explain the antibacterial effectiveness of OCS in previous reports [15, 16]. The most widely accepted means for antimicrobial activity of chitosan is the interaction between the positively charged amino groups of chitosan and the negatively charged microbial cell membranes. This interaction resulted in the extensive alterations to the cell surface and membrane barrier properties, leading to leakage of intracellular contents that results in cell death. In this study, for against gram-negative (*P. aeruginosa*), OCS.2% had good antibacterial efficiency by the appearance of apparent inhibitory zones. However, OCS.2% showed a narrow inhibitory zone against gram-positive (*S. aureus*). Therefore, in the case of this study, the higher amount of OCS in term of concentration or layer coating times need to be concerned.

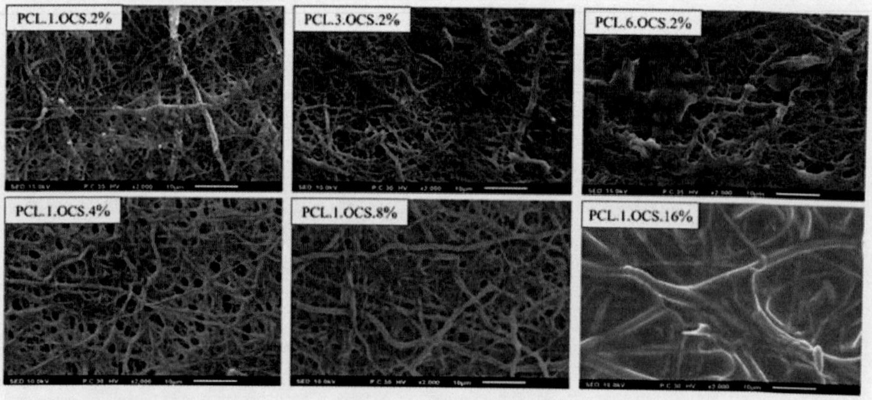

Fig. 3 SEM images of multi-coated membranes (PCL.1.OCS.2%, PCL.3.OCS.2%, and PCL.6.OCS.2%) and mono-coated membranes (PCL.1.OCS.2%, PCL.1.OCS.4%, PCL.1.OCS.8%, PCL.1.OCS.16%)

3.2 Morphological and Physicochemical Characterization of PCL.OCS Membrane

Figure 3 displays the morphology of PCL surfaces after OCS coating via SEM. For the multi-coated membrane, the increase in the number of coating times provided a more irregular morphology surface. The low content of OCS.2% corresponded to the fragmented distribution of OCS molecules onto the PCL membrane. At the first coat (PCL.1.OCS.2%), the OCS molecules were randomly stuck to PCL membranes due to chemical bonding between the hydroxyl and amide groups in OCS and plasma PCL. Then, the subsequent adhesion of OCS layers once again attached to both the residual PCL surface and the previous OCS layer fragment. As a result, the six-time coating membrane gained heterogeneous morphology with OCS clot and PCL mat structure. For the mono-coated membranes, their surface morphology became smoother as the higher OCS concentration. With the highest concentration of OCS, the surface of PCL.1.OCS.16% was overlaid entirely with the coating layer and hence lost the original PCL fibrous structure.

All the tensile test results are presented in Fig. 4a. Compared to the coated samples, uncoated membrane (PCL) gained a stress point or breaking strength of 4.15 MPa at a high strain of 81%. However, in the case of the coated membrane, there are various graphical stress–strain curves recorded during the testing process.

For the multi-coated samples, PCL.1.OCS.2% with the first coating layer reached a full stress point of 3.5 MPa at a strain of 30%. Nevertheless, the elongation was still progressing, although the maximum stress point had been reached. The PCL.1.OCS.2% membrane only fractured until the percent elongation went up to 71% parallel with a downgrade of the pulling forces. This stress–strain stage in PCL.1.OCS.2% associated with a form of plastic deformation that occurred in the polymer material, called necking when the stress throughout the sample or the surface

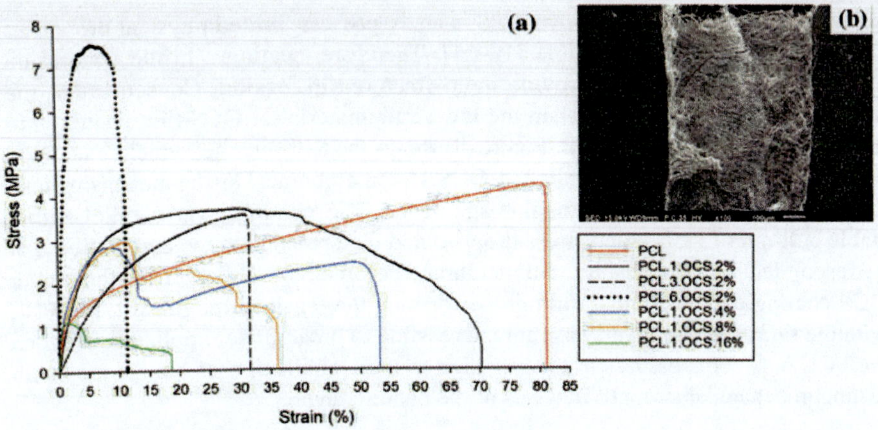

Fig. 4 **a** The stress–strain curve of multi-coated membranes (PCL.1.OCS.2%, PCL.3.OCS.2%, and PCL.6.OCS.2%) and mono-coated membranes (PCL.1.OCS.2%, PCL.1.OCS.4%, PCL.1.OCS.8%, PCL.1.OCS.16%); **b** The SEM image visualized the coating ridge cracks of PCL.1.OCS.16% membrane in the tension test

of the material was not uniform [17]. The SEM revealed that PCL.1.OCS.2% with one-layer coating was not completely overlaid the whole membrane; therefore, it quickly resulted in the necking stage.

Not like PCL.1.OCS.2%, the stress–strain curves of PCL.3.OCS.2% with three coating layers gained its stress point of 3.5 MPa at a strain of 32% without any residual force. Compared to the others, PCL.6.OCS.2% with six coating layers reached the highest stress point of 7.5 MPa; however, at the lowest strain of 12%. Therefore, raising the OCS content by increasing coating times made the PCS.OCS was rigid and less flexible. Compared to the mono-coated samples not regarding the OCS concentration, the multi-coated technique did not advantage in time, also the mechanical properties.

In the case of the mono-coated group, the stress–strain curves exhibit double stress maxima for PCL.1.OCS.4%, the stress reached its first maximum of 2.89 MPa at a strain of 9% before dropping to 1.48 MPa, but the membrane had not been wholly destructive. The stress increased again until it gained the second stress maximum of 2.4 MPa with 53% total elongation at break. The resulting stress–strain of PCL.1.OCS.8% and PCL.1.OCS.16% nearly exhibit the same geometric behavior that of PCL.1.OCS.4% with the steadily lower stress and strain to failure. Compared to the others, PCL.1.OCS.16% reached all weak stress point under 1.5 MPa and gained the lowest percent elongation of 12%.

To explain the various geometrical stress–strain, the typically mechanical property of PCL and OCS was considered. PCL is an elastomeric biomaterial polymer with high failure strain and has a long-time degradation. Therefore, PCL was usually incorporated with another polymer as a basement to toughen the inelasticity of low strain materials and modify the rate of degradation [18]. In contrast, the film-forming only by chitosan, or its derivative OCS, usually possess limited physical properties such as brittleness and inflexible [19–21]. Therefore, coating a brittle coating on a high strain substrate can provide multi-fractured in tension. Consequently, the double stress maxima occur when the low strain material of OCS film coated onto the high strain material of PCL. Figure 4b shows the SEM image, which visualized the first coating ridge cracks of PCL.1.OCS.16% membrane. During the tension test, the early fracture referred to the damage of the OCS coating layer parallel with a stable pull-out of PCL. Such stress dropped and then reached to the destruction that corresponded to PCL. Besides, the membrane became more inflexible as the higher OCS coating content. Thus, through performing the tensile strength test, the PCL with the OCS coating layer was more accessible to break under applied stress than the PCL only. This assures a warning sign to the combination of biomaterial with distinct mechanical strength in terms of the coating layer.

4 Conclusion

In this study, novel electrospun PCL membranes coating OCS were successfully fabricated by mono-coating and multi-coating techniques. The antibacterial pretest indicated that the coating OCS.2% solution had antibacterial efficiency against both gram-negative (*P. aeruginosa*) and gram-positive (*S. aureus*) despite the lowest OCS concentration. For the coating membrane, SEM evaluation displayed increased OCS coating content by several coating times provided a more irregular surface, while by one-layer coating with higher OCS concentration induced the smoother morphology. However, the stress–strain curve of PCL.OCS membrane became more rigid and less flexible as the more significant OCS content in both coating method. Besides, due to the distinct mechanical strength between PCL substrate and OCS coating film, the multi-fractured in tension occurred with the first stress point for the OCS layer and later for PCL.

Acknowledgements This research is funded by International University, VNU-HCM, under grant number T2019-03-BME.

Conflicts of Interest The authors have no conflict of interest to declare.

References

1. Broussard KC, Powers JG (2013) Wound dressings: selecting the most appropriate type. Am J Clin Dermatol 14(6):449–459
2. Guo SA, DiPietro LA (2010) Factors affecting wound healing. J Dent Res 89(3):219–229
3. Matica MA, Aachmann FL, Tøndervik A, Sletta H, Ostafe V (2019) Chitosan as a wound dressing starting material: antimicrobial properties and mode of action. Int J Mol Sci 20(23):5889

4. Loan Khanh L, Thanh Truc N, Tan Dat N, Thi Phuong Nghi N, Van Toi V, Thi Thu Hoai N et al (2019) Gelatin-stabilized composites of silver nanoparticles and curcumin: characterization, antibacterial and antioxidant study. Sci Technol Adv Mater 20(1):276–290
5. Thanh NT, Hieu MH, Phuong NTM, Thuan TDB, Thu HNT, Do Minh T et al (2018) Optimization and characterization of electrospun polycaprolactone coated with gelatin-silver nanoparticles for wound healing application. Mater Sci Eng C 91:318–329
6. Croisier F, Jérôme C (2013) Chitosan-based biomaterials for tissue engineering. Eur Polymer J 49(4):780–792
7. Schmitz C, González Auza L, Koberidze D, Rasche S, Fischer R, Bortesi L (2019) Conversion of chitin to defined chitosan oligomers: current status and future prospects. Mar Drugs 17(8):452
8. Naveed M, Phil L, Sohail M, Hasnat M, Baig MMFA, Ihsan AU et al (2019) Chitosan oligosaccharide (COS): an overview. Int J Bio Macromol
9. Thapa B, Narain R (2016) Mechanism, current challenges, and new approaches for non-viral gene delivery. In: Polymers and nanomaterials for gene therapy. Woodhead Publishing, pp 1–27
10. Shokrollahi M, Bahrami SH, Nazarpak MH, Solouk A (2020) Multilayer nanofibrous patch comprising chamomile loaded carboxyethyl chitosan/poly (vinyl alcohol) and polycaprolactone as a potential wound dressing. Int J Bio Macromol
11. Abrigo M, McArthur SL, Kingshott P (2014) Electrospun nanofibers as dressings for chronic wound care: advances, challenges, and future prospects. Macromol Biosci 14(6):772–792
12. Okur ME, Karantas ID, Şenyiğit Z, Okur NÜ, Siafaka PI (2020) Recent trends on wound management; new therapeutic choices based on polymeric carriers. Asian J Pharm Sci
13. Ho MH, Do TBT, Dang NNT, Le ANM, Ta HTK, Vo TV, Nguyen HT (2019) Effects of an acetic acid and acetone mixture on the characteristics and scaffold-cell interaction of electrospun polycaprolactone membranes. Appl Sci 9(20):4350
14. Xuan Du D, Xuan Vuong B (2019) Study on preparation of water-soluble chitosan with varying molecular weights and its antioxidant activity. Adv Mater Sci Eng
15. Zou P, Yang X, Wang J, Li Y, Yu H, Zhang Y, Liu G (2016) Advances in characterisation and biological activities of chitosan and chitosan oligosaccharides. Food Chem 190:1174–1181
16. Kumar S, Ye F, Dobretsov S, Dutta J (2019) Chitosan nanocomposite coatings for food, paints, and water treatment applications. Appl Sci 9(12):2409
17. Roylance D (2001) Stress-strain curves. Massachusetts Institute of Technology study, Cambridge
18. Brown DA, Lee EW, Loh CT, Kee ST (2009) A new wave in treatment of vascular occlusive disease: biodegradable stents—clinical experience and scientific principles. J Vasc Interv Radiol 20(3):315–324
19. Hamrang A, Howell BA (2013) Foundations of high-performance polymers: properties, performance and applications. Apple Academic Press
20. Nunthanid J, Puttipipatkhachorn S, Yamamoto K, Peck GE (2001) Physical properties and molecular behavior of chitosan films. Drug Dev Ind Pharm 27(2):143–157
21. Sarasam A, Madihally SV (2005) Characterization of chitosan–polycaprolactone blends for tissue engineering applications. Biomaterials 26(27):5500–5508

Effect of In Vitro Formed Fibrillar Fibronectin on Fibroblast Cellular Migration

Nghi Mai, Phong Le, and Khon Huynh

Abstract Fibronectin (FN) is an important adhesive extracellular matrix protein. It forms a fibrillar matrix on cell surface which controls cell morphology, adhesion, migration and proliferation. This study aimed to analyze the effect of in vitro cell-free formed fibrillar fibronectin with different morphologies on mouse fibroblasts L929 cells migration. To evaluate fibrils morphology on cell function, fibroblast L929 were cultured on surface pre-coated with FN and FN fibrils at concentration of 0.25 and 0.75 mg/mL at 4 °C overnight, then cell migration assay were performed. Migration speed of cells monolayers grown on different fibronectin surfaces was measured by using snapshot pictures with a regular inverted microscope at 0, 12, 48 and 72 h. Our cell-free FN fibrillogenesis indicated that at different concentrations, fibronectin organized into various structural matrix. Our data showed that large fibronectin fibrils decreased the rate of cell migration while small fibronectin fibrils (0.25 mg/mL) did not, compared with native plasma fibronectin. These data demonstrate that there is a strong relationship between structure and function of fibronectin. Taken together, our data draws new attention towards controlling biological function of fibronectin by its fibrillar structure.

Keywords Fibronectin · Fibronectin fibrils · Cell function

1 Introduction

Fibronectin (FN) mediates a wide variety of cellular interactions within the extracellular matrix (ECM) and plays important roles in cell adhesion, migration, growth and differentiation [1]. Fibronectin composes of three general types of homologous repeating units or modules, termed types I, II and III [2]. These modules comprise functional domains that mediate interactions with other ECM components, with

N. Mai · P. Le · K. Huynh (✉)
School of Biomedical Engineering, International University, Ho Chi Minh 700000, Vietnam
e-mail: hckhon@hcmiu.edu.vn

Vietnam National University-Ho Chi Minh Cityity (VNU-HCMC), Ho Chi Minh 700000, Vietnam

V. Van Toi et al. (eds.), *8th International Conference on the Development of Biomedical Engineering in Vietnam*, IFMBE Proceedings 85,
https://doi.org/10.1007/978-3-030-75506-5_35

heparin domains, with cell surface receptors and with FN itself [3, 4]. Studies with synthetic peptides focused on these areas identified a series of three Arg-Gly-Asp (RGD) peptides, also known as RGD chains, that are essential sequences for interaction with the receptor protein on the cell surface [5]. Depends on interactions between the cell-binding domain of FN and cell surface integrin receptors—usually α5β1 binding to the RGD and synergy site [5]. Moreover, FN has functional regions on the structural chain that can interact well with different components in the extracellular matrix such as collagen, fibrinogen, and receptor proteins on the cell surface.

The interaction of cells with the extracellular matrix via cell surface integrins generates a series of complex signaling events that serve to regulate several aspects of cell behavior, including growth, differentiation, adhesion, and motility. Integrins are a major class of transmembrane receptors that mediate cell adhesion to fibronectin [6, 7] as well as to other extracellular matrix proteins. Binding of extracellular matrix proteins to integrins leads to the generation of intracellular signals, many of which are similar to intracellular signals generated by growth factor stimulation [8]. Integrin-extracellular matrix interactions are also important for cell survival, as disruption of integrin-mediated attachment to the extracellular matrix induces apoptosis in epithelial and endothelial cells [9]. Receptor binding stimulates FN self-association mediated by the N-terminal assembly domain and organizes the actin cytoskeleton to promote cell contractility.

When FN is structurally stretched, the domains of FN are clearly exposed and interact with the cell and other components of the ECM better [4, 10]. Therefore, FN conformational changes expose additional binding sites that participate in fibril formation and in conversion of fibrils into a stabilized, insoluble form [10]. Once assembled, the FN matrix impacts tissue organization by contributing to the assembly of other ECM proteins [10]. It forms a fibrillar matrix at the cell surface which controls cell morphology, migration, proliferation, and other important cellular processes. Thus, the structure and composition of the fibrillar fibronectin matrix is also very significance for cell function. Our goal is to analyze the effect of in vitro cell-free formed fibrillar fibronectin with different morphologies on fibroblast L929 behavior towards the use of fibronectin to control migration of mouse fibroblast L929.

2 Material and Method

2.1 Materials

Fibronectin was in-house purification from frozen human plasma (Blood Transfusion and Hematology Hospital, HCM City, Vietnam) by Lê Qu´ôc Phong. Fibronectin fibrils obtained from Lê Qu´ôc Phong (International University, Vietnam National Universities, Vietnam). Integrin α5β1 and αvβ3 antibody (Millipore, Germany). Mitomycin C and Cytochalasin D (Sigma-Aldrich, USA). Phosphate buffered saline

(Sigma-Aldrich, Singapore). Tissue culture plate supplied by SPL Life Sciences, Korea.

The mouse L929 fibroblasts was purchased from the Cell Bank at ATCC. The cells were cultured in Dulbecco's modified Eagle's medium (DMEM; Gibco; Thermo Fisher Scientific, Inc., Waltham, MA, USA) and supplemented with 10% Fetal Bovine Serum (FBS) (Sigma-Aldrich), 1% Penicillin Streptomycin (PS) (Sigma-Aldrich).

2.2 *Methods*

Cell culture

L929 were cultured in DMEM containing 10% fetal bovine serum (FBS) and 1% penicillin/streptomycin. Cells were grown in T75 tissue culture flasks in a humidified atmosphere containing 5% CO_2, 37 °C and passaging either every 4–5 days or before reaching 90% confluent. For passaging, cells were rinsed with phosphate buffered saline (PBS) and incubated with 1 mL trypsin/EDTA at 37 °C for 3 min until detachment. Trypsin was inactivated by adding 5 mL of growth medium, and cells were diluted according to their growth rate.

Cell migration assay

An in vitro wound repair assay was used to measure cell migration [11]. Tissue culture plates (96-well plate) were coated with native fibronectin and fibronectin fibrils at 0.25 and 0.75 mg/mL in PBS1x for overnight at 4 °C. Wells were washed with PBS 1x to remove unbound protein, then blocked with 1% BSA in PBS 1x for 1 h at 37 °C, 5% CO_2. L929 were seeded at 2×10^5 cells/ml in culture media. Cells were allowed to adhere and spread 24 h at 37 °C, 5% CO_2 to establish a confluent monolayer. Then, the cells were treated with 10 μg/mL mitomycin for 2 h in order to inhibit cell proliferation in cell migration. For control, cells were treated with 10 μg/mL cytochalasin D to inhibit actin cytoskeleton of cells. Wells were washed with PBS 1x before a thin section of the monolayer was carefully removed with a sterile, pipette tip 200 μL. The media and cell debris were aspirated. Culture media were slowly added against the well wall to cover the bottom of the well and avoid additional cells from detaching. Five non-overlapping regions of the wound were viewed immediately after wounding with an inverted microscope using a 10x objective (time = 0 h). The tissue culture plate was placed in an incubator which was set at the appropriate temperature and CO_2 concentration (typically 37 °C and 5% CO_2). At several time points, the plate was taken out from the incubator and placed under an inverted microscope to take a snapshot picture and to check for wound closure. To analyze the results of snapshot pictures, we measure the distance of one side of the wound to the other using a scale bar. Cell migration was calculated according to the following formula:

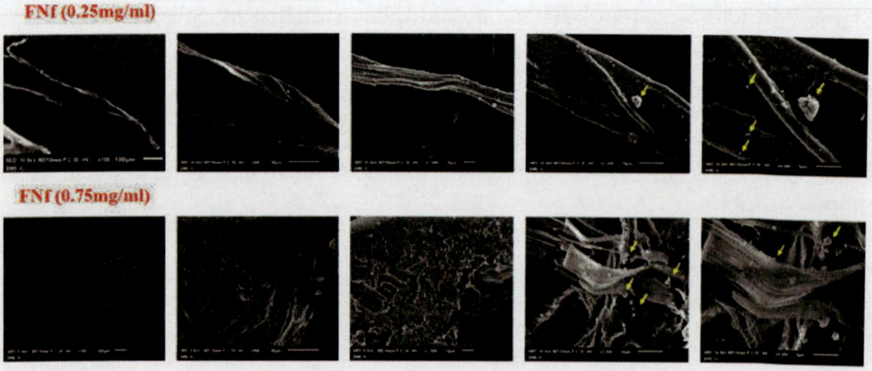

Fig. 1 SEM image of fibronectin fibrils with various morphology (FNf 0.25 mg/mL and FNf 0.75 mg/mL) at magnification 100, 500, 1000, 2000, 4000 (scale bar: 100 μm, 50 μm, 10 μm, 10 μm, 5 μm respectively). Multiple nodular FN attached on fibrils surface (yellow arrow)

$$\text{Wound Closure } (\mu\text{m/hour}) = \frac{\text{Initial Wound Width } (\mu\text{m}) - \text{Final Wound Width } (\mu\text{m})}{\text{Time (hour)}}$$

SEM analysis of cell-FN interaction

Cells were washed with PBS 1x, then fixed in 4% glutaraldehyde (GA) for 1 h at room temperature. Afterward, samples were progressively dehydrated in a 50, 60, 70, 80, 90, 95 and 100% ethanol series for 3 min each. After that, fixed cells were keep in clean bench 1 day to completely dry, then samples were sputter-coated to examine under SEM (JEOL JSM-IT100) at 15 and 20 kV, with magnifications of 100, 1000, 4000 and 8000.

Statistical analysis

Experiments were replicated at least 3 times. Statistical comparisons were performed using either one-way analysis of variance (ANOVA) followed by Student's *t*-test using GraphPad Prism software. Results were considered statistically significant when $p < 0.05$.

3 Results

3.1 Morphologies of FN Fibrils

SEM image showed that FN concentration have significant effect on fibronectin fibrils morphology (Fig. 1). FN morphology varied respectively to increase FN concentration. In particular, when FN concentration achieved 0.25 mg/mL, cross-linked FNf have small size and scattered and discreted. Surface of FNf 0.25 mg/ml were smooth and multiple nodular FN attached on fibrils surface (yellow arrow). As FN concentration increased to 0.75 mg/mL, FNf achieved an even distribution with great size and interlacing of FN fibers. Comparing to FNf 0.25, FNf 0.75 mg/mL surface were rougher since many thin fibers overlapped and more nodular FN attached on (yellow arrow). From two highest magnification SEM image of FNf, it can be seen that many FN fibrils intertwined with each other to form protein matrix and FN molecules stretched out and parallel interact to each other to form thin and "sheet" fiber.

3.2 Cell Migration

The wound closure assay was performed using fibroblasts L929 as a model system. In the wound closure assay, L929 cells were seeded in a 96-well tissue culture plate and grown to 100% confluence in 24 h. A wound of approximately 1300 μm wide was generated using a pipette tip and the wound closure (cell migration) was recorded using the time-lapse microscopy. The pictures from the time-lapse series at 0, 24, 48 and 72 h time points are shown in Fig. 2a. Based on the width of the wound, surfaces coated with the original Fibronectin promoted cell migration into the denuded area to a greater extent than surfaces coated with Fibronectin fibrils (0.25 mg/mL) (Fig. 2b). Substituting the Fibronectin fibrils (0.75 mg/mL) decreased the rate of cell migration (original Fibronectin). The rate of cell migration on the surfaces coated with the Fibronectin fibrils (0.25 mg/mL) did not reduce the rate of cell migration of original Fibronectin.

4 Discussion

Our cell migration experiments may be influenced by the difference in fibrils structure at the two concentrations of FN (0.25 and 0.75 mg/mL). FN fibrils sample were synthesized from FN (0.25 mg/mL) with only small fibrils. These fibrils which consisted of smaller fibrils bundled together have a smooth surface. In contrast, FN fibrils samples synthesized from FN (0.75 mg/mL) have many larger fibrils due to the combination of multiple small fibrils to form large fibrils. FN particles sticked to the surface of the large fibrils, which made the fibrils surface become rough.

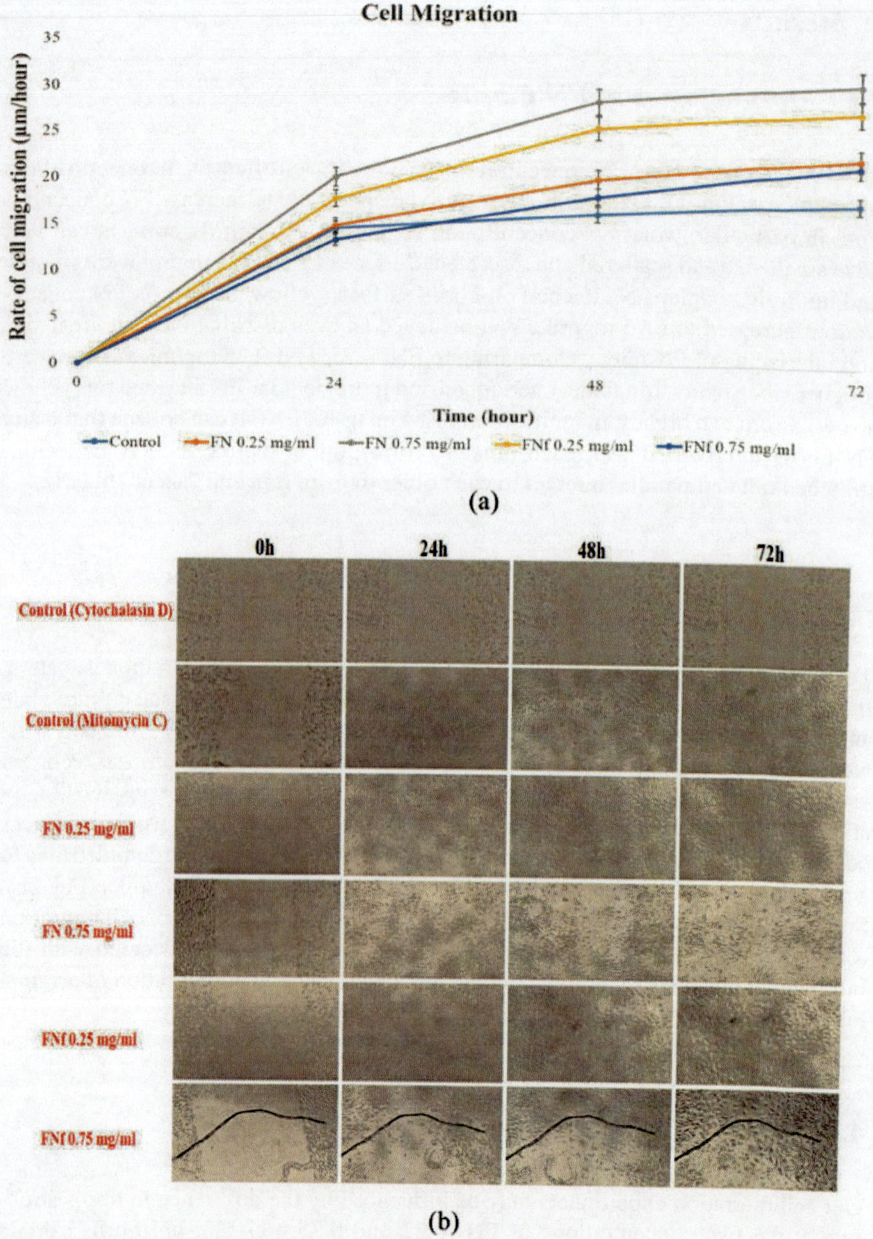

Fig. 2 Migration of fibroblasts to fibronectin. The rate of migration cells **a** was determine after 3 days. **b** Phase contrast images were obtained after 3 days (Scale bar: 100 µm). *$p < 0.05$ evaluated by unpaired *t*-test. Results are representative of three experiments

These fibrils, which were originally made from FN 0.75 mg/mL and were made of many smaller fibrils and FN particles, can stick together to form a layer between the small fibrils. Therefore, the size and structure of the fibrils' surface have the potential to interfere with the rate of cell migration. These data indicate that as adhesive substrates, the fibronectin fibrils (0.75 mg/mL) are able to decrease cell migration rate. Thus, there is considerable cell movement that is dependent on the synthesis of fibrils morphogenesis, changes in cell-surface receptor, and matrix remodeling.

5 Conclusion

We conclude that this study elucidates a key player in the mechanical interaction between cells and substrate, namely the presence of fibronectin fibrils. The results of cell migration assay with mouse fibroblast L929 and Fibronectin show that large fibronectin fibrils (0.75 mg/mL) have the ability to interfere with the migration rate of cancer cells. These data illustrate that fibronectin structure and its function can effect the cell migration rate.

Acknowledgements We gratefully acknowledge the financial support provided to this study by the Vietnam National Foundation for Science and Technology Development (NAFOSTED) under Grant Number 108.05-2017.328.

Conflicts of Interest The authors have no conflict of interest to declare.

References

1. Pankov R, Yamada KM (2002) Fibronectin at a glance. J Cell Sci 115(20):3861–3863
2. Schwarzbauer JE, DeSimone DW (2011) Fibronectins, their fibrillogenesis, and in vivo functions. Cold Spring Harb Perspect Biol 3(7):a005041
3. Lenselink EA (2015) Role of fibronectin in normal wound healing. Int Wound J 12(3):313–316
4. Mao Y, Schwarzbauer JE (2005) Fibronectin fibrillogenesis, a cell-mediated matrix assembly process. Matrix Biol 24(6):389–399
5. Schwarzbauer JE, Sechler JL (1999) Fibronectin fibrillogenesis: a paradigm for extracellular matrix assembly. Curr Opin Cell Biol 11(5):622–627
6. Pytela R, Pierschbacher MD, Ruoslahti E (1985) Identification and isolation of a 140 kd cell surface glycoprotein with properties expected of a fibronectin receptor. Cell 40(1):191–198
7. Hynes RO (1992) Integrins: versatility, modulation, and signaling in cell adhesion. Cell 69(1):11–25
8. Schwartz MA (1997) Integrins, oncogenes, and anchorage independence. J Cell Biol 139(3):575–578
9. Frisch S, Francis H (1994) Disruption of epithelial cell-matrix interactions induces apoptosis. J Cell Biol 124(4):619–626
10. Singh P, Carraher C, Schwarzbauer JE (2010) Assembly of fibronectin extracellular matrix. Annu Rev Cell Dev Biol 26:397–419
11. Small JV et al (2002) The lamellipodium: where motility begins. Trends Cell Biol 12(3):112–120

12. Jonkman JEN et al (2014) An introduction to the wound healing assay using live-cell microscopy. Cell Adhes Migr 8(5):440–451. https://doi.org/10.4161/cam.36224

Noninvasive System for Tracking Naïve Induced Pluripotent Stem Cells During Reprogramming

Tran Thi Hai Yen, Aya Fukuda, and Koji Hisatake

Abstract Naïve induced pluripotent stem (iPS) cells are useful for drug development, disease model and regenerative medicine. In female cells, X chromosome reactivation (XCR) is closely linked to the acquisition of naïve iPSCs. To visualize the status of X chromosome in living cells and further analyze XCR mechanism, we established a novel noninvasive detection system of XCR. Here we established female mouse embryonic stem cell (mESC) lines whose allele of both X chromosomes carry each of different fluorescent protein genes [Kusabira Orange (KO) and Enhanced green fluorescent protein (EGFP)]. These mESC clones, initially displaying yellow fluorescence owing to two active X chromosomes, generate chimeric cluster of cells showing either orange or green fluorescence (KO^+ or $EGFP^+$), indicating that X chromosome inactivation (XCI) occurs in these mESC clones during differentiation. When the single colored cells were reprogrammed, the iPS colonies displayed yellow ($KO^+/EGFP^+$) fluorescence, indicating XCR upon reprogramming. Our work reveals generated female $KO^+/EGFP^+$ mESC lines can be used for monitoring X chromosome status during differentiation and/or reprogramming. This novel system is also potential for tracking acquisition of pluripotency of iPSCs during reprogramming and screening for factors or culture conditions which promote this event.

Keywords iPSC · XCI · XCR · *Xist*

1 Introduction

In 2006, Yamanaka and his colleague succeeded in directly generating iPSCs (Induced Pluripotent Stem Cells) from somatic cells by co-expressing fours factors

---

T. T. H. Yen (✉)
School of Biotechnology, International University, Ho Chi Minh City, Viet Nam
e-mail: tthyen@hcmiu.edu.vn.com

Viet Nam National University, Ho Chi Minh City, Viet Nam

A. Fukuda · K. Hisatake
Faculty of Medicine, University of Tsukuba, 1-1-1 Tennodai, Tsukuba 305-8575, Ibaraki, Japan

V. Van Toi et al. (eds.), *8th International Conference on the Development of Biomedical Engineering in Vietnam*, IFMBE Proceedings 85,
https://doi.org/10.1007/978-3-030-75506-5_37

Oct4, Sox2, c-Myc and Klf4 [1]. Although they have been discovered recently, iPSCs have been already useful for the disease model, drug development, and regenerative medicine [2]. iPSC-derived tissues nearly match to cell donor, thus, probably prevent the transplant rejection. Moreover, bioethical issues are not applied to iPSC production; thus, generation of patient-specific stem cell lines is potential. However, a small number of high pluripotent iPSCs in reprogrammed population will limit the applicable potential [3]. For their applications in cell therapy, these high pluripotent cells need to be isolated.

Up to date, most of the high pluripotent iPSCs are male [4–11] although high pluripotent female cells are also essential for clinical research and applications such as models for X chromosome-linked human diseases. The reprogramming efficiency of female mouse embryonic fibroblasts (MEFs) is also known to be lower than that of male MEFs [12]. Since one X chromosome is inactivated in female somatic cells in mammals, the inactive X chromosome has to be reactivated during reprogramming to generate iPSCs with high pluripotency. Moreover, X chromosome reactivation (XCR) is rarely achieved in human iPSCs, and their pluripotency is lower than mouse iPSCs, suggesting that XCR is critical for high quality of iPSCs [13]. It has also been proven that XCR is closely linked to pluripotency [14]. In other words, XCR is used as an indicator of high pluripotent iPSCs.

In previous researches, some factors such as PRDM14 have been proven to be necessary for XCR process. However, the detailed molecular mechanism of XCR remains unclear [15]. Understanding of molecular mechanism that underlies XCR will enable scientists to overcome the barrier of iPSCs production and move it closer to clinical application. To clarify this mechanism, reactivated X chromosome iPSCs need to be distinguished and isolated from mixture of reprogrammed cells. H3K27me3 immunostaining is a common method. However, it requires fixation on living cells. Hence, it is impossible to investigate the cells' characteristics further. A live detection method is essential for these researches. Unfortunately, at present, there is no tool to monitor XCR in living cells. To visualize the status of X chromosome and further analyze the XCR mechanism, we established a novel noninvasive detection system of XCR. We also tested this system was utilized for monitoring of XCI by the fluorescent signals as well. Thus, our system provides a convenient and simple method for tracking dynamic changes by two directions in the epigenetic status of X chromosome in living cells.

2 Materials and Methods

2.1 Plasmid Preparation

To construct targeting vectors, 1000–2000 bp homologous fragments to X chromosome were placed into phEF1-EGFP-IRES-Puro^r and pEF1-hKO-IRES-Zeo^r vectors.

2.2 mESCs Transfection

Mitomycin C-treated SNLs (Riken Bioresource Center) on 6-well dish were cultured for 12 h, at 37 °C, 5% CO_2 in SNL media (DMEM + 10% FBS). Then, 5 × 10^5 female mESCs derived from inbred C57BL/6 (*B6*) (Riken Bioresource Center) were seeded and cultured for 5 h on one well of feeder-layered dish, at 37 °C, 5% CO_2 in mESC medium (DMEM; Nacalai tesque) with 15% Knockdown Serum Replacement (KSR), 0.1 μM 2-mercaptoethanol (2-ME) (Life Technologies), 1 mM sodium pyruvate (Nacalai tesque), 0.1 mM nonessential amino acids (NEAA) (Wako), 1000 U/ml LIF (Yeast Oriental). Two micrograms of pX330 vectors (with/without gRNA) were mixed with 2 μg targeting vectors and introduced into mESCs by Lipofectamine 2000. Forty-eight hours after transfection, EGFP and hKO expression was observed. Transfected cells were collected after Puromycin and Zeocin selection since candidate $EGFP^+/hKO^+$ cells with both inserted genes on X chromosome are resistant to these two antibiotics.

2.3 Genotyping Assay

To select candidate $EGFP^+/hKO^+$ mESCs, genomic DNA was extracted from transfected cells and used as a template for PCR (KOD-plus-Neo (Toyobo)) to evaluate genome editing efficiency. Prior to the DNA extraction, transfected cells were cultured under the condition of feeder-free to remove feeder cells.

The primer sets used for PCR-based genotyping assay were designed with one primer on the intergenic region of X Chromosome closed to the insertion site and another primer on reporter gene cassette (antibiotic resistance gene or fluorescence coding gene). Each cell was genotyped with 2 different primer sets on 2 reporter gene cassettes (EGFP-IRES-$Puro^r$ and hKO-IRES-Zeo^r).

2.4 Differentiation of mESCs

Candidate mESCs were grown on gelatin-coated plates with feeder layer to 80% confluence. These cells were collected, trypsinized and suspended in EB medium (DMEM supplemented with 20% FBS (Life Technologies), 0.1 μM 2-ME, 0.1 mM NEAA (Wako)). To remove feeder cells, the cell suspension was transferred to a culture dish for 20 min. Then, the supernatant was collected and cultured into bacterial dishes for embryoid body (EB) formation. After 5 days, EBs were collected, trypsinized and filtrated by cell strainers. The resulting single cells were then cultured on Collagen Type I coated plates with the addition of Puromycin or Zeocin to select single colored cells. During differentiation time, the fluorescent signal was observed to monitor XCI.

2.5 iPSCs Production

To generate miPSCs, 2.5 $\times$ 10^4 hKO^+ cells were cultured on one well of 24-well dish in DMEM + 10% FBS medium for 12 h, at 37 °C, 5% CO_2. Cells were infected with viral vectors expressing 4 Yamanaka factors for 16 h at 32 °C before seeding onto feeder cells. The infected cells were cultured in Knockout-DMEM (Life Technologies) supplemented with 15% KSR, 55 μM 2-ME, 2 mM GlutaMAX (Life Technologies), 0.1 mM NEAA, 1% Pen-Strep, 1000 U/mL LIF for 7 days. Then, the culture was replaced with mESC medium. Fluorescent expression was monitored. Reactivated X chromosome iPSCs were picked up and analyzed.

2.6 Quantitative Reverse Transcription (RT)-PCR

Total RNA of cultured cells was extracted by Sepasol-RNA I Super G (Nacalai tesque). Prior to the RNA extraction, miPSCs were cultured under the feeder-free condition to remove the contamination of feeder cells. Superscript III First-Strand Synthesis System (Life Technologies) was used to perform reverse transcription. mRNA levels were determined by quantitative PCR (qPCR) (GoTaq qPCR Master Mix (Promega)). The levels of g-tubulin and 45S rRNA were used for the normalization of the obtained data.

3 Results and Discussion

3.1 Generation of Female Dual-Color-Fluorescent-Labeled mESCs

To visualize the status of X chromosome in living cells, we generated female mouse embryonic stem cell (mESC) lines which carry fluorescent reporter genes on X chromosomes. The basic principle of the detection system was shown in Fig. 1a. By using the CRISPR/Cas9 system, we inserted reporter gene cassettes encoding two different fluorescent proteins, humanized Kusabira orange (hKO) and enhanced green fluorescent protein (EGFP), into downstream of EF1 promoter, into specific loci (the intergenic region that is close to *Syap1*) in the X chromosomes (Fig. 1b).

During transfection, we used Puromycin and Zeocin for selection. After drug selection, mESC clones that expressed yellow fluorescence were collected. PCR analysis with different primer sets was used to confirm their recombinant genome DNA. Among them, five clones showed expected insert at the targeted locus on both X chromosomes. In addition, further confirmation of single-copy integration at the targeted region was performed (data not shown). Two out of them had a single copy of reporter genes at the targeted locus. These results indicated heterozygous mESCs

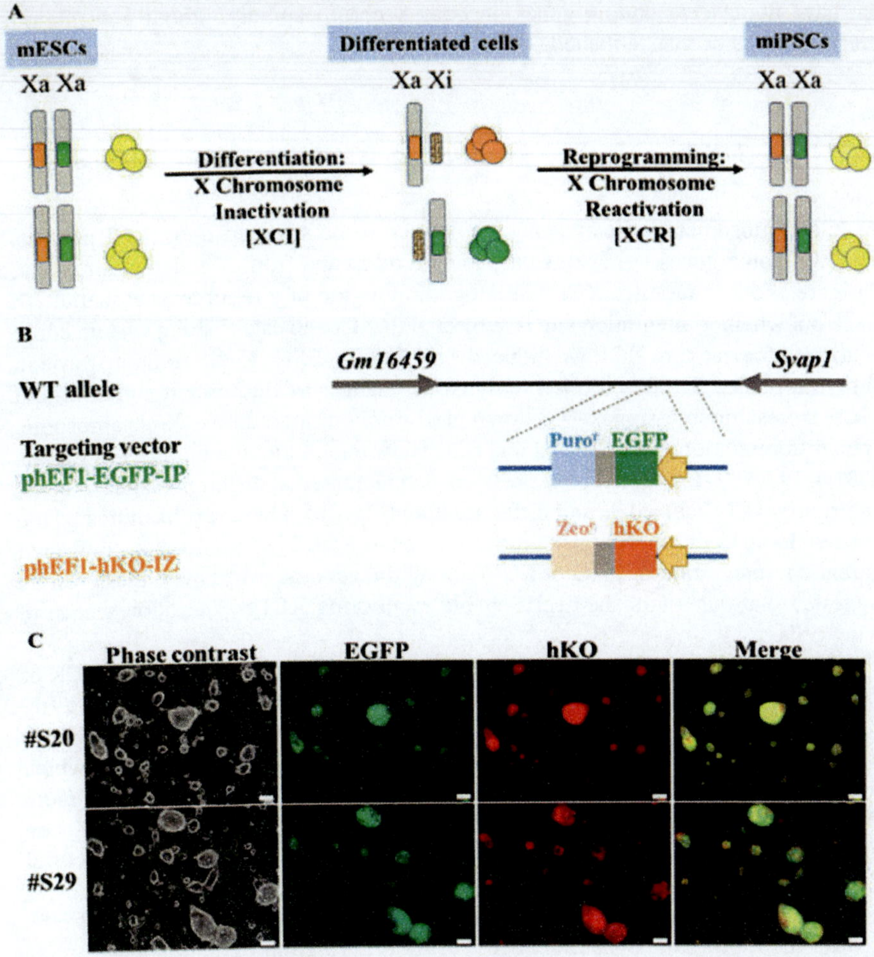

Fig. 1 Establishment of EGFP$^+$/KO$^+$ female mouse ESCs. **a** The detection system of X chromosome status. Two fluorescent reporter genes (hKO–EGFP) are genetically inserted into X chromosomes of ESCs. Initially, ESCs express dual fluorescence (hKO and EGFP) by owning two active X chromosomes (XaXa). During differentiation, by the random XCI (XaXi), they express either orange or green signal. During reprogramming, by the process of XCR, the inactive X chromosome is reactivated (XaXa), iPSCs display double fluorescence. **b** Structure of targeting vectors for the insertion locus. Between the homologous sequences in the targeted locus, each targeting vector contains one fluorescent protein gene, a drug-resistant gene, which was regulated by the EF1α promoter. **c** Morphology and fluorescent pattern of candidate clones. Scale bars: 100 μm

that have fluorescent coding genes on both X chromosomes without any random integration had been established (#S20 and #S29).

3.2 *Live Cell Imaging for Monitoring X Chromosome Status*

On X chromosomes, most of genes are known to be silenced upon XCI process. However, some genes (~15–20%) are known to escape from X inactivation event. If we are about to monitor XCI, the integration region of a reporter is essential. To check out whether integration site is subjected to XCI and establishing system could be utilized for tracking XCI, we induced $KO^+/EGFP^+$ ESCs to differentiate through embryoid bodies (EBs) and observe dynamic changes of fluorescent signal. Initial ESCs express double signals—yellow color, implying two active X chromosome. Upon differentiation, we observed that cells show one of the two fluorescent signals (KO^+ or $EGFP^+$) (Fig. 2a). *Xist* expression also increased at different stages of cells, confirming XCI occurred during differentiation (Fig. 2b). These results indicated the inserted fluorescent reporter genes are subjected to XCI. This results also confirmed no random integration of hKO or EGFP along the genome. Moreover, these results suggested that our established cells enable monitoring XCI by the fluorescence in living cells.

Afterwards, to examine whether our establishing system is able to detect XCR or not, we isolated single colored hKO^+ cells for reprogramming by infecting SeVdp (KOSM). Following reprogramming, XCR was monitored by fluorescence (EGFP reactivation). EGFP signal started to appear around day 15 after infection, which suggests the time point of XCR (Fig. 2c). These colonies gradually became more homogenous. These $EGFP^+/hKO^+$ colonies showed a reduction of *Xist* expression (Fig. 2d) and induction of stem cell marker genes (*Oct4*, *Rex1*) to a similar level of mESCs (Fig. 2e). The reduction of *Xist* expression is signed for XCR; thus, it indicated XCR occurs during reprogramming. Conclusively, our system can also be used for monitoring the dynamic process of XCR during reprogramming.

3.3 *Detection of Heterogeneous Population of iPSCs*

In a population of pluripotent stem cells, it is challenging to maintain homogeneity between cells or colonies. It is also challenging to detect heterogeneity of these population by the noninvasive way. In addition, obtaining naïve pluripotent stem cells is essential for research or transplantation medicine. In our experiment, some colonies maintained orange signal; some showed heterogeneous or homogenous pattern of yellow fluorescence, some of them. It may indicate that the XCR process has not simultaneously occurred in all formed colonies. Thus, this system may provide a simple method to detect heterogeneity of reprogrammed cells.

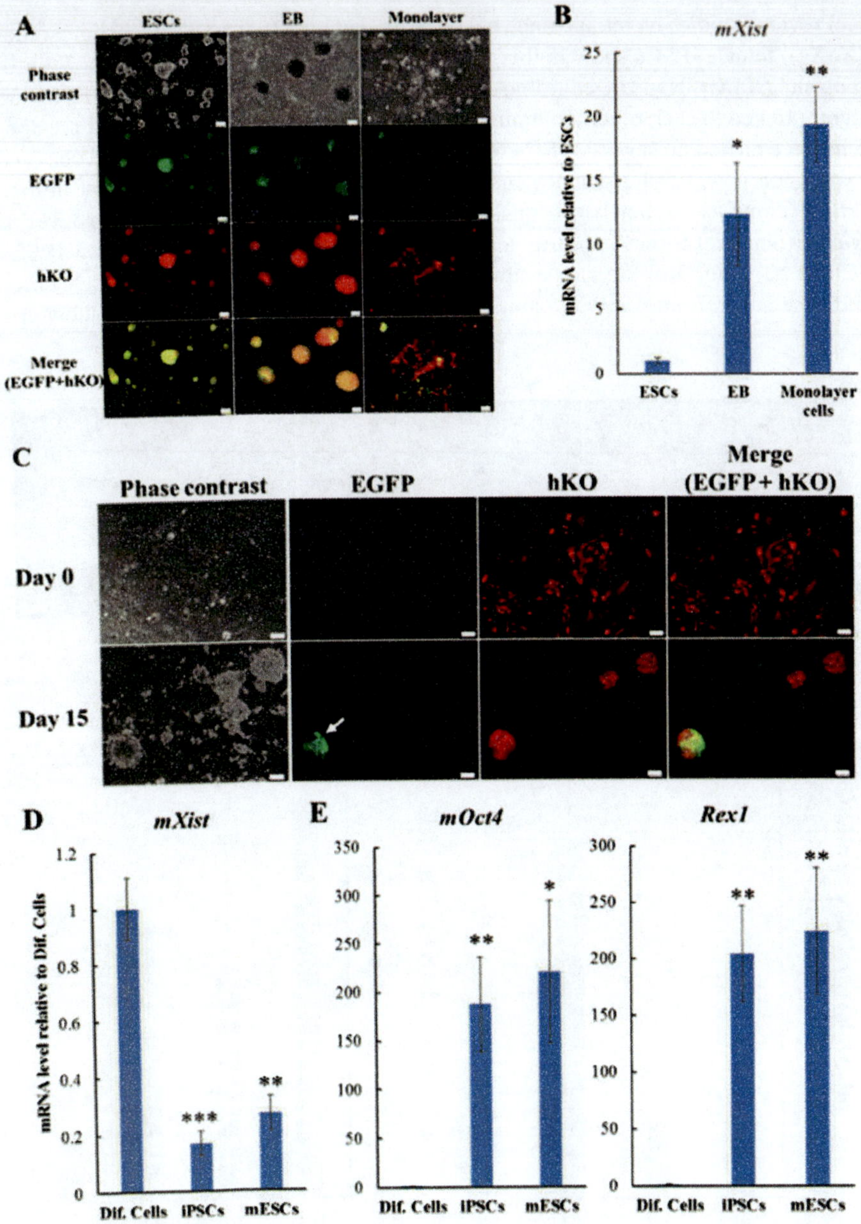

Fig. 2 Live cell imaging for monitoring X chromosome status. **a** Fluorescent patterns of representative clone during differentiation. **d** Expression of *mXist* of generated iPSCs. **e** Expression of *mOct4*, *Rex1* (pluripotency markers) of generated iPSCs. Scale bars: 100 μm. * $p < 0.05$, ** $p < 0.01$, *** $p < 0.001$

Previous studies on female mice have shown that with two active X chromosomes (XaXa), female iPSCs show higher pluripotency than cells with one active X chromosome (XaXi). Hence, X chromosome reactivation and pluripotency of iPSCs are tightly linked to each other. To figure out the correlation between XCR and pluripotency, we picked up several iPSC colonies, observed fluorescent signal and checked expression level of pluripotency marker genes. Notably, we found out that orange iPSCs (clone #1) or heterogeneously yellow (clone #2) expressed a lower level of *Nanog* (a marker for naïve pluripotent stem cells) than homogenously yellow iPSCs (Clone #3) (Fig. 3b). We also evaluated the expression of *Xist* to confirm X chromosome status in these iPSC clones. Clone #3 and #2 but not clone #1 showed a

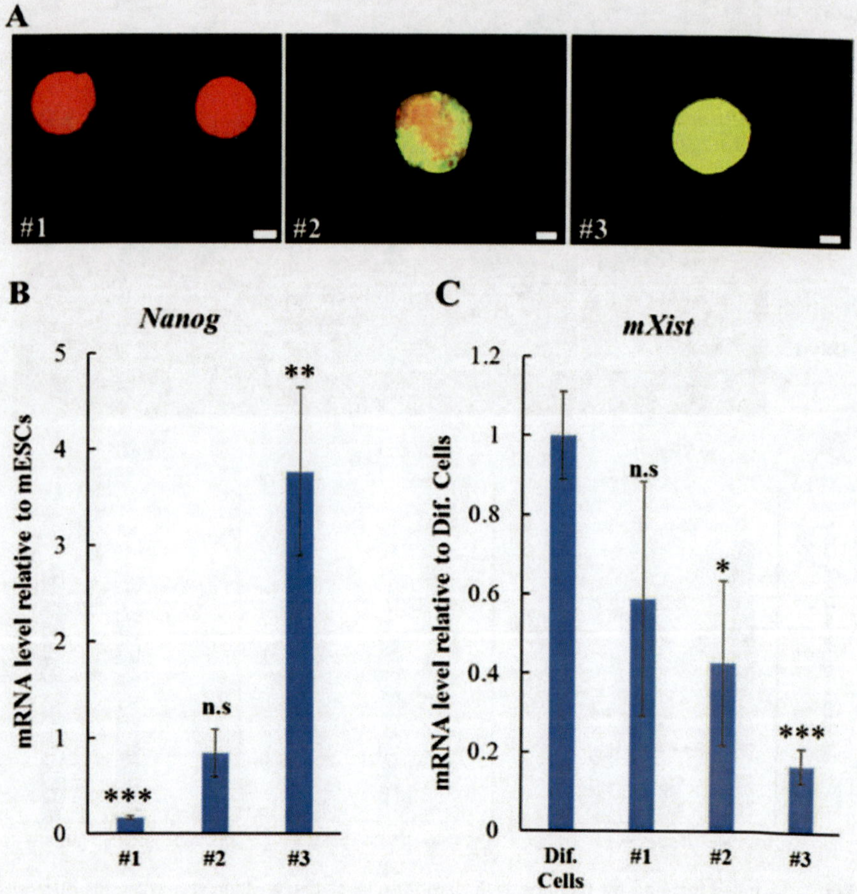

Fig. 3 Detection of reprogrammed population's heterogeneity. **a** Different fluorescent patterns of reprogrammed iPSC clones. Three clones (#1, #2, #3) were shown as representatives. **b** Expression of *Nanog* in different fluorescent patterned clones. **c** Expression of *Xist* in in different fluorescent patterned clones. Scale bars: 100 μm. [n.s] not significant, * $p < 0.05$, ** $p < 0.01$, *** $p < 0.001$

significant reduction of *Xist* expression. Since reactivation of X chromosome occurs after silencing of *Xist* RNA, it indicated XCR occurred in clone #3 and #2 but not in #1. These data reflect a correlation between the acquisition of pluripotency and reactivation of X chromosome. Although further in vivo experiments should be proceeded to confirm pluripotency of produced yellow iPSCs or erosion of X chromosome in orange iPSCs, this system provides simple methods for tracking naïve pluripotent stem cells by monitoring XCR during reprogramming.

3.4 High Throughput Screening for Culture Conditions that Promote Acquisition of Pluripotency

Colony morphology is a well-known marker for pluripotent stem cells (PSCs). However, based on our experiment, it is obviously difficult to distinguishing naïve from primed PSCs by evaluating morphology. Additionally, the expansion and maintenance of PSCs are relied on culture conditions. For their applications in medicine, these cells need to be at naïve state. For that purpose, researchers have been actively searching for compounds or culture conditions which promote the acquisition of pluripotency. In order to evaluate whether our established system is utilized for high throughput screening for potential factors or culture conditions, we initially tested two popular media for culturing reprogramming cells: 2i medium (Serum-free) and ES medium (Serum and LIF). Following the reprogramming process, to track XCR (EGFP reactivation, the fluorescent signal is monitored. Around day 15 of reprogramming, iPSC colonies in 2i medium started to show dual signal in a heterogeneous pattern, indicating initial time point of XCR. These colonies became more homogenous around day 17, indicating completed XCR (Fig. 4a). On the other hand, XCR initial time point of iPS colonies cultured in ES medium was slightly delayed. This medium also could not facilitate completed XCR; only heterogeneous colonies were formed (Fig. 4b). 2i medium assisted generated iPS colonies in maintaining rounder shape than ES medium. Conclusively, our system provides a simple tool for screenings factors or culture conditions that promote acquisition of pluripotency.

4 Conclusion

Conclusively, our established model reveals a novel noninvasive system for in vitro visualizing the epigenetic state of X chromosome in living cells. Our established female hKO^+/$EGFP^+$ mESC lines are utilized for monitoring two directions of X chromosome activity (XCI–XCR) during differentiation and/or reprogramming. This novel noninvasive detection system provides a simple method for sorting high pluripotent stem cells in the population of reprogrammed iPSCs, which are potential materials for regenerative therapy research. This system also facilitates the tracing of

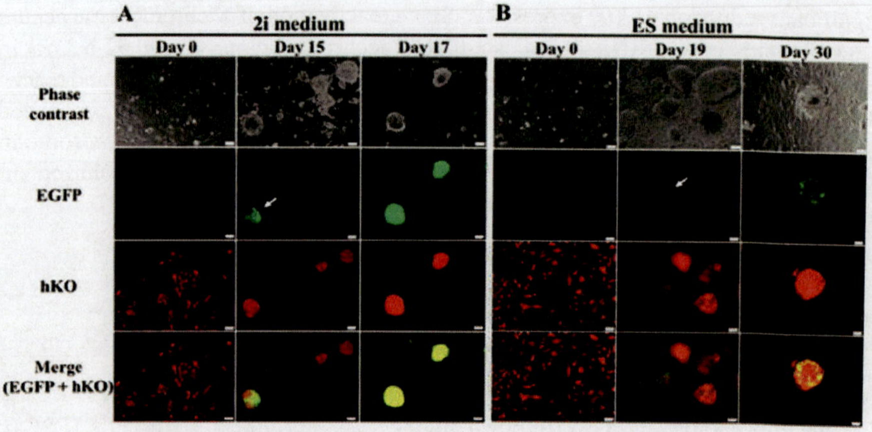

Fig. 4 Effect of cultured media on X chromosome reactivation detection. **a** 2i medium (Serum-free). **b** ES medium (Serum and LIF). Scale bars: 100 μm

the reprogramming process; consequently, it will support analysis of XCR mechanism, how to rescue naïve cells from primed cells. Understandings on these processes will help us in naïve human iPSCs establishment from primed ESCs effectively. Our results enable researcher to step closer to clarification of molecular mechanism that underlies reprogramming although they have just initially pointed out the XCR time point. Additionally, for the first time, our established cell lines, which are genetically engineered by revolutionised technique CRISPR/Cas9, has contributed into the storage of new cell lines for XCI or XCR researches.

Conflicts of Interest The authors have no conflict of interest to declare.

References

1. Takahashi K, Yamanaka S (2006) Induction of pluripotent stem cells from mouse embryonic and adult fibroblast cultures by defined factors. Cell 126(4):663–676
2. Singh VK (2015) Induced pluripotent stem cells: applications in regenerative medicine, disease modeling, and drug discovery. Front Cell Dev Biol 3:2
3. Polo JM (2012) A molecular roadmap of reprogramming somatic cells into iPS cells. Cell 151(7):1617–1632
4. Boland MJ (2009) Adult mice generated from induced pluripotent stem cells. Nature 461(7260):91–94
5. Carey BW (2011) Reprogramming factor stoichiometry influences the epigenetic state and biological properties of induced pluripotent stem cells. Cell Stem Cell 9(6):588–598
6. Di KQ (2015) Generation of fully pluripotent female murine-induced pluripotent stem cells. Biol Reprod 92(5):123
7. Kang L (2009) iPS cells can support full-term development of tetraploid blastocyst-complemented embryos. Cell Stem Cell 5(2):135–138

8. Okita K, Ichisaka T, Yamanaka S (2007) Generation of germline-competent induced pluripotent stem cells. Nature 448(7151):313–317
9. Tong M (2011) Mice generated from tetraploid complementation competent iPS cells show similar developmental features as those from ES cells but are prone to tumorigenesis. Cell Res 21(11):1634–1637
10. Wernig M (2007) In vitro reprogramming of fibroblasts into a pluripotent ES-cell-like state. Nature 448(7151):318–324
11. Zhao XY (2009) iPS cells produce viable mice through tetraploid complementation. Nature 461(7260):86–90
12. Do JT, Scholer HR (2009) Regulatory circuits underlying pluripotency and reprogramming. Trends Pharmacol Sci 30(6):296–302
13. Pasque V, Plath K (2015) X chromosome reactivation in reprogramming and in development. Curr Opin Cell Biol 37:75–83
14. Pasque V (2014) X chromosome reactivation dynamics reveal stages of reprogramming to pluripotency. Cell 159(7):1681–1697
15. Payer B (2013) Tsix RNA and the germline factor, PRDM14, link X reactivation and stem cell reprogramming. Mol Cell 52(6):805–818

Fabrication of Injectable Hydrogel from Decellularized Adipose Tissue

My Nguyen, Truong Do, and Ha Tran

Abstract The decellularized extracellular matrix (ECM) has been established as an ideal biomaterial in tissue engineering applications with potential clinical translation. In the present study, we aimed to generate a hydrogel from decellularized adipose tissue. Adipose tissue samples were collected from the liposuction procedure and subjected to undergo a series of delipidation and decellularization. Oil Red O staining illustrated the complete removal of lipid, whereas the absence of cellular nuclei in histology staining and low DNA residual content indicated an acellular matrix. Decellularized adipose tissue was further fabricated into a hydrogel. Histological evaluation and scanning electron images demonstrated a fibrous ultrastructure of the hydrogel. Furthermore, short-term in vivo observation detected the gelation after 2 h being implanted subcutaneously, which demonstrated dAT hydrogel was capable of self-assembling under physiological condition. Taken together, these results indicated the potential approach of hydrogel from decellularized adipose tissue in tissue engineering.

Keywords Extracellular matrix · Decellularization · Hydrogel · Biomaterial · Tissue engineering

M. Nguyen (✉) · H. Tran
Laboratory of Tissue Engineering and Biomedical Materials, University of Science, Ho Chi Minh City, Vietnam
e-mail: ntnmy@hcmus.edu.vn

Department of Human Physiology and Animal Biotechnology, Faculty of Biology–Biotechnology, University of Science, Ho Chi Minh City, Vietnam

Vietnam National University, Ho Chi Minh City, Vietnam

T. Do
Xuan Truong Paradise Plastic Aesthetic, Ho Chi Minh City, Vietnam

V. Van Toi et al. (eds.), *8th International Conference on the Development of Biomedical Engineering in Vietnam*, IFMBE Proceedings 85,
https://doi.org/10.1007/978-3-030-75506-5_38

1 Introduction

The extracellular matrix (ECM) is composed of macromolecules, including polysaccharide glycosaminoglycans (GAGs), collagens, laminins, and fibronectin [1, 2]. Bioactive scaffolds derived from ECM have been demonstrated to provide a suitable microenvironment for cell proliferation, migration, and differentiation; therefore, it is considered an ideal choice for the repair or reconstruction of damaged organs and tissues. Extracted ECM can be made into tissue engineering scaffolds as sheet [3, 4], sponge [5, 6], particles [7, 8] or hydrogel [9, 10]. Among the multiple these scaffold types, hydrogels have been shown for their various desired biomedical applications. The ECM hydrogel was fabricated from an enzymatic solubilization and neutralization procedure [10–12]. ECM hydrogel is characterized as a pre-gel viscous fluid that can self-polymerize into a hydrogel conforming to physiologic temperature. This special characteristic of ECM hydrogel is conveniently tunable to suit for in cell encapsulation and minimally invasive injection with greater ease and for the reconstruction of tissue defects.

Adipose tissue is a highly specialized connective tissue composed of a variety of ECM components, including collagen, elastin, glycosaminoglycan (GAG), and bioactive cytokines, which was shown for their benefits in tissue engineering [1, 13, 14]. Recently, ECM isolated from adipose tissue has been examined for biomaterial fabrication and potential as a stem cell-supporting scaffold in regenerative therapies [6, 7, 15, 16]. Adipose tissues in the human body can be obtained as a medical waste product of liposuction surgery. In this study, adipose tissue was processed into an acellular adipose matrix for the preparation of hydrogel. The production of hydrogel from adipose tissue was divided into two main periods: delipidation and decellularization of adipose tissue, enzymatic digestion and neutralization. The hydrogel was evaluated the in vitro biosafety and the ability to form a hydrogel in vivo in order to explore whether it can be used as an injectable material for soft tissue regeneration.

2 Materials and Methods

2.1 Preparation of Decellularized Adipose Tissue (dAT)

Fresh human adipose tissue was obtained from healthy female patients who underwent an abdominal liposuction procedure at Xuan Truong Paradise Plastic Aesthetic (Ho Chi Minh City, Vietnam). All protocols using human samples were approved by the Ethical Review Board of Xuan Truong Paradise Plastic Aesthetic (No. 2604/QĐ-TMXT), and the sample was obtained with written informed consent.

Adipose tissue was undergone a mechanical homogenization (Bead Rupter, Omni, USA) to remove the major oil contents. Further delipidation was carried out by incubating in Isopropanol for 6 h. The tissue suspension was decellularized in 10 mM

Tris–HCl (Invitrogen, USA) for 24 h, followed by the incubation in 0.1% Sodium Dodecyl Sulfate (SDS, Sigma, USA) for 12 h [4]. Finally, treated samples were rinsed in 1× Phosphate Buffer Saline (PBS, Gibco, USA) for 24 h to remove excess detergents and frozen at 80 °C overnight for lyophilization.

2.2 *Preparation of Hydrogel from dAT*

dAT was further processed into powder by milling (Bead Rupter, Omni, USA) and sterilized by ethylene oxide. DAT powder was digested 1 mg/mL pepsin in 0.01 N HCl solution (Sigma, USA) at room temperature for 24 h [10]. One-ninth, the digest volume of 10× PBS was added, followed by the addition of one-tenth of the digest volume of 0.1 N NaOH (Merck, Germany) to accomplish the neutralization. The mixture was kept in a refrigerator at 4 °C overnight. The gelation of hydrogel was determined by the inversion test after incubating at 37 °C.

2.3 *Oil Red O Staining*

Oil Red O staining was used to evaluate the presence of lipid residue in treated adipose tissue. The samples were fixed in 10% Formalin (Merck, Germany) for 30 min. Before staining, the samples were quickly incubated with 60% Isopropanol for 5 min. Oil Red O solution (Sigma, USA) was used to stain the samples for 1 h.

The samples were washed in distilled water and observed using a phase-contrast microscopy microscope (CKX53, Olympus, Japan) equipped with a digital camera.

2.4 *Hematoxyline and Eosin (H&E) Staining*

For histological examinations, treated samples were fixed in 10% Formalin (Merck, Germany) overnight, paraffin embedded, and cut into 7 μm thick sections. Sample sections were stained with the Hematoxyline (Thermo Scientific, USA) and the counterstain Eosin (Thermo Scientific, USA). The samples were observed using a phase-contrast microscopy microscope (CKX53, Olympus, Japan) equipped with a digital camera.

2.5 *DNA Extraction and Quantification*

DNA extraction from freeze-dried native tissue and treated samples were conducted using GenElute™ Mammalian Genomic DNA Kits (Sigma, USA) in accordance with the manufacturer's instructions. The DNA contents were quantified by measuring absorbance at 260/280 nm in a NanoDrop™ spectrophotometer (Thermo Fisher Scientific, USA). All measurements were performed in triplicate; the amount of DNA was averaged and expressed as ng per mg of dry weight.

2.6 Scanning Electron Microscope (SEM) Analysis

Hydrogel samples were fixed overnight in 10% formalin and then dehydrated using sequential ethanol series (70, 80, 90, 96, and 100% (v/v), 10 min in each). The structure of hydrogel samples were determined by scanning electron microscope (SEM) (JSM-6510 JEOL, Japan).

2.7 In Vitro Cytotoxicity Testing

The available human Fibroblasts (hFs) at the fourth passage were obtained from the Laboratory of Tissue Engineering and Biomedical Materials and were used in this cytotoxicity testing. hFs were grown in complete medium (DMEM-F12, Sigma, USA) supplemented with 10% Fetal Bovine Serum (FBS, Sigma, USA), 1X antibiotics (Sigma, USA)). The cells were detached from culture flasks and seeded in a 96-well plate (10^4 cells/well), and cultured overnight.

Hydrogel (10 mg/mL) was prepared by incubating at 37 °C for 30 min. The hydrogel was immersed in a complete medium (surface area 1.25 cm^2/mL) for 24 h to prepare a liquid extract of hydrogel (according to ISO 10993 part 12). The cells were exposed to the liquid extract of hydrogel for the next 24 h. Complete medium and liquid extract from the latex membrane were used as the blank and positive control, respectively ($n = 3$).

The MTT assay was performed to measure the relative growth rate of hGFs. After incubation with liquid extracts, the cells were treated with 0.5 mg/mL MTT solution at 37 °C for 4 h. MTT solution was aspirated, and a mixture of DMSO/Ethanol (Sigma, USA) was added to dissolve the formazan crystals and measured at 570 nm (OD570) using a spectrophotometer (ELISA READ 400, Biochrom, USA).

Cell viability was represented as the relative growth rate (RGR), which was calculated based upon OD570 values as RGR (%) = (OD test group/OD blank group) × 100%. If the RGR value is lower than 70%, the tested sample has potential cytotoxicity (ISO 10993 part 5).

2.8 *Examination of In Vivo Gelation*

To investigate the in vivo gelation of dAT hydrogels, a pre-gel was prepared as the above description and contained in 1 cc syringe. 0.5 mL of DAT pre-gel was injected subcutaneously into the backs of female mice (Mus musculus var. albino, 6–8 weeks old, 25–30 g, n = 3) using a 25-gauge needle. After 2 and 24 h after subcutaneous injection, the mice were sacrificed. The skin tissues surrounding the injection site were surgically removed for histology examination.

3 Results

3.1 *Preparation of dAT*

Native adipose tissue (Fig. 1a) maintains a bright yellowish appearance due to carotenoids dissolved in the lipid droplets, which stained positively with Oil Red O (Fig. 1c). After being processed, treated adipose samples exhibited a whitish appearance (Fig. 1b). Complete lipid removal from the tissue was confirmed by the absence of lipid droplets after staining with Oil Red O (Fig. 1d). In comparison with native adipose tissue (Fig. 1e), the absence of intact nuclei in histological sections indicated the removal of cellular content (Fig. 1f). Additionally, DNA quantification was as 1054 ± 4.681 ng DNA/mg lyophilized native adipose samples and 27.67 ± 1.528 ng DNA/mg in postprocessed samples, which show a DNA clearance ratio of 97.4%. These results demonstrated that decellularized adipose tissue was successfully prepared.

3.2 *Macroscopic and Ultrastructural Evaluation of dAT Hydrogel*

dAT was prepared, milled, and sterilized for use in hydrogel preparation (Fig. 2a). dAT concentrations of 1, 2.5, 5, 8, and 10 mg/mL were investigated for hydrogel formation. The solubilization of dAT powder in the pepsin-HCl solution was detected and yielded a higher turbidity mixture when increasing dAT concentration. After incubation at 37 °C for 30 min, the vials were inverted to determine gelation (Fig. 2b). The inversion test revealed that the higher dAT concentration of 8–10 mg/mL could result in the stable hydrogel. 2.5–5 mg/mL dAT concentration was also able to form hydrogel with an obvious shape; however, it could not remain at the vial bottom after inversion, which indicated a softer structure. H&E staining (Fig. 2c) and Scanning electron microscopy (Fig. 2d, white arrows) of 10 mg/mL dAT hydrogel visualized an oriented fibrillar structure and inter-connective pores.

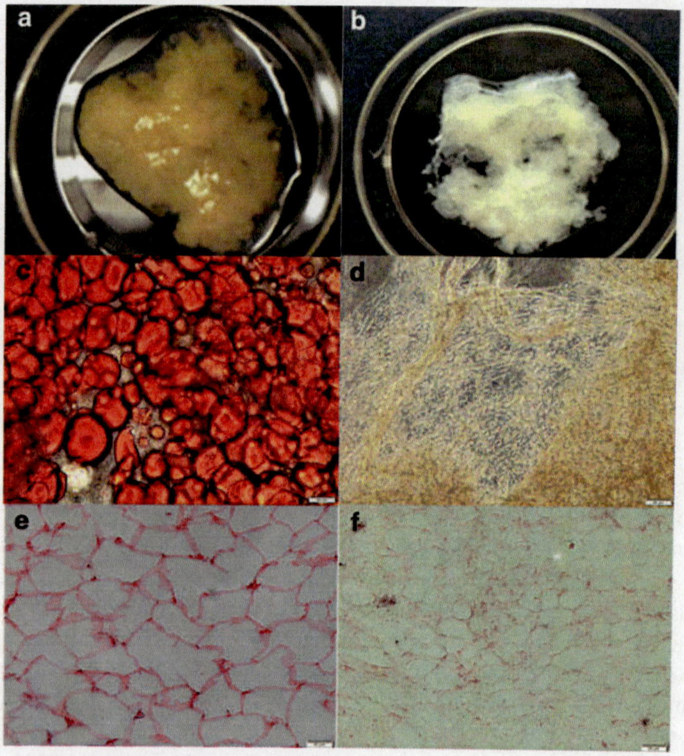

Fig. 1 Preparation of decellularized adipose matrix. **a** Native adipose tissue. **b** Treated adipose sample. **c** Oil red staining of native adipose tissue. **d** Oil red staining of treated sample. **e** H&E staining of native adipose tissue. **f** H&E staining of treated sample. Scale bar: 50 μm

3.3 In Vitro Cytotoxicity Testing

Liquid extract of dAT hydrogel did not present a cytotoxic effect on hFs viability indicated by normal cell morphology (Fig. 3c), which was found to be similar to hFs incubated in complete medium (Fig. 3a). In MTT assay, a dense formazan crystals formation was detected in both complete medium and dAT hydrogel extraction groups (Fig. 3d, f). Consequently, the RGR value of dAT hydrogel extraction was determined as 95.06%, which confirmed its potential biocompatibility. On the other hand, liquid extract from the latex membrane as positive control expectedly caused cytotoxic to cells, as shown in abnormal cell morphology and death. Thus, there was a lack of formazan crystals in the Latex extract group, resulting in the RGR value as 1.4% (Fig. 3b, e).

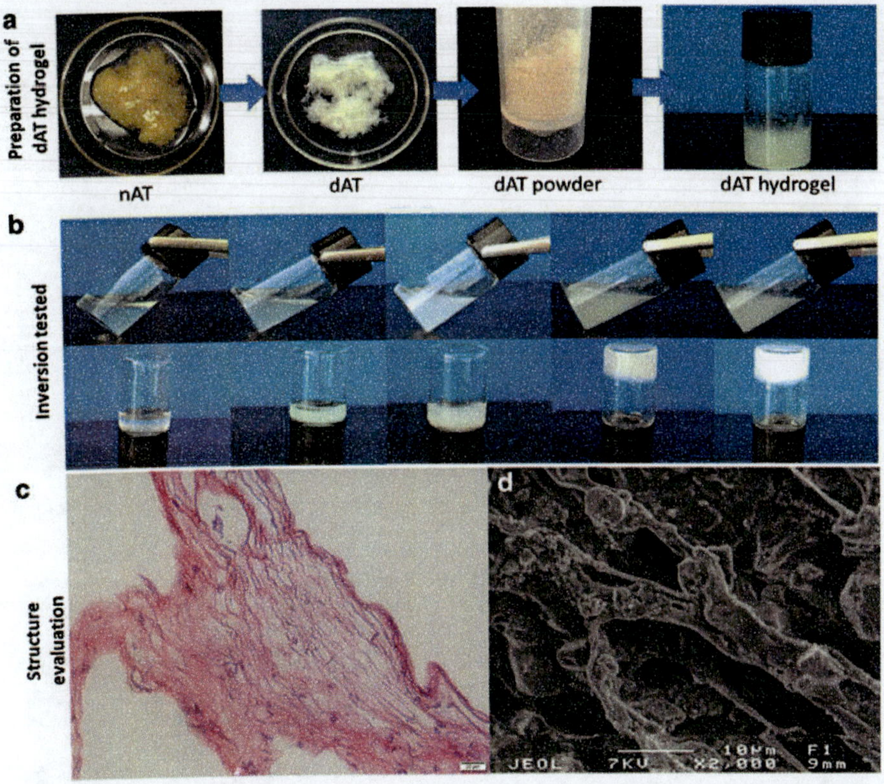

Fig. 2 Preparation and evaluation of hydrogel from decellularized adipose tissue. **a** Procedure for preparing dAT hydrogel. **b** Inversion test for gelation examination. **c** H&E staining of dAT hydrogel. **d** SEM evaluation of dAT hydrogel (white arrows indicate pores formed within the hydrogel sample). Scale bar: 20 μm

3.4 In Vivo Gelation

Pre-gel of 10 mg/mL dAT was able to pass through 25G needles. 2 and 24 h after injection, the implantation site was collected for H&E staining. At a 2-h time point, a fibrillar network was detected under the dermis (Fig. 4a, c). This structure appeared to be dispersed, which became more compact after 24 h (Fig. 4b, d). The infiltration of immune cells into the dAT hydrogel was also visualized, which is considered a normal host response early stage of grafting [17]. This histological evaluation showed that dAT hydrogel was capable of self-assembling under physiological condition. Therefore, the fabricated dAT exhibited the potential material to be injected and deformed to accommodate irregular defects.

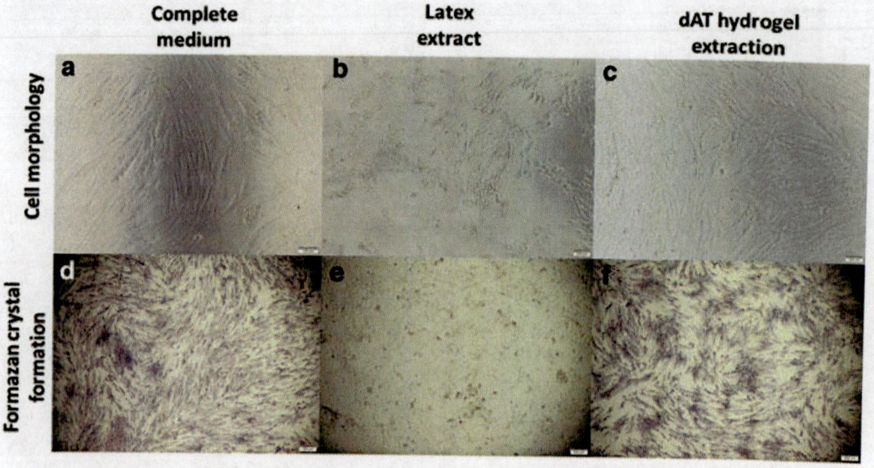

Fig. 3 Cytotoxicity testing of dAT hydrogel indirectly via liquid extract. **a–c** Cell morphology evaluation after 24 h incubation. **d–f** Formazan crystals formation in MTT assay. Scale bar: 200 μm

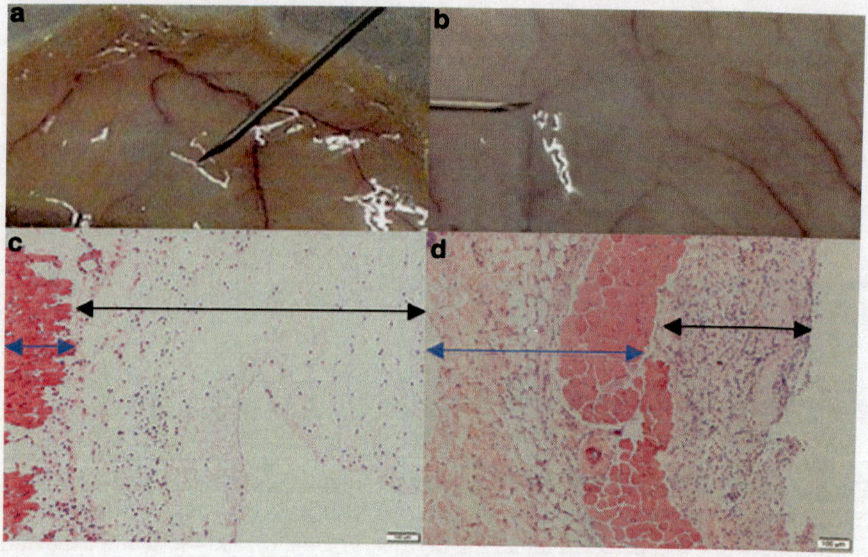

Fig. 4 In vivo formation of hydrogel and H&E staining for examining in vivo gelation. **a, c** 2 h post-injection. **b, d** 24 h post-injection. Blue arrows indicate skin layers. Black arrows indicate the implanted hydrogel. Scale bar: 100 μm

Acknowledgements This research is funded by Vietnam National University Ho Chi Minh City (VNU-HCM) under grant number C2019-18-20.

Conflicts of Interest The authors have no conflict of interest to declare.

References

1. Sano H et al (2014) Acellular adipose matrix as a natural scaffold for tissue engineering. J Plast Reconstr Aesthet Surg 67(1):99–106
2. Hinderer S, Layland S, Schenke-Layland K (2016) ECM and ECM-like materials—biomaterials for applications in regenerative medicine and cancer therapy. Adv Drug Deliv Rev 97:260–269
3. Nguyen MTN, Tran HLB (2018) Effect of modified bovine pericardium on human gingival fibroblasts in vitro. Cells Tissues Organs 206(6):296–307
4. Tran HLB (2016) Preparation and characterization of acellular porcine pericardium for cardiovascular surgery. Turk J Biol 40:1–8
5. Kokai LE et al (2020) Clinical evaluation of an off-the-shelf allogeneic adipose matrix for soft tissue reconstruction. Plast Reconstr Surg Glob Open 8(1)
6. Choi YC et al (2012) Decellularized extracellular matrix derived from porcine adipose tissue as a xenogeneic biomaterial for tissue engineering. Tissue Eng Part C Methods 18(11):866–876
7. Brown CF et al (2015) Effect of decellularized adipose tissue particle size and cell density on adipose-derived stem cell proliferation and adipogenic differentiation in composite methacrylated chondroitin sulphate hydrogels. Biomed Mater 10(4):045010
8. Choi JS et al (2009) Human extracellular matrix (ECM) powders for injectable cell delivery and adipose tissue engineering. J Control Release 139(1):2–7
9. Wu J et al (2015) An injectable extracellular matrix derived hydrogel for meniscus repair and regeneration. Acta Biomater 16:49–59
10. Wolf MT et al (2012) A hydrogel derived from decellularized dermal extracellular matrix. Biomaterials 33(29):7028–7038
11. Saldin LT et al (2017) Extracellular matrix hydrogels from decellularized tissues: structure and function. Acta Biomater 49:1–15
12. Duran P, Alperin M, Christman KL (2019) Decellularized extracellular matrix hydrogels: fabrication, properties, characterization, and current applications. In: Decellularized extracellular matrix, pp 116–138
13. Bora P, Majumdar AS (2017) Adipose tissue-derived stromal vascular fraction in regenerative medicine: a brief review on biology and translation. Stem Cell Res Ther 8(1):145
14. Song M, Liu Y, Hui L (2018) Preparation and characterization of acellular adipose tissue matrix using a combination of physical and chemical treatments. Mol Med Rep 17(1):138–146
15. Brown BN et al (2011) Comparison of three methods for the derivation of a biologic scaffold composed of adipose tissue extracellular matrix. Tissue Eng Part C Methods 17(4):411–421
16. Flynn LE (2010) The use of decellularized adipose tissue to provide an inductive microenvironment for the adipogenic differentiation of human adipose-derived stem cells. Biomaterials 31(17):4715–4724
17. Spahn JH, Li W, Kreisel D (2014) Innate immune cells in transplantation. Curr Opin Organ Transplant 19(1):14–19

Effect of the Grafting of Conjugated Linoleic Acid on Crystallinity and Thermal Properties of Electrospun Polymeric Membranes and Its Anticoagulation Mechanism

Nam Minh-Phuong Tran, An Nguyen-My Le, Nhi Thao-Ngoc Dang, Ha Thi Thanh Huong, Phuoc Long Truong, Phu Dai Huynh, and Thi-Hiep Nguyen

Abstract Grafting of conjugated linoleic acid (CLA) was an effective approach to improve surface hemocompatibility while preserving the initial mechanical properties and morphology. However, its effect on the crystallinity and thermal properties of synthetic polymeric materials were not investigated. Thus, this study aimed to evaluate and compare these properties of the non-grafted and the CLA-grafted polymeric membranes. The membranes were subjected to X-ray diffraction analysis, thermal gravimetric analysis, and differential scanning calorimetry. Moreover, the coagulation of blood plasma exposure to the membranes was evaluated to understand the antithrombotic mechanism of CLA on the membrane surface. The results showed that generally, the grafting caused no substantial impact on the properties being assessed, successfully preserved the initial state of the base materials. Moreover, the CLA exhibited slight obstruction on plasma coagulation cascade, specifically on the intrinsic pathway. Thus, CLA grafting could be a prospective approach to improve the hemocompatibility of a surface while preserving other properties.

Keywords Electrospinning · Thermal properties · Crystallinity · Conjugated linoleic acid · Anticoagulation mechanism

N. M.-P. Tran · A. N.-M. Le · N. T.-N. Dang · H. T. T. Huong · P. L. Truong · T.-H. Nguyen (✉)
Department of Tissue Engineering and Regenerative Medicine, School of Biomedical Engineering, International University, Ho Chi Minh City 70000, Vietnam
e-mail: nthiep@hcmiu.edu.vn

N. M.-P. Tran · A. N.-M. Le · N. T.-N. Dang · H. T. T. Huong · P. L. Truong · P. D. Huynh · T.-H. Nguyen
Vietnam National University, Ho Chi Minh City, Vietnam

P. D. Huynh
Faculty of Materials Technology, Ho Chi Minh City University of Technology, Ho Chi Minh City, Vietnam

V. Van Toi et al. (eds.), *8th International Conference on the Development of Biomedical Engineering in Vietnam*, IFMBE Proceedings 85,
https://doi.org/10.1007/978-3-030-75506-5_39

1 Introduction

Hemocompatibility is crucial for biomaterials in various biomedical applications, especially in blood-contacting devices such as vascular graft, stents, blood containers, etc. Unfortunately, the biomaterials—which satisfied these applications in terms of physicochemical properties and biocompatibility—commonly lack suitable blood compatibility. Thus, various methods have been investigated to improve this essential characteristic, most notably the loading or grafting of antithrombotic agents such as heparin, albumin, or nanomaterials [1]. Previously, our group has investigated the use of conjugated linoleic acid (CLA) for this purpose with the loading method [2]. CLA is an unsaturated fatty acid with antithrombotic properties [3], formerly applied to fabricate the hemodialysis membrane [4–6]. The loading of CLA improved the hemocompatibility of polymeric surface, mitigating platelet adhesion, and delaying blood coagulation. However, this method required an excessive amount of CLA and the intruding acid deteriorated the structural integrity of the membrane and its fibers.

Therefore, the following studies have focused on grafting CLA on the outer surface of the membranes to maximize its activities and eliminate undesirable effects [1, 7]. Compare to other common anticoagulation agents like heparin or albumin, the CLA molecule is significantly smaller and simpler. This characteristic, with low steric hindrance, allows CLA to be chemically grafted at denser concentrations. The results demonstrated effective grafting, which significantly improved the hemocompatibility of the underlying surface while preserving its initial characteristics like morphology and mechanical properties. However, the studies have not evaluated the impact of this treatment on the crystallinity and thermal behaviors of the main materials, influencing their performance. Moreover, CLA was proven to inhibit platelet function, but its anticoagulation mechanism on a surface was not fully understood.

Thus, in this study, we assessed the crystallinity and thermal properties of the electrospun membranes made from polyurethane (PU) and polycaprolactone (PCL) before and after they were grafted with CLA. PU and PCL, two FDA-approved polymers, were common electrospinning materials and potential candidates for biomedical applications [8, 9]. Moreover, human blood plasma was exposed to the membranes, and its coagulation times were evaluated to identify abnormal changes. Understanding the effects of this treatment would allow further studies into its application in blood-contacting materials and devices.

2 Materials and Methods

2.1 *Fabrication of CLA-Grafted PU/PCL Membrane*

2.1.1 Material

Poly[4,4′-methylenebis(phenyl isocyanate)-alt-1,4-butanediol /di(propylene glycol)/polycaprolactone] (PU), and polycaprolactone (PCL) were obtained from Sigma-Aldrich. Conjugated linoleic acid (CLA), tetrahydrofuran (THF), N,N-dimethyl formamide (DMF), glutaraldehyde and N-(3-Dimethylaminopropyl)-N′-ethyl carbodiimide (EDC) were also purchased from Sigma-Aldrich.

2.1.2 Electrospinning Procedure

PU/PCL pellets (1:1 w/w) was added into mixed DMF:THF (1:1 v/v) and stirred at room temperature overnight to create a 12% w/w polymer solution. The transparent solution was then contained in a glass syringe and pumped at the flow rate of 0.5 ml/h using a peristaltic pump (Harvard Apparatus, Infusion Syringe Pump 980638). A 20 kV voltage was supplied using a high voltage power supply designed by the School of Biomedical Engineering. A stationary disk was set 12 cm away from the needle tip to collect the fibers.

2.1.3 CLA Grafting Procedure

PU/PCL membranes were first treated with plasma under low pressure. Each 1 × 1 cm piece of the membrane was submerged in 500 μL of 0.4 mM CLA solution (dissolved in ethanol) and different concentrations of EDC (4, 8, and 12 mM, named C2-E4, C2-E8, and C2-E12 respectively) for 24 h at 4 °C. After this period, the samples were rinsed two times with absolute ethanol and thrice with distilled water. Finally, the membranes were dried in a vacuum. A sample cover with CLA but not treated with EDC was fabricated as a comparison, named C2-E0.

2.2 *Characterization of Crystallinity and Thermal Properties*

2.2.1 X-ray Diffraction Analysis (XRD)

XRD spectra of the membranes were obtained using the X-ray diffractometer (CuKα, D8 Advance, Bruker). The diffraction angle was varied from 5° to 90° (0.02°/s/step).

2.2.2 Thermalgravimetric Analysis (TGA) and Differential Scanning Calorimetry (DSC)

Thermogravimetric properties of the samples were analyzed by using TA Instrument—Series Q500. Membranes weighted from 20 to 30 mg were evaluated with the temperature heated from 30 to 600 °C at a rate of 10 °C min^{-1}. Nitrogen flow of 60 ml/minute was used in the procedure.

The membrane's thermal behaviors were studied using a Labsys EVO (Setaram, France) thermal analyzer. The DSC scan was made from 30 to 600 °C at a heating rate of 10 °C min^{-1} in a nitrogen atmosphere. The mass of each sample ranged from 10 to 15 mg.

2.3 Investigation of Blood-Material Interaction

The study was conducted according to the ethical regulations of the Institutional Review Board of School of Biomedical Engineering. Fresh human whole blood drew from 3 consented volunteers was mixed with 3.2% w/v sodium citrate solution (blood:sodium citrate v/v ratio = 9:1) in special blood tubes to stop blood coagulation. The experiment was conducted at most 2 h after blood collection to avoid inactivation.

2.4 Determination of Clotting Times of Blood Plasma

Alteration in clotting times of blood plasma—including activated partial thromboplastin time (aPTT), prothrombin time (PT), and fibrinogen time (FT)—after its exposure to the samples of this study were investigated as described elsewhere [10, 11]. Briefly, platelet-poor plasma (PPP) was prepared by centrifuging whole human blood at 3500 rpm for 10 min. Then, a 1 cm × 1 cm membrane was immersed and incubated in 0.5 mL of PPP for 1 h at 37 °C. After the incubation time, the membrane was removed, and the coagulation parameters of the exposed PPP were evaluated with an automated coagulation analyzer (CA-540, Sysmex, Japan). Unexposed PPP was used as control. The experiment was repeated three times. Data were reported as mean ± standard deviation.

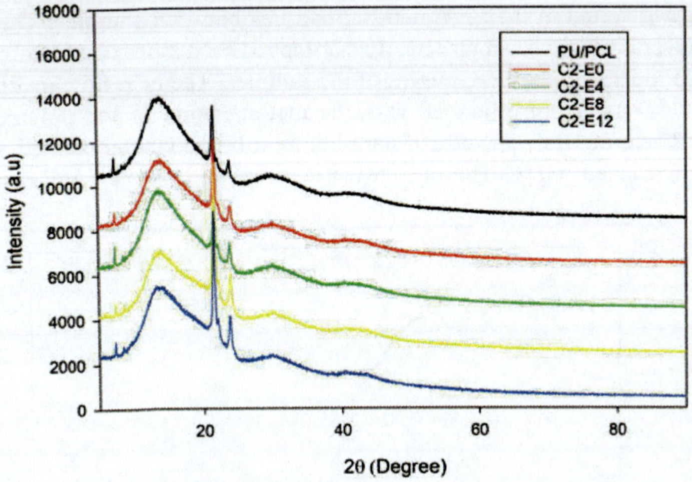

Fig. 1 XRD spectra of PU/PCL and treated samples. The spectra were translated vertically with a fixed y value for better comparison

3 Results

3.1 *XRD Spectra*

XRD spectra of the nid PU/PCL and its treated samples were analyzed to identify the undesirable effect of the process on their structure and crystallinity. As shown in Fig. 1, the spectra were highly similar, with no spectrum showed an outstanding peak. The spectra all exhibited two diffraction peaks at the 2θ angle of 21.71° and 23.91°, and a broad diffraction peak is ranging from 12° to 17°. Although there was no new signal, the intensities ratio between the broad peak and the other two peaks varied notably among the samples.

3.2 *Thermal Behavior of the PU/PCL Membrane and CLA-Grafted Membranes*

The thermal behavior of the samples was investigated using TGA and DSC. In TGA, the change in weight of each sample was monitored as the temperature gradually rose. The weight of all samples remained constant from the beginning to around 350 °C then dropped precipitously, indicated the start of thermal degradation. The drop slowed at 450 °C and decreased gradually until complete degradation at 600 °C. The degradation at these temperatures was typical for PU and PCL, and no significant difference was found in the gravimetric properties between samples. The weight-temperature curves of the samples nearly overlapped each other completely (Fig. 2).

Figure 3 illustrated the DSC curves of the samples. The heat flow in all samples dipped at 75–80 °C, corresponded with the melting point of the polymeric base materials. Then, the flow stayed constant before a broad change started when the temperature reached 300 °C. This change in heat flow was by the TGA data, indicated

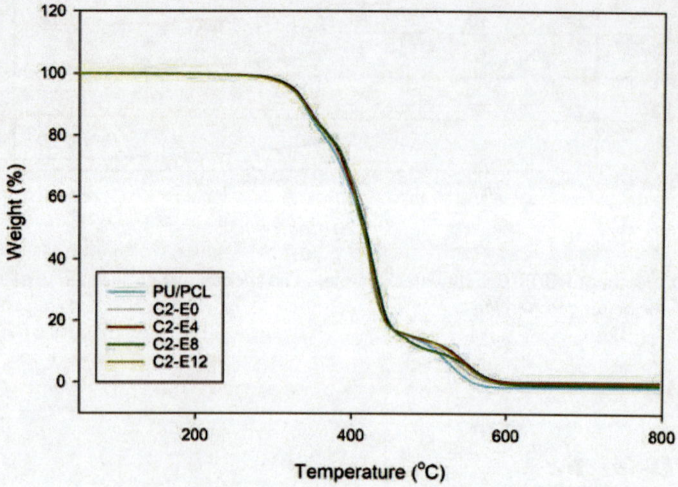

Fig. 2 Weight-temperature curves of the membranes

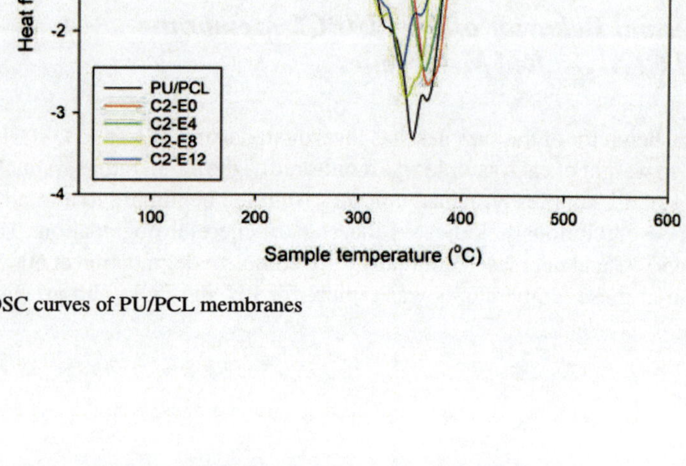

Fig. 3 DSC curves of PU/PCL membranes

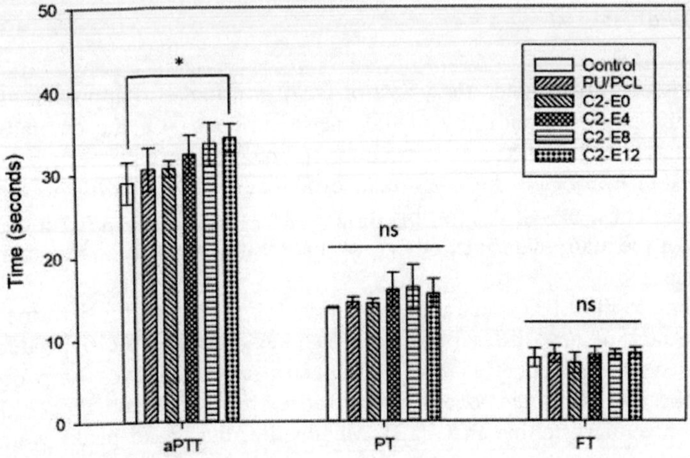

Fig. 4 aPTT, PT, and FT of PPP after exposure to the membranes. N = 3; error bars indicated standard deviations. ns: $p > 0.05$; *: $p < 0.05$

the degradation of the polymers. However, in this degradation range, the heat flow of each sample varied strongly, but no definitive trend or difference was noted.

3.3 Alteration in Clotting Times of Blood Plasma After Exposure to the Membranes

The clotting times of normal PPP and PPP exposed to the membranes were illustrated in Fig. 4. Overall, there was no substantial difference in the PT and FT of the samples, but results suggested that the presence of CLA prolonged aPTT values.

Particularly, the aPTT test showed prolongation in clot formation time according to the increase of CLA density of the membranes. The plain PU/PCL membrane yielded aPTT of approximately 29.5 s, similar to the value of control PPP. With the highest density of CLA, C2-E12 induced an aPTT value of 35.4 s, whereas C2-E4 and C2-E8 had aPTT of 32.4 and 33.6, respectively. Even though the mean value showed a definitive rising trend, the samples exhibited considerable deviation, which made the difference between samples not statistically significant ($p > 0.05$).

Similar elongation was not observed in the PT and FT test. The PU/PCL and C2-E0—with its negligible quantity of CLA—had their PT approximated the control value of 13.8 s. However, the PT of the CLA-grafted samples were from 15.5 to 16.2 s. However, the rise in PT was not following the CLA density of each sample, as C2-E12 showed shorter PT compared to which of C2-E8. Moreover, large deviations were also observed in the PT values of the CLA-grafted samples, rendered their difference statistically insignificant ($p > 0.05$). FT of the samples was also comparable to which of the control, indicated no substantial effect of CLA on this parameter.

4 Discussion

In this study, we investigated the effect of CLA grafting—an approach effectively improved the hemocompatibility of polymeric surfaces—on the crystallinity and thermal behaviors of PU/PCL electrospun membranes. These properties could have a considerable impact on the performance of materials, including in biomedical applications. Thus, the CLA grafting should not cause notable alteration to these properties of the main materials; otherwise, the treatment could induce undesirable compromise.

According to the XRD spectra, the treatment left the crystallinity of the surfaces intact. The lack of new diffraction peaks also suggests that the byproduct of the crosslinker was removed efficiently. All spectra presented two sharp diffraction peaks, which we attributed to the diffraction of the (110) and the (200) lattice plane of the semicrystalline PCL [12]. Meanwhile, the broad peaks were typical of polyurethane. There was a notable difference in the intensity ratio between these peaks among the samples. However, since they were attributed to different materials, these differences were likely due to the uneven distribution of PU and PCL on specific locations of the membrane, rather than the effect of CLA treatment.

TGA and DSC experiments also drew a similar conclusion. The minor density of grafted CLA and its benign treatment preserved the initial state of the polymeric biomaterials. The heat flow in DSC varied strongly between the samples when the temperature reached the degradation threshold. However, in biomedical applications, biomaterials would not be subjected to such extreme heat so that the alteration could be inconsequential.

As reported by previous studies [1, 7], CLA effectively hampered the activity of platelets, delaying the blood coagulation. However, besides the activity and aggregation of platelets (also known as primary hemostasis), the production of fibrin by a cascade of reactions (secondary hemostasis) is the essence of the hemostatic process. This cascade had two triggering pathways: the intrinsic pathway and the extrinsic pathway. The extrinsic pathway occurs when there is damage on tissue or the endothelium, which exposes the bloodstream with tissue factor, TF. On the other hand, trauma inside the vascular system initiates the intrinsic pathway, involving factors VIII, IX, XI, and XII. Both pathways end at the activation of factor X, which then activates prothrombin into thrombin in the common pathway [13]. Since the grafted CLA has demonstrated antithrombotic and anti-adhesion on whole blood and platelets, we investigated its effect on the clotting times of blood plasma to determine whether it interferes with the coagulation cascades. The aPTT is an indicator of the intrinsic pathway; the PT identifies abnormalities in the extrinsic pathway, whereas the FT detects changes in the common pathway. A prolonged clotting time indicates possible interference in the corresponding clotting pathway.

As the assays demonstrated, there were prolonged aPTT in CLA-grafted samples, but the PT and FT values showed no distinct alteration. C2-E8 yields an aPTT elongation of 8 s compared to the control values. Thus, CLA might also obstruct the coagulation cascade on intrinsic pathways besides its inhibitory effect on platelets. Since platelets were also secretors of factors that participate in the coagulation cascade (such as factor XI and factor V) [14], the inactivation of platelets certainly impeded the cascade.

However, the magnitude of these prolongations was minuscule by comparison with other studies. For instance, Kung et al. demonstrated that CLA-grafted polyacrylonitrile membranes extended aPTT value by 140 s, whereas a CLA-grafted polysulfone membrane induced and extension of 20 s in PT [4, 6]. Since the PU/PCL-CLA samples have demonstrated antithrombotic activity against whole blood, PRP and contained similar CLA content to these studies, these differences in plasma clotting times could stem from the difference in release kinetic of grafted CLA. Based on their reported aPTT–CLA concentration calibration curve, it can be inferred that the CLA-grafted polysulfone and polyacrylonitrile membranes had released their total CLA content into the exposed plasma in the incubation period of 30 min [4, 6]. Thus, the small increase in plasma clotting times by the PU/PCL membranes might occur due to the slower release kinetics of CLA. To confirm this assumption, further studies on its release in a different environment and various period are needed.

5 Conclusions

This study has provided evidence suggested that the grafting of CLA using carbodiimide crosslinker did not notably impact the crystallinity and thermal behavior of polymeric biomaterials, namely polyurethane and polycaprolactone. Moreover, the results indicated that CLA grafted on polymeric membranes could improve its hemocompatibility by inhibiting the intrinsic pathway of the coagulation cascade. Combined with published works, CLA grafting could be a prospective antithrombotic approach, specifically for polymeric surfaces with suitable chemistry.

Acknowledgements This research was supported by the National Foundation for Science and Technology Development (NAFOSTED, Vietnam) under grant number 108.06-2018.18.

Conflicts of Interest The authors declare that they have no conflict of interest.

References

1. Tran NM-p et al (2020) Conjugated linoleic acid grafting improved hemocompatibility of the polycaprolactone electrospun membrane. Int J Polym Sci 2020
2. Minh HH et al (2017) Fabrication of polycaprolactone/polyurethane loading conjugated linoleic acid and its antiplatelet adhesion. Int J Biomater 2017
3. Yu L, Adams D, Gabel M (2002) Conjugated linoleic acid isomers differ in their free radical scavenging properties. J Agric Food Chem 50(14):4135–4140
4. Kung FC, Yang MC (2006) Effect of conjugated linoleic acid grafting on the hemocompatibility of polyacrylonitrile membrane. Polym Adv Technol 17(6):419–425

5. Kung F-C, Yang M-C (2006) effect of conjugated linoleic acid immobilization on the hemocompatibility of cellulose acetate membrane. Colloids Surf B 47(1):36–42
6. Kung F-C, Yang M-C (2006) The effect of covalently bonded conjugated linoleic acid on the reduction of oxidative stress and blood coagulation for polysulfone hemodialyzer membrane. Int J Biol Macromol 38(3–5):157–164
7. Tran N et al (2020) Polyurethane/polycaprolactone membrane grafted with conjugated linoleic acid for artificial vascular graft application. Sci Technol Adv Mater 2020 (accepted)
8. Kim J, Kim S (2002) PEO-grafting on PU/PS IPNs for enhanced blood compatibility—effect of pendant length and grafting density. Biomaterials 23(9):2015–2025
9. Williamson MR, Black R, Kielty C (2006) PCL–PU composite vascular scaffold production for vascular tissue engineering: attachment, proliferation and bioactivity of human vascular endothelial cells. Biomaterials 27(19):3608–3616
10. Liu Y, Yang Y, Wu F (2010) Effects of L-arginine immobilization on the anticoagulant activity and hemolytic property of polyethylene terephthalate films. Appl Surf Sci 256(12):3977–3981
11. Amarnath LP, Srinivas A, Ramamurthi A (2006) In vitro hemocompatibility testing of UV-modified hyaluronan hydrogels. Biomaterials 27(8):1416–1424
12. Mi H-Y et al (2014) Poly (ε-caprolactone)(PCL)/cellulose nano-crystal (CNC) nanocomposites and foams. Cellulose 21(4):2727–2741
13. Antovic JP, Blombäck M (2013) Essential guide to blood coagulation. Wiley, Hoboken, NJ
14. Golebiewska EM, Poole AW (2014) Secrets of platelet exocytosis—what do we really know about platelet secretion mechanisms? Br J Haematol 165(2):204–216

Molecular and Cellular Biology in Medicine

Optimization of High Resolution Melting (HRM) Method for Genotyping and Estimating the Frequency of SNP rs751402 on *ERCC5* Gene in Vietnamese Breast Cancer Patients

Tran Thai Minh Nguyen, Luan Huu Huynh, Thanh Thi Ngoc Nguyen, Hoang Ngo Phan, and Hue Thi Nguyen

Abstract Breast cancer (BC) is the leading cancer-causing death for women worldwide. Among BC-causing genes, *ERCC5* plays an important role in DNA damage repair. The cancer risk can be modified by Single Nucleotide Polymorphisms (SNPs), and thus, SNPs may be considered as potential markers of carcinogenesis. Among SNPs of the *ERCC5* gene, rs751402 (A/G) had been found related to BC risk in the Chinese and Thailand populations. This study aims to develop the genotyping condition for rs751402 and then to predict the relation to BC risk in a Vietnamese population. 50 BC blood samples were examined. High Resolution Melting (HRM) genotyping method was optimized for rs751402. The allelic and genotypic frequencies of rs751402 in this study were compared with its data of Kinh Vietnamese in 1000Genomes as controls to predict the association with BC risk. The HRM genotyping assay was constructed with high sensitivity, specificity, and stability. The results revealed the minor allele frequency was 31.91%, indicating rs751402 is high polymorphism. The predicted association with BC showed rs751402 A allele tends to confer protection against BC (OR = 0.75) but not statistically significant (P-value = 0.28). Even so, rs75140 is still considered as a potential marker for further association studies with BC risk in the Vietnamese population due to its high polymorphism. 1000Genomes has certain reference value; however, some concern remains. Consequently, a further genotyping assay is proposed to conduct in a proper control group and larger sample size to confirm the relationship between rs751402 and BC risk in Vietnamese women.

Keywords Breast cancer · High resolution melting (HRM) · rs751402 · Single nucleotide polymorphisms (SNPs)

T. T. M. Nguyen · L. H. Huynh · T. T. N. Nguyen · H. N. Phan · H. T. Nguyen (✉)
Department of Physiology and Animal Biotechnology, Faculty of Biology and Biotechnology, University of Science, Ho Chi Minh City, Vietnam
e-mail: nthue@hcmus.edu.vn

Vietnam National University, Ho Chi Minh City, Vietnam

V. Van Toi et al. (eds.), *8th International Conference on the Development of Biomedical Engineering in Vietnam*, IFMBE Proceedings 85,
https://doi.org/10.1007/978-3-030-75506-5_40

1 Introduction

According to GLOBOCAN 2018, while the mortality to incidence ratio of breast cancer (BC) in the world was just close to 0.28, in Vietnam, this number showed a significant difference (around 0.40, greater than 0.12). This problem might be due to the late diagnosis [1]. Roughly 85% of Vietnamese patients were diagnosed with BC in the late stages, making the treatment less effective and thus increasing mortality [2]. BC is a disease characterized by genetic variation in different genes [3]. Therefore, it is hoped that the discovery of new biomarkers contributing to the BC development will enhance early diagnosis BC in the Vietnamese population.

DNA damage repair systems help repair endogenous and exogenous DNA damage [4]. Therefore, they are essential to maintain the genomic DNA integrity and protect from cancer forming mutations [5]. Their deficiency leads to genomic instability, DNA damage, increased mutagenesis, accumulation of potentially disadvantageous mutations, and increased cancer susceptibility, including BC [6–8]. Among DNA repair pathways, the NER pathway plays pivotal roles in repairing DNA damage responsible for DNA adducts, from bulky helical distortions to subtle modifications on sugar residues [9]. Excision repair cross-complementary group 5 (*ERCC5*) gene, also known as Xeroderma Pigmentosum complementation group G (*XPG*) [10], plays a key role in nucleotide excision repair (NER) pathway by making the 3'-incision in DNA excision repair following UV-induced damage [11]. Single Nucleotide Polymorphisms (SNPs) in *ERCC5* may influence the function of DNA repair and, in turn, affect the BC risk [12, 13].

Among SNPs of *ERCC5*, SNP rs751402 (A/G) was reported to decrease BC risk in the Han population of northwest China (AA vs. GG, OR = 0.052, 95% CI = 0.006–0.411, $P = 0.005$; AG vs. GG, OR = 0.145, 95% CI = 0.067–0.315, $P < 0.001$) [14]. On the other hand, in the Thailand population, this polymorphism was showed to increase BC risk (AG vs. GG, OR = 2.30, 95% CI = 1.22–4.35, $P = 0.01$) [15]. However, there have not yet been any reports on its role in relation to BC risk in the Vietnamese population. Therefore, in this study, we aimed to develop the genotyping condition for rs751402 and then to predict the association between this SNP and BC risk in the Vietnamese population.

High Resolution Melting (HRM) method is a close-tube, rapid, highly sensitive, reliable and cost-effective molecular method used for distinguishing specimens on the basis of SNP [16]. Hence, in this study, the HRM method was optimized for genotyping rs751402.

2 Materials and Methods

2.1 Materials

A total of 50 blood samples were collected from BC patients who were confirmed to have a tumor in their breast at Oncology Hospital Ho Chi Minh city. The consent forms of the sample donor were under the approval decision of the Medical Council of Oncology Hospital Ho Chi Minh city (No. 177/ĐĐĐ—CĐT November 18, 2014). Then, genomic DNA from these blood samples was extracted by the Salting–out method, followed by the protocol of Hue et al. [17] with some modifications. To determine the concentration and purity of DNA extracted samples, spectrophotometry using the NanoDrop 1000 Spectrophotometer (Thermo Fisher Scientific, USA) was performed. Finally, DNA samples having high purity (A 260/A 280 = 1.7–2.0) and high concentration ($\geq$ 30 ng/μl) were used for this study.

2.2 Methods

2.2.1 HRM Method Development

Primer design. The DNA sequence region of SNP rs751402 was identified using the SNP database on NCBI. Primers for real-time PCR HRM were designed and tested its specificity by NCBI Primer-BLAST. Oligo Analyzer was used to check the secondary structure of primers. uMelt HETS was used to predict HRM curves from the designed primer-based PCR products at different concentrations of $MgCl_2$ (1.5–3.5 mM), and the best $MgCl_2$ concentration would be chosen for the initial optimization HRM method.

HRM optimization. Annealing temperature (T_a) was optimized by the temperature gradient method. Components for one PCR reaction included: 1X Toptaq Mastermix, 0.2 μM forward primer, 0.2 μM reverse primer, 20 ng DNA and molecular H_2O. The result was analyzed by electrophoresis. The optimal gel electrophoresis procedure was performed on 2% agarose gel, at 90 V in 30 min, and 50 bp DNA Ladder (ThermoFisher) was used as a reference.

With the optimal T_a, the $MgCl_2$ concentration selected from uMelt HETS, primers and DNA concentration according to previous research of our group [18], HRM analysis was performed for several randomized samples by LightCycler® 96 Instrument to identify three control samples. HRM components in one reaction consisted of 1X LightCycler® 480 High Resolution Melting Master, $MgCl_2$ concentration selected from the uMelt HET software, 0.2 μM forward primer, 0.2 μM reverse primer, 20 ng DNA and molecular H_2O. The genotypes of samples selected for control samples were confirmed by Sanger automatic sequencing using The BigDye Terminator v1.1 Cycle Sequencing Kit, Applied Biosystems, USA.

After identifying the control samples, the post-optimization of $MgCl_2$ concentration might be conducted in a range of 1.5–3.5 mM to achieve the best discrimination among identified controls.

Genotyping. The optimal HRM's condition for rs751402 was applied to genotype 50 BC cases. Three positive controls and one negative control (molecular H_2O) were included in each running time. The results were analyzed using Light Cycler® 96 Application Software Version 1.1.

Evaluation of the optimization HRM method. Validation HRM method based on three criteria including the sensitivity (the ratio between successful genotyping samples and a total sample), the stability (The T_m stability among samples and distinct runs, using T-test to analyze) and the specificity (the ability to distinguish three genotypes through their T_m characteristics, using ANOVA test to analyze).

Association prediction. The allelic and genotypic frequencies of 50 BC cases were compared with data of rs751402 in Kinh Vietnamese in 1000Genomes as controls to predict the association with BC risk. Rstudio software was applied to analyze using the 'SNPassoc' package. The ORs and 95% CIs were calculated to assess BC risk. The threshold to determine a statistically significant association was set at $P = 0.05$.

3 Result

The most specific primer pair for the HRM method (Table 1) was selected. In addition, uMelt HETS prediction showed that the $MgCl_2$ concentration at 3.0 mM had the most distinct T_m of two types of homozygotes ($\Delta T_m = 0.6$ °C) and the clear shape of the melting curves of the given primer pair (Fig. 1). Therefore, 3.0 mM $MgCl_2$ was chosen for the initial optimization HRM method.

To optimal T_a, PCR reactions with gradient temperature ranged from 58 to 68 °C were run, and the result was analyzed using agarose gel electrophoresis (Fig. 2). The result showed that the band at 62 °C is brightest indicating that the PCR in this temperature synthesizes the most specific products. Therefore, 62 °C was selected as T_a for the HRM method.

With the optimal T_a of 62 °C and the $MgCl_2$ concentration at 3.0 mM, HRM analysis was performed for several randomized samples. The result showed that melting curves of randomized samples clustered into three groups (Fig. 3), with the distinct T_m of AA and GG homozygotes ($\Delta T_m = 0.6$ °C) being similar to uMelt HETS prediction. One sample of each group was selected for the control sample and confirmed by sequencing. The sequencing results were consistent with our prediction.

Table 1 The designed primer pair for HRM

Sequence (5′→3′)	T_m (°C)	Product size
F-GCCCTGCGCGTCGTATTA	60.28	71 bp
R-CTTCGCTGGTGGGTCCG	60.09	

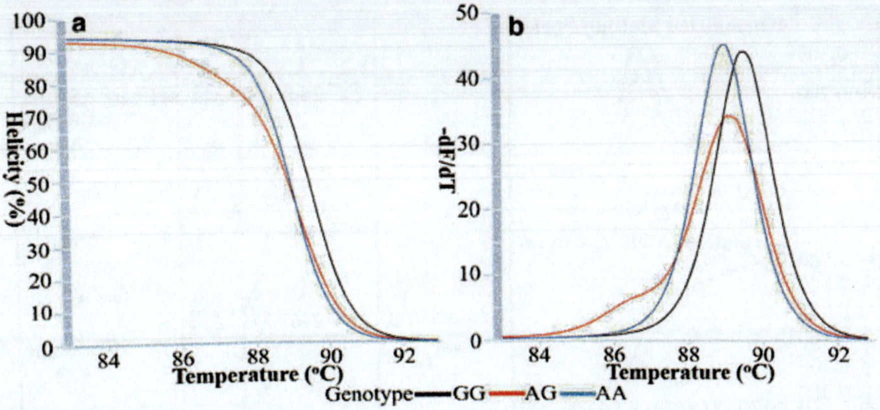

Fig. 1 The predicted melting curves (**a**) and melting peaks (**b**) of the given primer pair at $[Mg^{2+}] = 3.0$ mM

Fig. 2 Electrophoresis of PCR products with different T_a

Because of the clearly distinct melting curves of 2 homozygotes, there is no need for the post-optimization of MgCl2 concentration, and so the HRM condition for genotyping rs751402 was completely optimized (Table 2). This condition, then, was used for genotyping rs751402 in a total of 50 BC samples (case group). The result (Fig. 4) showed that while the melting curves of 47 samples were fit perfectly into the melting curves of three control groups and the Ct value of these samples came in the range of 25–30, there were still three samples that had unusual melting curves or did not express any amplification signal. Therefore, only 47 out of 50 samples were successfully identified genotype and able to produce data for the case group to predict the association (Table 3).

After genotyping, HRM method validation based on three criteria was performed. For the sensitivity, with 3/50 samples were unable to determine genotype, the sensitivity of the optimization HRM method was 94%. For the stability, the change in Tm of three genotypes among HRM distinct runs was not statistically significant (P-value > 0.05) in both control and case samples (Table 4). Therefore, this optimization HRM method is highly stable. For the specificity, the Tm characteristic of three genotypes

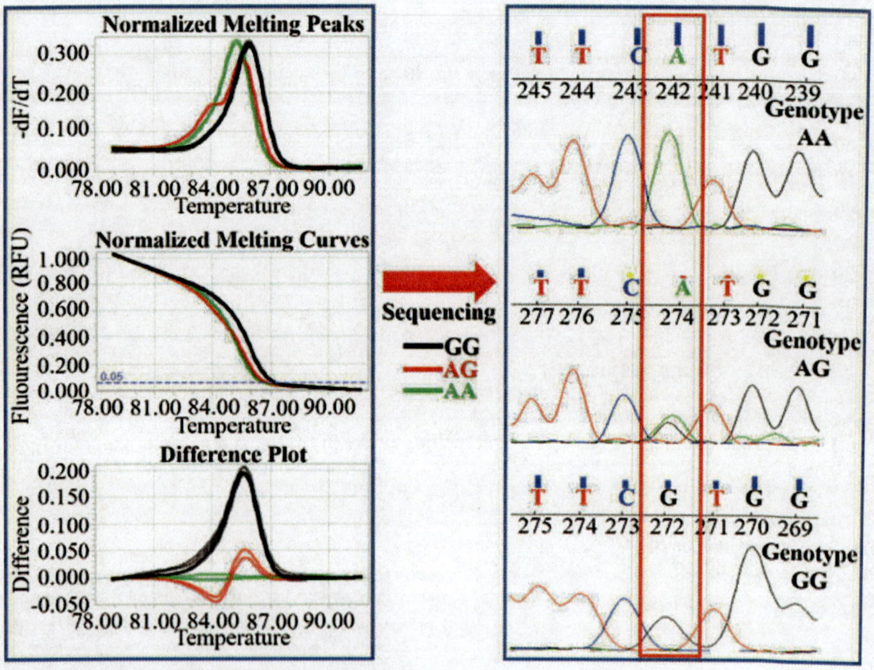

Fig. 3 Identification of three controls by sequencing

Table 2 HRM optimal protocol

	Final concentration	Stage	Temperature (°C)	Duration
Master mix	1X	Pre-incubation	95	5 min
$MgCl_2$	3.0 mM	Amplification (40 cycles)	95	30 s
Forward primer	0.2 μM		62	30 s
			72	30 s
			95	90 s
Reversed primer	0.2 μM	HRM	40	60 s
			65	30 s
			95	1 s
DNA	20 ng	Cooling	37	30 s

was statistically significant (P-value < 0.05) in both two groups (Table 5); thus, this optimization HRM method is highly specific.

For association prediction, the result (Table 3) showed that the frequency differences of alleles or genotypes between the case and control groups were not statistically significant (P-value > 0.05). This suggests that the polymorphism of rs751402

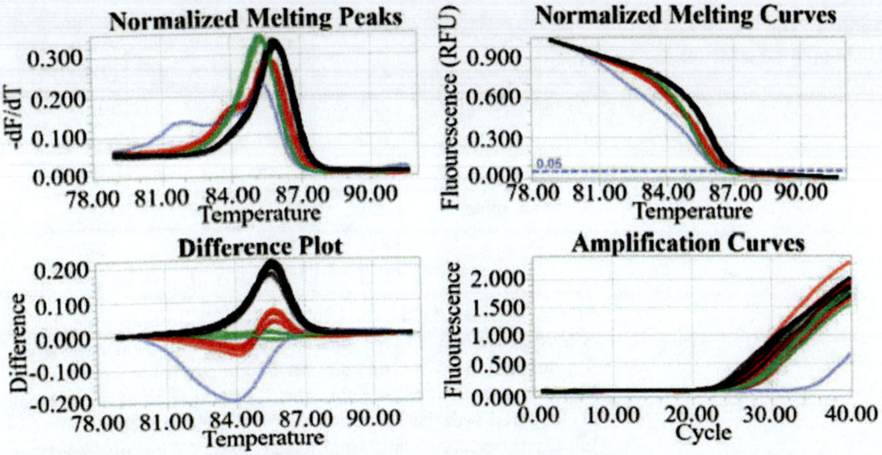

Fig. 4 The melting curves analysis of rs751402 by HRM. The green, red, black and blue curves presented AA, AG, GG genotype and the failed sample, respectively

Table 3 Genotypes and alleles frequency for rs751402

Group	Genotype			Allele	
	AA	AG	GG	A	G
Case	5 (10.64%)	20 (42.56%)	22 (46.80%)	30 (31.91%)	64 (68.09%)
Control (1000Genomes)	12 (11.88%)	53 (52.48%)	36 (35.64%)	77 (38.12%)	125 (61.88%)
P-value (Chi-squared)[a]	0.43			0.28 OR [95% CI] = 0.75 [0.43–1.28]	

[a]Chi-squared test was conducted to predict the association between SNP and BC. The difference between the frequency of alleles and genotypes in BC cases and "online" controls is tested. Based on the P-value of the Chi-squared test, the predicted association between a certain allele or genotype with the disease is indicated

is predicted not related to the BC risk in Vietnamese. However, it seems that the A allele has a protective effect on the disease (OR = 0.75).

4 Discussion

In this study, SNP rs751402 in the *ERCC5* gene was genotyped and predicted the relationship with BC risk. To our knowledge, this is the first study of this SNP with BC in the Vietnamese population. A genotyping method was successfully developed for this SNP with high sensitivity (94%), specificity, and stability. In order to identify the genotype of a sample, four criteria based on amplification curves, normalized melting

Table 4 The stability of the HRM optimal protocol

Genotype		$\overline{Tm} \pm SD$ (°C)	*P* value (T-test)[a]
Control samples	AA	84.98 ± 0.00	1.00
	AG	84.97 ± 0.29	0.99
	GG	85.53 ± 0.12	0.71
Case samples	AA	84.94 ± 0.19	0.69
	AG	85.03 ± 0.21	0.20
	GG	85.48 ± 0.14	0.76

[a]T-test was conducted to evaluate the stability of the HRM condition by analyzing the fluctuation of T_m within each genotype. T_m fluctuation of control samples is compared with the median T_m over 3 repetitions. Besides, the T_m fluctuation of all samples is compared with the T_m of the control sample in each run. In case this test is not statistically significant ($P > 0.05$), it is indicated that the samples fluctuation between runs and within each run is not significantly different. It means that the HRM method is stable

Table 5 The specificity of the HRM optimal protocol

Genotype		T_m range (°C)	*P* value (ANOVA)[a]
Control samples	AA	84.98–84.98	0.04
	AG	84.59–85.35	
	GG	85.38–85.61	
Case samples	AA	84.76–85.12	< 0.001
	AG	84.72–85.35	
	GG	85.14–85.67	

[a]ANOVA test was conducted to evaluate the ability to determine a specific genotype of the optimized HRM by analyzing the T_m difference between three genotypes. T_m of control samples in each genotype are compared together over 3 repetitions. Besides, T_m of all samples in each genotype are also compared together. In case this test is statistically significant ($P < 0.05$), it is indicated that the T_m between three genotypes are significantly different. It means that this HRM method can determine the genotype-specificity of all samples

curves, normalized melting peaks, and different plots were taken into consideration (Fig. 4): In amplification curves, the Ct value must be lower than 30; In normalized melting peaks, the different T_m of the two homozygotes must be higher than 0.5; In normalized melting curves, the heterozygote curve cuts the homozygote with the lower T_m in the middle and does not cut the other homozygote curve; In different plots, the lower-T_m homozygote is set as the baseline, and the heterozygote curve must divide into two sides of the baseline. The homozygote with the higher T_m curve must be on the upper side. Furthermore, the validation result (Tables 4 and 5) showed that the T_m of similar genotype samples were stable among HRM distinct runs and easily distinguished between genotypes, indicating that the developed HRM condition (Table 2) can be used for genotyping rs751402 on large sample size.

The genotype and allele frequency are summarized in Table 3. This SNP was a highly variable polymorphism, with the minor allele frequencies equal to 31.91% in BC cases and 38.12% in controls from the 1000Genomes database. The genotype frequencies of this polymorphism were all in agreement with the HWE in cases and controls with all P values being greater than 0.05, indicating that there was a normal distribution between genotypes inside this SNP in the tested population. This means the research sample represented the Vietnamese population, and the genotyping for this SNP was reliable.

As also shown in Table 3, using genotyping data from 1000Genome as control group, the Chi-squared test indicated that there was no significant difference between BC cases and controls for both allelic and genotypic frequency of rs751402 ($P > 0.05$). It is predicted that rs751402 is unrelated to BC risk in the Vietnamese population. However, the power of this association analysis is just 12.25% indicating that this prediction is unsure. This might be due to the small sample size. In studies on the Chinese and Thai populations, sample sizes were 202 samples and 277 samples, respectively. However, in this study, there were only 148 samples. Another reason is that the data of controls were taken from Kinh Vietnamese in 1000Genomes, which might not match the case group data in some criteria such as gender and age. Therefore, it is necessary to carry out further studies that conduct in a proper control group and larger sample size to find a more accurate association between rs751402 and the risk of BC in the Vietnamese population.

Further predicted association analysis of allelic models of rs751402 was carried out. The A allele of this SNP tends to confer protection against BC (OR = 0.75) (Table 3). This result is consistent with the other case–control study in Chinese women. The AA and AG genotype of this SNP were reported to decrease BC risk in the Han population of northwest China (AA vs. GG, OR = 0.052, 95% CI = 0.006–0.411, $P = 0.005$; AG vs. GG, OR = 0.145, 95% CI = 0.067–0.315, $P < 0.001$) [14]. On the other hand, in the Thailand population, AG genotype of this polymorphism was showed to increase BC risk comparing to GG genotype (AG vs. GG, OR = 2.30, 95% CI = 1.22–4.35, $P = 0.01$) [15]. This indicates that the association between this SNP and BC risk may be related to the ethnic specificity. Therefore, it is necessary to conduct studies to find an association between this SNP and BC risk among the Vietnamese people.

For the mechanism investigation, the *ERCC5* gene, also known as *XPG*, plays a key role in excision repair DNA damage induced by UV in the NER pathway [11]. ERCC5 protein is required for forming a stable complex with TFIHH possessed a structure-specific 3′ endonuclease, which then recruits XPF-ERCC1 to the incision complex produces the 5′-incision in the NER pathway [19, 20]. The disorder of *ERCC5* leads to impair transcription, decrease gene expression, decrease the function of certain NER-associated pathways, cause DNA repair defects, genomic instability, aggravate DNA damage [20–23], and so contributing to tumorigenesis, including breast carcinogenesis [24]. Furthermore, *ERCC5* amplification may limit the tumor mutational burden in metaplastic breast carcinoma [25]. Therefore, any variant present in *ERCC5* may affect the function of DNA repair and, in turn, influence BC risk [12, 13].

SNP rs751402 (A/G) locates in the *ERCC5* 5′-untranslated region (UTR), where variants might affect different splicing regulatory elements, leading to aberrant allele-specific splicing of ERCC5 pre-mRNA [26]. Also, this SNP is in the E2F1/YY1 binding, where it plays a key role in transcription. Evidently, rs751402 might influence ERCC5 expression [27] and, in turn, affect the BC risk. SNP rs751402 might play crucial roles in the development of BC and is expected as a potential biomarker for early diagnosis of breast cancer risk.

5 Conclusion

This study has been successful in developing the HRM method's condition for genotyping SNP rs751402 with high sensitivity, specificity, and stability. This condition can be used to genotyping rs751402 in a proper control group and larger sample size to confirm the association between this SNP and BC risk in the Vietnamese population. Despite the predicted association was not statistically significant, due to the high polymorphism and the role in the regulation of gene relating breast cancer, this SNP is still considered as a potential biomarker for association studies with BC risk in the Vietnamese population in the future.

Conflicts of Interest The authors have no conflict of interest to declare.

Acknowledgements This research is funded by Vietnam National University, Ho Chi Minh City (VNU-HCM) under grant number 562-2020-18-02. The authors would like to thank the Oncology Hospital HCMC for their contribution to collecting samples.

References

1. Lan NH, Laohasiriwong W, Stewart JF (2013) Survival probability and prognostic factors for breast cancer patients in Vietnam. Glob Health Action 6:1–9
2. Trieu PDY, Mello-Thoms C, Brennan PC (2015) Female breast cancer in Vietnam: a comparison across Asian specific regions. Cancer Biol Med 12(3):238–245
3. De Silva S, Tennekoon KH, Karunanayake EH (2019) Overview of the genetic basis toward early detection of breast cancer. Breast cancer (Dove Medical Press) 11:71–80
4. Chatterjee N, Walker GC (2017) Mechanisms of DNA damage, repair, and mutagenesis. Environ Mol Mutagen 58(5):235–263
5. Torgovnick A, Schumacher B (2015) DNA repair mechanisms in cancer development and therapy. Front Genet 6:157–157
6. Mohrenweiser HW et al (2002) Identification of 127 amino acid substitution variants in screening 37 DNA repair genes in humans. Cancer Epidemiol Biomark Prev 11(10):1054

7. Clancy S (2008) DNA damage & repair: mechanisms for maintaining DNA integrity. Nat Educ 1(1):103
8. Shen MR, Jones IM, Mohrenweiser H (1998) Nonconservative amino acid substitution variants exist at polymorphic frequency in DNA repair genes in healthy humans. Cancer Res 58(4):604–608
9. Wood RD (1999) DNA damage recognition during nucleotide excision repair in mammalian cells. Biochimie 81(1–2):39–44
10. O'Donovan A et al (1994) Isolation of active recombinant XPG protein, a human DNA repair endonuclease. J Biol Chem 269(23):15965–15968
11. Fagbemi AF, Orelli B, Scharer OD (2011) Regulation of endonuclease activity in human nucleotide excision repair. DNA Repair. 1568-7856 (Electronic)
12. Shakil Malik S et al (2019) Genetic polymorphism in ERCC5 and breast cancer risk. Mol Biol Res Commun 8(1):27–31
13. Malik SS et al (2018) An insight into clinical outcome of XPG polymorphisms in breast cancer. Mol Biol Rep. 1573-4978 (Electronic)
14. Wang H et al (2016) Association analysis of ERCC5 gene polymorphisms with risk of breast cancer in Han women of northwest China. Breast Cancer. 1880-4233 (Electronic)
15. Pongsavee M, Wisuwan K (2018) ERCC5 rs751402 polymorphism is the risk factor for sporadic breast cancer in Thailand. Int J Mol Epidemiol Genet. 1948-1756 (Print)
16. Jin Y et al (2014) High-resolution melting (HRM) analysis: a highly sensitive alternative for the identification of commercially important Crassostrea oysters. J Molluscan Stud 81(1):167–170
17. Hue NT et al Extraction of human genomic DNA from dried blood spots and hair roots. Int J Biosci Biochem Bioinform 2:21–26
18. Nhan BT (2015) Development of HRM method for genotyping and studying the presence of SNP rs2046210 in breast cancer Vietnamese patients. International University
19. Mu D, Hsu DS, Sancar A (1996) Reaction mechanism of human DNA repair excision nuclease. J Biol Chem 271(14):8285–8294
20. Ito S et al (2007) XPG stabilizes TFIIH, allowing transactivation of nuclear receptors: implications for Cockayne syndrome in XP-G/CS patients. Mol Cell 26(2):231–243
21. Schmitz K-M et al (2009) TAF12 recruits Gadd45a and the nucleotide excision repair complex to the promoter of rRNA genes leading to active DNA demethylation. Mol Cell 33(3):344–353
22. Walsh CS et al (2008) ERCC5 is a novel biomarker of ovarian cancer prognosis. J Clin Oncol 26(18):2952–2958
23. Nouspikel TP, Hyka-Nouspikel N, Hanawalt PC (2006) Transcription domain-associated repair in human cells. Mol Cell Biol 26(23):8722–8730
24. Smith TR et al (2003) DNA damage and breast cancer risk. Carcinogenesis. 0143-3334 (Print)
25. Febres-Aldana CA et al (2020) Understanding the histogenesis of a HRAS-PIK3R1 co-driven metastatic metaplastic breast carcinoma associated with squamous metaplasia of lactiferous ducts. Pathol Int 70(2):101–107
26. Zhang X et al (2016) Haplotype and diplotype analyses of variation in ERCC5 transcription cis-regulation in normal bronchial epithelial cells. Physiol Genomics 48(7):537–543
27. Blomquist TM, Crawford EL, Willey JC (2010) Cis-acting genetic variation at an E2F1/YY1 response site and putative p53 site is associated with altered allele-specific expression of ERCC5 (XPG) transcript in normal human bronchial epithelium. Carcinogenesis 31(7):1242–1250

Purification of Alpha-Fetoprotein from Human Umbilical Cord Plasma

Huu-Hung Nguyen and Thi-Phuong-Thao Le

Abstract Alpha-fetoprotein (AFP) is a sera protein that involves in fetal development, produced by the fetal liver and yolk sac. Human AFP (hAFP) is not detected in adults but it is increased in cases of hepatocellular carcinoma (HCC), cirrhosis and other cancers like germ cell cancer, stomach and intestine cancer. hAFP has been used as an important marker for HCC serological diagnosis and it has recently become a target for HCC immunization studies. Umbilical cord blood has approximately 150–250 μg/ml hAFP but it is usually discarded in hospitals. In order to take advantage of human umbilical cord blood as hAFP nature material, this study have established a three-step chromatography process for hAFP purification including removal of albumin by Hitrap blue HP column, hAFP captured chromatography using rabbit anti-hAFP IgG column, and filtration chromatography through Sephacryl S100 HR column. By native-PAGE, western blotting and bradford assays, the hAFP purity was showed more than 95% and 80 μg of hAFP per mL of initial cord plasma was recovered. Our established hAFP purification process can be performed on a larger scale for later applications in the serodiagnosis of HCC and other hAFP related diseases.

Keywords Alpha-fetoprotein · Chromatography · Hepatocellular carcinoma · Purification · Umbilical cord blood

H.-H. Nguyen (✉)
Faculty of Technology, Van Lang University, 45 Nguyen Khac Nhu Street, Co Giang Ward, District 1, Ho Chi Minh City, Vietnam
e-mail: hung.nh@vlu.edu.vn

T.-P.-T. Le
Humasis Co., Ltd, Lot D.01, Tan Thuan Street, Tan Thuan Export Processing Zone, Tan Thuan Dong Ward, District 7, Ho Chi Minh City, Vietnam

Faculty of Biotechnology, Nguyen Tat Thanh University, 298A Nguyen Tat Thanh Street, Ward 13, District 4, Ho Chi Minh City, Vietnam

V. Van Toi et al. (eds.), *8th International Conference on the Development of Biomedical Engineering in Vietnam*, IFMBE Proceedings 85,
https://doi.org/10.1007/978-3-030-75506-5_41

1 Introduction

Hepatocellular carcinoma (HCC) is in the top ten of common cancer in the world and of the highest in mortality [1]. In Vietnam, liver cancer has a relatively high incidence in the Ho Chi Minh City and tends to increase. Up to now, the cause of high mortality in cancers in general and HCC in particular are mainly due to the late detection of the disease. Therefore, the research and development of techniques for early detection of cancer is extremely urgent.

Alpha-fetoprotein is a member of the protein family which has 4 like-albumin members, including: ALB (albumin), DBP (vitamin-D binding protein), AFP and α-ALB (alpha-albumin). In particular, both AFP and ALB have 609 amino acids and 39% homologous molecule structure [2]. hAFP has a molecular weight of about 67–72 kDa depending on the percentage of carbohydrates attached to this molecule [3]. According to that point, the molecular weight of mature hAFP is 66.3 kDa (without carbohydrates) and 71 kDa with 4% carbohydrates.

hAFP has long been known to be an useful cancer marker, especially in the diagnosis of liver cancer. hAFP is an embryo-specific protein, intensively produced in the early embryonic stage by the liver but less produced by a gastrointestinal tract [4]. After birth, hAFP is discontinued, so it only exists in our circulation in amounts < 20 ng/mL in adults [3]. However, some clinical cases show a return of hAFP in the blood related to liver abnormalities such as hepatocellular carcinoma, metastatic liver cancer, chronic hepatitis. On the other hand, there are also cases where liver disease is not present but hAFP is higher than normal levels, including non-seminomatous germ cell tumors [5], submucosal cell tumors (gastric adenocarcinoma) [6].

Recently, determining the ratio of glycosylated hAFP present in total plasma hAFP allows distinguishing HCC from cirrhosis and chronic hepatitis [7–9]. In Southeast Asia, people who are negative for hepatitis B virus markers ($HbsAg^-$) with total plasma hAFP > 25 ng/mL are considered to have a 100% specific diagnostic value for HCC [10]. For patients with $HbsAg^+$, HCC specificity reached 79.8% when hAFP < 200 ng/mL and up to 91.5% if hAFP is in the range of 200–400 ng/mL [11]. A recent study showed that the threshold of hAFP with 400 ng/mL is better than that of 200 ng/mL in HCC [12]. In addition, monitoring hAFP levels in HCC patients blood after surgery helps to detect early HCC recurrence [13].

As a basis for developing diagnostic applications, it is essential to have refined hAFP as an important material. hAFP can be found in umbilical cord blood at concentrations of about 150–250 μg/mL [14]. This is a common source of discarded materials in maternity hospitals. Therefore, the aim of this study was to establish the hAFP purification procedure from umbilical cord blood donated from postpartum women.

2 Materials and Methods

2.1 *Source of hAFP*

Cord blood was donated by pregnant women (from 25 to 32 years old) and was centrifuged to obtain plasma. Normal blood plasma was donated by normal volunteers (from 20 to 28 years old) who are working in our lab, Nguyen Tat Thanh University.

2.2 *Removal of Albumin*

Hitrap Blue HP chromatographic column, which contains cibacron ligands, was used to remove ALB according to manufacturer's instructions (Citiva, formerly GE Healthcare Life Sciences). Cibacron ligands attached to the gel have an affinity for ALB and retain it when the plasma sample passes through the column. Buffer solutions used were binding buffer (50 mM KH_2PO_4, 0.05% NaN_3, pH 7.0), elution buffer (50 mM KH_2PO_4, 1.5 M KCl, 0.05% NaN_3, pH 7.0). The rate of sample loading and running of buffer solutions is 1 mL/min. Mixed plasma samples were diluted in binding buffer before loading into columns.

2.3 *Isolating hAFP*

The sample containing hAFP was concentrated and buffer-exchanged with 20 mM sodium phosphate buffer by centrifugal filtration method using amicon filter unit 10 kDa cut-off according to the manufacturer's instructions (Merck). Centrifugal speed was 3000 $\times$ *g* for 20 min. The sample was then passed through a rabbit anti-hAFP IgG chromatographic column. The rate of sample loading was 1 mL/min. Buffer solutions used were binding buffer (20 mM sodium phosphate buffer, 0.05% NaN_3, pH 7.0), elution buffer (0.1 M Glycine-HCl, 0.3 M NaCl, pH 2.7), neutral buffer (1 M Tris-HCl, 0.05% NaN_3, pH 9.0). The collected fractions were analyzed by native-PAGE which is described later.

2.4 *Gel Filtration Chromatography*

Sephacryl S-100 resin and XK26/60 column were used to separate proteins in the segment containing hAFP. The sample was concentrated and buffer-exchanged with 50 mM sodium phosphate buffer by centrifugal filtration using amicon filter unit 10 kDa cut-off (speed 3000 $\times$ *g* for 20 min) before loading into the column.

50 mM sodium phosphate buffer, 0.15 M NaCl, 0.05% NaN_3, pH 7.0 are used in the chromatography and flow rate of 1.3 mL/min.

2.5 *Native-PAGE*

Cord plasma and protein samples after purification were analyzed by non-denatured electrophoresis using polyacrylamide at 7.5% concentration. Proteins after electrophoresis are detected by dyeing with $AgNO_3$. Electrophoresis images were recorded by HP G4050 image scanner. Purified hAFP (69 kDa) purchased from Lee Biosolutions was used as positive control marker.

2.6 *Western Blotting*

The protein in the gel was transferred to a nitrocellulose film with a pore size of 0.45 μm in 1.5 h using a semi-dry film transfer system (Hoefer). The membrane was then blocked with 1% casein in TBST buffer (20 mM Tris, 150 mM NaCl, 0.1% Tween 20, pH 7.5) for 1 h and then washed 3 times with TBST buffer for 10 min. The membrane was incubated for 1 h with rabbit plasma after inducing an immune response to AFP (diluting plasma in TBST buffer to reach 0.73 μg/mL) and rinsed with TBST buffer. The membrane is then incubated with goat anti-rabbit IgG antibodies conjugated with HRP (Santa Cruz Biotechnology) for 1 h (dilute antibody in TBST buffer to reach 0.04 μg/mL) and wash with TBST buffer. To detect an antigen-antibody hybrid, the film is incubated with luminol photochemicals and scanned by a C-DiGit Blot Scanner (LI-COR).

2.7 *Protein Quantification*

The total plasma proteins are determined by spectrophotometric method of measuring the adsorption of protein at 280 and 260 nm [15]. The protein concentration is calculated according to the formula:

$$C(\text{mg/mL}) = (1.55 * A280 - 0.76 * A260) * D$$

in which:

A280 and A260 are the OD values of the corresponding plasma samples at wavelengths of 280 nm and 260 nm

D is the dilution factor of the plasma sample.

The amount of AFP after purification were determined by the Bradford method, bovine serum albumin (BSA) was used as standard curve.

3 Results

3.1 *Determine the Presence of hAFP in Human Cord Plasma Samples*

In order to identify hAFP in human cord blood samples, we analyzed human cord plasma by native-PAGE and western blotting. hAFP and ALB have approximately the same molecular weight (~ 69 kDa for hAFP and ~ 66.5 kDa for ALB in human cord blood). By native-PAGE and compared with commercial purified hAFP, the position of hAFP in human cord blood plasma samples in electrophoresis gel can be predicted (Fig. 1a). To confirm the presence of hAFP, we conducted a western blotting using rabbit anti-hAFP IgG antibody generated by rabbit immunization with purified hAFP (data not shown). Figure 1b shows that the normal human plasma sample gives a line at the position of ALB and the human cord plasma sample has an additional line with a strong intensity just above the position of ALB and corresponding to the position

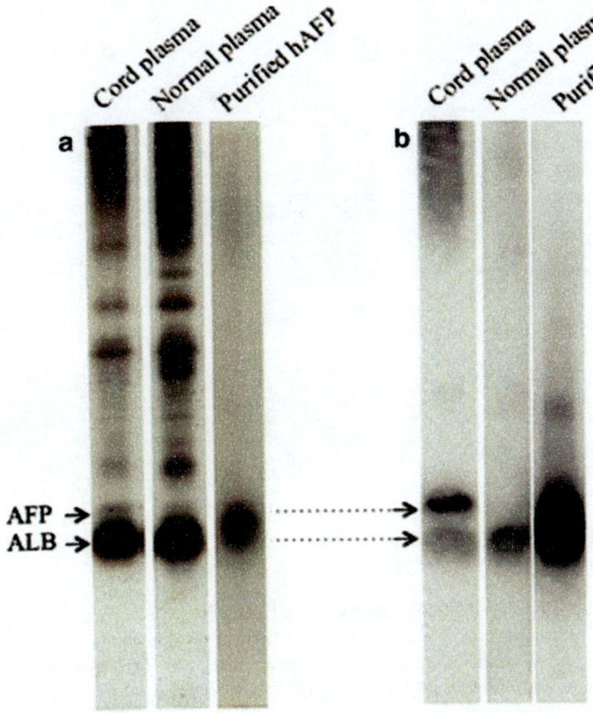

Fig. 1 Identification of hAFP in cord blood samples. **a** Native-PAGE of cord blood plasma and normal human plasma in 7.5% polyacrylamide. Protein was detected by $AgNO_3$ staining. **b** Western blotting of plasma samples used rabbit anti-hAFP antibody. Purified hAFP is used as positive control marker

of hAFP. This is consistent that hAFP and ALB have a 39% structural similarity [2], so that anti-hAFP antibodies may cross-react with ALB. The data presented indicate that there are hAFP present in the human cord blood sample and it could be used to purify hAFP in subsequent experiments.

3.2 *Removal of ALB from Human Cord Plasma*

ALB in the cord plasma was depleted by chromatography using Hitrap Blue column. Figure 2a shows two Peaks formed on the chromatogram. Peak 1A contains hAFP which is non-bound in the column and Peak 2A contains mainly ALB which is bound in the column. Similarity, as show in Fig. 2b, there are two Peaks (1B and 2B) were formed on the chromatogram in case of normal human plasma samples. Chromatographic products were analyzed by native-PAGE (Fig. 2c) and western blotting (Fig. 2d). The results show that most of ALB was eliminated from Peak 1 in both cord and normal plasma samples and hAFP was only detected in Peak 1A.

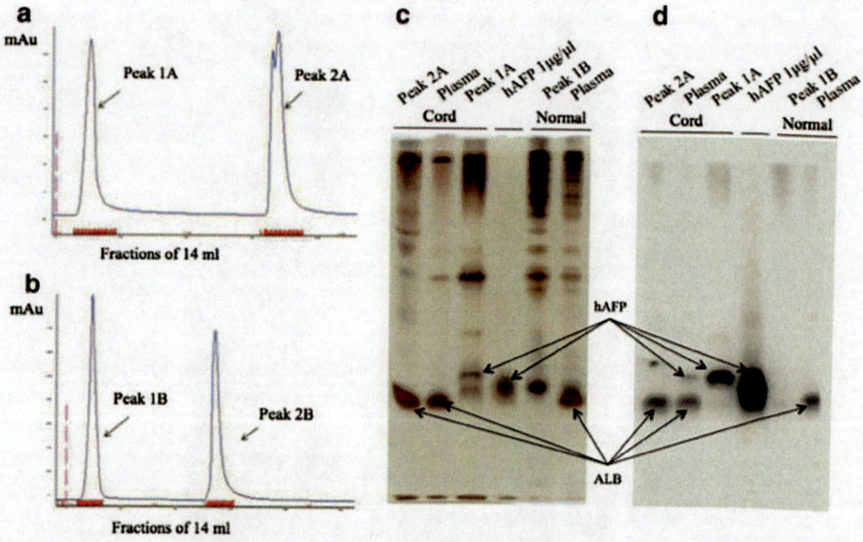

Fig. 2 ALB removal by affinity chromatography using Hitrap Blue column. **a** Chromatogram of removing ALB from umbilical cord blood plasma and **b** from normal human plasma. **c** Native-PAGE samples before and after chromatography, 2 µg protein/well. **d** Western blotting of samples before and after chromatography with rabbit anti-hAFP IgG antibodies

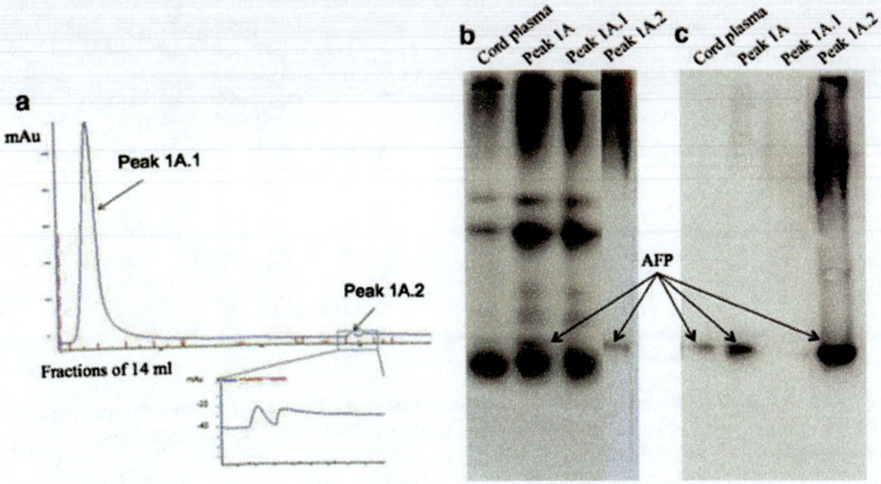

Fig. 3 Capturing of hAFP by affinity chromatography. **a** Chromatogram shows separation of the bound (Peak 1A.2) and non-bound (Peak 1A.1) proteins. **b** Native-PAGE in 7.5% polyacrylamide, 2 μg protein/well. **c** Western blotting of samples before and after chromatography

3.3 *Capturing of hAFP by Avidity Chromatographic Column Using Anti-AFP Rabbit IgG Antibodies*

Proteins in Peak 1A were then applied to hAFP binding column which contain rabbit anti-hAFP IgG. Figure 3a shows that Peak 1A.1 and 1A.2 were formed in the chromatogram, and that Peak 1A.2 is expected to contain hAFP. The presence of hAFP was confirmed by electrophoresis (Fig. 3b) and Western blotting using rabbit anti-hAFP IgG (Fig. 3c). However, an amount of high molecular weight proteins was found and required to be further depleted. This shows that the anti-hAFP IgG antibodies used in the experiment not only cross-reacts with ALB but also reacts to some extent with large molecular weight proteins in plasma.

3.4 *Purify hAFP by Gel Filtration Chromatography*

Peak 1A.2 was applied in gel filtration chromatography. The chromatogram shows three Peaks (1A.2.1, 1A.2.2 and 1A.2.3) (Fig. 4a). Analysis of these Peaks by native-PAGE show that hAFP was detected in all 3 Peaks but Peak 1A.2.3 contains hAFP with high purity estimated > 95%. AFP quantitative resulted by Bradford assay shows that 80 μg of hAFP per mL of initial cord plasma was recovered.

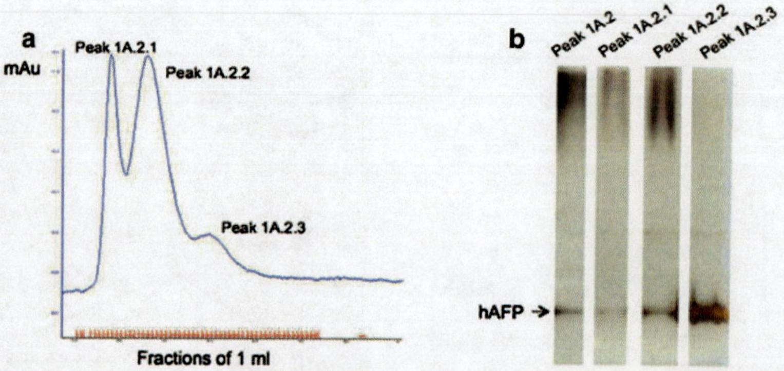

Fig. 4 Analysis of hAFP after filtration chromatography. **a** Chromatogram shows separation of proteins through the Sephacryl S-100 column. **b** Native-PAGE in 7.5% polyacrylamide, 2 μg protein/well

4 Discussion

AFP was first detected in human serum in 1956 [16]. To date, there have been numerous reports regarding AFP purification of many different species (rats, rodents, humans) from various sources of raw materials such as amniotic fluid, fetal blood, fetal liver and HepG2 liver cancer cell line. Among the methods used to purify AFP, the immuno-approach using anti-AFP IgG antibodies is the most effective. Gel filtration chromatography is a crucial step to obtaining purified AFP with high recovery efficiency. In addition, a number of other methods have been used to purify AFP. AFP from rat amniotic fluid was purified by chromatography through anti-AFP IgG antibodies column combined with DEAE negative ion exchange column [17]. Avidity chromatography through anti-AFP antibody column and Concavalin A (Con-A) Sepharose column was performed to purify AFP from human cord blood for 20% recovery efficiency [18]; However, Con-A has a predominant affinity for the glycoform of AFP in germ cell tumors and liver diseases such as HCC [19]. Affinity chromatography through E2 estrogens affinity column was performed to capture AFP. However, E2 has a strong affinity for mouse AFP and a weak affinity for human AFP [20]. Therefore, these methods showed above are unlikely effective when used to purify AFP from human cord bloods. The most difficult problem in refining AFP with traditional physical-chemical techniques is how AFP can be separated from ALB because these two proteins are shared in molecular weight, isoelectric point and composition of amino acid chains. Avidity chromatography via Blue Sepharose column with Cibacron ligand is currently the best method to remove ALB before refining AFP.

Our present three-steps purification process allowed us to obtain hAFP with high purity. The hAFP recovery efficiency is equivalent to some previously published results [14, 21].

5 Conclusions

We successfully purified hAFP from human cord blood plasma via the three-steps chromatographic procedure (affinity chromatography removing ALB, affinity chromatography binding hAFP via column conjugates rabbit anti-hAFP IgG antibodies and filtration chromatography). The obtained hAFP was > 95% purity and 80 μg of hAFP was recovered from 1 mL of initial cord plasma. This purification procedure can be potentially up-scaled in order to take advantage in using of the human cord blood and creating an important starting nature material for further investigations of HCC diagnosis and other hAFP associated diseases.

Conflicts of Interest No potential conflict of interest was reported by the authors.

References

1. Bray F, Ferlay J, Soerjomataram I, Siegel RL, Torre LA, Jemal A (2018) Global cancer statistics 2018: GLOBOCAN estimates of incidence and mortality worldwide for 36 cancers in 185 countries. CA Cancer J Clin 68(6):394–424
2. Morinaga T, Sakai M, Wegmann TG, Tamaoki T (1983) Primary structures of human alpha-fetoprotein and its mRNA. Proc Natl Acad Sci USA 80(15):4604–4608
3. Mizejewski GJ (2004) Biological roles of alpha-fetoprotein during pregnancy and perinatal development. Exp Biol Med (Maywood) 229(6):439–463
4. Gitlin D, Perricelli A, Gitlin GM (1972) Synthesis of -fetoprotein by liver, yolk sac, and gastrointestinal tract of the human conceptus. Cancer Res 32(5):979–982
5. Johnson PJ, Poon TC, Hjelm NM, Ho CS, Ho SK, Welby C, Stevenson D, Patel T, Parekh R, Townsend RR (1999) Glycan composition of serum alpha-fetoprotein in patients with hepatocellular carcinoma and non-seminomatous germ cell tumour. Br J Cancer 81(7):1188–1195
6. Chen J, Rocken C, Treiber G, Jentsch-Ulrich K, Malfertheiner P, Ebert MP (2003) Clinical implications of alpha-fetoprotein expression in gastric adenocarcinoma. Dig Dis 21(4):357–362
7. Kudo M (2013) Alpha-fetoprotein-L3: useful or useless for hepatocellular carcinoma? Liver Cancer 2(3–4):151–152
8. Gan Y, Liang Q, Song X (2014) Diagnostic value of alpha-L-fucosidase for hepatocellular carcinoma: a meta-analysis. Tumour Biol 35(5):3953–3960
9. Li Z, Li H, Deng D, Liu R, Lv Y (2020) Mass spectrometric assay of alpha-fetoprotein isoforms for accurate serological evaluation. Anal Chem 92(7):4807–4813
10. Lee HS, Chung YH, Kim CY (1991) Specificities of serum alpha-fetoprotein in $HBsAg^{+}$ and $HBsAg^{-}$ patients in the diagnosis of hepatocellular carcinoma. Hepatology 14(1):68–72
11. Tan CK, Law NM, Ng HS, Machin D (2003) Simple clinical prognostic model for hepatocellular carcinoma in developing countries and its validation. J Clin Oncol 21(12):2294–2298
12. Zhang J, Chen G, Zhang P, Zhang J, Li X, Gan D, Cao X, Han M, Du H, Ye Y (2020) The threshold of alpha-fetoprotein (AFP) for the diagnosis of hepatocellular carcinoma: a systematic review and meta-analysis. PLoS ONE 15(2):
13. Shirabe K, Takenaka K, Gion T, Shimada M, Fujiwara Y, Sugimachi K (1997) Significance of alpha-fetoprotein levels for detection of early recurrence of hepatocellular carcinoma after hepatic resection. J Surg Oncol 64(2):143–146
14. Czokalo M (1979) Human alpha-fetoprotein/AFP/ I. Isolation of homogenous AFP from cord blood serum. Mol Cell Biochem 25(3):179–185

15. Alastair A, Michèle P (2009) Protein determination by UV absorption. The protein protocols handbook. Humana Press
16. Bergstrand CG, Czar B (1956) Demonstration of a new protein fraction in serum from the human fetus. Scand J Clin Lab Invest 8(2):174
17. Kuhlmann WD (1975) Purification of mouse alpha1-fetoprotein and preparation of specific peroxidase conjugates for its cellular localization. Histochemistry 44(2):155–167
18. Chudy D, Zizkovsky V (1987) A simple and rapid method for the isolation of human alpha-fetoprotein from human cord serum. Neoplasma 34(4):491–496
19. Govindarajan S, Fong TL, Ashcavai M (1987) Concanavalin A affinity of alpha-fetoprotein. Its use in differentiating tumors. Am J Clin Pathol 88(6):722–724
20. Allen SH, Bennett JA, Mizejewski GJ, Andersen TT, Ferraris S, Jacobson HI (1993) Purification of alpha-fetoprotein from human cord serum with demonstration of its antiestrogenic activity. Biochim Biophys Acta 1202(1):135–142
21. Chaturvedi R, Agarkar V, Sharma GL, Sarma PU (1998) Purification of alpha feto protein from human cord blood. Prep Biochem Biotechnol 28(4):293–303

The Influence of High D-glucose Concentrations on Increasing the Expression of EGR-1, PTEN and GGPS-1 Involved in Insulin Resistance of AT-MSCs

Huu-Phuong Mai, Nhu-Thuy Trinh, Vong Binh Long, Nguyen Trong Binh, Dang-Quan Nguyen, and Hoa-Xo Duong

Abstract Type 2 diabetes mellitus (T2DM) is increasing worldwide, characterized by insulin resistance and hyperglycemia. The results of previous studies have demonstrated that high D-glucose concentrations alter the characteristic and function of adipose tissue-derived mesenchymal stem cells (AT-MSCs). Besides, *early growth response factor-1 (EGR-1)* and insulin resistance mediators (*PTEN* and *GGPS-1*) were highly upregulated in diabetic AT-MSCs (dAT-MSCs) compared with non-diabetic AT-MSCs (nAT-MSCs). In this research, we examined the effect of high glucose concentrations on nAT-MSCs in comparison to dAT-MSCs on the expression of *EGR-1*, *PTEN,* and *GGPS-1* involved in insulin resistance of human AT-MSCs. The expression of insulin resistance-related genes and *EGR-1* protein were analyzed by reverse transcription-quantitative polymerase chain reaction (RT-qPCR) and Western blot, respectively. The results show that the high D-Glucose concentrations (25, 50, and 100 mM) enhanced the expression of *EGR-1* and insulin resistance-related genes in nAT-MSCs compared with non-treated nAT-MSCs and dAT-MSCs. Notably, the increase of high D-Glucose concentration remarkably upregulated the expression of *EGR-1*, *PTEN*, and *GGPS-1* in both nAT-MSCs and dAT-MSCs. The effect of high D-glucose concentration (100 mM) remarkably increased the expression of *EGR-1*, *PTEN*, and *GGPS-1* in human AT-MSCs. The results of this study will expand our knowledge about the impact of high glucose concentration on insulin resistance in human AT-MSCs for the improvement in diabetic treatment.

Keywords Type 2 diabetes · Human adipose tissue-derived mesenchymal stem cells · High D-glucose concentrations · Metformin · *EGR-1* · *PTEN* · *GGPS-1*

H.-P. Mai · N. T. Binh · D.-Q. Nguyen · H.-X. Duong
Biotechnology Center of Ho Chi Minh City, Ho Chi Minh City 700000, Vietnam

N.-T. Trinh (✉) · V. B. Long
International University, Ho Chi Minh City 700000, Vietnam
e-mail: tnthuy@hcmiu.edu.vn

Vietnam National University of Ho Chi Minh City, Ho Chi Minh City 700000, Vietnam

V. Van Toi et al. (eds.), *8th International Conference on the Development of Biomedical Engineering in Vietnam*, IFMBE Proceedings 85,
https://doi.org/10.1007/978-3-030-75506-5_42

Abbreviations

EGR-1	Early growth response-1
PTEN	Phosphatase and tensin homologue
GGPS-1	Geranylgeranyl diphosphate synthase-1
IRS-1	Insulin receptor substrate-1

1 Introduction

Type 2 diabetes mellitus is related to insulin resistance and hyperglycemia, which causes serious complications, including microvascular and macrovascular diseases [1, 2]. Adipose tissue is an endocrine organ that suffered from insulin resistance. Previous reports showed that high glucose concentration increased insulin resistance in human adipocytes through the serine phosphorylation of insulin receptors [3, 4]. In addition, the early growth response-1 *(EGR-1),* which is a transcription factor, was highly expressed in abdominal fat of diabetic patients and in adipocytes of *db/db* mice associated with increased expression of insulin resistance-related genes such as *PTEN* and *GGPS-1* based on the characteristics of adipose tissue in type 2 diabetic patients [5–7].

Recent researches have clarified the benefits of mesenchymal stem cells (MSCs) due to its regeneration ability, which has shown promises in the prevention and treatment of diabetes and its complications [8–10]. AT-MSCs are considered to be an important source of stem cell collection for tissue technology and regenerative medicine, highlighted by their characteristics such as easy to be isolation, less invasive, and the high percentage of stem cells [2, 11]. However, our previous study demonstrated that AT-MSCs from diabetic donors (dAT-MSCs) accelerated the expression of insulin resistance-related genes, including *PTEN*, *GGPS-1*, and *EGR-1*. The phosphorylation of IRS-1 at serine 636/639 was elevated in dAT-MSCs under both normoxia and hypoxia conditions [7]. Cramer et al. also gave an idea of the high glucose concentrations alter the regenerative potential of AT-MSCs [1]. Thus, the aim of this study is to evaluate the impact of high D-glucose concentrations on the expression of insulin resistance-related factors, including the *EGR-1* at transcription and translation levels as well as *PTEN* and *GGPS-1* expression in nAT- MSCs, high D-glucose treated nAT-MSCs, and dAT-MSCs. This finding suggests EGR-1 may be the key factor for the therapeutic target of AT-MSCs under diabetic conditions.

2 Material and Methods

2.1 *Stem Cell Culture*

Non-diabetic AT-MSCs (nAT-MSCs) and diabetic AT-MSCs (dAT-MSCs) are provided by the Laboratory of Regenerative Medicine and Stem Cell Biology, University of Tsukuba, Japan [7, 12]. Cells were cultured in Iscove's Modified Dulbecco's medium (Thermo, USA), supplemented with 10% fetal bovine serum (Thermo, USA), 1% antibiotics (Sigma, USA) and 5 ng/ml basic fibroblast growth factor (bFGF, Sigma, USA) at 37 °C and 5% CO_2. The medium will be renewed every three days. Cells are frozen with cell banker solution (ZENOAQ, Japan) and preserved in liquid nitrogen. All AT-MSCs are at passage 5 to 8 for the experiments of this study.

2.2 *The Treatment of High D-glucose Concentrations on AT-MSCs*

AT-MSCs were plated on 12-well plate with a density of 3×10^4 cells/well and culture stable for 24 h. We then added D-glucose at different concentrations (25, 50, 100, and 200 mM) to the culture medium. The impact of high D-glucose concentrations on the cell morphology and the expression of insulin resistance-related genes will be determined at day 3.

2.3 *Quantitative Reverse Transcription Polymerase Chain Reaction*

nAT-MSC and dAT-MSC were cultured in D-glucose (25, 50 and 100 mM) supplementation medium to examine the impact of those on the expression of genes related to insulin resistance. The cell lysis buffer was Sepasol-RNA I Super G (Nacalai tesque, Japan). Total RNA was reversely transcribed by reverse transcription polymerase chain reaction (RT-PCR) Kit (TOYOBO, Japan). cDNA was used for detecting the expression of *EGR-1*, *PTEN*, and *GGPS-1* by real-time PCR with Maxima SYBR Green/ROX qPCR Master Mix (2X) Kit (Thermo, USA) and the reaction in a LightCycler 96 System (Roche, Switzerland). The target genes were analyzed by the $\Delta\Delta C_t$ method. The β-actin gene was used as the internal control for the experiments. The primer sets used for the PCR reactions are shown in Table 1.

Table 1 The primers used for quantitative polymerase chain reaction [7]

Function	Gene	Primer	Sequence
Internal control	*β-actin*	5′-primer	GTGCGTGACATTAAGGAGAAGCTGTGC
		3′-primer	GTACTTGCGCTCAGGAGGAGCAATGAT
Transcription factors	*EGR-1*	5’-primer	AGTCTTTTCCTGACATCTCTCTGAA
		3′-primer	ACTAGGCCACTGACCAAGCTGAA
		5′-primer	CTGGTAGGCGATGTCCTTA
Mediators of insulin resistance	*PTEN*	3′-primer	TTGGCGGTGTCATAATGTCT
		5′-primer	GCAGAA AGACTTGAAGGCGTA
	GGPS-1	3′-primer	ACTGTTTGGATTAGCAGTAGGTCTC
		5′-primer	GGAGTGTAGATTAGCATAATCATCC

2.4 Western Blotting

Cells were collected in RIPA buffer for the whole cell lyses, as described in our previous report [7]. The supernatant will be used for western blotting analyses. The same amount of protein in each group will be electrophoresed on 10% sodium dodecyl sulfate polyacrylamide gel electrophoresis (SDS-PAGE) gel and then transferred to PVDF membrane (Thermo, USA). The membranes are then incubated with a Rabbit anti-EGR1 antibody (Cell Signaling, 15F7). HRP-conjugated goat anti-rabbit IgG (Abcam, ab97040) and goat anti-mouse IgG (Abcam, ab7090) were used as secondary antibodies and was detected by an enhanced chemiluminescence Pierce ECL Western blotting substrate. Internal control was the Mouse anti-b-Actin antibody (Abcam, ab49900).

2.5 Statistical Analysis

The one-way ANOVA (Tukey post-hoc test; SPSS 20 software, IBM Corp.) has been used to identify significant differences among many tested groups. $P < 0.05$ value was considered as statistical significance. The data is presented as the mean ± standard deviation (SD).

3 Results

3.1 The Effect of High D-glucose Concentrations on the Morphology of AT-MSCs

D-glucose is one of the main sources of energy for many cellular activities [13]. In this study, we evaluated the impact of high D-glucose concentrations (25, 50, 100 and 200 mM D-glucose) on the morphology of nAT-MSCs that were captured and compared with nAT-MSCs and dAT-MSCs.

Our results showed that the morphology of all AT-MSCs groups did not change under 25, 50, or 100 mM D-glucose. However, nAT-MSCs with 200 mM D-glucose began to have the sign of spreading and elongating on the culture surface from day 3 (Fig. 1). This result indicates that high D-glucose concentration alters cell morphology, which can lead to changes in the characteristics and functions of AT-MSCs. Therefore, we chose 25, 50, and 100 mM for investigating the expression of genes and protein related to insulin resistance.

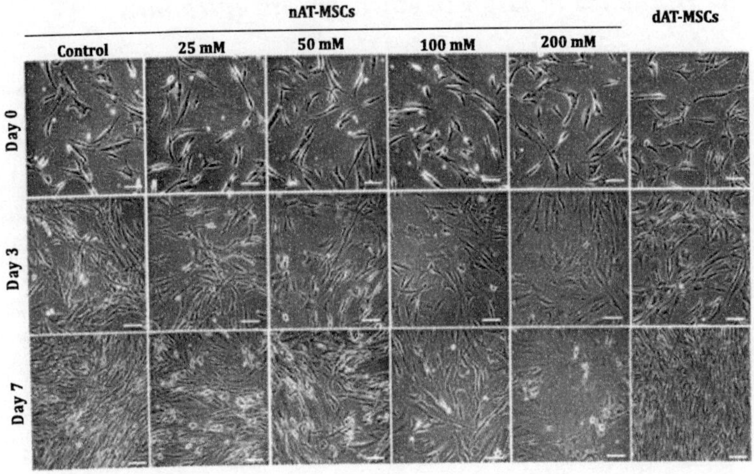

Fig. 1 The effect of high D-glucose concentrations on the morphology of AT-MSCs. High D-glucose concentrations make a nAT-MSCs spread on the culture surface compare with the control sample and dAT-MSCs. Scale bar: 100 μm

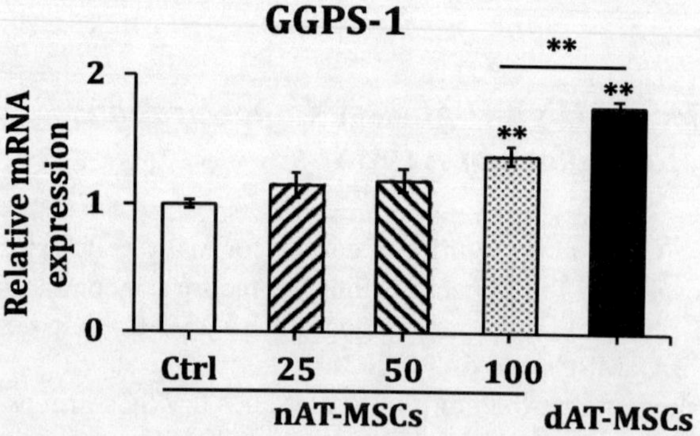

Fig. 2 The high D-glucose concentrations increased GGPS-1 expression in AT-MSCs. mRNA expression was determined by qRT-PCR with β-actin as the internal control. 100 mM D-glucose increased the expression of *GGPS-1* in nAT-MSCs. The data indicated the mean of the three independent experiments (mean ± SD); $P < 0.01$ (**); Ctrl: control

3.2 The Expression of GGPS-1 in AT-MSCs Under the Influence of High D-glucose Concentrations

Geranylgeranyl diphosphate synthase-1 (GGPS-1) activates MAPK/EKRK1/2 signal, which causes serine phosphorylation on IRS-1 and inhibits the activity of IRS-1, leading to insulin resistance [10, 14].

This result showed that the gene expression level of *GGPS-1* in dAT-MSCs were 1.76 ± 0.05 fold higher than in nAT-MSCs ($P < 0.01$, n = 3). Under the conditions of 25 mM and 50 mM D-glucose, the expression of the *GGPS-1* gene is not a significant difference between treated-nAT-MSCs and non-treated nAT-MSCs (the control). On the other hand, the expression of *GGPS-1* of nAT-MSCs under the influence of 100 mM D-glucose increased 1.38 ± 0.07 fold ($P < 0.01$, n = 3) compared to control (Fig. 2), suggesting that 100 mM D-glucose significant increased the gene expression involved in insulin resistance in nAT-MSCs.

3.3 The Expression of PTEN in AT-MSCs Under the Influence of High D-glucose Concentrations

In addition, *phosphatase and tensin homologue (PTEN)*, an enzyme that catalyzes phosphorylation, inhibits the PI3K/AKT pathway and decreases the insulin signal from insulin receptors substrate (IRS) [15, 16].

Our result showed that *PTEN* expression in dAT-MSCs were 1.77 ± 0.04 fold higher than in nAT-MSCs ($P < 0.01$, n = 3). Under influence of high D-glucose concentrations (50 and 100 mM), the expression of PTEN in nAT-MSCs was increased 1.22 ± 0.05 fold (50 mM D-glucose, $P < 0.05$, n = 3) and 1.58 ± 0.11 fold

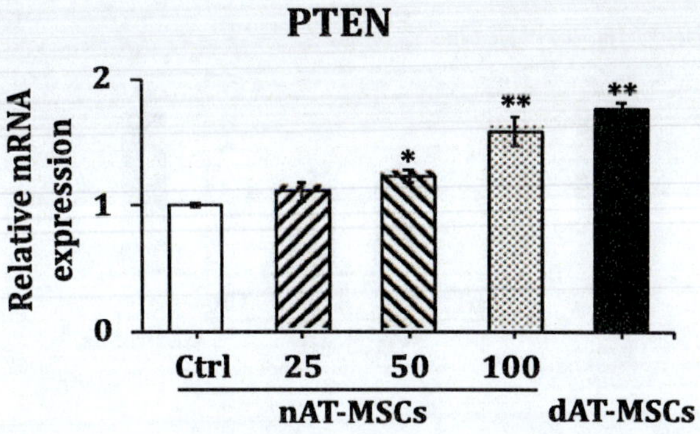

Fig. 3 The high D-glucose concentrations increased *PTEN* expression in AT-MSCs. mRNA expression was determined by qRT-PCR with β-actin as the internal control. 50 and 100 mM D-glucose increased the expression of *PTEN* in nAT-MSCs. The data indicated the mean of the three independent experiments (mean ± SD); $P < 0.01$ (**); $P < 0.05$ (*); Ctrl: control

(100 mM D-glucose, $P < 0.01$, n = 3) compared to control (Fig. 3), indicating that the high D-glucose concentrations (50 and 100 mM D-glucose) decrease the insulin signal from insulin receptors substrate in nAT-MSCs.

3.4 *The Expression of EGR-1 in AT-MSCs Under the Influence of High D-glucose Concentrations*

EGR-1, a transcription factor that plays an important role in the growth and development of many types of cells [17]. Remarkably, *EGR-1* regulates the expression of many genes by binding to target sequences, including those related to insulin resistance, such as *PTEN* and *GGPS-1* [5, 6]. Therefore, *EGR-1* was considered to be a factor regulating cellular insulin response. In this study, we investigated the impact of high D-glucose concentrations on *EGR-1* expression in AT-MSCs.

RT-qPCR results show that the gene expression of *EGR-1* in dAT-MSCs was 2.34 ± 0.14 fold higher than the control and D-glucose treated nAT-MSCs ($P < 0.01$, n = 3). *EGR-1* expression significantly upregulated in nAT-MSCs treated with 100 mM D-glucose compared to that of the control (*EGR-1*, 1.5 ± 0.12 fold increase, $P < 0.01$, n = 3) Specifically, *EGR-1* expression significantly upregulated in nAT-MSCs treated with 50 and 100 mM D-glucose compared to that of the control (1.29 ± 0.05 fold increase at 50 mM D-glucose, $P < 0.05$, n = 3; and 1.59 ± 0.07 fold increase at 100 mM D-glucose, $P < 0.01$, n = 3) (Fig. 4).

Consistent with the expression of *EGR-1* at the transcriptional level, western blot results showed that EGR-1 protein in dAT-MSCs and treated-nAT-MSCs (100 mM D-glucose) were significantly elevated compared to the control ($P < 0.01$, n = 3).

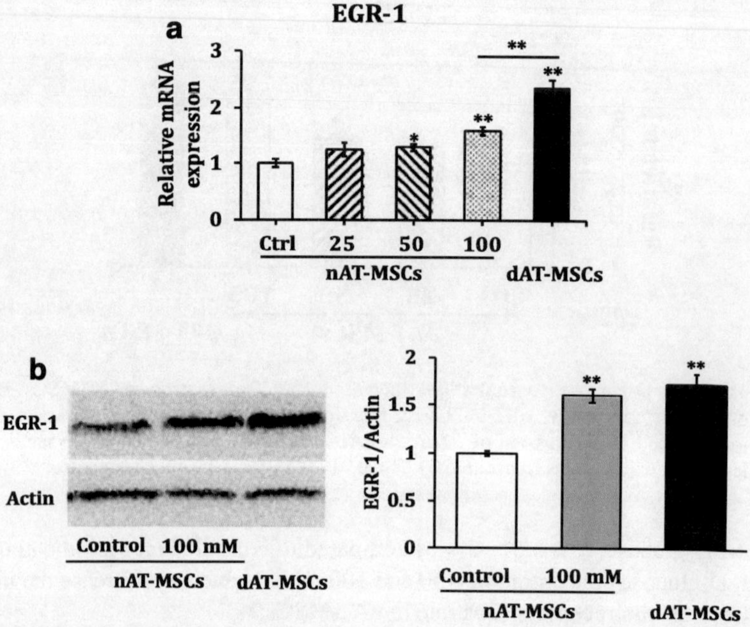

Fig. 4 *EGR-1* expression at transcription and translation levels. **a** The mRNA expression of *EGR-1* was increased due to the effects of high D-glucose concentrations; **b** the effect of high D-glucose concentration on *EGR-1* protein expression in AT-MSCs. The data indicated the mean of the three independent experiments (mean ± SD); $P < 0.01$ (**); $P < 0.05$ (*); Ctrl: control

Notably, EGR-1 protein expression is not significantly different between treated-nAT-MSCs and dAT-MSCs (Fig. 4).

The results suggest that nAT-MSCs under high D-glucose concentrations are similar as dAT-MSCs in the expression of *EGR-1* gene and protein related to insulin resistance.

4 Discussion

EGR-1, *GGPS-1,* and *PTEN* have been known to be involved in signal degradation from the insulin receptor substrate [5, 6, 15]. This study demonstrated that *EGR-1, PTEN* and *GGPS-1* expressed in dAT-MSCs and in nAT-MSCs under high D-glucose concentrations much higher than that in nAT-MSCs. Besides, high D-glucose concentrations enhanced the expression of *EGR-1*, *GGPS-1,* and *PTEN* in nAT-MSCs similar to those in dAT-MSCs.

Although D-glucose is known to provide energy for many cellular activities, high D-glucose levels have been shown to increase free radicals and increase the aging level of many types of cells, such as umbilical vein endothelial cells and human bone-delivered stem cells [13, 18, 19]. Our results show that 200 mM D-glucose has changed the morphology of nAT-MSCs (Fig. 1). In addition, previous studies on human and mouse adipocytes have shown that high glucose concentrations increased insulin resistance in these cell lines by inhibiting signals from the insulin receptors substrate [4, 17, 19]. Moreover, our results have also shown that 200 mM D-glucose strongly altered the morphology of AT-MSCs without proliferation. Therefore, we focused on lower D-glucose concentrations (25, 50 and 100 mM) for examining the expression of the insulin resistance-related factors for evaluating the effect of high D-glucose concentrations on human nAT-MSCs, mimicking the effect of persistent hyperglycemia in type 2 diabetic patients.

EGR-1 is often highly expressed in response to cellular stress caused by hyperglycemia [17–19]. In addition, *EGR-1* regulates the expression of many genes through the specific binding to target sequences, including insulin-mediated genes. Its overexpression has been shown to be associated with increase expression of genes related to insulin resistance [5, 20]. Our results show that *EGR-1* expression in nAT-MSCs lower than dAT-MSCs. *EGR-1* expression in nAT-MSCs was significantly increased when exposed to high D-glucose concentration (50 and 100 mM) in transcription and translation level (Fig. 4). These results were similar to those of other studies on the effect of high D-glucose concentrations (20–25 mM) on *EGR-1* expression in beta pancreatic cells, endothelial epithelial and adipocytes in mice [5, 17, 21]. Previously, *PTEN* and *GGPS-1* (target genes of *EGR-1*) have been associated with insulin resistance at the cellular level [14]. *PTEN* inhibits the PI3K/AKT pathway, which decreases the insulin signal from insulin receptors substrate-1 (IRS-1) [15, 19]. Whereas *GGPS-1* activates the MAPK/EKRK1/2 signal, causes serine phosphorylation on IRS-1 and inhibits the activity of IRS-1, leading to insulin resistance [10, 14]. In our study, along with an increase in *EGR-1* expression, its target genes (*PTEN* and *GGPS-1*) also increased expression in nAT-MSCs under the influence of 50 and 100 mM D-glucose (Figs. 2 and 3). In addition, expression levels of *EGR-1*, *PTEN* and *GGPS-1* were more upregulated in dAT-MSCs than nAT-MSCs (Figs. 2, 3 and 4). Thus, these results show the similarity of dAT-MSCs and nAT-MSCs under high D-glucose concentrations in the expression of gene and factors related to insulin resistance.

Taken together, our study demonstrated that high D-glucose concentrations, especially at 100 mM D-glucose, have an effect on human nAT-MSCs by increased the expression of *EGR-1*, *PTEN* and *GGPS-1*, relating to insulin resistance, similar as those of dAT-MSCs. The results suggest the potential of mimicking hyperglycemia effect on human stem cells for cellular diabetic disease modeling in the research and application of stem cells for treating diabetes mellitus or preventing its complications.

Acknowledgements The authors thank to the financial support by the grant from the Biotechnology Center of Ho Chi Minh City (DV01/17-19). We also thank professor Osamu Ohneda for providing stem cells.

Conflict of Interest The authors declare *no* conflict of interest.

References

1. Cramer EFC, Jones RK, Slakey DP, Dupin CL, Newsome ER, Alt EU, Izadpanah R (2010) Persistent high glucose concentrations alter the regenerative potential of mesenchymal stem cells. Stem Cells Dev 19:1875–1884. https://doi.org/10.1089/scd.2010.0009
2. Groop L, DeFronzo RA, Ferrannini E, Henry RR, Herman WH, Holst JJ, Hu FB, Kahn CR, Raz I, Shulman GI, Simonson DC (2015). Type 2 diabetes mellitus. Nat Rev. https://doi.org/10.1038/nrdp.2015.19
3. Xiao S, Pillay TS, Olefsky JM (1996) Glucose-induced phosphorylation of the insulin receptor—functional effects and characterization of phosphorylation sites. Am Soc Clinx Invest 97:613–620. https://doi.org/10.1172/JCI118457
4. Renström JBF, Svensson M, Eriksson JW (2007) Insulin resistance induced by high glucose and high insulin precedes insulin receptor substrate 1 protein depletion in human adipocytes. Metab Clin Exp 56:190–198. https://doi.org/10.1016/j.metabol.2006.09.012
5. Ning XY, Shen F-Y, Gao X, Xue B, Li C-J (2011) An early response transcription factor, Egr-1, enhances insulin resistance in type 2 diabetes with chronic hyperinsulinism. J Biol Chem 286:14508–14515. https://doi.org/10.1074/jbc.M110.190165
6. Ning Shen Xiao Yu, Zhang M-L, Pan F-Y, Wang C, Jia W-P, Liu C, Gao Q, Gao X, Xue B, Li C-J (2011) Egr-1 decreases adipocyte insulin sensitivity by tilting PI3K/Akt and MAPK signal balance in mice. EMBO J 30:3754–3765. https://doi.org/10.1038/emboj.2011.277
7. Trinh T-T, Ohneda K, Kimura K, Salazar GT, Sato F, Ohneda O (2016) Increased expression of EGR-1 in diabetic human adipose tissue-derived mesenchymal stem cells reduces their wound healing capacity. Stem Cells Dev 25:760–773. https://doi.org/10.1089/scd.2015.0335
8. Kern HES, Stoeve J, Klüter H, Bieback K (2006) Comparative analysis of mesenchymal stem cells from bone marrow, umbilical cord blood, or adipose tissue. Stem Cells 24:1294–1301. https://doi.org/10.1634/stemcells.2005-0342
9. Benisch P, Schilling T, Li YM, Zeck S, Meissner-Weigl J, Schneider D, Limbert C, Seufert J, Kassem M, Schütze N, Jakob F (2007) Effects of high glucose on mesenchymal stem cell proliferation and differentiation. Biochem Biophys Res Commun 363:209–215. https://doi.org/10.1016/j.bbrc.2007.08.161
10. Kimura MNK, Salazar G, Yamashita T, Tsuboi I, Mishima H, Matsushita S, Sato F, Yamagata K, Ohneda O (2014) The role of CCL5 in the ability of adipose tissue-derived mesenchymal stem cells to support repair of ischemic regions. Stem Cells Dev 23:488–501. https://doi.org/10.1089/scd.2013.0307
11. Seeberger KL, Yeung TY, Kin T, Adesida A, Jomha N, James Shapiro AM, Korbutt GS (2012) Human mesenchymal stem cells protect human islets from pro-inflammatory cytokines. PLoS One 7. https://doi.org/10.1371/journal.pone.0038189
12. Trinh TYNT, Tran Cam Tu, Kato T, Ohneda K, Sato F, Ohneda O (2016) Microvesicles enhance the mobility of human diabetic adipose tissuederived mesenchymal stem cells in vitro and improve wound healing in vivo. Biochem Biophys Res Commun 473:1111–1118. https://doi.org/10.1016/j.bbrc.2016.04.025
13. Aguiari BZSLP, Vindigni V, Rimessi A, Bianchi K, Franzin C, Cortivo R, Rossato M, Vettor R, Abatangelo G, Pozzan T, Pinton P, Rizzuto R (2008) High glucose induces adipogenic differentiation of muscle-derived stem cells. PNAS 105:1226–1231. https://doi.org/10.1073/pnas.0711402105
14. Jianming HH, Guo JG, Bai H, He H, Assi R, Isaji T, Wang T, Setia O, Lopes L, Yongquan Gu, Dardik A (2018) Adipose-derived mesenchymal stem cells accelerate diabetic wound healing in a similar fashion as bone marrow-derived cells. Am J Physiol Cell Physiol 315:885–896. https://doi.org/10.1152/ajpcell.00120.2018
15. Rosivatz E (2007) Inhibiting PTEN. Biochem Soc Trans 35:257–259. https://doi.org/10.1042/BST0350257
16. Dey CS, Gupta A (2012) PTEN, a widely known negative regulator of insulin/PI3K signaling, positively regulates neuronal insulin resistance. Mol Biol Cell 23:3882–3898. https://doi.org/10.1091/mbc.E12-05-0337

17. SørensenLR, Josefsen K, Buschard K, Birkenbach M (1999) Glucose induces early growth response gene (Egr-1) expression in pancreatic beta cells. Diabetologia 42:195–203. https://doi.org/10.1007/s001250051139
18. Castoldi M, Horn P, Wagner W, Diehlmann A, Bork S, Saffrich R, Benes V, Blake J, Pfister S, Eckstein V, Ho AD (2008) Replicative senescence of mesenchymal stem cells: a continuous and organized process. PLoS ONE 3. https://doi.org/10.1371/journal.pone.0002213
19. Munsey TS, Abuarab N, Jiang L-H,Li J, Sivaprasadarao A (2017) High glucose-induced ROS activates TRPM2 to trigger lysosomal membrane permeabilization and Zn^{2+}-mediated mitochondrial fission. Sci Signal 10. https://doi.org/10.1126/scisignal.aal4161
20. Song G, Cao X, Shao X, Zhao Y, Shi B (2014) Metformin rescues the MG63 osteoblasts against the effect of high glucose on proliferation. J Diabetes Res 2014. https://doi.org/10.1155/2014/453940
21. Rukhsana SP, Hasan N, Harada S (2003) Differential regulation of early growth response gene-1 expression by insulin and glucose in vascular endothelial cells, arteriosclerosis. Thromb Vasc Biol 23:988–993. https://doi.org/10.1161/01.ATV.0000071351.07784.19

Antioxidant Effect of Syringic Acid Grafted Chitooligosaccharides in RAW264.7 Cells

Van-Hoai Bui, Hong-Tham N Vo, and Dai-Nghiep Ngo

Abstract Reactive oxygen species (ROS) that attack biomolecules leading to many health disorders. Chitooligosaccharides (COSs) is a natural antioxidant that has a positive effect against ROS. It may be applied in the functional food and pharmaceutical industries. In recent years, importance has been given to synthesize functionalized COSs by chemical modifications to improve the bioactivities of COSs while keeping intact the basic chemical properties. Therefore, this work aims to develop a novel COSs derivatives by conjugation of syringic acid onto plain COSs to improve the antioxidant activity of COSs. In this study, the syringic acid was grafted onto COSs by a free radical mediated grafting method with an ascorbic acid/hydrogen peroxide redox catalysis pair. Antioxidant effects of the novel COSs derivatives were determined by reducing power, free radical scavenging ability, and the inhibition of cellular macromolecules' oxidation, such as lipid and DNA in RAW264.7 cells. The result showed that free radical scavenging effects and inhibitory effects of the novel derivatives of biomacromolecules oxidation were significative improved more than free COSs. These results suggest that the novel COSs derivatives can be potentially used as a free radical scavenger to protect biomacromolecules in cellular systems. Besides, the research is a foundation for further study of other activities such as the anti-inflammatory properties of these derivatives.

Keywords Chitooligosaccharides (COSs) · Antioxidant · RAW264.7

Van-Hoai Bui and Hong-Tham N Vo contributed equally.

V.-H. Bui · H.-T. N. Vo · D.-N. Ngo (✉)
Department of Biochemistry, Faculty of Biology-Biotechnology, University of Science, Ho Chi Minh City 700000, Vietnam
e-mail: ndnghiep@hcmus.edu.vn

Vietnam National University, Ho Chi Minh City 700000, Vietnam

V.-H. Bui
Ho Chi Minh City University of Food Industry (HUFI), Ho Chi Minh City 700000, Vietnam

V. Van Toi et al. (eds.), *8th International Conference on the Development of Biomedical Engineering in Vietnam*, IFMBE Proceedings 85,
https://doi.org/10.1007/978-3-030-75506-5_43

1 Introduction

Free radicals contain atoms, molecules, ions that have unpaired electrons in an atomic orbital. In this case, unpaired electrons are the reason to increase the ability to react to free radicals. They can afford to activate and accept different unpaired electrons to be a reducing agent and an oxidizing agent [1, 2]. Oxidative stress is the accumulation of excessive free radical formation that beyond the control of the antioxidant systems in human. Oxidative stress occurs at the time that the concentration of antioxidants is too low to eliminate free radicals that are formed during the physiological and biochemical processes of the body. Free radicals have two common types are reactive oxygen species (ROS) and reactive nitrogen species (RNS). In previous reports, oxidative stress is indicated to lead to different diseases, such as rheumatoid arthritis disease [3, 4], Alzheimer's disease [5], Parkinson disease [6], amyotrophic lateral sclerosis disease [7], cardiovascular disease [8], allergic and respiratory disease [9], disease relating to the immune system [10], diabetes and cancer disease [11, 12].

Chitin is partly deacetylated in an alkaline medium to give a hydrolyzed product calling chitosan. The deacetylation process converts the N-acetyl groups to the amino group at the C2 position in the structure. In this case, the deacetylation is not complete, and the product which has the deacetyl degree larger than 50 percent is called chitosan as well as the deacetyl degree less than 50 percent is called chitin [13]. Chitosan products have many research and diverse applications in different fields [14]. However, chitosan products remain limited like insoluble in water, and the human has no enzymes to decompose chitosan to metabolism type in the gastric.

Chitooligosaccharides (COSs) is the solution to solve the problem of chitosan products. COSs is the chitosan that has low molecular weight containing from 3 to 11 glucosamine units and N-acetylglucosamine units being linked by β-1,4-O-glycoside bonds [15]. Because of being the oligomer of chitosan, COSs carries most of the biological activity of chitosan likes anti-oxidant activity [16, 17], anti-microbial activity [18, 19], anti-cancer activity [20, 21], immuno-stimulating activity [22], anti-inflammatory [23], anti-HIV [24], and Alzheimer's disease [25].

Phenolic acids are natural compounds that have strong antioxidant capacity; they are found in many plants which supplied daily diets of human. Many studies have recently shown that grafted compounds of phenolic acid onto polysaccharide enhance the activity such as starch, oligosaccharide, inulin, and chitosan [26]. Syringic acid is a natural compound of phenolic acid obtained from many plants and has a high antioxidant effect that plays an important role in control oxidative stress. However, this compound has some limitations; its solubility in water is limited and is rapidly metabolized in the body. In addition, syringic has a functional group like carboxyl that can attach to the functional group of COSs material to improve the antioxidative capacity of COSs and slow release of syringic acid [27].

Macrophages are stimulable cells that can significantly increase oxygen and nitrogen species in a short time when they are stimulated. Free radicals from these species may oxidize to lipid peroxidation, protein oxidation and DNA damage. Therefore, we used a mouse macrophage model such as RAW264.7 to evaluate the antioxidant capacity of novel derivatives. In addition, in recent years, the RAW264.7 cells have also been used as models for evaluating antioxidant and anti-inflammatory activity such as Cho, Y. S., et al. (2011), Ngo, D. H., et al. (2011) [28, 29].

In recent years, great interest has been given functional group modifications to improve COS's bioactivities and keep intact the background characteristics of COSs. Notably, phenolic acid coherence linking functional groups on the COSs chain achieved interest in functional food, pharmaceutical and medicinal fields [30, 31]. Till now, there have been no reports on antioxidant activities of syringic acid conjugated onto chitooligosaccharides (SA-g-COSs) in live cells. Therefore, the aim of this study, we investigate the antioxidative effect of SA-g-COSs in RAW264.7 cells. In this work, COSs derivatives are synthesized by grafting syringic acid onto COSs backbone. The free radical mediated grafting method is used to synthesize the derivative and the derivatives' structure confirmed by the 1H-NMR spectrum. The antioxidative activities of derivatives are determined by the inhibitory effects of the bio-macromolecules injury.

2 Materials and Methods

2.1 *Materials*

Syringic acid (SA), lactic acid, sodium bicarbonate, ethanol, thiobarbituric acid (TBA), 3-(4,5-dimethylthiazol2-yl)-2,5-diphenyl tetrazolium bromide (MTT) were purchased from HiMedia (Mumbai, India). Dimethyl sulfoxide (DMSO), sodium hydroxyl, hydrogen peroxide, folin, TLC silica gel 60 F254 and agarose were from Merck (Darmstadt, Germany). Mouse macrophages RAW264.7 (KCLB, Seoul, Korea). Dulbecco's Modified Eagle Medium (DMEM), penicillin/streptomycin, fetal bovine serum (FBS), trypsin, and the other materials required for culturing of cells were purchased from Sigma–Aldrich (St. Louis, MO, USA). Ascorbic acid (VWR, Leuven, Belgium) and all other chemicals were of analytical grade or the highest grade available commercially.

Preparation of COSs powder from our previous research. Briefly, 0.8% chitosan solution was prepared in 0.8% lactic acid solution. The chitosan solution was hydrolyzed by cellulase with optimal parameters. The COSs, after hydrolysis, was filtered through an ultrafiltration membrane with membranes ranged in molecular weight cut-offs (MWCO). COSs fractionation was carried out spray drying (SD-06AG, LabPlant, UK) to created COSs powder. This COSs are packed by vacuum conditions for storage for the next experiments.

2.2 Methods

Synthesis of SA-g-COSs. In a previous study, we reported the method to synthesis the SA-g-COSs [32]. Briefly, the COSs (0.5 g) was dissolved in 25 mL of water, and then 1 mL of 1.0 M H_2O_2 that contains 0.054 g of ascorbic acid was added. After 30 min, the syringic acid (0.25 g) was added to the mixture. The mixture's pH value changed to 5, and the mixture was stirred at room temperature for 6 h. After the reaction, the mixture was centrifuged with filter membrane 1 kDa in 50 mL centrifuge tube in order to remove ingredients like H_2O_2, ascorbic acid and unreactive syringic acid. Finally, the resulting solution was lyophilized by using a freeze-drying system (FDU 2110, Eyela, Japan) to obtain SA-g-COSs solid samples.

Characterization Assay of SA-g-COSs. To identify whether Syringic acid was successfully grafted onto COSs backbones, TLC analysis was applied. Syringic acid, ascorbic acid, and SA-g-COSs were developed on a TLC silica gel 60 F254 (Merck, Darmstadt, Germany) with chloroform–ethyl acetate–acetic acid (50:50:1) as mobile phase, heating at 100 °C for 5 min. The performed TLC plate was recognized under UV light; The SA-g-COSs was characterized by the proton nuclear magnetic resonance (1H NMR) spectra. Samples were determined by the NMR-500 MHz spectrometer (Bruker, Germany) at 500 MHz; samples were dissolved in D_2O.

Cytotoxicity Determination in RAW264.7 Cells. DMEM medium containing 5% (v/v) FBS, 100 μg/mL penicillin–streptomycin and 5% CO_2 is the medium for the RAW264.7 cells growing at 37 °C. Cytotoxicity levels of COSs and derivatives were determined by the MTT method as the previously described report [33]. Briefly, the RAW264.7 cells were grown in 96-well plates at the concentration of 10^4 cells/well. The cells were grown for 24 h and treated with the fresh medium and supplemented COSs or derivatives with different concentrations (samples were dissolved in deionized water and filtered using a 0.22 μm syringe filter). After incubation for 24 h, the cells were rewashed. Subsequently, 50 μL of MTT solution (0.5 mg/mL in PBS) was added into every well and incubated for 4 h. Finally, 100 μL of DMSO was added to solubilize the formazan salt formed and was measured the absorbance at 540 nm by using a microplate reader (PerkinElmer, US). The live degrees of cells were calculated as a percentage of control.

Reducing Power Assay. Procedure determined based on previously reported procedure [34]. Briefly, 1 mL sample of COSs and derivatives were mixed with 2.5 mL of 0.2 M phosphate buffer pH 6.6 and 2.5 mL of 1% $K_3[Fe(CN)_6]$ was incubated at 50 °C for 20 min. The mixture was added 2.5 mL of 10% TCA, centrifuged at 12,000 rpm for 10 min, collected the supernatant. 1 mL the supernatant was added 2 mL of distilled water and 0.5 mL of 1% $FeCl_3$, mixed and measured the absorbance of the solution at 700 nm.

DPPH Free Radical Scavenging Assay. The DPPH free radical scavenging ability of COSs and its derivatives was based on the previously reported method with slight modifications [35]. Briefly, the mixture included 1.5 mL of DPPH solution (40 μg/mL) and 1 mL of sample (10–100 μg/mL) were vortexed in 15 s and incubated at room temperature for 30 min. The absorbance of the mixture was measured at 517 nm, the control was prepared by replacing the sample with deionized water, and meanwhile, the sample blank (sample added in ethanol without DPPH) was also measured to avoid any interference of background color. The DPPH radical scavenging activity was calculated based on Eq. (1).

$$\text{Freeradical scavenging effect } (\%) = (1 - \text{Asample}/\text{Acontrol}) * 100 \quad (1)$$

wherein: Asample and Acontrol are the absorbance of the sample and control, respectively.

Hydroxyl Free Radical Scavenging Assay. Hydroxyl radical scavenging method was performed based on the previous method with slight modifications [36]. Briefly, 0.5 mL sample of COSs and derivative (100–1000 μg/mL, dissolved in distilled water) mixed ethanol (0.5 mL; 95%), deoxyribose sugar (0.5 mL; 3.75 mM), H_2O_2 (0.5 mL; 1 mM), $FeCl_3$ (0.5 mL; 100 mM), EDTA (0.5 mL; 100 mM), ascorbic acid (0,5 mL; 100 mM) and phosphate buffer (2 mL; 20 mM; pH 7.4). The mixture was incubated at 37 °C for 60 min. 1 mL supernatant was mixed with TCA (1 mL; 2% w/v), TBA (1 mL; 2% w/v), heated with boiling water for 15 min and then cooled. The mixture was determined by the absorbance at 535 nm. The free radical scavenging effect was calculated based on Eq. (1).

Genomic DNA Isolation. The DNA sample was isolated from RAW264.7 cells using a standard phenol/proteinase K procedure following the previous study with minor modifications [37]. 10 cm culture dishes containing cells were washed twice with PBS and scraped into 1 mL of PBS containing 10 mM EDTA. Then the cells were removed PBS by centrifuge at 12,000 rpm for 5 min. After, cells were dissolved in sodium acetate (350 μL; 0.2 M), proteinase K (20 μL; 10 mg/mL), SDS (25 μL; 10% w/v), RNase (25μL; 1 mg/mL). The mixture was incubated at 37 °C for 30 min and 55 °C for an hour. After incubation, the mixture was added a solution of phenol: chloroform: isoamyl alcohol (25:24:1) to 1:1 ratio and centrifuged at 12,000 rpm for 5 min. Following centrifugation, the supernatant was added cool ethanol to 1:1.5 ratio and incubated for 30 min at −20 °C. After that mixture was centrifuged at 12,000 rpm for 5 min to remove the supernatant. The pellet was dissolved in TE buffer, and the purified DNA had a 260/280 nm absorbance ratio of 1.8–2.0. Further, the quality of isolated DNA was evaluated using 1% agarose gel electrophoresis.

DNA Oxidation Assay. The DNA oxidation assay indicated the DNA protection ability from RAW264.7 of derivatives samples and COSs. The method was performed according to the previously described report that being adjusted [38]. The mixture included DNA (5 μL; 1.0 μg), samples (4 μL; concentration ranged from 10 to 100 μg/mL), distilled water (3 μL), $FeSO_4$ (14 μL; 600 μM), H_2O_2 (14 μL; 3 mM) was incubated at room temperature for 10 min and stopped the reaction by adding EDTA (4 μL; 130 mM). The mixture was stained with GelRed and electrophoresed on a 1% agarose gel for 20 min at 100 V. Gels were then visualized by UV light using VisionCapt gel image analysis software (Bio-Print TX4, Vilber, Wembley, WA 6014, Australia).

Membrane Lipid Peroxidation Assay. The method was performed based on the previous report [37]. Briefly, 50 μL samples of COSs and derivatives (concentration ranged 10–100 μg/mL) were added to eppendorf tube containing cell solution (50 μL) or replace by distilled water as a control and incubated at room temperature for 10 min. The mixture was added H_2O_2 (50 μL; 2 mM) and $FeSO_4$ (50 μL; 0.1 M) and incubated at room temperature for 30 min. After that, the mixture was added twofold volume of TCA (10% w/v, cooled) to precipitated protein, which was removed by centrifugation at 12,000 rpm for 5 min. After centrifuging, the supernatant was collected and added an equal volume of TBA (500 μL; 1%) and incubated at 90 °C for 30 min; after cooling (4 °C), the mixture centrifuged at 12,000 rpm for 5 min and absorbance of the supernatant was determined at 528 nm.

Statistical Analysis. The result was displayed as the mean ± standard deviation (SD) of triplicates. The assay of variance was performed by statgraphics centurion XVI software. The difference was recognized to be statistically significant when $p < 0.05$.

3 Results and Discussion

3.1 Structural Characterization of SA-g-COSs

The TLC result showed that free syringic acid in derivatives was not observed. Whereas the free syringic acid and ascorbic acid appeared on the developed plate. This demonstrated that syringic acid was successfully grafted onto the COSs and not being a mixture between syringic acid and COSs. The 1H-NMR spectra showed resonance peaks of SA-g-COSs that occurred at 2 ppm (H-Ac), 3.09 ppm (H2), 3.32–3.91 ppm (H3 to H6), 5.41 ppm (H1) and 7.17 ppm belonging to phenyl protons. However, resonance peaks of COSs have not appeared phenyl protons at 6–8 ppm. Therefore, the results of 1H-NMR spectra confirmed the successful synthesis of SA grafted onto the COSs chain. This result is similar to the previously reported result when conjugation of syringic acid onto plain COSs by a peak at 7.278 ppm that represented for phenyl protons [30]. The proton nuclear magnetic resonance spectra of COSs and SA-g-COSs were presented in Fig. 1.

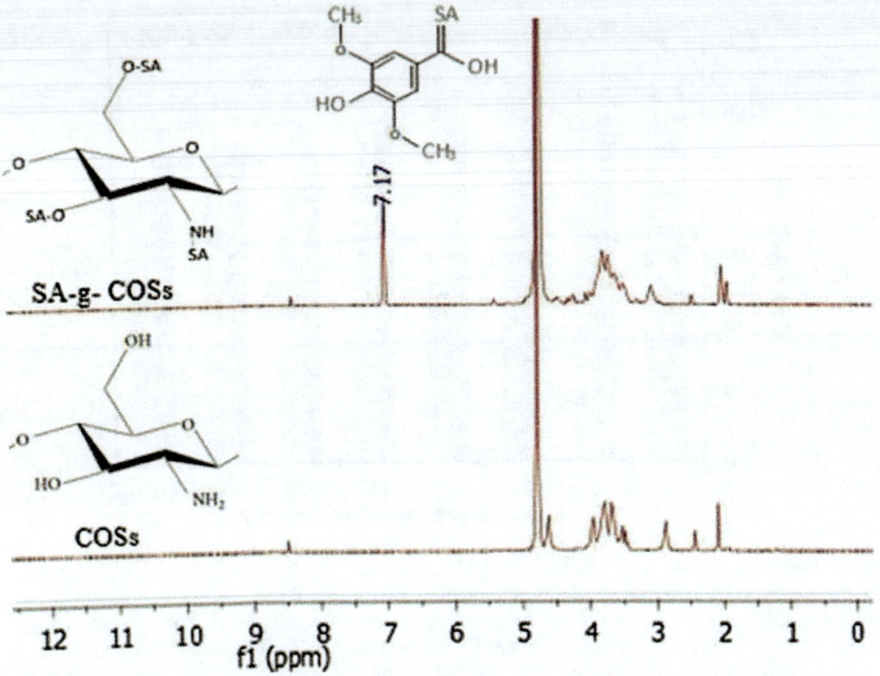

Fig. 1. 1H-NMR spectra of COSs and SA-g-COSs

3.2 *Assessment of Cell Viability in RAW264.7*

The levels that are toxic for cells were measured by the MTT method [39]. MTT was metabolized to formazan salt by enzymes in the cells system. The formazan salt was dissolved by DMSO and measured the absorbance at 540 nm. The results in Fig. 2 showed that no significant toxic effects in RAW264.7 cells at 100 μg/mL. At the concentration of 250–1000 μg/mL, the results showed the light significant toxic effects in RAW264.7 cells. In this case, COSs and SA-g-COSs are the safe materials that are not lead to cytotoxicity on RAW264.7. Therefore, nontoxic concentrations of COSs and SA-g-COSs can apply in further experiments. Till now, there has been no report confirmed the nontoxic of SA-g-COSs derivatives. However, in the previous report, GA-COSs (GA: gallic acid that is a kind of phenolic acids) was confirmed the nontoxic on RAW264.7 cells [28].

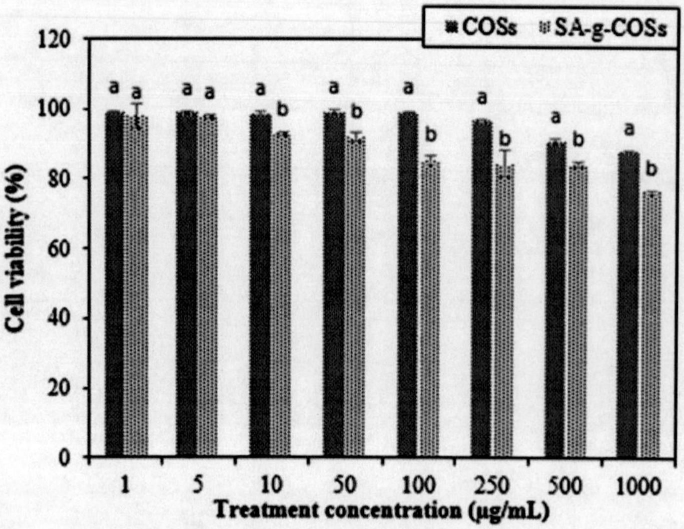

Fig. 2 Cytotoxic effects of COSs and SA-g-COSs in RAW264.7 cells. The cell viability was determined by MTT assay, as showed in the method. Firstly, the RAW264.7 cells were grown in DMEM medium at a density of 1×10^4 cells/well and treated test samples (1–1000 μg/mL). After incubation for 24 h, cells were treated with MTT solution for 4 h. Finally, the DMSO solution was added to solubilize formazan crystals, and absorbance was recognized at 540 nm. The experiments are run three times, and results are presented as means ± SD. Different letters a-b indicate significant difference, $p < 0.05$

3.3 Reducing Power

The reduced ability shows the antioxidative activity of a compound. The antioxidants potentially of COSs and SA-g-COSs subtract Fe^{3+} of potassium ferricyanide to Fe^{2+} of potassium ferrocyanide, which then reacts with ferric chloride to create a ferric ferrous complex that has a green color and an absorption maximum at 700 nm [40]. Figure 3 indicated that reducing the power of SA-g-COSs was an increased dose dependent manner and was significantly higher than COSs. Previously reported results also indicated that syringic acid grafted onto chitooligosaccharides improving the reducing power of chitooligosaccharides [30].

3.4 DPPH Free Radical Scavenging Effect

The antioxidative activity of COSs and its derivatives was evaluated by DPPH free radical scavenging ability. In the DPPH reaction, DPPH free radicals reacted with an antioxidant to achieve the stable status of the DPPH molecule by accepting an electron or hydrogen radical [35]. The DPPH free radicals are widely used for testing

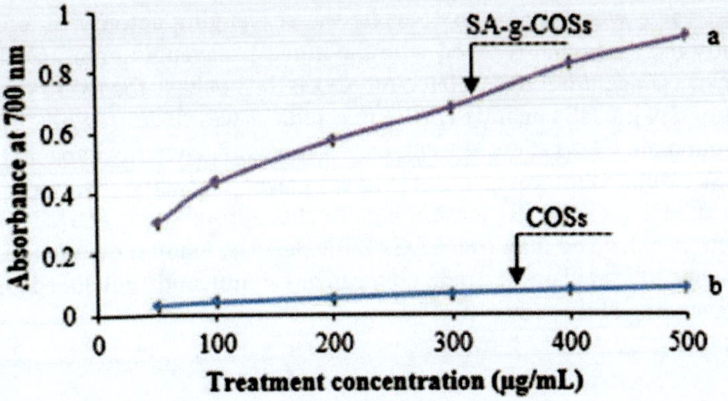

Fig. 3 Reducing power of COSs and SA-g-COSs. Test samples have potential reducing power reacted potassium ferricyanide (Fe^{3+}) to create potassium ferrocyanide (Fe^{2+}), after reacting with ferric chloride to form ferric ferrous complex that has green color and absorbances were read at 700 nm. Values are displayed as means ± SD of three repeated experiments. Different letters a-b indicate significant difference, $p < 0.05$

the preliminary plant extracts or compounds that they have antioxidants potentially [41]. As shown in Fig. 4, with the increase of concentration in the range of 10 to 100 μg/mL, COSs showed a slightly increasing trend of scavenging activity, but

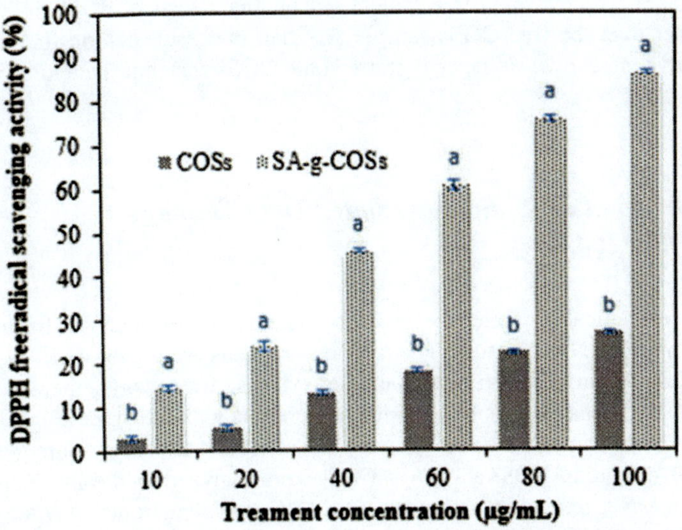

Fig. 4 DPPH scavenging activity of COSs and SA-g-COSs derivatives. The test samples reacted DPPH free radicals in the dark at room temperature for 30 min, and absorbances were recognized at 517 nm. All statistical analyses were performed with three independent experiments, and the data are shown as means ± SD. Different letters a-b indicate significant difference, $p < 0.05$

the highest value was only 27.06%, while the scavenging activity of SA-g-COSs significantly increased and reached with the highest scavenging ability of 86.15%. Moreover, at a concentration of 100 μg/mL COSs did not have the EC50 value, while the EC50 of SA-g-COSs against DPPH free radical was 56.75 μg/mL. According to this result, plain COSs showed weak DPPH radical scavenging capacity, whereas SA-g-COSs exhibited improved DPPH radical scavenging activities. In the previous study indicated that the DPPH scavenging effect of syringic acid grafted onto COSs had been more enhanced than free COSs [30]. Besides, many previous reports have shown that phenolic acid grafted onto chitosan has significantly enhanced DPPH free radical scavenging ability [29, 42].

3.5 *Hydroxyl Free Radical Scavenging Effect*

The hydroxyl radical is an oxidizing product relating to Fenton and Haber Weiss reaction. The ability of hydroxyl free radicals scavenging is applied to determine the potential antioxidant of compounds, including COSs and their derivatives. Potential antioxidants could reduce the pink color of TBA and malonaldehyde (MDA) complex, which at the high temperature condition and acid environment, MDA is formed between deoxyribose sugar and hydroxyl radicals that generating by the Fenton reactions [36]. The data displayed in Fig. 5. The results show that hydroxyl free radical scavenging effects of samples were enhanced when concentration increases. The hydroxyl free radical scavenging effects of derivatives samples were higher than the free COSs sample. So that, the result demonstrated that the antioxidative effect of syringic acid grafted onto COSs was improved more than free COSs.

3.6 *Inhibition of Radical Mediated DNA Damage in Raw264.7*

The hydroxyl radical generated by the Fenton reaction is considered to be the reason for breaking the DNA structure. The hydroxyl radicals are highly when exposed to DNA to destroy purines, pyrimidine and deoxyribose frameworks, thereby causing genomic DNA damage. This DNA damage increases with the increasing density of free radicals and can be the cause of many diseases such as cancer, mutation, aging, etc. [38]. In this study, COSs and SA-g-COSs were analyzed for the ability to protect DNA from DNA oxidation using genomic DNA isolated from RAW264.7 cells. Hydroxyl generation via the Fenton reaction started the oxidation of isolated DNA. COSs and SA-g-COSs were treated to determine the effect on the hydroxyl induced DNA damage was observed after electrophoresing the DNA on the agarose gel. The results demonstrated SA-g-COSs significantly inhibited oxidative damage of DNA.

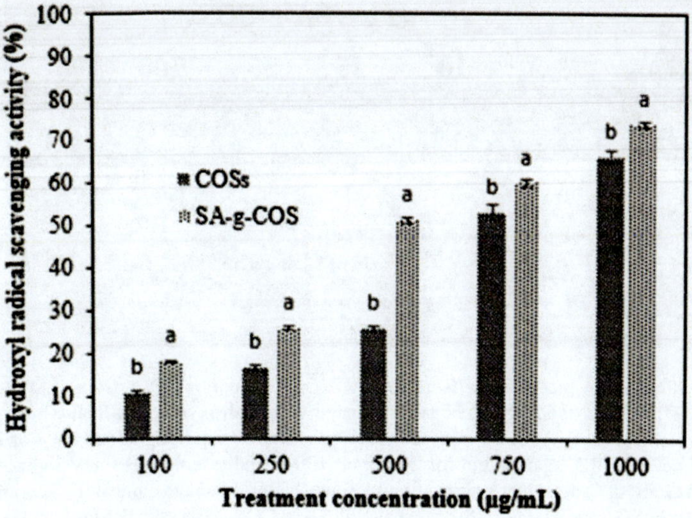

Fig. 5 OH radical scavenging by COSs and SA-g-COSs derivatives. Test samples as potential antioxidants could reduce the pink color of TBA and malonaldehyde (MDA) complex, which at the high-temperature condition and acid environment, MDA is formed between deoxyribose sugar and hydroxyl radicals generated by the Fenton system. Absorbances were read at 535 nm. The experiments are run three times, and results are presented as means ± SD. Different letters a-b indicate significant difference, $p < 0.05$

At 100 µg/mL, the band of SA-g-COSs shows the presence of DNA that is most clearly protected by the test substance and decreases gradually at 50 to 10 µg/mL. The protective effect of derivatives was 82.53%, while COSs was 26.26%. The results also show that the derivatives' DNA protection has improved significantly compared to the free COSs and dose dependent manner. Data were shown in Fig. 6.

No reports have confirmed the ability to protect the DNA damage in RAW264.7 cells of SA-g-COSs derivatives before. However, few previous reports have been shown that gallic acid grafted COSs displayed the protective effect of DNA oxidation more than 90% at 100 µg/mL [28], while this study results showed the protective effect of DNA oxidation more than 82.53% at the same concentration. In addition, other COSs derivatives also showed the protective ability of DNA damage, such as chitooligosaccharide-4-hydroxybenzyl derivatives that protected DNA damage from Chang liver cells [43]. The results of the study clarified that the derivatives formed, which could prevent oxidation from damaging the genomic DNA when the DNA is exposed to the hydroxyl free radical generated in the live cell system.

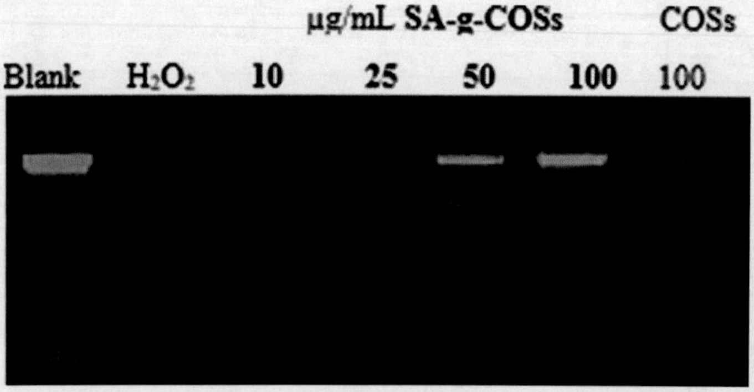

Fig. 6 DNA oxidative protection effects in RAW264.7. Genomic DNA from RAW264.7 cells was treated with COSs or SA-g-COSs and incubated with hydroxyl free radical which generated from the Fenton reaction. After 10 min, the reaction mixture containing about 1 µg of DNA was electrophoresed on a 1% agarose gel for 20 min at 100 V and visualized by UV light after being stained with GelRed. Lane 1: DNA alone (blank); Lane 2: DNA damage control ($FeSO_4$ and H_2O_2); Lanes 3 to 6: $FeSO_4$ and H_2O_2 in the presence of SA-g-COSs; Lane 7: $FeSO_4$ and H_2O_2 in the presence of COSs. Data are represented by three repeated experiments

3.7 Inhibitory Effect of Membrane Lipid Peroxidation in RAW264.7

Reports of many previous studies had shown that membrane lipid peroxidation is related to many pathologies in the human body [44]. In this experiment, the hydroxyl free radical was released from the Fenton reaction, and the OD value was recorded through the color product of the reaction between malondialdehyde and 2-thiobarbituric acid. The study results showed that the cell membrane lipid protection of CA-COSs was much better than the original COSs. At 100 µg/mL, the lipid protection activity of SA-g-COSs derivatives is 75.08%, while that of COSs is 38.00%. The results are detailed in Fig. 7. The previously reported result confirmed that gallic acid grafted COSs could decrease TBARS in the SW1353 cells and indicated that the protective effects of membrane lipids were 80% [37].

4 Conclusion

This study has demonstrated that the covalently linking syringic acid and COSs can afford to create a novel antioxidant. SA-g-COSs were found to be nontoxic and had protective effects of membrane lipid and DNA injury in RAW264.7 cell stronger than free COSs. Besides, derivatives could scavenge cellular radicals more enhanced than free COSs. Based on these results, we conclude that SA-g-COSs have potential free radical scavenging effects by both indirect and direct ways to inhibit and prevent molecular biological damage of free radicals in the live cells system.

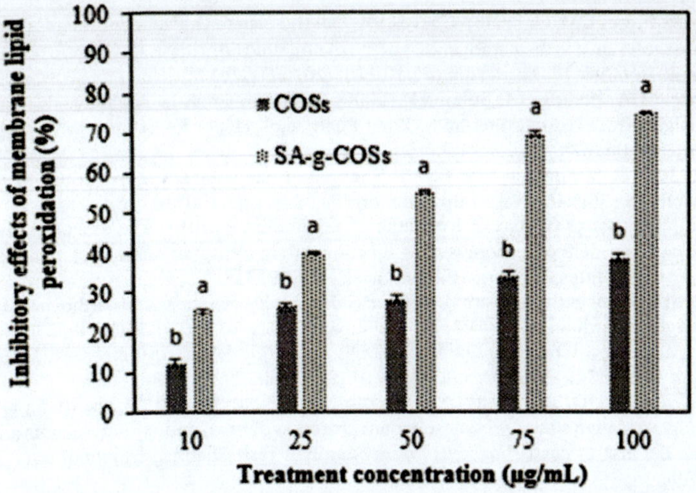

Fig. 7 The protective effects of membrane lipid peroxidation of COSs and SA-g-COSs in RAW264.7. The cells were exposed to hydroxyl radicals generated via Fenton's reaction, and peroxidation products of membrane lipids reacted with TBA to form a pink color complex which was measured spectroscopically at 528 nm. The experiments are replicated three times, and data are shown as means ± SD. Different letters a-b indicate significant difference, $p < 0.05$

The study is fundamental for antioxidant activity research in other cell models as well as other activities such as anti-inflammatory activity on the macrophage model. Furthermore, SA-g-COSs may promise used as additives in the pharmaceutical and functional food industry in order to the prevention or treatment of disease in which free radical formation.

Acknowledgements This work was supported by Vietnam National Foundation for Science and Technology Development (NAFOSTED) under no. 106.02-2019.47.

Conflicts of Interest The authors declare that they have no conflict of interest.

References

1. Cross CE, Halliwell B, Borish ET, Pryor WA, Ames BN, Saul RL, Mccord JM, Harman D (1987) Oxygen radicals and human disease. Ann Intern Med 107(4):526–545. https://doi.org/10.7326/0003-4819-107-4-526
2. McCord JM (2000) The evolution of free radicals and oxidative stress. Am J Med 108(8):652–659. https://doi.org/10.1016/S0002-9343(00)00412-5
3. Veselinovic M, Barudzic N, Vuletic M, Zivkovic V, Tomic-Lucic A, Djuric D, Jakovljevic V (2014) Oxidative stress in rheumatoid arthritis patients: relationship to diseases activity. Mol Cell Biochem 391(1–2):225–232. https://doi.org/10.1007/s11010-014-2006-6

4. Quinonez-Flores CM, Gonzalez-Chavez SA, Del Rio Najera D, Pacheco-Tena C (2016) Oxidative stress relevance in the pathogenesis of the rheumatoid arthritis: a systematic review. Biomed Res Int 2016:6097417. https://doi.org/10.1155/2016/6097417
5. Butterfield DA, Perluigi M, Sultana R (2006) Oxidative stress in Alzheimer's disease brain: new insights from redox proteomics. Eur J Pharmacol 545(1):39–50. https://doi.org/10.1016/j.ejphar.2006.06.026
6. Dias V, Junn E, Mouradian MM (2013) The role of oxidative stress in Parkinson's disease. J Parkinsons Dis 3(4):461–491. https://doi.org/10.3233/JPD-130230
7. Pollari E, Goldsteins G, Bart G, Koistinaho J, Giniatullin R (2014) The role of oxidative stress in degeneration of the neuromuscular junction in amyotrophic lateral sclerosis. Front Cell Neurosci 8:131. https://doi.org/10.3389/fncel.2014.00131
8. Pignatelli P, Menichelli D, Pastori D, Violi F (2018) Oxidative stress and cardiovascular disease: new insights. Kardiol Pol 76(4):713–722. https://doi.org/10.5603/KP.a2018.0071
9. Bowler RP, Crapo JD (2002) Oxidative stress in allergic respiratory diseases. J Allergy Clin Immunol 110(3):349–356. https://doi.org/10.1067/mai.2002.126780
10. Biller-Takahashi JD, Takahashi LS, Mingatto FE, Urbinati EC (2015) The immune system is limited by oxidative stress: dietary selenium promotes optimal antioxidative status and greatest immune defense in pacu *Piaractus mesopotamicus*. Fish Shellfish Immunol 47(1):360–367. https://doi.org/10.1016/j.fsi.2015.09.022
11. Hamada Y, Fujii H, Fukagawa M (2009) Role of oxidative stress in diabetic bone disorder. Bone 45:35–38. https://doi.org/10.1016/j.bone.2009.02.004
12. Khandrika L, Kumar B, Koul S, Maroni P, Koul HK (2009) Oxidative stress in prostate cancer. Cancer Lett 282(2):125–136. https://doi.org/10.1016/j.canlet.2008.12.011
13. Lodhi G, Kim YS, Hwang JW, Kim SK, Jeon YJ, Je JY, Ahn CB, Moon SH, Jeon BT, Park PJ (2014) Chitooligosaccharide and its derivatives: preparation and biological applications. Biomed Res Int 2014:654913. https://doi.org/10.1155/2014/654913
14. Hamed I, Özogul F, Regenstein JM (2016) Industrial applications of crustacean by-products (chitin, chitosan, and chitooligosaccharides): a review. Trends Food Sci Technol 48:40–50
15. Liaqat F, Eltem R (2018) Chitooligosaccharides and their biological activities: a comprehensive review. Carbohydr Polym 184:243–259. https://doi.org/10.1016/j.carbpol.2017.12.067
16. Phil L, Naveed M, Mohammad IS, Bo L, Bin D (2018) Chitooligosaccharide: an evaluation of physicochemical and biological properties with the proposition for determination of thermal degradation products. Biomed Pharmacother 102:438–451. https://doi.org/10.1016/j.biopha.2018.03.108
17. Park PJ, Je JY, Kim SK (2003) Free radical scavenging activity of chitooligosaccharides by electron spin resonance spectrometry. J Agric Food Chem 51:4624–4627. https://doi.org/10.1021/jf034039
18. Lillo L, Alarco'n J, Cabello G, Céspedes C, Caro C (2008) Antibacterial activity of chitooligosaccharides. J Biosci 63:644–648. https://doi.org/10.1515/znc-2008-9-1005
19. Jeon YJ, Kim SK (1999) Production of chitooligosaccharides using an ultrafiltration membrane reactor and their antibacterial activity. Carbohydr Polym 41:133–141. https://doi.org/10.1016/S0144-8617(99)00084-3
20. Shen KT, Chen MH, Chan HY, Jeng JH, Wang YJ (2009) Inhibitory effects of chitooligosaccharides on tumor growth and metastasis. Food Chem Toxicol 47:1864–1871. https://doi.org/10.1016/j.fct.2009.04.044
21. Li X, Wang J, Chen X, Tian J, Li L, Zhao M, Jiao Y, Zhou C (2011) Effect of chitooligosaccharides on cyclin D1, bcl-xl and bcl-2 mRNA expression in A549 cells using quantitative PCR. Chin Sci Bull 56(15):1629–1632. https://doi.org/10.1007/s11434-010-4501-9
22. Yang Y, Xing R, Liu S, Qin Y, Li K, Yu H, Li P (2019) Immunostimulatory effects of chitooligosaccharides on RAW 264.7 mouse macrophages via regulation of the MAPK and PI3K/Akt signaling pathways. Mar Drugs 17(1). https://doi.org/10.3390/md17010036
23. Ma P, Liu HT, Wei P, Xu QS, Bai XF, Du YG (2011) Chitosan oligosaccharides inhibit LPS-induced over-expression of IL-6 and TNF-α in RAW264.7 macrophage cells through blockade of mitogen-activated protein kinase (MAPK) and PI3K/Akt signaling pathways. Carbohydr Polym 84(4):1391–1398. https://doi.org/10.1016/j.carbpol.2011.01.045

24. Artan M, Karadeniz F, Karagozlu MZ, Kim MM, Kim SK (2010) Anti-HIV-1 activity of low molecular weight sulfated chitooligosaccharides. Carbohydr Res 345(5):656–662. https://doi.org/10.1016/j.carres.2009.12.017
25. Yoon NY, Ngo DN, Kim SK (2009) Acetylcholinesterase inhibitory activity of novel chitooligosaccharide derivatives. Carbohyd Polym 78:869–872. https://doi.org/10.1016/j.carbpol.2009.07.004
26. Liu J, Wang X, Bai R, Zhang N, Kan J, Jin C (2018) Synthesis, characterization, and antioxidant activity of caffeic-acid-grafted corn starch. Starch—Stärke 70(1–2). https://doi.org/10.1002/star.201700141
27. Cikman O, Soylemez O, Ozkan OF, Kiraz HA, Sayar I, Ademoglu S, Taysi S, Karaayvaz M (2015) Antioxidant activity of syringic acid prevents oxidative stress in l-arginine-induced acute pancreatitis: an experimental study on rats. Int Surg 100(5):891–896. https://doi.org/10.9738/INTSURG-D-14-00170.1
28. Ngo DH, Qian ZJ, Vo TS, Ryu B, Ngo DN, Kim SK (2011) Antioxidant activity of gallate-chitooligosaccharides in mouse macrophage RAW264.7 cells. Carbohydr Polym 84:1182–1288. https://doi.org/10.1016/j.carbpol.2011.01.022
29. Cho YS, Kim SK, Ahn CB, Je JY (2011) Preparation, characterization, and antioxidant properties of gallic acid-grafted-chitosans. Carbohydr Polym 83(4):1617–1622. https://doi.org/10.1016/j.carbpol.2010.10.019
30. Eom TK, Senevirathne M, Kim SK (2012) Synthesis of phenolic acid conjugated chitooligosaccharides and evaluation of their antioxidant activity. Environ Toxicol Pharmacol 34:519–527. https://doi.org/10.1016/j.etap.2012.05.004
31. Vo TS, Ngo DH, Bach LG, Ngo DN, Kim SK (2017) The free radical scavenging and anti-inflammatory activities of gallate-chitooligosaccharides in human lung epithelial A549 cells. Process Biochem 54:188–194. https://doi.org/10.1016/j.procbio.2017.01.001
32. Bui VH, Ngo DN (2019) The research determines appropriate parameters in the synthesis process of syringic acid grafted chitooligosaccharides. Sci Technol Dev J 22(3):317–323. https://doi.org/10.32508/stdj.v22i3.1289
33. Ngo DN, Kim MM, Kim SK (2012) Protective effects of aminoethyl-chitooligosaccharides against oxidative stress in mouse macrophage RAW 264.7 cells. Int J Biol Macromol 50:624–631. https://doi.org/10.1016/j.ijbiomac.2012.01.036
34. Lee DS, Woo JY, Ahn CB, Je JY (2014) Chitosan–hydroxycinnamic acid conjugates: preparation, antioxidant and antimicrobial activity. Food Chem 148:97–104. https://doi.org/10.1016/j.foodchem.2013.10.019
35. Blois MS (1958) Antioxidant determination by the use of a stable free radical. Nature 181:1199–1200
36. Halliwell B, Gutteridge JMC, Aruoma OI (1987) The deoxyribose method: a simple "test-tube" assay for determination of rate constants for reactions of hydroxyl radicals. Anal Biochem 165:215–219. https://doi.org/10.1016/0003-2697(87)90222-3
37. Ngo DH, Qian ZJ, Ngo DN, Vo TS, Wijesekara I, Kim SK (2011) Gally chitooligosaccharides inhibit intracellular free radical-mediated oxidation. Food Chem 128:974–981. https://doi.org/10.1016/j.foodchem.2011.03.128
38. Ngo DN, Kim MM, Kim SK (2008) Chitin oligosacchride inhibit oxidative stress in live cells. Carbohydr Polym 74:119–123. https://doi.org/10.1016/j.carbpol.2008.02.005
39. Ngo D-H, Qian Z-J, Ngo D-N, Vo T-S, Wijesekara I, Kim S-K (2011) Gallyl chitooligosaccharides inhibit intracellular free radical-mediated oxidation. Food Chem 128(4):974–981. https://doi.org/10.1016/j.foodchem.2011.03.128
40. Canabady-Rochelle LL, Harscoat-Schiavo C, Kessler V, Aymes A, Fournier F, Girardet JM (2015) Determination of reducing power and metal chelating ability of antioxidant peptides: revisited methods. Food Chem 183:129–135. https://doi.org/10.1016/j.foodchem.2015.02.147
41. Aytekin AO, Morimura S, Kida K (2011) Synthesis of chitosan-caffeic acid derivatives and evaluation of their antioxidant activities. J Biosci Bioeng 111(2):212–216. https://doi.org/10.1016/j.jbiosc.2010.09.018

42. Chatterjee NS, Panda SK, Navitha M, Asha KK, Anandan R, Mathew S (2015) Vanillic acid and coumaric acid grafted chitosan derivatives: improved grafting ratio and potential application in functional food. J Food Sci Technol 52(11):7153–7162. https://doi.org/10.1007/s13197-015-1874-4
43. Trinh MDL, Dinh MH, Ngo DH, Tran DK, Tran QT, Vo TS, Ngo DN (2014) Prevention of H_2O_2-induced oxidative stress in Chang liver cells by 4-hydroxybenzyl-chitooligomers. Carbohydr Polym 103:502–509. https://doi.org/10.1016/j.carbpol.2013.12.061
44. Liu J, Pu H, Liu S, Kan J, Jin C (2017) Synthesis, characterization, bioactivity and potential application of phenolic acid grafted chitosan: a review. Carbohydr Polym 174:999–1017. https://doi.org/10.1016/j.carbpol.2017.07.014

Identification of Plant Disease Based on Multi-feature Extraction

Kien Trang, Long TonThat, Khanh-Linh Nguyen, and Gia-Huy Tran

Abstract Plant disease is known as the largest intimidation, which directly affects on production and causes economic loss. Traditionally, human experience is used in defining the class of plant disease, but this often leads to wrong recognition. Thanks to the rapid development of modern technology, many machine-based methods with high efficiency and accuracy can be easily developed. Recently, computer vision has been widely used in various applications which have become intelligent supporters of human. This paper proposes an image-based method using Gray Level Co-occurrence Matrix (GLCM) and Local Binary Pattern (LBP) feature extraction for plant disease recognition. Contrast enhancement is used to increase the visual quality in the pre-processing stage. Finally, the classification is done by Support Vector Machine (SVM), and different schemes are used to justify the performance. Overall, the obtained result demonstrates that the proposed approach can identify six classes of plant-based on leaves.

Keywords Plant disease · Gray level co-occurrence matrix · Local binary pattern · Computer vision

1 Introduction

Agriculture is not only one of the essential sources of human livelihood but also one of the important economic fields aiming to produce food for our human existence. High quality of food is able to bring variety of nutrition for human sustenance. According to [1], a large portion of the population is facing with hunger. This may lead to the shortages in nutrients directly affecting the health. Besides, the growth of the population also becomes one of the burdens for agriculture manufacturing.

K. Trang · L. TonThat (✉) · G.-H. Tran
School of Electrical Engineering, International University, VNU-HCMC, Ho Chi Minh City, Vietnam
e-mail: ttlong@hcmiu.edu.vn

K.-L. Nguyen
Faculty of Mechatronics and Electronics, Lac Hong University, Ho Chi Minh City, Vietnam

V. Van Toi et al. (eds.), *8th International Conference on the Development of Biomedical Engineering in Vietnam*, IFMBE Proceedings 85,
https://doi.org/10.1007/978-3-030-75506-5_44

According to the Food and Agricultural Organization (FAO) report, the world population is estimated to reach 9.1 billion by the year 2050, leading to the fact that the demand for agricultural food will rise by 70%. Therefore, the process of agricultural manufacturing needs to be improved to serve the multi-national market. One of the solutions to increase the quality and quantity of crop production is to minimize the impact of plant disease. Several approaches are proposed for dealing with these kinds of problems. Nevertheless, not all the solutions are treated as safety in some cases. For instance, despite the fact that pesticide has many advantages in terms of increasing productivity and controlling quality, the abuse of pesticide may bring a large impact on the agricultural environment and human [2]. Indeed, the early identification of disease may benefit from taking action to control the spread of disease and pets. Thus, the identification of plant disease has attracted the attention of many scholars.

Thanks to the advances in image processing and computer vision in over the last few years, many image-based applications are widely used in agriculture. Image-based recognition for plant disease is used to automatically classify and identify the types of diseases based on appearance. Compared with human labour, this approach can easily monitor the plant's large field and gives high efficiency and reliability. Normally, the leaf is considered as the essential part expressing symptoms of disease in the early-stage; thus, images of the leaf are used to analyze in a huge number of approaches to disease identification. Thus, many research studies have been done in this field. Wang et al. [3] proposed that Principal Component Analysis (PCA) is employed to reduce the dimensionality of different types of features, while the classifier, in this case, is Neural Networks (NNs). Bhange et al. [4] introduced a system based on a web-based tool to classify the diseases of fruit from the uploaded images. The authors used color, morphology, and color coherence vector as feature extraction process and the clustering algorithm is then employed by K-mean. The classification is performed by Support Vector Machine (SVM), which reached an accuracy of 82%. Results obtained in [5] proposed the approach to deal with tomato leaves disease identification. In order to increase reliability for further processes, some techniques are applied, such as leaf image extraction, image rescaling, and background removal in the pre-processing stage.

Gabor Wavelet Transform (GWT) is used as a feature extraction, and SVM with different kernels is used in the classification phase. The study of Thilagavathi and Abirami [6] used approaches for guava leaf identification. Region growing segmentation is applied to highlight the disease regions, while color transformation YCbCr and CIELAB to detect the infected regions. Then, feature extraction is processed by Scale-Invariant Feature Transform (SIFT) algorithm based on these regions. After experiments with two types of classifiers, SVM gives a better performance than the k-nearest neighbors (k-NN) algorithm. The work in [7] performs a combination in feature extraction (color, shape, and textures) after using Otsu's algorithm to segment the leaf images. Thus, the classifier in this case used a classification tree to identify six different types of diseases. The authors in [8, 9] used an effective method based on a convolutional neural network with a multilayer to classify the plant disease and diagnosis. Thanks to CNN's advantages, this can work with a massive dataset and achieve an impressive performance.

In this paper, we introduce an image-based method that combines two feature extraction algorithms Gray Level Co-occurrence Matrix (GLCM) and Local Binary Pattern (LBP) to classify the diseases based on the leaf images. Then, different kernels of Support Vector Machine (SVM) are used as classifiers in order to compare the performances. In this work, a small set of Plant Village dataset is used. The structure of this paper is organized as follows. The proposed method is described in Sect. 2 in detail. The results and discussions are given in Sect. 3, while a conclusion and recommendations for future work are described in Sect. 4.

2 Proposed Method

The proposed model for plant disease identification is illustrated in Fig. 1, which contains four stages. Initially, the leaf image dataset is prepared and rescaled. Then, for the pre-processing stage, contrast enhancement is applied to all images to achieve improvement in terms of visual quality. Afterward, Gray Level Co-Occurrence Matrix (GLCM) and Local Binary Patterns (LBP) algorithm are used for feature extraction. A combination of two types of features is employed before feeding into Support Vector Machine (SVM) as a classifier.

2.1 *Plant Village Dataset*

The dataset of Plant Village is recognized as a large dataset of plant leaf diseases created by Hughes and Salathe [10]. It comprises of approximately 56,000 leaf pictures of 19 crops of 38 kinds of diseases in the resolution of 256 × 256 pixels. In this study, a subset of the Plant Village dataset is used, which only consists of 1152 images of 3 types of plants, as shown in Table 1. A limited dataset can be judged easily in case that the techniques are able to apply in the real-life environment.

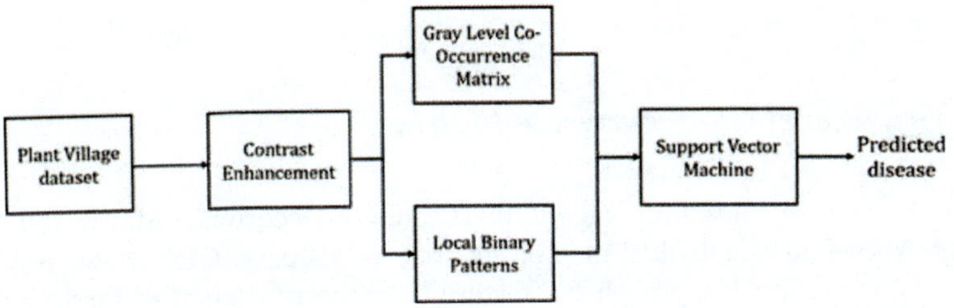

Fig. 1 Proposed approach

Table 1 Description for dataset

Plant	Status	Total images
Grape	Black rot	200
	Healthy	200
Peach	Bacterial spot	200
	Healthy	200
Potato	Early blight	200
	Healthy	152

Fig. 2 The visualization of contrast enhancement stage

2.2 *Contrast Enhancement*

Contrast is an essential element in determining image quality which is known as the luminance difference between adjacent regions. In terms of visualization, there are some pictures or images that may fall into the case of low contrast. This means that many areas in the image will not contribute to feature extraction or human perception. There are several methods for dealing with contrast enhancement, which has already been introduced in [11]. In this work, the approach described in [12] is applied to increase the quality of the image in the pre-processing stages. This approach's key concept is to keep the mean brightness of an input image when changing the contrast in the local areas. The visualization of the contrast enhancement stage is shown in Fig. 2.

2.3 *Gray Level Co-occurrence Matrix*

Firstly, the feature extraction stage, Gray Level Co-Occurrence Matrix (GLCM), which is known as a statistical method, is used in this case. GLCM also has been treated as an essential approach for feature extraction in terms of textures. This approach obtains the co-occurrence matrix by calculating the gray image and then obtains some features based on the co-occurrence matrix to represent some image's texture features. The GLCM method gives the image textures by computing the spatial relationship, which can reflect the gray image-level comprehensive information, including direction, adjacent interval, variation, etc... GLCM is obtained by calculating how often the pixel's grayscale intensity value occurs grayscale intensity value of adjacent pixels. This is computed by averaging four different angles: 0, 45, 90, and 135. Based on the GLCM, the formula for obtaining statistical texture features is defined, by contrast, correlation, energy, and homogeneity as follows.

Contrast implies a measure of the difference of intensity between neighbor pixels as calculated in

$$Contrast = \sum_{i,j} |i - j|^2 p(i, j) \tag{1}$$

where $p(i, j)$ is a GLCM, i and j are the indexes of $p(i, j)$.

Correlation implies a measure of how correlated between neighbor pixels as

$$Correlation = \sum_{i,j} \frac{(i - \mu_i)(j - \mu_j)p(i, j)}{\sigma_i \sigma_j} \tag{2}$$

where μ is the mean, and σ is the standard deviation of GLCM.

Energy implies the sum of squared elements of GLCM as defined in

$$Energy = \sum_{i,j} (p(i, j))^2 \tag{3}$$

Homogeneity implies the adjacency of the distribution of elements of GLCM as computed in

$$Homogeneity = \sum_{i,j} \frac{p(i, j)}{1 + |i - j|} \tag{4}$$

2.4 Local Binary Patterns

Simultaneously, Local Binary Patterns (LBP) features are also extracted from an input image proposed by Ojala et al. [13] since 1996. LBP is a texture descriptor that has the advantages of rotation invariance and simple computational process. The first step to building an LBP descriptor is to convert the input image to grayscale. For each pixel in the grayscale image, comparisons between the central pixel and other adjacent pixels occur. After thresholding, there will be a matrix that is used for storing the new value. In case that the value of the adjacent pixel is greater than or

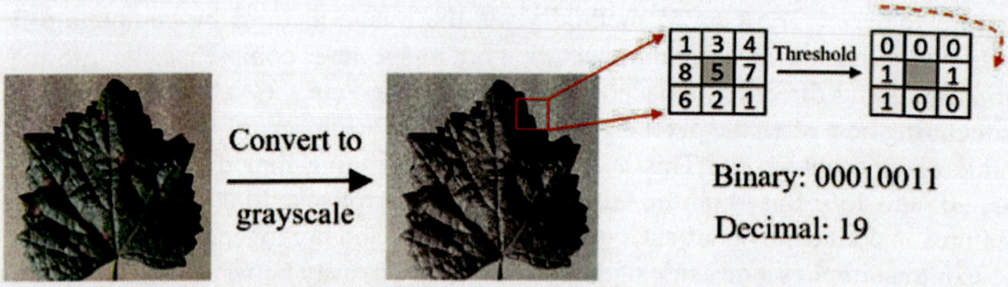

Fig. 3 The visualization of LBP features

equal to the value of the central pixel, it will return to 1. Otherwise, it will return to zero. The threshold formula for obtaining LBP value is calculated as

$$LBP(\chi_c, \gamma_c) = \sum_{p=0}^{P-1} 2^p s\left(i_p - i_c\right)$$

$$s(\chi) = \begin{cases} 1, \text{ if } \chi \geq 0 \\ 0, \text{ else} \end{cases} \tag{5}$$

where (χ_C, γ_C) is the central pixels; i_p and i_c represent the intensity value of neighbourhood pixels and the intensity value of the central pixel, respectively, while P is the number of neighbourhood pixels. The histogram statistics are then performed on the encoding of all pixels in a block to obtain LBP features. The visualization calculation of the LBP features is depicted in Fig. 3.

2.5 *Support Vector Machine*

SVM (Support Vector Machine) is a supervised machine learning algorithm commonly used in classification or regression problems which is proposed by [14]. Besides, SVM usage is applied in a variety of applications such as computer vision, image processing with various feature extraction algorithms. This method can reach better classification results on small sample data sets compared to other methods, leading to the widespread use of this classifier. The idea of SVM is to find an optimal hyperplane to separate data points so that the margins between the two classes are maximized. The learning algorithm of SVM is the optimal algorithm for solving convex quadratic programming. In addition, kernels are also used for nonlinear separable cases. In this study, after concatenating two types of features (GLCM and LBP), the polynomial kernel is applied for multi-class classification, which is defined as

$$K_{Poly}(\chi_i, \chi_j) = \left(\chi_i^T \chi_j + \eta\right)^{\rho} \tag{6}$$

where χ_i, χ_j are the feature vector and ρ represent the degree of the kernel.

3 Results and Discussions

To evaluate the proposed work, the dataset is divided into two sets, training and test sets. The training set accounts for 80% of the dataset using for training perimeters, while the remaining 20% is treated as a test set using for performance evaluation. This work is implemented on a computer with processor Intel(R) Core(TM) i5-7500, 32 GB for RAM. GPU used is GTX 1070Ti with 8 GB VRAM.

We test the work in three cases based on the SVM classifier. For the first scheme, only the LBP feature extraction method is used, while a combination of LBP and GLCM is used in the second scheme. The confusion matrices of the first and second schemes are shown in Tables 2 and 3, respectively. It is noticed that using the combined LBP and GLCM methods improves the accuracy in terms of potato while the bacterial spot of peach decreases 2.5% compared with scheme 1. The first and second schemes' average accuracy are 90.83% and 91.38% over the experiments, respectively. Especially for scheme 3, as shown in Table 4, we apply a contrast enhancement method in the pre-processing stage, which improves for almost all cases. The average accuracy in this 3rd scheme reaches the highest accuracy—92.08%.

Table 2 Confusion matrix for only LBP method (Scheme 1)

%	Grape_ Black Rot	Grape_ Healthy	Peach_ Bacterial Spot	Peach_ Healthy	Potato_ Early Blight	Potato_ Healthy
Grape_ Black Rot	**97.5**	0	2.5	0	0	0
Grape_ Healthy	0	**97.5**	2.5	0	0	0
Peach_ Bacterial Spot	0	0	**72.5**	17.5	10	0
Peach_ Healthy	0	0	0	**100**	0	0
Potato_ Early Blight	0	0	0	2.5	**97.5**	0
Potato_ Healthy	0	3.33	6.67	0	10	**80**

The bold elements (diagonal elements) represent the accuracy for which the predicted label is equal to the true label

Table 3 Confusion matrix for the combination of LBP and GLCM methods (Scheme 2)

%	Grape_ Black Rot	Grape_ Healthy	Peach_ Bacterial Spot	Peach_ Healthy	Potato_ Early Blight	Potato_ Healthy
Grape_ Black Rot	**97.5**	0	2.5	0	0	0
Grape_ Healthy	0	**97.5**	2.5	0	0	0
Peach_ Bacterial Spot	0	5	**70**	20	2.5	2.5
Peach_ Healthy	0	0	0	**100**	0	0
Potato_ Early Blight	0	0	0	0	**100**	0
Potato_ Healthy	0	0	6.67	0	10	**83.33**

The bold elements (diagonal elements) represent the accuracy for which the predicted label is equal to the true label

Table 4 Confusion matrix for the combination of LBP and GLCM methods with contrast enhancement step (Scheme 3)

%	Grape_ Black Rot	Grape_ Healthy	Peach_ Bacterial Spot	Peach_ Healthy	Potato_ Early Blight	Potato_ Healthy
Grape_ Black Rot	**97.5**	0	2.5	0	0	0
Grape_ Healthy	0	**100**	0	0	0	0
Peach_ Bacterial Spot	0	5	**75**	15	2.5	2.5
Peach_ Healthy	0	0	0	**100**	0	0
Potato_ Early Blight	0	0	0	0	**100**	0
Potato_ Healthy	0	0	6.67	0	13.33	**80**

The bold elements (diagonal elements) represent the accuracy for which the predicted label is equal to the true label

4 Conclusion

This paper has introduced an image-based approach using the combination of GLCM and LBP feature extraction methods to recognize the plant leaf disease with a limited dataset. Each of the features has different characteristics; thus, the concatenating two features may bring an improvement in classification rate. Besides, using the pre-processing step, such as contrast enhancement, also contributes to increasing visualization quality. Overall, the testing results show that the proposed approach reaches the average accuracy of 92.08%. The method can be improved in future studies by testing different combinational feature extraction methods with new training algorithms.

Conflicts of Interest The authors have no conflict of interest to declare.

References

1. Wise TA (2019) World hunger is on the rise. Chain React 136:42–43
2. Aktar MW, Sengupta D, Chowdhury A (2009) Impact of pesticides use in agriculture: their benefits and hazards. Interdiscip Toxicol 2(1):1–12. https://doi.org/10.2478/v10102-009-0001-7
3. Wang H, Li G, Ma Z, Li X (2012) Image recognition of plant diseases based on principal component analysis and neural networks. In: 2012 8th international conference on natural computation. Chongqing, pp 246–251. https://doi.org/10.1109/ICNC.2012.6234701
4. Bhangea M, Hingoliwala HA (2015) Smart farming: pomegranate disease detection using image processing. Procedia Comput Sci 58:280–288. https://doi.org/10.1016/j.procs.2015.08.022
5. Mokhtar U, Ali MAS, Hassenian AE, Hefny H (2015) Tomato leaves diseases detection approach based on Support Vector Machines. In: 2015 11th international computer engineering conference (ICENCO). Cairo, pp 246–250. https://doi.org/10.1109/ICENCO.2015.7416356
6. Thilagavathi M, Abirami S (2017) Smart farming: application of image processing in diagnosing guava leaf diseases. Int J Sci Res Manag 5(7):5927–5933. https://doi.org/10.18535/ijsrm/v5i7.19
7. Sabrol H, Satish K (2016) Tomato plant disease classification in digital images using classification tree. In: 2016 international conference on communication and signal processing (ICCSP). Melmaruvathur, pp 1242–1246. https://doi.org/10.1109/ICCSP.2016.7754351
8. Militante SV, Gerardo BD, Dionisio NV (2019) Plant leaf detection and disease recognition using deep learning. In: 2019 IEEE Eurasia conference on IOT, communication and engineering (ECICE), Yunlin, Taiwan, pp 579–582. https://doi.org/10.1109/ECICE47484.2019.8942686.
9. Singh UP, Chouhan SS, Jain S, Jain S (2019) Multilayer convolution neural network for the classification of mango leaves infected by anthracnose disease. IEEE Access 7:43721–43729. https://doi.org/10.1109/ACCESS.2019.2907383
10. Hughes DP, Salathe M (2016) An open access repository of images on plant health to enable the development of mobile disease diagnostics. arXiv preprint arXiv:1511.08060
11. Maragatham G, Roomi SMM (2015) A review of image contrast enhancement methods and techniques. Res J Appl Sci Eng Technol 9:309–326. https://doi.org/10.19026/rjaset.9.1409
12. Rahman MA, Liu S, Lin SCF, Wong CY, Jiang G, Kwok N (2015) Image contrast enhancement for brightness preservation based on dynamic stretching. Int J Image Process 9:241–253
13. Ojala T, Pietikainen M, Harwood D (1996) A comparative study of texture measures with classification based on featured distributions. Pattern Recogn 29(1):51–59. https://doi.org/10.1016/0031-3203(95)00067-4
14. Boser BE, Guyon IM, Vapnik VN (1992) A training algorithm for optimal margin classifiers. In: Proceedings of the fifth annual workshop on computational learning theory, pp 144–152. https://doi.org/10.1145/130385.130401

Cyclodipeptides Isolated from a Marine-Derived Fungus *Penicillium chrysogenum* M612 of Bai Tu Long Sea, Quang Ninh, Vietnam

Thi Nhu Quynh Bach, Duc Tuan Cao, Thi Hong Lien Hoang, Anh Duy Do, Mai Anh Nguyen, Thi Quyen Vu, Thi Dao Phi, Thi Hong Minh Le, Thi Mai Huong Doan, Van Cuong Pham, Trong Quan Khong, Quoc Tuan Nguyen, Jung-Woo Chae, Hwi-yeol Yun, Min-Kyun Na, Young-Ho Kim, Van Chuc Dang, Truong Son Nguyen, and Van Hung Nguyen

Abstract Bacterial resistance is getting more and more serious, and Vietnam is one of the high bacterial resistance prevalent countries in the world. Hence, the development of new antimicrobial agents, domestically, is urgently needed. During the course of our screening program, the ethyl acetate extract of a marine-derived fungus strain (*Penicillium chrysogenum* M612), isolated from sediment collected in Bai Tu Long sea, showed good inhibition activities against tested clinical important microbial pathogens. Subsequent analysis of the extracts prepared from the strain M612's culture broth (50L) led to the isolation of seven cyclodipeptides. Structure of isolated compounds was determined, based on MS; NMR spectroscopic data analysis and compared with literature, as cyclo-(L-Pro-L-Leu) (**1**), cyclo-(L-Pro-L-Val) (**2**), cyclo-(L-Ala-L-Pro) (**3**), cyclo-(L-Pro-L-ILe) (**4**), cyclo-(L-Pro-L-Phe) (**5**), cyclo-(D-Pro-D-Phe) (**6**) and cyclo-(L-Pro-L-Tyr) (**7**). All the compounds were previously

Bach Thi Nhu Quynh and Cao Duc Tuan contributed equally to this work as first author.

T. N. Q. Bach · D. T. Cao (✉) · V. C. Dang · T. S. Nguyen · V. H. Nguyen (✉)
Hai Phong University of Medicine and Pharmacy, Haiphong, Viet Nam
e-mail: cdtuan@hpmu.edu.vn

V. H. Nguyen
e-mail: nvhung@hpmu.edu.vn

T. H. L. Hoang
Buon Ma, Thuot University, Dak Lak, Viet Nam

A. D. Do
Research Institute for Marine Fisheries, Hai Phong, Viet Nam

M. A. Nguyen · T. Q. Vu · T. D. Phi · T. H. M. Le · T. M. H. Doan · V. C. Pham
Institute of Marine Biochemistry, Vietnam Academy of Science and Technology, Hanoi, Viet Nam

T. Q. Khong · Q. T. Nguyen · J.-W. Chae · H-y. Yun · M.-K. Na · Y.-H. Kim
Institute of Drug Research and Development and College of Pharmacy, Chungnam National University, Daejeon, Republic of Korea

V. Van Toi et al. (eds.), *8th International Conference on the Development of Biomedical Engineering in Vietnam*, IFMBE Proceedings 85,
https://doi.org/10.1007/978-3-030-75506-5_45

reported to have antimicrobial activities. The obtained results suggested that strain M612 can be further studied to develop an antimicrobial agent.

Keywords Antimicrobial · Cyclodipeptides · Fungus · Marine · *Penicillium chrysogenum* · Vietnam east sea

1 Introduction

Microorganisms (bacteria, fungi, virus and parasites) are the main causes of infectious diseases, a major public health problem all over the world [1]. The discovery of penicillin from fungus *Penicillium chrysogenum* in 1928 marked the new era of antibiotics [2]. Since then, many efforts have been made to develop new antimicrobial agents, especially from natural sources. During the period of 1981 to 2014, there were about 80 natural derived approved antibiotics [3]. However, the antibiotic resistance increases rapidly worldwide, and it is estimated to cause 10 million death by 2050 [4]. Therefore, there is an ongoing demand for the search for new antibiotics to deal with antibiotic resistance problem.

Marine environment, accommodate about 70% of the earth surface, is proven to be a valuable source for drug discoveries, especially marine microorganisms. In 2017, of 1490 new compounds from the sea, the marine microorganism was accounted for 57% [5]. Vietnam, situated in the west of the Pacific Ocean, with very high marine biodiversity, but the research on antimicrobial compounds from marine fungi has started recently, and there are not many data published [6].

During the course of our project in searching for antimicrobials from Vietnam marine-derived fungi, the EtOAc extract of a marine fungus strain *Penicillium chrysogenum* M612 isolated from sediment of Bai Tu Long sea, Quang Ninh, Vietnam showed inhibition activities against three Gram positive bacteria *Bacillus cereus* ATCC 14579, *Enterococcus faecalis* ATCC 29212, *Staphylococcus aureus* ATCC 25923 and one yeast strain *Candida albicans* ATCC 10231 with MIC values of 256, 256, 256 and 128 μg/mL, respectively. This paper described the isolation and structure characterization of seven cyclodipeptides (**1–7**) from the fungus *Penicillium chrysogenum* M612 (Fig. 1).

2 Material and Methods

2.1 General

A Polax-2L Polari-meter was used to measure optical rotations. Electron Spray Ionization Mass Spectroscopy (ESI–MS) experiment was done on an Agilent spectrometer (1100 LC-MSD Trap). A Bruker 300 MHz instrument (Avance 300) was

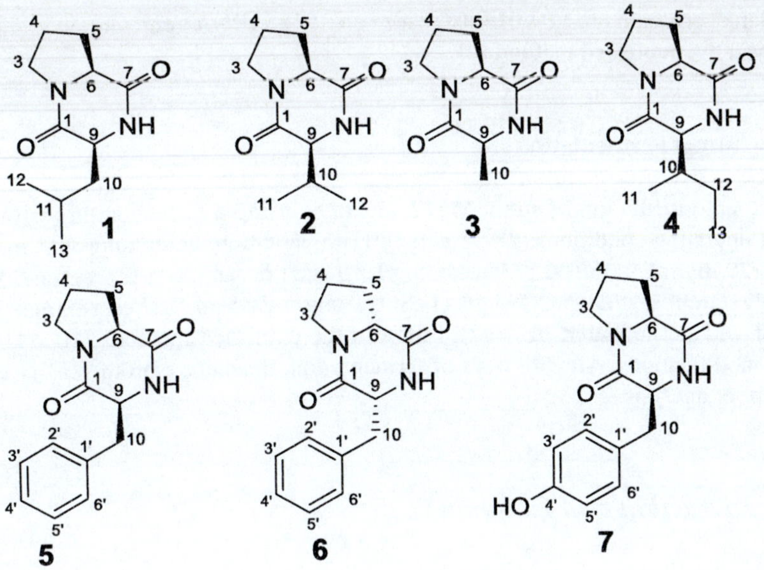

Fig. 1 Compounds 1–7 isolated from strain M612

used to record Nuclear Magnetic Resonance (NMR) spectra. Thin layer chromatography (TLC) was done on TLC silica gel (Merck 60 F_{254}). Sephadex LH-20 (25–100 μm) or silica gel (Kieselgel 60, 40–63 μm) was used as stationary phase for column chromatography (CC).

2.2 *Marine Materials*

2.2.1 Fungi Isolation and Identification

The marine sample (sediment) was collected in Bai Tu Long sea (Quang Ninh, Vietnam) in August 2019. Fungi isolation was done, followed published methods with modifications [7, 8]. First, 0.5 g of sediment was transferred to 4.5 mL of de-ionized sterile water, after homogenization, the suspension was heat-treated at 60 °C for 6 min. Next, 0.5 mL of the heat-treated suspension was used for serial dilutions in sterile distilled water to achieved 10^{-3} ratio. 50 μL of the final diluent were spread on agar dish pre-filled with A1 solid medium, including 15 g Agar; 4 g Yeast extract; 30 g Instant ocean; 2 g Peptone; 10 g Soluble starch and de-ionized sterile water to 1 L. The dishes were kept in the incubator at 28 °C for 15 days. Further purifications were done by transferring single colonies of fungi to new petri dishes of PDA medium (Agar: 15 g; Dextrose: 20 g; Potato extract: 30 g and de-ionized sterile water: to 1 L of total volume). The strain M612 was later identified as *Penicillium chrysogenum*,

GenBank accession no. **MW015801**, based on its 18S rRNA gene sequence analysis, followed the published method [9].

2.2.2 Fungi Fermentation

The 50 L fermentation of strain M612 was done using a fermentation instrument, preset operating conditions: 28 °C and 100 rpm agitation, containing $A1^+$ medium (KBr (20 mg/mL): 5 mL/L; Glucose 5 g/L; Instant ocean: 30 g/L; Peptone: 3 g/L; Soluble starch: 15 g/L; $CaCO_3$ 1 g/L; $FeSO_4$ (8 mg/mL): 5 mL/L; Yeast extract: 5 g/L; De-ionized sterile water; pH = 7), followed the published methods [10, 11], with some modifications. After 14 days of fermentation, the culture broth was harvested for further analysis.

2.3 *Extraction and Isolation*

The 50 L culture broth obtained from the fermentation of strain M612 was firstly filtered, then, allowed to pass slowly through a column (Ø = 10 cm) packed with Amberlite-XAD 16 (15 kg). After that, the column was eluted with MeOH (40 L), the eluent was collected, removed solvent under reduced pressure to obtain 20.1 g of methanol extract (MM612). The MM612 was re-suspended in 700 mL of distilled water and successively partitioned with EtOAc and CH_2Cl_2. Solvents were removed in vacuo to obtain the EtOAc (EM612, 2.92 g) and CH_2Cl_2 (CM612, 10 g) extracts. The EM612 (2.92 g) was separated by CC (CH_2Cl_2/MeOH gradient; silica gel) to afford seven fractions. Fraction 3 (458 mg) was purified by CC (silica gel; *n*-hexane/EtOAc, 10–70% EtOAc in *n*-hexane) to give compound 1 (9 mg). Fraction 4 (280 mg) was firstly chromatographed on CC column (Sephadex LH-20; MeOH), then undergone preparative thin layer chromatography (p-TLC) to afford compound 4 (12 mg). Fraction 5 (280 mg) was separated by CC (silica gel; *n*-hexane/acetone, 10–50% acetone in *n*-hexane) to give 6 sub-fractions, named F5.1-F5.6, respectively. Sub-fraction F5.3 (78 mg) was recrystallized in EtOAc to provide compound 2 (12 mg). Fraction 6 (1.42 g) was subjected to CC (silica gel, CH_2Cl_2/MeOH gradient) to achieve six sub-fractions, named F6.1-F6.6, respectively. Sub-fraction F6.3 (165 mg) was separated on silica gel CC, using 5% CH_2Cl_2 in methanol, providing compound 3 (8 mg). Sub-fraction F6.4 (185 mg) was separated by CC (Sephadex LH-20; MeOH), in combination with p-TLC to give compound 5 (9 mg). Recrystallizing sub-fraction F5.5 (70 mg) in CH_2Cl_2 brought about compound 6 (7 mg). Sub-fraction F6.6 (95 mg) was separated by CC (CH_2Cl_2/acetone gradient; silica gel) affording compound 7 (15 mg).

2.4 *Physical and 1-D NMR Data of Isolated Compounds (1–7)*

Cyclo-(L-Pro-L-Leu) (1): White solid, ESI–MS: *m/z* 249 [M+K]$^+$, $[\alpha]_D^{25}$ −87 (*c* 0.25, MeOH); ^{1}H-NMR (500 MHz, $CDCl_3$) and ^{13}C-NMR (125 MHz, $CDCl_3$): see Tables 1 and 2.

Cyclo-(L-Pro-L-Val) (2): Amorphous solid, ESI–MS: *m/z* 197 [M+H]$^+$, $[\alpha]^{25}_D$ = −97 (*c* 0.33, MeOH); ^{1}H-NMR (300 MHz, $CDCl_3$) and ^{13}C-NMR (75 MHz, $CDCl_3$): see Tables 1 and 2.

Cyclo-(L-Ala-L-Pro) (3): White solid, ESI–MS: *m/z* 169 [M+H], $[\alpha]^{25}_D$ = −88.6 (*c* 0.6, MeOH); ^{1}H-NMR (300 MHz, CD_3OD) and ^{13}C-NMR (75 MHz, CD_3OD): see Tables 1 and 2.

Table 1 ^{1}H-NMR data of compounds (**1–7**) isolated from strain M612

Proton	δ_H^{a,e} mult. (*J*, Hz)						
	1^{a,c}	**2**^{a,d}	**3**^{b,d}	**4**^{a,d}	**5**^{b,d}	**6**^{b,d}	**7**^{b,d}
3	3.55 m 3.60 m	3.57 m	3.54 m	3.58 m	3.57 m 3.40 m	3.57 m 3.22 m	3.57 m 3.38 m
4	1.90 m 2.05 m	2.34 m 2.00 m	2.00 m	2.02 m 1.93 m	1.83 m	1.68 m	1.83 m
5	2.34 m 2.13 m	2.00 m	2.33 m 2.00 m	2.37 m 1.99 m	2.12 dt (5.5, 6.0) 1.26 m	2.08 m 1.92 m	2.12 m 1.25 m
6	4.12 t (8.0)	4.23 t (7.2)	4.20 m	4.18 m	4.09 m	4.08 m	4.38 m
9	4.02 m	4.06 s	4.28 m	4.29 m	4.47 t (7.0)	4.23 t (4.5)	4.07 m
10	1.53 m 2.05 m	2.52 m	1.40 d (7.0)	2.32 m	3.19 d (5.0)	3.02 dd (4.6, 13.6) 2,65 m	3.08 m
11	1.77 m	0.95 d (7.2)		1.13 d (7.0)			
12	1.00 d (6.5)	1.12 d (7.2)		1.49 m 1.57 m			
13	0.95 d (6.5)			1.00 t (7.0)			
2′,6’					7.28 m	7.22 m	7.06 d (8.5)
3’, 5’					7.28 m	7.33 m	6.72 d (8.5)
4’					7.28 m	7.33 m	–

Measured in ^{a}$CDCl_3$, ^{b}CD_3OD, ^{c}500 MHz, ^{d}300 MHz

Table 2 ^{13}C-NMR data of compounds (**1–7**) isolated from strain M612

C	**1**[a,c]	**2**[a,d]	**3**[b,d]	**4**[a,d]	**5**[b,d]	**6**[b,d]	**7**[b,d]
1	167.52	167.56	167.65	167.51	166.90	167.43	166.96
3	45.03	46.17	45.03	46.14	45.95	46.14	45.92
4	22.87	23.25	22.22	22.21	22.77	22.48	22.72
5	27.66	29.52	27.78	29.51	29.37	29.83	29.40
6	58.87	61.51	59.07	59.95	60.07	59.77	60.06
7	171.39	172.58	171.20	172.24	170.92	168.69	170.78
9	53.24	60.03	50.69	61.27	57.68	59.11	57.90
10	38.01	29.87	14.30	37.01	38.18	40.98	37.67
11	24.37	16.66		15.54			
12	23.24	18.84		25.72			
13	20.79			12.59			
1'					128.06	128.52	127.63
2'					129.45	129.65	132.11
3'					131.03	131.28	116.19
4'					137.36	136.72	157.68
5'					131.03	131.28	116.19
6'					129.45	129.65	132.11

Measured in [a]$CDCl_3$, [b]CD_3OD, [c]125 MHz, [d]75 MHz

Cyclo-(L-Pro-L-ILe) (4): White solid, ESI–MS: *m/z* 211 $[M+H]^+$, $[\alpha]_D^{25}$ −79.7 (*c* 0.15, MeOH); ^{1}H-NMR (300 MHz, $CDCl_3$) and ^{13}C-NMR (75 MHz, $CDCl_3$): see Tables 1 and 2.

Cyclo-(L-Pro-L-Phe) (5): White solid, ESI–MS: *m/z* 245 $[M+H]^+$, $[\alpha]^{15}_D$ = −79.2 (*c* 0.6, MeOH); ^{1}H-NMR (300 MHz, CD_3OD) and ^{13}C-NMR (75 MHz, CD_3OD): see Tables 1 and 2.

Cyclo-(D-Pro-D-Phe) (6): White solid, ESI–MS (*m/z*): 245 $[M+H]^+$, $[\alpha]^{25}_D$ = + 59.6 (*c* 0.12, MeOH,); ^{1}H-NMR (300 MHz, CD_3OD) and ^{13}C-NMR (75 MHz, CD_3OD): see Tables 1 and 2.

Cyclo-(L-Pro-L-Tyr) (7): White solid, ESI–MS: *m/z* 261 $[M+H]^+$, $[\alpha]^{15}_D$ = −83.6 (*c* 0.6, MeOH); ^{1}H-NMR (300 MHz, CD_3OD) and ^{13}C-NMR (75 MHz, CD_3OD): see Tables 1 and 2.

3 Results and Discussion

3.1 *Structure and Biological Activities of Isolated Compounds (1–7)*

Compound **1** was separated and purified in the form of a white solid. It was optically active with $[\alpha]_D^{25} = -87$ (*c* 0.25; MeOH). The pseudo-molecular ion $[M+K]^+$ of 1 has appeared in its ESI (positive) mass spectrum at 249 (*m/z*). Analysis of the ^{13}C-NMR and DEPT spectra of 1 indicated that there were 11 carbons presented, including two carbonyl carbons at δ_C, 171.39 (C=O); 167.52 (C=O), three sp^3 methine groups at δ_C 58.87 (C-6), 53.24 (C-9) and 24.37 (C-11), four sp^3 methylene groups at δ_C 45.03 (C-3), 38.01 (C-10), 27.66 (C-5) and 22.87 (C-4), two methyl groups at δ_C 23.24 (C-12) and 20.79 (C-13). In the ^{1}H NMR spectrum, signals of 2 methyl groups at δ_H 1.00 (3H, d, $J = 6.5$ Hz, CH_3-12) and 0.95 (3H, d, $J = 6.5$ Hz, CH_3-13), two methine groups bearing nitrogen at δ_H 4.02 (1H, m, H-9); 4.12 (1H, t, $J =$ 8.0 Hz, H-6) were observed. There were also eight aliphatic protons signals shown in the δ_H region of 1.53 to 3.60 ppm. The chemical shifts of CH-9, CH-6 and CH_2-3 suggested their direct connection to nitrogen atoms. NMR spectra indicated that 1 was a cyclodipeptide, with the combination of proline and a leucine unit in its structure. Compared the obtained data of this compound with reference [12–14], **1** was confirmed to be cyclo-(L-Leu-L-Pro). Compound 1 shows weak antifungal activity against *Candida albicans* [15].

Compound **2** was separated and purified as an amorphous solid. It had negative optical rotation, $[\alpha]^{25}{}_D = -97$ (*c* 0,33, MeOH). The protonated ion $[M+H]^+$ of compound **2** appeared in its ESI (positive) mass spectrum at 197 (*m/z*). The ^{1}H-NMR and ^{13}C NMR spectra of compound **2** revealed that its structure was similar to **1**, except, there is an absence of a methylene group in the structure of **2**. The NMR spectra of **2** were analyzed entirely and compared with reported data [16–18], confirming that compound **2** was cyclo-(L-Pro-L-Val). This compound could inhibit the free radical (OH$^\bullet$) with the inhibition level of 64.9% [19]. In another study, compound **2**, isolated from a microorganism associated with a sponge from the South China Sea, also, inhibits *Aspergillus parasiticus* to produce aflatoxin [20].

Compound **3** was separated and purified as a white solid. It was active optically with $[\alpha]^{25}{}_D = -88.6$ (*c* 0.6, MeOH). The pseudo-molecular ion peak $[M+H]^+$ of compound **3** was shown in its ESI (positive) mass spectrum at $m/z = 169$. The ^{13}C-NMR and DEPT spectra of **3** indicated that there were eight carbons, including two carbonyl carbons at δ_C 167.65 (C=O) and 171.20 (C=O), three sp^3 methylene groups at δ_C 22.22 (C-4), 27.78 (C-5) and 45.03 (C-3), two sp^3 methine groups at δ_C 50.69 (C-9) and 59.07 (C-6), one methyl group at δ_C 14.30 (C-10). The chemical shifts of carbons at position 3, 6 and 9 suggested their direct connection to oxygen or nitrogen atom. The ^{1}H-NMR data of **3** indicated the appearance of a methyl group at δ_H 1.40 (3H, d, $J = 7.0$ Hz, CH_3-10), two sp^3 methine groups connecting to nitrogen at δ_H 4.20 (1H, m, H-6); 4.28 (1H, m, H-9) and one sp^3 methylene group connecting to nitrogen at δ_H 3,54 (2H, m, H-3) and 4 protons at δ_H 2.00–2.33. Comparison with the reference [21], compound **3** was confirmed as cyclo-(L-Ala-L-Pro). **3** has been reported to induce growth inhibition of *A. citreus* SBUG 321, *Bacillus subtilis* subsp. *spizizenii*, *Micrococcus luteus* and *Staphylococcus aureus* at the concentration range of 25–100 μg/mL [16]. Besides, compound **3** was shown as a macrophage cytokines stimulator, by inducing J774A.1 cells to increase secretion of IL-10; MCP-1; IFN-γ and reduce the secretion of TNF-α [22].

Compound **4** was separated and purified in the form of a white solid. It was optically active with $[\alpha]_D^{25} = -79.7$ (*c* 0.15, MeOH). The pseudo-molecular ion $[M+H]^+$ of **4** was observed in its ESI (positive) mass spectrum at 211 (*m/z*). Analysis of the ^{13}C NMR and DEPT spectra of **4** revealed that 11 carbons are presented in its structure, including two methyl groups at δ_C 15.54 (C-11) and 12.59 (C-13), three sp^3 methine groups at δ_C 37.01 (C-10), 59.95 (C-6) and 61.27 (C-9), four sp^3 methylene groups at δ_C 22.21 (C-4), 29.51 (C-5), 37.01 (C-10) and 46.14 (C-3), and two carbonyls at δ_C 167.51 (C=O) and 172.24 (C=O). There were signals of two methine groups bonded with nitrogen at δ_H 4.18 (1H, m, H-6) and 4.29 (1H, m, H-9), a nitrogen bearing methylene group at δ_H 3.58 (2H, m, CH_2-3), 2 methyl groups at δ_H 1.00 (3H, t, $J = 7.0$ Hz, CH_3-13) and 1.13 (3H, d, $J = 7.0$ Hz, CH_3-11) shown in the ^{1}H NMR spectrum of **4**. Signals of seven aliphatic protons ranging from δ_H 1.49 to 2.32 ppm were also noted. Comparison of obtained data led to the confirmation of the structure of **4** as cyclo-(L-Pro-L-ILe) [23, 24]. Compound **4** strongly inhibits the growth of *Vibrio anguillarum* (MIC = 0.12 μg/mL) [25, 26].

Compound **5** was separated and purified as a white amorphous solid. It is optically active with $[\alpha]_D^{25} = -79.2$ (*c* 0.15, MeOH). The ESI (positive) mass spectrum of **5** presented a pseudo-molecular ion $[M+H]^+$ peak at 245 (*m/z*). The ^{13}C NMR and ^{1}H-NMR data of **5** indicated signals of the proline unit, similar to compound **1**. However, in comparison with **1**, instead of signals of the isopropyl group, there is a phenyl group in the structure of **5**. This difference suggested the leucine moiety was replaced by a phenylalanine unit. Detail analysis of compound **5**'s ESI–MS and 1D-NMR data, in combination with the literature spectral data matching, confirmed compound **5** as cyclo-(L-Pro-L-Phe) [27]. Compound **5** showed a cytotoxic protective effect on INT-407 cells (human intestinal epithelial) induced by *Vibrio vulnificus* [28] and weak antifungal effects against *Aspergillus fumigatus* and *Penicillium roqueforti* (MIC = 20 mg/mL) [29].

Compound **6** was separated and purified in the form of a white solid. At the position of 245 (*m/z*) in its positive ESI mass spectrum, the pseudo-molecular ion $[M+H]^+$ of compound **6** was noted. The resonances in ^{13}C-NMR and ^{1}H-NMR spectra of compound **6** appeared very close to these of **5**. Besides, the optical rotation activities of compound **6** ($[\alpha]^{25}_D = +59.6$ (*c* 0.12, MeOH) and compound **5** ($[\alpha]_D^{25}$ -79.2 (*c* 0,15; MeOH) was opposite. Comparison of the NMR spectra with reported reference [20, 25, 26, 30] determined the structure of **6** as cyclo-(D-Pro-D-Phe). Compound **6** is a potent anti-tumour agent, inducing apoptosis in HT-29 colon cancer cell line and inhibiting the development of two cancer cell lines (MCF-7 and HeLa) [31]. Similar to compound **4**, compound **6** also showed strong inhibition (MIC = 0.03 μg/mL) against *Vibrio anguillarum* [18].

Compound **7** was separated and purified as a white solid with the negative optical rotation of $[\alpha]^{25}{}_D = -83.6$ (MeOH, *c* 0.6)., The $[M + H]^+$, pseudo-molecular ion, of **7** has appeared in its ESI (positive) mass spectrum at 261 (*m/z*). The signals appeared in ^{13}C-NMR, and ^{1}H-NMR spectra of compound **7** were close to these of **5**, except for the presence of an A_2B_2 system instead of the phenyl ring. This data strongly suggested that the phenyl fragment in **5** was replaced by a 4-hydroxyphenyl moiety in **7**. Matching recorded NMR data with Ref. [28] revealed that **7** was cyclo-(L-Pro-L-Tyr). Compound **7** inhibits in vitro and in vivo platelet aggregation induces by U46619 and ADP [32], and active against *Ralstonia solanacearum* and *Xanthomonas axonopodis* pv. citri at moderate level (MIC = 31.25 μg/mL) [33].

4 Conclusion

Seven cyclodipeptides had been isolated from the fermentation broth of the marine fungus strain *Penicillium chrysogenum* M612. Analyzing their spectra (MS and NMR) and comparing with published data led to the identification of these compounds as cyclo-(L-Leu-L-Pro) (**1**), cyclo-(L-Val-L-Pro) (**2**), cyclo-(L-Ala-L-Pro) (**3**), cyclo-(L-Ile-L-Pro) (**4**), cyclo-(L-Phe-L-Pro) (**5**), cyclo-(D-Pro-D-Phe) (**6**), cyclo-(L-Tyr-L-Pro) (**7**). All isolated compounds were previously reported to have antimicrobial activities, suggested that strain M612 can be further studied to develop an antimicrobial agent.

Acknowledgements This work was co-funded by Ministry of Science and Technology, Vietnam (Grant No. HNQT/SPĐP/11.19) and Institute of Drug Research and Development, College of Pharmacy, Chungnam National University, Republic of Korea under "The Vietnam Bilateral/Multilateral Joint International Research Program in Science and Technology to 2020".

Conflicts of Interest It is confirmed that there will be no conflict of interest to arise.

References

1. Choudhary A et al (2017) Current status and future prospects of marine natural products (MNPs) as antimicrobials. Mar Drugs 15(9):272
2. Dougherty TJ, Pucci MJ (2012) Antibiotic discovery and development. Springer, Boston, MA, USA
3. Newman DJ, Cragg GM (2016) Natural products as sources of new drugs from 1981 to 2014. J Nat Prod 79(3):629–661
4. Singer AC et al (2016) Review of antimicrobial resistance in the environment and its relevance to environmental regulators. Front Microbiol 7:1728
5. Carroll AR et al (2019) Marine natural products. Nat Prod Rep 36(1):122–173
6. Quang TH et al (2018) Macrolide and phenolic metabolites from the marine-derived fungus *Paraconiothyrium* sp. VK-13 with anti-inflammatory activity. J Antibiotics 71(9):826–830

7. Helmholz H, Etoundi P, Lindequist U (1999) Cultivation of the marine basidiomycete *Nia vibrissa* (Moore & Meyers). In: Osinga R (ed) Progress in industrial microbiology. Elsevier, pp 203–206
8. Cao ĐT et al (2019) Nghiên cứu phân lập vi nấm biển từ trầm tích khu vực biển Cát Bà, thành phố Hải Phòng, Việt nam. Vietnam Med J 484:570–576
9. Le THM et al (2018) Isolation, screening antimicrobial activity and identification of fungi from marine sediments of the area Thanh Lan, Co To, Vietnam. Vietnam J Biotechnol 16:721–728
10. Basilio A et al (2003) Patterns of antimicrobial activities from soil actinomycetes isolated under different conditions of pH and salinity. J Appl Microbiol 95(4):814–823
11. Cách NV(2004) Công nghệ lên men các chất kháng sinh. Nhà xuất bản khoa học kỹ thuật Hà Nội
12. He R et al (2013) Cyclodipeptides from metagenomic library of a Japanese Marine Sponge. J Braz Chem Soc 24:1926–1932
13. Yan PS et al (2004) Cyclo(L-leucyl-L-prolyl) produced by *Achromobacter xylosoxidans* inhibits aflatoxin production by *Aspergillus parasiticus*. Appl Environ Microbiol 70(12):7466–7473
14. Pedras MS et al (2005) Metabolites produced by the phytopathogenic fungus *Rhizoctonia solani*: isolation, chemical structure determination, syntheses and bioactivity. Z Naturforsch C J Biosci 60(9–10):717–722
15. Wang Y, Mueller UG, Clardy J (1999) Antifungal diketopiperazines from symbiotic fungus of fungus-growing ant *Cyphomyrmex minutus*. J Chem Ecol 25(4):935–941
16. Brack C, Mikolasch A, Schauer F (2014) 2,5-Diketopiperazines produced by *Bacillus pumilus* during bacteriolysis of *Arthrobacter citreus*. Mar Biotechnol 16(4):385–395
17. Kalinovskaya NI, Romanenko LA, Kalinovsky AI (2017) Antibacterial low-molecular-weight compounds produced by the marine bacterium *Rheinheimera japonica* KMM 9513(T). Antonie Van Leeuwenhoek 110(5):719–726
18. Huayue LI et al (2008) Bioactive cyclic dipeptides from a marine sponge-associated bacterium, *Psychrobacter* sp. Biomol Ther 16(4):356–363
19. Takaya Y et al (2007) Antioxidant constituents in distillation residue of Awamori spirits. J Agric Food Chem 55(1):75–79
20. Gao Y et al (2010) Diketopiperazines from two strains of South China Sea sponge-associated microorganisms. Biochem Syst Ecol 38(5):931–934
21. Mehnaz S et al (2013) Lahorenoic acids A-C, ortho-dialkyl-substituted aromatic acids from the biocontrol strain *Pseudomonas aurantiaca* PB-St2. J Nat Prod 76(2):135–141
22. Chen J-H et al (2012) The effects of diketopiperazines from *Callyspongia* sp. on release of cytokines and chemokines in cultured J774A.1 macrophages. Bioorg Med Chem Lett 22(9):3177–3180
23. Ye F et al (2018) Absolute configuration of (2R,3R,6S,8R)-methyl homononactate, a polyketide from actinomycetes *Streptomyces* sp. R-527F of the Arctic region. Chem Nat Compd 54(4):821–825
24. Wang N, Cui C-B, Li C-W (2016) A new cyclic dipeptide penicimutide: the activated production of cyclic dipeptides by introduction of neomycin-resistance in the marine-derived fungus *Penicillium purpurogenum* G59. Arch Pharm Res 39(6):762–770
25. Fdhila F et al (2003) dd-Diketopiperazines: antibiotics active against *Vibrio anguillarum* Isolated from marine bacteria associated with cultures of *Pecten maximus*. J Nat Prod 66(10):1299–1301
26. Fdhila F et al (2006) dd-Diketopiperazines: antibiotics active against *Vibrio anguillarum* isolated from marine bacteria associated with cultures of *Pecten maximus*. J Nat Prod 69(7):1120–1120
27. Vázquez-Rivera D et al (2015) Cytotoxicity of cyclodipeptides from *Pseudomonas aeruginosa* PAO1 leads to apoptosis in human cancer cell lines. BioMed Res Int 2015:197608
28. Li H et al (2008) Bioactive cyclic dipeptides from a marine sponge-associated bacterium, *Psychrobacter* sp. Biomol Ther 16:356–363

29. Ström K et al (2002) *Lactobacillus plantarum* MiLAB 393 produces the antifungal cyclic dipeptides cyclo(l-Phe-l-Pro) and cyclo(l-Phe-trans-4-OH-l-Pro) and 3-phenyllactic acid. Appl Environ Microbiol 68(9):4322–4327
30. Adamczeski M, Reed AR, Crews P (1995) New and known diketopiperazines from the Caribbean Sponge, Calyx cf. podatypa. J Nat Prod 58(2):201–208
31. Brauns SC et al (2004) Selected cyclic dipeptides inhibit cancer cell growth and induce apoptosis in HT-29 colon cancer cells. Anticancer Res 24(3a):1713–1719
32. Lee W et al (2017) Novel direct factor Xa inhibitory compounds from Tenebrio molitor with anti-platelet aggregation activity. Food Chem Toxicol 109:19–27
33. Wattana-Amorn P et al (2016) Antibacterial activity of cyclo(L-Pro-L-Tyr) and cyclo(D-Pro-L-Tyr) from *Streptomyces* sp. strain 22–4 against phytopathogenic bacteria. Nat Prod Res 30(17):1980–1983

Evaluation of a Protein-Depletion Kit for Enriching Low Abundant Proteins from Human Sera

Bao Chau Duong, Dung Van Nguyen, Nguyen Dinh Song Huy, Lim Teckwang, Lin Qifeng, Lin Qingsong, and Thi Thu Hoai Nguyen

Abstract The identification of low-abundance serum proteins in human sera can be severely affected by high-abundance proteins (HAPs), so many protein-depletion kits have been developed for enhancing the efficiency of identifying them. The ProteoSpin™ Abundant Serum Protein Depletion Kit is a fast, simple, and budget-friendly kit for the depletion of HAPs including albumin, alpha-antitrypsin, transferrin, and haptoglobin from serum and plasma samples but its depletion efficiency has not been investigated. Our study aimed to evaluate the efficiency of this kit in processing a pooled serum of hepatocellular carcinoma patients. We conducted a thorough evaluation following criteria: (i) protein recovery rate after depletion, (ii) efficacy of depletion, (iii) specificity of depletion using ProteoSpin™ Kit. Quantitative determination of total proteins, depleted proteins, and HAPs was performed. SDS-PAGE was used to analyze the serum samples before and after depletion while Shotgun LC–MS/MS was conducted to identify proteins from the depleted serum. The results then were compared with previous research to evaluate the efficiency of depletion. Our results indicated that the ProteoSpin™ Kit was efficient to deplete HAPs by revealing a protein recovery rate of 5.73% and the identification of 251 serum proteins. It reduced albumin, IgG, transferrin, fibrinogen, haptoglobin, and transthyretin by approximately 60%. The depletion efficacy of ProteoSpin™ Kit was comparable or even superior to some other commercial kits. It has significantly eliminated HAPs and enhanced the number of identified proteins in a serum sample. Our

---

B. C. Duong · D. Van Nguyen · T. T. H. Nguyen (✉)
School of Biotechnology, International University, Vietnam National University of Ho Chi Minh City, Ho Chi Minh City, Vietnam
e-mail: ntthoai@hcmiu.edu.vn

D. Van Nguyen
Institute of Biotechnology and Food Technology, Industrial University of Ho Chi Minh City, Ho Chi Minh City, Vietnam

N. D. S. Huy
Oncology Centre, Cho Ray Hospital, Ho Chi Minh City, Vietnam

L. Teckwang · L. Qifeng · L. Qingsong
Department of Biological Sciences, National University of Singapore, Singapore, Singapore

V. Van Toi et al. (eds.), *8th International Conference on the Development of Biomedical Engineering in Vietnam*, IFMBE Proceedings 85,
https://doi.org/10.1007/978-3-030-75506-5_46

study can help to facilitate the choice of optimal products for HAP depletion in studying serum proteins.

Keywords High-abundance proteins · Kit evaluation · Protein depletion · Shotgun proteomics

1 Introduction

Human serum is derived from the blood via the removal of all blood cells and clotting factors. It is a complex body fluid that contains 60–80 mg of protein per mL and diverse molecules such as salts, lipids, amino acids, and carbohydrates. The serum components, particularly serum proteins have been using as high potential and useful indicators of personal pathophysiological conditions [1]. The serum proteins consist of many functional polypeptides such as low-molecular-weight apolipoproteins, cytokines, hormones, antimicrobial peptides, and several peptides resulted from the degradation of serum proteins and other tissues as well as some oncoproteins [2]. For years, many studies have been carried out to search for proteinaceous biomarkers of different diseases especially cancers such as colorectal carcinoma [3] and hepatocellular carcinoma in the human sera [4]. However, as most of the potential disease-associated markers only present in trace amount in sera, they are often covered by serum high abundant proteins (HAPs), and it is very hard to detect them and their disease-associated subtle change.

Today even though the enhancement of mass spectrometry (MS)-based methods has provided modern and effective approaches to study serum proteome, it is still challenging to analyze the obtained data to generate useful diagnostic information due to the large dynamic concentration range of candidate proteins. Additionally, as previously mentioned, the presence of HAPs such as albumin or immunoglobulins (normally greater than mg/mL) in serum interferes with the detection of most low abundant proteins (often at ng/ml or less). There are almost 300 identified proteins in human serum but twenty-two of them including albumin, immunoglobulin, haptoglobin, transferrin, and lipoproteins already account approximately 99% of all the serum protein content [1]. Therefore, it is cumbersome to effectively find and characterize proteinaceous serum without removal or separation of HAPs.

At current, HAP depletion has become an increasingly essential step for the most analytical strategy of plasma or serum. It is not surprising that various commercial depletion kits have been developed to reach scientists' demand ranging from simple depletion of albumin and IgG to sophisticated depletion using centrifugal ultrafiltration or specific antibodies. Several studies comparing different depletion products have been performed but major pay attention to evaluate the removal of albumin or albumin in combination with IgG [5]. In recent years, ion-exchange based columns and immunoaffinity columns have been introduced to remove multiple HAPs simultaneously. They are coupled with other fractionation strategies and MS analysis to deeply explore the serum proteome, bringing in very sophisticated and promising approaches. While comparative studies on immunoaffinity technique have been carried out, the depletion efficacy of ion-exchange based columns has not been thoroughly evaluated. In this respect, our study aimed to evaluate The ProteoSpin™ Abundant Serum Protein Depletion Kit, an ion-exchange based column for its depletion efficacy of HAPs from human serum. We also compare our obtained data with similar previously published data using immunoaffinity-based columns.

2 Materials and Methods

2.1 *Sample Preparations*

Blood samples were collected from 24 hepatocellular carcinoma patients one-month post tumour section under the approval of the research ethics committee of Cho Ray Hospital (740/ CN- HĐĐĐ) with appropriate informed consents. The blood samples were let clotted at room temperature for at least 60 min then centrifuged at 4000 g for 10 min at room temperature to obtain serum samples. Serum was stored at − 80 °C until being used. In this study, 20 μL of each serum was taken and pooled together to obtain pooled serum E4. This process helps to minimize person-to-person variation in the serum protein profile.

2.2 *Removal of High Abundant Proteins from Serum*

60 μL of the pooled serum was subjected to HAP depletion using ProteoSpin™ Abundant Serum Protein Depletion Kit (Norgen Biotek, Product # 17,300) following the manufacturer's instruction. Subsequently, protein quantification was performed using the Bradford Protein Assay (Bio-Rad Laboratories, Inc.).

2.3 *SDS-Page*

10 μg protein of each crude, flow-through and depleted pooled serum was mixed with 2X sample buffer (120 Tris–HCL pH 6.8, 4% sodium dodecyl ulphate (SDS), 20% glycerol, 750 mM β-mercaptoethanol and 0.05% bromophenol blue), loaded on 10% polyacrylamide gel and run on a Bio-Rad system (Bio-Rad Laboratories, Inc.) with running buffer containing 25 mM Tris-base, 192 mM glycine and 0.1% SDS. The gel was run under a constant 150 V for 60 min and visualized with Coomassie staining [6].

2.4 Two-Dimensional LC/MS Analysis

100 μg of protein sample was separately mixed with S-Trap Lysis Buffer containing 50 mM TEAB and 5% SDS to form an equal starting volume that was used for S-Trap™ Mini Spin Column Protocol. The samples were incubated with five mM TCEP at 55 °C for 10 min and ten mM MMTS for 10 min at room temperature for reducing and alkylating respectively following acidified to 1.2% phosphoric acid. S-Trap binding buffer (90% MeOH, 100 mM TEAB final; pH 7.1) was added in a 1:7 (v/v) ratio to form the solution added into the spin column. The centrifugation for the30s at 4,000 g was performed until all solution has flown through. The column was washed four times using 400 μL S-Trap binding buffer for each time. Then, trypsin was added to the column at a ratio of 1:25 (w/w, trypsin: protein) and incubated for 1 h at 47 °C. 80 μL of 50 mM TEAB and 0.2% aqueous formic acid were added to each spin column and centrifuged at 1000 g for 60 s to elute the peptides. The final elution was carried out with 50% aqueous ACN containing 0.2% formic acid and centrifuged at 4000 g for 60 s. The eluted peptides were dried using an evaporator.

The separation of peptides was conducted on an Eksigent nanoLC Ultra and cHiPLC-nano flex system (Eksigent) in trap-elute configuration. The samples were desalted with Sep-Pak TC18 96-well μElution Plate (Waters) and reconstructed with 20 μl of the diluent containing 98% Water, 2% Acetonitrile, and 0.05% Formic acid. The next step is loading 5 μl of the sample on a 200 μm × 0.5 mm trap column and then separating on an analytical 75 μm × 150 mm column (ChromXP C18-CL, 3 μm (Eksigent)). The separation of peptides was approached by a gradient formed by 2% ACN, 0.1% FA (mobile phase A) and 98% ACN, 0.1% FA (mobile phase B) at a flow rate of 300 nl/min.

A TripleTOF 5600 system (AB SCIEX, Foster City, CA, USA) was used for mass spectrometry in Information Dependent Mode. The mass range of MS spectra was acquired among 400–1250 m/z in high-resolution mode (>30,000) using 250 ms accumulation time per spectrum. The fragmentation of maximum ten chosen precursors per cycle from each MS spectrum was applied with 100 ms minimum accumulation time for each precursor and dynamic exclusion for 8 s. Tandem mass spectra were noted in the resolution higher than 15,000 with rolling collision energy on.

2.5 Data Analysis

The ProteinPilot 5.0 software Revision 4769 (AB SCIEX) using the Paragon database search algorithm (5.0.0.0.4767) and the integrated false discovery rate (FDR) analysis function was applied for the identification and quantification of peptides. The results were searched against a database of 20190821_SwissProt_Human_with CRAP (total 20,350 entries) following this setting: Sample Type—Identification;

Cys; Alkylation—MMTS; Digestion—trypsin; Special Factors—None; Species—None. The processing was specified as follows: ID Focus—Biological Modifications; Search Effor—Thorough; FDR Analysis: Yes; Detected Protein Threshold—0.05 (10.0%). Data were searched against a concatenated database within silico on-the-fly reversal for FDR determination. Peptides/proteins identified with confidence interval ≥ 95% were selected. The estimated abundance of each protein (protein content in percentage) in the depleted serum was obtained based on its emPAI score (exponentially modified Protein Abundance Index) and its molecular weight.

In order to demonstrate HAP depletion efficacy, depletion percentage of each HAP was obtained via dividing its calculated percentage in the depleted serum to its reference percentage in crude serum [7]. The formula for calculating efficacy of depletion was shown below.

$$Efficacy\,of\,depletion\,(\%) = \left(1 - \frac{\Pr otein\,abundance\,in\,depleted\,serum\,(\%)}{Corresponding\,percentage\,in\,\mathrm{standardan}\,crude\,serum\,(\%)}\right) \times 100$$

3 Results and Discussion

3.1 *Quantitative Analysis of Serum Protein Samples After Depletion*

The total protein content before and after depleting the crude serum was presented in Table 1. With 60 μL of crude serum loaded into the ProteoSpin column, 200 μL of eluted samples was obtained with the concentration decreased by approximately 70 times, from 69.26 to 1.19 μg/μL. The result indicated a recovery rate of 238 μg out of 4155.6 μg or 5.73%.

A comparison was made on the protein recovery rate after depletion between our study and previous studies using either chemicals such as Dithiothreitol (DTT, GE Healthcare) [8] or other kits such as Top 12 Abundant Protein Depletion Spin Columns (Thermo Fisher Scientific) [9] and Qproteome Albumin/IgG Depletion kit (Qiagen) [9] (Table 2). As can be seen from the table, the recovery rate was highest if using DTT which can remove disulfide-rich HAPs including albumin and transferrin. However, applying unspecific removal like DTT can result in the large

Table 1 Quantitative analysis of serum protein before and after depletion

Sample	Serum loading capacity (μL)	Total protein content in crude serum (μg)	Total protein content in eluted sample (μg)	Recovery rate (%)
Pooled serum	60	4155.60	238.00	5.73

Table 2 Comparative analysis on the recovery rate and total protein content after one-stage HAP depletion of using ProteoSpin, DTT and alternative kits

Depletion method	Loading volume (μL)	Recovery rate (%)	Total protein content in the eluted sample (μg)	Depletion mechanism	Targeted proteins	References
Dithiothreitol (DTT, GE Healthcare)	20	26.40	~ 300.00[a]	Reducing agents	Albumin and transferrin	[8]
Top 12 Abundant Protein Depletion Spin Columns (Thermo Fisher Scientific)	10	25.70	223.59	Immunoaffinity	Albumin, IgG, α-1-Acid glycoprotein, α-1-Antitrypsin, Alpha-2-Macroglobulin, Apolipoproteins A-I and A-II, Fibrinogen, Haptoglobin, IgA, IgM, Transferrin	[9]
Qproteome Albumin/IgG Depletion kit (Qiagen)	25	14.70	319.73	Immunoaffinity	Albumin, IgG	[9]
ProteoSpin (Norgen)	60	5.73	238.00	Ion-exchange	Albumin, α-antitrypsin, transferrin, and haptoglobin	This study

[a]The amount of protein was estimated based on the concentration of initial serum represented in the square curve of Fig. 1a from corresponding research [8]

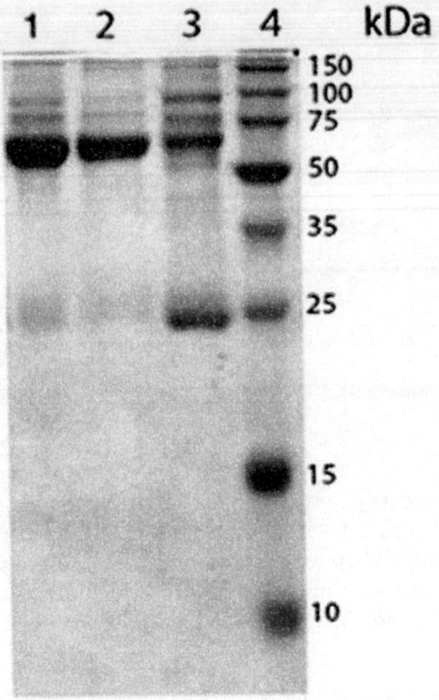

Fig. 1 SDS-PAGE analysis of crude serum, flow-through, and eluted samples. 10 μg of crude pooled serum, flow-through and eluted samples from ProteoSpin column were loaded on 10% SDS PAG, then stained with Coomassie Brilliant Blue. Lanes: 1, crude serum; 2, flow-through sample; 3, eluted sample (depleted serum); 4, Broad Range Protein Molecular Weight Markers (Promega)

retention of HAPs as well as the removal of low abundance disulfide-rich ones. Thus, this method is not recommended for serum biomarker investigation unless targets are well-known. With the other two kits using targeted antibodies to remove HAPs, the recovery rates were relatively good. Still their low loading capacity and high cost would hinder their usage when multiple samples are needed to be analyzed in one study. ProteoSpin, despite having low recovery rate, still generated comparable total protein content as higher crude serum volume can be loaded to the column (Table 2).

3.2 *Analysis of HAP Depletion Efficacy Using SDS-PAGE*

Protein quality of crude pooled serum, flow-through, and eluted samples were compared using SDS-PAGE (Fig. 1). Result showed that, the majority of the most abundant serum protein, albumin (66.5 kDa) was successfully depleted. Some HAPs with molecular weight approximately 50 kDa (probably, IgG heavy chain, ~ 55 kDa) were also successfully removed. Besides, more protein bands became visible in the eluted sample compared to the crude serum indicating the enrichment of low abundance proteins (Fig. 1).

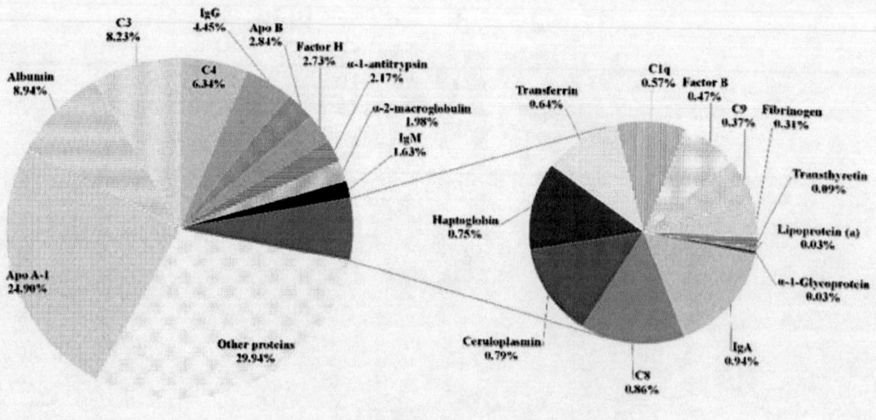

Fig. 2 Protein content in percentage of the pooled serum sample depleted with ProteoSpin. The abundance of each protein was calculated by multiplying emPAI score with its molecular weight and dividing to the sum of emPAIxMW of all detected proteins. Left pie chart shows proteins that account for more than 1% of total protein content (10 highly dominant HAPs). Right pie chart shows proteins that account for less than 1% of total protein content (12 dominant HAPs)

3.3 Identification and Absolute Quantification of Serum Protein After Depletion

HPLC–MS/MS analysis of the depleted pooled serum sample (eluted sample) resulted in the identification of 251 unique serum proteins. The estimated abundance of these proteins (protein content in percentage) was obtained by multiplying each protein's emPAI score with its molecular weight and then dividing to the sum of multiplication between each emPAI value and molecular weight for all identified proteins. The distribution of 22 dominant HAPs in the depleted serum was presented in Fig. 2. It can be observed that the percentage of the most abundant serum protein, albumin, accounted for only 8.94% of total protein content in the depleted serum. This indicates that this protein was quite well depleted with the ion-exchange ProteoSpin. Besides, apolipoprotein A-1 seemed not to be removed, thus became highly dominant in the depleted serum sample (Fig. 2).

Twenty-two most dominant HAPs accounting for 99% of total protein in crude serum [1] were selected for depletion efficacy analysis (Table 3). In order to demonstrate HAP depletion efficacy, depletion percentage of each HAP was obtained via dividing its calculated percentage in the depleted serum to its reference percentage in crude serum [7]. The formula for calculating efficacy of depletion was shown below (Table 4).

$$Efficacy\,of\,depletion\,(\%) = \left(1 - \frac{\Pr otein\,abundance\,in\,depleted\,serum\,(\%)}{Corresponding\,percentage\,in\,s\tan dard\,crude\,serum\,(\%)}\right) \times 100$$

Table 3 Protein abundance of 22 most dominant HAPs obtained using ProteoSpin depletion kit in comparison to previously published data

Protein (%)	Standard crude serum [7]	Protein abundance in depleted serum (%)				
		ProteoSpin (this study)	Top 12 (ThermoFisher Scientific) [9]	Qproteome (Qiagen) [9]	IgY12 spin column (Beckman Coulter) [10][1]	IgY14 spin column (Sigmaaldrich) [10][1]
α-1- Acid glycoprotein	1.00	0.03	2.73	3.33	n.d.	0.05
Lipoprotein (a)	0.60	0.03	0.0045	0.006		0.13
Fibrinogen	3.00	0.15	0.28	0.22		n.d.
Transthyretin	0.60	0.09	0.62	0.44		n.d.
Transferrin	4.00	0.64	1.50	4.6	4.52	1.97
Albumin	53.00	8.94	3.26	2.3	53.55	35.52
IgG total	16.00	4.45	3.33	3.19	2.17	1.63
IgA total	3.00	0.94	1.10	3.66	n.d.	1.03
Haptoglobin	2.00	0.75	3.83	3.63	9.50	0.42
C9 complement	0.60	0.37	0.26	0.25	n.d.	n.d.
α-2-macroglobulin	3.00	1.98	2.02	2.21	2.46	1.21
Complement factor B	0.60	0.47	0.89	0.53	n.d.	n.d.
IgM total	2.00	1.63	3.57	3.56	1.30	2.85
C1q complement	0.60	0.57	0.24	0.24	n.d.	n.d.
α-1-antitrypsin	2.00	2.17	2.92	3.88		0.45
Ceruloplasmin	0.60	0.79	1.26	0.7		2.72
C8 complement	0.60	0.86	0.78	0.53		n.d.
Apolipoprotein B	1.00	2.84	0.26	0.23		
Factor H	0.60	2.73	0.46	0.51		
C3 complement	1.00	8.23	3.17	1.79		
C4 complement	0.60	6.34	1.12	0.74		
Apolipoprotein A-1	1.60	24.90	2.40	3.38	5.22	0.54

[a]The measurement of protein abundance based on the percentage of peak area (%); *n.d.* not determined

Obtained data showed that 14/22 HAPs were significantly removed. Among them, 9 proteins which were depleted by more than 60% included α-1-Glycoprotein (97%), lipoprotein (a) (95%), fibrinogen (94.9%), transthyretin (85%), transferrin (84.06%), albumin (83.13%), IgG total (72.02%), IgA total (68.67%), and haptoglobin (62.50%). The total percentage of these 14 significantly depleted HAPs decreased from 90% of total protein in crude serum to only 21.04% in the depleted serum (Table 3). The left 8 HAPs accounting for 9% of total protein in crude serum, on the other hand, were enriched in the depleted serum to 48.86%. Interestingly, the

Table 4 Efficacy of depleting 22 most dominant HAPs using ProteoSpin depletion kit in comparison to previously published data

Protein (%)	Efficacy of depletion (%)[a]				
	ProteoSpin (this study)	Top 12 (ThermoFisherScientific) [9][b]	Qproteome (Qiagen) [9][b]	IgY12 spin column (Beckman Coulter) [10][c]	IgY14 spin column (Sigmaaldrich) [10][c]
α-1- Acid glycoprotein	97.00	Enriched	Enriched	n.d.	95.00
Lipoprotein (a)	95.00				78.33
Fibrinogen	94.90				n.d.
Transthyretin	85.00				n.d.
Transferrin	84.06	42.31		Enriched	50.75
Albumin	83.13	80.36	86.14	Enriched	32.98
IgG total	72.20	72.78	73.93	86.44	89.81
IgA total	68.67	37.63	Enriched	n.d.	65.67
Haptoglobin	62.50	Enriched		Enriched	79.00
C9 complement	38.33			n.d.	n.d.
α-2-macroglobulin	34.00			18.00	38.89
Complement factor B	21.67			n.d.	n.d.
IgM total	18.50			35.00	Enriched
C1q complement	5.00			n.d.	n.d.
α-1-antitrypsin	Enriched				77.50
Ceruloplasmin					Enriched
C8 complement					n.d.
Apolipoprotein B					
Factor H					
C3 complement					
C4 complement					
Apolipoprotein A-1				Enriched	66.25

[a]Efficacy of depletion for each HAP was obtained based on the its reference percentage in crude serum [7] and its calculated percentage in the depleted serum
[b]The efficacy retrieved directly from the research
[c]The efficacy obtained based on its abundance and corresponding percentage in crude serum
n.d. not determined

low-abundance proteins which generally account for only 1% in crude serum were enriched to 29.94% in the depleted serum (Fig. 2, Table 3).

In comparison to similar data obtained via using Top 12 (Thermo Scientific), Qproteome (Qiagen), IgY12 Spin column (Beckmann Counter), IgY14 Spin column (Sigma-Aldrich), it looked like that ProteoSpin worked quite well or even better (Table 3). Top 12 kit was indicated to remove 12 HAPs (Table 2) but only 4 HAPs were found to be depleted, with highest depletion efficacy for albumin (80.36%), followed by IgG total (72.78%), transferrin (42.31%) and IgA total (37.63%). In case of Qproteome kit, it was stated to remove albumin and IgG (Table 2) and the result indeed showed very high depletion efficacy of these two proteins, 86.14% for albumin and 73.93% for IgG total. However, other than the mentioned depleted HAPs, remaining HAPs in the group of 22 most dominant HAPs in serum were not efficiently removed by these two kits representing via their increased percentages in the depleted samples (Table 3). Regarding IgY12 and IgY14 spin column which target 12 and 14 HAPs respectively due to the presence of specific IgY antibodies against those proteins, the depletion efficacy was very high for IgG total, 86.44% for IgY12 and 89.81% for IgY14. However, while the IgY14 Spin column worked quite well for most HAPs, IgY12 spin column was not so efficient in removing HAPs other than IgG total. Also, both columns could not effectively deplete albumin (Table 3).

On overall, although the analyzed data showed quite superior efficacy of ProteoSpin in compared to previously published data of other kits, it should be taken into account that the calculated data were affected not only by the applied kits themselves but also by the downstream techniques used to detect, identify and quantify the serum proteins. Nonetheless, there were still relative similarities in downstream analysis between the methodologies of our study and the reference research [9] and [10]. After performing protein depletion, the samples were both separated by 1D—PAGE and the digestion of gel bands were carried out overnight using trypsin. However, there were the applications of other methods such as S-Trap mini spin column (this study) or Filter-aided Sample Preparation [9] which have been used to optimize digestion efficiency and proteomic coverage since it is vital to provide protease access to cleavage sites by establishing adequate solubilization and unfolding of all protein in a complete mixture [11]. However, although the main active agent in S-Trap mini spin column—SDS is one of the best protein solubilizers, it has been reported to be detrimental to LC–MS/MS peptide sensitivity [11]. Different concentration of formic acid was used for peptide reconstitution in this study and reference research.

Furthermore, reverse phase liquid chromatography including C18 column was applied as the trapping column and analytical column for peptide pre-concentration and fractionation. At the same time, the following shotgun strategy both included nano LC—MS/MS. Peptides entered the mass spectrometer which was operated in Data-dependent acquisition (DDA) mode in this study and in [10] whereas in [9], they did not show the specific detail in this setting. In general, the bioinformatics analysis both based on UniProt Swissprot Database, but each research applied the different p-value as well as the algorithm for protein significance and for peptide validation. Consequently, this setting dominantly affected the number of identified proteins. More in detail, a p-value ≤ 0.05 and FDR < 0.001 (FDR: false discovery rate) were chosen to be statistically significant in our study and in [9] while on the contrary only p-value ≤ 0.001 was used in [10] and no information about the FDR. So, the observation of fewer proteins in protein samples treated by IgY12 spin column might be caused by choosing this cut-off value. Finally, the protein samples were all conducted with label-free approach as protein abundance quantitative method. Techniques and software programs used in the downstream analysis of depleted serum samples in our study and two published research including compared commercial kits were shown in a detailed manner in Table 5.

In comparison on the usage convenience between commercial kits, ProteoSpin reportedly provides the largest size for performing the protein depletion (Table 6) with relatively short time to process samples. Among those kits, the capacity of

Table 5 Techniques and software used in the downstream analysis of depleted serum protein

Stage	This study	Reference [9]	Reference [10]
Protein separation	1D SDS-PAGE	1D SDS-PAGE	1D SDS-PAGE
Digestion of gel bands	Overnight trypsin digestion including S-Trap mini spin column	Overnight trypsin digestion coupling with Filter-aided Sample Preparation	Overnight trypsin digestion
Protein reconstitution	0.05% formic acid	0.2% formic acid	0.1% formic acid
Protein pre-concentration and fractionation	200 μm × 0.5 mm trap column and 75 μm × 150 mm analytical column (Eksigent)	Acclaim PepMap C18, 75 μm x 2 cm trap column and Acclaim PepMap RSLC C18, 75 μm × 50 cm nanoViper, 2 μm, 100 Å (Thermo Fisher Scientific) fractionated column	Dionex C18 Pepmap (150 mm length, 0.075 mm id)/Biobasic C18 PicofritTM column (100 mm length, 75 mm id)
Peptide ionization, separation, and fractionation methods	Eksigent nanoLC Ultra and cHiPLC-nanoflex system (Eksigent) A TripleTOF 5600 system (AB SCIEX, Foster City, CA, USA)	UltiMate 3000 RSLCnanoLC system Q Exactive™ Hybrid Quadrupole-Orbitrap™ Mass Spectrometer (Thermo Fisher Scientific, San Jose, CA, USA)	Dionex Ultimate 3000 (RSLCnano) chromatography system LTQ ORBITRAP XL mass spectrometer (Thermo Fisher Scientific, San Jose, CA, USA)
Bioinformatics analysis	Protein Pilot 5.0 (AB SCIEX) UniProt Swissprot Database p-value ≤ 0.05 FDR analysis function	Proteome Discoverer (version 1.4) (Thermo Fisher Scientific) UniProt Swissprot Database p-value ≤ 0.05 FDR analysis function	TurboSEQUEST (Bioworks Browser 3.3.1 SP1; Thermo Finnigan, UK) UniProt Swissprot Database p-value ≤ 0.001
Quantitative method	Label-free approach		

Table 6 Comparison of the usage convenience between commercial kits

	ProteoSpin (this study)	Top 12 (Thermo Fisher Scientific)	Qproteome (Qiagen)	IgY12 spin column (Beckman Coulter)	IgY14 spin column (Sigma Aldrich)
Size (preps)	25	24	6	2	2
Protein max capacity (μg)	150	600	1700-2000[a]	20000[c]	1400[e]
Protein min capacity (μg)	50	n.d.	n.d.	n.d.	1000
Protein min elution volume (μl)	30	n.d.	n.d.[b]	n.d.[d]	n.d.
Estimated time to process 10 samples (min)	30	40–60	15	n.d.	30

[a]The result based on 25 μl recommended sample volume
[b]Amount of protein recovered is 0.5–0.8 mg according to the product handbook
[c]The result based on 250 μl sample volume (protein concentration is 80 mg/ml)
[d]2 mg of LAPs was reported according to the manufacture
[e]The result based on 14–20 μl recommended sample volume (protein concentration is 70 mg/ml)

ProteoSpin seems to be quite low when compared with Qproteome or IgY12 spin column but it could be suitable for clinical samples with limited amount. An obvious drawback to this methodology is that the samples need to be diluted prior.

4 Conclusion

Even though it is often challenging to analyze serum proteome, this field still develops rapidly due to its promising achieved results in discovering potential and reliable disease-associated biomarkers. Through the development of this field, various HAP depletion kits have been introduced to reduce the dynamic range as well as the complexity of human serum protein samples to ease the complication of serum proteome analysis. At the same time, this also led to the difficulty for researchers to select the most appropriate depletion kit for conducting experiments. Our study results can be served as a reference to facilitate the choice of optimal products for HAP depletion in serum proteomics studies. Among the disposable columns, ProteoSpin™ Abundant Serum Protein Depletion Kit can be one of the good choices as it is a fast, simple, and budget-friendly kit for the depletion of major HAPs. Our data showed that ProteoSpin does not yield a high protein recovery rate but a wide-range elimination of HAPs with adequate depletion efficacy for follow-up analysis. This indicates the reasonable usage of ProteoSpin when the volume of samples is not extremely limited and the desired outcome of the experiment is to reduce multiple HAPs, making the samples feasible for further MS analysis.

Acknowledgements We would like to sincerely thank staffs at Oncology centre, Cho Ray hospital, Ho Chi Minh City for their assistance in collecting samples.

Conflicts of Interest The authors have no conflict of interest to declare.

References

1. Chan KC et al (2004) Analysis of the human serum proteome. Clin Proteomics 1(2):101–225
2. Chertov O et al (2014) Enrichment of low-molecular-weight proteins from biofluids for biomarker discovery. Expert Rev Proteomics 2(1):139–145
3. Pham TTH et al (2019) Isobaric tags for relative and absolute quantitation-based proteome analysis of Vietnamese colorectal carcinoma tissues. J Cancer Res Ther 15(1):96–103
4. Chauhan R, Lahiri N (2016) Tissue- and serum-associated biomarkers of hepatocellular carcinoma. Biomark Cancer 8(Suppl 1):37–55
5. Polaskova V et al (2010) High-abundance protein depletion: comparison of methods for human plasma biomarker discovery. Electrophoresis 31(3):471–482
6. Brunelle JL, Green R (2014) One-dimensional SDS-polyacrylamide Gel electrophoresis (1D SDS-PAGE). In: laboratory methods in enzymology: protein part C, pp 151–159
7. Ciordia S, Ríos V, Albar J (2006) Contributions of advanced proteomics technologies to cancer diagnosis. Clin Transl Oncol 8:566–580 (Official publication of the Federation of Spanish Oncology Societies and of the National Cancer Institute of Mexico)
8. de Jesus JR et al (2017) Simplifying the human serum proteome for discriminating patients with bipolar disorder of other psychiatry conditions. Clin Biochem 50(18):1118–1125
9. Pisanu S et al (2018) Comparative evaluation of seven commercial products for human serum enrichment/depletion by shotgun proteomics. Talanta 185:213–220
10. Corrigan L et al (2011) Evaluation and optimization of IgY spin column technology in the depletion of abundant proteins from human serum. Proteomics 11(16):3415–3419
11. Zhang Y et al (2013) Protein analysis by shotgun/bottom-up proteomics. Chem Rev 113(4):2343–2394

Ongoing Initiatives in Bringing the Preemptive Panel—Based Pharmacogenetic Testing to Clinical Decisions in Vietnam

Nguyen Thi Thu Phuong, Nguyen Thanh Hoi, and Jae-Gook Shin

Abstract Pharmacogenetics has become one of the leading areas of personalized medicine to increase both the prescribed drug's efficacy and safety. To implement pharmacogenetics into routine clinical practice, genotyping assay needs provide a rapid, simple, reliable and cost-saving method. A number of the kits commercially available contain limited coverage, omitting significant variations relevant for Asian populations. In response to this problem, we aimed to develop in-house SNaPshot multiplex assay for genotyping multiple SNP loci in the Vietnamese population and translate the pharmacogenetic results into clinical action. Polymerase chain reaction (PCR) with the extracted DNAs from 10 dried blood spot (DBS) samples across were analyzed the major polymorphisms of including both pharmacokinetic and pharmacodynamic genomic panels such as CYP2D6, CYP2C9, VKORC1, CYP4F2, CYP2C19, HLA-B, HLA-A, HLA-DRB1, HLA-DOA1, TPMT, G6PD, NUDT15, DPYD, UGT1A1, C11orf65, SLCO1B1, ABCG2, ACE, COQ2, KCNJ11, NEDD4L, SLC14A2, ADRA1A, AGTR1, GNB3, MMP3, PTGS2. The variants were selected using PharmGKB. Gen-Drug interaction is leveled on drug labels, clinical guidelines, and literature. We found that 100% of subjects carry one or more variants. The major alleles detected include HLA-B*1502, HLA*5701, HLA*5801, HLA-A*3303, CYP2C19*1, CYP2C9*10 and VKORC1 G/A. The genotypes were translated to inform the prescribing decision by creating the patient-centred results report. In this study, we developed not only an appropriate panel relevant to the prescription patterns and genotype prevalence but also an easy-to-understand report of the patient's drug—gene interactions information to recommend drug dosing.

N. T. T. Phuong (✉) · N. T. Hoi
Hai Phong University of Medicine and Pharmacy, Haiphong, Vietnam
e-mail: nttphuong@hpmu.edu.vn

Hai Phong International Hospital, Haiphong, Vietnam

J.-G. Shin
Department of Clinical Pharmacology, Inje University Busan Paik Hospital, Busan, Republic of Korea

SPMED Co., Ltd. Busan, Busan, Republic of Korea

V. Van Toi et al. (eds.), *8th International Conference on the Development of Biomedical Engineering in Vietnam*, IFMBE Proceedings 85,
https://doi.org/10.1007/978-3-030-75506-5_47

Keywords Pharmacogenetic · Pharmacogenomic · Preemptive panel · Drug therapy

1 Introduction

Pharmacogenomics is the use of genomic and other "omic" information to individualize drug selection and drug use to avoid adverse drug reactions and to maximize drug efficacy. Almost half of all adverse drug reactions are manifested in the skin. Although most are not serious, severe cutaneous adverse reactions (SCARs) are potentially lethal caused by antibiotics [1], allopurinol [2], harmacogenom [3]. Asians as a group comprise > 60% of the world's population. Pharmacogenetic differences in Asian populations represent an emerging area of research that stipulates certain adverse drug reactions (ADRs) or predictable alterations in drug metabolism due to genetic variation can be associated with Asian or other racial/ethnic backgrounds [4, 5]. Increasing evidence revealed that the prevalence of pharmacogenetics variants in different ethnic groups could be significant [6]. Evidence-based guidelines and summary statistics from the US Food and Drug Administration and the Clinical Pharmacogenetics Implementation Consortium were analyzed for ethnic differences in outcomes. Implicated drugs included commonly prescribed drugs such as warfarin, clopidogrel, Carbamazepine, and allopurinol. The majority of these associations are due to Asians more commonly being poor metabolizers of cytochrome P450 (CYP) 2C19 and carriers of the human leukocyte antigen (HLA)-B*15:02 allele [7]. We observed that harmacogenomics research in Asia was dominated by larger developed countries, notably Japan and Korea, and mainly driven by local researchers. All studies conducted on Cambodian and Vietnamese populations were performed in collaboration between the local and researchers from other non-Asian countries [8]. A decade ago, due to the limited pharmacogenomics information and the scarcity of drugs requiring such a special genetic attention, the clinical use of harmacogenomics tests in the community was also limited. Several commercial DNA isolation kits are available for extracting the genomic DNA from the ethylene diamine tetra-acetic acid (EDTA) whole blood samples. To obtain DNA from whole blood these DNA isolation procedures require quite some hands on time and are rather expensive. An alternative technique could be dried blood spot (DBS) sampling, with which DNA isolation is faster, cheaper and logistics are easier [9]. Therefore, we conducted this study to (i) initial develop and validate our in-house SnaPshot multiplex assay using DBS sample for genotyping multiple SNP loci in diagnosed ADR cases in Haiphong International Hospital (ii) propose the solution to translate the pharmacogenetic results into clinical action.

2 Materials and Methods

The study protocol conformed to the principles of the Declaration of Helsinki, and was approved by the Ethics Committee of Haiphong International Hospital. Blood spots were obtained from 10 patients in Haiphong International Hospital, Vietnam who fulfilled the diagnostic criteria of severe allergic drug reactions. The variants were selected using PharmGKB. Gen-Drug interaction is level on drug labels, clinical guidelines and literature. This study included 27 of the most common variants in Asians selected using the number of resources including The Pharmacogenomics Knowledgebase (PharmGKB, www.pharmgkb.org) and genetic-based dosing guidelines from the Clinical Pharmacogenetic Implementation Consortium (CPIC, www.pharmgkb.org/page/cpic) and the Dutch Pharmacogenetics Working Group (DPWG, www.pharmgkb.org/page/dpwg), which can also be accessed via the PharmGKB website. DNA was extracted from the dried blood spots collected on a filter paper as previously reported [10]. A one-tube multiplex polymerase chain reaction (PCR) was performed with pairs of primers amplifying DNA fragments containing common mutations by using the Complete PCR Reagent Set (Agena Bioscience). PCR primers were then eliminated using ExoSAP-IT reagent (Thermo Fisher Scientific) and used as the template for the subsequent primer extension reactions. Multiplex primer-extension reactions were performed using a SNaPshot Multiplex Kit.

3 Results and Discussion

Our study consisted of 10 patients comprising 6 SJS (Stevens–Johnson syndrome), 4 TEN (toxic epidermal necrolysis) (Table 1). The mean age was 50.6 ± 13.7 years. Overall, the HLA-B*58:01 allele were detected in 4/4 patients with allopurinol-induced SCARs (severe cutaneous adverse reactions). HLA-B*58:01 has been documented to strongly associate with the risk of allopurinol-induced SCARs in Han Chinese [4], Thai [11], European [12] (Table 2). The Clinical Pharmacogenetics Implementation Consortium (CPIC) recommended that allopurinol should be contraindicated to patients who have tested positive for HLA-B*58:01 to preventing SCAR [13, 14]. While other Asian countries have implemented the HLA-B*58:01 genotypic test in routine practice before being started on allopurinol, currently we have not been able to conduct it in Vietnam [15]. Simultaneously, the HLA-A*33:03 allele were detected in 4/4 patients with SJS during allopurinol treatment. HLA alleles A*3303 was reported in linkage disequilibrium and formed an extended haplotype with HLA-B*58:01 that could increase risk of severe cutaneous adverse reactions when treated with allopurinol [2]. SNapShot histogram of patient number 6 was shown in Fig. 1. The genotype of this patient was detected positively with HLA-A*33:03/*33:03 and HLA-B*58:01.

Table 1 Pharmacogenetics of drug hypersensitivity and related drug-induced syndromes

Patient no (n = 10)	Co-disease	Drug-induced hypersensitivity syndrome	Drug toxicity syndrome	Genotype	Gene—drug interaction
1	Hypertension, Gout	Carbamazepine	SJS	*CYP2D6*10/*10*	Decreased metabolism/decreased clearance of paroxetine, codein as compared to patients with the CYPD6*1/*1 or *2/*2 genotype (level 1A)
				*HLA-B*15:02*	High risk of SJS –TEN, DRESS due to carbazepine, oxacabazepine, phenytoin (level 1A)
2	Lipoprotein disorders	Abacavir	TEN	*CYP2D6*10/*10*	Normal metabolism. This genotype is common among Asians (level 1A)
				*CYP2C9*2/*2*	Decreased metabolism, increased plasma concentration, when treated with phenytoin, piroxicam, ibuprofen, warfarin, acenocoumarol when compared to patients with*1/*1 genotypes (level 1A)
				VKORC1 G/A	Patients with the G/A genotype who are treated with warfarin may require a lower dose as compared to patients with the G/G genotype, and a higher dose as compared to patients with the A/A genotype (level 1B)

(continued)

Table 1 (continued)

Patient no (n = 10)	Co-disease	Drug-induced hypersensitivity syndrome	Drug toxicity syndrome	Genotype	Gene—drug interaction
				*HLA-B*57: 01*	High risk of abacavir allergy. Before initiating treatment with abacavir, HLA-B*5701 status should be determined if possible (level 1A)
3	Lipoprotein disorders Reflux esophagitis	NSAIDs	SJS	*CYP2C19*1/*2*	poor metabolism of clopidogrel and decreased formation of active drug metabolite, resulting in decreased response and may have an increased risk for secondary cardiovascular events when treated with clopidogrel (level 1A)
				*HLA-B*15: 02*	High risk of SJS—TEN, DRESS due to carbazepine, oxacabazepine, phenytoin (level 1A)
				*UGT1A1*1/*28*	increased levels of bilirubin leading to an increased likelihood for hyperbilirubinemia when treated with atazanavir (in most studies boosted with low dose of ritonavir) (level 1A)
				*SLCO1B1*1B/*15*	higher risk of simvastatin-related myopathy (level 1A)

(continued)

Table 1 (continued)

Patient no (n = 10)	Co-disease	Drug-induced hypersensitivity syndrome	Drug toxicity syndrome	Genotype	Gene—drug interaction
4	Stomachache	NSAIDs	SJS	*VKORC1 G/A*	Patients with the G/A genotype who are treated with warfarin may require a lower dose as compared to patients with the G/G genotype, and a higher dose as compared to patients with the A/A genotype (level 1B)
				*CYP2C9*1/*2*	Decreased metabolism, increased plasma concentration, when treated with phenytoin, piroxicam, ibuprofen, warfarin, acenocoumarol when compared to patients with *1/*1 genotypes (level 1A)
				*CYP2D6*10/*10*	decreased metabolism/decreased clearance of paroxetine, codein as compared to patients with the CYPD6*1/*1 or*2/*2 genotype (level 1A)
5	Hypertension Renal failure	Allopurinol	SJS	*HLA-B*58: 01*	Patients with one or two copies of the HLA-B*58:01 allele have an increased risk of SCARs, such as SJS and TEN, when treated with allopurinol, carbamazepin, lamotrigine, antiepileptics (level 1A)

(continued)

Table 1 (continued)

Patient no (n = 10)	Co-disease	Drug-induced hypersensitivity syndrome	Drug toxicity syndrome	Genotype	Gene—drug interaction
				*HLA-A*33:03*	Patients with one or two copies of the HLA-A*33:03 allele who are treated with allopurinol may have an increased risk of severe cutaneous adverse reactions (SCAR)
6	Pneumoniae	TMP-SMX Allopurinol	TEN	*HLA-B*58: 01*	Patients with one or two copies of the HLA-B*58:01 allele have an increased risk of SCARs, such as SJS and TEN, when treated with allopurinol, carbamazepin, lamotrigine, antiepileptics (level 1A)
				*HLA-A*33: 03*	Patients with one or two copies of the HLA-A*33: 03 allele who are treated with allopurinol may have an increased risk of SCAR (level 2B)
7	Hypertension, Hemorrhoids	TMP-SMX Allopurinol	TEN	*HLA-B*58:01*	Patients with one or two copies of the HLA-B*58:01 allele have an increased risk of SCARs, such as SJS and TEN, when treated with allopurinol, carbamazepin, lamotrigine, antiepileptics (level 1A)

(continued)

Table 1 (continued)

Patient no (n = 10)	Co-disease	Drug-induced hypersensitivity syndrome	Drug toxicity syndrome	Genotype	Gene—drug interaction
				*HLA-A*33:03*	Patients with one or two copies of the HLA-A*33:03 allele who are treated with allopurinol may have an increased risk of SCAR (level 2B)
				*CYP2D6*10/*10*	decreased metabolism/decreased clearance of paroxetine, codein as compared to patients with the CYPD6*1/*1 or*2/*2 genotype (level 1A)
8	Atrial fibrillation, Heart failure	Allopurinol	TEN	*HLA-B*58: 01*	Patients with one or two copies of the HLA-B*58:01 allele have an increased risk of SCARs, such as SJS and TEN, when treated with allopurinol, carbamazepin, lamotrigine, antiepileptics (level 1A)
				*HLA-A*33: 03*	Patients with one or two copies of the HLA-A*33:03 allele who are treated with allopurinol may have an increased risk of SCAR (level 2B)
				*CYP2D6*10/*10*	decreased metabolism/decreased clearance of paroxetine, codein as compared to patients with the CYPD6*1/*1 or*2/*2 genotype (level 1A)

(continued)

Table 1 (continued)

Patient no (n = 10)	Co-disease	Drug-induced hypersensitivity syndrome	Drug toxicity syndrome	Genotype	Gene—drug interaction
9	Pneumoniae	Carbamazepine	SJS	*HLA-B*15:02*	High risk of SJS—TEN, DRESS due to carbazepine, oxacabazepine, phenytoin (level 1A)
10	Hypertension, Gout	Colchicin, Carbamazepine	SJS	*HLA-B*15: 02*	High risk of SJS—TEN, DRESS due to carbazepine, oxacabazepine, phenytoin (level 1A)
				*CYP2C9*1/*2*	Decreased metabolism, increased plasma concentration, when treated with phenytoin, piroxicam, ibuprofen, warfarin, acenocoumarol when compared to patients with*1/*1 genotypes (level 1A)

SJS Stevens–Johnson syndrome; *DRESS* Drug reaction with eosinophilia and systemic symptoms; *TEN* toxic epidermal necrolysis; *NSAIDs* Nonsteroidal anti-inflammatory drug; *TMP-SMX* Trimethoprim—sulfamethoxazole

Table 2 Frequency of alleles by ethnicity

Genotype/Phenotype	Target SNPs	Frequency (n (%))						References
		America	Caucasian	Asian	Korea	Thailand	Vietnam	
CYP2D6*10/*10 Decrease function	100 C > T	251(0.016)–104,509(14.9)	208–672(0.4–14.5)	98–206(8.6–64.1)	3417(46.2–54.3)	134 (53.7)	136(0.4–14.5)	[26–31]
CYP2C9*1/*3 Intermediate metabolizer	1075A > C	300 (3.3)	106(9.4)	102 (6.9)	105 (9.10%)	15 (4.53%)	100 (0.07)	[6, 32–34]
UGT1A1*1/*28	7-TA insertion in promoter; rs8175347	0.42–0.56	0.26 –0.31	0.09–0.16	390 (2.31)	96 (15.6)	NA	[35–38]
VKORC1 G/A	1639G > A	300(17.7)	106(45.3)	102 (21.6)	152 (13.77)	111 (33.53)	49 (14.3)	[6, 32, 39]
CYP2C19*1/*2 Intermediate Metabolizer	681 G > A	250 (0.4)	250 (16.1)	250 (37.2)	3587 (7.75)	331 (4.53)	157 (4.45)	[5]
HLA-B *1502	rs2844682C > T and rs3903184C > G	250 (0.2)	1133 (0.06)	358 (4.87)	5802 (0.3–0.5)	99 (6.1); 986 (15.2)	170 (13.5)	[40–45]
HLA-B *5801	rs9263726 A > G	251 (0.8–6.37)	265 (1.13)	358 (7.38)	5802 (7)	986 (16.3)	170 (6.5)	[40, 42–44]
HLA-A *3303	rs6921921(A) & rs2523945(G)	252 (0.53–3.97)	265 (0.57)	358 (11.7)	5802 (16.8)	183 (9.52)	170 (11.5)	[40, 42, 43, 46]

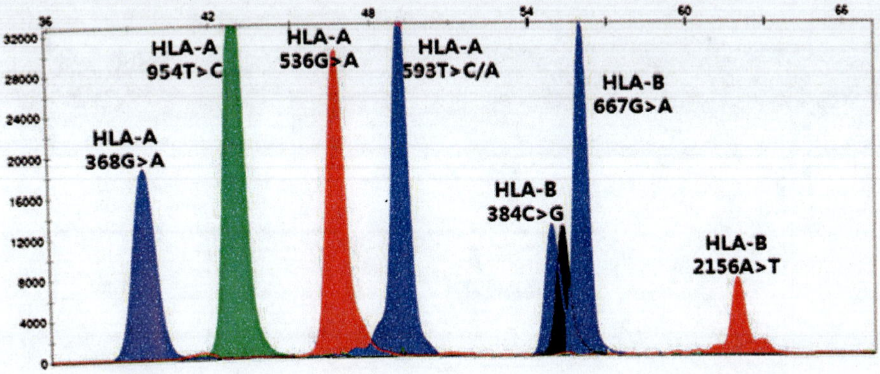

Fig. 1 SNapShot histogram of patient number 6. The genotype was detected positively with HLA-A*33:03/*33:03 and HLA-B*58:01

The HLA-B*15:02 allele were presented in 3/3 patients Carbamazepine (CBZ)-induced SCARs in our study. It has been documented that HLA-B*15:02 is the major genetic risk variant for CBZ-induced hypersensitivity reactions in Asian countries such as the Thais [11], Singaporeans [16], Malaysians [3], Indians [17], Vietnamese [18] and Chinese [19]. These findings could alert clinicians to individualize CBZ therapy for Asian CBZ users.

The CYP2D6 enzyme is responsible for the hepatic metabolism of up to 25% of commonly used medications [20]. The *CYP2D6* gene is highly polymorphic; over 100 allelic variants and subvariants have been designated to date by the Pharmacogene Variation Consortium (PharmVar) at www.PharmVar.org [20]. There are over 100 reported alleles of CYP2D6 with some variants more common in different ethnicities in which CYP2D6*10 is the most common CYP2D6 allele in Asians [21]. In our study, we also found that three patient have CYP2D6*10/*10 that should be optimize the selection and dosing of tamoxifen [22, 23] and paroxetine [24].

The *VKORC1* G/A frequencies ranged from 14.3 (Vietnam) to 45.3 (Caucasian) (Table 2). In our study, the *VKORC1* g.-1639G > A allele was detected in 2/10 patients. The Caucasian and Asians had significantly different in the distributions of VKORC1 G/A compared to other ethnic populations ($p < 0.0001$).

CYP2C19 contributes to the metabolism of a large number of clinically relevant drugs and drug classes such as antidepressants, benzodiazepines, mephenytoin, proton pump inhibitors, and the antiplatelet prodrug clopidogrel [25]. The 2017 FDA-approved drug label for clopidogrel includes a boxed warning concerning the reduced antiplatelet effect of clopidogrel in CYP2C19 poor metabolizers. The warning states that tests are available to identify patients who are CYP2C19 poor metabolizers, and to consider the use of another platelet P2Y12 inhibitor in patients identified as CYP2C19 poor metabolizers. Also, patient who own CYP2C19 intermediate metabolizer to leading to reduce platelet inhibition and increased risk for adverse cardiovascular events. Interestingly, 1/10 patient in our study carried CYP2C19*2/*2 (poor metabolizer) and 3/10 patients owned *1/*2 (intermediate metabolizer).

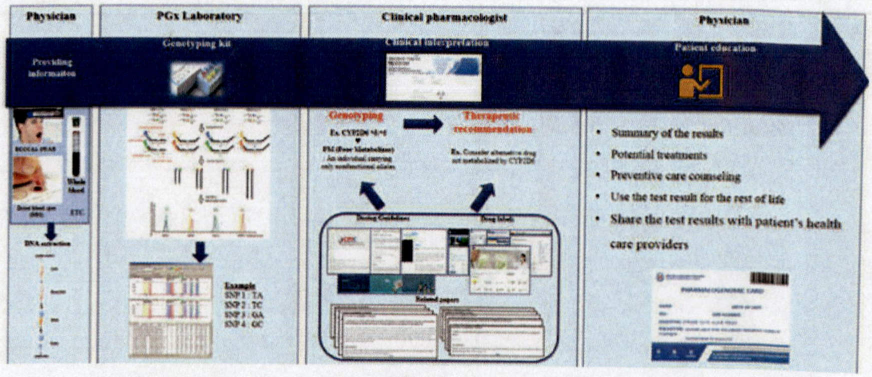

Fig. 2 Steps driving pharmacogenomics implementation into routine practice

Nowadays, ongoing programs of PGx implementation have been performed in America, Europe, and Asia for over ten year [47, 48]. However, the preemptive interpretation of PGx successes into the routine in Vietnam remains several barriers, including scientific, educational, ethical, legal, and social issues, information technology, and reimbursement [48]. Therefore, our hospital has built a system of the information infrastructure that integrates electronic medical records with the appropriate language to help translate genotype data at hand into phenotype information. The process of screening information to answering results is described in the following diagram (Fig. 2).

To promote the use of PGx in routine practice, comprehensive coordination between patients, physicians, engineers and pharmacists is required. First, the doctor should provide information about the test, including risk, benefits, economy and necessity. After genotyping results, clinical pharmacologists nalyse the interaction between the gene and the drug, producing simple, straightforward results. Finally, the doctor will talk to the patient, summarize the results, educate on risk prevention and issue a warning card. Our doctors also teach patients how to use these results for the rest of their lives. To improve comprehension, our clinician uses effective risk communication strategies by tailoring the information to the patient's needs and concerns.

4 Conclusion

Although our study is only the first step to validate our in-house SNaPshot multiplex assay using DBS sample in the clinical practice, it has also shown that the Vietnamese population with high risk genotypes needs to be screened to ensure safe and effective of drugs. However, the clinical application of PGx in Vietnam also faces challenges such as the support of national insurance, the knowledge of people and health care providers, the availability of testing nationwide. The following studies should be performed on a larger sample size to confirm the convenience and cost savings of pharmacogenomic test.

Acknowledgements We thank all physicians and nurses of Haiphong International Hospital for their contributions to this study.

Conflicts of Interest The authors have no conflict of interest to declare.

References

1. Aung AK et al (2014) Pharmacogenomics of antimicrobial agents. Pharmacogenomics 15(15):1903–1930
2. Hung SI et al (2005) HLA-B*5801 allele as a genetic marker for severe cutaneous adverse reactions caused by allopurinol. Proc Natl Acad Sci U S A 102(11):4134–4139
3. Chang CC et al (2011) Association of HLA-B*1502 allele with carbamazepine-induced toxic epidermal necrolysis and Stevens-Johnson syndrome in the multi-ethnic Malaysian population. Int J Dermatol 50(2):221–224
4. Lonjou C et al (2006) A marker for Stevens-Johnson syndrome …: ethnicity matters. Pharmacogenomics J 6(4):265–268
5. Martis S et al (2013) Multi-ethnic distribution of clinically relevant CYP2C genotypes and haplotypes. Pharmacogenomics J 13(4):369–377
6. Scott SA et al (2010) Combined CYP2C9, VKORC1 and CYP4F2 frequencies among racial and ethnic groups. Pharmacogenomics 11(6):781–791
7. Lo C et al (2020) Pharmacogenomics in Asian subpopulations and impacts on commonly prescribed medications. Clin Transl Sci
8. Ang HX et al (2017) Pharmacogenomics in Asia: a systematic review on current trends and novel discoveries. Pharmacogenomics 18(9):891–910
9. Wijnen PA et al (2008) Genotyping with a dried blood spot method: a useful technique for application in pharmacogenetics. Clin Chim Acta 388(1–2):189–191
10. Spooner N, Lad R, Barfield M (2009) Dried blood spots as a sample collection technique for the determination of pharmacokinetics in clinical studies: considerations for the validation of a quantitative bioanalytical method. Anal Chem 81(4):1557–1563
11. Tassaneeyakul W et al (2009) Strong association between HLA-B*5801 and allopurinol-induced Stevens-Johnson syndrome and toxic epidermal necrolysis in a Thai population. Pharmacogenet Genomics 19(9):704–709
12. Lonjou C et al (2008) A European study of HLA-B in Stevens-Johnson syndrome and toxic epidermal necrolysis related to five high-risk drugs. Pharmacogenet Genomics 18(2):99–107
13. Hershfield MS et al (2013) Clinical pharmacogenetics implementation consortium guidelines for human leukocyte antigen-B genotype and allopurinol dosing. Clin Pharmacol Ther 93(2):153–158
14. Saito Y et al (2016) Clinical pharmacogenetics implementation consortium (CPIC) guidelines for human leukocyte antigen B (HLA-B) genotype and allopurinol dosing: 2015 update. Clin Pharmacol Ther 99(1):36–37
15. Ko TM et al (2015) Use of HLA-B*58:01 genotyping to prevent allopurinol induced severe cutaneous adverse reactions in Taiwan: national prospective cohort study. BMJ 351:h4848
16. Chong KW et al (2014) Association of carbamazepine-induced severe cutaneous drug reactions and HLA-B*1502 allele status, and dose and treatment duration in paediatric neurology patients in Singapore. Arch Dis Child 99(6):581–584
17. Aggarwal R et al (2014) HLA-B*1502 is associated with carbamazepine induced Stevens-Johnson syndrome in North Indian population. Hum Immunol 75(11):1120–1122

18. Nguyen DV et al (2015) HLA-B*1502 and carbamazepine-induced severe cutaneous adverse drug reactions in Vietnamese. Asia Pac Allergy 5(2):68–77
19. Wu XT et al (2010) Association between carbamazepine-induced cutaneous adverse drug reactions and the HLA-B*1502 allele among patients in central China. Epilepsy Behav 19(3):405–408
20. Zhou SF, Liu JP, Chowbay B (2009) Polymorphism of human cytochrome P450 enzymes and its clinical impact. Drug Metab Rev 41(2):89–295
21. Johansson I et al (1994) Genetic analysis of the Chinese cytochrome P4502D locus: characterization of variant CYP2D6 genes present in subjects with diminished capacity for debrisoquine hydroxylation. Mol Pharmacol 46(3):452–459
22. Xu Y et al (2008) Association between CYP2D6 *10 genotype and survival of breast cancer patients receiving tamoxifen treatment. Ann Oncol 19(8):1423–1429
23. Kiyotani K et al (2012) Dose-adjustment study of tamoxifen based on CYP2D6 genotypes in Japanese breast cancer patients. Breast Cancer Res Treat 131(1):137–145
24. Nishimura M et al (2016) Influence of the cytochrome P450 2D6 *10/*10 genotype on the pharmacokinetics of paroxetine in Japanese patients with major depressive disorder: a population pharmacokinetic analysis. Pharmacogenet Genomics 26(9):403–413
25. Scott SA et al (2012) PharmGKB summary: very important pharmacogene information for cytochrome P450, family 2, subfamily C, polypeptide 19. Pharmacogenet Genomics 22(2):159–165
26. Gaedigk A et al (2002) Unique CYP2D6 activity distribution and genotype-phenotype discordance in black Americans. Clin Pharmacol Ther 72(1):76–89
27. Del Tredici AL et al (2018) Frequency of CYP2D6 alleles including structural variants in the United States. Front Pharmacol 9:305
28. Bradford LD (2002) CYP2D6 allele frequency in European Caucasians, Asians Africans and their descendants. Pharmacogenomics 3(2):229–243
29. Charoenchokthavee W et al (2016) Prevalence of CYP2D6*2, CYP2D6*4, CYP2D6*10, and CYP3A5*3 in Thai breast cancer patients undergoing tamoxifen treatment. Breast Cancer (Dove Med Press) 8:149–155
30. Byeon JY et al (2018) CYP2D6 allele frequencies in Korean population, comparison with East Asian, Caucasian and African populations, and the comparison of metabolic activity of CYP2D6 genotypes. Arch Pharm Res 41(9):921–930
31. Veiga MI et al (2009) Pharmacogenomics of CYP2A6, CYP2B6, CYP2C19, CYP2D6, CYP3A4, CYP3A5 and MDR1 in Vietnam. Eur J Clin Pharmacol 65(4):355–363
32. Gaikwad T, Ghosh K, Shetty S (2014) VKORC1 and CYP2C9 genotype distribution in Asian countries. Thromb Res 134(3):537–544
33. Kuanprasert S et al (2009) Prevalence of CYP2C9 and VKORC1 mutation in patients with valvular heart disease in northern Thailand. J Med Assoc Thai 92(12):1597–1601
34. Vu NP et al (2018) Polymorphic analysis of CYP2C9 gene in Vietnamese population. Mol Biol Rep 45(5):893–900
35. Hall D et al (1999) Variability at the uridine diphosphate glucuronosyltransferase 1A1 promoter in human populations and primates. Pharmacogenetics 9(5):591–599
36. Barbarino JM et al (2014) PharmGKB summary: very important pharmacogene information for UGT1A1. Pharmacogenet Genomics 24(3):177–183
37. Boyd MA et al (2006) Relationship between hyperbilirubinaemia and UDP-glucuronosyltransferase 1A1 (UGT1A1) polymorphism in adult HIV-infected Thai patients treated with indinavir. Pharmacogenet Genomics 16(5):321–329
38. Bae JW et al (2011) Effects of UDP-glucuronosyltransferase polymorphisms on the pharmacokinetics of ezetimibe in healthy subjects. Eur J Clin Pharmacol 67(1):39–45
39. Lee MT et al (2009) VKORC1 haplotypes in five East-Asian populations and Indians. Pharmacogenomics 10(10):1609–1616
40. Cao K et al (2001) Analysis of the frequencies of HLA-A, B, and C alleles and haplotypes in the five major ethnic groups of the United States reveals high levels of diversity in these loci and contrasting distribution patterns in these populations. Hum Immunol 62(9):1009–1030

41. Ferrell PB Jr, McLeod HL (2008) Carbamazepine, HLA-B*1502 and risk of Stevens-Johnson syndrome and toxic epidermal necrolysis: US FDA recommendations. Pharmacogenomics 9(10):1543–1546
42. Hoa BK et al (2008) HLA-A, -B, -C, -DRB1 and -DQB1 alleles and haplotypes in the Kinh population in Vietnam. Tissue Antigens 71(2):127–134
43. Park HJ et al (2016) HLA allele frequencies in 5802 Koreans: varied allele types associated with SJS/TEN according to culprit drugs. Yonsei Med J 57(1):118–126
44. Puangpetch A et al (2014) HLA-B allele and haplotype diversity among Thai patients identified by PCR-SSOP: evidence for high risk of drug-induced hypersensitivity. Front Genet 5:478
45. Gonzalez-Galarza FF et al (2011) Allele frequency net: a database and online repository for immune gene frequencies in worldwide populations. Nucleic Acids Res 39(Database issue):D913-9
46. Nakkam N et al (2018) HLA pharmacogenetic markers of drug hypersensitivity in a Thai population. Front Genet 9:277
47. Volpi S et al (2018) Research directions in the clinical implementation of pharmacogenomics: an overview of US programs and projects. Clin Pharmacol Ther 103(5):778–786
48. Klein ME, Parvez MM, Shin JG (2017) Clinical implementation of pharmacogenomics for personalized precision medicine: barriers and solutions. J Pharm Sci 106(9):2368–2379

Optimization of HRM Method for Genotyping and Estimating the Frequency of SNP rs12325489 on *Lincrna* Gene in Vietnamese Breast Cancer Patients

Thao Le Thi Mai, Thanh Nguyen Thi Ngoc, Hanh Duong Thi Hong, Luan Huynh Huu, Hoang Ngo Phan, and Hue Nguyen Thi

Abstract Breast cancer (BC) is the most dangerous cancer in women worldwide. Among well—known BC penetrant genes, *LincRNA* plays an important role in the cellular processes. Single Nucleotide Polymorphisms (SNPs) may be regarded as potential markers of carcinogenesis because its appearance can change gene's function. Among SNPs of *LincRNA*, rs12325489 was demonstrated to have an association with BC in Chinese. This study aims to develop the genotyping method for rs12325489 in 50 BC cases and predict whether it relates to BC risk in Vietnamese. This genotyping condition was then evaluated based on its sensitivity, specificity, and stability. The predicted association between rs12325489 and BC risk was conducted, using the allelic and genotypic frequencies of this SNP in Kinh-Vietnamese from 1000Genomes as controls. The results showed that the genotyping assay was successfully developed with high sensitivity, specificity, and stability. The minor allele of rs12325489 is 41.11% indicating that this SNP is high polymorphism in Vietnamese. The prediction of this SNP was increased with the BC risk by 1.43 times (OR (Odds Ratio) = 1.43), not statistically significant ($P = 0.17$). Because rs12325489 has high polymorphism in both patient and control group, it is considered as a potential marker for further association studies between this SNP and breast cancer in Vietnamese. Although the online database has certain reference values in this study, there are remaining concerns. Thus, further studies need to be conducted in proper healthy samples and a larger sample size to confirm the association of this SNP with BC risk in Vietnamese.

Keywords Breast cancer (BC) · High resolution melting (HRM) · LincRNA · Single nucleotide polymorphisms (SNPs) · rs12325489

T. Le Thi Mai · T. N. T. Ngoc · H. D. T. Hong · L. H. Huu · H. N. Phan · H. N. Thi (✉)
Faculty of Biology and Biotechnology, University of Science, Vietnam National University, Ho Chi Minh City, Vietnam
e-mail: nthue@hcmus.edu.vn

V. Van Toi et al. (eds.), *8th International Conference on the Development of Biomedical Engineering in Vietnam*, IFMBE Proceedings 85,
https://doi.org/10.1007/978-3-030-75506-5_48

1 Introduction

Breast cancer (BC) has been becoming the most common disease in women throughout the world. According to the statistics of the World Health Organization (WHO), in 2018, there were over 2 million new cases of BC diagnosed (accounted 24.2%) and 626 679 death cases (accounted 15%), ranked highest in all types of cancer in women around the world [1]. In developing countries like Vietnam, BC had a lower incidence than in developed countries but it had a higher mortality rate. Particularly, mortality/ incidence of Germany, Russia, England, Vietnam are 26.95%, 32.45%, 21.37%, 40.07%, respectively [2]. One of the main causes of this problem was because of diagnosis at a late stage (approximately 85% cases are detected in II, III, IV) [3], which could reduce the effective treatment leading to increase the mortality. Therefore, diagnosis at early stage is extremely necessary and can help resolve this problem.

Breast cancer is a multifactorial disease. Among that, unfavorable environmental exposure, lifestyle and genetic factors are closely related to the incidence of BC [4–6]. However, only a small percentage of individuals developed BC, suggesting that genetic factors, such as genetic variants play a more important role in the individual's risk of diseases [7–9]. Therefore, the discovery of new biomarkers can support diagnose and detect this disease for the Vietnamese early as well as give information into a big picture of the mechanism of causing BC.

In high-penetrance genes such as *BRCA 1* (breast cancer 1), *BRCA 2* (breast cancer 2), mutations appear rarely (< 0.1%) in the population. By comparison, variants in low penetrance genes (single nucleotide polymorphisms, SNPs) are more common (> 0.1%) and they play a vital role in the etiology [10]. Combinations of variants of multiple low penetrance loci/ genes may pose a more significant effect of the risk of BC [11]. Therefore, SNP, which has a higher frequency and association with the other risk SNPs in low-penetrant genes, can make potential genetic markers in the prediction of the BC risk. To contribute more information to literature, in this study, a SNP rs12325489 in exon region of *LincRNA* was investigated. Long intergenic non-coding RNAs (LincRNAs) are non-coding RNAs longer than 200 nucleotides that do not overlap coding genes. Among various functions, *LincRNA* especially participates in RNA-RNA interactions to conduct their regulatory roles in the cytoplasm [12, 13]. The exon region of *LincRNA* obtaining rs12325489 binds site for *microRNA 370* which acts as an oncogene by downregulating *WNK2* (with no lysine (K) 2 in BC) [14, 15]. Genetic variants in exon regions *LincRNA* may alter microRNA—mediated regulation, leading to abnormal cell processes, and they are linked to BC risk. In eastern, southern, and northern Chinese populations, the pooled analysis for association study between rs12325489 and BC was an increased risk (OR (Odds Ratio) = 1.79, CI (confidence interval) 95%: 1.50–2.12, $p = 10^{-7}$) [16]. However, in Vietnam, association studies in Vietnamese populations still lacks reports. Hence, in this study, we aimed to develop an HRM method for genotype rs12325489 and predict the relation between this SNP and BC risk in a Vietnamese population.

2 Materials and Methods

2.1 Materials

Our study used fifty blood samples which were collected from female patients diagnosed with breast cancer and approved by the ethic committee in the Oncology.

Hospital of Ho Chi Minh City (No. 177/ĐĐĐ—CĐT November 18, 2014). Then, genomic DNA was extracted from these blood samples, used by the salting-out method by Hue et al. which had some changes to suit whole blood samples [17]. To ensure suitable samples for HRM assay, concentration, and purity values of DNA were determined by using Nanodrop Spectrophotometer (Thermo Fisher Scientific, USA). A sample was regarded as being suitable when its concentration was above 30 ng/ μL and its OD value of 260/280 ratio was in the range of 1.7–2.0.

2.2 Methods

HRM method development

Primer design. The DNA sequence region and information around the SNP rs12325489 were determined by using the SNP database on NCBI. Primers for real-time PCR HRM were designed and tested specificity by Primer-BLAST. After choosing several potential pairs of primers, PCR desired products were identified by UCSC In-silico PCR, and the secondary structures of primers were run to examine by the Oligo Analyzer tool. Finally, the melting curve of amplifying product was predicted by the uMelt HETS tool at different concentrations of $MgCl_2$ (1.5–3.5 mM) to choose the best $MgCl_2$ concentration for the initial optimization HRM assay.

HRM optimization. Annealing temperature (Ta) is an important factor in the PCR reaction. The suitable annealing temperature will give the best target product and restrict any PCR undesired products. Based on the actual melting temperature (Tm) of the primer (provided by the manufacturer), Ta of the primer could be optimized by the temperature gradient method. Components for each PCR reaction included 1X Toptaq Mastermix, 0.2 μM forward primer (sigma), 0.2 μM reverse primer (sigma), 20 ng DNA and molecular H2O. After that, PCR products were analyzed by electrophoresis with 2% agarose gel, at 90V in 30 minutes, and 50bp DNA Ladder (ThermoFisher) was used as a reference.

After optimizing the Ta, to finding three control samples, we conducted PCR and HRM for some random DNA samples with selected Ta, Mg^{2+} concentration based on the In-silico results from the uMelt HET software, and primer concentration follow the recommendation from the manufacturer. The components in each reaction included LightCycler® 480 High Resolution Melting Master master (Roche) 1X, 0.2 μM forward primer (sigma), 0.2 μM reverse primer (sigma), 3 mM $MgCl_2$, 20 ng DNA and molecular H_2O. These were run on Lightcycler® 96 (Roche), analyzed on Lightcycler® 96 Application Software Version 1.1. Then, suspected DNA samples that represent for different genotypes were confirmed by Sanger automatic sequencing using The BigDye Terminator v1.1 Cycle Sequencing Kit, Applied Biosystems, USA. The results were analyzed by the BioEdit Sequence Alignment application. Three positive controls for the three genotypes of SNP would be identified and used as standardize genotype screening methods as well as performed genotype for the entire study sample.

During identifying the control samples, if Ct (cycle threshold) results were bad (> 30 cycles), Ct modifications would be conducted as the Ct results reflected the amplified ability. We would add 0.1 μM primer and 10 ng DNA each time depending on Ct results to increase the amplified ability of PCR product. After identifying the control samples, based on discrimination of melting curves controls, $MgCl_2$ concentration was considered conducting optimization of $MgCl_2$ concentration in a range of 1.5–3.5 mM.

Genotyping. 50 DNA samples were carried out with three positive controls and one negative control (molecular H_2O) in each running time, applied the optimal HRM's condition. The results were analyzed by using Light Cycler® 96 Application Software Version 1.1.

Evaluation of the optimization HRM method. After genotyping, the result of 25 random samples would be analyzed to evaluate the genotyping condition, based on its sensitivity, specificity, and stability. The sensitivity was the ratio between successful genotyping samples and a total sample. The stability was analyzed by the t-Test algorithm using statistical software R (R i386 5.2.0 and Rstudio) to evaluate the Tm stability of samples among different runs. And the specificity was the ability to distinguish three genotypes through the peak shape of curves and the different Tm of two homozygotes.

Evaluate the potential of SNP rs12325489

Polymorphism. The polymorphism of rs12325489 was given via the minor allele frequency. Frequencies of two alleles were calculated by the following formula: f(C) = f(CC) + ½ f(CT) and f(T) = f(TT) + ½ f(CT) in which f (CC, CT or TT) was the frequency of individuals having CC, CT and TT genotype, respectively.

Prediction of association. To predict the correlation between rs12325489 and BC risk, the frequencies of allele and genotype of 50 BC samples were compared with Kinh Vietnamese data of this SNP in 1000Genomes as controls. Rstudio software was applied to analyze using the 'SNPassoc' package and OR with 95% CI to determine the association between rs12325489 and Vietnamese patients. P= 0.05 was the threshold to identify a statistically significant association.

Table 1 The designed primer pairs for HRM

Sequence(5′–3′)	Tm (°C)	Product size (bp)
F-CACACATGACAACCTGGACTTTC	60.00	78
R-CCCCACAGGTTAAGAGCTTGAT	60.03	

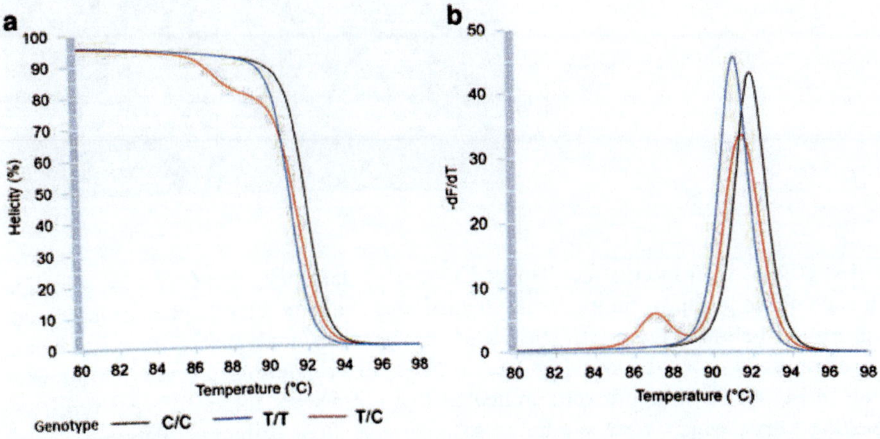

Fig. 1 The predicted normalize melting curves (**a**) and melting peaks (**b**) on uMelt HETS at 3.0 mM $MgCl_2$

3 Results

3.1 HRM Method Development

Primer design. The primer pairs which had a specificity and weak secondary structures were chosen (Table 1). In Silico prediction, at 3.0 mM $MgCl_2$, the homozygotes had the most distinction with ΔTm= 0.9 and the shapes of three melting curves in different genotypes were clearly distinguished (Figure 1). Therefore, 3.0 mM $MgCl_2$ was chosen for the initial optimization HRM method.

3.2 HRM Optimization

Optimization for Ta. After examining gradient temperature from 55 to 65 °C, among bright bands on gel electrophoresis results, to increase the specificity of primers, 61 °C was chosen as Ta for further HRM analysis (Fig. 2).

Controls identification. To find controls, the HRM assay with the optimal Ta of 61 °C and 3.0 mM $MgCl_2$ was applied to several random DNA samples. The result showed there was the shape of three curve groups as predicted melting curves on

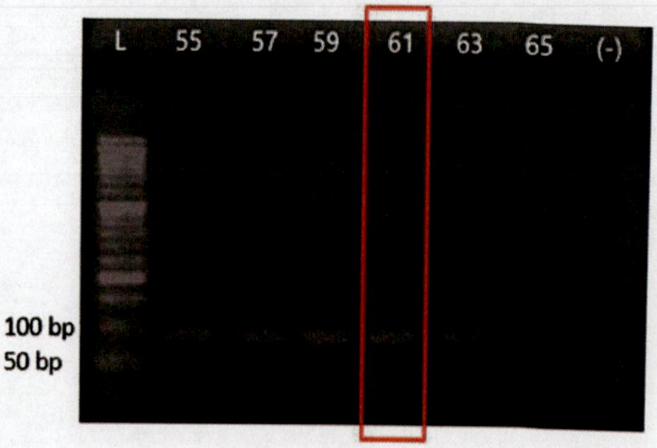

Fig. 2 The annealing temperature for HRM

uMelt HETS, with the distinct Tm of CC and TT homozygotes (ΔTm = 0.6 °C). In each curve group, a sample was selected and confirmed by sequence, indicated the results below (Figure 3). The HRM results were analyzed by three analysis channels as Normalized Melting Peaks, Normalized Melting Curves, Difference Plot in Figure 3a. These results indicated that there were three different types of melting curve which were predicted to represent three different genotypes. One sample in each genotype group was sequenced and the result in Figure 3b showed the corrected genotype as the prediction. These three samples were continued to be used as controls in subsequent HRM assay. *Final optimization conditions of HRM assay.* At 0.4 μM primer and DNA: 30 ng, the amplify curves were improved (Cq (quantification cycle) < 30 cycles), and the rates of failed samples were decreased. In addition, at 3.0 mM $MgCl_2$, the melting curves of two homozygotes were clearly differentiated in three groups. Therefore, there was no need for the post-optimization of $MgCl_2$ concentration. These HRM optimal protocol below were final optimization conditions of HRM and they were used for genotyping on 50 samples (Table 2).

Evaluation of the optimization HRM method. After genotyping, the HRM method was validated by three parameters: sensitivity, specificity, and stability on 25 random samples.

Sensitivity. Among the results of 25 samples, 4/25 failed samples which did not group into control samples or amplified poorly, the others were successful samples (Fig. 4). With the successful samples, the curves that were identified by clustering its respective controls in three charts of HRM results such as normalized melting peaks, normalized melting curves, difference plot. At the successful/total samples ratio, the sensitivity of this HRM method is 84%.

Specificity. The heterozygote had a unique pattern with two peaks, easily identified with one peak of homozygotes in normalized melting peaks (Fig. 4). With one peak of homozygotes, the Tm difference (ΔTm) of two homozygotes was greater than 0.5 °C at all times. Particularly, ΔTm = 0.5 °C at control samples and ΔTm = 0.59 °C at all successful samples (Table 3). Therefore, this optimization HRM method is high specificity.

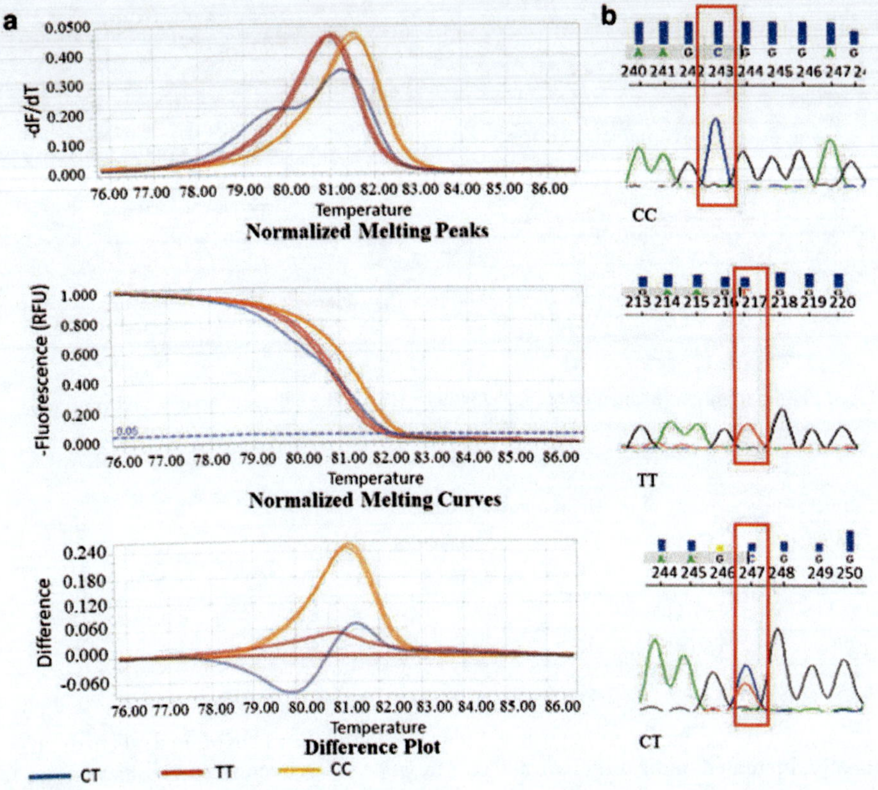

Fig. 3 Determining the genotypes of the controls. A. The melting curves of three control groups and B. the results of sequence

Table 2 HRM optimal protocol

	Final Concentration	Stage	Temperature (°C)	Duration
Master Mix	1X	Pre-incubation	95	5 min
$MgCl_2$	3.0 mM	Amplification	95	30 s
Forward Primer	0.4 μM	(40 cycles)	61	30 s
			72	30 s
		HRM	95	90 s
Reversed Primer	0.4 μM		40	60 s
			65	30 s
			95	1 s
DNA	30 ng	Cooling	37	30 s

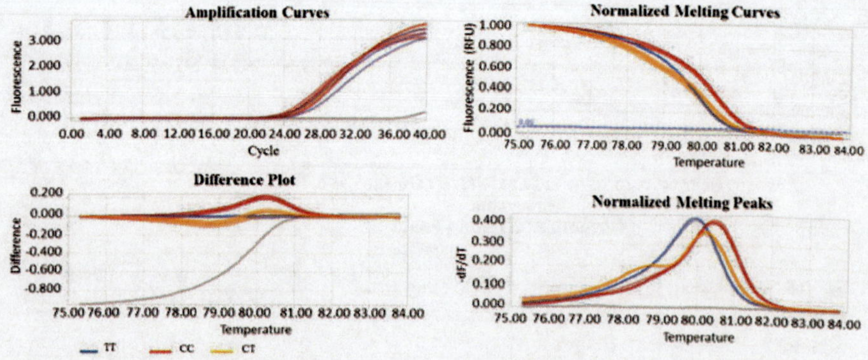

Fig. 4 The melting curves analysis of rs12325489 by HRM. The blue, red, orange, and grey curves presented TT, CC, CT genotype, and the failed sample, respectively

Table 3 The specificity of the HRM optimal protocol

Genotype	Control samples	All samples		
	Tm (°C)	ΔTm (°C)	Tm (°C)	ΔTm (°C)
CC	81.11	0.5	81.09	0.59
TT	80.61		80.54	

The stability: The Tm of control samples through 3 times HRM analysis insignificantly fluctuated, indicating by $p > 0.05$ (Table 4). Moreover, in each running time, all samples were also grouped into three respective control samples ($p > 0.05$) (Table 5). Therefore, this optimization HRM method had high stability.

Table 4 The stability of control samples

Genotype	Repeated times	Tm range	Tm (°C)	*P*-value (T-Test)
CC	3	80.27–81.73	81.11	0.83
CT	3	79.84–81.27	80.65	0.84
TT	3	79.79–81.12	80.61	0.82

Table 5 The stability of successful samples

Running times		*p*-value (T.Tes	t)
	TT	CC	CT
Once	0.20	(*)	(**)
Second	0.12	0.20	0.31
Third	0.11	0.24	0.16

(*) T. test cannot calculate because of few samples. (**) not having samples were recorded result

Table 6 Genotypes and alleles frequency for rs12325489

Group	Genotype			Allele	
	CC	CT	TT	C	T
Case	17 37.80%	19 42.20%	9 20.00%	53 58.89%	37 41.11%
Control 1000 Genomes	45 44.60%	46 45.50%	10 9.90%	53 58.89%	66 32.67%
P-value (chi-square test)	0.24			0.17 OR [95% CI]: 1.43 [0.86–2.38]	

3.3 *Evaluate the Potential of SNP Rs12325489*

Polymorphism. The frequencies of CC, CT, TT were 37.8%, 42.2%, 20.0% respectively; and the frequency of minor allele T at 41.11% in our study was higher than that in 1000 Genomes as control (T= 32.67%) (Table 6). Thus, SNP rs12325489 had high polymorphism in the Vietnamese population.

Prediction of association. The frequency difference of alleles or genotypes between the case and control groups was not statistically significant (*P*-value >0.05) (Table 6). The initial prediction stated that this SNP did not relate to the BC risk in Vietnamese. Nevertheless, the minor allele of rs12325489 increased the BC risk by 1.43 times (OR = 1.43).

4 Discussion

This is the first study investigated the genotype frequency of rs12325489 by the HRM method in the Vietnamese population. As the result, the genotyping HRM condition was successfully developed for this SNP with high sensitivity, specificity, and stability. Based on four charts of HRM results, the genotype of samples would be identified. In amplification curves, the Ct values of all samples were lower than 30 cycles, indicated that was good amplifying results. For normalized melting peaks, the heterozygote curve had two peaks, and two homozygotes had one peak which were distinguished by ΔTm over 0.4 °C. In normalized melting curves, the heterozygote curve cut the lower Tm homozygote curves. In different plots, based on the baseline was set by the lower Tm homozygote, other homozygote curves above that, and heterozygote curves were divided into two sides of that (Fig. 4). Besides, the same genotype samples clustered round respective controls and they stabilized mutually. The Tm of control samples themselves had stability among distinct HRM running times. Thus, all samples had stability and easy distinction (Tables 4 and 5), indicating that the developed HRM condition could be continued using for genotyping rs12325489 on a larger sample size.

In BC cases, the frequency of minor allele T was 41.11% which was higher than that of in both 1000 Genomes as controls (T = 32.67%) and the published data of

Na li in 2014 (T = 28%), indicated that SNP rs12325489 was a highly variable polymorphism.

This means the appearance of minor allele T of rs12325489 and the feasibility of association study between this SNP and breast cancer were quietly high in the Vietnamese population. In addition, results of P HWE (Hardy–Weinberg Equilibrium) for both genotype and allele frequencies were greater than 0.05 indicating that the genotypic distribution of this polymorphism followed by Hardy-Weinberg principle in both cases and controls. This means the research sample represented the Vietnamese population and the genotyping for this SNP was reliable. With P chi-square > 0.05, that statistics also showed no significant difference between BC cases and controls for both allelic and genotypic frequency of rs12325489 (Table 6). The polymorphism of this SNP is predicted to be irrelevant to the risk of BC in the Vietnamese population. However, the power of association analysis was just 4.86%, indicates an unsure result. This reason might be due to the small sample size. Whereas, according to the Na Li et al., the pooled analysis association for this SNP and BC risk in the Chinese population was significantly statistics ($p = 10^{-7}$), but it was carried out in a large size (over 2500 on each cases/ controls group), archived 90% power [16]. Thus, the future association study in the Vietnamese population should be analyzed in a larger size. Additionally, we used the online database on 1000Genomes as healthy control, which had certain reference values in this study, but this lacked available information such as aging and gender. Hence, the proper matching healthy samples should be conducted in further studies.

Among numerous functions, *LincRNA* is specially joined and regulated in RNA-RNA interactions. SNP rs12325489 is located on the exon region of *LincRNA* which was bound site for *microRNA-370* [16]. The expression of *microRNA-370* acts as an oncogene. Higher levels of *miR-370* were found in higher tumor stages (III), significantly expressed higher with larger tumor size, HER2 + , ER/P − status, P53 − sample in comparing case—healthy samples [18], linked to tumor progression, worse disease-free survival [15], and promotion of cell proliferation [14]. Recently, Huang & Liu reported the pathway of *miR-370* in BC, it had oncogenic function by downregulating the tumor suppressor *WNK2* [14]. *WNK2* belonged to the With-no-lysine (K) (WNK) serine-threonine kinase family and was a key regulator of cellular physiology, such as chloride ion transport [19]. Knockdown of *WNK2* promoted G1/S cell cycle transition, rose the influences of epidermal growth factor stimulation, and impaired cell proliferation [20]. Thus, on *LincRNA*, rs12325489 function could disrupt the binding site for *miRNA-370* and make dysregulation of set *LincRNA-Micro370−WNK2* and leading to abnormal in cell processes as well as BC. Therefore, SNP rs12325489 was expected as a potential biomarker for early diagnosis of breast cancer risk.

5 Conclusion

The HRM conditions for genotyping SNP rs12325489 were successfully developed with high sensitivity, specificity, and stability. Therefore, this HRM condition should be continued to examine for larger cases and healthy samples to confirm this association research results. Although the predicted results have not found the association between this SNP and BC risk yet ($p < 0.05$), the SNP had high polymorphism and the role in the regulation of gene relating breast cancer. Hence, SNP rs12325489 should be considered as a potential biomarker for further association studies with BC risk in the Vietnamese population.

Acknowledgements This research is funded by Vietnam National University, Ho Chi Minh City (VNU-HCM) under grant number 562-2020-18-02. The authors would like to thank the Oncology Hospital HCMC for their contribution to collecting samples.

References

1. Bray F, Ferlay J, Soerjomataram I, Siegel RL, Torre LA, Jemal A (2018) Global cancer statistics 2018: GLOBOCAN estimates of incidence and mortality worldwide for 36 cancers in 185 countries. CA Cancer J Clin 68(6):394–424. https://doi.org/10.3322/caac.21492
2. Global Cancer Observatory (2020). Available from: http://gco.iarc.fr/
3. Trieu PD, Mello-Thoms C, Brennan PC (2015) Female breast cancer in Vietnam: a comparison across Asian specific regions. Cancer Biol Med 12(3):238–245. https://doi.org/10.7497/j.issn.2095-3941.2015.0034
4. Fenga C (2016) Occupational exposure and risk of breast cancer. Biomed Rep 4(3):282–292. https://doi.org/10.3892/br.2016.575
5. Lichtenstein P, Holm NV, Verkasalo PK, Iliadou A, Kaprio J, Koskenvuo M, Pukkala E, Skytthe A, Hemminki K (2000) Environmental and heritable factors in the causation of cancer-analyses of cohorts of twins from Sweden, Denmark, and Finland. N Engl J Med 343(2):78–85. https://doi.org/10.1056/NEJM200007133430201
6. Nkondjock A, Ghadirian P (2005) Risk factors and risk reduction of breast cancer. Med Sci (Paris) 21(2):175–180. https://doi.org/10.1051/medsci/2005212175
7. Apostolou P, Fostira F (2013) Hereditary breast cancer: the era of new susceptibility genes. Biomed Res Int 2013:747318. https://doi.org/10.1155/2013/747318
8. Fachal L, Aschard H, Beesley J, Barnes DR, Allen J, Kar S, Pooley KA, Dennis J, Michailidou K, Turman C, Soucy P (2020) Fine-mapping of 150 breast cancer risk regions identifies 191 likely target genes. Nat Genet 52(1):56–73. https://doi.org/10.1038/s41588-019-0537-1
9. Lee KM, Choi JY, Kang C, Kang CP, Park SK, Cho H, Cho DY, Yoo KY, Noh DY, Ahn SH, Park CG, Wei Q, Kang D (2005) Genetic polymorphisms of selected DNA repair genes, estrogen and progesterone receptor status, and breast cancer risk. Clin Cancer Res 11(12):4620–4626. https://doi.org/10.1158/1078-0432.CCR-04-2534
10. Shaker OG, Senousy MA (2019) Association of SNP-SNP interactions between RANKL, OPG, CHI3L1, and VDR genes with breast cancer risk in Egyptian women. Clin Breast Cancer 19(1):e220–e238. https://doi.org/10.1016/j.clbc.2018.09.004
11. Pharoah PD, Antoniou AC, Easton DF, Ponder BA (2008) Polygenes, risk prediction, and targeted prevention of breast cancer. N Engl J Med 358(26):2796–2803. https://doi.org/10.1056/NEJMsa0708739

12. Guttman M, Rinn J (2012) Modular regulatory principles of large non-coding RNAs. Nature 482(7385):339–346. https://doi.org/10.1038/nature10887
13. Ransohoff J, Wei Y, Khavari P (2018) The functions and unique features of long intergenic non-coding RNA. Nat Rev Mol Cell Biol 19:143–157. https://doi.org/10.1038/nrm.2017.104
14. Huang L, Liu X (2019) microRNA-370 promotes cell growth by targeting WNK2 in breast cancer. DNA Cell Biol 38(6):501–509
15. Sim J et al (2015) High MicroRNA-370 expression correlates with tumor progression and poor prognosis in breast cancer. J Breast Cancer 18(4):323–328. https://doi.org/10.1089/dna.2018.4602
16. Li N, Zhou P, Zheng J, Deng J, Wu H, Li W, Li F, Li H, Lu J, Zhou Y, Zhang C (2014) A polymorphism rs12325489C>T in the lincRNA-ENST00000515084 exon was found to modulate breast cancer risk via GWAS-based association analyses. PLoS ONE 9(5):e98251. https://doi.org/10.1371/journal.pone.0098251
17. Nguyen TH, Nguyen DHC, Phan TP, Nguyen TTL, Nguyen DTG (2012) Extraction of human genomic DNA from dried blood spots and hair roots. Int J Biosci Biochem Bioinf 21–26. https://doi.org/10.17706/IJBBB
18. Mollainezhad H, Eskandari N, Pourazar A, Salehi M, Andalib A (2016) Expression of microRNA-370 in human breast cancer compare with normal samples. Adv Biomed Res 5:129. https://doi.org/10.4103/2277-9175.186987
19. Kahle KT, Ring AM, Lifton RP (2008) Molecular physiology of the WNK kinases. Annu Rev Physiol 70:329–355. https://doi.org/10.1146/annurev.physiol.70.113006.100651
20. Moniz S, Verissimo F, Matos P, Brazao R, Silva E, Kotevelets L, Chastre E, Gespach C, Jordan P (2007) Protein kinase WNK2 inhibits cell proliferation by negatively modulating the activation of MEK1/ERK1/2. Oncogene 26(41):6071–6081. https://doi.org/10.1038/sj.onc.1210706

The High D-Glucose Concentration Reduces the Ability of Wound Healing in Vitro of Human Adipose Tissue-Derived Mesenchymal Stem Cells

Huu-Phuong Mai, Nhu-Thuy Trinh, Vong Binh Long, Nguyen Trong Binh, Dang-Quan Nguyen, and Hoa-Xo Duong

Abstract Persistent hyperglycemia causes many microvascular and macrovascular complications in type 2 diabetics. Chronic wounds are one of the most common complications. Adipose tissue-derived mesenchymal stem cells (AT-MSCs) shows great potential for chronic wound healing because of their ability to secrete many factors to stimulate migration, growth, angiogenic. In previous research, we were determined that diabetic AT-MSCs (dAT-MSCs) impaired the ability of wound healing compared with non-diabetic AT-MSCs (nAT-MSCs). In this study, we aim to examine the effect of high D-glucose concentration on the mobility of nAT-MSCs and wound healing in vitro. The expression of migratory genes was determined by reverse transcription quantitative polymerase chain reaction (RT-qPCR). In vitro scratch assay was used to examine the healing ability of nAT-MSCs under the influence of high D-glucose concentration. The results show that high D-glucose treated-nAT-MSCs were reduced the expression of migration factors (*SDF-1* and *CXCR-4*), which suggested the high D-glucose treated-nAT-MSCs may decrease their migration mobility compared with those of non-treated nAT-MSCs. Moreover, the wound healing ability in vitro was impaired high D-glucose treated-nAT-MSCs compared with those of non-treated nAT-MSCs. Interestingly, the migration mobility of nAT-MSCs under the effect of high D-glucose condition was similar to those of dAT-MSCs. Our study provides further insight into the impact of high D-glucose levels on reducing wound healing ability of nAT-MSCs in supporting of the improving AT-MSC function in the treatment of diabetic complications.

Keywords Diabetes · Human mesenchymal stem cells · High D-glucose concentrations · Wound healing

---

H.-P. Mai · N. T. Binh · D.-Q. Nguyen · H.-X. Duong
Biotechnology Center of Ho Chi Minh City, Ho Chi Minh City 700000, Vietnam

N.-T. Trinh (✉) · V. B. Long
International University, Ho Chi Minh City 700000, Vietnam
e-mail: tnthuy@hcmiu.edu.vn

Vietnam National University of Ho Chi Minh City, Ho Chi Minh City 700000, Vietnam

V. Van Toi et al. (eds.), *8th International Conference on the Development of Biomedical Engineering in Vietnam*, IFMBE Proceedings 85,
https://doi.org/10.1007/978-3-030-75506-5_49

Abbreviations

SDF-1	Stromal cell-derived factor-1
CXCR-4	SDF-1 receptor

1 Introduction

Type 2 diabetes mellitus is a global health problem, characterized by insulin resistance and hyperglycemia [1]. The persistent hyperglycemia causes many complications in diabetic patients [1]. A chronic wound is a hard-to-heal wound, one of the most common complications, causing many difficulties in the lives of type 2 diabetes [2]. Nowadays, stem cell therapy shows great potential for chronic wound healing because of their ability to secrete many factors to stimulate migration, growth, angiogenic and immunomodulation [1, 3, 4].

However, our previous research showed that diabetic AT-MSCs (dAT-MSCs) were impaired wound healing ability compared with non-diabetic AT-MSCs (nAT-MSCs) in the flap mouse model and in vitro wound healing model [5, 6]. *Stromal cell-derived factor-1 (SDF-1)* and *CXCR-4* (*SDF-1 receptor)* have been shown to direct the movement of stem cells related to repairing damage in many cells and tissues [7]. Besidesesides, the expression of migration factors (*SDF-1, CXCR-4*) in dAT-MSCs was lower than those of nAT-MSCs [6]. Furthermore, previous studies have shown that high D-glucose concentrations impaired proliferation and migration of human fibroblast, keratinocyte [8, 9]. In vitro scratch assay is an inexpensive and easy method to assess cell movement in vitro of many cells such as fibroblasts, keratinocytes, and mesenchymal stem cells in chronic wound healing [6, 10, 11].

Therefore, we examined the impact of high D-glucose concentration on the mobility and wound healing ability in vitro of nAT-MSCs compared with dAT-MSCs.

2 Material and Methods

2.1 Stem Cell Culture

Non-diabetic AT-MSCs (nAT-MSCs) and diabetic AT-MSCs (dAT-MSCs) are provided by Laboratory of Regenerative Medicine and Stem Cell Biology, University of Tsukuba, Japan. These cells were characterized in the previous report [6]. Briefly, AT-MSCs were cultured in Iscove's Modified Dulbecco's medium (Thermo, USA), with 10% fetal bovine serum (Thermo, USA), 1% antibiotics (Sigma, USA) and five ng/ml basic fibroblast growth factor (bFGF, Sigma, USA) at 37 °C and 5% CO_2. The medium will be renewed every three days. Cells are frozen with cell banker solution (Sigma, USA) and preserved in liquid nitrogen for further experiments. The AT-MSCs at passage 5–8 were used for the experiments.

2.2 *Cell Proliferation Assay*

AT-MSCs were cultured on 24-well plate at a density of 1.5×10^4 cells/well and cultured under high D-glucose concentrations (100 and 200 mM). The cell culture medium was changed every three days. The morphology and growth of AT-MSCs will be monitored for 11 days after a supplement with D-glucose. Trypan blue staining method was used to determine cell proliferation every 48 h.

2.3 *Quantitative Reverse Transcription-Polymerase Chain Reaction*

Stem cell movement plays an important role in the healing of chronic wounds. Therefore, we cultured nAT-MSCs and dAT-MSCs in D-glucose (100 mM) supplementation medium to examine the influence of those on the expression of migratory genes. After adding D-glucose for three days, RNA was extracted with Sepasol-RNA I Super G (Nacalaitesque, Japan). Total RNA (300 ng) was reverse transcribed by transcription-polymerase chain reaction (RT-PCR) Kit (TOYOBO, Japan) to converse to cDNA. cDNA was analyzed using a LightCycler 96 System (Roche, Switzerland) using Maxima SYBR Green/ROX qPCR Master Mix (2X) Kit (Thermo, USA). The expression levels of migration genes were analyzed using the $2^{-\Delta\Delta C_t}$ methods. β-actin gene was used as an internal control. The primer sequences for the PCR reactions are shown in Table 1.

Table 1 The primers used for the quantitative polymerase chain reaction

Function	Gene	Primer	Sequence
Internal control	*β-actin*	5′-primer	GTGCGTGACATTAAGGAGAAGCTGTGC
		3′-primer	GTACTTGCGCTCAGGAGGAGCAATGAT
Migration factors	*SDF-1*	3′-primer	AACGCCAAGGTCGTGGTCGTGCTGG
		5′-primer	CTACAATCTGAAGGGCACAGTTTGG
	CXCR-4	3′-primer	CTGTGACCGCTTCTACCCCAAT GACTT
		5′-primer	CCAAGGAAAGCATAGAGGATGGGGTTC

2.4 In Vitro Scratch Assay

In vitro, wound-healing assay as described previously [12, 13]. Briefly, nAT-MSCs and dAT-MSCs were cultured on 4-well plates with a density of 4×10^4 cells/well. After 24 h, AT-MSCs were treated with 100 mM D-glucose. When cells had reached confluent monolayers, a scratch was created on the surface of the culture dish by a p1000 pipette tip (width 1 mm). Images of wound areas were recorded and analyzed by Wimasis software (https://mywim.wimasis.com) at 0 h and after 24 h. Data have presented the average of five measurements from wound areas.

2.5 Statistical Analysis

The significant differences among many test groups have been used to identify one-way ANOVA (Tukey posthoc test; SPSS 20 software, IBM Corp.). $P < 0.05$ value is statistical significance. The data indicated the mean of the three independent experiments (mean $\pm$ SD).

3 Results

3.1 The Effect of High D-glucose Concentrations on Growth Ability of AT-MSCs

D-glucose is the main source of energy for many cellular activities [14]. Despite this, high D-glucose levels have been shown to increase free radicals and increase the level of ageing of many types of cells [15, 16]. Hence, we evaluated the impact of high D-glucose concentrations on the growth of nAT-MSCs by adding 100 mM and 200 mM D-glucose to the culture medium. Cell proliferation of nAT-MSCs was monitored for 11 days after D-glucose supplementation and compare with dAT-MSCs.

Our results showed that the doubling times of nAT-MSCs (35.67 $\pm$ 0.49 h) and dAT-MSCs (34.9 $\pm$ 1.26 h) were not statistically different. However, nAT-MSCs were cultured with high D-glucose concentrations had slower growth rates than the control sample and dAT-MSCs. Differences in growth rates are most evident since day 3 of the culture process. At the log phase, we found that the doubling time of nAT-MSCs treated with 100 mM D-glucose was 60.95 $\pm$ 4.18 h (1.70 fold decrease, $p < 0.01$, n = 3) compared to the control sample. It is worth noting that nAT-MSCs treated with 200 mM D-glucose hardly proliferated with the doubling time was 201.07 $\pm$ 34.92 h (Fig. 1). Therefore, this result proves that high D-glucose concentration inhibits the growth of nAT-MSCs, which can lead to reduced wound healing ability. According to the results of cell proliferation, we considered that the concentration of 200 mM D-glucose is not suitable for further experiments, so we only examined the effect of

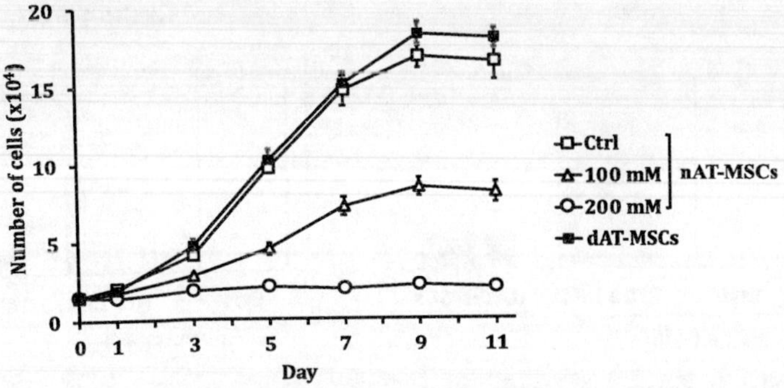

Fig. 1 Growth curve of AT-MSCs under the impact of high D-glucose concentrations. Several cells were determined every 48 h by using the trypan blue exclusion method for 11 days after D-glucose supplementation. High D-glucose concentrations inhibit the growth of nAT-MSCs. Data indicated the average values of three independent experiments (mean ± SD); Ctrl: control

100 mM D-glucose on nAT-MSCs for the expression of the migration factors (SDF-1 and CXCR-4) and wound healing in vitro.

3.2 *The High D-glucose Concentration Decreased Migration Factors Expression in AT-MSCs*

Chronic wounds are one of the common complications in diabetic patients [2]. *SDF-1* and *CXCR-4* factors are known to promote acute and chronic wound healings [7]. Previous studies have demonstrated that the wound healing ability of dAT-MSCs was impaired in vitro wound healing model and the flap mouse model [5, 6].

The results showed that nAT-MSCs have higher expression of migration factors than dAT-MSCs. Specifically, the expression of *SDF-1* (5.32 ± 0.35 fold, $P < 0.01$, n = 3) and *CXCR-4* (8.99 ± 0.60 fold, $P < 0.01$, n = 3) were higher than dAT-MSCs (Fig. 2). Notably, the expression of *SDF-1* and *CXCR-4* in 100 mM D-glucose treated-nAT-MSCs were reduced compared with control nAT-MSCs (*SDF-1*, 1.24 ± 0.08 fold decrease and *CXCR-4*, 1.94 ± 0.13 fold decrease, $P < 0.01$, n = 3). However, the expression of these factors in nAT-MSCs were still higher than dAT-MSCs (*SDF-1*, 4.30 ± 0.15 fold increase, $P < 0.01$, n = 3; *CXCR-4*, 4.63 ± 0.18 fold increase, $P < 0.01$, n = 3) (Fig. 2). Thus, the results suggest that high D-glucose concentration reducing the expression of migration factors in AT-MSCs. This may reduce the ability of nAT-MSCs to heal the wound under hyperglycemia or in diabetic patients.

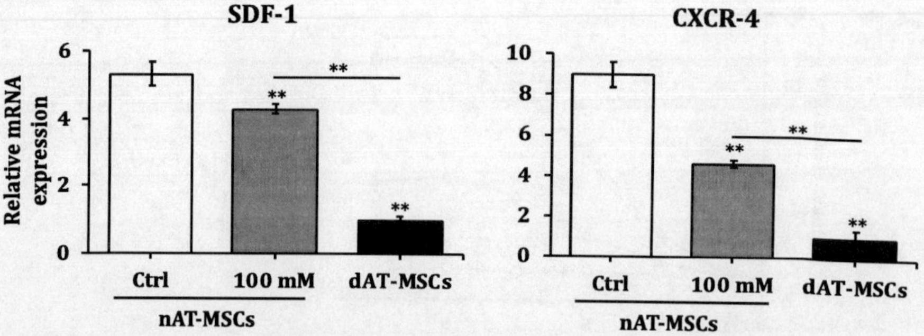

Fig. 2 The high D-glucose concentration decreased migration factors expression in nAT-MSCs. The mRNA expression level was examined by qRT-PCR and normalized to β-actin. The expression *of SDF-1* and *CXCR-4* were decreased in nAT-MSCs under the influence of 100 mM D-glucose. The data indicated the mean of the three independent experiments (mean ± SD); $P < 0.01$ (**); Ctrl: Control

3.3 The High D-glucose Concentration Reduced the Ability of AT-MSCs to Heal Chronic Wounds in Vitro

Previous results have elucidated that high D-glucose concentration altered the expression of genes involved in the ability of wound healing of AT-MSCs (Fig. 2). We performed the scratch assay to examine the effect of high D-glucose concentration on the ability of AT-MSCs to treat chronic wounds. This is an appropriate method to study the effect of cell–matrix and interaction, mimicking cell migration during wound healing in vivo [13].

The results demonstrated that the wound areas at 24 h of nAT-MSCs were significantly recovered in comparison with those of dAT-MSCs (wound area: nAT-MSCs, 12.80 ± 0.88% vs. dAT-MSCs, 22.95 ± 3.22%; $P < 0.01$; n = 5) (Fig. 3). Notably, the ability to cover wounds of nAT-MSCs has been significantly reduced when exposed to high D-glucose concentration (wound area: 100 mM D-glucose treated-nAT-MSCs, 23.48 ± 3.68% vs. nAT-MSCs, 12.80 ± 0.88%; $P < 0.01$; n = 5) (Fig. 3). The wound areas in 100 mM D-glucose treated-nAT-MSCs and dAT-MSCs were not statistical differences (wound area: 100 mM D-glucose treated-nAT-MSCs, 23.48 ± 3.68% vs. dAT-MSCs, 22.95 ± 3.22%, $P > 0.05$, n = 5) (Fig. 3). These results indicate that the ability of wound healing in nAT-MSCs is impaired by the effects of high D-glucose concentration, possibly due to the inhibition of factors that promote cell migration.

4 Discussion

The stem cell migration to the damaged area plays a very important role in wound healing [7, 12]. Previous studies have shown that the wound healing ability of dAT-MSCs is impaired due to an increase in the number of adhesion molecules and a

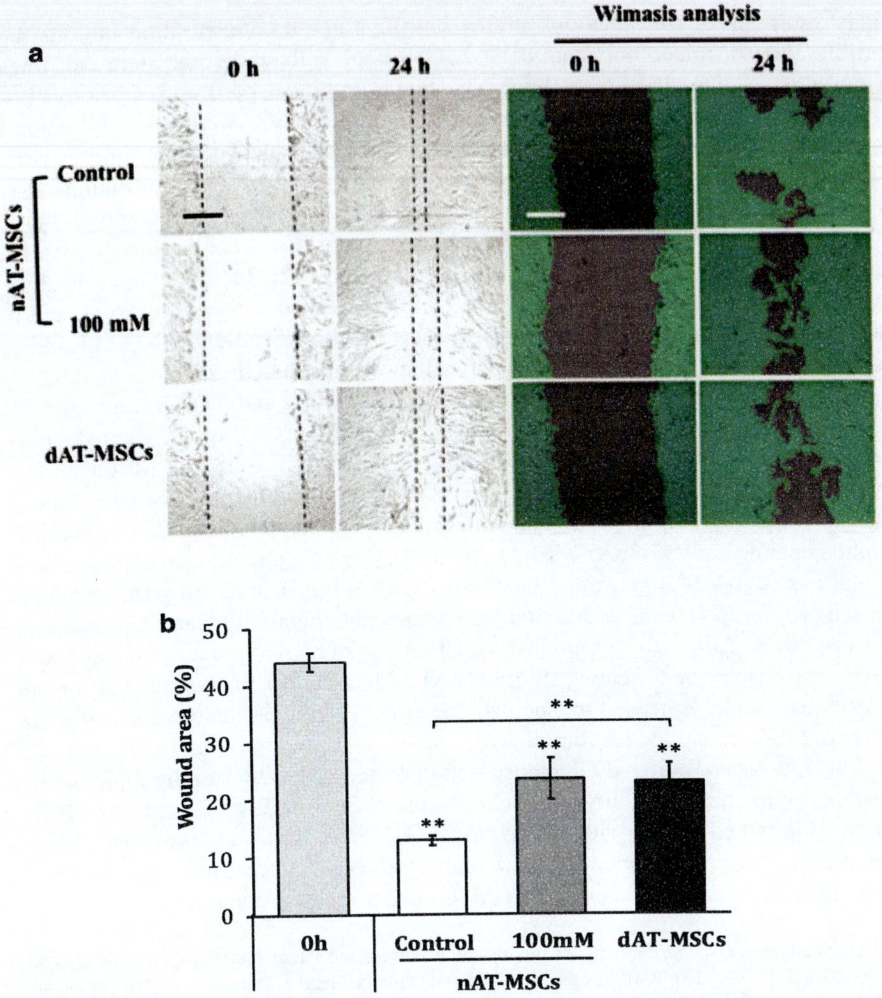

Fig. 3 High D-glucose concentration inhibits the ability of in vitro cell migration. **a** The image was taken using a microscope and analyzed by Wimasis software at 0 and 24 h. **b** Percentage of wound areas of nAT-MSCs (control and 100 mM) and dAT-MSCs at 0 and 24 h. Data were taken from the average of five different wound areas and three replications (mean value ± SD); **, $P < 0.01$. *Scale bar* 500 μm

prolonged inflammation stage [5, 6]. Here, we demonstrated that the high D-glucose concentration reduced the expression of migration factors (*SDF-1* and *CXCR-4*) and impaired the ability to heal wounds of AT-MSCs in vitro.

Previous studies have elucidated that the high D-glucose concentrations increase free radicals and the ageing level of any types of cells [15, 16]. Our results show that 100 and 200 mM D-glucose decreased the proliferation of nAT-MSCs (Fig. 1).

This result agrees with previous studies high D-glucose concentration inhibits the proliferation of mice bone stem cells, MG63 bone cells and some stem cell lines from other sources in human (bones, placenta, umbilical cord and chorion) [17–19]. Also, our previous publication showed that dAT-MSCs were impaired wound healing ability is compared with nAT-MSCs in the flap mouse model and in vitro wound healing model [5, 6]. Furthermore, the high D-glucose concentrations also impaired the proliferation and migration of human fibroblast, keratinocyte [8, 9].

The interaction between *SDF-1* and *CXCR-4* recruit stem cells into the wound area, promoting wound repair and neovascularization [2]. In vivo wound healing model, the concentration of SDF-1 remarkedly increased at the wound site, and it was manifested throughout the healing process. Blocking of SDF-1/CXCR4 leads to a significant reduction in epidermal stem cell movement and impairs wound healing in vivo [20]. Our results found that the expression of *SDF-1* and *CXCR-4* have significantly downregulated in dAT-MSCs than those in nAT-MSCs (Fig. 2). The reduction of these factors is expected to reduce the migration ability of nAT-MSCs during wound healing. Therefore, the scratch assay was performed to mimic cell movement during wound healing in vivo. The result of in vitro scratch assay shows that the wound healing ability of nAT-MSCs treated with high D-glucose concentration was similar as dAT-MSCs in wound healing in vitro (Fig. 3), indicating the reducing in cell proliferation, gene expression, and wound healing ability under high glucose concentration. Although we demonstrated the effect of high D-glucose concentration on impairing the wound healing ability of nAT-MSCs in vitro, the further study on the flap mouse model is needed to elucidate the effect of high D-glucose concentration on nAT-MSCs in wound healing in vivo.

Taken together, our study demonstrated that the expression of migration factors and the wound healing ability in high D-glucose treated-nAT-MSCs and dAT-MSCs were inhibited compared with non-treated nAT-MSCs. This study provides a better understating in the adverse effect of high D-glucose concentration on AT-MSC functions for future application of AT-MSCs under diabetic conditions.

Acknowledgements The funding for this research is from the Biotechnology Center of Ho Chi Minh City (DV01/17-19). We thank professor Osamu Ohneda (Head of Department of Regenerative Medicine and Stem Cell Biology at the University of Tsukuba, Japan) for providing the AT-MSC samples.

Conflict of Interest The authors declare that there is no conflict of interest in this study.

References

1. Ralph EF, DeFronzo A, Groop L, Henry RR, Herman WH, Holst JJ, Hu FB, Ronald Kahn C, Raz I, Shulman GI, Simonson DC, Testa MA, Weiss R (2015) Type 2 diabetes mellitus. Nat Rev 1:1–22. https://doi.org/10.1038/nrdp.2015.19
2. Jude E, Blakytny R (2006) The molecular biology of chronic wounds and delayed healing in diabetes. Diabet Med 23:594–608. https://doi.org/10.1111/j.1464-5491.2006.01773.x

3. Kern HES, Stocvc J, Klüter H, Bieback K (2006) Comparative analysis of mesenchymal stem cells from bone marrow, umbilical cord blood, or adipose tissue. Stem Cells 24:1294–1301. https://doi.org/10.1634/stemcells.2005-0342
4. Seeberger KL,Yeung TY, Kin T, Adesida A, Jomha N, James Shapiro AM, Korbutt GS (2012) Human mesenchymal stem cells protect human islets from pro-inflammatory cytokines. PLoS One 7. https://doi.org/10.1371/journal.pone.0038189
5. Trinh T-T, Ohneda K, Kimura K, Salazar GT, Sato F, Ohneda O (2016) Increased expression of EGR-1 in diabetic human adipose tissue-derived mesenchymal stem cells reduces their wound healing capacity. Stem Cells Dev 25:760–773. https://doi.org/10.1089/scd.2015.0335
6. Trinh TYNT, Tran Cam Tu, Kato T, Ohneda K, Sato F, Ohneda O (2016) Microvesicles enhance the mobility of human diabetic adipose tissuederived mesenchymal stem cells in vitro and improve wound healing in vivo. Biochem Biophys Res Commun 473:1111–1118. https://doi.org/10.1016/j.bbrc.2016.04.025
7. Chai L, Guo R, Chen L, Chen W, Ge L, Li X, Li H, Li S, Cao C (2015) Stromal cell-derived factor 1 (SDF-1) accelerated skin wound healing by promoting the migration and proliferation of epidermal stem cells. In Vitro Cell Dev Biol Anim. https://doi.org/10.1007/s11626-014-9862-y
8. Mizutani K, Buranasin P, Iwasaki K, Pawaputanon Na Mahasarakham C, Kido D, Takeda K, Izumi Y (2018) High glucose-induced oxidative stress impairs proliferation and migration of human gingival fibroblasts. PLoS One 12. https://doi.org/10.1371/journal.pone.0201855
9. Zhang J, Li L, Zhang Q, Zhang D, Xiang F, Jia J, Wei P, Zhang J, Hu J, Huang Y (2019) High glucose suppresses keratinocyte migration through the inhibition of p38 MAPK/autophagy pathway. Front Physiol 10. https://doi.org/10.3389/fphys.2019.00024
10. Wright KT, Waltera MNM, Fuller HR, MacNeil S, Johnson WEB (2010) Mesenchymal stem cell-conditioned medium accelerates skin wound healing: an in vitro study of fibroblast and keratinocyte scratch assays. Exp Cell Res 316:1271–1281. https://doi.org/10.1016/j.yexcr.2010.02.026
11. Gilger BC, Sherman AB, Berglund AK, Schnabe LV (2017) Effect of bone marrow-derived mesenchymal stem cells and stem cell supernatant on equine corneal wound healing in vitro. Stem Cell Res Ther 8. https://doi.org/10.1186/s13287-017-0577-3
12. Babaheydari FM, Dehkordi AN, Chehelgerdi M, Dehkordi SR (2019) Skin tissue engineering: wound healing based on stem-cell-based therapeutic strategies. Stem Cell Res Ther 10. https://doi.org/10.1186/s13287-019-1212-2
13. Park AY, Liang C-C, Guan J-L (2007) In vitro scratch assay: a convenient and inexpensive method for analysis of cell migration in vitro. Nat Protoc 2:329–333. https://doi.org/10.1038/nprot.2007.30
14. Aguiari BZSLP, Vindigni V, Rimessi A, Bianchi K, Franzin C, Cortivo R, Rossato M, Vettor R, Abatangelo G, Pozzan T, Pinton P, Rizzuto R (2008) High glucose induces adipogenic differentiation of muscle-derived stem cells. PNAS 105:1226–1231. https://doi.org/10.1073/pnas.0711402105
15. Castoldi M, Horn P, Wagner W, Diehlmann A, Bork S, Saffrich R, Benes V, Blake J, Pfister S, Eckstein V, Ho AD (2008) Replicative senescence of mesenchymal stem cells: a continuous and organized process. PLoS ONE 3. https://doi.org/10.1371/journal.pone.0002213
16. Munsey TS, Abuarab N, Jiang L-H, Li J, Sivaprasadarao A (2017) High glucose-induced ROS activates TRPM2 to trigger lysosomal membrane permeabilization and Zn^{2+}-mediated mitochondrial fission. Sci Signal 10. https://doi.org/10.1126/scisignal.aal4161
17. Tantrawatpan C, Manochantr S, Hankamolsiri W, Tantikanlayaporn D, Tapanadechopone P, Kheolamai P (2016) The effects of high glucose on adipogenic and osteogenic differentiation of gestational tissue-derived MSCs. Stem Cells Int https://doi.org/10.1155/2016/9674614

18. Wang XZW, Zheng J, Yang J (2010) High glucose stimulates adipogenic and inhibits osteogenic differentiation in MG-63 cells through cAMP/protein kinase A/extracellular signal-regulated kinase pathway. Mol Cell Biochem 338:115–122. https://doi.org/10.1007/s11010-009-0344-6
19. Song G, Cao X, Shao X, Zhao Y, Shi B (2014) Metformin rescues the MG63 osteoblasts against the effect of high glucose on proliferation. J Diab Res. https://doi.org/10.1155/2014/453940
20. Fangqiang Zhu Xiang Xu, Zhang M, Zeng D, Luo D, Liu G, Cui W, Wang S, Guo W, Xing W, Liang H, Li L, Xiaobing Fu, Jiang J, Huang H (2013) Stromal cell-derived factor-1 enhances wound healing through recruiting bone marrow-derived mesenchymal stem cells to the wound area and promoting neovascularization. Cells Tissues Organs 197:103–113. https://doi.org/10.1159/000342921

Evaluate and Optimize Cell-Free RNA Extraction Methods to Apply for Alzheimer's Disease Biomarkers Detection

Anh Phuc Hoang Le, Tai Tien Tran, Thien Hoang Minh Cao, Thao Mai Le, Phuc Truong Le, and Ha Thi Thanh Huong

Abstract Circulating cell-free RNA (cf-RNA) holds great potential as effective diagnostic and prognostic biomarkers, especially for neurodegenerative disorders such as Alzheimer's disease (AD). However, low yield, integrity, and contaminants can alter the final RNA quality and concentration, which hinders the application of cf-RNA. This study aims to: (1) evaluate different cf-RNA extraction methods, (2) optimize the protocols to achieve the optimal cf-RNA quality and quantity, and (3) evaluate hemolysis, which is the main cause of cellular RNA contamination to the final product. Two different types of blood-containing tubes (Norgen and Streck) were compared to evaluate their effects on hemolysis and RNA products. cf-RNA samples were extracted from plasma using Trizol and Zymo cf-RNA extraction kits. We also tested whether the addition of glycogen (20 mg/ml) or combination with RNeasy mini kit to the Trizol protocol will increase the yield of cf-RNA. cf-RNA was then converted to cDNA and examined with PCR and qPCR. The original Trizol protocol resulted in almost no detectable cf-RNA. Adding glycogen significantly increased the yield of cf-RNA to 4–15 ug/ml. Zymo kit's RNA yield was about 5–18 ug/ml. No trace of genomic DNA contamination was identified. Minimal hemolysis was still detected by spectrometer measurement and visual inspection; contamination of cellular RNA has not been determined. Interestingly, despite the high cf-RNA yield, we were not able to consistently detect the housekeeping gene *beta-actin (ACTB)* or Alzheimer-related transcripts using Trizol and Trizol-glycogen cDNA. Reliable detection of housekeeping gene ACTB and inconsistent detection of Alzheimer's related transcripts, namely *CNTF* and *MAG* were achieved when using cDNA samples from Zymo-kit. Concentration of cDNA is highly correlated with the ΔCt values of real-time PCR reactions for *ACTB* primers, suggesting the high integrity of cDNA products from Zymo-kit. Our preliminary data show that the Zymo

A. P. H. Le · T. H. M. Cao · T. M. Le · P. T. Le · H. T. T. Huong (✉)
School of Biomedical Engineering, International University, Ho Chi Minh City Vietnam National University, Ho Chi Minh City, Vietnam
e-mail: htthuong@hcmiu.edu.vn

T. T. Tran
Pham Ngoc Thach University of Medicine, Ho Chi Minh City, Vietnam
e-mail: trantientai@pnt.edu.vn

V. Van Toi et al. (eds.), *8th International Conference on the Development of Biomedical Engineering in Vietnam*, IFMBE Proceedings 85,
https://doi.org/10.1007/978-3-030-75506-5_50

outperformed the Trizol original and modified methods in both yield and integrity measures.

Keywords Cf-RNA · Alzheimer's disease · Molecular diagnostic

1 Introduction

Alzheimer's disease (AD) is the most prevalent type of neurodegenerative disorder in the world, affecting approximately 47 million people [1, 2]. In most forms of neurodegenerative diseases, molecular and cellular changes were found to precede the symptoms [3]. Clinical diagnosis for Alzheimer's disease (AD) mainly bases on neuropsychological tests and exclusion of other age-related dementias. The disease's progression and elevated severity of manifestation could reinforce the identification. At the moment, the most common AD clinical diagnostic methods are dependent on physiological signs such as brain lesions, amyloid plaques, and neurofibrillary tangles [3]. Additionally, there is no treatment for AD [4, 5]. This requires an early intervention before the structural changes happen in the brain to slow down the progression of AD. These contribute to the need for an early AD detection method. Most early diagnostic studies for AD focus on ignallin the pre-symptomatic changes at the molecular, cellular, or whole-brain structural and physiological levels. The two widely accepted biomarkers for AD include cortical thinning and an excessive accumulation of amyloid-β and tau in the central nervous system (CNS). These changes could be assessed with either imaging techniques such as magnetic resonance imaging (MRI), positron emission tomography (PET) of the whole brain, or biochemical assays with cerebrospinal fluid (CSF) as input [6]. Current diagnostic methods for AD are either highly invasive with CSF collection or expensive, time-consuming, and labor-intensive with neuroimaging techniques. Furthermore, the pathogenic inconvertibility is the primary obstruction with recent clinical trials [7]. These features are making them unsuitable for primary care, diagnostic setting, and drug efficacy testing in clinical trials. In addition, CSF biomarkers and neuroimaging are still challenging to achieve the ultimate sensitivity and specificity as well as minimizing variability [3]. To overcome these deficiencies, the key solution is to find a less invasive, more economical, and extensively applicable method for diagnosis of AD, for which blood biomarkers are the potential candidates.

Cf-RNA or cf-protein blood can serve as low invasive, economic and more approachable biomarkers for AD diagnosis. Cf-RNA is RNA existing outside the cells, which are released from the cells as packaged signalling molecules in communication between cells and tissues [8] or as the disposal from apoptosis [9]. Cf-RNAs, which can be used to detect Alzheimer's include *miR-29a, miR-29b-1,* and *miR-9* [10]. However, the association between miRNA and Alzheimer's pathogenesis has not been clarified, unlike mRNA. The mRNA related to *APP, PSD3, MOBP, MAG, SLC2A1, TCF7L2, CDH22, CNTF,* and *PAQR6*—the genes proven to relate to AD, can be identified in plasma and hold the great potential as AD biomarkers [11]. A few of these transcripts, MAG, and CNTF, are utilized as targets in our study for evaluating cf-RNA usability for future diagnostic applications. *MAG* (Myelin Associated Glycoprotein) encodes the protein family that participates in the myelination of neuronal cells. *CNTF* (Ciliary Neurotrophic Factor) encodes for a neuronal survival factor.

The amount of mRNA in an extracellular environment can be limited as it only accounts for 2% of the total extracellular RNA population, which sets a great challenge for any follow-up procedures to quantify or identify RNA [12]. Hence, the extraction method should be optimized to ensure the yield and integrity of final RNA products. Endogenous contaminants like Ca^{2+}, lipids, proteins, genomic DNA, cellular debris, and environmental contaminants such as glove powder, phenolic compounds can hinder or reduce the efficiency of reverse transcription and amplification [13, 14]. In this study, plasma is the subjected sample, which leads to another source of contamination: hemolysis. Hemolysis is the state when red blood cells are lysed and released a great number of cellular products, including proteins and nucleic acids [15, 16]. This event can interfere with the yield and component in the cf-RNA final product.

In our study, we aim to (1) evaluate different cf-RNA extraction methods, [17] optimize the protocols to achieve the optimal cf-RNA quality and quantity, and [18] evaluate hemolysis. This study has successfully enhanced the RNA yield in samples extracted with the Trizol method by adding Glycogen as the carrier and detected the AD-related mRNA in plasma samples. Additionally, the differences in the RNA yield using four different extraction methods, and the ssDNA yield using two different reverse transcription kits were recorded as significant. The difference in the hemolysis level in two types of blood collection tubes was significant as well.

2 Materials and Methods

2.1 Experimental Design

This study was approved by the Internal Review Board, School of Biomedical Engineering, International University, Ho Chi Minh City Vietnam National University. Blood samples were collected from six healthy volunteers in our laboratory, aging between 20 and 35, with no medical records on dementia or other forms of neurodegenerative disorders. Donor samples were drawn into two different blood collection tubes: Norgen Biotek's cf-DNA/cf-RNA Preservative Tubes and Streck's Cell-Free RNA BCT® (Fig. 1).

Two products for isolating cf-RNA were tested: Quick-cfRNA Serum and Plasma Kit (Zymo Research, Irvine, CA) and Trizol LS Reagent (Thermo Fisher Scientific, Waltham, MA). For several extractions using Trizol LS, the protocol was modified in two ways: with the addition of glycogen, and with the use of spin columns [19]. We chose these two products because we wanted to compare a commercial RNA isolation

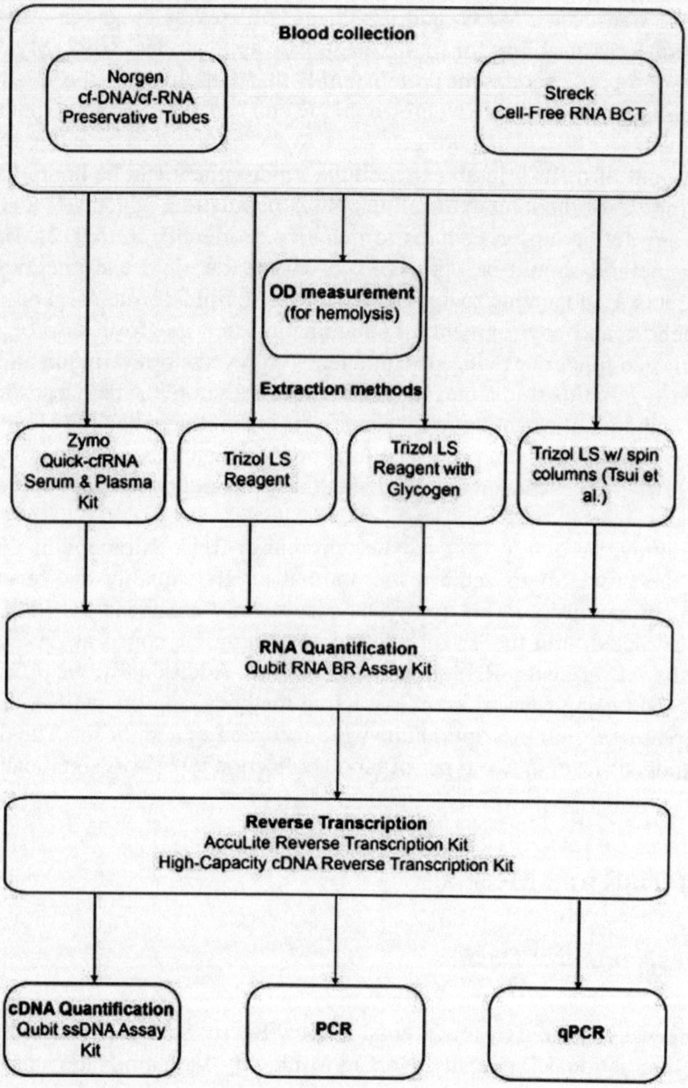

Fig. 1 Study design

kit to the traditional guanidinium thiocyanate-phenol–chloroform extraction method. The modified protocols of Trizol Reagent are our attempts to improve its yield. 30 blood samples were collected in total, 12 samples were treated with Zymo (n = 12), six samples were treated with Trizol (n = 6), eight samples were treated with Trizol w/glycogen (n = 8) and four samples w/column (n = 4).

Isolated RNA samples were quantified by staining with Qubit™ RNA BR Assay Kit and then measured using Qubit™ Fluorometer. Afterwards, a DNase treatment was performed (Dnase I, Rnase-free, Thermo Scientific) and samples were converted to cDNA with the AccuLite Reverse Transcription Kit A (KT Biotech, Ho Chi Minh City, Vietnam). For different extractions, the High-Capacity cDNA Reverse Transcription Kit (Applied Biosystems, Foster City, CA) was used for comparison purposes. cDNA product was quantified using the Qubit™ ssDNA Assay Kit. Qubit fluorometric assays were chosen to quantify RNA and single-stranded DNA (ssDNA) for their sensitivity to low nucleic acid concentrations, which is typical for cf-RNA extraction, and specificity to RNA or ssDNA. Finally, a combination of PCR and qPCR targeting to identify and quantify Alzheimer's-related transcripts was used to evaluate the cf-RNA quality as material for diagnostic application.

2.2 Plasma Preperation

10 mL of whole blood was collected from each donor using a BD Vacutainer® Safety-LokTM blood collection set into the Norgen and Streck tubes. The tubes were then inverted according to each manufacturer's protocol (5 times for Norgen, and 10 times for Streck). It was found that in many cases with Norgen tubes, the donor's blood did not fill the tube completely, so PBS 1X had to be added according to Norgen's instructions before proceeding to plasma separation. To isolate plasma, the Norgen and Streck tubes were centrifuged for 20 min, respectively, at 325 g or 300 g. The plasma layer was transferred to a new 15 ml conical tube and centrifuged at 12,000 g for 15 min to remove any cell debris. Finally, the processed plasma was aliquoted for immediate cf-RNA isolation.

OD measurement. To assess hemolysis, optical density (OD) of plasma samples ($n > 3$) was measured using a spectrophotometer (Vis Spectrophotometer Vinmax 721 LDC Digital Lab Visible Spectrophotometer) at 414 nm [20]. A series of dilutions comprising 50, 25, and 10% were prepared by diluting the plasma samples with PBS 1X. Absorbance measurements were taken at each dilution point.

2.3 RNA Extraction

Extraction of RNA from buccal cells. For the extraction of RNA using Accupid HCV Quantification Kit/Catalog Numbers: Q01HCV02.2A kit, buccal cells were collected and resuspended in PBS 1X solution. First of all, KTS1 was added to the test tube, and a strong vortex was required for complete lysis of the sample. Secondly, KTS2 and KTS3 were added respectively, following which centrifugation was required. Subsequently, the supernatant was removed. KTS4 was then added to wash the pellet and was removed afterwards by centrifuging. Then, the pellet was suspended in DEPC water so that further DNAse and Reverse transcriptase could take place. The resulted cDNA was utilized as positive controls in PCR and qPCR experiments.

Extraction of RNA from plasma

Trizol. The sample was lysed in Trizol (TRIzol™ LS Reagent/Catalog Numbers 10296010 and 10296028) and incubated for complete dissociation of the nucleoproteins complex. Chloroform was then added to allow the phasing process to take place; the aqueous phase was later transferred to a new test tube and mixed with isopropanol. After centrifugation, the supernatant was removed carefully by a white gel-like pellet (RNA precipitate) was not visible at the bottom of the tube. Then ethanol 70% be added for RNA washing. Afterwards, the supernatant was discarded, then the pellet was eluted by mixing with DEPC water. As a result, RNA samples were able to be utilized for further procedures.

Trizol glycogen. For the extractions with the glycogen-Trizol method, after the aqueous phase separation step, 1/10 volume of 5 M ammonium acetate and 0.05 volume of 20 mg/ml glycogen were added to the aqueous phase. The glycogen was RNA-grade, purchased from Oyster. A few different glycogen concentration levels were used (range from 0.4 to 20 mg/ml) in order to assess any possible effects of glycogen on downstream applications. After the addition of glycogen, 1 volume of isopropanol was added, and the mixture was incubated for 1 h at −20 °C. The mixture was then centrifuged at 10,000 g, and the resultant pellet was rinsed with cold ethanol, air-dried, and dissolved in 20 μl of DEPC-treated water.

Zymo. For the Zymo Quick-cfRNA Serum and Plasma Kit, the manufacturer's protocol was followed. For each reaction, we used the maximum sample input of 3 ml. It is worth-noting that the procedure requires the use of a vacuum pump, vacuum manifold, and a high-speed microcentrifuge.

The detailed protocol of each method is described in Fig. 2.

2.4 PCR and qPCR

Materials. The primer sets were designed by Primer-BLAST and synthesized by PHUSA Biochem., Vietnam. The set included the forward and reverse primers, which were, in turn, AAGATCATTGCTCCTCC and GTTAGGTTTTGTCAAGAAAGG for β-*actin (ACTB)*, GCTTCACATAGGATTCCG and GCTTCACATAGGATTCCG for *ciliary neurotrophic factor (CNTF)*, CCAAGTAGTCCACGAGA and TGCTCTGAGAAGGTGT for *myelin-associated glycoprotein (MAG)*.

PCR. Polymerase chain reaction (PCR) was performed with the Master Mix 2x (PHUSA Biochem., Vietnam) on the system of Mastercycler Nexus Gradient (Eppendorf, USA). Corresponding primer pairs of genes of interest were used. Gel images were generated by the chemiluminescent Western blot imaging system C300 (Biosystem Azure, USA).

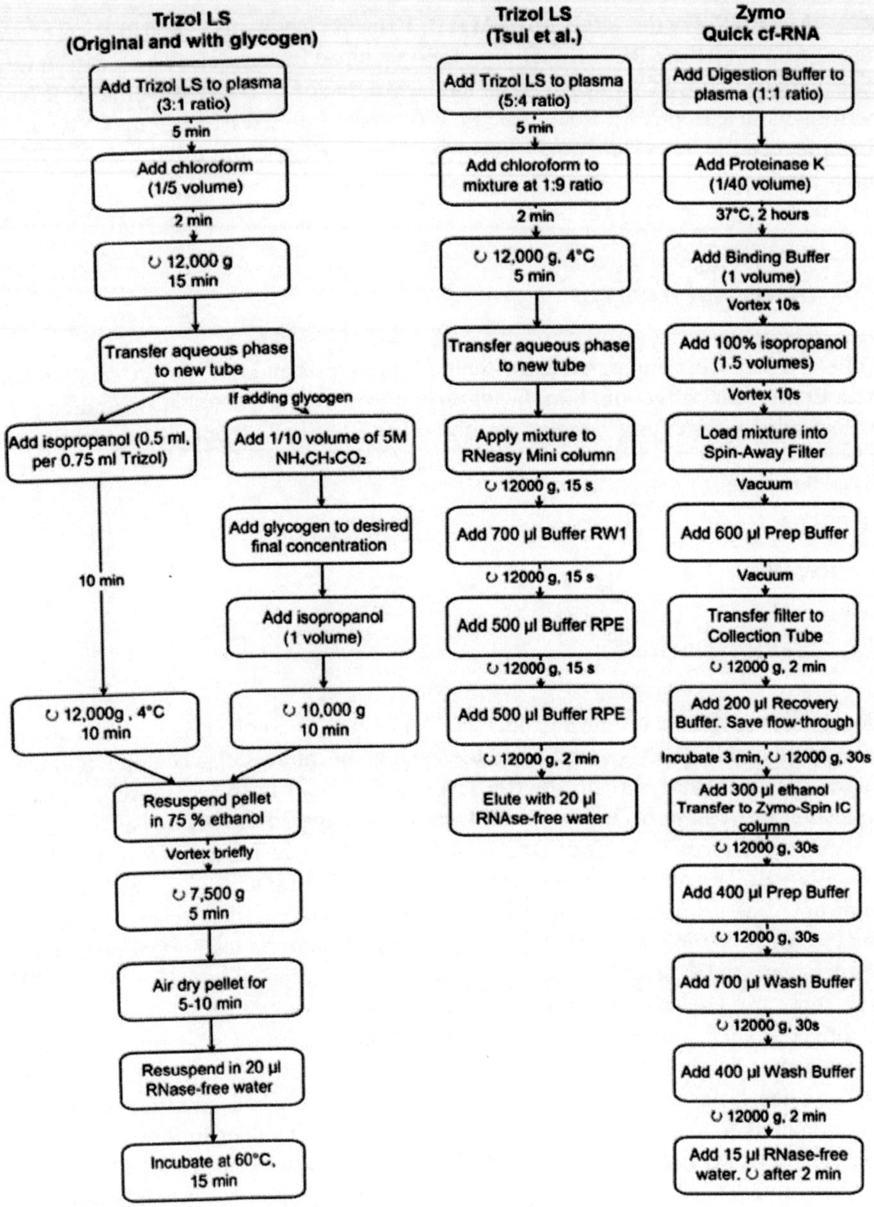

Fig. 2 RNA extraction protocols

Real-time PCR. For the detection of ACTB, 10 ng of cDNA was added to 10 μl of the SensiFAST SYBR Hi-ROX Kit (Biolin, USA) with added primers. The reaction was then added with DEPC-water to reach a final volume of 20 μL. Real-time PCR was performed on a StepOne™ Real-Time PCR System (Applied Biosystem, USA) with recommended PCR cycling conditions from the SYBR kit and modified annealing temperature.

2.5 Statistical Analysis

Collected data were compared using the multiple-t tests and one-way ANOVA tests with Bonferroni correction. Significant differences were expected at two-tailed $p < 0.05$. Statistical Analyses were performed using SPSS v.23.0 and GraphPad Prism 8.0.

3 Results

3.1 Hemolysis Status

Hemolysis usually occurs during blood collection [20]. Hemolysis is the breakdown of red blood cells (RBCs), leading to the leakage of intracellular contents into the plasma. Severe hemolysis can affect the levels of plasma cf-RNA [21]. Several blood-collection tubes have been developed. Here, we compared the hemolysis levels of plasma from Norgen Biotek's cf-DNA/cf-RNA Preservative Tubes and Streck's Cell-Free RNA BCT® collection tubes (abbreviated as Norgen sample and Streck sample from now on) using visual inspection and spectrophotometric. According to the color chart of hemolysis level (Fig. 3a) [17], the hemolysis level of plasma from Norgen was graded at 250 mg/dl Hgb, and that from Streck was nearly 500 mg/dl Hgb (Fig. 3b). However, in most situations, the color difference is not significant to compare. Therefore, we need to utilize an additional method, spectrophotometric assessment, which is more objective and accurate. 10–50% diluted samples were used, so the differences in hemolysis level between two collection tubes can be more obvious to assess. At 10% and 25% dilutions, OD values in Streck samples were significantly higher than those in Norgen (Fig. 3c, $p = 0.011$ and $p = 0.003$, respectively). Thus, all results indicated that Norgen samples had lower hemolysis levels than Streck samples did.

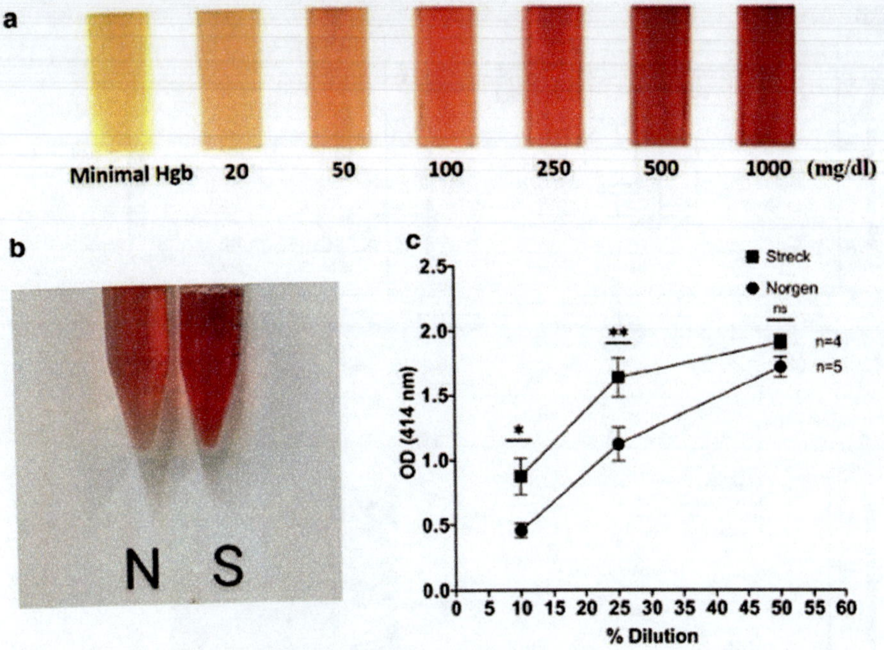

Fig. 3 Hemolysis status of Norgen and Streck samples. **a** Color scale to evaluate hemolysis [17]. **b** Plasma samples extracted from blood collected with Norgen or Streck tubes. c Measured absorbance at 414 nm of plasma samples from Norgen or Streck tubes

3.2 *RNA and ssDNA Yield*

Glycogen is a carrier of RNA precipitation and thus, could increase RNA yield [22]. However, when using the recommended concentration of glycogen (20 μg/μl) from the producer, we found that it influences Qubit measurement (Data not shown). Hence, in this study, a low concentration of glycogen (4 mg/ml) was used to minimize the direct effect of glycogen on Qubit output. At that concentration, white pellets appeared after adding ethanol (Fig. 4a). When compared with the Trizol method, the Trizol column did not yield more RNA (Fig. 4a, $p > 0.99$) whereas Trizol glycogen and Quick cf-RNA Serum and Plasma Kit resulted in higher RNA concentration (Fig. 4a, $p = 0.02$ and $p = 0.01$, respectively). No significant differences in RNA yield were found between Quick cf-RNA Serum and Plasma Kit and Trizol glycogen (Fig. 4a, $p > 0.99$). In conclusion, the addition of glycogen in the Trizol method could increase the yield of RNA concentrations up to the level of the Quick cf-RNA Serum and Plasma Kit method.

Next, Thermo Reverse Transcriptase Kit (Thermo samples) and AccuLite Reverse enzymes (AccuLite samples) were also compared to their efficiency in converting

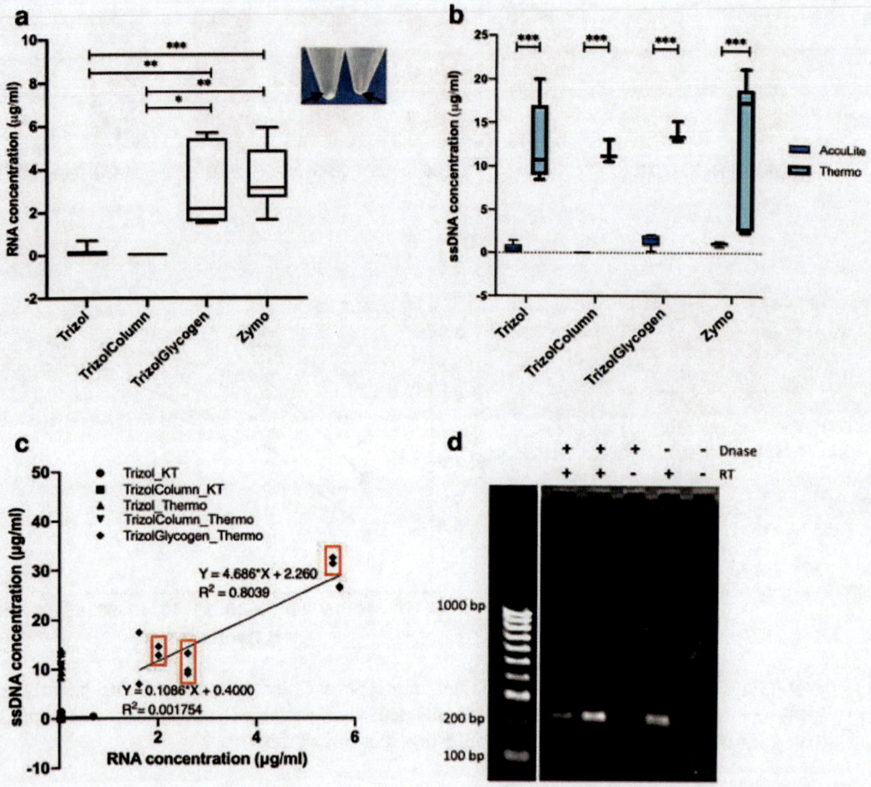

Fig. 4 RNA and ssDNA concentration. **a** RNA concentration of different extraction methods and image of RNA pellets (left islet). **b** ssDNA concentration in various extraction methods. **c** The correlation between RNA and ssDNA concentrations. **d** DNA gel result to evaluate genomic DNA contamination

RNA into cDNA (ssDNA). ssDNA concentration of Thermo samples was significantly higher than that of AccuLite samples with $p < 0.001$ (Fig. 4b). Therefore, Thermo Reverse Transcriptase Kit has higher RNA to cDNA conversion efficiency.

To evaluate the correlation between ssDNA concentration and RNA concentration, a linear regress was applied on data of different extraction kits and reverse transcription enzymes. In samples from Trizol Glycogen protocol with Thermo Reverse Transcriptase Kit, ssDNA concentration is highly-correlated with RNA level (Fig. 4c, $p < 0.001$, $R^2 = 0.8039$). A similar correlation was not found in any other group of samples. Identical replications of RNA sample (same kit, same extraction methods and reverse transcription enzyme) have variable ssDNA concentration (Fig. 4c, red square), suggesting the reverse transcriptase reaction is not highly stable. This result indicated that RNA concentration is not a reliable factor to estimate the final ssDNA level.

Residual genomic DNA in RNA samples can abnormally increase transcript levels and lead to incorrect RT-qPCR results [23]. The level of genomic DNA in RNA samples was assessed using two methods in this study: (1) Double-stranded DNA (dsDNA) concentration in RNA samples and (2) presence of the *ACTB* band in non-reverse samples. The level of dsDNA in the collected samples was too low to be detected by Qubit Fluorometer 2.0. Only samples treated with Reverse transcriptase showed *ACTB* amplicon bands; no amplification of *ACTB* from other samples was detected (Fig. 4d). This result suggested that there is no significant level of genomic DNA contamination was detected in the collected samples. Comparison of commercial RNA isolation kits and reverse transcription enzyme based on operational factors.

3.3 *Detection of Housekeeping and Alzheimer's Disease-Related Transcripts in Plasma Cell-Free cDNA*

Assessment of house-keeping gene beta-actin (*ACTB*) expression

Real-time PCR was used to quickly examine the cell-free existence of the housekeeping gene *ACTB*. We first assessed *ACTB* with SYBR-based assay. Reactions with plasma samples generated an obvious Ct distance from the negative reactions (Fig. 5a). To compare inter-experiments, ΔCt was computed. The ΔCt value was generated by subtraction of Ct from the negative groups of real-time PCR batch, with the baseline set at 0.1. ΔCt showed fluctuations between extraction methods. Detectable but low amount of *ACTB* was found with multiple sources of plasma cell-free cDNA (Fig. 5b). The melt curve of every real-time PCR experiment was also assessed in order to evaluate the product specificity. As shown in Fig. 5c, the melt curve includes the peak of *ACTB* at 86 °C and a secondary structure of primers at 72 °C. Besides, primer variation has two predominant types with a melting temperature of about 72 and 77 °C (Fig. 6). The peak 77 °C infrequently appears among experiments (Figs. 5c and 6). Occasionally, there was an unspecific product with the T_m about 80 °C. To clarify these characteristic waves, we conducted electrophoresis and then sequenced with the same primer set. Resulted image from the electrophoresis of real-time PCR products was found to be consistent with the melting curves. The molecular weight of the structure other than *ACTB* was low and near the area of secondary primers structures (Fig. 5d). Bands a, b, c, and d are corresponding to the peaks 72 °C, 77 °C, 86 °C and 80 °C respectively. Sequencing results confirm that predominant bands over 200 bp are indeed *ACTB* (Fig. 5e). Hereby, we have demonstrated the unstable presence of *ACTB* among extraction methods.

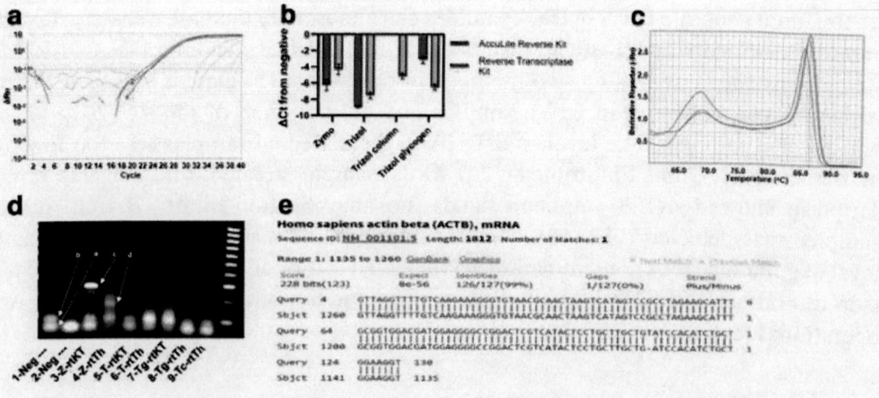

Fig. 5 Detection of house-keeping beta-actin. **a**. Amplification plot of real-time PCR with primers set of beta-actin gene (*ACTB*). Gray and blue lines represent the amplification of negative and target samples, respectively. Each sample was triplicated. **b** ΔCt value of real-time PCR products using plasma cDNA from different extraction methods (mean ± SEM). **c** Melting curve of products from panel A (n = 3) with two clear peaks at 71 and 86 °C. (D) Gel image shows the presence of *ACTB* in plasma cDNA from different extraction methods [Quick cf-RNA Serum and Plasma Kit (Z), Trizol (T), Trizol using glycogen (Tg), Trizol using column (Tc) in combination with reverse transcriptase AccuLite Reverse (KT) and Thermo Reverse Transcriptase Kit (Th)]. (E) Sequencing result of the band a from Fig. 5d. The resulting sequence was aligned by the Nucleotide BLAST to confirm the identification

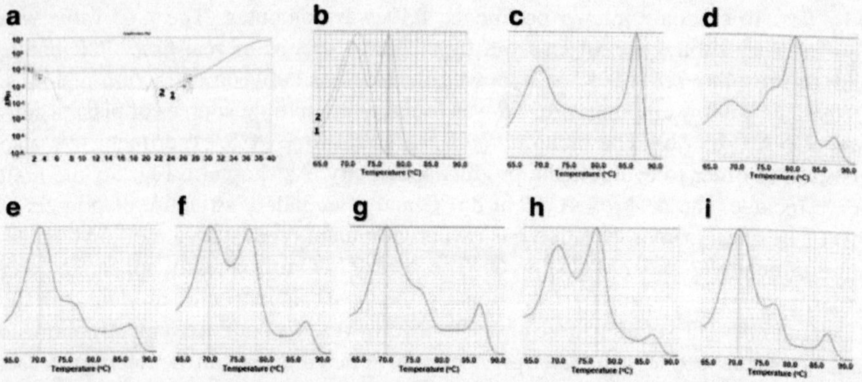

Fig. 6 The melting curves of real-time PCR products from different extraction methods. **a** Amplification plot of real-time PCR with primers set of beta-actin gene (*ACTB*) and negative samples. Gray lines (1 and 2) are replication. **b** Melting curves responding to lines from panel A. (C-I) Melting curves represent real-time PCR products of sample from different methods **c** Quick cf-RNA Serum and Plasma Kit—AccuLite Reverse Kit (kt), **d** Quick cf-RNA Serum and Plasma Kit—Thermo Reverse Transcriptase Kit (th), **e** Trizol—kt, **f** Trizol—th, **g** Trizol using glycogen—kt, **h** Trizol using glycogen—th, **i** Trizol using column—th, matching in tune with lanes 3–9 of Fig. 5d

4 Correlation Between ssDNA and Ct Value

To indirectly estimate the integrity of cf-RNA, the correlation between the concentration of cDNA and the ΔCt values of real-time PCR reactions with *ACTB* primers was analyzed (Fig. 7a). The assumption is if the cDNA product is fragmented, its concentration will not reflect the numbers of *ACTB*-containing DNA fragments. There was generally no correlation between the concentration of cDNA and the ΔCt value. The ΔCt being used to compare among multiple experiments was generated by the subtraction of the mean negative Ct of each experiment. The only significant correlation was exclusively recorded from samples using the Quick cf-RNA Serum and Plasma Kit and Reverse Transcriptase Kit (Fig. 7b).

Examination of Alzheimer's disease (AD)-related RNA in plasma. Aside from the house-keeping gene, we also assessed brain-related ones by conventional PCR and electrophoresis. In detail, we focused on the two previously identified biomarkers of AD, including ciliary neurotrophic factor (*CNTF*) and myelin-associated glycoprotein (*MAG*) [11]. The gel image demonstrated the existence of AD-related genes from cDNA samples of Quick cf-RNA Serum and Plasma Kit and Reverse Transcriptase Kit (Fig. 8). Specifically, *CNTF* and *MAG* respectively appeared in turn with the bandsize of 119 and 162 bp. However, in several other experiments using similar extraction methods or using Trizol methods did not show any visible band (Data not shown). To sum up, the present study detected the *CNTF* and *MAG*, AD-related genes, in plasma.

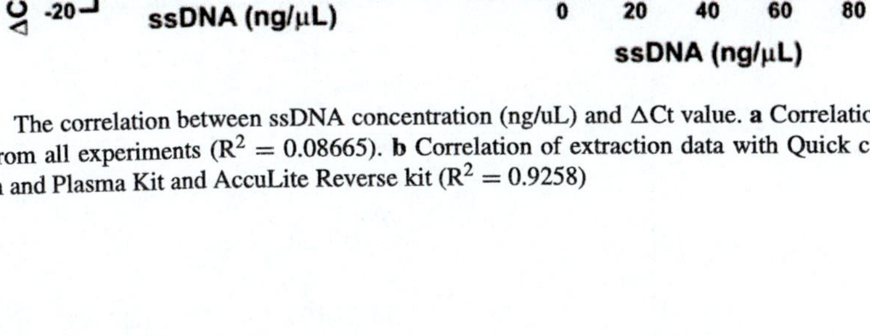

Fig. 7 The correlation between ssDNA concentration (ng/uL) and ΔCt value. **a** Correlation with data from all experiments ($R^2 = 0.08665$). **b** Correlation of extraction data with Quick cf-RNA Serum and Plasma Kit and AccuLite Reverse kit ($R^2 = 0.9258$)

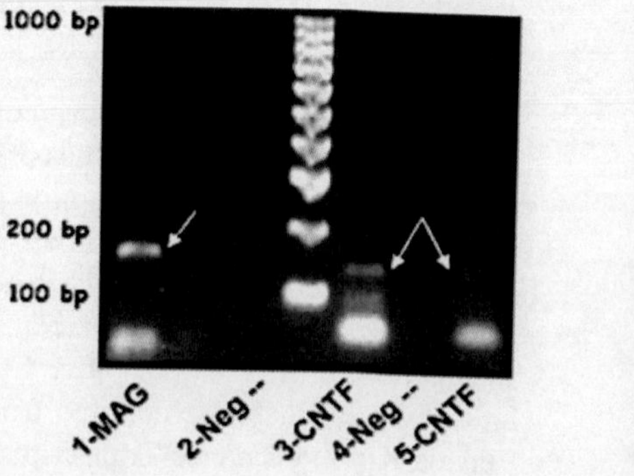

Fig. 8 The expression of Alzheimer's disease (AD) related gene in plasma cDNA extracted with Quick cf-RNA Serum and Plasma Kit

4.1 Operational Factors

Besides the concentration of both RNA and ssDNA, several operational factors of different kits should be considered carefully. They consist of subjective aspects associated with the use, application, cost, and health effects. Among several isolation methods, Trizol glycogen and Quick cf-RNA Serum and Plasma Kit gained the highest amount of RNA (Fig. 4b). Although Quick cf-RNA Serum and Plasma Kit are considerably expensive and inconvenient, they have several advantages such as shorter time input and harmless to use. Trizol and related methods are substantially more affordable yet require a longer time input. Additionally, the organic solvents in Trizol and related methods could be harmful to the respiratory system (Table 1).

Furthermore, although the cost of AccuLite enzymes is higher than Thermo, its runtime is shorter and fewer chemical ingredients quantity. This indicates that the enzymes of AccuLite are more significantly convenient and time-efficient (Tables 2 and 3).

Table 1 PCR and qPCR experimental conditions

Gene	Product size (bps)	Annealing temp. (°C)	Thermal cycles
ACTB	195	55.5	95 °C-1 min (95 °C-15 s, 55.5 °C-15 s, 72 °C-20 s) × 35 cycles 72 °C-5 min, 4 °C
CNTF	119	55.6	95 °C-1 min (95 °C-15 s, 55.6 °C-15 s, 72 °C-20 s) × 35 cycles 72 °C-5 min, 4 °C
MAG	162	58.1	95 °C-1 min (95 °C-15 s, 58.1 °C-15 s, 72 °C-20 s) × 35 cycles 72 °C-5 min, 4 °C

Table 2 Comparison of operational factors between different RNA isolation methods

Extraction kit	Ease of use	Time input (h)	Sample input	Cost in VND/sample	Harmfulness
Trizol	**	7 h	Flexible (2–3 ml)	1,556,524	Toxic
Trizol column	***	6.5 h	Flexible (2–3 ml)	1,362,580	Toxic
Trizol glycogen	*	8 h	Flexible (2–3 ml)	1,606,524	Toxic
Quick cf-RNA Serum and Plasma Kit	**	6.5 h	Flexible (2–3 ml)	2,367,203	–

*: low, **: medium, ***: high

Table 3 Comparison of different reverse transcriptase enzymes

Enzyme	Time input (min)	Cost in VND/reaction
AccuLite	45	52,000 [24]
Thermo	135	49,225 [25]

5 Discussion

5.1 *Hemolysis Status*

The significantly elevated levels of hemolysis in Streck tubes compared to Norgen tubes (Fig. 3c) suggest that differences in preservation technology might be the cause. Another study by Ward Gahlawat et al. (2019) also showed elevated hemolysis in Streck tubes comparing to other blood preservation products, including Norgen [26]. However, it is uncertain how much hemolysis has affected cf-RNA expression in our case. According to our correspondence with Norgen Biotek, they claimed that our current level of hemolysis with Norgen tubes was acceptable. Further testing is necessary to determine the correlation between hemolysis and cf-RNA levels.

Although the RNA samples showed undetectable levels of genomic DNA, the levels of cellular RNA released from the hemolysis process have not yet been assessed, for which further experiments are required. The ratio between miR-451a and miR-23a was used to assess the hemolysis status at a low level of contamination in previous studies: low hemolysis has miR-ratio below five, moderate has miR-ratio between five and seven or severe has miR-ratio higher than seven [18]. The yield of *miR-15b-3p* and *miR-16-5p* are also highly related to the level of hemolysis. They display significant differences in three hemolysis levels [20]. Considering miRBase, *miR-16-5p* also exists in some exosomes, which cannot act as a specific cellular RNA marker. Apart from erythrocytes, RNAs from other blood cell types including lymphocytes, white blood cells, etc., can also be used as detection markers for the levels of cellular RNA contamination in cf-RNA samples [27]. qPCR using the primers of these suggested miRNAs can be used to evaluate the level of cellular nucleic acid contamination and minimal hemolysis.

5.2 RNA and ssDNA Results

Given that the extraction efficiency of Quick cf-RNA Serum/Plasma Kit outperformed Trizol protocols in yield (Fig. 4a), integrity (Figs. 5, 6) and convenience (Table 1), it is recommended for future studies of cf-RNA. The addition of glycogen to Trizol protocol significantly improved the yield (Fig. 4a), yet the high concentration of Glycogen can affect the Qubit measurements. Henceforth, future experiments should be conducted to clarify. The cf-RNA concentrations varying amongst different trials can be explained by the severity of hemolysis differing due to (1) the blood withdrawal velocity and (2) subjects' conditions such as fever or inflammation. It is well-known that hemolysis can lead to false readings in cf-RNA studies as it releases an additional amount of cellular RNA. Optimization in the blood collection step, asides from limiting hemolysis, might improve the yield of final cf-RNA samples. The following optimization suggestions can be considered to further optimize the protocols: (1) Trying blood collection tubes with different preservatives to increase the RNA yield and the samples' stability throughout storage period [28]; (2) Adding ultracentrifugation and ultrafiltration step to enrich the exosomes contained before proceeding RNA extraction[29]; (3) Treating the samples with proteinase K after the lysis step or adding GITC-based denaturing buffer to increase RNA yield[30]. In another outlook, previous studies have proven that some extraction methods outperform the others and their efficiencies are not the same when applied to different biofluids [31]. Hence, further studies using other extraction methods and a replicate of this study on other biofluids are required to complete the evaluation of cf-RNA extraction methods.

The inconsistency in the correlation between RNA concentration and ssDNA concentration implies weak dependency of ssDNA concentration on RNA concentration (Fig. 4c). However, the ssDNA concentration increases significantly when altering the reverse transcription kit, showing strong effects of reverse transcriptase on the ssDNA yield (Fig. 4b). To explain the great differences in ssDNA concentration yielded by two Reverse Transcriptase kit, primers are one of the key variables to be concerned. Two main types are mainly served for the reverse transcription purposes: oligo (dT)N and random primers. Oligo (dT)N is a 2-step cDNA and specific for mRNA. Oligo (dT)N also only targets large mRNA; therefore, the yield is low. On the other hand, random primers target a variety of RNA subpopulations [32]. In this study, both Thermo and Acculite Reverse Transcriptase Kit used random hexamer primers yet still yielded different results. However, the difference in their productivity might be due to the ratio between the primer concentrations and the template concentrations. The higher this ratio, the better yield the reverse transcription might get. The optimal concentration of random primers is suggested to be approximately

60 μM, and an addition of 2.5 μM can also be beneficial. On the other hand, the reverse transcriptase in different kits is also recorded to have different efficiency [33].

Integrity of RNA product. Despite the similar input amount of ssDNA (10 ng/reaction), Ct value was not stable among experiments, batches, and groups (Fig. 6). Presuming the variation of relativity concentration of primers and samples was random and equal among experiments, the fluctuation of Ct value could originate from variable integrity of the extracted RNA. To evaluate the integrity, one could use specific techniques such as examination of average RNA integrity (RIN) [34]. Another approach to indirectly estimate the integrity of cf-RNA is to use the correlation between the concentration of cDNA and the ΔCt values of real-time PCR reactions with *ACTB* primers. In our experiments, a low correlation was detected with samples from Trizol and Trizol-glycogen or Trizol-column kit (Fig. 7a) yet a very high correlation was seen with samples from Zymo kit (Fig. 7b). Additionally, the electrophoresis results usually showed a clear product band of *ACTB* with Quick cf-RNA Serum and Plasma Kit and AccuLite Reverse Kit instead of the one from ThermoFisher (Fig. 5d). Both results suggest that cDNA products from Zymo-kit combining with AccuLite Reverse Kit have the highest integrity.

Acknowledgements This research is funded by Vietnam National University Ho Chi Minh City (VNU-HCM) under grant number NCM2020-28-01, by International University, VNU-HCM under grant number SV2019-BME-02 and Vietnam Alzheimer Network grant, organized by the Vietnam National Geriatric Hospital and the University of California, Davis, US.

We thank Dr. Khoi Nguyen, Mr. Thuc Pham, Mr. Lam Nguyen for providing technical support and advise; Mr. Luan Tran and Ms. Quynh Nguyen for revising the manuscript; RCHAA Center (University of Science, VNU-HCM) for allowing us to use the molecular lab facility and the Molecular Biology Lab, Genetics Department, Faculty of Biology and Biotechnology (University of Science, VNU-HCM) for provid- ing us the glycogen reagent.

Conflicts of Interest The authors have no conflict of interest to declare.

References

1. Dos Santos Picanco LC et al (2018) Alzheimer's disease: a review from the Pathophysiology to diagnosis, new perspectives for pharmacological treatment. Curr Med Chem 25(26):3141–3159
2. Dylan Wint M, Jinny Tavee MD, Patrick Sweeney MD (2014) Alzheimer's Disease. https://www.clevelandclinicmeded.com/medicalpubs/diseasemanagement/neurology/alzheimers-disease/. Accessed Apr 2020.
3. Khan TK, Alkon DL (2015) Alzheimer's disease cerebrospinal fluid and neuroimaging biomarkers: diagnostic accuracy and relationship to drug efficacy. J Alzheimers Dis 46(4):817–836
4. Cummings JL, Morstorf T, Zhong K (2014) Alzheimer's disease drug-development pipeline: few candidates, frequent failures. Alzheimers Res Ther 6(4):37
5. Schott JM et al (2019) Unsuccessful trials of therapies for Alzheimer's disease. Lancet 393(10166):29

6. Oliver P et al (2019) Serum neurofilament dynamics predicts neurodegeneration and clinical progression in presymptomatic Alzheimer's disease. Nat Med 25(2):277–283
7. Patrick RE et al (2019) The limitations of using cognitive cutoff scores for enrollment in Alzheimer trials. Am J Geriatr Psychiatry 27(10):1153–1158
8. Lässer C (2012) Exosomal RNA as biomarkers and the therapeutic potential of exosome vectors. Expert Opin Biol Ther 12(Suppl 1):S189–S197
9. Kubiritova Z, Radvanszky J, Gardlik R (2019) Cell-free nucleic acids and their emerging role in the pathogenesis and clinical management of inflammatory bowel disease. Int J Mol Sci 20(15)
10. Maciotta Rolandin S, Meregalli M, Torrente Y (2013) The involvement of microRNAs in neurodegenerative diseases. Front Cellular Neurosci 7(265)
11. Koh LCWQ, Stephen R, Fan HC, Pan W (2018) Methods for profiling and quantitating cell-free RNA, in Google Patents
12. Sadik N et al (2018) Extracellular RNAs: a new awareness of old perspectives. Methods Mol Biol 1740:1–15
13. Fleige S, Pfaffl MW (2006) RNA integrity and the effect on the real-time qRT-PCR performance. Mol Aspects Med 27(2–3):126–139
14. Spornraft M et al (2014) Optimization of extraction of circulating RNAs from plasma--enabling small RNA sequencing. PLoS One 9(9):e107259
15. Kirschner MB et al (2013) The Impact of Hemolysis on Cell-Free microRNA Biomarkers. Front Genet 4:94
16. Pritchard CC et al (2012) Blood cell origin of circulating microRNAs: a cautionary note for cancer biomarker studies. Cancer Prev Res (Phila) 5(3):492–497
17. Arora SK, Dhupia JS (2014) Hemolyzed Samples Should be Processed for Coagulation Studies: The Study of Hemolysis Effects on Coagulation Parameters. Ann Med Health Sci Res 4
18. Blondal T et al (2013) Assessing sample and miRNA profile quality in serum and plasma or other biofluids. Methods 59(1):S1-6
19. Tsui NBY, NEKO, Lo YMD (2006) Molecular analysis of circulating RNA in plasma. In: Lo YMD, Chiu RWK., Chan K.C.A. (eds) Clinical applications of PCR, in Humana Press. Humana Press, pp 123–134
20. Shah JS, Soon PS, Marsh DJ (2016) Comparison of methodologies to detect low levels of hemolysis in serum for accurate assessment of serum microRNAs. PLoS One 11(4): e0153200
21. Ostendorf APPPW (2019) Nursing interventions and clinical Skills 7th. Mosby.
22. Kosamo S, Peterson K (2001) Sensitive ribonuclease protection assay employing glycogen as a carrier and a single inactivation/precipitation step. BioTechniques 30, 1198–200, 1202, 1204
23. Hashemi-Petroudi S et al (2018) Assessment of DNA contamination in RNA samples based on ribosomal DNA. J Visualized Exp (131):55451
24. KT Biotech (2020) https://kt-biotech.com/san-pham/acculite-reverse-transcription-kit-a/. Accessed Apr 2020
25. Vitech Development (2020) http://vitechltd.vn/vn/sinh-hoc-phan-tu-1595426701.html. Accessed Apr 2020
26. Gahlawat AW et al (2019) Evaluation of storage tubes for combined analysis of circulating nucleic acids in liquid biopsies. Int J Mol Sci 20(3):704
27. Juzenas S et al (2017) A comprehensive, cell specific microRNA catalogue of human peripheral blood. Nucleic Acids Res 45(16):9290–9301
28. Fernando MR et al (2012) Stabilization of cell-free RNA in blood samples using a new collection device. Clin Biochem 45(16–17):1497–1502
29. Lobb RJ et al (2015) Optimized exosome isolation protocol for cell culture supernatant and human plasma. J Extracell Vesicles 4:27031
30. Danielson KM et al (2017) High throughput sequencing of extracellular RNA from human plasma. PLoS One 12(1):0164644
31. Srinivasan S, Yeri A, Cheah PS, Chung A, Danielson K, De Hoff P, Filant J, Laurent CD, Laurent LD, Magee R, Moeller C (2019) Small RNA sequencing across diverse biofluids identifies optimal methods for exRNA isolation. Cell 177(2):446–462.e16

32. Garcia-Milian R (2012) Is a random primer or oligo (dT) better for RT-PCR reaction? https://www.researchgate.net/post/Is-a-random-primer-or-oligodT-better-for-RT-PCR-reaction. Accessed Apr 2020
33. Miranda JA, Steward GF (2017) Variables influencing the efficiency and interpretation of reverse transcription quantitative PCR (RT-qPCR): an empirical study using Bacteriophage MS2. J Virol Methods 241:1–10
34. Brown RAM et al (2018) Total RNA extraction from tissues for microRNA and target gene expression analysis: not all kits are created equal. BMC Biotechnol 18(1):16
35. Agamanolis DP (2020) Neuropathology. In: Chap 9: Alzheimer's disease. http://neuropathologyweb.org/chapter9/chapter9bAD.html. Accessed Apr 2020
36. El Kadmiri N, Said N, Slassi I, El Moutawakil B, Nadifi S (2018) Biomarkers for Alzheimer disease: classical and novel candidates' review. Neuroscience 370:181–190
37. Huynh RA, Mohan C (2017) Alzheimer's Disease: Biomarkers in the Genome, Blood, and Cerebrospinal Fluid. Front Neurol 8
38. Kolte S, Dhupia JS, Arora S (2014) Hemolyzed samples should be processed for coagulation studies: The study of hemolysis effects on coagulation parameters. Ann Med Health Sci Res 4(2):233
39. Phelan MP, Reineks EZ, Berriochoa JP, Schold JD, Hustey FM, Chamberlin J, Kovach A (2017) Impact of use of smaller volume, smaller vacuum blood collection tubes on hemolysis in emergency department blood samples. Am J Clin Pathol 148(4):330–335

Determination of Structure and Anticancer Activity of MM_1 Compound Isolated from Endophytic Fungus *Aspergillus terreus-RTN3* of *Alpinia chinensis* Rosc.

Vo Thi Ngoc My and Nguyen Van Thanh

Abstract *Alpinia chinensis* Rosc. was widely used as spices and medicines. Leaves, stems and rhizomes of the plant contain essential oils, provide a useful habitat for endophytic fungi groups differently. This study has isolated endogenous *A. terreus*—RTN3 strain from young stems of *Alpinia chinensis* Rosc.-Zingiberaceae family. Cultivating and extracting compounds containing metabolites that are resistant to *Staphylococcus aureus* and Methicillin-resistant *Staphylococcus aureus* (MRSA). Determining the number of antibacterial compounds in the crude extract by thin-layer chromatography and vacuum liquid chromatography. The research has isolated pure MM_1 compound with high antioxidant activity and spectral resolution to find the structure compound MM_1 (methyl 2-acetyl-5-methoxy-4-oxo-4*H*-chromen-7-carboxylate). This is a new compound that has not been announced in any works before. MM_1 compound has a high antioxidant activity, the ability to catch α,α-diphenyl-β-picrylhydrazyl (DPPH) free radicals increases linearly by concentration, at concentration 400 ppm of MM_1 is capable of capturing over 80% of DPPH free radicals. The results showed that MM_1 compound's ability to inhibit cancer cell lines increased linearly, at concentrations of 100 μg/mL of MM_1 compound, which inhibited nearly 80% of cancer cell lines and was highest in liver cancer cell line HepG2 (inhibited 86.82%). The MM_1 compound has also been shown to inhibit cancer cells through the ability to induce apoptosis on the NCI-H460 lung cancer cell line under the investigated conditions.

Keywords *Aspergillus terreus* · Endophytic fungi · Cancer cell

V. T. N. My (✉) · N. V. Thanh
Faculty of Pharmacy, Nguyen Tat Thanh University, 300A, Nguyen Tat Thanh Street, 4 District, Ho Chi Minh City, Vietnam
e-mail: vtnmy@ntt.edu.vn

V. Van Toi et al. (eds.), *8th International Conference on the Development of Biomedical Engineering in Vietnam*, IFMBE Proceedings 85,
https://doi.org/10.1007/978-3-030-75506-5_51

1 Introduction

In plants, endophytic fungi and plant are closely related, endophytic fungi can use substances in plants as nutrients to survive. In return, they bring many benefits to the plant, playing an essential role in protecting the host plant against the harmful effects of insects, harmful microorganisms or environmental disadvantages. Recently, secondary fungi metabolites, especially endophytic fungi, are gaining interest because they can produce many bioactive metabolites with antibacterial, anticancer and antioxidant properties. Some endophytic fungi are noted as *Aspergillus, Penicillium, Fusarium* due to the production of many metabolites for biological effects such as antibacterial, antiviral, anticancer … in which *A. terreus* for some compounds have properties antibacterial such as terremid A, terremid B, terrein … [1, 2]. *A. terreus*—RTN3, isolated from young stems of *A. chinensis* Rosc.-Zingiberaceae. In culture media, *A. terreus*—RTN3 is a metabolite that is resistant to *S. aureus* and MRSA bacteria. In addition, *A. terreus* from various sources can give metabolites such as acetylaranotin and related natural products; aspernolid D, and furandion, (+)—Geodin asperteron, along with four butenolid and aspernolid B. *A. terreus* also produces enzymes such as invertase and gluc-glucosidase when grown in submerged fermentation containing barley black as carbon source; β-xylanase production; cis-aconitic decarboxylase acid (CAD)—an important enzyme in the production of itaconic acid; production of beta-glucosidase, … [3, 4].

2 Materials and Methods

2.1 Materials

A. terreus—RTN3 strain has extracted from *A. chinensis* Rosc.-Zingiberaceae. Subjects include testing activity: *Escherichia coli* ATCC 25,922, *Pseudomonas aeruginosa* ATCC 27,853, *Staphylococcus aureus* ATCC 29,213, *MRSA* ATCC 43,300, *Streptococcus faecalis* ATCC 29,212, *Candida alibicans* ATCC 10,231.

2.2 Investigation of Antibacterial, Antifungal Effects of Crude Extracts from A. terreus Culture Solution

After being inoculated in the TSA plate, incubated at 37 °C for 24 h, take 3–5 groups of bacteria with diameter ≥ 1 mm to form a suspension in TSB solution. Distribute evenly with vortex, increased in 4–6 h at 37 °C. The amount of bacteria in suspension is determined by measuring the absorbance with a spectrophotometer at 625 nm; the blank is the TSB solution.

2.3 Investigation of Antibacterial and Antifungal Activities of Crude Extracts

The extract by ethyl acetate is collected and brought to bite by either a rotary evaporator or a stove on a stove at 50 °C.

The bite is redissolved with methanol, diluted to the proper concentration, taking 10 μL of diluent impregnated on a paper plate, leaving in a sterile petri dish to evaporate the solvent. Spread the test bacteria on a cotton swab to 20 mL of TSA medium. Place the dried solvent paper plate on a freshly spread petri dish. Incubate at 37 °C for 24 h.

2.4 Split the Segments for Antibacterial Activity from the Crude Extract

Thin-layer chromatography (TLC). Explore the best separating solvent system for crude extracts from *A. terreus* culture medium. Raw extracts of *A. terreus* culture medium were dissolved in MeOH at a concentration of 8 mg/ml. Develop with solvent system: $CHCl_3$–CH_3COOH (9:1). Record results detected by UV254, UV365 and VS reagent. Evaluate the separability of the solvent system for the test substance based on the number and characteristics of traces on the chromatogram.

Rapid column chromatography (VLC). Separates a complex mixture into simpler, different polarized segments so that more pure fractions can be obtained containing antibiotic active ingredients. The stages of vacuum column chromatography include mobile phase survey, column preparation and deployment, evaporation of solvents resulting in dry bites or concentrated solutions.

Mobile phase survey for VLC. Exploring many different solvent systems on TLC and selecting the most appropriate system as the mobile phase for VLC. Requirements for the mobile phase of the solvent system after one-time deployment are: traces need to be separated with $Rf > 0.3$ and traces that move, separate and do not drag into long streaks.

The different segments are determined biological activity:

- Determination of segments for antibacterial activity by the self-mapping method.
- Test the antibacterial activity of the segment by diffusion method through paper disc.
- Determine the segment for antioxidant activity by fast qualitative method using DPPH.

2.5 Methods for Determining the Chemical Structure of a Compound

The substances obtained through column chromatography and thin-layer chromatography are processed to determine the following criteria: Infrared spectrum (IR), visible ultraviolet (UV–vis), mass spectrometry (MS), resonance spectrum from nuclear (NMR).

2.6 Identification

If a foreign substance appears in the sample, the mass spectrometry can identify its unique chemical structure. The structure of this substance is then compared to a structural library of known substances. Suppose the corresponding substance cannot be found in the library. In that case, we can obtain new data and contribute to the structural library after taking further measures to identify the new compound correctly.

Determine the structure of the isolated compounds based on spectral methods: Electronic atomizing mass spectrometry (ESI–MS); One-way magnetic resonance spectrum (^{1}H-NMR, ^{13}C-NMR, DEPT) and bidirectional (HMBC, HSQC, COZY, NOESY).

2.7 Cell Toxicity Test

The in vitro cytotoxicity test has been confirmed by the US National Cancer Institute as a standard cytotoxicity test to screen and detect substances that can inhibit or kill cancer cells in in vitro conditions [5]. The test was performed to determine the total cellular protein content based on the OD optical density measured when the protein component of the cell was stained with SRB. The test was repeated 3 times. Ellipticine (Sigma) is used as a positive control; DMSO 10% is used as a negative control. IC_{50} values are determined by table Curve computer software. Any reagent with an $IC_{50} < 20$ µg/mL (with crude extract, or with chemical fractions) or $IC_{50} \leq 4$ µg/mL (with pure active substance) is considered to have a cellular inhibitory activity and is capable of ability to inhibit growth or kill cancer cells.

2.8 Statistical Analysis of Data

Data is processed statistically by Microsoft Excel software in Microsoft Office.

3 Results

3.1 Investigation of Antibacterial Activity of Extracts from Fungal and Biomass Media

Extracts from the culture medium of *A. terreus* R-TN3 strain and from biomass. The extracts were investigated for antibacterial activity by diffusion method through paper disc.

Extracts from *A. terreus* R-TN3 media have antibacterial effects on *S. aureus* and MRSA. Biomass extracts of *A. terreus* R-TN3 strain did not have an antibacterial effect on *S. aureus* and MRSA. This indicates that the antibacterial active ingredients from *A. terreus* R-TN3 strain are extracellularly produced, so extracting from biomass of fungi with the same solvent does not give antibacterial effect.

3.2 Cultivation and Extraction of Antibacterial Active Ingredients from A. terreus R-TN3 Strain

From the results in Table 1, we used the method of extracting antibacterial agents from *A. terreus* R-TN3 strain.

Culture after 7 days: gray-brown color-extracted with ethyl acetate to give coarse extract-Ethyl acetate (CT-EtOAc). CT-EtOAc has hygroscopic to constant weight: Dark brown, scented, plastic. Culture results from 1 L of the optimum culture medium, yielding approximately 415 mg of crude high. In order to have enough raw material to isolate the pure compound for the structural determination spectrum, we cultured over 50 L of medium.

Table 1 Antibacterial effect of extracts of strain *A. terreus* R-TN3

Reagent	Suppressive ring diameter (mm)	
	S. aureus	MRSA
Extract from culture solution	17.33 ± 0.93^a	17.33 ± 0.93^a
Extracted from biomass	0.0^b	0.0^b

[a] Notes are referenced using alpha superscripts
[b] Self-supporting

Table 2 Antibacterial activity of spots on TLC of CT-EtOAc

Stain	Trace inhibition ring diameter (mm)	
	S. aureus	MRSA
6	11.33 ± 1.15[b]	11.33 ± 1.15[b]
12	7.33 ± 1.15[c]	7.33 ± 1.15[c]
13	15.33 ± 1.15[a]	15.33 ± 1.15[a]

[a] Notes are referenced using alpha superscripts
[b]Self-supporting
[c]Deposited over Al backing

3.3 Locate Antibacterial Active Substance on the Thin Plate

On the TLC chromatographic chromatography chart, the $CHCl_3$–CH_3COOH solvent system (9:1) was detected by UV scanning at 254, 365 nm and VS reagent spray for a total of 3/17 spots for antibacterial activity (Table 2).

CT-EtOAc contains 3 stain for antibacterial activity, in which the stain 13 has the highest activity against *S. aureus* and *MRSA*. Stain 6 gives a moderate antibacterial effect and stain 6 shows the weakest antibacterial effect on *S. aureus* and *MRSA* strains.

3.4 Refine the Sample by Column Chromatography

Perform high-chromatographic column chromatography with chloroform–methanol solvent system with the rate of change in the direction of increasing methanol, dot and combine the segments and proceed to crystallize the compound. The total high mass of all nine fractions obtained through column chromatography was 10.6 g. From segment 1, it is purified to be light yellow, amorphous MM_1 compound, with Rf of 0.692. Conducting qualitative antioxidant ability of MM_1 compound by DPPH, the results showed that MM_1 compound has strong antioxidant activity.

3.5 Determination of Antioxidant Ability by DPPH of MM_1 Compound

The results showed that MM_1 has a relatively high antioxidant activity, the ability to catch DPPH free radicals increases linearly with concentration (Fig. 1).

The results showed that MM_1 has high antioxidant activity, the ability to catch DPPH free radicals increases linearly with concentration. At a concentration of 100 ppm, MM_1 has the ability to catch more than 80% of DPPH free radicals.

+ MM_1 compound

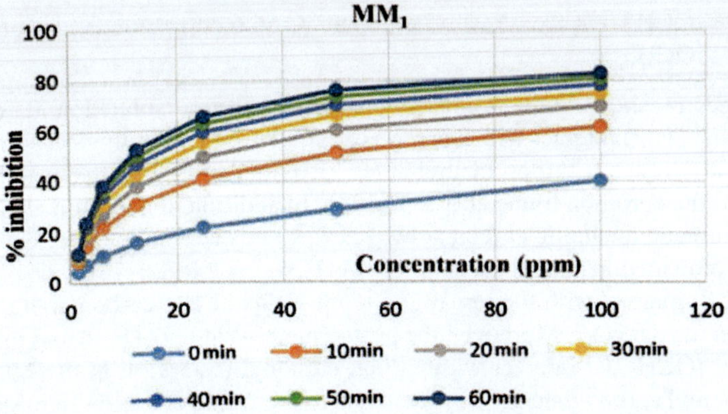

Fig. 1 Diagram showing the antioxidant capacity by DPPH test of MM_1 substance

MM_1 compound obtained from column chromatography, structure is determined by spectroscopic techniques such as ^1H, ^{13}C, HRMS and 2-dimensional spectra to determine the structure. The isolated MM_1 compound is in the form of powder, light yellow, and soluble in methanol and DMSO, but insoluble in n-hexane. HRMS spectrum: positive m/z: 277.07123. 13C NMR (125 MHz, CD_3OD, δ ppm): 167.9 (C-1), 146.2 (C-2), 109.7 (C-3), 173.6 (C-4), 111.8 (C-4a), 158.7 (C-5), 56.6 (OCH_3-5), 108.9 (C-6), 127.7 (C-7), 104.3 (C-8), 159.9 (C-8a), 202.2 (C-9), 21.0 (C-10), 167.9 (C-1′), 52.6 (OCH_3-1′).

^1H NMR Spectrum (500 MHz, CD_3OD, δ ppm): 6.23 (1H, d, J = 0.5 Hz, H-3), 6.97 (1H, d, J = 2.0 Hz, H-6), 6.66 (1H, d, J = 2.0 Hz, H-8), 2.29 (1H, d, J = 0.5 Hz, H-10), 3.71 (3H, s, OCH_3-5), 3.69 (3H, s, OCH_3-1′).

HR-ESI–MS spectra for pseudo-molecular ion tip with m/z: [M + H] + = 277.07123 corresponding to $C_{14}H_{13}O_6$ (theory 277.07067, deviation of 0.56 mmass), allowing determination of CTPT of MM_1 is $C_{14}H_{12}O_6$.

^{13}C-NMR spectrum (125 MHz, CD_3OD, δ ppm) combined with DEPT90, DEPT135 shows that MM_1 has 14 carbon: 3 carbonyl carbon, 3 carbon quadrant aromatic oxygen adjacent, 2 carbon quadrant aromatic ring, 3 aromatic aromatic methine, 2 carbon oxymethyl and 1 carbon methyl.

The presence of 1 carbonyl carbon at δC 173.6 (C-4), 3 carbon quadrant aromatic adjacent oxygen, 2 carbon aromatic quadruple carbon, 3 aromatic aromatic methine corresponding to proton at δH 6.23 (1H, d, J = 0.5 Hz, H-3), 6.97 (1H, d, J = 2.0 Hz, H-6), 6.66 (1H, d, J = 2.0 Hz, H-8), characteristic of the chromon potential frame at position 2, 5, 7 [5, 6].

In addition, MM_1 also shows three substituents: 1 acetyl group with carbonyl carbon (ketones) at δC 202.2 and carbon methyl at δC 21.0 corresponding to protons at δH 2.29 (1H, d, J = 0,5 Hz, H-10); 1 group of methyl carboxylate with carbon carbonyl (ester) at δC 167.9 and 1 carbon oxymethyl at δC 52.6 correspond to protons at δH

3.69 (3H, s, OCH3-1′); 1 oxymethyl group at δC 56.6 corresponds to protons at δH 3.71 (3H, s, OCH3-5).

HMBC spectra, showing the aromatic ring-shaped proton isolated at δH 6.23 (1H, d, J = 0.5 Hz, H-3) interact with 2 carbon of acetyl group including carbon carbonyl (ketones) at δC 202.2, carbon methyl at δC 21,0, indicating that the acetyl group is attached to the chromon frame at C-2 position. In addition, this proton also interacts with an aromatic quadratic carbon at δC 111.8, so this carbon is C-4a. On the other hand, 2 proton methin pairs meta at δH 6.97 (1H, d, J = 2.0 Hz, H6), 6.66 (1H, d, J = 2.0 Hz, H-8) interact with the remaining aromatic quadratic carbon at δC 127.7, so this carbon must be C-7. Moreover, the proton methine at 6.97 (H-6) and the protons at δH 3.69 (OCH_3-1′) interact with carbon carbonyl (esters) at δC 167.9, proving that the methyl carboxylate group is attached to the frame chromon in position C-7. In addition, the remaining oxymethyl proton at δH 3.71 interacts with the aromatic quadratic aromatic carbon adjacent to δC 158.7 (C-5), indicating that C-5 has been oxidized (Tables 3 and 4).

From the HRMS, ^{1}H, ^{13}C-NMR, spectral data, combined with DEPT, HSQC, HMBC spectra and compared with Refs. [5, 6]; we identified MM_1 as methyl-2-acetyl-5-methoxy-4-oxo-4H-chromen-7-carboxylate (methyl 2-acetyl-5-methoxychromon-7-carboxylate) (Fig. 2).

Table 3 Spectral interpretation of MM_1

Carbon number	CH_n			Carbon number	CH_n		
2	C	159.68	0.2	9	CH	106.69	0.8
3	C	116.86	0.5	10	C	161.31	4.6
4	C	177.73	2.7	12	C	192.40	2.4
5	CH	112.48	–	14	C	167.12	1.0
6	C	155.60	1.9	17	CH_3	56.20	1.1
7	CH	104.95	4.4	18	CH_3	25.90	–
8	C	136.35	1.6	20	CH_3	52.51	0.4

Table 4 Spectral results of MM_1

J	Group 1	Group 2	Value	Error	Group	nH	Shift	Error
4J	7	9	2.00	–	5	1	6.95	0.06
5J	9	17	0.25	0.10	7	1	7.61	0.37
2J	17	17	9.40	2.80	9	1	7.48	0.21
2J	18	18	2.88	8.40	17	3	3.85	0.15
2J	20	20	9.40	2.80	18	3	2.65	0.17
					20	3	3.89	0.02

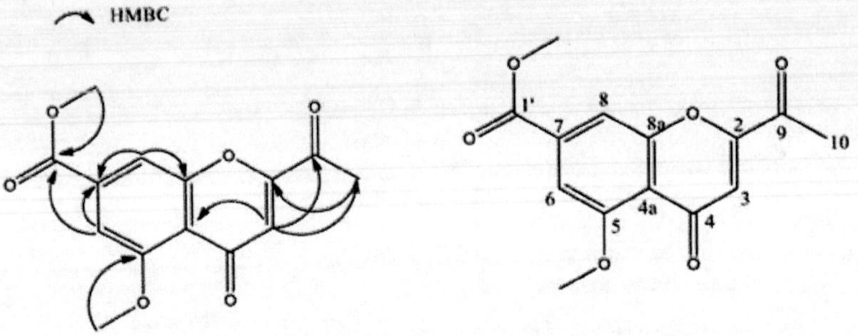

Fig. 2 Primary HMBC interactions and structural formulas of MM_1

3.6 *Survey of Inhibition of Cancer Cell Lines Ability of MM_1 Compound*

Determination of MCF-7 inhibition of breast cancer cell lines, Hela cervical cancer, HepG2 and NCI-H460 lung cancer of MM_1 compound. Results are presented in Table 5.

The results showed that MM_1 compound's ability to inhibit the test cancer cell lines increased linearly with concentration. At a concentration of 100 µg/mL, MM_1 compound inhibits nearly 80% of cancer cell lines and is highest in Hela cervical cancer cell line (at a concentration of 100 µg/mL, inhibits 85.36% cells Hela cancer cells) (Fig. 3).

The results showed that the compound MM_1 was able to inhibit the proliferation of in vitro cancer cell lines. MM_1 compound is not inhibitory to somatic cells, proving it is a highly safe compound.

Table 5 Results of inhibiting cancer cell lines of compound MM_1

Concentration (µg/mL)	Percentage of cell suppression (%)			
	MCF-7	Hela	Hep G2	NCI-H460
100	80.65	85.36	79.62	78.34
75	75.67	65.37	75.41	71.75
50	68.47	59.33	65.68	62.87
25	33.01	23.35	33.79	30.68
10	9.96	– 0.01	13.94	19.88

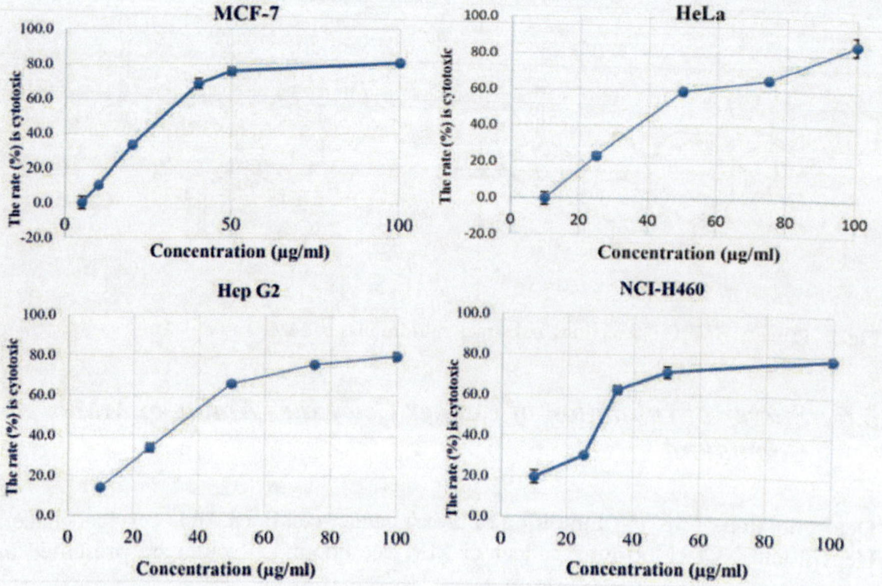

Fig. 3 Results of inhibiting cancer cell lines of compound MM_1

3.7 Mechanism of Inhibiting Cancer Cells of MM_1 Compound

The mechanism of action of MM_1 compound through the ability to induce apoptosis by fragmented DNA tests, fluorescence microscopy with dual AO/EB dyes, and caspase-3 on cancer cell lines lungs NCI-H460. Select NCI-H460 lung cancer cell line do apoptosis because this cell line can perform caspase-3 test; the other cancer cell lines mentioned in the topic did not meet the caspase-3 test.

3.8 DNA Fragmentation Test Results

During apoptosis in mammalian cells, activated endonucleases cleave the chromatin structure at the link between nucleosomes to form small fragments whose sizes are multiples of 180–200 bp. When electrophoresis on the agarose gel these segments form a characteristic "DNA scale". DNA fragmentation is considered a "marker" of apoptosis. We processed NCI-H460 cells with MM_1 at a concentration of 100 µg/mL for 48 and 69 h.

Results showed that fragmented DNA appeared clearly in the NCI-H460 cell population treated with MM_1 at a concentration of 100 µg/mL for 48 and 69 h (wells 1 and 3, Fig. 4), while it was not significant this phenomenon was observed in cell

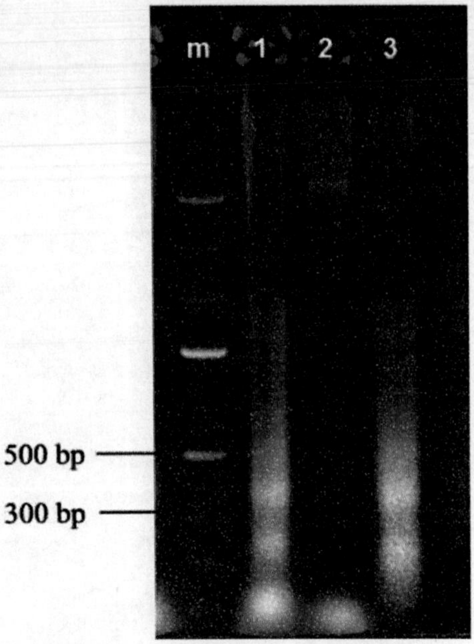

Fig. 4 DNA fragmentation test results; m: standard scale 100 bp. 1. NCI-H460 cells were treated with MM_1 at a concentration of 100 μg/mL for 48 h. 2. Control NCI-H460 cell (not processed). 3. NCI-H460 cells were treated with MM_1 at a concentration of 100 μg/mL for 69 h

samples NCI-H460 (untreated, well 2). This proves that MM_1 compound induced apoptosis in NCI-H460 lung cancer cells under the survey conditions.

3.9 *Testing Fluorescence Microscope with AO/EB Dual Dyes*

The fluorescence microscope method combines the different absorption capacity of Acridine orange (AO) and Ethidium bromide (EB) dyes with the chromatin concentration in the nucleus of the stained cells when entering the apoptosis process. The method allows detection of living cells, apoptosis and necrosis.

Living cells with intact cell membranes will fluoresce blue throughout the nucleus due to the selective permeability of the living cell membrane only. Early apoptosis cells have intact cell membranes but concentrated, or fragmented chromatin will glow very bright green (due to strong color AO) in the nucleus (solid red arrow). The apoptosis cell at a later stage has a broken cell membrane, so AO and EB simultaneously permeate into the cell, the nucleus will emit an orange color which is a combined signal of AO and EB (dashed red arrow). We performed the control NCI-H460 (untreated cells) and incubated cells with MM_1 compound at 100 μg/mL, for 36, 48 and 60 h with dye solution AO: EB.

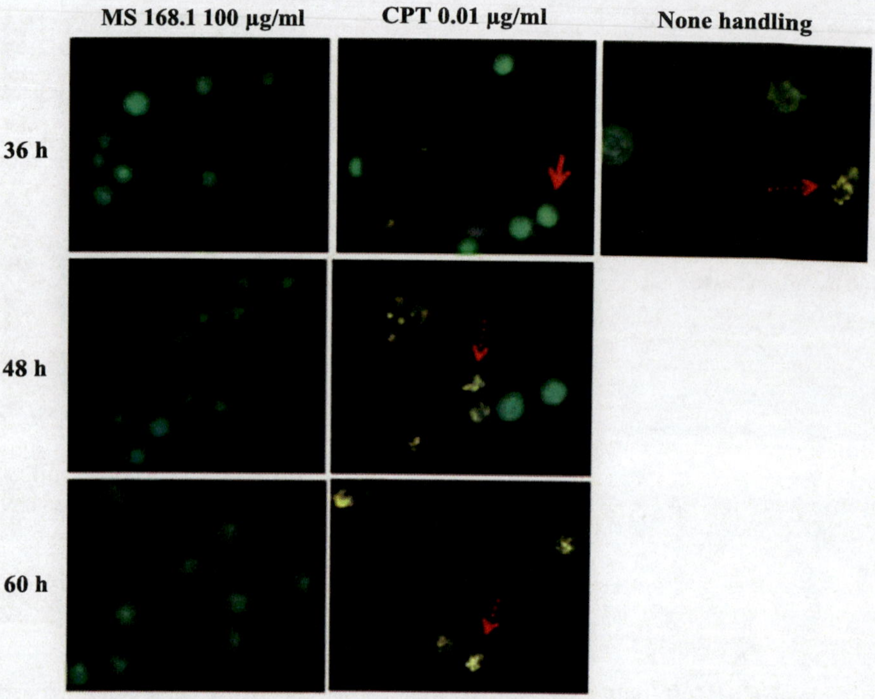

Fig. 5 Fluorescence microscope test with AO/EB dual dyes. (*Solid red arrow:* Early apoptosis cells have a nucleus that emits a very bright green color; *Red dashed arrow:* Late Apoptosis cells that have a nucleus glow orange)

Observation results under a fluorescence microscope (Fig. 5) show a change in morphology of the incubated cell with MM_1 compared to the control cell. The NCI-H460 cell sample treated with Camptothecin standard (CPT) at a concentration of 0.01 µg/mL for 36 h was a positive test of the procedure.

4 Discussion

The results showed that the compound MM_1 (ethyl 2-acetyl-5-methoxy-4-oxo-4H-chromen-7-carboxylate) was able to inhibit many different cancer cell lines, through the ability to inhibit 4 experimental cell lines: MCF-7 breast cancer, Hela cervical cancer, HepG2 liver cancer and NCI-H460 lung cancer. In normal cell lines (Soma), this compound has a higher IC_{50} value and is outside the cell suppression threshold (> 4 mg/mL). This proves that this compound is highly safe, does not inhibit normal cells. This new compound has not been announced on any works in the country and around the world. This study is completely in line with previous studies on the ability to inhibit cancer cells of *A. terreus* fungi. However, the source of *A. terreus* isolates of previous studies is from soil, water, mangroves, sea … For example, terrein with antifungal activity, known from 1935, was born by *A. terreus*, it was also found to inhibit breast cancer cells by inducing apoptosis pathway with IC_{50} of 1.1 nM in MCF-7 cell line, which is 100 times more likely to be taxol in inhibits this cell line. Furthermore, terrein was also found to be active against pancreatic and hepatic cancer cell lines PANC-1 (IC_{50} = 9.8uM) and HepG2 (IC_{50} = 66.8uM) [7, 8]. Two butyrolactone I derivatives in *A. terreus* isolated from desert sands with antifungal activity *A. fumigatus* [9]. Compound 6-(4′-hydroxy-2′-methyl phenoxy)-(-)-(3R)-mellein isolated from *A. terreus* fungal extract exhibits antioxidant activity. In addition, in the fungal epidemic there are terremide A and B and terrelactone, in which terremide A and B have antibacterial activity [9]. Suja et al. 2014 also demonstrated that *A. terreus* mushroom extract isolated from the ocean was resistant to cancer. Lovastatin is used as a medicine to lower total and LDL cholesterol, treat high cholesterol … In a 2010 study, the study also showed that lovastatin induced apoptosis in 10 ovarian cancer cell lines with IC_{50} from 2 to 39 μM, lovastatin was also recently detected to inhibit cell flow of MCF-7 cancer, Hep G2 and cervical cancer with an IC_{50} value of 0.7; 1.1 and 0.6 μM [10, 11].

5 Conclusion

After conducting research, we cultured and extracted a substance containing a crude extract that acts against *S. aureus* and MRSA of *A. terreus* R-TN3 strain. By the method of TLC and VLC, the study isolated the pure MM_1 compound with high antioxidant activity and spectra to find the structure of MM_1 compound with the name: Methyl 2-acetyl-5-methoxy4-oxo-4H-chromen-7-carboxylate. This is a new compound that has not been published in any works in the world. MM_1 compound has a high antioxidant activity, the ability to catch DPPH free radicals increases linearly with the concentration, at a concentration of 400 ppm MM_1 has the ability to catch more than 80% of DPPH free radicals. In addition, the compound MM_1 also has the ability to inhibit 4 test cell lines: MCF-7 breast cancer, Hela cervical cancer, HepG2 liver cancer and NCI-H460 lung cancer. The results showed that MM_1 compound's ability to inhibit the cancer cell lines increased linearly, at the concentration of 100 μg/mL of MM_1 compound, which inhibited nearly 80% of the cancer cell lines and was high. Most in liver cancer cell line HepG2 (at a concentration of 100 μg/mL inhibits 86.82% of HepG2 liver cancer cells). The MM_1 compound has also been shown to inhibit cancer cells through the ability to induce apoptosis on the NCI-H460 lung cancer cell line under the investigated conditions. On normal cell lines (Soma), this compound has a higher IC_{50} value and is outside the cell suppression threshold (> 4 mg/mL). This proves that this compound is highly safe, does not inhibit normal cells and has high potential for application.

Acknowledgements We would like to thank the Department of microbiology-parasitic, Faculty of Pharmacy, University of Medicine and Pharmacy Ho Chi Minh city; Faculty of Pharmacy, University of Nguyen Tat Thanh have supported and created favorable conditions to complete this research.

Conflicts of Interest The authors have no conflict of interest to declare.

References

1. DewiR T, Tachibana S, Itoh K (2012) Isolation of Antioxidant compounds from *Aspergillus terreus* LS01. J Microb Biochem Technol (4)
2. Gu W, Qiao C (2012) Furandiones from an endophytic *Aspergillus terreus* Residing in *Malus halliana*. Chem Pharm Bull 60:1474–1477
3. Jain P, Pundir R (2010) Effect of different carbon and nitrogen sources on *Aspergillus terreus* antimicrobial metabolite production. Int J Pharm Sci Rev Res 5(3):72–76
4. Shurygina MP, Kurskii YA, Druzhkov NO, Chesnokov SA (2008) Photolytic decarbonylation of obenzoquinones. Tetrahedron 64:9784–9788
5. Gaspar A, Cagide F, Quezada E (2013) Synthesis and NMR studies of novel chromone-2-carboxamide derivatives. Magn Reson Chem 51:251–254
6. Reis J, Gaspar A, Borges F (2014) Synthesis, spectroscopic characterization and X-ray structure of novel 7methoxy-4-oxo-N-phenyl-4H-chromene-2-carboxamides. J Mol Struct 1056–1057:31–37
7. Liu JL, Li XH (2015) 4-nor-b-patchoulene sesquiterpenoids from the essential oil of *Pogostemon cablin*. Phytochem Lett 12:27–30
8. Macedo FC, Andre LM, Anita JM (2004) Terreinol-a novel metabolite from *Aspergillus terreus*: structure and 13C labeling. Tetrahedron Lett 45:53–55
9. Rao KV, Sadhukhan AK (2000) Butyrolactones from *Aspergillus terreus*. Chem Pharm Bull (Tokyo) 48(4):559–562
10. Pawlak M, Marcin B (2012) Kinetic modelling of lovastatin biosynthesis by *Aspergillus terreus* cultivated on lactose and glycerol as carbon sources. Chem Process Eng 33(4):651–665
11. Sun ZH, Zhang CF, Zhang M (2010) A new benzoic acid derivative from *Eclipta prostrata*. Chin J Nat Med 8(4):0244–0246

Identification of Bacterial Profile in Root Canals of Teeth with Chronic Periapical Lesions in Vietnam

Huy An Thi Tran, Trung Vu Nguyen, and Hai Thanh Pham

Abstract The effectiveness of the treatment of endodontic infections with antiseptics is highly dependent on the occurrence of bacterial etiologies associated with endodontic pathology. However, the distribution of bacterial etiologies associated with endodontic pathology varies across regions of the world. This study aimed to determine the bacteria profile in root canals of teeth with chronic periapical lesions in order to help physicians select suitable irrigating agents and intra-canal medicament for endodontic treatment. Forty-seven teeth with chronic periapical lesions were collected from patients for this study. Specimens used for microbiological culture were collected using a paper cone. Bacteria growing on blood and chocolate medium were then identified by 16S rRNA gene sequencing. Forty-five different cultivable bacteria were successfully cultured and identified by 16S rRNA gene sequencing in all samples, in which each canal found one to nine bacteria species. Facultative anaerobic bacteria were the highest proportion (79.31%) including *Streptococcus, Neisseria, Haemophilus, Actinomyces* (proportion: 78.7%, 27.66%; 17.02% and 17.02% respectively. Followed by obligate anaerobic bacteria distributed for 11.72% with *Prevotella, Veillonella, Campylobacter, Oribacterium, Fusobacterium, Dialister*. The number of aerobic bacteria was only 8.97%, of which mainly *Bacillus* species. *Bacillus, Haemophilus, Actinomyces, Streptococcus* were presented in most of the teeth with chronic apical periodontitis showing pain and swelling. This is the first data on the microbiology species in root canals of teeth with chronic periapical lesions of Vietnamese people. Our data identified facultative anaerobic as the most common species in root canals of teeth with chronic periapical lesions, and four main microorganisms notably occurred in case with pain and swelling. This data

H. A. T. Tran · H. T. Pham (✉)
Haiphong University of Medicine and Pharmacy, Faculty of Dentistry, 72 A Nguyen Binh Khiem, Haiphong, Vietnam
e-mail: pthai@hpmu.edu.vn

H. A. T. Tran
e-mail: tahuy@hpmu.edu.vn

T. V. Nguyen
Hanoi Medical University, Faculty of Microbiology, No 1 Ton That Tung, Đong Đa, Hanoi, Vietnam

V. Van Toi et al. (eds.), *8th International Conference on the Development of Biomedical Engineering in Vietnam*, IFMBE Proceedings 85,
https://doi.org/10.1007/978-3-030-75506-5_52

has important implications for the effective treatment of root canal infections with antiseptics.

Keywords Microorganisms in endodontic · Microbiology · Anaerobic · 16S rRNA gene

1 Introduction

Chronic apical periodontitis (CAP) appears to be a chronic oral disease. The presence of the microorganism in the root canal system seems to be a significant cause of periapical pathology [1]. There are two methods for treatment of the teeth with chronic apical periodontitis, including surgical endodontic and non-surgical endodontic. Previously, surgical endodontic treatment was commonly used in all cases of chronic apical periodontitis. But the method revealed major disadvantages like anxiety before surgery and postoperative pain for the patient [2].

Currently, advanced technology and their application in dental short out the limitations of previous methods, the conservative treatment of apical periodontitis by nonsurgical management of periapical lesions has been widely used. However, to get a better treatment outcome, it is necessary to remove the bacterial factors to achieve optimal healing for the periapical tissues [3].

Regarding treatment protocol, irrigation and shaping process have been shown to decrease the number of bacteria ranges from 100 to 1000 times [4]. However, it is still struggling to eliminate bacteria and their toxins by irrigation and canal shaping at unreachable areas [5]. Therefore, to remove the remaining bacteria in the root canal, medicaments are additionally applied to eliminate all surviving microorganisms from the root canal after shaping and irrigation [5, 6]. In practice, it has shown that each antiseptic has the superiority of killing on several different bacteria [7]. Thus, Identification of bacterial etiologies associated with endodontic pathology in the canal and choosing appropriate antiseptic is crucial.

With the advanced technologies of medical microbiology, the technique of cultivating anaerobic bacteria and especially molecular biology methods have been used widely to determine microorganisms involved in endodontic pathology. Recently, *polymerase chain reaction* (PCR) and nucleotide sequencing techniques have also been used hugely to identify microorganisms precisely by using the 16S rRNA [8–10].

The detection of bacterial involved in endodontic pathology is currently gained great attention worldwide. Previous studies of microorganisms in root Canal-Treated teeth had been reported by Pinheiro et al., Isabela, Molven et al., Mindere et al., [11–14]. However, the distribution of bacterial etiologies associated with endodontic pathology varies across regions of the world.

In VietNam, to the best of our knowledge, there is no study relating to microorganism profile in the root canal of teeth with chronic periapical lesions. Thus, this study aimed to identify the bacteria profile in root canals of teeth with chronic periapical lesions helping physicians select appropriate irrigants or intra-canal medicaments.

2 Materials and Methods

2.1 Patients and Samples Setting

Forty-seven one-legged teeth with chronic periapical lesions were obtained from patients that were diagnosed by clinical and X-ray. Patients with full eligibility criteria for nonsurgical endodontic treatment were selected. The tooth with CAP has a vertical fracture or broken crown more than one half, abnormal root, calcified root canal, internal resorption, external resorption, open apex, severe periodontitis, endodontic treatment or apical lesions on X-ray film with a diameter more than 10 mm or root canal treatment were excluded.

All patients were examined and imaged with periapical X-rays. The sample was performed at the School of Odonto-Stomatology, Department of Odonto-Stomatology, Hanoi Medical University Hospital. The clinical features (clinical symptoms: sinus, swelling, etc.) and radiographic findings of the patient were recorded. All root-canal samples were proceeded at the Clinical Laboratory of National Hospital for Tropical Diseases in VietNam from December 2013 to December 2016.

2.2 Sampling Procedure

Tartar was removed by ultrasonic. Each tooth was isolated with a rubber dam and disinfected with Betadine. After cavities were prepared and cleaned by sterile Saline 0.9%, the pulp chamber was exposed with sterilized high-speed bur. *Explored* the root canal with file K. Inserted a sterile paper cone into the full length of the root canal as size as the root canal for 60 s. After that took the paper cone out and immediately inserted in eppendorf 1.5 mL sterilized contains BHI (Brain heart infusion) broth with Glycerol 30%. Eppendorfs containing the sample was transferred to the Clinical Laboratories of National Hospital for Tropical Diseases within 1 h. Samples were kept in a box containing dry ice during transportation.

2.3 Isolation and Evaluation of Bacteria

50 μL of the specimen was used to incubate on the surface of anaerobic blood agar (BA), and anaerobic chocolate agar medium and all of these were incubated at 37 °C

for 48 h in an anaerobic cabinet (Don Whitley Scientific VA500 Workstation—Don Whitley Scientific Limited—UK): 10% H_2, 10% CO_2, 80% N_2, pressure from 2 to 4 bar. The colonies growing on the BA and SC medium were selected and assessed for morphology and properties. Colony characteristics were evaluated: shape, color, blood melting, etc.; the number of each colony was counted to measure the t of bacteria (CFU/mL). Selected each representative colony of each type and subcultured it on new media to obtain a pure colony. Finally, picked up a pure colony for Gram staining observe bacterial morphology via microscope, the images of the bacteria morphology and conducted the proliferation of each colony. International standard Propionibacterium acnes ATCC 6919 was used as a standard strain for media testing.

2.4 PCR Amplification and 16S rRNA Gene Sequencing

Bacterial DNA was isolated from the colonies using Qiagen kits (QIAampDNAMini Kit, Cat No. 51304; Qiagen Sciences Germantown, MD, USA) in accordance with the manufacturer's protocol. The DNA was then used as a template for PCR. The PCR amplification of the 16S rRNA gene was performed in a 50 μL reaction volume using HotStarTaq® Master Mix Kit (Qiagen). A PCR mix was prepared in 0.2 mL, thin-walled tube; reaction contained 25 μL 2 × Master mix, 2 μL MgCl2 (25 mM) 1 μL DMSO: 20 pmol primer 16SF (10 μM), 20 pmol primer 16SR (10 μM), and 10 ng template ADN. The final volume was made up to 50 μL with distilled water. The following primers [14] were used in the PCR reaction: 16S-F (5′-AGAGTTTGATCCTGGCTCAG-3) and 16S-R (5′-TACGGTTACCTTGTTACGAC-3′). The PCR was performed using the ProFlex PCR System (Applied Biosystem). The amplification parameters included 1 cycle of denaturation at 95 °C for 15 min and 35 cycles of initial heat activation at 95 °C for 30 s, annealing at 58 °C for 30 s, elongation at 72 °C for 1 min, and a final extension at 72 °C for 10 min.

The PCR products were then visualized using gel electrophoresis; their size and concentration were measured and standardized. Procedure to perform 16S rRNA gene sequencing for bacteria identification using BigDye™ Terminator Chemistry v3.1 (Applied Biosystem, Foster City, CA, USA) in accordance with the manufacturer's instructions. Both forward (16S-F) and reverse (16S-R) for bacterial identification were used as sequencing primers, the samples were placed into the ABI 3130 sequencer (Applied Biosystem) to read the sequences. The sequences obtained from ABI 3130 sequencer were then analyzed by using ATGC software v7.02, the completed nucleotide sequences were blasted to the GenBank database to identify similar microorganisms.

2.5 Data Analysis

Data is stored and processed on SPSS 20.0 software. Data are analyzed and presented in frequency, percentage, Chi-square test were used in data analysis.

3 Results

The result of cultivable anaerobic microbiota, using PCR and sequencing of genes of 47 samples were taken from root canal determined 45 bacteria species. All root canals were positive for cultivable bacteria.

Identification of colonies obtained from the culture by 16S rRNA gene sequencing showed the presence of nine species of facultative anaerobic bacteria belonged to the *Streptococcus* genus, of which *Streptococcus sanguinis* presented in 24 samples (51.1%), followed by *Streptococcus salivarius* (21.27%), other Streptococcus species appeared to range between 2.1 to 12.77%. Other facultative anaerobic bacteria such as *Enterococcus, Corynebacterium, Lactobacillus, Staphylococcus, Acinetobacter, Morococcus* … also presented with rates ranging between 2.1 and 14.79%. The obligated anaerobic bacteria were mainly presented by *Veillonella parvula* (17.02%), *Prevotella bucca* (6.38%), *Campylobacter gracilis* (4.2%), and a minor proportion of *Oribacterium sinus, Fusobacterium nucleatum, Dialister pneumosintes, Prevotella denticola* were also obtained. Furthermore, aerobic bacteria were also found in both media, including Bacillus species such as *B. licheniformis, B. polyfermenticus, B. amyloliquefaciens, B. cereus.* There was also a presentation of *Pseudomonas reactans, Pseudomonas flourescens*, and *Enterobacter cloacae. Enterococcus faecalis was* found in 2 root canals (4.2%) (Table 1).

The results of the data analysis showed that a total of 25 genera of bacteria were identified among 47 root canal specimens. Of which, six genera of obligate anaerobes (including *Prevotella, Veillonella, Campylobacter, Oribacterium, Fusobacterium* and *Dialister*); 2 genera of aerobic bacteria (*Bacillus, Pseudomonas*) and the rest are genera of facultative anaerobes. The prevalence of Streptococcus was highest (78.7%). Prevalence of *Neisseria, Haemophylus, Actinomyces*: 27.66%; 17.02% and 17.02%, respectively (Fig. 1).

There were one to nine isolated microorganism species from each tooth. One to two species were identified in 28 teeth; 3 species and more than 3 species were identified in 19 teeth. Cultivation results of forty-seven specimens collected from the root canal indicated that 145 colonies were growing on both types of media (blood and chocolate agar). Of these, facultative anaerobes accounted for the highest proportion with 115 colonies (79.31%). Followed by obligated anaerobes with 17 colonies (11.72%), and the lowest proportion was aerobic bacteria with only 13 colonies (8.97%) (Table 2).

The additional data revealed that the *Streptococcus* genus was found in twenty-seven root canal specimens collected from the tooth with pain and swelling. While

Table 1 Molecular identification of cultivable bacteria from infected root canals of patients

Bacterial species	Oxygen condition	Number of teeth contains bacteria (%) (n = 47)
Streptococcus gordonii	Facultative anaerobe	6 (12.77)
Streptococcus oralis	Facultative anaerobe	5 (10.64)
Streptococcus sanguinis	Facultative anaerobe	24 (51.1)
Streptococcus mutans	Facultative anaerobe	4 (8.5)
Streptococcus mitis	Facultative anaerobe	3 (6.38)
Streptococcus	Facultative anaerobe	6 (12.77)
Streptococcus salivarius	Facultative anaerobe	10 (21.27)
Streptococcus oligofermentans	Facultative anaerobe	1 (2.1)
Streptococcus australis	Facultative anaerobe	1 (2.1)
Pseudomonas reactans	Aerobe	1 (2.1)
Pseudomonas fluorescens	Aerobe	1 (2.1)
Paenibacillus masssiliensis	Facultative anaerobe	1 (2.1)
Aggregatibater segnic	Facultative anaerobe	1 (2.1)
Oribacterium sinus	Obligated anaerobe	1 (2.1)
Fusobacterium nucleatum	Obligated anaerobe	1 (2.1)
Klebsiella pneumoniae	Facultative anaerobe	2 (4.2)
Enterococcus faecalis	Facultative anaerobe	2 (4.2)
Enterobacter cloacae	Aerobe	2 (4.2)
Capnocytophaga ochracea	Facultative anaerobe	1 (2.1)
Corynebacterium	Facultative anaerobe	1 (2.1)
Eikenella corrodens	Facultative anaerobe	1 (2.1)
Morococcus cerebrosus	Facultative anaerobe	1 (2.1)
Acinetobacter schindleri	Facultative anaerobe	1 (2.1)
Proteus mirabilis	Facultative anaerobe	1 (2.1)
Lactobacillus salivarius	Facultative anaerobe	1 (2.1)
Lactobacillus plantarum	Facultative anaerobe	2 (4.2)
Lactobacillus fermentum	Facultative anaerobe	2 (4.2)
Veillonella parvula	Obligated anaerobe	8 (17.02)
Bacillus licheniformis	Aerobe	4 (8.5)
Bacillus polyfermenticus	Aerobe	1 (2.1)
Bacillus amyloliquefaciens	Aerobe	3 (6.38)
Bacillus cereus	Aerobe	1 (2.1)
Campylobater gracilis	Obligated anaerobe	2 (4.2)
Neisseria flavescens	Facultative anaerobe	6 (12.77)
Neisseria elongata	Facultative anaerobe	6 (12.77)

(continued)

Table 1 (continued)

Bacterial species	Oxygen condition	Number of teeth contains bacteria (%) (n = 47)
Neisseria sicca	Facultative anaerobe	1 (2.1)
Haemophylus haemolyticus	Facultative anaerobe	3 (6.38)
Haemophylus parainfluenzae	Facultative anaerobe	5 (10.64)
Actinomyces oris	Facultative anaerobe	7 (14.89)
Actinomyces naeslundii	Facultative anaerobe	3 (6.38)
Dialister pneumosintes	Obligated anaerobe	1 (2.1)
Staphylococcus aureus	Facultative anaerobe	2 (4.2)
Staphylococcus epidermidis	Facultative anaerobe	5 (10.64)
Prevotella bucca	Obligated anaerobe	3 (6.38)
Prevotella denticola	Obligated anaerobe	1 (2.1)

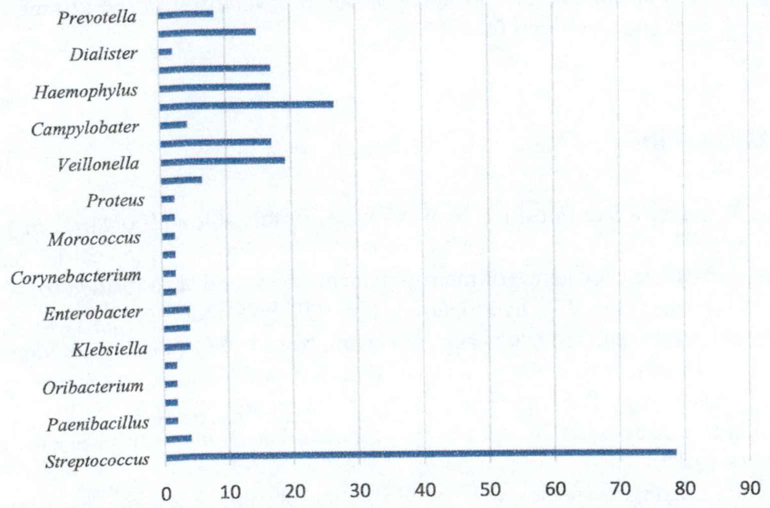

Fig. 1 The distribution of bacterial genera found in 47 teeth with chronic apical periodontitis

Table 2 Prevalence of isolated microorganisms in root canals of teeth with chronic periapical lesions

Microorganism	No. of colony	Percentage of colony
Aerobes	13	8.97
Obligated anaerobes	17	11.72
Facultative anaerobes	115	79.31

Table 3 Distribution of bacterial genus in root canal of teeth with *swelling and* pain

Bacteria	Teeth with *swelling and* pain		Teeth without *swelling and* pain	
	n	%	n	%
Streptococcus	27	81.81	6	18.19
Bacillus	8	100	0	0
Haemophylus	8	100	0	0
Actinomyces	8	100	0	0
Neisseria	10	71.43	3	28.57

there were only six specimens collected from the teeth without pain and swelling. Moreover, genera of *Bacillus, Haemophilus*, and *Actinomyces* were only found in root canal specimens of the tooth with pain and swelling. Still, they did not present in the tooth without pain and swelling. However, the *Neisseria* genus was found in both root canal specimens but presented a higher proportion in the groups of the tooth with pain and swelling (Table 3).

4 Discussion

Cultivable bacteria were positive in all 47 cases in this study (100%), from 1 to 9 species were isolated from root canals of teeth with chronic periapical lesions. In similar studies, the prevalence of microorganism was found at root-filled teeth with periapical lessions were 85% by Pinheiro [11] or 97% by Mindere [14]. Regarding the number of species isolated from each root canal, from 1 to 6 species were identified in Mindere's study [14], while in study of Pinheiro, 46.7% cases presented a single microorganism [10]. The higher prevalence of microorganism species in our data compared to other studies is explained by the fact that all the teeth in our research were not treated.

45 microorganism species were found in the present research. There were 9 bacteria species of *Streptococcus* genus. *Streptococcus sanguinis* is the highest percentage 55.1% (24 samples). *Enterococcus faecalis was* found in 2 teeth (4.2%) (Table 1). Li-Wan Lee et al. found 34 bacterial species in the apical root canals of 50 teeth with apical periodontitis, *Prophyromonas endodontalis* was detected in 10 teeth (20%), the other bacterial species were detected in less than 10 teeth [15]. *The number of bacterial species* and species of *bacterial* with the high rate was different in studies may be depend on the country for research and teeth that had been treatment endodontic or not. *Enterococcus faecalis* was also found in the canal at a rate of 4.2% in the present study. But *Enterococcus faecalis* were found up to 38% in root canal-treated teeth with CAP of Qian-Qian Wang's study (2012) [16] and 52.94% in reporting of Pinheiro et al. [11]. *Enterococcus faecalis* is higher percentage in

the canal of root-filled teeth with apical periodontitis may be explained by the difficulty of treating this microorganism. *Enterococcus faecalis* was considered as an endodontic challenge [9, 17].

Twenty-five genera belonging to 45 species of bacteria were isolated. In which, 6 genera of obligate anaerobes, 2 genera of aerobic bacteria that is *Bacillus, Pseudomonas, Enterobacter* and the rest are genera of facultative anaerobes. The result of our study indicated that the number species of facultative anaerobes were largest, most commonly for *Streptococcus* species (78.7%). The following species were identified: *Neisseria,* 27.66%; *Haemophilus,* 17.02%. *Actinomyces,* 17.02%, *Bacillus,* 17.02%. Our result was similar to Anda Mindere's study on the teeth with CAP in 2010. He showed that *Streptococcus* and *Actinomyces* accounted for the highest percentage 27%; 27% and *Haemophilus* and *Neisseria* were not isolated in his study. *Streptococcus* and *Actinomyces* were less responsive to endodontic treatment as they were capable of adapting to environmental changes. They were often found in teeth with failed endodontic treatment [14, 18].

Veillonella, Campylobacter, Fusobacterium, Dialister, Prevotella, Oribacterium were obligate anaerobes that found in this study. Pinheiro E.T, Gome B.P. found *Veillonella, Prevotella, Fusobacterium* in the canal of root-filled teeth with periapical lesions [10]. *Dialister invisus* or *Fusobacterium nucleatum* were isolated in 9 teeth (18%) at Li-Wan Lee's study. This proved that these bacteria were removed in endodontic treatment difficultly [8]. Therefore, in order to succeed in the treatment of teeth with CAP, a modern canal shaping technique should be applied in combination with a suitable cleaning solution and antiseptic to purify the bacteria in the canal.

According to Table 2, Facultative anaerobes were the highest rate (79.31%), obligate anaerobes were 11.72%. Facultative anaerobes always were isolated in the root canal with high prevalence. Similar to some research: Anda Mindere (2010): 56% facultative anaerobes were in root-filled teeth with apical periodontitis in Latvian patients; Molander et al. discovered 69% facultative anaerobes in 100 root-filled teeth with apical periodontitis [14, 19]. Facultative anaerobes were pathogens in primary endodontic and its involvement in the *failure* of *the endodontic* treatment [17]. So, in order to be successful in endodontics treatment, it is necessary to apply modern root canal shaping techniques in combination with appropriate irrigation and disinfectant to clean bacteria in the root canal.

Five bacterial genera were found in teeth with swelling and pain from 71.43% to 100% (Table 3). Almost these microorganisms were facultative anaerobes [20]. These indicators that *Streptococcus, Neisseria. Haemophilus, Actinomyces, Bacillus* have usually been found in symptomatic teeth. Pinheiro et al. and Shaikha Al-Samahi, Mohammad Al-Omari (2003), (2012) had found *Streptococcus, Actinomyces, Bacillus* associated with teeth having pain and swelling [11, 21]. Irrigation and disinfectant were chosen for clean root canal of symptomatic teeth depends on bacterial species that were identified.

5 Conclusion

This is the first data on the microbiology species in root canals of teeth with chronic periapical lesions of Vietnamese people. Our data identified facultative anaerobic as the most common species in root canals of teeth with chronic periapical lesions, and four main microorganisms notably occurred in case with pain and swelling: *Haemophilus, Actinomyces, Streptococcus, Bacillus,* respectively.

Conflicts of Interest The authors have no conflict of interest to declare.

References

1. Kakehashi S, Stanley HR, Fitzgerald RJ (1965) The effects of surgical exposures of dental pulps in germ-free and conventional laboratory rats. Oral Surg Oral Med Oral Pathol 20:340–349
2. Trinh HTT (2013) Cavity filling and endodontics. Education publisher, VietNam
3. Lin LM, Ricucci D, Lin J, Rosenberg PA (2009) Nonsurgical root canal therapy of large cyst-like inflammatory periapical lesions and inflammatory apical cysts. J Endod 35:607–615
4. Byström A, Sundqvist G (1981) Bacteriologic evaluation of the efficacy of mechanical root canal instrumentation in endodontic therapy. Scand J Dent Res 89:321–328
5. Van Peters LB, Winkelhoff AJ, Buijs JF (2002) Effects of instrumentation, irrigation and dressing with calcium hydroxide on infection in pulpless teeth with periapical bone lesions. Int Endod J 35(1):13–21
6. Nisha G, Amit G (2013) Textbook of endodontics, 2nd edn. Paypee Brothers Medical Publishers, New Delhi
7. Kalchinov V, Dimitrov SL, Belcheva M (2009) In vitro study of bactericidal effect of antimicrobial agents used in modern endodontics. J IMAB Ann Proc (Sc Papers) 2:79–82
8. Franco-Duarte R (2019) Advances in chemical and biological methods to identify microorganisms—from past to present. Microorganisms 7(5):130
9. Alberts B, Johnson A, Lewis J, Raff M, Roberts K, Walter P (2002) Molecular biology of the Cell, 4th edn. Garland Science, New York
10. Suzuki T, Iihara H, Uno T (2007) Suture-related keratitis caused by Corynebacterium macginleyi-case reports. J Clin Microbiol 45(11):3833–3836
11. Pinheiro ET, Gome BP (2003) Microorganisms from canal of root-filled teeth with periapical lessions. Int Endod J 36:1–11
12. Isabela N, Roças RF, Siqueira (2012) Characterization of microbiota of root canal-treated teeth with posttreatment disease. J Clin Microbio 50(5):1721–1724
13. Molven O, Halse A, Fristad I (2002) Periapical changes following root-canal treatment observed 20–27 years postoperatively. Int Endod J 35:784–789
14. Mindere A, Kundzina R, Nikolajev V (2010) Microflora of root filled teeth with apical periodontitis in Latvian patients. Stomatologija Baltic Dent Maxillofacial J 12:116–121
15. Lee LW, Lee YL, Hsiao SH, Lin HP (2016) Bacteria in the apical root canals of teeth with apical periodontitis. J Formosan Med Assoc 1–9
16. Wang QQ, Zhang CF, Chu CH, Zhu XF (2012) Prevalence of *Enterococcus faecalis* in saliva and filled root canals of teeth associated with apical periodontitis. Int J Oral Sci 4(1):19–23
17. Peciuliene V, Maneliene R, Balcikonyte E, Drukteinis S, Rutkunas V (2008) Microorganisms in root canal infections: a review. Stomatologija, Baltic Dent Maxillofacial J 10(1):4–9
18. Stuart CH, Schwartz SA, Thomas J (2006) Its role in root canal treatment failure and current concepts in retreatment. J Clin Microbiol 32(2):93–98

19. Molander A, Reit C, Dahlén G, Kvist G (1998) Microbiological status of root-filled teeth with apical periodontitis. Ent Endod J 31(1):1–7
20. George M (2007) Root canal microflora. Acta Medica 50:7–15
21. Al-Samahi S, Al-Omari M (2012) Detection of bacteria in endodontic samples and its association with defined clinical signs and symptoms of endodontic infection. J Clin Microbiol 50(5):1721–1724

The Role of Cell-Free RNA in Clinical Diagnosis and Evaluation of Cell-Free RNA Extraction Methods

Anh Phuc Hoang Le and Ha Thi Thanh Huong

Abstract Cell-free RNA (cf-RNA) is released from cells and circulated in the body via the circulatory and endocrine systems. Subsequently, cf-RNA can be extracted from different biofluids by various extraction methods. This review aims to: (1) clarify the clinical role of cf-RNA and methods of extraction, (2) introduce the evaluation parameters and methods, (3) summarize the evaluation of several common extraction methods and propose optimizing solutions for them. Eighty articles on the application of cf-RNA in diagnosing diseases and cf-RNA extraction published in journals with good impact factor (higher than 2) were then summarized and sorted into groups based on their targeted disease: (1) Cancers and (2) Neurodegenerative diseases since the cf-RNA is being studied mainly in these two fields. The general design of these studies is comparing cf-RNA to the current means of diagnosis, which are considered as standards such as endoscopy, biopsy, CSF analysis. By quantifying cf-RNA using qRT-PCR, previous studies were able to detect several cancers and neurodegenerative diseases. The key advantage of using the cf-RNA as a biomarker is reducing the risks and invasiveness without losing sensitivity since it can be extracted in the peripheral area. Furthermore, cf-RNA expression levels vary according to disease stages, suggesting its potential as prognostic markers. The main challenge in using cf-RNA is achieving the optimal yield, purity, integrity with minimal contamination. The current methods like Trizol LS, miR-Neasy, mirVANA have yielded unsatisfying results; therefore, they require modifications to optimize their results. miRCURY kit has the highest RNA yield but has been discontinued. In conclusion, cf-RNA holds excellent promises as a promising biomarker in the future. However, due to certain challenges, the use of cf-RNA is currently limited to clinical studies due to the instability in yield and integrity, requiring supporting studies and extraction methods evaluation.

Keywords Cf-RNA · Molecular biomarker · Molecular diagnostic

A. P. H. Le · H. T. T. Huong (✉)
School of Biomedical Engineering, International University, Ho Chi Minh City, Vietnam
e-mail: htthuong@hcmiu.edu.vn

Vietnam National University, Ho Chi Minh City, Vietnam

V. Van Toi et al. (eds.), *8th International Conference on the Development of Biomedical Engineering in Vietnam*, IFMBE Proceedings 85,
https://doi.org/10.1007/978-3-030-75506-5_53

1 Introduction

Previous studies had stated that the term "Biological markers" or "Biomarkers" only covers the cellular, biochemical or molecular compounds that can be extracted from biofluids and tissues, which display alterations according to diseases [1]. Later on, the term was also broadened to cover any biological characteristics which can be evaluated [2]. For example, a positron emission tomography (PET) scan shows the blood flow. There are two major types of biomarkers: Biomarkers of exposure which are used for risk prediction and can be applied in clinical trials, and biomarkers of disease, which are used in screening, diagnosis, and prognosis [3]. An ideal biomarker is proposed to have the following characteristics: specificity, sensitivity, robustness, non-invasiveness, ability to provide early diagnosis, and connect preclinical with clinical statements [4]. To be qualified as a (1) diagnostic method, the biomarker must be able to detect at least one disease specifically; (2) prognostic method, the biomarker must be able to specify each stage of the disease; (3) screening method, the biomarker must be able to detect the tumor with high-sensitivity, and the difference between normal cohort and patient cohort must be significant. Amongst the available biomarkers, applications of molecular biomarkers are having innovative discoveries [5–8]. They are applicable and multifunctional: they can help pathological studies, clinical trials, disease prevention, management, and treatment [3]. Molecular biomarkers include proteins and nucleic acids such as DNAs and RNAs. This review will focus on cell-free RNA (cf-RNA), a subclass of RNA molecular biomarkers.

Cell-free RNA (cf-RNA) or extracellular RNA can be defined as any RNA that exists outside a cell [9]. They consist of several subpopulations such as miRNAs (40.4%), piwiRNAs (40.0%), pseudo-genes (3.7%), lncRNAs (2.4%), tRNAs (2.1%), and mRNAs (2.1%) [10]. These cf-RNAs are encapsulated in exosomes, microvesicles, oncosomes, or other lipid/protein complexes [11]. Thanks to the protection layer, cf-RNA can be circulated in different biofluids such as plasma, serum, saliva, urine and CSF,[1] without being degraded by ribonucleases [12]. They are released from the cells to the extracellular matrix, mainly function as packaged signaling molecules in communication between cells and tissues [11] or released from apoptosis [9].

Cf-RNA biomarkers possess many advantages of a nucleic acid biomarker: they can differentiate the control group from the disease-carrier group in both early and late stages of the disease, which allows screening and monitoring of diseases [13]. This shows promises in providing more information, specificity, and accuracy compared to other biomarkers [14]. Additionally, protein composition in the blood is complicated, and the post-translational structures of proteins are sophisticated. Another point cf-RNA detection has high success-rate, and cf-RNA stay homogenous most of the time [4]. Moreover, cf-RNA can give information about not only mutation but also the level of expression through different diseases' stages [14]. Hence, cf-RNA has the potential of a less-invasive biomarker for clinical diagnosis of various types of

[1] Cerebrospinal fluid.

diseases compared to other available biomarkers, including cancers like lungs [15], prostate, breast cancer [4, 16]; cardiovascular diseases [17]; autoimmune diseases like Lupus [18]; prenatal diagnosis to neurodegeneration diseases like [19, 20].

As it has been discussed, cf-RNA has the potential of a promising biomarker. To apply it to clinical practice, methods such as RT-qPCR, RNA sequencing, and microarray are used [14]. However, to ensure reliable results, there are challenges to overcome. Low-quality RNAs can negatively affect follow-up procedures such as RT-qPCR, microarray, or Next Generation Sequencing (NGS) [21, 22]. For instance, low-yield or low-integrity-RNA can hinder reverse transcription reaction, amplification reaction, and RNA sequencing [23].

Additionally, endogenous contaminants such as Ca^{2+}, fat, genomic DNA, proteins, cell debris, and exogenous contaminants like glove powder and phenolic components from extraction can hinder reverse transcription or PCR and reduce the efficiency. However, these contaminants can be easily controlled by strictly following the protocols and maintain the lab sanitary. On the contrary, hemolysis is a contaminant that is difficult to supervise. In case the biofluid used to extract cf-RNA is plasma/serum, hemolysis is a cautious issue needed to be avoided. It can alter the level of RNA concentration in plasma/serum, affect the accuracy of the final results by contributing not only a high-level of cellular RNA but also a significant amount of cellular protein and cell debris [24, 25]. For example, hemolyzed plasma samples show *miR-16-5p* level 20–30 times higher than those of non-hemolyzed, which questioned miR-16-5p's role as an internal control for cf-miRNA quantification [10]. These problems occur moderately frequently in cf-RNA extraction due to inaccurate experimental operation or environmental/technical factors which are unable to control entirely. As plasma/serum is currently one of the most favorable biofluids, hemolysis needed to gain more consideration. Consequently, it is necessary to evaluate and compare different protocols to find the most optimal, suitable one while keeping the procedure contaminants-free.

2 Review Method

This review based on 80 articles on the application cf-RNA in diagnosing diseases and cf-RNA extraction published in journals with good impact factor (higher than 2). The studies are sorted into groups based on their targeted disease: (1) Cancers and (2) Neurodegenerative diseases. Other non-clinical studies were used to provide supporting statements or explanations. The general design of these studies is comparing cf-RNA to the current means of diagnosis which are considered as standards such as endoscopy, biopsy, CSF analysis and compare different extraction methods altogether, evaluate them and the optimization methods. The cited papers were assessed based on ROBIS Generic assessment in five main aspects: Random Sequence Generation refers to the randomness in selecting participants; Allocation concealment refers to the strategy which prevents the participants from picking their groups; Blinding of Participants and Personnel ensures that the participants within

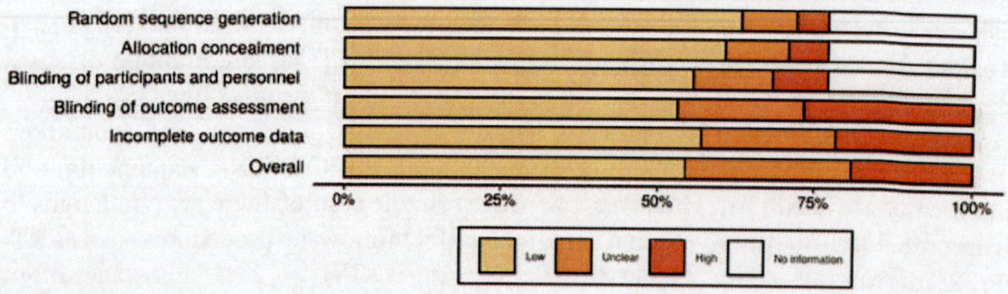

Fig. 1 Risk of bias summary

a study are unaware of the study details (aside from the main points included in the IRB[2]); Blinding of outcome assessment prevents the bias coming from the data collector; Incomplete outcome data refers to the bias created from lacking data, incomplete study. In case no concerns are available in a study, it is marked as a "low risk of bias". For a few concerns that appeared in a study, it is labelled as "unclear". If the risk is unavoidable and noticeable, the study would be marked as "high risk". In the case a study does not possess a criterion to be assessed (a review study can no recruit subjects), it is labelled as "no information" for that criteria. The risk of bias test ensures the reliability of a systematic review [26] (Fig. 1).

In general, there is a low risk of bias in more than 50% of the included studies. Most of the bias risk came from the results analysis and the lack of supporting studies. Many of the included studies (about 25%) lack of information to assess the Random Sequence Generation, Allocation Concealment, and Blinding of Participants variables (Fig. 3). This is due to some of the cited articles are literature review papers and textbooks. In the remaining experimental studies, subjects were recruited randomly, sorted into groups based on published criteria and not excessively understand the details of the study design. The risk of bias summary was generated by Robvis Visualization Tool [27].

3 Cell-Free RNA in Clinical Diagnosis

As mentioned in the introduction, cf-RNA holds great clinical potential. In the following section, the authors will discuss the clinical role of cf-RNA in cancer and neurodegenerative diseases' diagnosis.

---

[2] Institutional Review Board—Ethics and Informed Consent.

3.1 Cell-Free RNA in Cancer Diagnosis

Currently, qRT-PCR using primers targeting either non-specific cf-RNA or tumor-related cf-RNA in biofluids can support the screening, clinical diagnosis and prognosis of cancer. For instance, mRNA related to tumor suppressor genes like *p53, RB, MCC, APC, DCC, NF1, WT* or mRNA related to tumor-specific antigens like *MAGE1, MAGE2, MAGE3, MAGE4, GP-100,* and *MAGE 6, MUC 18, P97* [28]. On the other hand, reduction in expression of miR-125a-5p and miR-1343-3p also involves many cancer types tests with the ability to classify the cancer types (False Discovery Rate (FDR) < 0.05) [10].

Breast cancer

Cell-free *MAGE-A1, -A2, -A3, -A12, BORIS* mRNAs; microRNAs *let-7b, miR-202* in serum has been proven to relate to breast benign/malignant tumor [29]. The general accuracy for these biomarkers was reported to be 82%. Even though the potential of these cf-RNAs as breast cancer biomarkers have been proven in the study (Table 1), further conformational experiments are required to support the reliability of the biomarkers. Studies on the relationship between these cf-RNAs and the pathogenesis are also encouraged. Other cf-RNA such as *miR-34a, -93,* and *-373* in serum plasma were also able to detect cancer and proven to involve changes in *HER2* expression, especially with *miR-373* whose sensitivity is 96.8% and specificity is 94.1% [30].

Table 1 Capabilities and potential application of signature cf-RNAs biomarkers for breast cancer

Cf-RNA	Abilities	Application
MAGE-A1	– Identify normal, benign and malignant states – Differentiate pre- and post-surgery groups	– Screening test – Diagnostic and Prognostic method
MAGE-A2	– Identify normal, benign and malignant states	– Screening test – Diagnostic method
MAGE-A3, MAGE-A12, let-7b	– Differentiate malignant from benign and normal	– Diagnostic method
BORIS	– Identify normal, benign and malignant states – Differentiate pre- and post-surgery groups – Highly-correlated with observation survival	– Screening test – Diagnostic and Prognostic method
miR-202	– Differentiate malignant from benign and normal – Differentiate pre- and post-surgery groups – Highly-correlated with observation survival	– Diagnostic and Prognostic method

Urinary cancer

An experiment by Zhao et al. (2015) used the cf-RNA integrity index in urine to differentiate the renal carcinoma group from the control group, which was defined as the ratio between 416 and 106 bp RT-PCR amplicons of the *GAPDH* housekeeping gene. The index was recorded to be approximately 0.07 in the disease group, significantly lower than the control group (0.25, $p < 0.001$). This potential diagnosis meth-od has a sensitivity of 72.0% and a specificity of 92.63% in distinguishing between the patients and the normal subjects. Another potential biomarker suggested by the same study is *VEGF*-mRNA (Vascular Endothelial Growth Factor), displaying a significant difference ($p < 0.001$) between two experimental groups. In another research by [31], cf-RNAs of five genes existing in urine: *UBEC2, TOP2A, CDC20, ESM1,* and *CA9. UBEC2* were found to hold great potential as bladder carcinoma biomarkers for diagnostics and prognostics. This is due to sensitivity for different stages (the number of qPCR transcripts increases in later stages) ($p < 0.001$). Compared with cystoscopy and urinary cytology which are set as the gold standard diagnostic methods for bladder carcinoma, urinary biomarkers reduce the invasiveness yet still have a similar sensitivity.

Lung cancer

In an experiment conducted by Rabinowits et al. [15], total plasma cf-RNA concentration in the lung cancer group was 158.6 ng/mL on average, significantly higher than the control group which was 68.1 ng/mL ($p < 0.001$). In another study, the significance was also recorded in total serum cf-RNA and was able to distinguish each stage of the disease: recurrent tumor group (~ 64.9 ng/μL; $p = 0.045$), primary tumor samples (~ 25.6 ng/μL; $p = 0.049$), comparing to normal group (~ 16.6 ng/μL) [32]. This significance can be explained by the release of cf-RNAs into the bloodstream from various sources, including the (circulating) tumor cells, and increasing apoptosis [33]. Twelve promising miRNAs with upregulated behavior (confirmed with the standard diagnosis method: Analysis using miRNAs extracted from tumor biopsy) were also detected: *miR-17-3p, miR-21, miR-106a, miR-146a, miR-155, miR-191, miR-192, miR-203, miR-205, miR-210, miR-212, and miR-214* [15]. Additionally, expression levels of *miR-21* and *miR-155* can alter specifically to the primary and recurrent tumors [32]. However, both total RNA yield and miRNA expression level still need further studies gathering a larger subject population before publishing any standards that can be used as a diagnostic or a prognostic test for lung cancer.

Other cancers

Based on an experimental design by Yuan et al. [10] conducted on colorectal, prostate and pancreatic cancer, it was concluded that six transcripts including *miR-125a-5p, miR-1343-3p, miR-708-5p, miR-381-3p, miR-543,* and *piR_019825* could detect colon cancer specifically at different stages with 72% sensitivity and 80% specificity. Additionally, five transcripts, which include *miR-125a-5p, miR-1343-3p, let-7d-5p, miR-191-5p, miR-204-5p, miR-7-5p* and *piR_016658*, were able to distinctively distinguish between two subclasses of prostate cancer: hormone-sensitive prostate cancer (HSPC) and castration-resistant prostate cancer (CRPC) with 94% sensitivity and 88% specificity. Three other cf-RNAs that specifically correlate with pancreatic cancer are *piR-001311, -016658,* and *miR-7-5p.* Other types of cf-RNA were also found in the plasma of the disease cohort: *C6orf226, C-CAS09594, LOC388692* showed significantly different expression in colon cancer patients stages II–IV. Two other mRNAs *CDHR1* and *PAQR5* were also able to differentiate between the prostate and pancreatic cancer groups with normal groups. According to another study, *CEA* and *CK19* mRNAs were exceedingly expressed in plasma samples from the colorectal carcinoma cohort, comparing to the control cohort [34]. *CEA* displays higher specificity and lower sensitivity (96 and 32%) than *CK19* (80 and 73.6%). All of the suggested cf-RNAs hold great clinical potential. Yet, each still has certain limitations including difficult and expensive detection, low correlation with pathogenesis (miRNAs), and unsatisfying sensitivity (mRNA), which can be explained by the low abundance of mRNA in biofluids mentioned in the introduction.

3.2 *Cell-Free RNA in Neurodegenerative Diseases Diagnosis*

Neurodegenerative diseases are a set of diseases that involve the progressive loss of neuronal function, which is the consequence of neuronal cell death [35]. The common diseases are Alzheimer's Disease (AD), Parkinson's disease (PD), and Amyotrophic Lateral Sclerosis (ALS). The diseases' pathogenesis can be either polygenic, monogenic, or environmental [35]. Studying the RNA expression, or specifically, the cf-RNA expression can contribute to the knowledge background of the pathogenesis, therefore, contribute to the development of early clinical diagnosis and prognosis methods. There were significant differences in the expression level of multiple cf-RNA between the control group and the disease group (Table 2). Some in this group are highly correlated to the disease progression: formation of Lewy body, neurofibrillary tangles, plaque density and Braak stages (Table 3). The details about the potential cf-RNAs will be discussed in the following sections, along with their clinical challenges.

Alzheimer's disease (AD)

AD is the most popular type of dementia, accounting for two-thirds of the overall dementia cases. The patients suffer from cognition disability, due to the neurodegeneration [41]. There is currently no cure available and delaying the progress of the disease is the most optimal treatment. The most precise and early-staged diagnosis method is CSF protein biomarkers, including amyloid-β (Aβ42), total Tau (t-Tau), and phosphorylated tau (P-tau), with 85–90% sensitivity and specificity [42, 43]. A study by Müller et al. [65] evaluating new AD biomarkers candidates has found that *miR-146a* level in CSF decreased significantly in AD patients, which can act as a biomarker with high specificity yet moderate sensitivity. As the cf-RNA concentration was not measurable using Nanodrop, input for the qPCR process was equivalent in volume. Most of the miRNAs were not detectable except for *miR-16* (95%) and

Table 2 Expressions of some signature cf-miRNAs as biomarkers for common neurodegenerative diseases can be used for diagnosis [36–39]

Disease	Cf-RNAs	Source	Changes
Parkinson's disease [36]	miR-181c, -331-5p, -193a-p, -196b, -454, -125a-3p, -137	Plasma	Up-regulation
Alzheimer's disease [37, 38]	let-7f, miR-105, -125a, -135a, -138, -141, -151, -186, -191, -197, -204, -205, -216, -302b, -30a-5p, -30a-3p, -30b, -30c, -30d, -32, -345, -362, -371, -374, -375, -380-3p, -429, -448, -449, -494, -501, -517, -517b, -518b, 518f, 520a, 526a	Cerebrospinal fluid	Up-regulation
	miR-10a, -10b, -125, -126, -127, 142-5p, -143, -146b, -154, -15b, -181a, -181c, -194, -195, -199a, -214, -221, -328, -422b, -451, -455, -497, -99a	Cerebrospinal fluid	Down-regulation
	miR-92a-3p, miR-181c-5p, miR-210-3p, and miR-125b	Plasma/Serum	Up-regulation
Amyotrophic lateral sclerosis [39]	miR-142-5p, -150-5p, -378a-3p, Let-7f-5p	Cerebrospinal fluid	Down-regulation
	miR-124-3p, -127-3p, -9-5p, -27b-3p	Cerebrospinal fluid	Up-regulation

Table 3 Some cf-RNAs can be used as prognostic methods for AD and PD due to a high correlation with the disease progression [40]

Source	Cf-RNA	Progression type	Correlation
Cerebrospinal fluid	miR-34-5p, 37a-5p	Lewy body	+
	miR-9-3p, 181a-5p, 181b-5p, 181d	Neurofibrillary tangle	−
	miR-195-5p, 30c-5p	Plaque density	−
	miR-708-3p, 9-3p	Braak stages	−
Serum	miR-130-3p,181-5p	Lewy body	+
	Let-7i-3p, miR-10a-5p	Neurofibrillary tangle	±
	miR-30b-5p, 106a-5p	Plaque density	−
	miR-166-5p, 183b-5p	Braak stages	−

miR-146a (87.5%). Howcver, thc hemolysis still triggered noise in the collected data and required careful control. Another study by Burgos et al. [40] aimed to profile the cf-RNA database in CSF and plasma samples to differentiate AD from Parkinson's Disease (PD), Parkinson's Disease with Dementia (PDD) and control groups. They found the potential of cf-RNAs in not just diagnosis but also prognosis. Forty-one miRNAs in CSF samples and 20 miRNAs in serum samples show significant differences between normal and AD groups. Among the set, *miR-127-5p* and *873-5p* in CSF; *miR-1285-3p* and *21-5p* are also able to differentiate AD from PD groups. They also are potential prognosis biomarkers (*miR-127, -21-5p* have a negative correlation with neurofibrillary tangle development, and *miR-1285-3p* has a positive correlation with the development of Braak stages) (Table 4). According to another study, *miR-34c* (AUC = 0.99) and *-34a* (AUC = 0.81), are potential screening and diagnostic markers for AD from the early stages, since they are highly sensitive and specific [44]. Despite the peripheral RNAs are less-invasive than CSF biopsy, CSF RNAs are more reliable with higher sensitivity and specificity [40].

Other cf-mRNAs such as ACHE, ESR1, PHGDH, MRAS, APP, PSEN1, PSEN2, APOE-ε4, BIN1, clusterin, PICALM, TREM2, PTK2B, SORL1, SLC24A4, DSG2, INPP5D, MEF2C, NME8, ZCWPW1, CELF1, FERMT2, CASS4, PSD3, MOBP, MAG, SLC2A1, TCF7L2, CDH22, CNTF, and PAQR6, et cetera, can also be considered and studied as future candidates for Alzheimer's biomarkers [1, 45, 46].

Parkinson's Disease (PD)

Parkinson's Disease is the second-most common neuron degenerative disease, occurs in 2–3% of people aged above 65 [47]. The disease's most apparent symptom is abnormal movements: the patients are incapable of controlling their actions [48]. The disease damages mostly the motor neurons but also affects the non-motor cells

Table 4 Pros and Cons of other integrity assessment methods [13, 57, 58, 60, 61]

Technique	Advantages	Limitations
RNA integrity number (Bioanalyzer) [58]	High accuracy Has scoring system for easier evaluation	High cost
RT-PCR [57, 60, 61]	Low cost Easy procedure	Require many primers to be designed
Microarray [13]	High throughput Relative Low Cost	Not suitable for accurate quantification Can detect a wide range of target RNAs Low dynamic range
Deep sequencing [13]	Detection of novel miRNA Detection of Splicing variants (isomiRs) Ability to distinguish similar sequences	Sequence-specific bias High Cost Need for Special Equipment and bioinformatician A relative high amount of starting material

later-on [49]. In the discussed study by Burgos et al. [40], *miR-34c* was stated to be connected with PD, along with 17 CSF miRNAs and five peripheral cf-miRNAs. *miR-34c* expression levels reduce by 40–65% in the amygdala, substantia nigra, cerebellum, and frontal cortex of PD patients by previous studies [50]. This miRNA shows a strong connection with *DJ-1, Parkin* expression levels, and activities [51].

Other cf-RNAs related to *SNCA, Parkin, PINK1, DJ-1, LRRK2, ATP13A2, UCH-L1, NR4A2* are also positive options for the future PD biomarker studies and applications [52, 53].

Amyotrophic Lateral Sclerosis (ALS)

ALS is a fatal disease that causes the loss of motor neurons, affecting the motor cortex and the spinal cord. According to a study by Waller et al. [39], eight CSF miRNAs were found to be significantly expressed in ALS patients compared to the normal cohort ($p < 0.05$). They included the increase of *miR-9, miR-124* (which were reported to involve in brain injuries) and *miR-127-3p, miR-143-3p* (which were suspected to relate to the neuronal apoptosis) (Table 3). Despite that the sequencing results implicated significant changes between patients and normal cohorts, qPCR results were not as positive: The qPCR showed the same directional changes in the collected data, but the significance is inadequate to conclude. It is then concluded by the investigators that the current quality of the detected biomarkers is not sufficient for qPCR due to their integrity, length, the cross-reactivity with primers and not ideally-designed primers. This is a challenge for future applications of the biomarkers as RNA-sequencing is high cost, complicated protocols, and equipment.

Other potential cf-mRNAs include ALS2, mGLUR2, NEFH, SOD1, C9orf72, FUS, TARDBP, ERBB4, TUBA4A, CHMP2B, MATR3, SQSTM1, OPTN, ANXA11, HNRNPA1, TBK1, ANG, SPG11, PFN1, VAPB and UBQLN2 can be considered for further ALS biomarker studies [52, 54].

4 Extraction of cf-RNA

4.1 Extraction Methods

As discussed above, the clinical potential of cf-RNA has been clarified. To apply the cf-RNA into clinical practice, an optimal extraction procedure is necessary. There are various extraction methods available in the market. According to a study by Srinivasan et al. (2019), depending on which type cf-RNA is targeted and which biofluid source is used, there will be optimal extraction methods. For instance, ExoRNeasy and ExoQuick were the most suitable methods for small RNAs from serum/plasma samples. On the other hand, miRNeasy targeted RNAs larger than 200 nucleotides. Another study by McAlexander et al. [74] showed that miRCURY RNA Isolation Kit—Biofluids (Exiqon) outperformed four other extraction methods: mirVana (Life Technologies), miRCURY RNA Isolation Kit—Cell and Plant, miRNeasy

Serum/Plasma Kit (Qiagen), and TRIzol LS reagent (Life Technologies). miRCURY Biofluids exhibited more stable results through trials with higher abundance and integrity, tested with qRT-PCR. Their efficiency also slightly improved when added Glycogen as a carrier, which will be discussed in detail later. mirVana also yields similarly high results as miRCURY Biofluids but could not be optimized by Glycogen. miRNeasy is also a good option for plasma samples as the study suggested it had similar efficiency but less stability. Hence, miRNeasy and mirVana methods can be considered as alternatives for miRCURY Biofluids since this method has been discontinued. In general, the efficiency of extraction methods varies due to biofluids and its capability. Methods that are highly recommended are miRNeasy, miRVana, and miRCURY Biofluids (discontinued). All of the mentioned methods have been applied in many studies and proven to work. In another remark by Kuang et al. (2018), the extraction methods based on Trizol-protocols usually have lower cost and yield higher RNA concentration than column-based methods whilst the spin columns methods provide RNA with higher purity. Trizol-based methods precipitate the nucleic acid molecules from biofluids and then collect it (TRIZOL RNA Isolation protocol). Column-based methods use filtration and centrifugation based on size and molecular weight. This allows more intact and specific extraction than ultracentrifugation [55].

4.2 *Evaluation Criteria and Methods*

As discussed above, certain challenges occurred in cf-RNA extraction methods such as low yield, integrity, and purity in plasma/serum samples. Based on the given challenges, this study proposed a set of criteria and methods to evaluate the results of future cf-RNA extraction experiments (Fig. 2). These criteria can be applied to assess the available methods, methods in trial or methods under development.

RNA yield

RNA yield is defined as the concentration of RNA samples. It can be measured using a spectrometer, measuring the emitted light intensity at certain wave-length. To measure the RNA yield, we can use Qubit Fluorometer 2.0, Nanodrop [22], and Bioanalyzer [56]. Qubit's mechanism based on the chemistry of fluorochromes: the fluorescent dye will bind specifically onto the RNA strand. This method has high accuracy, portable, cost-effective. However, the methods of preparation can be time and chemical-consuming. Nanodrop's mechanism is based on the absorbance of nucleic acids at 260 nm. This method only requires a small amount of sample with no additional preparation and can measure the relative purity of the samples. On the other hand, this method cannot differentiate between types of nucleic acids and can be costly. Bioanalyzer uses the principles of electrophoresis, combining with the fluorescent dye to analyze the RNA samples. This method can quantify the RNA yield and assess the RNA integrity with high accuracy using only a small sample volume. Unfortunately, this method is time-consuming and costly. However, the readings are

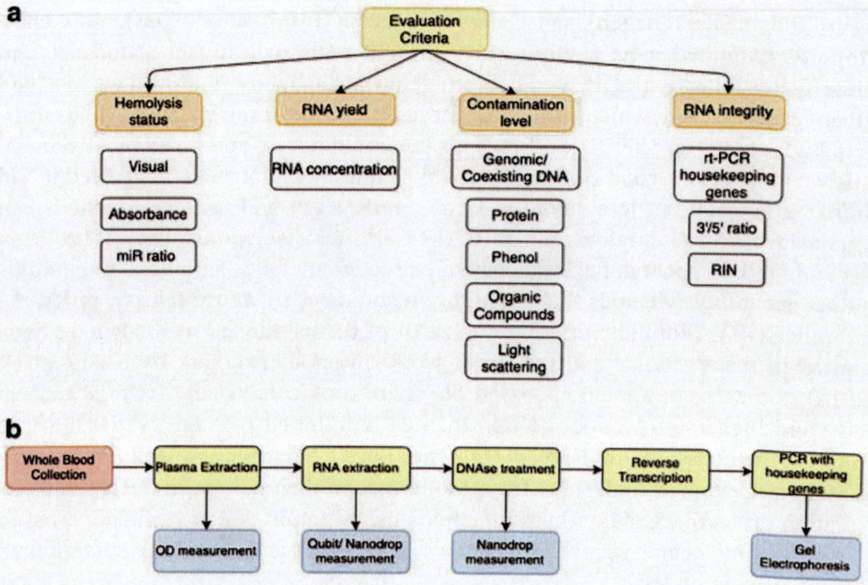

Fig. 2 Schematic representation of the assessment process. **a** A diagram summarizing the evaluation criteria can be applied for the cf-RNA extraction procedure. **b** Diagram suggesting experimental design which can be applied to evaluate a cf-RNA extraction method

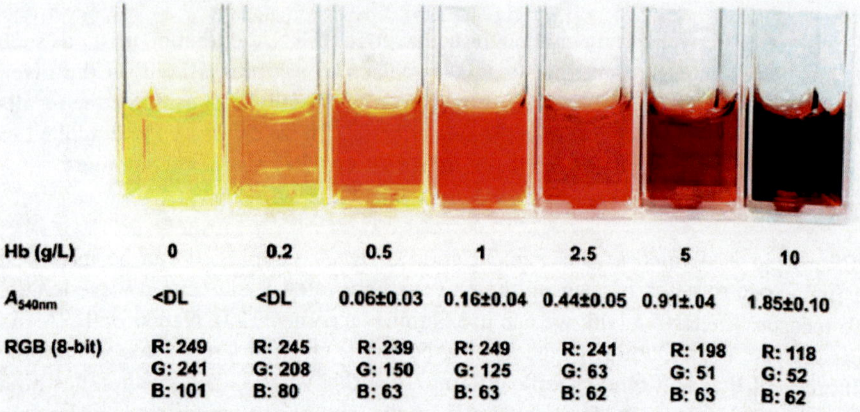

Fig. 3 Hemolysis ladder for visual assessment and the sample absorbance at 540 nm [68]. The visible colors can also be converted into digital data and represented in 8-bit RGB color

not always accurate due to excessive amounts of other nucleic acids such as small DNA fragments and ssDNA in the sample, and hence, require further evaluation of purity.

RNA integrity

Bioanalyzer is a microfluidic capillary electrophoresis assay, combining with a fluorometer, based on fluorescent dyes binding to RNA to measure not only RNA quality but also RNA quantity [57]. Similar to the gel electrophoresis method, the bioanalyzer is also able to detect the concentration of 28S and 18S rRNA, but the difference is that the system uses the fluorescent intensity to assess which is more accurate [58]. The results will then be displayed as an electropherogram for analysis and RNA integrity number—RIN, an important parameter in any RNA extraction evaluation until now, will be provided. RIN range from 1 to 10, with 1 stand for totally degraded RNA, and 10 is for perfect RNA integrity [59]. qRT-PCR and microarray experiments require nearly intact RNA (RIN 8) [21].

The integrity of extracted RNAs can also be evaluated using RT-PCR [57, 60]. The method uses housekeeping genes like *Beta-actin* or *GAPDH* as targeted genes in PCR or real-time PCR. According to the same paper's result, using this method to compare the integrity of the fresh frozen sample and Formalin-fixed Paraffin-embedded (FFPE) samples, the gel band of fresh frozen sample is more visible while FFPE band is blurred. Another advanced approach of this method discussed by the same author is 3'/5' mRNA Integrity Assay: A real-time PCR reaction is carried out using three differently labelled TaqMan probes designed to pair with the transcript at three different locations, usually are 100, 400–500, and 800–1,000 regions counting from the 3′ end of the gene [60, 61]. These regions are called 3′, intermediate, and 5′ segments. The ratio between the Ct values of the 3′ and 5′ amplicons reflects the intact level of the sequences and the efficiency of the reverse transcription. The RNA fragmentation is inversely proportional to the RT efficiency, and subsequently, the PCR yield. A 3′/5′ ratio of approximately 1 indicates high RNA integrity, whereas gradual results can imply considerable degradation occurring in the collected sample [61]. Other methods used for integrity assessment are mentioned in Table 4.

Contamination level

As mentioned in the previous section, contaminants can intrude into extracted RNA samples during extraction or storage. Levels of contaminants existing in extracted RNA samples can be measured based on absorbance wavelength using the NanoDrop, UV–Vis, NanoPhotometer, or other sensitive spectrophotometers. The optical density (OD) of each biological molecule is different: 230 and 320 nm are the optimal wavelengths for contaminants and background, 260 nm—the optimum wavelength for nucleic acids, and 280 nm—maxima wavelength of proteins. 260/280 ratio higher than 1.8 is assumed as acceptable protein residue while 260/230 should be at least 1.8 to represent the independence of the RNA samples from back-ground contaminants [62] (Table 5). Other contaminants like genomic DNA or cf-DNA can also cause infatuation to the final results readings of cf-RNA [21, 22]. However, as they have the same absorption wavelength (260 nm), Nanodrop may not be a suitable option to evaluate the contamination level. In this case, Qubit Fluorometer is a better option as it is based on the chemistry of fluorochromes: the fluorescent dye will bind specifically onto the RNA strands and DNA strands, which allows researchers to differentiate

Table 5 Absorbance criteria to evaluate the purity of RNA samples

Absorbance (nm)	Indicates the presence of	Target measurements
230	Organic compounds, sugars, urea, salts	$A_{260}/A_{230} > 1.8$
260	All Nucleic Acids	$A_{260} \approx 0.1–1.0$
270	Phenol	$A_{260}/A_{270} > 1.0$
280	Protein	RNA: $A_{260}/A_{280} \approx 2.0$ DNA: $A_{260}/A_{280} \approx 1.8$
330	Light scattering	$A_{330} = 0$

between these two types of nucleic acids. If the DNA readings return too low, then the DNA contamination level in the sample is not considerable. Another method is to run PCR and qPCR using primers that target ribosomal DNA (rDNA) [63]. This, however, only benefits the genomic DNA contamination evaluation but not cf-DNA.

Hemolysis status

Hemolysis is the state that red blood cells break down under chemical or mechanical effects [64]. Due to technical/mechanical effects, hemolysis can occur commonly during the withdrawal step. Besides the drawbacks caused by hemolysis introduced in the previous section, erythrocytes also have a negative impact on the cf-RNA preservation (erythrocytes population has a negative correlation with cf-RNA population) [65]. Hemolysis also alters the components of the collected cf-RNA samples by contributing some blood-cells-specified RNAs. Based on these understandings, one can evaluate hemolysis status using (1) Coulter AcT diffTM Analyzer hematology analysis, (2) visual assessment (Fig. 2), (3) measurement of absorbance of hemoglobin at 414 nm via spectrometer and (4) the miR-ratio of miR-451a (signature miRNAs of erythrocytes) to miR-23a (miRNA that exists only in serum) (miR-ratio). The Coulter1 AcT diffTM Analyzer could detect as low as 1% hemolysis but still unable to differentiate between 0.25% hemolyzed and the non-hemolyzed sample [66, 67]. Visual inspection can detect 0.25% hemolysis [67]. On the other hand, spectrometers could detect as low as 0.004% hemolysis [67]. Finally, miR-ratio is the most sensitive indicator and can detect 0.001% hemolysis [66, 67]. Based on this grading, low hemolysis severity has miR ratio < 5, moderate has miR ratio between 5 and 7 or severe has miR ratio > 7 hemolysis [66]. miR-15b-3p and miR-16-5p are also indicative of hemolysis and increase proportionally with miR-ratio [67]. Their levels are different among low, moderate, and severe hemolyzed samples. However, miR-16-5p is also found in exosomes. Therefore, it is not a specific blood cell marker. Despite the sensitivity to hemoglobin lysis, the mentioned methods do not cover all blood cell types such as lymphocytes and white blood cells. Based on a database provided by Juzenas et al. (2017), confirmed using miRBase, some potential miRNA families can be used as detection markers for the listed blood cells: miR-17 (Natural Killer cells: hsa-miR-106a, monocytes: hsa-miR-18a, neutrophils: hsa-miR-93); miR-30 (B-cells: hsa-miR-30a, Neutrophils: hsa-miR-30e); let-7 (helper T-cells: hsa-let-7e, neutrophils: hsa-let-7d); miR-10 (Natural Killer cells: hsa-miR-10a, helper T-cells: hsa-miR-99a) and miR-196 (cytotoxic T-cells: hsa-miR-196a-1/2). These miRNAs' yield can be assessed using qPCR, Bioanalyzer, and microarray to conclude hemolysis status (Fig. 3).

5 Discussion and Conclusion

The current extraction methods still yield unsatisfying results. In order to optimize cf-RNA extraction process, many steps can be interfered and adjusted to maximize the quality and quantity of RNA samples like blood collecting step (reduce hemolysis by using methods such as blood preservation tubes and anticoagulants); exosomes extraction (can be modified as suggested previously); RNA extraction (try different methods and protocols) and RNA purification. The suggested methods that can be combined with the traditional protocols are ultracentrifugation, exosome precipitation and immunoaffinity capture [69, 70] (Table 6).

Commercialized reagents are also able to reduce the difficulties in cf-RNA extraction as well. According to Prendergast et al. (2018), ExoQuick (a column-based extraction method) gave better results than ultracentrifugation. Additionally, the study also showed that a combination of different methods with the same purpose could increase efficiency. In exceptional cases, assistant molecules can be added to provide a beneficial environment for the process—such as carrier molecules (Table 6). A comparison between extraction methods with/ without the addition of Glycogen as a carrier molecule showed that methods with glycogen performed slightly better [58, 74]. Optimization modifications can also be added from the beginning step—blood collection step: Blood drawn into K3EDTA tubes shows stable preservation of cf-RNA over long-time storage, even at room temperature, which serves excellent benefits for future clinical applications [73].

According to the studies discussed in the previous sections, the potential of cf-RNA in diagnosis and prognosis is undeniable. However, there are remaining challenges that keep this biomarker from being publicized for clinical uses. The expression levels of cf-RNA in biofluids are not stable, and some RNA subpopulations get affected more than the others. The additional amounts of cellular RNA due to hemolysis during sample collection step and pathogenic RNA due to infection can cause background noises, leading to false readings. The variation in expression levels amongst different biofluids needs to be considered before clinical execution. Furthermore, during extraction, RNA is removed from their protection layers, requiring the following steps to proceed rapidly or the addition of preservatives beforehand. The reproducibility and credibility of cf-RNA are also questioned: their yield varied in different experiments, causing contradictions. There has not been any uniform, tangible guideline published for the use of cf-RNA in clinical procedures [12, 75]. For

Table 6 Analysis of some optimization methods can be used to modify the original protocols

Technique	Based techniques	Advantages	Disadvantages
Ultracentrifugation [69, 70]	Based on the vesicles' density and sizes	Reduce reagents' cost, contamination Increase yield and capacity of exosomes	Time-consuming Labor-intensive Not portable Might impeding downstream analysis by damaging the nucleic acids at a high spin rate
Filtration [69]	Filtration based solely on the size difference between exosomes and other components	Fast Portable Direct RNA extraction possible	Possible clogging Moderate purity Time-consuming
Exosome precipitation [69]	Changing solubility and dispersibility by chemical reagents	Easy	Need additional chemicals Time-consuming
Immunoaffinity capture [69]	Apply ELISA mechanism, catching exosomes by pairing the antigens localized in exosomes and the immobilized antibodies	High specificity Can be portable by integrating on microfluidics-based system Works well with cell-free samples	Expensive Low yield
Carrier molecules [71, 72]	Aggregate the nucleic acid molecules in the collected samples, trap them then form visible precipitate after centrifugation	Increase the yield significantly	Additional steps If not adding the right amount can lead to false readings Not RNA-specific
Sample collection tubes with preservatives [73]	Contain stabilizing agents which aid the storage process and keep the RNA intact	Increase the integrity and yield from the beginning step The samples can be stored long-termed	Needs purification step as an excessive amount of preservatives can affect the results

neurodegenerative diseases, peripheral biofluids cf-RNA are not as credible as there are not many studies supporting the relationship between these biomarkers and the pathogenesis [40]. Hence, the current applications of cf-RNA are limited to scientific and clinical studies. Further studies about the molecular background of cf-RNA, its correlation with pathogenesis, and comparisons amongst the extraction methods are also required before this type of biomarker can be used in clinical diagnosis.

Acknowledgements This research is funded by Vietnam National University Ho Chi Minh City (VNU-HCM) under grant number NCM2020-28-01 and Vietnam Alzheimer Network grant, organized by the Vietnam National Geriatric Hospital and the University of California, Davis, US.

Conflicts of Interest The authors have no conflict of interest to declare.

References

1. Huynh RA, Mohan C (2017) Alzheimer's disease: biomarkers in the genome, blood, and cerebrospinal fluid. Front Neurol 8:102
2. Naylor S (2003) Biomarkers: current perspectives and future prospects. Expert Rev Mol Diagn 3(5):525–529
3. Mayeux R (2004) Biomarkers: potential uses and limitations. NeuroRx 1(2):182–188
4. Etheridge AaL, Inyoul, Hood L, Galas D, Wang K (2011) Extracellular microRNA: a new source of biomarkers. Mut Res 717:85–90
5. Galasko D (2001) New approaches to diagnose and treat Alzheimer's disease: a glimpse of the future. Clin Geriatr Med 17(2):393–410
6. Reiber H, Peter JB (2001) Cerebrospinal fluid analysis: disease-related data patterns and evaluation programs. J Neurol Sci 184(2):101–122
7. Rohlff C (2001) Proteomics in neuropsychiatric disorders. Int J Neuropsychopharmacol 4(1):93–102
8. Verbeek MM, De Jong D, Kremer HP (2003) Brain-specific proteins in cerebrospinal fluid for the diagnosis of neurodegenerative diseases. Ann Clin Biochem 40(Pt 1):25–40
9. Kubiritova Z, Radvanszky J,Gardlik R (2019) Cell-free nucleic acids and their emerging role in the pathogenesis and clinical management of inflammatory bowel disease. Int J Mol Sci 20(15)
10. Yuan T et al (2016) Plasma extracellular RNA profiles in healthy and cancer patients. Sci Rep 6:19413
11. Lässer C (2012) Exosomal RNA as biomarkers and the therapeutic potential of exosome vectors. Exp Opin Biol Ther 12(Suppl 1):S189–S197
12. Sadik N et al (2018) Extracellular RNAs: a new awareness of old perspectives. Methods Mol Biol 1740:1–15
13. Schwarzenbach H et al (2011) Evaluation of cell-free tumour DNA and RNA in patients with breast cancer and benign breast disease. Mol Biosyst 7(10):2848–2854
14. Xi X et al (2017) RNA biomarkers: frontier of precision medicine for cancer. Noncoding RNA 3(1)
15. Rabinowits G et al (2009) Exosomal microRNA: a diagnostic marker for lung cancer. Clin Lung Cancer 10(1):42–46
16. Papadopoulou E et al (2006) Cell-free DNA and RNA in plasma as a new molecular marker for prostate and breast cancer. Ann N Y Acad Sci 1075:235–243
17. Cheng Y et al (2010) A translational study of circulating cell-free microRNA-1 in acute myocardial infarction. Clin Sci (Lond) 119(2):87–95
18. Wang G et al (2010) Serum and urinary cell-free MiR-146a and MiR-155 in patients with systemic lupus erythematosus. J Rheumatol 37(12):2516–2522
19. Jessica A, Dietz KLJ, Lauren JM, Schaper J, Horlitz M, Cowan J, Bianchi DW (2011) Comparison of extraction techniques for amniotic fluid supernatant demonstrates improved yield of cell-free fetal RNA. Prenatal Diagnosis
20. Kiko T et al (2014) MicroRNAs in plasma and cerebrospinal fluid as potential markers for Alzheimer's disease. J Alzheimers Dis 39(2):253–259
21. Fleige S et al (2006) Comparison of relative mRNA quantification models and the impact of RNA integrity in quantitative real-time RT-PCR. Biotechnol Lett 28(19):1601–1613
22. Spornraft M et al (2014) Optimization of extraction of circulating RNAs from plasma—enabling small RNA sequencing. PLoS One 9(9):e107259
23. Becker C et al (2010) mRNA and microRNA quality control for RT-qPCR analysis. Methods 50(4):237–243
24. Kirschner MB et al (2013) The impact of hemolysis on cell-free microRNA biomarkers. Front Genet 4:94
25. Pritchard CC et al (2012) Blood cell origin of circulating microRNAs: a cautionary note for cancer biomarker studies. Cancer Prev Res (Phila) 5(3):492–497

26. Higgins JP, Thomas J, Chandler J, Cumpston M, Li T, Page MJ, Welch VA (eds) (2019) Cochrane handbook for systematic reviews of interventions. J.W. Sons
27. McGuinness LAR (2019) An R package and web application for visualising risk-of-bias assessments. 2019 [cited 2020 May 13]. Available from: https://github.com/mcguinlu/robvis
28. Kopreski MS, Christopher D. Gocke (2005) Method of detecting tumor-associated DNA in plasma or serum with a premalignant solid tumor (U.S. Patent No. US 8,361,726 B2), U.S. Patent and Trademark Office, Washington, DC.
29. Joosse SA et al (2014) Circulating cell-free cancer-testis MAGE-A RNA, BORIS RNA, let-7b and miR-202 in the blood of patients with breast cancer and benign breast diseases. Br J Cancer 111(5):909–917
30. Eichelser C et al (2013) Deregulated serum concentrations of circulating cell-free microRNAs miR-17, miR-34a, miR-155, and miR-373 in human breast cancer development and progression. Clin Chem 59(10):1489–1496
31. Kim WT et al (2016) UBE2C cell-free RNA in urine can discriminate between bladder cancer and hematuria. Oncotarget 7(36):58193–58202
32. Munagala R, Aqil F, Gupta RC (2016) Exosomal miRNAs as biomarkers of recurrent lung cancer. Tumour Biol 37(8):10703–10714
33. Schwarzenbach H (2013) Circulating nucleic acids as biomarkers in breast cancer. Breast Cancer Res 15(5):211
34. Silva JM et al (2002) Detection of epithelial tumour RNA in the plasma of colon cancer patients is associated with advanced stages and circulating tumour cells. Gut 50(4):530–534
35. Maciotta S, Meregalli M, Torrente Y (2013) The involvement of microRNAs in neurodegenerative diseases. Front Cell Neurosci 7:265
36. Cardo LF et al (2013) Profile of microRNAs in the plasma of Parkinson's disease patients and healthy controls. J Neurol 260(5):1420–1422
37. Cogswell JP et al (2008) Identification of miRNA changes in Alzheimer's disease brain and CSF yields putative biomarkers and insights into disease pathways. J Alzheimers Dis 14(1):27–41
38. Tielking K et al (2019) Extracellular RNA in central nervous system pathologies. Front Mol Neurosci 12:254
39. Waller R et al (2017) Small RNA sequencing of sporadic amyotrophic lateral sclerosis cerebrospinal fluid reveals differentially expressed miRNAs Related to neural and glial activity. Front Neurosci 11:731
40. Burgos K et al (2014) Profiles of extracellular miRNA in cerebrospinal fluid and serum from patients with Alzheimer's and Parkinson's diseases correlate with disease status and features of pathology. PLoS One 9(5):e94839
41. Nussbaum RL, Ellis CE (2003) Alzheimer's disease and Parkinson's disease. N Engl J Med 348(14):1356–1364
42. Shaw LM et al (2009) Cerebrospinal fluid biomarker signature in Alzheimer's disease neuroimaging initiative subjects. Ann Neurol 65(4):403–413
43. Welge V et al (2009) Combined CSF tau, p-tau181 and amyloid-beta 38/40/42 for diagnosing Alzheimer's disease. J Neural Transm (Vienna) 116(2):203–212
44. Bhatnagar S et al (2014) Increased microRNA-34c abundance in Alzheimer's disease circulating blood plasma. Front Mol Neurosci 7:2
45. KOH LCW, Stephen R. Quake, Hei-Mun Christina Fan,Wenying Pan (2013) Methods for profiling and quantitating cell-free rna (World Patent Application No. WO2015069900A1), WIPO (PCT).
46. Yan Z et al (2020) Presymptomatic increase of an extracellular RNA in blood plasma associates with the development of alzheimer's disease. Curr Biol 30(10):1771-1782.e3
47. Poewe W et al (2017) Parkinson disease. Nat Rev Dis Primers 3:17013
48. Hoehn MM, Yahr MD (1967) Parkinsonism: onset, progression and mortality. Neurology 17(5):427–442
49. Pollanen MS, Dickson DW, Bergeron C (1993) Pathology and biology of the Lewy body. J Neuropathol Exp Neurol 52(3):183–191

50. Miñones-Moyano E et al (2011) MicroRNA profiling of Parkinson's disease brains identifies early downregulation of miR-34b/c which modulate mitochondrial function. Hum Mol Genet 20(15):3067–3078
51. Junn E, Mouradian MM (2012) MicroRNAs in neurodegenerative diseases and their therapeutic potential. Pharmacol Ther 133(2):142–150
52. Agrawal M, Biswas A (2015) Molecular diagnostics of neurodegenerative disorders. Front Mol Biosci 2(54)
53. Yang YX, Wood NW, Latchman DS (2009) Molecular basis of Parkinson's disease. NeuroReport 20(2):150–156
54. Ferrara D et al (2018) Role of extracellular vesicles in amyotrophic lateral sclerosis. Front Neurosci 12:574
55. Enderle D et al (2015) Characterization of RNA from exosomes and other extracellular vesicles isolated by a novel spin column-based method. PLoS One 10(8):e0136133
56. Eldh M et al (2012) Importance of RNA isolation methods for analysis of exosomal RNA: evaluation of different methods. Mol Immunol 50(4):278–286
57. Dotti IaB (2011) Serena, integrity assessment of nucleic acids, pp 81–85
58. Cerkovnik PaP, Andraz, Zgajnar J, Novakovic S (2007) Optimization of an RNA isolation procedure from plasma samples. Int J Mol Med 20:293–300
59. Lightfoot S, Salowsky R, Buhlmann C (2005) RNA integrity number: towards standardization of RNA quality assessment for better reproducibility and reliability of gene expression experiments. Breast Cancer Res 7(Suppl 2):P7 05
60. Wong BC, Lo YM (2006) Plasma RNA integrity analysis: methodology and validation. Ann N Y Acad Sci 1075:174–178
61. Wong BC et al (2006) Reduced plasma RNA integrity in nasopharyngeal carcinoma patients. Clin Cancer Res 12(8):2512–2516
62. Manchester KL (1996) Use of UV methods for measurement of protein and nucleic acid concentrations. Biotechniques 20(6):968–970
63. Hashemipetroudi SH et al (2018) Assessment of DNA contamination in RNA samples based on ribosomal DNA. J Vis Exp (131)
64. Heireman L et al (2017) Causes, consequences and management of sample hemolysis in the clinical laboratory. Clin Biochem 50(18):1317–1322
65. Müller M et al (2014) MicroRNAs in Alzheimer's disease: differential expression in hippocampus and cell-free cerebrospinal fluid. Neurobiol Aging 35(1):152–158
66. Blondal T et al (2013) Assessing sample and miRNA profile quality in serum and plasma or other biofluids. Methods 59(1):S1-6
67. Shah JS, Soon PS, Marsh DJ (2016) Comparison of methodologies to detect low levels of hemolysis in serum for accurate assessment of serum microRNAs. PLoS One 11(4):e0153200
68. Killilea DW et al (2017) Identification of a hemolysis threshold that increases plasma and serum zinc concentration. J Nutr 147(6):1218–1225
69. Li Q et al (2015) Plasma long noncoding RNA protected by exosomes as a potential stable biomarker for gastric cancer. Tumour Biol 36(3):2007–2012
70. Patel T (2018) MARTON, extracellular RNA. Springer Nature, ed. L. Springer Science+ Business Media
71. Hengen PN (1996) Carriers for precipitating nucleic acids. Trends Biochem Sci 21(6):224–225
72. Moret I et al (2013) Assessing an improved protocol for plasma microRNA extraction. PLoS One 8(12):e82753
73. Fernando MR, Luna KK, Lechner JM, Qin J (2012) Stabilization of cell-free RNA in blood samples using a new collection device. Clin Biochem 45(16–17):1497–1502
74. McAlexander MA, Phillips MJ, Witwer KW (2013) Comparison of methods for miRNA extraction from plasma and quantitative recovery of RNA from cerebrospinal fluid. Front Genet 4:83
75. Zaporozhchenko IA et al (2018) The potential of circulating cell-free RNA as a cancer biomarker: challenges and opportunities. Exp Rev Mol Diagn 18(2):133–145

76. Zhao An, Péoc'h M, Cottier M, Genin C, Mottet N, Li G (2015) Cell-free RNA content in urine as a possible molecular diagnostic tool for clear cell renal cell carcinoma. International Journal of Cancer 136 (11):2610–2615
77. Srinivasan S, Yeri A, Cheah PS, Chung A, Danielson K, De Hoff P, Filant J, Laurent CD, Laurent LD, Magee R, Moeller C, Murthy VL, Nejad P, Paul A, Rigoutsos I, Rodosthenous R, Shah RV, Simonson B, To, C, ... Laurent LC (2019) Small RNA Sequencing across Diverse Biofluids Identifies Optimal Methods for exRNA Isolation. Cell, 177(2), 446–462.e16.
78. Kuang J, Yan X, Genders A J, Granata C, & Bishop D J (2018) An overview of technical considerations when using quantitative real-time PCR analysis of gene ex-pression in human exercise research. PloS One, 13(5), e0196438.
79. Juzenas S, Venkatesh G, Hübenthal M, Hoeppner MP, Du, Z G, Paulsen M, Rosenstiel P, Senger P, Hofmann-Apitius M, Keller A., Kupcinskas L, Franke A., Hemmrich-Stanisak G (2017) A comprehensive, cell specific microRNA catalogue of human peripheral blood. Nucleic Acids Research, 45(16), 9290–9301.
80. Prendergast E N, de Souza Fonseca M A., Dezem F S, Lester J, Karlan B Y, Noushmehr H, Lin X, Lawrenson K (2018) Optimizing exosomal RNA isola-tion for RNA-Seq analyses of archival sera specimens. PloS One, 13(5), e0196913.

Study on Extracting Crude Phycocyanin from Spirulina Algae and Determining Its Ability in Protecting Fibroblasts from Oxidative Stress of Hydroxyl Peroxide

Quan Minh To, Nhan Dinh Tran, Phuc Thi-Kim Pham, My Truong-Nhu Ho, Bien Dinh Lai, Tri Quang Le, Long Thanh Le, and Son Nghia Hoang

Abstract Phycocyanin is a powerful natural antioxidant from *Arthrospira platensis*. In this study, crude phycocyanin (crPC) was extracted, and its ability to protect fibroblast from hydroxyl peroxide (H_2O_2) in vitro was determined. crPC was extracted from Spirulina by a step-by-step method: 1/cell breaking by a repeated freezing/thawing method; 2/contaminant removal by 8% active charcoal and ammonium sulfate 15%; 3/phycocyanin obtaining by ammonium sulfate 50% and dialysis. Abilities of crPC 6, 12.5, 25, 50 and 75 μg/mL in inducing cell proliferation and migration were tested on fibroblasts. The protective ability of crPC against oxidative stress of H_2O_2 was determined on fibroblasts as the following method: fibroblasts were pretreated with crPC 25 μg/mL for 24 h before exposing to H_2O_2 175 μM for 90 min (experimental group). Then cell viability, cell phase and senescence β-galactosidase expression were evaluated. The results showed that: PI of crPC was 2.79 ± 0.03 and concentration was 1.13 ± 0.05 ($n = 5$). crPC below 150 μg/ml was non-toxic to fibroblasts and crPC above 25 μg/mL stimulated cell proliferation (doubling time was 44.5 ± 1.2 h, $n = 3$) and migration (24 h, $n = 3$). Compared to control group (cells were treated only with H_2O_2 175 μM), cell viability in the experimental group was higher ($79.3 \pm 5.4\%$ vs. $53.4 \pm 2.3\%$, $n = 3$) and cell ratio in G0/G1 and senescence β-galactosidase expression was lower ($78.7 \pm 4.1\%$ vs. $85.8 \pm 6.1\%$) ($22.8 \pm 4.6\%$ vs. $68.4 \pm 6.8\%$). Conclusion: crPC was extracted successfully from *Arthospira platensis,* and crPC 25 μg/mL was able to protect fibroblast against oxidative stress of H_2O_2.

Q. M. To (✉) · N. D. Tran
University of Science, HCM National University, Ho Chi Minh City, Vietnam
e-mail: tomquan@hcmus.edu.vn

P. T.-K. Pham · M. T.-N. Ho · B. D. Lai
HCMC University of Food Industry, Ho Chi Minh City, Vietnam

T. Q. Le
7A Military Hospital, Ho Chi Minh City, Vietnam

L. T. Le · S. N. Hoang
Institute of Tropical Biology, Vietnam Academy of Science and Technology, Ho Chi Minh City, Vietnam

V. Van Toi et al. (eds.), *8th International Conference on the Development of Biomedical Engineering in Vietnam*, IFMBE Proceedings 85,
https://doi.org/10.1007/978-3-030-75506-5_54

Keywords Phycocyanin · Hydroxyl peroxide · Fibroblast

1 Introduction

Arthrospira platensis (Cyanobacteria) is acknowledged as one of the most superfoods that contain many necessary nutritious elements, especially a powerful antioxidant, phycocyanin (PC) [1]. PC is a blue accessory pigment which transfers light energy to chlorophyll. Phycocyanin is water-soluble, non-toxic, antioxidant, anti-inflammatory, anti-cancer. Today, PC is mainly employed as a coloring agent or nutritional ingredient in fields of food, cosmetics [2]. Moreover, many researches focus on its antioxidant, anti-inflammatory and anti-cancer function. These findings showed that PC could effectively eliminate Reactive Oxygen Species (ROS) involved in some diseases, for example, atherosclerosis, liver fibrosis, neurodegenerative diseases [3]. In the cosmetic field, PC was used as a natural dye for some commercial products composed of lipsticks, eye-shadows ... but its antioxidant characteristics have the most value. UV radiation is the main reason for photoaging. UV-light affects the skin by generating ROS in the dermis of the skin. An excessive amount of ROS leads to cell damage, alternation of structure and function of the protein in the skin, including collagen, elastin ... [4, 5]. Photoaging can be curbed by antioxidants such as PC which has the ability to scavenge or inhibit ROS such as hydroxyl peroxide (H_2O_2) [2]. Moreover, PC shows its proliferation and migration effects on skin cells, for example, fibroblast and keratinocyte [6, 7]. In 2008 and 2013, Madhyastha proved PC (75 μg/mL and 75 μM) induced migration of skin cells to full-thickness wounds in Sprague–Dawley male rats after 6-day treatment, so wound healing process is faster than control [6, 8]. Consequently, PC has a large potential to be an anti-aging agent in skincare. In the market, the value of PC is depended on the purity of PC (PI) which determined by OD_{620}/OD_{280}. It is divided into food grade (PI $> = 0.7$) and analytical grade (PI $> = 4$). It costs from 500 USD/kg to 15 USD/mg [9]. In cosmetic products, food graded PC is frequently utilized; however, the quality is different among companies. Even though the effect of PC on various diseases has been proved, such as cancer, its anti-aging was neglected [10]. Furthermore, the use of over-purified PC could be costly to be adapted for industrial products. In this study, we want to extract crude PC (crPC) by a simple protocol to make a high-quality and low-cost material used in cosmetics.

2 Materials and Methods

2.1 Materials and Reagents

Arthrospira platensis was received from the University of Science, Vietnam National University of Ho Chi Minh City.

2.2 Extraction and Purification of C-Phycocyanin

The extraction and purification process was performed according to Safaei (2019) with some modification [10]. The biomass of *Arthrospira platensis* was added to lysis buffer (Na phosphate or $CaCl_2$) with a ratio of 1 g dried biomass/20 mL solution. The solution was subjected to repeated freezing/thawing cycle (−20 °C for 3 h and 4 °C for 8 h) with 3 replications for cell breaking. Then, the solution was centrifuged at 8000 rpm for 10 min to obtain supernatant (crude Arthrospira extraction, CRE).

Next, activated charcoal was applied to CRE (8 g/100 mL), and the mixture was centrifuged at 8000 rpm for 15 min to discard pellet (1st step). Then, the crude phycocyanin (crPC) was purified by 2-step salting-out method using ammonium sulfate ($(NH_4)_2SO_4$). $(NH_4)_2SO_4$ was supplemented to CRE to reach saturation of 15% (2nd step, to remove contaminants) and 50% (3rd step, to obtain phycocyanin precipitation). Then, the precipitation was resuspended in 10 mL sodium phosphate buffer (10 mM, pH 7.0), dialyzed against distilled water overnight at 4 °C and filtered stepwise with 0.8, 0.45 and 0.22 μm membrane. The absorbance spectra were obtained to determine the concentration and purity of C-PC by using the following equation of Safaei, 2019. Purity index: OD_{620}/OD_{280} and concentration: (OD620—0.474 × A650)/5.34.

2.3 Evaluation of Cytotoxicity of CrPC In Vitro [11]

Human fibroblasts were cultured in DMEM/F12 (Dulbecco's Modified Eagle Medium) medium containing 10% fetal bovine serum (FBS). The cells were seeded at a density of 5×10^4 cells/well in 24-well plate. After 24 h, crPC was added to wells with concentrations: 150, 100, 50, 25, 12.5 and 0 μg/mL and incubated for 24 h. One day later, the medium was changed, and the MTT assay was performed to assess the cytotoxicity of crPC at various concentrations.

2.4 Evaluation of the Proliferation Effect of CrPC [12]

Human fibroblasts were seeded in 24-well plate with a concentration of 5×10^3 cells/well and exposed to medium containing crPC (75, 50, 25, 12.5 and 0 μg/mL) for 7 days, and the medium was exchanged 2 days per time. After 2, 4, 6 culturing days, cell number in each group was determined by hemocytometer.

2.5 *Evaluation of the Migration Effect of CrPC by Wound Healing Assay [13]*

Human fibroblast was seeded into a 24-well plate with concentration 5×10^4 cells/well. After 24 h, a scratch was made on a cell monolayer in each well. Then medium containing crPC at concentrations of 75, 50, 25, 12.5 and 0 μg/mL was supplemented to wells. The migration of fibroblasts was observed by microscope until 24 h.

2.6 *Evaluation of the Protective Ability of CrPC Against H_2O_2*

The H_2O_2-induced oxidative stress was performed according to Dimozi [14] with some modifications [14]. The experiment was divided into 3 groups: normal group (fibroblasts were cultured in DMEM/F12 medium), control group (fibroblasts were treated with H_2O_2 175 μM for 90 min), experimental group (fibroblasts were pretreated with medium containing crPC 25 μg/mL for 1 day, followed by treating with H_2O_2 175 μM for 90 min). Cell viability, cell ratio in the G0/G1 phase and senescence β-galactosidase (SβGA) expression were determined in each group.

2.7 *Statistical Analysis*

All experiments were repeated at least three times, and the results were shown as mean ± S.D. One-way ANOVA using Statgraphics 18 was employed to analyze the results statistically. A value of $P < 0.05$ was considered as significant.

3 Result

3.1 *Extraction of CrPC*

As shown in Table 1, the highest purity index (PI) of CRE was obtained in the group using $CaCl_2$ as lysis buffer (1.01 ± 0.07), and its concentration was 1.77 ± 0.04 mg/mL. The result obtained from the salting method showed that the previous adding of salt to 15% increased the purity of C-PC in the solution to 1.29 mg/mL.

Table 1 The result of extraction and purification of crPC from *Arthrospira platensis*

Phase	Variants	Concentration (mg/mL)	PI
CSE	1% Calcium chloride	1.77 ± 0.04^{a}	1.01 ± 0.07^{a}
	2% Calcium chloride	1.68 ± 0.06^{b}	1.00 ± 0.02^{a}
	10 mM Na phosphate buffer	1.56 ± 0.03^{c}	0.42 ± 0.04^{b}
	100 mM Na phosphate buffer	1.79 ± 0.04^{a}	0.48 ± 0.04^{b}
	Distilled water	1.62 ± 0.06^{b}	0.45 ± 0.05^{b}
1st purification	8% Activated charcoal	1.65 ± 0.08	1.52 ± 0.08
2nd purification	15% Ammonium sulfate	1.60 ± 0.03	1.84 ± 0.04
crPC	45% Ammonium sulfate	1.13 ± 0.05	2.79 ± 0.03

^{a,b,c}Were different statistically, $\alpha = 0.05$

After the pellet was discarded, the further supplement of ammonium sulfate precipitated C-PC, C-PC was precipitated the most by 50% of the salt solution. The concentration of crPC group was 1.13 ± 0.05 mg/mL, and the purity index was 2.79 ± 0.25 (Fig. 1).

Fig. 1 Crude phycocyanin. **a** Arthrospira, **b** crPC

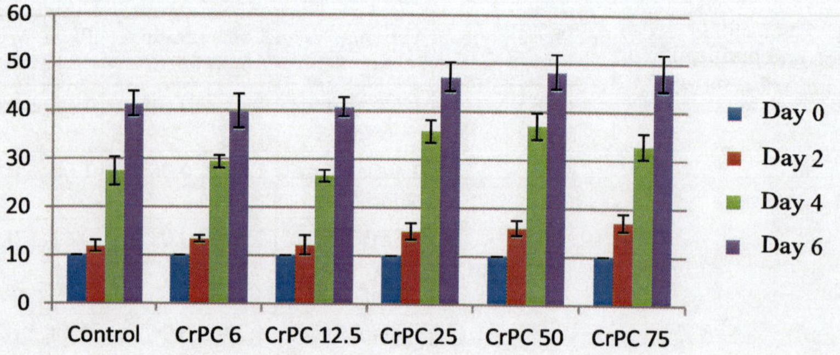

Fig. 2 The results of the proliferation effect test

3.2 Evaluation of Cytotoxicity of CrPC

This test is according to ISO 10993 standard. The fibroblasts are seeded in the plate and contacted the agent (crPC). Later, MTT assay is carried out to determine the OD of cells in experiment group (exp) and control group (con). The cytotoxic level is based on OD_{exp}/OD_{con}. The result of MTT assay showed that OD absorption in all experimental groups (medium containing crPC at 12.5–150 μg/mL) was higher than the control group (fibroblasts were cultured in DMEM/F12), the ratio of OD597 ex/con was higher than 1. Cytotoxicity degree is 0, crPC was not toxic to fibroblasts at 200 μg/mL and lower.

3.3 Evaluation of Proliferation Effect of CrPC

Fibroblasts were cultured in medium containing crPC 6, 12.5, 25, 50, 75 μg/mL for 6 days. After 6-day in culture, cells in control group covered about 90% surface area and reached a density of $41.3 \pm 2.5 \times 10^4$ cells/well. In experimental groups, cells in crPC reached the full confluence with concentration from 47.1×10^4 to 47.9×10^4 cells/well and cell proliferation in these groups was faster than crPC 6, 12.5 μg/mL (Fig. 3 and Table 2). The results showed that crPC stimulated proliferation of fibroblast in vitro at concentrations above 25 μg/mL (Fig. 2).

3.4 Evaluation of Migration of CrPC

After 24 h, cells in all groups migrated into the scratch with different levels. In control and crPC6 groups, cells reached to the scratch center but did not cover total wound area (close to 50%). In groups of crPC 12.5, crPC 25, crPC 50 and crPC 75, cell invaded completely into the wound and covered over 90% of the scratch (Fig. 3).

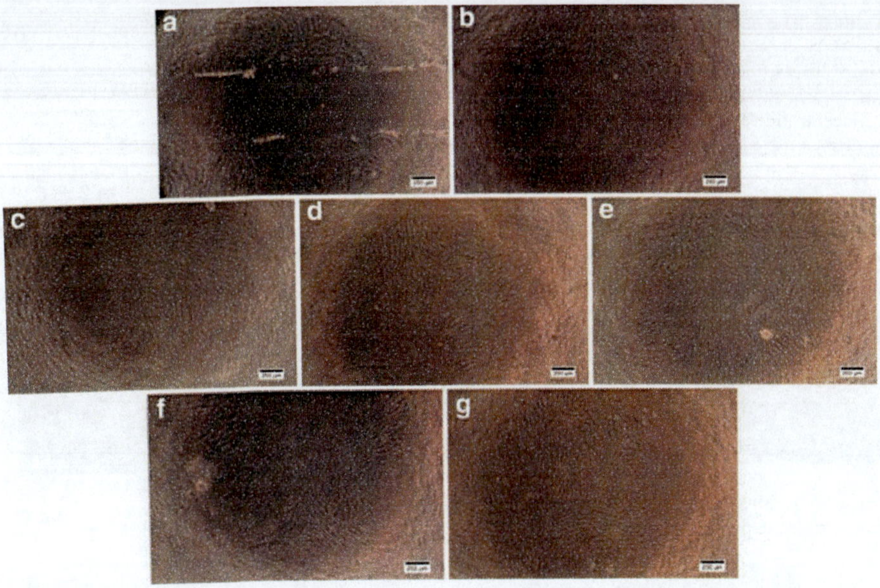

Fig. 3 Wound Healing test. **a** the scratch on cell monolayer; **b–g** cell migration on crPC 0, 6, 12.5, 25, 50, 75 μg/mL

Table 2 The results of the cytotoxicity test

	Control	crPC 12.5	crPC 25	crPC 50	crPC 100	crPC 150
Aver	0.30 ± 0.02	0.36 ± 0.02	0.32 ± 0.05	0.33 ± 0.04	0.37 ± 0.01	0.36 ± 0.03
Ratio		1.2 ± 0.06	1.09 ± 0.15	1.11 ± 0.14	1.25 ± 0.04	1.2 ± 0.11

The results showed that crPC at concentrations above 12.5 μg/mL enhanced the cell migration effectively.

3.5 *Evaluation of the Protective Ability of C-PC Against H_2O_2*

Table 3 shows the results of this protective effect test. The result showed that in the normal group, most cells were in the G0/G1 phase (72.5 ± 4.8%) and a few cells expressed Sβ-GA (4.4 ± 2.1%). In the control group, cell viability declined to 53.4 ± 2.3%, cell ratio in G0/G1 increased to 85.8 ± 6.1%, percentage of Sβ-GA expressed cells increased to 68.4 ± 6.8% (about 17 times). These results showed that H_2O_2 affected negatively on cell: it led to cell death and senescence. In the experimental group, cells were pretreated with crPC at a concentration of 25 μg/ml before exposing

Table 3 The number of cells in all groups after 6 days in culture [a] and [b] were different statistically, $\alpha = 0.05$

	Control	crPC 6	crPC 12.5	crPC 25	crPC 50	crPC 75
Concentration ($\times 10^4$)	41.3 ± 2.5^a	39.8 ± 3.7^a	40.7 ± 2.0^a	47.1 ± 3.0^b	48.2 ± 3.5^b	47.9 ± 3.8^b
DT	47.3 ± 1.4^a	48.2 ± 2.1^a	47.6 ± 1.2^a	44.5 ± 1.2^b	44.1 ± 1.4^b	44.2 ± 1.5^b

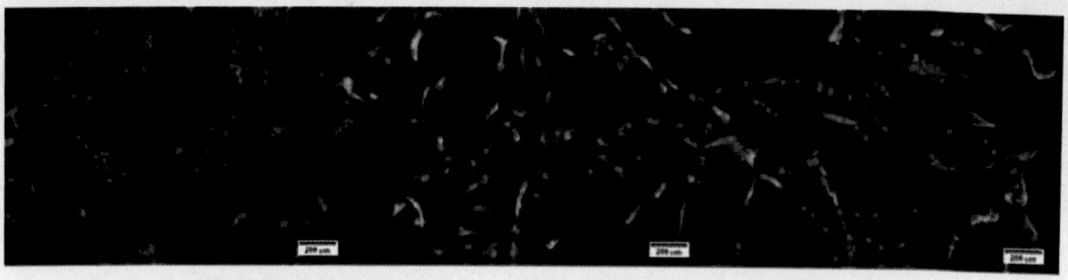

Fig. 4 Fibroblasts were stained with Hoetch and actin staining. Left, normal group; center, control group; right: experimental group

to H_2O_2 (Figs. 4, 5, 6 and Table 4). The results showed that crPC alleviated oxidative stress of H_2O_2 on fibroblasts: viability was higher than the control group and the ratio of G0/G1 cells, Sβ-GA expressed cells was lower than the control group.

4 Discussion and Conclusion

Phycocyanin is a powerful natural antioxidant. It costs from 500 to 15 USD/mg, which depend on its purity. Purity index (PI) is calculated by formulation of A620/A280 and divided to food-grade (PI > = 0.9) and analysis grade (PI > = 4) [15]. In the cosmetic field, PC is used as a natural dye. PC was proved to have positive effects on proliferation and migration of skin cells such as fibroblasts and keratinocytes in vitro [8, 10, 16]. However, affection of PC on protecting skin cells from oxidation agent in vitro is limited. Most commercial PC products belong to food grade. PC used in the experiment is analysis grade, but it is too expensive to apply in commercial products. In this study, we want to extract crude PC whose purity is in a range of 2.5–3 with low price and determine its ability in protecting skin cells (fibroblasts) against oxidative stress. Thus, we used a simple and cheap process to extract the PC from Spirulina. Crude Spirulina extract (CSE) was obtained by repeated freezing/thawing method with 3 replications. Then, contaminants were discarded by using activated charcoal and ammonium sulfate 15% combined with centrifugation. Final, PC was precipitated by charcoal and ammonium sulfate 50%, dialyzed against sodium phosphate overnight and cryopreserved (crude phycocyanin, crPC). The crPC had a concentration of 1.13 mg/mL and a PI of 2.89.

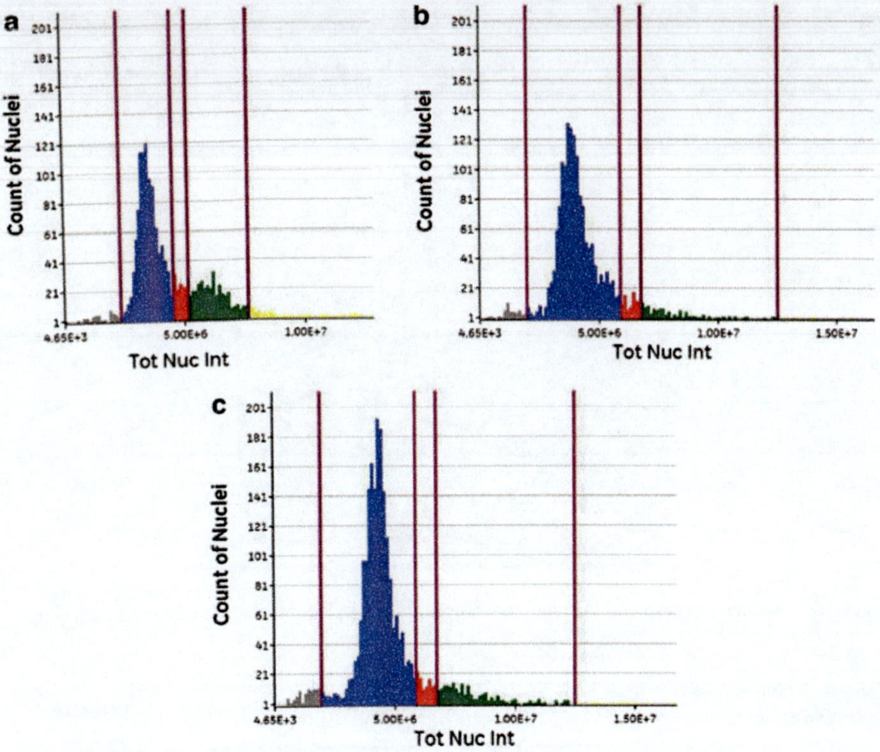

Fig. 5 Flow cytometry results of cells in protective effect test. **a** normal group; **b** control group, **c** experimental group

Some of its biological characteristics were determined in vitro, including cytotoxicity, proliferation and migration induction on fibroblasts and ability in protecting fibroblasts. Fibroblasts are the major cells in the skin's dermal layer. In physiological condition, fibroblasts synthesize collagen and elastin which retain the strength and elasticity of skin [17]. When being hurt by flame or chemicals, fibroblasts move into the wound and proliferate to regenerate the dermis [18]. In aging skin, the number of fibroblasts is diminished, and they produce less collagen and elastin. The skin becomes thinner, weaker and looser [19, 20]. In cytotoxicity, proliferation and migration test, crPC in concentrations of 6, 12.5, 25, 50, 75, 150 μg/mL was used. Cytotoxicity test was performed according to ISO 10993 standard. OD597 obtained from MTT assay in all experimental groups (cells were cultured in medium supplemented with crPC) was higher than control (cells were cultured in culturing medium). This result showed that crPC was not toxic to fibroblasts from 6 to 150 μg/mL. Migration test was performed according to Wound Healing assay. After 24 h, fibroblasts cultured in crPC 12.5—75 μg/mL moved faster and covered more scratch area than in control and crPC 6 μg/mL. In proliferation effect test, after 6 days, cells in groups

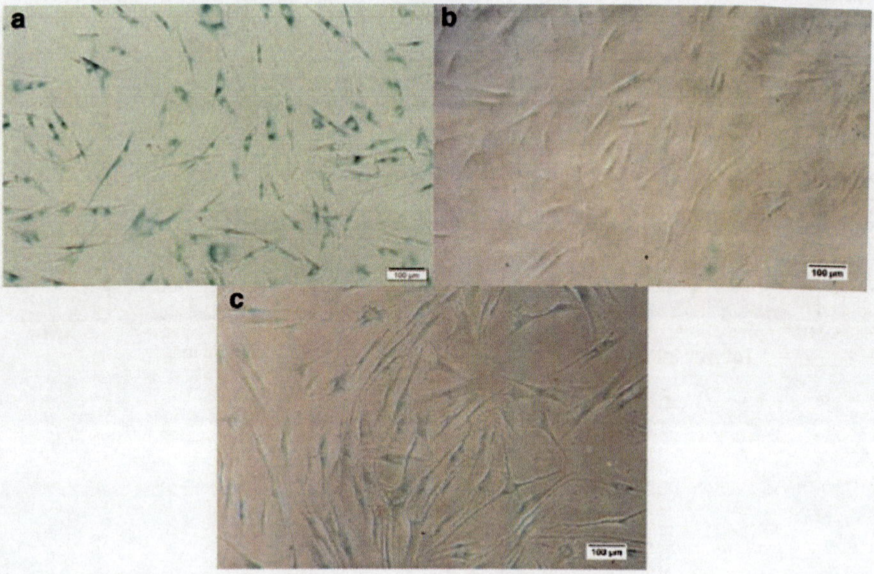

Fig. 6 Senescence—β galactosidase staining. **a** normal group; **b** control group; **c** experiment group

Table 4 The result of protective effect test

	Normal group (%)	Control group (%)	Experimental group (%)
Cell viability	100[a]	53.4 ± 2.3^{b}	79.3 ± 5.4^{c}
Ratio cells in the G0/G1 phase	72.5 ± 4.8^{a}	85.8 ± 6.1^{b}	78.7 ± 4.1^{c}
Sβ-GA expression	4.4 ± 2.1^{a}	68.4 ± 6.8^{b}	22.8 ± 4.6^{c}

[a,b,c]Were different statistically, $\alpha = 0.05$

of crPC 25–75 µg/mL reached 100% confluence and had doubling times about 41 h. While cells in groups of control and crPC 6–12.5 µg/mL covered nearly 90% the surface and had DT longer. Based on these results, we concluded that crPC was not toxic to fibroblasts with concentration under 150 µg/mL. Moreover, 25 µg/mL was the minimum concentration of crPC, which can stimulate both proliferation and migration of fibroblast, so this concentration was used in the following experiments.

Next, we assessed affection for its ability to protect fibroblasts from hydroxyl peroxide (H_2O_2). H_2O_2 is a strong oxidant which could damage cell compartments including membrane, nucleus … Oxidants such as H_2O_2 were produced in the body as the by-products of metabolisms. Oxidants were overproduced in aging skin or by the affection of UV light. H_2O_2 can destroy the skin cells or make the cells become senescence and break up collagen and elastin, and this symptom occurs in photoaging

[21–23]. PC can inhibit H_2O_2—induced injuries on hepatocyte, porcine embryos but its protective ability on fibroblast in vitro was not concerned [24–26]. So, we tried to evaluate crPC protective ability on fibroblast from H_2O_2. The cells were pretreated with crPC 25 μg/mL for 24 h and exposed to H_2O_2 175 μM for 90 min as our previous protocol (experimental group). Three aspects were evaluated: cell viability (MTT assay), cell ratio in G0/G1 phase (Hoetch staining and using Cytell machine) and senescence—β galactosidase expression (staining). This result showed H_2O_2 175 μM destroyed fibroblasts (cell viability decreased to 50%) and most remain cells had signs of senescence: 85% of cells were arrested in G0/G1 phase and 60% of cells expressed SA—β galactosidase. In the experimental group (crPC—pretreated cells), viability was higher than control, and cell rate was arrested in G0/G1 was lower and fewer cells expressed SA—β galactosidase. We concluded crPC 25 μg/mL inhibited oxidative stress of H_2O_2 175 μM on fibroblasts.

5 Conclusion

crPC was successfully extracted from *A. platensis,* and crPC inhibited oxidative stress of H_2O_2 on fibroblasts.

Conflicts of Interest The authors have no conflict of interest to declare.

References

1. Jung F, Krüger-Genge A et al (2019) Spirulina platensis, a super food? J Cell Biotech 5:43–54
2. de Morais MG, da Fontoura Prates D et al (2018) Phycocyanin from microalgae: properties, extraction and purification, with some recent applications. Ind Biotech 14:30–37
3. İlter I, Akyıl S et al (2018) Optimization of phycocyanin extraction from *Spirulina platensis* using different techniques. J Food Compos Anal 70:78–88. https://doi.org/10.1016/j.jfca.2018.04.007
4. Tu Y, Quan TJC (2016) Oxidative stress and human skin connective tissue aging 3:28. https://doi.org/10.3390/cosmetics3030028
5. Wang B (2011) Photoaging: a review of current concepts of pathogenesis. J Cutan Med Surg 15(Suppl 1):S374–S377. https://doi.org/10.2310/7750.2011.00006
6. Madhyastha HK, Radha KS et al (2008) uPA dependent and independent mechanisms of wound healing by C-phycocyanin. J Cell Mol Med 12:2691–2703. https://doi.org/10.1111/j.1582-4934.2008.00272.x
7. Gur CS, Erdogan DK et al (2013) In vitro and in vivo investigations of the wound healing effect of crude Spirulina extract and C-phycocyanin. J Med Plants Res 7:425–433
8. Madhyastha H, Madhyastha R et al (2012) Regulation of growth factors-associated cell migration by C-phycocyanin scaffold in dermal wound healing. Clin Exp Pharmacol Physiol 39:13–19. https://doi.org/10.1111/j.1440-1681.2011.05627.x
9. Chaiklahan R, Chirasuwan N et al (2018) Stepwise extraction of high-value chemicals from Arthrospira (Spirulina) and an economic feasibility study. Biotechnol Rep (Amst) 20:e00280–e00280. https://doi.org/10.1016/j.btre.2018.e00280

10. Safaei M, Maleki H et al (2019) Development of a novel method for the purification of C-phycocyanin pigment from a local cyanobacterial strain Limnothrix sp. NS01 and evaluation of its anticancer properties. Sci Rep 9:9474. https://doi.org/10.1038/s41598-019-45905-6
11. Aslantürk ÖS (2018) In vitro cytotoxicity and cell viability assays: principles, advantages, and disadvantages. https://doi.org/10.5772/intechopen.71923
12. Hao S, Li S et al (2019) C-phycocyanin suppresses the in vitro proliferation and migration of non-small-cell lung cancer cells through reduction of RIPK1/NF-κB Activity. Mar Drugs 17:362. https://doi.org/10.3390/md17060362
13. Jonkman JEN, Cathcart JA et al (2014) An introduction to the wound healing assay using live-cell microscopy. Cell Adh Migr 8:440–451. https://doi.org/10.4161/cam.36224
14. Dimozi A, Mavrogonatou E et al (2015) Oxidative stress inhibits the proliferation, induces premature senescence and promotes a catabolic phenotype in human nucleus pulposus intervertebral disc cells. Eur Cell Mater 30:89–102; discussion 103. https://doi.org/10.22203/ecm.v030a07
15. Liu Q, Huang Y et al (2016) Medical application of *Spirulina platensis* derived C-phycocyanin. Evid Based Complement Altern Med 2016:7803846. https://doi.org/10.1155/2016/7803846
16. Gunes S, Tamburaci S et al (2017) In vitro evaluation of Spirulina platensis extract incorporated skin cream with its wound healing and antioxidant activities. Pharm Biol 55:1824–1832. https://doi.org/10.1080/13880209.2017.1331249
17. Rodrigues M, Kosaric N et al (2018) Wound healing: a cellular perspective. Physiol Rev 99:665–706. https://doi.org/10.1152/physrev.00067.2017
18. Gonzalez ACdO, Costa TF et al (2016) Wound healing—a literature review. Anais brasileiros de dermatologia 91:614–620. https://doi.org/10.1590/abd1806-4841.20164741
19. Lago JC, Puzzi MB (2019) The effect of aging in primary human dermal fibroblasts. PLoS ONE 14:e0219165. https://doi.org/10.1371/journal.pone.0219165
20. Tigges J, Krutmann J et al (2014) The hallmarks of fibroblast ageing. Mech Ageing Dev 138:26–44. https://doi.org/10.1016/j.mad.2014.03.004
21. Bladier C, Wolvetang EJ et al (1997) Response of a primary human fibroblast cell line to H2O2: senescence-like growth arrest or apoptosis? Cell Growth Differ 8:589–598
22. Sander CS, Chang H et al (2002) Photoaging is associated with protein oxidation in human skin in vivo. J Invest Dermatol 118:618–625. https://doi.org/10.1046/j.1523-1747.2002.01708.x
23. Shin MH, Rhie GE et al (2005) H2O2 accumulation by catalase reduction changes MAP kinase signaling in aged human skin in vivo. J Invest Dermatol 125:221–229. https://doi.org/10.1111/j.0022-202X.2005.23823.x
24. Ou Y, Zheng S et al (2010) Protective effect of C-phycocyanin against carbon tetrachloride-induced hepatocyte damage in vitro and in vivo. Chem Biol Interact 185:94–100. https://doi.org/10.1016/j.cbi.2010.03.013
25. Niu Y-J, Zhou W et al (2017) C-Phycocyanin protects against mitochondrial dysfunction and oxidative stress in parthenogenetic porcine embryos. Sci Rep 7:16992. https://doi.org/10.1038/s41598-017-17287-0
26. Liang S, Guo J et al (2018) C-Phycocyanin supplementation during in vitro maturation enhances pre-implantation developmental competence of parthenogenetic and cloned embryos in pigs. Theriogenology 106:69–78. https://doi.org/10.1016/j.theriogenology.2017.09.001

Isolation of Antimicrobial Probiotic Bacteria from Sour Shrimps in Hue City-Vietnam and Optimization for Biomass and Acid Production

Hoang Thi Kim Hong, Nguyen Thi Thu Thuy, Vo Thi Bich, and Nguyen Thi Ngoc Hanh

Abstract This study isolated antibacterial probiotic strains from sour shrimp, a Vietnamese traditional fermented food, and optimized culturing conditions for the chosen strain. Strain V101 was chosen for optimization experiments for its high antagonistic activity and production of strong antibacterial compounds against indicator harmful bacteria – *Escherichia coli*, and *Staphylococcus aureus*. Molecular identification of 16S rRNA sequencing showed that V101 strain was *Lactobacillus paracasei*. Optimization of culturing conditions found that this strain produces the highest biomass and acid accumulation on modified MRS at pH 5.5; temperature 37 °C in 60 h. Under optimum condition, biomass and acid production increased 76.61% and 10.69% respectively, and antibacterial activities against *E. coli* improved up to 20%. Pilot fermentation of L. *paracasei* V101 on production Vietnamese pickles, sour shrimp showed that the fermentation process was quicker, tastier, and fermented products could be preserved longer. These findings will be a potential prerequisite for further study on *L. paracasei* V101 as a strain not only good for food fermentation and preserving but also a helpful probiotic bacteria bring benefits to human health.

Keywords Antibacterial · *Lactobacillus paracasei* · Probiotics · LAB

1 Introduction

Lactic acid bacteria (LAB probiotics) are well-known for their benefits on health for their application as probiotics – good bacteria for the human digestive system and a popular bacteria for fermentation [1]. A wide range of antimicrobial compounds (organic acids, hydrogen peroxide, diacetyl, low molecule weight compounds, and bacteriocin) enable probiotics to inhibit both Gram (+) and Gram (–) bacteria [2].

H. T. K. Hong (✉) · N. T. T. Thuy · V. T. Bich
Hue University of Sciences, Hue University, Ho Chi Minh City, Vietnam
e-mail: hkhong@hueuni.edu.vn

N. T. N. Hanh
Hong Bang International University, Ho Chi Minh City, Vietnam

V. Van Toi et al. (eds.), *8th International Conference on the Development of Biomedical Engineering in Vietnam*, IFMBE Proceedings 85,
https://doi.org/10.1007/978-3-030-75506-5_55

Among organic acids produced by probiotics, lactic acid was produced in a noticeable amount contributing to antimicrobial activities. On the other hand, bacteriocins from LAB Probiotics were well-known for their antimicrobial activities. Bacteriocins' antimicrobial mechanisms were first studied on Nisin, a bacteriocin Gram (+). They can kill bacteria due to the capacity of changing membrane permeability by forming channels on bacteria's membrane. Moreover, bacteriocins can break down DNA and RNA and weaken the peptidoglycan layer of cell walls. These antimicrobial activities make it possible to extend the shelf life of fermented products without adding artificial preservatives [3–5].

Fermented products, such as (yogurt, sour cherry, pickled noodles, sour bamboo shoots, and sour shrimp, etc.) represented an irreplaceable part of Vietnamese cuisine. However, the quality of fermented foods fluctuated from batch to batch due to microorganisms in raw materials and the surrounding environment; skills, and workmanship of ones carrying out the fermentation process. Moreover, food preservatives added to extend the preservation period could cause bad impacts on consumer's health [1]. As a result, there was an urgent need for microorganisms to perform high antimicrobial activities against rotting microbes to improve fermented food quality and extend preservation time.

In this study, we isolated high antimicrobial probiotics strains from traditional fermented products. Optimal culture conditions for isolated strains would be screened and applied to the production of traditional fermented products.

2 Materials and Methods

2.1 Materials

A number of 25 samples of sour shrimps were collected in Hue City, Vietnam. *E. coli*, *S. aureus*, from the Department of Applied Biology, Faculty of Biology, University of Science, Hue University, Vietnam, were used as indicator strains for screening antimicrobial activity of isolated LAB probiotics strains.

2.2 Isolation of Potential Probiotics Strains

An amount of 100 μL of serial dilutions 10^{-1}–10^{-5} of sour shrimp samples was widely spread on MRS agar plates, followed by 48 h incubation at 30 °C. Separated colonies forming a surrounding clear zone would be transferred onto slant agar for keeping stock at 4 °C [6].

2.3 Primary Screening for Potential Probiotics Strain

Isolated colonies were transferred to $CaCO_3$ added MRS agar and incubated for 60 h at 30 °C. Deduction of the size of inoculum (d) from the resolution zone (D) represented the total acid production abilities of probiotics. Strain producing the highest total acid would be selected for the next experiments.

2.4 Screening for Probiotics Strains Strongly Inhibiting Indicator Bacteria

Bacteria isolates were transferred to liquid MRS medium and cultured at 30 °C for 48 h, centrifuged at 7000 rpm to obtain supernatant to apply to antibacterial assays.

2.5 Agar-Well Diffusion Assay

Indicator strains were mixed with a nutrient agar at 50 °C and shaken well before pouring into plates. An amount of 50 μL of supernatant collected from each isolate would be pumped into 10 mm wells on plates. Plates were incubated at 4 °C for 12 h in the fridge and followed by 18–24 h incubation at 35 °C. Diameters of inhibition zones were calculated as a deduction of diameters of the wells (d) from diameters of inhibition areas (D) [7].

2.6 Dual Culture Overlay Assay

LAB probiotics strains were streaked on MRS plates and incubated for 48 h. Then, the indicator strains were cultured in a nutrient agar medium at 50 °C and shaken well before overlaying MRS agar. Plates were incubated at 4 °C for 12 h in the fridge followed by a period of 18–24 h incubation at 35 °C. Diameters of inhibition zones were calculated as a deduction of diameters of colonies (d) from diameters of inhibition areas (D) [8].

2.7 Molecular Identification of Potential Strain

The probiotics strain named V101 chosen from the previous experiment was identified by 16s rRNA sequencing. After being observed for morphology, it was sent to

Nam Khoa Service and Trading Co., Ltd. to analyze 16s rRNA sequence and aligned sequences on Blast Search [9].

2.8 Optimization of Culture Conditions for V101 Growth and Total Acid Production

V101 strain was cultured in MRS broth with 10% culturing rate and shaken at 120 rpm. Optimum culturing conditions found from previous experiments would be applied to later on optimization experiments. Biomass and total acid production of the culture would be evaluated. Firstly, incubation time were optimized from 12; 24; 36; 48; 60; 72; 84; 96 h. Then initial pH were optimized: 4.0; 4.5; 5.0; 5.5; 6.0; 6.5; 7.0; 7.5; 8.0; next to NaCl concentrations 3; 4; 5; 6; 7; 8%. Optimum carbon sources were screened on mannitol, lactose, glucose, CMC, starch, molasses, and saccharose. Finally, KNO_3, NH_4NO_3, peptone, meat extract, gelatin, urea were used to screen for optimal Nitrogen sources for V101 culturing. Mc Ilvaine buffer was used for preparing medium and adjusting pH.

2.9 Growth and Antimicrobial Activities of Probiotics Strain Under Optimal Culture Conditions

V101 strain was incubated under optimal culture conditions found from previous experiments to evaluate biomass, total acid accumulation, and antimicrobial activities.

2.10 Building the Biomass Standard Curve of V101 Strain

An amount of 1 g of biomass collected from 72 h culture of V101 in MRS by centrifugation at 4000 rpm for 10 min was desiccated to constant weight. The absorbance of serial dilutions from another 1 g of biomass in sterile water was measured at 600 nm. Linear regression equation between optical density: dry biomass (g/L) was plotted by Microsoft Excel 2016 [10].

2.11 *Quantitative Determination of Total Acid Concentration by Titration*

After having centrifuged, 10 mL of each supernatant was transferred to a flask, then 20 mL of distilled water and 1–2 drops of phenolphthalein was added to the flask for a titration with NaOH 0.1 N. The amount of total acid in 10 mL of the supernatant was equivalent to the multiplication of the volume of NaOH 0.1 N in mL used in titration with 0.009 (1 mL of NaOH is equal to 0.009 g of total acid) [11].

2.12 *Data Analysis*

All experiments were triplicated and analyzed by Excel 2016 and SPSS with $p < 0.05$.

3 Results and Discussions

3.1 *Isolation Potential Probiotics Strains*

Bacterial densities in 25 studied samples were significantly different, from 10.71×10^6 to 43.47×10^6 CFU/mL. Around 110 LAB Probiotics strains, named from V1, V2, V3 … to V110, were isolated. Most colonies were 0.5–1.5 mm in diameter; round shape; smooth surface; round edge or lobe edge; opaque, white, cream, or milky color. Some colonies were 2–3 mm, round, smooth, opaque, or milky. There were only a few large colonies with a diameter of around 3.5–4 mm; round shape; opaque or clear white; irregular and dry edge. Among 110 isolates, 20 strains with large resolution lines were selected for the next experiments, including V9, V11, V21, V37, V38, V40, V48, V50, V51, V52, V54, V55, V62, V76, V78, V95, V96, V101, V103, and V104. Results of screening for the antimicrobial activities of selected strains by diffusion method using indicator bacteria were shown in Table 1.

Table 1 demonstrated that most strains were assumed to produce antimicrobial compounds as they could inhibit indicator bacteria. V52 and V101 performed outstanding capability of producing antimicrobial compounds and stronger inhibitory activities than other strains. Moreover, the V101 strain showed higher inhibitory ability against *Salmonella* than V52. Thus, it was selected for further experiments.

Table 1 Sizes of inhibition zones

Strain	Agar-well diffusion assay		Dual culture overlayer assay	
	E. coli	*E. coli*	*S. aureus*	*S. aureus*
V9	54.5 ± 0.3	31.5 ± 0.5	–	9.1 ± 0.8
V11	43.5 ± 0.3	–	27.5 ± 0.5	–
V21	–	–	–	10.5 ± 0.1
V37	65.5 ± 0.2	22.8 ± 0.1	32.5 ± 0.3	–
V38	37.4 ± 0.6	–	–	–
V40	34.5 ± 0.9	–	–	19.5 ± 0.4
V48	61.3 ± 0.2	–	28.8 ± 0.8	23.8 ± 0.1
V50	72.0 ± 0.5	33.5 ± 0.5	12.0 ± 0.1	25.3 ± 0.9
V51	31.0 ± 0.8	–	–	11.6 ± 0.2
V52	72.8 ± 0.6	36.7 ±.0.8	31.5 ±.0.5	32.5 ± 0.5
V54	–	–	–	11.8 ± 0.3
V55	37.0 ± 0.1	12.5 ± 0.6	–	–
V62	–	–	–	–
V76	29.5 ± 0.1	–	–	–
V78	50.5 ± 0.7	–	16.2 ± 0.7	–
V95	–	28.0 ± 0.5	–	30.5 ± 0.1
V96	21.2 ± 0.8	–	–	–
V101	63.3 ± 0.7	33.5 ±.0.3	29.6 ±.0.4	29.2 ± 0.1
V103	14.3 ± 0.9	–	–	–
V104	35.5 ± 0.7	–	22.7 ± 0.5	–

3.2 Colony Morphological Analysis and Sequencing 16S rRNA of V101 Strain

Colony morphological analysis. V101 strain was cultured on MRS plates in 48 h. Morphology of colonies was observed. V101 colonies were convex, 1.0–1.5 mm in diameter, ivory-white. Gram staining showed that V101 was a Gram-positive strain. Under the microscope, V101 cells were scattered, long rod-shaped (Fig. 1).

Sequencing analysis. Sequencing analysis of 16 s rRNA, shown in Fig. 2 was carried out in Nam Khoa Service and Trading Co., Ltd. Molecular identification resulted was analyzed similarities with sequences in Genbank by BLAST.

It was found that the V101 16S rRNA sequence having 99% similarity with *Lactobacillus paracasei* Strain IIA, ID CP014985.1. Thus it was identified as *L. paracasei* (Table 2).

In Thai fermented shrimp or Kung-Som, isolated LAB strains were *L. futsaii* CS3 and *L. futsaii* CS5 [12]. LAB Probiotics strain found on meat traditional product

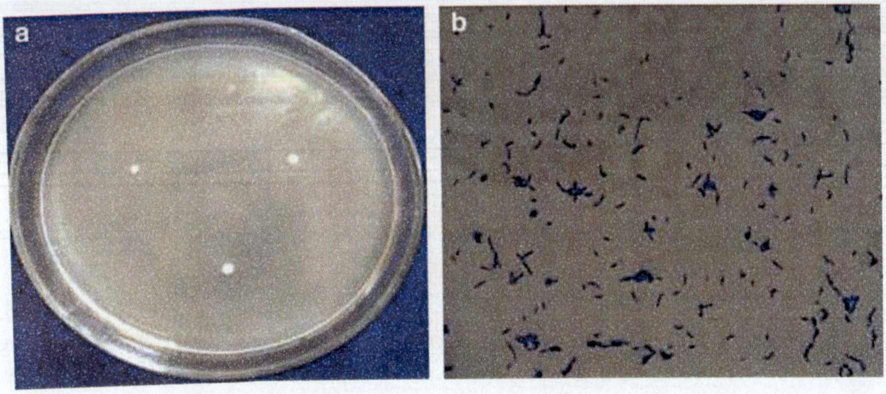

Fig. 1 **a** V101 colonies on MRS plates. **b** V101 Gram staining under the microscope

of Portugal was *Lactobacillus sakei* [13]. *L. plantarum* was used to ferment and improve the nutritional values of fermented silage from Korea [14]. *Lactobacillus futsaii* CS3 was found as the main fermentation bacteria for Kung-Som (a Thailand fermented shrimp) [12]. Potential LAB probiotics isolated from Nem Chua, another Vietnamese fermentation food were found as *Lactobacillus plantarum* and *Pediococcus pentosaceus* [1]. It was found that *L. plantarum* could prevent the occurrence of yeast, molds, and other harmful microorganisms in silage. It was also proven as a good probiotic strain for its characteristics including bile salt tolerance, non-virulent and non-hemolysis characteristics, high survival rate, or ability to grow in low pH from 2.5 to 4.0 [15].

3.3 Biomass Standard Curve

The standard curve of biomass was built against biomass and OD at 600 nm. The graph was plotted in Fig. 3. From the standard curve, R^2 is 0.9988, which is near 1. Thus this standard curve was reliable to be used for interpreting data from OD to biomass content.

3.4 Optimization of Appropriate Culturing Conditions for Bacterial Growth and Total Acid Accumulation

Factors affecting the fermentation process including incubation time, pH, NaCl concentrations, carbon sources, and nitrogen sources were optimized. Overall, biomass accumulation and acid production correlated with each other in which they peaked up at the same culturing conditions in 4 out of 5 optimization experiments.

```
Features:
Query  1       GATCCTGGCTCAGGATGAACGCTGGCGGCGTGCCTAATACATGCAAGTCGAACGAGTTCT  60
               ||||||||||||||||||||||||||||||||||||||||||||||||||||||||||||
Sbjct  983608  GATCCTGGCTCAGGATGAACGCTGGCGGCGTGCCTAATACATGCAAGTCGAACGAGTTCT  983549

Query  61      CGTTGATGATCGGTGCTTGCACCGAGATTCAACATGGGAACGAGTGGCGGACGGGTGAGT  120
               ||||||||||||||||||||||||||||||||||||| ||||||||||||||||||||||
Sbjct  983548  CGTTGATGATCGGTGCTTGCACCGAGATTCAACATGG-AACGAGTGGCGGACGGGTGAGT  983490

Query  121     AACACGTGGGTAACCTGCCCTTAAGTGGGGGATAACATTTGGAAACAGATGCTAATACCG  180
               ||||||||||||||||||||||||||||||||||||||||||||||||||||||||||||
Sbjct  983489  AACACGTGGGTAACCTGCCCTTAAGTGGGGGATAACATTTGGAAACAGATGCTAATACCG  983430

Query  181     CATAGATCCAAGAACCGCATGGTTCTTGGCTGAAAGATGGCGTAAGCTATCGCTTTTGGA  240
               ||||||||||||||||||||||||||||||||||||||||||||||||||||||||||||
Sbjct  983429  CATAGATCCAAGAACCGCATGGTTCTTGGCTGAAAGATGGCGTAAGCTATCGCTTTTGGA  983370
Query  241     TGGACCCGCGGCGTATTATCTAGTTGGTGAGGTAATGGCTCACCAAGGCGATGATACGTA  300
               |||||||||||||||||| |||||||||||||||||||||||||||||||||||||||||
Sbjct  983369  TGGACCCGCGGCGTATTAGCTAGTTGGTGAGGTAATGGCTCACCAAGGCGATGATACGTA  983310

Query  301     TCCGAACTGAGAGGTTGATCGGCCACTTTGGGACTGAGACACGGCCCAAACTCCTACGGG  360
                ||||||||||||||||||||||||| |||||||||||||||||||||||||||||||||
Sbjct  983309  GCCGAACTGAGAGGTTGATCGGCCACATTGGGACTGAGACACGGCCCAAACTCCTACGGG  983250

Query  361     AGGCAGCAGTAGGGAATCTTCCACAATGGACGCAAGTCTGATGGAGCAACGCCGCGTGAG  420
               ||||||||||||||||||||||||||||||||||||||||||||||||||||||||||||
Sbjct  983249  AGGCAGCAGTAGGGAATCTTCCACAATGGACGCAAGTCTGATGGAGCAACGCCGCGTGAG  983190

Query  421     TGAAGAAGGCTTTCGGGTCGTAAAACTCTGTTGTTGGAGAAGAATGGTCGGCAGAGTAAC  480
               ||||||||||||||||||||||||||||||||||||||||||||||||||||||||||||
Sbjct  983189  TGAAGAAGGCTTTCGGGTCGTAAAACTCTGTTGTTGGAGAAGAATGGTCGGCAGAGTAAC  983130

Query  481     TGTTGTCGGCGTGACGGTATCCAACCAGAAAGCCACGGCTAACTACGTGC  530
               ||||||||||||||||||||||||||||||||||||||||||||||||||
Sbjct  983129  TGTTGTCGGCGTGACGGTATCCAACCAGAAAGCCACGGCTAACTACGTGC  983080
```

Fig. 2 16S sequence of V101 strain

Table 2 Identification of V101 strain

Species name	Strain	ID	Similarity
Lactobacillus paracasei	IIA	CP014985.1	99%

Effects of fermentation time. Firstly, the effect of incubation time was screened from 12 to 96 h. Both biomass and acid production rose from 0.65 and 10.17 g/L to their peaks at 60 h – biomass 2.01 g/L and acid content 17.25 g/L then gradually declined. As a result, 60 h was chosen as the optimum incubation time for V101 for further experiments (Fig. 4).

Fig. 3 Biomass standard curve of V101

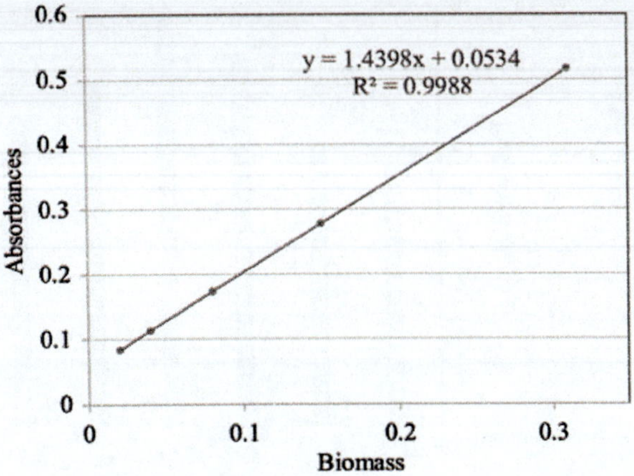

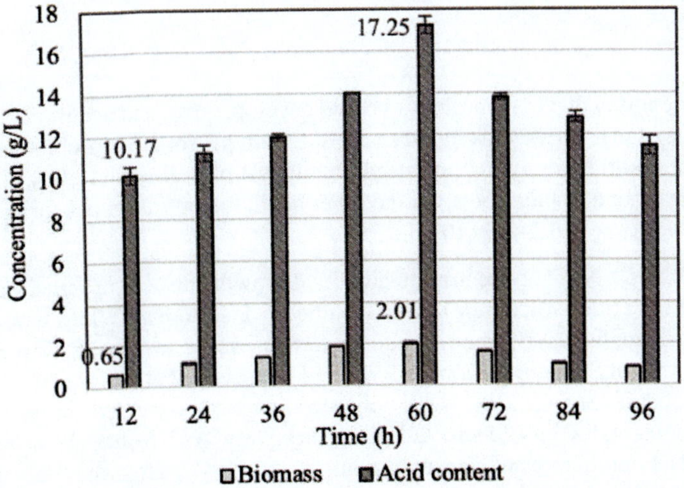

Fig. 4 Effects of fermentation time

In a previous study, it took only 19 h for LAB probiotics *Lactobacillus sakei* to produce the highest bacteriocin ST22Ch at 1600AU/mL and remained unchanged for the next 5 h before dropping down [13].

Effects of pH. Biomass and acid content were optimized against initial pH, from acidic conditions to slightly alkaline of pH range – from 4 to 8. Biomass accumulation, acid content produced by V101 strain climbed up to a high at pH 5.5, with biomass at 17.46 g/L and acid content at 2.31 g/L. pH higher than 5.5 then reduced biomass and acid accumulation of V101, to a low at pH with biomass at 9.66/L and acid content at 0.69 g/L. Hence, pH 5.5 was chosen as optimum pH and would be applied for the next experiments (Fig. 5).

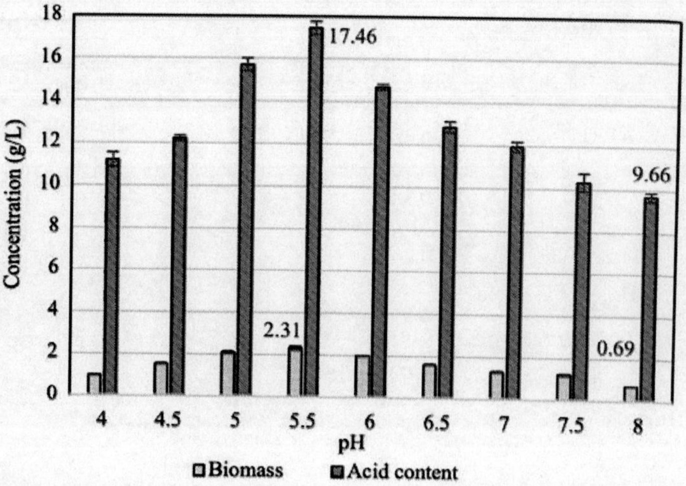

Fig. 5 Effects of pH

It was reported that LAB probiotics could develop, grow, and accumulate acid in a wide range of pH. pH 5.5 was found as the optimum pH for LAB probiotics. *L. sakei* from meat traditional product produced the highest bacteriocin at pH 4 [13], lower than results found in this study. On the other hand, the optimum pH for bacteriocin LAB probiotics was 6.2–8.5 [10].

Effects of NaCl. NaCl concentration did affect growth and acid production of V101 strain and the effect of its on V101 was studied from 3 to 8%. While biomass fluctuated, acid content performed an old pattern of rising to a peak then declined. However, they still exhibited correlation when both peaked at 6% of NaCl – 17.64 g biomass/L and 2.36 g acid/L. NaCl concentration of 8% yielded minimal growth and acid content, 9.66 g/L, and 1.22 g/L, respectively. As a result, 6% of NaCl was chosen as an optimum NaCl concentration.

It was reported that *Lactobacillus plantarum* from the shrimp gut performed great antibacterial activities against many harmful bacteria. It produced the highest bacteriocin at 0.9% NaCl [16], less than 1/6 of optimum NaCl concentration found in this study. Strain V101 here was isolated from quite salty fermented food such as Vietnamese pickles, sour shrimps. Thus this strain might prefer a higher NaCl concentration for its growth (Fig. 6).

Effects of carbon sources. Effects of various carbon sources on growth and acid production of V101 were screen on a medium containing mannitol, lactose, glucose, CMC, starch, molasses, and saccharose (Fig. 7).

Glucose beat other carbon sources to induce the best growth and acid production of the strain, with 3.08 g biomass/L and 18.21 g acid/L. CMC was the worst studied carbon source, yielded 4.13 g/L for acid content, only less than ¼ of that with medium prepared by glucose, and biomass only 1.68 g/L – nearly a half of that in medium

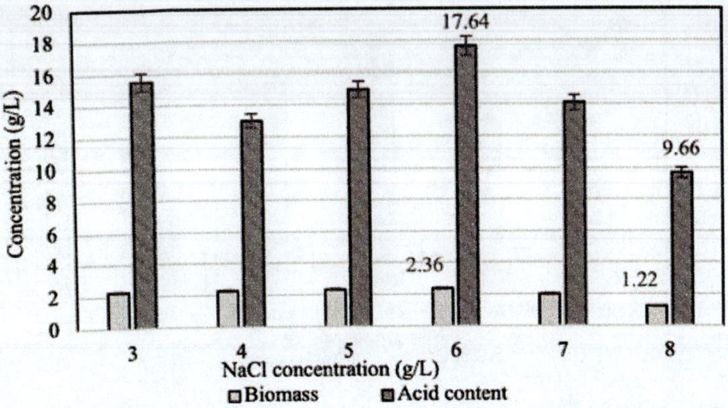

Fig. 6 Effects of NaCl concentrations

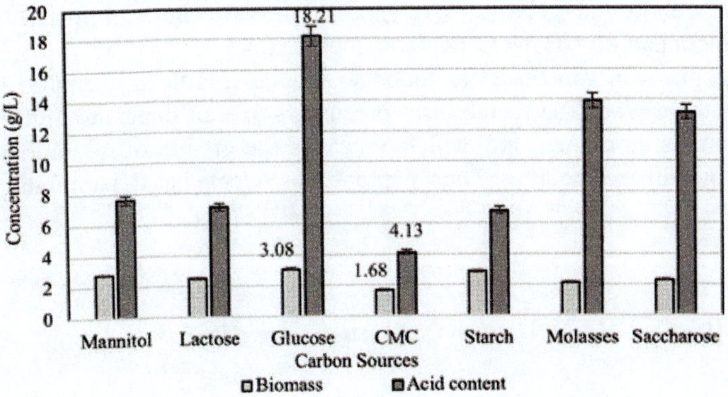

Fig. 7 Effects of carbon sources

with glucose. It could be explained that glucose was simple and ready to use carbon source as it was a monosaccharide and the main material for many main biochemical reactions in the cells.

CMC, a kind of cellulose, on the other hand, was not only a polysaccharide but also has a low conversion rate or slow breaking down reaction speed. *L. sakei* from meat traditional product produced the highest bacteriocin – 1600AU/mL on medium containing glucose, fructose, lactose, and saccharose [13]. This finding consistent with our results, as LAB PROBIOTICS, tended to use simple carbon sources as monosaccharides and disaccharides for its growth. All in all, glucose was chosen as an optimum carbon source for growing V101 for outstanding biomass and acid yield.

Effect of nitrogen sources. The effects of various nitrogen sources on growth and acid production of V101 were screened on a medium containing KNO_3, NH_4NO_3, peptone, meat extract, gelatin, and urea (Fig. 8).

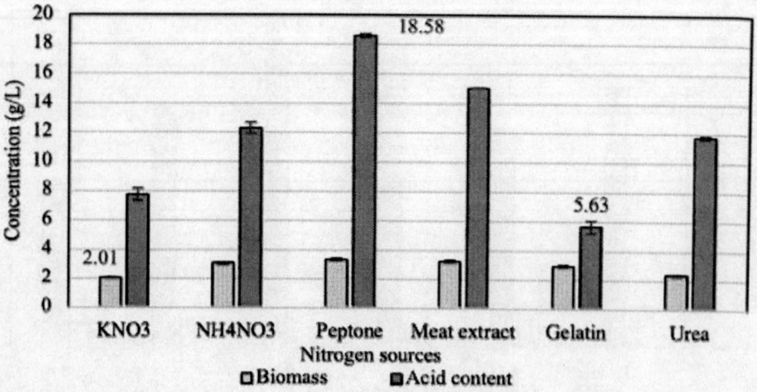

Fig. 8 Effects of nitrogen sources

Total acid content was highest on medium containing peptone, 5.63 g/L, and lowest on medium with gelatin, 18.58 g/L. Biomass was not significantly different in medium containing NH_4NO_3, peptone, meat extract.

Organic nitrogen sources were found to induce significantly higher inorganic sources as there would be metabolism precursors or trait concentration of growth factor, vitamins, hormones, etc. which enhanced the growth of microorganisms in dried organic nitrogen sources. Thus, peptone was selected as the optimum nitrogen source.

3.5 Culturing V101 Under Optimum Conditions

All in all, optimum conditions for culturing V101 strain were found as pH 5.5, 6% NaCl on medium using glucose and peptone as carbon and nitrogen sources in 60 h. Under optimum conditions, the V101 strain produced 3.55 g/L biomass and 19.44 g/L acid content, 76.61%, and 10.69% higher than previous non-optimal conditions, respectively (Table 3).

V101 acid production doubled that of L. *helveticus* and L. bulgaricus, which were 10.1 g/L and 9.6 g/L, respectively [17]. However, total acid content produced by the V101 strain was a little lower than that ability of *L. pentosus* ATCC 8041 – 21.8 g/L [18], or nearly a half of lactic acid produced by *L. bulgaricus* NRRL B-548-38.7 g/L [19]. Differences in acid production could be caused by different LAB strains, fermentation conditions, and substrates.

Table 3 Dry biomass and acid content in optimum culture condition

	Dry biomass (g/L)	Acid content (g/L)
Optimum conditions	3.55 ± 0.01	19.14 ± 0.16
Traditional MRS	2.01 ± 0.12	17.51 ± 0.13

The culture supernatant of V101 under optimum conditions was collected by centrifugation and applied to antibacterial assays. Large and clear inhibition zones were observed by both dual culture overlay assay and agar-well diffusion assay on agar plates against *E. coli* and *S. aureus* (Figs. 9 and 10).

Table 4 showed that most inhibition zones against both *E. coli* and *S. aureus* were moderately improved. Sizes were statistically significantly different.

Among 2 indicating bacteria, optimization of culturing conditions improved antibacterial of V101 against E. *coli* better and more impressive than *S. aureus.*

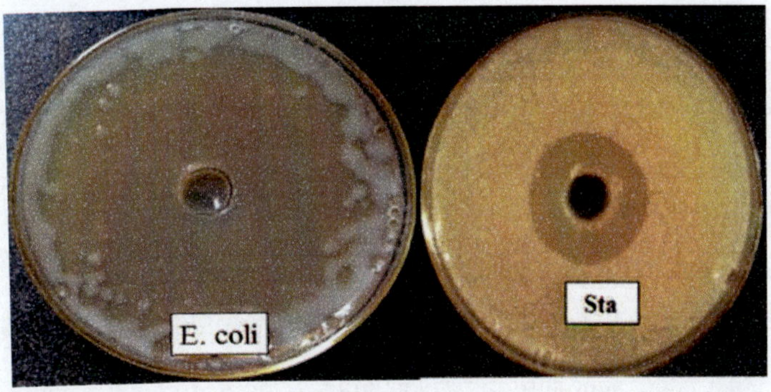

Fig. 9 Inhibition zones by agar-well diffusion assay

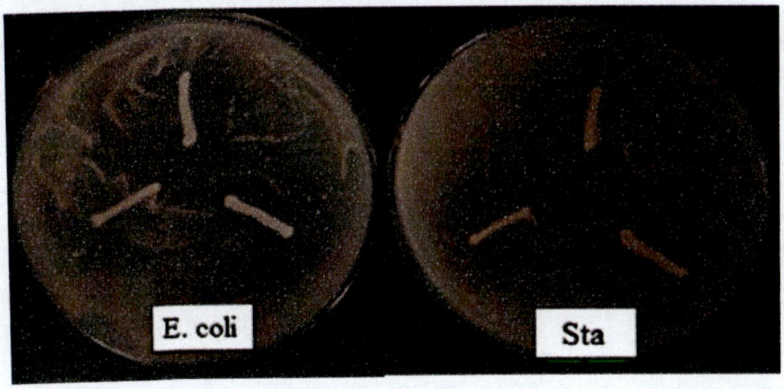

Fig. 10 Inhibition zones by dual culture overlayer assay

Table 4 Inhibition zones of indicator bacteria inhibited by V101 on optimum medium

	Dual culture overlay assay		Agar-well diffusion assay	
	E. coli	*S. aureus*	*E. coli*	*S. aureus*
Optimum conditions	65.5 ± 0.9	31.2 ± 0.5	35.5 ± 0.6	31.5 ± 0.8
Traditional MRS	63.3 ± 0.7	33.5 ± 0.3	29.6 ± 0.4	29.2 ± 0.1

Inhibition zones by agar-well diffusion assay against *E. coli* increased 19.93%, from 29.6 mm to 35.5 mm.

It could be explained that a 76.61% increase in biomass accumulation might lead to the secretion of more bacteriocins and other antibacterial bioactive compounds. Together 10.69% higher acid produced and secreted, supernatant from culture under optimum conditions performed significant improvement on antibacterial activities.

3.6 Pilot Fermentation

L. paracasei V101 was applied to ferment sour shrimp under pilot productions. Shrimps were prepared as the same protocol for traditional sour shrimp fermentation. Samples inoculated with 3% *L. paracasei* V101 was fermented and compared with control without adding *L. paracasei* V101. Results of changes in sensory assessments were summarized in Table 5.

It took less time for *L. paracasei* V101to ferment Vietnamese pickles and sour shrimp. When both cultures were not different on Day 1, fermentation was quicker in the sample with *L. paracasei* V101 with the appearance of broth and slightly good smell.

On the other hand, sour shrimp fermented by *L. paracasei* V101 could be well preserved almost double the preservation period of control samples. More impressively, during preservation time, the texture of sour shrimps fermented by *L. paracasei* V101 was firmer. It was until Day 35 that shrimps *L. paracasei* V101 start their degrading phase while control samples appeared degrading signals from Day 20. Moreover, the shrimp body was intact and remain their shape for a longer time (up to 35 days) while control shrimp lost their heads after 20 days and shrank their bodies at 35 days.

Acid accumulation in the culture of *L. paracasei* V101 fermented and control sour shrimps was recorded in Fig. 11.

The acid content in culture was recorded from Day 1 to Day 35. In general, the acid content in both cultures climbed up to a peak then declined. Acid contents in culture fermented with *L. paracasei* V101 were significantly higher than that of control culture in all studied periods. In *L. paracasei* V101 fermented culture, acid content rose to a high on Day 25, at 39.90 g/L, and decreased to around 32 g/L on Day 35. The acid content in the control batch was much lower. It peaked earlier on Day 20, only reached 29.16 g/L then declined. Lactic acid bacteria added to culture would increase the LAB probiotics population and thus, increase the capacity of total acid-producing of the culture. Higher acid content could explain why fermentation by *L. paracasei* V101 was quicker with longer preservation time than non-L. *paracasei* V101 adding samples.

As a result, products were more delicious and more impressive, could be preserved longer. Fermentation processes induced by potential *L. paracasei* V101 should be studied more carefully in the future to be applied as a helpful probiotics strain to

Table 5 Quality of sour shrimp fermented by *L. paracasei* through time

Time (day)	Sample	Sensory assessments				Shrimp status
		Color		Flavor		
		Shrimp	Broth	Shrimp	Broth	
1	Control	Transparent grey	Transparent grey	Fishy	No broth	Intact, firm texture
	L. paracasei	Transparent grey	Transparent grey	Fishy	No broth	Intact, firm texture
2	Control	Transparent grey	Coral	Fishy	Slight flavor	Intact, firm texture
	L. paracasei	Coral	Coral	Slight flavor	Slight flavor	Intact, firm texture
3	Control	Orange	Coral red	Sour, good flavor	Sour, good flavor	Intact, firm texture
	L. paracasei	Coral red	Coral red	Sour, good flavor	Sour, good flavor	Intact, firm texture
5	Control	Coral red	Coral red	Sour, good flavor	Sour, good flavor	Intact, firm texture
	L. paracasei	Coral red	Coral red	Sour, good flavor	Sour, good flavor	Intact, firm texture
20	Control	Coral red	Coral red	Too sour, good flavor	Too sour, good flavor	Head detached, loose texture
	L. paracasei	Coral red	Coral red	Too sour, good flavor	Too sour, good flavor	Intact, firm texture
35	Control	Coral red	Coral red	Too sour, good flavor	Too sour, good flavor	Head detached, loose texture
	L. paracasei	Coral red	Coral red	Too sour, good flavor	Too sour, good flavor	Head detached, firm texture

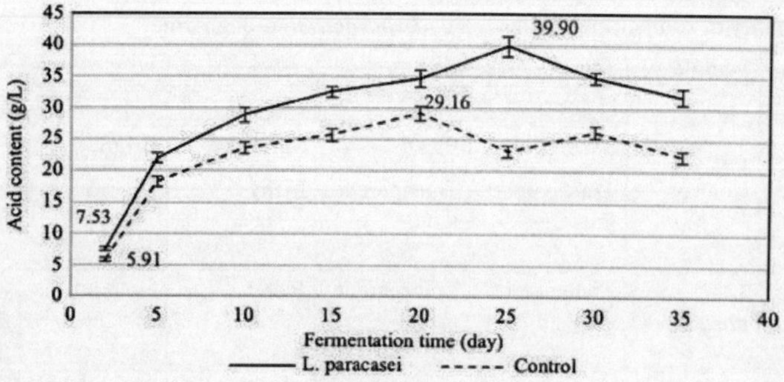

Fig. 11 Acid content in sour shrimp culture through time

bring good benefits for human health, traditional fermentation food improvement, and industrialization.

4 Conclusions

V101 strain isolated from traditional fermented food was identified as *L. paracasei.* It exhibited optimum cell growth and biomass, total acid production on modified MRS with 10 g/L meat extract, 5 g/L yeast extract, 10 g/L peptone, 20 g/L glucose, 5 g/L CH_3COONa, 2 g/L K_2HPO_4, 0.2 g/L $MgSO_4$, 0.05 g/L $MnSO_4$, 1 mL/L Tween 80, pH 5.5 and 60 h. Under optimal culture conditions, biomass and total acid accumulation increased by 76.61% and 10.69% respectively, and antimicrobial activities increased by 20%. *L. paracasei* V101 was a potential LAB probiotic for the improvement of fermented food and benefits for human health.

Conflicts of Interest The authors have no conflict of interest to declare.

References

1. Phong HX, Van TH, Thanh NN, Long BHD, Dung NTP (2017) Antifungal activity of lactic acid bacteria isolated from nem chua. Chem Biomol Eng 1(2):49
2. Kaškonienė V, Stankevičius M, Bimbiraitė-Survilienė K, Naujokaitytė G, Šernienė L, Mulkytė K, Malakauskas M, Maruška A (2017) Current state of purification, isolation, and analysis of bacteriocins produced by lactic acid bacteria. Appl Microbiol Biotechnol 101(4):1323–1335
3. Thomas LV, Ingram RE, Bevis HE, Davies EA, Milne CF, Delves-Broughton J (2002) Effective use of nisin to control *Bacillus* and *Clostridium* spoilage of a pasteurized mashed potato product. J Food Prot 65(10):1580–1585
4. Reis J, Paula A, Casarotti S, Penna A (2012) Lactic acid bacteria antimicrobial compounds: characteristics and applications. Food Eng Rev 4(2):124–140

5. Gautam N, Sharma N (2009) Bacteriocin: safest approach to preserve food products. Indian J Microbiol 49(3):204–211
6. Zhang B, Wang Y, Tan Z, Li Z, Jiao Z, Huang Q (2016) Screening of probiotic activities of lactobacilli strains isolated from traditional Tibetan Qula, a raw yak milk cheese. Asian Aust J Anim Sci 29(10):1490
7. Gharib SA (2020) Antimicrobial activity and probiotic properties of lactic acid bacteria isolated from traditional fermented dairy products. J Mod Res
8. De Las Cardoso M, Manzo RM, Tonarelli GG, Simonetta AC (2012) Characterisation of a cell-free supernatant obtained from cultures of *Enterococcus faecalis* DBFIQ E24 with antagonistic activity against bacteria, yeasts and moulds. Int J Dairy Technol 65(4):568–577
9. Magnusson J, Ström K, Roos S, Sjögren J, Schnürer J (2003) Broad and complex antifungal activity among environmental isolates of lactic acid bacteria. FEMS Microbiol Lett 219(1):129–135
10. Yang E, Fan L, Yan J, Jiang Y, Doucette C, Fillmore S, Walker B (2018) Influence of culture media, pH and temperature on growth and bacteriocin production of bacteriocinogenic lactic acid bacteria. AMB Exp 8(1):10
11. Khusniati T, Choliq A, Rahayu RD (2020) The organoleptic and physic characteristics and lactic acid contents of yoghurt with commercial starter added *Bifidobacteria bifidum*. J Bio Res 17:87–90
12. Sanchart C, Rattanaporn O, Haltrich D, Phukpattaranont P, Maneerat S (2017) Lactobacillus futsaii CS3, a new GABA-producing strain isolated from Thai fermented shrimp (Kung-Som). Indian J Microbiol 57(2):211–217
13. Todorov SD, Oliveira R, Vaz-Velho M (2012) Media optimization of bacteriocin ST22CH production by *Lactobacillus sakei* ST22CH isolated from salpicao, a traditional meat-product from Portugal. Chem Eng Trans 27:283–288
14. Arasu MV, Jung MW, Kim DH, Ilavenil S, Jane M, Park HS, Al-Dhabi NA, Jeon BT, Choi KC (2014) Enhancing nutritional quality of silage by fermentation with *Lactobacillus plantarum*. Indian J Microbiol 54(4):396–402
15. Arasu MV, Jung M-W, Kim DH, Ilavenil S, Jane M, Park HS, Al-Dhabi NA, Jeon BT, Choi KC (2014) Enhancing nutritional quality of silage by fermentation with Lactobacillus plantarum. Indian J Microbiol 54(4):396–402
16. Karthikeyan V, Santosh S (2009) Isolation and partial characterization of bacteriocin produced from *Lactobacillus plantarum*. Afr J Microbio Res 3(5):233–239
17. Plessas S, Bosnea L, Psarianos C, Koutinas A, Marchant R, Banat I (2008) Lactic acid production by mixed cultures of *Kluyveromyces marxianus*, *Lactobacillus delbrueckii* ssp. bulgaricus and *Lactobacillus helveticus*. Bioresour Technol 99(13):5951–5955
18. Bustos G, Moldes AB, Cruz JM, Domínguez JM (2004) Production of fermentable media from vine-trimming wastes and bioconversion into lactic acid by *Lactobacillus pentosus*. J Sci Food Agric 84(15):2105–2112
19. Burgos-Rubio CN, Okos MR, Wankat PC (2000) Kinetic study of the conversion of different substrates to lactic acid using *Lactobacillus bulgaricus*. Biotechnol Prog 16(3):305–314

Investigation Fermentation of Oolong for Stress Suppressing Drink and Enriching GABA—A Neurological Nourishing Compound

Ngan Thi-Thuy Truong, Khanh Tu My Tuong, Minh Thai Nguyen, Hanh Thi-Ngoc Nguyen, and Oanh Ngoc Huynh

Abstract Nowadays, the modern living lifestyle made the number of stressful people increasing day by day. This strategy brought up a problem of overusing stress managing pills, whose side effects are weight gains, changes of appetite and sleep habits, frustration, etc. As a result, there is a need for natural antidepressant remedies for the community to reduce such stressful situations. Tea (*Camellia sinensis*) becomes one of the most popular drinks worldwide because of its effective medicinal properties such as antioxidant, antitumor, antidepression, cancer and cardiovascular diseases prevention, etc. However, fresh tea might be quite bitter due to high tannin content, hard to preserve and insomnia causing. Fortunately, fermented tea could be another better choice as it reduced tannin content, increased fragrant and eye-catching color, together with the richness of γ-aminobutyric acid (GABA), a neuro-benefit compound, could overcome those downsides. In this study, tea was screened for enzyme inactivation conditions, then fermented for good color ratios, lower tannin content, and higher GABA content. Oolong tea inoculated with 8% *L. brevis* and fermented in 4 days yielded better GABA content than fermentation of Oolong tea with and without Viscozyme L. Under these conditions, fermented tea had a good sensory assessment with color, flavor almost like Pu-erh tea. Other attributions of fermented tea would be TF/TR at 1/9.7, TC at 1.53%, tannin content at 13.72%, and GABA content at 302.38 mg/kg, almost 4 times higher than GABA content in raw material. This finding is a requisite for further production of GABA rich tea to create widely consuming and naturally stress suppressing drinks.

Keywords GABA · Oolong · Tea · *Camellia sinensis* · *L. brevis*

N. T.-T. Truong · K. T. M. Tuong · O. N. Huynh (✉)
Department of Biotechnology, Faculty of Chemical Engineering, Bach Khoa University, Vietnam National University Ho Chi Minh City, Ho Chi Minh City 70000, Vietnam
e-mail: ngocoanh_cnsh@hcmut.edu.vn

M. T. Nguyen · H. T.-N. Nguyen
Hong Bang International University, Ho Chi Minh City 70000, Vietnam

V. Van Toi et al. (eds.), *8th International Conference on the Development of Biomedical Engineering in Vietnam*, IFMBE Proceedings 85,
https://doi.org/10.1007/978-3-030-75506-5_56

1 Introduction

In modern life, 10–20% of the population are affected by mental health diseases, mostly anxiety and depression [1]. Depression, known as a psychiatric disease characterized by mood disorders, caused by disturbed brain activity or a psychological factor that triggers abnormal changes in behavioral thinking [2], affects 264 million people worldwide [3]. Long-term use of sedatives or antidepressants will cause many side effects such as withdrawal, committing suicide, emotional blunting, etc. [4], or could reduce the secretion of regulators or nerve stabilizers in the body [5, 6].

Tea, water broth from *Camellia sinensis*, is the most popular drink worldwide [7]. Tea has many groups of bioactive compounds such as catechins, phenols, tannins, alkaloids, etc., which bring various benefits to human health thanks to their antioxidant, antitumor, antidepression, cancer and cardiovascular diseases prevention, etc. activities [8, 9]. In fermented tea such as Pu-erh and Oolong tea, Theaflavin (TF) and Thearubigin (TR) are critical compounds responsible for the beautiful color of the tea. TF and TR are yellow and reddish-brown compounds from the oxidation of polyphenol-catechin L-epigallocatechin (L-EGC) and L-epigallocatechin gallate (L-EGCG) during tea fermentation [7]. Recent researches showed that TF/TR ratio should be between 1/8.8 to 1/11.4 for the tea broth to have shiny transparent specific for fermented tea [10, 11]. TR and TF were proved to in vitro inhibit HCT 116 colon cancer cells and HT 460 lung cancer cells and could induce apoptosis in both cultures [8]. The fermented tea is also rich in γ-aminobutyric acid (GABA). It is synthesized from glutamic acid in our brains, helps maintain normal activities of our brains, regulates the activities of neurons. It could reduce stress, prevent insomnia, premenstrual syndrome, etc. [12–14].

In this study, Oolong tea was fermented to produce Pu-erh tea-like product together with enriching GABA content to create widely consumed natural antidepressant drinks.

2 Materials and Methods

2.1 Materials

Oolong tea was from CJ Cau Tre Company.

Viscozyme L (arabanase, cellulase, β-glucanase, hemicellulase, and xylanase from *Aspergillus aculeatus*) was purchased from Brenntag.

Lactobacillus brevis was from Hochiminh City University of Agriculture and Forestry.

MRS medium was from HiMedia.

Phenol, $KMnO_4$, Na_2HPO_4, boric acid, NaCl, Sodium tetraborate were from Xilong. Isobutyl mcthyl ketone (IBMK), n-butanol were from Chemsol.

2.2 *GABA Enriching Fermentation of Oolong Tea*

Oolong tea was adjusted to 45% moisture. Then it was spread to bulk with a thickness of 25 cm, covered with a sheet to maintain moisture and heat, and fermented at 43 °C in 7 days as bulk fermentation.

2.3 *GABA Enriching Bulk Fermentation of Oolong Tea Supplied with Viscozyme L*

Oolong tea was supplied with 10% of Viscozyme L and adjusted to 45% moisture. Then it was fermented under previous conditions. Volumes of enzyme and water needed for moisture adjustment were calculated by:

$$45\% = \frac{x\% \times m + y}{m + y} \times 100\%$$

m: the initial weight of tea (g).
x: the initial moisture of tea (%).
y: the weight of Viscozyme solution (g).

2.4 *GABA Enriching Bulk Fermentation of Oolong Tea Inoculated with* L. brevis

***L. brevis* culture**. *L. brevis* was kept on the MRS slant. A single colony of *L. brevis* was cultured on MRS broth in 48 h at 38 °C for seeding tea fermentation.

GABA enriching bulk fermentation of Oolong tea inoculated with *L. brevis*. Oolong tea was adjusted to 45% moisture. Then it was inoculated with 2, 4, 6, 8, 10% of *L. brevis* liquid MRS culture, mixed well, and fermented under previous conditions.

2.5 Determination of Tannin Content

Tannin content was determined by titration with $KMnO_4$ 0,1 N followed by Lowenthal Permanganate methodology with Indigo carmine as a color indicator. Briefly, 2 g tea was extracted in 80 ml boiling water, then adjusted to 250 ml to make tea extract. An amount of 1 ml of tea extract was added to Erlene with 75 ml distilled water and 2.5 ml indigo carmine 0.1% in acid. The solution was then titrated with $KMnO_4$ 0.1 N until the blue color evaded. The control sample was prepared with 1 ml of tea extract boiled with activated coal. Under acidic conditions, tannin was oxidized by $KMnO_4$ by the below reaction.

$$\text{Tannin} + KMnO_4 \rightarrow CO_2 + H_2O$$

Tannin content was calculated by the following formula:

$$X = \frac{V \cdot 0.00582 \cdot (a - b)}{v \cdot m} \cdot 100$$

X (%): Tannin content (percentage on dry biomass).
a: Volume of $KMnO_4$ 0.1 N used for titration of sample.
b: Volume of $KMnO_4$ 0.1 N used for titration of control;
V: Volume of tea broth from 2 g tea (250 ml);
v: Volume of analyzed tea sample (1 ml);
m: Weight of tea sample;
0.00582: Tannin constant (1 ml $KMnO_4$ could oxidize 0.00582 g tannin compounds).

2.6 Determination of Total Color (TC)

An amount of 1 ml of tea extract was diluted with 9 ml of distilled water and measured at 460 nm. The result was recorded as the below formula [15]:

$$TC(\%) = \frac{A_{460} \times 10}{m} \times 100\%$$

A_{460}: Absorbance at 460 nm.
M: the weight of tea sample (g).

2.7 Determination of Thearubigins (TR) and Theaflavin (TF)

TF and TR content was determined by the modified Takeo and Oosawa method. TF could be dissolved in isobutyl methyl ketone (IBMK) and Na_2HPO_4, while TR was water-soluble but could not be dissolved in IBMK. Tea extract was mixed with the same amount of IBMK at room temperature to obtain the upper (IBMK) phase containing TF and water phase containing TR. The upper phase (MIBK) was diluted 10 times by 45% ethanol and measured OD 380 nm (A). Then 10 ml of upper phase was extracted with the same amount of 2.5% Na_2HPO_4 to collect the IBMK phase, again was diluted 10 times and measured OD at 380 nm (B). The lower phase or water phase was extracted with the same volume of n-butanol. Then n-butanol layer was collected, diluted 10 times with ethanol, and measured OD at 380 nm (C) [16]. TF and TR were calculated by the following formulas:

$$TF = 4.313 \times B$$
$$TR = 13.643(A + C - B)$$

2.8 *Determination of GABA Content*

In alkaline solution, ($-NH_2$) group in GABA reacted with (OCl–) of HClO to form monochloramine, which reacted with 2 phenol groups to form indophenols, having a specific color of the reaction. An amount of 0.6 ml tea extract was added to 0.4 ml borate buffer and 2 ml phenol 6%, followed by 1 min vigorous vortex and 5 min cooling at 20 °C. Then 1.6 ml NaOCl 8% was added to the solution, followed by 1 min vigorous vortex and 5 min cooling at 20 °C. OD was measured at 630 nm [12].

2.9 *Sensory Assessment*

All sensory experiments were carried out in Sensory Laboratory, Cau Tre Food Inc. Tea was tasted and tested based on standard sensory assessment criteria.

2.10 *Data Analysis*

All experiments were triplicated and analyzed by SPSS with $P < 0.05$.

3 Results and Discussions

3.1 *Quality of Oolong Tea Sample Prepared for Fermentation*

The Oolong tea samples prepared for fermentation were initially recorded with a moisture content of 10.67% and a Tannin level of 19.4%.

3.2 *Investigate the Fluctuation of Physical and Chemical Components of Oolong Tea in the Bulk Fermentation*

Temperature fluctuation in the bulk fermentation process. According to Fig. 1, the temperature increased gradually from 30.3 to 33.6 °C in the first two days of the fermentation before getting a peak at 45.7 °C on day 3.

After that, the temperature was decreased to 36.6 °C and 32.7 °C on day 4 and day 5, respectively. It was recognized that fermentation got the most intense reaction at day 3, known as the exponential phase of the reaction, and then turned to the decline phase after 4 days.

The fluctuation of TF, TR, and TC (total color) content in bulk fermentation. In Table 1, the result TF/TR on day 3 at 1/11.9 was quite similar to the standard rate of the Pu-erh tea, so it can be concluded that the time for stopping the fermentation would be on the 3rd day of processing.

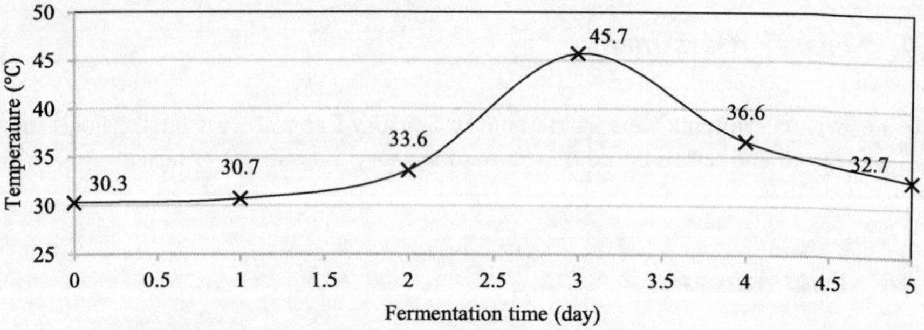

Fig. 1 Changes in temperature of bulk fermentation of Oolong tea

Table 1 TF/TR ratio of enzyme inactivated Oolong tea

Fermentation time (day)	0	1	2	3	4	5
TF/TR	1/64.3	1/54.5	1/15.3	**1/11.9**	1/12.2	1/21.3

bold indicates color of pu-erh tea

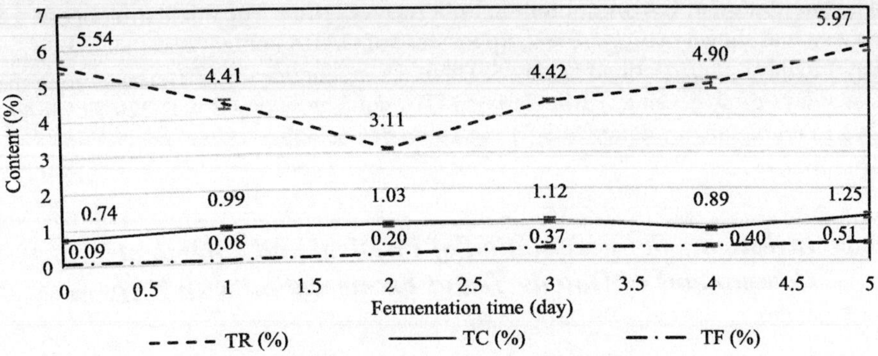

Fig. 2 TR, TF, TC content of bulk fermentation of Oolong tea

As showed in Fig. 2, the TF content was increased continuously from 0.086% to 0.515% corresponding to day 0 to day 5 of the process, while TR content has fluctuated unstably. The ratio TF/TR was slightly uptrend from 1/64.3 at day 0 to 1/11.9 on day 3 of the process. The rate TF/TR was not significantly different among day 3, day 4, and day 5. As a result, the total color (TC) content of tea samples increased gradually from the first day of incubation to day 3 and it quite fluctuated on day 4 and day 5.

The correlation between the total color TC and TF/TR ratio was recognized: when the TC value was high, the ratio of TF/TR would be low and vice versa. The higher the TF/TR ratio, the clearer and brighter the color of the tea water.

The fluctuation of tannin content in bulk fermentation. Figure 3 showed that the tannin content was decreased continuously from 19.4% on the first day till 8.7% after 4 days of fermentation. There was no clear difference between tannin contents on day 3 and day 4 in the tea after fermented.

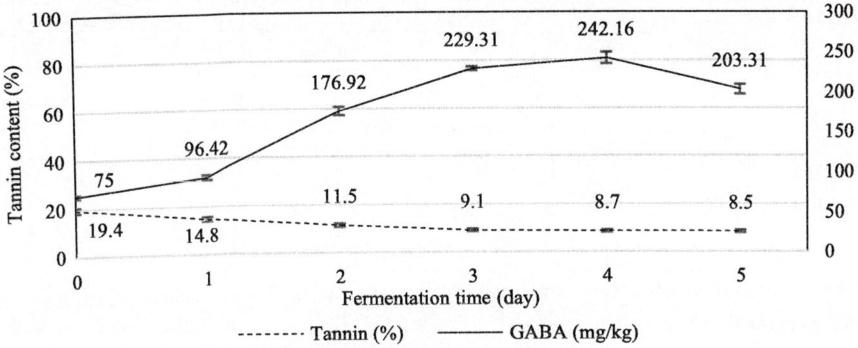

Fig. 3 Tannin and GABA content of bulk fermentation of Oolong tea

The fluctuation of GABA content in bulk fermentation. The fluctuation of GABA content was shown in Fig. 3, the sharp increase of GABA content was recorded from day 1 to day 4 then declined on day 5. There was not noticeably different among thc last 3 days of processing. Particularly, GABA content on day 3 was 229.31 mg/kg, 242.16 mg/kg on day 4, and 203.31 mg/kg on day 5.

3.3 *Investigate the Fluctuation of Physical and Chemical Components of Oolong Tea in Fermentation with Enzyme*

The fluctuation of TF, TR, and TC content in fermentation with the enzyme. The color of the tea after fermented with enzyme was shown in Fig. 4. The pattern was in order from left to right: from day 2 to day 7 of fermentation (Table 2).

As showed in Fig. 4, the TF content as well as total color content increased constantly from the first day to the 7th day of fermentation with the enzyme.

The ratio TF/TR between day 6 and day 7 was not significantly different (1/9.1 and 1/10.9, respectively), so the fermentation process can stop at day 6 for time-saving. Comparing to the bulk fermentation time—only 3 days, the fermentation with enzyme took longer due to the characteristic of the tea that has been stored for a long time.

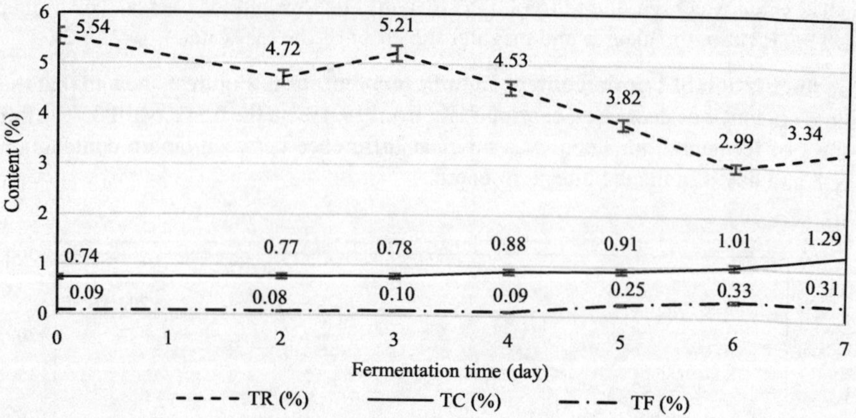

Fig. 4 TR, TF, TC content of bulk fermentation of Oolong tea with enzyme

Table 2 TF/TR ratio of enzyme inactivated Oolong tea in 7 days of fermentation with enzyme

Fermentation time (day)	0	2	3	4	5	6	7
TF/TR	1/64.3	1/60.8	1/54.9	1/51	1/15.2	**1/9.1**	**1/10.9**

bold indicates color of pu-erh tea

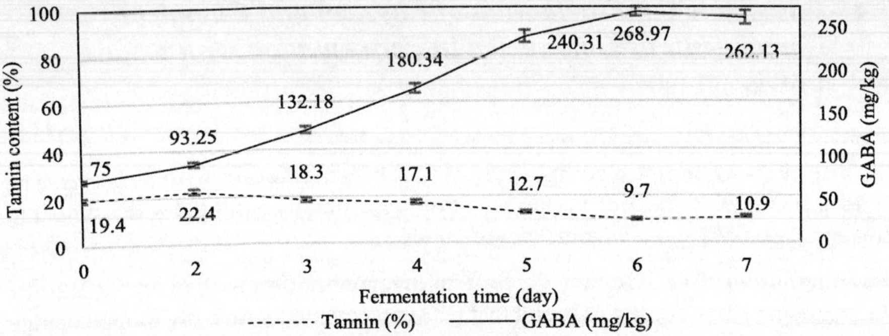

Fig. 5 Tannin and GABA content of bulk fermentation of Oolong tea with enzyme

The fluctuation tannin content in fermentation with the enzyme. Tannin content was reduced constantly after 6 days from 22.4% at day 2 down to 9.7% on day 6 when fermented with the enzyme. There was a quite upturn in the level of tannin to 10.9% on day 7, but it was not significantly different from that of day 6. The information was shown in Fig. 5.

Comparing to the tannin content in bulk fermentation at 9.1% after 3 fermented days while the tannin content in fermentation with enzyme was still high at 18.3%.

The fluctuation of GABA content in fermentation with the enzyme. GABA level climbed from Day 0 to its peak, 268.97 mg/kg, on Day 6 then remained unchanged d in day 7 as there was no significant difference in GABA content between days 6 and 7.

Comparing Oolong tea in bulk fermentation to fermentation with the enzyme. The GABA content in tea that fermented with enzyme supplement reached a high level similar to in bulk fermentation and also got nearly 4 times higher when compared to the raw tea sample. Moreover, the tannin content in raw tea samples significantly decreased from 19.4% to 9.1% and 9.7% in bulk fermentation and enzyme supplemented fermentation, respectively. This shows that the tea's acridness is reduced as well as the broth's color is brighter when fermented.

It was reported that the enzyme had been using in tea processing before tea fermentation, before tea extraction, and during extraction to enhance solubility as well as extraction yields [7]. Viscozyme, an enzyme cocktail containing hemicellulase, cellulase, xylanase weakened cell walls of tea leaves, promotes releasing of bioactive compounds inside tea leaves and speeded up fermentation time. GABA content in tea added with 10% Viscozyme L was 17% higher than that in fermented non-enzyme-added tea.

3.4 *Investigate the Fluctuation of Physical and Chemical Components of Oolong Tea in Fermentation with* L. brevis

Quality of the Oolong tea sample prepared for fermentation with *L. brevis*. The Oolong tea samples prepared for fermentation were initially recorded with a moisture content of 10.67% and a Tannin level of 19.4%.

The fluctuation of TF, TR, and TC content in fermentation with *L. brevis*. Results were shown in Table 3. It was found that tea supplemented with a higher percentage of microorganisms resulted in a TF/TR ratio closer to the requirement. Particularly, on the 4th day of fermentation, samples of 8 and 10% microorganism supplement have reached close to the ratio of 1/12 ÷ 1/10 and there was not significantly different to the results on the 5th day.

Compared to samples that fermented with enzyme supplementation (6 days), the fermentation time of the microbial supplement samples was faster (4 days), due to the assistance of microorganisms in metabolizing the color compounds faster.

The fluctuation of TF and TR content was closely related to the activity of microorganisms. By the activation of the microbial enzyme system concurrently with the PPO enzyme, tea leaf cells were broken down, as the result, these enzymes would come into and contact with the substance in tea to convert tannins into TF and TR pigments.

The total color (TC) variation correlates, in Table 4, with the variation of the TF/TR ratio. On the 4th day of fermentation with microorganisms, the TC contents were 1.53 and 1.74 in samples in cases of 8 and 10% microorganisms supplement. There were no significant differences between day 4 and day 5 in samples with different microorganisms rate. As the recording, it was found that the TC content reached great value and was higher than other achieved samples in both bulk fermentation and fermentation with the enzyme.

The fluctuation of tannin content in fermentation with *L. brevis*. The tannin content decreased gradually every 24 h of fermentation and reached 12.17% after 4 days treated with an 8% microorganisms supplement. The tannin content trend was still down on day 5, but it was not significantly different among samples on day 4

Table 3 TF/TR ratio of Oolong supplied with *L. brevis* in 5 days of fermentation

Percentages of microorganisms (%)	Day 1	Day 2	Day 3	Day 4	Day 5
2	1/42.7	1/81.3	1/17.5	1/13.6	1/12.8
4	1/55.2	1/22.6	1/18.6	1/16.8	1/14.9
6	1/37.6	1/36.3	1/34.4	1/23.9	1/22.5
8	1/77.7	1/25.1	1/13.9	**1/9.7**	1/10.2
10	1/46.4	1/27.2	1/17.1	**1/12.2**	1/11.6

bold indicates color of pu-erh tea

Table 4 TC (%) of Oolong supplied with *L. brevis* in 5 days of fermentation

Percentages of microorganisms (%)	Day 1	Day 2	Day 3	Day 4	Day 5
2	1.24 ± 0.03	1.24 ± 0.02	1.53 ± 0.03	1.53 ± 0.03	1.58 ± 0.04
4	0.96 ± 0.02	1.33 ± 0.04	1.52 ± 0.01	1.50 ± 0.02	1.53 ± 0.02
6	1.36 ± 0.03	1.35 ± 0.02	1.45 ± 0.02	1.75 ± 0.03	1.84 ± 0.04
8	1.24 ± 0.02	1.23 ± 0.03	1.28 ± 0.03	**1.53** ± 0.03	1.56 ± 0.05
10	1.57 ± 0.03	1.33 ± 0.02	1.34 ± 0.04	**1.74** ± 0.06	1.82 ± 0.04

bold indicates color of pu-erh tea

and day 5. On day 4, 10% of microorganisms supplement could bring the result of 13.54% tannin content and it was rather higher when compared to the group of 8% microorganisms supplement. Therefore, tea was fermented with 8% microorganisms supplement in 4 days could bring an efficient result. The reduction of tannin content during fermentation could be explained by the microbial enzyme system to transfer the tannin into TF and TR color compounds. The fluctuation of tannin content was shown in Table 5.

The fluctuation of GABA content in fermentation with *L. brevis*. GABA content was presented in Table 6. Based on the results of the TF/TR ratio in Table 3, corresponding to the ratio of 1/9.7 of the 4th day of fermentation using the microbiological ratio of 8%, the GABA content was quantified as 302.38 mg/kg. After that, on the 5th day of fermentation, GABA values were significantly reduced to 272.13 mg/kg and still high when compared to other samples on that day with various percentages of microorganisms supplement. GABA content in sample inoculated with 10% of *L. brevis* on day 4 was a bit higher—313.21 mg/kg. However, this increment was small in comparison to the cost of increasing the microbial inoculum, together with extended time for fermentation.

Table 5 Tannin content (%) of Oolong supplied with *L. brevis* in 5 days of fermentation

Percentages of microorganisms (%)	Day 1	Day 2	Day 3	Day 4	Day 5
2	25.46 ± 0.76	20.07 ± 0.54	18.81 ± 0.58	13.64 ± 0.45	14.22 ± 0.51
4	22.15 ± 0.65	18.81 ± 0.63	18.33 ± 0.42	13.64 ± 0.57	13.76 ± 0.53
6	22.54 ± 0.71	18.33 ± 0.49	17.20 ± 0.67	16.67 ± 0.69	15.23 ± 0.66
8	22.54 ± 0.68	18.81 ± 0.66	13.72 ± 0.53	**12.17** ± 0.38	11.87 ± 0.48
10	20.07 ± 0.37	17.06 ± 0.59	16.42 ± 0.75	**13.54** ± 0.62	13.15 ± 0.61

bold indicates color of pu-erh tea

Table 6 GABA content (mg/kg) of Oolong supplied with *L. brevis* in 5 days of fermentation

Percentages of microorganisms (%)	Day 1	Day 2	Day 3	Day 4	Day 5
2	102.17 ± 3.06	104.23 ± 3.23	124.74 ± 5.03	132.23 ± 4.61	97.33 ± 3.15
4	103.64 ± 3.15	123.21 ± 5.72	138.12 ± 4.71	142.27 ± 4.77	158.29 ± 4.22
6	83.33 ± 4.21	129.33 ± 4.63	154.27 ± 3.82	192.54 ± 5.06	234.17 ± 5.82
8	102.68 ± 4.01	142.86 ± 5.51	231.42 ± 5.83	**302.38** ± 7.21	272.13 ± 5.49
10	101.37 ± 3.68	156.28 ± 6.29	259.87 ± 6.44	**313.21** ± 5.18	264.86 ± 6.13

bold indicates color of pu-erh tea

In the research of R. Dhakal and colleagues, the microorganisms that can produce GABA mostly were lactic acid bacteria, their growth could inhibit food spoilage pathogens and act as probiotics in the digestive tract [17].

The GABA content in raw tea samples (raw materials) was 75 mg/kg. After adding 2, 4, 6, 8, and 10%, microorganisms, the GABA content increased and during the survey, the additional sample of 8% microorganisms seem to be the best as it has a good TF/TR ratio, together with almost optimum GABA content of all surveyed samples. GABA content reached and stabilized at 302.38 mg/kg in that group.

The strain *L. brevis*, which belonged to the lactic acid bacteria group, has been used to produce GABA and is known as the most sufficient method. According to Namura M., lactic acid bacteria have special physiological activity, which contributes to components of intestinal bacteria and helps to fight germs in the body. This strain of Lactic bacteria is quite safe for use in purposes of human health care [18]. The application of these microorganisms in the tea production process would create a new medicinal product or also be a good probiotic drink for the immune system. In the study of Yukihiro Nobuta et al., *L. brevis* KB290 was tested for effectiveness and safety as a probiotic [19].

3.5 *Sensory Assessment the Oolong Fermented with* L. brevis

The Oolong tea which fermented in bulk was not much different in flavor as well as color when compared to the raw tea production. However, the tea when fermented with an 8% *Lactobacillus* supplement can bring both the flavor and color similar to the famous Pu-erh tea. The color of the tea samples was showed in Fig. 6.

In a previous study, Pu-erh fermentation time can last from 42 to 180 days [20], which yielded longer the fermenting time, lighter the taste and darker color of tea water. According to the results in this study, the fermentation time lasts for about 3 or 4 days and there may be a huge difference to other researches. There may be due

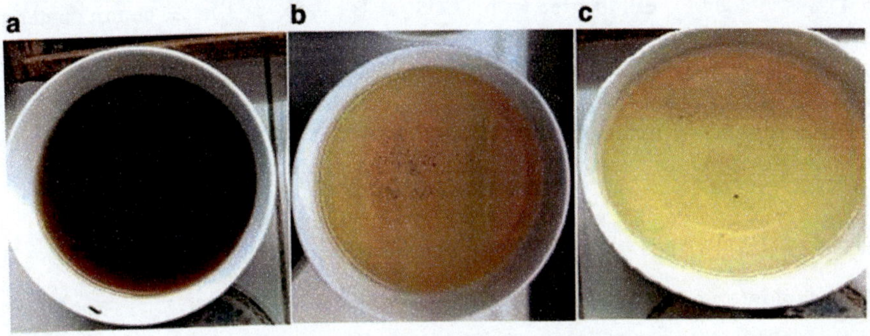

Fig. 6 Cau Tre Pu-erh tea (**a**), Oolong supplied with 8% *L. brevis* (**b**) and bulk fermented Oolong (**c**)

Fig. 7 Oolong supplied with 8% *L. brevis* fermented in 4 days

to many reasons such as the humidity in the experimental area was quite high, the ambient temperature was different from the locations of the above studies, and also differences in tea types. Sensory evaluation of the tea sample which supplemented 8% microorganisms can bring close similarities to Pu-erh tea both in color and flavor. Consequently, fermentation with *Lactobacillus* supplement should be suggested in creating probiotic tea products for health benefits (Fig. 7).

4 Conclusions

Oolong tea fermentation of tea added with microorganisms was better than one added with Viscozyme L and non-enzyme nor microorganism supported sample. It was found that *L. brevis* speed up the fermentation of Oolong tea, with fermentation time only 4 days to reach good TF/TR ratio, low tannin content, and high GABA content. Amongst all studied samples, Oolong tea inoculated with 8% *L. brevis* fermented in 4 days yielded Pu-erh like tea with TF/TR at 1/9.7, TC at 1.53%, tannin content at 12.17%, and GABA content at 302.38 mg/kg, almost 4 times higher than GABA content in raw material, around 75 mg/kg. However, mechanisms on how *L. brevis* could speed up and improve the fermentation, nor other factors affecting the process were not studied. It could be suggested that this GABA enriching process should be studied more and level up to a larger scale heading to creating good health products.

Acknowledgements We would like to thank Ho Chi Minh City University of Technology (HCMUT), VNU-HCM for the support of time and facilities for this study.

Conflicts of Interest The authors have no conflict of interest to declare.

References

1. Keles B, McCrae N, Grealish A (2020) A systematic review: the influence of social media on depression, anxiety, and psychological distress in adolescents. Int J Adolesc Youth 25(1):79–93
2. Kok RM, Reynolds CF (2017) Management of depression in older adults: a review. JAMA 317(20):2114–2122
3. James SL, Abate D, Abate KH, Abay SM, Abbafati C, Abbasi N, Abbastabar H, Abd-Allah F, Abdela J, Abdelalim A (2018) Global, regional, and national incidence, prevalence, and years lived with disability for 354 diseases and injuries for 195 countries and territories, 1990–2017: a systematic analysis for the Global Burden of Disease Study 2017. Lancet 392(10159):1789–1858
4. Read J, Williams J (2018) Adverse effects of antidepressants reported by a large international cohort: emotional blunting, suicidality, and withdrawal effects. Curr Drug Saf 13(3):176–186
5. Wichniak A, Wierzbicka A, Walęcka M, Jernajczyk W (2017) Effects of antidepressants on sleep. Curr Psychiatry Rep 19(9):63
6. Fuchs T, Jefferson SJ, Hooper A, Yee P-H, Maguire J, Luscher B (2017) Disinhibition of somatostatin-positive GABAergic interneurons results in an anxiolytic and antidepressant-like brain state. Mol Psychiatry 22(6):920–930
7. Chandini SK, Rao LJ, Gowthaman M, Haware D, Subramanian R (2011) Enzymatic treatment to improve the quality of black tea extracts. Food Chem 127(3):1039–1045
8. Imran A, Butt MS, Xiao H, Imran M, Rauf A, Mubarak MS, Ramadan MF (2019) Inhibitory effect of black tea (*Camellia sinensis*) theaflavins and thearubigins against HCT 116 colon cancer cells and HT 460 lung cancer cells. J Food Biochem 43(5):e12822
9. Khan N, Mukhtar H (2007) Tea polyphenols for health promotion. Life Sci 81(7):519–533
10. Someswararao C, Srivastav P (2012) A novel technology for production of instant tea powder from the existing black tea manufacturing process. Innov Food Sci Emerg Technol 16:143–147
11. Teshome K (2019) Effect of tea processing methods on biochemical composition and sensory quality of black tea (*Camellia sinensis* (L.) O. Kuntze): a review. J Hortic For 11(6):84–95
12. Karladee D, Suriyong S (2012) γ-aminobutyric acid (GABA) content in different varieties of brown rice during germination. Sci Asia 38(1):13–17
13. Liao J, Wu X, Xing Z, Li Q, Duan Y, Fang W, Zhu X (2017) γ-aminobutyric acid (GABA) accumulation in tea (*Camellia sinensis* L.) through the GABA shunt and polyamine degradation pathways under anoxia. J Agric Food Chem 65(14):3013–3018
14. Rashmi D, Zanan R, John S, Khandagale K, Nadaf A (2018) γ-aminobutyric acid (GABA): biosynthesis, role, commercial production, and applications. Stud Nat Prod Chem 57:413–452
15. Joshi R, Babu GK, Gulati A (2013) Effect of decaffeination conditions on quality parameters of Kangra orthodox black tea. Food Res Int 53(2):693–703
16. Jayabalan R, Marimuthu S, Swaminathan K (2007) Changes in content of organic acids and tea polyphenols during kombucha tea fermentation. Food Chem 102(1):392–398
17. Dhakal R, Bajpai VK, Baek K-H (2012) Production of GABA (γ-aminobutyric acid) by microorganisms: a review. Braz J Microbiol 43(4):1230–1241
18. Nomura M, Kimoto H, Someya Y, Furukawa S, Suzuki I (1998) Production of γ-aminobutyric acid by cheese starters during cheese ripening. J Dairy Sci 81(6):1486–1491
19. Nobuta Y, Inoue T, Suzuki S, Arakawa C, Yakabe T, Ogawa M, Yajima N (2009) The efficacy and the safety of Lactobacillus brevis KB290 as a human probiotics. Int J Probiotics Prebiotics 4:263–270
20. Jeng K-C, Chen C-S, Fang Y-P, Hou RC-W, Chen Y-S (2007) Effect of microbial fermentation on content of statin, GABA, and polyphenols in Pu-Erh tea. J Agric Food Chem 55(21):8787–8792

Enhanced Activation of Integrin $\alpha_{IIb}\beta_3$-Dependent Signaling in the Pro^{33} (HPA-1b) Variant

Khon C. Huynh

Abstract The major platelet integrin $\alpha_{IIb}\beta_3$ is polymorphic at residue 33, and the Pro^{33} isoform exhibits prothrombotic character. We explored via which pathways the Leu^{33}/Pro^{33} polymorphism modulates integrin $\alpha_{IIb}\beta_3$ mediated outside-in signaling. We have investigated Src Y418 phosphorylation in human platelets adherent on fibrinogen in the presence or absence of divalent cation Mn^{2+} and of the soluble form of the CD40 ligand (sCD40L). Whereas HPA-1a/1a platelets did not show Src activation on 10 μg/ml immobilized fibrinogen even after 40 min, we observed a significant Src phosphorylation increase in the HPA-1b/1b isoform. Mn^{2+} alone also increased the Src activity in the platelets and addition of Mn^{2+} to platelets adherent on fibrinogen resulted in a synergism of Src Y418 phosphorylation in both isoforms, whereas after 40 min adhesion HPA-1b/1b platelets demonstrated a substantially (sixfold) higher Src kinase activation than HPA-1a/1a. Our observation that in the presence of Mn^{2+} the Src pY418 phosphorylation increases in both isoforms was independent of the fibrinogen concentration suggesting a dominant role of the Mn^{2+}. Whereas sCD40L, another $\alpha_{IIb}\beta_3$ ligand, did not enhance the Src activity in HPA-1a/1a platelets adherent on fibrinogen, we observed a threefold Src kinase activation in HPA-1b/1b platelets in the presence of sCD40L.

Keywords HPA-1b/1b · Integrin polymorphism · Src Y418

1 Introduction

The major platelet integrin $\alpha_{IIb}\beta_3$ plays a crucial role in platelet adhesion and aggregation. $\alpha_{IIb}\beta_3$ activation and "outside-in" signaling can also be achieved either via engagement of adhesive proteins to the integrin [1–3] or by utilizing Mn^{2+} cations. $\alpha_{IIb}\beta_3$ integrins activated by Mn^{2+} also display a higher affinity to soluble fibrinogen

K. C. Huynh (✉)
School of Biomedical Engineering, International University, Ho Chi Minh City, Vietnam
e-mail: hckhon@hcmiu.edu.vn

Vietnam National University, Ho Chi Minh City, Vietnam

V. Van Toi et al. (eds.), *8th International Conference on the Development of Biomedical Engineering in Vietnam*, IFMBE Proceedings 85,
https://doi.org/10.1007/978-3-030-75506-5_57

[4–7]. Another potential ligand of integrin $\alpha_{IIb}\beta_3$ is the tumor necrosis factor (TNF) family member soluble CD40L (sCD40L). With its KGD-sequence, it binds and activates the integrin receptor [8, 9]. A high number of molecules and pathways are involved in $\alpha_{IIb}\beta_3$-mediated outside-in signaling [1, 10–12]. To the most important biological responses belong tyrosine phosphorylation procedures. Src family kinases have been reported to play an important role in outside-in signaling [11, 13–16]. Src kinase is directly associated with the β_3 cytoplasmatic tail of $\alpha_{IIb}\beta_3$ and becomes fully activated through trans-autophosphorylation at the Y418 residue in human platelets after the fibrinogen engagement to the receptor. Via its substrates it plays a pivotal role in the regulation of various cellular processes between integrin and cytoskeleton, such as cell adhesion, spreading, cytoskeletal rearrangement, cell growth and migration [11, 16–18].

The β_3 subunit of $\alpha_{IIb}\beta_3$ is polymorphic at residue 33: Leu (HPA-1a) or Pro (HPA-1b). Pro^{33} platelets exhibit enhanced aggregation, α-granule secretion, and they demonstrate greater affinity to immobilized fibrinogen [19–22]. The HPA-1b/1b genotype could be a potential association with premature myocardial infarction by patients with coronary artery disease [23]. It is known from previous works, that the substitution of Leucine to Proline at the residue 33 enhances signaling to ERK2 and MLCK in human platelets and CHO cells and increases phosphorylation of focal adhesion kinase (FAK) at Y125 [10, 21, 24, 25]. Moreover, as these signal molecules play an important role in cytoskeletal reorganization, adhesion, shape change and spreading, this observation correlates well with the increased $\alpha_{IIb}\beta_3$ function observed in the HPA-1b/1b isoform. To find out, which signaling pathways may be involved in the modulation of the $\alpha_{IIb}\beta_3$ function by the polymorphism, we have studied the proximate vicinity of the receptor. In the presence of Mn^{2+} fibrinogen adherent HPA-1b/1b platelets exhibited substantially greater Src activation than HPA-1a/1a platelets. sCD40L also demonstrated an enhanced Src Y418 phosphorylation in fibrinogen adherent HPA-1b1b platelets but not in HPA-1a1a.

2 Materials and Methods

Antibodies and Reagents

Anti-Src pY418 was from Invitrogen (Darmstadt, Germany), anti-v-Src were purchased from Calbiochem (Darmstadt, Germany), PE-conjugated anti-rabbit secondary antibody from Biozol (Eching, Germany), secondary antibodies rabbit-HRG, mouse-HRG from GE Healthcare (München, Germany), FITC-conjugated non-specific mouse-IgG from Becton–Dickinson (Heidelberg, Germany), sCD40L from PeproTechInc. (Offenbach, Germany), abciximab from Lilly ("ReoPro", Gießen, Germany and Bio-Rad Staining kit from Bio-Rad (München, Germany). All other reagents were from Sigma (Taufkirchen, Germany).

Platelet Adhesion
Twenty-four-well plates were coated with 10 μg/ml or 100 μg/ml human fibrinogen or 1% heat-denatured BSA (as negative control) in PBS buffer and incubated at 37 °C in 5% CO_2 for 1 h. For adhesion experiments, 60 μl resting human platelets of 10^9/ml were added to immobilized fibrinogen or BSA at 37 °C, in 5% CO_2. In some experiments platelets were previously incubated for 3 min with 0.5 mM $MnCl_2$ or 1 μg/ml sCD40L prior to the adhesion. After incubation times of 2.5 min, 5 min, 10 min, 20 min, and 40 min non-adherent platelets were removed and adherent ones were solubilised with ice-cold lysis buffer containing 0.5% Triton X-100, 10 mM Tris (pH 7.4), 150 mM NaCl and protease and phosphatase inhibitors (1 mM Na_3VO_4, 0.5 mM NaF, 2 mg/ml pepstatin, 50 mg/ml AEBSF, 5 mg/ml chymostatin, 5 mg/ml antipain, 5 mg/ml leupeptin, 1% phosphatase inhibitor mixture) and incubated on ice for 30 min. Lysates were then clarified by centrifugation at 16.000 × g and 4 °C for 30 min. Protein concentration was determined by Bradford method. For negative control, 10 μg/ml abciximab was added to the platelets.

Fluorescent Microscopy
Washed human platelets were adhered in Tyrode's buffer on 100 μg/ml fibrinogen-coated coverslips for 20 min in the presence of 0.5 mM Mn^{2+} at room temperature. Adherent platelets were fixed with 2% paraformaldehyde and incubated with Src pY418 antibody (1:100) or PE anti-rabbit-IgG (1:100) or FITC-P2 antibody (1:10). Single images were taken with an Axiovert 100 M microscope (Carl-Zeiss, Jena, Germany).

3 Results

Ligand engagement and conformational change are of capital importance in $\alpha_{IIb}\beta_3$ activation and in the regulation of the integrin-mediated outside-in signaling. Src kinase is known to bind directly to the fibrinogen receptor [16, 17, 26]. To analyze in which extent ligand binding or activation by Mn^{2+} cations contribute to the outside-in signaling, we investigated Src Y418 phosphorylation in human HPA-1a/1a and HPA-1b/1b platelets either allowing them to adhere onto fibrinogen or maintaining them in suspension over a BSA substrate in the presence or absence of Mn^{2+}.

Immobilized Fibrinogen Induces Higher Src Activation in HPA-1b/1b than in HPA-1a/1a Platelets
To study the effect of the ligand occupation and to assess the role of the ligand concentration in the signaling process, we plated resting washed human platelets on 10 μg/ml or 100 μg/ml immobilized fibrinogen. On 10 μg/ml fibrinogen, the HPA-1a/1a isoform did not show a significant Src activation even after 40 min incubation time, whereas the HPA-1b/1b platelets showed a significantly higher Src activation (Fig. 1). On 100 μg/ml immobilized fibrinogen, HPA-1b/1b platelets exhibited an enhanced Src pY418 activity comparing to HPA-1a/1a at as early as 2.5 min

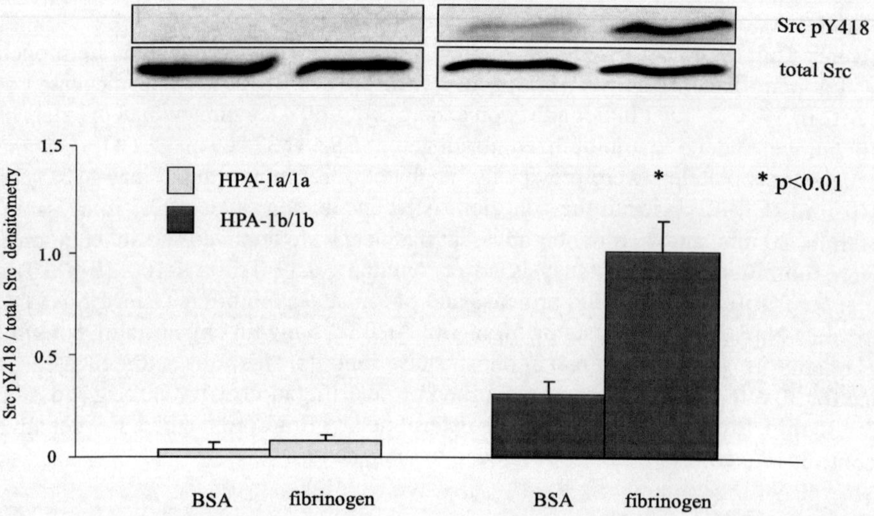

Fig. 1 Src pY418 activity in fibrinogen adherent platelets. Washed human platelets of known HPA-1a/1a and HPA-1b/1b were allowed to adhere onto 10 µg/ml fibrinogen or kept in suspension over 1% BSA for 40 min. Adherent platelets were then solubilized and analyzed by Western blot with antibodies to Src pY418 (*upper level*) or total Src (*lower level*) and quantified by densitometry (Src pY418/total Src in arbitrary units). *$p < 0.01$, evaluated unpaired t test. Results are representative of three experiments

and they maintained approximately constant extent of Src activity without a significant increase over the whole incubation time of 40 min. The HPA-1a/1a isoform, however, showed less Src activation in the first 20 min, but after 40 min adhesion time both isoforms displayed approximately equal extents of activation under these experimental conditions (Fig. 2).

The Combination of Mn^{2+} and Immobilized Fibrinogen Results in the Synergism of Src Activation

To examine whether Mn^{2+} alone stimulates Src kinase, we added 0.5 mM Mn^{2+} to HPA-1a/1a platelets maintained in suspension on 1% BSA. As shown in Fig. 3 Mn^{2+} alone induced Src activation, and the extent of the Src pY418 activity was comparable with Src pY418 in platelets adhering to 100 µg/ml fibrinogen. To determine whether Mn^{2+} modifies Src Y418 phosphorylation in adherent platelets, we plated the HPA-1a/1a isoform on 100 µg/ml immobilized fibrinogen after the addition of various concentrations of Mn^{2+}. In the presence of 0.5 mM Mn^{2+} platelets exhibited a fourfold increase in the Src activity comparing to platelets adhering to fibrinogen in the absence of Mn^{2+}. Performing the adhesion experiments using various concentrations of Mn^{2+}, we observed considerably less enhancement upon increasing the Mn^{2+} concentrations (0.5 mM and 1 mM Mn^{2+} presented approximately equivalent Src Y418 phosphorylation and 2 mM Mn^{2+} displayed 1.6-fold increase than 1 mM Mn^{2+}) than between the Src kinase activities in adherent platelets prior to and after

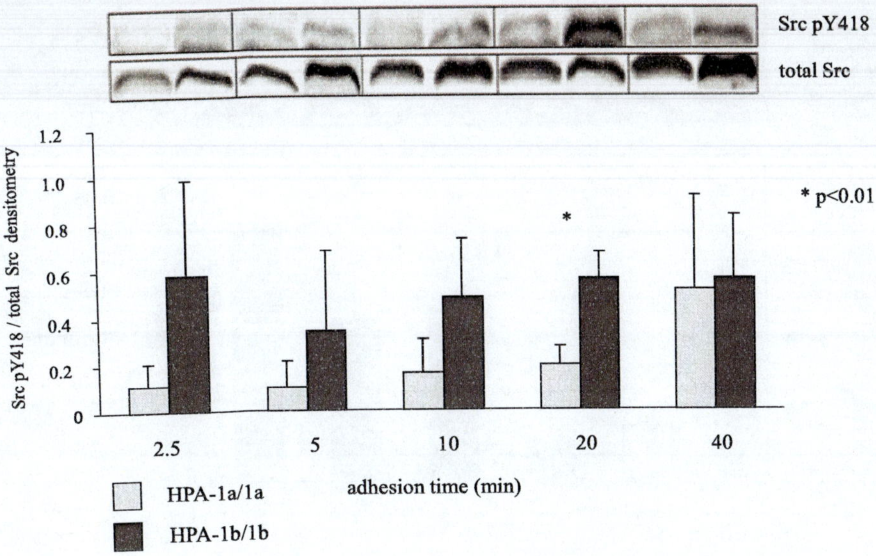

Fig. 2 Time course of Src pY418 activation in fibrinogen adherent platelets

0.5 mM Mn^{2+} addition, respectively (Fig. 3a). $\alpha_{IIb}\beta_3$ specific inhibitor abciximab significantly reduced the Src activity induced by the combination of immobilized fibrinogen and Mn^{2+} indicating an $\alpha_{IIb}\beta_3$ mediated signaling (Fig. 3b).

On the basis of these results, we analyzed the time course of the outside-in signaling to Src in the HPA-1a/1a and HPA-1b/1b platelets utilizing the synergetic effect of the combination of immobilized fibrinogen and Mn^{2+} cations. Since 1 mM and 2 mM Mn^{2+} concentrations resulted in artificial morphological changes in resting platelets after longer incubation times (data not shown), we allowed adhering the human platelets on 10 μg/ml or 100 μg/ml fibrinogen in the presence of 0.5 mM Mn^{2+}. As shown in Fig. 4a, b, at 2.5 min, both isoforms showed approximately equivalent Src pY418 phosphorylation on both 10 μg/ml and 100 μg/ml fibrinogen matrix. After 5 min adhesion time HPA-1b/1b adhering to 100 μg/ml fibrinogen exhibited a two-fold increase comparing to HPA-1a/1a, whereas on 10 μg/ml immobilized fibrinogen the two isoforms did not significantly differ from each other. At 10 min and 20 min, HPA-1b/1b demonstrated approximately two-fold Src activities independently of the fibrinogen concentration as compared HPA-1a/1a. At 40 min, whereas the Src activity in HPA-1a/1a remained constant, the amount of Src pY418 in HPA-1b/1b substantially raised on both 10 μg/ml and 100 μg/ml fibrinogen matrix (twofold and threefold, respectively) resulting in HPA-1b/1b Src activities fivefold (10 μg/ml) and sixfold (100 μg/ml) higher than HPA-1a/1a.

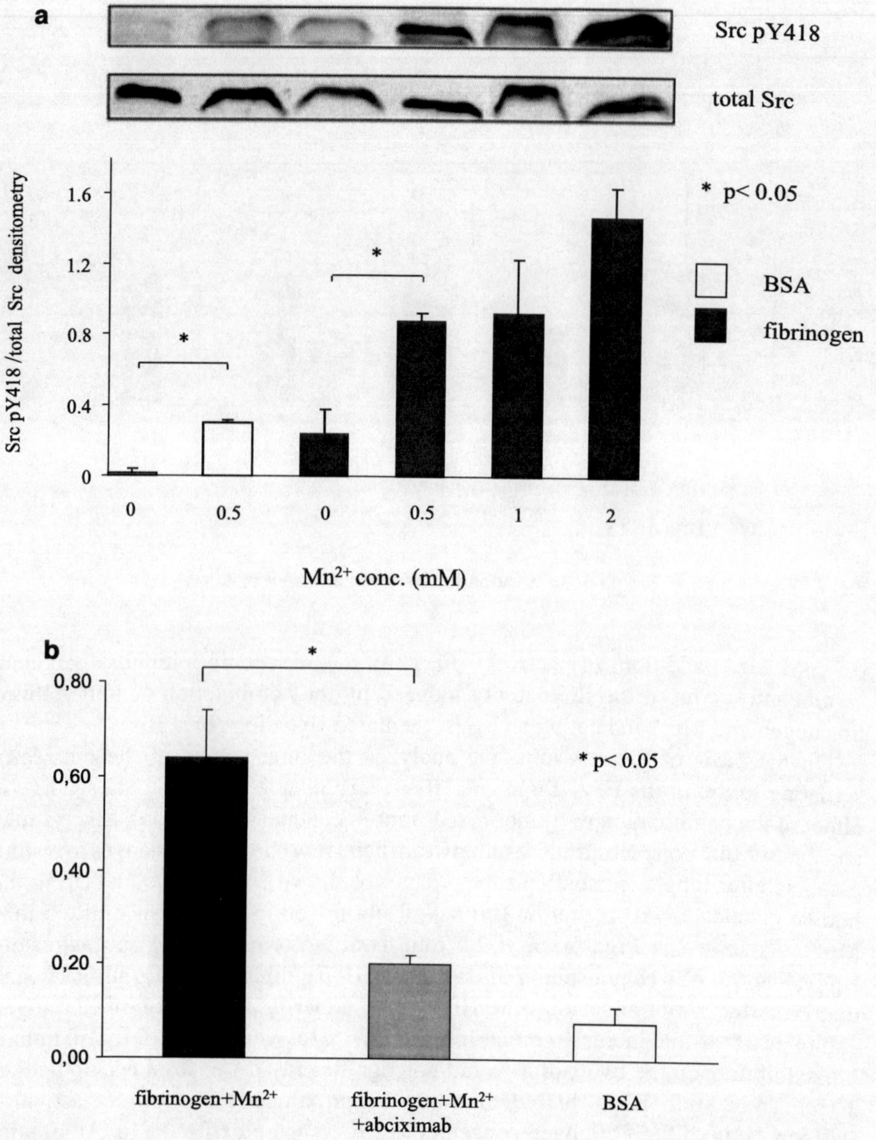

Fig. 3 Effect of Mn^{2+} on Src activity in human platelets adherent on immobilized fibrinogen or in suspension. Adhesion assays were repeated in the presence or absence of Mn^{2+}. **a** 10 μg/ml abciximab significantly reduced Src activity in HPA-1a/1a fibrinogen adherent platelets in the presence of Mn^{2+}. **b** Lysates were processed as described in Fig. 1

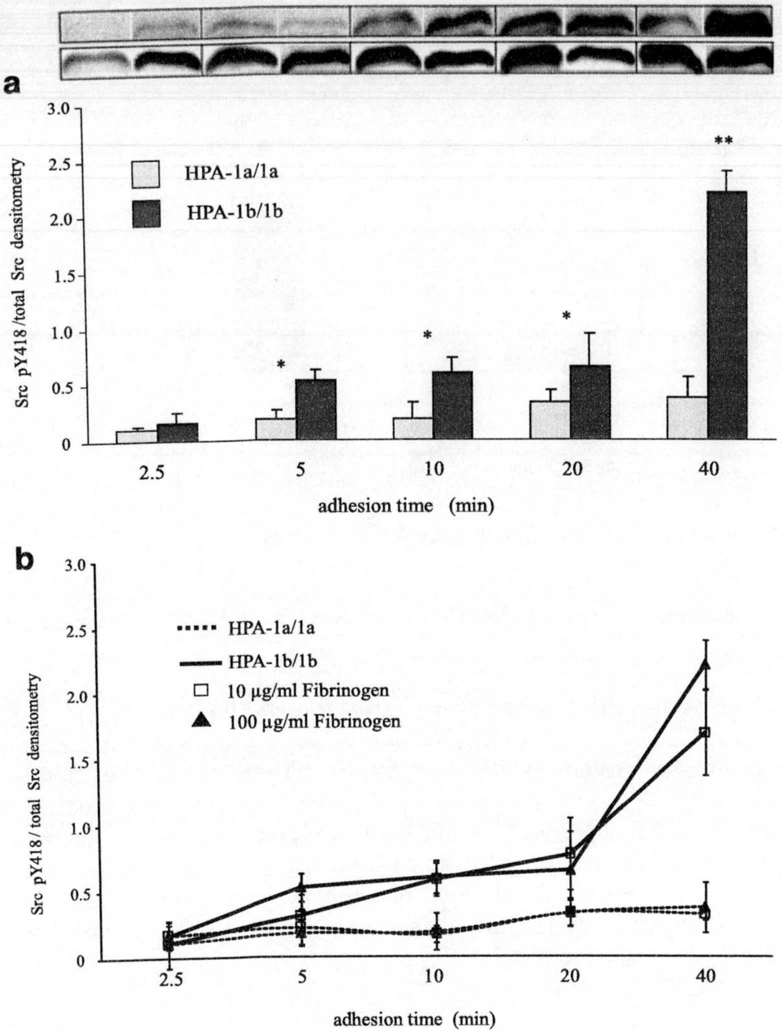

Fig. 4 **a** Time course of Src pY418 activity in fibrinogen adherent platelets in the presence of 0.5 mM Mn^{2+}. **b** Comparison of Src pY418 in platelets adhering onto different concentrations of fibrinogen in the presence of 0.5 mM Mn^{2+}

sCD40L Increases Src Activation Only in Fibrinogen Adherent HPA-1b/1b Platelets

It is known from recent works (38, 39) that sCD40L plays an essential role in thrombogenesis via integrin $\alpha_{IIb}\beta_3$ activation. We could demonstrate that the HPA-1 polymorphism modulates the sCD40L-induced $\alpha_{IIb}\beta_3$-mediated outside-in signaling to Src in human platelets. HPA-1b/1b platelets exhibited a threefold increase of Src kinase activation in the presence of sCD40L compared to adhesion on 100 µg/ml

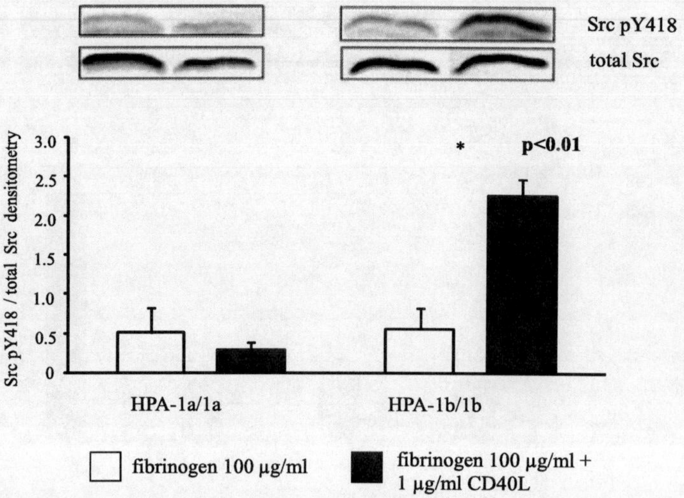

Fig. 5 Effect of sCD40L on platelets adherent on immobilized fibrinogen

fibrinogen alone, whereas no effect of sCD40L was observed in HPA-1a/1a platelets (Fig. 5).

Colocalisation of Active Src Kinase and $\alpha_{IIb}\beta_3$ in Human Platelets

To study the distribution of activated Src kinase in fibrinogen-adherent platelets fluorescent microscopy was utilized using the anti-Src pY418 antibody. As shown in Fig. 6, anti-Src pY418 was mainly confined to the edge of the spreading platelets. $\alpha_{IIb}\beta_3$ labeled with FITC-conjugated specific antibody P2 (green), and the merged image (the combination of the red-labeled Src pY418 and green-labeled $\alpha_{IIb}\beta_3$ giving a yellow staining) indicate a colocalization between the $\alpha_{IIb}\beta_3$ integrin and Src kinase. This observation confirms previous results reported by other groups that Src is associated with $\alpha_{IIb}\beta_3$ receptor [15, 16].

4 Discussion

The purpose of our study was to investigate the role of the Leu^{33}/Pro^{33} polymorphism in the modulation of the $\alpha_{IIb}\beta_3$ receptor-mediated outside-in signaling. It is known from previous works of various groups that Pro^{33} isoform demonstrates an enhanced $\alpha_{IIb}\beta_3$ function [10, 16, 19, 20, 27, 28].

It has been shown that ERK2 and MLCK are regulated by polymorphism (9). Therefore, we wanted to find out which pathways are involved in this process. Here we focused on the Src kinase. This tyrosine kinase plays an essential role in integrin signaling and is directly associated with the $\alpha_{IIb}\beta_3$ integrin [11, 16, 17].

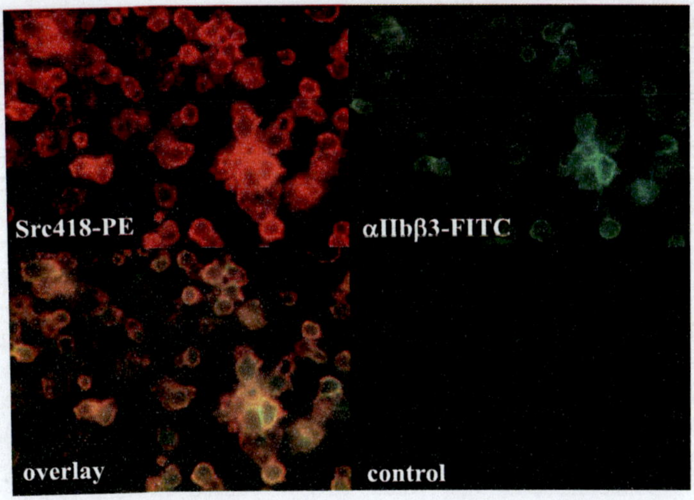

Fig. 6 Distribution of Src pY418 and $\alpha_{IIb}\beta_3$ in HPA-1b/1b human platelets adherent on immobilized fibrinogen. HPA-1b/1b washed human platelets were preincubated with 0.5 mM Mn^{2+} for 3 min, plated out on 100 μg/ml fibrinogen-coated coverslips for 20 min and prepared for fluorescence microscopy. Platelets immunoreacting with Src pY418 are red and FITC-conjugated P2 antibody specific against $\alpha_{IIb}\beta_3$ is green. In merged image the combination of the two staining is yellow. In the control experiment platelets were only with the conjugated second antibody incubated

Whereas on 10 μg/ml immobilized fibrinogen only the HPA-1b/1b platelets showed a significant Src stimulation, the employment of a fibrinogen concentration of 100 μg/ml induced Src activation in both isoforms. HPA-1b/1b platelets thereby reached a maximal activity rapidly indicating a higher sensitivity for Src kinase stimulation than HPA-1a/1a (Fig. 2). However, the Src pY418 values were significantly variable between the individual experiments indicating that experimental procedures and mechanical stress prior to the adhesion (i.e. plating out) might significantly contribute to the activation values. On the other hand, the high Src kinase activity in adherent HPA-1b/1b platelets on 10 μg/ml fibrinogen after 40 min adhesion (Fig. 1) also indicates a greater sensitivity for Src kinase stimulation by the prothrombotic isoform.

Another possibility to activate integrins is shifting them from their inactive to their active states by divalent cations [4, 29]. Litvinov et al. have reported a direct correlation between the ligand-binding affinity of $\alpha_{IIb}\beta_3$ and the conformational rearrangement of the receptor using Mn^{2+} cations [6]. Although both ligand occupation and Mn^{2+} maintain the integrin in an active conformational structure, they interact with $\alpha_{IIb}\beta_3$ in a different way. Whereas fibrinogen engages to the dodecapeptide (γ400–412) binding site in the α_{IIb} subunit and to the RGD binding sequence in the β_3 subunits, Mn^{2+} can bind to the three metal-binding sites in the β_3 subunits: ADMIDAS (adjacent to the metal ion-dependent adhesion site), MIDAS (metal ion-dependent adhesion site) and LIMBS (ligand-induced metal-binding site) [6, 30].

We raised the question to what extent Mn^{2+} ions alone regulate $\alpha_{IIb}\beta_3$ mediated outside-in signaling comparing to ligand engagement. In the absence of a ligand, only ADMIDAS is occupied by Mn^{2+}. In the presence of fibrinogen, however, MIDAS and LIMBS are engaged by Mn^{2+} as well, and stabilize the ligand–occupation confirmation of the $\alpha_{IIb}\beta_3$ integrin [6, 30]. To analyze how this different way of activation reflects in the receptor signaling, we examined outside-in signaling induced by immobilized fibrinogen or Mn^{2+} alone and by combining both. Furthermore, we analyzed how these processes are modified by the Leu^{33}/Pro^{33} polymorphism.

Our results show that 0.5 mM Mn^{2+} alone stimulates Src Y418 phosphorylation to an approximately equivalent degree to the Src kinase activation triggered by 100 μg/ml immobilized fibrinogen (Fig. 3). Moreover, the addition of 0.5 mM Mn^{2+} to platelets adherent on fibrinogen generated a considerable augmentation in Src activity. The large synergetic effect in the Src kinase activity induced by the combination of fibrinogen and Mn^{2+} may reflect the stabilized activated conformation of $\alpha_{IIb}\beta_3$ raised by the two additionally metal bindings at MIDAS and LIMBS.

As Mn^{2+} concentrations higher than 0.5 mM caused artificial morphological changes in platelets during adhesion times longer than 10 min, we used 0.5 mM Mn^{2+} for analyzing the modulation of the Leu^{33}/Pro^{33} polymorphism to Src signaling in fibrinogen adherent platelets. In the presence of Mn^{2+} HPA-1b/1b showed a significantly enhanced Src Y418 phosphorylation (Fig. 4a, b) than HPA-1a/1a demonstrating that Leu^{33}/Pro^{33} polymorphism exerts its modulating effect on intracellular processes via pathways involving the Src kinase signaling. After 20 min HPA-1b/1b platelets exhibited a huge increase in Src activity, whereas the amount of Src pY418 in HPA-1a/1a did not change (Fig. 4a, b). As we know that Src activation is one of the earliest biochemical responses after fibrinogen engagement [1, 11] and this observation also accords to our results of detectable Src kinase activities after already 2.5 min adhesion, we can only speculate that this rate increase is at least partially due to generation of further activating processes in the platelets, such as the release of fibrinogen or ADP [31]. Why only the HPA-1b/1b isoform displays, this phenomenon raises further questions and could be the topic of future studies. Our observation that Src kinase activation is independent of the fibrinogen concentration if Mn^{2+} is present (Fig. 4b), suggests a dominant role of Mn^{2+} in this process.

CD40L is not only an important modulator of the immune system, but it is also expressed in human platelets and released to the blood plasma in small concentrations (soluble form) by activated platelets [8]. sCD40L binds $\alpha_{IIb}\beta_3$ integrin in its active conformation via a KGD sequence [9]. We focused on the effect of sCD40L on platelets adherent onto fibrinogen of both isoforms. In the presence of sCD40L HPA-1b/1b fibrinogen-adherent platelets exhibited a threefold increase of Src kinase activation, whereas no effect was observed in HPA-1a/1a platelets (Fig. 5). This indicates that the prothrombotic isoform is more susceptible to alternative ligands such as sCD40L. As it is known that sCD40L plays an important role in stabilizing thrombi during myocardial infarction [32], this might give a hint, why HPA-1b/1b carriers (who suffer from coronary artery disease) exhibit their myocardial infarction earlier than carriers of the HPA-1a/1a genotype [23].

In conclusion, we provided evidence that the modulating effect of the Pro^{33}/Leu^{33} polymorphism is mediated via the Src kinase signaling and that the Pro^{33} isoform exhibits an enhanced signaling to Src kinase. A higher activation of Src kinase in response to the conformational changes induced by Mn^{2+} or sCD40L can probably be due to its directly associated position on the β_3 cytoplasmatic tail. Until now the enhanced functionality of the Pro^{33} isoform is not cleared. At present, only information about the crystal structure of Leu^{33} β_3 of $\alpha_{IIb}\beta_3$ is available [30], and one can only speculate that the substitution of a helix-favouring residue Leu with a helix-disrupting residue Pro might modify the local conformation into a rigid structure that prefers an "open state" activated conformation of the receptor [10]

Acknowledgements This research is funded by Vietnam National University Ho Chi Minh City (VNUHCM) under grant number C2018-28-01.

Conflict of Interest The authors declare that they have no conflict of interest.

References

1. Shattil SJ, Newman PJ (2004) Integrins: dynamic scaffolds for adhesion and signaling in platelets. Blood 104(6):1606–1615
2. Takagi J et al (2002) Global conformational rearrangements in integrin extracellular domains in outside-in and inside-out signaling. Cell 110(5):599–611
3. Liddington RC, Ginsberg MH (2002) Integrin activation takes shape. J Cell Biol 158(5):833–839
4. Bazzoni G, Hemler ME (1998) Are changes in integrin affinity and conformation overemphasized? Trends Biochem Sci 23(1):30–34
5. Shimaoka M, Takagi J, Springer TA (2002) Conformational regulation of integrin structure and function. Annu Rev Biophys Biomol Struct 31:485–516
6. Litvinov RI et al (2004) Functional and structural correlations of individual alphaIIbbeta3 molecules. Blood 104(13):3979–3985
7. Craig D et al (2004) Structural insights into how the MIDAS ion stabilizes integrin binding to an RGD peptide under force. Structure 12(11):2049–2058
8. Andre P et al (2002) CD40L stabilizes arterial thrombi by a beta3 integrin–dependent mechanism. Nat Med 8(3):247–252
9. Prasad KS et al (2003) Soluble CD40 ligand induces beta3 integrin tyrosine phosphorylation and triggers platelet activation by outside-in signaling. Proc Natl Acad Sci U S A 100(21):12367–12371
10. Vijayan KV, Bray PF (2006) Molecular mechanisms of prothrombotic risk due to genetic variations in platelet genes: Enhanced outside-in signaling through the Pro^{33} variant of integrin beta3. Exp Biol Med (Maywood) 231(5):505–513
11. Shattil SJ (2005) Integrins and Src: dynamic duo of adhesion signaling. Trends Cell Biol 15(8):399–403
12. Zou Z et al (2007) Structure-function analysis reveals discrete beta3 integrin inside-out and outside-in signaling pathways in platelets. Blood 109(8):3284–3290
13. Frame MC (2004) Newest findings on the oldest oncogene; how activated src does it. J Cell Sci 117(Pt 7):989–998
14. Totani L et al (2006) Src-family kinases mediate an outside-in signal necessary for beta2 integrins to achieve full activation and sustain firm adhesion of polymorphonuclear leucocytes tethered on E-selectin. Biochem J 396(1):89–98

15. Roskoski R Jr (2005) Signaling by Kit protein-tyrosine kinase—the stem cell factor receptor. Biochem Biophys Res Commun 337(1):1–13
16. Obergfell A et al (2002) Coordinate interactions of Csk, Src, and Syk kinases with [alpha]IIb[beta]3 initiate integrin signaling to the cytoskeleton. J Cell Biol 157(2):265–275
17. Arias-Salgado EG et al (2003) Src kinase activation by direct interaction with the integrin beta cytoplasmic domain. Proc Natl Acad Sci U S A 100(23):13298–13302
18. Fincham VJ et al (2000) Active ERK/MAP kinase is targeted to newly forming cell-matrix adhesions by integrin engagement and v-Src. EMBO J 19(12):2911–2923
19. Michelson AD et al (2000) Platelet GP IIIa Pl(A) polymorphisms display different sensitivities to agonists. Circulation 101(9):1013–1018
20. Feng D et al (1999) Increased platelet aggregability associated with platelet GPIIIa PlA2 polymorphism: the Framingham offspring study. Arterioscler Thromb Vasc Biol 19(4):1142–1147
21. Vijayan KV et al (2000) The Pl(A2) polymorphism of integrin beta(3) enhances outside-in signaling and adhesive functions. J Clin Invest 105(6):793–802
22. Loncar R et al (2007) HPA-1 polymorphism of alphaIIbbeta3 modulates platelet adhesion onto immobilized fibrinogen in an in-vitro flow system. Thromb J 5:2
23. Zotz RB et al (2005) Association of polymorphisms of platelet membrane integrins alpha IIb(beta)3 (HPA-1b/Pl) and alpha2(beta)1 (alpha807TT) with premature myocardial infarction. J Thromb Haemost 3(7):1522–1529
24. Vijayan KV et al (2005) The Pro^{33} isoform of integrin beta3 enhances outside-in signaling in human platelets by regulating the activation of serine/threonine phosphatases. J Biol Chem 280(23):21756–21762
25. Vijayan KV et al (2003) Enhanced activation of mitogen-activated protein kinase and myosin light chain kinase by the Pro^{33} polymorphism of integrin beta 3. J Biol Chem 278(6):3860–3867
26. de Virgilio M, Kiosses WB, Shattil SJ (2004) Proximal, selective, and dynamic interactions between integrin alphaIIbbeta3 and protein tyrosine kinases in living cells. J Cell Biol 165(3):305–311
27. Vielreicher M et al (2007) Dynamic interaction between Src and C-terminal Src kinase in integrin alphaIIbbeta3-mediated signaling to the cytoskeleton. J Biol Chem 282(46):33623–33631
28. Kamm KE, Stull JT (2001) Dedicated myosin light chain kinases with diverse cellular functions. J Biol Chem 276(7):4527–4530
29. Smith JW, Piotrowicz RS, Mathis D (1994) A mechanism for divalent cation regulation of beta 3-integrins. J Biol Chem 269(2):960–967
30. Xiao T et al (2004) Structural basis for allostery in integrins and binding to fibrinogen-mimetic therapeutics. Nature 432(7013):59–67
31. Salsmann A et al (2005) A new functional role of the fibrinogen RGD motif as the molecular switch that selectively triggers integrin alphaIIbbeta3-dependent RhoA activation during cell spreading. J Biol Chem 280(39):33610–33619
32. Aukrust P et al (1999) Enhanced levels of soluble and membrane-bound CD40 ligand in patients with unstable angina. Possible reflection of T lymphocyte and platelet involvement in the pathogenesis of acute coronary syndromes. Circulation 100(6):614–620

AI and Data Science for Health

Object Contour in Medical Images Based on Saliency Map Combined with Active Contour Model

Vo Hoang Thien, Vo Thi Hong Tuyet, and Nguyen Thanh Binh

Abstract Medical imaging is useful in the diagnosis and treatment of diseases. A wide range of pathologies have been discovered by identifying abnormalities of the object's boundary in the medical image. Almost tokens of illness are shown in salient regions of objects, such as the intensity of color, connection together, etc. Therefore, the contour is a pressing-issue in the cutting-edge diagnose. This paper proposed a method for detecting contour of the object in medical images based on the saliency map combined with the active contour model. The proposed method detected salient regions of the medical image and activated the boundary of objects by energy reducing. The proposed method includes four steps: firstly, the pre-processing is applied for input images by the median filter. Secondly, the threshold of intensity is divided by super-pixels. Thirdly, the outstanding map is made by clustering the above super-pixels. Finally, the active contour model is built in salient regions of concerned about. We test results from DICOM images dataset, which includes high-quality and low-quality medical images. The accuracy of the proposed method is 97.52%, more than the results of the other methods, 5.0%.

Keywords Active contour model · Contour detection · Salient map

V. H. Thien · V. T. H. Tuyet (✉)
Faculty of Information Technology, Ho Chi Minh City Open University, Ho Chi Minh City, Vietnam
e-mail: tuyet.vth@ou.edu.vn

V. H. Thien
e-mail: 1651012179thien@ou.edu.vn

N. T. Binh (✉)
Faculty of Computer Science and Engineering, Ho Chi Minh City University of Technology, VNU-HCM, 268 Ly Thuong Kiet Street, District 10, Ho Chi Minh City, Vietnam
e-mail: ntbinh@hcmut.edu.vn

Vietnam National University Ho Chi Minh City, Linh Trung Ward, Thu Duc District, Ho Ho Chi Minh City, Vietnam

V. Van Toi et al. (eds.), *8th International Conference on the Development of Biomedical Engineering in Vietnam*, IFMBE Proceedings 85,
https://doi.org/10.1007/978-3-030-75506-5_58

1 Introduction

Medical image has an important role in detecting the disease. The boundary of each part in the body can determine the severity of the pathology. The concept of the contour can be explained by edge detection or edge map. This approach is depended on the color intensity of each pixel. Many types of researches had used threshold for dividing the level of the neighbor-pixels. All of them must fit with a wide range of parameters or level of decomposition. Therefore, these algorithm complexities are high.

Another approach for contour detection is the detection for distinctive regions. The differences between areas can create the best choice for object recognition. If the highlights in the input image are created, the inside and outside of the target will be very clear. The edge map can be easily created, and the connection will be stronger [1]. K-means and colony optimization are the combination for image segmentation, and this solution depends on the edge map [1]. Canny method and mean shift technique are applied in circle segmentation by Nurpur [2]. Therefore, the collection of edges is the popular methods in object detection.

Saliency is the edge map of objects in the frame. The differences between the previous and current frame are the feature for boundary detection. Saliency is also a powerful tool for a wide range of fields, such as detection and segmentation [3, 4], edge saliency map detection depends on texture suppression [5]. The other solutions about edge map not only based on saliency but also context-aware [6], using saliency as a filter in contrast region detection [7]. Xu [8] used the power of energy which call active contour model—Snakes for computing as diffusion of binary edge map derived from the image. Saadatmand [9] used self-affine snake for medical image segmentation which is an inherited task from edge and contour detection. The concept of gradient vector flow [8, 10] depended on the snake model in convex function. That was a significant shift of the active contour model. In recent years, the snake comes back forcefully with image segmentation. The local and global intensity information are a parent for the active contour model [11, 12]. Another study, the local similarity factor, was connected to the energy function for robust noise region [13]. The employment of transform, Nageswara [14] overcomes the drawbacks of energy contraction by contourlet domain, and Qing [15] uses frequency tuned. From these methods, we can conclude that the active contour model will be used to the best of its ability if there is a clear contrast. Therefore, edge map or a clear area is extremely useful with contour detection by minimum or maximum energy [16]. The feature extraction by convolution neural network classifier [17] also gives valuable information for the recognition process. However, deep learning is complex, and a wide range of execution time. Besides, Sen [18] also used salient regions for the dividing area of objects with a background in images. In medical images processing, to avoid losing information in images after processing is a hard task. The case is true that the special regions are applied in the edge map depended on the Sobel mask as a condition for retinal vessels segmentation [19]. From the above researches, we can see that salient map is very advantageous for object detection.

In this paper, we propose an approach for contour detection based on active contour model in the edge map. This map is created by the level of salient regions. The proposed method includes four steps: firstly, the pre-processing is applied for input images by the median filter. Secondly, the threshold of intensity is divided by super-pixels. Thirdly, the outstanding map is made by clustering the above super-pixels. Finally, the active contour model is built in salient regions of concerned about. The paper's organization consists of five sections: the current section introduced about the concern about, the background of the salient map and active contour model is presented in Sect. 2. The proposed method is indicated clearly in Sect. 3, and our experiments are presented in Sect. 4. The final section is conclusions and future works.

2 Salient Map and Active Contour Model

2.1 Salient Map

The salient map is a similar basis with edge map based on the quality of pixel in the image. The simplification and convenience analysis are the aims of saliency. This way responses for object's region detection [3] and includes five steps:

(i) Calculating the superpixel for clustering by Simple Linear Iterative Clustering (SLIC) algorithm. This step bases on a local version of K-mean and each pixel is connected to vector $\Psi_{(x,y)}$ which is defined as Eq. (1):

$$\Psi_{(x,y)} = \begin{bmatrix} \lambda_x \\ \lambda_y \\ I_{(x,y)} \end{bmatrix} \tag{1}$$

(ii) Creating a grid for input image with size M × N, where M and is defined as Eq. (2):

$$M = \left[\frac{imageWidth}{regionSize}\right], \quad N = \left[\frac{imageHeight}{regionSize}\right] \tag{2}$$

(iii) From grid which created in the previous step (i), the center point (x_i, y_j) of the grid is manifested as Eq. (3):

$$x_i = round\, i = \frac{imageWidth}{regionSize}, \quad y_j = round\, j = \frac{imageWidth}{regionSize} \tag{3}$$

(iv) The center points substitute length by 3 × 3 to minimize image intensity. The edge (x, y) is defined as Eq. (4):

$$edge(x, y) = ||I(x+1, y) - I(x, -1, y)||^2 + ||I(x, y+1) - I(x, y-1)||^2 \quad (4)$$

(v) From centers (x_i, y_j), each region which creates by superpixel is calculating by Eq. (5):

$$C = \{\Psi(x_i, y_j), i = 0, 1, \ldots, M-1\ j = 0, 1, \ldots, N-1\} \quad (5)$$

The dividing between object region with the background is not simple. However, the most prominent object can be appeared by the level of saliency.

2.2 *Active Contour Model*

The aim of the active contour model (snake) is the finding boundary of objects in the image. The energy, which creates from the snake, is the minimizing [18]. This is an approach which calculates the approximate shape of ground truth is known. The object shape is a spline. It is defined by v(s) = [x(s), y(s)], with s is the arc length. The energy of the snake is defined as Eq. (6):

$$E_{snake} = \int_0^1 E_{snake}(v(s))ds \int_0^1 \left(E_{internal}(v(s)) + E_{image}(v(s)) + E_{con}(v(s))\right)ds \quad (6)$$

where $E_{internal}$ is the internal energy of the spline due to bending, E_{image} gives rise to the image forces, and E_{con} gives rise to the external constraint forces. From Eq. (6), the active contour model will extract the object contour depended on local energy minimizing.

The minimizing process is the benefits package for finding boundary exactly. These have four advantages of this method which can be mentioned. Firstly, the flexibility adapts with minimum plight. That reason is that the search does not stop for finding the optimal value. Secondly, this process begins with outside of the image. Thirdly, the appearing of the denoising period in the image energy function. The smoothing joins in here for creating the scale sensitivity. Finally, continuous updates are extremely great for tracking moving objects. Any whimsy will prevail in some frame of video or pixels with ground-truth.

Even though its advantages, the snake still have many shortcomings, as following: the restriction in the local minima states constitutes no seamless in contour results, the cluster contraction of the entire contour overlooks some features, and the evaluation absolutely relies on the convergence policy of features in global.

The combination with other feature extraction, edge map or other solution is proposed for improving the traditional snake [8]. Rui [9] proposed a self-affine snake for medical image segmentation. This algorithm overcomes the local regions based on energy minimizing. Another direction and gradient have retrieved the frontier based on the convex function [10].

3 Contour Detection Based on Active Contour Model in Salient Regions

As the above section, the traditional snake needs to improve by local regions in each image. In this section, we present a method to detect contour for medical images.

Figure 1 presents the general diagram of the proposed method. It includes two periods: pre-processing and contour detection. In the pre-processing period, we reduce the noise to improve the region quality of objects by the median filter. The contour detection period is also divided into three steps: thresholding for superpixels, calculating the salient map and active contour model for the output. The detail of steps presents as below.

3.1 Improving the Region's Quality of Objects

The quality images can be affected by bad pixels which have blur or noise information. Sometimes, in a blur, details also have noise in its [20]. The bad pixels will create unclear regions for objects. Thereupon, improving quality images is frequently concern about. On the other hand, this amelioration must keep information in input images. The case is true that a wide range of these algorithms has to cope with a severe lack of information. The median filter is a non-linear digital filtering technique, and it is one kind of smoothing algorithm as linear Gaussian filtering.

The median filtering depends on the sliding window, entry by entry, and only applies on one colour channel. Each edge from a half-part of window width will be calculated in each edge from a half-part of the window height. In size width x height of images, the subtraction of each direction and index pixels of a half-part calculates the value for window array. Each value will demonstrate the same with an average of

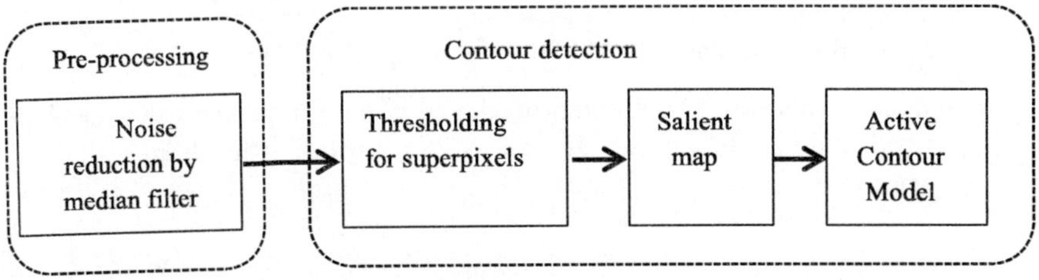

Fig. 1 The process of the proposed method

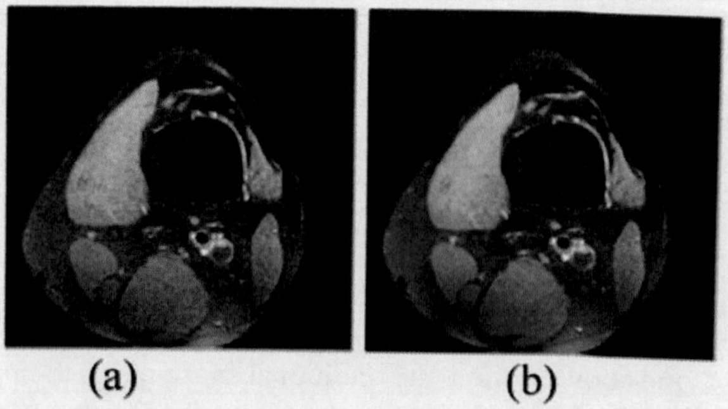

Fig. 2 The result of improving the image quality by median filter

each direction in vector. Then, replacing in each entry based on neighbouring entries. Figure 2 presents an example of the result of improving the image quality by median filtering.

In Fig. 2a, the medical image has noise in around pixels. Therefore, the quality of the input images is reduced. Figure 2b is the output of pre-processing by the median filter. All of the pixels are clearer than the previous medical image. This is a convenience for detecting the strong pixels by intensity or object areas. The aim of this step is supporting the results of salient map step.

3.2 Superpixels for Regions

The previous step, improving the quality by medial filter gives the uniform values in medical images. In the thresholding step, the average value of the entire pixel will determine by comparison. Any pixel with a value less than the threshold will be assigned 0 (black), whereas the pixel value has a greater value threshold will be set to a value of 255 (white). The process of superpixels for regions presents as follows:

i. The read-data process.
ii. The calculating a centre-to-centre distance value between pixels.
iii. The performing for superpixel bases on the linear iteration clustering algorithm SLIC. Firstly, the medical image is divided into a grid of size M × N. Secondly, and clusters will be calculated by K-mean clustering. This process begins from the centers and finds the same areas depended on the features. Finally, the clustering thresholds are divided into two values 0 and 255.

The output of this step is presented in Fig. 3. Figure 3a presents the result of creating a grid, and Fig. 3b presents the result of the clustering threshold.

Fig. 3 The result of superpixel for regions

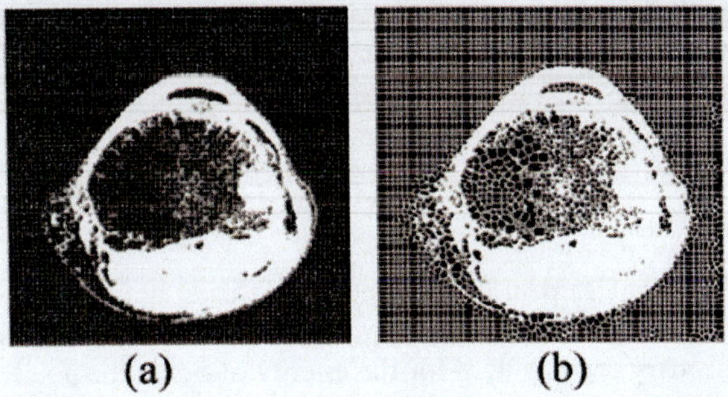

3.3 Salient Map for Medical Images

The diagram of a superpixel, which is the result of Sect. 3.2 has fitting parameters for creating the saliency map:

$$SALS(I_k) = \sum_{i=1}^{N} |I_k - I_i| \tag{7}$$

where, I_i is the value of pixel i in the range [0, 255]. $SALS_{(Ik)}$ shows the nature of the histogram of the object.

However, superpixel in the previous step has a key to level. Therefore, the beginning for saliency map must calculate the distance between the average feature vector which extraction by the threshold. If we call D that is the Euclidean distance between two points (p and q), then D will set up by Eq. (8):

$$D(p, q) = \sqrt{\sum_{i=1}^{n} (q_i - p_i)^2} \tag{8}$$

where n is the number of spaces.

Then, this process continues with the contrast saliency value for creating salient regions. The contrast saliency value is the distance between the average feature vector of the pixel with its neighbourhood.

$$c_{i,j} = D\left[\left(\frac{1}{N_1}\sum_{p=1}^{N_1} v_p\right), \left(\frac{1}{N_2}\sum_{p=1}^{N_2} v_q\right)\right] \tag{9}$$

where, $c_{i,j}$ is the contrast saliency value for the pixel at position (i, j), N_1 and N_2 are the number of pixels in R_1 (inner region) and R_2 (outer region), respectively; v is the vector of feature elements corresponding to a pixel, D is shown as Eq. (8). The output of this step is the salient map for objects. These regions are differences with the background or other parts. The proposed method will use the active contour model for contour detection.

3.4 Active Contour Model in Salient Regions

The final step in contour detection used active contour model in salient regions. Each medical image only has the object's area to present in the input. Therefore, the primary regions help for the energy of neighbor pixels stronger. The active contour model is defined as Eq. (6), and each part is explained as follows:

(i) Calculating the internal energy: with v(s) = [x(s), y(s)] is the spline of object shape, $\alpha(s)$ and $\beta(s)$ are user-defined weights. The internal energy of the snake is composed as:

$$E_{internal} = E_{con} + E_{curu} \tag{10}$$

E_{con} is the continuity of the contour. E_{curu} is the smoothness of the contour. Another way for expanding as:

$$E_{internal} = \frac{1}{2}\left(\alpha(s)|v_s(s)|^2 + \beta(s)|v_s(s)|^2\right) \tag{11}$$

(ii) Calculating the energy in the image:

$$E_{image} = w_{line}E_{line} + w_{edge}E_{edge} + w_{term}E_{term} \tag{12}$$

where $w_{line}, w_{edge}, w_{term}$ are weights of these salient features; $E_{line}, E_{edge} and E_{term}$ are lines, edges and terminations in the images respectively of an image I(x, y). In other approaches, the line function can be appeared as:

$$E_{line} = filter(I(x, y)) \tag{13}$$

(iii) Expanding the external constraint forces E_{con}.
(iv) The summing process of Eq. (6).

In here, this model will apply by the loop. The continued with Sect. 3.3 is the appearing of a boundary around the medical image. Then, the value of Eq. (6) is done for identifying the contour step-by-step of loops. The number of loops is 2000 and removing excess contour is the purpose. Figure 4 shows the processing of the active contour model in salient map. Figure 4a is the result of the salient map of Sect. 3.3, Fig. 4b is the original boundary with blue color in a box shape. Figure 4c is the result of energy minimizing with 2000 loops. The contour detection for the medical image is presented in Fig. 4d.

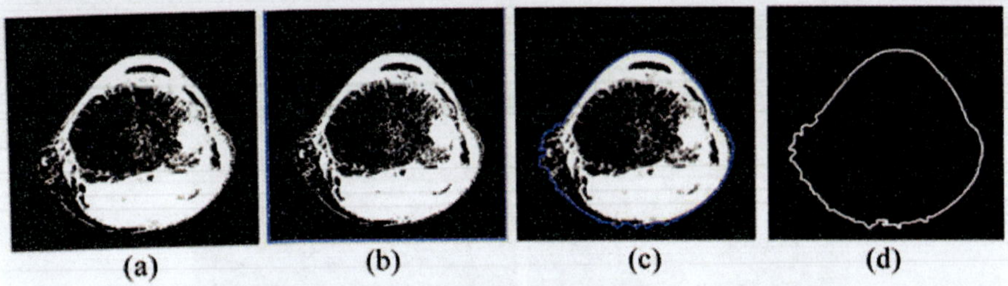

Fig. 4 The processing of the active contour model in salient map

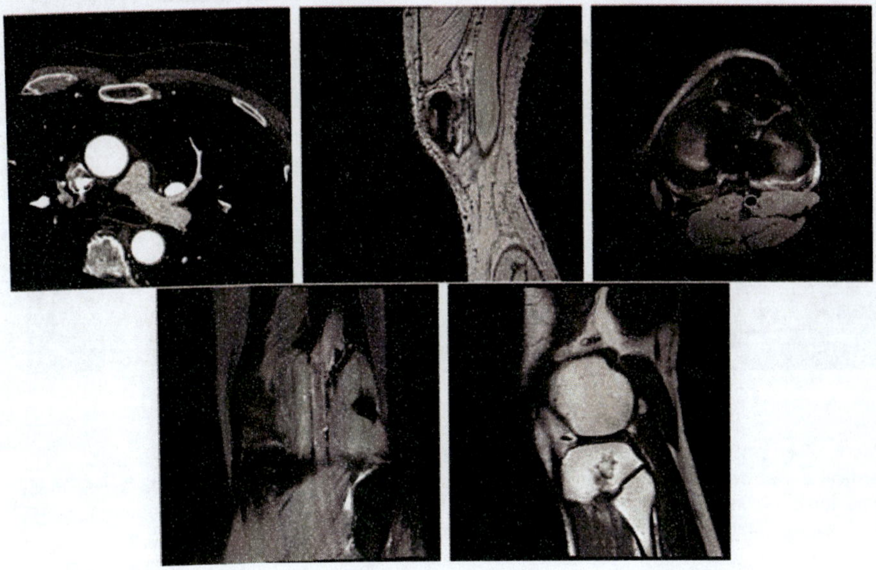

Fig. 5 Some images in the above dataset

4 Experiments and Results

Images can have blur or noise or pair [20]. The medical images quality engenders a wide range of mistake in edge or contour detection. The objects in medical images of these datasets include the parts of the body, such as femur, abdominal, kidney, skull, backbone, etc. To evaluate the results of the proposed method, we compare the results of the proposed method with the result of the other methods, such as Canny method [21], Sobel method [22] and Active contour model—Snakes [9] through the number of nonzero matrix elements (NNZ) value. To experiments and evaluate the results of the proposed method, we use DICOM dataset [23] for testing in this case. The experiments are built in Matlab R2018a language, and we use 800 medical images

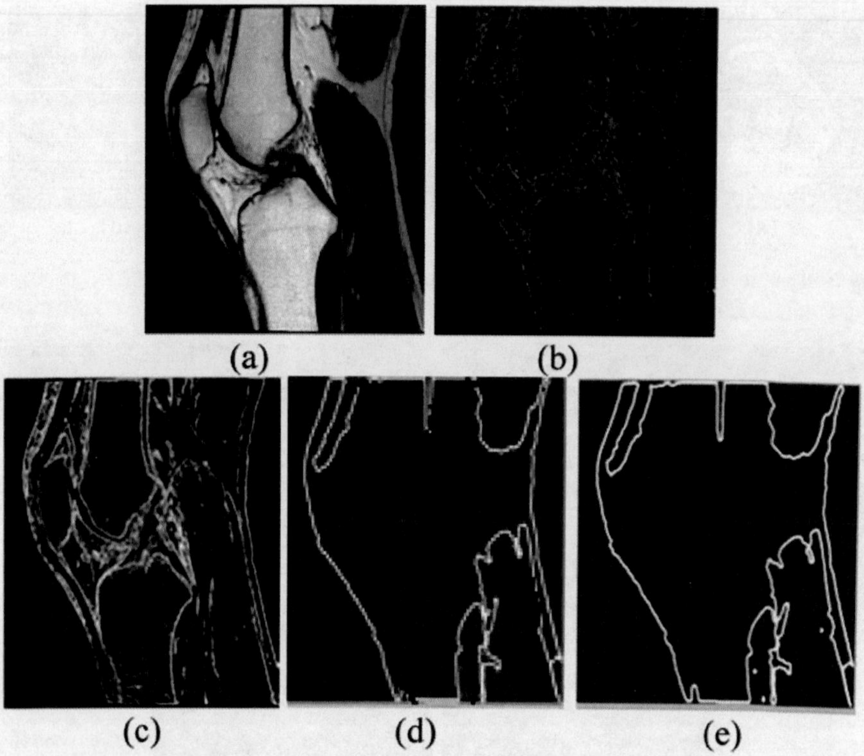

Fig. 6 The results of contour detection by the proposed methods and the other methods. **a** The medical image original. **b** Contour detection by Canny method [21] (NNZ = 15,037). **c** Contour detection by Sobel method [22] (NNZ = 62,304). **d** Contour detection by active contour model [9] (NNZ = 79,054). **e** Contour detection by the proposed method (NNZ = 94,023)

which are exported from DICOM dataset. The quality of each medical images in the dataset is unequal. Figure 5 presents some medical images of these datasets.

We count non-zero number and record non-border values. In fact, the background may contain several pixels that do not belong to the border. Therefore, they create some mistakes. We test with 800 images from the above dataset. In here, we only show the result of the proposed method with the results of the other methods as Fig. 6. Figure 6 shows the results of the medical images contour detection with the other methods. Figure 6a is the medical image original. The results of Canny method [21], Sobel method [22], active contour model method [9] present as Fig. 6b–d, respectively. Figure 6e presents the result of the proposed method. From the NNZ values, we can conclude that the results of the proposed method have better than the results with the other methods.

The average NNZ value of all images in this dataset is presented in Table 1. From Table 1, the result of the proposed method is better than the results of the other

Table 1 The average NNZ value of the above four methods

Total of medical images	The average NNZ value of Canny method [21]	The average NNZ value of Sobel method [22]	The average NNZ value of active contour model [9]	The average NNZ value of the proposed method
800 images	9729.81	40,350.7	43,378.64	45,915.26

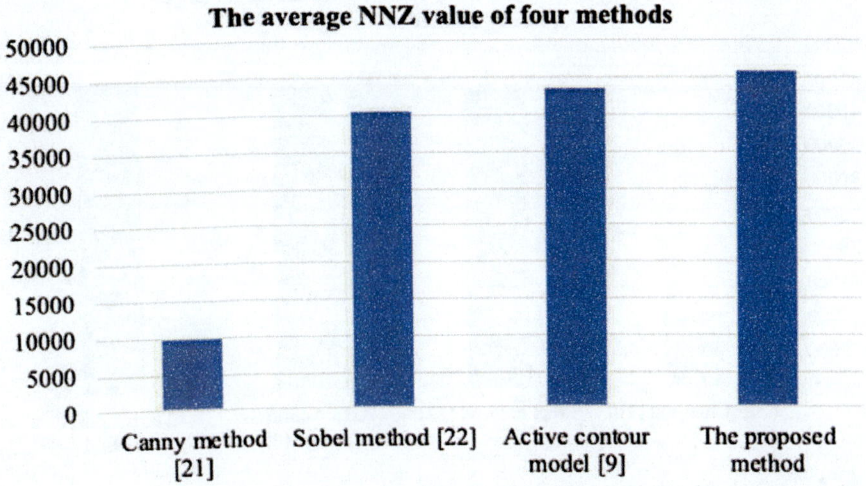

Fig. 7 The column chart for comparison between the above four methods by the average NNZ value

Table 2 The average NNZ value of the above four methods with high-quality medical images

Total of medical images	The average NNZ value of Canny method [21]	The average NNZ value of Sobel method [22]	The average NNZ value of active contour model method [9]	The average NNZ value of the proposed method
479 images	11,080.01	43,738.23	44,745.05	45,993.95

methods. This positive result shows that our solution adapts with contour detection in medical images.

We can compare clearly the average NNZ value of the above four methods in Fig. 7. This is a column chart for comparison between each method, and the proposed method is the best result.

On the other hand, we divide the above dataset into two groups: high-quality images and low-quality images by peak signal to noise ratio (PSNR). Tables 2 and 3 are the average NNZ value of the above four methods in high-quality and low-quality medical images, respectively. We use the PSNR is 23 db for the threshold between two groups.

Table 3 The average NNZ value of the above four methods with low-quality medical images

Total of medical images	The average NNZ value of Canny method [21]	The average NNZ value of Sobel method [22]	The average NNZ value of active contour model method [9]	The average NNZ value of the proposed method
321 images	8379.61	30,963.16	42,678.93	45,836.57

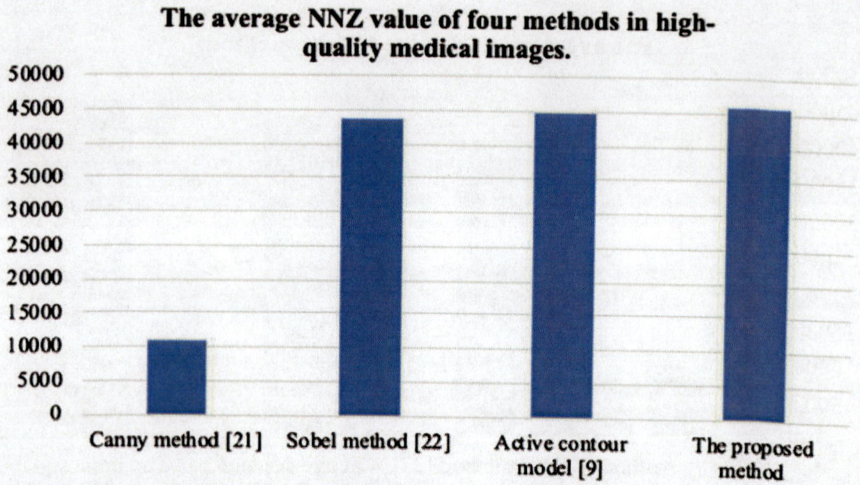

Fig. 8 The column chart for comparison between four methods by the average NNZ value in high-quality medical images case

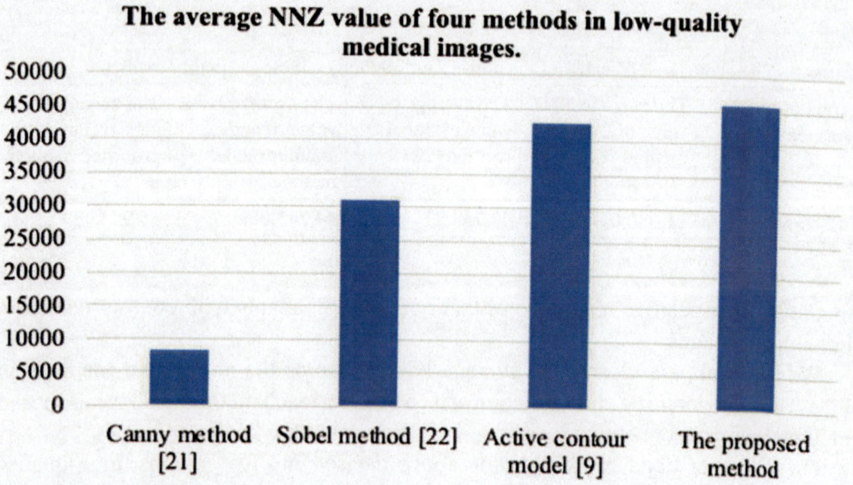

Fig. 9 The column chart for comparison between four methods by the average NNZ value in low-quality medical images case

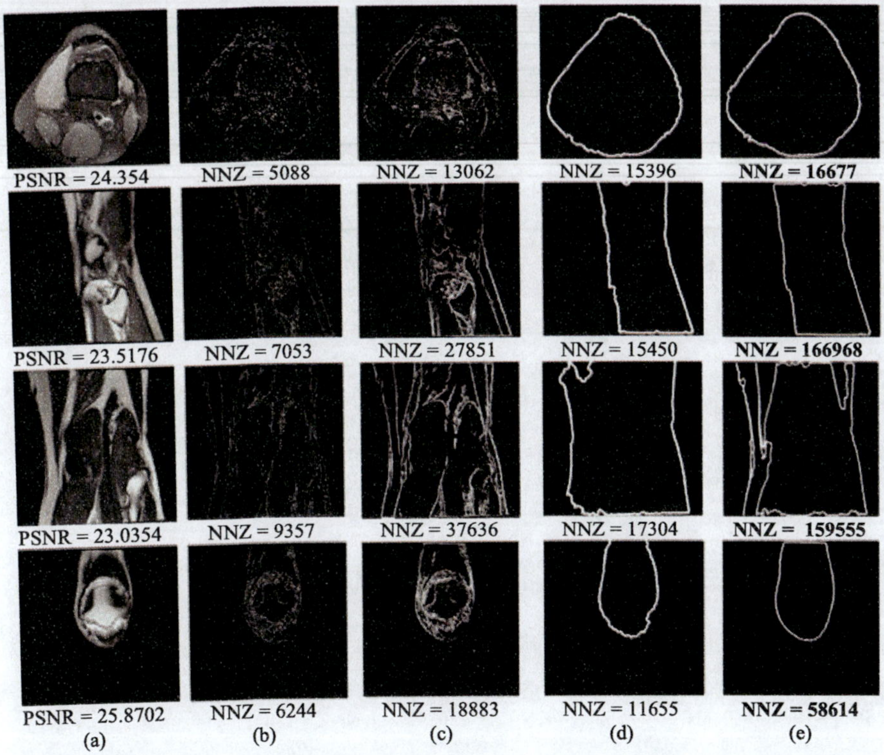

Fig. 10 The results of contour detection by the proposed methods and the other methods in high-quality medical images case. **a** The original high-quality medical images. **b** Contour detection by Canny method [21]. **c** Contour detection by Sobel method [22]. **d** Contour detection by active contour model [9]. **e** Contour detection by the proposed method

Table 2 has the average PSNR is 24.5458 db and Table 3 has 20.4612 db. The higher PSNR is better for quality. We can compare easily in Figs. 8 and 9.

Some high-quality and low-quality medical images cases are presented in Figs. 10 and 11. In Fig. 10a presents the original high-quality medical images, Fig. 10b presents the result of Canny method [21], Fig. 10c presents the result of Sobel method [22], Fig. 10d presents the result of active contour model method [9] and the proposed method presents as Fig. 10e.

In Fig. 11, the low-quality medical image of the dataset is showed in Fig. 11a, the contour detection by Canny method [21] is shown in Fig. 11b, the contour detection by Sobel method [22] is shown in Fig. 11c, the contour detection by active contour model is shown in Fig. 11d, and the Fig. 11e is the result of our way. From these values, the NNZ of snake in the salient map, the proposed method, is the higher.

The evaluation for the above solutions with the accuracy by percent is presented in Table 4. We compare with the number of nonzero matrix elements of ground-truth.

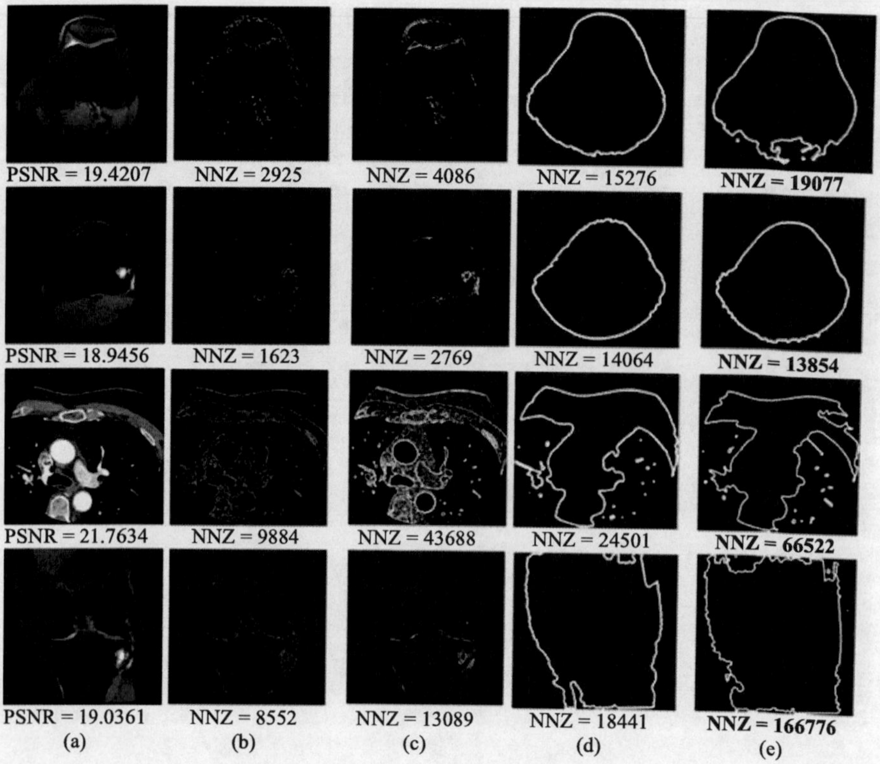

Fig. 11 The results of contour detection by proposed methods and the other methods in low-quality medical images case. **a** The original low-quality medical image. **b** Contour detection by Canny method [21]. **c** Contour detection by Sobel method [22]. **d** Contour detection by active contour model method [9]. **e** Contour detection by the proposed method

Table 4 The accuracy of four methods

Total of medical images	The accuracy of Canny method [21]	The accuracy of Sobel method [22]	The accuracy of active contour model [9]	The accuracy of the proposed method
800 images	20.67%	85.7%	92.13%	97.52%

The above figure and table, the NNZ values of the proposed method are higher than of Canny method [21], Sobel method [22], and active contour model method [9]. In two cases, high-quality and low-quality medical images, the proposed method gives good results in the average value. The proposed method has better results with other solutions due to the fact that energy minimizing in salient regions is easy for object detection.

5 Conclusions

The contour in the medical image is very vital for the medical field. In recent years, many researchers have focused on contour in medical images. However, this task has many challenges. In this paper, we proposed a new method for contour detection in medical images. The proposed method has four steps: pre-processing, the threshold of intensity, outstanding map and the active contour model to detect object contour. This method uses the concept about salient regions for the input of snake that the superpixel calculation for dividing the level of the edge map. We apply the median filter for pre-processing to balance color intensity between regions and to make clearer differences between background and subject. The result of the proposed method is compared with other methods by NNZ value and gives better results. In the future, we will be calculated with the execution time for minimizing. The case is true that the boundary is approximated in a long time by the loop. However, the cost of time is sometimes a pressing issue.

Conflicts of Interest The authors have no conflict of interest to declare.

References

1. Ju Z-W, Chen J-Z, Zhou J-L (2013) Image segmentation based on edge detection using K-means and an improved ant colony optimization. In: International conference on machine learning and cybernetics, pp 297–303
2. Gandhi NJ, Shah VJ, Kshirsagar R (2014) Mean shift technique for image segmentation and modified canny edge detection algorithm for circle detection. In: 2014 international conference on communication and signal processing. IEEE, pp 246–250
3. Achanta R, Estrada F, Wils P, Süsstrunk S (2008) Salient region detection and segmentation. In: Proceedings of international conference on computer vision systems, vol 5008. LNCS. Springer, pp 66–75
4. Paul L (2009) Rosin: a simple method for detecting salient regions. Pattern Recogn 42(11):2363–2371
5. Ao H, Yu N (2011) Edge saliency map detection with texture suppression. In: Proceedings of sixth international conference on image and graphics, pp 309–313
6. Goferman S, Zelnik-Manor L, Tal A (2012) Context-aware saliency detection. IEEE Trans Pattern Anal Mach Intell 34(10):1915–1926
7. Perazzi F, Krähenbühl P, Pritch Y, Hornung A (2012) Saliency filters: contrast based filtering for salient region detection. In: Proceedings of IEEE conference on computer vision and pattern recognition, pp 733–740
8. Xu C, Prince JL (1998) Snakes, shapes, and gradient vector flow. IEEE Trans Image Process 7(3):359–369
9. Saadatmand-Tarzjan M, Ghassemian H (2015) Self-affine snake for medical image segmentation. Pattern Recogn Lett 59:1–10
10. Zhang R, Zhu S, Zhou Q (2016) A novel gradient vector flow snake model based on convex function for infrared image segmentation. Sensors 16(10):1–17
11. Zhou S, Wang J, Zhang S, Liang Y, Gong Y (2016) Active contour model based on local and global intensity information for medical image segmentation. Neurocomputing 186:107–118

12. Chen Y-W, Shih C-T, Cheng C-Y, Lin Y-C (2017) The Development of skull prosthesis through active contour model. J Med Syst 41(10):1–10
13. Niu S, Chen Q, de Sisternes L, Ji Z, Zhou Z, Rubin DL (2017) Robust noise region-based active contour model via local similarity factor for image segmentation. Pattern Recogn 61:104–119
14. Nageswara Reddy P, Mohan Rao CPVNJ, Satyanarayana C (2017) Brain MR image segmentation by modified active contours and Contourlet transform. ICTACT J Image Video Process 8(2):1645–1650
15. Guo Q, Sun S, Ren X, Dong F, Gao BZ, Feng W (2018) Frequency-tuned active contour model. Neurocomputing 275:2307–2316
16. Zhu Y, Hao B, Jiang B, Nian R, He B, Ren X, Lendasse A (2017) Underwater image segmentation with co-saliency detection and local statistical active contour model. In: Proceedings of the OCEANS 2017—Aberdeen. IEEE, pp 1–5
17. Kasinathan G, Jayakumar S, Gandomi AH, Ramachandran M, Fong SJ, Patan R (2019) Automated 3-D lung tumor detection and classification by an active contour model and CNN classifier. Exp Syst Appl 134:112–119
18. He S, Pugeaulty N (2018) Salient region segmentation. Comput Vis Pattern Recogn 1–6
19. Binh NT, Tuyet VTH, Hien NM, Thuy NT (2019) Retinal vessels segmentation by improving salient region combined with Sobel operator condition. In: International conference on future data and security engineering, vol 11814. LNCS. Springer, pp 608–617
20. Binh NT, Tuyet VTH (2015) Enhancing the quality of medical images containing blur combined with noise pair. Int J Imag Graph Sig Process 7(11):16–25
21. Zahorodnia D, Pigovsky Y, Bykovyy P (2016) Canny-based method of image contour segmentation. Int J Comput 15(3):200–205
22. Halim MAA, Ruhaiyem NIR, Fauzi ERI, Jamil MSC, Mohamed ASA (2016) Automatic laser welding defect detection and classification using sobel-contour shape detection. J Telecommun Electron Comput Eng 8(6):157–160
23. http://www.osirix-viewer.com/resources/dicom-image-library/. Last accessed 15 Mar 2020

Visual Object Tracking Method of Spatio-temporal Context Learning with Scale Variation

An Hoang Nguyen, Linh Mai, and Hung Ngoc Do

Abstract In the development of image processing and computer vision field, visual tracking is considered as an attractive research field in regard to its practical characteristic in security surveillance, computer–human based interaction, motion and activity recognition in health care or control systems, etc. In a typical visual tracking model, the most difficult task is to handle the changes in the target objects' appearances and their surrounding backgrounds. As a matter of fact, if the changes are severe, information extracted to detect the object in interest will be limited. In this paper, we enhance the robustness of a tracking model to adapt to these changes and increase the tracking accuracy level by exploiting local context information. In particular, the study implements an efficient tracking model that utilizes the spatio-temporal context information. The context model relation of the tracking target with its surrounding background is generated by computing a devolution task due to its spatial correlation. Then, the analyzed relationship is exploited to update a spatio-temporal context in subsequent frames. The tracking process is computed using a confidence map by integrating the information within the spatio-temporal context. The model also implements the exhaustive scale estimation method to calculate the target's scale characteristic changes while maintaining computational efficiency. Finally, the TB-100 dataset is applied to evaluate the performance of the model.

Keywords Visual tracking · Spatio-temporal context · Scale variation · Confidence map · TB100 dataset

A. H. Nguyen (✉) · L. Mai · H. N. Do
School of Electrical Engineering, International University, Vietnam National University Ho Chi Minh City (VNU-HCMC), Ho Chi Minh City 700000, Vietnam
e-mail: nhan@hcmiu.edu.vn

L. Mai
e-mail: mlinh@hcmiu.edu.vn

H. N. Do
e-mail: dnhung@hcmiu.edu.vn

V. Van Toi et al. (eds.), *8th International Conference on the Development of Biomedical Engineering in Vietnam*, IFMBE Proceedings 85,
https://doi.org/10.1007/978-3-030-75506-5_59

1 Introduction

In the visual tracking field, the well-known methods are classified into three classes, which are (1) Deep Learning tracker DLT [1], (2) Correlation Filter trackers, such as CSK [2], STC [3], and KCF [4], and (3) others representative tracking algorithms using a single or multiple online classifiers, such as MIL [5], Struck [6], CT [7], LSHT [8], TLD [9], SCM [10], MEEM [11], and TGPR [12].

In general, no matter how the above algorithms work, they either follow generative or discriminative principles. The generative principle represents a learning model that studies and generates a representative model for the target object. The algorithm detects the object in interest in the following frames by searching for regions with the best matching scores with the representative model. However, this principle often discards the regions surrounding the object that may contain useful information, and the computation process involved is generally complex. On the contrary, the discriminative principle considers the tracking process as a binary classification task, which estimates the target object's borderline with its background. Due to its characteristics, the discriminative principle requires the algorithm to process through the feature extraction for classification, thus resulting in extensive computation.

In a more specific definition, a local context includes an object and its surrounding scene. Since the local context around the target's position can be assumed to be mostly stable, as the changes between consecutive frames are usually smooth within a limited time interval of approximately 30 frames per second. Thus, the characteristic results in a strong spatio-temporal context between the current frame with its consecutive ones. Moreover, the local context also provides the necessary information regarding the structure of the scene, results in an ability to distinguish the object with its background if the changes in appearance are significant.

Therefore, in this study, the researched model treats the local region of the object's surrounding background as potential context regions, which contain useful information to determine the location of the object in the visual tracking process. Furthermore, the correlation between the object in motion and its local context within a series of consecutive frames could be computed well using FFT. The study also provides a method to enhance the flexible characteristic to adapt to changes in the tracking target's size, resulting in the loss of the target trajectory during the tracking process.

2 Methods of Visual Tracking

The mentioned algorithm exploits a fast and effective visual tracking method that utilizes the dense spatio-temporal context information. First, a spatial context model represents the relationship of the object and its background in frame t, is generated by computing a devolution task. Secondly, the analyzed relationship is exploited to update a spatio-temporal context model for frame $t + 1$. The tracking process is

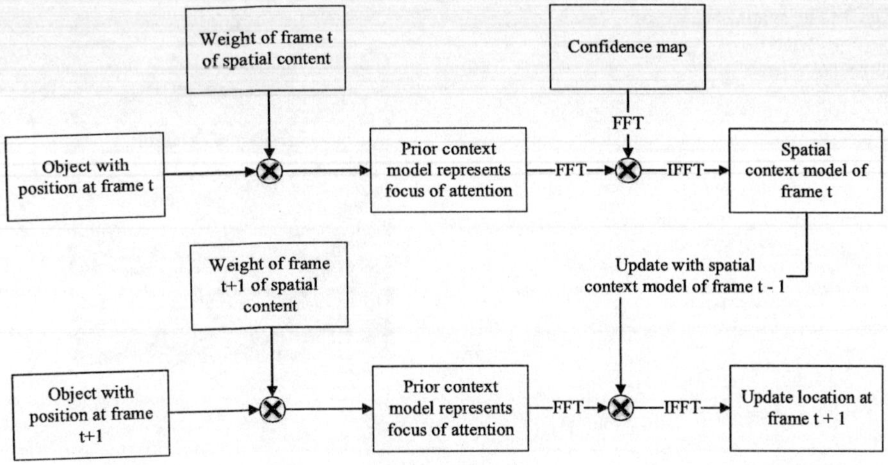

Fig. 1 The spatio-temporal context (STC) algorithm [3]

then computed using a confidence map by integrating the information within the spatio-temporal context (STC) by performing a convolution step (Fig. 1).

2.1 *Spatio-Temporal Context Algorithm*

Formula establishment. The tracking algorithm is executed by starting with a confidence map to estimate the coordinates of the object in the current frame:

$$f(\boldsymbol{x}) = P(\boldsymbol{x}|e) \tag{1}$$

In (1), $\boldsymbol{x} \in \mathbb{R}^2$ indicates the target position, whereas e denotes the existence of the target in the current frame. Figure 2 illustrates the relationship between the object with its surrounding background, which creates a dense spatio-temporal local context.

The local context feature set is defined in Eq. (2).

$$X^s = \{s(\boldsymbol{c}) = (I(\boldsymbol{c}), \boldsymbol{c})|\boldsymbol{c} \in A_s(\boldsymbol{x}^*)\}, \tag{2}$$

where $I(\boldsymbol{c})$ is the image gray-scale level at coordinate point $\boldsymbol{c}$., and $A_s(\boldsymbol{x}^*)$ is the area twice as much as the target object's size, centered on the location $\boldsymbol{x}^*$ (i.e., the centroid of the object). By summing up all the probability of the target to be appeared, i.e., marginalizing the joint probability $P(\boldsymbol{x}, \boldsymbol{c}(z)|e)$, the existence of the object location can be estimated by the formula).

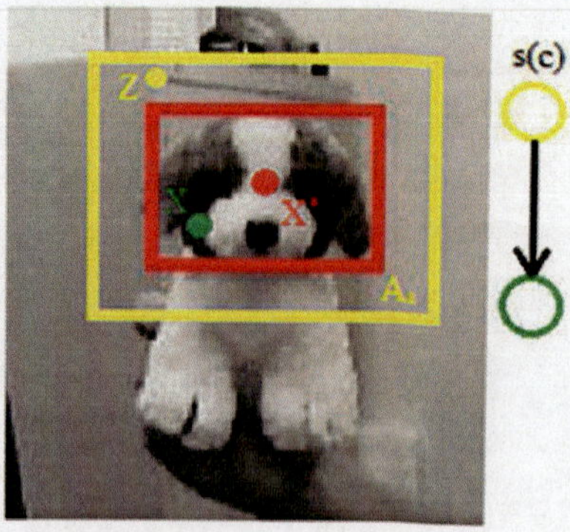

Fig. 2 The spatial context of the object and its surrounding area

$$f(\boldsymbol{x}) = P(\boldsymbol{x}|e) = \sum_{s(c)\in X^s} P(\boldsymbol{x}, s(\boldsymbol{c})|e)$$
$$= \sum_{s(c)\in X^s} P(\boldsymbol{x}|, s(c), e)P(s(c)|e) \quad (3)$$

The term $P(\boldsymbol{x}|, s(\boldsymbol{c}), o)$ illustrates the spatial relationship of the target object with its local context information, while $P(\boldsymbol{c}(z)|e)$ is a context prior to the probability that describes the model appearance of the target local context.

Spatial context model. As described in Eq. (3), the conditional probability can be obtained by applying Eq. (4).

$$P(\boldsymbol{x}|, s(\boldsymbol{c}), e) = h^{sc}(\boldsymbol{x} - \boldsymbol{c}) \quad (4)$$

The term $h^{sc}(\boldsymbol{x} - \boldsymbol{c})$ represents the relative vector of the distance between the target object location $\boldsymbol{x}$ with respect to its local context coordinate point $\boldsymbol{c}$, which handles the role of encoding the spatial relationship between the object and its spatial context. Equation (4) also helps to solve the problem of ambiguity in the case there is presented within the examining frame, a similar object with the object in interest, such as the right and left eyes of a human with a typical surrounding background.

Context prior model. The context prior mentioned in Eq. (3) is used to describe the spatial context's appearance simply by applying Eq. (5).

$$P(s(c)|e) = I(\boldsymbol{c})w_\theta(\boldsymbol{c} - \boldsymbol{x}^*), \quad (5)$$

where $I(\boldsymbol{c})$ is the image intensity that indicates the appearance of the local context, while $w_\theta(\boldsymbol{c} - \boldsymbol{x}^*)$ is the Gaussian weight formula. Equation (5) explains how important the distance between coordinate point $\boldsymbol{c}$ of spatial context with currently tracking location for the target object $\boldsymbol{x}^*$. As the distance getting smaller, the more important it becomes for the model to predict the object location in the following frame; thus, the weight function should be adjusted to be more significant.

Fast Learning Spatial Context model. The confidence map takes into account the problem of location ambiguity in the tracking process by considering the prior information of the target location. On the contrary, with the weight function, the ambiguity problem becomes easier to occur the closer the object's location is getting into with respect to the current tracking location. The confidence map can be achieved by (6)

$$f(x) = ke^{-\left|\frac{x-x^*}{\sigma}\right|^{\lambda}}, \tag{6}$$

Based on the above equations, the problem of tracking the object in interest from Eq. (2) can be rewritten as a convolution operation:

$$\begin{aligned} f(x) &= \sum_{c \in A_s(x^*)} h^{sc}(x-c)I(c)w_\theta(c-x^*) \\ &= h^{sc}(x) \otimes \left(I(x)w_\theta(x-x^*)\right) \end{aligned} \tag{7}$$

Applying the Fast Fourier Transform (FFT) algorithm for extracting target feature in frame t, then updating the spatial context model h^{sc} for relative spatial relations between different pixels in the object's consecutive frames, by following a Bayesian framework in equation number (8) [13].

$$h^{sc}(x) = \mathcal{F}^{-1}\left(\frac{\mathcal{F}\left(ke^{-\left|\frac{x-x^*}{\sigma}\right|^{\lambda}}\right)}{\mathcal{F}(I(c)w_\theta(x-x^*))}\right) \tag{8}$$

2.2 *Size Variation Problem and Solution*

The basis of the above algorithm is centered on the declaration in [3], in which the variation of the object and its surrounding background does not change drastically over a certain number of consecutive frames, which make it possible for the model to adapt to changes over time. However, due to certain circumstances, such as object scale rapid variation, motion blur, the object being covered completely or partially, illumination or brightness variation, etc. result in the target area of the object would be extracted inaccurately.

In [13], the authors propose a method combined with two processes of translation and scale estimation method for visual tracking purposes by learning a 3-dimensional scale-space correlation filter. The size of the filter is $M \times N \times S$, in which M, N are the height and width of the filter, whereas S is the number of scales, respectively.

For the purpose of updating the filter for the consequence frames, a feature pyramid in a rectangular area around the target is computed in the first frame. The feature

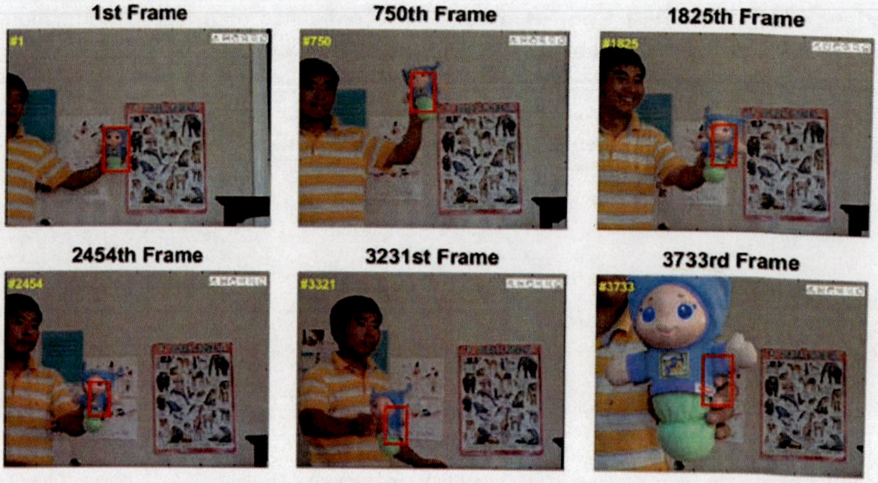

Fig. 3 Tracking result under the heavy effect of target scale variation

pyramid is also constructed such that the size of the target is also M × N at its respective scale S. The training sample f, in which there is only one frame, is then set into a rectangular cuboid of the pyramid set. The cuboid is of size $M \times N \times S$ and is located at the center of the target has calculated position and scale, in which we are only interested in the scale estimation of the object in consecutive frames (Fig. 3).

Since the model only considers one training sample, which is the very first frame of the image sequences, we can generalize the formula as described in Eq. (9).

$$\varepsilon = \left\| \sum_{j=1}^{d} h^j \odot f^j - g^2 \right\| + \lambda \left\| \sum_{j=1}^{d} h^j \right\|^2 \tag{9}$$

The value of $\lambda \geq 0$ indicates the weight of the regularization term. Thus, the solution to (9) for updating the object's scale is obtained in Eq. (10).

$$H^j = \frac{\overline{G} F^j}{\sum_{k=1}^{t} \overline{F^k} F^k + \lambda} \tag{10}$$

3 Results and Discussions

In this study, the image sequences are originally created from [14], and accepted as benchmark tracking dataset for performance evaluation in the visual tracking field, referred to as TB100 dataset with 100 image sequences within the dataset in total.

Originally, the authors in [3] applied the STC into 18 image sequences. In this study, for the purpose of emphasizing the effect of applying the exhaustive scale estimation model, we focus only on applying the STC and the combination model onto five image sequences selected as: 'Car2', 'Dancer', 'Dog1', 'Doll', and 'Shaking'.

The two methods used for the performance evaluation stage are called the Precision plot and the Successive plot [14]. The precision plot is used to measure the ability to track down the implemented model's target object, while the Successive plot indicates the efficiency in measuring the scale and shape of the model toward tracking the target.

The precision plot measures the distance between the centroid of the current tracking result of the evaluating model with the centroid of the target provided by the dataset creator. The threshold to indicate whether the tracking process is successive or not is within 20 pixels. Throughout the tracking period, the model estimates the percentage of frames in which the model is successfully tracking the target object. In Fig. 4, the results of the precision plot of the five image sequences are presented.

The results in Fig. 4 indicate 4 out of 5 image sequences are enhanced in terms of tracking precisely. In the case of 'car2', the effect does not present clearly, as the centroid of the target is located down perfectly. The model slightly increases the precision level from 90.26 up to 93.46% in the' doll' sequence. On the other hand, for the remaining three sequences, the model enhances more than 11% in terms of accurate tracking down the object positions. In more detail, the 'dancer' increases from 86.67 to 97.77%. Moreover, the 'shaking' rises up 24.4 to 94.52%. Last but not least, the 'dog1' sequence enhances more than 26% from 73.48%.

On the other hand, the precision plot only indicates the centroid position without taking into account the size and shape of the tracking target. The successive plot estimates the overlap region of bounding boxes, defined as Intersection over Union (IoU), of the currently evaluating model, with the position of the target provided in advance. The result of the overlap region is normalized in the 0 to 1 range. Throughout the tracking period, the model will calculate the average of overlap region to present the applied model's effectiveness.

The results in Fig. 5 indicate 4 out of 5 image sequences are tracking successfully. In the case of 'shaking', the model does not enhance the result, as the nature of this sequence is affected by other characteristics such as motion blur and brightness variation. However, as indicated by the Precision plot, the tradeoff in the tracking effect is worth consideration. On the other hand, of 4 remaining sequences, the effect is significant. The success level increases at least 19% overall such as in the 'doll' case, the range of enhancing is from 42.07 to 61.32%. For the remaining cases, 'dog1' increases around 34–74.166%, 'dancer' rises to 76.56% from 32.66%. The most significant affected sequence is 'car2' by increasing approximately 51–89.75%.

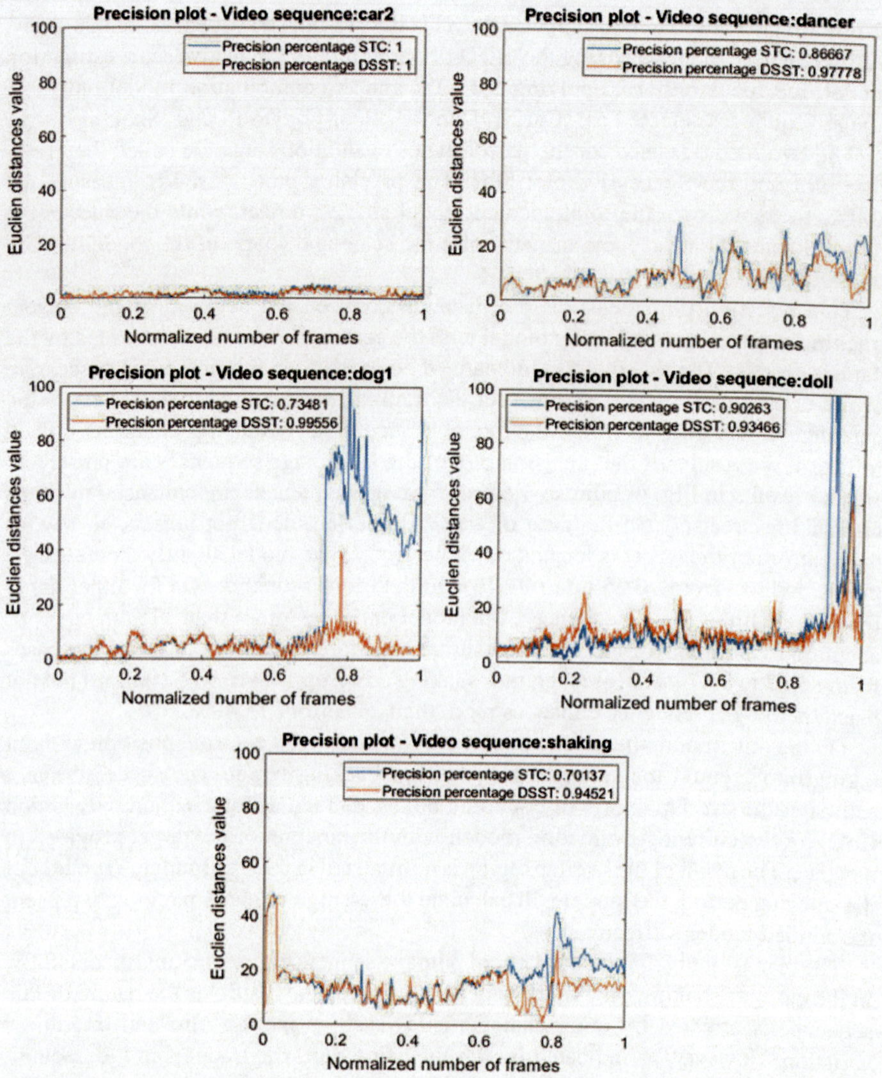

Fig. 4 Results of precision plot in performance evaluation

4 Conclusions

The study initially implements a visual tracking model using spatio-temporal context learning by exploiting the feature extracted from the object in interest as well as the information from the target surrounding background. The exhaustive scale estimation application into the existing model also provides positive outcomes by enhancing the robustness of the model against scale variation characteristic in general and the

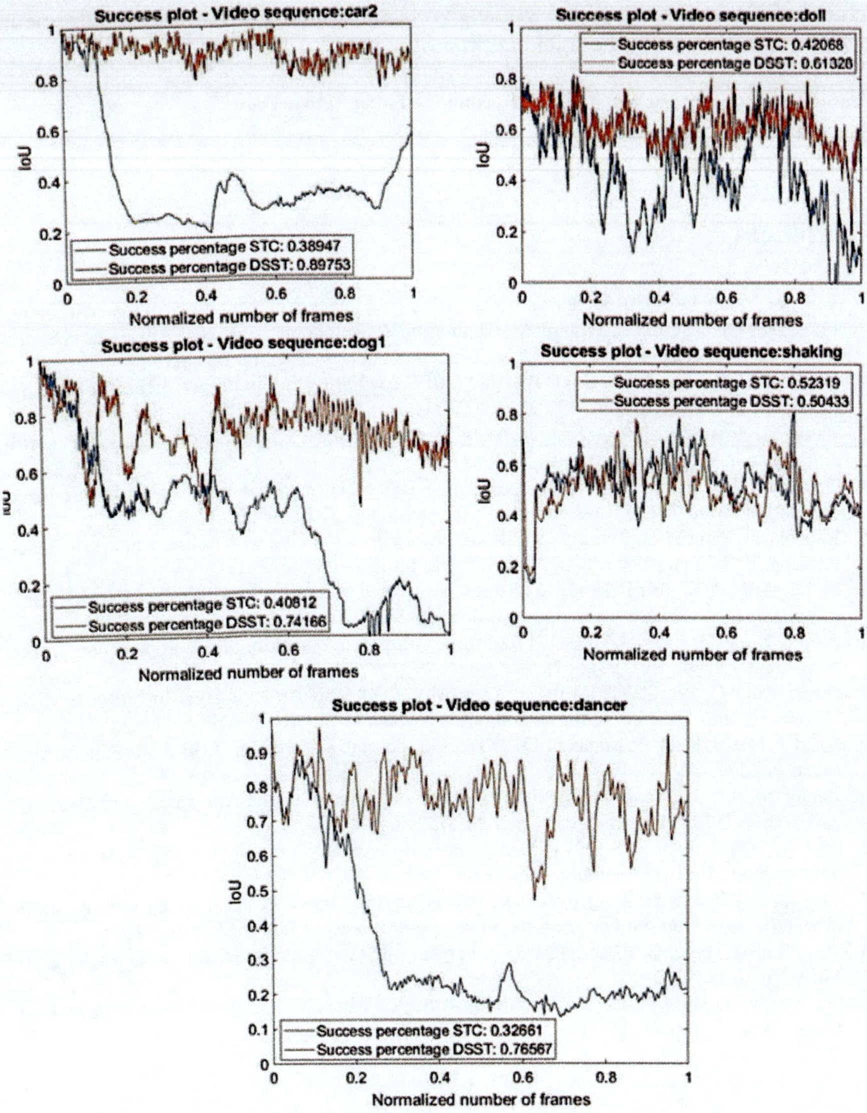

Fig. 5 Results of success plot in performance evaluation

tracking results of different image sequences as a whole. However, by applying another estimation process into the model, the computation process becomes taxing and, subsequently, time performance in total. Moreover, the effect of each character such as illumination or brightness variation, motion blur, partial occlusion, etc. on the combination model is still open for further investigation and research in order to achieve an ideal tracking system with high robustness, less pre-learning requirements and computation during the tracking process.

Conflicts of Interest The authors have no conflict of interest to declare.

References

1. Wang N, Yeung D-Y (2013) Learning a deep dompact image representation for visual tracking. In: Burges CJC, Bottou L, Welling M, Ghahramni Z, Weinberger KQ (eds) Advances in neural information processing systems, vol 26. Curran Associates, Inc., pp 809–817
2. Henriques J, Caseiro R, Martins P, Batista J (2012) Exploiting the circulant structure of tracking-by-detection with kernels. LNCS 7575:702–715
3. Zhang K, Zhang L, Liu Q, Zhang D, Yang M-H (2014) Fast visual tracking via dense spatio-temporal context leraning. In: ECCV 2014
4. Henriques J, Caseiro R, Martins P, Batista J (2015) High-speed tracking with kernelized correlation filters. IEEE Trans Pattern Anal Mach Intell 37:583–596
5. Babenko B, Yang M-H, Belongie S (2011) Robust object tracking with online multiple instance learning. IEEE Trans Pattern Anal Mach Intell, pp 983–990
6. Saffari A, Hare S, Torr PHS (2011) Struck: structured output tracking with kernels. In: IEEE international conference on computer vision, pp 263–270
7. Kaihua Z, Lei Z, Yang M-H (2012) Real-time compressive tracking. In: European conference on computer vision ECCV, vol 7574, pp 864–877
8. He S, Yang Q, Lau RWH, Wang J, Yang M (2013) Visual tracking via locaility sensitive histogram. In: IEEE conference on computer vision and pattern recognition, pp 2427–2434
9. Kalal Z, Mikolajczyk K, Matas J (2010) Tracking-learning-detection. IEEE Trans Pattern Anal Mach Intell 6
10. Zhong W, Lu H, Yang M-H (2014) Robust objective tracking via sparse collaborative appearance. IEEE Trans Image Process 23(5):2356–2368
11. Zhang J, Ma S, Sclaroff S (2014) MEEM: robust tracking via multiple experts using entropy minimization. In: Lecture note on computer science—ECCV 2014, vol 8694
12. Gao J, Ling H, Hu W, Xing J (2014) Transfer learning based visual tracking with gaussian processes regression. In: Lecture note on computer science—ECCV 2014, vol 8691
13. Danelljan M, Hager G, Khan F (2014) Accurate scale estimation for robust visual tracking. Br Mach Vis Conf, 1–11
14. Wu Y, Lim J, Ming-Hsuan Y (2015) Object tracking benchmark. IEEE Trans Pattern Anal Mach Intell 37:1–14

Weighted Box Fusion Ensembling for Lung Disease Detection

Huy Tran, Long TonThat, and Kien Trang

Abstract Lung disease is a kind of infection that cannot be observed by human eyes, but it can be detected by X-ray images. Nevertheless, diagnostic accuracy from the look on these X-ray images is always a challenging task because it frequently relies on the experiences of the doctors. Hence, many computer vision techniques are developed to support medical staffs to perform lung disease detection. So as to encourage different detectors to identify different kinds of lung disease, this paper proposes a novel post-processing technique, a Weighted Box Fusion (WBF) ensembling algorithm, which may improve the efficiency in lung disease detection. This algorithm is applied to an available benchmarking dataset of X-ray chest in literature for justifying its performance.

Keywords Weighted box fusion · Lung disease detection · Post-processing · Image processing

1 Introduction

Every year, more than 250,000 people in the US seek care in hospitals due to pneumonia, and around 20% of them die [1]. Chest X-ray imaging becomes an important part of diagnosing pneumonia. Thus, pneumonia identification task, especially pneumonia localization, also plays an important role in identifying lung diseases. Pneumonia identification in chest X-rays is a difficult task because it needs to be analyzed by expert radiologists. Deep learning models can be developed to learn this knowledge through numerous of annotated data. The most recent model is designed to automatically detect pneumonia from chest X-rays [2–6]. The optimal goal of these approaches is to increase the performance of the identification task, i.e. pneumonia classification. Through years, the performance of the pneumonia identification task is improved, and the deep learning models get an acceptable accuracy to be applied in hospitals.

H. Tran (✉) · L. TonThat · K. Trang
International University, VNU-HCMC, Ho Chi Minh City, Vietnam

V. Van Toi et al. (eds.), *8th International Conference on the Development of Biomedical Engineering in Vietnam*, IFMBE Proceedings 85,
https://doi.org/10.1007/978-3-030-75506-5_60

Comparing to pneumonia identification, localizing exactly the disease region in chest X-rays is not always an easy task for doctors or even expert radiologists. It is also a challenging task for deep learning models because they need to deal with vague and noise in X-ray images which can mimic other diagnoses. Although many different approaches and proposed deep networks are designed to solve the pneumonia localization [7, 8], the performance is not good enough.

In this research, we introduce a method that can increase the performance of pneumonia localization by leveraging the effectiveness of different localization and detection models. This method is called Weighted Boxes Fusion (WBF) which can re-calculate the coordinates of the available boxes to give a more precise box. In pneumonia localization, areas of diseases are identified and bounded with rectangles accompanied by their pneumonia labels. There are multiple predictions of each model for the image. Every prediction consists of 6 numbers: 4 coordinates of the rectangle, class ID of pneumonia (for examples, *Atelectasis-1*, *Cardiomegaly-2*, *Effusion-3*, etc. ...) and a probability from 0 to 1—the feasibility of this pneumonia in the region [9]. Suppose that each image is analyzed with several different models which leads to many predictions. These predictions can be theoretically combined to obtain the higher quality of predictions. For computer vision imaging, NMS (Non-maximum Suppression) method and its extension—Soft-NMS [10] are known to have good performance in combining multiple detection results between different object detectors. Inspired by this idea, Weighted Boxes Fusion (WBF) is introduced to deal with the pneumonia localization in X-ray chest imaging. Note that WBF can significantly improve the accuracy of localization rectangles because it uses all the predicted rectangles instead of simply removing parts of the predictions like NMS [9].

The paper is described as follows. Firstly, WBF is introduced and used to deal with the pneumonia localization in X-ray chest imaging by leveraging multiple predictions of different localization and detection models. Secondly, the proposed WBF, an optimized version of the former approaches, uses all the predicted rectangles instead of simply remove part of the predictions. Finally, WBF outperformed baseline CheXNet by a large margin (0.08328 in IoU) on the pneumonia localization task of NIH Chest X-ray dataset by using only CheXNet models (i.e. name of a convolutional neural network normally used for X-ray imaging analysis) with different average boxes setup.

2 Proposed Method

2.1 Dataset

The ChestX-ray14 dataset contains 112,120 frontal-view X-ray images of 30,805 unique patients. The dataset, released by Wang et al. [11], used automatic extraction methods on radiology reports to annotate 14 different thoracic pathology labels in each image. However, due to the limitation of hardware resources, only images from the first eight classes {*Atelectasis, Cardiomegaly, Effusion, Infiltration, Mass, Nodule, Pneumonia, Pneumothorax*} are selected for training and reproducing. Note that there is no patient overlapping between the train set and the test set. Images are normalized and downscaled to the size of 224 × 224 before inputting into the network. Training data are augmented with horizontal flipping, shifting, rotating, and center cropping.

2.2 *Weighted Boxes Fusion*

WBF is a post-processing method. To achieve the effectiveness of it, a full process framework is introduced: (i) Disease Prediction; (ii) Heat Map Computation; (iii) Bounding boxes Localization, and (iv) WBF Application using bounding boxes outputted from different localization models.

Disease Prediction (Step 1)
Since pneumonia localization is developed upon pneumonia identification, it is necessary to train a high-performance pneumonia model. CheXNet is a 121-layer convolutional neural network trained on NIH ChestX-ray dataset. It is the current state-of-the-art model which is trained to identify diseases in X-ray Chest images. Thus, we leverage the architecture of CheXNet and train on data given in Sect. 2.1. At the end of this step, disease prediction models with different average boxes setups are produced.

Heat Map Computation (Step 2)
The disease prediction models got from Step 1 can output the distribution over pneumonia, i.e., confident scores. These scores are then passed through class activation mappings (CAMs) module introduced in [12] to produce heat maps. This is shown in Fig. 1.

Bounding Boxes Localization (Step 3)
After calculating the heat map, the global peak value with the scale of 0.9 will become the threshold to select local maximum points with qualified intensity. Regions with local maximal centroids are localized by calculating the difference between resulting heat maps and embedding maximum filter and a minimum filter in each heat map. Dilation is applied to all the local maxima, which are greater than the threshold in the image to accumulate multiple candidate points. The center of a predicted bounding box is the centroid of the accumulated components.

From observation, the similar disease tends to have the equally bounding box size. A box covering each local maximal centroid with boundaries higher than the threshold is firstly constructed for each image with its corresponding prediction and heat map. The box size of every distinct disease is the average of all these boxes. After each experiment, the threshold is decreased by 0.05 to get more bounding boxes. These boxes are then assembled to produce new boxes which have the most

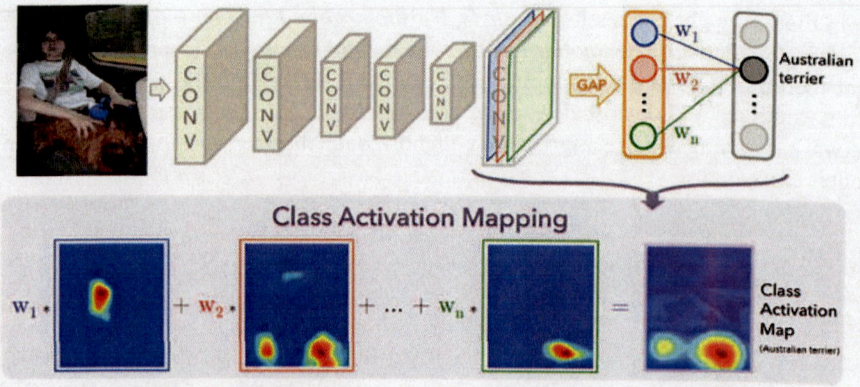

Fig. 1 Class activation mapping [12]

similar size with the ones in the validation set. The box of the centroid for each distinct disease is then re-calculated using the mean width and height.

WBF Application Using Bounding Boxes Outputted from Different Localization Models (Step 4)
Multiple bounding boxes from Step 3 output from different models are then passed through WBF to achieve a better area of pneumonia. Note that because this method only targets the localization of diseases in Chest, pneumonia prediction performance is not affected.

2.3 *Weighted Boxes Fusion Algorithm Description*

Suppose that there are several bounding box predictions from N different models for each image, and model predictions are considered as independent for the mirrored images [9].

The description of the WBF algorithm can have the following steps. Firstly, a single list B is created to contain all predicted boxes from each model. Each box has confidence scores C and is sorted in decreasing order. It is shown in Eq. (1) as follows

$$C = \frac{C_1 + C_2 + \cdots + C_T}{T} \quad (1)$$

Next, two empty lists L and an empty list F (Fusion) are created. The boxes in list B are circularly iterated from top to bottom. For each class, box has maximum Intersection over Union (IoU) which is larger than threshold (THR) value is considered to be 'matched box' and put into list F (THR = 0.55 was mostly closed to an optimal threshold in this experiment). If the box does not match, it is put to the end of list L and list F. The matched bounding box will be then added in L, including its position Z in F. This results in the situation where multiple bounding boxes can access the same stored position that L currently holds. A clustering method is then formed. Finally, we re-calculate the bounding box with its corresponding position stored in F by using all T bounding box at L[Z]. All of the descriptions above can be found via the following formulae written as follows:

$$X1 = \frac{C_1 * BX1_1 + C_2 * BX1_2 + \cdots + C_T * BX1_T}{C_1 + C_2 + \cdots + C_T} \tag{2}$$

$$X2 = \frac{C_1 * BX2_1 + C_2 * BX2_2 + \cdots + C_T * BX2_T}{C_1 + C_2 + \cdots + C_T} \tag{3}$$

$$Y1 = \frac{C_1 * BY1_1 + C_2 * BY1_2 + \cdots + C_T * BY1_T}{C_1 + C_2 + \cdots + C_T} \tag{4}$$

$$Y2 = \frac{C_1 * BY2_1 + C_2 * BY2_2 + \cdots + C_T * BY2_T}{C_1 + C_2 + \cdots + C_T} \tag{5}$$

The confidence of a fused box is a mean value of all confidence of boxes that form it. Weighted sums of coordinates of all boxes are the coordinates of the fused box. In this case, the confidence of boxes is weight; therefore, boxes with a large value of confidence will be dominant in the fused box coordinates. Finally, the confidences in F need to be re-scaled. The initial confidence value is multiplied with the ratio of T and N, where T is the number of boxes in a cluster, and N is a number of models. They need to be re-scaled because when the number in a cluster is low, that means just a few models predict the disease, then the confidence threshold needs to be decreased for those case. In some cases, a number of models could be smaller than a number of boxes in one cluster [9]. The following two formulae can be used:

$$C = C * \frac{\min(T, N)}{N} \tag{6}$$

$$C = C * \frac{T}{N} \tag{7}$$

3 Results and Discussions

The proposed method in this paper is implemented by using Py-Torch [13]. The experiments are conducted on a single NVIDIA Titan V with 12 GB RAM. For all of the experiments, the learning rate is set to 0.0002 and batch size is set to 64. Adam optimizer [14] is used to train the models.

To evaluate the contribution of WBF, we conduct ablation studies when training on the train set and testing on the test set of NIH X-ray Chest dataset.

3.1 *The Effectiveness of WBF Over Baseline*

Starting with the CheXNet model [2] as the baseline, we show the effectiveness of proposed WBF when they are integrated into the baseline. As shown in Table 1, with using WBF, the single CheXNet models achieve lower performances about ~0.04805 to ~0.08328 in IoU metric. The obtained results indicate that by leveraging predictions of different pneumonia detectors, the bias in each deep learning model is reduced and thus, increases the confidence of identified disease areas. Details about the visualization between with/without using WBF are shown in Fig. 2. The red box in each image is the ground-truth bounding box while the predicted box is the blue one.

Note that all of the models use DenseNet121 as the backbone and WBF is computed using all of these models. An average box is added for each prediction.

Table 1 Result of different CheXNet models with Bounding Box Algorithms and the introduced WBF

Model	IoU
CheXNet + Local Maxima	0.21780
Local Maxima + Average Box 1st setup	0.25075
Local Maxima + Average Box 2nd setup	0.25303
WBF	0.30208

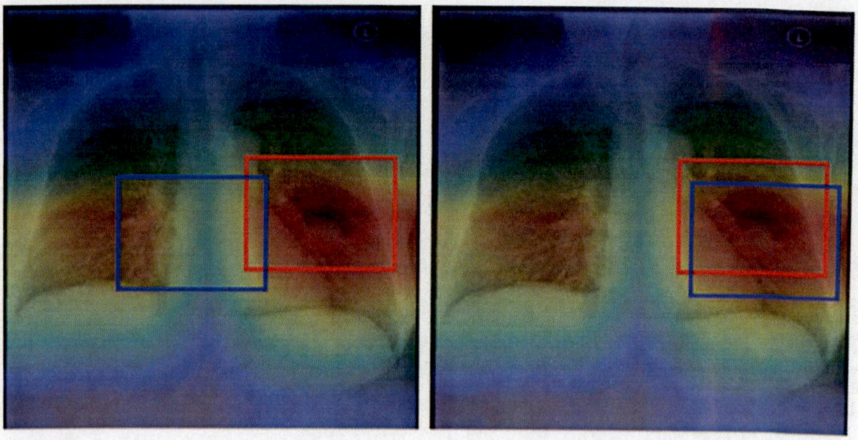

Fig. 2 The visualization between using WBF (right side) and without using WBF (left side)

Table 2 Result of different localization algorithms with their best thresholds

Model	IoU
Non-maximum suppression (NMS) (Best THR IoU = 0.1)	0.27012
Soft-NMS Gaussian (Best THR IoU = 0.45, Best Sigma = 0.05)	0.27778
Non-maximum suppression (NMW) (Best THR IoU: 0.2)	0.28477
Weighted box fusion (WBF) (Best THR IoU: 0.65)	0.30208

The 1st setup of Average Box is as follows. The average coordinate is calculated by the same method for the average box size, as described in step 3 of Sect. 2.2. The 2nd setup is the revised version of the 1st setup to reduce predictions with relatively lower likelihoods.

3.2 *The Effectiveness of WBF When Comparing with Different Approaches*

We compare the results of four methods NMS, Soft-NMS Gaussian, Non-maximum Suppression (NMW) [15] and the proposed WBF result in NIH X-ray Chest dataset for pneumonia localization task. Grid search is applied to find optimal parameters for each of the three methods, which is illustrated in Table 2.

For a fair comparison, all four methods use models following the CheXNet architecture as components. The results in Table 2 have qualified the effectiveness of introduced WBF over other methods.

4 Conclusion

We have proposed a post-processing method Weighted Boxes Fusion (WBF) accompanied by a full process framework to improve the performance of pneumonia localization by leveraging predicted bounding boxes of different detection and localization models. The extensive experimental results show that the proposed method yields the best performance with the most competitive approaches on the NIH X-ray chest dataset in pneumonia localization.

Conflicts of Interest The authors have no conflict of interest to declare.

References

1. Centers for Disease Control and Prevention. https://www.cdc.gov/dotw/pneumonia/
2. Rajpurkar P, Irvin J, Zhu K, Yang B, Mehta H, Duan T, Ding D, Bagul A, Langlotz C, Shpanskaya K, Lungren MP (2017) Radiologist-level pneumonia detection on chest X-rays with deep learning. arXiv preprint arXiv:1711.05225
3. Lakhani P, Sundaram B (2017) Deep learning at chest radiography: automated classification of pulmonary tuberculosis by using convolutional neural networks. Radiology 284(2):574–582
4. Huang G, Liu Z, Van Der Maaten L, Weinberger KQ (2017) Densely connected convolutional networks. In: Proceedings of the IEEE conference on computer vision and pattern recognition, pp 4700–4708
5. Islam MT, Aowal MA, Minhaz AT, Ashraf K (2017) Abnormality detection and localization in chest X-rays using deep convolutional neural networks. Computer Science, Cornell University, arXiv:1705.09850v3 [cs.CV]
6. Yao L, Poblenz E, Dagunts D, Covington B, Bernard D, Lyman K (2017) Learning to diagnose from scratch by exploiting dependencies among labels. Computer Science, Cornell University, arXiv:1710.10501v2 [cs.CV]
7. Li Z, Wang C, Han M, Xue Y, Wei W, Li LJ, Fei-Fei L (2018) Thoracic disease identification and localization with limited supervision. In: Proceedings of the IEEE conference on computer vision and pattern recognition, pp 8290–8299
8. Cai J, Lu L, Harrison AP, Shi X, Chen P, Yang L (2018) Iterative attention mining for weakly supervised thoracic disease pattern localization in chest X-rays. In: International conference on medical image computing and computer-assisted intervention. Springer, Cham, pp 589–598
9. Solovyev R, Wang W (2019) Weighted boxes fusion: ensembling boxes for object detection models. Image Vis Comput 107, Article 104117
10. Bodla N, Singh B, Chellappa R, Davis LS (2017) Soft-NMS—improving object detection with one line of code. In: Proceedings of the IEEE international conference on computer vision, pp 5561–5569
11. Wang X, Peng Y, Lu L, Lu Z, Bagheri M, Summers RM (2017) ChestX-ray8: hospital-scale chest X-ray database and benchmarks on weakly-supervised classification and localization of common thorax diseases. In: Proceedings of the IEEE conference on computer vision and pattern recognition, pp 2097–2106
12. Zhou B, Khosla A, Lapedriza A, Oliva A, Torralba A (2016) Learning deep features for discriminative localization. In: Proceedings of the IEEE conference on computer vision and pattern recognition, pp 2921–2929
13. Paszke A, Gross S, Massa F, Lerer A, Bradbury J, Chanan G, Killeen T, Lin Z, Gimelshein N, Antiga L, Desmaison A (2019) PyTorch: an imperative style, high-performance deep learning library. In: Advances in neural information processing systems, pp 8024–8035
14. Kingma DP, Ba J (2014) Adam: a method for stochastic optimization. 3rd International Conference on Learning Representations (ICLR) (Poster section), arXiv:1412.6980v9 [cs.LG]
15. Zhou H, Li Z, Ning C, Tang J (2017) CAD: Scale invariant framework for real-time object detection. In: Proceedings of the IEEE international conference on computer vision workshops, pp 760–68

Epileptic Seizure Detection Based on Electroencephalography Signals and One-Dimensional Convolutional Neural Network

Quynh Vu Nguyen Phuong, Minh Hiep Do Tran, and Huong Nguyen Thi Minh

Abstract Epilepsy is a group of neurological disorders characterized by recurrent epileptic seizures. According to the World Health Organization, there are about 50 million people who have epilepsy. Electroencephalography (EEG) is used as a powerful tool for doctors in diagnosis. By visualizing the EEG recordings, experts can initiate antiepileptic drug therapy and reduce the risk of future seizure. However, this current method is time-consuming and inflexible. With the development of deep learning, the problems can be solved. In this study, a 25-channel EEG data recorded at Neurology Department of 115 Hospital was converted into images after filtering, and a one-dimensional convolutional neural network (1-D CNN) model was applied to classify seizure states as seizure or non-seizure accurately. The accuracy of the model is about 90%. With the development of deep learning, it is more convenient to distinguish between different data of seizure and non-seizure without difficulties and consumption of time.

Keywords CNN · Seizure detection · Deep learning

1 Introduction

Epilepsy is a group of neurological disorders characterized by seizures. As a clinical tool, EEG plays an important role in the diagnosis of epilepsy. With the electrodes placed along patients' scalp, it recorded the voltage fluctuations produced by the neurons in the brain [1]. Traditionally, EEG signals are used to diagnosis by visual inspection of the doctors. This is such a slow process which may lead to disagreements between experts.

Deep learning, including deep neural network, deep belief networks, recurrent neural network and convolutional neural network, has been applied to many fields and

Q. Vu Nguyen Phuong (✉) · M. H. Do Tran · H. Nguyen Thi Minh
Ho Chi Minh City University of Technology—VNU-HCM, Ho Chi Minh City, Viet Nam

H. Nguyen Thi Minh
e-mail: nguyentmhuong@hcmut.edu.vn

V. Van Toi et al. (eds.), *8th International Conference on the Development of Biomedical Engineering in Vietnam*, IFMBE Proceedings 85,
https://doi.org/10.1007/978-3-030-75506-5_61

produced results comparable to human experts [2]. In this experiment, we introduce the Convolutional Neural Network (CNN) system to classify the images into healthy and epileptic.

2 Related Work

In our understanding of seizures detection, there are various ways that have been proposed in the past decade, but the main theme of them can be seen as extracting features by hand-engineered and using deep learning method to extract the features of EEG signal. In the paper [3], Yash Paul has done a great job in reviewing most of the state-of-the-art technology that has been used in decomposing EEG signal. In his review, the EEG signal can be separated into five domains, thereby using classification to detect a seizure. The five domain is time domain, frequency domain, wavelet domain, rational domain and EMD domain. Each domain has a different approach and also have an advantage as well as a disadvantage than the others. Overall, they all give a good result with the accuracy of over 90%, which is very promising.

Another approach is using Deep Learning method. In the last several years, there are many types of research has been conducted using CNN to detect seizure based on EEG signal. The study of Emami et al. [4] using the VGG-16 as a CNN architecture applied on the plot images of the EEG data. Ullah et al. [5] built a pyramid one-dimensional convolutional neural network (P-1D-CNN) which consist of three main modules:

1. Input module, which takes an input EEG signal, splits it into sub-signal using fixed-size overlapping windows and passes them to base P-1D-CNN models
2. An ensemble where the sub-signals are classified by base P-1D-CNN models
3. Fusion and decision.

Their work had process beautifully and give the result of 0.99 accuracies in classification. Another work is by Park et al. [6] his team has developed a model using 1D CNN, and 2D CNN in considers of the temporal-spatial correlation. Their work gave a promising result at 90.5% prediction accuracy with SNUH-HYU EEG dataset.

Bouaziz et al. [7] give a new perspective in data construction as they extract a fixed temporal window and transform it into a set of intensity images, the result approaches the accuracy over 99% give us great encouragement as our ideas of import data is relevant. Their research consists of two main steps: dataset construction and a convolutional neural network. In the first step, they extract a fixed temporal window from continuous EEG signals issued from the annotated dataset of 23 electrodes. The obtained windows are then transformed to a spatial representation by producing a set of intensity images. The convolutional neural network they are using is 2D CNN and consists of 8 layers comprising one initial input layer and five hidden layers followed by one fully connected layer and ending with the output classifier layer. Through these works, we can see that the deep learning approach have a high potential to investigate in.

The first idea of our team is to discover a new and simple model to detect the pattern of the person with epilepsy. The model that we chose is one dimensional convolutional neural network (1D CNN). The convolutional neural network is a model developed for image classification, in which the model accepts a two-dimensional input representing an image's pixels and colour channels, in a process called feature learning. With the same process, a 1D CNN extracts features from sequences data and maps the internal features of the sequence. This model is very effective for deriving features from a fixed-length segment of the overall dataset, where it is not so important where the feature is located in the segment.

While two-dimensional convolutional network (2D CNN) had been widely used for image classification, it may not be a suitable option in various applications over 1D signals, especially when the training data is scarce or application-specific. To overcome this problem, 1D CNNs have been proposed and immediately achieved the state of the art performance levels in several applications [8]. Real-time Electrocardiogram (ECG) Monitoring is an adaptive implementation of 1D CNN as it fuses two major blocks of ECG classification into a single learning body: feature extraction and classification. The results over the MIT-BIH arrhythmia benchmark database demonstrate that the proposed solution achieves a superior classification performance than most of the state-of-the-art methods for the detection of ventricular ectopic beats (VEB) and supraventricular ectopic beats (SVEB) [9].

3 Method

Two groups are considered in these experiments: (a) seizure and (b) non-seizures individuals. The recorded EEG data of these groups were converted into the two-dimensional images. Then, CNN was built using 1D-kernels to provide high classification rates.

3.1 Dataset

This step consists of building the input data for CNN architecture. The EEG obtained with 25 channels, and the sampling rate is 128 Hz was collected in 115 Hospital with diagnosis by doctors. The training data is divided into two groups, corresponded to the classes label—seizure and non seizure and was obtained from 4 patient. The patient is in good mental alertness and has a good attitude in cooperating with the doctor. Data were filtered with 1-Hz high pass and 40 Hz-low pass filters. Then those filtered EEG signals ware segmented to a 1-s epoch image, which means they are 128×25 in size (Fig. 1).

Fig. 1 The intensity image

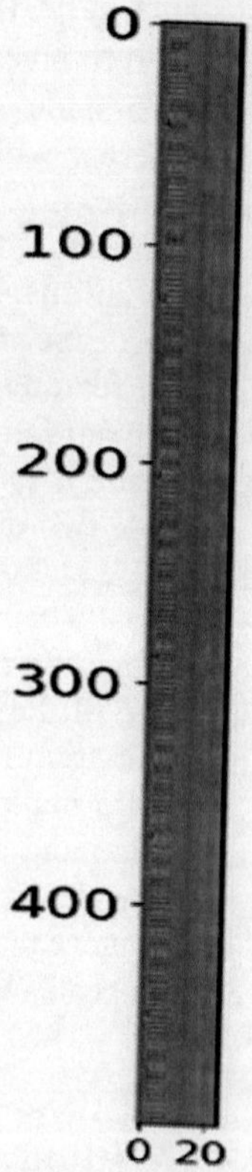

There are about 4696 images used as training data. The testing data consisted of over 15,000 images and was collected from 11 patient. The reasons for this imbalance in training and testing data are:

- While there are only two ictal data files in the training dataset, these two files have the longest ictal time and have no non-ictal epoch in the recording. The other files used for test dataset have a mixed between ictal and non-ictal epoch so it would be not proficient when using for training.
- As the proportion of the ictal and non-ictal data in training must be relatively equal, the non-ictal in the training data could not be too long. Therefore, the remain training data is relatively short due to the testing data.

Here is an illustration of our intensity image, presented as a gray-scale image with the x-axis is the number of channels, the y-axis is the sampling frequency and each pixel present for the intensity as each point.

3.2 Convolutional Neural Network Architecture

In our research, the process consists of 2 main step, that is:

1. Construct the data to create an input for the model
2. Create a 1D CNN model to extract features and classify seizure pattern.

In the first step, we are inspired by Bouaziz et al. work and construct the data into a set of intensity images. We first import the data as a picture with sizes of 128 × 25 pixel, as the sampling frequency is 128 Hz, each image contains the information of 1 s EEG signal. By using this technique, each pixel will consist of the information of the data and construct a condensed picture depicting how EEG signal varies. Another advantage is the small size of image input which will help reduce the complexity of the model to extract the feature.

The next step is to construct a model for detecting seizure pattern. Our model is composed by 7 layers with the input are 128 × 25 pixel pictures, and the output is seizure or non-seizure.

The CNN architecture and hyperparameters that have been used in our research was referred to the post of Singh [10]. This model was first used for speech recognition by using EEG signal sent to the larynx with binary classification. The result received in this post is quite promising, and regarding that, the required training input used for the model is relatively small consider to other models such as VGG16. By these reasons, we have chosen this model for our research as there is a limitation of the dataset we used. The result in our paper demonstrated that this model is highly effective in seizure detection (Table 1).

The activation function that we used to define the output of a node are Rectified Linear Unit function (ReLU), Softmax function and the loss function to optimize the parameters of the neural networks is the cross entropy (Table 2).

Table 1 Layers of the CNN architecture

Layer	Parameter
Convolutional 1D	40 kernels with the size of each are 10 pixels and use ReLU activation function
Drop out	0.2 (randomly subtract 20% of the previous node)
Max pooling 1D	The size of this layer is 3 (reduce 66% data from the previous layer)
Global average pooling 1D	Default
Fully connected 1	50 (Turn the previous matrices into a 50 nodes vector)
Drop out	0.2 (randomly subtract 20% of the previous node)
Fully connected 2	2 (Turn the vector into two node corresponding with two labels) using the softmax activation function

Table 2 Activation function and loss function

Function	Equation
Rectified linear unit	$f(z) = \max(0, z)$
Softmax	$a_i = \frac{\exp(z_i)}{\sum_{j=1}^{C} \exp(z_j)}, \forall i = 1, 2, \ldots, C$
Cross entropy	$H(p, q) = -\sum_{i=1}^{C} p_i \log q_i$

4 Result

4.1 Training Efficiency

After 200 training times, it can be seen that the accuracy of the training data is 99.78%, of the validation data is 97.8%. This is considered reasonable when compared to previous research, which has a similar accuracy [11]. The output of this method is the percentage of the images belong to seizure or non-seizure class. At first, we classify by which percentage of the label is higher; the image will belong to that label. Then we increased the threshold from 50 to 90% to classify the label for those images (Fig. 2).

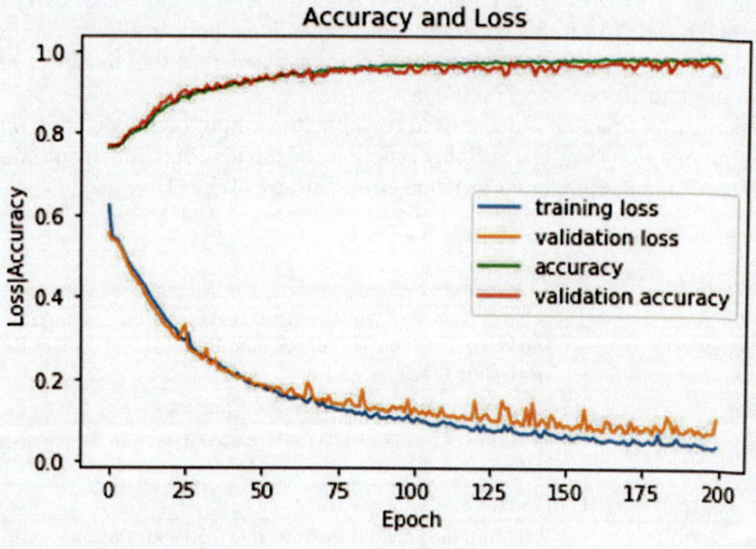

Fig. 2 Accuracy and loss value

4.2 *Prediction Result*

After the training process, the EEG data of 11 test subjects which is the seizure patients in 115 Hospital located in Ho Chi Minh City was converted into 14,000 image epochs in total have been used to evaluate the model. It is clear that the total percentage of correctly identifying the presence of seizure in patients is 81.8% (9/11) and the percentage of misidentification is 18.2% (2/11) (Table 3).

It is clear that the total percentage of correctly identifying the presence of seizure in patients is 81.8% (9/11), and the percentage of misidentification is 18.2% (2/11).

In order to avoid the 50/50 chance in the prediction result, we increase the threshold from 50 to 90% (Tables 4 and 5).

Table 3 Prediction result

Subject	Prediction results	Prediction results compared with the doctor's conclusions
CC121009	Seizure	Incorrect
HN161009	Non seizure	Correct
HS011009	Non seizure	Correct
VS080909	Non seizure	Correct
NA061109	Non seizure	Correct
VC261009	Non seizure	Correct
KP220709	Non seizure	Correct
TA030309	Seizure	Correct
NH121009	Seizure	Incorrect
TB300709	Non seizure	Correct
TK101109	Non seizure	Correct

Table 4 The number of images predicted to have epilepsy of the model

Subject	50% threshold	60% threshold	80% threshold	90% threshold
CC121009t	1057	1042	1008	987
HN161009	47	46	43	41
HS011009	552	491	431	393
VS080909	208	140	59	25
NA061109	435	416	383	348
VC261009	161	153	130	117
KP220709	965	959	932	907
TA030309	3464	3429	3339	3251
NH121009	338	324	293	264
TB300709	73	71	66	60
TK101109	384	299	278	255

Table 5 The number of images predicted not to have epilepsy of the model

Subject	50% threshold	60% threshold	80% threshold	90% threshold
CC121009	105	94	68	50
HN161009	411	409	405	402
HS011009	882	850	759	688
VS080909	603	525	337	205
NA061109	1079	1049	1005	940
VC261009	1084	1077	1042	1009
KP220709	351	343	314	292
TA030309	311	280	221	180
NH121009	55	48	21	13
TB300709	766	764	754	741
TK101109	1044	1020	986	951

In general, except for subject VS080909 and NH121009, the number of predicted number saw a trivial decrease.

5 Discussion

The final outputs of the model contain two numbers considered as the probability of classification. A higher numbers identified as the image label. The setting of threshold value can make effects on the accuracy of classification. Therefore, we examine the parameters of the ANOVA single factor to guaranteed that probability distribution does not have an impact on the result prediction (Tables 6 and 7).

When the threshold increases from 50 to 90%, the percentage of images classified as epilepsy decrease from 100 to 75.55% and those of images classified as epilepsy decrease from 100 to 71.83%. In general, we can see that changing the threshold result in a minor decrease in the prediction result. In the bar chart (Fig. 3), it can be seen that except for subject VS080909 and NH121009, there is no significant change in the number of the images which are predicted to be seizure or non-seizure.

In order to guaranteed the final result, we had used ANOVA-single factor method to assess the correlation between threshold and classification results. The ANOVA is an analysis tool commonly used in statistic; it allows us to determine if one factor is systematic or random. The systematic factors have a statistical influence on the given data set, while the random factors do not [10] (Tables 8 and 9).

The F value (0.020349 for epilepsy label and 0.1874 for the non-epilepsy label) is lower than the criteria value (2.838745 for epilepsy label and 2.838745 for the non-epilepsy label), which means that there is no evidence that the expected values in the four groups differ when we change the threshold. Thereby, we can say that the

Table 6 Attenuation of number of images predicted to have epilepsy when changing the threshold

Subject	50% threshold	60% threshold	80% threshold	90% threshold
CC121009	100	98.581	95.36	93.377
HN161009	100	97.872	91.49	87.234
HS011009	100	88.949	78.08	71.196
VS080909	100	67.308	28.37	12.019
NA061109	100	95.632	88.05	80
VC261009	100	95.031	80.75	72.671
KP220709	100	99.378	96.58	93.99
TA030309	100	98.99	96.39	93.851
NH121009	100	95.858	86.69	78.107
TB300709	100	97.26	90.41	82.192
TK101109	100	77.865	72.4	66.406
Average	100	92.06	82.23	75.55

Table 7 Attenuation of number of images predicted not to have epilepsy when changing the threshold

Subject	50% threshold	60% threshold	80% threshold	90% threshold
CC121009	100	89.52	64.76	47.61
HN161009	100	99.51	98.54	97.81
HS011009	100	96.37	86.05	78.00
VS080909	100	87.06	55.88	33.99
NA061109	100	97.21	93.14	87.11
VC261009	100	99.35	96.12	93.08
KP220709	100	97.72	89.45	83.19
TA030309	100	90.03	71.06	57.87
NH121009	100	87.27	38.18	23.63
TB300709	100	99.73	98.43	96.73
TK101109	100	97.70	94.44	91.09
Average	100	94.68	80.55	71.83

probability distribution varies just a random factor and does not catch any influence on the precision of the model.

Another result to be discussed is the model accuracy. The term of our research is to create a simple and effective model to detect epilepsy seizure. There are many types of research that had been using the CNN model, such as VGG16, ANN, state-of-the-art technology to detect seizure. These researches give a great result but also very complicated to proceed and have a high requirement with the processor. On the other hand, 1D CNN is simple to approach and do not require much in the processor. The result has shown us that 1D CNN can give a great detection in seizure signal. By

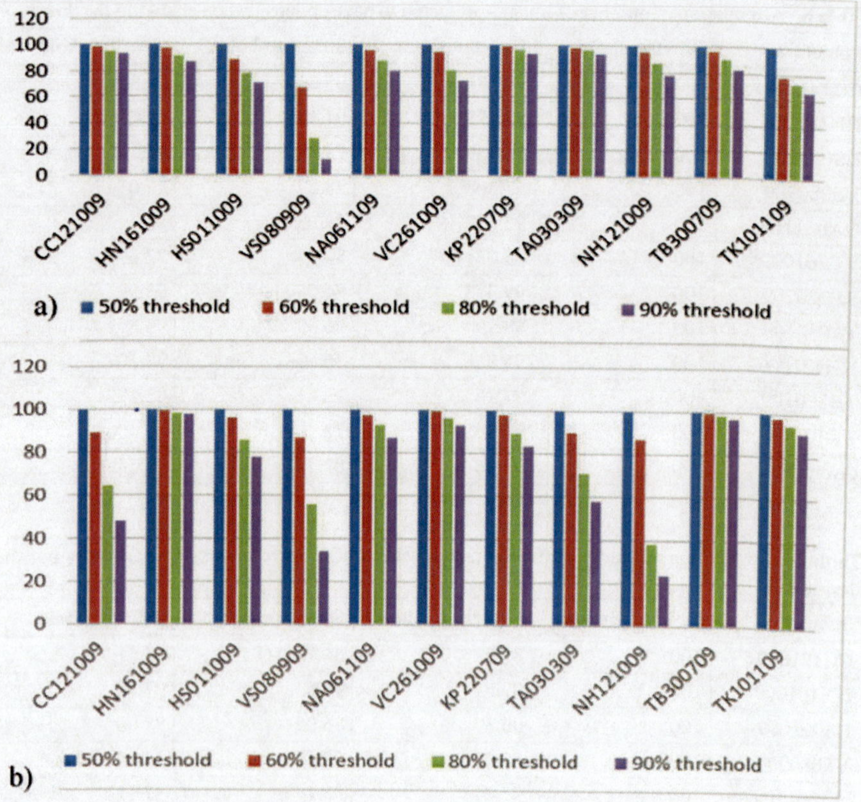

Fig. 3 Attenuation of the number of images. **a** Predicted to have seizure. **b** Predicted not to have seizure

Table 8 ANOVA analysis of images predicted to have epilepsy

Source of variation	SS	df	MS	F	*P*-value	F crit
Between groups	56,352.727	3	18,784.24242	0.020349	0.995993	2.838745
Within groups	36,924,108	40	923,102.7045			
Total	36,980,461	43				

comparing the result of our work with Park et al. [6], as his model using a combination of 1D CNN and 2D CNN in consider of spatial and temporal correlation, we can see that by using the 1D CNN only we still got an accuracy on the test set very close to the result of C.Park et al. (90.5% on SNUH-HYU EEG database, 85.6% prediction accuracy on CHB-MIT EEG database). Another result to be mentioned is Bouaziz

Table 9 ANOVA analysis of images predicted not to have epilepsy

Source of variation	SS	df	MS	F	*P*-value	F crit
Between groups	82,247.7	3	27,415.9	0.187411	0.904347	2.838745
Within groups	5,851,515	40	146,287.9			
Total	5,933,763	43				

Note SS Sum of square, *MS* Mean of square

et al. [7] as their work using 2D CNN and gave the validation accuracy and precision over 99% which is our result very close to.

6 Conclusion

By applying the CNN architecture to the EEG data, it becomes easier to distinguish between different data of seizure and non-seizure. To further improve the performance of the model, additional work should be considered; for instance, analyzing the real-time data.

Acknowledgements This research is funded by Vietnam National University Ho Chi Minh City (VNU-HCM) under grant number C2020-20-11.

Conflicts of Interest Vu Nguyen Phuong Quynh, Nguyen Thi Minh Huong and Do Chan Minh Hiep declare that there is no conflict of interest.

References

1. Tatum WO (2014) Handbook of EEG interpretation. Demos Medical Publishing, New York, pp 155–190
2. Bengio Y (2009) Learning deep architectures for AI. Foundations and trends in machine learning, vol 02, pp 1–55
3. Paul Y (2018) Various epileptic seizure detection techniques using biomedical signals: a review. Brain Inf 5:6. https://doi.org/10.1186/s40708-018-0084-
4. Ullah I, Hussain M, Emad-ul-Haq Qazi, Aboalsamh H (2018) An automated system for epilepsy detection using EEG brain signals based on deep learning approach. Expert Syst Appl 107. https://doi.org/10.1016/j.eswa.2018.04.021
5. Ullah I, Hussain M, Qazi E, Aboalsamh H (2018) An automated system for epilepsy detection using EEG brain signals based on deep learning approach. https://doi.org/10.1016/j.eswa.2018.04.021
6. Park C et al (2018) Epileptic seizure detection for multi-channel EEG with deep convolutional neural network. In: 2018 international conference on electronics, information, and communication (ICEIC), Honolulu, HI, pp 1–5
7. Bouaziz B, Chaari L, Batatia H, Rincon Q (2019) Antonio epileptic seizure detection using a convolutional neural network. In: Digital health approach for predictive, preventive, personalized and participatory medicine. (Advances in predictive, preventive and personalized medicine). Springer, Cham, pp 79–86. ISBN 978-3-030-11799-3

8. Kiranyaz S, Avci O, Abdeljaber O, Ince T, Gabbouj M, Inman D (2019) 1D convolutional neural networks and applications: a survey. Mech Syst Signal Process 151:107398
9. Kiranyaz S, Ince T, Gabbouj M (2015) Real-time patient-specific ECG classification by 1D convolutional neural networks. IEEE Trans Bio-med Eng 63. http://doi.org/10.1109/TBME.2015.2468589
10. Singh J (2019) Merging with AI: how to make a brain-computer interface to communicate with Google using Keras and OpenBCI. Internet: https://towardsdatascience.com/merging-with-ai-how-to-make-a-brain-computer-interface-to-communicate-with-google-using-keras-and-f9414c540a92. Sept 5, 2019
11. Cun YL (1995) Convolutional networks for images, speech, and time series. In: Arbid A (ed) Handbook of brain theory & neural networks

Classification of Four-Class Motor-Imagery Data for Brain-Computer Interfaces

Ngan Thanh Luu, Anh Hoang Lan Duong, Dung Xuan Nguyen, Khiet Thi Thu Dang, Chau Nu Ngoc Pham, Thuong Hoai Nguyen, and Ha Thi Thanh Huong

Abstract Brain-computer interface (BCI) provides opportunities for patients with severe motor impairment to regain their motor controls by decoding and converting Electroencephalogram (EEG) signals into the motor control signal. Existing processing schemes, including artifact removal and band-pass filtering, usually cause leakage of EOG artifacts. Hence, the preprocessing order in this paper was optimized for artifact removal in event-related desynchronization. To address this challenge, we implemented an algorithm to classify four different motor imagery tasks, including the left hand, right hand, both feet, and tongue movements. Here, we use Graz dataset A published in the BCI competition IV in 2008. Band-pass filter (7–30 Hz) with 5th order Butterworth IIR combined with linear regression was applied to remove the artifact of the signal. Mutual Information Best Individual Features (MIBIF) and Linear Discriminant Analysis (LDA) algorithms were used for selecting features and classifying the classes. The mean training kappa of our classifier is 0.57, and the mean evaluated kappa is 0.41. Besides, the application of the Common Spatial Pattern enhanced the kappa score (from 0.33 to 0.41). Those results are comparable to those from the BCI competition (range from 0.29 to 0.57). Hence, our approach could be the foundation for further development of motor imagery based BCI research. For future works, we propose to create an algorithm for the automatic procedure by automatically choosing proper frequency bands for band power features.

Keywords EEG · BCI · Motor imagery · Common spatial pattern · Linear discriminant analysis · Mutual information

---

N. T. Luu · A. H. L. Duong · D. X. Nguyen · K. T. T. Dang · C. N. N. Pham · T. H. Nguyen · H. T. T. Huong (✉)
School of Biomedical Engineering, International University, Ho Chi Minh City, Vietnam
e-mail: htthuong@hcmiu.edu.vn

Vietnam National University-Ho Chi Minh City (VNU-HCMC), Ho Chi Minh City, Vietnam

V. Van Toi et al. (eds.), *8th International Conference on the Development of Biomedical Engineering in Vietnam*, IFMBE Proceedings 85,
https://doi.org/10.1007/978-3-030-75506-5_62

1 Introduction

With the development of non-invasive techniques, electroencephalography (EEG) plays an important role in the research of neuroscience, cognitive science, and cognitive psychology, especially in the Brain-Computer Interface (BCI). EEG-based BCI system mainly relies on modulating the responses of large populations of neurons, either through a training period time or using external stimuli [1]. An important advantage of EEG is the extremely low risk to the subjects, compared to other invasive and semi-invasive techniques, such as electrocorticography (ECoG) [2]. Besides, EEG has an extremely high temporal resolution on milliseconds level, and its signals can be divided into several different frequency patterns (e.g. mu, beta) which can present different characteristics, patterns and access different functional brain states [3, 4]. Many factors are contaminating EEG signals, including both objective factors (e.g. environmental noises, faulty electrodes) and subjective factors [e.g. electrooculography (EOG)]. These factors result in various undesired artifacts, which reduce the classification accuracy in the BCI system, alter the characteristics of neurological phenomena, and make significant mistakes in controlling the BCI system [5, 6]. Therefore, it is important to apply the artifact removal and band-pass filtering to avoid causing leakage of EOG artifacts and gains meaningful EEG signals.

Through modulation of brain signals, such as motor imaginary (MI), BCI can be applied to both medical and non-medical fields. Applications in medicine, including cochlear implants for the deaf, deep brain stimulation for Parkinson's disease has become more and more popular in medical treatments [7, 8]. Besides, scientists and engineers have also investigated various non-medical BCI products such as lie detector, alertness monitor, games, e-learning system, etc. [9–12]. As an important role in BCI, EEG-based BCI systems have strong opportunities to help patients with severe motor impairment regain their motor controls such as cursor control, prosthetic control, and speller [13–15]. EEG in BCI also has applications in interacting with a Web browser, controlling robots as well as lie detector [16, 17]. Commonly, a BCI system works based on detecting changes in brain activity when the brain responds to voluntary or involuntary mental commands. Motor Imagery Brain-Computer Interface (MI-BCI) is the most popular used to detect EEG signals through imagined movements. When subjects imagine moving a particular part in the body, the neural activity in the sensorimotor cortex will be spatiotemporally similar to the activity during performing the real movement. Therefore, this type of neural response allows us to discriminate and match the EEG signals with each imagined body movement [18]. The biggest challenge in MI-BCI system is the difficulty in feature extraction, which needs making the selective features robust, informative, and discriminative. The motor imagery frequencies have both inter- and intra-subject variabilities, leading to low accuracy in this system [19, 20].

Related to EEG recording and motor imaginary system, Hans Berger has already described and discovered the oscillations in alpha waves during closed and open eyes in the 1930s [21]. This is a common phenomenon observed in the EEG experiments. A phenomenon like event-related desynchronization (ERD) and synchronization (ERS) was introduced to investigate the dynamics of EEG oscillations [22]. By imagining the movement with EEG and EMG recording, these phenomena can be able to change the power of the mu (8–13 Hz) and beta (13–30 Hz) bands [23, 24].

Event-related desynchronization (ERD) is a neurological phenomenon that was introduced by Gert Pfurtscheller and colleagues in the 1970s [25]. This can be used in BCI to classify different motor imagery classes. In the case of a subject imagine about a limb movement, event-related desynchronization will happen. It leads to a drop in the mu and beta waves' power. Otherwise, the event-related synchronization (ERS) increases the power of mu and beta waves when the imaginary movement stops.

The band power is the most common methodology used to determine event-related desynchronization and synchronization. First, the signal is filtered in distinct bands such as 8–14 Hz, 24–30 Hz..., then the influence level of the ERD/ERS effect was represented with different frequency components (e.g. in 8–14 Hz, the ERD effect occurs the most clearly). In the same frequency bands, the effect of ERD/ERS does not perform similarly for each subject, so it is recommended to select the frequency bands for individual participants. Then, the signal will be converted from frequency bands into power values. It means that the ERD/ERS alters when the power changes [26]. In an experiment, we perform various trials to get the final results, so the resulting signal is the average value over trials. This above information describes how to detect and quantify the effect of ERD/ERS by the band power method.

In this paper, we used the BCI Competition IV 2008 dataset IV 2a that contains the EEG signal recording of 9 subjects. This dataset includes 18 files of data recordings (nine files for the training set and nine files for the evaluation set) and 18 files of true labels for the evaluation step. The dataset was divided into four classes that are four different motor imaginary tasks (left hand, right hand, both feet, and tongue) [27]. This work is aimed to build an algorithm that is utilized to distinguish four above different motor imagery tasks. Normally, multichannel electroencephalogram (EEG) signals give fuzzy images of brain activity due to its low signal-to-noise ratio (SNR) [28]. This is the reason why it is required to filter signals before usable in BCI applications. In our proposed algorithm, we applied numerous methods to detect and remove all unwanted factors to get clean EEG signals. First, EEG signals were filtered by bandpass filter in the frequency band between 7 and 30 Hz with 5th order Butterworth IIR. Besides, we also applied linear regression and down-sampling 62.5 Hz in preprocessing.

In comparison with other techniques such as wavelet-based filter, linear regression is a simpler method that allows us to easily predict the effect of the filtering [29]. Due to its efficiency and reliability, these methods are extremely helpful to the signal-to-noise ratio improvement. In the feature extraction, the common spatial pattern (CSP) method was introduced. CSP is the most efficient and common method in BCI that is used to extract the discriminability classes from EEG signals [30, 31]. This algorithm is a spatial filtering method that can be able to optimize the discriminability of two classes [30].

Furthermore, the number of features CSP components can be reduced by the CSP method; it leads to optimize differences between motor imagery classes. After that,

Mutual Information Best Individual Features (MIBIF) was used to select the best features for classification. The combination of CSP and MIBIF can raise the classification accuracy and Kappa score. In classification algorithms, linear discriminant analysis (LDA) was used to classify features and reduce the dimension of the feature vector.

2 Materials and Methods

In this work, we aim to translate the signal from the brain into a typical movement. To reach this goal, a set of processing phases is required to create the control signal, which comprises pre-processing, feature extraction, feature selection, and classification. An illustration of the processing scheme has been shown in Fig. 1 and the detail of each stage is represented in the following sections.

2.1 Data Acquisition and Dataset

The dataset used for this work is the dataset 2a from BCI Competition IV in 2008 [27]. This data set is provided by the Institute for Knowledge Discovery (Laboratory of Brain-Computer Interfaces), Graz University of Technology, (Clemens Brunner, Robert Leeb, Gernot Müller-Putz, Alois Schlögl, Gert Pfurtscheller).

The dataset 2a includes EEG data of 9 subjects. For each subject, two sessions were recorded on different days. There are six runs for each session, and each run consisted of 48 trials for four classes which means a single session during the experiment consisted of 288 trials. The cue of each trial was related to four different motor imaginary tasks of the left hand (class 1), right hand (class 2), foot (class 3), and tongue (class 4). The trial begins with a warning tone and a fixation cross displayed on the computer screen at the same time. After two seconds, a cue was shown as a small arrow pointing to the left, right, up, and down, and lasted for 1.25 s. The subjects were prompted to perform the imaginary movement task until the fixation cross disappeared from the screen at $t = 6$ s. After that, there was a short break. Figure 2 illustrates the paradigm of a single trial.

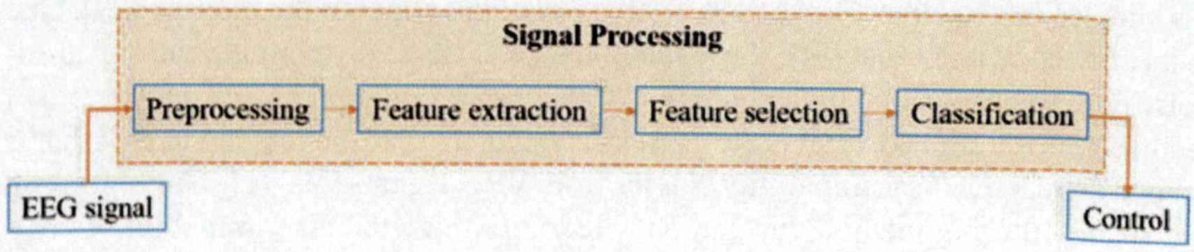

Fig. 1 The workflow of the experiment

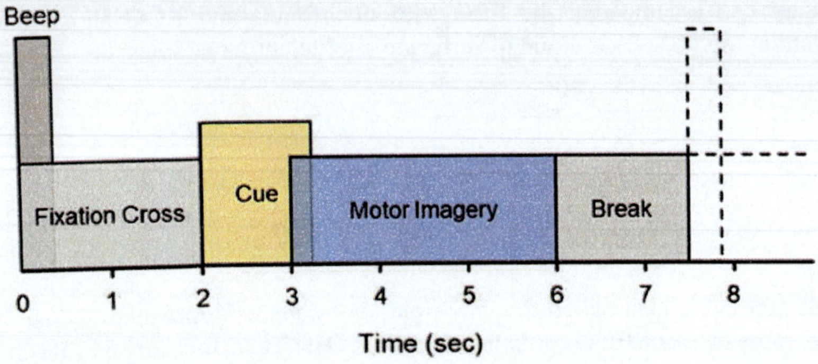

Fig. 2 Timing scheme of the experimental paradigm

Twenty-two EEG channels were recorded using Ag/AgCl electrodes. The specific challenge of this dataset is the eye movement artifact, so three monopolar electro oculographic (EOG) channels were added for artifact processing purposes. The signals were sampled at 250 Hz and bandpass filtered between 0.5 and 100 Hz with 50 Hz notch filtered.

2.2 *Preprocessing*

Artifact Removal

As the output of this proposed method is the four different motor imagery classes, the frequency band of brain rhythm is within the ERD/ERS information. A 5th order Butterworth infinite impulse response (IIR) filter with the passband in the range 7–30 Hz to concentrate on the mu (8–13 Hz) and beta (13–30 Hz) rhythms. Additionally, three EOG components were used to utterly "subtract" the EOG artifacts in the EEG signals by linear regression [32]. We assume the regression model as follow:

$$X = E + kO \tag{1}$$

Accordingly, X is the recorded EEG which presented as a matrix of N channels $X = [x_1, x_2, x_3, \ldots x_N]$, E denotes the uncontaminated EEG signal without eye movement artifact, O denotes the pure EOG channel, and k denotes the weight of EOG artifact affecting on N component of EEG signal.

The dataset includes EOG channels for each subject (which electrodes are placed adjacent to the eye to avoid other artifacts). To compute the unknown k, a multiplication of (1) with O^T, where T denotes the recording time-points, yields:

$$\langle O^T.X\rangle = \langle O^T.E\rangle + k\langle O^T.O\rangle \tag{2}$$

As we can assume that the EEG without artifacts and EOG signal have no correlation, the $\langle O^T . E \rangle$ is equal to zero, and coefficient k becomes:

$$k = \langle O^T.X \rangle = \langle O^T.O \rangle^{-1} \quad (3)$$

And the uncontaminated EEG signal can be found according to:

$$E = X - kO \quad (4)$$

This method is also called the "multiple least-squares approaches" since it can remove more than one EOG component. In Graz Dataset 2a, there are 3 EOG channel recordings for each subject, so the algorithm was applied respectively to each channel in order to choose k at the minimum mean square of E.

Resampling

This step lessens the time spent on processing the signal and limits small variations in the data. Regarding the Nyquist theorem [33], the highest frequency of the input signal accounts for half of the sampling rate. Thus, the filtered signal must be sampled more than 60 Hz. One-fourth of the original sampling rate was applied for our work.

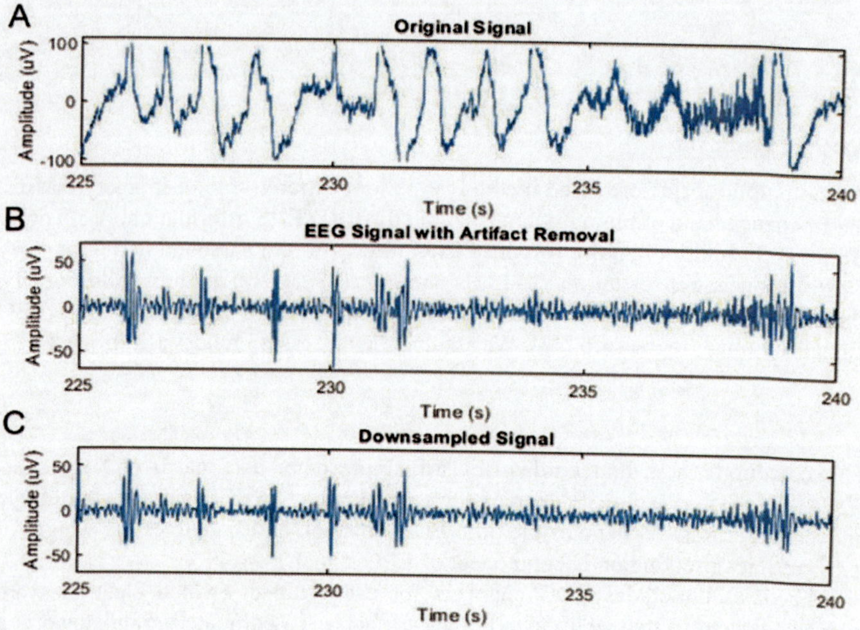

Fig. 3 An illustration of the preprocessing method. **a** Original EEG signal of the channel from subject 1. **b** The filtered signal. **c** The downsampled signal

A visualization of the preprocessing step is presented in Fig. 3. This figure shows the period time from 225 to 240 s of the first channel of subject 1.

2.3 *Feature Extraction*

Common Spatial Pattern (CSP)

This algorithm can magnify the discriminative of classes by optimizing the variances of the filtered data [34]. Let $\boldsymbol{W}^{\boldsymbol{T}}$ is a $\boldsymbol{N} \times \boldsymbol{M}$ matrix (M spatial filters for N samples) with T denotes recording time-points and $\boldsymbol{E}(\boldsymbol{t})$ is the input EEG signal, the CSP model is as follow:

$$E_{CS}(t) = W^T E(t) \tag{5}$$

In terms of the spatial filter, covariance matrices $\boldsymbol{R}_c$ under the $c \in \{1,2\}$ conditions have been shown in the below equation. Though multiple conditions were used for this proposed method, two conditions were applied for the simplicity of the illustration.

$$R_c = \frac{1}{\mathrm{K}} \sum_{\mathrm{i}} \mathrm{Z}_{\mathrm{c}}^{\mathrm{i}} \left(\mathrm{Z}_{\mathrm{c}}^{\mathrm{i}}\right)^{\mathrm{T}} \tag{6}$$

$Z_c{}^i$ denotes the EEG data in the *i-th* trial at the condition *c*. Moreover, the CSP also estimates:

$$W^T R_c W = \tau_c \tag{7}$$

$$\tau_1 + \tau_2 = I \tag{8}$$

The sum of two diagonal matrices τ_1 and τ_2 is the identity matrix I. This result illustrates the sum variances must be 1. Thus, if the variance of class 1 is high, the variance of class 2 is accordingly low and vice versa. The results of CSP are estimated by the mean of diagonalization of covariance matrices for two classes.

Band Power and Time Domain Parameters

Band power features were extracted in this method by applying the *BioSig library* [35] to calculate the target bandwidths by bandpass filtering the signal. Our target frequency bands are 8–14, 19–24, and 24–30 Hz. The signal was filtered by the 5th order Butterworth IIR bandpass filter. The bandpass frequencies of this filter are our target bands. To prevent leakage effects that reduce the quality of the processing signal, we apply the smoothing window of 2 s by continue implementing the filter. Finally, natural logarithms of output signals are calculated to enhance to performance of linear classification.

2.4 Feature Selection

Feature selection is a critical step in the classification problem. Some extracted features are irrelevant to the output variable. They do not contribute much to model performance but take time for training. Furthermore, the high dimensionality of data will lead to complicated computation and overfitting results. Therefore, feature selection methods are applied to reduce dimensions of input while maintaining the essential information for classification.

Mutual Information Best Individual Features

Mutual information is a quantity that computes the relationship or correlation between two random variables [36]. The formula of mutual information between two variable X and Y is given by:

$$I(X, Y) = \sum_{x \in X} \sum_{y \in Y} P(x, y) \frac{P(x, y)}{P(x)P(y)} \tag{9}$$

In which, P(X) and P(Y) are the marginal distributions of X and Y.

The Mutual Information (MI) Function

The MI algorithms were developed by Räsänen [37].

The function MI estimates the mutual information and uses the number of variables ranking to score the weight of features based on their individual mutual information with four output classes. The more weight the feature has, the better the performance is when applied to the model.

From the previous step, 24 features corresponding to 3 bandwidths 8–14, 19–24, 2430 Hz are extracted. Mutual information of 24 features is calculated and weighted. Through the experiment, 15 largest-score features are collected.

2.5 Classification

Linear discriminant analysis (LDA) is implemented in our experiment.

Although multiple category classification was done in this work, an illustration for the principle of the LDA algorithm is binary classification due to its simplicity. Specifically, in order to discriminate two classes X and Y, LDA algorithms define a line $\boldsymbol{w}$ such that when the projection of each value of both classes onto the line satisfy two following conditions:

- Maximize the distance between the means of given classes.
- Minimize the variation within each class.

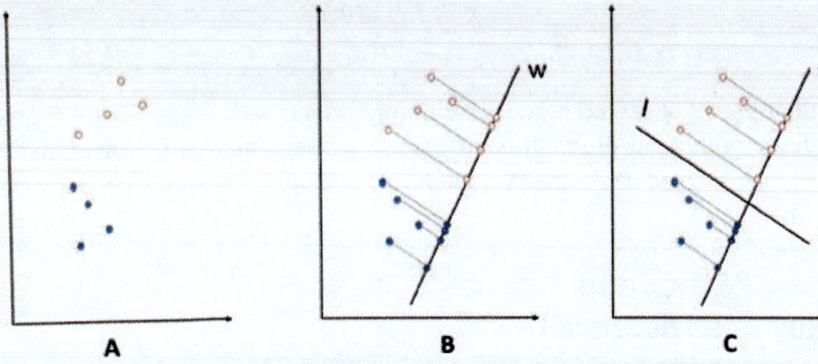

Fig. 4 The projection of red and blue data points representing two classes of the label in the dataset onto different vectors ***w***, is done by ***the LDA classifier***. **a** Scatter plot representing data of the two classes. **b** Projection of each data point onto new vector w. **c** Separation of data points of the two classes done by line ***l***

By projecting each value onto line ***w***, data of the two classes are presented in a new axis which is easier to establish the decision boundary ***l*** to discriminate them (Fig. 4).

A good model will induce a large distance between the two classes when projecting these data points two classes onto vector ***w***. Therefore, a threshold can be defined to separate the two classes and used to predict new data points.

Regarding the implementation of LDA algorithm on the classification of four mentioned imaginary motor tasks, function *fitcdiscr* of MATLAB is used in classification tasks with:

- Bayesian optimization.
- Iterations: 30.

Through each time of iteration, the optimizer algorithm automatically changes the shape of the objective function model by changing the hyperparameters to obtaining the best fit to the observed data points. Therefore, the classification performance of the model is improved.

3 Evaluation Metric

Cohen's kappa coefficients were calculated to evaluate the performance of the classification model. Kappa scores are considered a robust evaluation of categorized discrimination against the problem as it not only takes into account the percentage of the agreement but also the possibility of chance agreement. Kappa statistics focus on the object-class distribution and ignore the number of features used to find out this distribution, so this measure is useful for imbalanced datasets.

Cohen's kappa equation [38] is defined as:

$$k = \frac{\Pr(a) - \Pr(e)}{1 - \Pr(e)} \tag{10}$$

Pr(a) the relative observed agreement among raters.

Pr(e) the hypothetical probability of chance agreement that is calculated as the probabilities of each observer randomly seeing each category using observed data.

4 Result and Discussion

As mentioned in the evaluation metrics part, the kappa value is suggested to address several accuracy validation problems in BCI research. The different trials can result in different kappa values, so we took the average results of several trials to get a more representative kappa; also, the kappa value will be more general. This paper reached a 0.57 score of mean kappa when using the Graz dataset 2a to train this model. Additionally, we added the testing set from a similar dataset to evaluate our model, and we observed that the mean of evaluation kappa was 0.41, which approaches the training result. Consequently, our model is considered reliable. Table 1 summarizes the kappa coefficient of the training and evaluating set are represented.

Improvement. To compare the efficiency of our method with traditional methods, we propose two other models. The first model is artifact removal without linear regression, and another is feature extraction without CSP. Figure 5 illustrates that the best model is applying our method (k = 0.41), while the model applying linear regression algorithm in artifact removal is less efficient (k = 0.39). This experiment also proves that the CSP which is used to extract the features improved the kappa score (from 0.33 to 0.41).

To recognize the role of linear regression and CSP on improving the kappa score, the Kruskal–Wallis test [39] was utilized to compare the difference between the mean

Table 1 The final training and evaluation kappa scores for each subject

Subjects	Training kappa	Evaluating kappa
1	0.72	0.70
2	0.50	0.11
3	0.78	0.52
4	0.35	0.21
5	0.38	0.10
6	0.38	0.18
7	0.76	0.66
8	0.74	0.59
9	0.47	0.61
Mean ± SEM	**0.57 ± 0.061**	**0.41 ± 0.084**

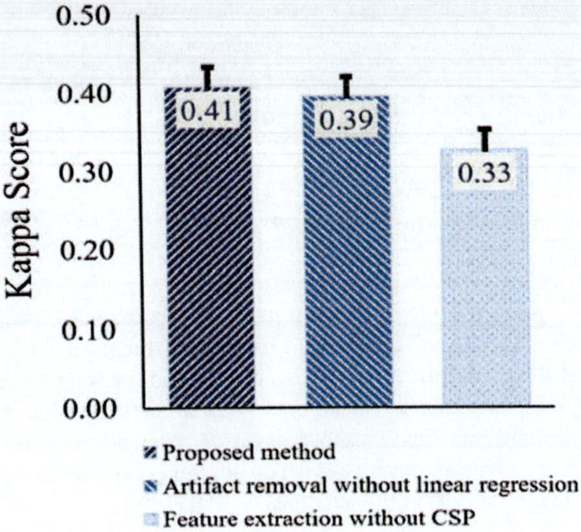

Fig. 5 Comparison of the effects of linear regression and CSP to kappa score

value of 3 statistical models. The first reason is that the population data has a small size with nine subjects per each group method and another reason is the non-normal distribution of this data, according to the Levene test ($p = 0.025$) [39, 40]. In the average ranks (see Table 2), the feature extraction without CSP was worth ranked bottom, which proves that the CSP plays an important role in improving our model. On the other hand, using the linear regression method just makes a small impact on the mean value compared to the method without using CSP. Furthermore, Fig. 5 shows that when applying both linear regression and CSP, the mean kappa score was enhanced. Nevertheless, no significant difference in the mean is observed among the three approaches ($p = 0.641$).

Confusion Matrix

To describe the effect of a 4-class classification problem, we use the confusion matrix which illustrates the relationship between the predicted classes (intended labels by user) and the true classes (known labels). Table 3 shows the proportion of samples from evaluation data predicted by the proposed model based on the ground truth, in which the bold ones illustrate the correct classifier. The higher the bold values of the confusion matrix, the more efficient model.

Table 2 The Kruskal–Wallis test summarizes the mean rank of 3 methods

Method	Mean rank
Proposed method	15.33
Artifact removal without linear regression	14.67
Feature extraction without CSP	12.00

Table 3 Confusion matrix of the evaluation data

		Predicted class			
		Left hand	Right hand	Both Feet	Tongue
True class	Left hand	**0.64**	0.25	0.08	0.03
	Right hand	0.133	**0.851**	0.009	0.007
	Both feet	0.03	0.04	**0.79**	0.14
	Tongue	0.004	0.002	0.116	**0.878**

According to Table 3, tongue motor imagery got the best result with 87.8%. In contrast, the left-hand motor imagery just reached 64% accurate prediction; this class still was significantly confused back and forth with the right-hand class when being predicted (25% left-hand class was labeled right-hand motor imagery). Meanwhile, feet and right-hand motor imagery got considerable results, 79% and 85.1% sequentially. Overall, our model is considered good working in classifying four-class motor imagery (79% in the average).

Comparison with the BCI Competition IV (2008)

Compare with the BCI Competition IV (2008) on the same Graz dataset 2a, our work with removing eye artifact and applying Mutual Information Best Individual Features (MIBIF) and LDA algorithm has yielded promising results. Our kappa is higher than that of the 3rd prize in the competition, which is respectively 0.41 and 0.31 [41] (see Table 4). By using the Filter Bank CSP and Naive Bayes Parzen Window technique, K. K. Ang et al. won the first prize with the 0.57 kappa value [42]. Similar to the proposed method, Liu Guangquan et al. also applied the LDA algorithm, however, the combination of it with the Log variance of the best eight components and Bayesian is the reason why this kappa value holds more weight than the proposed method (k = 0.52) [43].

Table 4 The comparison kappa values of 9 subjects with the three best competitors

Subjects	Kai Keng Ang	Liu Guangquan	Wei Song	Proposed method
1	0.68	0.69	0.38	0.70
2	0.42	0.34	0.18	0.11
3	0.75	0.71	0.48	0.52
4	0.48	0.44	0.33	0.21
5	0.40	0.16	0.07	0.10
6	0.27	0.21	0.14	0.18
7	0.77	0.66	0.29	0.66
8	0.75	0.73	0.49	0.59
9	0.61	0.69	0.44	0.61
Mean	**0.57**	**0.52**	**0.31**	**0.41**

Code Deposit

Our code has been uploaded on the Github: https://github.com/nganluu0903/BCI_project.

Limitation and Future Steps

In this proposed method, the utilized techniques in signal processing procedure (including feature extractions, feature selection, and classification) are considered to be very common. Therefore, the future steps are to select and apply recent advantage techniques (e.g. recurrent neuronal network) to improve the quality of this study. Furthermore, there are various limitations in the evaluated dataset—Graz dataset A which was published in 2008. So, recent datasets or a self-constructed dataset are highly recommended to approach in the next work.

5 Conclusion

The EEG-based BCI is known as the non-invasive method that is growing significantly and bringing many advantages as well as challenges in EEG signal processing [44]. From recent studies, motor imagery based BCI (MI-BCI) has become a common approach; however, it is challenging to classify imaginary motor tasks [32].

In this paper, the proposed algorithm works on the event-related resynchronization (ERD) phenomenon to categorize four different motor imagery classes. We implemented band power features, time-domain parameters, and the common spatial patterns (CSP) to extract features while Mutual Information Best Individual Features was considered in feature selection. Besides, the linear discriminant analysis (LDA) classifier was utilized to classify four classes. The proposed algorithm was evaluated using the kappa score. The kappa value was obtained from the confusion matrix that evaluates the effectiveness of the algorithm and supports to compare the accuracy of models with other ones easily. Besides, the spatial filter is recommended to enhance the kappa score and increase the reliability of the model.

For future work, it is recommended to focus on raising the performance of the MIBCI algorithm by utilizing different techniques in encoding the EEG signal. In this paper, some factors influence the performance of the proposed algorithm. For instance, frequency band selection for each subject is necessary since the effect of the ERD/ERS phenomenon is different for each individual. To optimize this system quality, an automatic method for selecting the frequency bands should be considered before performing preprocessing. Furthermore, another method named the filter bank common spatial patterns (FBCSP) algorithm should be considered for feature extraction in separated smaller frequency bands.

Acknowledgements The Ministry of Science and Technology funds this research, Grant # KC-4.007/19-25.

Conflicts of Interest The authors have no conflict of interest to declare.

References

1. Rao RPN (2013) Brain-computer interfacing: an introduction. Cambridge University Press, Cambridge
2. Panoulas KJ, Hadjileontiadis LJ, Panas SM (2010) Brain-Computer Interface (BCI): types, processing perspectives and applications. In: Tsihrintzis GA, Jain LC (eds) Multimedia services in intelligent environments: integrated systems. Springer, Berlin, Heidelberg, pp 299–321
3. Gu X, Cao Z, Jolfaei A, Xu P, Wu D, Jung T-P, Lin C-T (2020) EEG-based Brain Computer Interfaces (BCIs): a survey of recent studies on signal sensing technologies and computational intelligence approaches and their applications
4. Zhang X, Yao L, Huang C, Gu T, Yang Z, Liu Y (2019) DeepKey: an EEG and gait based dual-authentication system. arXiv:1706.01606 [cs]
5. Jiang X, Bian G-B, Tian Z (2019) Removal of artifacts from EEG signals: a review. Sensors 19:987. https://doi.org/10.3390/s19050987
6. Mannan MMN, Kamran MA, Kang S, Jeong MY. Effect of EOG signal filtering on the removal of ocular artifacts and EEG-based brain-computer interface: a comprehensive study. https://www.hindawi.com/journals/complexity/2018/4853741/
7. Deep brain stimulation: BCI at large, where are we going to? https://www.researchgate.net/publication/51596032_Deep_brain_stimulation_BCI_at_large_where_are_we_going_to
8. Roy D, Hash L (2016) Advisor: study of brain computer interfacing (BCI) with cochlear implant as an example
9. Wang H, Chang W, Zhang C (2016) Functional brain network and multichannel analysis for the P300-based brain computer interface system of lying detection. Expert Syst Appl 53:117–128. https://doi.org/10.1016/j.eswa.2016.01.024
10. Maksimenko VA, Runnova AE, Zhuravlev MO, Makarov VV, Nedayvozov V, Grubov VV, Pchelintceva SV, Hramov AE, Pisarchik AN (2017) Visual perception affected by motivation and alertness controlled by a noninvasive brain-computer interface. PLoS ONE 12:e0188700. https://doi.org/10.1371/journal.pone.0188700
11. Finke A, Lenhardt A, Ritter H (2009) The MindGame: a P300-based brain–computer interface game. Neural Netw 22:1329–1333. https://doi.org/10.1016/j.neunet.2009.07.003
12. Marchesi M, Riccò B (2013) BRAVO: a brain virtual operator for education exploiting braincomputer interfaces. Presented at the Apr 27
13. Li Y, Long J, Yu T, Yu Z, Wang C, Zhang H, Guan C (2010) An EEG-based BCI system for 2-D cursor control by combining mu/beta rhythm and P300 potential. IEEE Trans Biomed Eng 57:2495–2505. https://doi.org/10.1109/TBME.2010.2055564
14. Guger C, Harkam W, Hertnaes C, Pfurtscheller G (1999) Prosthetic control by an EEG-based brain-computer interface (BCI). In: Proceedings AAATE 5th European conference for the advancement of assistive technology
15. Chen X, Chen Z, Gao S, Gao X (2014) A high-ITR SSVEP-based BCI speller. Brain-Comput Interfaces 1:181–191. https://doi.org/10.1080/2326263X.2014.944469
16. Bensch M, Karim AA, Mellinger J, Hinterberger T, Tangermann M, Bogdan M, Rosenstiel W, Birbaumer N (2007) Nessi: an EEG-controlled web browser for severely paralyzed patients. Comput Intell Neurosci 2007:e71863. https://doi.org/10.1155/2007/71863
17. Chae Y, Jeong J, Jo S (2012) Toward brain-actuated humanoid robots: asynchronous direct control using an EEG-based BCI. IEEE Trans Rob 28:1131–1144. https://doi.org/10.1109/TRO.2012.2201310
18. dos Santos EM, Cassani R, Falk TH, Fraga FJ (2020) Improved motor imagery brain-computer interface performance via adaptive modulation filtering and two-stage classification. Biomed Signal Process Control 57:101812. https://doi.org/10.1016/j.bspc.2019.101812
19. A review on motor imagery signal classification for BCI. https://www.cscjournals.org/library/manuscriptinfo.php?mc=SPIJ-284
20. Hersche M, Benini L, Rahimi A (2020) Binary models for motor-imagery brain–computer interfaces: sparse random projection and binarized SVM. Presented at the 2nd IEEE international conference on artificial intelligence circuits and systems (AICAS 2020)

21. Lemm S, Müller K-R, Curio G (2009) A generalized framework for quantifying the dynamics of EEG event-related desynchronization. PLoS Comput Biol 5:e1000453. https://doi.org/10.1371/journal.pcbi.1000453
22. Schlögl A, Supp G (2006) Analyzing event-related EEG data with multivariate autoregressive parameters. Prog Brain Res 159:135–147. https://doi.org/10.1016/S00796123(06)59009-0
23. Toro C, Deuschl G, Thatcher R, Sato S, Kufta C, Hallett M (1994) Event-related desynchronization and movement-related cortical potentials on the ECoG and EEG. Electroencephalogr Clin Neurophysiol/Evoked Potentials Sect 93:380–389. https://doi.org/10.1016/0168-5597(94)90126-0
24. Percio CD, Infarinato F, Iacoboni M, Marzano N, Soricelli A, Aschieri P, Eusebi F, Babiloni C (2010) Movement-related desynchronization of alpha rhythms is lower in athletes than non-athletes: a high-resolution EEG study. Clin Neurophysiol 121:482–491. https://doi.org/10.1016/j.clinph.2009.12.004
25. Klimesch W, Pfurtscheller G, Mohl W, Schimke H (1990) Event-related desynchronization, ERD-mapping and hemispheric differences for words and numbers. Int J Psychophysiol 8:297–308. https://doi.org/10.1016/0167-8760(90)90020-E
26. Pfurtscheller G, Lopes da Silva FH (1999) Event-related EEG/MEG synchronization and desynchronization: basic principles. Clin Neurophysiol 110:1842–1857. http://doi.org/10.1016/S1388-2457(99)00141-8
27. Brunner C, Leeb R, Müller-Putz G, Schlögl A, Pfurtscheller G (2008) BCI competition 2008—Graz data set A. Institute for Knowledge Discovery (Laboratory of Brain-Computer Interfaces), Graz University of Technology, 16
28. Singla R, Devgan A, Gogna P, Batra A (2014) Fixation of delayed union or non-union posterior cruciate ligament avulsion fractures. J Orthop Surg. https://doi.org/10.1177/230949901402200118
29. Fogarty DP, Deering AL, Guo S, Wei Z, Kautz NA, Kandel SA (2006) Minimizing image-processing artifacts in scanning tunneling microscopy using linear-regression fitting. Rev Sci Instrum 77. https://doi.org/10.1063/1.2390633
30. Lu H, Plataniotis KN, Venetsanopoulos AN (2009) Regularized common spatial patterns with generic learning for EEG signal classification. In: 2009 annual international conference of the IEEE engineering in medicine and biology society, pp 6599–6602
31. Lotte F, Guan C (2011) Regularizing common spatial patterns to improve BCI designs: unified theory and new algorithms. IEEE Trans Biomed Eng 58:355–362. https://doi.org/10.1109/TBME.2010.2082539
32. Wang D, Miao D, Blohm G (2012) Multi-class motor imagery EEG decoding for brain-computer interfaces. Front Neurosci 6:151
33. Srinivasan R, Tucker DM, Murias M (1998) Estimating the spatial Nyquist of the human EEG. Behav Res Methods Instrum Comput 30:8–19. https://doi.org/10.3758/BF03209412
34. Rao RPN, Scherer R (2010) Statistical pattern recognition and machine learning in brain–computer interfaces. In: Statistical signal processing for neuroscience and neurotechnology. Elsevier, Amsterdam, pp 335–367
35. Vidaurre C, Sander TH, Schlögl A (2011) BioSig: the free and open source software library for biomedical signal processing. Comput Intell Neurosci 2011
36. Learned-Miller EG (2013) Entropy and mutual information
37. Pohjalainen J, Räsänen O, Kadioglu S (2013) Feature selection methods and their combinations in high-dimensional classification of speaker likability, intelligibility and personality traits. Comput Speech Lang 29. http://doi.org/10.1016/j.csl.2013.11.004
38. Cohen J (2016) A coefficient of agreement for nominal scales. Educ Psychol Meas. https://doi.org/10.1177/001316446002000104
39. Stephanie: non parametric data and tests (Distribution Free Tests). https://www.statisticshowto.com/parametric-and-non-parametric-data/
40. Stephanie: Levene test for equality of variances. https://www.statisticshowto.com/levenetest/
41. Tangermann M, Müller K-R, Aertsen A, Birbaumer N, Braun C, Brunner C, Leeb R, Mehring C, Miller KJ, Müller-Putz GR, Nolte G, Pfurtscheller G, Preissl H, Schalk G, Schlögl A, Vidaurre

C, Waldert S, Blankertz B (2012) Review of the BCI competition IV. Front Neurosci 6. http://doi.org/10.3389/fnins.2012.00055
42. Ang KK, Chin ZY, Zhang H, Guan C (2008) Filter bank common spatial pattern (FBCSP) in brain-computer interface. In: 2008 IEEE international joint conference on neural networks (IEEE world congress on computational intelligence), pp 2390–2397
43. Liu G. BCI competition IV: results. http://www.bbci.de/competition/iv/results/ds2a/LiuGuangquan_desc.txt
44. Machado S, Araújo F, Paes F, Velasques B, Cunha M, Budde H, Basile LF, Anghinah R, Arias-Carrión O, Cagy M, Piedade R, de Graaf TA, Sack AT, Ribeiro P (2010) EEG-based brain-computer interfaces: an overview of basic concepts and clinical applications in neurorehabilitation. Rev Neurosci 21:451–468. https://doi.org/10.1515/REVNEURO.2010.21.6.451

Segmentation of Brain Tumour Using UNET Architecture

Ngo Hoang Anh Vy, Le Thi Thu Uyen, and Huynh Quang Linh

Abstract Gliomas is the most widely recognized essential brain cancer with particular degrees of aggressive, variable prediction and different heterogeneous histological sub-regions including: edema, necrotic center, enhancing and non-enhancing tumour core. Exact tumour division of sub-areas plays a crucial role for diagnosis, treatment planning and risk factor identification. This study was conducted by using the MultiResUnet architecture for computerized segmentation of gliomas in multimodal MRI scans (FLAIR, T1CE, T2). Previously, the images were enhanced through various preprocessing stages, then they were fed by the MultiResUnet network. From the 3D database gave by BraTS 2019, we extracted 2176 images to train and 1056 images to assess our network. The segmentation model performance was qualitatively evaluated with the accuracy and under the Dice score metrics. Our study suggested a model which Dice score accomplished 0.78, 0.82 and 0.88 on the training set and 0.63, 0.73 and 0.86 on the validation set for the enhancing tumour, the tumour core, and the entire tumour, separately. In general, the system sectioned the entire tumour more precisely than it accomplishes for the core of tumour or enhancing tumour. They are significantly harder to segment because of the closeness between all areas. We plan to further our network to improve the testing phase results and use sub-regions tumour on Radiomic Features extraction for tumour investigation and survival expectation.

Keywords Gliomas tumour · Sub-regions segmentation · UNet

N. H. A. Vy (✉) · L. T. T. Uyen · H. Q. Linh
Department of Biomedical Engineering, Ho Chi Minh City University of Technology, Vietnam National University Ho Chi Minh City (VNU-HCMC), Ho Chi Minh City 700000, Vietnam
e-mail: 1614194@hcmut.edu.vn

V. Van Toi et al. (eds.), *8th International Conference on the Development of Biomedical Engineering in Vietnam*, IFMBE Proceedings 85,
https://doi.org/10.1007/978-3-030-75506-5_63

1 Introduction

1.1 A Subsection Sample

Gliomas (GLMs) is a tumour developed from the glial organization of cells within the brain, which give support capabilities to neurons all through the brain. According to the World Health Organization's classification based on the tumour's histologic features, there are four grades in the malignancy scale [1]. The increase of the serious level of lesions follows the increase of every grade. Malicious gliomas are the most widely recognized essential brain cancers, where glioblastomas account for the largest proportion of gliomas [2]. This study focuses on glioblastoma (GLM), also known as high-grade glioma (HGG).

It is considered a grade IV gliomas and is the most combative diffuse glioma of astrocytic genealogy. GBM may be primary (arising de novo) or secondary (arising from a lower-grade tumor). Primary glioblastomas are more regular, as a rule happen in patients at more seasoned matured, and have a worse forecast than patients with secondary GBMs [3]. GBMs arise from not only glial cells but also from various cell types with neural stem cell characteristic. It's ordinarily found within the supratentorial region (frontal–temporal, parietal and occipital lobes) and barely in the cerebellum [4]. Standard treatment for newly determined patients with glioblastoma includes the subsequent approaches: Operation with radiation therapy and chemotherapy, treatment with radiation therapy, Carmustine-impregnated polymer implanted during the initial surgery, radiation therapy and simultaneous chemotherapy [5]. Whichever method is chosen to treatment, it is still extremely necessary to section the brain growth and sub-regions in that growth exactly. This step is essential for treatment navigation (support for plan radiotherapy plan or surgical navigation).

Magnetic resonance imaging (MRI), the most commonly applied method, has important functions in detecting and treating brain neoplasms. MRI has a superior soft-tissue resolution. Almost GBMs on MRI enhance gadolinium contrast and show a dense region of enhancement and hypo-intense necrosis center. Necrosis is a characteristic of GBM, and the existence of necrosis is illustrated by the World Health Organisation classification scheme for GBM. Surrounding vasogenic edema (which can induce a mass effect), hemorrhage and the change in the shape of the ventricle (ventricular distortion) may also be performed for diagnostic imaging [4]. In short, the MR sequences provide distinct informative regions within the GBM, this information is tremendously valuable for not only pre-treatment but also prognostic. Although the fact that the quality, flexibility and resolution of MRI technology has significantly developed in the past, the comprehension of images remains largely personal and non-quantitative. Radiomics are studied to advance image analytic procedures rely on the extraction of objective, quantitative data from MRI scans. The correlation between them and underlying tissue dynamics can boost cancer imaging research's scope like detecting automatically grade of the tumour, predicting survival time or further studying [6]. Thus, accurate and timely detection of HGG tumor may improve patient survival duration and be a reliable base for the aforementioned research.

In this day and age, many modern or advanced techniques are utilized to automatically segment brain tumours based on machine learning and deep learning. Convolution Neural Networks (CNNs) are especially proposed and gave good performances in image semantic segmentation tasks. So far, there are a number of brain tumour segmentation methodologies focused on CNNs or mixing CNNs with traditional methods are also suggested [7]. However, CNN architecture usually requires the massive volumes of training data. U-Net is a type of fully convolutional neural network (FCN) which can solve that problem. When it comes to biomedical imaging, we usually face with the lack of data, so due to skip connection, the most advanced aspect of U-Net and U-Net like models, they have been successfully used in segmenting biomedical images of neuronal structures [8].

Our investigation developed a recent architecture called MultiResUNet based on U-Net architecture inspired by Nabil Ibtehaz and M. Sohel Rahman for segmentation of brain tumour [9]. This model can solve some drawbacks of UNet models; one of them is the ability to analyze objects at different scales. The work used data obtained from multi-institutional MRI scans made available as part of BraTS 2019 to segment three sub-regions of glioblastoma consisting: the whole tumour, enhancing tumor and tumour core. We also apply some pre-processing stages to boost the performance.

2 Method

2.1 Database

This study used data from the BraTS 2019 challenge [10–12]. To be more detail, this dataset includes pre-operative multimodal MRI scans (T1, T1CE, T2, FLAIR) of glioblastoma (GBM/HGG) and lower-grade glioma (LGG) from many institutes. All the imaging datasets were segmented manually by professional neuroradiologists. In our study, we extracted 2D images which are slices with clear tumour region from HGG set. Thus, in turn, the training dataset includes 2176 cases; the validation dataset includes 1056 cases.

In provided dataset, labels are signed as: 1 for non-enhancing tumor (NET) and necrotic (NCR), 2 for edema (ED), 4 for enhancing tumor (ET) [10]. However, in this study, we segmented only three sub-regions: whole tumour (label 1 + 2 + 4), enhancing tumour (label 4), tumour core TC (label 1 + 4). FLAIR, T2, T1ce sequences were used to segment the whole tumour, tumour core and enhancing tumour respectively (Fig. 1).

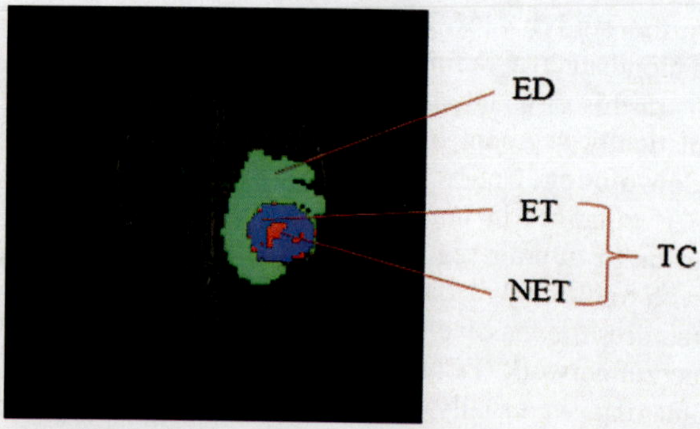

Fig. 1 Glioma sub-regions

Table 1 Enhancement used in each MRI modalities

Modalities	Enhancement
FLAIR	Mean filter, increasing brightness potential pixels[a]
T1CE	Gaussian filter, intensity scaling, increasing brightness potential pixels[a], mean filter
T2	Mean filter, local histogram equalization, Laplacian filter

[a]We extract the histogram of all image and assess intensity of potential pixels statistically

2.2 Pre-processing

Firstly, all MRI images were normalized by subtracting the mean intensity and dividing by the standard deviation. Secondly, we remove 0 ground part by cropping to focus on only the brain. The same pre-processing steps were also applied to both the validation and training data set before segmenting. Finally, we enhance the data following these approaches (Table 1).

2.3 MultiResUNet

The model used in this research has a similar architecture as that of U-net [13]. The network architecture was illustrated in Fig. 2. It consists of an encoder (left side) and a decoder (right side).

Both paths consist of MultiRes blocks, instead of the sequence of two convolutional layers in the conventional U-net. This replacement aims to improve the performance when it comes to the case that these ROIs are of irregular and different scales. This multiRes block is inspired by Inception blocks [14]. Therefore, Ibtehaz and Rahman [9] supposed to utilize convolutional layers of varying kernel sizes 3×3, 5×5, 7×7 follow the approach of Inception model. However, this way will increase the memory requirement. So, firstly, they replaced the bigger and more costly 5×5 and 7×7 kernels as a series of 3×3 kernels (see Fig. 3). The outputs of

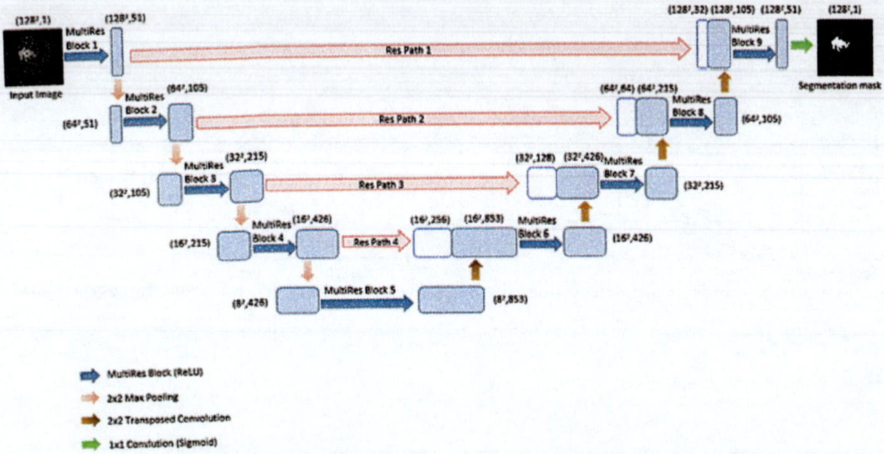

Fig. 2 Proposed MultiResUNet architecture

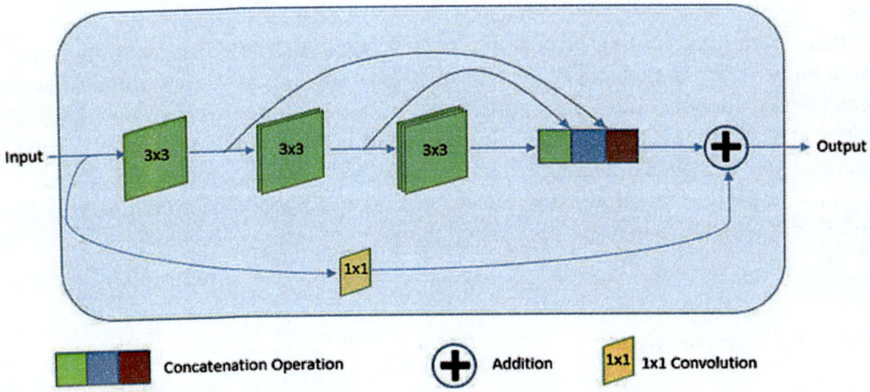

Fig. 3 Structure of MultiRes block

2nd and 3rd 3 × 3 filters from three consecutive convolutional blocks, each of them have increasing filters in sort instead of keeping them the same. Then three blocks are concatenated together to extract features. In addition, a residual connection (along with 1 × 1 kernel for conserving dimensions) is added to modify weights efficiently (Fig. 4).

The ordinary concatenations corresponding to the contracting path are replaced with the proposed Res paths. Because of a possible semantic gap between features in two paths (encoder and decoder), the contracting features are passes through a series of convolutional layers. It could be seen that the intensity of the semantic gap between the contracting and expanding feature maps are likely to fall down because of the succeeding shortcut connections. Therefore, the number of convolutional blocks are gradually reduced along the Res paths. Particularly, 4; 3; 2 and 1 convolutional

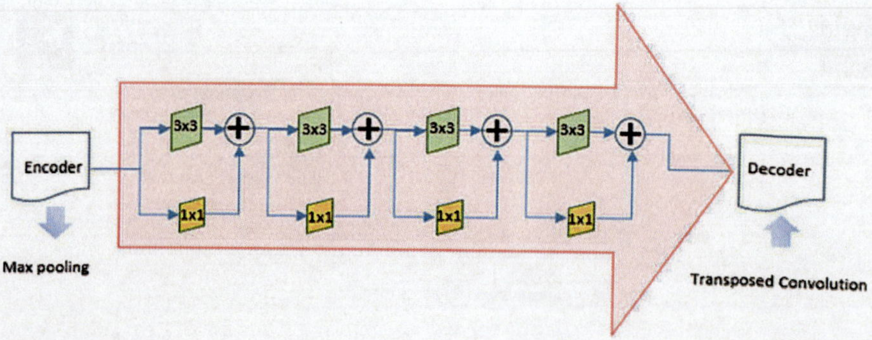

Fig. 4 Structure of Res path

blocks are used respectively along the four Res paths. Then, we use 256; 128; 64 and 32 filters in the four Res paths blocks respectively to suitable with the numbers of filters in encoder-decoder. Residual connections are applied to make the learning easier. Finally, they were concatenated with the decoder features.

After each max-pooling layer in the contracting path, the down-sampling operation halves each dimension of feature maps. On the contrary, each dimension of feature maps doubles after passing transposed convolution. All the convolutional layers used in this network, except for the output layer, comprise a Batch Normalization layer, a ReLU activation function, as used in the original architecture. A Sigmoid activation function is used in the last layer of the encoder path to map the multi-channel feature maps to the desired number of classes.

2.4 *Post-processing*

Different types of MRI modalities were passed through the MultiResUNet model to segment each sub-region type, specifically as in Table 2. Due to the Sigmoid activation function in the last layer, the model's output is a matrix whose elements are in the range of [0, 1]. Then, we converted this matrix into a binary image by the threshold of 0.5. To visually show the results, we coloured the pixels in T1-weighted image corresponding to each type sub-regions' location based on the segmented binary images.

Table 2 The used modalities for the segmentation of different sub-regions

Modalities	Sub-region
FLAIR	Whole tumour
T1CE	Enhancing tumour
T2	Tumour core

2.5 Evaluation Metric

In brain tumour segmentation task, the examination zone occupies a low percentage of the whole image. Metrics such as recall and accuracy are thus insufficient and therefore contribute to a false sense of superiority, distorted with background detection excellence. Dice score measures the spatial overlap between two segments. This factor is often used to evaluate reliability when the number of background pixels in the image is much larger than the target pixels [15]. Dice score is defined as:

$$Dice\,score = \frac{2|A \cap B|}{A + B} \tag{1}$$

When A illustrates the ground truth binary segmentation mask, and B corresponds to the predicted binary segmentation output.

3 Results and Discussion

We developed the MultiResUNet by using Tensorflow, with Keras in Python. Experiments were conducted with Intel ® CoreTM i7-4790 CPU @3.60 GHz (14.0 GB).

The dataset was divided into training and testing datasets with 2176 and 1056 images respectively. The size of both input and output images are 128 × 128 pixels. The loss function applied in our model is Binary cross-entropy (Eq. 2) [9]. The cross-entropy function evaluates each pixel's prediction and then averages all the pixels in the output image.

$$H_p(q) = -\frac{1}{N}\sum_{i=1}^{N} y_i.\log(p(y_i)) + (1 - y_i).\log(1 - p(y_i)) \tag{2}$$

Here are hyper-parameters used in MultiResUNet model:

- Learning rate: 10^{-4}
- Epochs: 40
- Batch size: 32.

The results are shown in Fig. 5. In this analysis, the Dice coefficient was the metric chosen to measure the efficiency of our model. In the training data set, the Dice coefficient for WT, TC and ET were 0.78, 0.82 and 0.88. The mean loss for WT, TC and ET was 0.0122, 0.0091 and 0.006 in the training set. In the test data set, the Dice coefficient for WT, TC and ET were 0.63, 0.73 and 0.86. The mean loss for WT, TC and ET was 0.179, 0.0363 and 0.032 in the test set. Table 3 indicates the estimation results of test set obtained by the proposed model on data after training 40 epochs.

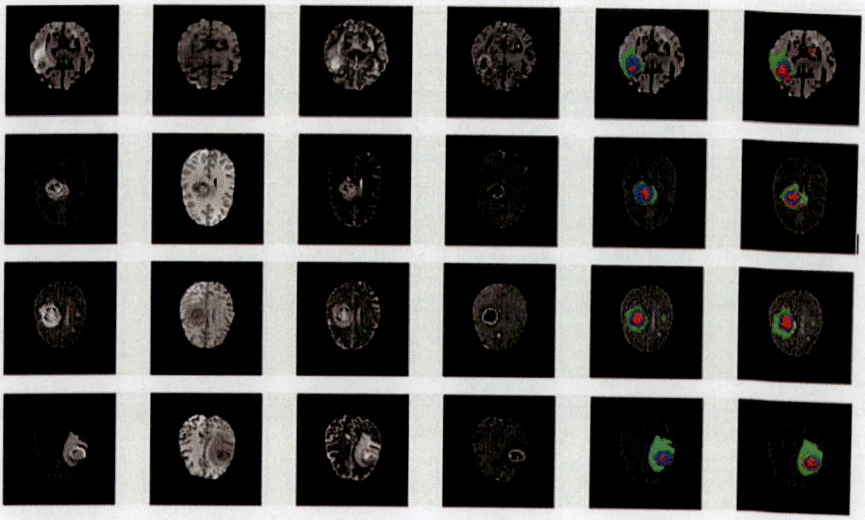

Fig. 5 Segmentation results on several examples. Colours: whole tumour (green), tumour core (blue), enhancing tumour (red)

Table 3 Results for validation dataset

	Dice WT	Dice TC	Dice ET
Mean	0.881	0.727	0.629
Std	0.119	0.257	0.277
Median	0.891	0.782	0.749
25 Quantile	0.827	0.604	0.369
75 Quantile	0.925	0.881	0.848

The enhancing tumour was segmented less accurately than the tumour core and whole tumour. This trend is normal when considering the results in other studies. The reasons are likely that the enhancing tumour has a similar intensity and the smallest area than other parts.

Our segmentation findings were very positive, and the assessment criteria indicate that our outcomes are consistent with other studies. Serrano-Rubio et al. used Extremely Randomly Trees algorithms to segment brain tumour with 57.19%, 62.85%, 79.92% of Dice score for ET, TC and WT [16]. Hyung Eun Shin et al. using 2D UNet to segment achieved Dice score with 61.5%, 70% and 76.8% for ET, TC and WT [17].

4 Conclusion

In this research, we utilized the 2D conventional MultiResUNet model to subdivide brain tumour into different sub-regions (whole tumour, tumour core and enhancing tumour) with the dataset provided by the BRATS 2019 challenge. The dataset was obtained by extracting 2D MRI images (FLAIR, T1CE and T2) from 3D data of BRATS 2019 and dividing into training and testing set. The results achieved 0.63, 0.73 and 0.86 for ET, TC and WT respectively. The sub-regions segmentation can be utilized to forecast the overall survival of patients. Future work will focus on extracting features from segmentation parts and establishing a model to forecast the overall survival.

Acknowledgements The authors would like to thank the anonymous reviewers for valuable comments and suggestions that have greatly improved the quality of this manuscript.

Conflict of Interest Ngo Hoang Anh Vy, Le Thi Thu Uyen and Huynh Quang Linh declare that they have no conflict of interest.

References

1. Louis DN, Perry A, Reifenberger G, von Deimling A, Figarella-Branger D, Cavenee WK, Ohgaki H, Wiestler OD, Kleihues P, Ellison DW (2016) The 2016 World Health Organization classification of tumors of the central nervous system: a summary. Acta Neuropathol 131(6)
2. Jovčevska I, Kočevar N, Komel R (2013) Glioma and glioblastoma—how much do we (not) know? Mol Clin Oncol 1(6):935–941
3. Bralten LB, French PJ (2011) Genetic alterations in glioma. Cancers 3:1129–1140
4. Davis M (2016) Glioblastoma: overview of disease and treatment. Clin J Oncol Nurs 20(5):S2–S8
5. NCBI Homepage. PDQ cancer information summaries. National Cancer Institute (US), Bethesda (MD). https://www.ncbi.nlm.nih.gov/books/NBK82221/. Last accessed 2020/4/5
6. Zhou M, Scott J, Chaudhury B, Hall L, Goldgof D, Yeom KW, Iv M, Ou Y, Kalpathy-Cramer J, Napel S, Gillies R, Gevaert O, Gatenby R (2017) Radiomics in brain tumor: image assessment, quantitative feature descriptors and machine-learning approaches. Am J Neuroradiol 39(2):208–216
7. Wadhwa A, Bhardwaj A, Verma VS (2019) A review on brain tumor segmentation of MRI images. Magn Reson Imaging 61:247–259
8. Drozdzal M, Vorontsov E, Chartrand G, Kadoury S, Pal C (2016) The importance of skip connections in biomedical image segmentation. Lecture notes in computer science, pp 179–187
9. Ibtehaz N, Rahman MS (2019) MultiResUNet: rethinking the U-Net architecture for multi-modal biomedical image segmentation. Neural Netw 121:74–87
10. Menze BH, Jakab A, Bauer S, Kalpathy-Cramer J, Farahani K, Kirby J et al (2015) The multimodal brain tumor image segmentation benchmark (BRATS). IEEE Trans Med Imaging 34(10):1993–2024
11. Bakas S, Akbari H, Sotiras A, Bilello M, Rozycki M, Kirby JS, Freymann JB, Farahani K, Davatzikos C (2017) Advancing the cancer genome atlas glioma MRI collections with expert segmentation labels and radiomic feature. Nat Sci Data 4:170117

12. Bakas S, Akbari H, Sotiras A, Bilello M, Rozycki M, Kirby JS, Freymann JB, Farahani K, Davatzikos C (2017) Segmentation labels and radiomic features for the pre-operative scans of the TCGA-LGG collection. The Cancer Imaging Archive
13. Ronneberger O, Fischer P, Brox T (2015) U-Net: convolutional networks for biomedical image segmentation. In: Medical image computing and computer-assisted intervention—MICCAI 2015, pp 234–241
14. Christian S et al (2015) Going deeper with convolutions. In: Proceedings of the IEEE conference on computer vision and pattern recognition, pp 1–9
15. Zou KH, Warfield SK, Bharatha A, Tempany CMC, Kaus MR, Haker SJ, Wells WM, Jolesz FA, Kikinis R (2004) Statistical validation of image segmentation quality based on a spatial overlap index1. Acad Radiol 11(2):178–189
16. Serrano-Rubio JP, Everson R, Hutt H (2018) Brain tumour segmentation method based on sparse feature vectors. In: Pre-conference proceedings of the 7th MICCAI BraTS challenge, Granada, Spain, pp 420–427
17. Shin HE, Park MS (218) Brain tumor segmentation using 2D U-net. In: Pre-conference proceedings of the 7th MICCAI BraTS challenge, Granada, Spain, pp 428–437

Deep Learning-Based Automatic Detection of Defective Tablets in Pharmaceutical Manufacturing

Huynh Thanh Quan, Dong Duc Huy, Ngo Thanh Hoan, and Nguyen Thanh Duc

Abstract With many tablets produced everyday in manufacturing plants, the pharmaceutical industry needs automatic, highly accurate methods for inspection of tablet quality. Detecting defective tablets is of importance to reduce unqualified products to consumers. In this paper, we propose a deep learning method combining image processing and deep convolutional neural networks (DCNN) for detection of defective tablets using images captured by a multiple-camera inspection system. A dataset of 6000 images of tablets labelled either GOOD or NOT-GOOD were collected at a pharmaceutical factory using commercial camera inspection systems. After collecting and labelling, the images were preprocessed to normalize intensity values. The entire dataset was split into a training set (50%, 3000 images), a validation set (16.6%, 1000 images) and a testing set (33.3%, 2000 images). We trained DCNN ResNets (ResNet50, ResNet101) and DenseNets (DenseNet169, DenseNet201) models on the training set and validated them on the validation set. We applied transfer learning techniques by using pre-trained models that had been trained on the ImageNet dataset in combination with data augmentation and training strategies such as learning rate rescheduling overtime. We compared our deep learning methods with various machine-learning ones such as Support Vector Machine (SVM), K-Nearest-Neighbors (KNN), AdaBoost that used intensity histograms as features. Tuning hyperparameters were performed to seek the best hyper-parameters and algorithms. We achieved best performances using the deep learning models as

H. T. Quan
Department of Electrical Engineering and Information Technology, Vietnamese German University, Thu Dau Mot, Vietnam

D. D. Huy · N. T. Hoan
School of Biomedical Engineering, International University, Ho Chi Minh City, Vietnam

Vietnam National University-Ho Chi Minh City, Ho Chi Minh City, Vietnam

N. T. Duc (✉)
Department of Biomedical Science and Engineering, Integrated of Integrated Technology, Gwangju Institute of Science and Technology, Gwangju, South Korea
e-mail: ducnguyen@gm.gist.ac.kr

Montreal Neurological Institute, McGill University, Montreal, Canada

V. Van Toi et al. (eds.), *8th International Conference on the Development of Biomedical Engineering in Vietnam*, IFMBE Proceedings 85,
https://doi.org/10.1007/978-3-030-75506-5_64

the ResNet50, and DenseNet169 obtained 96.60% ± 4.9% and 94.13% ± 4.2% accuracies (ACC), respectively. In contrast, SVM achieved 87.75% ACC, KNN achieved 76.09% ± 7.7% ACC while AdaBoost achieved 81.25% ACC, respectively.

Keywords AI · Tablet visual inspection · CNN network · ResNet · Pharmaceutical industry

1 Introduction

Medicine is of great importance to treat human's diseases. Nowadays, to meet the world's demand, the pharmaceutical industry focuses on manufacturing a large quantity of medicine in the form of tablets. In the mass production of tablets, various defects can arise during the production processes or even during transportation processes [1]. Tablet defects impart financial burden on pharmaceutical companies and may result in a therapeutic failure as well as reducing patient compliance [1, 2]. Therefore, manufactured medicine needs to be carefully inspected and classified before being shipped to customers. With the number of tablets produced nowadays, the manual visual inspection is subjective, unreliable, tedious, and even harmful to the operators. Thus, an automated inspection and verification system that can detect defective tablets is required [3].

Traditional image analysis techniques such as real-time image segmentation [3, 4], edge detection [5], image registration [6, 7] have been applied to detect defective tablets. With segmentation method, tablet region and background are separated. The tablet region is compared to the sample to determine whether the tablet defects. Edge detection method [5] is a method that extracts the edge feature of the objects in the image. Morphology opening then is used to remove noise from the binary image. After this process, area and perimeter are determined, and then the metric is calculated. Metric closer to one indicates that the tablet is not broken or is completely round. Image registration [7] based on the direct matching of circular profile matching and pixel intensities. A devised "gold standard" is used for comparing against registration; circular profile matching method proved to be a powerful image processing tool. However, tablets from the different company are produced in different sizes, shapes, colours, surface property and have many varieties of possible defects. The limitation of traditional image analysis techniques is it needs to be specific for one type of tablet. Automated visual tablet inspection systems on the current market are robust and general enough to be able to be used for detecting various defection from a large variety of tablets. However, its weakness is of lower sensitivity. This trade-off means that these inspection systems have advantages of being able to conduct well-large defects on smooth tablets without imprints or texture but have the suboptimal performance on tablets with texture and imprints [3].

In recent years, convolutional neural networks (CNNs) for deep learning data analytics have become a promising area of research and have applications towards visual inspection and appearance testing [8–10]. CNNs are capable of matching or even exceeding human performance [3, 11–13] and hold the ability to be quickly adapted to a new product. CNNs have no limitations when using a large-scale inspection. The industries can generalize and apply CNNs in production line rapidly. Deep learning has higher capacity to represent complex structures compared to classical machine-vision methods, and capacity to learn more complex features, thus deep learning focus on using an automated learning process instead of reply on hand engineering of features [3, 11, 14]. In recent years, CNNs become the most common approach in the field of computer vision and outperform previous state-of-the-art machine learning techniques in several fields [6].

In this paper, we trained ResNets (ResNet50, ResNet101) and DenseNets (DenseNet169, DenseNet201) models to detect defective tablets and compared the result to machine-learning ones such as Support Vector Machine (SVM), Nearest-Neighbors (NN), AdaBoost. The ability of the trained CNN for classifying sample images in a training test set into "GOOD" or "NOT-GOOD" category was tested and evaluated using the dataset of 6000 images collected at a pharmaceutical factory using commercial camera inspection systems.

The remainder of this paper is organized as follows. In Sect. 2, we provide an overall framework for intelligent visual inspection and data acquisition, image preprocessing and feature extraction. Deep learning models ResNet, DenseNet are also reviewed. The basic architectures, training processes of each group are presented. In Sect. 3, we review the result of Deep learning models and compare the results to machine-learning models. Finally, the discussion and conclusion are presented in Sect. 4.

2 Material and Methods

2.1 The Overall Framework for Intelligent Visual Inspection

Figure 1 depicts our proposed all-in-one framework, for which state-of-the-art artificial intelligence concepts are based, for automated tablet visual inspection in pharmaceutical manufacturing. In this paper, we specifically aim at identifying the defective tablets, which we define NOT-GOOD class, among the normal ones, which we define the GOOD class. Examples of acquired images of two defined classes are provided in Fig. 2 (labelling block). The first block in the framework is the hardware and high-resolution camera system that is used to capture the high-resolution two-dimensional images of the inspective tablets. For quality assurance inspection, the hardware architecture with multiple cameras acquires the tablet on different inspected sides. Then, various advanced image preprocessing approaches are performed to enhance the quality of the raw images before they are fed into AI algorithms for automatic inspection.

Additionally, these image processing methods are expected to extract the salient features which powerfully discriminate images of defective tablets compared to

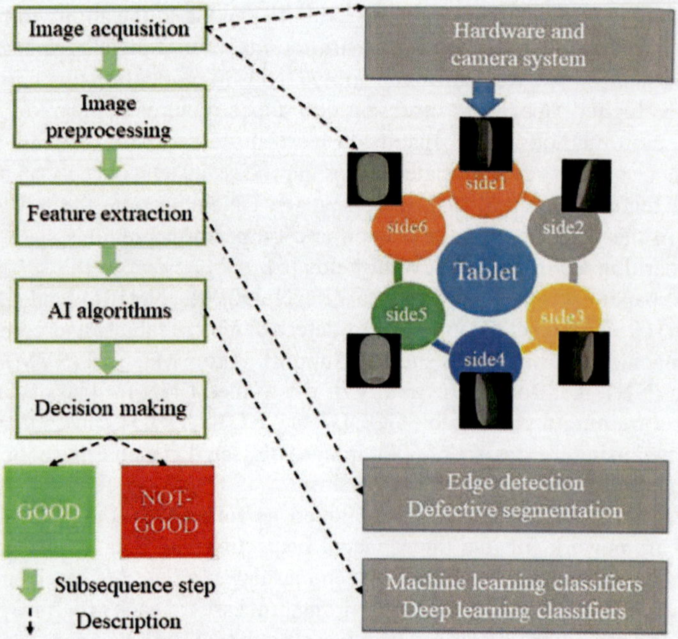

Fig. 1 Illustrations of the overall process of automatic tablet visual inspection in the pharmaceutical industry using artificial intelligence technology

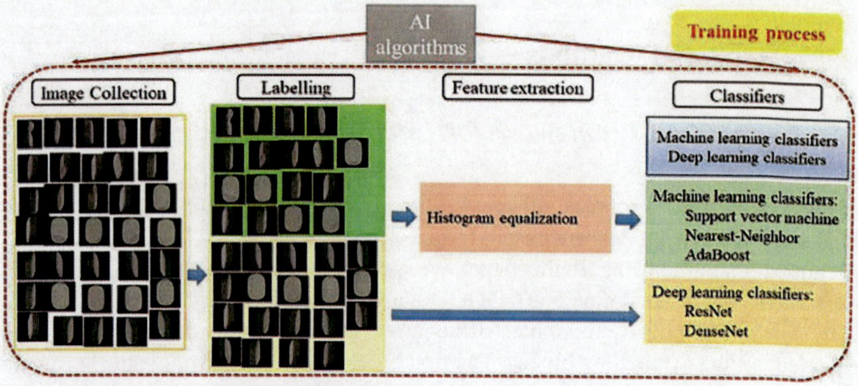

Fig. 2 Illustrations of artificial intelligence algorithms and training processes

normal ones. The core block of intelligent visual machines is AI algorithms which are used to make a probability-based decision to which class (either GOOD or NOT-GOOD) new unseen tablets belong to. In this paper, we examined the efficacies of the classical machine learning classifiers; that is Support Vector Machine (SVM), Nearest-Neighbor (NN), Decision Tree, Random Forest, AdaBoost; as well as the most advanced deep learning classifier; that are ResNet, DenseNet, InceptionNet. Detailed algorithms for the classical machine learning classifiers and state-of-the-art deep learning ones can be seen in Fig. 2.

2.2 *Data Acquisition, Image Preprocessing and Feature Extraction*

The data collection was carried out at a pharmaceutical factory site using an existing industrial machine vision system which is based on imaging processing supported by high-performance software algorithms and mechanical structures with high-sensitive measurement. Next, the collected datasets were further inspected and then labelled by five quality control engineers who have more than five years with industrial visual inspection to assure the quality of collected images. We discarded the images that were not qualified for further analysis. Finally, about 6000 raw images were acquired, about 3000 images for each class; i.e., GOOD and NOT-GOOD. The balanced binary class datasets were further divided into training, validation and testing groups for training, validating and testing the artificial intelligence algorithms, respectively, by random selection. As a result, the number of images for training, validation and testing was 3000, 1000, and 2000 images, respectively. Detailed distributions of the data are provided in Table 1.

Image processing, which is a driven step for the success of AI algorithms, has a vital role in the automation of visual inspection. Image processing algorithms were used at preprocessing and feature engineering phases. In the preprocessing phase, the raw images were cropped to have same-size datasets (trimmed size: 224 × 224) while keeping the useful image pixels, normalized their intensities between ranges of zero and one, smoothed and scaled to convert all collected raw images into a standard form. The normalization ensures the classification results to be invariant of intensities, thereby enabling the model to make predictions solely based on the shape configurations of the images. Subsequently, the feature extraction phase is expected to be useful in discriminating, for example, relevant components in defective tablets and normal tablets. Note that the feature extraction phase is needed only for the classical machine learning classifiers, not applicable for deep learning classifiers. There are numbers of imaging features existing in the literature that one can employ for such intelligent visual inspections. And so is the case with the type of classifiers. However, the question here is how to determine the precise 'feature-classifier' combination that

Table 1 The distribution of images for training, validation and testing sets

Dataset	No. of images in the GOOD class	No. of images in NOT-GOOD class
Training	1500	1500
Validation	500	500
Testing	1000	1000

works for a specific application. In this study, a larger number of histogram features constructed out of image pixel values was represented in a feature vector. Histogram features were used to observe the range of intensity variation in images from class to class. The information derived from histogram plots may provide salient features for visual inspection. The feature vector was used to train the classical machine learning classifiers, i.e., Support Vector Machine (SVM), Nearest-Neighbor (NN), Random Forest, AdaBoost.

2.3 Deep Learning Models

The availability of extensive training data and computing advancements are the critical factors to drive the success of deep learning CNN models, allowing ones to utilize the parallel processing capabilities of GPU into faster training complex CNN architecture for higher performances. Several CNN models have proved their effectiveness on public image databases, for example, ImageNet Large Scale Visual Recognition Challenge (ILSVRC) [15], PASCAL 2010 VOC [16], and COCO dataset of Microsoft Research team [17], and therefore are expected to provide promising results on our classification task. Table 2 highlights the achievements of some featured CNN-based deep learning models on public datasets. In this present study, we employed four CNN-based deep learning models, which had previously demonstrated high performance in a classification task for solving our similar problem; and those models include ResNet50, ResNet101, DenseNet169 and DenseNet201. The detailed descriptions of these models, which include the error rate trained on the well-established ImageNet dataset, are given in Table 2.

Residual Network (ResNet)

The process of training deep CNN models usually requires the computational expense and is more difficult when the network increases in depth. Typically, the problem that

Table 2 Highlight features of the most well-known models in ILSVRC, PASCAL and VOC competitions of recent years

Architecture	Year	Error rate	Main points
VGG	2014	ImageNet: 7.3	Use small size kernels
Inception-V3	2015	ImageNet: 3.5 Multi-Crop: 3.58 Single-Crop: 5.6	Handle the representational bottleneck
ResNet	2016	ImageNet: 3.6 CIFAR-10: 6.43	Residual connections identity kernel
DenseNet	2017	CIFAR-10+ :3.46 CIFAR100+ :17.18 CIFAR-10: 5.19 CIFAR-100: 19.64	Cross-layer information flow

has arisen in deep neural networks is the backpropagation of error getting a smaller gradient value when approaching low-level layers. Gradient vanishing arises when the network introduces too many non-linear mapping feature layers. These stacked layers reduce the gradient extremely small (approximately to zero) when it reaches the very first layers. In theory, these gradient vectors still can be recognized. However, in practice, when float numbers are declared mostly in 32bit or 64bit (FLOAT32 or FLOAT64), the floating points still can be approximated to become zero when it is too small. This leads to all subsequent values to become zero or gradient vanishing [18].

Residual network (ResNet) [19] exploited the idea of bypass pathways used in Highway Networks. The identity layer or "shortcut" helps to address gradient vanishing by letting the gradient values flow directly to upper layers when back-propagation. Their idea was not new; in fact, there were many model designs such as "shortcut connection", "highway" or adding linear layers connected to auxiliary classifiers to solve vanishing/exploding gradients [20].

ResNet has five versions (ResNet-18, 34, 50, 101, and 152) depending on the depth of the convolution. In this study, we employed two ResNet architectures for the classification task. These were ResNet50 and ResNet101; which have 50 layers and 101 layers, respectively. Illustration example of the ResNet50-based deep learning architecture is shown in Fig. 3. ResNet50 has a total of 50 layers, and its configuration is as follows [21]. The set consists of a convolution layer, batch normalization, and ReLU (activation function) with 49 layers, and a fully connected last layer. Only the first convolution layer is set to a 7×7 kernel with stride two and padding 3, and the kernel size of all subsequent convolution layers is 3×3.

DenseNet

Similar to ResNet, DenseNet also uses the concept of residual connection. However, unlike ResNet, that preserves information through additive identity mapping on a subset of layers, DenseNet uses cross-layer connectivity more densely. DenseNet

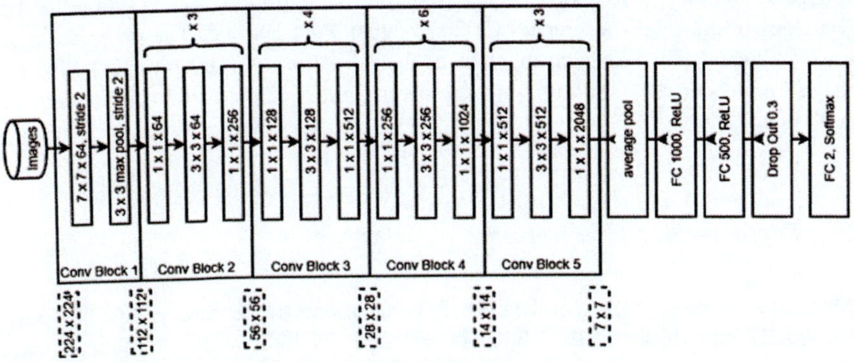

Fig. 3 The architecture of ResNet50 convolutional neural network

connects all preceding layers to the next coming layer in the feed-forward mechanism. All feature-maps of previous layers have been used on all subsequent layers. Therefore, DenseNet is considered as a very narrow structure and is parametrically expensive by the increase of the number of feature maps [22]. DenseNet has three versions (DenseNet-121, 169, and 201 with 32 channels). In this work, we use two DenseNet architectures; these are DenseNet169 that has 169 layers and DenseNet201 that consists of 201 layers.

2.4 *Training Strategy and Experiment Setup*

Several CNN-based deep learning networks, as well as conventional machine learning, were used for classification tasks in this work. We used TensorFlow (ver. 2.2.0, https://tensorflow.org/) as the deep-learning library to train, validate, and test the CNN. In this work, we implemented the transfer learning network, which had been pre-trained on the well-established large-scale ImageNet database for image classification. However, to be adapted for our application, we re-configured the fully connected last layer which was at 2 (output dimension is changed from 1000 to 2 as we have two classes: GOOD and NOT-GOOD). The parameters of the convolution layers in pre-trained networks were used as initial values. The weights in the pre-trained deep learning networks had learned well through a large number of images in the ImageNet dataset [15]. Therefore, the transfer learning using the pre-trained weights was expected to learn faster than the scratched networks, and the better performances are promised. To increase the number of images for training and obtain sufficient robustness, we applied on-the-fly data augmentation which each training image was flipped to create a new training image. However, the data augmentation was not used in the validation and testing phases.

We used the binary cross-entropy function as a loss function for binary classification. Adam was used as an optimizer with a scheduled learning rate strategy for precise training. One epoch is defined as performing backpropagation once for all images in the total training-set (3000 images) were learned. The deep learning networks were verified for each epoch using the validation-set (1000 images). Finally, the best models on the validation set were saved for testing on the left-out testing-set (2000 images).

2.5 *Performance Evaluation*

To evaluate the performance of the classifiers, we report the Accuracy (ACC), Precision (PRE), and Recall (REC). TP, TN, FP, and FN indicate the number of true positives, true negatives, false positives, and false negatives, respectively. In terms of these numbers, ACC, PRE, and REC can be computed as follows:

$$ACC = (TP + TN)/(TP + TN + FP + FN) \quad (1)$$

$$PRE = TP/(TP + FP) \quad (2)$$

$$REC = TP/(TP + FN) \quad (3)$$

To minimize the effect of random weight initializations and maximize the reliability of the results, we performed the training 20 times. The means and standard errors of the classification results for each algorithm are reported.

3 Classification Results

The classification results of our proposed artificial intelligent approach for automated inspection of tablets in the pharmaceutical industry are shown in Table 3. The means and standard errors of the evaluation metrics, including classification accuracy (ACC), precision (PRE), and recall (RECALL) of about 20 reruns are reported. The performances of four CNN-based deep learning classifiers as well as six machine learning classifiers are presented. Illustrative plots of multiple comparative methods are provided in Fig. 4. In overall, as we can see from Table 3, the deep learning networks significantly outperformed the machine learning ones. Specifically, the

Table 3 The classification results of the proposed artificial intelligence approaches which include conventional machine learning and CNN-based deep learning methods of automatic tablet inspection

Algorithms	Evaluation metrics		
	Accuracy (ACC)	Precision (PRE)	Recall (REC)
Classical machine learning classifiers			
SVM (Support Vector Machine)	87.75% ± 0.0%	87.75% ± 0.0%	87.75% ± 0.0%
SGD (Stochastic Gradient Descent)	69.03% ± 0.7%	70.01% ± 0.5%	69.04% ± 0.7%
KNN (K-Nearest-Neighbor)	76.09% ± 7.7%	78.70% ± 5.8%	76.09% ± 7.7%
Decision Tree	75.08% ± 0.2%	75.25% ± 0.2%	75.08% ± 0.2%
Random Forest	73.58% ± 2.7%	74.62% ± 2.5%	73.58% ± 2.7%
AdaBoost	81.25% ± 0.0%	81.58% ± 1.1%	81.25% ± 0.0%
CNN-based deep learning classifiers			
ResNet50	**96.60% ± 4.9%**	**94.33% ± 3.3%**	**99.37% ± 4.9%**
ResNet101	93.26% ± 1.7%	92.51% ± 0.9%	94.11% ± 1.1%
DenseNet169	94.13% ± 4.2%	93.56% ± 1.7%	95.02% ± 1.3%
DenseNet201	89.27% ± 2.8%	91.49% ± 5.0%	87.68% ± 0.4%

Bold indicates the best performance model

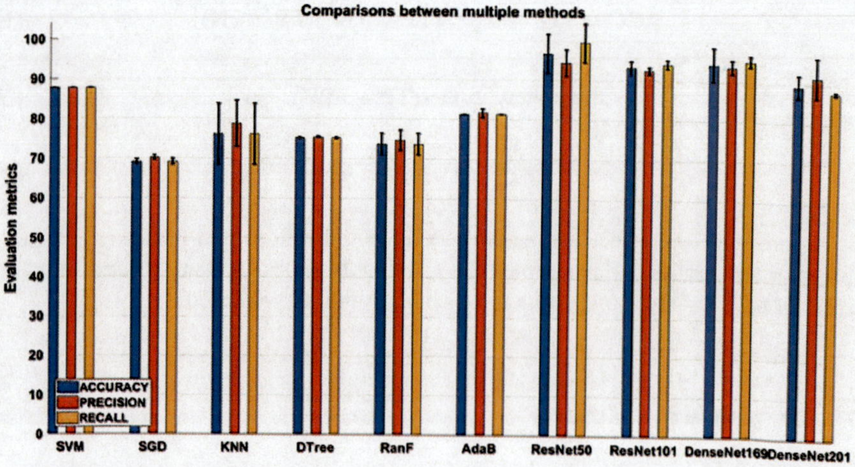

Fig. 4 The comparison of multiple methods described in this study. Abbreviation: SVM: Support Vector Machine; SGD: Stochastic Gradient Descent; KNN: K-Nearest Neighbor; DTree: Decision Tree; RanF: Random Forest; AdaB: AdaBoost

ACC of deep learning classifiers were more than 90% (only DenseNet201' ACC is slightly below 90% at 89.27% ± 2.8%).

On the other hand, the ACCs of all machine learning algorithms were deeply below 90%, ranging from a minimum value of 69.03% ± 0.7% with SGD algorithm to a maximum value of 87.75% ± 0.0% with SVM algorithm. All in all, the best performance among all studied architectures was achieved using ResNet50 model in which up to an ACC of 96.60% ± 4.9%, a PRE of 94.33% ± 3.3% and a RECALL of 99.37% ± 4.9% are obtained, followed by a DenseNet169 model in which we can achieve up to an ACC of 94.13% ± 4.2%, a PRE of 93.56% ± 1.7% and a RECALL of 95.02% ± 1.3%.

4 Discussion and Conclusion

4.1 Comparison with the Previous Studies

Deep learning models have shown promising results on automatic visual inspection applications [2, 12, 23]. In this paper, we introduce an end-to-end deep learning framework to automatically inspect (detect) defective tablets in the manufacturing process of the pharmaceutical industry. For the classification problem, we applied four fine-tuning transfer learning on deep learning networks, i.e., ResNet50, ResNet101, DenseNet169, DenseNet201, which had been pre-trained on the large-scale ImageNet dataset. A total of 6000 224 × 224 images were used as our dataset. Out of the 6000 images, a total of 3000 images (50% of the total images), 1000 images, and 2000 images were randomly chosen for training, validation, and testing sets, respectively. With more than 100 epochs, the best performance was reported in which up to an accuracy of 96.60% ± 4.9% a PRE of 94.33% ± 3.3% and a RECALL of 99.37% ± 4.9% on the unseen testing dataset were achieved. The robustness of our network was evaluated on a separate left-out dataset of 2000 images, 1000 images for each class. Illustrative examples of how our proposed deep learning models predict the images belonging to NOT-GOOD class are provided in Fig. 5. The overall performances of our proposed deep learning models have shown high reliability and robustness in the identification of defective tablets. The designed CNN-based deep learning approach is expected to be applied or combined with the existing mechanical actual inspection machines to increase the inspection results.

In recent years, many studies have been carried out for visual inspection of defective products using artificial intelligence approaches in the manufacturing industry. Nagata et al. [12] developed a deep convolutional neural network to inspect defects such as crack and burr phenomena occurred in the manufacturing process of resin moulding. The proposed model showed the ability to classify sample images in a training test set into "OK" (good products) and "NG" (defective products) category

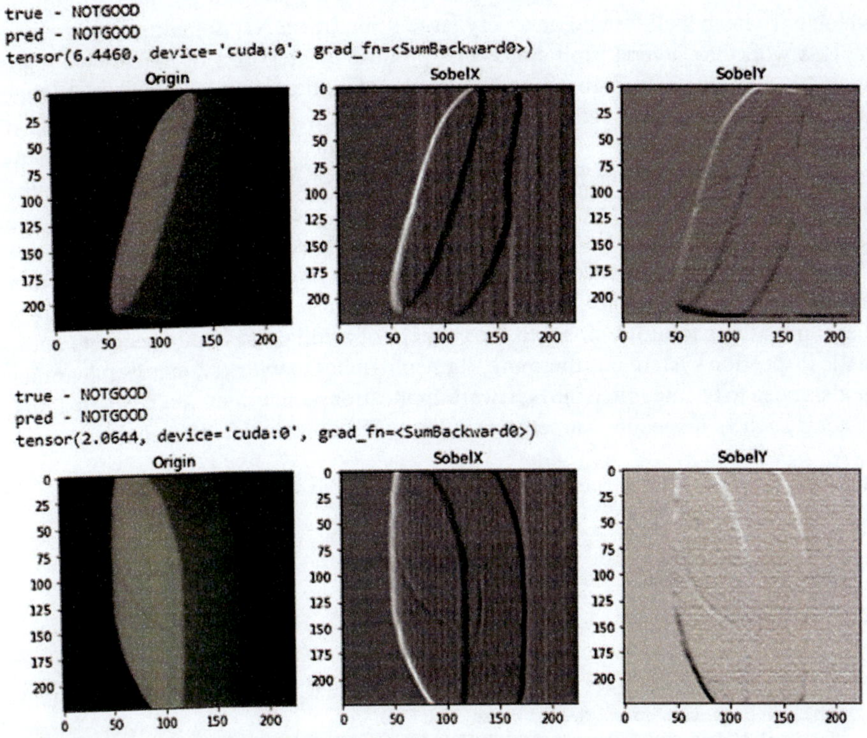

Fig. 5 Examples of the true and predicted images predicted by our proposed deep learning models. These provided images are from the NOT-GOOD class

with high accuracy (96%). Similarly, Perez et al. [23] introduced an application of CNN networks towards an automated detection and localization of key building defects such as mould, deterioration, and stain using images. The introduced approach was based on pretrained CNN-based VGG-16, ResNet50, and Inception networks, with class activation mapping for object localization. They achieved a maximum performance of 97.83% with 0.0572 loss on the testing set. However, direct performance comparison with other studies would not be fair because of the different datasets, preprocessing pipelines, feature measures, and classifiers.

4.2 *Limitations and Future Works*

Several challenges are remaining in this work. One of the primary challenges was the availability of the large labelled datasets, which can be used to train the CNN-based deep learning models for this type of problems. To deal with this problem, we used image augmentation techniques to generate synthetic data for largely enough dataset to train our models. On the other hand, transfer learning can be a good choice for fewer label data problems as in this case. Indeed, we employed pre-trained models which have been well trained on a very large-scale ImageNet domain.

This work has several limitations. First, multiple categories of defects were not considered as we only focused on the binary classification problem, which aims at detecting defective tablets among the normal ones. It would be more robust if the deep learning model can detect different defective categories with high accuracy [23]. Second, only the images with visible defects were considered. Third, the localizations of defective areas in the defect tablets were not addressed. Four, the ensemble learning in which multiple deep learning classifiers are combined was not studied. In the literature, ensemble learning has been proved to be a wise choice that significantly boosts the classification performance. In future works, we will address these limitations to move closer to the concept of a fully and highly accurate automatic inspection system. Furthermore, for deployment in industry, our deep learning models need to be integrated into software applications, once combined with imaging systems such as inspection cameras, can detect defective tablets in real-time.

Conflicts of Interest The authors have no conflict of interest to declare.

References

1. Albion K, Briens L, Briens C, Berruti F (2006) Detection of the breakage of pharmaceutical tablets in pneumatic transport. Int J Pharm 322(1–2)
2. Sabri AH, Hallam CN, Baker NA, Murphy DS, Gabbott IP (2018) Understanding tablet defects in commercial manufacture and transfer. J Drug Deliv Sci Technol 46:1–6. https://doi.org/10.1016/j.jddst.2018.04.020

3. Možina M, Tomaževič D, Pernuš F, Likar B (2013) Automated visual inspection of imprint quality of pharmaceutical tablets. Mach Vis Appl 24(1):63–73. https://doi.org/10.1007/s00138-011-0366-4
4. Možina M, Tomaževič D, Pernuš F, Likar B (2011) Real-time image segmentation for visual inspection of pharmaceutical tablets. Mach Vis Appl 22(1):145–156. https://doi.org/10.1007/s00138-009-0218-7
5. Manzoor H, Randhawa YS (2014) Edge detection in digital image using statistical method. IOSR J Electron Commun Eng 9(3):15–19. https://doi.org/10.9790/2834-09311519
6. Voulodimos A, Doulamis N, Doulamis A, Protopapadakis E (2018) Deep learning for computer vision: a brief review. Comput Intell Neurosci 2018:7068349
7. Špiclin Ž, Bukovec M, Pernuš F, Likar B (2011) Image registration for visual inspection of imprinted pharmaceutical tablets. Mach Vis Appl 22(1):197–206. https://doi.org/10.1007/s00138-007-0104-0
8. Nguyen DT, Ryu S, Qureshi MNI, Choi M, Lee KH, Lee B (2019) Hybrid multivariate pattern analysis combined with extreme learning machine for Alzheimer's dementia diagnosis using multi-measure rs-fMRI spatial patterns. PLoS One 14(2):e0212582
9. Duc NT, Lee B (2019) Microstate functional connectivity in EEG cognitive tasks revealed by a multivariate Gaussian hidden Markov model with phase locking value. J Neural Eng 16(2):026033
10. Duc NT, Ryu S, Choi M, Iqbal Qureshi MN, Lee B (2019) Mild cognitive impairment diagnosis using extreme learning machine combined with multivoxel pattern analysis on multi-biomarker resting-state FMRI. In: Conference proceedings IEEE engineering in medicine and biology society, vol 2019, pp 882–885
11. Kato N, Inoue M, Nishiyama M, Iwai Y (2020) Comparing the recognition accuracy of humans and deep learning on a simple visual inspection task. Lecture notes in computer science, pp 184–197. http://doi.org/10.1007/978-3-030-41299-9_15
12. Nagata F et al (2018) Basic application of deep convolutional neural network to visual inspection. In: Proceedings of the 6th IIAE international conference on industrial application engineering 2018. http://doi.org/10.12792/iciae2018.004
13. Duc NT, Ryu S, Qureshi MNI, Choi M, Lee KH, Lee B (2020) 3D-deep learning based automatic diagnosis of Alzheimer's disease with joint MMSE prediction using resting-state fMRI. Neuroinformatics 18(1):71–86
14. Tabernik D, Šela S, Skvarč J, Skočaj D (2019) Segmentation-based deep-learning approach for surface-defect detection. J Intell Manuf. https://doi.org/10.1007/s10845-019-01476-x
15. Russakovsky O et al (2015) ImageNet large scale visual recognition challenge. Int J Comput Vision 115(3):211–252. https://doi.org/10.1007/s11263-015-0816-y
16. Everingham M, Van Gool L, Williams CKI, Winn J, Zisserman A (2010) The pascal visual object classes (VOC) challenge. Int J Comput Vision 88(2):303–338. https://doi.org/10.1007/s11263-009-0275-4
17. Lin T-Y et al (2014) Microsoft COCO: common objects in context. In: Computer vision—ECCV 2014, pp 740–755. http://doi.org/10.1007/978-3-319-10602-1_48
18. On the difficulty of training recurrent neural networks. https://arxiv.org/pdf/1211.5063.pdf. Accessed 27 Apr 2020
19. He K, Zhang X, Ren S, Sun J (2016) Deep residual learning for image recognition. In: 2016 IEEE conference on computer vision and pattern recognition (CVPR). http://doi.org/10.1109/cvpr.2016.90
20. Highway Networks. https://arxiv.org/abs/1505.00387. Accessed 27 Apr 2020
21. Kim YD et al (2020) Effects of hypertension, diabetes, and smoking on age and sex prediction from retinal fundus images. Sci Rep 10(1). http://doi.org/10.1038/s41598-020-61519-9
22. Huang G, Liu Z, Van Der Maaten L, Weinberger KQ (2017) Densely connected convolutional networks. In: 2017 IEEE conference on computer vision and pattern recognition (CVPR). http://doi.org/10.1109/cvpr.2017.243
23. Perez H, Tah JHM, Mosavi A (2019) Deep learning for detecting building defects using convolutional neural networks. Sensors 19(16). http://doi.org/10.3390/s19163556

Adaptive Lung Diseases Images Classification Technique Based on Deep Learning

Nguyen Huu The, Nguyen Thi Hong Nhung, and Nguyen Thanh Binh

Abstract Medical images have made an important contribution to improving the accuracy and effectiveness of disease diagnosis, such as diseases related to lung, heart, liver, kidney, etc. Pneumonia has increased rapidly in the world in recent years. Chest X-ray image analysis is a common method for detecting lung diseases. An advanced artificial intelligence system will help doctors have accurate conclusions, timely treatment for patients and reducing mortality. Using machine learning on X-ray images is of great interest, but research results are still limited in accuracy. This paper proposed an adaptive technique for lung diseases image classification based on the deep learning method. We improved the convolutional neural network for lung diseases image classification, created a training model with a suitable number of hidden network layers and optimal algorithms to detect pneumonia images. As a result, the rate of correct detection of pneumonia image was 98.72%. We used chest X-ray images dataset that published by Kaggle, including 5863 chest X-ray images. The results of the proposed method are better than the other methods.

Keywords Pneumonia · Lung diseases · Convolutional neural network · Chest X-ray pneumonia

---

N. H. The · N. T. H. Nhung
Faculty of Information Technology-Communication, University of Cuu Long, Vinh Long, Viet Nam
e-mail: nguyenhuuthe@mku.edu.vn

N. T. H. Nhung
e-mail: nguyenthihongnhung@mku.edu.vn

N. T. Binh (✉)
Faculty of Computer Science and Engineering, Ho Chi Minh City University of Technology, VNU-HCM, 268 Ly Thuong Kiet Street, District 10, Ho Chi Minh City, Viet Nam
e-mail: ntbinh@hcmut.edu.vn

Vietnam National University Ho Chi Minh City, Linh Trung Ward, Thu Duc District, Ho Chi Minh City, Viet Nam

V. Van Toi et al. (eds.), *8th International Conference on the Development of Biomedical Engineering in Vietnam*, IFMBE Proceedings 85,
https://doi.org/10.1007/978-3-030-75506-5_65

1 Introduction

Detect pneumonia on chest X-rays is a challenging task even for X-ray specialists. Pneumonia on X-rays is difficult to distinguish from some pathologies about the other chest area [1]. This work requires knowledge of chest-related pathologies, as well as expertise and experience in viewing X-rays to make an initial diagnosis. The intelligent computer systems to support doctors became essential to help doctors diagnose pneumonia from X-ray images more effectively.

Many recent researches in the field of image processing have achieved good results in image classification, especially using Convolutional Neural Networks (CNN), a deep learning model that is heavily used in image recognition, because CNN is capable of extract key features from image fast and efficiently [2]. In the field of medicine, there have been many scientific papers that have applied the CNN model [3–9] such as breast cancer detection [3], skin cancer [4], pneumonia detection [9], etc.

To assist doctors in detecting pneumonia, Ayan [10] studied the diagnosis of pneumonia with the CNN model when tested on pretrained models of Xception and Vgg16. To adjust the parameters for the model in the process of training the model, the authors combined the method of transfer learning and fine-tuning. The result was the accuracy of Xception with 82% and Vgg16 with 87%. Ponnada [11] presented a system for detecting pneumonia. The author and collaborators suggested a CNN method consisting of seven hidden layers to train their model. The result with accuracy is 86%, precision is 79%, and recall is 98% with LIDC-IDRI and Mendeley dataset. Kadam [12] have developed a deep neural network, which helped predict the presence of pneumonia using chest X-rays, a CNN model has been employed to increase efficiency and accuracy, to further add to the performance, optimum differential learning rates have been selected using the techniques of cosine annealing and stochastic gradient with restarts, and the result with accuracy is 92.90%, precision is 90.88%, and recall is 99.27%.

The detection of pneumonia on chest X-ray is an image classification problem to determine whether the presence of pneumonia in the chest X-ray image. The system input is a chest X-ray image; the output is a normal X-ray image or pneumonia image. In this paper, we proposed an adaptive technique for lung diseases image classification based on the deep learning method. We improved the convolutional neural network for lung diseases image classification, created a training model with a suitable number of hidden network layers and optimal algorithms to detect pneumonia images. The rest of this paper is organized as following: in Sect. 2, we describe deep learning for classification. The proposed method for lung diseases image classification is presented clearly in Sect. 3. The experiment results and conclusion are presented in Sects. 4 and 5.

2 Background of Deep Learning for Classification

Deep learning is the neural network with multiple hidden layers. A neural network may have few hidden layers, but with deep learning, the number of hidden layers is very large, up to hundreds of layers. More layers will increase the accuracy of the neural network. In deep learning technique, the more the input data, the more accurate the output [5]. Deep learning is a sub-branch of machine learning; deep learning algorithms will perform a task several times, each time adjusting it a little to improve results.

CNN is a class in deep learning. CNN has applications in image and video identification [13], recommendation systems [14], image classification, medical image analysis [10–12], natural language processing [15], and financial time series [16]. CNN performs images classification by receiving input images, processing them, and classifying them as labels. The computer receives the input as an array of pixels based on the resolution of the image, from which the computer recognizes the image as height × width × dimension (h × w × d). An image with (150 × 150 × 3) means there are three color channels RGB.

In the CNN model, each input image scans through convolutional layers, also known as hidden layers. Each convolutional layer includes filters (kernel), activation function, max pooling, and the result is the feature map. To classify the output data, the feature map will switch to the fully connected layer (FC) and apply an activation function.

Figure 1 illustrates the process of receiving input image, extracting features of the image through layers, and classifying the output object.

The convolutional layer is used to extract features of the image. The input image of each convolutional layer is a 3-dimensional tensor of size (H × W × D) that will slide over by a kernel of size (Wk × Hk × Dk) with stride S and padding P. Each convolutional layer has N kernel; the output is a 3-dimensional tensor with the formula shown in the lecture [17]:

$$\left(\frac{W - Wk + 2P}{S} + 1\right) * \left(\frac{H - Hk + 2P}{S} + 1\right) * N \quad (1)$$

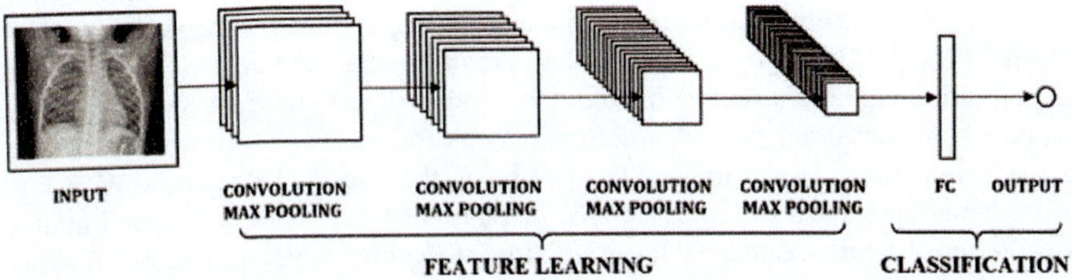

Fig. 1 A neural network with many convolutional layers

$$\underset{X}{\begin{bmatrix} 0 & 1 & 1 & 0 & 0 \\ 0 & 1 & 1 & 1 & 0 \\ 0 & 0 & 1 & 1 & 1 \\ 0 & 0 & 1 & 1 & 0 \\ 0 & 1 & 1 & 0 & 0 \end{bmatrix}} \otimes \underset{K}{\begin{bmatrix} 1 & 0 & 1 \\ 0 & 1 & 0 \\ 1 & 0 & 1 \end{bmatrix}} = \underset{Y}{\begin{bmatrix} 3 & 3 & 4 \\ 2 & 4 & 3 \\ 2 & 3 & 4 \end{bmatrix}}$$

Fig. 2 Features map

Consider a matrix X [5 × 5] containing image pixels with values 0 or 1, and a filter K [3 × 3] slides up to X with stride = 1. Each time K slides over X, extract from X a matrix A [3 × 3], perform the element-wise calculation on two matrices by the formula:

$$Y = X \otimes K \tag{2}$$

For each element x_{ij} in a matrix X, take out a matrix equal to the size of kernel K, and element x_{ij} in the central, called matrix A. Then, sum the elements of the element-wise calculation of matrix A and matrix K, write the result into matrix Y. Figure 2 illustrates the element-wise calculation to get the features map [18].

The value of y_{11} is calculated: $y_{11} = \text{sum}(A \otimes K) = x_{11} \times k_{11} + x_{12} \times k_{12} + x_{13} \times k_{13} + x_{21} \times k_{21} + x_{22} \times k_{22} + x_{23} \times k_{23} + x_{31} \times k_{31} + x_{32} \times k_{32} + x_{33} \times k_{33} = 3$. And perform the same calculation with the remaining elements in the matrix.

However, color images have three color channels: red, green, blue. The image is represented as a 3-dimensional tensor, and the filter is also a 3-dimensional tensor of size (k × k × 3).

The output of the convolutional layer will apply the activation function, and the input of the next convolutional layer will be received from the output of the previous layer as in the lecture [17]. Each kernel has dimensions (Wk × Hk × Dk) and has a bias coefficient, the total parameter of a kernel is (Wk × Hk × Dk + 1). With N kernels in the layer, the total number of parameters in this layer is (N × (Wk × Hk × Dk)).

Pooling layers are used between convolutional layers, to reduce data size, but retain important attributes. The reduced data size helps reduce computations in the model. If the pooling layer is size = (2 × 2), stride = 2, padding = 0, the output size of width and height will be halved, the depth will not change. After the image has been transferred to many convolutional layers and pooling layers, the model has learned the features of the image. The last layer, the tensor of size (H × W × D), will be converted to a size vector (H × W × D) in the fully connected layer. Finally, we combine the fully connected layer with image features to get the model's output.

3 Lung Diseases Image Classification

In this section, we proposed the method for lung diseases images classification based on deep learning. The proposed method is presented in Fig. 3.

In Fig. 3, the proposed method includes six steps. The detail of the above steps presents as follows:

1. *Get training data from the dataset*: perform reading chest X-ray image data from dataset "train" directory. Images, in turn, are put into the system, the total number of photos in training is 5218. The parameters are configured as follows, batch_size = 32, epoch = 32, image_dimension = 150 × 150 px. Therefore, corresponding to one epoch, the training dataset will be put into the system with 163 batches (5218/32 batch_size).
2. *Data augmentation*: deep learning techniques require a huge amount of data to get good results. In many cases, the amount of data is not large enough, such as medical data. In machine learning, the solution often chosen to solve this problem is data augmentation, help avoid redundancy and improve algorithm accuracy [19]. In this work, we have used different data augmentation techniques such as rescaling, image change adjust brightness, shear intensity, range for random zoom (rescale = 1./255, brightness_range = [0.7, 1.0], shear_range = 0.2, zoom_range = 0.2). Data Augmentation is an effective way to increase training data.
3. *Image analysis using CNN*: the model was built using five convolutional blocks with kernel_size = 3 × 3, activation function is "Relu", pool_size = 2 × 2 and uses "Dropout" to reduce overfitting with rate = 0.2. The last layer uses the activation function is "Sigmoid" because of binary classification. The optimization chosen was "Adam". Using Callbacks including ModelCheckpoint to save a copy of the best model, and EarlyStopping to stop model training when the difference between "training" and "validation error" starts to increase, instead of decreasing. The detail of this step is presented in Fig. 4.
4. *Compile and build the model:* implementing the model training process.
5. *Test model on test data:* use the "confusion matrix" and get the results for precision, recall, accuracy.

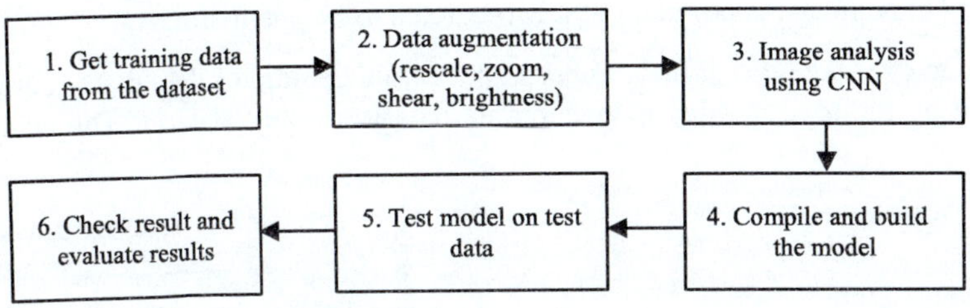

Fig. 3 Process flow of the proposed method

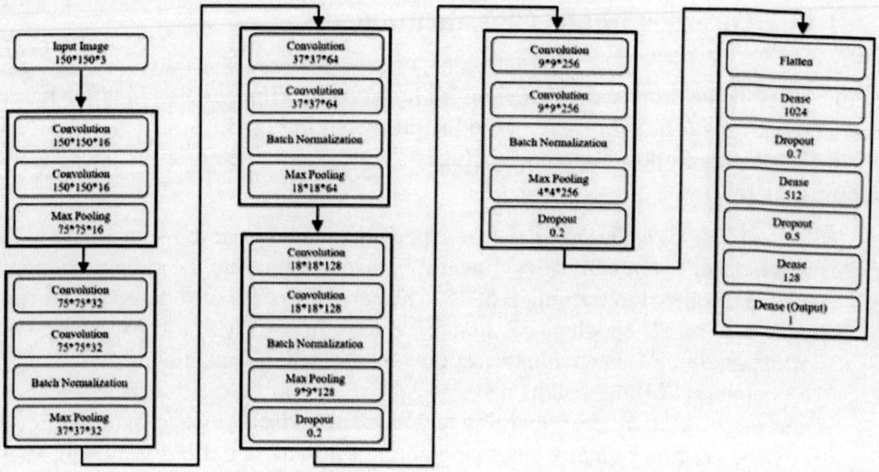

Fig. 4 Process flow of image analysis using convolutional neural network

6. *Check and evaluate results:* in this step, we check and evaluate the results of the proposed method.

To evaluate the results of the model, we use evaluation scales such as accuracy, precision and recall. A confusion matrix is used to present the values obtained after the model testing process, with these values, we can calculate the effectiveness of the training model. The components of the confusion matrix are described in detail in Fig. 5.

We apply the diagnosis of pneumonia to explain these four indicators. In the diagnosis of pneumonia, we have two layers: the pneumonia layer and the normal layer. The grade of pneumonia was marked as positive, and the normal layer was marked as negative. We define the parameters for the evaluation results such as True Positive (TP), False Negative (FN), True Negative (TN) and False Positive (FP).

TP: Actual image is pneumonia, and the predicted model is pneumonia.
FN: Actual image is pneumonia, but it is predicted to be normal.
TN: Actual image is normal, and it is predicted to be normal.
FP: Actual image is normal, but it is predicted to be pneumonia.

Accuracy is the ratio of the number of pneumonia predictor data that is correctly divided by the total number of pneumonia data in the test dataset. The accuracy

Fig. 5 Confusion matrix [20]

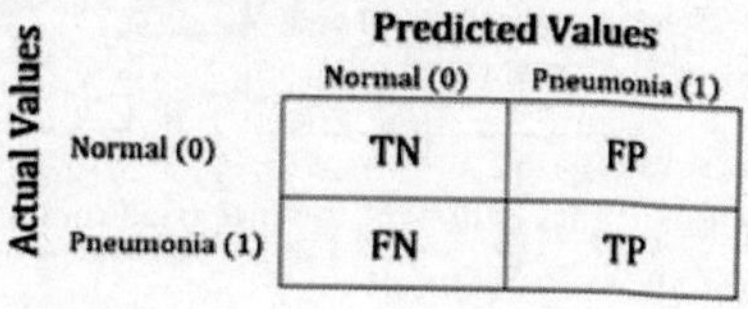

method is only suitable for general assessment. If we need to get the number of details of each type of prediction is true or false, this method has not provided enough information. For example, a model with 95% accuracy can be either very good or bad depending on actual data classification requirements [20, 21]. The value of accuracy is calculated as Eq. (3):

$$Accuracy = \frac{\text{TP} + \text{TN}}{\text{TP} + \text{TN} + \text{FP} + \text{FN}} \tag{3}$$

The precision is the ratio of the number of cases that the pneumonia image prediction model was true to the actual dataset divided by the total number of cases the model predicted pneumonia images. The value of precision is calculated as Eq. (4)

$$Precision = \frac{\text{TP}}{\text{TP} + \text{FP}} \tag{4}$$

The recall is the ratio of the number of cases that correctly predicted pneumonia divided by the total number of pneumonia image cases in the real dataset. The value of recall is calculated as Eq. (5):

$$Recall = \frac{\text{TP}}{\text{TP} + \text{FN}} \tag{5}$$

In the diagnosis of pneumonia, if the recall rate is very high, the efficiency of the model will be very good, because the model can detect most cases of pneumonia.

4 Experiments and Results

In this section, we experiment and evaluate the results of the proposed method. We compare the results of the proposed method with the results of the other methods, such as Ayan method [10] and Kadam method [12].

4.1 The Dataset for Training and Testing

We use X-ray dataset provided by Kaggle [22]. This dataset contains 5863 chest X-ray (JPEG) images with different resolutions. The dataset is structured into three parts, such as: training, testing and validation. Each part containing subfolders for each category is pneumonia or normal. X-ray images have been manually classified by doctors and divided into two categories. The images in the dataset have different resolutions, such as from 407 × 178 to 2916 × 2583 pixel. There are 1583 normal cases, 4273 cases of pneumonia images in the dataset. In our study, all images were

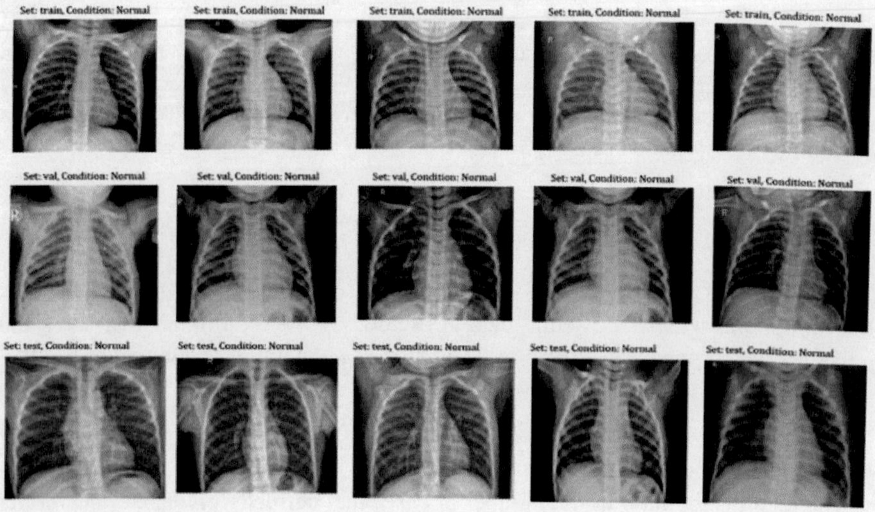

Fig. 6 X-ray image of the lungs normal case

scaled down to 150 × 150 pixels to maintain the same aspect ratio between all images. Some images from the dataset are shown in Fig. 6 including fifteen normal images, and Fig. 7 including fifteen images of pneumonia. The number of medical images for training, testing and validation present in Table 1.

Now, we test the proposed method with the above dataset and compare with the other methods, such as Ayan method [10] and Kadam method [12].

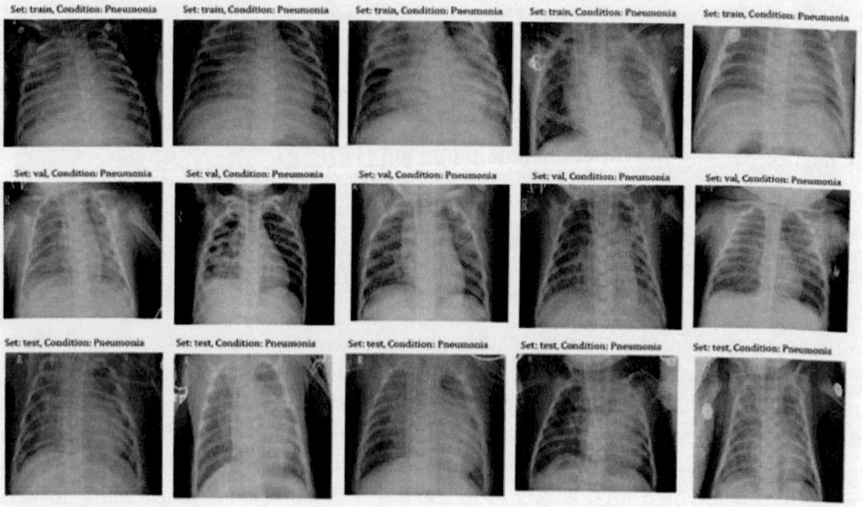

Fig. 7 X-ray image of lungs pneumonia case

Table 1 Illustrates the number of images in three parts

	Train	Validation	Test	Total
Normal	1342	9	234	1583
Pneumonia	3876	9	390	4273

Table 2 The configuration of CNN networks

Configuration	Values
Training data	All
Input size	150 × 150
Batch size	32
Epoch	32
Step per epoch	163
Kernel size	3 × 3
Activation function	Relu
Activation function of the last layer	Sigmoid
Optimizer	Adam
Loss function to use for classification	binary_crossentropy

4.2 Experiments

Our experiments developed on Google Colab, including resources such as 13 GB of RAM, 2 CPUs of 2.2 GHz, GPU and TPU 12 GB. Dataset X-ray images stored on Google Drive cloud. Methods to build training models and classify diseases through six steps as in Sect. 3. Table 2 present the configuration information for the parameters in the model.

Table 3 describes the layers in CNN, including one input layer, five convolution blocks, fully connected layer and output layer.

Information on model test results is presented in "confusion matrix" in Fig. 8.

Now, we compared the results of the proposed method with the results of the other methods. The results comparisons are presented in Table 4.

According to the Table 4 presented the above, our experimental results are better than the experimental results of Ayan method [10] and Kadam method [12] on the same dataset.

With the results of Ayan method [10], the results of the proposed method are better than this method on all three indicators: accuracy, precision, and recall. With the results of Kadam method [12], the value of recall of Kadam method [12] is 99.27% while the value of the proposed method is 98.72%. However, both the precision and accuracy values of the proposed method are 91.23% and 93.27%, which are greater than the precision and accuracy values of the Kadam method are 90.88% and 92.90%.

Table 3 Convolutional neural network architecture

Layer (type)	Output shape	Parameters
InputLayer	(None, 150, 150, 3)	0
Conv2d_1_1	(None, 150, 150, 16)	448
Conv2d_1_2	(None, 150, 150, 16)	2320
Max_pooling2d_1	(None, 75, 75, 16)	0
Separable_conv2d_2_1	(None, 75, 75, 32)	688
Separable_conv2d_2_2	(None, 75, 75, 32)	1344
Batch_normalization_2	(None, 75, 75, 32)	128
Max_pooling2d_2	(None, 37, 37, 32)	0
Separable_conv2d_3_1	(None, 37, 37, 64)	2400
Separable_conv2d_3_2	(None, 37, 37, 64)	4736
Batch_normalization_3	(None, 37, 37, 64)	256
Max_pooling2d_3	(None, 18, 18, 64)	0
Separable_conv2d_4_1	(None, 18, 18, 128)	8896
Separable_conv2d_4_2	(None, 18, 18, 128)	17,664
Batch_normalization_4	(None, 18, 18, 128)	512
Max_pooling2d_4	(None, 9, 9, 128)	0
Dropout_4	(None, 9, 9, 128)	0
Separable_conv2d_5_1	(None, 9, 9, 256)	34,176
Separable_conv2d_5_2	(None, 9, 9, 256)	68,096
Batch_normalization_5	(None, 9, 9, 256)	1024
Max_pooling2d_5	(None, 4, 4, 256)	0
Dropout_5	(None, 4, 4, 256)	0
Flatten_6	(None, 4096)	0
Dense_7	(None, 1024)	4,195,328
Dropout_8	(None, 1024)	0
Dense_9	(None, 512)	524,800
Dropout_10	(None, 512)	0
Dense_11	(None, 128)	65,664
Dropout_12	(None, 128)	0
Dense_13 (Output)	(None, 1)	129

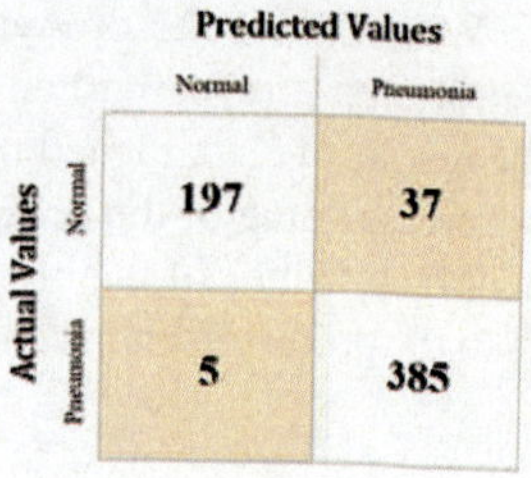

Fig. 8 Confusion matrix result

Table 4 The table compares the results of researches on the chest X-ray image

Method	Dataset chest X-ray	Accuracy (%)	Precision (%)	Recall (%)
Ayan method [10]	Kaggle	87.00	91.00	89.00
Kadam method [12]	Kaggle	92.90	90.88	99.27
Proposed method	Kaggle	93.27	91.23	98.72

The reason which the results of the proposed method has better results than the other method because the classification results of pneumonia are higher in the evaluation methods.

5 Conclusions

Lung diseases image classification is hard work and challenge. In this paper, we developed an automated method of checking chest X-ray images with high accuracy, using the dataset Chest X-ray images. These images have been pre-classified by radiologists and divided into two types: pneumonia and normal images. We use deep learning methods, build models starting from steps such as pre-processing, image acquisition, image layering, setting appropriate parameters for each layer such as image resolution, number of training times, optimal algorithms, threshold changes, and state transitions during training to get the best and most appropriate model. The test results are highly effective image classification model. The results of the proposed method have high precision and better than those recently published on the same dataset of chest X-ray images. In the future work, we will extend the experiment to update the number of layers, the time of training or testing on the other pre-train models such as: InceptionNet, ResNet, InceptionResNet, MobileNet and CheXNet. Those tasks aim to reduce the training time and increase the level accuracy of the model compared to the above results.

Acknowledgements This research is funded by University of Cuu Long.

Conflicts of Interest The authors have no conflict of interest to declare.

References

1. Neuman MI et al (2012) Variability in the interpretation of chest radiographs for the diagnosis of pneumonia in children. J Hosp Med 7(4):294–298
2. Krizhevsky A, Sutskever I, Hinton GE (2012) Imagenet classification with deep convolutional neural networks. In: Advances in neural information processing systems, pp 1097–1105
3. Selvathi D, Aarthy Poornila A (2017) Deep learning techniques for breast cancer detection using medical image analysis. In: Biologically rationalized computing techniques for image processing applications. Springer, Berlin, pp 159–186

4. AliKadampur M, Al Riyaee S (2020) Skin cancer detection: applying a deep learning based model driven architecture in the cloud for classifying dermal cell images. In: Informatics in medicine unlocked, vol 18. Elsevier, Amsterdam
5. Le QV, Ranzato M, Monga R, Devin M, Chen K, Corrado GS, Dean J, Ng AY (2012) Building high-level features using large scale unsupervised learning. In: Proceedings of the 29th international conference on international conference on machine learning, Edinburgh, Scotland, UK, pp 507–514
6. Katamreddy S, Doody P, Riordan D (2018) Visual udder detection with deep neural networks. In: 12th international conference on sensing technology. IEEE, Ireland, pp 166–171
7. Lujan-Garcia JE, Yanez-Marquez C, Villuendas-Rey Y, Camacho-Nieto O (2020) A transfer learning method for pneumonia classification and visualization, artificial intelligence for medical image analysis. Appl Sci 10(8)
8. Sirazitdinov I, Kholiavchenko M, Mustafaev T, Yixuan Y, Kuleev R, Ibragimov B (2019) Deep neural network ensemble for pneumonia localization from a large-scale chest x-ray database. In: Computers and electrical engineering, vol 78. Elsevier, Amsterdam, pp 388–399
9. Sirish Kaushik V, Nayyar A, Kataria G, Jain R (2019) Pneumonia detection using convolutional neural networks. In: Proceedings of first international conference on computing, communications, and cyber-security, vol 121. Springer, Berlin, pp 471–483
10. Ayan E, Unver HM (2019) Diagnosis of pneumonia from chest X-ray images using deep learning. In: Scientific meeting on electrical-electronics & biomedical engineering and computer science. IEEE, Turkey
11. Ponnada VT, Naga Srinivasu SV (2019) Edge AI system for pneumonia and lung cancer detection. Int J Innov Technol Exploring Eng 8(9)
12. Kadam K, Ahirrao S, Kaur H, Phansalkar S, Pawar A (2019) Deep learning approach for prediction of pneumonia. Int J Sci Technol Res 8(10):86–89
13. Ji S, Xu W, Yang M, Yu K (2013) 3D convolutional neural networks for human action recognition. IEEE Trans Pattern Anal Mach Intell 35(1):221–231
14. Adiyansjah, Sgunawan AA, Suhartono D (2019) Music recommender system based on genre using convolutional recurrent neural networks. Procedia Comput Sci 157:99–109
15. Ramaswamy S, DeClerck N (2018) Customer perception analysis using deep learning and NLP. Procedia Comput Sci 140:170–178
16. Xu Z, Zhang J, Wang J, Xu Z (2020) Prediction research of financial time series based on deep learning. Soft Comput 24:8295–8312
17. CNN lectures. https://cs231n.github.io/convolutional-networks. CS231n, Stanford. Last accessed 2020/04/28
18. Classification Convolutional Neural Network (CNN). https://www.thelearningmachine.ai/cnn. Last accessed 2020/04/28
19. Ayan E, Unver HM (2018) Data augmentation importance for classification of skin lesions via deep learning. In: 2018 electric electronics, computer science, biomedical engineerings' meeting. IEEE, Turkey, pp 1–4
20. Powers DMW (2011) Evaluation: from precision, recall and F-measure to ROC, informedness, markedness & correlation. J Mach Learn Technol 2(1):37–63
21. Confusion Matrix in Machine Learning. https://www.geeksforgeeks.org/confusion-matrix-machine-learning. Last accessed 2020/04/12
22. Kaggle Chest X-Ray Images. https://www.kaggle.com/paultimothymooney/chest-xray-pneumonia. Last accessed 2020/04/24

A Machine Learning Framework Based on Extreme Gradient Boosting for Intelligent Alzheimer's Disease Diagnosis Using Structure MRI

Hong Ong, Hoang Le, Hoang Nguyen, Dong Nguyen, Huong Ha, Hoan Thanh Ngo, and Nguyen Thanh Duc

Abstract Alzheimer's Disease (AD) is difficult to diagnose even with recent advanced diagnostic methods. Other mental disorders, such as frontotemporal lobe dementia or vascular dementia, could be misdiagnosed as AD. Various deep learning models for early detection of AD using MRI data have demonstrated promising results. However, the results of these methods are difficult to interpret as they did not identify specific structural changes that are related to the disease. Additionally, they require a large amount of data and computational resources. In this study, we proposed a machine learning framework for diagnosis of AD. Our framework employed FreeSurfer library to extract insightful features such as volumetric measures and voxel- or vertex-wise atrophy from structural MRI brain scans collected from Alzheimer's Disease Neuroimaging Initiative. These extracted features were then fed into Extreme Gradient Boosting (XGBoost) which is an ensemble learning algorithm with a decision tree as a base learner to distinguish AD patients from cognitively normal subjects (CN). XGBoost also provides the concept of variable importance which was evaluated by various criteria such as information gain or feature frequency to give an insight on which structural features have critical impacts on the final diagnosis. Our model was trained on 144 features extracted from 924 sMRI images (462 images of AD and 462 images of CN) and achieved 91% Area Under the Curve (AUC) on average using 5-fold cross-validation. Based on feature ranking, we observed that the 3rd ventricle, intracranial and supratentorial volume were

H. Ong · H. Nguyen · D. Nguyen · H. Ha · H. T. Ngo (✉)
School of Biomedical Engineering, International University, Ho Chi Minh City, Vietnam
e-mail: nthoan@hcmiu.edu.vn

Vietnam National University, Ho Chi Minh City, Vietnam

H. Le
Department of Electrical Engineering and Computer Science, York University, Toronto, ON, Canada

N. T. Duc (✉)
Department of Biomedical Science and Engineering, Institute of Integrated Technology, Gwangju Institute of Science and Technology, Gwangju, South Korea
e-mail: ducnguyen@gm.gist.ac.kr

Montreal Neurological Institute, McGill University, Montreal, Canada

V. Van Toi et al. (eds.), *8th International Conference on the Development of Biomedical Engineering in Vietnam*, IFMBE Proceedings 85,
https://doi.org/10.1007/978-3-030-75506-5_66

some of the most crucial brain structural features affected in AD. This information could assist doctors and experts in AD diagnosis. We propose a framework for AD diagnosis using XGBoost and brain structural atrophy. Thc feature extraction phase which consumed significant computational resources is currently the bottleneck of our diagnosis pipeline. In future, we plan to improve the feature extraction efficiency to reduce computational cost, yet maintaining diagnosis accuracy.

Keywords AD diagnostic · Gradient boosting · Feature extraction · Feature selection

1 Introduction

One of the most common yet severe types of dementia is Alzheimer's disease (AD), which could be described as a progressive neurodegenerative disorder. Patients who contract the disease witness a rapid and irreversible decline in cognitive functions such as loss of memory and reasoning [1] and eventually loss of motor functions [2]. There were 5.8 millions of people diagnosed with AD in 2019, and 14 millions will catch AD by 2050 [3]. Up to date, there is still no cure for this disease [4]. However, several measures that can delay the disease progress are available. That is the reason why early diagnosis of AD has a crucial role to play in saving those patients.

There are various neuroimaging modalities, such as magnetic resonance images (MRI) and positron emission tomography (PET) that could provide useful yet precise information to help doctors and researchers understand and monitor the anatomical and functional neural changes related to Alzheimer's disease. Recently, machine learning [5, 6] or computer vision techniques have been employed extensively, combined with multimodality neuroimages, to explore further biomarkers and develop computer-aided system for early AD diagnosis [7–12].

Many AD diagnosis studies based on classical machine learning get features from neuroimaging modalities by Freesurfer [13] and feed them into classification models. In Oliveira et al. [14] study, they analyze cortical thickness and perform volume measurement of cortical and subcortical structures using Freesurfer in order to train a Support Vector Machine for AD detection [14]. Frenzel et al. [15] develops a method for AD detection using supervised learning, and Freesurfer plays its role in their preprocessing part of cortical reconstruction and volumetric segmentation [15].

Ensemble learning has also been widely employed to build a machine learning model for AD diagnosis. This technique is based on building a stronger classifier from weaker ones. Liu et al. [16], instead of extracting features from each structural and functional brain image and training a global classifier, randomly partition such an image into many small local patches and then use sparse coding based models to learn from those patches. Although these models are weak on their own as they only learn spatial features from local patches, when combined together to get a final decision, they form a state-of-the-art AD classifier.

In this work, we investigate the use of boosting machines (one of the ensemble methods) during the two most important phases in the proposed framework, namely feature selection and AD classification. We will also present the and interpretability of our proposed framework which is suitable to use in the hospital to aid the clinician in the diagnosis process.

The rest of the paper is organized as follows. Section 2 presents the detail on how we prepare the dataset for the experiments and the summary of our proposed framework. Results including comparison of classification score from our models with previous works and the effects on classifiers' performance of feature selection will be reported in Sect. 3. Finally, Sects. 4 and 5 close with the discussion on the limitation of this work and perspectives for future studies.

2 Materials and Methods

2.1 Data Acquisition

All data in this work were collected from the Alzheimer's Disease Neuroimaging Initiative (ADNI) database, which was available at http://adni.loni.usc.edu/. We collected the T1-weighted magnetic resonance (MR) imaging data that belonged to two groups, namely AD, and CN. AD was a group of those who were diagnosed with Alzheimer's disease, and CN was cognitively normal, representing those who did not contract the disease. The mean ages of AD and CN were 75.30 ± 7.71 and 75.02 ± 5.58, respectively. General criteria for categorizing AD and CN were well explained on the ADNI web site (http://adni.loni.usc.edu/). We only chose baseline scans for this study. The demographic details of the subjects are presented in Table 1, in which MMSE stands for Mini-Mental State Examination score, and CDR denotes the Clinical Dementia Rating (CDR). Functional assessments of AD/CN by using cognitive variables such as MMSE and CDR were independently performed by the research institutions and the assessments met the National Institute of Neurological and Communicative Disorders and Stroke and the Alzheimer' disease and Related Disorders Association (NINCDS/ADRDA) criteria for probable AD. In this study, a dataset with a balanced number of subjects (231 subjects for each group) were used.

Table 1 Demographic details of all participants used in this study

Group	Number of subjects	Gender (M/F)	Age (years)	MMSE score	CDR score
AD	231	126/105	75.30 ± 7.71	23.34 ± 1.99	0.75 ± 0.27
CN	231	123/108	75.02 ± 5.84	29.22 ± 0.93	0.00 ± 0.04

MMSE mini-mental state examination; *CDR* clinical dementia rating

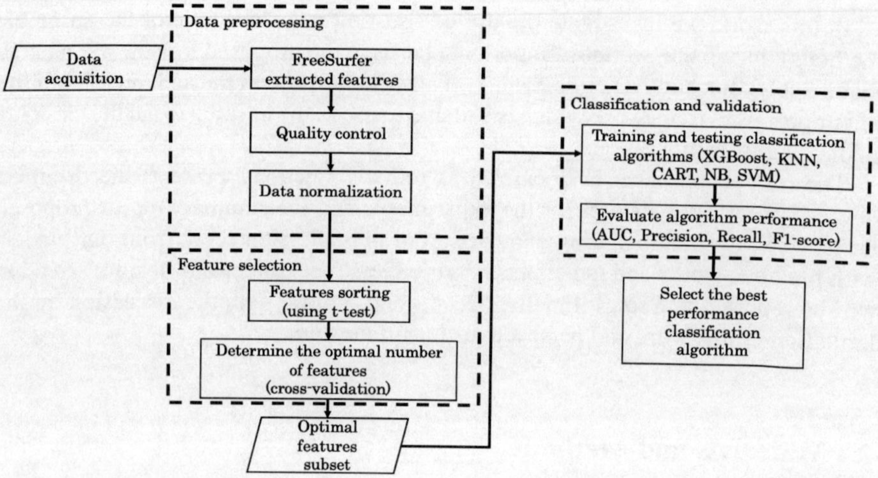

Fig. 1 Illustrations of the proposed overall framework. *Abbreviations* XGBoost = Extreme Gradient Boosting; KNN = K-Nearest Neighbors; CART = Classification And Regression Tree; NB = Naive Bayes; SVM = Support Vector Machine; AUC = Area Under receiver operating Curve (increase font size)

2.2 Overall Framework

The overall framework of the proposed approach for automatic Azheimer's disease diagnosis is shown in Fig. 1. Firstly, the raw sMRI images from subjects are fed into the FreeSurfer open-source toolbox for data pre-processing which includes: (1) skull-stripping that extracts only brain areas and removes non-brain areas, (2) bias correction that enhances the image quality, and (3) data normalization which registers individual brains to a standard MNI152 template. The next step is feature extraction which we extract the cortical volume features. In this study, we focused on 72 cortical volume features covering the entire cortex. Subsequently, we performed the feature selection by sorting features using their corresponding statistical significance t-test values. By doing so, the optimal number of features and performance of different machine learning algorithms were determined. Finally, we evaluated the proposed models with AUC, Precision, Recall and F1-score and the best performance classification algorithm was reported.

2.3 Preprocessing of Structural MRI and Feature Extraction

Preprocessing on the structural MRI scans is expected to extract the patterns of informative knowledge. In this work, we performed preprocessing using FreeSurfer 5.3 software with a fully automatic pipeline. FreeSurfer software is a well-known tool for

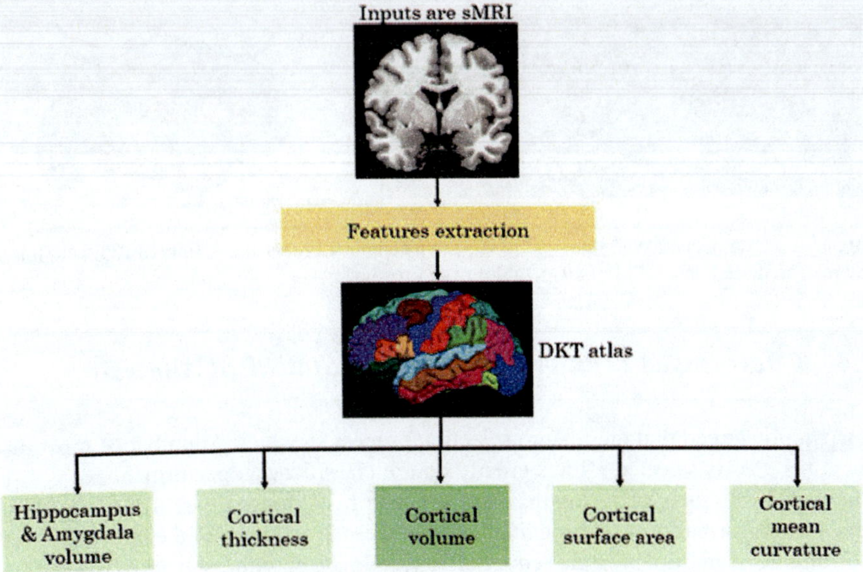

Fig. 2 Various typical cortical measures extracted by FreeSurfer. In this work, we only focus on cortical volumes for further analysis

neuroimaging image processing and has earned its reputation for obtaining high performance in brain segmentation and measurement [17]. The software proceeds with motion correction, T1-weighted image averaging, skull stripping to obtain cortical areas, followed by registering individual brains to the standard MNI152 brain space with a deformable template model. Next, a cortical surface-based atlas (DKT atlas) was mapped to a sphere aligning the cortical patterns, which provides accurate matching of the morphologically homologous cortical locations across subjects [18]. This atlas is produced by using 40 subjects which include 14 male and 26 female whose ages between 18 and 87 years old in which 30 non-demented and 10 demented are all captured by Siemens 1.5T Vision (Wash U). For each of the DKT-based segments, FreeSurfer computes nine different measures which include gray matter volume, average cortical thickness, cortical thickness standard deviation, cortical mean curvature, cortical Gaussian curvature, surface area, surface vertices, cortical folding index, and cortical curvature indices. In Fig. 2, we highlight several typical measures made by FreeSurfer that one can extract for their analysis. In this study, we use grey matter volume as input features for classification algorithms (Fig. 3).

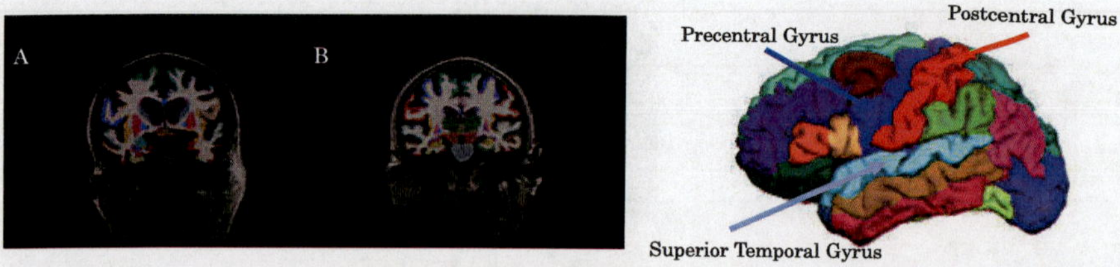

Fig. 3 **A** AD subject male 75 years old; **B** CN subject male 73 years old. Colors of ROI (Region of Interest) are based on LUT (Look Up Table) from FreeSurfer

2.4 T-Test-Based Feature Selection and Statistical Analysis

It is widely agreed that the feature selection process in which a number of most relevant features is selected affects greatly on the classification performances [5, 12]. The reason for such impact is that, in data with high dimensions, there is a lot of redundant information which significantly reduces the accuracy of the discriminative algorithms. In addition, feature reduction may help the classifiers to avoid overfitting problems, especially in the neuroimaging domains where the number of sample sizes is limited as compared to the number of features. Therefore, selecting meaningful features before training data plays a crucial role in modeling. Importantly, feature selection should be performed using training data only. Once identified, the same brain regions identified during training are used to access the classifier predictive accuracy on the leave-out testing data. In this study, we perform an univariate statistical t-test between two studied groups to: (1) find the cortical volumes that show significant changes in AD patients as compared to normals ones, and (2) select the most salient features which could be used to improve machine learning classifiers. For statistical analysis, generally, univariate statistical t-tests are used to show abnormalities, at the level of group average, in one or more brain features in a diseased group as compared to a control one. Here, we report several cortical regions where we observe the significant changes in their volume due to AD (p-value < 0.05). As for feature selection utilizing statistical t-test results, the key results of the t-test-based feature selections are expressed by means of p-values. By performing statistical analysis of each independent brain feature between the two groups, we obtain the p-values sorted in descending manner. Then in each modeling and evaluation we add features incrementally and see the changes in performance reported by AUC metric.

2.5 Machine Learning Classifier

In this present work, we examine the performances of different machine learning classification techniques, i.e., K-nearest neighbors, classification and regression tree, Gaussian Naive Bayes, Support Vector Machine, and Extreme Gradient Boosting

(XGBoost), for automatic identifications of the AD patients using their gray matter volumes computed from baseline sMRI scans. The details of those above-mentioned algorithms are as follows:

Chen and Guestrin proposed a novel implementation of gradient boosting decision trees, which is so-called XGBoost [19]. Decision trees (DTs) are supervised learning techniques that are non-parametric. They are utilized for not only classification but also regression problems. The goal of DTs is to generate a model which is capable of predicting target variables through learning simple decision rules from features. Boosting is an ensemble technique in which models are combined sequentially until no further improvements or no decrease in error metrics. Gradient boosting is the way that latest models could be constructed and added together to narrow the errors of previous ones by using a gradient descent algorithm for reducing the objective function. In this way, we could take advantage of all weak models to form a stronger one. In addition, XGBoost could help to optimize the computation resources and speed by using a combining predictive and regularization terms as well as compressing data, sharding and cache access techniques.

K-Nearest Neighbors (KNN) [20] is a simple yet powerful machine learning algorithm which could classify new cases based on the similarity metric between existing data points. KNN is an instance-based learning type, meaning that it does not build an internal model, but simply stores data points. KNN assigns values for a query point based on a majority vote of its nearest neighbors.

Classification and Regression Trees (CART) [21], on the other hand, is the technique that could support numerical target variables (regression) and does not compute rule sets, unlike DTs. CART takes features and threshold into consideration so that it can construct binary trees generating the largest information gain at each node.

Gaussian Naive Bayes is implemented using the Gaussian Naive Bayes algorithm for classification in which the likelihood of the features is assumed to be Gaussian:

$$P(x_i|y) = \frac{1}{\sqrt{2\pi\sigma_y^2}} \exp\left(-\frac{(x_i - \mu_y)^2}{2\sigma_y^2}\right) \tag{1}$$

Support Vector Machine (SVM) [22] algorithm is another supervised machine learning algorithm, which is usually adopted for classification problems. By taking advantage of kernel tricks (converting low dimensional data to higher ones), this type of algorithm could conduct both linear and non-linear mapping. SVM creates hyperplanes between classes, and these planes could be considered as decision boundaries. SVM is generally used in such a case that the number of dimensions are greater than the number of sample data.

2.6 Performance Evaluation

To evaluate the performance of the classifiers, we report the Area Under receiver operating Curve (AUC), Precision (PRE), Recall (REC), and F1-score. TP, TN, FP, and FN indicate the number of true positives, true negatives, false positives, and false negatives, respectively. In terms of these numbers, AUC, PRE, and REC, F1-score can be computed as follows:

$$PRE = TP/(TP + FP)$$
$$REC = TP/(TP + FN)$$

3 Results

3.1 Preprocessing Results

After completing preprocessing and features extraction, we export a data list of subjects where each one is accompanied by 72 features for evaluation. Table 2 is a portion of data

3.2 Classification Results

Table 3 and Fig. 4 show the performance of each classifier on different evaluation criteria where the results shown in the table are the average of the 5-fold cross-validation. We see that the best performance on AUC is XGBoost classifier obtains 0.91 on average and overall F1-score is 0.81. The next is Gaussian Naive Bayes, its AUC is 0.88 and F1-score is 0.79. The third is SVM, its AUC is 0.84 and F1-score is 0.76.

Table 2 Subject information in form of cortical volume measurements

subj_id	rh_BrainSeg VolNotVent	rh_eTIV	rh_ bankssts	rh_caudalan terior cingulate	rh_caudal middlefrontal	…	rh_cuneus
941_S_4376	949,104	1,419,527.897	2091	1325	4049	…	2542
941_S_4365	1,057,067	1,712,397.88	2233	1559	4956	…	3336
941_S_4292	1,019,224	1,464,614.757	1780	1149	5308	…	3323
941_S_4255	1,059,422	1,600,066.421	1746	2100	5998	…	2756
…	…	…	…	…	…	…	…
941_S_1203	922,204	1,444,002.244	1687	1444	4139	…	2781

Table 3 Performance comparisons between different classifiers

Model name	Approach	Abbreviation	AUC (mean ± SD)	Precision (mean ± SD)	Recall (mean ± SD)	F1-score (mean ± SD)
Extreme gradient boosting	Tree-based	XGBoost	0.91 ± 0.02	0.83 ± 0.06	0.79 ± 0.04	0.81 ± 0.02
K-Nearest Neighbors	Instance-based	KNN	0.79 ± 0.01	0.86 ± 0.03	0.43 ± 0.06	0.57 ± 0.05
Classification and regression trees	Tree-based	CART	0.77 ± 0.04	0.84 ± 0.06	0.62 ± 0.05	0.71 ± 0.03
Gaussian Naive Bayes	Probabilistic	NB	0.88 ± 0.03	0.86 ± 0.01	0.74 ± 0.09	0.79 ± 0.06
Support-vector machine	Linearity	SVM	0.84 ± 0.03	0.76 ± 0.05	0.76 ± 0.06	0.76 ± 0.04

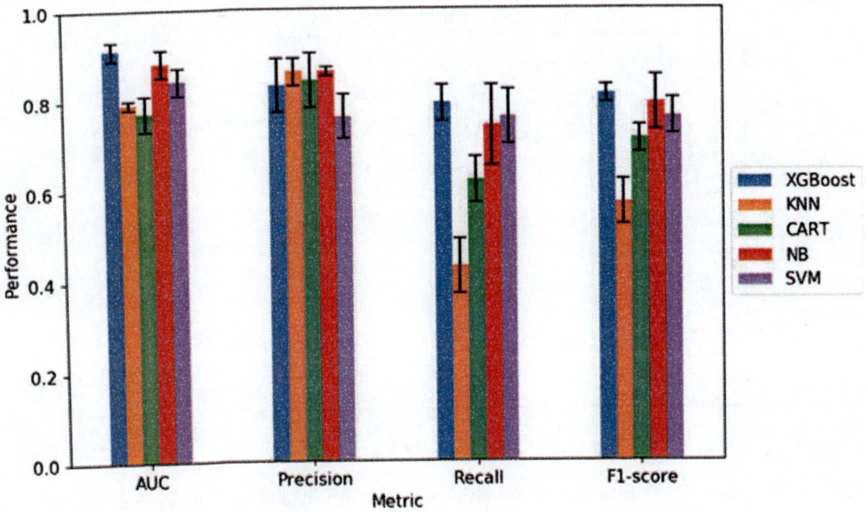

Fig. 4 Performance comparisons between different classifiers

3.3 *Affected Cortical Regions (Feature Ranking and Feature-Based Classification Results)*

Figure 5 shows the ranking of statistical (t-test p-value) of cortical volume. Left hemisphere posterior cingulate has most significant value (0.078).

We also test classifiers by adding features incrementally, and record their average AUC along the way. Overall, most classifiers show its robustness and stability

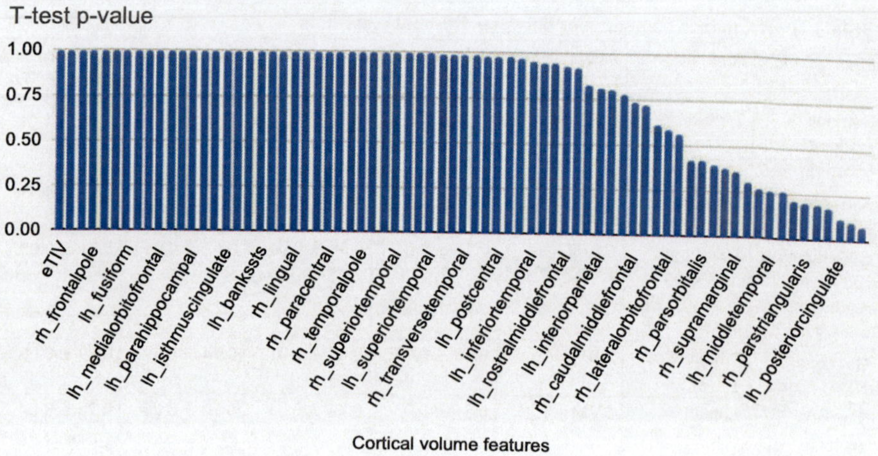

Fig. 5 T-test *p*-value of cortical volume features

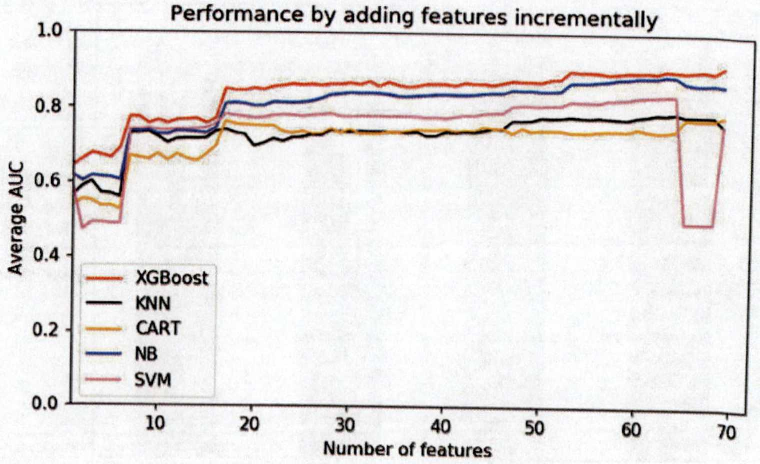

Fig. 6 Performance comparisons between different classifiers by adding features incrementally

throughout the experiment as can be seen in Fig. 6, and XGBoost still tops the chart even with the least or most number of features. However, after adding more than 60 features, only XGBoost and CART could improve their average AUC further, while SVM, NB, and KNN witnessed their AUC to decrease. More remarkably, when adding around 65 features, SVM sees its average AUC plummet from 0.8 to 0.5, but soar again to 0.7 when adding around 68–70 features.

4 Discussion

We found some significant cortical volume features from 72 areas that play an important role in our discriminative results. For all classifiers, the trend is that the more number of volume features used, the higher classification accuracy achieved. One exception is SVM, when accuracy dropped significantly after adding more than 65 features. One possible hypothesis is that if we keep adding insignificant features to the SVM, SVM's discriminative power may be reduced. In general, the more features fed to the classifiers, the higher classification accuracy we can get.

Our work shows the performance of classifiers when adding features extracted from 72 volume areas incrementally. Only XGBoost could utilize all the features to reach the highest average AUC, while SVM gets worse when adding more than 65 features. Oliveira Jr. et al., in contrast, show that their SVM classifier could achieve very high accuracy using features from 45 volume areas [14]. In the work of Alam S. et al., they do not use volumetric features but extract features by transforming MRI images using dual-tree complex wavelet; and instead of using one SVM as usual, they combine two SVM models for their classifiers [23]. They show that not only volumetric features but also features from wavelet transforms could be used to increase classifiers' performance. It is quite obvious that some classifiers are more sensitive than others. While SVM considers insignificant features to be noises, XGBoost could use them effectively to contribute to its final decision, as expected from an ensemble method.

One of the critical limitations of our approach is that it takes us a significant amount of time to preprocess the dataset. Specifically, to preprocess 462 subjects in the training phase, we needed roughly three weeks to run Freesurfer. But fortunately this process may take up to a couple of hours in the testing phase as we can only deal with one patient at a time. Thus, it is still applicable when we use it in hospitals to aid doctors diagnose AD. Another limitation of our work is the small dataset. As the dataset is not well presented for the whole patient population, our model's generalization could not be guaranteed.

5 Conclusion

In this study, we proposed an ensemble framework for Alzheimer disease detection. We also showed how classifiers' performance is affected by adding features incrementally. In the future, we need to improve our pipeline to make it less dependent on Freesurfer. We could try to implement a separate deep-learning model for extracting features automatically with the goal that it will help to speed up our pipeline and improve the accuracy. Also in this way, we may discover some hidden yet significant features for our XGBoost model.

Acknowledgements This research is funded by International University, VNU-HCM under grant number T2019-05-BME. This research is also supported by Smart Health-care Research Grant through the Daewoong Foundation (DS188).

Conflicts of Interest The authors have no conflict of interest to declare.

References

1. Collie A, Maruff P (2000) The neuropsychology of preclinical Alzheimer's disease and mild cognitive impairment. Neurosci Biobehav Rev 24(3):365–374
2. Hubert RJ, VanMeter KC (2017) Gould's pathophysiology for the health professions. Elsevier
3. Alzheimer's Disease International (ADI) (2018) World Alzheimer report 2018. London. https://www.alz.co.uk/research/WorldAlzheimerReport2018.pdf
4. Mayeux R, Sano M (1999) Treatment of Alzheimer's disease. N. Engl. J. Med. 341(22):1670–1679 PMID: 10572156
5. Nguyen D, Ryu S, Qureshi MNI, Choi M, Lee K, Lee B (2019) Hybrid multivariate pattern analysis combined with extreme learning machine for Alzheimer's dementia diagnosis using multi-measure RS-FMRI spatial patterns. PLOS ONE 14:e0212582
6. Duc NT, Ryu S, Choi M, Iqbal Qureshi MN, Lee B (2019) Mild cognitive impairment diagnosis using extreme learning machine combined with multivoxel pattern analysis on multi-biomarker resting-state FMRI. In: 2019 41st annual international conference of the IEEE engineering in medicine and biology society (EMBC), pp 882–885
7. Cheng B, Liu M, Suk H-I, Shen D, Zhang D (2015) Multimodal manifold-regularized transfer learning for MCI conversion prediction. Brain Imag Behav 9:913–926
8. Suk H-I, Shen D (2013) Deep learning-based feature representation for AB/MCI classification. Med Image Comput Comput Assist Interv 16:583–90
9. Suk H-I, Lee S-W, Shen D (2014) Hierarchical feature representation and multimodal fusion with deep learning for AD/MCI diagnosis. NeuroImage 101:07
10. Suk H-I, Lee S-W, Shen D (2013) Latent feature representation with stacked auto-encoder for AD/MCI diagnosis. Brain Struct Funct 220
11. Zhang D, Wang Y, Zhou L, Yuan H, Shen D (2011) Multimodal classification of Alzheimer's disease and mild cognitive impairment. NeuroImage 55:856–67
12. Nguyen D, Ryu S, Qureshi MNI, Choi M, Lee K, Lee B (2019) 3d-deep learning based automatic diagnosis of Alzheimer's disease with joint MMSE prediction using resting-state fMRI. Neuroinformatics 18
13. Fischl B (2012) Freesurfer. NeuroImage, 62(2):774–781 (20 years of fMRI)
14. Oliveira P, Nitrini R, Busatto G, Buchpiguel C, Sato J, Amaro E (2010) Use of SVM methods with surface-based cortical and volumetric subcortical measurements to detect Alzheimer's disease. J Alzheim Dis JAD 19:1263–72
15. Frenzel S, Wittfeld K, Habes M, Klinger-König J, Bülow R, Völzke H, Grabe HJ (2020) A biomarker for Alzheimer's disease based on patterns of regional brain atrophy. Front Psychiatry 10:953
16. Liu M, Zhang D, Shen D (2012) Ensemble sparse classification of Alzheimer's disease. NeuroImage 60:1106–16
17. Fischl B (2012) Freesurfer. NeuroImage 62(2):774–781 (2012) (20 years of fMRI)
18. Qureshi MNI, Min B, Jo H, Lee B (2016) Multiclass classification for the differential diagnosis on the ADHD subtypes using recursive feature elimination and hierarchical extreme learning machine: structural MRI study. PLOS ONE 11:e0160697
19. Chen T, Guestrin C (2016) Xgboost: a scalable tree boosting system. In: Proceedings of the 22nd ACM SIGKDD international conference on knowledge discovery and data mining. KDD'16. Association for Computing Machinery, New York, pp 785–794

20. Altman NS (1992) An introduction to kernel and nearest-neighbor nonparametric regression. Am Stat 46(3):175–185
21. Breiman L, Friedman J, Stone CJ, Olshen RA (1984) Classification and regression trees. CRC Press
22. Cortes C, Vapnik V (1995) Support-vector networks. Mach Learn 20:273–297
23. Alam S (2017) Twin SVM-based classification of Alzheimer's disease using complex dual-tree wavelet principal coefficients and lDA. J Healthc Eng 2017

Automatic White Blood Cell Classification Using Deep Learning Network

Thinh Tran Nguyen and Hoan Thanh Ngo

Abstract Changes in white blood cell, a.k.a. leukocyte, subtypes count are often indicator of medical conditions. Counting each of white blood cell subtypes is thus essential in blood tests. To do this, it is necessary to classify white blood cell images into different subtypes, namely, Eosinophils, Lymphocytes, Monocytes, Neutrophilic, etc. Traditional image processing techniques have been utilized to classify white blood cell images. However, in general, accuracy and stability of these techniques are sensitive to variations in image size, staining technique, and image quality. Recently, application of deep learning networks in white blood cell classification has drawn a lot of interest. In this work, we present two deep learning networks based on VGG16 and SqueezeNet convolutional neural networks (CNNs) that can automatically classify white blood cell images into different subtypes. The two networks were trained on a public dataset of leukocytes named Blood Cell Count and Detection (BCCD) and a self-built dataset named the Extended BCCD dataset. Although having relatively light-size, 16,005,884 and 35,858 parameters for VGG16 and SqueezeNet respectively, both models achieved ~ 97% accuracy in the BCCD dataset. In the extended BCCD dataset, the VGG16-based model achieved 97% accuracy, and the SqueezeNet-based model achieved 95% accuracy. These two models are thus potential candidates for deployment on computers and IoT devices with limited computational power for white blood cell classification task.

Keywords Leukocyte classification · Convolutional neural network · Hematological diseases

1 Introduction

White blood cells (WBCs), also called leukocytes, play a crucial role in human immune system. Hematologists usually utilize shape and granular information of

T. T. Nguyen · H. T. Ngo (✉)
School of Biomedical Engineering, International University, Vietnam National University—Ho Chi Minh City (VNU-HCMC), 700000 Ho Chi Minh City, Vietnam
e-mail: nthoan@hcmiu.edu.vn

V. Van Toi et al. (eds.), *8th International Conference on the Development of Biomedical Engineering in Vietnam*, IFMBE Proceedings 85,
https://doi.org/10.1007/978-3-030-75506-5_67

leukocytes to classify WBCs into granular cells (including neutrophil, eosinophil, and basophil) or agranular cells (containing lymphocyte and monocyte). Each of the WBC subtypes has its own function in the human immune system.

Leucopenia and leukocytosis are respectively characterized by the reduction and elevation of the absolute WBCs count. These WBCs count changes can occur in response to infection, inflammation, primary bone marrow disorders, medication, environmental exposure (i.e., heavy metals or radiation), etc. Furthermore, changes in count of WBC subtypes may also be indicator of diseases. For example, an abnormally low number of lymphocytes count could relate to HIV infection. A high neutrophil count could be sign of bacterial infection. Therefore, counting each of WBC subtypes is essential, and WBC classification is needed [1].

Acute Lymphoblastic Leukemia (ALL), a type of cancer of the blood and bone marrow, could progress rapidly to a fatal condition within a few months if not treated. In 2015, it caused 111,000 deaths globally and profoundly affected lives of at least 876,000 patients. ALL is characterized by the development of large numbers of immature lymphocytes [2]. WBC classification is important to ALL diagnosis.

Manual WBC classification using hematology microscopic images has some drawbacks such as labor intensive, time consuming, and requires highly trained personel. In manual WBC classification, pathologists often investigate thin blood smear samples under light microscopes to identify each kind of WBCs. In this process, pathologists have to look at thousands of blood cells in the blood smear. Analysis of blood smear film remains a challenge globally not only because of the long time needed to analyze a blood smear image but also due to the need of highly trained pathologists, which is usually in shortage.

These issues raise the necessity of an automated blood smear analyzing system. Such a system is not only meaningful in reducing workload of pathologists but also advantageous in medicine since it is time-saving, labor-saving, and can enhance diagnosis accuracy diagnosis by giving doctors a second or reference opinion.

Traditional automated WBC classification systems are usually built based on classical image processing techniques (e.g., the Diffmaster Octavia and Cella-Vision DM96 systems) and have overall classification accuracy from 87 to 92% [3]. In these systems, features of WBC subtypes have to be identified by domain experts and extracted by traditional computer vision methods such as morphology, thresholding, watershed, etc. Then the extracted features are fed into learning algorithms to provide classification result. This process has several limitations since it may perform well on one dataset but often perform inadequately in another dataset with different lighting condition, image size, background intensity, and so on.

Deep learning, however, provides an end-to-end approach by automatically extracting relevant features and learning how to make use of those features for classification. Recent works have shown that, in many cases, deep learning achieves higher performance than traditional image processing techniques given the same task.

In this paper, we developed two state-of-the-art deep-learning models for leukocyte subtypes classification using blood smear microscopic images. Both two models achieved accuracy of 95%–97% and are light-size, making them potential candidates for deployment on computers and IoT devices with limited computational power for WBC classification.

2 Methodology

2.1 Dataset

BCCD (Blood cell count and detection) dataset [4] is a small-scale dataset for blood cell detection. It contains 410 images of blood cells with subtype labels and bounding boxes. The distribution for eosinophil, lymphocyte, monocyte, and neutrophil are 88, 33, 20, and 206 images respectively (Fig. 1). Size of each image is 640 pixel width and 480 pixel height (640 × 480).

On the other hand, Extended BCCD Dataset (Fig. 2) is a self-built dataset which combines the BCCD Dataset and WBC images from Jiangxi Tecom Science Corporation, CellaVision's Blog, histology-world.com website, imagebank.hematology.org website, etc. To be specific, three hundred 120 × 120 WBC images with 24 bits color depth were from Jiangxi Tecom Science Corporation. One hundred 300 × 300 WBC color images were from CellaVision's blog. The remaining images were from open-source websites such as histology-world.com [5], imagebank.hematology.org [6], etc.

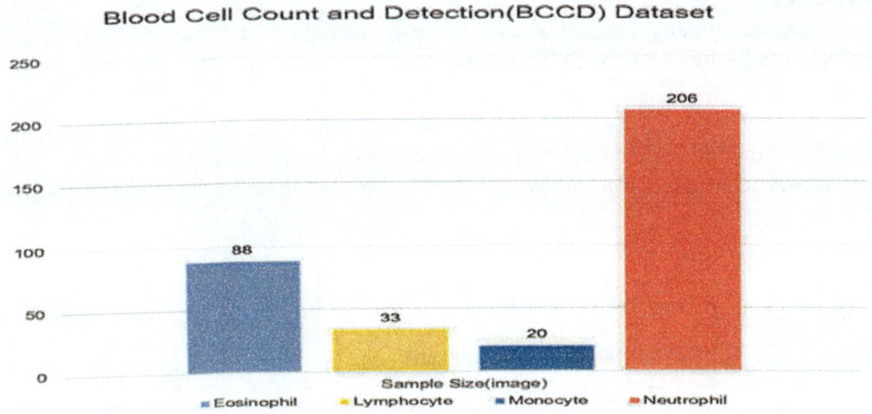

Fig. 1 Distribution of samples in the BCCD Dataset

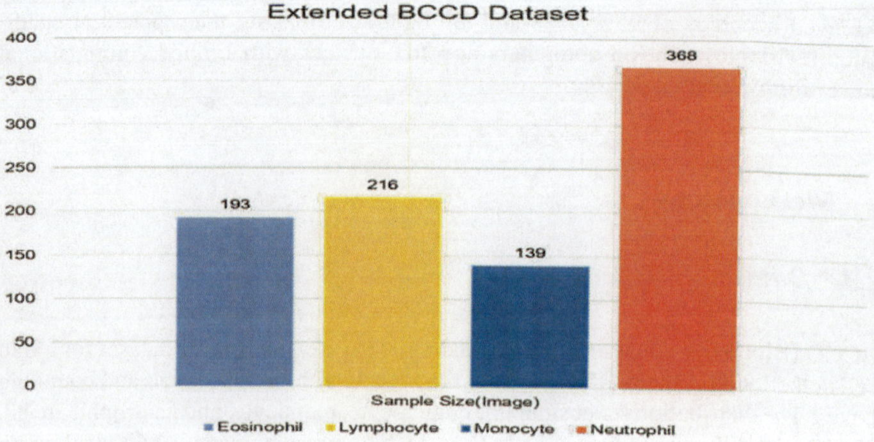

Fig. 2 Distribution of samples in the Extended BCCD Dataset

2.2 *Experimental Framework*

To perform leukocyte classification using blood smear microscopic images, a three-stage classifier system was built. The working process of the system can be found in Fig. 3.

In the first stage of the system, data preprocessing was conducted. In this stage, images were rescaled from 0–255 to 0–1 and resized to fit with CNN models in the second stage.

In the second stage of the system, the preprocessed WBC image was fed into a CNN model (VGG16 or SqueezeNet), which outputs extracted features of the image.

In the last stage of the system, a classifier consisted of full connected layers analysed the extracted features and output probability for each class. These probabilities are passed to a perceptron network, which gave classification label that predicts

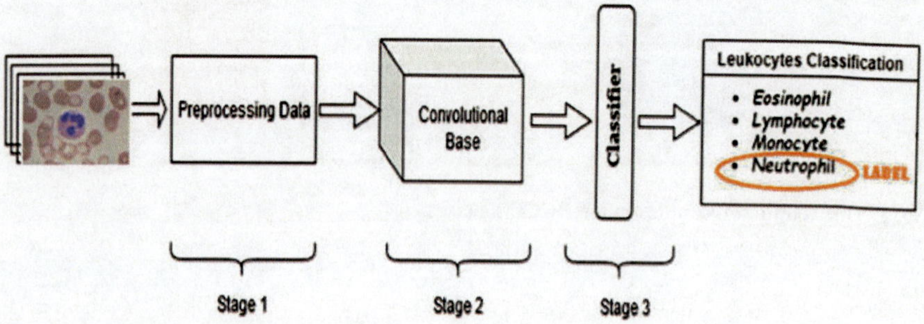

Fig. 3 Working process for the proposed classifier

whether the input image is Neutrophils, Eosinophils, Monocytes, or Lymphocytes. For training of the models, splitting ratio of 80 training: 20 testing was used for both the BCCD dataset and the Extended BCCD dataset. Data augmentation techniques that don't remove WBCs from the images were applied to increase the number of training examples. Transfer learning were also applied by using pre-trained VGG16 and SqueezeNet models.

3 Result and Discussion

Performance of both VGG16 based model and SqueezeNet based model reach the accuracy of ~97% in the BCCD dataset (Table. 1). As for the Extended BCCD dataset, the accuracy of SqueezeNet based model is 95%, lower than that of the VGG16 based model (97%).

Example image of eosinophil and neutrophil are shown in Fig. 4. In classification techniques using traditional image processing, these two kinds of leukocyte are classified based on the number of lobes in nucleus as well as characteristics of granules [7]. To be specific, eosinophil usually have a nucleus with two lobes while neutrophil

Table 1 The classification results of author's AI models

		VGG16	SqueezeNet
BCCD dataset	Accuracy	0.972	0.97
	Precison	0.9725 ± 0.02625	0.975 ± 0.02
	Recall	0.9725 ± 0.04125	0.975 ± 0.025
Extended BCCD	Accuracy	0.975	0.95
	Precison	0.9775 ± 0.01375	0.95 ± 0.03
	Recall	0.975 ± 0.0175	0.95 ± 0.02

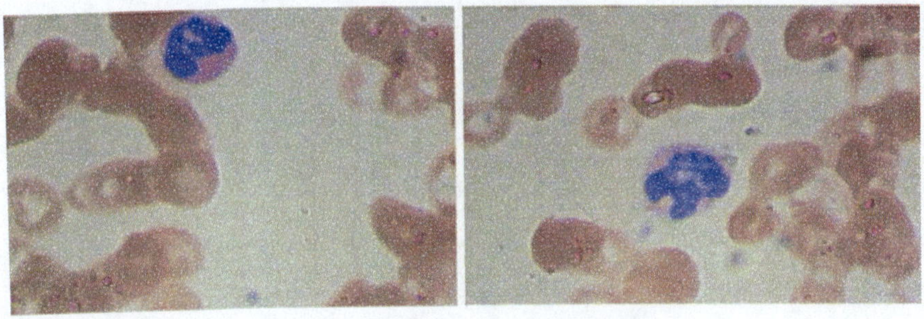

Fig. 4 Images of the eosinophil and neutrophil from the BCCD dataset

nucleus is multi-lobed (two to five lobes). Besides, the presence of pink/red granules is only in eosinophils whereas neutrophils' granules are very thin and with low visibility.

4 Conclusion

In this work, we developed two deep learning networks based on VGG-16 and SqueezeNet CNN for automatic WBC classification. Two WBC datasets, the BCCD dataset and the Extended BCCD dataset, which consist of four types of WBCs including Eosinophil, Neutrophil, Lymphocyte, and Monocyte, were utilized to train and test the two networks. The results showed that both the VGG16-based deep learning network and the SqueezeNet-based deep learning network achieved ~97% accuracy in the BCCD dataset. As for the Extended BCCD dataset, the VGG16-based network performed slightly better (97%) than the SqueezeNet-based network (95%). With relatively light-size, the two deep learning networks are potential candidates for deployment of computers and IoT devices with limited computational power for automatic WBC classification.

Conflicts of Interest The authors have no conflict of interest to declare.

References

1. AL-Dulaimi K et al (2018) Classification of white blood cell types from microscope images:techniques and challenges. Microsc Sci Last Approaches Educ Programs Appl Res 8:17–25
2. Putzu L, Caocci G, Di Ruberto C (2014) Leucocyte classification for leukaemia detection using image processing techniques. Artif Intell Med 62:179–191. https://doi.org/10.1016/j.artmed.2014.09.002
3. Ceelie H, Dinkelaar RB, Van Gelder W (2007) Examination of peripheral blood films using automated microscopy; evaluation of Diffmaster Octavia and Cellavision DM96. J Clin Pathol 60:72–79. https://doi.org/10.1136/jcp.2005.035402
4. BCCD Object Detection Dataset. https://public.roboflow.ai/object-detection/bccd. Accessed 21 May 2020
5. Histology-World! http://histology-world.com/. Accessed 21 May 2020
6. ImageBank|Home|Regular Bank. http://imagebank.hematology.org/. Accessed 21 May 2020
7. Chan YK, Tsai MH, Huang DC, Zheng ZH, Hung KD (2010) Leukocyte nucleus segmentation and nucleus lobe counting. BMC Bioinform 11:558. https://doi.org/10.1186/1471-2105-11-558

Automatic Foveal Avascular Zone Segmentation Using Hessian-Based Filter and U-Net Deep Learning Network

Quang Ngoc Nguyen, Vinh Tuong-Quang Nguyen, Tammy Hsu, Lejla Vajzovic, and Hoan Thanh Ngo

Abstract Accurate segmentation of Foveal Avascular Zone (FAZ) in Optical Coherence Tomography Angiography (OCTA) images is important for OCTA images analysis. In this work, we developed an algorithm for automatic segmentation of FAZ in OCTA images using a Hessian-based filter and an U-Net deep learning network. A total of 260 OCTA images were used to train and test the algorithm. The images were first enhanced by a Hessian-based filter and then fed into a U-Net deep learning network. Eighty percent of the dataset was used for training and twenty percent was used for testing. Our method achieved 87.8% Jaccard Index (Intersection over Union metric) with 6% false-negative error and 5% false-positive error. The results showed that U-Net deep learning network could achieve good accuracy in automatically segmenting FAZ in OCTA images despite a small training set. The study also showed that image preprocessing techniques such as Hessian-based filtering helped to improve accuracy of U-Net deep learning network.

Keywords Deep learning · OCT-A · U-Net · Forvea avascular zone (FAZ)

1 Introduction

The creation of Deep Convolutional Neural Networks (DCNN) has had a significant impact on recent developments in all major computer vision tasks such as image classification [1, 2], object detection [3–5], and semantic segmentation [6, 7]. In particular, there have been numerous researchers who leveraged the power of

Q. N. Nguyen
Cao Thang International Eye Hospital, Ho Chi Minh City, Vietnam

T. Hsu · L. Vajzovic
Duke Eye Center, Duke University, Durham, NC 27705, USA

V. T.-Q. Nguyen · H. T. Ngo (✉)
School of Biomedical Engineering, International University, Vietnam National University—Ho Chi Minh City (VNU-HCMC), Ho Chi Minh City, Vietnam
e-mail: nthoan@hcmiu.edu.vn

V. Van Toi et al. (eds.), *8th International Conference on the Development of Biomedical Engineering in Vietnam*, IFMBE Proceedings 85,
https://doi.org/10.1007/978-3-030-75506-5_68

DCNN in the medical field to support disease detection [8, 9] or lesions segmentation [10, 11].

Besides the emergence of Deep Learning, the development of medical devices and equipment have an equally important role in medical diagnosis and treatment. Recently, OCT (Optical Coherence Tomography)-based angiography has been researched as a new method for diagnosis in ophthalmology. OCTA has emerged as a prominent imaging tool due to its high resolution, speed, and non-invasiveness. OCTA is able to capture high-resolution volumetric angiographic images in just several seconds. Thus, OCTA have been considered a promising imaging tool in evaluating and detecting common retinal diseases such as age-related macular degeneration, diabetic macular edema, and choroidal neovascularization [12–14].

Segmentation of FAZ areas in OCTA images is important for OCTA images analysis. Several researchers have attempted to use traditional computer vision algorithms [15] to extract FAZ area in OCTA image. Other researchers used deep learning to solve FAZ segmentation [16]. Although showing great promises, application of deep learning networks in FAZ segmentation needs more investigation regarding their effectiveness and generalization ability.

This paper studied a new method that combines traditional computer vision techniques with a U-Net-based deep learning model to automatically segment FAZ area in OCTA image. In brief, blood vessels in OCTA images will be enhanced by a Hessian-based filter. Then we fed those enhanced images into a U-shape semantic segmentation model to extract FAZ area. Figure 1 shows the overall process that combines traditional computer vision algorithm, a Hessian-based filter, with deep convolutional neural networks, a U-Net deep learning model. We also compared our approach with a more traditional approach called Level Set, which has been widely used in segmentation of biomedical images before the onset of Deep Learning. We demonstrated that our approach outperformed the Level Set by a large margin.

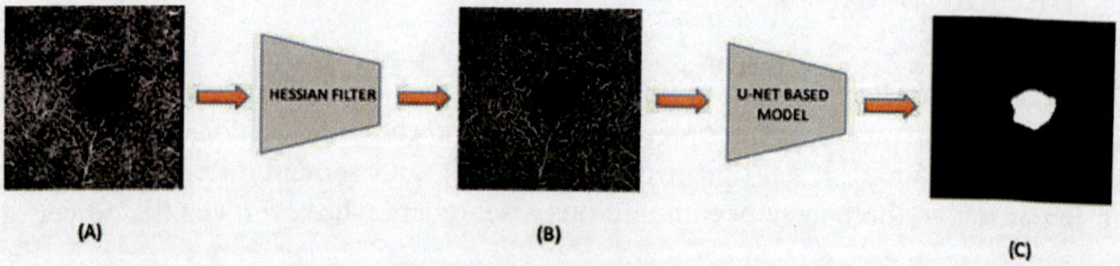

Fig. 1 Enhancing blood vessels in OCTA image (**a**) with a Hessian-based filter. The enhanced image (**b**) is fed to a U-net based model to generate a FAZ mask (**c**)

2 Method

2.1 Dataset

In total, there were 260 anonymous OCTA images used in this study in which 184 images in the training set, 30 images in the validation set, and 46 images in the test set. All images had been collected with IRB approval.

The images were first enhanced by a Hessian-based filter and then fed into a U-Net deep learning network.

2.2 Hessian Filter

The first step of our approach is to enhance the input image so that vascular and avascular regions could be effectively distinguished. The vessel enhancement process was achieved by performing the Hessian multi-scale filtering [17]. In [17], Frangi et al. demonstrated that the eigenvalues and eigenvectors of Hessian matrix are closely related to vascular intensity and direction.

In our study, a Hessian based filter was firstly applied to enhance vascular structures and to suppress non-vascular structures and image noise. The filter that we used was developed by Tim Jerman [18–20]. The code [21] is based on Dirk-Jan Kroon's implementation of Frangi's vesselness filter. We selected sigma values σ range to be from 0.5 to 2.5.

2.3 U-Net Deep Learning Network

Segmentation model was modified from U-Net [22]. Different from U-Net, on the left-branch, we leveraged a state-of-the-art architecture called SeResNext-50 to encode the whole context of the image. Then based on the backbone, we redesigned the decoder branch for up-sampling the encoded feature to receive FAZ segmentation mask. There are many deep neural network architectures that could be used as backbone of U-Net. But in our experiences, SeResNext-50 gave the best result which laveraged the advantages of Squeeze-and-Excitation block [23] in Resnext50. We trained the model for 70 epochs, batch-size = 8, resolution = 256×256, learning rate = 0.0002, and using Adam optimizer.

2.4 Level Set

In order to demonstrate the robustness of our deep learning-based approach, we compared its performance with that of a traditional method called Level Set.

Introduced in 1988 by Oscher and Sethian [24], Level Set method has been widely used in optical image segmentation due to its ability to capture dynamic, complex shapes and interfaces, especially in biomedical images. Basically, the Level Set represents an object's contour as the zero level set of a higher dimensional function, called a level set function (LSF). Then the algorithm constructs the motion of the contour as the evolution of the LSF.

The level set method that we used in this study is new type of level set evolution called distance regularized level set evolution [25] (DRLSE), developed by Chunming Li et al. It removes the requirement for reinitialization and thereby avoids its induced numerical errors. DRLSE also allows the use of more general and robust initialization of the level set function.

We developed an intensity-based method to locate the initial point automatically, and then applied the Level Set method on filtered images to identify the FAZ area.

2.5 Jaccard Coefficient:

Jaccard Coefficient. We compared FAZ segmented by our proposed automatic method and the manual grading. The similarity in FAZ is measured by the Jaccard coefficient, defined as the ratio between the area of intersection and that of union.

$$J = \frac{|A \cap M|}{|A \cup M|} \times 100 \tag{1}$$

where A = area of automated segmentation and M = area of manual segmentation. The value ranging from 0, meaning no overlap, to 100%, meaning complete congruence.

False Positive and False Negative. A false positive indicates a predicted object mask had no associated ground truth object mask.

$$\text{False Positive Error (FP)} = \frac{false\ positive\ area}{Total\ area\ of\ predicted\ mask} \times 100 \tag{2}$$

A false negative indicates a ground truth object mask had no associated predicted object mask.

$$\text{False Negative Error (FN)} = \frac{false\ negative\ area}{Total\ area\ of\ ground - truth\ mask} \times 100 \tag{3}$$

Table 1 Performance of three methods

Methods	Jaccard		FP		FN	
	Mean	STD	Mean	STD	Mean	STD
Hessian + Level Set	54.59	37.12	27.37	41.53	44.59	38.64
Hessian + U-Net	**87.76**	**5.49**	**3.59**	**5.26**	8.83	**4.31**
U-Net	86.49	9.63	9.17	9.53	**5.17**	4.62

3 Results

We compared performance of our proposed method with that of the Level Set. In both scenarios, the input image had been processed by a Hessian filter. We also tested the performance of the U-Net without the Hessian filter in order to examine the effect of filtering. The result is summarized in Table 1.

The combination of Hessian filter and U-Net yielded the best result in Jaccard coefficient, much better than the traditional Level Set. In addition, we also see that the Hessian filter helped improve the overall performance of U-Net with a higher averaged Jaccard score. Also, with the presence of the Hessian preprocessing, we observed a more stable performance seen from the smaller variance of Jaccard coefficient (5.49%), as compared to that of U-Net alone (9.63%). The proposed method also had the least FP's mean, FP's standard deviation, and FN's standard deviation. Some segmentation results are presented in Fig. 2.

In most cases, performance of the Level Set was not very good, probably due to the low resolution (304 × 304 pixels) of the input images and the heavy noise inside FAZ. Level Set must have mistaken the noise as border of FAZ and could not extend the contour further. Another drawback of the Level Set-based method is that it requires initial point setting. We developed an algorithm that can automatically locate initial point for Level Set. However, in some cases, the algorithm failed, especially in OCTA images with low image quality or with severe motion artifacts.

4 Conclusion and Discussion

This study demonstrated the effectiveness of U-Net deep learning network in segmenting the FAZ in OCTA image. With a relatively small training set, our proposed method outperformed a more traditional and widely used segmentation technique Level Set. In addition, our study also showed that combining image preprocessing using Hessian-based filter with U-Net deep learning network helped to improve and stabilize U-Net deep learning network's performance.

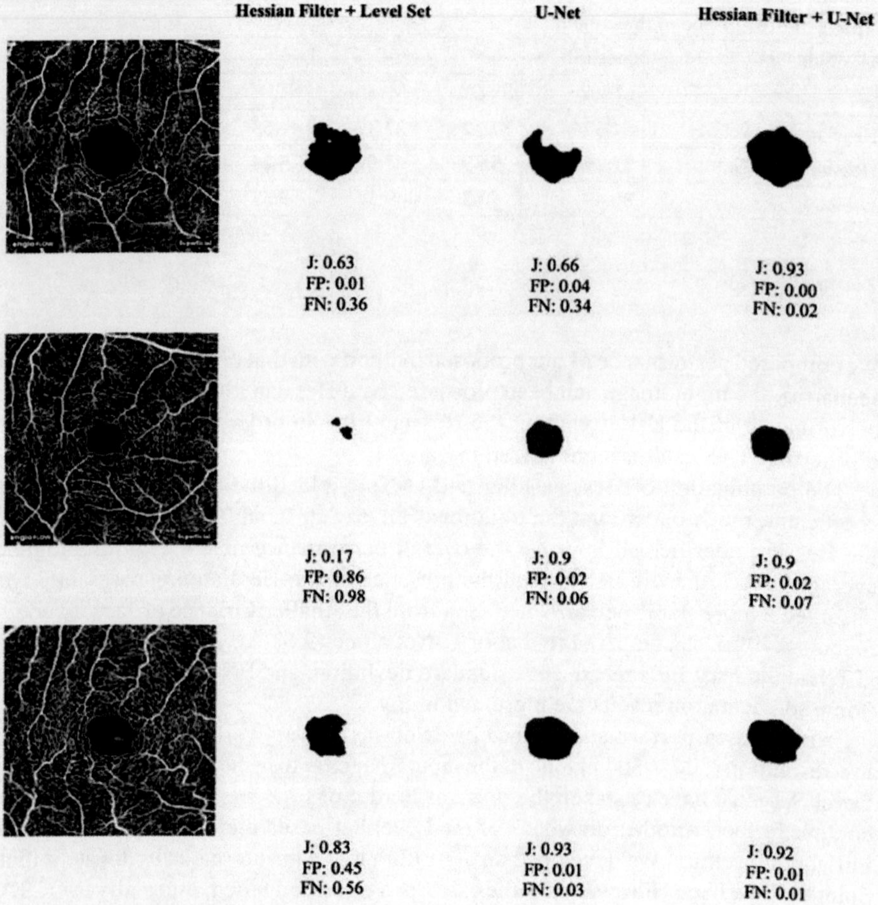

Fig. 2 Sample output of examined methods. first column: Hessian filter + Level Set, second: U-Net, third: Hessian filter + U-Net

Conflicts of Interest The authors have no conflict of interest to declare.

References

1. Krizhevsky A, Sutskever I, Hinton GE (2017) ImageNet classification with deep convolutional neural networks. Commun ACM 60(6):84–90. https://doi.org/10.1145/3065386
2. He K, Zhang X, Ren S, Sun J (2016) Deep residual learning for image recognition. In: Proceedings of the IEEE Computer Society Conference on Computer Vision and Pattern Recognition, Dec 2016, vol. 2016, pp 770–778. https://doi.org/10.1109/CVPR.2016.90

3. Girshick R, Donahue J, Darrell T, Malik J (2014) Rich feature hierarchies for accurate object detection and semantic segmentation. In: Proceedings of the IEEE Computer Society Conference on Computer Vision and Pattern Recognition, Sep. 2014, pp 580–587. https://doi.org/10.1109/CVPR.2014.81
4. Ren S, He K, Girshick R, Sun J (2017) Faster R-CNN: towards real-time object detection with region proposal networks. IEEE Trans Pattern Anal Mach Intell 39(6):1137–1149. https://doi.org/10.1109/TPAMI.2016.2577031
5. Lin T-Y, Dollár P, Girshick R, He K, Hariharan B, Belongie S (2018) Feature pyramid networks for object detection, Dec 2016. Accessed: Nov 27 [Online]. Available: https://arxiv.org/abs/1612.03144
6. Long J, Shelhamer E, Darrell T (2014) Fully convolutional networks for semantic segmentation. IEEE Trans Pattern Anal Mach Intel 39(4):640–651. [Online]. Available: http://arxiv.org/abs/1411.4038
7. Chen LC, Papandreou G, Kokkinos I, Murphy K, Yuille AL (2018) DeepLab: semantic image segmentation with deep convolutional nets, atrous convolution, and fully connected CRFs. IEEE Trans Pattern Anal Mach Intell 40(4):834–848. https://doi.org/10.1109/TPAMI.2017.2699184
8. Poplin R et al (2018) Prediction of cardiovascular risk factors from retinal fundus photographs via deep learning. Nat Biomed Eng. https://doi.org/10.1038/s41551-018-0195-0
9. Han Z, Wei B, Zheng Y, Yin Y, Li K, Li S (2017) Breast Cancer multi-classification from histopathological images with structured deep learning model. Sci Rep 7(1):1–10. https://doi.org/10.1038/s41598-017-04075-z
10. Orlando JI, Prokofyeva E, del Fresno M, Blaschko MB (2018) An ensemble deep learning based approach for red lesion detection in fundus images. Comput Methods Programs Biomed 153:115–127. https://doi.org/10.1016/j.cmpb.2017.10.017
11. Yuan Y, Chao M, Lo YC (2017) Automatic skin lesion segmentation using deep fully convolutional networks with Jaccard distance. IEEE Trans Med Imaging 36(9):1876–1886. https://doi.org/10.1109/TMI.2017.2695227
12. de Carlo TE, Romano A, Waheed NK, Duker JS (2015) A review of optical coherence tomography angiography (OCTA). Int J Retin Vitr 1(1):5. https://doi.org/10.1186/s40942-015-0005-8
13. Virgili G et al (2015) Optical coherence tomography (OCT) for detection of macular oedema in patients with diabetic retinopathy. Cochrane Database Syst Rev 1:2015. https://doi.org/10.1002/14651858.CD008081.pub3
14. Jia Y et al (2014) Quantitative optical coherence tomography angiography of choroidal neovascularization in age-related macular degeneration. Ophthalmology 121(7):1435–1444. https://doi.org/10.1016/j.ophtha.2014.01.034
15. Díaz M, Novo J, Cutrín P, Gómez-Ulla F, Penedo MG, Ortega M (2019) Automatic segmentation of the foveal avascular zone in ophthalmological OCT-A images. PLoS One 14(2):e0212364. https://doi.org/10.1371/journal.pone.0212364
16. Guo Y, Camino A, Wang J, Huang D, Hwang TS, Jia Y (2018) MEDnet, a neural network for automated detection of avascular area in OCT angiography. Biomed Opt Express 9(11):5147. https://doi.org/10.1364/boe.9.005147
17. Frangi AF, Niessen WJ, Vincken KL, Viergever MA (1998) Multiscale vessel enhancement filtering. Springer, Berlin, pp 130–137
18. Jerman T, Pernus F, Likar B, Spiclin Z (2016) Enhancement of vascular structures in 3D and 2D angiographic images. IEEE Trans Med Imaging 35(9):2107–2118. https://doi.org/10.1109/TMI.2016.2550102
19. Jerman T, Pernus F, Likar B, Spiclin Z (2016) Blob enhancement and visualization for improved intracranial aneurysm detection. IEEE Trans Vis Comput Graph 22(6):1705–1717. https://doi.org/10.1109/TVCG.2015.2446493
20. Jerman T, Pernuš F, Likar B, Špiclin Ž (2015) Beyond Frangi: an improved multiscale vesselness filter, vol 9413, pp 94132A. https://doi.org/10.1117/12.2081147

21. “Jerman Enhancement Filter—File Exchange—MATLAB Central.” https://www.mathworks.com/matlabcentral/fileexchange/63171-jerman-enhancement-filter. Accessed 30 Aug 2020
22. Ronneberger O, Fischer P, Brox T (2015) U-net: convolutional networks for biomedical image segmentation. In: Lecture Notes in computer science (including subseries Lecture notes in artificial intelligence and lecture notes in bioinformatics), May 2015, vol 9351, pp 234–241. https://doi.org/10.1007/978-3-319-24574-4_28
23. Hu J, Shen L, Albanie S, Sun G, Wu E (2020) Squeeze-and-excitation networks. IEEE Trans Pattern Anal Mach Intell 42(8):2011–2023. [Online]. Available: http://arxiv.org/abs/1709.01507
24. Osher S, Sethian JA (1988) Fronts propagating with curvature-dependent speed: algorithms based on Hamilton-Jacobi formulations. J Comput Phys 79(1):12–49. https://doi.org/10.1016/0021-9991(88)90002-2
25. Li C, Xu C, Gui C, Fox MD (2010) Distance regularized level set evolution and its application to image segmentation. IEEE Trans Image Process. 19(12):3243–3254. https://doi.org/10.1109/TIP.2010.2069690

Healthcare Informatics and Public Health

Ascorbic Acid as a Radiation-Protective Agent Against Ionizing Radiation

Tran Thi Nhan, Youichirou Matuo, Maradi Abdillah, Lukas Wisnu Wicaksono, and Yoshinobu Izumi

Abstract Ascorbic acid (AA) is well known as a good radical scavenger and antioxidant. Here, the protection effects of AA against free radicals (mainly hydroxyl radicals) caused by low and high linear energy transfer radiation were examined. In this study, cobalt-60 gamma rays and helium ion beams were used as the sources of ionizing radiation. DNA strand breaks of PUC118 plasmid in the presence and absence of AA were monitored using agarose gel electrophoresis. Moreover, the survival rates of *Saccharomyces cerevisiae* yeast cells irradiated with different AA concentrations in growing medium were analysed using the linear quadratic survival model. Based on the results of DNA strand break experiments, the yield of strand breaks increased with an increase in absorbed dose. In the presence of AA, the yield of strand breaks decreased with an increase in AA concentration. In other words, the greater the additive concentration, the more effective the radiation-protective agent. On the other hand, in the survival rate experiments, the dependence on additive concentration was not similar to that of the DNA experiments. We consider that the main reason for our observations is the difference in additive concentration between the outside and inside of the cell membrane of yeast cells. These results suggest that AA plays an effective role as a radiation-protective agent for DNA and living cells against free radicals formed by radiation.

Keywords Radiation protection · Free radical · Ascorbic acid · DNA strand breaks · Cell survival

T. T. Nhan (✉)
Electric Power University, Hanoi, Vietnam
e-mail: nhantt@epu.edu.vn

T. T. Nhan · Y. Matuo · M. Abdillah · L. W. Wicaksono · Y. Izumi
Research Institute of Nuclear Engineering, University of Fukui, Fukui, Japan

T. T. Nhan
Vietnam Atomic Energy Institute, Hanoi, Vietnam

M. Abdillah · L. W. Wicaksono
Indonesia Nuclear Energy Regulatory Agency, Jakarta, Indonesia

V. Van Toi et al. (eds.), *8th International Conference on the Development of Biomedical Engineering in Vietnam*, IFMBE Proceedings 85,
https://doi.org/10.1007/978-3-030-75506-5_69

1 Introduction

Ionizing radiation may stem from the decay process of unstable nuclei or by de-excitation of atoms and their nuclei in nuclear reactors, X-ray machines, cyclotrons, and other devices. Both natural cosmic and terrestrial or man-made sources of ionizing radiation contribute to human exposure and constitute a hazard for human health. In the case of nuclear accidents, large amounts of radioactive material are released into the environment. In particular, since the Fukushima Daiichi nuclear accident, the risks of radiation exposure have garnered more public attention. Ionizing radiation generates free radicals, which damage cellular targets like DNA and cell membranes as well as cause cell death or mutations. There are two interaction mechanisms of ionizing radiation on DNA. One involves the ionization of atoms in the DNA (direct effect) while the other involves attacks by free radicals produced by the radiolysis of surrounding water molecules (indirect effect) [1]. Radiation causes a wide range of lesions in DNA, such as single-strand breaks (SSBs) in phosphodiester linkages, double-strand breaks (DSBs) on opposing sites or displaced, base damage, protein-DNA crosslinks, and protein–protein crosslinks. Depending on the amount of damage, type of damage, and damage location, protein oxidation and lipid peroxidation could lead to cell death given that damage to DNA play a critical role in cell killing, DSB effects of DNA-DSB [1]. Moreover, the biological effects of ionizing radiation are generally thought to depend on linear energy transfer (LET). High LET radiation is likely to cause more DSBs. Low LET radiation interacts with cellular water to generate reactive intermediate species, such as hydroxyl radicals (·OH), which then attack the DNA. This process is called indirect action. Most of the energy from high LET radiation is absorbed directly by the DNA to create DNA lesions (via direct action).

Antioxidant compounds can prevent the uncontrolled formation of free radicals or inhibit their reaction with biological sites. Antioxidants can be categorized in two types based on their activity: enzymatic and non-enzymatic. Enzymatic antioxidants work by breaking down and removing free radicals. This type of antioxidant converts dangerous oxidative products to hydrogen peroxide (H_2O_2) and then to water via a multi-step process in the presence of cofactors, such as copper, zinc, manganese, and iron. Non-enzymatic antioxidants operate by interrupting free radical chain reactions [2]. Ascorbic acid (AA) is considered to be a strong antioxidant that is found naturally in some fruits, including citrus fruits, and many vegetables and is also available as a dietary supplement [3, 4]. Given the chemical structure of AA with four hydrogen bond donors and six hydrogen bond acceptors, AA is an excellent source of electrons for free radicals that are seeking out an electron to regain their stability [5, 6]. AA (vitamin C) is well known as a protective agent [7, 8]. The radioprotection of epigallocatechin gallate (EGCG) for gamma irradiation in mice has also been studied [10–12]. Moreover, it was reported that ECGC is effective in suppressing DNA strand breaks caused by ionizing radiation [13]. However, there are few studies in which the protective effect of radical scavengers was carried out at the cellular and DNA levels simultaneously. Further experimental reports about the radiation protection of

DNA in the presence of AA through direct or indirect reactions of radiation via the capture of free radicals are required.

In this study, we examined the capacity of AA to protect DNA and cell survival from cobalt-60 (^{60}Co) gamma rays and helium ion beams. While gamma rays emitted from ^{60}Co are classified as low LET radiation, helium ion beams appear to have higher LET. The number of DNA strand breaks were monitored using agarose gel electrophoresis. The survival rate of living cells was also analysed by the linear quadratic (LQ) survival model [14]. We clarified how radical scavengers have a protective effect on cells and DNA. In summary, we studied the contribution of radical scavengers to the field of radiation protection.

2 Materials and Methods

2.1 Samples

DNA (pUC118, 3162 bp) in tris-ethylenediaminetetraacetic acid (TE) buffer was extracted from *Escherichia coli* JM109 competent cells (Takara Bio, Shiga, Japan) using a Plasmid Midi Kit (QIAGEN, Hilden, Germany).

Saccharomyces cerevisiae yeast cells were incubated for 24 h at 30 °C in yeast extract-peptone-dextrose (YPD) liquid medium. After growing, the yeast cells were counted, diluted, and filtered through nitrocellulose membranes. Each membrane containing 200 yeast cells was placed in a 50 mm petri dish for all cell experiments. Five samples were utilized for each condition.

To investigate the protective effects of AA on DNA and living cells, AA was added to the TE buffer and YPD liquid medium with different concentrations from 0 to 1000 μM. AA was purchased from Nacalai Tesque, Inc. (Kyoto, Japan) with a purity ≥ 99.5%.

2.2 Irradiation

DNA and living cell samples were irradiated with ^{60}Co gamma rays with a LET of 0.2 keV/μm in the Radiation Laboratory of ISIR, Osaka University (Osaka, Japan) and helium ion beams with a LET of 5.5 keV/μ at the Wakasa-Wan Energy Research Center (Tsuruga, Japan). The absorbed dose (Gy; J/kg) of DNA and living cells ranged from 0 to 100 Gy with each type of ionizing radiation in which the absorbed dose is a dose quantity, which is the measure of the energy deposited in the matter by ionizing radiation per unit mass (SI base unit: J/kg).

2.3 Analysis of Strand Breaks

The analysis of plasmid DNA for SSBs and DSBs was performed using agarose gel electrophoresis. When irradiated, plasmid DNA molecules containing SSBs and DSBs migrate independently from the intact supercoiled form on agarose gels because mobility in an agarose gel depends on the DNA structure, thereby enabling quantification. The faster band represents the original form or form I. The open circle or form II indicates that the plasmid was cleaved in one of the strands (SSB), resulting in a wider and slower molecule in the agarose gel. The linear or form III plasmid molecule is slightly faster than form II and represents a DSB (Fig. 1) [15].

Agarose gel was prepared using Agarose S from Nippon Gene Co., Ltd. (Toyama, Japan) in a 1% Agarose S concentration. Electrophoresis was performed at 100 V for 30 min. Then, the gel was stained with SYBR Green (LONZA, Tokyo, Japan). The DNA bands were visualised under ultraviolent light, and the result was captured with imaging equipment (AS One, Osaka, Japan). We used Image Lab™ software version 6.0.0 (BioRad Laboratories, Hercules, CA, USA) to measure the intensity of the gel image produced from electrophoresis to determine the percentage of DNA strand breaks. Fluorescent images of electrophoresis were analysed for each of the three lanes. To analyse the role of AA as a radiation-protective agent on DNA, we used the relative damage function, which is calculated according to the following Eq. (1):

$$Relative\ damage = \frac{G\ value\ of\ SBs}{G\ Value\ of\ SBs\ No\ Additive} \quad (1)$$

where G value represents the number of molecules damaged by 100 eV of radiation energy absorption [16].

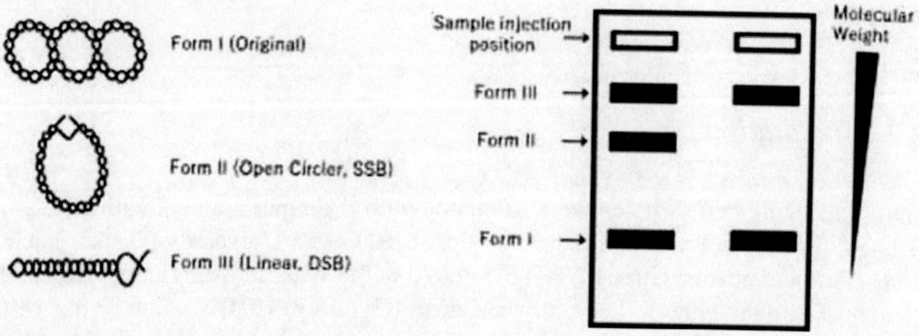

Fig. 1 Principle of DNA strand break evaluation using agarose gel electrophoresis. DNA that has undergone strand breaks can be distinguished from form I (original) by form II (open circle, SSB) and form III (linear, DSB)

2.4 *Survival Rate (SR)*

Membranes containing yeast cells were irradiated and then transferred to solid YPD medium to grow the cells at 30 °C for two days. The number of remaining living cells after irradiation was compared to the initial number of cells (before irradiation) to obtain the SR, as in Eq. (2):

$$SR = \frac{Number\ of\ colonies_{irradiated}}{Number\ of\ colonies_{non\text{-}irradiated}} \tag{2}$$

The SR of yeast cells was analysed using the LQ survival model, which shows that cell survival after absorbing a dose of ionizing radiation follows a combination of single-hit or linear kinetics (representing DSBs) and double-hit kinetics represented by a quadratic term (representing SSBs) [17]. The LQ survival model is given by Eq. (3):

$$S = e^{-(\alpha D+\beta D2)} \tag{3}$$

where S is the SR following the absorbed dose (D) while the constants α and β are coefficients for the respective single-hit and double-hit components. For high LET irradiation, the quadratic component (β) is small or non-existent [1]. The higher the α/β rate, the higher death cell rate.

3 Results

3.1 *Protective Effect of AA on DNA*

DNA strand breaks of the pUC118 plasmid in the absence and presence of AA were monitored using agarose gel electrophoresis. The data collected with a digital camera and band observation showed only the original form (form I) and open circle (form II) for gamma irradiation and helium ion beam irradiation. In other words, no significant DSBs were observed (Fig. 2).

The percentage of SSB images is shown in Fig. 3. This finding indicates that all studied concentrations of AA have protective effects on DNA, and the yield of SSBs decreased as the concentration of AA increased. The protective effect of AA was clearer when DNA treated with different concentration of AAs was irradiated by higher LET radiation. The SSB percentage of DNA irradiated by helium ion beams is indicated in Fig. 4.

For both gamma ray irradiation and helium ion beams, the percentage of SSBs with 1000 μM of AA treatment was always the smallest, followed by 500 and 300 μM. This suggests that the higher the AA concentration, the fewer the number of DNA strand breaks.

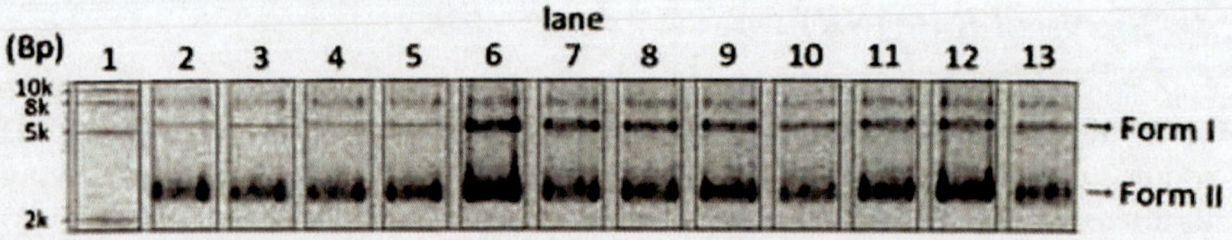

Fig. 2 Gel electrophoresis pattern of plasmid pUC118 exposed to gamma rays. Lane 1, molecular weight marker; lanes 2–4, 0 Gy; lanes 5–7, 25 Gy; lanes 8–10, 50 Gy; and lanes 11–13, 100 Gy. Arrows indicate the structure of plasmid DNA (form I or form II). Form 3 is not displayed because DSBs could not be detected by gamma irradiation up to 100 Gy

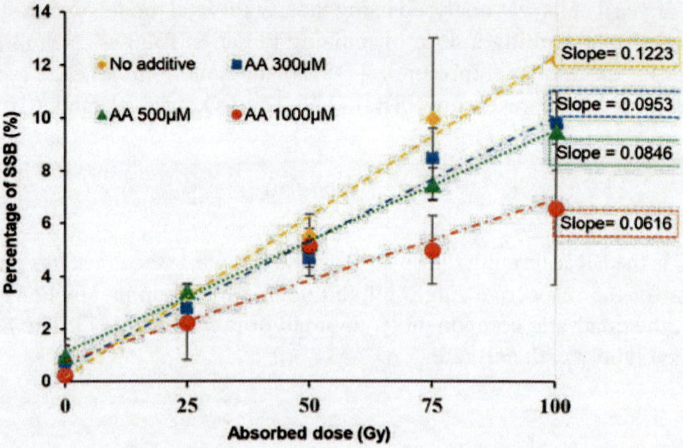

Fig. 3 The SSB percentage of DNA irradiated by gamma rays with different concentrations of AA

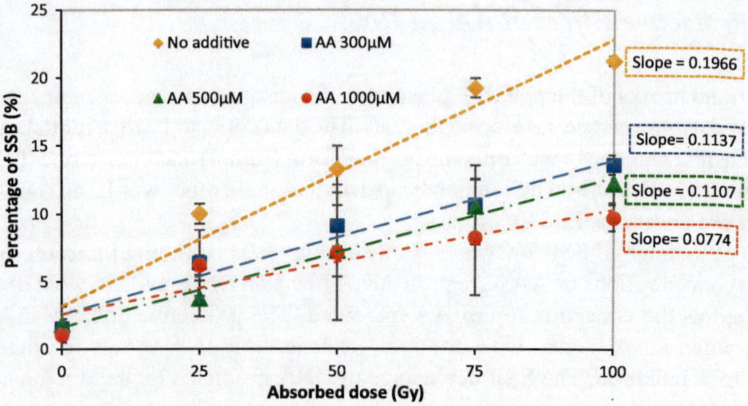

Fig. 4 The SSB percentage of DNA irradiated by helium ion beams with different concentrations of AA

The G values of SSB formation were calculated from the SSB slopes (Figs. 3 and 4) and are summarized in Fig. 5. The higher the G value, the higher the number of damaged molecules and vice-versa. The SSB G value of DNA under helium ion beam irradiation was higher than that of gamma ray irradiation. This finding suggests that the damage to DNA from helium ion beams was more significant than that of gamma rays. However, the G values tended to decrease when the concentration of AA increased for both low and high LET radiation. The scavenging effect of AA at different LETs is demonstrated by this relative damage, which is represented by Eq. (1).

Figure 6 demonstrates that AA is able to protect DNA from damage caused by ionizing radiation. In other words, DNA damage is reduced clearly in the presence of AA. Compared to the 300 and 500 μM AA concentrations, the amount of relative DNA damage in the 1000 μM concentration was the lowest at approximately 0.5 and

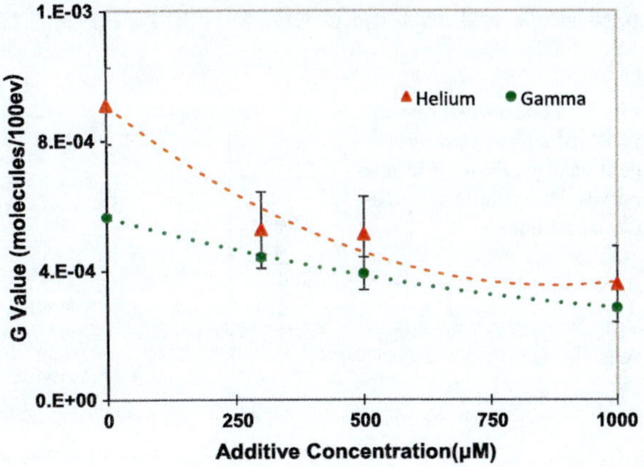

Fig. 5 The DNA SSB G value is correlated with the additive AA concentration

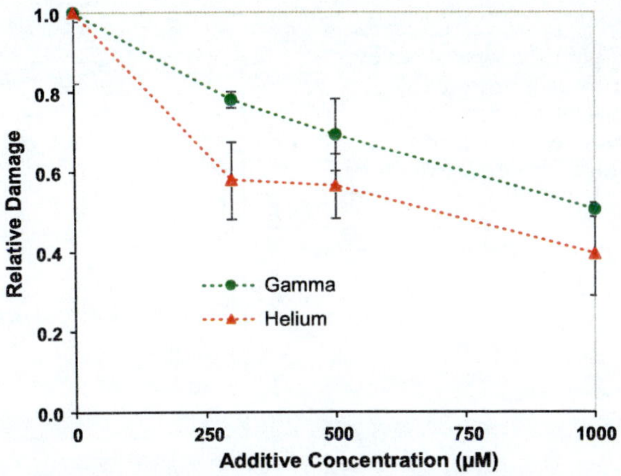

Fig. 6 Relative damage is correlated with the additive AA concentration

0.4 for gamma rays and helium ion beam radiation, respectively. The DNA strand break analysis results showed that AA plays an effective role as a radiation-protective agent for DNA against radiation, and the strand break yield decreased with increases in AA concentration.

3.2 *Protective Effect of AA on Living Cells*

The survival rates of yeast cells as a function of the absorbed dose in the presence and absence of AA for gamma rays and helium ion beams are shown in Figs. 7 and 8, respectively.

The survival rate of yeast cell growth in YPD medium containing AA was always higher compared to cell growth in YPD medium without AA. For example, at a 100 Gy absorbed dose of gamma rays, the cell survival rate increased from 36% to 62%, 45%, and 56% the presence of 300, 500, and 1000 μM of AA, respectively.

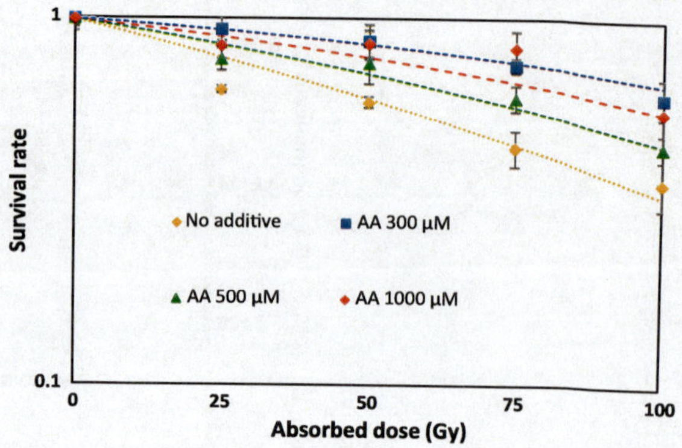

Fig. 7 The survival rate of yeast cells irradiated by gamma rays. Dose–response curves were fitted using the LQ equation

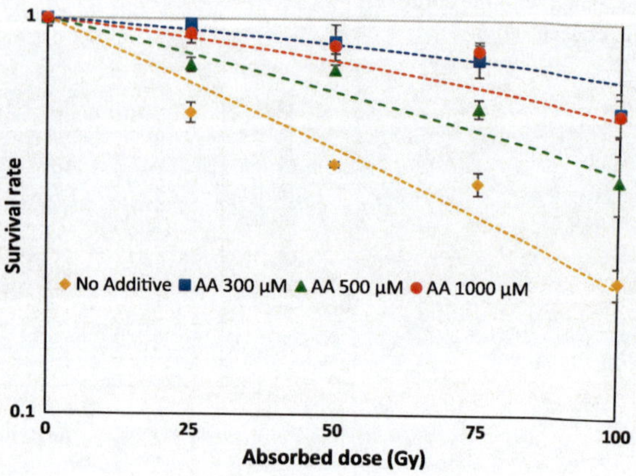

Fig. 8 The survival rate of yeast cells irradiated by helium ion beams. Dose–response curves were fitted using the LQ equation

However, the dependence on additive concentration was not similar to that of the DNA experiments. An AA concentration of 300 μM displayed the greatest effect of protecting cells against ionizing radiation for both low and high LETs (Figs. 7 and 8).

The α and α/β values of the LQ survival model were estimated using Eq. (3) and are presented in Figs. 9 and 10. The α constant deceased in the presence of AA in medium cell liquid for both low high LETs (Fig. 9). This observation indicates the effects of AA in reducing the number of DNA strand breaks and cell death upon cell irradiation. An AA concentration of 300 μM was the most effective for protecting cells from ionizing radiation: gamma rays ($\alpha = 0.0023$ and $\alpha/\beta = 147$) and helium ion beams ($\alpha = 0.024$ and $\alpha/\beta = 240$).

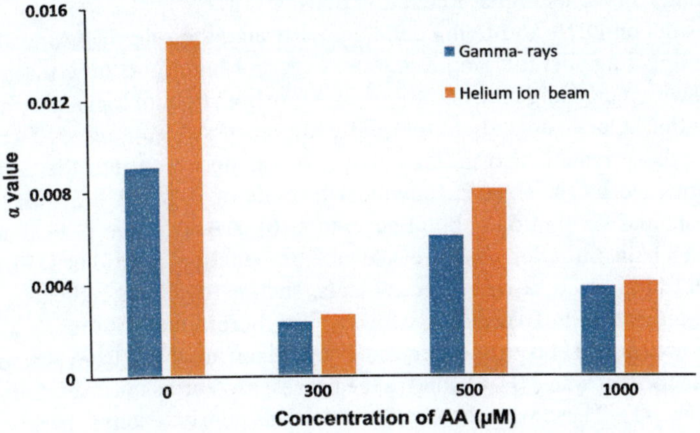

Fig. 9 The α constant follows the LQ survival model

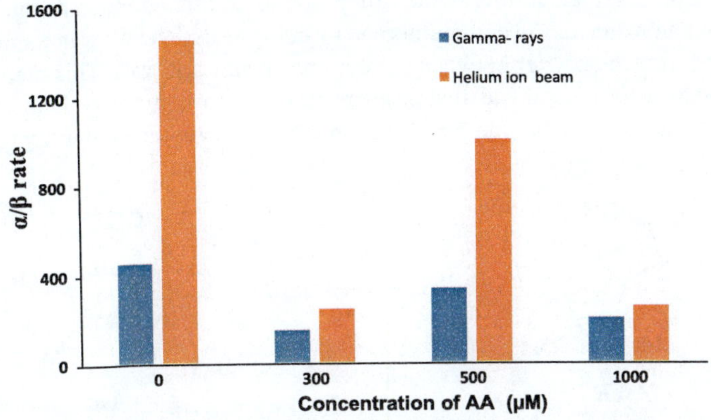

Fig. 10 The α/β constant α/β follows the LQ survival model

4 Discussion

Our study showed that high LET radiation is able to cause more DSBs. The critical difference between low LET photons and high LET charged particles is the density of ionizations along a track. In terms of physicochemical aspects, ionizations (chiefly of H_2O) induced by different types of radiation are otherwise quite similar. Clustered lesions are more difficult to repair, hence the increased cytotoxicity of high LET radiation simply reflects the intrinsic weakness of DNA repair systems to process clustered DNA damage. It has been argued that the cytotoxic effects of low LET ionizations are largely due to occasional clustered lesions rather than the far more numerous isolated lesions [18]. Thus, the greater "relative biological effectiveness" of high LET radiation can be explained by its greater propensity to induce clustered DNA damage due to its higher ionization density [18].

Our results on DNA and living cells demonstrated the role of AA as a radiation protective agent against ionizing radiation. Figure 9 highlights how the α constant decreased with increasing concentrations of AA in both helium ion beam and gamma ray irradiation. Our results show that the addition of AA suppresses DNA DSBs. Of course, those results also indicate that AA was absorbed into the cells, which can be explained by the $C_6H_8O_6$ chemical formula of AA. AA has four hydrogen bond donors and six hydrogen bond acceptors [6]. AA as a free radical scavenger can inhibit chain initiation and break chain propagation reactions [19]. AA can also protect biomolecules from free radicals, such as hydroxyl radicals (·OH), by donating an electron to free radicals for stability before they can reach and attack the biomolecules. ·OH takes one electron or hydrogen atom (H) from AA to create a more stable form of water (H_2O), and in the process, AA turns into ascorbate radicals ($\cdot Asc^-$) (Fig. 11). These new ascorbate radicals are poorly reactive, harmless to the surroundings, and easily neutralized by other reactions [20, 21].

However, in the survival rate experiments, the dependence on additive concentration was not similar to that of the DNA experiments. We suggest that the main reason for our observed results is the difference in additive concentration between the outside and inside of the cell membrane of living cells. The plasma membrane of yeast cells forms the barrier between the cytoplasm and cell wall. This membrane is surrounded by a porous cell wall that protects the cell from osmotic lysis and excludes larger macromolecules. AA can pass through the cell membrane either by passive or

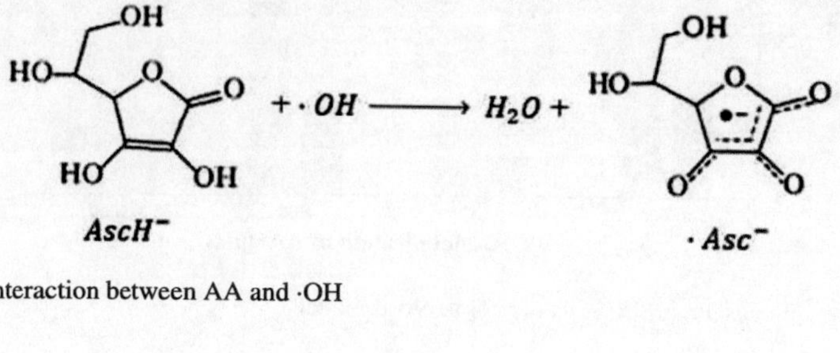

Fig. 11 Interaction between AA and ·OH

active transport. Passive transport means that the cell does not need to spend energy to absorb substances or molecules from the outside of the cell, such as by diffusion (movement of a substance from an area of high concentration to an area of low concentration) and channel proteins (open pores in the membrane that allow small molecules of the appropriate size and charge to pass freely through the lipid bilayer). Active transport means that the cell needs to spend energy to absorb substances or molecules, such as by carrier proteins (which bind specific molecules to be transported on one side of the membrane), endocytosis (bringing a molecule "into the cell" by enveloping it in a portion of its cell membrane), phagocytosis ("cell eating") for large particles, and pinocytosis ("cell drinking") for fluid containing dissolved substances. Several factors affect the rate of transport because of the various transport mechanism [22]. One of the reasons given is the molarity of AA. The higher the AA molarity, the thicker the AA density, thus the rate of AA transport is slower because it is more difficult to pass through the cell membrane. We also assume that there is a maximum capacity for vitamin C inside of a single yeast cell, and, in our experiment, 300 μM was the highest capacity for AA inside of a single cell relative to 500 and 1000 μM. Thus, in the present study, we believe that AA serves as a protective agent against ionizing radiation.

For the given irradiated cells, the α/β value was significantly higher than that of living animal cells (from 0.4 to 36) [23]. As mentioned by Ballarini [24], many experimental data have shown that the semilogarithmic plot is different between haploid yeasts, some bacteria and diploid yeasts, and almost all mammalian cells. In this case, haploid yeasts are characterized by a purely exponential dose–response curve, which becomes a straight line in the usual semilogarithmic representation. Here, we used haploid yeast as a sample for irradiation. Thus, the α and α/β values for irradiated cells in our study are higher than those of diploid yeasts and mammalian cells.

5 Conclusion

This study suggests that AA plays an effective role as a radiation-protective agent for DNA and living cells against free radicals formed by radiation. The results of the DNA strand break experiments demonstrated that strand break yields decrease with increases in the AA concentration. The survival rate of cells increased in the presence of AA. The α and α/β values calculated using the LQ model highlighted the differences in doseresponse curves between haploid and diploid yeasts.

Acknowledgements The authors would like to thank the institute of Scientific and Industrial Research of Osaka University and the facilities of Radioisotope Research Center at the Suita campus of Osaka University.

Conflicts of Interest The authors have no conflict of interest to declare.

References

1. Dance DR, Christofides S, Maidment ADA, McLean AD, Ng KH (2014) Diagnostic radiology physics. handbook for teachers and students. International Atomic Energy Agency
2. Nimse SB, Palb D (2015) Free radicals, natural antioxidants, and their reaction mechanisms. Royal Soc Chem 5:27986–28006
3. Yamamoto T, Kinoshita M, Shinomiya N, Hiroi S, Sugasawa H, Matsushita Y, Majima T, Saitoh D, Seki S (2010) Pretreatment with ascorbic acid prevents lethal gastrointestinal syndrome in mice receiving a massive amount of radiation. J Radiat Res
4. Yanagisawa A, Michie Iwata F, Akiyama S (2011) Effect of Vitamin C and anti- oxidative nutrition on radiation induced gene expression in Fukushima Nuclear Plant workers—a pilot study. Japanese College of Intravenous Therapy (JCIT)
5. Bindhumol V, Chitra KC, Mathur PP (2003) Bisphenol A induces reactive oxygen species generation in the liver of male rats. Toxicology 188:117–124
6. NCBI homepage, https://pubchem.ncbi.nlm.nih.gov/. Last accessed 27 Dec 2018
7. Verma RS, F., Mhta A, Srivastava T (2007) In vivo chlorpyrifos induced oxidative stress: attenuation by antioxidant vitamins. Pesticide Biochem Physiol 88:191196
8. Cai L, Koropatnick J, Cherian MG (2001) Roles of vitamin C in radiation-induced DNA damage in presence and absence of copper. Chem Biol Interact 137:75–88
9. Smith TA, Kirkpatrick DR, Smith S, Smith TK, Pearson T, Kailasam A, Herrmann KZ, Schubert J, Agrawal DK (2017) Radioprotective agents to prevent cellular damage due to ionizing radiation. J Transl 15:232
10. Yi J, Chen CC, Liu X, Kang Q, Hao L, Huang J, Lu J (2020) Radioprotection of EGCG based on immunoregulatory effect and antioxidant activity against 60-Co γ radiation-induced injury in mice. Food Chem Toxicol 161:175–186
11. Matuo Y, Nishijima S, Ikeda T, Shimizu K (2008) Radical scavenging effect of naturally-occuring epigallocatechin gallate against oxidative damage caused by gamma-ray radiation. Radioisotopes
12. Gudkov SV, Popova NR, Bruskov VI (2015) Radioprotective substances: history, trends and prospects. Biophysics. Pleiades Publishing, Inc.
13. Xie LW, Cai S, Zhao TS, Li M, Tian Y (2020) Green tea derivative (−)-epigallocatechin3-gallate (EGCG) confers protection against ionizing radiation-induced intestinal epithelial cell death both in vitro and in vivo. Free Radical Biol Med 161:175–186
14. Brenner DJ (2008) The linear-quadratic model is an appropriate methodology for determining iso-effective doses at large doses per fraction. NIH Public Access
15. De Mattos JCP, Dantas FJS, Caldeira-de-Araújo A, Moraes MO (2004) Agarose gel electrophoresis system in the classroom. Biochem Mol Biol Educ 32(4):254–257
16. Rashid RA, Ariffin NA, Dollah N, Rahman WN (2019) Application of linear quadratic ans multi-target model to single and fractionated radiotherapy irradiation. Jurnal Sains Nuklear Malaysia 31(2):23–31
17. Flickinger JC, Kondziolka D, Lunsford LD (2003) Radiobiological analysis of tissue responses following radiosurgery. Technol Cancer Res Treat 2(2). ISSN 1533-0346
18. Nickoloff JA, Sharma N, Taylor L (2020) Clustered DNA double-strand breaks: biological effects and relevance to cancer radiotherapy. Genes (Basel) 11(1):99. https://doi.org/10.3390/genes11010099
19. Ighodaro OM, Akinloye OA (2017) First line defence antioxidants-superoxide dismutase (Sod), catalase (Cat) and glutathione peroxidase (Gpx): their fundamental role in the entire antioxidant defence grid. Alexandria J Med
20. Narra VR, Howell RW, Sastry KS, Rao DV (1992) Vitamin C as a radioprotector against iodine-131 in vivo. J Nucl Med 34:637–640
21. Dabbagh HA, Azami F, Farrokhpour H, Chermahini AN (2014) UV-VIS, NMR and FTIR spectra of tautomers of vitamin C. Experimental and DFT calculations, J Chil Chem Soc 59(3):2248–2254

22. Prescianotto-Baschong C, Riezman F (1998) Morphology of the yeast endocytic pathway. Mol Biol Cell 9:173–189
23. Hawkins RB (2017) Biophysical models, microdosimetry and the linear quadratic survival relation. Ann Radiation Therapy Oncol
24. Ballarini F (2010) From DNA radiation damage to cell death: theoretical approaches. J Nucl Acids

The Effectiveness Validation of Novel UVC Air Disinfection Device in a Hospital Room Using Computational Fluid Dynamic (CFD) Simulation

Thai Hien Nguyen, Ngoc An Dang Nguyen, Quoc Khai Le, Anh Tu Tran, Thanh Nha Nguyen, Trung Nghia Tran, and Tich Thien Truong

Abstract Airborne-mediated diseases like chickenpox, tuberculosis, influenza, influenza, measles, and rhinovirus represent major public health challenges. In the process of examining the causes related to the transmission of infection, studies on the airborne pathogen and the way it spread were carried out mainly in the medical field. In low-middle income countries, studies on the diffusion of air pollutants and virus diffusion in the hospital were limited. Short-wavelength ultraviolet C (UVC) light has an airborne antimicrobial great potential for preventing airborne transmission by inactivating the airborne pathogens. However, its widespread use in public areas is limited because conventional UVC light sources may risk skin cancers and eye damages. To overcome these problems, we propose a novel UVC air disinfection device and utilize the computational fluid dynamics (CFD) method to simulate the dispersion of airborne droplets in a double-bed hospital room. This paper reports the airborne pathogens spreading patterns according to the location of the proposed device. The transient simulation results concerning the airborne pathogen tracing

Thai Hien Nguyen, Thanh Nha Nguyen, Trung Nghia Tran, Tich Thien Truong contributed equally to this chapter.

T. H. Nguyen (✉) · N. A. D. Nguyen · Q. K. Le · A. T. Tran · T. N. Nguyen (✉) ·
T. N. Tran (✉) · T. T. Truong (✉)
Ho Chi Minh City University of Technology (HCMUT), 700000 Ho Chi Minh City, Vietnam
e-mail: thnguyen@hcmut.edu.vn

T. N. Nguyen
e-mail: nhanguyen@hcmut.edu.vn

T. N. Tran
e-mail: ttnghia@hcmut.edu.vn

T. T. Truong
e-mail: tttruong@hcmut.edu.vn

T. H. Nguyen · Q. K. Le · A. T. Tran · T. N. Nguyen · T. N. Tran · T. T. Truong
Vietnam National University Ho Chi Minh City (VNUCM), 700000 Ho Chi Minh City, Vietnam

N. A. D. Nguyen
University of Medicine and Pharmacy At Ho Chi Minh City, 700000 Ho Chi Minh City, Vietnam

V. Van Toi et al. (eds.), *8th International Conference on the Development of Biomedical Engineering in Vietnam*, IFMBE Proceedings 85,
https://doi.org/10.1007/978-3-030-75506-5_70

combined with their concentration and distance provided evidence of the contaminant zones that should be checked and cleaned. The effectiveness of the proposed device was confirmed through the airborne pathogen tracing simulation and device's location analyses in a double-bed hospital room.

Keywords Hospital room · Computational fluid dynamic · Novel UVC air disinfection · Airborne · Ventilation

1 Introduction

The outbreaks or pandemic like the severe acute respiratory syndrome (SARS) outbreaks in 2003, the human avian influenza A (H5N1) infections, the pandemic influenza A (H1N1) in 2009, and the recent ongoing pandemic coronavirus disease 2019 (COVID-19) cases in this year 2020, represent major public health challenges. The concerns about the aerosol or airborne transmission of infection have become important [1–5].

Following the large-scale COVID-19 out-break this year, these experts reported evidence of the likelihood of airborne disease transmission. The COVID-19 has highlighted the potential for airborne transmission not only in indoor environments but also in outdoor environments and crowded spaces. The airborne transmission of infection in hospitals or clinical treatment units occurs even frequent and more significant threats than before [6, 7].

The airborne-mediated diseases (tuberculosis, influenza, etc.) were often not diagnosed before patients were admitted to the clinical treatment units. The patients may be accompanied by undiagnosed infectious diseases, such as chickenpox, tuberculosis, influenza, measles, and rhinovirus, that are spread primarily through airborne transmission. The risk of cross-infection from these airborne respiratory diseases can be higher among vulnerable hospital patients and the healthcare staff who care for them [8]. Hence, the virus spread silently among other patients and healthcare staff, then even redirected back into the community.

In developing countries, studies on the dispersion of airborne in the hospital were limited. Despite that the airborne transmission becomes essential and getting more attention, there is little research on strategies to control airborne transport and environmental contamination. The hospital room is usually not equipped with indoor air disinfection to handle these forms of airborne infection.

Short-wavelength ultraviolet C (UVC) light has an airborne antimicrobial great potential for preventing airborne transmission by inactivating the airborne pathogens [9–12]. However, its wide-spread use in public areas is limited because conventional UVC light (254.0 nm) sources may risk both skin cancers and eye damages. To overcome these problems, we propose a novel UVC air disinfection device and simulate with Computational Fluid Dynamics (CFD) for evaluating the airborne pathogens spreading patterns in a double-bed hospital room according to the location of the proposed device.

2 Materials and Method

2.1 Computational Fluid Dynamics (CFD) Modeling Method

Computational fluid dynamics (CFD) is the numerical analysis approach utilizing computer-based simulation. It provides detailed insight into the large and complex fluid flow, heat transfer, and associated phenomena like chemical reactions or particle spreading with the assistance of high-speed computers. This method plays a vital role in numerous fields in engineering and spans a broad range of application areas (both in industrial and non-industrial) including aerodynamics and aerospace analysis, weather simulation, civil and environmental engineering, industrial system design and analysis, biological engineering, engine, and combustion analysis, and optimize airflow pattern, ventilation in hospital or facilities.

The results of CFD analyses can help reduce the effort or cost required for experimentation, conceptual studies of latest designs, detailed development, troubleshooting, or redesign.

CFD modelling has the great potential to provide detailed insight into the airborne dispersion and therefore, the effectiveness of disinfection strategies. Many studies focus on hospital ventilation design and applications included pathogen dispersion and assessing air disinfection device performance within single-bed space, multi-beds space, operating room, clinical treatment room, and isolation room [13–22].

In this work, the CFD analyses of the pathogen dispersion and assessing air disinfection device performance within a double-bed hospital room by applying the finite volume method with the assistance of ANSYS CFX Solver. The fluid domain is decomposed into a finite set of control volumes. The conservation equations for mass, momentum, energy, species are solved on this set of finite volumes. The governing equations of the flow of a Newtonian fluid are continuity equation, momentum equations, energy equation, and equation of state. These continuous partial differential equations are discretized into a system of linear algebraic equations which can be solved by ANSYS software. The discretized solutions for a CFD problem velocity, pressure, temperature, volume fraction, etc., that are distributed on the model.

To simulate the behavior of flow, CFD solver is set up to find the converged solution of the system of governing equations such as the continuity, momentum conservation and turbulence equations [23]. In this study, the standard k-ε model is used for the turbulence flow because it offers a good compromise between numerical effort and computational accuracy [24]. The governing equations are listed below:

- continuity equation:

$$\frac{\partial \rho}{\partial t} + \nabla \cdot (\rho \mathbf{v}) = 0, \tag{1}$$

where ρ is the density and $\boldsymbol{v}$ is the velocity vector,

- momentum conservation equations:

$$\frac{\partial(\rho \mathbf{v})}{\partial t} + \nabla \cdot (\rho \mathbf{v} \otimes \mathbf{v}) = \rho \mathbf{g} - \nabla p + \mu \nabla^2 \mathbf{v}, \tag{2}$$

where p denotes the pressure, $\boldsymbol{v}$ is the velocity vector, $\boldsymbol{g}$ stands for the accelerate vector and μ is the dynamic viscosity.

2.2 *Airborne Dispersion in Double-Bed Hospital Room Equipped with UVC Air Disinfection Devices*

The procedure for a CFD simulation is shown in Fig. 1. There are four main steps needed for a CFD problem, including problem identification, pre-processing, solving, and post-processing.

The simulated double-bed hospital room has roughly 40.0 m^2 floor space (5.0 m × 8.0 m) and 2.6 m ceiling height. The double-bed hospital room in our model contains two patients, two beds, three small wardrobes, table for medical equipment. There is no inlet and exhaust, as the room was considered as a closed and isolated space since the typical hospital room in low and middle-income countries rarely equipped with the central ventilation system. This room is equipped with the UVC air disinfection device, and the models are shown in Fig. 2. There are three models of double-bed hospital room chosen for simulating in this paper (see Fig. 2). In the first model (see

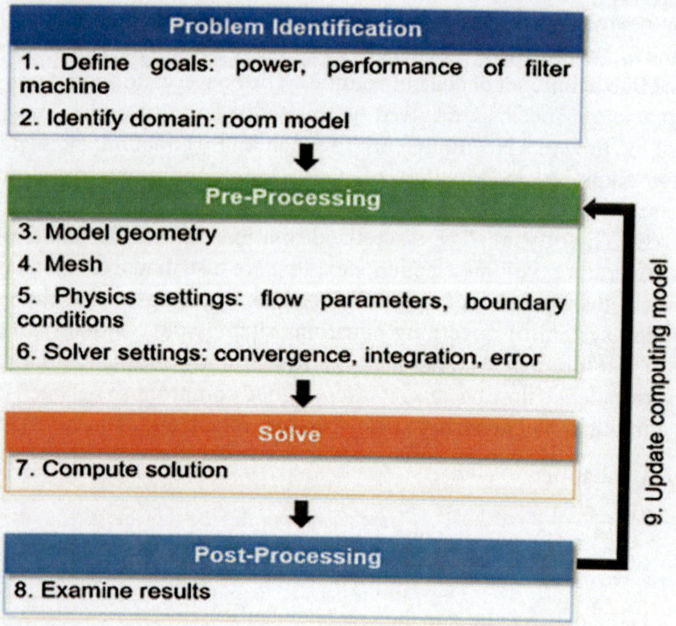

Fig. 1 CFD simulation procedure

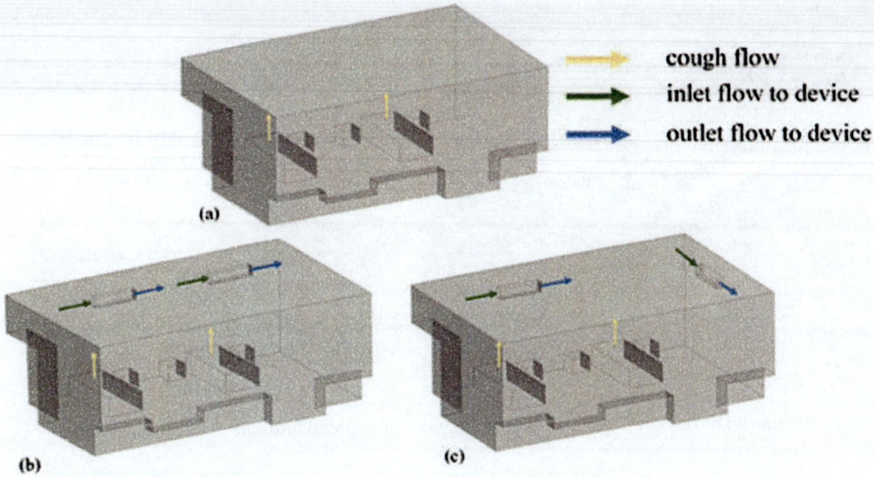

Fig. 2 3D models of the double-bed hospital room

Fig. 2a), there is no UVC air disinfection device in the room. In the second model (see Fig. 2b), there are two UVC air disinfection devices are set-up on the same wall and the top of the beds. Finally, in the last model (see Fig. 2c), each disinfection device is placed on different walls.

The simulation was conducted with the assumption that two persons had a heavy cough event continuously 04 times in 4.0 s. Coughing is a transient phenomenon with the characteristics obtained from "Flow dynamics and characterization of a cough" [25, 26]. The general airflow pattern of the room could be assumed that the coughing events do not affect because the air-flow rate from coughing and its momentum were negligible compared to the airflow rate inside the room while using disinfection device.

The cough flow rate profile with the highest maximum value (8.0 l/s) is chosen for this simulation; the flow direction is shown in Fig. 2, indicated by the yellow arrows. The total simulation time is 15.0 s. In order to improve the air quality by filtering dust and killing the airborne pathogens, the room was equipped with a novel UVC air disinfection device (1.20 m × 0.27 m × 0.22 m) designed by our group. Two UVC 254 nm lamps (G15T8OF 55 V 20/CS 1/SKU) with the total UV emission power equal to 16.0 W, placed at the exhaust gate of the UVC air disinfection device. The dose in the middle of the airflow approximately 2000 μW s/cm^2. The power of the UVC air disinfection device is 80.0 W. The inlet and outlet flows of the device are shown in Fig. 2, indicated by the green and blue arrows, respectively. It is assumed that these flows have the same velocity $v = 1.0$ m/s.

The room air was recirculated with a fan and air filtration system. The airflow is 240.0 m^3/h, and the speed is 2600 rpm. The ventilation rates studied were 4 Air-Change-Per-Hour (ACH) to 6 ACH (0.057–0.085 m^3/s), according to ASHRAE Handbook 2019 [27].

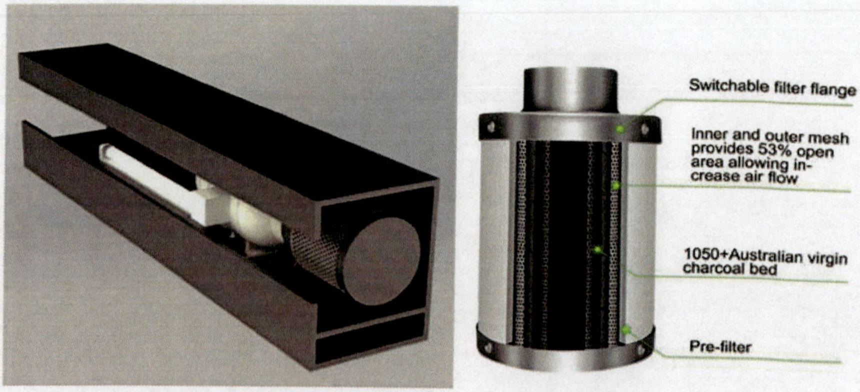

Fig. 3 3D model of the UVC air disinfection device (associated with air filter)

Figure 3 shows the 3D model of the UVC air disinfection device and the filtration module. The airborne pathogen usually attached to dust particles or water molecules [28]. Most of the airborne pathogens in hospitals could be isolated by 90–95% using air standard filters [29]. However, the viruses with their sizes less than 1.0 micron required a better filter to isolate. Hence, the high-efficiency particulate air (HEPA) filter could be added to improve air filtration reducing viral and bacterial particles for healthcare facilities [30–37]. HEPA type filter could provide higher filtering efficiency and usually recommended. On the other hand, due to their expensive cost, their use should be kept at a minimum.

3 Results and Discussion

In this paper, we report three models for the airborne pathogen spread tracing from two airborne infection sources in a double-bed hospital room. Figure 4 visualizes the distribution of volume fraction of airborne pathogen spreading in the room with three time-steps: 5.0, 10.0, and 15.0 s.

In case (a) as shown in Fig. 4a, without the UVC air disinfection device, the airborne pathogens spread in the room and they cannot run out because of the closed space. With time, the room will have filled with airborne pathogens. It can be harmful to other healthy people in the room.

In case (b) (see in Fig. 4b) and case (c) (see in Fig. 4c), the airborne pathogens were filtered out and the volume fraction of contaminated air is reduced effectively. However, as shown in simulation results, a small number of airborne pathogens could be presented on the surface of the furniture in the corner. It is recommended to clean and sterilizer the surface more often.

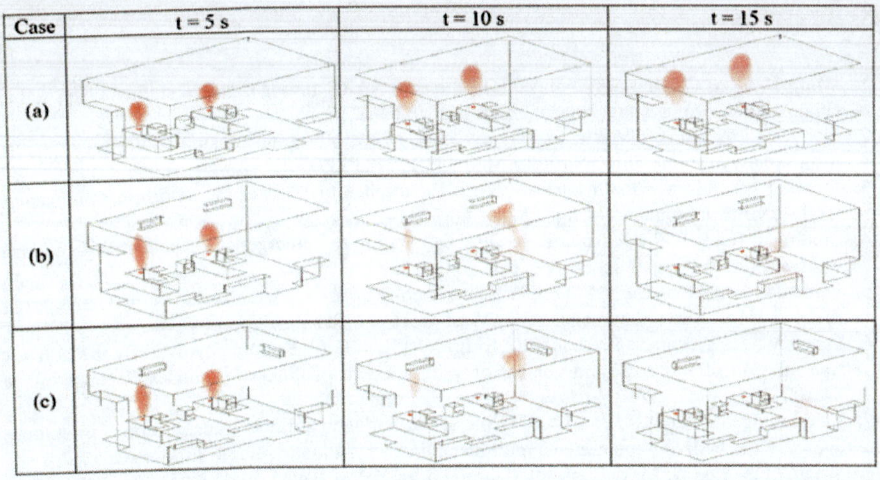

Fig. 4 Distribution of volume fraction of airborne pathogens spreading (red color)

4 Conclusion

With the view towards the realization of an air disinfection system to reduce the airborne pathogen dispersion, a novel UVC air disinfection device was proposed and designed. The computational fluid dynamics (CFD) method was used to simulate the spreading of airborne pathogens in a double-bed hospital room. The diffusion patterns, according to the location of the proposed device, were analyzed to find out the optimized the device's location. The analysis of transient simulation results concerning the airborne pathogens spreading with their concentration provided evidence of the contaminant zones that should be checked and cleaned. Through the results of this research, the feasibility and the effectiveness of the proposed device was confirmed by the CFD simulation.

It is recommended that we should do these simulations using the CFD method to understand and optimize the hospital design. The ventilation system and the furniture location should be carefully considered.

In order to improve the air quality by filtering dust and killing the airborne pathogens, the results provide an overview of evaluating the effectiveness of the proposed air disinfection device and various alternatives for optimizing the disinfection device placement in a double-bed hospital room. The effectiveness validation of the proposed device and model in the experiment is our next step.

Acknowledgements The authors acknowledge the support of time and facilities from Ho Chi Minh City University of Technology (HCMUT), Vietnam National University Ho Chi Minh City (VNUHCM) for this study.

Conflicts of Interest The authors have no conflict of interest to declare.

References

1. Wong TW, Lee CK, Tam W et al (2004) Cluster of SARS among medical students exposed to single patient. Hong Kong. Emerg Infect Dis 10:269–276
2. Yu ITS, Li Y, Wong TW et al (2004) Evidence of airborne transmission of the severe acute respiratory syndrome virus. N Engl J Med 350:1731–1739
3. Dawood FS, Jain S, Finelli L et al (2009) Emergence of a novel swine-origin influenza A (H1N1) virus in humans. N Engl J Med 360:2605–2615
4. Smith GJ, Bahl J, Vijaykrishna D et al (2009) Dating the emergence of pandemic influenza viruses. Proc Natl Acad Sci USA 106:11709–11712
5. Peiris JS, Poon LL, Guan Y (2009) Emergence of a novel swine-origin influenza A virus (S-OIV) H1N1 virus in humans. J Clin Virol 45:169–173
6. Tang JW, Li Y, Eames I, Chan PK, Ridgway GL (2006) Factors involved in the aerosol transmission of infection and control of ventilation in healthcare premises. J Hosp Infect 64:100–114
7. Li Y, Leung GM, Tang JW et al (2007) Role of ventilation in airborne transmission of infectious agents in the built environment e a multidisciplinary systematic review. Indoor Air 17:2–18
8. Wong BCK, Lee N, Li Y, Chan PKS, Qiu H, Luo Z, Lai RWM, Ngai KLK, Hui DSC, Choi KW, Yu ITS (2010) Possible role of aerosol transmission in a hospital outbreak of influenza. Clin Infect Dis 51:1176–1183
9. Szeto W, Yam WC, Huang H, Leung DYC. (2020) The efficacy of vacuum-ultraviolet light disinfection of some common environmental pathogens. BMC Infect. Diseases 20, Article number: 127
10. Eisenlöffel L, Reutter T, Horn M, Schlegel S, Truyen U, Speck S (2019) Impact of UVC-sustained recirculating air filtration on airborne bacteria and dust in a pig facility. PLoS ONE 14(11):e0225047
11. Do-Kyun K, Dong-Hyun K (2018) UVC LED irradiation effectively inactivates aerosolized viruses, bacteria, and fungi in a chamber-type air disinfection system. Appl Environ Microbiol 84(17):e00944
12. Escombe AR, Moore DAJ, Gilman RH, Navincopa M, Ticona E (2009) Upper-room ultraviolet light and negative air ionization to prevent tuberculosis transmission. PLoS Med 6(3)
13. Hathway EA, Noakes CJ, Sleigh PA, Fletcher LA (2011) CFD simulation of airborne pathogen transport due to human activities. Build Environ 46(12):2500–2511
14. Noakes CJ, Sleigh PA, Escombe AR, Beggs CB (2006) Use of CFD analysis in modifying a TB ward in Lima, Peru. Indoor Built Environ 15(1):41e7
15. Li Y, Huang X, Yu ITS, Wong TW, Qian H (2005) Role of air distribution in SARS transmission during the largest nosocomial outbreak in Hong Kong. Indoor Air 15(2):83e95
16. Beggs CB, Kerr KG, Noakes CJ, Hathway EA, Sleigh PA (2008) The ventilation of multiple-bed hospital wards: review and analysis. Am J Infect Control 36(4):250e9
17. Chao CYH, Wan MP (2006) A study of the dispersion of expiratory aerosols in unidirectional downward and ceiling-return type airflows using a multiphase approach. Indoor Air 16:296e312
18. Chow TT, Yang XY (2003) Performance of ventilation system in non-standard operating room. Build Environ 38(12):1401e11
19. Richmond BJ (2009) Transport of exhaled particulate matter in airborne infection isolation rooms. Build Environ 44(1):44e55
20. Kao PH, Yang RJ (2006) Virus diffusion in isolation rooms. J Hosp Infect 62(3):338e45
21. Yu ITS, Li YG, Wong TW, Tam W, Chan AT, Lee JHW et al (2004) Evidence of airborne transmission of the severe acute respiratory syndrome virus. New Engl J Med 350(17):1731e9
22. Noakes CJ, Sleigh PA, Fletcher LA, Beggs CB (2006) Use of CFD modelling to optimize the design of upper-room UVGI disinfection systems for ventilated rooms. Indoor Built Environ 15(4):347e56
23. Ferziger JH, Peric M (2002) Computational methods for fluid dynamics, 3rd ed. Springer, Berlin

24. Chung TJ (2002) Computational fluid dynamics. Cambridge University Press, Cambridge
25. Gupta JK, Lin CH, Chen Q (2009) Flow dynamics and characterization of a cough. Indoor Air 19:517–525
26. Kermani A (2015) CFD modeling for ventilation system of a hospital room. In: Proceedings of the COMSOL conference 2015, 6p. Available online: https://www.comsol.com/paper/download/257311/kermani_paper.pdf. Last accessed 11 May 2020
27. ASHRAE Handbook 2019, available online: https://www.ashrae.org/file%20library/technical%20resources/covid-19/si_a19_ch09healthcarefacilities.pdf. Last accessed 11 May 2020
28. Dascalaki EG, Lagoudi A, Balaras CA, Gaglia AG (2008) Air quality in hospital operating rooms. Build Environ 43:1945e52
29. Balaras CA. Dascalaki E, Gaglia A (2007) HVAC and indoor thermal conditions in hospital operating rooms. Energy Build 39:454e70
30. CDC (2005) Guidelines for preventing the transmission of Mycobacterium tuberculosis in health-care settings. U.S. Department of Health and Human Services Centers for Disease Control and Prevention (CDC), USA
31. Ulrich R, Zimring C, Quan X, Joseph A, Choudhary R (2004) The role of the physical environment in the hospital of the 21st century: a once-in-a-lifetime opportunity. The Center for Health Design for the Designing the 21st Century Hospital Project
32. Qian H, Li Y, Nielsen PV, Hyldgaard CE, Wong TW, Chwang ATY (2006) Dispersion of exhaled droplet nuclei in a two-bed hospital ward with three different ventilation system. Indoor Air 16:111e28
33. Memarzadeh F, Manning A (2000) Thermal comfort, uniformity, and ventilation effectiveness in patient rooms: performance assessment using ventilation indices. ASHRAE
34. Qian H, Lia Y, Nielsen PV, Hyldgaard CE (2008) Dispersion of exhalation pollutants in a two-bed hospital ward with a downward ventilation system. Build Environ 43:344e54
35. Atkinson J, Chartier Y, Pessoa-Silva C, Jensen P, Li Y, Seto WH (2009) Natural ventilation for infection control in health-care settings e WHO guidelines. World Health Organization
36. Escombe AR, Oeser CC, Gilman RH, Navincopa M, Ticona E (2007) Natural ventilation for the prevention of airborne contagion. PLoS Med 4(2)
37. Qian H, Li Y, Seto WH, Ching P, Ching WH, Sun HQ (2010) Natural ventilation for reducing airborne infection in hospitals. Build Environ 45:559e65

The Effect of Mindfulness-Based Stress Reduction Among Various Subject Groups: A Literature Review

Anh Hoang Minh An, Nghia Trung Nguyen, Thuong Hoai Nguyen, Khiet Thi Thu Dang, and Huong Thanh Thi Ha

Abstract Mindfulness-Based Stress Reduction (MBSR) is a non-invasive psychological therapy based on mindfulness developed by Dr. Jon Kabat-Zinn in 1979. The potential of MBSR includes improving mental health, enhancing working performance, and reducing mental stress. This literature review aims to: evaluate the effect of MBSR on stress reduction in different subject groups and examine the outcome of MBSR on various physiological and structural measurements. The literature review is based on 40 articles then narrowed down to 25 articles published from 2010 to 2019. The keywords to search for these articles were: MBSR, Electroencephalogram signal, stress, functional Magnetic Resonance Imaging, cortisol, functional Near-Infrared Spectroscopy. Subjects participated in the MBSR course. These studied populations, selected based on standard criteria DSM-IV, included one control group and one intervention group. Measurement stage comprises pre- and post-intervention by using three types of questionnaires: Mental Health Scale (MHS) and Mindfulness Scale and other scales. Besides, several studies combined common physiological and structural assessments. The reviewed studies show that MBSR is effective in several subject groups, including Social Anxiety Disorder patients, employees, and students. Other groups with more severe psychological symptoms have not benefited significantly. The questionnaire score reflects a significant improvement, whereas biological measurements as heart rate, brain imaging, or cortisol level do not show consistent change. The limitation of these studies is that there is a lack of a control group. In conclusion, MBSR has effects on mental health and mindfulness for a wide range of subjects. It is suggested that people who have mild psychological disorders should practice MBSR frequently to reduce symptoms. For future studies, the control group and repeated measure designs should be implemented for better results.

A. An · N. Nguyen · K. Dang · H. Ha (✉)
School of Biomedical Engineering, International University, Ho Chi Minh, Vietnam
e-mail: htthuong@hcmiu.edu.vn

Vietnam National University-Ho Chi Minh (VNU-HCM), Ho Chi Minh, Vietnam

N. Nguyen
University of Medicine and Pharmacy, Ho Chi Minh, Vietnam

V. Van Toi et al. (eds.), *8th International Conference on the Development of Biomedical Engineering in Vietnam*, IFMBE Proceedings 85,
https://doi.org/10.1007/978-3-030-75506-5_71

Keywords MBSR · Stress · Social anxiety disorder · EEG · Physiological measurement

1 Introduction

Mindfulness is an essential human ability that is intentionally paying attention to the current experiences in life, such as emotions, awareness, affective states, imagination, sensations, and then developing a stable and non-reactive awareness [1]. Mindfulness is one of the eight qualities cultivated in Buddhism to reduce mental disorders and promote well-being. Training exercises of mindfulness are designed to train the regulation of awareness via multiple forms of meditation: sitting, walking, lying down, and moving. The purposes of these training include improving the symptoms of mental health, enhancing working performance, reducing psychiatric symptoms, and distress. Because of an increasing number of people facing daily pressure and the rising demand in mental health services, nowadays, mindfulness becomes popular in society, and many studies are demonstrating the usefulness of mindfulness in stress reduction [2–5].

Mindfulness-Based Stress Reduction is an eight-week program that offers mainstream, intensive mindfulness coaching. The manualized MBSR program was developed by Dr. Kabat-Zinn, which based on the Mindfulness-Based program. MBSR instruction has three formal techniques: mindfulness meditation, body scanning, and simple yoga postures [6]. The course includes eight-week practice and one retreat day, sitting mindfulness in seven hours. The benefits of MBSR are: coping with anxiety, stress in short and long-term situations; increasing mindfulness; promoting well-being, and health-related quality of life [5]. Besides, it has shown to be useful for a variety of conditions and support to medical treatments, including chronic pain, hot flashes, asthma, fibromyalgia [7, 8]. Moreover, MBSR has positive effects on persisting well beyond the initial treatment period for promoting cancer recovery [9].

The goals of this literature review include to examine the effect of MBSR on stress reduction in different groups of subjects, and to evaluate the outcome of MBSR on various physiological and structural measurements whose results demonstrate the highest stress-reducing and mindfulness-improving effects.

2 Methodology

2.1 *MBSR Procedure*

Mindfulness-Based Stress Reduction program was developed by Kabat-Zinn in 1979, which consists of mainly an eight-week session [6]. Participants perform the lesson between 1.5 and 2.5 h a day per week and one meditation retreat day lasting six to seven hours. Subjects practice three formal practices: mindfulness meditation, breathing mindfulness, and yoga postures. To guarantee the effect of MBSR, specialists would ask participants to practice at home by providing a CD or online file of recorded lessons. Home practice includes formal practice and informal practice, which integrated into their daily life: eating, sitting or standing mindfulness; reading-related journals; and concentrating on their emotion and awareness. MBSR specialists and coaches will guide the whole MBSR practice. Participants are chosen based on the purpose of research and randomized into two different types of groups: an MBSR-intervention group and a waitlist control group, which means a non-MBSR- intervention group. In the intervention groups, studies divided subjects into small groups, which include from 8 to 16 participants who meet the criteria of such research, commit to following the study, and guarantee to attend the MBSR program fully. With a non-MBSR-intervention group, researches use this group to compare the effect of MBSR at the pre-intervention and post-intervention time points. Besides, many studies use other forms of meditation or intervention such as Yoga, Aerobic, Cognitive-Behavioral Therapy on the control group to compare with the MBSR group [10–12].

2.2 *Subject Selection Criteria*

Included Criteria

Almost all previous studies base on the Diagnostic and Statistical Manual of Mental Disorder IV (DSM-IV) as a standard criteria selection. There is no restriction on age, gender, ethnicity, and education in choosing subjects. Subjects volunteered to participate in MBSR research, and they attended to MBSR session without payment. After selection, they were asked to sign the informed consent, which guaranteed their benefits and their understanding of the purpose and potential risks of research. In several groups of subjects, there were specific included criteria, which will be mentioned in Table 1.

Excluded Criteria

Depending on the populations in the studies, researchers employed a specific set of exclusion criteria. These studies excluded participants who currently had pharmacotherapy or psychotherapy, a history of head trauma, generalized anxiety disorder, major depressive disorder, obsessive–compulsive disorder, agoraphobia without a history of panic attacks or specific phobia, mental disorders. Participants have endocrine disorders, history of using a psychotropic drug or substance abuse are also excluded [2, 11–15]. They also excluded participants having cardiovascular disorders, thought disorders, suicide thoughts, or bipolar disorder [14]. These studies also did not enroll participants whole previous participated in the MBSR course, long-term meditation retreats, practised meditation regularly, received intervention therapy such as Daily Yoga, Taichi, Qigong [14, 16].

Table 1 Summary results from previous studies of MBSR among various groups of subjects

Study	Type of subjects	Sample size (meditators/control)	Included criteria	Procedure description	Methodology	Result
Mental health disorders						
Bremner et al. [1]	PTSD male veteran	9/0 *Parallel group: Present Centered Group Therapy (n = 8)	Had no experienced significant head trauma or loss of consciousness for at least two minutes or the diagnosis of PTSD preceding military service	Standard MBSR program	Questionnaire: FFMQ for mindfulness assessment, the Clinician-Administered PTSD Scale (CAPS) for measuring PTSD symptom and PET for brain imaging	Comparing to pre- and post-intervention, the MBSR group decreased PTSD symptom significantly than PCGT ($p = 0.016$) and increased mindfulness ($p = 0.04$) Brain imaging was collected at six-month follow-up measurement, MBSR group increased anterior cingulate and inferior parietal lobule and decreased insula and precuneus function in response to traumatic reminders
Dessa et al. [18]	PTSD veteran	21/0 *Parallel group: Primary Brief care Mindfulness program (n = 21)	Met no more criteria subthreshold PTSD or diagnostic-level PTSD related to military service, as measured by the CAPS	A Primary Brief care Mindfulness Program (PCbMP), based on the MBSR standard program, involved four weeks of practice	Questionnaire: data were collected at baseline, week four, and week 12 Cortisol level: samples were collected at baseline and four-week follow-up assessment. Five saliva samples each day at specified times for two consecutive days for a total of ten samples	The PCbMP showed significant changes in cortisol awakening response ($p < 0.05$) compared to the usual treatment group
Wolmer et al. [44]	PTSD students studying in fourth and fifth grades	782 / 666		Participants implemented a standard MBSR program three months before the traumatic exposure by the MBSR group	Questionnaire: UCLA-PTSD Reaction Index and the Stress/Mood Scale three months after the end of rocket attacks	The intervention group showed significantly lower symptoms of post-trauma and stress/mood than the control group ($p < 0\ 0.001$)

(continued)

Table 1 (continued)

Study	Type of subjects	Sample size (meditators/control)	Included criteria	Procedure description	Methodology	Result
Goldin et al. [14]	Generalized SAD patients	36/36 *Parallel group: Cognitive Behavioral Therapy (n = 36)	The SAD patients endorsed greater than moderate social fear in five or more distinct social situations assessed by the ADIS-IV-L, and they had to achieve a score higher than 60 on the Liebowitz Social Anxiety Scale – Self-Report (LSAS-SR)	Both two intervention programs included 12 weeks of treatment. MBSR group practised standard MBSR programs	Measurement was assessed based on Diagnostic interviews using the ADIS-IV-L at baseline. Patients completed the weekly assessment online, which designed suitably to the study, at baseline, during treatment, at post-treatment completion, and every three months during a one-year follow-up	By comparison between baseline and post-intervention, the CBT group (43%) decreased in social anxiety than the MBSR group (40%). According to disputing (challenging anxious thought and feeling), CBT (32%) increased significantly than MBSR (18%). However, in mindfulness improvement, the MBSR group (53%) increased considerably than CBT (49%). By contrast, the waitlist control group did not show any changes
Goldin et al. [11]	SAD adult patients	31/0 *Parallel group: Aerobic (n = 25) (RCT)		Standard MBSR program	fMRI scanning at pre- and post-intervention in both groups	BOLD signal response in post-MBSR decreased significantly percent reduction in right posterior superior temporal gyrus (STG), and increased in left IPL, right anterior IPL, right posterior IPL and right SPL ($p = 0.05$, $p = 0.02$, respectively baseline and post-treatment)

(continued)

Table 1 (continued)

Study	Type of subjects	Sample size (meditators/control)	Included criteria	Procedure description	Methodology	Result
Goldin and Gross [46]	SAD adult patients	14/0	Patients had to pass an MRI scanned safety screen on three occasions	Standard MBSR program	They did clinical symptom self-reports and the emotion regulation task, examined two forms of attention deployment: breath-focused attention (the target regulation strategy), distraction-focused attention (a control regulation strategy), and fMRI scanning	Compared with baseline, post-measurement showed greater BOLD responses during breath-focused attention versus reactivity to negative self-beliefs in visual attention brain regions, including medial cuneus ($p = 0.03$), left cuneus ($p = 0.04$), and right middle occipital gyrus ($p = 0.01$), the more reduction of negative emotion. Additionally, according to self-report in clinical symptoms, participants improved stress, anxiety, and self-esteem ($p < 0.005$)
Faucher et al. [12]	Non-medication SAD adult patient	16 (SAD)/30 (Healthy Group) *Parallel group: Cognitive Behavioral Group Therapy (SAD n = 20)	No comorbid disorder	Standard MBSR program	Patients performed a speech task before and after treatment, while HC performed it only once Participants were asked to collect physiological responses that included salivary cortisol (collected five minutes before the speech task, and 15, 30, 45, and 60 min thereafter) and heart rate variability (recorded at baseline and throughout the speech task)	Reduction in anxiety was found from pre- to post-treatment with the mindfulness interventions, but the CBGT group improved significantly Cortisol level in post-MBSR was lower than post-CBGT, but it was not significant comparing to baseline ($p = 0.051$, $p = 0.440$, respectively) There had a significant difference between intervention groups and HC group in heart rate low-frequency and high-frequency ratio (LF/HF ratio) ($p = 0.010$) due to differences between MBSR and HC values ($z = -3.072$, $p = 0.003$) before treatment. During the speech task, MBSR reactivity was lower while the ratio in both HC and CBGT groups increased

(continued)

Table 1 (continued)

Study	Type of subjects	Sample size (meditators/control)	Included criteria	Procedure description	Methodology	Result
McIndoo et al. [30]	Depressed college students	18/12 (Randomized) *Parallel group: Behavioral Activation (n = 14)	Students had to report current depression by Beck Depression Inventory-II (BDI-II), which reported at least mild depression (>= 14) and had no current psychosis and substance abuse	Mindfulness-Based Therapy is relevant MBSR program. It involved four-weekly one-hour lessons and three main techniques: Body scan, mindfulness meditation-express feeling and mindfulness check-in	Questionnaire: – At post-treatment, all participants finished self-report measures: the ADIS-IV depression module and the Hamilton rating scale for depression (HRSD). Patients in two treatment groups completed the client satisfaction questionnaire (CSQ-8), FFMQ, Beck Anxiety Inventory (BAI), PSS, Ruminative response scale (RRS) – At the one-month follow-up measurement, they completed all questionnaires except the CSQ-8	At post-treatment, the majority of participants in the MBT group showed significant improvement in depression, stress, and mindfulness (87%) while BA group (71%) At a one-month follow-up, 73% of participants in MBT experienced clinically reduction in depression
Medical conditions						
Matousek et al. [13]	Breast cancer women finished medical treatment	33/0	Had no endocrine disorder and psychiatric disorder	Standard MBSR program	Questionnaire: Center for Epidemiologic Studies Depression Scale (CES-D), Medical Symptom Checklist (MSCL), and PSS Cortisol level was collected at three days -pre and three days post-treatment at 0, 30, 45 min after awakening in the morning	Outcome measures showed significant changes in pre- and post-intervention ($p = 0.0167$). Cortisol level indicated that stress levels, depressive symptomatology, and medical symptoms improved Cortisol levels showed a prolonged increase after awakening at the post-MBSR evaluating period, especially, cortisol level substantially increased at 45 min after awakening at the second session ($p = 0.012$)

(continued)

Table 1 (continued)

Study	Type of subjects	Sample size (meditators/control)	Included criteria	Procedure description	Methodology	Result
Andersen et al. [45]	Breast Cancer women underwent surgery for stage I–III in three to eighteen months previously	168/168 (RCT)		Standard MBSR program	Patients were required to complete "The validated Medical Outcome Study sleep scale" for measuring sleep quality and "SCL-90-R" to assess psychological distress four times: before and after the intervention, six-months, and 12-months follow-up	MBSR had a small effect on sleep quality over follow-up (all p values > 0.05) and had no changes in distress When compared with the control group, the MBSR group showed differences between baseline and post-intervention ($p = 0.05$), and at baseline and the six-month follow-up ($p = 0.02$) in sleep quality improvement
Caroline et al. [41]	Women with breast cancer in stage 0 to III	114/115	Women were diagnosed with stage 0 to III breast cancer within two months to two years after the completion of the surgery, chemotherapy, and radiotherapy, from 18 - 80 years old, and were able to complete questionnaires	Standard MBSR program included formal and informal practice at home with a two-week follow-up	Primary outcomes included the Profile of Mood States (POMS). Secondary outcomes included the Functional Assessment of Cancer Therapy-Breast (FACT-B) scale, Functional Assessment of Cancer Therapy-Endocrine Symptoms (FACT-ES) scale, FACT PWB, FACT SWB, FACT EWB, FACT FWB, WHO-5	Primary outcomes (POMS) had statistically significant differences between treatment groups. Secondary outcomes included the higher mean scores in the experimental group compared with the control group. Increased hours of practice in the experimental group were associated with improved scores in POMS and FACT-ES

(continued)

Table 1 (continued)

Study	Type of subjects	Sample size (meditators/control)	Included criteria	Procedure description	Methodology	Result
Gross et al. [40]	Adults with primary chronic insomnia	20/0 * Parallel group: Pharmacotherapy (PCT) (n = 10)	People were diagnosed with primary chronic insomnia without prescription, from 18 to 65 years old, and able to read and speak English	Standard MBSR program, included formal and informal practice at home with a five-month follow-up and pharmacotherapy (PCT)	Using actigraphy, sleep diary and questionnaires, including Standardized Sleep Scales (Insomnia Severity Index (ISI), Pittsburgh Sleep Quality Index (PSQI), Sleep Self - Efficacy Scale, Dysfunctional Beliefs and Attitudes about Sleep (DBAS)); Health-Related Quality of Life Measures (Activity impairment, State-Trait Anxiety Inventory (STAI), Center for Epidemiologic Studies Depression Scale, Short Form-12)	Sleep diary showed significant improvements in MBSR arm in a few measurements ($p < 0.01$). MBSR may impact SOL more than PCT Actigraphy Outcomes had a statistically significant in MBSR arm ($p = 0.04$) The questionnaire reported statistically significant improvements in MBSR arm in standardized sleep scales and activity impairment. Besides, DBAS scores were significantly associated with home practice ($p < 0.02$) One half the MBSR subjects had recovered, based on the ISI and PSQI. The recovery rates from poor sleep were considerably higher following MBSR than PCT at eight weeks and five months

(continued)

Table 1 (continued)

Study	Type of subjects	Sample size (meditators/control)	Included criteria	Procedure description	Methodology	Result
Adults						
Hölzela et al. [34]	Healthy Adults	16/17 (Controlled)	No contraindications for MRI scanning (i.e., metallic implants, claustrophobia) No Physically and psychological issues	Standard MBSR program, included formal and informal practice at home	Questionnaire: FFMQ MRI: MBSR participants were scanned during the two weeks before (Pre) and after (Post) participation in the program. Control participants were also scanned twice, approximately two months apart	In a questionnaire, MBSR changed significantly (In detail, awareness: $p = 0.003$; observing: $p = 0.001$; non-judging: $p = 0.003$), whereas the control group did not According to MRI result, the concentration of gray matter in the left hippocampus of the MBSR group increased gray matter concentration ($p = 0.014$), while the control group did not show a change from the Pre to Post time-point ($p = 0.736$) According to whole a brain imaging, MBSR group increasing gray matter concentration were significantly greater than the control group, For specific, Posterior cingulate gyrus ($p =$ 0.004), Cerebellum lobule ($p =$ 0.018), Cerebellum, Vermis, Lobule ($p = 0.001$), Temporo-parietal junction which peak in middle temporal gyrus: (p = 0.036)

(continued)

Table 1 (continued)

Study	Type of subjects	Sample size (meditators/control)	Included criteria	Procedure description	Methodology	Result
MacCoon et al. [42]	Healthy Adults	31/0 * Parallel group: Health Enhancement Program (HEP) (n = 32)	People from 18–65 years old, weighs under 300 lb, meets MRI safety standards, right-handed, no previous experience with meditation or other mind–body techniques, good general health, able to walk, understand and speak English, provide written consent before admission, see without glasses	Standard MBSR program and Health Enhancement Program (HEP), include formal and informal practice at home with a four-month follow-up	Thermal pain generated by a thermal stimulator used to establish a participant's pain threshold. There were 32 trials; after each trial, participants had to complete two Likert scales (measuring intensity and unpleasantness) Participant-report Outcomes (PROs) included SCL-90-R, Global Severity Index (GSI, overall psychological distress), Medical Symptoms Checklist (MSC), Experience Check Questionnaire (ECQ). Besides, participants needed to record time practice at home	Thermal pain ratings showed that mindfulness condition decreased pain ratings at T2 ($p < 0.001$) and at T3 ($p = 0.01$) compared to T1. Besides, MBSR had lower thermal pain ratings compared to HEP PROs included GSI (significant improvement in MBSR groups ($p = 0.04$)); SCL-90-R (showed increasing depressive symptoms in MBSR participants between T2 and T3 ($p = 0.002$), decreasing depressive symptoms in HEP participants between T2 and T3 ($p = 0.002$), decreasing anxious symptoms over time ($p = 0.02$)); MSC (reduced medical symptoms over time ($p = 0.001$)). Therefore, MBSR is effective but no more effective than other active interventions in HEP

(continued)

Table 1 (continued)

Study	Type of subjects	Sample size (meditators/control)	Included criteria	Procedure description	Methodology	Result
Rosenkranz et al. [43]	Healthy Adults	No clear	Age: 19–59 years old	Standard MBSR program and Health Enhancement Program (HEP), include formal and informal practice at home with a four-month follow-up	Using Trier Social Stress Test (TSST), a neurogenic inflammatory response which collected cytokine (tumor necrosis factor-alpha (TNF-alpha) and interleukin (IL)-8) and questionnaire SCL-90-R, Medical Symptoms Checklist (MSC), Global Severity Index (GSI)). Besides, flare size and cortisol were also collected	Flare size showed an increase from T1 to T2. MBSR show no significant ($p = 0.09$), while HEP showed a remarkable increase; the flare size of MBSR at T2 was substantially smaller than HEP ($p < 0.001$). Besides, flare size showed a significant increase from T1 to T3 ($p < 0.001$) in both groups Questionnaire had scores decreased in GSI ($p = 0.007$) and MSC ($p = 0.016$) Daily cortisol release (AUCg) showed a considerable relationship with practice. In contrast, diurnal cortisol rhythm experienced the trend toward a steeper slope in the MBSR group and less steep slope in the HEP group About neurogenic inflammatory response, the measurement showed that more considerable practice in MBSR, smaller TNF-alpha ($p = 0.028$), while greater practice in HEP, greater TNF-alpha. Besides, levels of both cytokines were lower at T3 than T1 ($p < 0.001$)

(continued)

Table 1 (continued)

Study	Type of subjects	Sample size (meditators/control)	Included criteria	Procedure description	Methodology	Result
Gundel et al. [3]	Meditator experts	14/16	No evidence of a current mental disorder	Mindfulness paradigm comprised four baseline blocks (two min each) and three mindfulness periods (three min each), which were conducted in an alternating fashion. A meditation bowl was presented with one new strike every 12 s to find group differences in the auditory system and adjacent cortical areas	NIRs: data individually averaged according to the task condition (rest periods and mindfulness). In the first 60 s of all mindful periods were discarded for practitioners to reach a meditative state, resulting in a 120 s average for each participant and condition	The control group decreased activation in higher auditory areas, whereas the meditation experts had a remarkable increase in those areas In mindfulness state, Intervention group increased significantly activation in areas of these areas than control group: primary somatosensory cortex([L] $p =$ 0.024), supramarginal gyrus ([R] $p = 0.002$, [L] $p = 0.085$) and Broca's area ([R] $p = 0.056$, [L] $p = 0.015$)
Van der Gucht et al. [25]	Low-income urban adults who were receiving social welfare benefits	42/0		The Mindfulness-Based Intervention was designed based on standard MBSR and MBCT manual	Participants were required to complete four types of questionnaires: DASS 21, Children's Acceptance and Mindfulness Measure (CAMM) to assess mindfulness skill; The revised Leiden Index of Depression Sensitivity (LEIDS-R) and The Attitudes Toward Self Revised (ATS-R) to assess cognitive vulnerability through three measurements: pre-, post- and three-months follow-up	Comparing to post- and three-month follow up, the reduction in symptoms of depression ($p = 0.293$), anxiety ($p = 0.187$), and stress ($p =$ 0.153) were not significant during the waitlist interval. The reductions in cognitive reactivity ($p = 0.048$) and overgeneralization ($p = 0.028$) were found but not enough, whereas mindfulness skills showed considerable improvement ($p = 0.002$)

(continued)

Table 1 (continued)

Study	Type of subjects	Sample size (meditators/control)	Included criteria	Procedure description	Methodology	Result
Moynihan et al. [35]	Older adults	100/101	People from 63 or older and had a stable medication regimen	Standard MBSR program, include formal and informal practice at home with 24-week follow-up	Using measurements from anti-KLH antibody response; Trail Making Test; questionnaire includes Center for the Epidemiologic Studies Depression Scale-Revised (CES-D-R), PSS, Mindful Attention Awareness Scale (MAAS); and brain electrical activity (EEG signal)	Trail Making Test showed a significantly lower ratio in the MBSR team (better executive control) ($p = 0.04$) Brain Electrical Activity (focused on F3/4) had a significantly greater leftward alpha asymmetry in MBSR subjects ($p = 0.03$) at T2. Otherwise, waitlist control group subjects exhibited a substantial rightward shift in alpha asymmetry ($p = 0.02$) at that time The questionnaire exhibited significantly higher MAAS scores in the MBSR group after treatment ($p = 0.023$) and at 24-week follow-up ($p = 0.006$) Antibody response to KLH showed a higher level in MBSR at T2 ($p = 0.007$) and lower at T4 ($p = 0.036$) compared to WLC (opposite of the expected direction)

(continued)

Table 1 (continued)

Study	Type of subjects	Sample size (meditators/control)	Included criteria	Procedure description	Methodology	Result
Students						
Hindman et al. [36]	University Student	13/10 *Parallel group: Mindful stress management-Informal (n = 11)		MSM is a six-week program based on MBSR and MBCT program, with formal mindfulness meditation and informal practice, and completed the post-test measures at the end of the final meeting. MSM-I involved informal mindfulness practice only	Questionnaire: they were required to fill in 13 questionnaires in total. MSM and MSM-I had to fill in one individual questionnaire: Daily Formal Mindfulness Log (MSM group only) and Daily Informal Mindfulness Log (MSM-I group only)	MSM group reported results more significant than MSM-I in reducing stress ($p = 0.001$), improving stress level ($p = 0.03$), and worrying through FFMQ, PSWQ In Mindfulness, MSM participants exhibited greater increases in related variables (psychological inflexibility, decentering, and self-compassion) while MSM-I increases self-compassion. However, both groups did not change in anxiety, depression, or positive mood variables Comparing to the Control group, both two intervention groups showed significantly in Stress reduction and mindfulness through FFMQ, SCS
Galante et al. [15]	University students	233/208 (RCT)	Had no serve mental health before	Standard MBSR program	Questionnaire: the Clinical Outcomes in Routine Evaluation Outcome Measure (CORE–OM) for distress assessment and Warwick–Edinburgh Mental Wellbeing Scale (WEMWBS) for well-being assessment The study had done in student's stressful week	MBSR group reduced distress but no negligible change. Bedsides, well-being in the MBSR group increased after the course and was maintained during the examination period, although less so than CORE–OM scores

(continued)

Table 1 (continued)

Study	Type of subjects	Sample size (meditators/control)	Included criteria	Procedure description	Methodology	Result
Yeoungsuk et al. [37]	Nursing students	21/23 (RCT)	Had no regular meditation and yoga practice within the past six months and no physical contraindications to exercise	Standard MBSR program	Participants were asked to complete questionnaires: DASS 21 for depression, stress and anxiety measurement, MAAS for mindfulness measurement	There were significantly greater differences between the MBSR and WL groups in stress ($p < 0.001$), anxiety ($p = 0.023$), depression ($p = 0.02$), and mindfulness ($p = 0.010$)
Michelle et al. [31]	Undergraduate pharmacy students	51/48	Undergraduate pharmacy students who were over 18 years old	A Brief care Mindfulness Program based on the MBSR standard program involved four weeks of practice	The questionnaires included Stress (PSS); mental distress (General Health Questionnaire (GHQ)); burnout (Maslach Burnout Inventory - Student Survey (MBI-SS) University Form); mindfulness (FFMQ); empathy (Jefferson Scale of Empathy - Health Professions Student version (JSE-HPS)); satisfaction (Likert scale)	Regarding mental distress, GHQ showed significant results ($p < 0.005$) In the mindfulness aspect, FFMQ showed a significant effect on observing ($p = 0.01$) and describing ($p = 0.01$) When results were split by gender, stress, distress remarkably improved only for females. Therefore, females were responding better to the intervention ($p = 0.042$) Attendance had significant positive effects on mental distress ($p < 0.005$) and observing facet ($p = 0.004$). Thus, higher attendance, bigger improvements. Participants reported high satisfaction (4 or 5)

(continued)

Table 1 (continued)

Study	Type of subjects	Sample size (meditators/control)	Included criteria	Procedure description	Methodology	Result
Sibinga et al. [33]	Fifth to eighth-grade middle school students who are low-income and minority	159/0 (RCT) *Parallel group: Health Topics group (n = 141)		12-week program based on standard MBSR instruction	Participants completed questionnaires at two sequential days at baseline and after program completion	Comparing to pre-intervention, MBSR participants reported lower levels in: depressive symptoms ($p = 0.02$), self-hostility ($p = 0.02$), somatization ($p = 0.03$), negative affect ($p = 0.003$), negative coping ($p = 0.04$) and rumination ($p = 0.03$). Besides, they showed impressive result at re-experiencing ($p = 0.008$), but this variable did not have baseline data to compare
Sibinga et al. [32]	Male students in seventh and eighth grade in a low-income urban area	22/0 (RCT) *Parallel group: Health Education (n = 19)	Have no significant psychopathology, developmental delay, substance abuse, or behavioral problems	MBSR and HT program included 12-week lessons; each lesson lasted 50 min	Cortisol level, sleep quality, and psychological functioning in three-stage measurements: baseline, post-intervention, and three-month following	Psychological functioning showed that the MBSR group had less anxiety ($p = 0.01$), rumination ($p = 0.02$), and negative coping ($p = 0.06$) than HT groups Sleep quality had no differences between the two groups Cortisol level had no significantly different between groups at baseline, post-program or follow-up

(continued)

Table 1 (continued)

Study	Type of subjects	Sample size (meditators/control)	Included criteria	Procedure description	Methodology	Result
Children						
Bennett and Dorjee [38]	Sixth-grade students	11/13 (non RCT)		Standard MBSR program	Participants completed questionnaires in pre-, post- and three-month follow up: Stress measurement based on DASS 21, Wellbeing assessment based on World Health Organisation Five-Item Well-Being Index (WHO-5)	Depression, anxiety, stress level measured by DASS-21 did not increase significantly between two groups in post-treatment and follow-up stages. However, depression level reduced more than stress and anxiety in the intervention group ($p = 0.08$ and $p = 0.09$, respectively). Although WHO-5 reflected significant well-being improvement ($p < 0.01$) in post-intervention, it did not reveal in follow-up measurement ($p > 0.12$)

Abbreviations: *ADIS-IV-L*—Anxiety Disorders Interview Schedule Lifetime Version, *ATS-R*—The Attitudes Toward Self Revised, *AUCg*—Area under the curve with respect to ground, *BAI*—Beck Anxiety Inventory, *BDI-II*—Beck Depression Inventory-II, *CAMM*—Children's Acceptance and Mindfulness Measure, *CAPS*—Clinician-Administered PTSD Scale, *CBT*—Cognitive Behavioral Therapy, *CBGT*—Cognitive Behavioral Group Therapy, *CES-D-R*—Center for Epidemiologic Studies Depression Scale Revised, *CORE–OM*—Clinical Outcomes in Routine Evaluation Outcome Measure, *CSQ-8*—Client satisfaction questionnaire, *DASS 21*—Depression, Anxiety and Stress Scale - 21 Items, *DBAS*—Dysfunctional Beliefs and Attitudes about Sleep, *ECQ*—Experience Check Questionnaire, *FACT-B*—Functional Assessment of Cancer Therapy-Breast scale, FACT-*ES*—Functional Assessment of Cancer Therapy-Endocrine Symptoms scale, *FACT EWB*—Functional Assessment of Cancer Therapy-Emotional Well-Being scale, *FACT FWB*—Functional Assessment of Cancer Therapy-Functional Well-Being scale, *FACT PWB*—Functional Assessment of Cancer Therapy-Physical Well-Being scale, *FACT SWB*—Functional Assessment of Cancer Therapy-Social/ family Well-Being scale, *FFMQ*—Five Facet Mindfulness Questionnaire, *GHQ*—General Health Questionnaire, *GSI*—Global Severity Index, *GAD*—Generalized Anxiety Disorder, *HC*—Healthy control, *HT*—Healthy topics, *HEP*—Health Enhancement Program, *HF*—High frequency, *HRSD*—Hamilton rating scale for depression, *IPL*—Inferior parietal lobule, *ISI*—Insomnia Severity Index, *SE-HPS*—Jefferson Scale of Empathy-Health Professions Student version, *KLH*—Keyhole limpet hemocyanin, *LEIDS-R*—Revised Leiden Index of Depression Sensitivity, *LF*—Low frequency, LSAS-SR—Liebowitz Social Anxiety Scale-Self-Report, *MAAS*—Mindful Attention Awareness Scale, *BCT*—Mindfulness-Based Cognitive Therapy, *MBSR*—Mindfulness-Based Stress Reduction, *MBI-SS*—Maslach Burnout Inventory-Student Survey, *MSCL/MSC*—Medical Symptom Checklist, *MSM-I*—Mindful stress management- Informal, *CT*—Pharmacotherapy, *PCbMP*—Primary Brief care Mindfulness Program, *PCGT*—Present Centered Group Therapy, *OMS*—Profile of Mood States, *PROs*—Participant-report Outcomes, *PSS*—Perceived Stress Scale, *PSQI*—Pittsburgh Sleep Quality Index, *PSWQ*—Penn State Worry Questionnaire, *PTSD*—Posttraumatic Stress Disorder, *RCT*—Randomized controlled trials, *RRS*—Ruminative response scale, *SCL-90-R*—Symptom Checklist-90-Revised, *SCS*—Self-Compassion Scale, *STAI*—State-Trait Anxiety Inventory, *STG*—Superior temporal gyrus, *TNF-alpha*—Tumor necrosis factor-alpha, *TSST*—Trier Social Stress Test, *UCLA-PTSD*—University of California at Los Angeles Posttraumatic Stress Disorder, *WEMWBS*—Warwick–Edinburgh Mental Wellbeing Scale, *WHO-5*—The World Health Organisation - Five Well-Being Index, *WLC*—Waiting list control

2.3 Assessment Methods

Questionnaire

Among different methods used to investigate MBSR effects, the questionnaire was the most popular. Each questionnaire was designed to evaluate specific psychological aspects from the answers of research participants. The advantage of this method is its low cost and ease of implementation; therefore, it is widely utilized by almost all researchers as the sole outcome. Other researches also used questionnaires to be a counterpart with biological data. However, the varieties of questionnaires make it challenging to compare the results between different studies. Questionnaires also need to be translated and culturally adapted before using non-English language populations. Two types of questionnaires were used in MBSR studies: the ones to evaluate mindfulness skills and those to monitor psychology symptoms or mental health levels.

Cortisol Level

Hypothalamic–pituitary–adrenal (HPA) axis is an interface between the central nervous system and the endocrine system, which is activated by stress stimuli and is responsible for the stress response [17]. Cortisol, the main glucocorticoid in humans, is considered a biomarker for stress levels [18]. According to the previous report, Matousek proposed that MBSR treatment could be associated with improvement in cortisol levels. Researchers have found that stress reduction is accompanied by a decrease in diurnal cortisol levels and contemporarily improved regulation of the HPA axis [13].

Brain Activity

Electroencephalography signal (EEG). EEG signal is a non-invasive, portable, and low-cost technique that records the electrical activity of large numbers of neurons from the scalp through electrodes. EEG waveforms that frequently use for diagnostics include alpha, delta, theta, beta wave [19, 20]. Among them, alpha and beta signals, associated with local and global brain networks, are most related to mental states in humans. The low power of alpha and the high power beta bands in the prefrontal cortex are indicators of high-level stress [2, 21]. In addition to alpha and beta power, increasing theta rhythm power in the frontal cortex is associated with the better task-solving ability [21]. The authors discovered that meditation helped increase the power of alpha, beta, and theta brainwaves, which then improved mental health and neurocognitive performances [2, 19].

Functional Near-Infrared Spectroscopy (fNIRS). fNIRS is the use of Near-Infrared Spectroscopy (NIRS), a non-invasive imaging method for measuring the cerebral hemodynamics associated with neural activity. It can distinguish the concentration of oxygenated (O_2Hb) and deoxygenated (HHb) hemoglobin levels. On the other hand, changes in oxygenated in hemoglobin link to mental stress. Many researchers demonstrated that mindfulness practice could lead to the decrease of O_2Hb concentration in the right prefrontal cortex and, in turn, reduces stress and improves memory, emotion, and decision making during practising [2].

Brain Imaging

Functional Magnetic Resonance Imaging (fMRI). fMRI technique measures brain activity via observing blood flow in the brain. The blood-oxygen-level-dependent (BOLD) contrast is a type of specialized brain and body measurement used to map neural occupation in the brain or spinal cord of humans related to brain cells by imaging the change in hemoglobin responses. The BOLD signal is considered as a biomarker of mental symptoms and psychological issues: stress, anxiety, and depression through specific brain regions [22]. The researchers found that the more significant BOLD responses improved attention, as demonstrated by increased activity of the parietal cortical regions; and affected emotion, as shown by decreased limbic activity [11].

Positron-emission tomography (PET). PET is a nuclear functional imaging test that uses radioactive tracers to observe metabolic processes in particular regions in the body as an aid to the diagnosis of the disorder. For brain disorders, PET detects glucose concentration in different brain regions related to certain types of conditions. There have been many studies using PET to demonstrate neural correlates of mindfulness on mental health through functional brain image [1, 11].

Other Physiological Metrics

Respiratory and heart rates are simple objective physiological signals that are responsive to meditation and could be used as physiological markers. Slow breathing decreased heart rate and blood pressure and reduced heart rate variability [12].

2.4 Subject Groups

Mental Health Disorder Groups

Based on the aims of self-awareness and self-management abilities, the MBSR program can be helpful in a wide range of mental disorders with the core of anxiety or depressive emotion. Meanwhile, there was no study conducted on a group of people who have mental health disorders because of the contraindication of psychological therapy on this specific group. Several mental disorders that were most common in MBSR studies were listed below.

Social Anxiety Disorder (SAD). The symptoms of SAD consist of fear, anxiety, or avoidant of social circumstances in which they are probably watched by others. The situations involve interactions with strangers or performance in front of others or even being watch while eating or drinking. Individuals with SAD fear being negatively evaluated, humiliated, or rejected by others, making themselves embarrassed or offending others [23].

Posttraumatic Stress Disorder (PTSD). PTSD is a mental health issue presenting in people who had experienced traumatic events in which others' or their life or body was in endanger. The situations consist of, but not limited to, wars, natural disasters, accidents, sexual abuse. The heterogeneous clinical picture of PTSD was contributed by intrusive, cognitive, emotional, and behavioral symptoms relating to the traumatic memories [23].

Generalized Anxiety Disorder (GAD). The clinical symptoms of GAD should be persistent at least six months and must include excessive anxiety and worry about several domains that are difficult to control. Besides, physical symptoms also are the primary concerns, which include restlessness or feeling keyed up or on edge; being easily fatigued; difficulty concentrating or mind going blank; irritability; muscle tension; and sleep disturbance [23].

Major Depressive Disorder (MDD). Individuals with MDD suffer from depressive mood or anhedonia nearly every day, and most of the day in at least two weeks. Other symptoms consist of changes in appetite and sleep, retard or irritable psychomotor, fatigue, difficulty concentrate, negative thought about self-image, and thoughts of death or suicide behaviors [23, 24].

Medical Condition Groups

This review decides to target Breast Cancer survivors groups and insomnia adult groups with mental health problems after clinical treatment and reduction of sleep quality. Sleep deprivation is a potential risk of stress since it affects mood that makes people irritable. Besides, short-tempered, caused by sleep deprivation, quickly lead to stressed-out. Our review based on three previous studies that had cooperated with these populations.

Other Groups

The other participant groups were categorized based on two main factors: age and occupation. We reviewed the effects of MBSR in the elder, who are above 50 years old, employees, university students, and children [24–26]. Due to the pressure of life, approximately 20% of adolescents before the age of 18 stressed-out worldwide, whereas college students have depressive symptoms in global reach to 30.6% [4, 27]. These subjects can easily have psychological issues due to social pressure and not having time to take care of their mental health. These issues lead to a remarkable reduction in their quality of life, physical health, and concentration in working and studying.

2.5 Studies Selection

This literature review based on a total of 40 articles published in English from 2010 to 2019 updated until present and downloaded from Google scholar, Pubmed Central, and Elsevier. The keywords to search for these articles were: MBSR, mindfulness, stress, EEG signal, fNIRS, fMRI, and cortisol. The GRADE instrument of Cochran is utilized to evaluate the "risk of bias" (https://www.riskofbias.info) of the collected articles [28]. Based on this evaluation, these articles were narrowed down to 25 for evaluating the quality of this scientific evidence. Each article carefully dissected and summarized the results in Table 1. Almost all the previous studies are before-after studies, whereas there have six Randomized Control Trial (RCT) studies.

3 Results

3.1 Risk of Bias

This literature review is based on a total of 40 articles published in English from 2010 to 2019 and updated until the present. Using the GRADE instrument of Cochran, which called the "risk of bias" tool, to assess the methodological quality [28]. The quality assessments include sequence generation, allocation concealment, blinding of participants and personnel, blinding of participant assessors, blinding outcome assessment, incomplete outcome data, and selective reporting [29]. The bias is evaluated as a judgment, includes a high, low, and unclear risk of bias. The lower risk of bias is, the more reliable evidence performs.

To be more precise, sequence generation and allocation concealment are considered as a bias selection. Sequence generation describes the technique used to generate the allocation sequence in adequate detail to permit an evaluation of whether it should produce a comparable group. In contrast, allocation concealment determines whether the intervention allocations have predicted before or during enrollment. Blinding of participants and personnel is a performing bias; used to prevent the participant from receiving the intervention's knowledge. Blinding outcome assessment is a detection bias, and it describes all methods used to blind outcome assessor from the information of the allocated information. Additionally, incomplete outcome data is an attrition bias, which represents the completeness of outcome data from the attrition and exclusion in analysis, demonstrates sufficient detail and reasonable. Selective reporting is considered as a reporting bias, shows how the report of the study was examined possibly and what was found.

The review considers four factors to assess the quality of evidence: (1) heterogeneity, (2) population, (3) intervention, comparison, and application, (4) reliability and validity. In several titles/articles, four full-texts of symmetric reviewed or meta-analysis and 11 full-texts were excluded due to the primary purposes of our review. They also cannot meet the Cochrane Collaboration so that the collection of articles was narrowed down to 25 for evaluating the quality of this scientific evidence.

Approximate 70% of the included studies had a low risk of bias in sequence generation and allocation concealment. Randomization methods of the remaining studies were unclear because they did not mention random sequence generation in their research and did not explain how they allocated participants. Besides, nearly half of the included studies mentioned using blinded personnel. Among the included studies, 80% had a low risk of bias in blinding outcome assessment and incomplete outcome data because they had reported using intention-to-treat analysis. No reporting bias was detected in the included studies. The risk of bias summary is presented in Fig. 1.

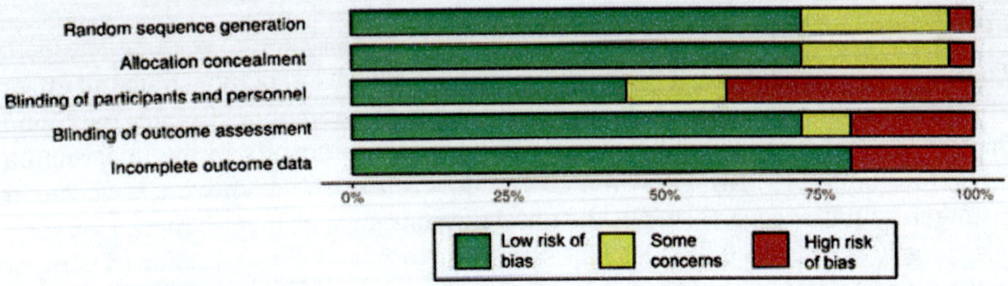

Fig. 1 Risk of bias summary

3.2 *Assessment Result*

Table 1 describes the characteristics of 25 studies in examining the effect of MBSR in various subjects and briefly describing some variations of the MBSR procedure. With the variations of MBSR, there had three studies applied Brief Care Mindfulness Program, which involved four-week practice [18, 30, 31], and four studies used a 12-week MBSR program as the central intervention [14, 31–33].

Psychological and Physiological Effects of MBSR

Questionnaires

Our first goal is to understand how MBSR effects on stress responses through different methods. First of all, the questionnaire score showed significant improvement ($p < 0.05$). There are two main types of questionnaires used in previous studies: Mindfulness Scales and Mental Health Scales. The Mindfulness assessments showed consistent improvement between pre- and post-intervention in almost all studies. Some of which showed shreds of evidence for long-term benefits up to 3 months [11, 25, 30, 31, 33–37]. More specifically, the Five Facet Mindfulness Questionnaire (FFMQ) and the Mindfulness Attention Awareness Scale (MAAS) indicated substantial changes in promoting mindfulness of the participants ($p < 0.01$) [31, 37]. Besides, the World Health Organization Five-Item Wellbeing Index (WHO-5) also reflected significantly in well-being improvement ($p < 0.01$) [38].

Meanwhile, other psychological aspects showed controversial results. MBSR and related techniques could help to improve stress, mental distress, anxiety, depression or depressive symptoms, mood state, well-being, medical symptoms or somatization, cognitive functions, PTSD, SAD [11, 14, 16, 25, 28, 30, 31, 33–37, 39–44]. Expressly, the outcome data of Depression, Anxiety, Stress Scale-21 (DASS-21), and Symptom Checklist-90-R (SCL-90-R), which used in many previous studies, indicated significant results in reducing stress and anxiety ($p < 0.05$) [37, 42]. However, several other studies showed no difference or no substantial change in depression, stress, anxiety levels, sleep quality, and distress [25, 31, 35, 36, 38, 45]. Therefore, although the questionnaire could be the most practical method to evaluate the MBSR intervention outcome, it should be noted that the result could be dubious and hard to repeat in different research. This method should be used in accompany with other indexes such as biological symptoms.

Cortisol level

The cortisol level changes at different time measurements to demonstrate the effect of MBSR on the stress response. However, the results are not negligible. We found that the level of cortisol awakening was decreased significantly in the intervention group ($p < 0.05$) [13, 18]. However, cortisol levels results, which were collected at distinguish time, in several studies did not show consistent changes between pre- and post-measurement ($p > 0.05$) [12, 32]. According to results, the regulation of cortisol in the morning after awakening is highest, and the level would decline throughout the day until midnight [13]. So that the results of studies that used cortisol awakening response showed a significant decreasing through the intervention and cortisol circadian cycle.

Brain activity and structure

Several studies used Electroencephalogram signal (EEG signal) and functional Near-Infrared Spectroscopy (fNIRS) to detect the response between brain electrical activity and the MBSR effect. First, the MBSR training leads to positive results in the participants from the EEG signal. The EEG signals at F3/4, which placed on the prefrontal cortex, noticed the more significant leftward alpha asymmetry in MBSR subjects ($p = 0.03$) at post-MBSR compared to the control group. The leftward activity approached motivation and positive emotion, while the rightward activity approached negative emotions, such as anger [35]. With the fNIRS data, the control group decreased activation in higher auditory areas, whereas the intervention group had a remarkable increase in those areas. In the mindfulness state, the intervention group increased activation significantly in these areas than the control group: primary somatosensory cortex, supramarginal gyrus, and Broca's area [3].

Previous studies also used functional Magnetic Resonance Imaging (fMRI) and Positron-emission tomography (PET) to demonstrate correlates of MBSR and functional brain image. According to brain image from fMRI, the BOLD signal response showed more considerable changes in different brain regions between pre- and post-MBSR. More specifically, BOLD responses in post-intervention changed significantly in the visual attention brain regions, including medial cuneus, left cuneus, and right middle occipital gyrus ($p < 0.05$). The results mean that the reduction of negative emotion was substantial in the MBSR group and increased attentional-related parietal cortical region [11, 46]. Besides, another study showed positive results between the intervention group and the control group by using an MRI image for detecting gray matter concentration in the brain. The gray matter concentration of the MBSR group increased greater than control groups in specific brain regions: Posterior cingulate gyrus, Cerebellum lobule, Cerebellum, Vermis, Lobule, and Temporo-parietal [34]. These brain regions involved emotional regulation and learning and memory processes and proved that MBSR affected these regions. Furthermore, a study from Emory University used PET to indicate the effect of MBSR on mental health. They demonstrated that MBSR leads to increased prefrontal response to stress, anxiety, and depression. The brain imaging collected at a six-month follow-up by PET indicated that the MBSR group increased the activation in anterior cingulate and inferior parietal lobule, and decreased insula and precuneus function in response to traumatic reminders [1].

More specifically, we summarized the positive effect of MBSR across multiple regions of the brain from brain imaging and brain activity. We clarified and characterized information represented in patterns of activity distributed in main brain regions based on previous studies in Fig. 2 and Table 2.

Other physiological metrics

A previous study used heart rate as the other biological metrics to show changes between pre- and post-intervention. The research of Faucher reported the difference in heart rate frequency between the MBSR group and the healthy group ($p < 0.05$). MBSR group's reactivity was lower, while the low-frequency and high-frequency ratio of heart rate in both the control group and the Cognitive Behavioral Group Therapy group increased [12]. Notably, during speech tasks, the ratio of heart rate frequency in the intervention group was lower than in other groups.

In addition to the main assessment methods above, there are other specific methods used in these articles, such as antibody response, neurogenic inflammation, and pain sensation. The study of Moynihan also measured the antibody responses in which the protein antigen Keyhole limpet hemocyanin (KLH) was injected into the arm of participants and determined by enzyme-linked immunoassay. However, the antibody response findings ran in the opposite of the expected direction, while significant levels of antibody were detected after the intervention [35]. The other study used neurogenic inflammation to investigate the effect of MBSR by flare size. The study performed the test of neurogenic inflammation was produced by using the topical application of capsaicin cream to forearm skin to investigate psychological stress in healthy individuals. In this study, flare response and the levels of TNF-alpha and (IL)-8 are the main measurements of neurogenic inflammation [43]. The results showed that the flare size of MBSR was substantially smaller than HEP ($p < 0.001$) at post-MBSR. Besides, levels of both cytokines were lower at follow-up measurement than baseline ($p < 0.001$). Therefore, it can be concluded that MBSR training might have reduced

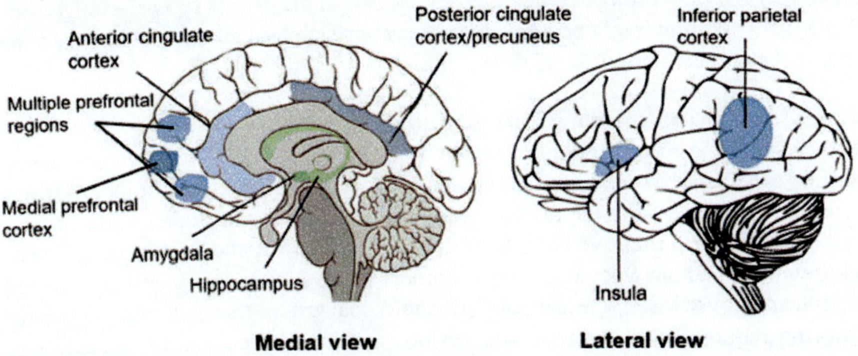

Fig. 2 Brain regions involved in the components of MBSR

Table 2 Functional outcome related to MBSR practice

Brain area	Functional outcome related to MBSR practice	MBSR-induced structural and physiological changes	Citation
Amygdala	Response to chronic stress and emotional regulation of fear	Activation in the brain decreased	[1]
Multiple prefrontal regions	Emotional regulation: the leftward approaches to motivation and positive emotion, whereas the rightward approaches negative emotion	Alpha signal increased in leftward and decreased in rightward	[35]
Medial prefrontal cortex	Involved in self-awareness	BOLD signal increased	[46]
Inferior parietal lobule	Increased the reorientation of attention, reduction of negative emotion	Activation in the brain and the BOLD signal increased	[1, 11]
Anterior cingulate cortex	Attention control	Activation in the brain increased	[1]
Posterior cingulate cortex/precuneus	Involved in self-awareness	Activation in the brain decreased while gray matter concentration increased	[1, 34]
Hippocampus	Involved in long-term memory	Gray matter concentration increased	[34]
Insula	Responses to stress as well as regulating a sense of time	Activation in the brain decreased	[1]

the sympathetic activity to the TSST relative to HEP, leading to less robust stress at post-training [43]. About the pain sensation, the mindfulness condition decreased the unpleasantness of pain ratings at post-intervention ($p < 0.001$) and at follow-up measurement ($p = 0.01$) compared to pre-intervention. Besides, the MBSR group had lower thermal pain ratings compared to other programs. The results also suggested that MBSR training might be a much greater approach to regulating pain than other approaches based on music and fitness [42].

The Effect of MBSR Among Various Subject Groups

The second goal of our literature review is to identify how MBSR affects mental health, especially in reducing stress and promoting mindfulness among various groups of subjects. With the mental health disorder groups, included studies found that MBSR helped improve clinical psychology symptoms and promote awareness. The review found that Social Anxiety Disorder (SAD) subjects are positively affected significantly by MBSR, whereas other psychological groups do not receive the benefit from the therapy. Based on a report of previous studies, SAD patients decreased stress and anxiety significantly between pre- and post-intervention according to self-report in clinical symptoms. Furthermore, they also improved mindfulness and awareness after practising MBSR [14]. Goldin's studies in 2010 and 2013 [11, 46] found impressive results when they used fMRIs to demonstrate the relationship between blood-oxygen-level-dependent (BOLD) in the brain and MBSR. Results from both previous studies showed an increasing level of BOLD in the left brain region and decreasing in the right that means the reduction of negative emotion was significant in the MBSR group and increased attentional-related parietal cortical region. By contrast, other groups with more severe psychological symptoms have not benefited significantly, especially for seniors who have Posttraumatic Disorder. The study of Bremmer in 2017 reported results from questionnaires and PET scanning. He stated the change in PET imaging in six-month follow-up measurement that the MBSR group increased anterior cingulate and inferior parietal lobule and decreased insula and precuneus function in response to traumatic reminders [1]. However, according to the questionnaire, mindfulness, stress, and PTSD symptoms did not improve significantly. From the result above, we can conclude that MBSR affected in mild psychological symptom groups substantially than serve symptom groups in reducing mental issues.

In the medical condition groups, previous studies indicated that besides reducing stress and anxiety level, the other effect of MBSR is improving sleep quality among survivor Breast Cancer groups and the elder. We found that the Breast Cancer patient groups decreased stress, improved mindfulness in few outcome measurements, whereas sleep quality had a small effect after the intervention. On the other hand, with old adults who had primary chronic insomnia, Gross reported that half of MBSR participants recovered from chronic insomnia and the other improved sleep quality significantly ($p < 0.01$) while comparing baseline and after the intervention [40]. Taken all together, MBSR showed clear advantages in stress reduction in comparison to improve sleep quality in both Breast Cancers groups and chronic insomnia elder group.

Finally, the reviewed studies discovered that the MBSR had positive effects in several non-disorder subject groups, including adults, employees, and students who stressed out easily due to pressure from daily life. According to the questionnaire results, all populations showed the MBSR had a strong effect on stress and anxiety reduction. Furthermore, Hölzela integrated MRI imaging into his study to support the effect of MBSR on improving emotional regulation and learning and memory processes in specific brain regions. In addition, in student groups, especially nursing students who are considered easily stressed, almost all participants responded that they reduced stress, anxiety, and improved attentional level, mindfulness, awareness significantly after the intervention, and follow-up measurement according to their self-report ($p < 0.01$). They reported that participating MBSR helped them increase study quality and improve life quality. From the results above, we can conclude that the MBSR helped support well-being and mental health in these groups of subjects efficiently.

Besides, all previous studies did not mention about the external factors as a potential risk for MBSR results, such as environment, duration, material. So, we can conclude that the results of the MBSR program of all subject groups cannot be affected by external factors.

4 Limitation and Recommendation

The main purpose of these studies above is to examine the effects of MBSR on various subject groups. MBSR helps reduce stress, depression, anxiety or promote mindfulness skills, memorize ability (in the elderly population), attentional skill, and sleeping quality (in Breast Cancer patients and old adults) and through various techniques and measurements. However, the impact of MBSR is controversial.

4.1 Limitation

According to the review above, each study has its limitations. First of all, a few populations had a small sample size average from nine to seventeen subjects in their study. The reason is these studies required specific groups of people who have mental health disorders or medical conditions; for instance, PTSD patients, especially elderly veterans, SAD patients, and breast cancer patients. Small size limits the ability to detect differences between MBSR and control groups [1, 3, 15, 46].

Secondly, several studies had no control group (non-intervention in the same population) [1, 11, 13, 18, 25, 40, 42, 46]. Despite having other parallel intervention groups such as Cognitive Behavioral Group Therapy (CBGT), Aerobic, Healthy Education, these results could not be persuasive enough to prove their hypothesis. Although the initial purpose of these studies was comparing two different therapies and proving the effect of MBSR, they should add a control group.

Thirdly, several studies did not design a long-term experiment. A few studies had a short follow-up measurement, approximately three-week or six-week post-MBSR [18, 41]. However, this duration is not long enough to give evidence of the long-lasting effect of MBSR.

Finally, the final results between pre-intervention and post-intervention showed less substantial changes in psychological symptoms. The first reason is previous studies used only questionnaires as outcome measurements, which are subjective and prone to error [14, 25, 30, 31, 33, 36, 38]. These results could not be convincing enough since it did not reflect brain response, and evidence related to brain response or other physiological measurements should be added. In studies combining questionnaires and other measures, results were not strong enough to demonstrate and distinguish before and after the intervention, especially in serve psychological symptoms groups. A few previous studies reported that psychological issues did not decrease negligibly in the intervention group, but MBSR helped improve awareness and mindfulness via questionnaire scores [15, 25]. This proves that MBSR acts as an improvement therapy rather than psychological treatment therapy.

4.2 *Recommendation*

Future studies should consider including the control group as well as using longitudinal and repeated measure designs. Besides, they should combine outcome measurement and other procedures such as common diagnostic procedure (EEG signal), and brain imaging (NIRs, MRI). The EEG signal is a potential method to demonstrate the effect of MBSR. Although a lack of previous studies used the EEG signal for the standard MBSR program, it was applied to the "Mediation Program" to reflect stress response in the brain. Notably, the EEG data showed significantly in stress response ($p < 0.05$) [35]. In addition to clinical devices, in the biomedical instrument market, there have developed many portable devices that can collect the EEG signal by giving high reliable results such as Emotiv, Muse Headband. We believe that these devices are convenient for use in a practical study. Moreover, MRI is a feasible method for demonstrating MBSR effectiveness through brain region imaging [47]. Design a repeated longitudinal measure, which lasts a three-month follow-up, is also critical. The standardized study helps the MBSR study's results be valuable and convincing. In addition, combining other therapies such as art, music, room color can help increase the effect of MBSR during the intervention.

5 Conclusion

Mindfulness-Based Stress Reduction is a non-pharmacological intervention in improving mental health and mindfulness for a wide range of subjects. This review aims to estimate the efficiency of MBSR in various populations, identify groups that need to practice the intervention and potential assessments for MBSR. Besides, MBSR has been proven to enhance self-awareness and mindfulness skills. The society currently has several stressful stimuli causing stress, anxiety, and depression for living people, regardless of age or gender. It is suggested that people with mild mental health disorders should practice MBSR frequently to reduce symptoms, for example, stress and anxiety, while depression symptoms reduce slightly. Future studies should include a control group, combine outcome and structural measurement, and longitudinal measure design to assess the effectiveness of MBSR on psychological disorders.

Acknowledgements This research is funded by **International University, VNU-HCM** under grant number T2019-04-BME. Conflicts of Interest: The authors have no conflict of interest to declare.

References

1. Bremner JD et al (2017) A pilot study of the effects of mindfulness-based stress reduction on post-traumatic stress disorder symptoms and brain response to traumatic reminders of combat in operation enduring freedom/operation Iraqi freedom combat veterans with post-traumatic stress disorder. Front Psychiatry 8:157
2. Al-Shargie F et al (2016) Mental stress assessment using simultaneous measurement of EEG and fNIRS. Biomed Opt Express 7(10):3882–3898
3. Gundel F et al (2018) Meditation and the brain - Neuronal correlates of mindfulness as assessed with near-infrared spectroscopy. Psychiatry Res Neuroimaging 271:24–33
4. Ibrahim AK et al (2013) A systematic review of studies of depression prevalence in university students. J Psychiatr Res 47(3):391–400
5. Virgili M (2015) Mindfulness-based interventions reduce psychological distress in working adults: a meta-analysis of intervention studies. Mindfulness 6(2):326–337
6. Niazi AK, Niazi SK (2011) Mindfulness-based stress reduction: a non-pharmacological approach for chronic illnesses. N Am J Med Sci 3(1):20–23
7. Carmody J et al (2011) Mindfulness training for coping with hot flashes: results of a randomized trial. Menopause (New York, NY) 18(6):611
8. Pbert L et al (2012) Effect of mindfulness training on asthma quality of life and lung function: a randomised controlled trial. Thorax 67(9):769–776
9. Kabat-Zinn J (2003) Mindfulness-based interventions in context: past, present, and future. Clin Psychol Sci Pract 10(2):144–156
10. Marchand WR (2012) Mindfulness-based stress reduction, mindfulness-based cognitive therapy, and Zen meditation for depression, anxiety, pain, and psychological distress. J Psychiatr Pract 18(4):233–252
11. Goldin P et al (2013) MBSR vs aerobic exercise in social anxiety: fMRI of emotion regulation of negative self-beliefs. Soc Cogn Affect Neurosci 8(1):65–72
12. Faucher J et al (2016) Effects of CBT versus MBSR treatment on social stress reactions in social anxiety disorder. Mindfulness 7(2):514–526
13. Matousek RH, Pruessner JC, Dobkin PL (2011) Changes in the cortisol awakening response (CAR) following participation in mindfulness-based stress reduction in women who completed treatment for breast cancer. Complement Ther Clin Pract 17(2):65–70
14. Goldin PR et al (2017) Trajectories of social anxiety, cognitive reappraisal, and mindfulness during an RCT of CBGT versus MBSR for social anxiety disorder. Behav Res Ther 97:1–13
15. Galante J et al (2018) A mindfulness-based intervention to increase resilience to stress in university students (the mindful student study): a pragmatic randomised controlled trial. Lancet Public Health 3(2):e72–e81
16. Hoge EA et al (2013) Randomized controlled trial of mindfulness meditation for generalized anxiety disorder: effects on anxiety and stress reactivity. J Clin Psychiatry 74(8):786–792
17. Miller GE, Chen E, Zhou ES (2007) If it goes up, must it come down? Chronic stress and the hypothalamic-pituitary-adrenocortical axis in humans. Psychol Bull 133(1):25
18. Bergen-Cico D, Possemato K, Pigeon W (2014) Reductions in cortisol associated with primary care brief mindfulness program for veterans with PTSD. Med Care 52(12 Suppl 5):S25-31
19. Hashemi A et al (2016) Characterizing population EEG dynamics throughout adulthood. eNeuro 3(6)
20. Seo S-H, Lee J-T (2010) Stress and EEG. Converg Hybrid Inform Technol 1(1):413–424
21. Ahani A et al (2014) Quantitative change of EEG and respiration signals during mindfulness meditation. J Neuroeng Rehabil 11:87
22. Goldstein JM et al (2010) Sex differences in stress response circuitry activation dependent on female hormonal cycle. J Neurosci 30(2):431–438
23. Association, A.P. (2013) Diagnostic and statistical manual of mental disorders (DSM-5®). American Psychiatric Publications
24. Chi X et al (2018) Effects of mindfulness-based stress reduction on depression in adolescents and young adults: a systematic review and meta-analysis. Front Psychol 9:1034

25. Van der Gucht K et al (2015) A mindfulness-based intervention for economically disadvantaged people: effects on symptoms of stress, anxiety, and depression and on cognitive reactivity and overgeneralization. Mindfulness 6(5):1042–1052
26. Wetherell JL et al (2017) Mindfulness-based stress reduction for older adults with stress disorders and neurocognitive difficulties: a randomized controlled trial. J Clin Psychiatry 78(7):e734–e743
27. Naicker K et al (2013) Social, demographic, and health outcomes in the 10 years following adolescent depression. J Adolesc Health 52(5):533–538
28. McGuinness LA (2019) robvis: an R package and web application for visualising risk-of-bias assessments. 07 Jan 2019. Available from: https://github.com/mcguinlu/robvis
29. Higgins JP et al (2019) Cochrane handbook for systematic reviews of interventions. Wiley, New York
30. McIndoo CC et al (2016) Mindfulness-based therapy and behavioral activation: a randomized controlled trial with depressed college students. Behav Res Ther 77:118–128
31. O'Driscoll M et al (2019) Impact of a mindfulness-based intervention on undergraduate pharmacy students' stress and distress: quantitative results of a mixed-methods study. Curr Pharm Teach Learn 11(9):876–887
32. Sibinga EM et al (2013) School-based mindfulness instruction for urban male youth: a small randomized controlled trial. Prev Med 57(6):799–801
33. Sibinga EM et al (2016) School-based mindfulness instruction: an RCT. Pediatrics 137(1)
34. Holzel BK et al (2011) Mindfulness practice leads to increases in regional brain gray matter density. Psychiatry Res 191(1):36–43
35. Moynihan JA et al (2013) Mindfulness-based stress reduction for older adults: effects on executive function, frontal alpha asymmetry and immune function. Neuropsychobiology 68(1):34–43
36. Hindman RK et al (2015) A comparison of formal and informal mindfulness programs for stress reduction in university students. Mindfulness 6(4):873–884
37. Song Y, Lindquist R (2015) Effects of mindfulness-based stress reduction on depression, anxiety, stress and mindfulness in Korean nursing students. Nurse Educ Today 35(1):86–90
38. Bennett K, Dorjee D (2016) The impact of a mindfulness-based stress reduction course (MBSR) on well-being and academic attainment of sixth-form students. Mindfulness 7(1):105–114
39. Berger R, Gelkopf M (2009) School-based intervention for the treatment of tsunami-related distress in children: a quasi-randomized controlled trial. Psychother Psychosom 78(6):364–371
40. Gross CR et al (2011) Mindfulness-based stress reduction versus pharmacotherapy for chronic primary insomnia: a randomized controlled clinical trial. Explore (NY) 7(2):76–87
41. Hoffman CJ et al (2012) Effectiveness of mindfulness-based stress reduction in mood, breast- and endocrine-related quality of life, and well-being in stage 0 to III breast cancer: a randomized, controlled trial. J Clin Oncol 30(12):1335–1342
42. MacCoon DG et al (2012) The validation of an active control intervention for mindfulness based stress reduction (MBSR). Behav Res Ther 50(1):3–12
43. Rosenkranz MA et al (2013) A comparison of mindfulness-based stress reduction and an active control in modulation of neurogenic inflammation. Brain Behav Immun 27(1):174–184
44. Wolmer L, Hamiel D, Laor N (2011) Preventing children's posttraumatic stress after disaster with teacher-based intervention: a controlled study. J Am Acad Child Adolesc Psychiatry 50(4):340–348
45. Andersen SR et al (2013) Effect of mindfulness-based stress reduction on sleep quality: results of a randomized trial among Danish breast cancer patients. Acta Oncol 52(2):336–344
46. Goldin PR, Gross JJ (2010) Effects of mindfulness-based stress reduction (MBSR) on emotion regulation in social anxiety disorder. Emotion 10(1):83–91
47. Parkinson TD, Kornelsen J, Smith SD (2019) Trait mindfulness and functional connectivity in cognitive and attentional resting state networks. Front Hum Neurosci 13:112

Reliability and Validity of Vietnamese Version of Patient Health Questionnaire 9 Items (PHQ-9) Among UMP Medical Freshmen

Nghia Trung Nguyen, An Pham Le, and Dong Phuong Tien Nguyen

Abstract Patient Health Questionnaire (PHQ–9) is a global tool that can be effective for depression screening, especially in the context of high prevalence of the disorder in medical students. The aims of this study are: (1) Translating and cultural adapting the Vietnamese version of PHQ-9; (2) Establishing reliability and (3) Identifying validity of PHQ-9 Vietnamese version within Vietnamese medical students. The process of establishing the PHQ-9 Vietnamese version strictly followed the international guideline for translation and cultural adaptation of psychometric instruments recommended by WHO. Four hundred two medical freshmen were involved in PHQ-9 study for internal reliability. After 10 days, 76 students from the former list were randomly selected for a structured interview with psychiatrists and underwent repeated PHQ-9 test. Records of the latter time were used to calculate test–retest reliability and criteria validity. ROC curve analysis was performed to evaluate diagnostic efficiency. High internal and test–retest consistency for PHQ-9 (Cronbach's alpha as 0.81 and Spearman correlation coefficient as 0.83) were identified. The sensitivity and specificity at the cut-off 12 were 71% and 93%, and 7 was 82% and 71%, respectively. The reliability and validity of PHQ-9 Vietnamese version are advisable for applying in mental health management for medical freshmen. The cut-off score is 12 for epidemiological purposes and is 7 for screening purposes. Further researches should assess PHQ-9 Vietnamese version's psychometric features towards other medical college groups or other forms of handling tests, listing as mobile apps or web-based systems.

N. T. Nguyen (✉)
Psychiatry Department, Medicine Faculty, University of Medicine and Pharmacy, Ho Chi Minh City, Vietnam
e-mail: ntnghiay11@ump.edu.vn

A. Pham Le
Family Medicine Training Center, University of Medicine and Pharmacy, Ho Chi Minh City, Vietnam

D. P. T. Nguyen
Foreign Language Department, University of Medicine and Pharmacy, Ho Chi Minh City, Vietnam

V. Van Toi et al. (eds.), *8th International Conference on the Development of Biomedical Engineering in Vietnam*, IFMBE Proceedings 85,
https://doi.org/10.1007/978-3-030-75506-5_72

Keywords PHQ-9 · Depression · Medical student · Reliability · Validity · Sensitivity · Specificity · Cultural adaptation

1 Introduction

Depression is one leading fatal cause globally, from which more than 322 million people are estimated to suffer. Proportions vary from 2.6 to 5.9% in different living areas [1]. In comparison to other normal populations, the proportion of medical students suffering from depression is much higher, 27% [2]. In comparison to other students of other majors sharing the same age, the proportion of medical students suffering from depression also exceeds [3, 4].

Medical students have to stretch themselves to deal with various hassles. Health Science is one specialized field with a massive sum of knowledge, stressful examinations, competitive pressures, and highly demanding professional requirements [5–7]. Additionally, medical students have been confronted by excessive workload and difficulties in balancing study—job—life and social interactions and peer relationships and health and financial dilemmas [8]. Besides, in clinical practice, medical students have to face ethical issues as well as witness their patients languishing in pain and death [9].

These chronic and severe stressors can be associated with depression, which leads to a decrease in academic performance and in sympathy to patients, as well as the increase in medical school dropping out [6, 10–14]. Furthermore, depression contributes to broken relationships, drug, and alcoholic addiction, along with committing suicide [9].

University of Medicine and Pharmacy at Ho Chi Minh City (UMP) is one leading medical training institute in Việt Nam with the number of students up to 10,000. From 2012–2019, the proportion of UMP students suffering from depressive disorder reaches approximately 27–50% [4, 15–17]. The situation replicates in other major medical schools across the country with the depressive prevalence ranging from 25 to 51% [18, 19]. Although these pioneering studies provided essential knowledge, they regrettably got weak points. The psychometric scales used in these studies were achieved by proceeding only the first step of Vietnamese translation without an opinion of experts' panel and lacking evidence of psychometric features within the particular group of medical students. Summing up, depression is a significant hassle in UMP and Vietnamese medical students, raising a prerequisite for reliable and validated instruments to access the problem accurately.

1.1 Patient Health Questionnaire 9 Items (PHQ-9)

From psychiatric disorder scoring at primary healthcare centers (PRIME-MD PHQ), in 2001, Dr. Robert L. Spitzer, Janet BW Williams, Kurt Kroenke et al. developed PHQ-9 to screen and evaluate depression severity [20]. PHQ-9 has been translated into more than 60 versions and assessed in terms of validity and reliability within various objects, including general populations, patients at the primary healthcare centers, out-patient clinics, inpatient departments, refugee groups, and low-income residents. Moreover, PHQ-9 is common in medical students' well-being researches [2, 4, 17, 21]. The scale shows the internal reliability (Cronbach's alpha) ranging from 0.74 to 0.89. The interval reliability (between 2 times of survey) is consistent at a high level [20, 22–26].

In terms of content value, the scoring items are constructed to evaluate the frequency of 9 depressive symptoms listed in DSM-IV criteria and remained unchanged in DSM-5 criteria. In Việt Nam, the Pfizer translated version or other self-translated versions of PHQ-9 seem to be more familiar with our researchers. These are simple renditions without the cross-culture adjustment step admitted by the professional consultant board, especially psychiatric experts' recommendations. According to official guidelines widely accepted by mental health professionals, translated versions of scales must be justified to cross the gap between the two languages and cultures as well as to maintain their content validity [27–29]. Otherwise, these versions' content validity may not be achieved, and the research outcomes may not exactly demonstrate the actual situation in target populations [29].

In terms of criteria value, PHQ-9 authors suggested 10 as the cut-off point for detecting depression disorders (sensitivity at 88% and specificity at 88%) [20]. However, studies in different populations recommended several other reasonable cut-offs. For example, 5 and 6 points score were chosen for patients in Uganda and Holland, respectively [30, 31]. Another analysis research combines the sorting list of the cut-offs from 8 to 10 [32]. This variety results from the differences in specific contexts, participants and aims of the researches [26]. Thus, it is a must to perform a study to identify the standard cut-off point of the PHQ-9 in the Vietnamese medical student group before applying the scale in screening or researching processes [33].

In terms of feasibility, the more widely the scoring benches are applied, the shorter time to collect scale's data is required [34]. The original research showed that the full 28-question-PHQ scale (3-page long for depressive disorder and other mental statuses) took patients approximately 3 min to finish. Consequently, the total length to complete only 9-question-depressive scale (PHQ-9) was estimated to be around 1 min [20]. This duration is appropriate for a short survey to screen depressive disorders on medical student participants.

1.2 This Study

Based on the need for a validated scale which is suitable for depression screening in medical students and the fact that PHQ-9 is a conventional scale with qualified psychometric features but have not been standardized in this specific group, our study was designed to achieve 3 aims: (1) Establishing the cultural adaptation of the Vietnamese version of PHQ-9 with the expert panel's opinions to warrant the content validity; (2) Estimating internal and test–retest reliability; (3) Identifying the optimal cut-off point for detecting depression in medical students in term of criteria validity. We hypothesized that both reliability coefficients would range from good to excellent level.

The study consisted of 2 stages: translation-adaptation with a step-by-step procedure and psychometric identification with repeated surveys and structured interviews. The Vietnamese PHQ-9 version resulting from our work was expected to be beneficial in healthcare profile recording, advance treatment support, and prompt community intervention to help depressive students return to normal life [35].

2 Methods

2.1 Linguistic Translation and Cultural Adaptation

In this study, the procedure of linguistic translation and standardization of the PHQ-9 scale was designed based on the 'Process of translation and adaptation of instruments' recommended by WHO [36]. The whole process comprised 5 steps: Translation from English to Vietnamese—Translation from Vietnamese to English—Presentation to Professional Board—Cognitive interview and pilot research—Completing the official Vietnamese final version. Further details regarding this process were as follows:

(1) (1) English-to-Vietnamese translation was performed by one Vietnamese specializing in English linguistics and one Vietnamese psychiatrist mastering the English language. After having the original English questionnaire translated into the Vietnamese language by two independent authors, the two translated versions were compared and contrasted to other available Vietnamese versions used in Việt Nam so as to unify the most accepted *English-to-Vietnamese translated version.*

(2) Vietnamese-to-English translation was carried out by two Vietnamese persons who have been living and teaching more than 10 years at Universities in Australia and America; primarily, none had ever known PHQ-9. The unified Vietnamese version obtained from step one went through the reverse steps in which two authors worked independently to translate it back to English. The

two *Vietnamese-to-English translated versions* then were also unified into one final logical version.

(3) Professional Board, consisting of one professor-psychiatrist, one experienced psychologist, one expert at scoring standardization, and one linguistic specialist, discussed the translated versions' appropriateness. The Professional Board's consideration based on 3 documents: (1) The unified English-to-Vietnamese version, (2) The unified Vietnamese-to-English version, and (3) The original English version. The panel meeting suggested revising the translation errors, adjusting cross-cultural issues, and warranting the content validity of PHQ-9 Vietnamese pre-final version.

(4) After granted by Professional Board, the pre-final translated version was tested in terms of the potential errors raised from 10 medical students' replies varied in gender and dialects (regional languages). These students conveyed content validity when they were asked to express their understanding of the questions. In the same way, students from different regions were asked about difficulties in understanding the questionnaire.

(5) These recommendations were noted for amending and completing the PHQ-9 Vietnamese final version.

2.2 *Participants in Surveys and Interviews*

402 medical freshmen at UMP, aging 18 years and older were involved in our study from October 2019 to February 2020. Sole criteria were as follow: Having inadequate capacity in the Vietnamese language; Being patients of systematical diseases (cancer, deteriorated immunity, endocrine diseases, and so on); Experiencing psychosis symptoms; Being diagnosed and treated, at least, with one psychosis disorder; Being in inappropriate conditions to participate in the survey or interview (listing as restlessness, irritable and aggressive states; Lack of concentration required to understand and respond to the survey and interview questions; Inadequate time to complete the procedure of PHQ-9 questions and MINI interview).

Sample size: The minimal sample size required to estimate the reliability of PHQ-9 was calculated using a method introduced by Bonett [37]. The PHQ-9 consisted of 9 items with 4 points Likert scale for every item. The coefficient of Cronbach's alpha in the null hypothesis and that of the alternative hypothesis were assumed to be equal to 0.0 and 0.8, respectively [26]. Based on the alpha value fixed at 0.05, the minimum sample size was 13 in order to achieve the power of 80.0%. Besides, the least sample size for performing Receiver Operating Characteristic (ROC) curve analysis was 72 students ($\alpha = 0.05$, $\beta = 0.20$, area under ROC curve [AUC] $= 0.70$, null hypothesis value $= 0.50$, ratio of sample sizes in negative/positive groups $= 2$; MedCalc Software bvba, 2019).

In the first phase of the research, the number of 402 students involving in the survey at a lecture hall in UMP achieved the needed sample size. Then, 90 students were randomly selected from 402 students in the previous survey by using random.org.

76 over 90 invited students accepted to participate in the second phase. Therefore, the sample size needed for ROC curves analysis was fulfilled.

In the second phase of our research, which occurred 10 days after the first phase, the participants answered the survey in the same setting for the first time and attended interviews with psychiatrists. The Mini International Neuropsychiatric Interview was used as the golden criteria for depressive disorders diagnosing, which was associated with determining the optimal cut-off values of PHQ-9 screening for depression in the medical students' population. In this step, half of the participants were interviewed first and conducted the survey afterward, and the other half did in the reverse direction.

2.3 *Measures*

The means of measurement in this study comprised several questionnaires and an interviewing structure, listing as: General Information File, PHQ-9 Vietnamese version (Patient Health Questionnaire 9 items), and MINI (Mini International Neuropsychiatric Interview) structured interview (Appendix 1).

Patient Health Questionnaire 9 Items

The PHQ-9 scale consisted of 9 items corresponding with 9 symptoms of major depressive disorder (DSM-IV criteria). DSM-5 version, updated in 2013, presented no change in the symptoms listed in DSM-IV. The survey participants used a Likert scale from 0 (not at all) to 3 (nearly every day) to report the frequency of the symptoms they had experienced within the last 2 weeks. The total score ranging from 0 to 27 was used to screen depressive disorders and evaluate depression severity: the higher, the worse. An example item was that 'Within the last 2 weeks, how often have you been bothered by thoughts that you would be better off dead or of hurting yourself?'.

Additionally, the 10th item asked the survey respondents to acknowledge how the symptoms influence them by choosing the ordinal level, from 'not difficult at all' to 'extremely difficult'. In the original study, PHQ-9 scale showed high internal consistency with Cronbach's alpha = 0.89 [20]. Currently, Pfizer owned this copyright. It was free of payment to download the original English version from https://www.phqscreeners.com/. Likewise, those interested could use this scoring for translation and research performance without prior permission. The Vietnamese version of this scoring had not been available on this website.

Mini International Neuropsychiatric Interview (MINI)

MINI was a structured diagnostic interview of psychiatric disorders [38]. The items were based on DSM-IV diagnostic criteria. However, diagnostic criteria of major depression disorder and generalized anxiety disorder in DSM-5 remained unchanged compared to that of the earlier version. Therefore, the structure was acceptably applicable. The psychiatrists interviewed medical students based on the structure of MINI so as to diagnose the major depressive disorder. This result was used as the golden standard to identify the sensitivity and specificity of PHQ-9. Structured interviews conducted with MINI scale comprised several distinctive parts for different psychiatric disorders. Only the interview outcomes related to major depressive disorder were further explained in this study duc to the research aims. The MINI presented high reliability and validity compared to the Composite International Diagnostic Interview questionnaire [39]. The questionnaire was translated and adapted to Vietnamese culture concurrently with PHQ-9 scale procedure.

2.4 Data Analysis

Raw data was recorded via Epidata 4.1 and analyzed via Stata 14 software. Chi-square test was used to compare the rate of categorical and ordinal variants. The internal reliability of the PHQ-9 questionnaire identified by using Cronbach's alpha value was considered acceptable from 0.7, good from 0.8, and excellent from 0.9 [40]. Meanwhile, test–retest reliability was evaluated by the correlation between 2-phases' PHQ-9 records. In terms of validity of PHQ-9, the ROC curve analysis was applied to demonstrate the criteria validity. The ROC curve was exhibited by plotting sensitivity against the false positive rate (1-specificity) across all possible threshold values along with Area Under the Curve (AUC). The AUC helped us to evaluate the ROC curve's diagnostic accuracy [41]. A curve ranging from 0.70 to 0.90 manifests moderate test accuracy [42]. The most advisable sensitivity, specificity, and AUC were considered to select the optimal cut-off score. Over our research, the difference was statistically significant when the p-value was < 0.05.

3 Results

3.1 Linguistic Translation and Scoring Standardization

The problems and the accompanied solutions occurring during linguistic translation and cultural adaptation of PHQ-9 were further presented in Table 1.

4 Overall Research Result

In the first phase, 402 students engaging in this research were aged 18–21 (M = 18.9, SD = 0.5). Male proportion accounted for 64.2%. Apart from 2 students being under generalized anxiety disorder treatment, no other student was suffering from psychiatric problems and other situations that should be eliminated. The median (IQR) total PHQ-9 score was 7 (4 to 10). The testing length of PHQ-9 ranged from 22 to 322 s (M = 122.7, SD = 42.3). The total score added by self and noted into

Table 1 Summary of linguistic translation and cultural adaption of PHQ-9

Original version	Former Vietnamese Version	Issues–Solutions	Final Vietnamese version
Over the last 2 weeks, how often have you been bothered by any of the following problems?	*Trong 2 tuần vừa qua, các vấn đề sau đây **có**thường xuyên làm phiền bạn?*	The Yes–No question did not reflect the desired point related to the frequency of the original question. It was relatively rare to be asked 'bạn thường xuyên ở mức độ nào' (meaning 'how often') in Vietnamese language ▶ The change into Imperative sentence requested the participants to identify the existing state as well as the frequency of each issue	*Bạn **hãy xác định bạn có bất kỳ vấn đề**nào sau đây và **mức độ các vấn đề này**làm phiền bạn trong 2 tuần vừa qua*
Item 1: Little interest or pleasure in doing things	Ít hứng thú và thỏa mãn khi làm việc gì đó	–	*Ít hứng thú hoặc thỏa mãn khi làm việc gì đó*
Item 2: Feeling down, depressed, or hopeless	Cảm thấy suy sụp, chán nản hay vô vọng	In terms of semantics, the Vietnamese language had no equivalent words or terms for 'feeling down,' and 'depressed' ▶ 3 emotional words in the original version were altered into other 3 words with the closet meanings in Vietnamese language and arranged in the order of increasing severity of symptoms	*Cảm thấy **chán nản, buồn rầu**và **vô vọng***

(continued)

Table 1 (continued)

Original version	Former Vietnamese Version	Issues–Solutions	Final Vietnamese version
Item 3: Trouble falling or staying asleep, or sleeping too much	Gặp vấn đề trong việc vào giấc ngủ khó duy trì giấc ngủ hoặc ngủ quá nhiều	It was rare to find 'gặp vấn đề trong việc vào giấc ngủ' (meaning 'trouble falling or staying asleep') expression in Vietnamese language ▶ 'trouble' was translated as 'khó' (meaning 'difficult'), instead of 'vấn đề' (meaning 'trouble' or 'problem') so as to be appropriate to Vietnamese expressive style. E.g., 'khó vào giấc ngủ' term was more easily understandable than 'gặp vấn đề trong việc ngủ' phrase	***Khó vào giấc ngủ, khó duy trì giấc ngủ hoặc ngủ quá nhiều***
Item 4: Feeling tired or having little energy	Cảm thấy mệt hoặc ít năng lượng	–	*Cảm thấy mệt hoặc ít năng lượng*
Item 5: Poor appetile or overeating	Chán ăn hoặc ăn quá nhiều	–	*Chán ăn hoặc ăn quá nhiều*

(continued)

Table 1 (continued)

Original version	Former Vietnamese Version	Issues–Solutions	Final Vietnamese version
Item 6: Feeling bad about yourself or that you are a failure or have let yourself or your family down	Cảm thấy bản thân mình tệ hại/ bản thân thất bại/ mình khiến bản thân hoặc gia đình thất vọng	This item consisted of many 'or' connectors that caused problems in understanding and miscommunication in terms of semantics. That the former translated version did not use 'hoặc' (meaning 'or') could not clarify the idea that there were 3 different cases. 'Khiến gia đình thất vọng' phrase sounds more serious than the target idea of 'let family down' ▶ Remained 'or' in the 3 opinions and replaced 'or' in 'yourself or family' phrase with the '/' quotation mark so as to avoid misunderstanding. Replaced 'thất vọng' with 'chán nản'	*Cảm thấy bản thân tệ hại hoặc bản thân thất bại hoặc mình khiến bản thân/gia* ***đình chán nản***
Item 7: Trouble concentrating on things, such as reading the newspaper or watching television	Gặp vấn đề trong việc tập trung, ví dụ: khi đọc báo hoặc xem tivi	There was no equivalent Vietnamese term for 'things' in this context ▶ Used 'việc đang làm' phrase (meaning 'things that you are carry on'), instead of 'các việc' word, in order to match the normal daily Vietnamese expression. Added 'đọc sách' word (meaning 'reading book') into the example list so as to be suitable for medical students	***Khó tập trung vào việc đang làm,*** *ví dụ đọc sách, đọc báo hoặc xem ti vi*

(continued)

Table 1 (continued)

Original version	Former Vietnamese Version	Issues–Solutions	Final Vietnamese version
Item 8: Moving or speaking so slowly that other people could have noticed. Or the opposite – being so figety or restless that you have been moving around a lot more than usual	Vận động hoặc nói rất chậm đến mức người khác nhận thấy sự khác biệt; hoặc ngược lại, cảm thấy bồn chồn, bứt rứt đến mức đi tới đi lui nhiều hơn bình thường	–	*Vận động hoặc nói rất chậm đến mức người khác nhận thấy sự khác biệt; hoặc ngược lại, cảm thấy bồn chồn, bứt rứt đến mức đi tới đi lui nhiều hơn bình thường*
Item 9: Thoughts that you would be better off dead, or of hurting yourself	Có suy nghĩ rằng tốt hơn mình nên chết đi cho rồi hoặc suy nghĩ về việc làm tổn thương bản thân	It was rather hard to translate 'better off dead' phrase into the equivalent Vietnamese expression. 'chết đi cho rồi' phrase was not formal enough to appear in this questionnaire ▶ The Professional Board agreed to use 'tốt hơn mình nên chết đi' phrase (closet meaning to 'better off dead'), excluded 'cho rồi' word	*Có suy nghĩ rằng tốt hơn mình nên chết đi hoặc suy nghĩ về việc tự làm tổn thương bản thân*
Item 10: If you checked off any problems, how difficult have these problems make it for you to do your work, take care of things at home, or get along [not difficult at all – somewhat difficult – very difficult – extremely difficult]	Nếu bạn trả lời có ở bất kỳ vấn đề nào, hãy cho biết các vấn đề đó đã gây khó khăn cho bạn như thế nào trong công việc, ở nhà, và trong mối quan hệ với người khác. [Không khó khăn chút nào cả - Hơi khó khăn – Rất khó khăn – Cực kỳ khó khăn]	These questions above were not Yes–No type, so it was inappropriate to use 'nếu bạn trả lời có' phrase (meaning 'if you said yes') ▶ 'nếu bạn đã xác định có bất kỳ vấn đề nào ở trên' phrase (meaning 'if you checked off any problems') was recommended to change YES–NO question into Confirmative type	*Nếu bạn* ***đã xác định có bất kỳ vấn đề nào ở trên,*** *hãy cho biết các vấn đề đó đã gây khó khăn cho bạn như thế nào trong công việc, học tập, ở nhà, và trong mối quan hệ với người khác. [Không khó khăn chút nào - Hơi khó khăn - Rất khó khăn - Cực kỳ khó khăn]*

Table 2 Internal and test–retest reliability of PHQ-9

Item	Cronbach's alpha	Test-rest score correlation	
PHQ-9	**0.81**	**0.83***	*Spearman's Rank Sum Correlation Test, $p < 0.05$
1	0.80	0.59*	
2	0.79	0.64*	
3	0.82	0.63*	
4	0.80	0.56*	
5	0.81	0.63*	
6	0.80	0.76*	
7	0.81	0.65*	
8	0.82	0.45*	
9	0.82	0.85*	

record papers (M = 7.17, SD = 4.55) was statistically significantly lower than the score automatically summarized by the computer after the data input step (M = 7.27, SD = 4.58); Wilcoxon signed-rank test, $p < 0.001$).

In the second phase, 76 students enrolling in the research were randomly selected from the first phase list. The mean (SD) age was 19.1 (0.6), ranging from 18 to 20. The male percentile was 63%. The median (IQR) total PHQ-9 score was 5 (3 to 10). The prevalence of major depressive disorder diagnosed by psychiatrists was 21%.

4.1 *Reliability*

Internal Reliability

Cronbach's alpha coefficient of PHQ-9 was 0.81; those of the 9 items ranged from 0.79 to 0.82; in which, the highest was of items 3, 8, and 9, whereas the lowest went to item 2 (Table 2).

Test–retest Reliability

PHQ-9 overall score in the first and second phase of this research was strongly correlated and statistically meaningful. The score of each item also shared similar results. Spearman's rank correlation coefficient between the PHQ-9 overall score of 2-time responses was 0.83 ($p < 0.05$) whereas, for items, the correlation coefficient ranged from 0.45 (item 8) to 0.85 (item 9) (Table 2).

4.2 *Criteria Validity*

Table 3 presented information related to specificity and sensitivity of PHQ-9 at different cut-off points. AUC = 0.88 (SE = 0,05; 95% CI [0.78; 0.98]).

Table 3 Criteria validity of PHQ-9 at different cut-off points

Cut-off point	Sensitivity (%)	Specificity (%)	Youden index	Depressive prevalence (%)
≥ 4	94.12	47.46	0.40	61.8
≥ 5	94.12	54.24	0.48	56.6
≥ 6	88.24	62.71	0.50	48.7
≥ 7	82.35	71.19	0.53	40.8
≥ 8	76.47	76.27	0.50	35.5
≥ 9	70.59	79.66	0.50	31.6
≥ 10	70.59	88.14	0.58	25.0
≥ 11	70.59	91.53	0.62	22.4
≥ 12	70.59	93.22	0.64	21.0
≥ 13	64.71	93.22	0.58	19.8

The optimal cut-off point with the best Youden index was 12 (specificity 93% and sensitivity 71%). At this point, the prevalence of depression detected by PHQ-9 scale was 21%, which was similar to the majority indicated by the gold standard. Meanwhile, 7 points were the highest score, which has sensitivity over 80% (sensitivity 82% and specificity 71%) (Fig. 1).

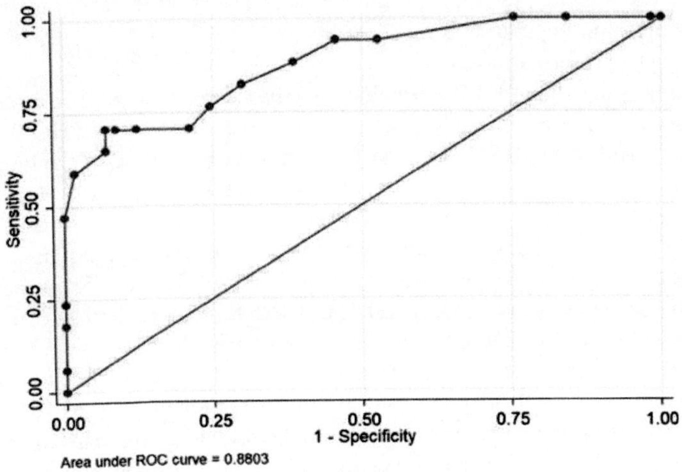

Fig. 1 Receiver Operating Characteristic (ROC) curve of PHQ-9

5 Discussion

Our Vietnamese version of PHQ-9 scale has some advantages over previous Vietnamese PHQ9 versions: the translation and cultural adaptation process strictly followed WHO-recommended procedure in addition to specific adjustments for the medical student group. Our finding shows that the changing of some pivotal Vietnamese words at the introducing sentence and several items (2, 3, 6, 7, 9, 10) are necessary because the words demonstrated in older versions are not semantically corresponding to the ordinal symptoms' severity and the Yes–No question type does not reflect the desired point related to the frequency of the original question. In our novel version, the change into imperative sentence requests the participants to identify the existing state as well as the frequency of each issue. As a result, our questionnaire minimizes the possibility that the survey participants could give wrong answers due to content misunderstanding. The content validity of our Vietnamese version of PHQ-9 is achieved by applying a crucial multi-dimension approach to hold by the Professional Board mentioned above. Among the Board, the psychiatrist and the psychologist's role was to guarantee that the content of the questionnaire was precise and targeted at depressive disorder screening, mainly in terms of DSM-5. The psychometrics and the linguistic expert recommended the more appropriate syntactic forms and expressions that match the specific survey subject and Vietnamese culture. Finally, we can exactly demonstrate the actual depression phenomena in our medical students, prevent systemic biases, and build the accuracy screening tools on depression in primary care.

Psychometric information of scale remains a big challenge for any international PHQ-9 version that includes the previous Vietnamese versions. The result above showed that the internal reliability of our Vietnamese version of PHQ-9 not only reached the good level (0.81) but also was similar to the PHQ-9's internal reliability in other studies, which ranged from 0.74 to 0.89 [20, 22–26]. Moreover, the requirement of test–retest reliability was achieved through a strong correlation with statistically significant between the records of two consecutive times answering the survey within a 10-day gap. In contrast to the high level of our Spearman's rank correlation coefficient ($\rho = 0.83$, $p < 0.05$), the test–retest reliability of PHQ-9 in Filipino female labor immigrants just reaches a average level ($\rho = 0.43$, $p < 0.05$) [26]. This outcome results from the difference in survey accomplishment of the latter study: At the first phase, their survey participants gather together to answer on tablets at the same place with supervision, whereas in the second phase, participants themselves answered the questionnaire by using various kinds of personal equipment at home without any supervision.

Meanwhile, medical students shared the same progress within the same testing condition in our research: answering on printed questionnaires and gathering in lecture halls under direct supervision. This result is meaningful in terms of clinical and communal surveys. For the clinical aspect, the comparison of PHQ-9 records completed by the same participant but at different time points can be more reliable if he or she answers the questions under the same conditions and the same apparent supervision. In terms of the epidemiological aspect, it is the elements of survey performance, environment, and equipment types that can influence the survey outcomes. Therefore, towards the survey subjcct of medical students, there should be more research that performs our new PHQ-9 Vietnamese versions on mobile applications or web pages to collect essential psychometric for each method.

In terms of criteria validity, which means the scale can discriminate between depressive and healthy medical students, the optimal cut-off score at 12 was depicted. This score not only has high specificity (93%) and acceptable sensitivity (71%) but also describes the most accurate depressive prevalence in comparison with the diagnosis of psychiatrists [43]. Meanwhile, the original 10-point score cut-off suggested by the PHQ-9 authors reaches the same sensitivity but lower specificity (88%). The situation may over-estimate the proportions of students suffering from depression in previous research [4]. Therefore, from the perspective of epidemiology, we recommend applying 12-point cut-off for studies using PHQ-9 Vietnamese version with medical students in order to attain the most factual data.

On the other hand, screening purpose requires a highly sensitive instrument [43]. Using a cut-off point at 7 allows us to achieve the sensitivity better than 80% and reasonable specificity (82% and 71%, respectively). Simultaneously, the cut-off at10 points indicated by the original research just established the sensitivity of aticalsensitivity students [20]. The difference may result from the context of researches. The original finding was conducted in the hospital and primary clinical care system while our study took place in the community where almost all participants belong to the healthy group. Because lower sensitivity means higher false negative, if we chose 10 points as the cut-off, more students who acquired depressive disorders were not recognized by the screening process. These students appeared to be disadvantaged because early depression detection and prompt treatment contributed to decreasing suicide risk, limiting deteriorated functions, and reducing medical treatment maintenance [44, 45]. Hence, to screen depressive disorder in medical students, we recommend the 7-point cut-off in order to reduce the missing out of students who may need help.

Testing length, another feature of scale that can affect psychometric quality, demonstrated approximately 2 min in our study group. This duration is acceptable for a quick survey in the community as well as logical towards the assumption from original research [20]. As a result, towards the medical student group in primary care, PHQ-9 can be combined with other brief scoring collections with mental healthcare administration.

The gap between calculating by self and automatic calculation by computers was discovered from our research with statistical significance. This meant the error of hand calculation did take place despite the concentrative environment with full supervision. Therefore, clinicians in practical screening usage are advised to add scores by themselves or use the automatic adding function available on the websites or mobile applications version of PHQ-9 instead of relying on their patients' work.

Strongpoints and Weak Points of This Research

In terms of strongpoints, this research, firstly, strictly followed the linguistic translation and cultural adaptation procedure recommended by WHO. The Professional

Board guaranteed the semantic appropriateness and content validity of the novel PHQ-9 Vietnamese version. Secondly, the criteria validity was conscientiously investigated. We used the MINI structured interview as the golden criteria, and all interviewers were experienced psychiatrists. We separated the second phase participants into two groups: one finished the PHQ-9 surveys before interviewing and underwent the reverse steps to exclude possible processing bias. The research results suggested 7 instead of 10 as the cut-off point with strong AUC and ROC curve evident. Last but not least, our research provided persuasive proofs for logically and feasibly applying the International PHQ-9 Vietnamese version into the context of Vietnamese medical students.

This research, nevertheless, reveal some weaknesses as below. Firstly, our study targeted medical freshmen, not all the remaining medical students. Secondly, the activity of gathering at lecture halls could result in some psychological effects; listing as: some students were ashamed of being revealed their mental health conditions, some students were stressed when others finished the questionnaire ahead of time. Further researches in the future can apply this Vietnamese version of PHQ-9 on websites or mobile phone applications to increase convenience and privacy. Thirdly, this research, due to finance and time limitation, could not evaluate inter-rater reliability of the structured interview.

6 Conclusion

This New International Vietnamese version of PHQ-9 has proved suitable psychometrics: the Professional Board's consensus achieved the content validity; the internal reliability (Cronbach's alpha 0.81), and the test–retest reliability after 10-day length remained good. The cut-off score recommended for epidemiological purposes is 12 (sensitivity 71% and specificity 93%), and for depression screening on medical students was 7 (sensitivity 82% and specificity 71%). The testing length took approximately 2 min. Therefore, our new International Vietnamese version of PHQ-9 was concluded reliable, valuable and feasible enough to be applied for depression assessment on medical students.

7 Appendix 1

Vietnamese version of PHQ-9 and MINI

THANG ĐIỂM PATIENT HEALTH QUESTIONAIRE 9 (PHQ-9)

Họ tên: **Ngày:**

	Bạn hãy xác định bạn có bất kỳ vấn đề nào sau đây và mức độ các vấn đề này làm phiền bạn trong 2 tuần vừa qua. (Sử dụng dấu "X" để đánh dấu vào câu trả lời của bạn).	Không chút nào	Vài ngày	Hơn một nửa số ngày	Gần như mọi ngày
1	Ít hứng thú hoặc thỏa mãn khi làm việc gì đó	0	1	2	3
2	Cảm thấy chán nản, buồn rầu và vô vọng	0	1	2	3
3	Khó vào giấc ngủ, khó duy trì giấc ngủ hoặc ngủ quá nhiều	0	1	2	3
4	Cảm thấy mệt hoặc ít năng lượng	0	1	2	3
5	Chán ăn hoặc ăn quá nhiều	0	1	2	3
6	Cảm thấy bản thân tệ hại hoặc bản thân thất bại hoặc mình khiến bản thân/gia đình chán nản	0	1	2	3
7	Khó tập trung vào việc đang làm, ví dụ đọc sách, đọc báo hoặc xem ti vi	0	1	2	3
8	Vận động hoặc nói rất chậm đến mức người khác nhận thấy sự khác biệt; hoặc ngược lại, cảm thấy bồn chồn, bứt rứt đến mức đi tới đi lui nhiều hơn bình thường	0	1	2	3
9	Có suy nghĩ rằng tốt hơn mình nên chết đi hoặc suy nghĩ về việc tự làm tổn thương bản thân.	0	1	2	3
	Cộng từng cột				
	Tổng cộng				

10	Nếu bạn đã xác định có bất kỳ vấn đề nào ở trên, hãy cho biết các vấn đề đó đã gây khó khăn cho bạn như thế nào trong công việc, học tập, ở nhà, và trong mối quan hệ với người khác	☐ Không khó khăn chút nào ☐ Hơi khó khăn ☐ Rất khó khăn ☐ Cực kỳ khó khăn

Vietnamese version of Mini International Neuropsychiatric Interview

A. GIAI ĐOẠN TRẦM CẢM CHỦ YẾU

A1	Em có ở trong trạng thái buồn rầu hoặc chán nản dai dẳng, trong hầu như cả ngày, gần như mọi ngày, trong hai tuần vừa qua?	KHÔNG	CÓ
A2	Trong 2 tuần vừa qua, - em có cảm thấy ít hứng thú hơn nhiều trong hầu hết mọi chuyện - hoặc cảm thấy ít vui thích hơn nhiều với những thứ từng trước đây thường làm mình vui thích?	KHÔNG	CÓ
	A1 hoặc A2 được ghi CÓ	▸ KHÔNG	CÓ
A3	**Trong 2 tuần vừa qua, trong khi em đang buồn rầu hoặc mất hứng thú:**		
	a. Em có bị giảm hay tăng cảm giác ngon miệng trong hầu như mọi ngày? Cân nặng của em có giảm hoặc tăng ngoài ý muốn (có nghĩa là tăng giảm hơn 5% khối lượng cơ thể hoặc 3.5kg ở một người 70kg trong 1 tháng) NẾU TRẢ LỜI CÓ Ở BẤT KỲ CÂU NÀO THÌ GHI CÓ	KHÔNG	CÓ
	b. Em có gặp vấn đề về giấc ngủ hầu như mỗi đêm (khó vào giấc ngủ, thức giấc giữa đêm, thức giấc sáng sớm hoặc ngủ quá nhiều?)	KHÔNG	CÓ
	c. Em có nói hoặc vận động chậm chạp hơn bình thường? hoặc em có bồn chồn, bứt rứt, khó ngồi yên một chỗ trong hầu như mọi ngày?	KHÔNG	CÓ
	d. Em có cảm thấy mệt mỏi hoặc mất năng lượng trong hầu như mọi ngày?	KHÔNG	CÓ
	e. Em có cảm thấy mình vô dụng hoặc tội lỗi hầu như mọi ngày?	KHÔNG	CÓ
	f. Em có cảm thấy khó tập trung hoặc ra quyết định hầu như mọi ngày?	KHÔNG	CÓ
	g. Em có suy nghĩ lặp lại nhiều lần về việc làm hại bản thân, cảm thấy muốn tự sát hoặc ước gì mình chết đi cho rồi?	KHÔNG	CÓ

CÓ ÍT NHẤT 5 CÂU HỎI (A1-A3) ĐƯỢC GHI CÓ?

CÓ	**KHÔNG**
GIAI ĐOẠN TRẦM CẢM CHỦ YẾU, HIỆN TẠI	

NẾU BỆNH NHÂN CÓ GIAI ĐOẠN TRẦM CẢM CHỦ YẾU TRONG HIỆN TẠI, TIẾP TỤC TỚI CÂU A4, NẾU KHÔNG THÌ TỚI MODULE TIẾP THEO:

A4	a. Từ đó đến giờ, em đã từng có giai đoạn nào kéo dài ít nhất 2 tuần, trong đó e cảm thấy buồn rầu hoặc mất hứng thú trong hầu hết mọi chuyện, và gặp phải hầu hết các vấn đề chúng ta vừa nói ở trên?	▸ KHÔNG	CÓ
	b. Em đã từng có một giai đoạn ít nhất 2 tháng mà không có bất kỳ trầm buồn hay mất hứng thú nào giữa 2 giai đoạn trầm cảm?	KHÔNG GIAI ĐOẠN TRẦM CẢM CHỦ YẾU, TÁI PHÁT	CÓ

B. DYSTHYMIA

NẾU CÁC TRIỆU CHỨNG CỦA BỆNH NHÂN ĐÃ THỎA TIÊU CHUẨN CỦA GIAI ĐOẠN TRẦM CẢM CHỦ YẾU THÌ KHÔNG HỎI MODULE NÀY

B1	Em có cảm thấy trầm buồn trong hầu hết thời gian trong 2 năm qua?	KHÔNG	CÓ
B2	Giữa giai đoạn trầm buồn này, em có giai đoạn nào cảm thấy ỔN trong ít nhất hai tháng?	KHÔNG	CÓ
B3	**Trong giai đoạn em cảm thấy trầm buồn trong hầu hết thời gian:**		
	a. Cảm giác ngon miệng của em có thay đổi đáng kể?	KHÔNG	CÓ
	b. Em có gặp vấn đề về giấc ngủ hoặc ngủ quá nhiều?	KHÔNG	CÓ
	c. Em có cảm thấy mệt mỏi hoặc mất năng lượng?	KHÔNG	CÓ
	d. Em có cảm thấy mất tự tin vào bản thân?	KHÔNG	CÓ
	e. Em có cảm thấy khó tập trung hoặc khó ra quyết định?	KHÔNG	CÓ
	f. Em có cảm thấy vô vọng?	KHÔNG	CÓ
	Ít nhất 2 câu trả lời ở B3 ghi **CÓ**	▸ KHÔNG	CÓ
B4	Các triệu chứng của trầm cảm gây ra sự đau khổ hoặc suy giảm đáng kể chức năng xã hội, học tập, nghề nghiệp, hoặc các lĩnh vực quan trọng khác?	**KHÔNG**	**CÓ** **LOẠN KHÍ SẮC HIỆN TẠI**

O. RỐI LOẠN LO ÂU LAN TỎA

O1	a. Em có lo lắng quá mức về nhiều vấn đề khác nhau trong 6 tháng vừa qua?	▸ KHÔNG	CÓ
	b. Tình trạng lo lắng này xuất hiện trong đa số các ngày?	▸ KHÔNG	CÓ
	TÌNH TRẠNG LO ÂU CỦA BỆNH NHÂN GIỚI HẠN TRONG HOẶC ĐƯỢC GIẢI THÍCH TỐT HƠN BỞI BẤT KỲ RỐI LOẠN NÀO ĐÃ ĐƯỢC PHÁT HIỆN Ở CÁC PHẦN TRÊN?	KHÔNG	▸ CÓ
O2	Em có cảm thấy khó khăn trong việc kiểm soát sự lo lắng?	▸ KHÔNG	CÓ
O3	TRONG CÁC CÂU HỎI DƯỚI ĐÂY, GHI **KHÔNG** NẾU CÁC TRIỆU CHỨNG GIỚI HẠN TRONG CÁC RỐI LOẠN ĐƯỢC PHÁT HIỆN Ở CÁC PHẦN TRÊN.	KHÔNG	CÓ
	Khi em cảm thấy lo lắng trong 6 tháng qua, em có trải nghiệm những điều sau trong hầu hết thời gian:		
	a. Cảm thấy bồn chồn, bứt rứt hoặc bất an?	KHÔNG	CÓ
	b. Cảm thấy căng cơ? (đau đầu, đau mỏi vai gáy, đau mỏi lưng…)	KHÔNG	CÓ
	c. Dễ cảm thấy mệt, yếu hoặc kiệt sức?	KHÔNG	CÓ
	d. Khó tập trung hoặc đầu óc trở nên trống rỗng?	KHÔNG	CÓ
	e. Cảm thấy bực bội cáu gắt?	KHÔNG	CÓ
	f. Khó ngủ (khó vào giấc ngủ, thức giấc giữa đêm, thức giấc sáng sớm, hoặc ngủ quá nhiều?)	KHÔNG	CÓ

CÓ ÍT NHẤT 3 CÂU HỎI Ở O3 ĐƯỢC GHI CÓ?

CÓ	**KHÔNG**
RỐI LOẠN LO ÂU LAN TỎA, HIỆN TẠI	

8 Appendix 2

Acknowledgement of Pfizer for PHQ-9's authorities

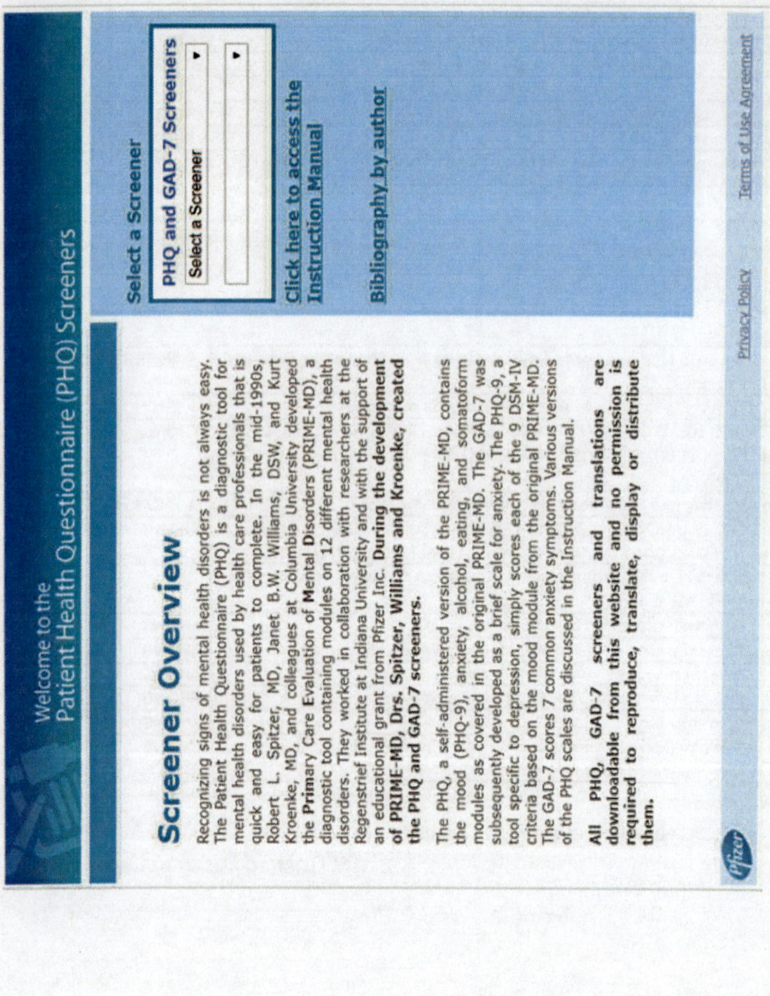
Welcome to the
Patient Health Questionnaire (PHQ) Screeners
Screener Overview
Recognizing signs of mental health disorders is not always easy. The Patient Health Questionnaire (PHQ) is a diagnostic tool for mental health disorders used by health care professionals that is quick and easy for patients to complete. In the mid-1990s, Robert L. Spitzer, MD, Janet B.W. Williams, DSW, and Kurt Kroenke, MD, and colleagues at Columbia University developed the Primary Care Evaluation of Mental Disorders (PRIME-MD), a diagnostic tool containing modules on 12 different mental health disorders. They worked in collaboration with researchers at the Regenstrief Institute at Indiana University and with the support of an educational grant from Pfizer Inc. During the development of PRIME-MD, Drs. Spitzer, Williams and Kroenke, created the PHQ and GAD-7 screeners.
The PHQ, a self-administered version of the PRIME-MD, contains the mood (PHQ-9), anxiety, alcohol, eating, and somatoform modules as covered in the original PRIME-MD. The GAD-7 was subsequently developed as a brief scale for anxiety. The PHQ-9, a tool specific to depression, simply scores each of the 9 DSM-IV criteria based on the mood module from the original PRIME-MD. The GAD-7 scores 7 common anxiety symptoms. Various versions of the PHQ scales are discussed in the Instruction Manual.
All PHQ, GAD-7 screeners and translations are downloadable from this website and no permission is required to reproduce, translate, display or distribute them.
Select a Screener
PHQ and GAD-7 Screeners
Select a Screener
Click here to access the Instruction Manual
Bibliography by author
Pfizer
Privacy Policy
Terms of Use Agreement

Acknowledgements We thank Dr. Ngô Tích Linh (University of Medicine and Pharmacy at Ho Chi Minh city) and Mr. Nguyễn Huỳnh Luân (University of Social Sciences and Humanities—Vietnam national university Ho Chi Minh city) for their essential roles in the Professional Board. We also wish to acknowledge the help provided by Mr. Trương Văn Đạt (UMP) in funding and initializing this project. This research was granted by the Ethical Board in terms of Bio-medical research at University of Medicine and Pharmacy at Ho Chi Minh City, Code 19452. ĐHYD. According to the announcement by Pfizer company patent ownership, PHQ-9 scoring is allowed to be used f'free of payment for clinical and researching purposes and linguistic translation without any prior permission (Appendix 2).

Conflicts of Interest The authors have no conflict of interest to declare.

References

1. WHO (2017) Depression and other common mental disorders: global health estimates. World Health Organization, Geneva
2. Rotenstein LS, Ramos MA, Torre M, Segal JB, Peluso MJ, Guille C, Sen S, Mata DA (2016) Prevalence of depression, depressive symptoms, and suicidal ideation among medical students: a systematic review and meta-analysis. JAMA 316(21):2214–2236. https://doi.org/10.1001/jama.2016.17324
3. Brazeau CM, Shanafelt T, Durning SJ, Massie FS, Eacker A, Moutier C, Satele DV, Sloan JA, Dyrbye LN (2014) Distress among matriculating medical students relative to the general population. Acad Med 89(11):1520–1525. https://doi.org/10.1097/acm.0000000000000482
4. Lê Minh T, Trần Thị Hồng N, Trần Quý Phương L (2018) Thực trạng trầm cảm trong sinh viên đại học. Y học Thành phố Hồ Chí Minh 22(1):166
5. Santen SA, Holt DB, Kemp JD, Hemphill RR (2010) Burnout in medical students: examining the prevalence and associated factors. South Med J 103(8):758–763
6. Linn BS, Zeppa R (1984) Stress in junior medical students: relationship to personality and performance. J Med Educ
7. Chang E, Eddins-Folensbee F, Coverdale J (2012) Survey of the prevalence of burnout, stress, depression, and the use of supports by medical students at one school. Acad Psychiatry 36(3):177–182
8. Hill MR, Goicochea S, Merlo LJ (2018) In their own words: stressors facing medical students in the millennial generation. Med Educ Online 23(1):1530558. https://doi.org/10.1080/10872981.2018.1530558
9. Dyrbye LN, Thomas MR, Shanafelt TD (2005) Medical student distress: causes, consequences, and proposed solutions. In: Mayo clinic proceedings, vol 12. Elsevier, Amsterdam, pp 1613–1622
10. Neumann M, Edelhäuser F, Tauschel D, Fischer MR, Wirtz M, Woopen C, Haramati A, Scheffer C (2011) Empathy decline and its reasons: a systematic review of studies with medical students and residents. Acad Med 86(8):996–1009
11. Dyrbye LN, Harper W, Durning SJ, Moutier C, Thomas MR, Massie FS Jr, Eacker A, Power DV, Szydlo DW, Sloan JA (2011) Patterns of distress in US medical students. Med Teach 33(10):834–839
12. Yamada Y, Klugar M, Ivanova K, Oborna I (2014) Psychological distress and academic self-perception among international medical students: the role of peer social support. BMC Med Educ 14(1):256
13. Rosal MC, Ockene IS, Ockene JK, Barrett SV, Ma Y, Hebert JR (1997) A longitudinal study of students' depression at one medical school. Acad Med J Assoc Am Med Coll 72(6):542–546
14. Stecker T (2004) Well-being in an academic environment. Med Educ 38(5):465–478

15. Trần Kim T (2012) Stress, lo âu và trầm cảm ở sinh viên y khoa. Y học TP Hồ Chí Minh 16(01):355–361
16. Tô Gia K, Lê Trường Vĩnh P, Huỳnh Ngọc Vân A (2019) Trầm cảm ở sinh viên khoa y tế công cộng. Y học Thành phố Hồ Chí Minh 23(2):120–126
17. Nguyen T, Kosik RO, Tran D, Ko Y-CI, Lu C, Fan AP (2016) Symptoms of depression and anxiety: a study of first and fifth year medical students in Vietnam. 醫學教育 20 (2):73–81
18. Huynh QHN, Tanasugarn C, Kengganpanich M, Lapvongwatana P, Khuong LQ, Thai TT (2020) Mental well-being, and coping strategies during stress for preclinical medical students in Vietnam. J Popul Social Stud [JPSS] 28(2):116–129
19. Quynh AT, Dunne MP, Ngoc HL (2014) Well-being, depression and suicidal ideation among medical students throughout Vietnam. Vietnam J Med Pharmacy 6(3):23–30
20. Kroenke K, Spitzer RL, Williams JB (2001) The PHQ-9: validity of a brief depression severity measure. J Gen Intern Med 16(9):606–613
21. Pham T, Bui L, Nguyen A, Nguyen B, Tran P, Vu P, Dang L (2019) The prevalence of depression and associated risk factors among medical students: an untold story in Vietnam. PLoS One 14(8):e0221432. https://doi.org/10.1371/journal.pone.0221432
22. Löwe B, Unützer J, Callahan CM, Perkins AJ, Kroenke K (2004) Monitoring depression treatment outcomes with the patient health questionnaire-9. Medical Care 1194–1201
23. Adewuya AO, Ola BA, Afolabi OO (2006) Validity of the patient health questionnaire (PHQ-9) as a screening tool for depression amongst Nigerian University students. J Affect Disord 96(1–2):89–93
24. Titov N, Dear BF, McMillan D, Anderson T, Zou J, Sunderland M (2011) Psychometric comparison of the PHQ-9 and BDI-II for measuring response during treatment of depression. Cogn Behav Ther 40(2):126–136
25. Yu X, Tam WW, Wong PT, Lam TH, Stewart SM (2012) The Patient Health Questionnaire-9 for measuring depressive symptoms among the general population in Hong Kong. Compr Psychiatry 53(1):95–102
26. Garabiles MR, Lao CK, Yip P, Chan EWW, Mordeno I, Hall BJ (2019) Psychometric validation of PHQ-9 and gad-7 in Filipino migrant domestic workers in Macao (SAR), China. J Pers Assess 1–12. https://doi.org/10.1080/00223891.2019.1644343
27. Eignor DR (2013) The standards for educational and psychological testing. In: APA handbook of testing and assessment in psychology, vol 1. APA handbooks in psychology®. American Psychological Association, pp 245–250
28. Muñiz J, Elosua P, Hambleton RK (2013) International Test Commission Guidelines for test translation and adaptation: second edition. Psicothema 25(2):151–157. https://doi.org/10.7334/psicothema2013.24
29. Guillemin F, Bombardier C, Beaton D (1993) Cross-cultural adaptation of health-related quality of life measures: literature review and proposed guidelines. J Clin Epidemiol 46(12):1417–1432. https://doi.org/10.1016/0895-4356(93)90142-n
30. Nakku JEM, Rathod SD, Kizza D, Breuer E, Mutyaba K, Baron EC, Ssebunnya J, Kigozi F (2016) Validity and diagnostic accuracy of the Luganda version of the 9-item and 2-item Patient Health Questionnaire for detecting major depressive disorder in rural Uganda. Glob Ment Health (Camb) 3:e20. https://doi.org/10.1017/gmh.2016.14
31. Zuithoff NP, Vergouwe Y, King M, Nazareth I, van Wezep MJ, Moons KG, Geerlings MI (2010) The patient health questionnaire-9 for detection of major depressive disorder in primary care: consequences of current thresholds in a crosssectional study. BMC Fam Pract 11:98. https://doi.org/10.1186/1471-2296-11-98
32. Manea L, Gilbody S, McMillan D (2012) Optimal cut-off score for diagnosing depression with the Patient Health Questionnaire (PHQ-9): a meta-analysis. CMAJ 184(3):E191-196. https://doi.org/10.1503/cmaj.110829
33. Ali G-C, Ryan G, De Silva MJ (2016) Validated screening tools for common mental disorders in low and middle income countries: a systematic review. PloS One 11(6):e0156939
34. Lambert MJ, Hawkins EJ (2004) Measuring Outcome in professional practice: considerations in selecting and using brief outcome instruments. Prof Psychol Res Pract 35(5):492

35. Hall BJ, Shi W, Garabiles MR, Chan EW (2018) Correlates of expected eMental Health intervention uptake among Filipino domestic workers in China. Global Mental Health 5
36. WHO (2019) Process of translation and adaptation of instruments. https://www.who.int/substance_abuse/research_tools/translation/en/. Accessed 19 Sept 2019
37. Bonett DG (2002) Sample size requirements for testing and estimating coefficient alpha. J Educ Behav Statistics 27(4):335–340
38. Sheehan DV, Lecrubier Y, Sheehan KH, Amorim P, Janavs J, Weiller E, Hergueta T, Baker R, Dunbar GC (1998) The mini-international neuropsychiatric interview (M.I.N.I.): the development and validation of a structured diagnostic psychiatric interview for DSM-IV and ICD-10. J Clin Psychiatry 59 Suppl 20:22–33;quiz 34–57
39. Lecrubier Y, Sheehan DV, Weiller E, Amorim P, Bonora I, Harnett Sheehan K, Janavs J, Dunbar GC (1997) The mini international neuropsychiatric interview (MINI). A short diagnostic structured interview: reliability and validity according to the CIDI. Eur Psychiatry 12(5):224–231. https://doi.org/10.1016/S0924-9338(97)83296-8
40. Gliem JA, Gliem RR (2003) Calculating, interpreting, and reporting Cronbach's alpha reliability coefficient for Likert-type scales. In: 2003. Midwest research-to-practice conference in adult, continuing, and community
41. Zou K, O'Malley A, Mauri L (2007) Statistical primer for cardiovascular research. receiver-operating characteristic analysis for evaluating diagnostic tests and predictive models. Circulation 115(5):654–657
42. Swets JA (1988) Measuring the accuracy of diagnostic systems. Science 240(4857):1285–1293. https://doi.org/10.1126/science.3287615
43. Streiner DL, Norman GR, Cairney J (2015) Health measurement scales: a practical guide to their development and use. Oxford University Press, Oxford
44. Kupfer DJ, Frank E, Perel JM (1989) The advantage of early treatment intervention in recurrent depression. Arch Gen Psychiatry 46(9):771–775. https://doi.org/10.1001/archpsyc.1989.01810090013002
45. Taylor DM, Barnes TR, Young AH (2018) Antidepressant Prophylaxis. In: The Maudsley prescribing guidelines in psychiatry. Wiley, New York, pp 306–307

Effect of Maxillary Skeletal Expansion on Upper Airway Patency and Respiratory Function: A Narrative Review

Sachin Kulkarni and **N. Doan**

Abstract Rapid maxillary skeletal expansion (RME) has been proposed as a treatment for Obstructive Sleep Apnea (OSA), and numerous studies exist on the effectiveness. However, conclusive results have not been obtained. The aim of this narrative review was to identify suitable studies to assess the effectiveness of maxillary skeletal expansion on airway and respiratory function. Studies were identified with the following search strategy: [Palatial expansion technique] AND [Respiratory function test OR pharyngeal volume] through four databases PubMed, Scopus, Cochrane and Google scholar. Inclusion criteria for studies included human in vivo studies and quantitative analysis. In the selected 9 studies, 6 studies primarily assessed function change with RME, and 3 studies primarily assessed dimensional change in the airway, most commonly the transverse dimension of the nasal valve. Not all studies were able to be accessed full-text, and only 4 studies were able to be included in the analysis. The significant increase was noted in the transverse dimension and nasal volume, as well as a decrease in inspiration and expiration resistance during breathing. We could conclude based on the available evidence that there was a positive increase in the volume of the airway, especially when an obstruction was caused due to nasal valve constriction, and reduction in nasal airflow resistance.

Keywords Rapid maxillary expansion · Nasal volume · Respiratory function · Inspiration resistance

1 Introduction

Obstructed nasal breathing, and/or mouth breathing is associated with a characteristic dentofacial deformity, especially maxillary hypoplasia, deviated septum, as well as hypertrophy of inferior turbinates [1]. Nasal breathing has been known to improve

S. Kulkarni
School of Dentistry, University of Adelaide, Adelaide, SA, Australia

S. Kulkarni · N. Doan (✉)
School of Dentistry and Oral Health, Griffith University, Gold Coast, QLD, Australia

V. Van Toi et al. (eds.), *8th International Conference on the Development of Biomedical Engineering in Vietnam*, IFMBE Proceedings 85,
https://doi.org/10.1007/978-3-030-75506-5_73

with individual and/or simultaneous treatment of the above three common deformities. This has come in favor with Oral and Maxillofacial surgery, as the complications associated with treating maxillary dysplasia (Le Fort 1 osteotomy), and intranasal procedures are not any higher when treated together compared to individually. The benefits have been known to be one admission for the patient and greater improvement in nasal breathing [1].

However, conservative treatment by an Orthodontist in growing individuals is also popular with Rapid maxillary expansion being the stapling procedure for improved nasal breathing and enlargement of the upper airway. This is backed by many studies begining with Eysel in 1886, who demonstrated that post RME, the width of the nasal cavity was increased resulting in larger intranasal capacity, and reduced resistance during breathing [2]. However, RME later became a treatment modality used by Orthodontists to manage dentofacial deformities, and it is only recently that the effect on nasal breathing and Obstructive sleep apnea (OSA) is becoming the reason for RME [3].

RME is a common distraction procedure that allows the growth of maxilla transversely by expanding at the mid-palatal suture. In growing individuals, this is considered to be a short period of treatment, typically with an expansion screw. It is typically done in patients from 4 years of age to an upper limit of 25 or 30 years of age. The results have been found to be better with younger patients. Indications typically include nasal septal deformity, dental crossbite, class III malocclusion, maxillary compression, and poor nasal airway. The original RME was a tooth-borne (TB) procedure where bands would be placed on permanent molars and expansion screws welded onto them [4]. Side effects include dental and alveolar bone tipping. Recently, Tooth-bone-borne (TBB) RME has been introduced with two mini-implants being anchored into the palate along with bands on molars to counteract the side effects [5]. Figure 1 illustrates and compares TB and TBB-RME.

However, currently, no systematic review is existing comparing and assessing the effects of TB and TBB RME on respiratory function and nasal dimensions.

With many studies now looking at the effect of RME on nasal cavity dimensions as well as respiratory function, this narrative review aimed at selecting studies based on inclusion criteria and comment on the current state of literature.

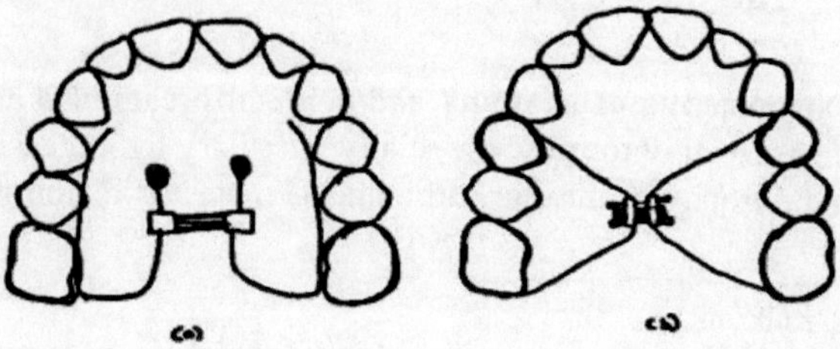

Fig. 1 Rapid maxillary expanders example **a** TBB-RME, **b** TB-RME

2 Methods

Upon definition of the topic, the following themes were identified:

1. Maxillary skeletal expansion
2. Airway volume assessment—nasal
3. Respiratory function test PFT—maximum inspiratory and expiratory pressure, oral peak expiratory flow and inspiratory nasal flow.

Based on the themes, Medical subject headings (MeSH) were selected:

1. (Palatial expansion technique) under corrective orthodontic covering all terms associated with Maxillary skeletal expansion
2. (Respiratory function test) covers airway resistance, lung volume, respiratory pressures, spirometry, pulmonary ventilation.

The search strategy was constructed: [Palatial expansion technique] AND [Respiratory function test OR pharyngeal volume]—and relevant studies were identified through three databases: PubMed, Cochrane, Google Scholar. Resulting articles were downloaded onto the Endnote reference manager. Duplicate articles were removed, and then they were first screened by title, and later the abstract. The inclusion criteria for the selected articles were:

- Outcome measure includes pharyngeal volume assessment with CBCT
- Assessment and diagnostic criteria used to assess for the indication of maxillary skeletal expansion
- Quantitative analysis conducted
- Human, in vivo study
- English language or translating available.

3 Results

4 Discussion

The functional matrix theory first introduced to the Orthodontic literature by Moss suggests that function such as nasal breathing is related to the growth of the dento-facial complex [6, 7]. There have been reports where a prolonged mouth breathing, especially in growing individuals, altered palatal vault to be narrower and higher [8]. The experimental interest in observing the reverse change, that is, the resulting change in function due to changes in structure produced by orthodontic treatment, in three dimensions, began as CBCT and CT became more readily available. Specifically, the method being tested was RME. This interest was compounded by the established fact that lateral displacement of the nasal cavity was associated with enlargement of the upper airway [8]. On average, this improvement occurred 6 months after RME, and stability of this was tested for up to 12 months. Clinically, this translates to decreased nasal resistance and possibly better nasal breathing [8]. In order to review the existing literature in support of this hypothesis, this narrative review was conducted.

In this narrative review, selected articles lacked this observation with only one out of four studies reviewing changes 6 months post-RME completion. Ottaviano et al. [9], conducted a randomized control trial with no reported blinding with TB-RME procedure and assessed for airflow change. Peak inspiratory flow significantly increased post-treatment; however, the study reporting mechanism seemed inappropriate—use of box plots to represent all data sets with no supporting tabulated values to read the actual results. In this study, it was not reported as to how large the data set was, which begs the question whether plotting individual points would have been a more accurate representation of the result, improving external credibility of the results. In line with this positive change in airflow, Bazargini conducted a single-blinded randomized control trial (RCT) with the comparison between TB and TBB-RME procedures for the outcome measures of airflow change as well as resistance change during inspiration and expiration. They concluded that TBB was significantly more effective in causing airflow change and had significantly less resistance [5]. No other study compared these two types of RME; hence no conclusion can be drawn from this comparison. This stand-alone study does, however, provide the framework for future studies in this evaluation and still proves that RME produces positive changes to airflow and reduced resistance in both cases [5].

Di Vece et al. [4] also conducted a non-randomized prospective longitudinal study to assess the effect of TB-RME on resistance change during inspiration and expiration. This was the only study that separated inspiration resistance and expiration resistance and reported on all data. In both cases, the resistance was almost halved and showed a significant change, with positive results like Bazargini [5]. Cappallete et al. [3] was the only study that reported the effect of TB-RME on nasal dimension change. It was also the only one of the two studies (other being Bazargini) to report the mean palatal expansion post-RME. Changes in nasal widths as well as area was reported in mm and mm^2. Significant changes were noted; however, no function test was done on those patients.

Essentially, wholistic measurements were not conducted whereby assessing mean expansion, nasal dimensions change, airflow change, and resistance change. The observation period was also deemed inadequate with only one of the four studies conducting follow up 6 months post RME treatment to allow for osseous changes to happen and for airflow changes to stabilize [3]. Additionally, only one study compared between male and female changes in RME and all others reported an approximate 50:50 male: female participation ratio (Table 1), adding a possible confounding variable. Only two out of four studies were randomized trials, and in order to achieve a high level of evidence, there is a need for rigorous protocols for RCT to be done with looking into all the above-mentioned limitations.

OSA has been identified as a unique challenge to the development of a child. It has been proposed to have varying effects on growth. Other than sleep fragmentation, and daytime fatigue and sleepiness, it has been suggested that the production of Growth hormone produced during slow-wave sleep is altered. In addition, a reduction in oxygen saturation due to OSA can place a child at a unique developmental hurdle.

Table 1 Characteristics and outcomes of included studies

Author(s)	Type of study	Gender ratio of participants [M:F]	Age range of participants	Blinding	Diagnosis	Tvpe of RME	Observation period	Mean expansion	Nasal dimension change	Airflow change	Resistance change
Bazargani et al. [5]	RCT	21:19	8–13, mean age 9.7 years	Single	Uni- or bilateral crossbite with the constricted maxilla	Tooth-borne and Tooth-bone-bome	Not reported	TB = 4.8 mm TBB = 5.48 mm	Not reported	TBB > TB $p = G.018$	TBB < TB $p = 0.005$
Di Vece et al. [4]	Prospective longitudinal study—nonrandomised	12:18	7–11, mean age 8.7 years	Not reported	Not specified	Tooth-borne	Not reported	Not reported	Not reported	Not reported	Inspiration:reduced by 48% $P < 0.001$ Expiration:reduced by 45% pO.OOL
Cappallete et al. [3]	Prospective longitudinal study—nonrandomised	35:2–6:00	6.5–13, Mean age 9.6 years	Not reported	Uni- or bilateral crossbite with the constricted maxilla	Tooth-borne	3 months after expansion	TB = 3.6 mm	Change in nasal width = 2.9 mm Change in nasal area 235.4 mm^2	Not reported	Not reported
Ottaviano et al. [9]	RCT			Not reported	Uni- or bilateral crossbite with the constricted maxilla	Tooth-borne	6 months after expansion	Not reported	Not reported	Peak inspiratory nasal flow = increased significantly (number not specified)	Not reported

RME has a place in orthopedically assisting nasal breathing and reducing the risk of OSA [8]. Results from the studies included in this narrative review are consistent with those proposed, whereby RME reduced resistance and improved volume of the nasal cavity and nasal breathing. However, limitations exist.

The purpose of this narrative review is also to evaluate existing studies to advise on the future direction of research. Future research should include blinding at the time of measurement, be randomized, compare the two mechanisms of RME (TB and TBB), and assess long-term effects beyond 6 months.

5 Conclusion

This narrative review suggests that rapid maxillary expansion, besides the limitations, can produce positive changes to nasal airflow and reduce resistance; however, the level of evidence is low. In order to produce evidence to support orthodontic practice, it will be necessary to address the limitations of outcome reporting and rigorous RCT protocols.

Conflict of Interest Authors declare no conflict of interest.

References

1. Posnick JC, Agnihotri N (2010) Consequences and management of nasal airway obstruction in the dentofacial deformity patient. Curr Opin Otolaryngol Head Neck Surg 18(4):323–331
2. Haas AJ (1961) Rapid Expansion of the maxillary dental arch and nasal cavity by opening the midpalatal suture. Angle Orthod 31(2):73–90
3. Cappellette M Jr, Nagai LHY, Goncalves RM, Yuki AK, Pignatari SSN, Fujita RR (2017) Skeletal effects of RME in the transverse and vertical dimensions of the nasal cavity in mouth-breathing growing children. Dental Press J Orthod 22(4):61–69
4. Di Vece L, Doldo T, Faleri GPM, Salerni L, Ugolini A, Goracci C (2018) Rhinofibroscopic and rhinomanometric evaluation of patients with maxillary contraction treated with rapid maxillary expansion. A prospective pilot study. J Clin Pediatr Dent 42:27–31
5. Bazargani F, Magnuson A, Ludwig B (2018) Effects on nasal airflow and resistance using two different RME appliances: a randomized controlled trial. Eur J Orthod 40:281–284
6. Moss ML (1962) The functional matrix. In: Kraus B, Reidel R (eds) Vistas in orthodontics. Lea and Febiger, Philadelphia, PA, pp 85–98
7. Moss ML (1997) The functional matrix hypothesis revisited. The role ofmechanotransduction. Am J Orthod Dentofacial Orthop 112:410–417
8. McNamara JA, Lione R, Franchi L, Angelieri F, Cevidanes LH, Darendeliler MA, Cozza P (2015) The role of rapid maxillary expansion in the promotion of oral and general health. Prog Orthod 16:33
9. Ottaviano G, Maculan P, Borghetto G, Favero V, Galletti B, Savietto E, Scarpa B, Martini A, Stellini E, De Filippis C, Favero L (2018) Nasal function before and after rapid maxillary expansion in children: a randomized, prospective, controlled study. Int J Pediatr Otorhinolaryngol 115:133–138

An Evaluation of the Correlation Between Temporomandibular Disorders Defined by Joint Vibration Analysis Both with Sleep Disorder Breathing and Patient Characteristics

A. Robinson, P. Reher, and N. Doan

Abstract Limited research exists on the correlation between sleep disorder breathing and TMD. The aim of this study is to assess the relation between TMJ noises generated using Joint Vibration Analysis (JVA) and the most common subtype of TMD; TMD and SDB, and variables amongst patients and TMD. This study is a 10-year retrospective analysis of 68 patients. Patients were selected for the study if they possessed a full set of dental records comprising of demographics, cone beam computed tomography (CBCT), and JVA data. Based on a clinical diagnosis of SDB, patients were divided into two groups, patients with SDB (n = 37) and those without SDB (n = 31). JVA was used to diagnose patients with TMD (n = 56) and those without TMD (n = 12) as well as to identify the most common subtype of TMD. SDB and non-SDB patient groups were compared to each other in order to identify whether a link concerning SDB and TMD and SDB versus other TMD subtypes. Finally, a multivariant analysis was done in order to identify a correlation between patient characteristics and TMD using Graph Pad Prism v8.2.0. The analysis of results indicated that ligamentous laxity was the most common diagnosis of TMD defined by JVA. No statistically significant association identified between SDB and TMD cohort ($p = 0.7598$); TMD subtypes in patients with SDB. However, multivariant analysis of patient characteristics below average patient oral hygiene was significantly associated with TMD ($p = 0.0309$). JVA successfully identified ligamentous laxity as the most common diagnosis of TMD. Due to limitations in this patient cohort, an association between TMD and SDB was not able to be established in this analysis. The correlation between TMD and SDB requires further research. If an association does exist, JVA represents a novel tool for the early identification of TMD, which may lead to earlier diagnosis and intervention for SDB.

Keywords JVA · CBCT · Sleep-disordered breathing · TMD

A. Robinson · P. Reher
School of Dentistry, Griffith University, Gold Coast, QLD, Australia

N. Doan (✉)
The University of Queensland, Griffith University, Gold Coast, QLD, Australia

V. Van Toi et al. (eds.), *8th International Conference on the Development of Biomedical Engineering in Vietnam*, IFMBE Proceedings 85,
https://doi.org/10.1007/978-3-030-75506-5_74

1 Introduction

SDB has become increasingly more common in the general community. Numerous illnesses are covered under the umbrella of SDB, and its gravity varies from snoring to sleep apnea. The more severe disorders are characterized by atypical displays of respiration and changes in oxygen saturation during sleep and are associated with adverse outcomes across multiple organ systems. SDB patients commonly report noises in their TMJ suggestive of underlying TMD. Multiple papers have identified an association between TMD and JVA. However, currently, little research has been done in identifying a link between SDB and TMD. The aim of this study is to identify the most common subtype of TMD diagnosed by BioPak software through JVA. Secondly, this study aims to investigate whether an association exists between SDB and TMD and whether certain TMD subtypes have a greater association with SDB. Finally, the study will look at which variables amongst patients are associated with a higher likelihood of TMD. If an association between SDB and TMD exists, the JVA device could be used to identify at-risk patients and offer early intervention.

2 Literature Review

2.1 Definition and Pathophysiology of Sleep Disordered Breathing

During sleep for healthy patients, the air is seen to move in and out of the lungs at a consistent rhythm. In patients who suffer SDB, this air movement is occasionally reduced or stopped completely. Sleep-disordered breathing indicates to brief, frequently cyclical, stops in breathing pattern (apnoeas) or temporary or persistent reductions in-breath amplitude (hypopneas) [1]. There are two main types of SDB which are demonstrated in sleep apnea (SA).

- Obstructive sleep apnoea (OSA) can be characterized by decreases or pauses of airflow in sleep, notwithstanding continuing respiratory effort. It is owing to upper airway impediment.
- Central sleep apnoea is characterised by recurring crescendo-decrescendo breathing exertion and airflow in alertness or sleep, exclusive of upper airway blockage.

The upper airway is a complicated arrangement needed to accomplish multiple intricate motor behaviours. Partly, the hyoid bone which is an important attaching location for pharyngeal dilator muscles is to blame for upper airway obstruction in sleep as it is not firmly attached to skeletal structures [1]. Hence the human pharynx has no firm backing excepting for the extreme upper and lower ends.

Reports by means of computer tomography (CT), nasal pharyngoscopy and magnetic resonance imaging (MRI) have revealed that the retropalatal region of the oropharynx is the usual utmost site of airway failure in OSA [1]. However, airway tapering is an active development that differs noticeably amongst patients and can also include the retroglossal and hypopharyngeal areas (see Fig. 1) [2].

Expansion of soft tissue, including the soft palate and tongue, around the upper airway (UA), renders drastically to pharyngeal airway contraction in the anterior–posterior plane in many cases of OSA (see Fig. 2) [5]. While thickening of the lateral pharyngeal partitions has been indicated to be the main location of airway concession in the lateral dimension (see Fig. 3) [3]. As well as this, OSA individuals exhibited retroposed and undersized lower jaws, inferiorly positioned hyoid bones, lengthier soft palates, broader uvulas, and greater slenderer hard palates [6].

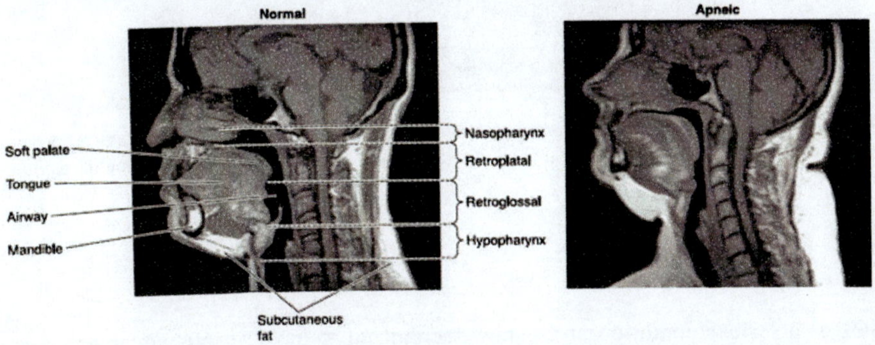

Fig. 1 Picture of midsagittal magnetic resonance (MRI) in a typical patient (left) and a subject with advanced OSA (right). Underlined are the four upper airway areas and upper airway soft tissue [3, 4]

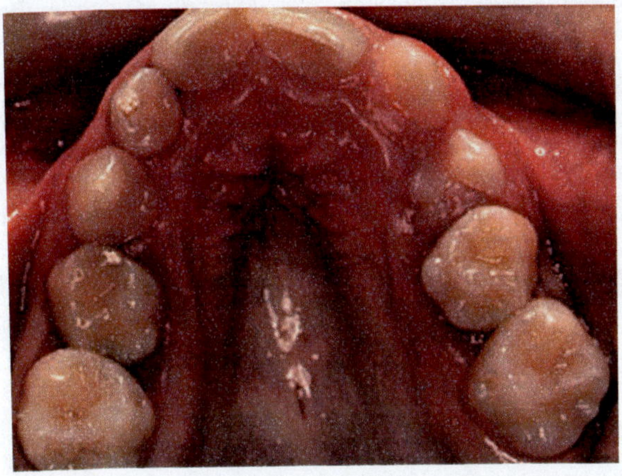

Fig. 2 Palate with a high and narrow maxillary arch in a subject has obstructive sleep apnoea. The tongue cannot fit into the narrow palate. Therefore, the tongue tends to move backwards when the patient is supine, blocking the airway [7, 8]

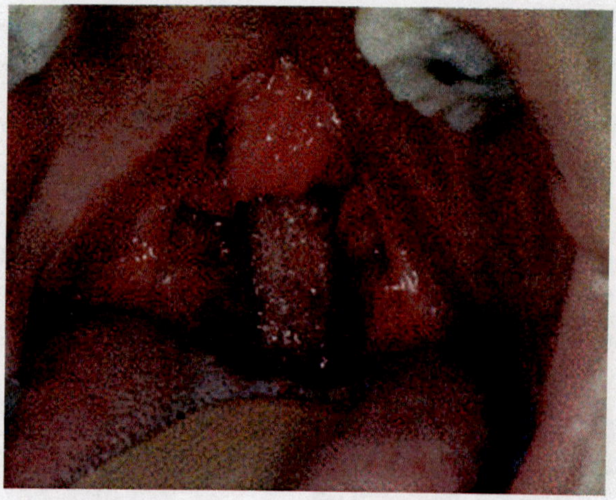

Fig. 3 This photograph is from a snoring patient in whom obstructive sleep apnea was excluded by polysomnography. The oropharyngeal narrowing is present in the lateral dimension due to residual tonsillar tissue [9, 10]

2.2 Epidemiology

SDB is a serious condition and is now recognized to have serious adverse consequences [11]. It can affect a person's ability to safely perform daily activities, including disproportionate daytime somnolence, behaviour and cardiovascular illness, and death. Approximately 25% of adults are at risk for sleep apnea of some grade [12].

OSA is currently the very usual sleep-related breathing disorder. It is more prevalent among older men and postmenopausal women; however, it also has been shown in children [13, 14]. The presence of OSA has also been shown to vary with race. OSA is farther widespread in African Americans age less than 35 when contrasted to Caucasians in the same age bracket [15]. It was also noted that OSA prevalence in Asia is similar to the United States, despite lower rates of obesity [16].

Multiple reports have shown that the occurrence of OSA seems to be on the rise, whether this is due to increased awareness and diagnosis of OSA or obesity is unclear [12, 17]. One study estimated the prevalence of OSA between 1990 and 2010 to have increased from 11 to 14% in adult males [12].

2.3 Risk Factors and Diagnosis

Many risk factors are linked with OSA and include Older age [18, 19], Male gender [15, 20], Obesity [1] and body mass index (BMI) [21], Craniofacial and upper airway anomalies [10, 22], Smoking [22], Big neck circumference [23].

In general, patients who suffer from OSA will complain of daytime sleepiness, or their partner will report loud snoring, choking, or interruptions in breathing while sleeping. The diagnosis of OSA is established both on the existence or lack of related symptoms, as well as the occurrence of respiratory incidents during sleep. The ideal standard diagnostic test for the rate of breathing incidents in sleep is in-laboratory polysomnography (PSG).

The most common variety of evaluation tool used in the diagnosis of OSA, which is readily available for clinicians is the STOP-Bang questionnaire (see Table 1). This tool is often used by clinicians preoperatively to assess the risk of undiagnosed OSA [24].

Table 1 STOP-Bang questionnaire [9, 25]

STOP-Bang questionnaire		
Yes	No	**Snoring?** Do you **snore loudly** (loud enough to be heard through closed doors, or bed partner elbows you for snoring at night?)
Yes	No	**Tired?** Do you often feel **tired, fatigued or sleepy** during daytime? (such as falling asleep during driving?)
Yes	No	**Observed?** Has anyone **observed** you **stop breathing** or **choking/gasping** during your sleep?
Yes	No	**Pressure?** Do you have or are being treated for **high blood pressure?**
Yes	No	**Body mass index more than 35 kg/cm^2?**
Yes	No	**Age older than 50 years old?**
Yes	No	**Neck size large? (measure around Adam's apple)** For male, is your shirt collar 17 inches or larger? For female, is your shirt collar 16 inches or larger?
Yes	No	**Gender = male?**
Scoring criteria:		
Low risk of OSA: Yes to 0–2 questions		
Intermediate risk of OSA: Yes to 3–4 questions		
High risk of OSA: Yes to 5–8 questions		

2.4 Understanding TMD and JVA

TMDs are common; however, they are not easily classified. Substantial morbidity results from patients who suffer from TMD. It is estimated in the US that for each 100 million working individuals, TMD will result in 17.8 million working days lost yearly [26].

The pathogenesis of TMDs is multifactorial with the greatest frequent being temporomandibular joint (TMJ) damage. Trauma to the TMJ ligaments, articular cartilage, disc, and bone leads to a disparity between free radicals and antioxidants in the body, recognized as oxidative stress. This, in turn, leads to free radical generation in the intra-radicular space and subsequent synovial fluid inflammation, producing cytokines that cause degenerative TMJ disease. The initial joint trauma leading to this degeneration can result from external injuries to the jaw, or bruxism [27].

The major clinical manifestations of TMD revolves around the two distinct elements which contribute to TMD. These being muscle dysfunction and specific problems within the TMJ. Thus, the major symptoms are a pain in the muscles of chewing or pre-auricular zone, joint noises, and limitation in jaw function [28].

JVA utilizes accelerometers to capture pulsations result from movement of two irregular sides, being in this instance the disc and condyle surfaces (see Fig. 4). In theory, joints in the right biomechanical connexion should make a very minute rubbing, and hence slight vibration which can be detected by JVA [29]. Changes in the two irregular surfaces, for example, those produced by degeneration, tear of the disc, or displacement of the disc, commonly create rubbing and hence vibration

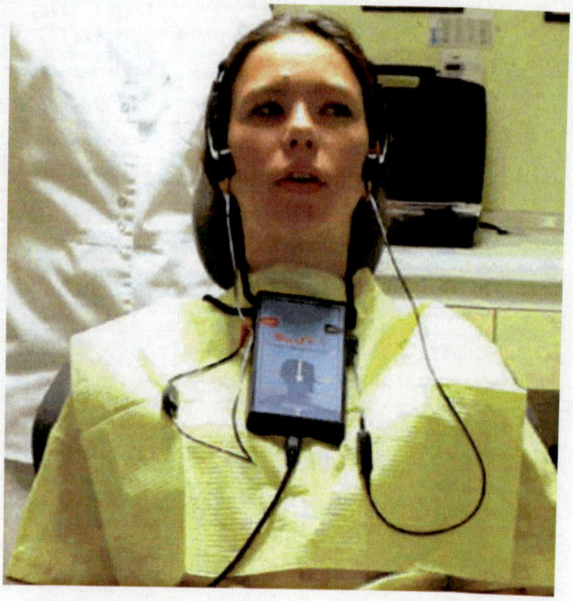

Fig. 4 Joint vibration analysis procedure [25, 34]

[30]. It was hypothesized that distinctive conditions might yield diverse patterns of vibration [31].

The most broadly applied tool for diagnosis of TMD, The Research Diagnostic Criteria aimed at temporomandibular disorder (RDC/TMD), uses TMJ sounds for patient diagnosis [32]. The TMJ noise kinds employed in RDC/TMD diagnosis are identified by particular measures of these noises by palpation and auscultation [30]. Traditionally the diagnostic accuracy of TMJ sounds is questionable due to the inherent inconsistency of joint noises and the trivial inter-examiner consistency. The advent of objective tools to measure and record TMJ sounds resulted in the development of the JVA.

The analysis of vibrations detected by the JVA of the TMJ is a measurable procedure that gauges the intensity and frequency dispersal of vibratory waves emitted from the joint during its full range of movement [33]. There are two companies, BioResearch and Myotronics, that presently promote devices that can be employed for appraisal of JVA data. The US Food and Drug administration accepts the BioResearch advertised device for the ensuing indications:

1. To trace and exhibit sound/noises from the TMJ
2. Help the practitioner in the study of joint sound/noise by permitting him/her to realise the wave-form in several typical plans
3. To aid the practitioner in equating a patients' existing normal graphs to preceding records before, during and after usage.

2.5 *TMD Diagnosis Subcategories as Defined by JVA*

The analysis of TMJ joint noises generated using JVA is the focus of this study, with the conditions of TMD diagnosed into the following subgroups as determined by the JVA flow chart (see Fig. 5).

1. Disc displacement with reduction (DDR)
2. Disc displacement without reduction (DD)
3. Degenerative joint disease (DJD)
4. Ligamentous laxity (LL)
5. Deliver mathematical standards that can be utilized to calculate the objective features of the sounds/vibrations, permitting between-patient contrasts by the practitioner [33]. The reliability of the vibrations captured by JVA has been evaluated in previous studies. All variables displayed decent to outstanding consistency throughout various sessions and through diverse dates [30]. Hence JVA may well provide a relatively quick, non-invasive, and easily repeatable technique to record the health of the TMJ.

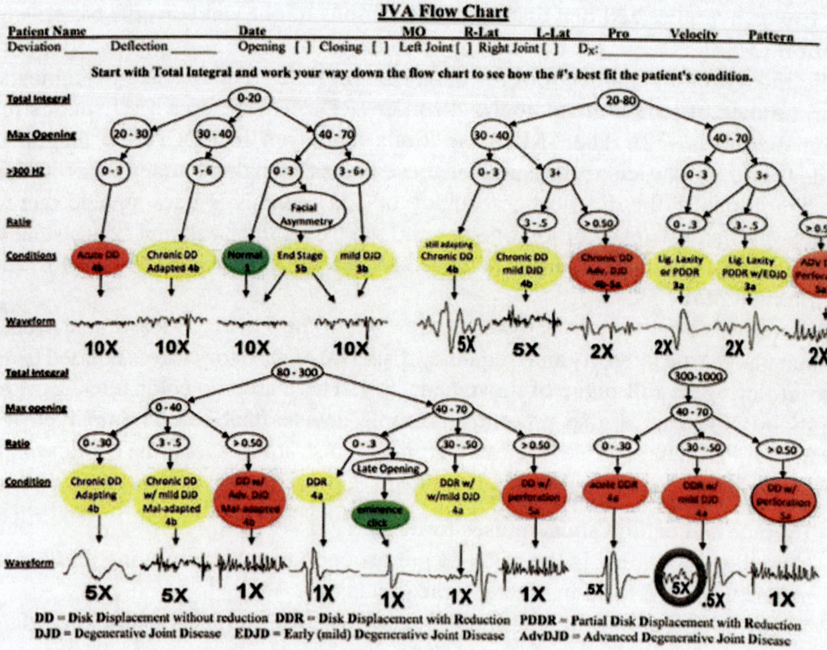

Fig. 5 Joint vibration analysis flow chart [17, 35]

2.6 Healthy Relationship Within TMJ and Association Between SDB and TMD

It is best to first understand a healthy relationship within the TMJ joint capsule on opening and closing before discussing dysfunction (see Fig. 6). There are two different activities, rotation and translation, which happen in the TMJ throughout opening and closing. The initial stage of mandibular opening, being turning of the mandibular condyle inside the glenoid fossa, is initiated by the suprahyoid muscles. The additional opening is because of a frontward translation, guided by the lateral pterygoid muscle, of the condyle alongside the posterior angle of the articular prominence. Finally, the action of mandibular closure is directed by the temporalis, masseter, and medial pterygoid muscle groups. During temporomandibular articulation, a fibrocartilaginous disc divides the condyle and glenoid fossa from straight interaction and apportions the joint area into inferior and superior zones [36].

The correlation between TMD and SDB is contentious. This is contributed by the lack of consistency in diagnostic aspects of both disorders. In support of this association, it has been found that a bulk (>50%) of subjects with TMD complain of deprived sleep characteristic [38]. Additionally, Smith et al. noted that 43% of the patients with TMD in their study had evidence of sleep disorders [39]. Furthermore,

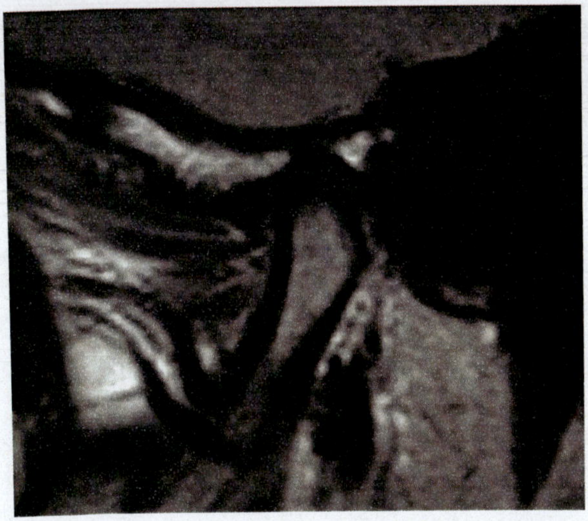

Fig. 6 Magnetic resonance images show the healthy TMJ disc position. The posterior band of the disc is located above the apex of the condylar head (at the 12 o'clock position) with the mouth closed [24, 37]

Dubrovsky et al. [40] using PSG on patients with TMD, found a pattern of sleep stage alterations as well as an increase in respiratory effort-related arousals [4, 40].

On the other hand, most previous research has focused only on the relationship between sleep bruxism and TMD. Bruxism is defined by the American Academy of Sleep Medicine, as an oral motor complaint categorized by sleep-related crunching and/or clenching of the teeth [8]. Alterations in the pattern of sleep, micro-arousal phenomena, that may be caused by SDB appear to show the main role in bruxism [41]. Of entire the parafunctional activities of the jaw system, bruxism is contemplated to be a major risk feature for TMD [42]. Manfredini et al. [42] carried out a systematic review noted that solid backing for the connection concerning bruxism and TMD only derived from reports which utilized a scientific bruxism analysis, whereas other studies which adopted diagnostic modalities failed to confirm the association [42]. Current evidence can therefore only support a weak association between TMD and bruxism, and hence SDB. This study aims to explore this gap in the literature.

3 Methodology

3.1 Study Design

Aims

The aims of this study are:

1. To analyze and quantify TMJ noises generated using JVA to determine the most common subtype of TMD diagnosed by BioPak software.
2. To analyze whether an association exists between TMD and SDB and whether certain TMD subtypes have a greater association with SDB.
3. To identify which variables amongst patients are associated with a higher likelihood of TMD.

Hypotheses

1. I hypothesize that JVA will be able to determine the most common subtype of TMD.
2. I hypothesize that there will be an association between TMD and SDB and that certain TMD subtypes will have a greater association with SDB.
3. I hypothesize that certain patient characteristics will be associated with TMD.

3.2 Methods

The design of this study is a cross-sectional retrospective study. Patients were selected from private dental practice in Brisbane, with data expanding from 2008 to 2018. The study sample was completed by methods of convenience sampling where subjects over the age of 18 that had an existing JVA and CBCT with adequate health records were randomly selected. Participants voluntarily gave informed consent in accordance with human ethics approval from Griffith University (Ref No: 2017/626). The randomization the indiscriminate consignment of study subjects to exposure groups to stopping any associations between exposure and confounders. Participants willingly provided informed consent in line with human ethics approval from Griffith University (Ref No: 2017/626). The study population consists of 68 individuals with age ranging from 22 to 85 years of age, of which 47 were female, and 21 were male. The attempt was made to match for age and sex. However, due to the small sample size, this was not possible. A power analysis has shown that a full sample of 270 participants would be required to identify sizeable outcomes (80% chance), built on a t-test at 5% level of significance. However, due to available data and randomization, no further participants were included.

In addition, the selected CBCT scans were acquired in a vertical direction only as this were the most common capturing position. This may affect the outcomes of the collected data.

In addition, patients gave informed consent for data collection to be used in further research. A full set of dental records comprising of demographics, CBCT, and JVA data. If this data was not available, then patients were excluded. Based on a clinical diagnosis of SDB as determined by the primary examiner, patients were divided into two groups, patients with SDB ($n = 37$) and those without SDB ($n = 31$). JVA was used to diagnose patients with TMD ($n = 56$) and those without TMD ($n = 12$) as well as to identify the most common subtype of TMD. SDB and non-SDB patient groups were compared to each other in order to identify whether an association between SDB and TMD exists. Further subgroup analysis of TMD subtypes was carried out to ascertain whether certain TMD subtypes were more strongly associated with SDB compared to other TMD subtypes. Finally, a multivariant analysis was done in order to identify a correlation between patient characteristics and TMD. The data set was analysed using Graph Pad Prism v8.2.0 (GraphPad Software, Inc., La Jolla, CA).

3.3 Data Collection and Statistical Analysis

The subsequent data (2008–2018) were gathered for entire participants from demographic information (age, gender, ethnicity), smoking standing, physical features and characteristics linking to sleep and temporomandibular joint disorders. These cofounding factors were inferred to take into account of SDB risk factors, for example, age, obesity, sex, family history, craniofacial anomalies, and smoking history.

Furthermore, we also gathered inspection statistics documented by an oral surgeon skilled in the assessment of sleep disorders. This comprised of vital signs, height, weight, and neck perimeter dimensions, palpitation findings of muscle and joint, joint vibration analysis (JVA), a TMD planned questionnaire, the incidence of micrognathia, tongue shape evaluation (scalloped, above occlusal plane), the occurrence of tori, uvula changes and Mallampati's score. To identify patients with possible TMD, the Diagnostic Criteria for Temporomandibular Disorders for Clinical and Research Applications was also employed.

Continuous variables are shown as mean ± standard error of the mean (SEM) if discovered to keep on a Gaussian distribution consistent with a D'Agostino-Pearson omnibus normality test, or as median ± lower and upper quartiles, if seen to track a non-Gaussian distribution. Normally distributed variables were equated by means of Student's t-test and non-normally distributed variables by means of Mann–Whitney U test. Categorical variables are expressed as percentages. A Chi-square analysis was done to equate categorical variables.

A *p*-value of ≤ 0.05 was regarded as statistically significant for all investigates. Of the statistically significant data set, an odds ratio and confidence interval using Baptista-Pike method was calculated.

Entire studies were accomplished by means of Graph Pad Prism v8.2.0 (GraphPad Software, Inc., La Jolla, CA).

3.4 Results

In patients with TMD, there is a statistically significant difference in diagnosis as defined by JVA. The most common diagnostic criteria in this patient population were LL, followed by DDR, followed by LL + DDR, followed by DD, with the least

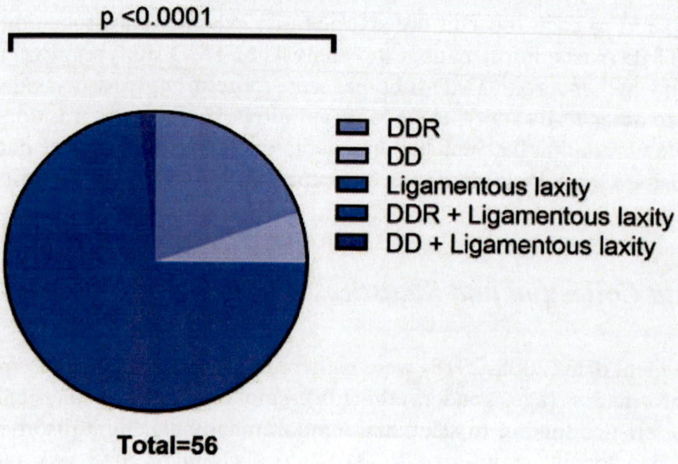

Fig. 7 Pie chart showing the most common subtype of TMD as defined by JVA

common diagnostic criteria being DD + LL (64.3% vs. 19.6% vs. 8.9% vs. 5.4% vs. 1.8%; $p < 0.0001$; see Fig. 7).

Analysis for an association exists between TMD and SDB and whether certain TMD subtypes have a greater association with SDB

In patients with SDB, there was no statistical significance between those diagnosed with TMD by JVA versus those without a diagnosis of TMD ($p = 0.7598$). Furthermore, in a subgroup analysis of TMD subtypes, there was no statistical significance between TMD subtypes and the likelihood of SDB ($p = 0.8010$; see Fig. 8).

Identification of which variables amongst patients are associated with a higher likelihood of TMD

In a multivariant analysis comparing patient characteristics between TMD and non-TMD groups, there were no statistically significant differences between TMD and non-TMD groups except for patient oral hygiene evaluation. There was a statistically significant percentage of patients with average or below average oral hygiene in the TMD group versus the non-TMD group (above average = 7.14%, average = 58.93%; below average = 33.93% in TMD group vs above average = 33.34%; average = 33.33%; below average = 33.33% in non TMD group; $p = 0.0309$; see Fig. 9). An odds ratio and confidence interval using Baptista-Pikes' method for oral hygiene was calculated (OR = 6.5; CI 95% = 1.569–25.08) (Table 2).

Highlights

SDB patients tend to have the following features: sizable neck perimeters, Mallampati Class II and III, the inadequacy of cervical motion, forward head posture, incompetent lips, nasal deviance, skeletal Class II, small jaw/micrognathia, pain on palpation of chewing muscles, TMJ noises and clinical indication of TMD.

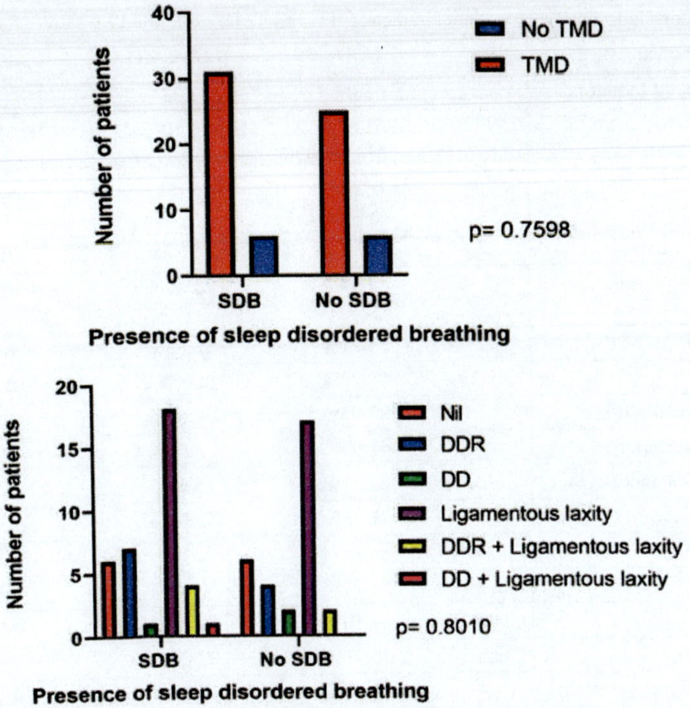

Fig. 8 No statistical significance between SDB and TMD. Furthermore, there was no statistical significance in the TMD patient population groups and SDB

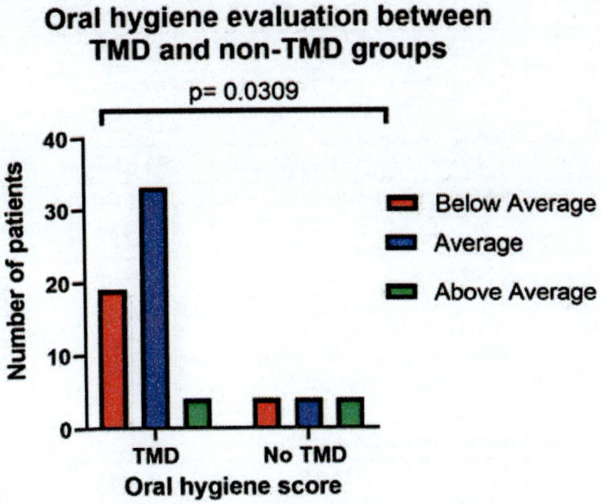

Fig. 9 Statistically significant percentage of patients with average or below average oral hygiene in the TMD versus non-TMD group

Table 2 Correlation between patient characteristics and TMD as defined by JVA

Patient characteristics		TMD	No TMD	*p*-value
Gender (% of female)		66.07	83.33	0.3169
Age		63.82 (53.98–69.68)	64.10 (54.71–74.72)	0.981
Weight		162.12 (54–81.25)	73 (52.75–76.5)	> 0.9999
Height		162.11 (±2.99)	161.75 (±2.99)	0.9052
Neck circumference > 40 cm		32.14%	25%	0.7416
BMI > 30		35.71%	33.33%	> 0.9999
Smoker		14.29%	8.33%	> 0.9999
Forward head posture		83.93%	66.67%	0.2223
Lip incompetence		41.07%	25.00%	0.3481
Dark circles under eyes		84.31%	15.69%	0.4773
Nasal deviation		71.43%	50.00%	0.182
Skeletal profile				
	Class 1	51.79%	50.00%	0.4316
	Class 2	37.50%	50.00%	
	Class 3	10.71%	0.00%	
Micrognathia		25.00%	16.67%	0.7171
Mn Range of Motion < 35 mm		1.79%	0.00%	> 0.9999
Dental classifications				
	Class 1	55.36%	66.67%	0.4767
	Class 2	33.93%	33.33%	
	Class 3	10.71%	0.00%	
Periodontal status				
	None	21.43%	33.33%	0.6765
	Mild	48.21%	41.67%	
	Moderate	30.36%	25.00%	
	Severe	0.00%	0.00%	
Oral removable prosthetics		5.36%	16.67%	0.2107
Limited cervical range of movement		35.71%	16.67%	0.3113
Professional dental oral hygiene evaluation				
	Below average	33.93%	33.33%	0.0309*
	Average	58.93%	33.33%	

(continued)

Table 2 (continued)

Patient characteristics		TMD	No TMD	*p*-value
	Above average	7.14%	33.33%	
Mallampti Classification	Class 1	39.29%	41.67%	0.9622
	Class 2	39.29%	41.67%	
	Class 3	19.64%	16.66%	
	Class 4	1.78%	0.00%	

4 Discussion

JVA is the first device that can absolutely quantify TMJ noises on opening and closing. JVA demonstrates several advantages over traditional methods of examination of the TMJ for symptoms of TMD. These advantages include not having to rely on inconsistent inter-examiner reliability, being able to pick up vibrations far below the frequency of human hearing, offering an absolute and simplified method of quantification, the ability to diagnose different signatures of vibratory waves within the joint capsule, and greater reproducibility of results. JVA is able to quantify vibration within the TMJ capsule and measure intensity and frequency distributions of vibratory waves that are stemming from the TMJ. Given JVAs proposed advantages, it is of particular interest in being able to more reliably diagnose patients with TMD in general dental surgery [33].

Determination of the most common subtype of TMD diagnosed by BioPak software

After acquiring data from all 68 patients involved in the report, 12 of the 68 subjects were not included in the analysis due to not being diagnosed into a subcategory of TMD defined by JVA BioPak software. Of the remaining 56 patients diagnosed with having vibratory waves indicative of TMD detected by JVA, the most common diagnosis was LL (64.3%).

LL of the TMJ is defined as larger than usual range of motion in the joint and suggests joint instability [43]. As indicated by the RDC/TMD, LL can be divided into two subcategories based on the severity of symptoms: these being subluxations and locations [44]. There are multiple etiological predictors of TMD development that have been postulated, one of which is joint hypermobility, also known as LL. In a clinical paper published in 2012, Ogren et al. indicated that general joint hypermobility is an imperative etiological aspect for the progress of TMD [45]. The findings in this study agree with this and suggest LL is the main pathological driver of TMD. This is an important correlation as LL may be considered as an early predictor and a significant factor in the development and advancement of TMD. Further research is warranted to identify the pathogenesis of LL as well as the role of therapeutic interventions.

Finding the association exists between TMD and SDB and whether certain TMD subtypes have a greater association with SDB

The data acquired from the patient cohort was analyzed to look for a correlation between patients with symptoms of SDB and prevalence of TMD diagnosed by JVA showed no statistical significance found between those patients with symptoms of SDB diagnosed with TMD defined by JVA versus those without a diagnosis of TMD ($p = 0.7598$).

Additionally, in a subgroup analysis of TMD subtypes, there was no statistical significance between TMD subtypes and the likelihood of SDB. It was hypothesized that varying degrees of severity of TMD would show a greater correlation in incidences of patients with symptoms of SDB, but this was not shown in this study.

The correlation between SDB and TMD is contentious. Smith et al. noted that 43% of patients that had diagnosed TMD had two or more sleep disorders [46]. Furthermore, [39] Dubrovsky et al. using polysomnographic data were able to determine that patients with TMD have a mild degree of sleep disturbances and a mild increase in upper airway resistance during sleep [39]. Instead, in a systematic review, Manfredini et al. were only able to establish a weak correlation between SDB and TMD [42]. In my patient cohort, I was not able to identify a correlation between SDB and TMD.

Results from this study indicate that an association between TMD and SDB does not exist. However, as discussed further below, this study had multiple limitations that impact this conclusion. Additionally, SDB is a complex process, with TMD possibly representing one part of multifactorial disease.

The importance of determining an association between TMD and SDB should not be understated. If a correlation is ascertained, then symptoms of TMD could lead the patient to a prompter diagnosis of SDB. This has ramifications in the overall health of the patient.

Further research is required to better ascertain a clear association between TMD and SDB. Patients with TMD and healthy controls should be tested using PSG as the standard gold diagnosis of SDB. Those patients with a confirmed diagnosis of SDB determined by PSG require JVA. A true healthy control group with no diagnosis of SDB and no TMD is required for adequate analysis.

Larger group analysis is warranted to not only establish whether a link does exist but also to use novel JVA technology in order to well comprehend TMDs character in the pathogenesis and treatment of SDB.

Identification of variables amongst patients that are associated with a higher likelihood of TMD.

The data acquired was analyzed in a multivariant analysis to look for correlations between different patient characteristics and TMD diagnosed by JVA. In this patient cohort, poor oral hygiene was significantly associated with TMD ($p = 0.0309$). This association was further confirmed by using Baptista-Pike method for odds ratio calculation; poor oral hygiene was again significantly associated with TMD (OR = 6.500, CI 95% = 1.569–25.08). The association between TMD and poor oral hygiene is an interesting finding of this study which, to my knowledge, has not been identified in prior literature. Poor oral health has significant ramifications on cardiovascular health and wellbeing, possibly through the elevation of cholesterol and resulting inflammation [47]. The current understanding of TMD is that of an inflammatory process driven by trauma to the TMJ. This initial trauma can result from external injuries and/or internal derangement (bruxism). The trauma leads to oxidative stress and hence free radical generation in the intra-radicular space and subsequent synovial fluid inflammation. This, in turn, produces cytokines which through a number of actions results in the degeneration of the TMJ and resulting TMD [48].

It is known that gingivitis and its more sinister infection periodontitis is also an inflammatory disease which tends to be a direct result of dental plaque build-up associated with poor oral hygiene. The inflammatory nature of gum disease is modulated through the production of interleukins, prostaglandins, and matrix metalloproteinases [49]. An isolated disturbance in the homeostatic equilibrium of the human body should not be considered as a secluded spectacle and hence limited in its area of impact [50]. Hence, the modulators of the inflammatory reaction of gum disease are not likely to remain confined to the initial area of onset. It could, therefore, be postulated that this inflammatory process could contribute to the pathogenesis of TMD. Further larger group analysis is warranted with more formal quantification of oral hygiene to assess if this result is replicable. If replicated, this could have significant ramifications to our understanding of TMD as well as the potential for preventative and targeted therapies.

Limitations

The data from this study do not back the original hypothesis. As such, it is important that the limitations of this study are looked at. Firstly, the report was constrained by a lesser subject size, particularly in the non-TMD group. Furthermore, the other limit of this report was the absence of a true healthy control group. Ideally, a true healthy control group would have been confirmed using PSG as the gold standard for diagnosing patients with symptoms indicative of SDB.

A further limitation is that the patient cohort that was analyzed had a wide range of dental disease which would affect occlusion and hence TMJ stability. The study was also limited by the method of diagnosing patients with SDB, which did not follow the gold standard of PSG. Moreover, finally, the study was limited by having only one examiner involved in taking the JVA data. As such, there was no inter-examiner agreement and hence the potential for information bias.

Confounding factors that affect the accuracy of JVA measurements can include the stress of the patient on the day of the measurement, ability to understand the language command given by the examiner, psychological factors of the patient, and experience of the operator. The rationale of randomization the indiscriminate consignment of study subjects to exposure groups to stopping any associations between exposure and confounders. This lessens the potential for confounding by creating clusters that are objectively analogous with regards to recognized and unrecognized confounding variables. Within the context of this study, matching does not completely solve the problem of confounding but might be useful in particular situations to the small sample controls.

A drawback of this study conveys to its sample size, as power analysis shown that a full sample of 270 participants would be required to identify sizeable outcomes (80% chance), built on a t-test at 5% level of significance. Additional limitation pertains to the information that the entire CBCT scans were acquired in a vertical direction. This may affect the collected data as the anteroposterior measurements of the oropharyngeal airway reduced respectively, with the highest contractions happening fat the soft palate.

5 Conclusion

In this study JVA was demonstrated as an innovative and objective tool in the diagnosis of TMD, JVA identified that LL (64.3%) was the most common disorder of the TMJ which is considered an early etiological factor in the deterioration of TMD. A relationship between SDB and TMD was not able to be identified, suggesting that TMD plays only a small role in the pathophysiology of SDB. However, this study has multiple limitations which include; a small patient cohort, the lack of a healthy control group, a wide range of patient dental health status, and other confounding factors. A significant link between TMD and poor oral hygiene was identified. This finding raises interesting questions regarding the pathophysiology and pathogenesis of TMD and warrants further research, especially in early identification of both SDB and TMD.

Conflict of Interest The authors have no conflict of interest to declare.

References

1. Anderson DM (2007) Dorland's illustrated medical dictionary. Saunders, Elsevier, Philadelphia
2. Babiec DF (2017) Temporomandibular pain caused by sleep disorders: a review and case report. Gen Dent 65:30–33
3. Baulkman J (2016) National Dentist's Day 2016: how poor dental care negatively affects health and well-being [Online]. Available: https://www.medicaldaily.com/dental-care-health-benefits-well-being-376680 [Accessed]
4. Schwab RJ, Gupta KB, Gefter WB, Metzger LJ, Hoffman EA, Pack AI (1995) Upper airway and soft tissue anatomy in normal subjects and patients with sleep-disordered breathing. Significance of the lateral pharyngeal walls. Am J Respir Crit Care Med 152:1673–1689
5. Bixler EO, Vgontzas AN, Lin HM, Ten Have T, Rein J, Vela-Bueno A, Kales A (2001) Prevalence of sleep-disordered breathing in women: effects of gender. Am J Respir Crit Care Med 163:608–613
6. Caton JG, Ciancio SG, Blieden TM, Bradshaw M, Crout RJ, Hefti AF, Massaro JM, Polson AM, Thomas J, Walker C (2000) Treatment with subantimicrobial dose doxycycline improves

the efficacy of scaling and root planing in patients with adult periodontitis. J Periodontol 71:521–532
7. Chiu HY, Chen PY, Chuang LP, Chen NH, Tu YK, Hsieh YJ, Wang YC, Guilleminault C (2017) Diagnostic accuracy of the Berlin questionnaire, STOP-BANG, STOP, and Epworth sleepiness scale in detecting obstructive sleep apnea: a bivariate meta-analysis. Sleep Med Rev 36:57–70
8. Paruthi S (2019) Evaluation of suspected obstructive sleep apnea in children [Online]. Available: https://www.uptodate.com/contents/evaluation-of-suspected-obstructive-sleep-apnea-in-children?search=palate%20obstructive%20sleep%20apnea&source=graphics_search&graphicRef=72817#graphicRef72817 [Accessed]
9. Chung F, Yegneswaran B, Liao P, Chung SA, Vairavanathan S, Islam S, Khajehdehi A, Shapiro CM (2008) STOP questionnaire: a tool to screen patients for obstructive sleep apnea. Anesthesiology 108:812–821
10. Rowley JA (2019) Snoring in adults [Online]. Available: https://www.uptodate.com/contents/snoring-in-adults?search=oropharynx%20of%20a%20snorer&source=graphics_search&graphicRef=67818#graphicRef67818 [Accessed]
11. Ciscar MA, Juan G, Martinez V, Ramon M, Lloret T, Minguez J, Armengot M, Marin J, Basterra J (2001) Magnetic resonance imaging of the pharynx in OSA patients and healthy subjects. Eur Respir J 17:79–86
12. Dempsey JA, Veasey SC, Morgan BJ, O'donnell CP (2010) Pathophysiology of sleep apnea. Physiol Rev 90:47–112
13. Dubrovsky B, Raphael KG, Lavigne GJ, Janal MN, Sirois DA, Wigren PE, Nemelivsky LV, Klausner JJ, Krieger AC (2014) Polysomnographic investigation of sleep and respiratory parameters in women with temporomandibular pain disorders. J Clin Sleep Med 10:195–201
14. Dworkin SF (2010) Research diagnostic criteria for temporomandibular disorders: current status & future relevance. J Oral Rehabil 37:734–743
15. Epstein LJ, Kristo D, Strollo PJ Jr, Friedman N, Malhotra A, Patil SP, Ramar K, Rogers R, Schwab RJ, Weaver EM, Weinstein MD (2009) Clinical guideline for the evaluation, management and long-term care of obstructive sleep apnea in adults. J Clin Sleep Med 5:263–276
16. Ferrazzo KL, Osório LB, Ferrazzo VA (2013) CT images of a severe TMJ osteoarthritis and differential diagnosis with other joint disorders. Case Rep Dentistry 5
17. Gözler S (2018) JVA, mastication and digital occlusal analysis in diagnosis and treatment of temporomandibular disorders
18. Guilleminault C, Partinen M, Hollman K, Powell N, Stoohs R (1995) Familial aggregates in obstructive sleep apnea syndrome. Chest 107:1545–1551
19. Harness DM, Donlon WC, Eversole LR (1990) Comparison of clinical characteristics in myogenic, TMJ internal derangement and atypical facial pain patients. Clin J Pain 6:4–17
20. Hasturk H, Kantarci A (2015) Activation and resolution of periodontal inflammation and its systemic impact. Periodontol 2000(69):255–273
21. Hwang IT, Jung DU, Lee JH, Kang DW (2009) Evaluation of TMJ sound on the subject with TMJ disorder by joint vibration analysis. J Adv Prosthodont 1:26–30
22. Jennum P, Riha RL (2009) Epidemiology of sleep apnoea/hypopnoea syndrome and sleep-disordered breathing. Eur Respir J 33:907–914
23. Jordan AS, Mcsharry DG, Malhotra A (2014) Adult obstructive sleep apnoea. Lancet 383:736–747
24. Koh KJ, Park HN, Kim KA (2013) Relationship between anterior disc displacement with/without reduction and effusion in temporomandibular disorder patients using magnetic resonance imaging. Imaging Sci Dent 43:245–251
25. Koval S (2016) Condylar movement assessment is severe attrition patient undergoing orthodontic treatment: Joint vibrant analysis procedure [Online]. Available: http://www.imedpub.com/articles/condylar-movement-assessment-is-severe-attrition-patient-undergoing-orthodontic-treatment-joint-vibrant-analysis-procedure.php?aid=17969 [Accessed]
26. Lavigne GJ, Khoury S, Abe S, Yamaguchi T, Raphael K (2008) Bruxism physiology and pathology: an overview for clinicians. J Oral Rehabil 35:476–494

27. Lechner M, Breeze CE, Ohayon MM, Kotecha B (2019) Snoring and breathing pauses during sleep: interview survey of a United Kingdom population sample reveals a significant increase in the rates of sleep apnoea and obesity over the last 20 years—data from the UK sleep survey. Sleep Med 54:250–256
28. Li KK, Kushida C, Powell NB, Riley RW, Guilleminault C (2000) Obstructive sleep apnea syndrome: a comparison between Far-East Asian and white men. Laryngoscope 110:1689–1693
29. Maixner W, Diatchenko L, Dubner R, Fillingim RB, Greenspan JD, Knott C, Ohrbach R, Weir B, Slade GD (2011) Orofacial pain prospective evaluation and risk assessment study--the OPPERA study. J Pain 12: T4–11.e1–2
30. Manfredini D, Lobbezoo F (2010) Relationship between bruxism and temporomandibular disorders: a systematic review of literature from 1998 to 2008. Oral Surg Oral Med Oral Pathol Oral Radiol Endod 109:e26-50
31. Medicine, A. A. O. S. (2005) International classification of sleep disorders. Diagnostic Coding Manual 51–55
32. Meyer RA (1990) Clinical methods: the history, physical, and laboratory examinations. Butterworths, Boston
33. Milam SB, Schmitz JP (1995) Molecular biology of temporomandibular joint disorders: proposed mechanisms of disease. J Oral Maxillofac Surg 53:1448–1454
34. Ogren M, Faltmars C, Lund B, Holmlund A (2012) Hypermobility and trauma as etiologic factors in patients with disc derangements of the temporomandibular joint. Int J Oral Maxillofac Surg 41:1046–1050
35. Peppard PE, Young T, Barnet JH, Palta M, Hagen EW, Hla KM (2013) Increased prevalence of sleep-disordered breathing in adults. Am J Epidemiol 177:1006–1014
36. Peppard PE, Young T, Palta M, Dempsey J, Skatrud J (2000) Longitudinal study of moderate weight change and sleep-disordered breathing. JAMA 284:3015–3021
37. Poluha RL, Canales GT, Costa YM, Grossmann E, Bonjardim LR, Conti PCR (2019) Temporomandibular joint disc displacement with reduction: a review of mechanisms and clinical presentation. J Appl Oral Sci 27:e20180433
38. Quintana-Gallego E, Carmona-Bernal C, Capote F, Sanchez-Armengol A, Botebol-Benhamou G, Polo-Padillo J, Castillo-Gomez J (2004) Gender differences in obstructive sleep apnea syndrome: a clinical study of 1166 patients. Respir Med 98:984–989
39. Redline S, Tishler PV, Hans MG, Tosteson TD, Strohl KP, Spry K (1997) Racial differences in sleep-disordered breathing in African-Americans and Caucasians. Am J Respir Crit Care Med 155:186–192
40. Schiffman E, Ohrbach R, Truelove E, Look J, Anderson G, Goulet JP, List T, Svensson P, Gonzalez Y, Lobbezoo F, Michelotti A, Brooks SL, Ceusters W, Drangsholt M, Ettlin D, Gaul C, Goldberg LJ, Haythornthwaite JA, Hollender L, Jensen R, John MT, De Laat A, De Leeuw R, Maixner W, Van Der Meulen M, Murray GM, Nixdorf DR, Palla S, Petersson A, Pionchon P, Smith B, Visscher CM, Zakrzewska J, Dworkin SF (2014) Diagnostic criteria for temporomandibular disorders (DC/TMD) for clinical and research applications: recommendations of the international RDC/TMD consortium network* and orofacial pain special interest groupdagger. J Oral Facial Pain Headache 28:6–27
41. Sharma S, Crow HC, Kartha K, Mccall WD Jr, Gonzalez YM (2017) Reliability and diagnostic validity of a joint vibration analysis device. BMC Oral Health 17:56
42. Sharma S, Crow HC, Mccall WD Jr, Gonzalez YM (2013) Systematic review of reliability and diagnostic validity of joint vibration analysis for diagnosis of temporomandibular disorders. J Orofac Pain 27:51–60
43. Smith MT, Wickwire EM, Grace EG, Edwards RR, Buenaver LF, Peterson S, Klick B, Haythornthwaite JA (2009) Sleep disorders and their association with laboratory pain sensitivity in temporomandibular joint disorder. Sleep 32:779–790
44. Somers VK, White DP, Amin R, Abraham WT, Costa F, Culebras A, Daniels S, Floras JS, Hunt CE, Olson LJ, Pickering TG, Russell R, Woo M, Young T (2008) Sleep apnea and cardiovascular disease: an American Heart Association/American College of Cardiology Foundation

Scientific Statement from the American Heart Association Council for High Blood Pressure Research Professional Education Committee, Council on Clinical Cardiology, Stroke Council, and Council On Cardiovascular Nursing. In collaboration with the National Heart, Lung, and Blood Institute National Center on Sleep Disorders Research (National Institutes of Health). Circulation 118:1080–1111

45. Sunwoo JS, Hwangbo Y, Kim WJ, Chu MK, Yun CH, Yang KI (2018) Prevalence, sleep characteristics, and comorbidities in a population at high risk for obstructive sleep apnea: a nationwide questionnaire study in South Korea. PLoS One 13:e0193549
46. Tanaka E, Kawai N, Tanaka M, Todoh M, Van Eijden T, Hanaoka K, Dalla-Bona DA, Takata T, Tanne K (2004) The frictional coefficient of the temporomandibular joint and its dependency on the magnitude and duration of joint loading. J Dent Res 83:404–407
47. Tasaki MM, Westesson PL, Isberg AM, Ren YF, Tallents RH (1996) Classification and prevalence of temporomandibular joint disk displacement in patients and symptom-free volunteers. Am J Orthod Dentofacial Orthop 109:249–262
48. Tuijt M, Parsa A, Koutris M, Berkhout E, Koolstra JH, Lobbezoo F (2018) Human jaw joint hypermobility: diagnosis and biomechanical modelling. J Oral Rehabil 45:783–789
49. Wetter DW, Young TB, Bidwell TR, Badr MS, Palta M (1994) Smoking as a risk factor for sleep-disordered breathing. Arch Intern Med 154:2219–2224
50. Wright EF (2014) manual of temporomandibular disorders. Wiley, Hoboken

Neuroengineering

Multi-Class Classification of Alzheimer's Disease Using Frontal Cortex Non-invasive fNIRS

Thi Kieu Khanh Ho, Minhee Kim, Harish Garg, Byeong C. Kim, Jang Jae Lee, Kyu Yeong Choi, Jae Gwan Kim, Kun Ho Lee, Jong-In Song, and Jeonghwan Gwak

Abstract Early and accurate diagnosis of Alzheimer's disease (AD) is especially important for neurodegenerative disorders allowing patients, who exhibit different patterns of severity and progression risks, to take prevention and treatments before brain damages are shaped. Functional Near-Infrared Spectroscopy (fNIRS), as a clinical test, can be employed in an ecological setting to support AD diagnosis at early stages. In this study, we aimed to comprehensively study the hemodynamic response patterns among four subject groups from 140 participants by visualizing the 4-class distributions based on t-SNE visualization techniques and showing the fNIRS topographic mapping to have a pictorial representation on the surface of the prefrontal cortex region. We demonstrated that the changes in hemodynamic concentrations of oxygenated hemoglobin (HbO) and deoxygenated hemoglobin (Hb) from healthy

T. K. K. Ho · J.-I. Song
School of Electrical Engineering and Computer Science, Gwangju Institute of Science and Technology, Gwangju 61005, Korea

M. Kim · J. G. Kim
Department of Biomedical Science and Engineering, Institute of Science and Technology, Gwangju 61005, Korea

H. Garg
School of Mathematics, Thapar Institute of Engineering and Technology, Deemed University, Patiala, Punjab 147004, India

B. C. Kim
Department of Neurology, Chonnam National University Medical School, Gwangju 61469, Korea

J. J. Lee · K. Y. Choi · K. H. Lee
Gwangju Alzheimer's Disease and Related Dementias (GARD) Cohort Research Center, Chosun University, Gwangju 61452, Korea

K. H. Lee
Department of Biomedical Science, Chosun University, Gwangju 61452, Korea

Aging Neuroscience Research Group, Korea Brain Research Institute, Daegu 41062, Korea

J. Gwak (✉)
Department of Software, Korea National University of Transportation, Chungju 27469, Korea

V. Van Toi et al. (eds.), *8th International Conference on the Development of Biomedical Engineering in Vietnam*, IFMBE Proceedings 85,
https://doi.org/10.1007/978-3-030-75506-5_75

controls to three stages of AD did exist during four stages of our experimental protocols. Results also revealed that various machine learning techniques were capable of both analyzing AD patterns and its related features and classifying multiple pathophysiologic mechanisms with the highest accuracies of $74.39 \pm 4.7\%$ and $84.91 \pm 4.01\%$ corresponding to raw and SMOTE-based features, respectively. These findings demonstrate the feasibility of fNIRS-based approaches to investigate the stages of AD progression.

Keywords fNIRS · Alzheimer's disease · Machine learning · Multi-class classification

1 Introduction

Alzheimer's disease (AD), the most common cause of dementia accounting for 60–80% cases worldwide [1], is an irreversible and chronic progressive neurodegenerative brain disease that is characterized by progressive impairments in the cognitive abilities and memory [2]. By a significant loss of neurons in the neurotransmitter systems, the present proteins such as amyloid-beta deposits and neurofibrillary tangles may disrupt the communication with nerve cells, damage cells and lead to the development of AD [3]. AD patients typically suffer from several symptoms of functional impairments, including memory, communication, reasoning, and behavior functions which interfere with them in daily life while there are currently no medication therapies or treatments being able to halt the disease's progression [4]. Hence, a significant effort has been made to develop the strategies for early detection, particularly at a presymptomatic stage, with the hope that the intervention can delay or even prevent those clinical symptoms.

Since the diagnosis of AD primarily relies on patients' medical history, neuropsychological examinations or clinical rating scores which require experienced clinicians and exhaustive clinical tests, it has been upgraded by adopting brain imaging techniques such as magnetic resonance imaging (MRI), single-photon emission computed tomography (SPECT), positron emission tomography (PET) and electroencephalography (EEG) introduced in recent years. Functional brain imaging tests are modified from a non-paced stage of administration and free doctor-patient interaction tasks to a paradigm event where an automated stimulation can replace doctors [5–7]. Despite the usefulness of these modifications from a neuroimaging standpoint, they still alter the ecology of the interaction between doctors and patients. With this in mind, Functional Near-Infrared Spectroscopy (fNIRS), which is a scalp-located non-invasive method recording neural activity and oscillation within the oxygenated (HbO) and deoxygenated (Hb) hemoglobin in the brain based on the Blood Oxygen Level Dependent (BOLD) effect [8], is one of the most suitable neuroimaging techniques that are feasible for an outpatient environment.

fNIRS offers various benefits over the aforementioned techniques by its higher temporal resolution, high portability, relatively low cost, lightweight measurement, lower susceptibility to motion artifacts, lack of ionizing radiation, and fewer constraints to the subjects during experiments. It can integrate with a mechanical structure resembling EEG [9] thus being appropriate for measurements during clinical administration tests. Furthermore, several studies have examined the feasibility of fNIRS to differentiate the hemodynamic responses between healthy controls and AD patients [10]. fNIRS is also able to successfully monitor AD treatment in the clinic which was reported by [11] showing a significant difference in cerebral blood flows and the effect of memantine on AD patients. These investigations showed that AD patients had lower levels of activation at specific regions in the brain compared to healthy controls during cognitive experiments. However, although these previous studies denote the potential of fNIRS on measuring the differences between healthy and AD groups, there are still various AD stages remaining unknown. It requires the recruitment of several participants at different stages of AD to comprehensively compare and evaluate the efficiency of fNIRS in therapeutic monitoring and diagnosis. Besides, the pathological mechanism of AD progression has not yet been thoroughly documented and investigated, which is expected to have an intact experimental design dealing with multiple subject groups.

To address the challenges, we comprehensively investigate the hemodynamic responses of healthy subjects and patients with three degrees of AD. In detail, the classification studies will be described, including seven machine learning methods that generally require four main components: feature extraction, feature selection, reduction of dimensionality, and processed feature-based classification algorithms. Due to our imbalanced datasets, a preprocessing step called Synthetic Minority Oversampling Technique (SMOTE) [12] was used to improve multi-class classification accuracies.

2 Materials and Methods

2.1 *Participants and Data Preprocessing*

In this study, 140 subjects dwelling in Gwangju city in South Korea and the adjacent cities were recruited from the Chonnam National University Hospital and the National Research Center for Dementia (Gwangju, South Korea). A set of medical examinations–Mini-Mental State Examination (MMSE), MRI, PET, and individual interview– was conducted to adequately analyze and diagnose different AD stages. Four categories were subsequently divided among them: HC class: Normal AD biomarkers, cognitively unimpaired (72.7 $\pm$ 5.3 years, 21 M/32F), asymptomatic AD (aAD) class: Very mild MCI with the cognitive decline on memory and executive functions (74.5 $\pm$ 4.3 years, 15 M/13F), prodromal AD (pAD) class: Explicit brain dysfunction symptoms (75.8 $\pm$ 3.9 years, 33 M/17F), AD Dementia (ADD) class:

Severe deterioration of memory, language, and social abilities (75.4 ± 6.8 years, 4 M/5F). The subject with a mental and behavioral disorder was disregarded in this cohort. All subjects underwent our experimental protocol without any previous experience. No subject had any previous experience with our experimental protocol. Each subject was fully informed of the purpose of research and consent forms prior to conducting experiments. Table 1 summarizes the demographic information of all subjects (SD: Standard Deviation).

To curtail any environmental disturbance, the experiments were carried out in a confined room. Subjects were requested to seat in a chair and rest by calmly watching a white cross that occurred on the monitor screen during resting periods. Then, they underwent a series of three tasks: (i) Oddball which is a cognitive ability test, (ii) 1-back which is a memory ability test, and (iii) Verbal fluency which is a language ability test. Six fNIRS channels measuring in the frontal cortex area were recorded. Each channel was visually inspected, and channels with large spikes were marked as noises and excluded from our studies. The concentration changes in hemoglobin, HbO, Hb, and total hemoglobin (THb), were calculated based on the Modified Beer-Lambert Law [13]. A low-pass filter with a cut-off frequency of 0.5 Hz was used to remove artifacts. Figure 1 shows the imbalanced datasets and the complexity of distributions

Table 1 Participant Information

	HC	aAD	pAD	ADD
Number of subjects	53	28	50	9
Age ± SD	72.7 ± 5.3	74.5 ± 4.3	75.8 ± 3.9	75.4 ± 6.8
Gender (M/F)	21/32	15/13	31/17	4/5
MMSE ± SD	27.0 ± 4.2	26.9 ± 2.5	26.0 ± 3.2	20.2 ± 4.8
Education ± SD	9.8 ± 4.7	10.2 ± 5.2	10.6 ± 5.2	8.5 ± 5.3

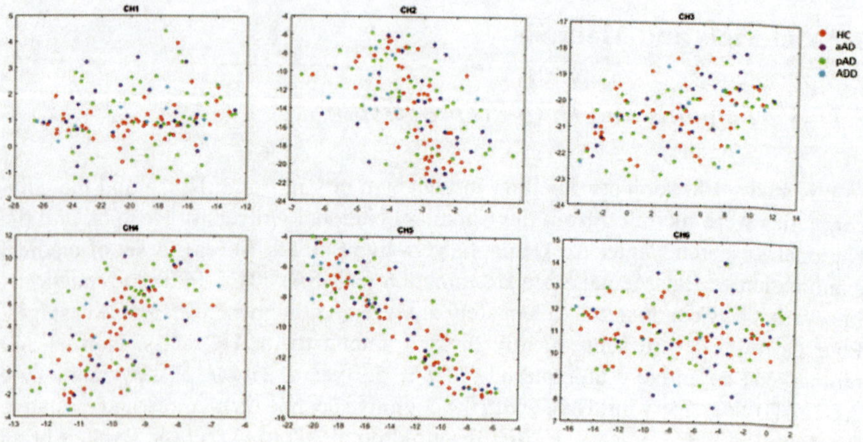

Fig. 1 t-SNE Visualization from 6 channels of HbO signals

among the four classes (i.e., HC, aAD, pAD and ADD) from HbO signals by using the t-Stochastic Neighbor Embedding (t-SNE) visualization technique.

2.2 *Experimental Protocol*

Figure 2a represents the fNIRS device setup. The transmitter (LED), the detector (photodiode), and the channel (CH) are denoted as the red circle, the black circle, and the green rectangular, respectively. The distance between the transmitter and the detector is 30 mm. Figure 2b displays the experimental protocol, and there are four main stages in our experiments. After finishing recording the fNIRS data, we segmented trial sections corresponding to the experimental stages. Each section was segmented by 60 s. To sum up all four sections, we had a resting section (60 s), the Oddball stage (300 s), the 1-back stage (270 s), and the Verbal stage (390 s). Therefore, the total experimental time for one trial was 1020 s.

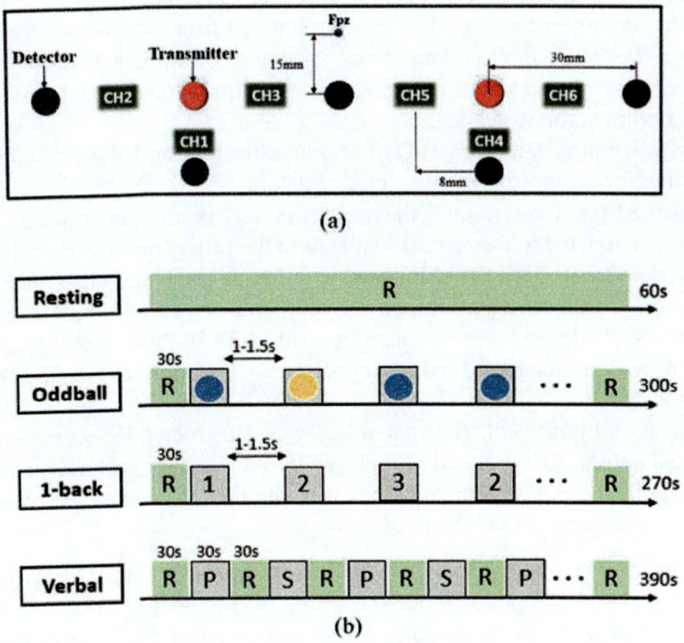

Fig. 2 **a** fNIRS device setup; **b** experimental protocol (R: Resting, P: phonemic, S: semantic)

2.3 Classification Algorithms

The following seven supervised machine learning classifiers was selected to classify the four categories:

- Gaussian Discriminant Analysis (GDA) [14]: With a set of samples (each belongs to each class), the intra-class and inter-class matrices were calculated, and the linear transformation was obtained based on solving the generalized eigenvalues. Then, the classification task was performed on the transformed space using Euclidean distance after 200 epochs.
- K-Nearest Neighbor (KNN) [15]: $K = 4$ was chosen as the most optimal value to reduce errors and allow the algorithm to calculate the distance between each data point and the cluster centroid by Euclidean distance. KNN was trained with 200 epochs.
- Gaussian Naïve Bayes (GNB) [16]: The conditional probability for each class was computed based on the prior and the posterior probability. The class showing the highest probability was considered as the final predicted class.
- Support Vector Machine (SVM) [17]: The decision hyperplane was constructed when the margin of the classifier was maximized from support vectors to separate four classes. A sigmoid kernel was specified for the non-linearly separation problem. SVM would stop training whenever the margin error was trivial, and no further optimization is needed.
- Adaptive Boosting (AdaBoost) [18]: A classifier was built by combining 1000 weak classifiers, and the number of estimators is 200. During training, if any misclassified point was found, the weight of that point was boosted. The new weight was updated to the next classifier, and the procedure was repeated.
- Neural Network (NN) [19]: Along with an input layer and an output layer (4 neurons), NN with different settings of hyperparameters was trained until the best optimal set of hidden neurons was obtained (100–80–50). A ReLU activation function, 1e-5 learning rate, and 300 epochs were manually chosen to achieve the highest accuracy.
- Random Forest (RF) [20]: A decision tree was constructed for every sample and generated output. RF consisted of multiple trees to vote for each predicted output and select the output with the most votes as the final prediction. RF was trained several times with the max depth of the tree as 8.

For a fair comparison, fivefold cross-validation was used for each classifier. The SMOTE step was also applied by over-sampling the minority classes (it drew a line between a set of neighboring samples in the feature space and over-sampled new synthetic points along the line) to improve accuracies in the latter part.

3 Experimental Results

The distributions of four classes based on t-SNE visualization techniques are depicted in Fig. 1. To have an apparently pictorial representation on the surface of the prefrontal cortex, we showed the fNIRS topographic mapping in Fig. 3 taken from the resting and three ability tests. The white digits being shown in Fig. 3 represent the channel numbers and their positions on the prefrontal cortex region. The bar graph on the right side in Fig. 3 denotes the fNIRS signal intensity. It was observed that channels 1, 2, and 3 had a significant impact on HC during Oddball and 1-back and aAD patients during Oddball and pAD during Verbal. Meanwhile, the activation spots located at channels 4, 5, and 6 were observed from pAD during 1-back and ADD patients during Verbal. The local activation regions in the same experimental task were totally opposite. For example, c hannels 4, 5, and 6 were highlighted in HC whilst not being spotted in aAD during the Oddball stage, or channels 1, 2, and 3 were dominantly punctuated in HC while those channels are inactivated in pAD. It induces that each channel owned peculiar characteristics and behaved differently counting on corresponding experimental tasks; we thereby utilized all channels for the classification tasks.

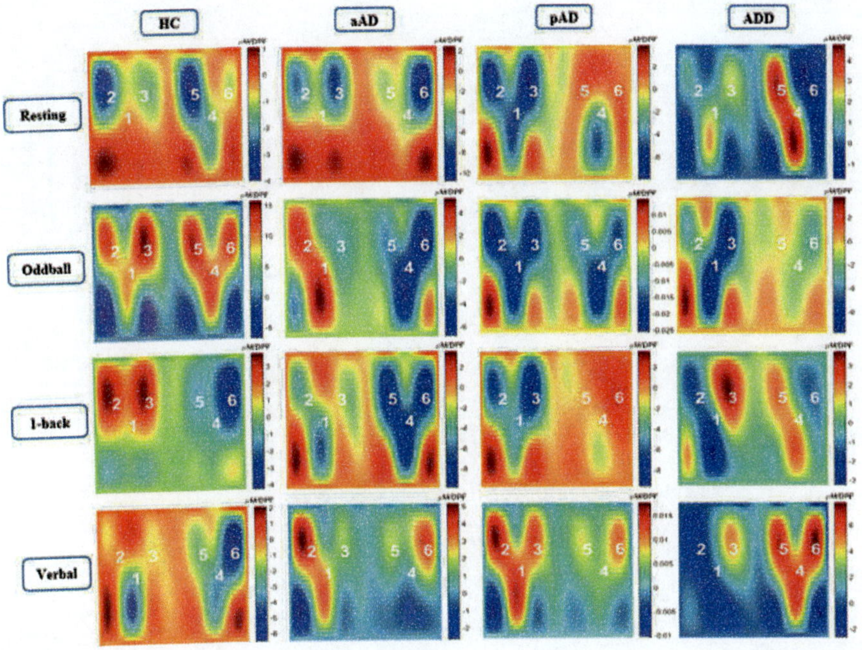

Fig. 3 Comparison of activation maps (topographic mapping) using HbO based on four experimental stages measuring from HC subjects and three AD-degree patients during four experimental stages

Regarding classification tasks, we evaluated the performances of different classifiers on each experimental task dealing with the original features from HbO, Hb, and THb. In general, all classifiers performed better on the THb input compared to the other hemoglobin types. In addition, NN classifiers with the accuracy of $74.39 \pm 4.7\%$ outperformed the others, followed by SVM, GDA, and RF (see Fig. 4). Compared to the other classifiers which require to identify features, break down to different parts, and recombine in the final stages, NN was more flexible since it adopted a sufficient amount of our original data size without any feature engineering steps and learned the high-level features in an end-to-end manner. The remaining classifiers obtained low accuracies. This indicates that the original data was not easily separated using a convoluted decision boundary in KNN, a probabilistic approach from GNB, or boosting algorithms from AdaBoost.

Due to the imbalance of our datasets that patients with three stages of AD were difficult to recruit, SMOTE was applied to magnify the number of small aAD samples and extremely small ADD samples to be balanced with two dominant HC and pAD classes. As expected, the mean classification accuracies considerably improved up to $84.91 \pm 4.01\%$, and the efficacy of each classifier on the resampled data behaved similarly as it applied on the original dataset (see Fig. 5).

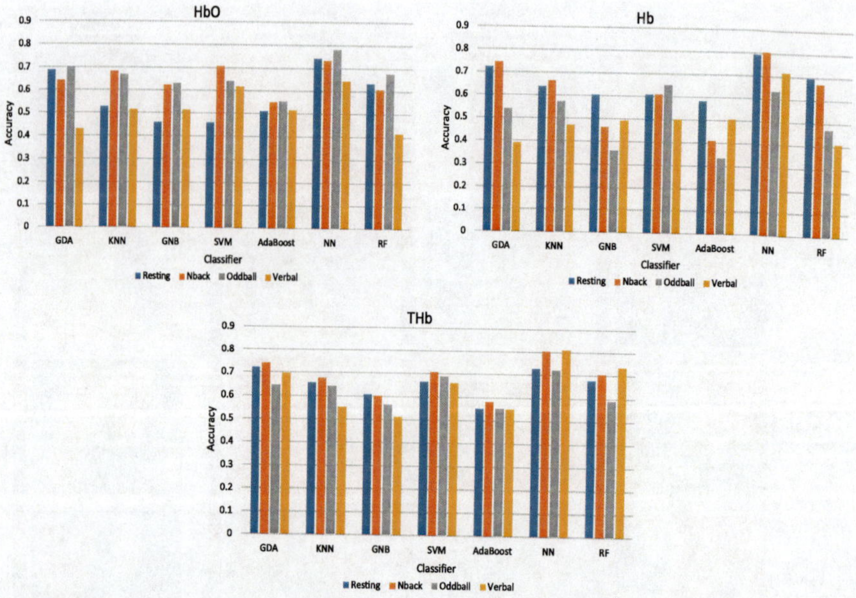

Fig. 4 Classification accuracies of 7 classifiers from three types of hemoglobin when the original data was used

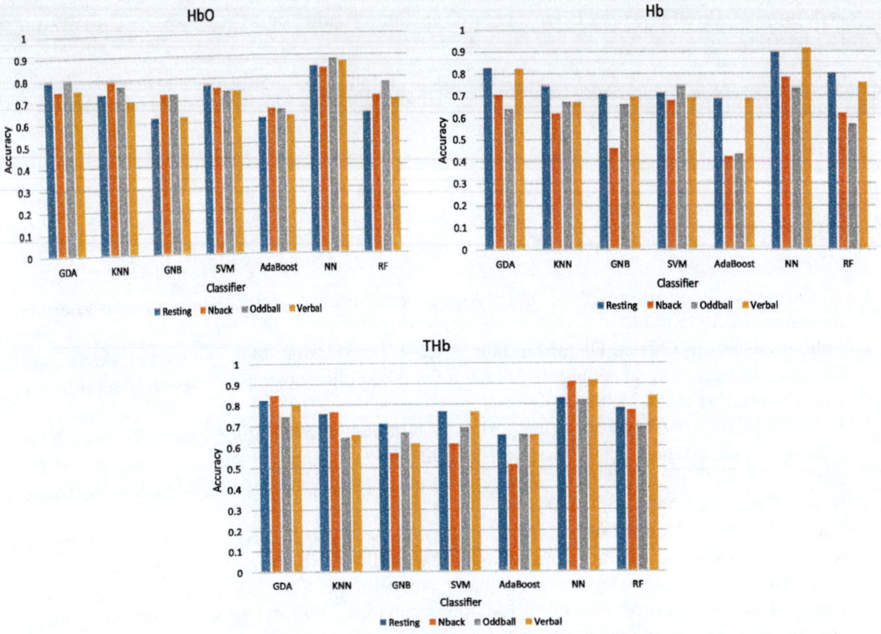

Fig. 5 Classification accuracies of 7 classifiers from three types of hemoglobin when SMOTE was applied

4 Conclusion and Future Works

Emerging evidence yields that exploring alterations of AD's progress has great importance in understanding and timely treatments of this cognitive deficit disease. In this study, we aimed to access the capability of employing fNIRS as a clinical test being commonly utilized to diagnose AD. We demonstrated that the changes in hemodynamic concentrations from healthy controls to the three stages of AD did exist. We evaluated and optimized the most representative features from three types of hemoglobin signals recorded from four stages of experiments. Thus, a set of machine learning classifiers were able to inexpensively and rapidly classify HC with three stages of AD patients and further supplement the diagnosis of the degree of dementia in AD. Due to the relatively small sample size of AD patients recruited, which could easily result in misclassification performances, a larger cohort should be carried out to validate our present findings. In addition, further advancement of non-linear classifiers, such as deep learning techniques is an indispensable topic to address the challenging fNIRS multi-class classification problems.

Acknowledgements This research was supported by the Brain Research Program through the National Research Foundation of Korea (NRF) funded by the Ministry of Science, ICT & Future Planning (NRF-2016M3C7A1905475, NRF-2019M3C7A1020406), and KBRI basic research program through Korea Brain Research Institute funded by Ministry of Science and ICT (20-BR-03-02).

Conflicts of Interest The authors have no conflicts of interest to declare.

References

1. Alzheimer's Association (2019) 2019 Alzheimer's disease facts and figures. Alzheimer's Dementia 15(3):321–387
2. Veitch DP et al (2019) Understanding disease progression and improving Alzheimer's disease clinical trials: recent highlights from the Alzheimer's disease neuroimaging initiative. Alzheimer's Dementia 15(1):106–152
3. Lowe VJ et al (2019) Neuroimaging correlates with neuropathologic schemes in neurodegenerative disease. Alzheimer's Dementia 15(7):927–939
4. Ibrahim MM, Gabr MT (2019) Multitarget therapeutic strategies for Alzheimer's disease. Neural Regen Res 14(3):437–440
5. Tian F, Yennu A, Smith-Osborne A, Gonzalez-Lima F, North CS, Liu H (2014) Prefrontal responses to digit span memory phases in patients with post-traumatic stress disorder (PTSD): a functional near infrared spectroscopy study. NeuroImage: Clin 4:808–819
6. Lancia S, Cofini V, Carrieri M, Ferrari M, Quaresima V (2018) Are ventrolateral and dorsolateral prefrontal cortices involved in the computerized Corsi block-tapping test execution? An fNIRS study. Neurophotonics 5(1):11–19
7. Ho TKK, Gwak J, Park CM, Song JI (2019) Discrimination of mental workload levels from multi-channel fNIRS using deep leaning-based approaches. IEEE Access 7:24392–24403
8. Ferrari M, Quaresima V (2012) A brief review on the history of human functional near-infrared spectroscopy (fNIRS) development and fields of application. Neuroimage 63(2):921–935
9. Fazli S, Mehnert J, Steinbrink J, Curio G, Villringer A, Müller KR, Blankertz B (2012) Enhanced performance by a hybrid NIRS–EEG brain computer interface. Neuroimage 59:519–529
10. Li R, Nguyen T, Potter T, Zhang T (2019) Dynamic cortical connectivity alterations associated with Alzheimer's disease: An EEG and fNIRS integration study. NeuroImage: Clin 21:101622:1–11
11. Araki T et al (2014) The effects of combine treatment of memantine and donepezil on Alzheimer's disease patients and its relationship with cerebral blood flow in the prefrontal area. Int J Geriatric Psychiatry 29(9):881–889
12. Chawla NV, Bowyer KW, Hall LO, Kegelmeyer WP (2002) SMOTE: synthetic minority oversampling technique. J Artif Intell Res 16:321–357
13. Scholkmann F, Kleiser S, Metz AJ, Zimmermann R, Pavia JM, Wolf U, Wolf M (2014) A review on continuous wave functional near-infrared spectroscopy and imaging instrumentation and methodology. Neuroimage 85:6–27
14. Ju J, Kolaczyk ED, Gopal S (2003) Gaussian mixture discriminant analysis and sub-pixel land cover characterization in remote sensing. Remote Sens Environ 84(4):550–560
15. Weinberger KQ, Saul LK (2009) Distance metric learning for large margin nearest neighbor classification. J Mach Learn Res 10(2009):207–244
16. Murphy KP (2006) Naive Bayes Classifiers. University of British Columbia, Department of Computer Science
17. Fung GM, Mangasarian OL (2005) Multicategory proximal support vector machine classifiers. Mach Learn 59(1):77–97
18. Sun Y, Liu Z, Todorovic S, Li J (2007) Adaptive boosting for SAR automatic target recognition. IEEE Trans Aerosp Electron Syst 43(1):112–125

19. Misra BB, Dehuri S (2007) Functional link artificial neural network for classification task in data mining. J Comput Sci 3(12):948–955
20. Pal M (2005) Random forest classifier for remote sensing classification. Int J Remote Sens 26(1):217–222

Retinal Vessels Segmentation Based on Histogram Equalization Combined with Improving Multi-Scale Line Detection

Nguyen Mong Hien, Nguyen Thanh Binh, Ngo Quoc Viet, and Ha Thi Thuy Vi

Abstract In the health care field, doctors usually analyse some details presented in medical images to detect some diseases. If the doctor sees some changes related to the shape, colour or size of the retinal blood vessel. The doctor will guess early stage of dangerous sickness like age-related macular degeneration, diabetes, hypertension, arteriosclerosis. It is a good time for the doctor to build a treatment plan as soon as possible. One of the useful retinal blood vessel analysis ways is segmentation. The task should be done before other works are deployed to cure the patient. The paper proposed a method to segment retinal blood vessels by using histogram equalization combined with improving multi-scale line detection. This method uses histogram equalization to improve the quality of the input image, and then improving multi-scale line detection (MLD) technique to enhance the accuracy of vessels detection. Particularly, using the MLD technique to detect large vessels and small ones with adaptive window sizes finally combined. As a result, our method can work effectively to segment more vessels. The method is tested in quality and quantity on publicly available DRIVE datasets with an average accuracy reaching to 0.9515. The result of the proposed method is better than these other methods.

Keywords Retinal vessels · Segmentation · Line detection · Histogram equalization

N. M. Hien (✉) · H. T. T. Vi
Tra Vinh University, Tra Vinh City, Viet Nam
e-mail: hientvu@tvu.edu.vn

N. M. Hien · N. T. Binh
Faculty of Computer Science and Engineering, Ho Chi Minh City University of Technology, VNU-HCM, 268 Ly Thuong Kiet Street, District 10, Ho Chi Minh City, Viet Nam

Vietnam National University Ho Chi Minh City, Linh Trung Ward, Thu Duc District, Ho Chi Minh City, Viet Nam

N. Q. Viet
Faculty of Information Technology, Ho Chi Minh City University of Education, Ho Chi Minh City, Viet Nam

V. Van Toi et al. (eds.), *8th International Conference on the Development of Biomedical Engineering in Vietnam*, IFMBE Proceedings 85,
https://doi.org/10.1007/978-3-030-75506-5_76

1 Introduction

Nowadays, medical image processing plays an important role in diagnosing some popularly dangerous sickness. Evidence of this, hypertension manifested by changes in vessel calibre [1, 2]. Diabetic retinopathy found out by the onset of neovascularization, which is the first sign of blindness [3]. The stroke often presented by appearing arteriovenous nicking [4, 5]. Therefore, it is very important to detect these retinal vessel changes to intervene immediately at the early stage. Thus, the patients have more chances to avoid losing a major vision. The effective performance of medical diagnosis depends on the accurate vessel segmentation. The result segmentation contributes to the success of the next tasks.

Thus, the medical image segmentation plays an important role in supporting with diagnosing. In fact, some various regions in the image are separated by the segmentation technique based on some features such as [6]: intensity, texture and colour. The segmented image is useful to identify regions of interest. The researchers have used many useful techniques such as threshold [7–9], joined between threshold and domain [10, 11], deep learning [12], filter [13, 14], clustering [15], etc. However, these methods have been very complex calculations. For example, the methods are the poor segmentation in some special cases such as vessel central light reflex, close vessels, small vessels, etc. These problems should be solved as soon as to enhance the quality of retinal vessel segmentation result. In another word, to avoid misunderstanding as two vessels when they consist of the central reflex pixels or prevent from merging two close vessels each other. These drawbacks are going to make vascular network analysis be low accuracy in some cases. Here, they are the identification of single vessel segments, vascular abnormality detection or vessel diameter measurement.

In fact, these disadvantages are easy to find out in some recent methods. Evidence of this, the disconnected vessels, and vessel merging presented by Staal [16] and Soares [17] result, respectively. Moreover, both Staal and Soares cannot avoid losing of objects located in a middle of vessels because of happening central reflection in a vessel. Nguyen [18] had used to improve salient region combined with Sobel operator condition to detect retinal blood vessel. The method is not effective to detect the vessels when the vessel pixels have various saliency levels. Nguyen [19] proposed an effective segmentation method to segment blood vessels in the retinal image. However, this segmentation technique also exists some drawbacks. It cannot detect both major vessels and thin ones at whole. This problem comes from using only one window size to segment all blood vessels. In fact, the line basic detection technique depends on the size of the window. In another word, if the window size is set very small, a major vessel will separate into two vessels.

In contrast, if the window size is set too large, the two close thin vessels will merge as one larger vessel. Mustafa [20] proposed the blood vessel extraction method, which works more effectively than Kirsch templates' detection technique. However, the method cannot detect small vessels.

A useful retinal blood vessel segmentation technique is proposed to solve the drawback mentioned above in the study. Our approach based on the using histogram equalization (HE) combined with improving the multi-scale line detection (IMLD). The proposed algorithm includes two stages: using HE to improve the quality of source image and applying MLD with adaptive window sizes to detect more vessels. The result of the proposed method dominates among recent approaches. This next sections in the paper are presented by following: presenting the basic line detector in section 2, describing the proposed method clearly in section 3, showing the experiment results and conclusion in section 4 and section 5 respectively.

2 Line Detector

To a server for line detector [19], the retinal image is separated into three channels: green, blue, and green. The green channel is the most suitable for line detector work. The vessels presented brighter than the background. A window with size W x W chosen at every pixel position. The grey level average at each pixel calculated as I_{avg}^{W}.

The number of lines passing through the centre pixel depends on the angular resolution. Each line has its average value of pixels. A line has the highest value considered as

'winning line' and this figure presented by $I_{\max}^{W}$. Each pixel has a line response calculated as Eq. (1):

$$R_W = I_{\max}^{W} - I_{avg}^{W} \tag{1}$$

A pixel will consider as the vessel pixel if the line response is large. The winning line is arranged along the vessel. In opposite, the pixel will be background pixel if the difference between the average grey level of the surrounding window and the winning line is tiny. The performance of the technique is depended on the size of window W. It is ideal for window size with a balance between the number of the vessel and background pixels. In another word, the window size should be double vessel calibre.

The line detector can work well in the case; there are the light reflection phenomena in the middle vessel. In fact, the middle pixel value is usually lower than the others. The performance of classification is not accurate because of the similar intensity values. However, these pixels recognized by the technique as a part of the vessel. The winning line consists only a few central reflex pixels. Therefore, the average value not reduced significantly by these pixels. As a result, these pixels considered as vessel pixels because of large responses. Additionally, most of the other pixels classified as background ones because of using long-length lines.

The basic line detector demonstrated its benefits. But it also consists of a few limitations such as merging close vessels, extending at crossover points and causing misclassification when the background pixels located near strong vessels.

3 IMLD for Retinal Vessels Segmentation

In the section, we proposed an efficient retinal blood vessel segmentation method that uses HE combined with IMLD. The generalized block diagram of our retinal blood vessel segmentation technique presented in Fig. 1.

The method includes two stages: histogram processing and improving multi-scale line detection. Each stage will be explained in more details in its subsection.

3.1 *Histogram Processing*

Firstly, the input image enhanced by using HE technique [21]. The task helps to improve the contrast between the vessels and others. This technique explained by the following:

1. Calculate the histogram of the source image by Eq. (2)

$$h(r_k) = n_k \tag{2}$$

   where r_k is the kth value of intensity, n_k is the number of pixels in the range of intensity r_k.
2. Calculate the Cumulative Distributive Function (CDF) of the source image. The probability $P(r_k)$ of intensity level r_k is calculated by Eq. (3):

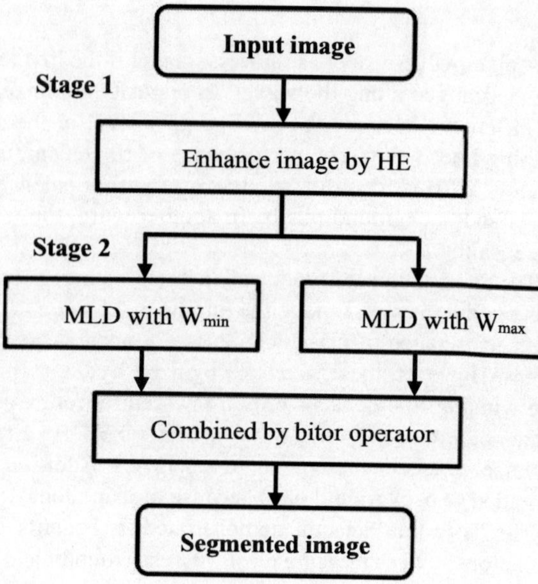

Fig. 1 Block diagram of the retinal blood vessel segmentation technique

$$P(r_k) = \frac{n_k}{MN} \quad k = 0, 1, 2, \ldots L - 1 \tag{3}$$

where MN is the sum of pixels, and L is the number of intensity levels.
A particular transformation function in image processing described by equation (4):

$$s = T(r) = (L - 1)\int_0^r \Pr(w)dw \tag{4}$$

where w is an integration dummy variable. The CDF of random variable r is computed by Eq. (4). The discrete form of previously mentioned transformation is described by Eq. (5):

$$s_k = T(r_k) = (L - 1)\sum_{j=0}^{k} \Pr(r_j) \quad k = 0, 1, 2, \ldots, L - 1 \tag{5}$$

Thus, each pixel in the source image will be mapped with intensity r_k into a corresponding pixel with level s_k in the output image. The transformation $(r_{k)}$ is called a histogram equalization.

3. The histogram of the source image computed by Eq. (6):

$$I_p = T(I_p) \tag{6}$$

where I_p is pixel intensity value.

3.2 *Improving Multi-scale Line Detection*

Nguyen [19] proposed the segmentation method in a retinal image using MLD. The result improved significantly compared with the basic line detection method. However, the improvement is not enough accuracy. In fact, this method can only detect all large vessels and some small ones. In another word, the method is false to detect a few smaller vessels. It is clear if we use only a window size, the result will get low accuracy. We, therefore, proposed IMLD by using MLD with various window sizes to detect better. The MLD also reuse the basic line detector with using various the length of the aligned lines. The line detector calculated by equation (7):

$$R_W^L = I_{\max}^L - I_{avg}^W \tag{7}$$

where $1 \leq L \leq W$, $I_{\max}^L$ and I_{avg}^W are the same role in the previous section. The line detectors at various scales are implemented by controlling the values of L.

In our proposed method, we use two window sizes to detect major vessels and minor ones by maximum and minimum window size. We define maximum window size and minimum ones as $W_{\max}$ and $W_{\min}$ in orderly. The $W_{\max}$ approximates double large vessel calibre. And the approximates double small vessel calibre.

Large vessels are detected by equation (7) with $W_{\max}$ and $1 \leq \mathrm{L} \leq W_{\max}$. Parallel, the small vessels are detected by equation (7) with $W_{\min}$ and $1 \leq \mathrm{L} \leq W_{\min}$. Moreover, the $W_{\max}$ is applied to avoid separating the large vessel into two small ones. Additionally, $W_{\min}$ also used to get more small vessels and reduce some drawbacks when a background pixel located at a special position. It is a fact that if line length reduced, the more background noise would appear. To solve this limitation, the line responses at various scales are linearly combined. We should note that line detector at each scale returns the raw response values that are in a limited range. As a result, the difference between the background and the vessels be very low. To improve the contrast in the images, the raw response image values standardized to create the unit standard deviation distribution and zero mean as equation (8):

$$R' = \frac{R - R_{mean}}{R_{sdt}} \tag{8}$$

where R is the value of the raw response, R' is the value of the standardized response, R_{mean} and R_{std} are the mean and standard deviation of the raw response values orderly. The standardization is useful for the distribution of the intensity values. It needs to apply to the response images produced by line detectors at different scales because its result is helpful for linearly combination process.

In the combination technique, each scale assigned the same weight. The line responses with various scales combined linearly to create the segmentation. This response at every pixel defined by equation (9):

$$R_{combined} = \frac{1}{n_L + 1}\left(\sum_{L} R_w^L + I_{igc}\right) \tag{9}$$

where I_{igc} is the inverted green channel value at the corresponding pixel, n_L is the number of scales used, and ${R_w}^L$ is the line detector response at scale L. The green channel integrated into the combination to distinguish similarity among the blood vessels and other structures easier.

Finally, we applied bitor operator to combine between the segmented image with $W_{\max}$ and segmented image with $W_{\min}$ for creating the final segmented image.

4 Experiment and Results

The experiment uses DRIVE dataset [22], which includes the colour images focusing on eyes and result of segmented images. It considered as a useful benchmark to appreciate the efficiency among the method and others.

Obviously, the segmentation method's result is the classification of pixels, which are either as a vessel or background tissue. Consequently, on the positive side, the events are true positive (TP) and true negative (TN) as a pixel accurately segmented as a vessel or non-vessel. In the negative side, the two misclassifications are false negative (FN), and false-positive (FP) as a pixel is incorrectly segmented.

To evaluate algorithms' performance, we use two widely known statistical measures, including sensitivity and specificity [23]. The sensitivity is a normalized measure of true positives as equation (10), while specificity measures the percentage of true negatives as equation (11):

$$\text{sensitivity} = \frac{\text{TP}}{\text{TP} + \text{FN}} \tag{10}$$

and,

$$\text{specificity} = \frac{\text{TN}}{\text{TN} + \text{FP}} \tag{11}$$

The binary classification accuracy described by equation (12):

$$\text{accuracy} = \frac{\text{TP} + \text{TN}}{\text{P} + \text{N}} \tag{12}$$

With N and P show the sum of negatives (non-vessel) and positives (vessel) pixels in during segmentation. The accuracy is the homogeneity level, which compares binary classification and the ground truth. Accuracy presents the degree of effective segmentation. Therefore, it used to evaluate and compare our method and others.

We implemented the proposed approach in Sect. 3. It is a truth that the original image includes three channels such as red, green, and blue. To prepare for the first stage, processing the green channel of an image extracted. In the green channel, the difference of intensity between the blood vessel and background is higher than the rest of channels. In this paper, the ground truths vessel extracted from DRIVE is considered a measurement standard to compare the method's effective performance and recent others [19, 20]. We test all images in DRIVE. In the study, we show two typical cases.

The segmented images presented in Figs. 2 and 3. Figure 2a presents the source image, segmented image by the method in Fig. 2b, the Nguyen [19] in Fig. 2c and the Mustafa [20] in Fig. 2d. Furthermore, in the Fig. 3a presenting the source image, segmented image by the method in Fig. 3b, the Nguyen [19] in Fig. 3c and the Mustafa [20] in Fig. 3d.

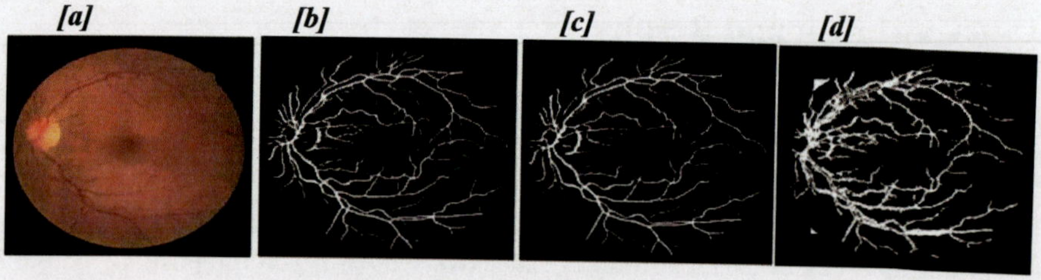

Fig. 2 The result images of the method and other methods

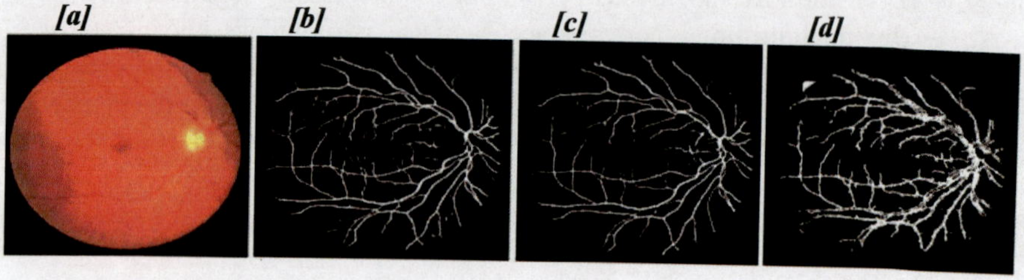

Fig. 3 The result images of the method and other methods

To look insight Figs. 2 and 3 show the quality of our result images (Figs. 2b and 3b) is more effective than results of other methods because of consisting more vessels. Thanks to using HE to improve the quality of source image and choose adaptive window sizes to detect all kinds of vessels.

The performance measures like sensitivity, specificity, and accuracy presented in Table 1. The segmented image is considered as better when these parameters value gets higher, which written by italics.

According to the number in Table 1, we can observe that the method's sensitivity and accuracy values are the highest in all cases. While in Table 2, our specificity values are higher than those in the Mustafa method [20]. In fact, our proposed method is less

Table 1 The first image

Measures	Sensitivity	Specificity	Accuracy
Nguyen et al. method [19]	67.90	*98.93*	95.50
Mustafa et al. method [20]	68.41	94.82	91.53
Proposed method	*74.40*	98.52	*95.85*

Table 2 The second image

Measures	Sensitivity	Specificity	Accuracy
Nguyen et al. method [19]	71.91	*98.37*	96.00
Mustafa et al. method [20]	65.02	95.54	92.44
Proposed method	*78.43*	98.06	*96.30*

Table 3 The average performance of various segmentation techniques regarding sensitivity, specificity, and accuracy on DRIVE dataset

Measures	The average performance		
	Sensitivity	Specificity	Accuracy
Nguyen et al. method [19]	62.90	*98.78*	94.86
Mustafa et al. method [20]	61.07	95.39	91.17
Proposed method	*69.35*	98.32	*95.15*

effective than Nguyen method [19] in specificity parameter. However, the difference is very tiny and related background pixels.

To evaluate the performance of these methods mentioned above, we test all images in DRIVE dataset. The average of sensitivity, specificity, and accuracy presented in Table 3. According to the figures in the table, we can say that our proposed method is more advantageous than the rest two methods. Ours can segment vessel pixels better. Evidence of this, in term of sensitivity and accuracy, our results are highest, and the difference of sensitivity is very great (69.35 compared to 62.90 and 61.07). However, ours is less effective than Nguyen method [19] in term of specificity. In fact, our drawback is very small because the difference is very light (98.32 compared to 98.78) and the figure related to background pixels. In our method, we focus on vessel pixels.

5 Conclusions

Retinal blood vessel segmentation is the main work in images processing. Because its result is very useful for doctors to diagnose some dangerous diseases. Based on the segmented image result, the doctor easily finds out some main symptoms of the disease. In our paper, a novel method proposed to segment retinal blood vessel using HE combined with IMLD. The proposed method includes two stages: using HE to enhance the input image and apply MLD with various window sizes to detect vessels. Thanks to the using the adaptive window size, the method can detect both large vessels and small ones. It concludes that our proposed method is more effective than recent other algorithms according to computing sensitivity, specificity and accuracy.

Acknowledgements This research is supported by scientific researching fund of Tra Vinh University.

Conflict of Interest The authors declare that they have no conflict of interest.

References

1. Wasan B, Cerutti A, Ford S, Marsh R (1995) Vascular network changes in the retina with age and hypertension. J Hypertens 13(12):1724–1728
2. Wong TT, McIntosh R (2005) Hypertensive retinopathy signs as risk indicators of cardiovascular morbidity and mortality. Br Med Bull 73(1):57–70
3. Sussman EJ, Tsiaras WG, Soper KA (1982) Diagnosis of diabetic eye disease. JAMA. J Am Med Assoc 247(23):3231–3234
4. Wong TY, Klein R, Couper DJ, Cooper LS, Shahar E, Hubbard LD, Wofford MR, Sharrett AR (1982) Retinal microvascular abnormalities and incident stroke: the atherosclerosis risk in communities study. The Lancet 358(9288):1134–1140
5. Wong TY, Klein R, Sharrett AR, Duncan BB, Couper DJ, Klein BEK, Hubbard LD, Nieto FJ (2004) Retinal arteriolar diameter and risk for hypertension. Ann Internal Med 140(4):248–255
6. Kaur D, Kaur Y (2014) Various image segmentation techniques: a review. Int J Comput Sci Mob Comput 3(5):809–814
7. Al-amri SS, Kalyankar NV, Khamitkar SD (2010) Image segmentation by using thereshod techniques. J Comput 2:83–86
8. Marín D, Aquino A, Gegundez-Arias ME, Bravo JM (2011) A new supervised method for blood vessel segmentation in retinal images by using gray-level and moment invariantsbased features. IEEE Trans Med Imag 30(1):146–158
9. Evelin Sujji G, Lakshmi YVS, Wiselin Jiji G (2013) MRI brain image segmentation based on thresholding. Int J Adv Comput Res 3:97–101
10. Hamdi MA (2011) Modified algorithm marker-controlled watershed transform for image segmentation based on curvelet threshold. Can J Image Process Comput Vis 2(8):88–91
11. Reddy PN, Mohan Rao CPVNJ, Satyanarayana Ch (2017) Brain MR image segmentation by modified Active contours and Contourlet transform. ICTACT J Image Video Process 8(2):1645–1650
12. Holger RR, Shen C, Oda H, Oda M, Hayashi Y, Misawa K, Mori M (2018) Deep learning and its application to medical image segmentation. Med Imaging Technol J 36(2):1–6
13. Li Q, You J, Zhang D (2012) Vessel segmentation and width estimation in retinal images using multiscale production of matched filter responses. Expert Systems with Application 39(9):7600–7610
14. Permuter H, Francos J, Jermyn I (2006) A study of Gaussian mixture models of color and texture features for image classification and segmentation. Pattern Recogn 39:695–706
15. Ng HP, Ong SH, Foong KWC, Goh PS, Nowinski WL (2006) Medical image segmentation using k-means clustering and improved watershed algorithm. In: Proceedings of IEEE Southwest symposium on image analysis and interpretation, pp 61–65
16. Staal J, Abràmoff MD, Niemeije M, Viergever MA, Ginneken BV (2004) Ridge-based vessel segmentation in color images of the retina. IEEE Trans Med Imag 23(4):501–509
17. Soares JVB, Leandro JJG, Cesar RM, Jelinek HF, Cree MJ (2006) Retinal vessel segmentation using the 2-d gabor wavelet and supervised classification. IEEE Trans Med Imag 25(9):1214–1222
18. Binh NT, Vo THT, Nguyen MH, Nguyen TT (2019) Retinal vessels segmentation by improving salient region combined with Sobel operator condition. In: International conference on future data and security engineering, pp 608–617
19. Nguyen UTV, Bhuiyan A, Park LAF, Ramamohanarao K (2013) An effective retinal blood vessel segmentation method using multi-scale line detection. Pattern Recogn 46:703–715
20. Mustafa WA, Mahmud AS, Khairunizam W, Razlan ZM, Shahriman AB, Zunaidi I (2019) Blood vessel extraction using combination OF Kirsch's templates and fuzzy C-means (FCM) on Retinal Images. IOP Conf Ser Mater Sci Eng 557:012009
21. Nguyen MH, Nguyen TB (2017) Efficient framework for magnetic resonance image fusion using histogram equalization combined with cross bilateral filter. In: 6th international conference on the development of biomedical engineering in Vietnam (BME6), IFMBE Proceedings 63, pp 351–356. https://doi.org/10.1007/978-981-10-4361-1_59

22. Digital Retinal Images for Vessel Extraction. https://www.isi.uu.nl/Research/Databases/DRIVE/. Last accessed 4 Apr 2020
23. Vlachos M, Dermatas E (2010) Multi-scale retinal vessel segmentation using line tracking. Comput Med Imaging Graph 34:213–227

SOREM and CAP Parameters in Narcolepsy Patients and Healthy Subjects

Huyen Nhung Hoang, Quoc Khai Le, and Quang Linh Huynh

Abstract This study aimed to compare sleep architects and CAP parameters between patients with Narcolepsy and the normal person using the electroencephalogram signal (EEG) analysis. In this study, we analyzed sleep data from five Narcolepsy patients and fourteen normal controls. Participants' macrostructure parameters and CAP parameters were calculated and compared to indicate the abnormal in narcoleptics' nocturnal sleep. A prime abnormal in Narcolepsy patient's sleep structures was the early onset of REM with significant shot REM latency; this phenomenon was observed in two out of five patients' hypnograms. The nocturnal sleep of Narcolepsy patients was usually disturbed by sleep arousals that appeared regularly at night, which reduced the quality of nocturnal slumber. In this study, Narcolepsy individuals had a lower CAP/NREM rate, shorter CAP total time, and a decreased in the A1 index during NREM sleep compared to healthy individuals. The obtained results suggested that people with Narcolepsy be likely to have a short REM latency with fragmented night sleep that reduced sleep quality. The increase in the CAP ratio could relate to the instability of night sleep in Narcolepsy patients. A further study on EEG wave activity should be conducted to assist the inequality between a group of Narcolepsy patients and another group of normal subjects.

Keywords Narcolepsy · CAP · Hypnogram · EEG · Electroencephalogram

H. N. Hoang (✉) · Q. K. Le (✉) · Q. L. Huynh
Department of Biomedical Engineering, Faculty of Applied Science, Ho Chi Minh City University of Technology (HCMUT), 268 Ly Thuong Kiet Street, District 10, Ho Chi Minh City, Vietnam
e-mail: 1612447@hcmut.edu.vn

Q. K. Le
e-mail: quockhai@hcmut.edu.vn

Vietnam National University Ho Chi Minh City, Linh Trung Ward, Thu Duc District, Ho Chi Minh City, Vietnam

V. Van Toi et al. (eds.), *8th International Conference on the Development of Biomedical Engineering in Vietnam*, IFMBE Proceedings 85,
https://doi.org/10.1007/978-3-030-75506-5_77

1 Introduction

Narcolepsy is unknown by the majority of the population in the world as it is a rare sleep disorder that is primarily characterized by four symptoms, including excessive daytime sleepiness (EDS), cataplexy, disturbed nocturnal slumber, and hallucinations [1]. There are two types of hallucinations that a Narcolepsy individual may face that happens when a person is waking up and when the person is falling asleep, call hypnopompic hallucinations and hypnagogic hallucinations [1]. Several articles report that Narcolepsy might occur for the first time at any time from an early age to under 50 years old [2]. Some research indicates around the age of 15 and the age of 36 as two peak time periods of narcolepsy onset [2]. In the United States, the disorder has been estimated to affect 0.05 percent of the population, affecting both males and females equally [1]. The prevalence seems to be higher or lower in some regions. For instance, Narcolepsy is commonest in the Japanese population, which affects at a rate of 160 cases per 100,000 individuals, but the percentage of Narcolepsy is lower in the Jewish and Saudi Arabian population, only 10 cases for every 100,000 persons [1]. However, determining its true frequency in the general population over the world with some difficulties since it takes years to be recognized and diagnosed.

In 1880, Gelineau first described Narcolepsy disorder in the medical literature. At an early age, Narcolepsy was diagnosed according to four main symptoms: excessive daytime sleepiness, sudden loss of muscular tonus, fragmented slumber, and vivid hallucinations [3]. Years later, the presence of sleep-onset rapid eye movement (SOREM) in narcolepsy sleep stages had been found as a shred of abnormal evidence for diagnosing by Vogel, and Honda et al. [4], had correlated the disease with the lack of Class II HLA-DR2 antigens in patients' spinal fluid in 1983 [4]. Many studies were conducted to find the relation between cerebrospinal fluid (CSF) hypocretin levels and Narcolepsy during the last decade. These studies indicate that 90% of the patients with Narcolepsy have lower hypocretin levels than in healthy individuals, which contributes to the evidence that Narcolepsy is an autoimmune disorder [5, 6].

Diagnosing Narcolepsy requires several tests, including polysomnography (PSG), Multiple Sleep Latency Test (MSLT). In sleep analysis, polysomnography (PSG) plays an important role as this is a non-invasive technique that makes it possible to collect data from eight hours of night sleep. Previous PSG studies on Narcoleptics mainly focus on sleep architectures, which refer to the basic structural organization of normal sleep. However, misdiagnosis is common in Narcolepsy since the symptoms are unclear; the low awareness of people about this disorder and the features extracted from these studies are not clear enough to diagnose the disorder. In this study, the main purpose is to find the differences in macrostructure and microstructure between Narcolepsy individuals and normal individuals using sleep data.

2 Methodology

This research applies the Declaration of Helsinki principles in human studies.

Macrostructure and microstructure are two major elements of sleep that are defined according to specific events of the EEG signals. For the macrostructure, the whole recording is divided into 30 s epochs then classified into rapid eye movement (REM) or non-REM (NREM) [7]. The microstructure is scored in 1 s epoch based on the transient and phases event in brain electrical activity. Terzano et al. [8] defined a way to analyze the microstructure through the cyclic alternating pattern (CAP). Analyzing the CAP in EEG activity is considered as a method to study sleep instability and sleep disturbance. The CAP includes transient periods (A-phase) that happen in NREM and interrupts the background activities of NREM sleep (B-phase) [8]. The duration of both phases is at least 2 s and no longer than 60 s. A succession of an A-phase and a B-phase form a CAP cycle. A CAP sequence consists of two or more CAP cycles [8, 9]. There are three sub-phases of A-phase (A1, A2, and A3) [8, 9] that can be examined for detailed sleep disturbance analysis or sleep instability.

In this work, we use 14 healthy individuals (31.64 ± 5.20 years) full-night recordings and 5 Narcolepsy patients (31.60 ± 10.33 years) recordings that are uploaded by the Sleep Disorders Center in Parma [8, 10]. The CAP sleep database was downloaded freely from the URL "http://physionet.org/physiobank/database/nstdb/". The database provides the EEG, ECG, EMG waveforms recorded throughout the night stored in .edf files. The scoring of sleep stages and the CAP were detected in agreement with Terzano's reference atlas of rules by expert neurologists trained at the Sleep Center. The scores for each recording are provided as.txt files. This information would be used to calculate the macro and micro sleep parameters of each subject.

2.1 *Marco-Structure Parameters*

Sleep architecture represents the cyclical sleep pattern when it shifts between sleep stages and provides a picture of what our sleep looks like over the night. Non-rapid eye movement (NREM) and rapid eye movement (REM) sleep are two major parts of sleep architecture. Sleep was usually staged base on the rules of the American Academy of Sleep Medicine (AASM) Manual for the Scoring of Sleep and Associated Events. The criteria that were focused on in this study were the duration of each sleep stage in sleep architecture [11]. Normally, sleep goes through NREM stages (N1, N2, N3) before getting into the REM stage. The first cycle often lasts from 90 to 120 min [12]. However, REM sleep does not last for the whole night; in fact, individual sleep architect includes 3–4 cycles between stages of NREM and REM [12]. NREM sleep constitutes most of the total time spent in sleep [12]. REM accounts for around one-fourth of slumber time [12]. Study the sleep architect could reveal sleep quality and other sleep disorders. In narcolepsy studying, the patient's condition was analyzed based on the symptoms, which made the patient fall asleep unpredictably [1]. Hypnogram visualized the alternating between sleep stages and was normally used to make a quick diagnose of sleep disorders [1].

Sleep architect's criteria were used to evaluate the differences between narcolepsy patients and normal controls. One of the most remarkable differences that distinguish Narcolepsy patients against healthy individuals is the significant short REM sleep latency [13]. Quinnell [14] conducted research to estimate the REM sleep latency in Narcolepsy patients compared to normal controls. The numbers indicated that the onset of sleep in 80% of Narcolepsy patients was around 8 min, followed by REM sleep occurred almost immediately within several seconds [14].

In this study, we focused on the duration of each sleep stage scored in hypnogram and the proportion of each stage to sleep period time. The quality of sleep was estimated through the total sleep period time, total actual sleep time, and the total time of arousals after sleep onset [11]. Figure 1 illustrates the time parameters that were calculated in this study. The hypnogram for each subject was built based on the annotations of sleep stages in the data. Time in bed was defined as the total recording time. The period time between the first time of falling asleep and the last time of waking up was called sleep period time and total actual sleep time was the sleep period time excluded all wakefulness time during the whole night. We calculated the duration time of each type of sleep stage and their percentage. The number of arousal per hour was calculated to estimate the fragmentation of narcolepsy sleep. After calculating all the macrostructure indices, the results were presented in Table 1 as median and standard deviation.

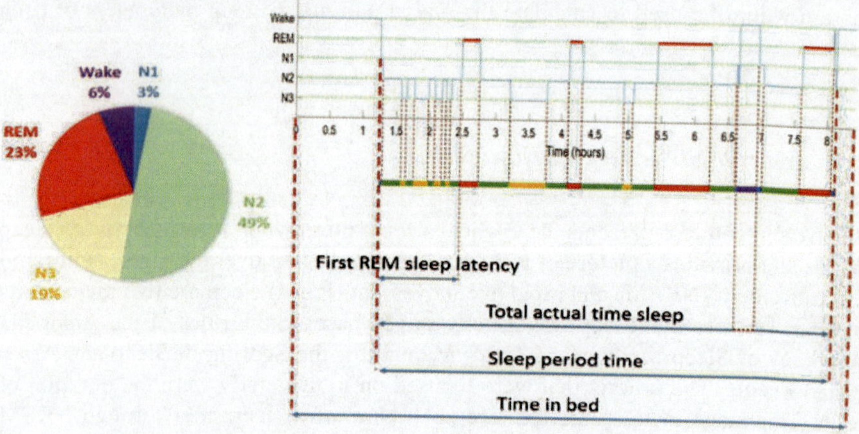

Fig. 1 Example of hypnogram, time and percentage parameters of sleep

Table 1 Comparison of sleep Macrostructural parameters between Narcolepsy group and Controls group (*p-value < 0.05)

	Narcolepsy	Controls	p-value
	Mean ± SD	Mean ± SD	
Sleep period time (min)	451.40 ± 102.80	481.40 ± 14.91	0.105
Total real sleep time (min)	431.10 ± 97.76	479.70 ± 14.16	0.853
N1 time (min)	30.10 ± 23.60	9.80 ± 7.61	0.246
N2 time (min)	170.80 ± 71.01	224.40 ± 20.74	0.165
N3 time (min)	104.40 ± 42.13	121.40 ± 16.46	0.405
REM time (min)	125.80 ± 39.74	124.10 ± 16.65	0.459
Wake time (min)	20.30 ± 7.22	1.70 ± 1.83	**0.033***
Percent of Wakefulness (%)	4.49 ± 1.12	0.35 ± 0.37	**0.042***
Percent of N1 stage (%)	6.58 ± 5.11	2.01 ± 1.50	0.267
Percent of N2 stage (%)	37.39 ± 11.04	46.65 ± 4.50	0.096
Percent of N3 stage (%)	22.43 ± 5.86	25.21 ± 3.14	0.116
Percent of REM stage (%)	29.10 ± 9.22	25.78 ± 3.42	0.869
REM latency (min)	26.00 ± 28.45	74.70 ± 11.01	**0.003***
REM/NREM	0.47 ± 0.21	0.35 ± 0.06	0.229
Awakening (arousals/h)	0.90 ± 0.15	0.15 ± 0.14	0.052

2.2 *Micro-structure Parameters*

In recent years, slumber instability has been analyzed through CAP parameters since the CAP distribution in NREM stages were considered to have a close relationship with sleep structure. According to Terzano et al. [8], phase A could be classified into three sub-phases (A1, A2, A3). The dominance of A1 subtypes in the N1 stage can be the expression of REM-off activity, while the REM-on activity is reflected by the increase in A2 and A3 subtypes in N2 and N3 [8, 12]. It is reported that in normal sleepers, CAP cycles occurred between 233 to 343 times during sleep [8]. The length of CAP cycles ranges between 25 and 31 s, with a high rate of B-phase [8]. Bruni et al. [12], the mean amount of CAP cycles in children aged between 6 and 10 years old is higher at 363. In children's sleep, the CAP usually lasts longer (30, 49 s) and B-phase consists of 81% of the total CAP time [8]. In this study, we calculated the following: total CAP time, CAP and three subtypes rate, the number of three subtypes and their duration according to Terzano rules [8]. The.txt files provided the

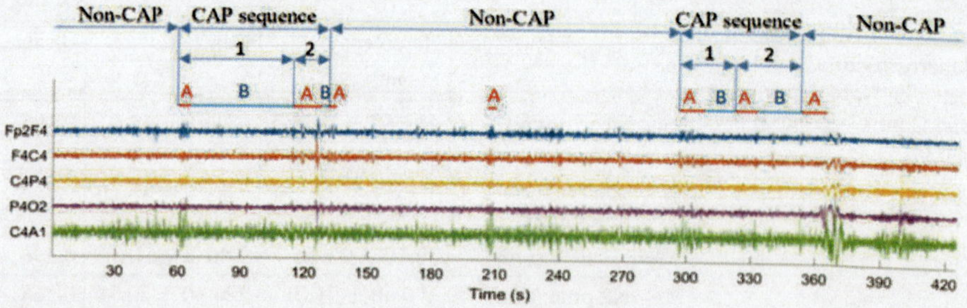

Fig. 2 Minimal requirements for the definition of a CAP sequence. The A phases are marked by red line and the B phases are marked by cyan line. The third A phase of each CAP sequence is not included in that CAP sequence

annotation of A phases and their subtypes. B phases are defined as periods between two A phases; any B phase that lasts longer than 60 s would be excluded. An A phase followed with a B phase form up a CAP cycle. Two or more CAP cycles make up a CAP sequence. Figure 2 provides an example for CAP sequences with at least 2 CAP cycles (numbered as 1, 2); the third A phase belongs to the non-CAP sequence and would be excluded when calculating the microstructure parameters. An isolated A phase, which is separated from other phases by more than 60 s is classified as non-CAP. CAP time (min) is the total time of NREM sleep occupied by CAP sequences and the CAP rate is the fraction of CAP time sleep during the total NREM sleep.

Statistical analysis

All comparisons between groups were carried out using the nonparametric Mann–Whitney test for independent data sets. Differences were considered significant when they reached a $p < 0.05$ level.

3 Results and Discussion

3.1 Macrostructural Variables

Vaňková et al. [15] conducted a research on sleep architects between two groups in order to compare and evaluate the differences that could easily be observed through sleep hypnogram. The researchers indicated that although the total time of sleep between two groups was nearly the same (425.9 ± 75.5 min) for Narcolepsy group and 419 ± 54.5 (min) for the healthy group) [15], the slumber latency was significantly shorter in comparison with the healthy group [15, 16]. In some cases, the onset of REM sleep appeared as soon as the patient falling asleep [15, 16], while in a normal person, REM onset appeared around 80 min after sleep [1].

The results were presented in Table 1. This study observed remarkable differences between the two groups, including early REM onset, wake time, and percentage of wake time. There were some slight differences found for the percentage of the N2 stage and the number of arousals per hour of sleep during nocturnal sleep. Other parameters such as sleep period time, total real sleep time, duration of sleep stages, percentage of REM and NREM's stages were not significantly different between the two groups. Narcoleptic patients often experience disturbed sleep, which expanded the wakefulness and light sleep duration (N1) while reduced deep sleep duration (N2 + N3) for bad sleep. Vivid dreams were one of the prime features in the nocturnal sleep of narcolepsy patients, which might relate to the early appearance of the REM stage in hypnogram. There are searches indicate some similar abnormalities in narcolepsy nocturnal sleep, such as reduced REM and NREM latency, increased micro-arousal in sleep, increased sleep–wake transition, and increased NREM first stage [17–19].

3.2 *Microstructural Variables*

Terzano et al. [16] reported that the CAP rate might reflect poor slumber quality or sleep disturbed by internal or external factors. Therefore, the CAP rate became one of the criteria for studying sleep disorders [16]. CAP criteria of narcoleptics and normal subjects are presented in Table 2. The result was presented as mean ± standard deviation (SD). In particular, there was no significant difference in the CAP rate in the Narcolepsy group. There were slight differences in the A1 rate and the total number of A3 between the two groups. The A1 rate in the Narcolepsy group was two-thirds of that in the healthy group ($p = 0.052$). In contrast, the total number of A3 in Narcolepsy individuals doubled the number of A3 in healthy individuals ($p = 0.052$). For other criteria, no remarkable differences were observed. Accordingly, phase A3-related indices were better than the CAP rate to distinguish normal sleep from the narcolepsy sleep.

The most well-known microstructure parameters index is CAP, as a higher CAP rate could reflect poor quality sleep [11]. The relationship between CAP sequences and arousals is that the subtypes A2 and A3 correlated positively with arousals, even though there is no statistically significant correlation between subtype A1 and arousals. Parrino et al. [20] indicated that the A1 subtypes dominate during deep NREM sleep and reflect neural mechanisms for maintaining NREM sleep. In contrast, A2 and A3 subtypes are the expressions of NREM disruption. Figure 3 shows the distribution of three sub-phases (A2 and A3 are presented as one bar because they both reflect the disruption of NREM sleep) marked on the sleep hypnogram.

Table 2 Comparison of Microstructural parameters between Narcolepsy group and Controls group (*p < 0.05)

	Narcolepsy	Controls	p-value
	Mean ± SD	Mean ± SD	
CAP time (min)	161.26 ± 83.98	140.17 ± 40.65	0.711
CAP rate (%)	52.73 ± 18.52	40.94 ± 9.64	0.079
Total number of A1	185.60 ± 109.36	192.21 ± 79.95	0.926
Total A1 rate (%)	49.21 ± 9.78	64.47 ± 13.25	0.052
Total number of A2	83.20 ± 58.03	56.57 ± 29.16	0.643
Total A2 rate (%)	17.27 ± 6.97	18.71 ± 7.85	0.781
Total number of A3	135.20 ± 74.35	67.40 ± 22.92	0.052
Total A3 rate (%)	33.52 ± 12.69	16.83 ± 7.26	**0.026***
A mean duration (s)	8.60 ± 1.02	7.79 ± 1.08	0.139
A total duration (min)	62.36 ± 30.94	50.75 ± 16.48	0.517

4 Conclusion

This study has some restrictions. First, there were only 19 recordings used (with only five samples for Narcolepsy groups) in this study and the samples were not matched by age and sex factors. The sample included in this study were from two different groups of people without age-match or gender-match, which could lead to limitations of the study results. Second, although there are significant differences in SOREM and A3 index, these features are not strong enough to recognize or diagnose every individual. In conclusion, the narcoleptic patients investigated in the study show early REM stage sleep. Compared to controls, there was a slight increase in CAP rate in narcolepsy patients which related to poor sleep quality. There were studies that showed contrary results [3, 13, 15]; in our study, some samples show a lower CAP rate but not for all the cases. However, the sample size is limited, and the p-value does not show a remarkable difference between the two groups. For better performance, it is needed to add more samples for patient group as well as match the samples by age and sex. Our study shows some noticeable features in narcolepsy patients that are different from normal individuals which may help detect the disorder. Other woths could be conducted to improve the diagnosing method, such as analyzing power spectrum of the EEG wave forms.

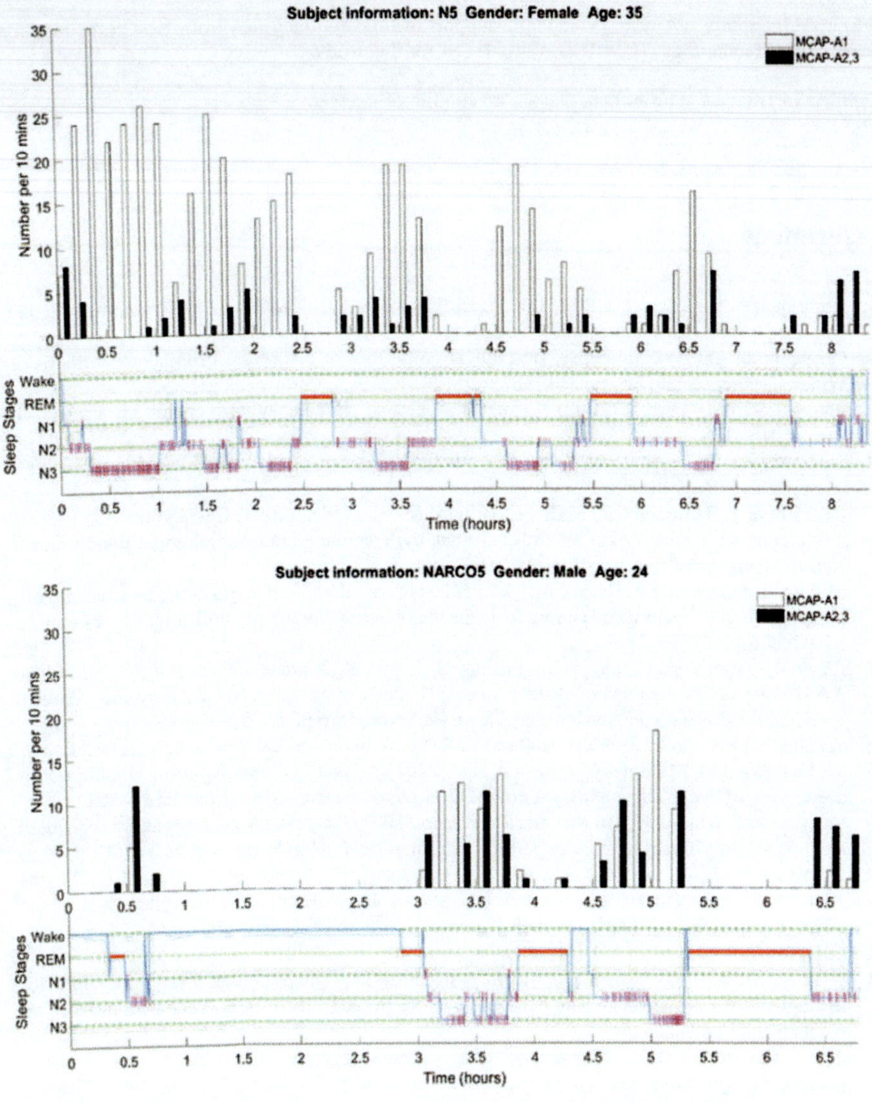

Fig. 3 Distribution of three sub-phases (white bars represent A1, black bars represent A2 and A3) marked on the sleep hypnogram. Each bar illustrated the number of events happen every 10 min over the recording night. The top panel indicated the hypnogram and the number of phase A in normal persons compared with the hypnogram and the number of phase A in a narcolepsy patient (bottom panel). Small magenta vertical lines on the hypnogram indicated the occurrence of each sub-phase event. Thick red lines represented the appearance and endurance of REM sleep stage. Sleep onset REM periods (SOREMs) in narcoleptic patient appeared very early (almost as soon as the patient fall asleep), while in a healthy individual subject SOREM appeared 2.5 h after the subject falling asleep. A significant difference can be observed through the graphs were the reduction in the number of phase A1 and the increase in the number of phase A2 and phase A3 in narcoleptic individual compared to healthy control

Acknowledgements We acknowledge the support of time and facilities from Ho Chi Minh City University of Technology (HCMUT), VNU-HCM for this study.

Conflicts of Interest The authors declare no conflict of interest.

References

1. Kornum BR, Knudsen S, Ollila HM, Pizza F, Jennum PJ, Dauvilliers Y, Overeem S (2017) Narcolepsy. Nat Rev Dis Primers 3:16100
2. Goswami M (2003) Narcolepsy. In: NORD guide to rare disorders. Lippincott Williams & Wilkins, Philadelphia, PA, pp 561–562
3. Plazzi P, Serra L, Ferri R (2008) Nocturnal aspects of Narcolepsy with cataplexy. Sleep Med Rev 12:109–128
4. Sturzenegger C, Bassetti C (2004) The clinical spectrum of Narcolepsy with cataplexy: a reappraisal. J Sleep Res 13:395–406
5. Dauvilliers Y, Baumann CR, Maly FE, Billiard M, Bassetti C (2003) CSF hypocretin-1 levels in Narcolepsy, Kleine-Levin syndrome, other hypersomnias and neurological conditions. J Neurol Neurosurg Psychiatry 74:1667–1673
6. Mignot E, Lammers GJ, Ripley B, Okun ML, Nevsimalova S et al (2002) The role of cerebrospinal fluid hypocretin measurement in the diagnosis of Narcolepsy and other hypersomnias. Arch Neurol 59:1553–1562
7. Berry R, Brooks R, Gamaldo C, Harding S, Lloyd R, Marcus C, Vaughn B (2017) The AASM manual for the scoring of sleep and associated events: rules, terminology and technical specifications. American Academy of Sleep Medicine, Darien, IL, SA
8. Terzano M, Parrino L, Sherieri A, Chervin R, Chokroverty S, Guilleminault C, Hirshkowitz M, Mahowald M, Moldofsky H, Rosa A et al (2001) Atlas, rules, and recording techniques for the scoring of cyclic alternating pattern (CAP) in human sleep. Sleep Med 2:537–553
9. Smerieri A, Parrino L, Agosti M, Ferri R, Terzano M (2007) Cyclic alternating pattern sequences and non-cyclic alternating pattern periods in human sleep. Clin Neurophysiol 118:2305–2313
10. Goldberger AL, Amaral LAN, Glass L, Hausdorff JM, Ivanov PCh, Mark RG, Mietus JE, Moody GB, Peng C-K, Stanley HE (2000) PhysioBank, PhysioToolkit, and PhysioNet: components of a new research resource for complex physiologic signals. Circulation 101(23):e215–e220
11. Tang X, Xu Z, Li Y, Shan S (2016) Macrostructural and microstructural sleep variables for distinguishing normal sleep from pathologic sleep. In: 2016 IEEE international conference on mechatronics and automation
12. Chāberová Bc J (2017), Analysis of sleep polysomnography data using advanced signal processing algorithms. Master Thesis, Czech Technical University in Prague, Part 1, January 2017
13. Bruni O, Novelli L, Miano S, Parrino L, Terzano MG, Ferri R (2010) Cyclic alternating pattern: a window into pediatric sleep. Sleep Med 11(7):628–636
14. Quinnell TG, Smith IE (2011) Narcolepsy, idiopathic hypersomnolence and related conditions. Clin Med 11(3):282–285
15. Vankova-Bouskova J, Busek P, Volna J, Sonka K, Nevsimalova S (2007) Abnormal sleep microstructure and autonomic response in narcolepsy. Cesk Slov Neurol N 70/103(2):158–162
16. Terzano MG, Smerieri A, Del Felice A, Giglia F, Palomba V, Parrino L (2006) Cyclic lternating pattern (CAP) alterations in narcolepsy. Sleep Med 7(8):619–626
17. Montplaisir J, Billiard M, Takahashi S, Bell IR, Guilleminault C, Dement WC (1978) Twenty-four-hour recording in REM-narcoleptics with special reference to nocturnal sleep disruption. Biol Psychiatry 13:73–89

18. Montplaisir J, Godbout R (1986) Nocturnal sleep of narcoleptic patients: revisited. Sleep 9:159–161
19. Broughton R, Dunham W, Newman J, Lutley K, Duschesne P, Rivers M (1988) Ambulatory 24 hour sleep-wake monitoring in narcolepsy-cataplexy compared to matched controls. Electroencephalogr Clin Neurophysiol 70:473–481
20. Terzano MG, Parrino L, Smerieri A, Carli F, Nobili L, Donadio S, Ferrillo F (2005) Cap and arousals are involved in the homeostatic and ultradian sleep processes. J Sleep Res 14(4):359–368

A Selective EOG Removal Method for EEG Signals: The Multi-thresholding Technique

Quoc Tuong Minh, Sieu Le Thi Be, Khai Le Quoc, and Linh Huynh Quang

Abstract EOG is one of the major artifacts in EEG signal processing. There are varieties of methods have been proposed that aim to eliminate the influence of Occular artifacts on the EEG signals. However, the problem is the trade-off between their performance of removing EOG artifact and their simplicity. In this study, we propose a simple and reliable method but giving a good performance. The idea of this method is to use a multi-threshold technique to target EOG contaminated parts in the signal then selectively subtract it out in order to get a corrected signal with a minimum alteration on the uncontaminated parts. In this study, we used triple-threshold, both in time and frequency domain, to target the contaminated parts (or EOG artifact component). The result shows that besides its simplicity, this method also reliable and effective when selectively removed some typical EOG artifacts like blinks or eye movements without altering other clean parts in the EEG signals. More than that, our method is also able to extract the estimated EOG artifact component from the EEG signal. The need for this method is only one single prefrontal EEG channel, no need for an EOG reference channel for the input. The source code of this method is freely available to download in the form of a MATLAB function by request. We encourage the researchers to give it a try.

Keywords Simple EOG artifact removal · Electroencephalogram (EEG) · Electrooculogram (EOG) · Multi-thresholding technique

1 Introduction

EOG artifact (or Ocular artifact) is very common and dramatically influences on EEG signal. Especially , eye blink and eye movement are two major sources of the ocular artifact that cause the most impact on the EEG signal, both in amplitude

Quoc Tuong Minh and Sieu Le Thi Be: These authors contributed equally.

Q. Tuong Minh (✉) · S. Le Thi Be · K. Le Quoc · L. Huynh Quang
Department of Biomedical Engineering, HCMUT-VNU, 268 Ly Thuong Kiet Street, Ho Chi Minh City 70000, Viet Nam

V. Van Toi et al. (eds.), *8th International Conference on the Development of Biomedical Engineering in Vietnam*, IFMBE Proceedings 85,
https://doi.org/10.1007/978-3-030-75506-5_78

and frequency perspective. Many researchers have proposed their methodologies to suppress the EOG artifact that all work so effectively. However, the problem of those methods is the trade-off between the performance of removing EOG artifact and their simplicity. For that reason, researchers in the BCI research community are spending their attention on finding a simple, fast, and reliable method to minimize the effect of those EOG noises without the need for the EOG reference channel instead [1]. For example, authors of [2] and have proposed a method that used only one prefrontal channel EEG.

There are many well-known algorithms to eliminate EOG artifact without EOG reference such as Wavelet + ICA hybrid [3], ICA – Independent Component Analysis [4], Regression + ICA hybrid [5], template matching based [6], Stationary Wavelet + Adaptive thresholding [7]. Those methods are all showed very good results in SNR increase, RMSE reduction, or increase in accuracy and sensitivity. However, they are such tough algorithms to understand and implement, especially with non-Computer-science-background students and researchers. Because of that reason, researchers and students with non-Computer-science-background are commonly using either coefficients threshold like wavelet coefficient in [8] or high-pass method like in [9] because of their simplicity and acceptable performance.

Since coefficient thresholding and high-pass filtering are simple and a little bit effective, we proposed a method that combines both of their main ideas to keep their simplicity while gaining a better performance. The combining is using thresholds in both the time and frequency domain. The uniqueness of our method is that we did not use a conventional coefficient like the wavelet coefficient. Instead, we have proposed using a cubed and scaled amplitude coefficient with multiple threshold values both in time and frequency domain. It helps target the contaminated parts in the signal and selectively subtracts it out to avoid altering other uncontaminated parts in the signal. We also compared our result with the wavelet coefficient thresholding method and high-pass filtering method when applying those on the semi-simulated signal taken from the EEG-EOG data set of Klados and Bamidis [10] to achieve a more objective evaluating. The result shows that our proposed method is simple and has better performance than the conventional thresholding and high-pass filtering method.

2 Materials and Methods

2.1 EEG-EOG Data Set

As mentioned in the introduction section, to obtain an objective evaluation for the outcomes, we have used the semi-simulated EEG-EOG data set of Klados and Bamidis [10]. They created the data set to help researchers evaluate their EOG artifact removal method performance objectively. They recorded the signal at 200 Hz sampling frequency, Notch filtered it at 50 Hz, and band-pass filtered it at 0.5–40 Hz. The data set contains samples from 27 subjects (males and females) with 19 electrodes placed according to the 10–20 International system layout. Each sample lasts about 30 s. The data set contains 4 files: "Pure_Data.mat" is EOG artifact-free EEG signal recorded during eye closed, "HEOG.mat" and "VEOG.mat" are respectively horizontal and vertical EOG data, and finally, "Contaminated_Data" is the semi-simulated EOG artifact contaminated data created by the following formula:

$$\text{Contaminated\_EEG}_{ij} = \text{Pure\_EEG}_{ij} + a_j\text{VEOG} + b_j\text{HEOG} \quad (1)$$

where index i indicates the subject's number and index j is for the electrode's number. Finally, a_j and b_j are the contamination coefficients of VEOG and HEOG, respectively [10].

We also would like to thank the authors of [10] have allowed us to use their data set.

2.2 A Summary of Influence of EOG on EEG Signal

Ocular (or EOG) artifact reflects the eye activities' influence on the EEG signal, mainly by the electrical field in anterior of the head be changed due to the polarity of the eye globes. It causes a big impact on the electrodes near the eyes and quickly decreases its impact degree in the anterior–posterior direction. Eyeblink and eye movement are 2 types of activities that affect the EEG signal greatest [11, 12].

A blink happens in a very short duration, about 0.1–0.5 s, or 2–10 Hz frequency. The amplitude of blink artifact varies from about 40 μV in the central plane, 80 μV in the frontal plane, and over 200 μV in the frontopolar plane. The authors of [11] also pointed out that blink artifact is bearing beneath the <5 Hz frequency range of the signal.

The rest type of EOG artifact is the eye movement or saccade, which can be seen both at horizontal and vertical EOG electrodes. The eye movement causes a longer duration, about 1.25 s (but just ~ 150 ms for Onset-to-peak duration), and with approximately the same peak amplitude with eyeblink on EEG signals [11]. See Fig. 1 for an example of typical waveforms of blink and saccade artifact on the EEG signal.

2.3 The Methodology

The idea of our method has 2 main parts:

- Locating the contaminated parts in the signal, both in time and frequency domain.

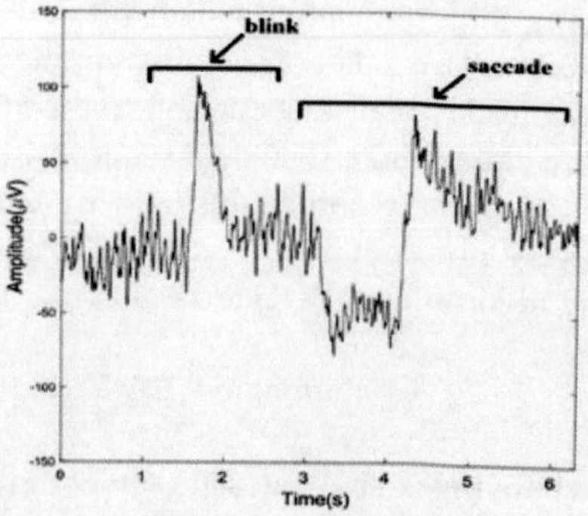

Fig. 1 Typical waveforms of a blink and a saccade we extracted from Klados and Bamidis [10] data set

- Selective subtracting that component out to get a corrected signal with as least alter as possible on the uncontaminated parts.

The diagram in Fig. 2 explains the stream of the signal's flow through steps in our method.

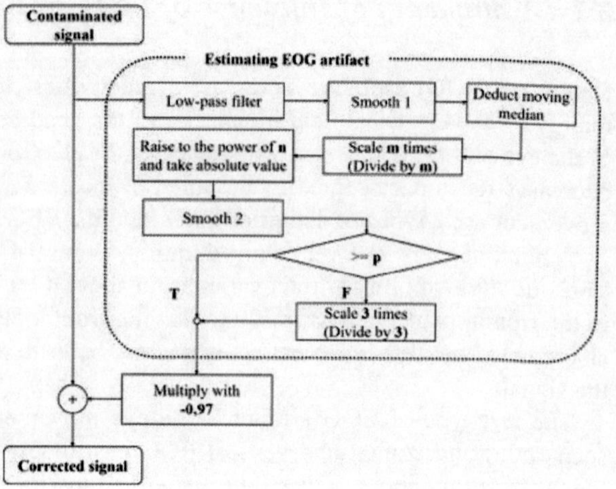

Fig. 2 Signal processing flowchart

Detail explanation

(Note: In this study, we use the signal of referential montage to process. We have not tested on other montages yet.)

- **Step 1**: Make out 2 copies of the original contaminated signal. One goes into the estimating process (denoted with a dotted frame), the other one is the reference signal for later subtracting step.
- **Step 2**: *Low-pass filter steps. This step is primarily to extract the rough EOG artifact out of the EEG signal. We* used **6 Hz** as a cut-off value because this value

is slightly higher than 5 Hz—the frequency that contains both blink and saccade artifacts, to be sure that all peak values are captured as correctly as possible.

- **Step 3**: *Smooth 1 step* helps attenuate the leakage frequency of the previous low-pass filter to gain better results in the final. Because the shortest duration of blink and saccade is about 100 ms long (or 10 Hz), we will take a bit shorter duration length ***0.08*Fs****(or 12.5 Hz, slightly higher than 10 Hz)* as the window width, where Fs is the sampling frequency, to retains peak values.
- **Step 4:** *Deduct moving median.* This step helps retain low-frequency characteristics of the signal in the contaminated part. To do this, subtract the resulting signal from *step 3* with the value of itself after going through the moving median filter to exclude the low-frequency oscillations out of estimated EOG. A duration of 3–5 s is optimal for window width.
- **Step 5 and 6**: Steps 4 and 5 can be call *coefficient calculating step.* First, scale the signal ***m*** times to make the sample points lower than m become lower than 1. Then, we raised them to the power of ***n*** to make a distinct gap for scaled sample points gained above. The value ***m*** was set to 50 (μV) as we found that this is the highest amplitude of below 5 Hz frequency oscillations in the awake adult's EEG signal. When choosing the value ***n***, we individually tried with a different value and got ***n*= *3*** as optimal when applying to the data, using the ROC curve as the evaluating tool. ***Remember to take the absolute value of the coefficient before going to the next step*** to avoid a flat and discontinuous error for the best result.
- **Step 7**: *Smooth 2 step.* Step 4 and step 5 above makes the artifact's waveform narrower than its actual width. This step helps expand the width of them to make locating step to be more accurate. We advise using **1*Fs to 1.8*Fs** window width because this is the average onset to end duration range of EOG artifact duration [11]. *Note: the bigger the window width, the more inclusive the artifact range we get accompany with more susceptible we cause to artifact-free trials. We empirically recommend value* ***1.5*Fs*** *as an optimal window width for natural eye open condition.*
- **Step 8 and 9**: *Comparing coefficient with threshold* ***p***. This step is to locate the contaminated trials (position and range). We have calculated the average maximum coefficient value for 50 clean trials (picked manually) and found that they are always below roughly 0.15. As a result, we have taken **p = 0.15** for the final threshold value. The value points below threshold ***p*** are considered as cleans and were scaled-down by 3 (30%) because these clean ones still include about 18–50% of EOG artifact in them (including type II error miss-detected trials) [11]. At the end of this step, we are now able to extract the estimated EOG artifact component in the contaminated EEG signal.
- **Step 10**: *Final step.* Scale the estimated EOG 97% and multiply it with (−1) to get a negative value. After that, plus the reference signal from **step 1** with the result. Due to the negative value, we can subtract (or in other word, eliminate) 97% of the estimated EOG when adding it with the reference signal. We use a 97% value because we do not want to strictly remove the estimated EOG since it is still just an estimated one, there are probably still have some components of

the EEG signal underlying in it. As 97% is just a semi-arbitrary value, in practice, researchers could use any value from 95 to 100%.

- For the rest of EEG channels on the head, use the result from step 8—the estimated EOG artifact, and remove it out from the desire channels, referencing their hemisphere. The only thing that changes between channels is the percentage value chosen to subtract artifact component out from the signal in step 8. Lins et al. [11] have described more detail about those percentages for different EEG channels.

This research applies the Declaration of Helsinki principles in human studies.

3 Result and Disscusion

In this study, we compare the result of our proposed with 2 most relevant method: Wavelet Coefficient Thresholding (WCT) and High-pass Filtering (HF) in 2 perspective, quality and quantity.

3.1 *Quality Evaluating*

The easiest way to evaluate the result of one EOG artifact removal method is by looking at how it altered our signal. A good filtering method will shift the EOG artifacts slipping down to the baseline and did as minimum as possible change on clean parts. Because we are using a semi-simulated data set, we have a pure signal as the reference for the corrected result.

For further comparison of the performance between the methods, besides the RMSE value, we also used an additional value ΔSNR in percentage unit. The ΔSNR was calculated using 3 simple steps. First, we calculated SNR of pure signal versus reference EOG signal and called it $\mathrm{SNR_{ref}}$. Then, we calculated SNR for each of the used methods and called it $\mathrm{SNR_{method}}$. Finally, the ΔSNR was calculated using the following formula:

$$\Delta SNR(\%) = \frac{SNR_{method} - SNR_{ref}}{SNR_{ref}} \times 100\% \tag{2}$$

As seen in Figs. 3, 4, and 5, applied on over a hundred of contaminated trials from 54 samples of the data set, in average, the proposed method gives the best corrected-data, which is stick very close to the reference (average RMSE $\approx$ 10 for amplitude value, average RMSE $\approx$ 2 for PSD value) and do very little impact on clean trials (see Fig. 5). Also, Fig. 6 shows us the estimated EOG artifact component extracted by the method from channel FP1-ref is also very close with the real VEOG.

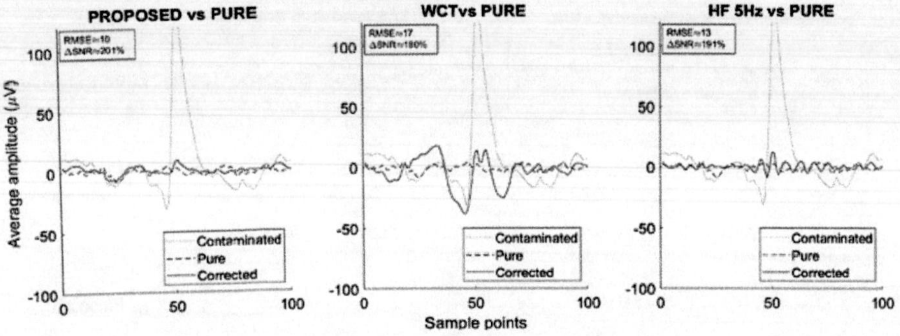

Fig. 3 Average amplitude before and after remove artifact of three methods

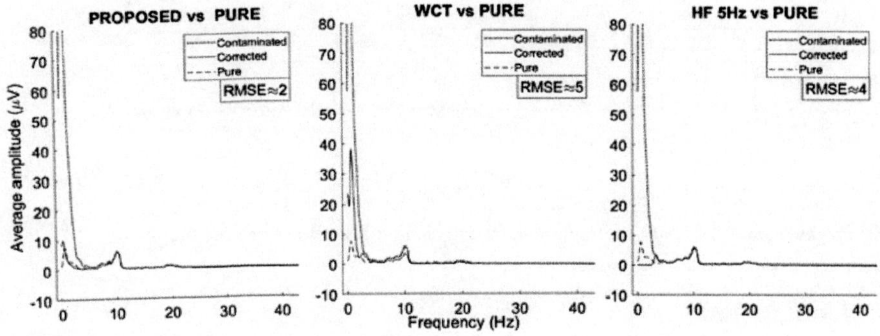

Fig. 4 The average value of PSD of before and after removing artifact result of three methods on contaminated trials

In short, our proposed method gives the most accurate corrected result compare with the reference pure signal in the visual look. It also gives a very accurate estimated EOG artifact component when compared with the real EOG signal.

3.2 *Quantity Evaluating*

Using T-test 2 sample with a significant level of 5%, we found out that the result of our method is better than the others in both time and frequency domain, but the same result in the effect of enhancing SNR value (see Fig. 7).

The proposed method gives RMSE value roughly 10 for amplitude and roughly 2 for PSD differences. It also helps double the signal-to-noise ratio (nearly 201%) (see Table 1).

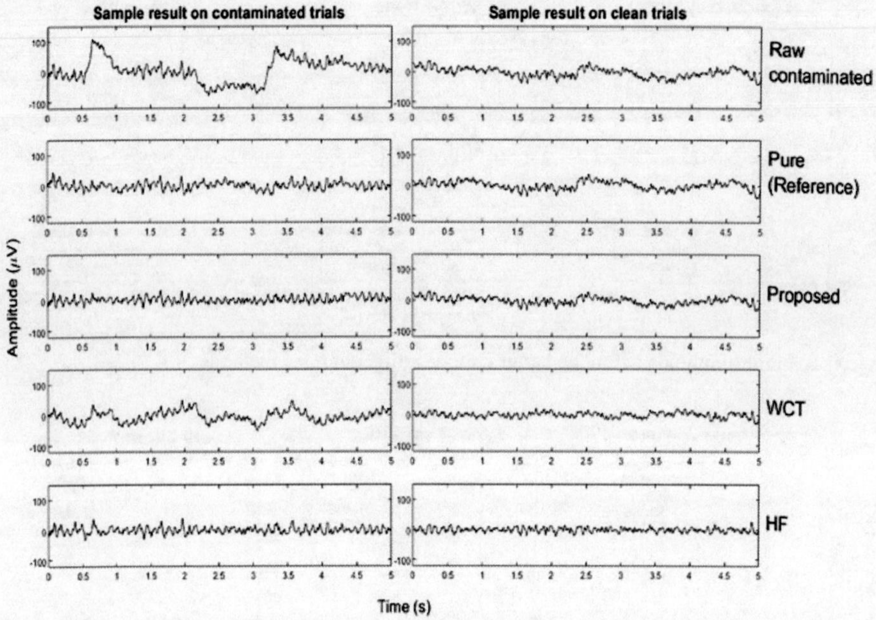

Fig. 5 One sample of result on contaminated and clean trials of the three methods

4 Conclusion

This study proposed a simple and effective method to remove the EOG artifact from the EEG signal, the Multi-Thresholding Technique. The first advantage of this method is that it was able to selectively remove the ocular artifact from contaminated trials without altering clean ones so much. Besides, only one frontopolar or frontal channel is required to detect contaminated trials. Also, FP1or FP2 is the recommended channel, if possible, or F3 and F4 are the alternatives in the case FP channels are not available. Furthermore, this method was able to extract the estimated EOG artifact component from the contaminated signal. Furthermore, the proposed method also has better performance than the coefficient thresholding and high-pass filtering method. Last but not least, this method is very easy to implement with just a few simple steps.

The result does not conflict with the authors' hypothesis. However, we should conduct a few more statistical studies to fully evaluate the effectiveness of this method or compare its effectiveness with other methods about the ocular artifact removal.

Researchers who want to download our source code, feel free to contact us for further instruction and getting a link to download the code.

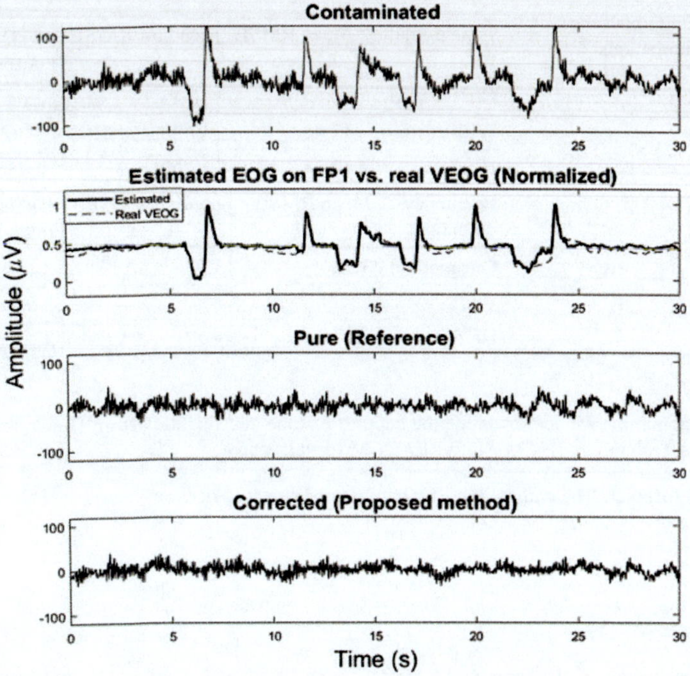

Fig. 6 The full result of our proposed method, including corrected data and estimated EOG artifact component

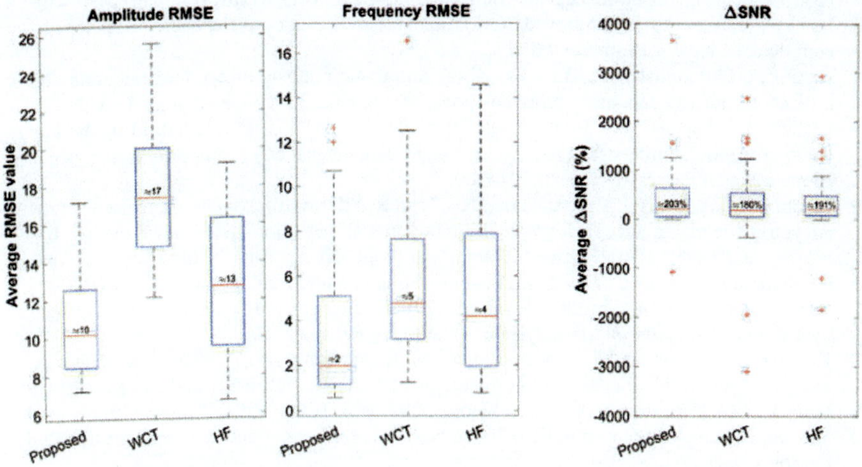

Fig. 7 The distribution of RMSE value in both time and frequency domain (left) and ΔSNR (right) of three methods after applying on the data set

Table 1 Detail about the RMSE of the methods

Time domain	Mean RMSE	Median RMSE	STD Include outliers
Our method	10.67	10.15	2.66
WCT	17.43	17.38	3.30
HF	12.68	12.69	3.75
Frequency domain	Mean RMSE	Median RMSE	STD Include outliers
Our method	3.32	1.98	3.15
WCT	5.87	4.77	3.67
HF	5.17	4.17	3.77

Acknowledgements We acknowledge the support of time and facilities from Ho Chi Minh City University of Technology (HCMUT), VNU-HCM for this study.

Conflicts of Interest The authors have no conflict of interest to declare.

References

1. Fatourechi M, Bashashati A, Ward RK, Birch GE (2007) EMG and EOG artifacts in brain computer interface systems: a survey. Clin Neurophysiol 118:480–494
2. Chang W-D, Cha H-S, Kim K, Im C-H (2016) Detection of eye blink artifacts from single prefrontal channel electroencephalogram. Comput Methods Programs Biomed 124:19–30
3. Issa MF, Juhasz Z (2019) Improved EOG artifact removal using wavelet enhanced independent component analysis. Brain Sci 9:355
4. Joyce CA, Gorodnitsky IF, Kutas M (2004) Automatic removal of eye movement and blink artifacts from EEG data using blind component separation. Psychophysiology 41:313–325
5. Klados MA, Papadelis C, Braun C, Bamidis PD (2011) REG-ICA: a hybrid methodology combining blind source separation and regression techniques for the rejection of ocular artifacts. Biomed Signal Process Control 6:291–300
6. Bizopoulos PA, Al-Ani T, Tsalikakis DG, Tzallas AT, Koutsouris DD, Fotiadis DI (2013) An automatic electroencephalography blinking artefact detection and removal method based on template matching and ensemble empirical mode decomposition. In: 2013 35th annual international conference of the IEEE engineering in medicine and biology society (EMBC)
7. Yong X, Fatourechi M, Ward RK, Birch GE (2012) Automatic artefact removal in a self-paced hybrid brain-computer interface system. J Neuroeng Rehabil 9:50
8. Krishnaveni V, Jayaraman S, Malmurugan N, Kandaswamy A, Ramadoss K (2004) Non adaptive thresholding methods for correcting ocular artifacts in EEG. Acad Open Int J 13
9. Mannan MMN, Kamran MA, Kang S, Jeong MY (2018) Effect of EOG signal filtering on the removal of ocular artifacts and EEG-based brain-computer interface: a comprehensive study. Complexity 2018
10. Klados MA, Bamidis PD (2016) A semi-simulated EEG/EOG dataset for the comparison of EOG artifact rejection techniques. Data Brief 8:1004–1006
11. Lins OG, Picton TW, Berg P, Scherg M (1993) Ocular artifacts in EEG and event-related potentials I: Scalp topography. Brain Topogr 6:51–63
12. Libenson MH (2012) Practical approach to electroencephalography E-Book. Elsevier Health Sciences

Automated EOG Removal from EEG Signal Using Independent Component Analysis and Machine Learning Algorithms

Tu Thanh Do, Thuong Hoai Nguyen, Tho Anh Le, Sinh Anh Thi Nguyen, Quyen Thao Ngoc Nguyen, Thinh Quang Vu Tran, Khoi Nguyen Khac Nguyen, Khiet Thu Thi Dang, Thien Hoai Luong, Chau Minh Ngoc Trinh, Hanh Hong Cao, Hieu Quang Le Truong, Khoa Anh Tran, Nhu Huynh Vo Nguyen, Phuong Nguyen Nguyen, Quyen Hoang Quoc Vo, and Huong Thanh Thi Ha

Abstract Contamination of EEG by ocular artifacts (EOG) is the major artifact that reduces the accuracy of applications using Electroencephalogram (EEG) signal. To resolve this issue, Independent Component Analysis (ICA) is a common method to remove EOG artifacts from EEG recordings, by decomposing multichannel EEG signals into maximally Independent Components (ICs). ICs representing ocular activities can be identified visually, then be eliminated to reconstruct EOG-free EEG signals. However, this approach requires prior domain knowledge, and hence, undermine reliability and reproducibility. To address this, our study proposed a method to remove EOG contamination by applying machine learning techniques. We acquired an EEG database of 20 healthy subjects using Alice 5 Polysomnography system to record signals from 12 electrodes. Randomly selected 15-s data segments from EEG channels were used to run ICA, which resulted in 10 ICs. For each IC, we plotted its topography map and labelled whether this IC is "EOG" or "non-EOG". A total of 612 labelled data points of ICs, topography maps and labels were collected. After applying several classifiers for model training and evaluation using cross-validation, the best classifier, Extremely Randomized Tree, achieved an average accuracy of 92%, precision of 83%, recall of 71%, and F1 score of 76%. In conclusion, the proposed method showed promising results in identifying EOG components and attenuating ocular activity on reconstructed EEG signals. Compared with existing

Tu Thanh Do, Thuong Hoai Nguyen and Tho Anh Le: These authors contributed equally to this paper.

T. T. Do · T. H. Nguyen · T. A. Le · S. A. T. Nguyen · Q. T. N. Nguyen · T. Q. V. Tran · K. N. K. Nguyen · K. T. T. Dang · T. H. Luong · C. M. N. Trinh · H. H. Cao · H. Q. Le Truong · K. A. Tran · N. H. V. Nguyen · P. N. Nguyen · Q. H. Q. Vo · H. T. T. Ha (✉)
School of Biomedical Engineering, International University, Ho Chi Minh City, Vietnam
e-mail: htthuong@hcmiu.edu.vn

Vietnam National University, Ho Chi Minh City, Vietnam

V. Van Toi et al. (eds.), *8th International Conference on the Development of Biomedical Engineering in Vietnam*, IFMBE Proceedings 85,
https://doi.org/10.1007/978-3-030-75506-5_79

automated solutions, our proposed method only used a small number of channels and had the potential to be applied in real-time applications due to its fast computation.

Keywords Independent component analysis · ICA · Ocular artifact · EOG · Classification models · Extra Trees Classifiers

1 Introduction

EEG signal (Electroencephalogram) has represented its unique role in neuroscience, clinical engineering, psychiatry studies as well as rehabilitation engineering with its non-invasive, inexpensive high temporal resolution technique [1–3]. Compared to other neuroimaging methods such as fMRI and PET, brain electrical signals have higher temporal resolution. This advantage of EEG signal enables various studies of cognitive processes. Different from conventional diagnostic tools used in mental and psychiatric studies such as questionnaires, EEG signals are more quantitative. Specifically, robust features from EEG signals that can be used for classification of different mental states or cognitive processes mainly fall in higher frequencies ranging from 30 to 80 Hz (gamma band) [4–6].

Due to its low amplitude, the EEG signal is sensitive to various noise sources coming from biological artifacts and the environment. This problem hinders doctors and researchers from obtaining good diagnostic information without excluding valuable EEG signals. For instance, power-line artifacts, as well as biological artifacts stemming from the subject including electrical signals from muscle tension, contractions of the heart and respiration, can also contaminate EEG signals [7]. EOG artifact is the most common source of artifacts that is affecting EEG signals and overlapping frequency spectrum. More specifically, Freeman and his colleagues have demonstrated that higher frequencies, e.g. gamma band or higher, are most easily overshadowed by EOG artifacts [8]. Hence, apart from rigorous experimental design for data collection, an algorithm for EOG artifact removal is imperative in most parts of EEG studies to eliminate undesired artifacts.

Many techniques have been proposed for EOG artifacts removal. These methods can be primarily separated into two categories: either by estimation of the artifact signals using reference channels or by decomposing the EEG signal into other domains [7]. Linear Regression is a method using reference channels that assumes that each EEG channel is the sum of the non-noisy source signal and a fraction of the source artifact that is available through a reference channel(s) [9]. While regression methods are simple and reduce computational demands, they still need good regression reference channels [9]. On the other hand, the Wavelet Transform algorithm decomposes the signal into a set of coefficients, for various scales, which represent the similarity of the signal with the wavelet at that scale. Nevertheless, it fails to identify EOG signals completely that overlap with the spectral properties [9]. Another decomposing method, Empirical mode Decomposition (EMD) is a fully data-driven method for decomposing multicomponent signals into a set of amplitude & frequency modulated (AM/FM) components known as intrinsic mode functions (IMFs) [10]. This method is sensitive to noise because it could not work effectively with multi-dimensional signals [11]. On the other hand, because mutually independent sources generate EOG artifacts such as eye movements, eye blinks, Blind Source Separation (BSS) methods, especially Independent Component Analysis (ICA) can remove EOG with great accuracy [7].

In our study, we used the MNE library implementation of ICA to decompose EEG signals to their independent components (ICs). From said components, MNE provides us with a scalp topography map of each IC. With the topography map, we were able to identify which IC represents ocular activity. However, ICA requires the need for visual inspection by experts to classify EOG and EEG. Hence, in this paper, we propose a new automatic EOG removal technique that uses ICA to decompose EEG signals into ICs then apply Machine Learning algorithms to detect EOG from these ICs. The algorithms will take the topography map of each IC as an input vector to predict whether the IC should be rejected. This new technique gives us the advantage of removing EOG artifacts without a reference channel, while requiring a low number of electrodes, and short computing time.

2 Materials and Methods

2.1 *Experiment and Database*

The full database of EEG signal was obtained from the Alice 5 Polysomnography system using 10 EEG channels including Fp1, Fp2, F3, F4, F7, F8, T5, T6, O1 and O2 with the ground electrodes Fpz at the forehead, M1 and M2 channels on the mastoid bones (Fig. 1).

20 Subjects were undergraduate students between the ages of 18–22 years at the time of the research study. All subjects were chosen based on exclusion criteria include (i) smokers, (ii) left-handers, (iii) native English speakers, (iv) those with a vision that was not corrected to normal, (v) antihistamine, glucocorticoid or asthma medication users, (vi) those with exposure to general anaesthesia in the last year, (vii) those with a personal or first degree family diagnosis of a DSM-IV, axis I disorder (a list of these disorders was given at the time of initial inquiry), and (viii) those with endocrine abnormalities. These exclusion criteria were self-affirmed by the prospective participants.

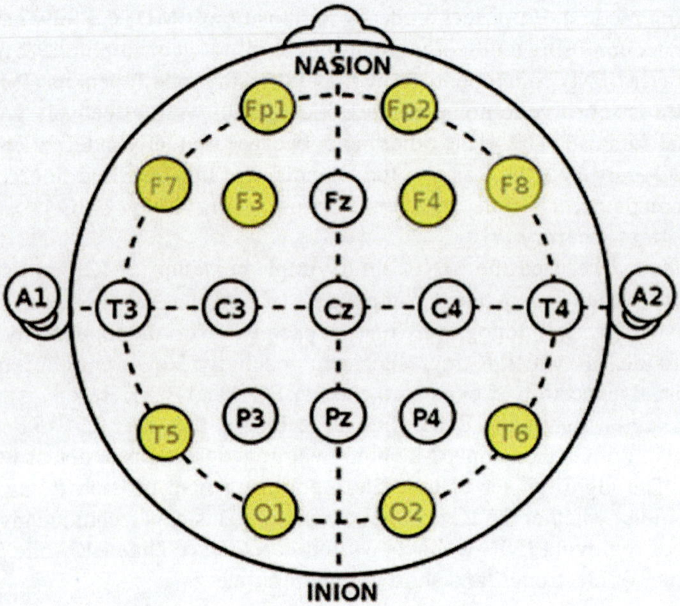

Fig. 1 Electrode mapping using the 10–5 system. The EEG database includes ten channels: Fp1, Fp2, F3, F4, F7, F8, T5, T6, O1, O2

2.2 *Pre-processing*

Since EEG signals have low amplitudes and are easily affected when processed, in this paper, baseline correction and bandpass filtering were used as standardized EEG preprocessing methods to avoid losing useful information. First, all original EEG recordings were bandpass filtered with cut-off frequencies at 0.5 and 45 Hz using one-dimension with an IIR or FIR filter. All functions were adapted from the Python library SciPy [12]. Second, baseline correction was applied in which the data after bandpass subtract to their average value to remove the baseline drift.

2.3 *Independent Component Analysis*

ICA is a generative model describing how the data are generated by the process of mixing the components x = As. ICA computes both mixing matrix A and independent components so that s is maximally independent. In this study, the goals of utilizing ICA are to calculate the independent components and topography map of each component across electrodes, and then to use them as input for classification models discussed in Sect. 2.4. This paper utilized the MNE library's implementation of ICA, using 'extended infomax' [13].

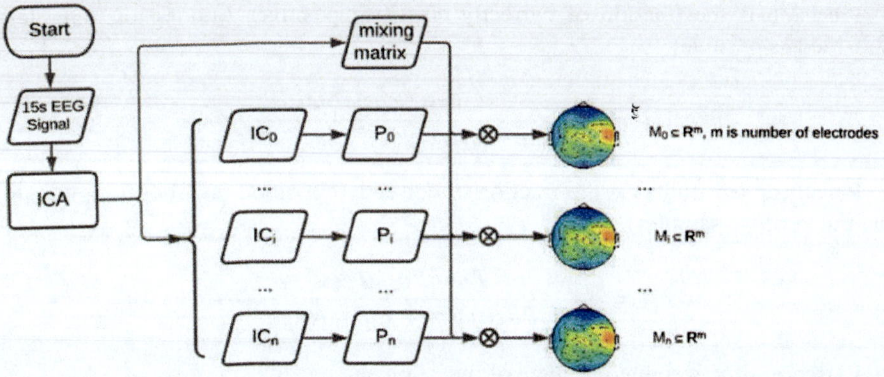

Fig. 2 Calculating the independent component's topography map. P_i is the signal power of component ith; M_i is topography map of component ith, representing activity level of the component on the scalp

After applying ICA for each 15-s chunk of preprocessed EEG signal, we had a matrix consisting of 10 ICs time series (s) and mixing matrix (A). From the matrix of ICs, we calculated power spectrum density (PSD) for each component. From the mixing matrix, we extracted the topography map for each component, as followed Fig. 2.

2.4 *Classification Models*

Four supervised learning models were used for comparison: support vector machine (SVM), random forest (RF), extremely randomized trees (ExTrees) and extreme gradient boosting (XGBoost). For the former three, the implementations from the Python library sci-kit-learn [14] are used, and for the latter, there is a dedicated library called xgboost [2].

In classification tasks, true positive refers to the number of correctly classified positive points (in this case, the number of correctly classified EOGs), false positive is the number of incorrectly classified EOGs. Similarly, true negative means the number of correctly classified non-EOGs and false-negative represents the samples that are incorrectly classified as non-EOGs. The metrics used in the experiment are solely based on these four elements.

Accuracy is calculated as the fraction of the labels that exactly match the ground truth.

$$accuracy = \frac{TruePositives + TrueNegatives}{TruePositives + TrueNegatives + FalsePositives + FalseNegatives}$$

Precision is the fraction of correctly classified positives (among the samples classified as positive)

$$precision = \frac{TruePositives}{TruePositives + FalsePositives}$$

Recall (or sensitivity) is the fraction of correctly classified positives (among all the true positive samples).

$$recall = \frac{TruePositives}{TruePositives + FalseNegatives}$$

F1-score is the weighted average of precision and recall

$$F1 = 2 \times \frac{precision \times recall}{precision + recall}$$

Accuracy can be a good metric for balanced datasets (where the number of positives and negatives are roughly equal). However, the metric suffers from imbalanced classes (where the class distribution is not uniform). If a dataset has 100 samples with only 10 out of them are EOGs then a model that only predicts non-EOG for all samples would still have an accuracy of 90%, but such model would not be considered 'good' since it fails to recognize any of the EOG samples (recall = 0). In our study, there were only a few EOG samples compared to the large number of non-EOGs. This is an example of an imbalanced dataset, where precision, recall and F1-score can be particularly useful for model evaluation.

These subsections give a brief overview of the methods we used.

Support vector machine

Support vector machine (SVM) [15] is one of the commonly-used machine learning algorithms in EEG classification. In classification context, SVM tries to find a hyperplane, which can be a line in 2-dimensional space or a plane in 3-dimensional space, that maximizes the margins—the distances between the hyperplane and the closest points to such hyperplane in each class. Since the dataset we needed to classify is not linearly separable, SVM with a non-linear kernel is used to map the data into a higherdimensional space where linear separability can be obtained. In our experiment, the radial basis function (RBF) kernel was used.

Random forest

Random forest (RF) is a type of ensemble learning model. The main idea of the method is to take advantage of many decision trees, where each tree is built from a bootstrap sample (random sample drawn with replacement) taken from the data, and to build each tree with all or a random subset of variables. The randomness introduced above will help decrease variance and thus prevent model overfitting, which is one of the main drawbacks of vanilla decision trees. The random forest implementation in scikit-learn calculates the predicted output by averaging the probabilistic predictions.

Since decision trees are non-linear as there is no formal equation to express the relationship between the features and the target, the random forest is expected to be able to solve the problem of non-linearly separability of the dataset.

Extremely randomized trees
Extremely randomized trees (ExTree) was first introduced in 2006 by Pierre Geurts, Damien Ernst and Louis Wehenkel [16]. Though the algorithm is similar to the random forest, the difference between these two ensemble learning models lies in the level of randomness. In node splitting, while the random forest model tries to find the best split, ExTrees chooses the variable splitting value randomly. This can normally reduce the variance of the model even more, but at the cost of increased bias, according to the authors. Like random forest and other tree-based models, ExTrees is non-linear and is expected to solve the problem of non-linearly separability.

Extreme gradient boosting
Extreme gradient boosting (XGBoost), is a scalable implementation of the gradient boosting algorithm [17]. Gradient boosting is, like random forest and extremely randomized trees, an ensemble learning method in a sense that the predicted output will be based on an ensemble of many models. The difference between boosting and bagging, which is the technique used in random forest and extremely randomized trees, is that the bootstrap samples are weighted so that the samples with which the model incorrectly predicted get higher weights and thus be sampled more often. The idea behind weighing samples is that the model would focus more on 'difficult' samples. The gradient is used when optimizing the training loss. Hence the name gradient boosting. XGBoost further improves the original boosting method by introducing second-order gradients and regularization that help prevent overfitting.

3 Results

3.1 Preprocessing of EEG signal

Bandpass-filter with cut-off frequencies at 0.5 and 45 Hz and baseline correction were applied for each chunk of 15 s original EEG signals.

To understand the changes in raw EEG signals after our preprocessing, we compared raw EEG data (Fig. 3a) and preprocessed EEG data (Fig. 3b). The noise was reduced by the bandpass filter as indicated by the reduced thickness of the data line, especially at channels Fp1, Fp2 (Fig. 3a, b). Baseline drifts were removed in data lines after the baseline correction (Fig. 3c).

Nevertheless, the general waveforms of processed EEG recordings still kept their origins, which proved that EEG signals do not lose their representative information after the preprocessing step.

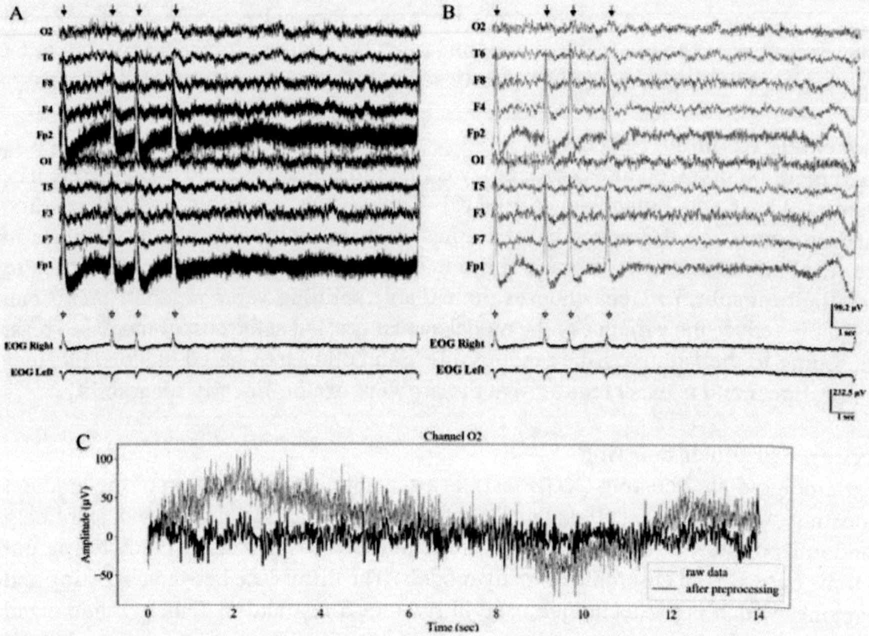

Fig. 3 Preprocessing of EEG signals. **a** A 15 s segment of raw EEG data (top) and EOG data (bottom). **b** the same segment of EEG data of figure **a** after band-pass filtering and baseline correction. Black arrows mark EOG peaks in EEG signal (above), white arrows mark EOG peaks in EOG reference's channel. **c** Another 15 s segment of raw EEG data—channel O2 after bandpass filtering and baseline correction

3.2 Independent Component Analysis of EEG Signal

To acquire the training dataset of ICs signal, its topography map, and its label, we divided our preprocessed EEG signal into chunks of 15 s. For each chunk of 15 s preprocessed EEG signal, ICA was used to calculate a matrix consisting of 10 ICs time series (s) and mixing matrix (A). From the matrix of ICs, the power spectrum density (PSD) for each component is calculated. From the mixing matrix, we extracted the topography map for each component. Upon visual inspection of the topography map and the IC itself, ICs that represent ocular activity are labelled 1, and other ICs were labelled 0. We observed that ICA did not always successfully isolate EOG artifacts from EEG signals. For successful cases, ICs were very distinguishable from each other (Fig. 4a, c, e). In this successful case, there was one IC (ICA000) with waveform resembling EOG artifacts when comparing with EOG reference channels (Fig. 4a). Each EOG peak was marked with a black arrow for the IC and white arrow for the EOG reference channels. The topography of this IC represents activity exclusively in the frontal lobe area (Fig. 4c), which is expected for eye-derived electrical activity. From the PSD (Fig. 4e), we could see these ICs carry very little

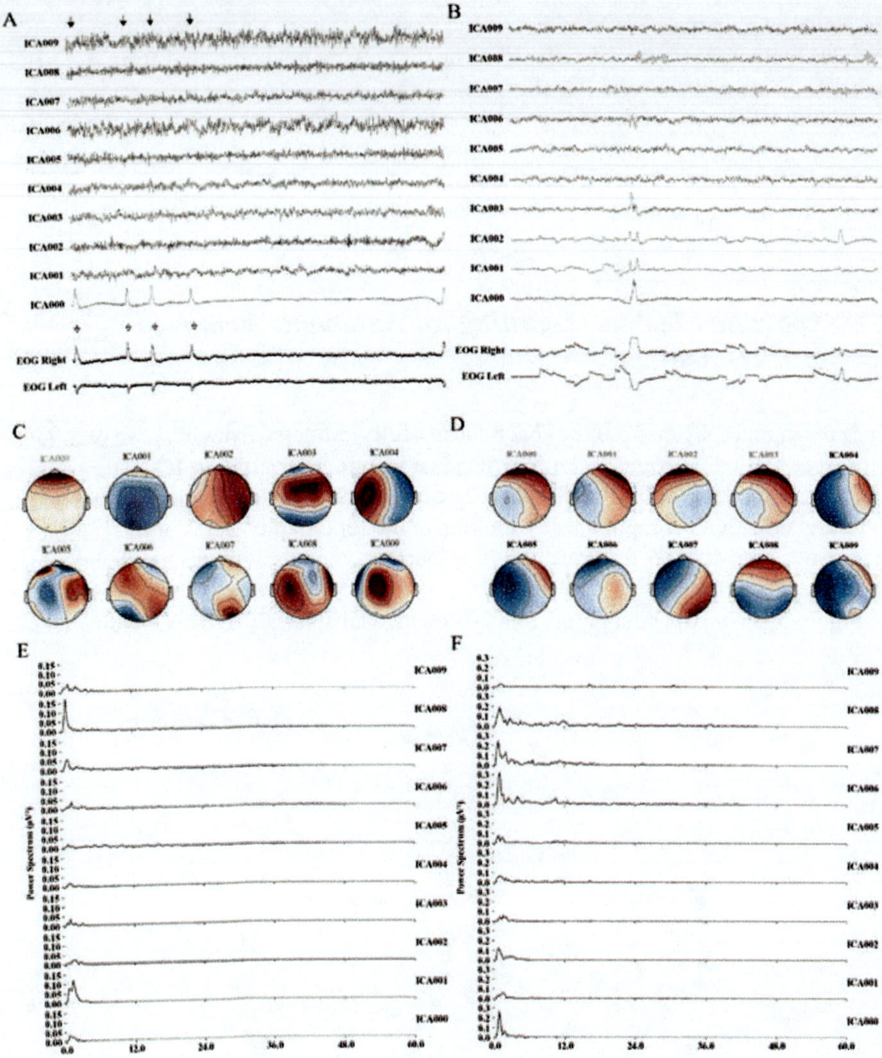

Fig. 4 Comparison of different ICA-related features. **a**, **b** Signal of independent components (ICs) from an example in which ICA successfully separated EOG artifact from EEG signal (**a**) or not (**b**). EOG reference channels are shown at the bottom. Black arrows mark EOG peaks in IC signal; white arrows mark EOG peak in EOG channels. **c**, **d** Topography maps corresponding to the ICs in A and B. **e**, **f** Power spectrum density plots of each ICs in A and B

bio-signal in the range 0–40 Hz. For unsuccessful cases, ICs were indistinguishable from each other. More specifically, EOG artifacts were not separated from the EEG signal and existed in several ICs (Fig. 4b).

Additionally, none of the topography exclusively represents activity in the frontal lobe area (Fig. 4d). For our training dataset, we only included cases in which ICA successfully separates EOG artifacts from the EEG signal. This training dataset was utilized for training several classifiers to detect EOG components in our ICs.

3.3 *Applying Machine Learning for Automatic Removal of EOG Artifact*

Once the topography map data has been successfully extracted from ICA, we obtained a dataset of 612 data points, each of which is a feature vector of raw IC features plus the map components we chose. Visually, one could notice a clear distinction between EOG and non-EOG components by looking at the topography maps of the samples. Still, we would like to find out how the learning models will perform with this particular dataset.

Figure 5 shows the data points in a 2-dimensional space. The map features were

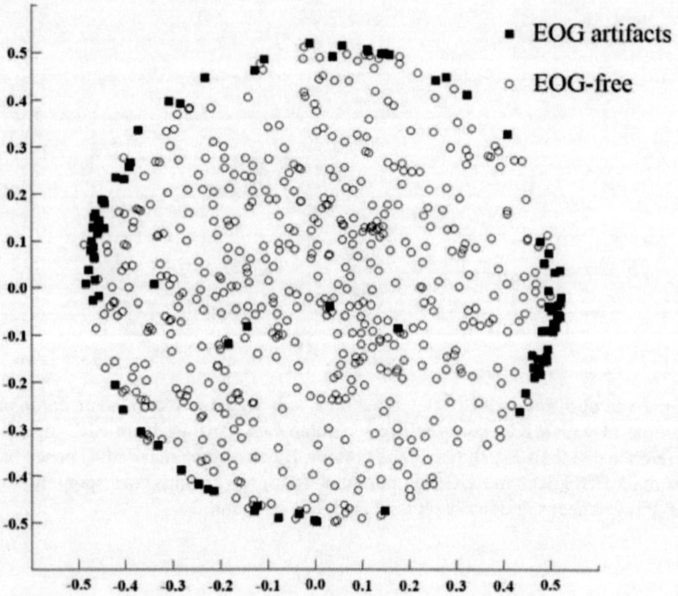

Fig. 5 Topographic map data visualized in 2-dimensional space after Principal Component Analysis (PCA). EOG components are presented as dark rectangles, and non-EOG components are presented as white circles

transformed from 10 dimensions into two dimensions using Principal Component Analysis (PCA). PCA is a widely used linear dimensionality reduction technique that aims to project multi-dimensional data into a lower-dimensional space and to retain maximum variance between data points [18]. Since Fig. 5 suggested our data is not linearly separable, we were tempted to use non-linear models for the dataset.

The data was standardized so that each component to have a mean of 0 and a standard deviation of 1 before being trained by the models. For each model, 3-fold cross-validation was used. The experiment on each model was repeated ten times with different random number generators for cross-validation splitting in order. The metrics were averaged across ten runs. The comparison boxplots for different metrics of the models with raw ICA features included along with map features used in training are shown in Fig. 5. The results in Fig. 5 suggested that all the models do not perform well when the raw ICA features are included along with topo map features in the training step. The best-performing model in this experiment was XGBoost with the top score in all metrics. While all models still managed to have accuracy above 0.8, only XGBoost had F1-score higher than 0.5 (0.59 ± 0.01). The rest failed to detect most EOGs, with the most extreme cases being ExTrees and SVM, which had precision, recall and F1-score of 0. We hypothesised that too many predictors as in the case of ICA features with 7500 dimensions would create the problem of high dimensionality, where the predictive power can at first increase along with more features, but then decreases when the number of observations is fixed [19].

To enhance the performance of the models, we selected another approach, which only included map features in training. From the results in Fig. 7, we observed that all models managed to have a high accuracy of over 0.9. ExTrees significantly outperformed other models in terms of F1-score and recall ($p = 0.001$ and $p = 0.022$) with the average F1-score of 0.77 ± 0.009 and the average recall of 0.71 ± 0.01. In terms of precision, Random Forest produced the results with the highest score (0.85 ± 0.01), but the score was not significantly better than that of ExTrees (0.84 ± 0.008) ($p > 0.05$). Both SVM and XGBoost fell behind RF and ExTrees with clearer differences in the precision score.

To reconstruct EOG free signal from preprocessed EEG signals, we used ICA to decompose ten channels of EEG signal into a matrix of 10 ICs. With the trained classifier mentioned earlier, we were able to detect ICs representing EOG activity. We then proceeded to set the value of this IC in the matrix to zero. With this new matrix, we were able to inverse transform to EOG free signal [20]. Figure 8 demonstrated the result of the algorithm successfully removing EOG peaks from the signal while preserving other bio-signals. The black arrows on Fig. 8a marked EOG peaks that were removed by the algorithm. EOG-free EEG signals were shown in Fig. 8b.

In addition to evaluating the performance of EOG classification, we were also interested in investigating the computation time, which is an important factor for a scalable pipeline. We executed the pipeline from initial processing to EOG removal of a segment ten times and took the average computation time. The pipeline script was run on a laptop with 16 Gb of memory and a Core i5 processor. As Table 1 suggested, the total pipeline takes around 5 s on average, with most of the computation time being from the ICA processing step.

Table 1 The computation time of the steps in the preprocessing pipeline. Computations are executed ten times on a single machine with 16 Gb of RAM and an Intel Core i5 processor

Task	Time (s) (mean $\pm$ SEM)
Bandpass filter	0.017 ± 0.002
Baseline correction	0.004 ± 0.0004
ICA	4.941 ± 0.251
EOG classification	0.115 ± 0.003
EOG removal	0.005 ± 0.001
Total	5.083

4 Discussion

To summarize, the proposed approach to EOG artifact removal consists of three steps: preprocessing signal, decomposing preprocessed signals into components, and using a classifier to detect components that represent EOG activity. Firstly, baseline correction and bandpass filtering were proven to be an effective preprocessing method to remove powerline noise while preserving EEG waveforms. Secondly, the independent component analysis showed the capacity to isolate EOG artifacts from EEG signals. However, for certain cases, EOG artifact and EEG signals were still mixed in one or many ICs. And finally, several machine learning classifiers were applied to detect components representing ocular activities. However, the classifier was not yet able to detect IC with mixed signals from EOG artifact and EEG signal, which left room for improvement in the future.

With the proposed method, we could automatically remove EOG artifacts from EEG signals without the need for reference channel and domain expertise. Also, by removing the manual step of determining EOG artifacts, it was more convenient to implement an online artifact removal implementation using ICA.

In our EEG signal, the numbers of sources were larger than the number of recordings, and the EOG artifacts had high magnitude. Therefore, ICA could be applied successfully to isolate EOG artifacts from EEG signals. However, there were several shortcomings in the proposed approach. First, our classifier could not determine components with mixed EOG artifacts and EEG signals from components that include purely EOG artifacts. This results from our training process in which we only included two classes: EOG components—consisting only EOG artifact and non-EOG components—consisting only EEG signals. We excluded components with mixed EOG artifacts and EEG signals from the training dataset. Second, our approach did not offer to remove EOG artifacts from a signal channel recording of EEG and required a large resource of computing power. [21] Finally, we would like to discuss the classification techniques used to determine components representing EOG artifacts. From the results shown in Figs. 6 and 7, ExTrees gave a significantly better performance in terms of F1-score and recall. Interestingly, raw ICA features made the models fail to recognize EOG samples, hypothetically due to the problem of high dimensionality. Compared to a previous study [22] which used SVM for eye-blink artifact detection and a fourfold CV, our best classification accuracy was lower (99.3% vs. 93%). One

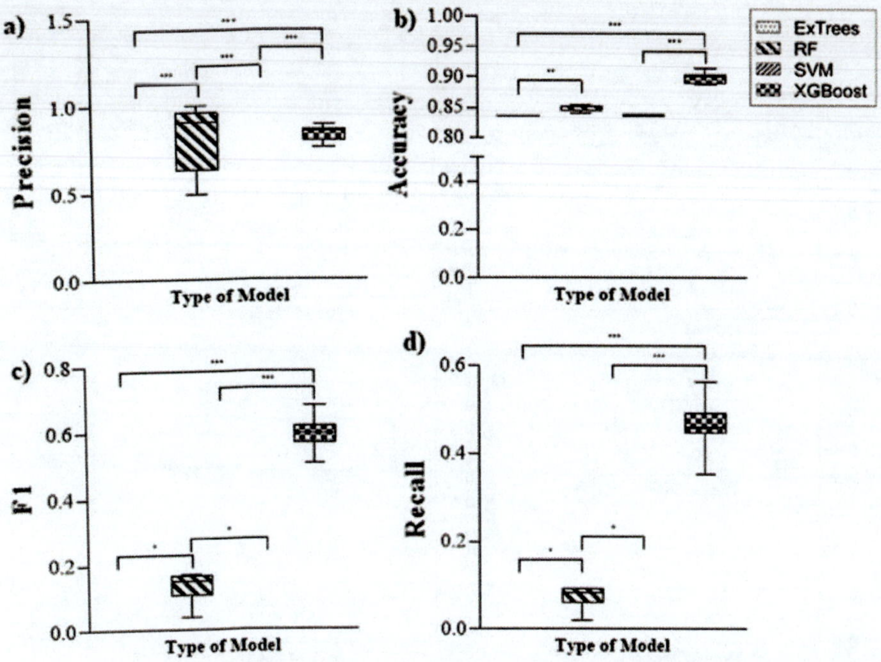

Fig. 6 Performance of different machine learning models using both signals from topography map and raw IC signals as input for training). **a** Precision. **b** Accuracy. **c** F1-score. D) Recall. One-way ANOVA followed by Kruskal–Wallis multiple comparisons (* $p < 0.05$; **, $p < 0.01$; ***, $p < 0.001$)

potential difference was that our study utilized an imbalance dataset while the dataset in [22] was perfectly balanced with 100 samples of each class. In another study [23] that used a similar classification approach, they managed to get high accuracy scores for eye-blink artifacts with a balanced dataset and more samples (99.39% for eye blink and 99.62% for eye movement). Given the limited number of samples and the imbalanced nature of the dataset we have, these results were encouraging.

Future Works
There exists certainly room for improvement in the aspect of F1-score by proper feature extraction for ICA data, using either statistical features (mean, median, kurtosis) or some sorts of signal transformations like discrete Fourier transform, or wavelet transform that might be able to capture the inner nature of the ICA components and the difference between EOGs and non-EOGs. Another topic that we would like to improve in the future is including mixed classes in our training dataset and curating a balanced dataset for the training. These approaches would help the classifier to determine which components consist of pure EOG artifacts and which components consist of both EOG artifacts and EEG signals and improve the accuracy of the models.

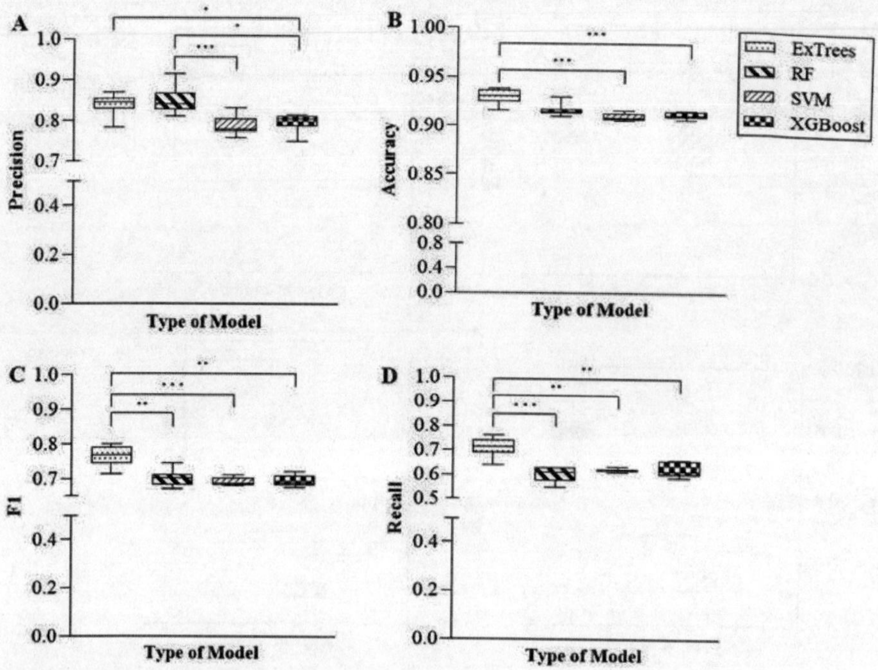

Fig. 7 Performance of different machine learning models using only the signal from topography maps as input for training. **a** Precision. **b** Accuracy. **c** F1-score. **d** Recall. One-way ANOVA followed by Kruskal–Wallis multiple comparisons (* $p < 0.05$; **, $p < 0.01$; ***, $p < 0.001$)

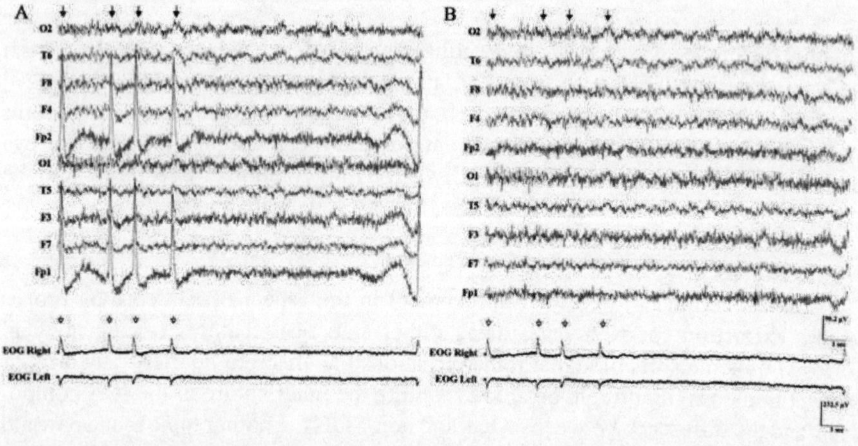

Fig. 8 Comparison of preprocessed EEG signal and EOG-free signal after using the algorithm. **a** 15 s of preprocessed EOG channels (top) and EOG channels (bottom). **b** The Same segment of EEG data after the EOG artifact was removed by the algorithm. Black arrows mark EOG peaks in the EEG channels. White arrows mark EOG peaks in the reference channels

Code deposit: https://github.com/Young1906/ica_paper

Acknowledgements This research is funded by Vietnam National University Ho Chi Minh City (VNU-HCM) under grant number C2020-28-06.

Disclosure of Potential The authors declare that they have no conflict of interest.

References

1. Light GA et al (2010) Electroencephalography (EEG) and event-related potentials (ERPs) with human participants. Curr Protocols Neurosci 52(1):6.25.1–6.25.24: https://doi.org/10.1002/0471142301.ns0625s52
2. Hughes JR, John ER (1999) Conventional and quantitative electroencephalography in psychiatry. J Neuropsychiatry Clin Neurosci 11(2):190–208
3. Loo SK, Makeig S (2012) Clinical utility of EEG in attention-deficit/hyperactivity disorder: a research update. Neurotherapeutics 9(3):569–587
4. Bucci P, Mucci A, Galderisi S (2011) Normal EEG patterns and waveforms. Standard Electroencephalog Clin Psychiatry, 33–57. https://doi.org/10.1002/9780470974612.ch4
5. Moffett SX, O'Malley SM, Man S, Hong D, Martin JV (2017) Dynamics of high frequency brain activity. Sci Rep 7(1). https://doi.org/10.1038/s41598-017-15966-615966-6
6. Muthukumaraswamy SD (2013) High-frequency brain activity and muscle artifacts in MEG/EEG: a review and recommendations. Front Hum Neurosci 7:138
7. Jiang X, Bian G-B, Tian Z (2019) Removal of artifacts from EEG signals: a review. Sensors 19(5):987. https://doi.org/10.3390/s19050987
8. Freeman WJ, Burke BC, Holmes MD (2003) Aperiodic phase re-setting in scalp EEG of beta-gamma oscillations by state transitions at alpha-theta rates. Hum Brain Mapp 19(4):248–272
9. Urigüen JA, Garcia-Zapirain B (2015) EEG artifact removal—state-of-the-art and guidelines. J Neural Eng 12(3):031001. https://doi.org/10.1088/1741-2560/12/3/0310012560/12/3/031001
10. Looney D, Li L, Rutkowski TM, Mandic DP, Cichocki A (2007) Ocular artifacts removal from EEG using EMD. In: Advances in Cognitive Neurodynamics ICCN 2007, pp 831–835. https://doi.org/10.1007/978-1-4020-8387-7_145
11. Xu X, Chen X, Zhang Y (2018) Removal of muscle artefacts from few-channel EEG recordings based on multivariate empirical mode decomposition and independent vector analysis. Electron Lett 54(14):866–868. https://doi.org/10.1049/el.2018.0191
12. Virtanen P et al (2020) SciPy 1.0: fundamental algorithms for scientific computing in Python. Nat Methods 17(3):261–272
13. Gramfort A, Luessi M, Larson E, Engemann D, Strohmeier D, Brodbeck C, Goj R, Jas M, Brooks T, Parkkonen L, Hämäläinen M (2020) MEG and EEG data analysis with MNE. Front Neurosci. https://mne.tools/dev/generated/mne.preprocessing.ICA.html. Accessed 26 Apr 2020
14. Varoquaux G, Buitinck L, Louppe G, Grisel O, Pedregosa F, Mueller A (2015) Scikitlearn. GetMobile: Mobile Comput Commun 19(1):29–33. https://doi.org/10.1145/2786984.2786995. http://paperpile.com/b/SPfWWe/pOxf
15. Cortes C, Vapnik V (1995) Support-vector networks. Mach Learn 20(3):273–297. https://doi.org/10.1007/bf00994018
16. Geurts P, Ernst D, Wehenkel L (2006) Extremely randomized trees. Mach Learn 63(1):3–42. https://doi.org/10.1007/s10994-006-6226-1
17. Chen T, Guestrin C (2016) XGBoost. In: Proceedings of the 22nd ACM SIGKDD international conference on knowledge discovery and data mining— KDD '16. https://doi.org/10.1145/2939672.2939785

18. Jolliffe IT (2013) Principal component analysis. Springer Science & Business Media
19. Trunk GV (1979) A problem of dimensionality: a simple example. IEEE Trans Pattern Anal Mach Intell 1(3):306–307
20. Djuwari D, Kant Kumar D, Palaniswami M (2005) Limitations of ICA for artefact removal. Conf Proc IEEE Eng Med Biol Soc 2005:4685–4688
21. Nguyen H-AT et al (2012) EOG artifact removal using a wavelet neural network. Neurocomputing 97:374–389. https://doi.org/10.1016/j.neucom.2012.04.016
22. Shoker L, Sanei S, Chambers J (2005) Artifact removal from electroencephalograms using a hybrid BSS-SVM algorithm. IEEE Signal Process Lett 12(10):721–724. https://doi.org/10.1109/lsp.2005.855539
23. Halder S et al (2007) Online artifact removal for brain-computer interfaces using support vector machines and blind source separation. Comput Intell Neurosci, 82069

An Algorithm for Removing Artifacts in Polysomnography Signals

Quyen Thao Ngoc Nguyen, Toan Le, Quan T. T. Vu, Khue D. Bui, and Hoan T. Ngo

Abstract Polysomnography (PSG) is considered the gold standard for sleep disorders diagnosis. However, its signals are difficult to read in the presence of artifacts. This study presents a biosignal processing method that can help to remove artifacts from PSG signals. The method consists of two main parts. Firstly, sleep diary and feature extraction using Fourier transform and Wavelet transforms were used to detect different artifacts in PSG signals including physiologic artifacts such as cardiac artifact, muscle artifact, movement artifact, ocular artifact...and non-physiologic artifacts. Secondly, ICA (Independent Component Analysis) and Wavelet transform were used to remove the detected artifacts and reconstruct the signal. The results indicate that our method could adequately remove ECG and EOG artifacts in 70%–80% of the cases and wandering baseline and 50 Hz artifacts in 100% of the cases. We hope this algorithm will become a useful tool in removing artifacts in PSG signals, thus helping to make PSG signal reading more straightforward.

Keywords Polysomnography · Signal processing

1 Introduction

In sleep study, PSG collects brain activity, heart activity, respiratory activity, oxygen saturation in blood, body position, and eye and leg movements during sleep [1]. An example of biosignal data recorded by PSG is shown in Fig. 1. The collected data can be used for determining patient's sleep stages, arousals, apneas, hypopneas, and possible leg movement disorder, etc. In addition, apnea index (number of apneas per hour), apnea–hypopnea index (AHI), snoring time, number of rapid eye movement (REM) sleep, sleep latency…can also be calculated [2]. PSG provides

Q. T. N. Nguyen · T. Le · H. T. Ngo (✉)
School of Biomedical Engineering, International University, Vietnam National University, Ho Chi Minh City, Vietnam
e-mail: nthoan@hcmiu.edu.vn

Q. T. T. Vu · K. D. Bui
University of Medicine and Pharmacy, Ho Chi Minh City, Vietnam

V. Van Toi et al. (eds.), *8th International Conference on the Development of Biomedical Engineering in Vietnam*, IFMBE Proceedings 85,
https://doi.org/10.1007/978-3-030-75506-5_80

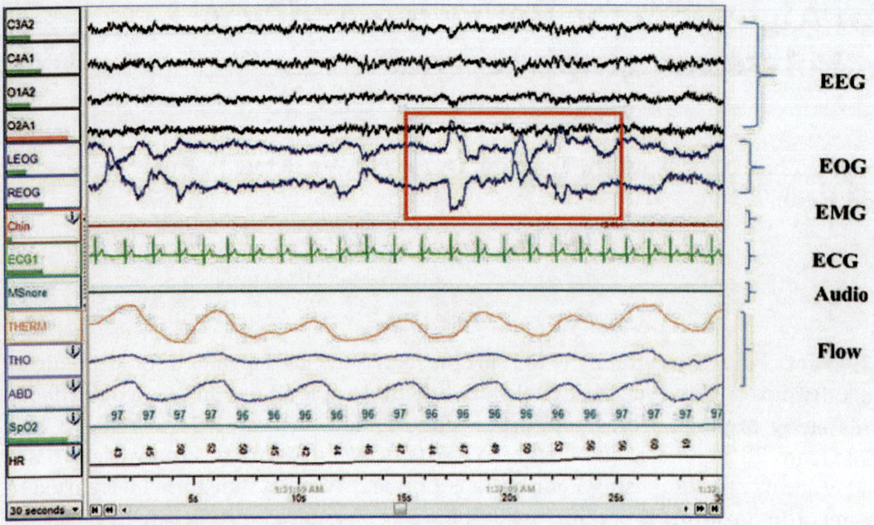

Fig. 1 Polysomnographic record of sleep

a large amount of information for diagnosing sleep disorders. And it is quite time-consuming for evaluators to read PSG signals due to the existence of different artifacts in PSG signals. In this paper, we propose an algorithm for removing artifacts in PSG signals.

2 Background

2.1 Characteristics of PSG Signals

In general, two main applications of PSG are sleep stages quantification and sleep disorders identification. There are various biosignals recorded by PSG: EEG, EOG, EMG, Respiratory events, SpO_2, etc. Below is a brief review of some biosignals which are often present in sleep reports:

Electroencephalogram (EEG) measures and records brain waves to determine sleep stages and detect seizure. Frequency of **EEG** signals ranges from 0.01 Hz to around 100 Hz with main EEG bands and their corresponding frequencies shown in Table 1 [3].

Electrooculogram (EOG) measures eye movement. These information are important for determining sleep stages, especially the REM stage [4].

Electromyogram (EMG) records muscle activity (e.g., teeth grinding, face twitching, limb movements). Chin EMG is important to differentiating REM from

Table 1 The frequency range of Sleep EEG bands and events

Freq. Band	Freq. range (Hz)
Alpha	8 –13
Beta	13–30
Delta	0.5–4
Theta	4–8
Sleep Spindles	12–14
K-Complex	0.5–1.5

Table 2 Characteristic of EOG and EMG in REM and Wake stage

EOG Chin EMG	REM	WAKE
	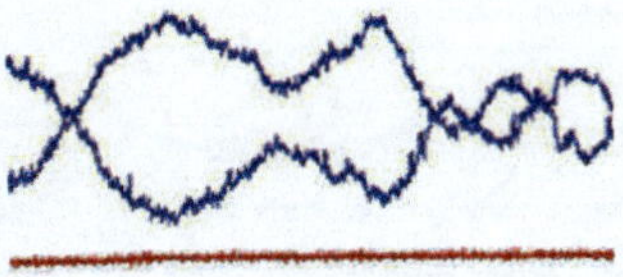	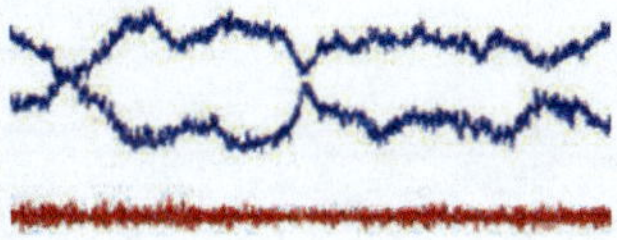

wakefulness (chin EMG tone remains low in REM). Limb EMG can detect periodic limb movement disorder (PLMD) [5].

Table 2 shows EOG and EMG signals during the REM and Wake stages, respectively. It illustrates that tone of EOG and EMG in Wake are higher than in REM due to higher muscle activities.

Electrocardiogram (ECG) records heart rate (HR) and rhythm. During sleep, analysis of HR and heart rhythm from ECG are useful for detecting apnea and phenotyping sleep sections besides respiratory patterns [6].

Pulse oximetry monitors oxygen saturation in blood (SpO_2). It can be used to detect hypopnea and respiratory disorders in human sleep (hypopnea is defined as reduced airflow for at least 10 s and a fall in SpO_2 of at least 4%) [7].

Respiratory monitor measures respiratory effort (thoracic and abdominal) during sleep interval. There are several types of sensor for monitoring respiratory effort including impedance sensor, inductance sensor, strain gauges sensor, etc. [8].

2.2 *Types of Artifacts*

Artifacts are extraneous signals interfering with the desired signals and can significantly affect quality of the desired signals. In general, a holistic knowledge about type or source of the artifact is needed to efficiently eliminate the artifact. Artifacts can be classified into physiologic artifacts (cardiac, muscle, movement, ocular, sweat, respirations...) and non-physiologic artifacts (including equipment artifacts: electrode impedance/50 Hz/60 Hz, electrode pop, overamplification, loose respiratory belt, etc. and environmental artifacts: nearby equipment, phones...) [9]. Main physiologic artifacts during sleep recording including ocular artifact, muscle artifact, cardiac artifact as well as non-physiologic artifacts will be explained below:

Ocular Artifacts are caused by eye movements and blinks which can propagate over the scalp and typically affects EEG signals (Fig. 2).

Muscle Artifacts is caused by increased muscle activity, subject talking, etc. (Fig. 3).

Cardiac artifacts exist when ECG signal is present in other channels such as EEG channels (Fig. 4).

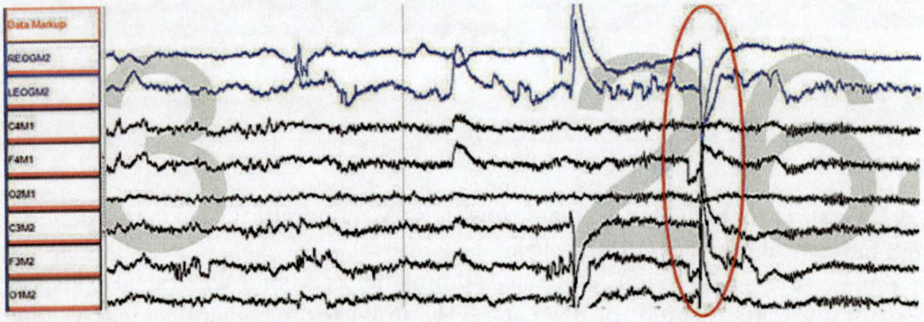

Fig. 2 Example of Ocular artifacts in EEG channels

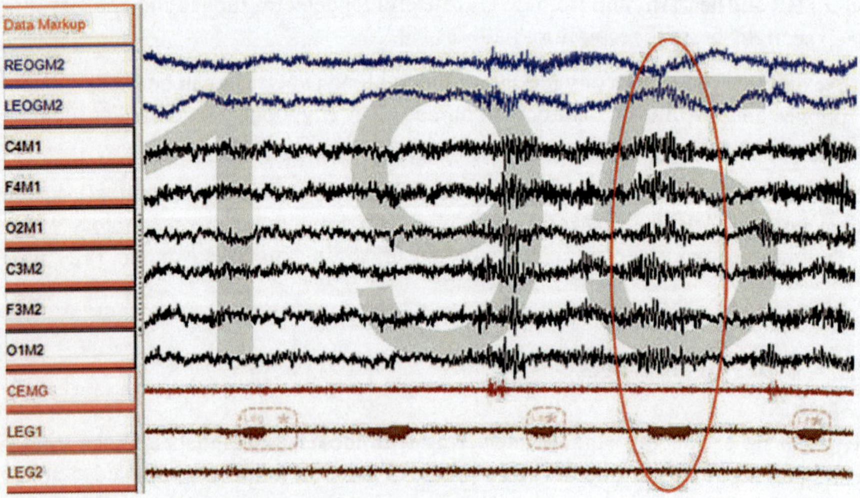

Fig. 3 Example of muscle artifacts in EOG and EEG channels

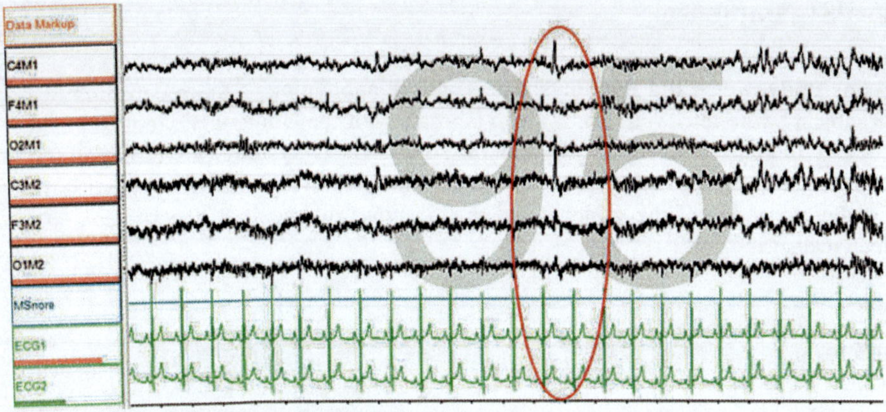

Fig. 4 Example of cardiac artifacts in EEG channels

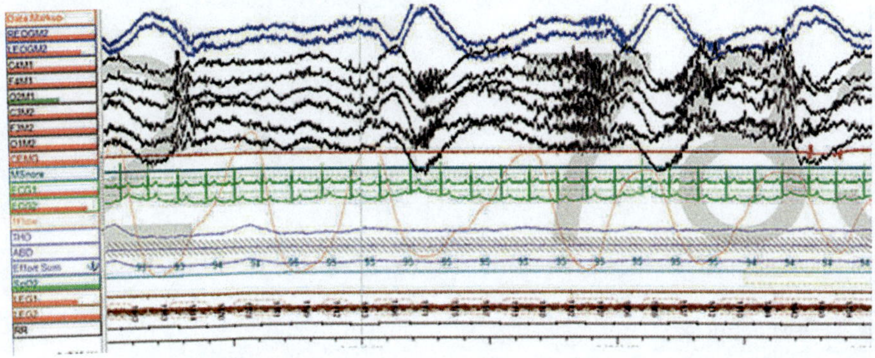

Fig. 5 Example of non-physiologic artifacts caused by poor electrode contact

Non-physiologic artifacts can be caused by poor electrode contact which results in variations in electrode impedance and can be seen in different channels such as EEG, ECG, EOG, EMG (Fig. 5).

3 PSG Database

In this study, we built a PSG database. Total 48 anonymous PSG recordings from 48 subjects were retrieved from a sleep laboratory of the University Medical Center Clinic 1 in Ho Chi Minh City, Vietnam. Alice 6 PSG system was used for PSG recordings. The study was approved by a institutional review board. Figure 6 shows distribution of age of the subjects and Fig. 7 shows diagnosis of the subjects after sleep measuring.

Fig. 6 Age distribution of the subjects

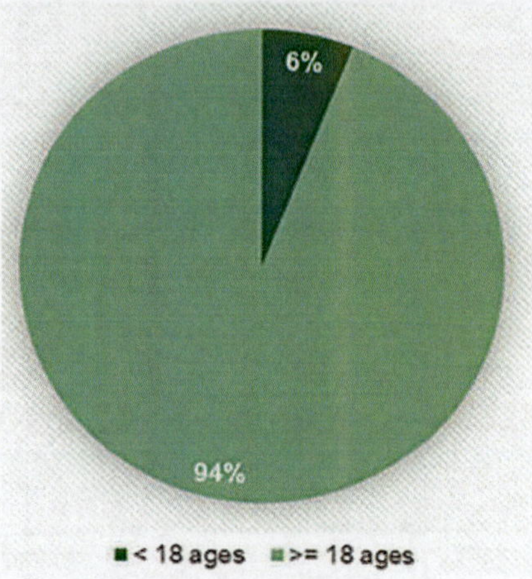

Fig. 7 Diagnosis of the subjects includes OSA (objective sleep apnea), CSA (Central sleep apnea), AHI (abnormal apnea–hypopnea index), HA (Hypopnea), and Normal

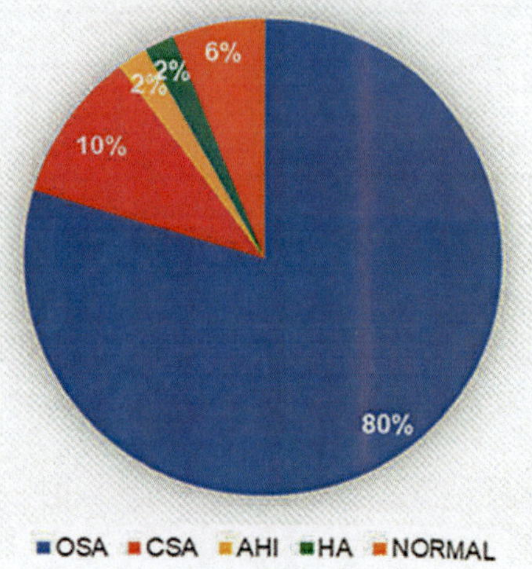

4 Methodology

As mentioned above, there are various PSG artifacts. We propose a framework for removing PSG artifacts as shown in Fig. 8.

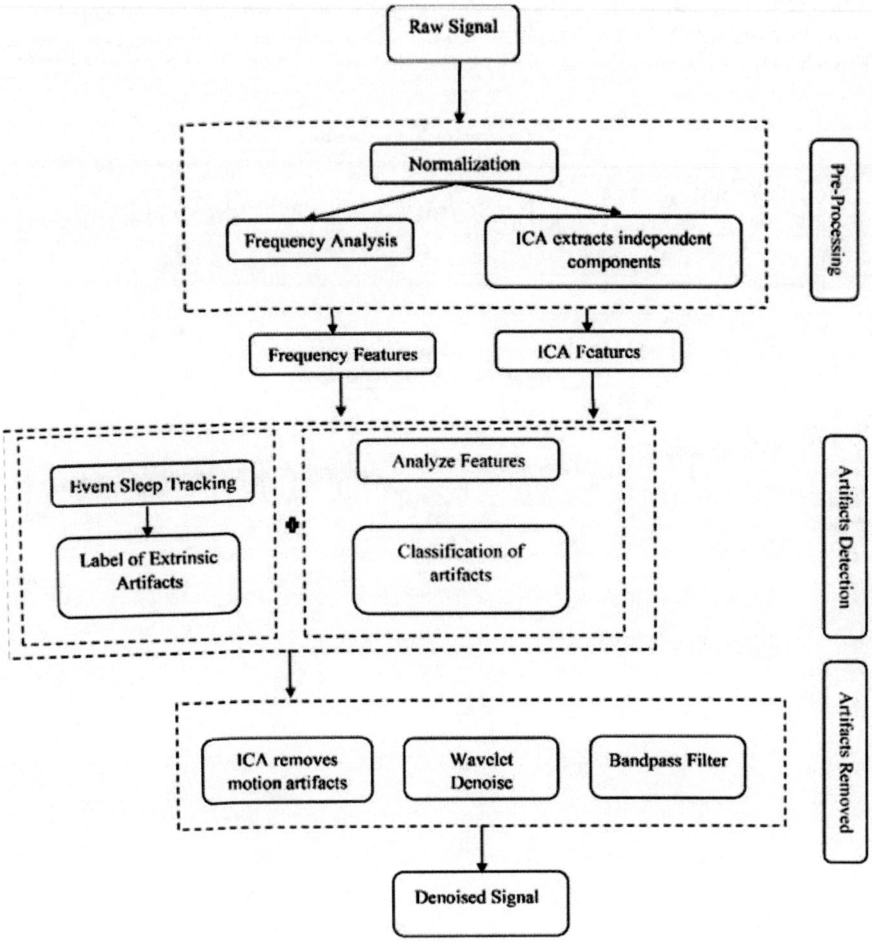

Fig. 8 Artifacts removal diagrams

4.1 Fourier Transform

Spectrum frequency is found by applying fast Fourier transform to detect extrinsic artifacts (50 Hz and 60 Hz impedances, white noise, etc.). Besides, the concept of Short Time Fourier Transform (STFT) is the localization of motion artifacts and sleep events (EEG signal) [10]. Figures 9 and 10 show the mixing of various un-known signal in EEG while the pure EEG which used to detect sleep stages has range from 0.5 to 30 Hz.

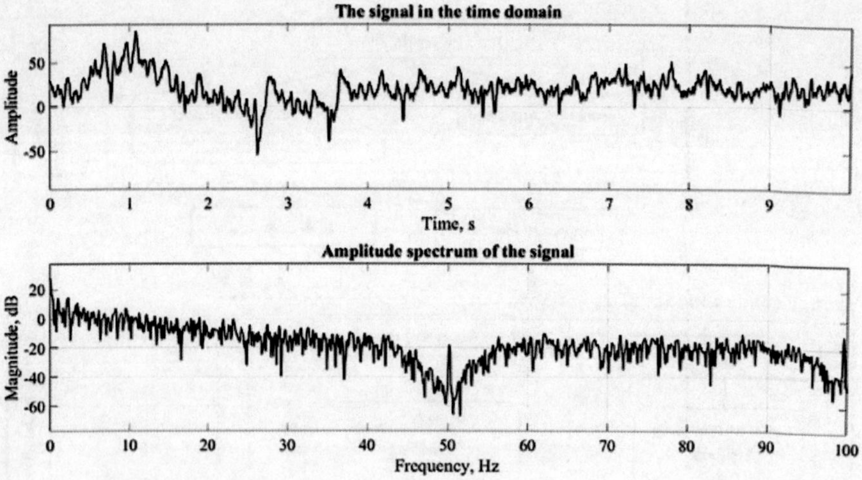

Fig. 9 Analyzing raw clinical EEG signal in frequency domain

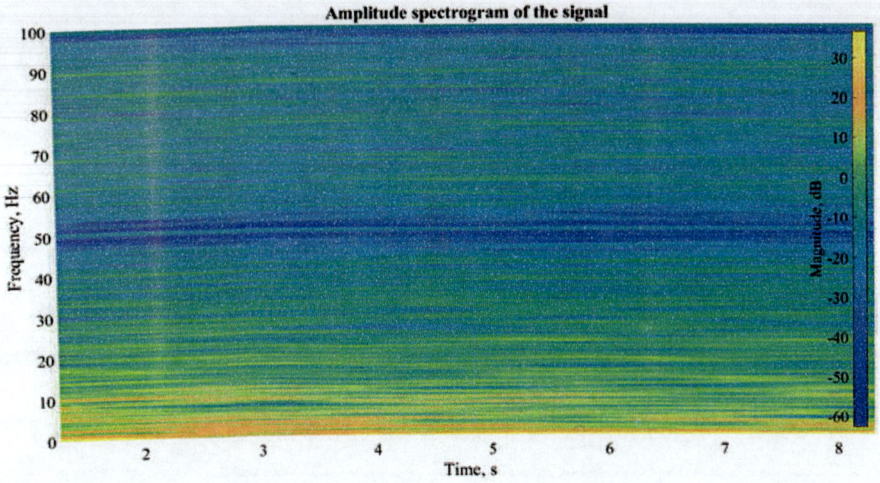

Fig. 10 Analyzing raw clinical EEG signal in time and frequency domain by using STFT

4.2 *Discrete Wavelet Transform:*

Discrete Wavelet Transform (DWT) is characterized by the main concept that includes two filters: low pass filter and high pass filter which are considered as a quadrature mirror filter. When the signal passed through DWT; then the output of the first filter are the approximation coefficients and the output of the second filter are the detail coefficients [11]. In summary, DWT (db4) (Fig. 11) is a potential method which can be used to remove EOG and ECG artifacts in EEG channels (Figs. 12 and 13).

Fig. 11 Discrete wavelet transform multilevel denoise algorithm

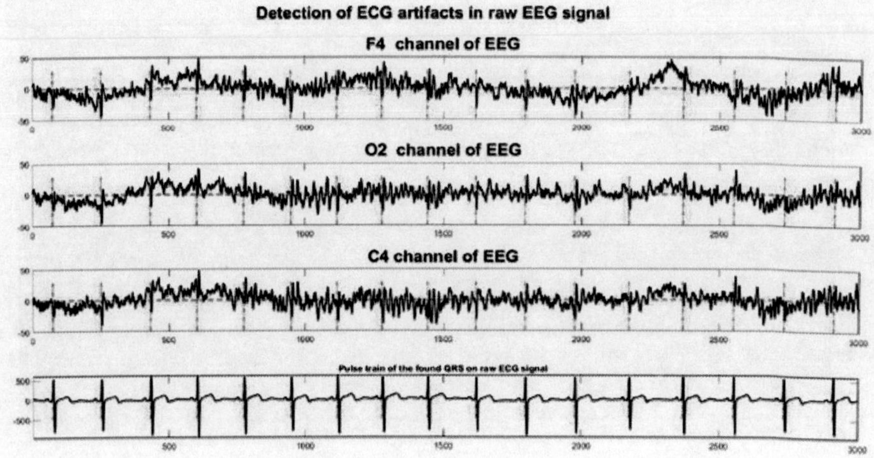

Fig. 12 Detection of ECG artifacts in clinical EEG signal

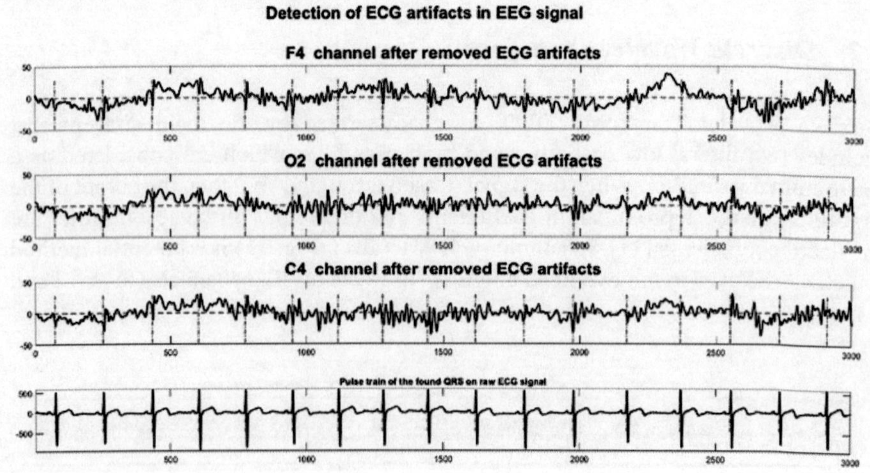

Fig. 13 Remove ECG artifact from clinical EEG signal using DWT

4.3 Independent Component Analysis (ICA)

ICA is a method for separating a multivariate signal into their independent components by exploiting their independence. This independence is found by focusing on finding sources that are the most non-Gaussian. The joint density of Gaussian sources is completely symmetric, making it impossible to estimate the individual independent sources [12]. By using the ICA, the list of artifacts is determined and inverse ICA which is used to remove those artifacts out of original signals. Figures 14, 15 and 16 presents the features extraction to detect the ocular artifact in EEG signal by using Fast-ICA and max-kurtosis algorithms [13, 14]. Fast-ICA find an orthogonal rotation of mixing matrix to maximizes a measure of non-Gaussianity by recursively maximizing both the kurtosis and negentropy of the signals. Kurtosis is the fourth order cumulant of a signal and is zero for a random Gaussian source. The kurtosis of a signal provided by Fast-ICA is defined as

$$kurt(y) = E\left\{y^4\right\} - 3. \tag{1}$$

Negentropy has been shown with statistical theory to be good measure of non-Gaussianity. Fast-ICA algorithm proposed an approximated function to estimate Negentropy (J)

$$J(y) = [E\{G(y) - E\{G(v)\}]^2 \tag{2}$$

where J is the negentropy approximation for any non-quadratic function G.

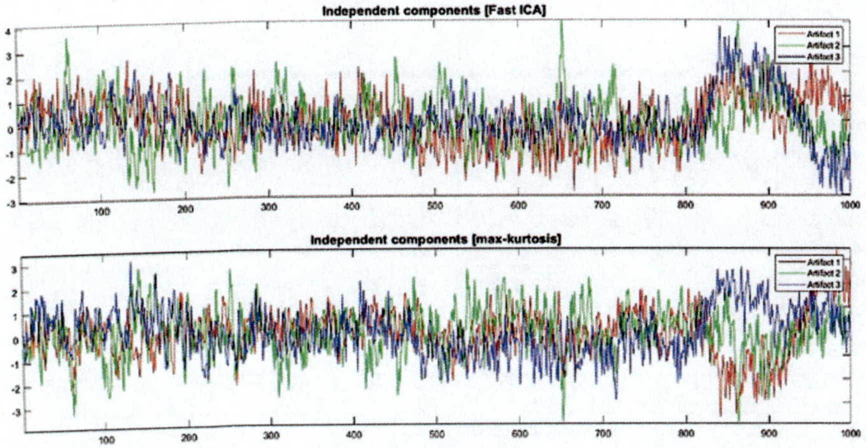

Fig. 14 Feature Extraction of the independent components from EOG signal in EEG signal

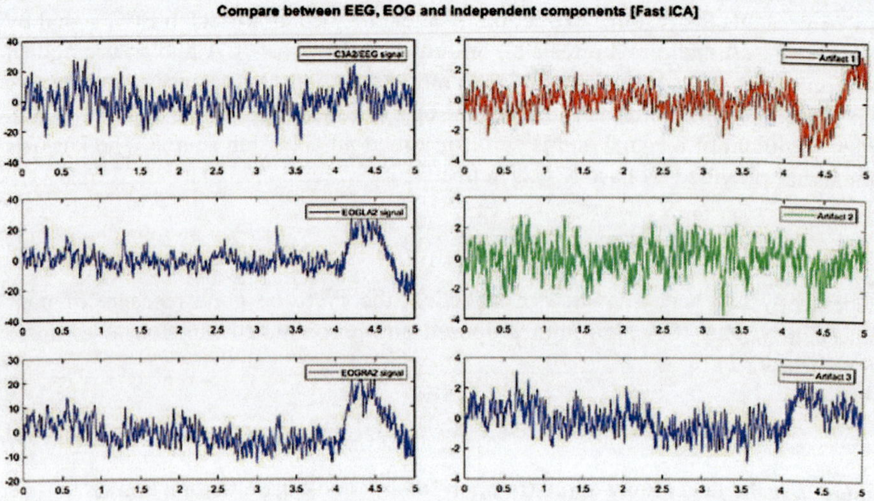

Fig. 15 Comparison between the independent components from Fast-ICA with EOG and EEG to detect ocular artifacts

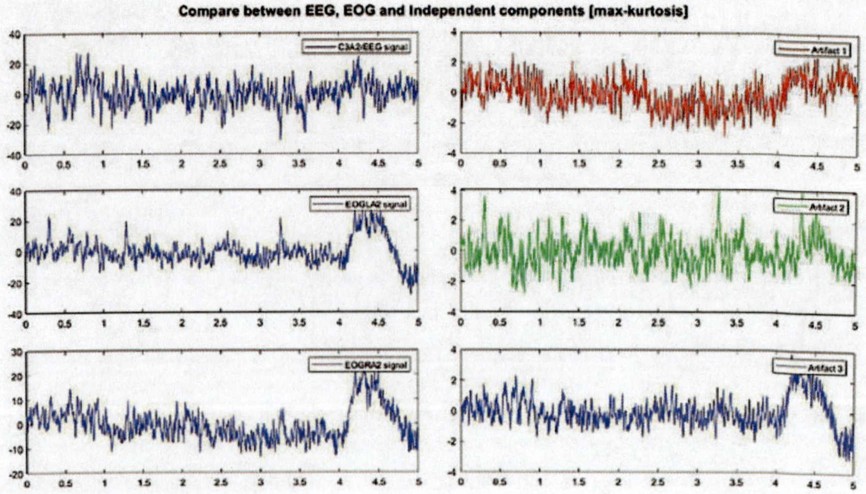

Fig. 16 Comparison between the independent component from max-kurtosis with EOG and EEG to detect ocular artifacts

5 Results

In this research, we aim to develop a method to remove artifacts out of EEG, EOG, EMG, ECG PSG signals since these signals are important in scoring sleep stage. According to this study protocol, clinical coordinators and physician were provided sleep diaries to record different events that occurs during sleep recording sessions. From this sleep diaries, artifacts in PSG signals were labeled. In addition, the PSG signals were also analyzed by our algorithm to automatically detect artifacts which could not be labeled using sleep diary. Artifacts existing in EOG, ECG, EEG, LEG EMG channels are shown in Fig. 17.

According to Fig. 17, EEG channels typically have numerous noise which affects the morphology of EEG waves such as alpha, sleep spindle. Muscle and Extrinsic artifacts are common reasons that change waveform of EOG signal because EOG originates from movement of the eye muscle so movement of other muscles could affect quality of EOG signal. Finally, in sleep measurement, ECG and LEG EMG channels usually have better quality compared to EOG and EEG channels because ECG and LEG EMG channels are usually only influenced by extrinsic/non-physiologic artifacts which are easier to eliminate than intrinsic/physiologic artifacts.

To evaluate the effectiveness of our proposed algorithm in removing artifacts in PSG signals, first, every epoch of sleep data was divided into 5–6 small windows based on R peak of ECG in each epoch. Second, labels of artifacts were generated for each window and denoise algorithm was applied. Number of windows that could be adequately denoised from each of wandering baseline, EOG, ECG, 50 Hz artifacts was counted. Finally, percentage of windows that could be adequately denoised from each of the aforementioned artifacts was calculated. Figure 18 shows that wandering baseline and 50 Hz artifacts caused by muscle and extrinsic artifacts could be adequately removed 100% of the cases because they have specific frequencies and space transition compared to other artifacts. On the other hand, EOG and ECG artifacts could be adequately removed 80.07% and 73.91% of the cases respectively. To compare performance of ICA and Wavelet transform in removing artifacts

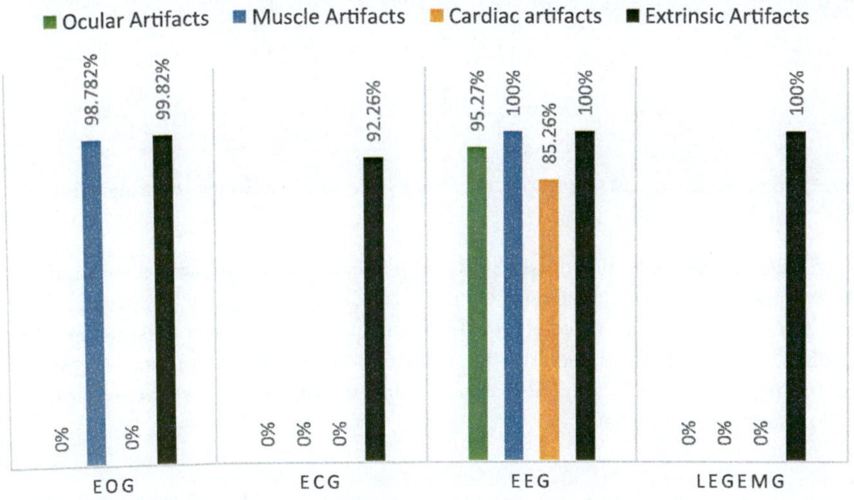

Fig. 17 Frequency of four artifacts including ocular artifact, muscle artifact, cardiac artifact, and extrinsic artifacts that exist in EOG, ECG, EEG, LEG EMG channels of PSG data

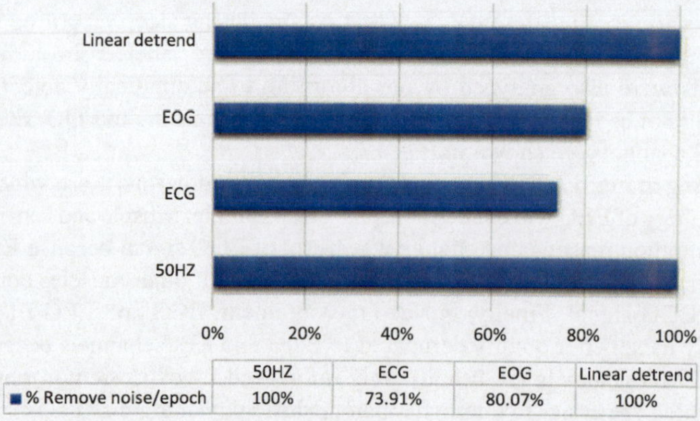

Fig. 18 Wandering baseline (linear detrend) and 50 Hz artifacts could be adequately removed 100% of the cases while EOG and ECG artifacts 80.07% and 73.91% of the cases respectively

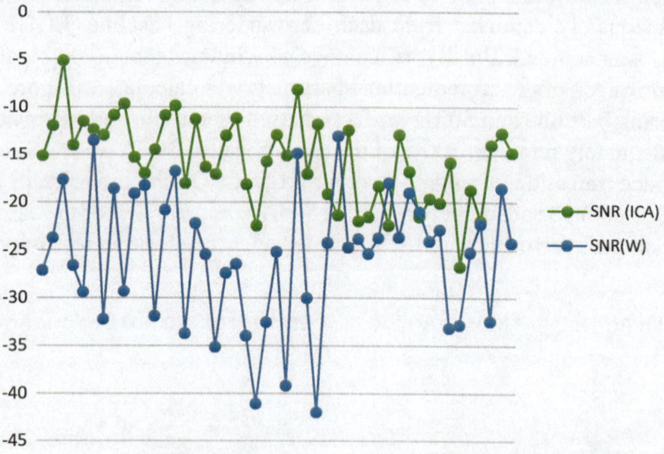

Fig. 19 SNR (in dB) of EEG signal after denoised by ICA and Wavelet transform algorithms

from EEG channel, signal to noise (SNR) ratio of EEG signal after artifact removal by the two methods was calculated (in dB unit) and shown in Fig. 19. The results show that ICA is better than Wavelet transform in removing artifacts existing in EEG channel. This can be explained by the existence of independent components, in this case independent artifacts, in EEG channel and ICA is better in recognizing these independent components.

6 Conclusion

There are different physiologic and non-physiological artifacts that affect quality of PSG signals. These artifacts make PSG signals reading more difficult, and removing them is challenging. In this study, we propose an algorithm that can effectively detect and remove artifacts such as EOG, ECG, 50 Hz from PSG signals. In addition, usage of sleep diary during sleep recording is useful for labeling artifacts in PSG signals. The proposed method has some limitations because it is not very stable and requires manual tuning to achieve the best performance. In future work, automatic tuning could be added, and we hope this algorithm would become a useful tool for PSG signal pre-processing.

Acknowledgements We thanks to School of Biomedical Engineering, International University, Vietnam National Universities, HCMC and UMC clinics for helping us collect data for this project.

Conflicts of Interest The authors have no conflict of interest to declare.

References

1. Collura TF (1993) History and evolution of electroencephalographic instruments and techniques. J Clin Neurophysiol 10(4):476–504
2. Aserinsky E, Kleitman N (1953) Regularly occurring periods of eye motility, and concomitant phenomena, during sleep. Science 118(3062):273–274
3. Campbell IG (2009) EEG recording and analysis for sleep research. Curr Protocols Neurosci 49(1):10.2.1–10.2.19
4. Berry RB, Wagner MH (2014) Sleep medicine pearls E-Book. Elsevier Health Sciences
5. Butkov N, Lee-Chiong TL (2007) Fundamentals of sleep technology. Lippincott Williams & Wilkins
6. Clifford GD (2002) Signal processing methods for heart rate variability. Oxford University, UK
7. Morillo DS et al (2009) Poincaré analysis of an overnight arterial oxygen saturation signal applied to the diagnosis of sleep apnea hypopnea syndrome. Physiol Meas 30(4):405
8. Nakajima K, Matsumoto Y, Tamura T (2001) Development of real-time image sequence analysis for evaluating posture change and respiratory rate of a subject in bed. Physiol Meas 22(3):N21
9. Carney PR, Berry RB, Geyer JD (2005) Clinical sleep disorders. Lippincott Williams & Wilkins
10. Van Hoey G, Philips W, Lemahieu I (1997) Time-frequency analysis of eeg signals. In: Proceedings of the ProRISC workshop on circuits, systems and signal processing
11. Akansu AN et al (2001) Multiresolution signal decomposition: transforms, subbands, and wavelets. Academic Press
12. Comon P (1994) Independent component analysis, a new concept? Signal Process 36(3):287–314
13. Stone JV (2004) Independent component analysis: a tutorial introduction. MIT Press
14. Ungureanu M et al (2004) Independent component analysis applied in biomedical signal processing. Meas Sci Rev 4(2):18

A Neuronavigation Toolkit for 3D Visualization, Spatial Registration and Segmentation of Brain Vessels from MR Angiography Images

Nguyen Thanh Duc and Boreom Lee

Abstract Neuronavigations are real-time approaches implemented to help neurosurgeons precisely localize different intracerebral pathologies by using multiple neuroimaging modalities. However, current systems suffer from several shortcomings including time-consuming, 3D registration and segmentation accuracies as well as difficulties transferring 3D vessel imaging to the neuronavigation systems. In this work, we introduce a standalone platform for image-fusion, and semantic segmentation of MR angiography images supported for neurovascular interventions. The full-stack toolkit consists of two parts: back-end and front-end. At back-end, Tensorflow library and Ajax for application programming interface server request-response interactions are embedded. As to front-end, we implement C++_based Qt platform for GUI (Graphic User Interface) in Visual Studio integrated development environment. The toolkit provides 3D volume and slice-based visualizations of brain multimodal images. All medical images are archived in DICOM standard, then converted into NIfTI formats. Visualization Toolkit was used to render 3D MR images. The implemented volume-rendering techniques allow the direct visualization of vascular structures, thus reveal vessel abnormalities more faithfully. We provide GUI interface of the modality brain image registration and segmentation functionality. There are two machine-learning based automatic registration categories: (1) intensity-based, (2) geometry-based. The registration error in mm is computed by using Hausdorff Distance metric. As for cerebrovascular segmentation, a method that utilizes an enhanced vesselness filter which bases on multiscale Hessian eigenvalues is performed to extract neurovascular trees. The introduced toolkit is expected to be a validated platform that allows researchers to apply insightful results into the operating room for clinical evaluations.

N. T. Duc · B. Lee (✉)
Department of Biomedical Science and Engineering, Institute of Integrated Technology, Gwangju Institute of Science and Technology, Gwangju, South Korea
e-mail: leebr@gist.ac.kr

N. T. Duc
Montreal Neurological Institute, McGill University, Montreal, Canada

V. Van Toi et al. (eds.), *8th International Conference on the Development of Biomedical Engineering in Vietnam*, IFMBE Proceedings 85,
https://doi.org/10.1007/978-3-030-75506-5_81

Keywords Deep learning · Neurovascular · Multimodality registration · Brain vessel segmentation · Neuronavigation system

1 Introduction

Neuronavigation systems, or image-guided neurosurgery tools, have become a standard practice and are used to allow preoperative phase of surgical trajectories to aid surgical procedures. Neuronavigation systems also allow objective real-time acquired neuroimaging in three-dimensional (3D) volumes, thus significantly reducing uncertainties throughout the surgical operation [1]. It is widely reported that the navigation systems help to guarantee the safeness of the surgery, reduce risks and provides tools for the inexperienced surgeons to self-practice and gain experiences. Several typical commercial neuronavigations including SonoWand system, Medtronic, and Brain Lab, have normally been manipulated in the surgery operations for multiple neuro-interventional purposes [2]. However, such neuronavigation systems experience several drawbacks regarding to visualization functions, registration, and segmentation accuracy. Furthermore, the recent developments in image-guided neurosurgery have moderately impact on the cerebrovascular surgery due to practical challenges moving 3D vessel volumes to the systems [3].

Advanced developments of 3D Computed Tomography Angiography (CTA) and Time-of-Flight (TOF) Magnetic Resonant Angiography (MRA) images have open the way for visualization and curing cerebral vascular diseases i.e., aneurysm, arteriovenous malformation (AVM), and brain tumor surgery [4]. Bringing the 3D MRA to image-guided neuronavigation to the real-time neurosurgical operation is a current state-of-the-art evolution to offer the surgeons to robustly localize brain vessels [5]. Particularly, it is crucial to determine correctly 3D interactions of an aneurysm with nearly vasculatures and penetrating arteries during the surgery of aneurysm, or correct drainage features of an arteriovenous malformation, or precise associations of a brain tumor with closed brain vessels and degree of vascular invasion [3].

There are several critical limitations of the current neuronavigation systems assisted for cerebrovascular interventions [6]. The most critical limitation is the low accuracy when segmenting small brain vessels using angiographic neuroimaging. Second limitation relates to the registration procedure that is still time-consuming especially when registering the rotated 3D angiographic images to the within-subject anatomical MRI. Giving the gaps in the exist literature, in this paper we introduce a preliminary version of our navigation toolkit that is specifically designed for neurovascular surgical interventions. Specifically, we describe several important techniques for visualization purposes of magnetic resonant angiography images, MRI-MRA image registrations of the same subjects as well as cerebral vascular structure segmentations, which are expected to be essential in the neurovascular interventions.

In the following sections, the core blocks, i.e., functionalities, architectures, methodologies, and interfaces of the proposed toolkit are presented. In section Materials and Methods, we first report various registration approaches implemented in our software and graphic user interface implementation for users to choose favored methods. Neuroimaging registration includes intensity-based methods that compare the intensity values of the assessed pixels, and geometry-based methods that use handcrafted feature descriptors. In addition, we describe a statistical method that employs an enhancement vesselness filter, which is based on the multiscale Hessian eigenvalues to extract cerebrovascular trees. In the Results section, we present the full-stack architecture of the toolkit. We also detail the Graphical User Interface (GUI) for 3D volume and slice-based visualization as well as volume-rendering of the vessels, spatial registration, and vascular-tree segmentation in which the users are able to select the preferred methods based on their prior-knowledge to perform the task. Finally, we discuss on the paper' limitations as well as several future works which should be done to improve the toolkit performance in order to move a step forward to the clinical application.

2 Materials and Methods

2.1 *Neuroimaging Registration in Neuronavigation Systems*

The registration of images is a growing research topic and forms an integral part in many medical image analysis tasks, and is a vital important task in neurosurgery interventions. Spatial registration refers to a process of finding corresponding structures within different images. Areas of clinical applications include alignments of datasets from multiple neuroimaging modalities, within-subject follow-up and baseline scans comparisons, pre-operative and post-operative within-subject examinations, therapeutic selections for radiotherapy, segmentation and parcellation based on template atlas, aligning training images for disease classification and clinical variable regression [7].

Various neuroimaging modalities are deployed in multiple phases of the neurosurgical interventions; for instance, several common types are CT brain images, T1- and T2-weighted anatomical MRI, metabolic screening using nuclear Positron Emission Tomography (PET), estimating the white matter structure using Diffusion Tensor Imaging (DTI), and angiographic images CTA, MRA for brain vessel diagnosis. Spatially aligning of multiple brain images of the same subject to a standard coordination is of importance to provide additional information of the patients and robust interpretations to the neurosurgeons throughout the interventions. Furthermore, spatial registration is needed for neuro-related clinical applications, for example, examining the clinical progression of a patient's health condition by using within-subject follow-up scans, and radiotherapy operations as well as image-guided neurosurgery [8].

2.2 Registration Methods

Medical image registration is a time-consuming process that spatial remodels coordinates of different medical images acquired from different acquisition sites into one standard coordinates while keeping the clinical contents unchanged. The primary task is basically to transform a so-called moving image, $\boldsymbol{I}_{M}(\boldsymbol{x})$, to make it aligned to the fixed image $\boldsymbol{I}_{F}(\boldsymbol{x})$. Specifically, registration procedure requires the finding of a transformed coordinate $\boldsymbol{T}(\boldsymbol{x})$ that allows $\boldsymbol{I}_{M}(\boldsymbol{x})$ spatially fit to $\boldsymbol{I}_{F}(\boldsymbol{x})$. The registration performance is assessed through a measure of match or cost function $\mathsf{C}(\boldsymbol{T}; \boldsymbol{I}_{F}(\boldsymbol{x}), \boldsymbol{I}_{M}(\boldsymbol{x}))$ that measures the differences or similarity between the transformed images and its reference. The transformation of coordinate can be predicted using an optimization method of the cost function based on $\boldsymbol{T}(\boldsymbol{x})$. The optimization algorithm is usually performed by iteratively updating the parameters of the transformation matrix such that the differences between these images are minimized.

There are multiple factors that should be borne in mind when selecting a precise registration method. For instance, these considered factors include models for coordinate transformation, optimization algorithms, multiresolution analyses, interpolation methods to evaluate $\boldsymbol{I}_{M}(\boldsymbol{T}(\boldsymbol{x}))$, and last but not least, a robust cost function. Regarding the coordinate transformation models, the degrees-of-freedom parameter, $\boldsymbol{T}(\boldsymbol{x})$, determine the deformations methods that can be used. There are several transformation models in the current literature covering from a very simple linear translation to more sophisticated linear rigid transforms and non-linear non-rigid or elastic models. Illustrations of various transformation models used for neuroimaging registrations are provided in Fig. 1.

The ordinary image registration approaches include intensity-based methods that compare the values of intensity at all pixels that represent the images, and geometry-based methods that use handcrafted image feature descriptors. In the existing literature, multiple cost functions $\mathsf{C}(\boldsymbol{T}; \boldsymbol{I}_{F}(\boldsymbol{x}), \boldsymbol{I}_{M}(\boldsymbol{x}))$ for intensity-based registration are commonly used including the mean squared difference (MSD, Equal. 1), or normalized correlation coefficience [5], and mutual information (MI). On the other hand, the regular geometrical descriptor for 3D brain volume registrations is a well-known

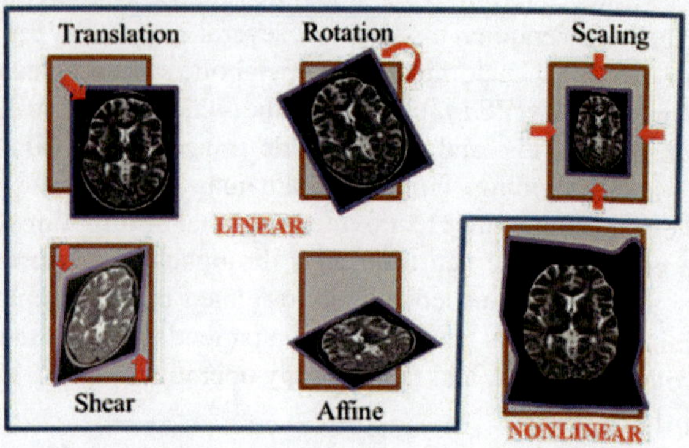

Fig. 1 Transformation methods for spatial neuroimaging registrations

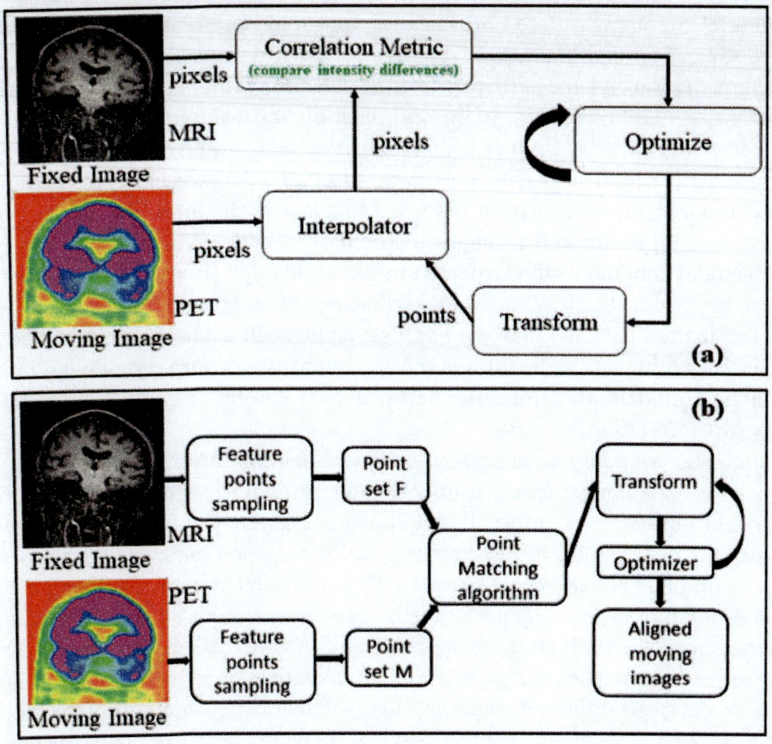

Fig. 2 Traditional neuroimaging (example of MRI and PET aligning) registration methods. **a** Intensity-based, **b** geometry-based pipelines

Iterative Closest Point (ICP) while Fast Point Feature Histogram (FPFH) is applied by several available researches. Figure 2a illustrates the intensity-based registrations while geometry-based neuroimaging registrations can be seen in Fig. 2b, respectively. Further details of iterative machine learning-based spatial registrations in neuroimaging applications can be seen in review papers.

$$\mathrm{MSD}(\boldsymbol{T}; \boldsymbol{I}_F(\boldsymbol{x}), \boldsymbol{I}_F(\boldsymbol{x})) = \frac{1}{N} \sum_{x \in \Omega_F} (\boldsymbol{I}_F(\boldsymbol{x}) - \boldsymbol{I}_M(\boldsymbol{T}(\boldsymbol{x})))^2 \quad (1)$$

2.3 *Cerebrovascular Structure Extractions*

Many traditional methods for cerebrovascular segmentation have been proposed: deformable models; statistical models; and recently developed deep neural networks (DNN). In deformable models, a classic method is a geodesic active contour that fits structural topologies of cerebrovascular appeared on MRA images. However, there are several critical limitations reported by using deformable models for brain vessel segmentation and one of them is leaking detection found around the edge of the vessels. During iteration process of the optimization, the leakage is wrongly detected in the nearly area which is right outside of brain vessels, which leads to poor performance. Statistical models perform cerebral vascular structure extraction by using statistical models such as Gaussian Mixture Models to fit the intensity distribution of multiple tissue types found the images. Particularly, Hidden Markov Random Field (HMRF) model integrated with Expectation–Maximization (EM) algorithm has been employed to separate brain White Matter with Gray Matter tissues and Cerebrospinal fluid in the anatomical MRI images. Their semantic segmentation performances are significantly contingent on the fitting between intensity histogram modelled in MRI and statistical models; therefore, the performances can be strongly limited by the intensity distortion in MRA data.

In this work, we apply a method, so-called vesselness filter proposed in [9, 10], to extract the vascular structure from MR angiography images. Vesselness filter employs multi-scale Hessian filter by implementing a newly enhanced function that overcomes the deficiencies of the currently well-developed ones and get properties close to a complete enhancement function. The proposed vesselness filter achieved the best evaluation scores compared to the existing methods in the literature and therefore it boosts a promising prospection and inspires the evolutions for better state-of-the-art approaches for separation and visualization of cerebral vasculature using MRA dataset. Further information and mathematic presentations of the method as well as proven assessment improvements of the vesselness filter over the current cutting-edge methods can be accessed in [9, 10].

2.4 Dataset

In this paper, we utilize a public available 3D time-of-flight angiography dataset of the TubeTK toolkit [11]. This dataset contains 100 (T1, MRA) image pairs acquired by a Siemens Allegra head-only 3 T MR system with voxel space of $0.53 \times 0.53 \times 0.83$ mm and 3D volume dimension of $448 \times 448 \times 128$ voxel for MRA images. The voxel spacing is 1 mm isotopic with a volume size of $176 \times 256 \times 176$ voxel for T1 images. The dataset has been collected with healthy subjects whom the ages are ranging from 18 to 74 years old, and without previous major cerebral vascular disease. This is the multimodality dataset in which our different neuroimaging scans are collected for each subjects, i.e., T1-weighted, T2-weighted MRI, MRA and DTI scans. However, for registration, we only utilize T1-weighted MRI and MRA in which the purpose is to align MRA image to corresponding MRI images of the same subject. For cerebrovascular segmentation, we utilize only MRA scans since MRA scans show the brain vascular structures.

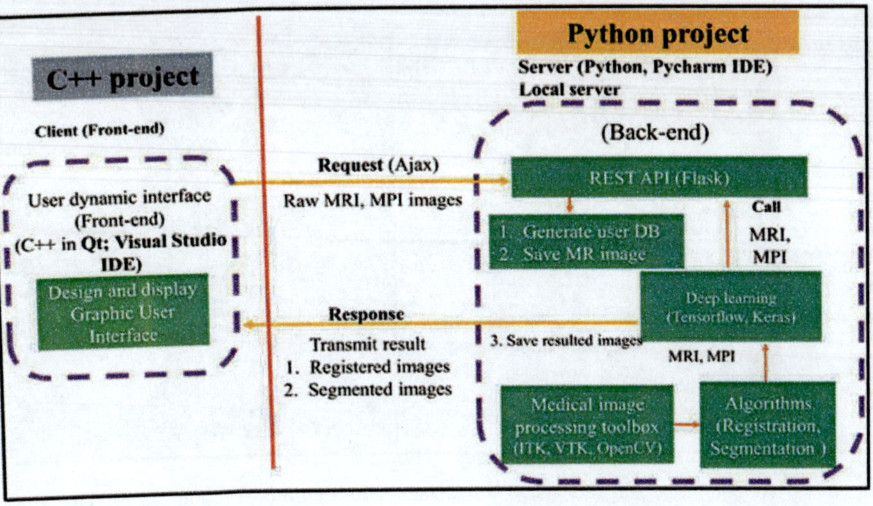

Fig. 3 An overview of the full-stack architecture of our toolkit. The toolkit consists of two parts: (i) Server-side (backend, right subfigure), and (ii) client-side (frontend, left subfigure)

3 Results

3.1 *A Toolkit Platform*

The objective of this sub-section is to implement a practical software for neuroimaging visualization and registration. The window form application consists of two parts: (i) Server-side (Backend), and (ii) Client-side (frontend). In case of the server-side, we used the Keras and Tensorflow for performing the deep neural network algorithms and Ajax for API server (Application programming Interface). All the registration methods including traditional machine-learning and deep convolutional neural network algorithms, which described in previous section, are implemented in this backend. In case of the client-side, we implemented C++_based Qt (https://www.qt.io/) platform for GUI (Graphic User Interface) in Visual Studio IDE (Integrated Development Environment) (https://visualstudio.microsoft.com/). Overall scheme for implementation of our software can be illustrated through Fig. 3.

3.2 *3D Neuroimaging Visualization and Volume Rendering*

Visualization of 3D medical images has a crucial role in the current development of the healthcare technology. Especially, 3D volume rendering techniques of neuroimaging provide various advantages and deep information needed in image-guided surgery, which offers neurosurgeons to see further the tissue level and brain

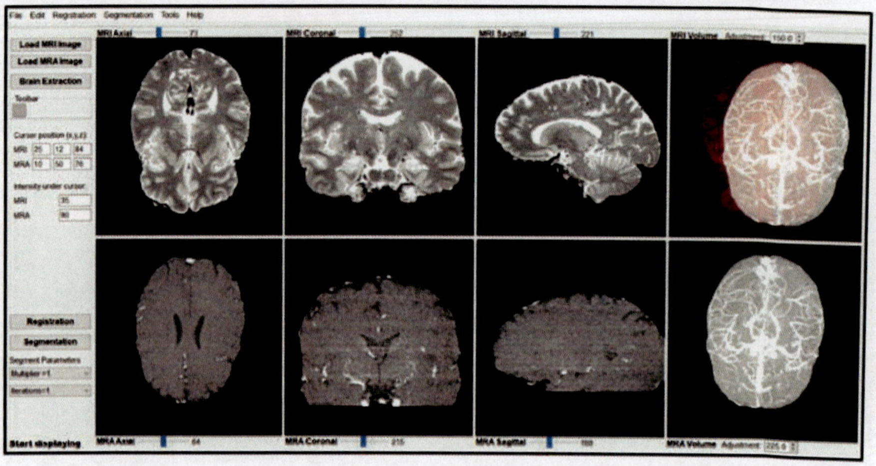

Fig. 4 3D volume-based and slice-based visualizations of multimodal brain images. The upper panel shows slice-based and 3D volume of MRI while the lower shows that of MRA images

vessels can be clearly seen by 3D visualization of multi-model MRI, MRA, CT and CTA, and Magnetic Particle Imaging (MPI) data. Visualizations of multimodal neuroimaging has significant role in brain surgical operation. In this study, we developed a software of 3D visualization of brain multimodal CTA, MRA and MRI images for optimal vascular neurosurgical planning. The input brain images are in DICOM format, then converted into NIfTI files. Visualization Toolkit (VTK) (https://vtk.org/) [12] is utilized to render MRA and MRI data. MRA images offer robust features of rigid structures and MRI provides better visualization of soft tissues. When volume rendering of MRA and MRI are performed, diverse information of the underlying features can be revealed respectively. Rendering implementation of 3D volumes unveils the straight view of the underlying contents which are hidden in the normal visualization, and help to disclose atypical vessels robustly.

In this software, we also implemented volume rendering functionality, which allows us to be able to divulge vascular structures. Figure 4 presents the GUI of the software, 3D MRI visualizations of brain images, and volume rendering with a focus on vessel visualization is depicted in Fig. 5.

3.3 Neuroimaging Registration Results

The Insight Segmentation and Registration Toolkit (ITK) which can be found at (https://itk.org/) [13, 14] is among the most common tools for researchers in the field of medical imaging analyses, and is already being implemented in various clinical applications. However, ITK is not designed for visualization and it does not have graphic user interface functionality for a dedicated development platform. Our

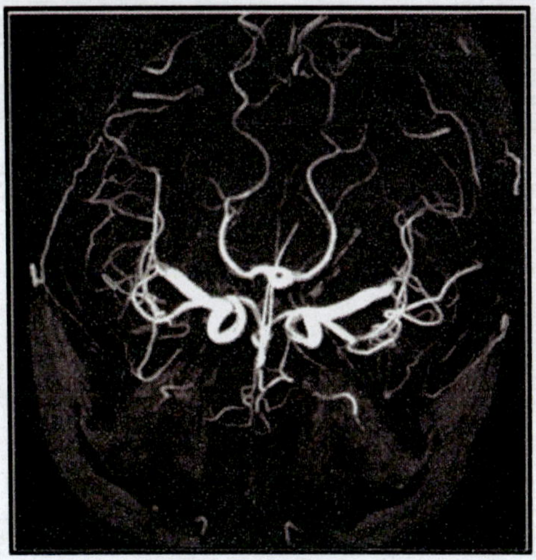

Fig. 5 Brain vascular tree visualization using volume-rendering technique implemented in our toolkit

software offers a visual programming interface with multiple necessary functionality of registration and visualization provide by VTK library. Figure 6 provides GUI interface of the modality brain image registration developed in our software package. There are three registration categories: (1) intensity-based, (2) geometry-based, and (3) landmark-based models. We compute and show the registration error in mm by using Hausdorff Distance metric. Figure 7a, b show the registration results, which have been performed by intensity-based and geometry-based registration algorithms, respectively. We are able to achieve a registration spatial error of 0.386 and 0.4266 mm by using intensity- and geometry-based algorithms, respectively. Figure 8 illustrates of the image differences before registration and after registration in which the robust alignments are achieved.

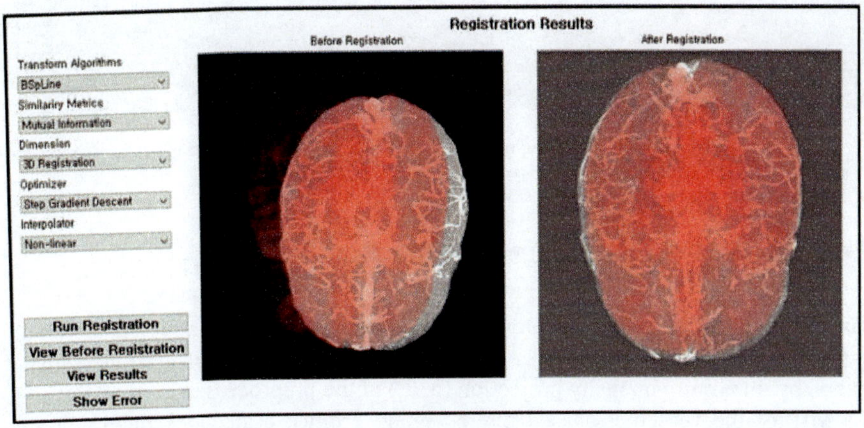

Fig. 6 GUI of 3D brain image registration implemented in our toolkit

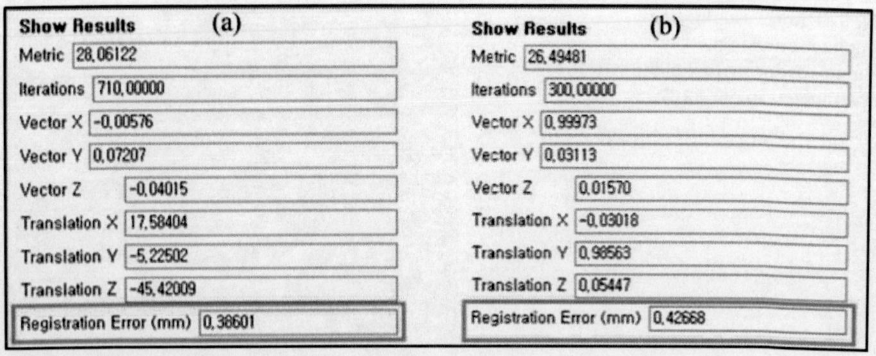

Fig. 7 3D registration errors using the intensity-based (**a**) and geometry-based (**b**) registrations

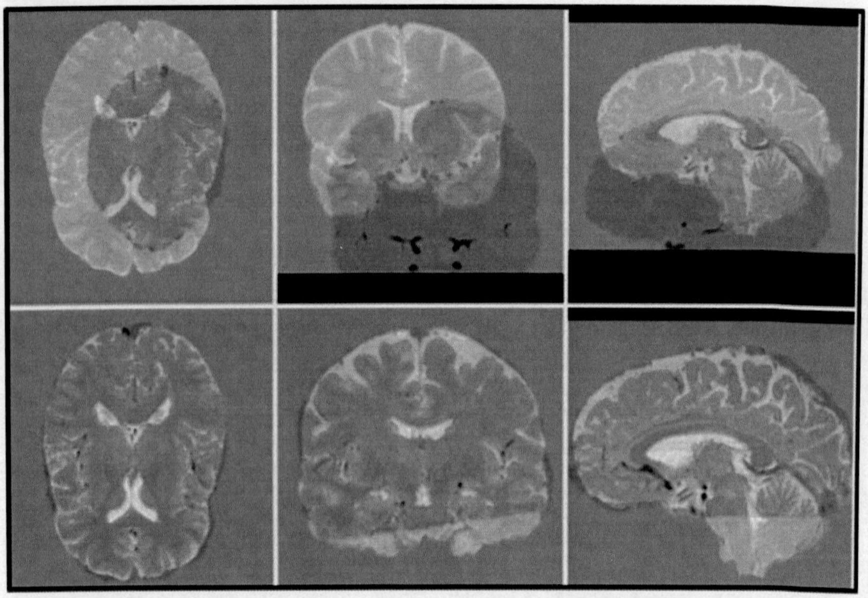

Fig. 8 Illustrative results of before (upper panel) and after (lower panel) registrations between in-subject MRI and MRA images

3.4 Cerebral Vascular Segmentation Results

Visualization and performance comparisons of the enhancement function for cerebral vascular trees using a 3D TOF-MRA data by applying the introduced vesselness filter, shown by maximum intensity projection (MIP) of the preprocessed image (left) and the MIP of the vesselness-based filtered image (right), is depicted in Fig. 9. As clearly seen in Fig. 9, the vesselness method results in high response values in all

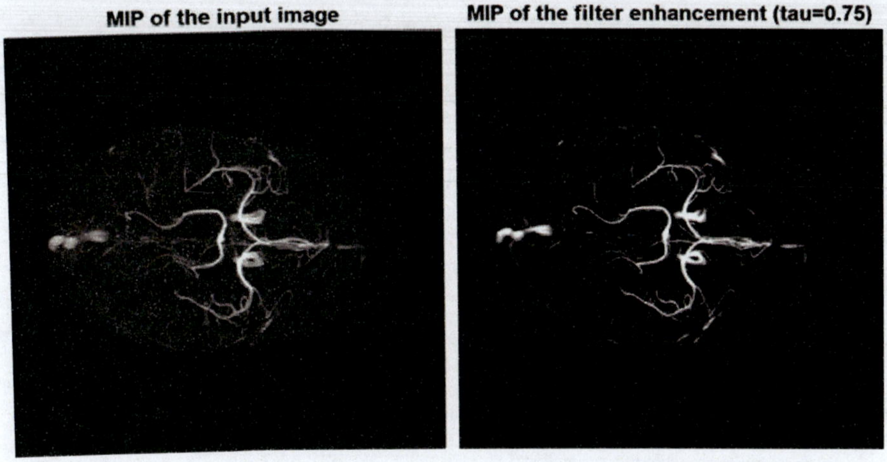

Fig. 9 Results of before (left) and after (right) segmentation using the vesselness filter

the cerebral vasculature including small and large vessels. Thus, the visualization of the segmented structures of the smallest brain vascular evidently proves with the effectiveness of the introduced method.

4 Discussion and Conclusion

In this work, we introduce a toolkit as an important feature that can be used to transform results of image-guided neurosurgery for neurovascular interventional research from the lab to the operation room. We have showed a full-stack toolkit, which includes several necessary components for implementation of an improved workflow for the image-guided neurovascular pathology. Our toolkit allows concurrent visualizations of 3D anatomical T1- and/or T2-weighted MRI images together with TOF MR Angiography image, which is spatially aligned to anatomical ones. Visualization features offer ones to view the 3D volumes and three slice-based (axial, sagittal, and coronal views) visualization functionalities. Moreover, we provide the volume rendering visualization, which allows researchers to visualize the 3D cerebrovascular structures from MRA images.

Our future works for improvements of the toolkit's performance focus on implementations of 3D deep learning approaches for registration and segmentation purposes. Whilst machine-learning approaches have long gained their reputations in implementation of registrations between within-subject pairwise images, recently proposed deep-learning-based frameworks directly estimate displacement fields without iterative optimisation for testing left-out images, using deep convolutional neural networks learned from a large amount of data [8, 15–20]. These recently developed deep learning methods are expected to solve various multiple critical drawbacks found in conventional pairwise machine learning based methods, i.e., time-consuming, robust generality [8, 17].

Methods using deep learning architectures are robustly appropriate for 3D high-dimension neuroimaging registration and segmentation, because these methods naturally learn to accumulate the information of diverse features that are applicable for the functionality. Moreover, deep learning approaches are expected to obtain high robust performance while reduce time processing. More importantly, deep learning architectures such as convolutional neural networks offer parallelism learning that boosts up implementation and learning mechanism using multiple GPUs fast and straightforward [15].

Acknowledgements This work was supported by GIST Research Institute (GRI) ARI grant funded by the GIST in 2020. This work was also supported by the Technology Innovation Program (or Industrial Strategic Technology Development Program-Development of Core Industrial Technology) (20003822, Development of Navigation System Technologies of MicroNano Robots with Drug for Brain Disease Therapy) funded By the Ministry of Trade, Industry and Energy (MOTIE, Korea). This research was also supported by Smart Healthcare Research Grant through the Daewoong Foundation (DS188).

Conflicts of Interest The authors have no conflict of interest to declare.

References

1. Mathiesen T et al (2007) Neuronavigation for arteriovenous malformation surgery by intraoperative three-dimensional ultrasound angiography. Neurosurgery 60(4, Suppl 2):345–50; discussion 350-1
2. Drouin S et al (2017) IBIS: an OR ready open-source platform for image-guided neurosurgery. Int J Comput Assist Radiol Surg 12(3):363–378
3. Jabbour P, Tjoumakaris S, Rosenwasser R (2009) Angiography, MRA in image guided neurosurgery. In: Lozano AM, Gildenberg PL, Tasker RR (eds) Textbook of stereotactic and functional neurosurgery. Springer, Berlin, pp 299–305
4. Leal PR, Hermier M, Froment JC, Souza MA, Cristino-Filho G, Sindou M (2010) Preoperative demonstration of the neurovascular compression characteristics with special emphasis on the degree of compression, using high-resolution magnetic resonance imaging: a prospective study, with comparison to surgical findings, in 100 consecutive patients who underwent microvascular decompression for trigeminal neuralgia. Acta Neurochir (Wien) 152(5):817–825
5. Zhang Q et al (2016) CBCT-based 3D MRA and angiographic image fusion and MRA image navigation for neuro interventions. Medicine (Baltimore) 95(32):e4358
6. Stidd DA et al (2014) Frameless neuronavigation based only on 3D digital subtraction angiography using surface-based facial registration. J Neurosurg 121(3):745–750
7. Klein S, Staring M, Murphy K, Viergever MA, Pluim JPW (2010) Elastix: a toolbox for intensity-based medical image registration (in English). IEEE Trans Med Imaging 29(1):196–205
8. de Vos BD, Berendsen FF, Viergever MA, Sokooti H, Staring M, Isgum I (2019) A deep learning framework for unsupervised affine and deformable image registration (in English). Med Image Anal 52:128–143
9. Jerman T, Pernus F, Likar B, Spiclin Z (2016) Enhancement of vascular structures in 3D and 2D angiographic images. IEEE Trans Med Imaging 35(9):2107–2118

10. Ourselin S, Styner MA, Jerman T, Pernuš F, Likar B, Špiclin Z (2015) Beyond Frangi: an improved multiscale vesselness filter 9413:94132A
11. Aylward SR, Bullitt E (2002) Initialization, noise, singularities, and scale in height ridge traversal for tubular object centerline extraction. IEEE Trans Med Imaging 21(2):61–75
12. Schroeder W, Martin K, Lorensen B (2006) The visualization toolkit: an object-oriented approach to 3D graphics. Kitware
13. McCormick M, Liu X, Jomier J, Marion C, Ibanez L (2014) ITK: enabling reproducible research and open science. Front Neuroinform 8:13
14. Yoo TS et al (2002) Engineering and algorithm design for an image processing Api: a technical report on ITK–the Insight Toolkit. Stud Health Technol Inform 85:586–592
15. Balakrishnan G, Zhao A, Sabuncu MR, Guttag J, Dalca AV (2019) VoxelMorph: a learning framework for deformable medical image registration. IEEE Trans Med Imaging
16. Duc NT, Lee B (2019) Microstate functional connectivity in EEG cognitive tasks revealed by a multivariate Gaussian hidden Markov model with phase locking value. J Neural Eng 16(2):026033
17. Duc NT, Ryu S, Qureshi MNI, Choi M, Lee KH, Lee B (2020) 3D-deep learning based automatic diagnosis of Alzheimer's disease with joint MMSE prediction using resting-state fMRI. Neuroinformatics 18:71–86
18. Duc NT, Ryu S, Choi M, Iqbal Qureshi MN, Lee B (2019) Mild cognitive impairment diagnosis using extreme learning machine combined with multivoxel pattern analysis on multi-biomarker resting-state FMRI. In: Conference on proceedings of IEEE engineering in medicine and biology society, vol 2019, pp 882–885
19. Nguyen DT, Ryu S, Qureshi MNI, Choi M, Lee KH, Lee B (2019) Hybrid multivariate pattern analysis combined with extreme learning machine for Alzheimer's dementia diagnosis using multi-measure rs-fMRI spatial patterns. PLoS One 14(2):e0212582
20. Livne M et al (2019) A U-net deep learning framework for high performance vessel segmentation in patients with cerebrovascular disease. Front Neurosci 13:97

Pre-processing Block Design for the Electroencephalography Signal by Using Notch Filter and Blind Source Separation Technique

Minh Bao Pham, Xuan Yen Nhi Phan, Quoc Khai Le, and Quang Linh Huynh

Abstract The digital filters play an indispensable role in recording and displaying EEG data. Without filters, many segments of EEG would be virtually unreadable. The main benefit of filters is that it makes the EEG tracing cleaner, easier to interpret and observe. Specific filter settings can also be used to accentuate particular types of EEG activity. Filters can, however, be misused, and at times their use can lead to unintended consequences. In this study, we present and compare the conventional filter design methods applied to significant types of noises. We built a program with a friendly interface with effective noise filtering.

Keywords EEG signal · Notch filter · BSS · ICA

1 Introduction

Research and application of biomedical signals, especially EEG signals, are always an attractive topic for scientists. Electroencephal signals are an exceptional type of signal with average voltage amplitude values only at the microvolt threshold, so there are many other bioelectric signals or causes from external signal sources that can affect the brain signal [1, 2]. There are two main approaches for the efficient exploitation of EEG signals in studies: the first is the use of original data obtained from electroencephalography (raw data) applications using machine learning algorithms [3] or various models of artificial neuron networks [4]; and the second direction is to extract the characteristics of the signal after it has been pre-processed, before including the corresponding character identification or identification steps [5, 6]. This study is based on the second approach and is also the most approach to the study of EEG signals. The main objective of the research is to establish a complete software

M. B. Pham (✉) · X. Y. N. Phan (✉) · Q. K. Le · Q. L. Huynh
Biomedical Engineering Department, HCMUT-VNU, 146 Ly Thuong Kiet Street, Ho Chi Minh City 70000, Viet Nam
e-mail: bao.pham1811539@hcmut.edu.vn

X. Y. N. Phan
e-mail: nhi.phanyennhi3107@hcmut.edu.vn

V. Van Toi et al. (eds.), *8th International Conference on the Development of Biomedical Engineering in Vietnam*, IFMBE Proceedings 85,
https://doi.org/10.1007/978-3-030-75506-5_82

with effective noise reduction processing for the EEG signal. The results of the project can be applied to complete the function of the preprocessing block in all EEG signal processing systems in real-time or in the form of file analysis archived (off-line). Many various algorithms and methods are used in previous studies when denoising in EEG signals. Algorithms can be divided into groups: time-domain processing [6, 7], z-domain processing [8, 9], random distribution processing [8, 10], processing to remove noise and artifacts caused by the other bioelectric signals affecting the brain signals [11–13] and blind source [14]. In this research, we focus on effectively exploiting various methods to remove noise with the two basic options of analysis and processing being fully automatic and semiautomatic. With the approach set out by the topic, the methods after being applied to real data that have noise will be evaluated and compared to the effectiveness of each other. Thereby creating an effective method to remove unwanted noises of EEG signals. The result of the research is a user interface software that allows import and export various types of EEG devices and is compatible with many programs when processing completed.

2 Materials

This research applies the Declaration of Helsinki principles in human studies.

All the experimental process was conducted in Biomedical Laboratory 204B4 at Ho Chi Minh University of Technology. The recording room was designed based on the characteristics of the Faraday cage. The data was recorded using NicoletOne V32 device (Natus Neurology Incorporated, WI, USA) providing 32 channels with the ability to amplify the original signal to gain high-quality signal. Signal gained from NicoletOne device is extracted into a digital signal and processed using Matlab software (The Mathworks, Natick, MA, USA).

3 Results and Discussions

Infinite Impulse Response (IIR) filters and Finite Impulse Response (FIR) filters are two common and basic types of digital, which have specific advantages and disadvantages. IIR filters work fast but easily lead to distorting the signal. In contrast, FIR filters can be designed without any distortion, but more compute intensity is required. FIR filters are preferable due to the development of technology.

3.1 The Low Pass Filter, High Pass Filter

EEG signals usually have a frequency range in 1–30 Hz, only gamma waves from 30 to 100 Hz. Table 1 describes the frequency bands of waves in the EEG signals [15].

Table 1 Comparison of EEG signal bands

Band	Frequency (Hz)	Characteristics	Location
Delta (δ)	0.2–4	Sleep	
Theta (θ)	4–8	Drowsiness	Frontal, temporal
Alpha (α)	8–13	Relaxed	Occipital, parietal
Beta (β)	13–30	Active, busy	Frontal, parietal, temporal
Gamma (γ)	30–100	Arousal	Frontal, parietal

To limit undesired signals, frequencies outside the range of 0.5–100 Hz should be removed by highpass and lowpass filters. A highpass filter needs to be implemented first to remove DC offset. It should be noted that the filter order increases, the sharpness of the attenuation characteristic in the transition band increases.

3.2 *The Notch Filter*

The notch filter is the band-stop filter with a narrow stopband. Band-stop filter is a filter attenuates the signal range at the specified frequency to the very low levels called the stopband, and pass the remain signals.

In this research, we use the *DSP system Toolbox* of Matlab to filter the powerline interference, the function of the notch filter is `d= fdesign.notch('N, F0, Q', value1, value2, value3, Fs)`. This function constructs a notch filter specification object while providing the sampling frequency `Fs` of the signal, as are all the other frequency values provided, with the specification set to '`N, F0, Q`' and `value` provided for all members of the '`N, F0, Q`'. The variable of the specification are listed as follows:

- `N`: Filter Order, in this case, we are processing the EEG signals, so `N`= 6.
- `F0`: Center Frequency. The signal we want to remove is the powerline interference so F0 = 50 Hz
- `Q`: Quality Factor of the filter is defined as the ratio of the centre frequency to the 3 dB bandwidth. The powerline interference is not only confined to 50 Hz but also affects neighbouring frequencies, particularly in the 47–53 Hz range. Therefore, we recommend the bandwidth is 6 Hz, and the suitable value of the quality factor is 8.3 [16].

Figure 1 is the result we use notch filter in Matlab to remove the power line interference of the raw EEG signals.

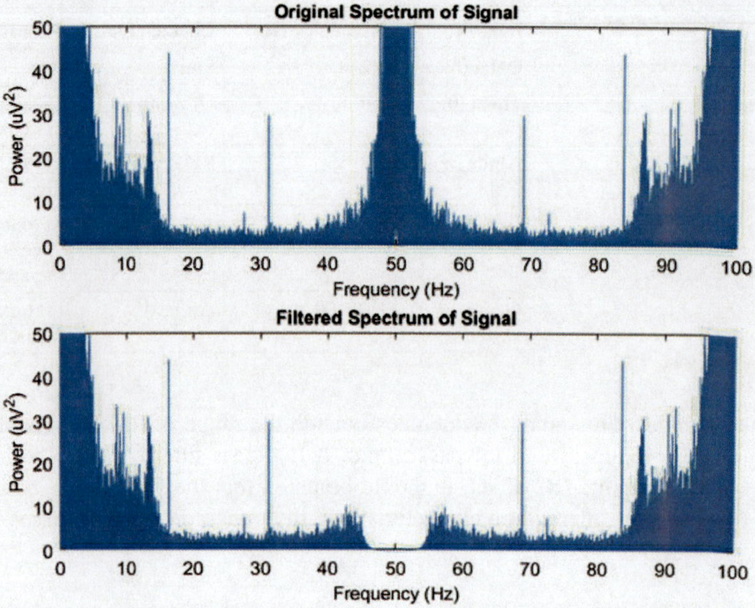

Fig. 1 Filtered data after using the notch filter

3.3 *Blind Source Separation*

The term *blind* refers to the fact that there is no specific information about the mixing process or about the existing source signals (Figs. 2, 3).

There are two distinct classes of present algorithms for solving BSS problem: the algorithms in the time domain and the algorithms in the frequency domain. The time-domain algorithms bring about better performance due to the reason that frequency-domain algorithms have to take account of a permutation and to scale ambiguity at each frequency, also, as many samples as possible are necessary to offer exact results. Moreover, the frequency domain approaches use a large number of frequency bins and, consequently, boost the computation cost. On the other side, the better independence assumption for full-band signals and possible high convergence near the optimal point is the salient advantages of time-domain approaches. But the degradation of convergence in the strong reverberant environment and the needs of many parameters to be adjusted for each iteration step have become the drawback of the time-domain algorithms. Comparisons of traditional BSS methods over an incremental learning scheme (an online mode), based on computational complexity & storage requirements and works, are based on Frequency-domain blind source separation (BSS) techniques to increase the speed and reduce the computational complexity of conventional algorithms, mainly for applications that involve sizeable

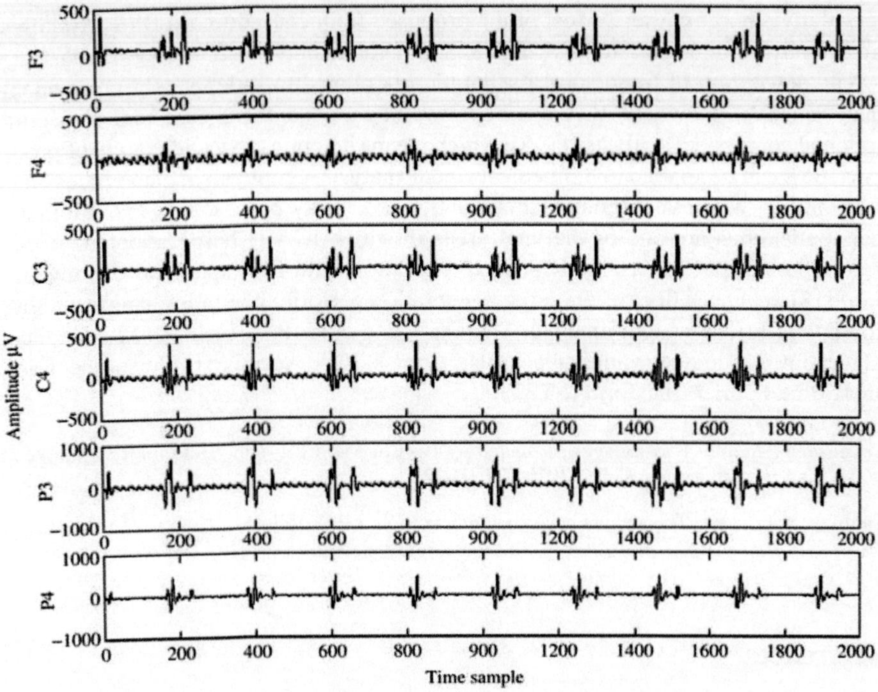

Fig. 2 A sample of an EEG signal simultaneously recorded with fMRI [1]

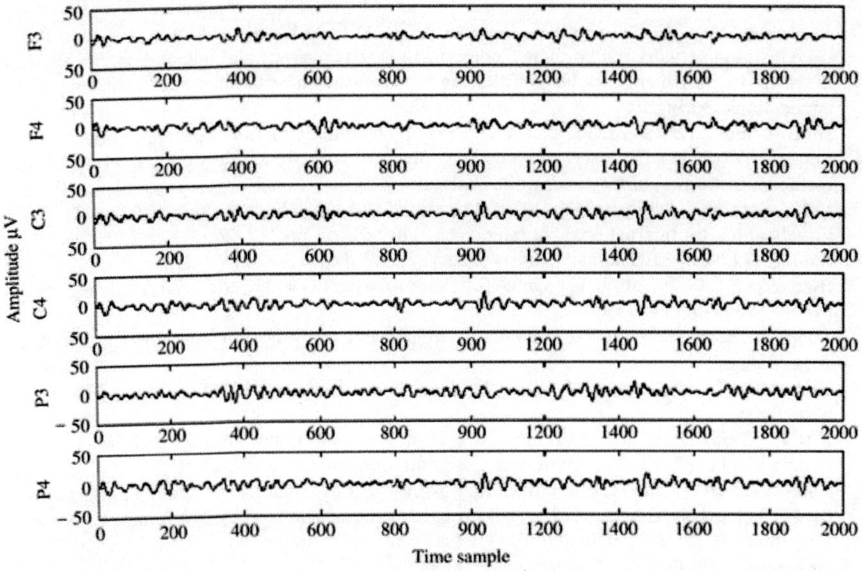

Fig. 3 The EEG signals after removal of the scanner artifacts [1]

convolutive mixture filters. Most of the proposed frequency-domain BSS structures are derived from the classic independent component analysis (ICA) algorithm.

The perception of blind source separation is related to independent component analysis (ICA). However, ICA can be viewed as a general-purpose tool replacing principal component analysis (PCA), which means it applies to a wide range of problems. BSS can be applied to biomedical signal analysis, geophysical data processing, data mining, wireless communications and sensor array processing [17]. Biomedical applications are one of the vital areas in which ICA is being adopted. It has been reported that the use of ICA in combination with SD (spherical deconvolution) [18] enhances fiber orientation regeneration performance in low-angle resolution DWI (diffuse weight images). There are also other BSS methods like Genetic Algorithm, Underdetermined Approach, Sparse ICA Approach, Noise scenario, Short-time Fourier transform (STFT).

Acknowledgements This research is funded by Ho Chi Minh City University of Technology - HCMC under grant number SVCQ-2019-KHUD-27.

Conflicts of Interest The authors have no conflict of interest to declare.

References

1. Sanei S, Chambers JA (2007) EEG signal processing. John Wiley & Sons Ltd, pp 93–94
2. Tong S, Thakor NV (2009) Quantitative EEG analysis methods and clinical applications. Artech House
3. Jirayucharoensak S, Pan-Ngum S, Israsena P (2014) EEG-based emotion recognition using deep learning network with principal component based covariate shift adaptation. Sci World J
4. Siuly S, Li Y, Zhang Y (2016) EEG signal analysis and classification: techniques and applications. Springer
5. Hu L, Zhang Z (2019) EEG signal processing and feature extraction. Springer, Singapore
6. Leong WY (2019) EEG signal processing: feature extraction, selection and classification methods. Institution of Engineering & Technology
7. Liu Y, Sourina O, Nguyen MK (2010) Real-time EEG-based human emotion recognition and visualization. In: International conference on cyberworlds, pp 262–269
8. Subha PD (2010) EEG signal analysis: a survey. J Med Syst 34(2):195–212
9. Blankertz B (2007) Optimizing spatial filters for robust EEG single-trial analysis. IEEE Signal Process Mag 25(1):41–56
10. Lopes R, Betrouni N (2009) Fractal and multifractal analysis: a review. Med Image Anal 13(4):634–649
11. Blinowska KJ, Zygierewicz J (2012) Practical biomedical signal analysis using MATLAB. CRC Press
12. Xue Z (2006) Using ICA to remove eye blink and power line artifacts in EEG. In: First international conference on innovative computing information and control
13. Zhou W, Zhou J, Zhao H, Ju L (2006) Removing eye movement and power line artifacts from the EEG based on ICA. In: IEEE engineering in medicine and biology 27th annual conference
14. Fitzgibbon SP (2007) Removal of EEG noise and artifact using blind source separation. J Clin Neurophysiol 24(3):232–243
15. Kim H, Hwang Y, Yoon D, Park CH (2013)An analysis of driver's workload in the lane change behavior. In: 2013 international conference on ICT convergence (ICTC)

16. Leske S, Dalal SS (2019) Reducing power line noise in EEG and MEG data via spectrum interpolation. NeuroImage
17. Jain SN (2012) Blind source separation and ICA techniques: a review. Int J Eng Sci Technol IV(04):1490–1501
18. Jing M, McGinnity MT, Coleman S, Zhang H, Fuchs A, Kelso JAS (2012) Enhancement of fiber orientation distribution reconstruction in diffusion-weighted imaging by single channel blind source separation. IEEE Trans Biomed Eng 59(2):363–373

Characteristics of Lateral Cephalometric by X-Ray in the Obstructive Sleep Apnea Syndrome Patients

Hoang Anh Tien and Nguyen Thi Thanh Duyen

Abstract Sleep apnea syndrome poses many dangers to the cardiovascular, nervous and respiratory systems. Evaluating changes in anatomical features of facial bones on head X-ray was an accessible and helpful tool in assessing the severity of obstructive sleep apnea syndrome. A case–control study of 49 patients (case group: 33, control group: 16) with symptoms of loud snoring and/or evidence of suspected obstructive sleep apnea syndrome (OSAS) admitted to the Hue University of Medicine and Pharmacy's Hospital, from May 2018 to May 2019. Sleep apnea syndrome was measured with StarDust II, Germany. X-ray was performed with Amrad ARM, USA. Average age of the study group: 71.47 ± 11.46 years. The average Epworth score of OSAS group (+) 7.64 ± 3.13 was higher than the OSAS group (−) 5.00 ± 2.13 ($p < 0.05$). High-risk Berlin scores accounted for 82.9% higher proportion in the OSAS group (+) than the OSAS group (−) ($p < 0.05$). SNA, SNB, ANB, MPH and PAS in OSAS group (+) differed from OSAS group (−) with $p < 0.05$. The average value of ANB and PAS were statistically significant, according to AHI with $p < 0.05$. There was a correlation between AHI and Epworth, SNB, ANB, MPH, PAS in the study group ($p < 0.05$). Among OSAS patients (+): Correlated between AHI and SNB, ANB, MPH, PAS was − 0.45 ($p < 0.05$), 0.58 ($p < 0.05$), 0.39 ($p < 0.05$), − 0.46 ($p < 0.05$), respectively. The lateral head X-ray provides SNB, ANB, MPH, PAS values correlated with AHI in obstructive sleep apnea syndrome patients.

Keywords Obstructive sleep apnea syndrome · Lateral cephalometric · X-ray

H. A. Tien (✉)
Hue University of Medicine and Pharmacy, Hue University, Hue, Vietnam
e-mail: hatien@hueuni.edu.vn; hatien@huemed-univ.edu.vn

N. T. T. Duyen
Danang C Hospital, Danang, Vietnam

V. Van Toi et al. (eds.), *8th International Conference on the Development of Biomedical Engineering in Vietnam*, IFMBE Proceedings 85,
https://doi.org/10.1007/978-3-030-75506-5_83

1 Background

Sleep apnea syndrome is a common sleep disorder [1] that has a negative impact on health and quality of life [2], which is the cause of drowsiness and fatigue during the day, leading to neurophysiological changes such as memory loss and concentration ability [3], and increase the risk of occupational accidents and traffic accidents [4]. This disease is common in both developed and developing countries. In Asia, the prevalence is about 8.5% [5].

Changes in the anatomical structure of the face affect the structure of the upper airway, thus affecting sleep apnea syndrome in patients with loud snoring and/or evidence of sleep apnea. The relationship between obstructive sleep apnea syndrome and facial bone features on the skull X-ray has been mentioned in many studies [6, 7].

Sleep apnea syndrome is a syndrome causing many dangers to the cardiovascular, nervous and respiratory systems. However, the diagnosis of sleep apnea syndrome is often overlooked, and until it is discovered, sleep apnea syndrome has left serious complications [8]. Assessing changes in facial anatomical features on a tilted skull X-ray is an accessible tool, useful in relation to the severity of obstructive sleep apnea syndrome [7].

2 Subjects and Methods

2.1 Subjects

Forty-nine patients (disease group 33, control group 16) who have been examined and treated at the Department of Internal Cardiology at Hue University of Medicine and Pharmacy have symptoms of loud snoring and/or evidence of suspected obstructive sleep apnea syndrome (OSAS).

2.2 Methods

A case–control study of 49 patients (case group: 33, control group: 16) with symptoms of loud snoring and/or evidence of suspected obstructive sleep apnea syndrome (OSAS) admitted to the Hue University of Medicine and Pharmacy's Hospital, from May 2018 to May 2019. Sleep apnea syndrome was measured with StarDust II, Germany. X-ray was done with Amrad ARM, USA. Data analysis used SPSS 20.0 software. Diagnosis of sleep apnea syndrome according to the American Sleep Association 2017 [9].

Some of the landmarks and measurements commonly used in X-ray film analysis [10]:

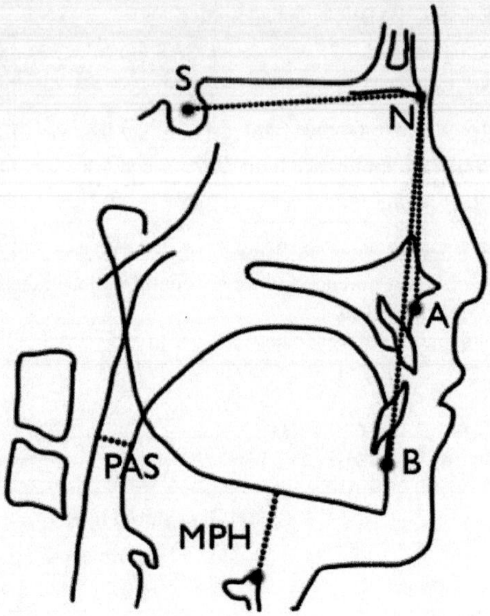

Fig. 1 The SNA, SNB angle and MPH, PAS distance

- Na point (Nasion): the first point on the frontonasal suture along the middle vertical line.
- S point (Sella Turcicca): the middle point of the hypophyseal fossa of alar bone.
- A point (Subspinale): the posterior point of the maxillary bone.
- B point (Submental): the posterior point of the mandibular bone.
- The angle and distance measurement:
- SNA angle: Angle created by straight lines connecting points S, Na and A.
- SNB angle: Angle formed by straight lines connecting point S, Na and B.
- ANB angle: Angle formed by the difference between the SNA angle and the SNB angle.
- MPH distance: The distance from the hyoid bone to the line parallel to the lower edge of the mandibular bone.
- PAS distance: Distance from the bottom of the tongue to the back of the throat (Fig. 1).

3 Results

3.1 Characteristics of the Study Group

Baseline characteristics of the study group

See Table 1.

Table 1 Baseline characteristics

Characteristics	OSAS (+) (n = 33)	OSAS (−) (n = 16)	Total (n = 49)	*p*
Male/female	19/14	9/7	28/21	> 0.05
Neck circumference (cm)	39.94 ± 3.63	37.50 ± 3.45	39.14 ± 3.72	< 0.05
Waist circumference (cm)	92.48 ± 8.5	82.44 ± 10.33	89.20 ± 10.21	< 0.05
BMI (kg/m^2)	23.97 ± 2.95	22.48 ± 2.79	23.42 ± 2.92	> 0.05
Age	71.85 ± 10.78	70.69 ± 13.10	71.47 ± 11.46	> 0.05

Neck circumference and waist circumference in the group OSAS (+) higher than in the group OSAS (−) with $p < 0.05$
No gender difference between the group OSAS (+) and the group OSAS (−)

Table 2 OR of the risk factors for OSAS

Risk factors	OR	95% CI	*p*
Obesity	7.5	0.874–64.37	0.039
Smoking	0.65	0.195–2.165	0.482
Hypertension	2.417	0.518–11.279	0.253
Diabetes	1.304	0.337–5.048	0.7
Dyslipidemia	1.056	0.316–3.522	0.93
Heart failure	0.774	0.16–3.739	0.749

Obese patients are 7.5 times more likely to have OSAS than those with normal weight with OR = 7.5 (95% confidence, 0.874–64.37, $p = 0.039$)

Odds ratio (OR) of the risk factors for OSAS

See Table 2.

Polysomnography parameters in the group OSAS (+) and group OSAS (−)

See Table 3.

Table 3 Polysomnography parameters

Parameters	OSAS (+) (n = 33)	OSAS (−) (n = 16)	Total	*p*
AHI (event/h)	22.73 ± 10.20	3.5 ± 0.632	16.45 ± 12.35	< 0.05
Maximum apnea duration (s)	24.12 ± 8.05	11.69 ± 3.4	20.06 ± 9.03	< 0.05
Minimum apnea duration (s)	14.64 ± 6.95	8.44 ± 2.68	12.61 ± 6.56	> 0.05
Average of apnea duration (s)	19.64 ± 6.48	9.94 ± 3.13	16.47 ± 7.22	< 0.05

AHI in the group OSAS (+) higher than in the group OSAS (−) $p < 0.05$
Maximum apnea duration and Minimum apnea duration in the OSAS (+) higher than in the group OSAS (−) $p < 0.05$

3.2 Relationship Between OSAS and Lateral Cephalometric X-Ray Characteristics

Lateral cephalometric X-ray characteristic and gender

See Table 4.

Lateral cephalometric X-ray characteristics and OSAS

See Table 5.

Lateral cephalometric X-ray characteristic and AHI

See Table 6.

The correlation between AHI, Epworth scale, BMI and SNA, SNB, ANB, MPH, PAS in the group OSAS (+)

See Table 7.

Table 4 Lateral cephalometric X-ray characteristic and gender

Parameters	Male (n = 19)	Female (n = 14)	p
SNA (°)	78.26 ± 3.53	77.64 ± 3.69	> 0.05
SNB (°)	71.26 ± 4.67	70.71 ± 3.95	> 0.05
ANB (°)	7.21 ± 1.72	7.14 ± 2.38	> 0.05
MPH (mm)	24.71 ± 4.12	20.39 ± 3.86	< 0.05
PAS (mm)	11.93 ± 2.05	13.42 ± 3.51	> 0.05

In the group OSAS (+), MPH in male higher than MPH in female $p < 0.05$

Table 5 Lateral cephalometric X-ray characteristics in the group OSAS (+) and group OSAS (−)

Parameters	OSAS (+) (n = 29)	OSAS (−) (n = 11)	p
SNA (°)	78 ± 3.55	81.69 ± 4.22	< 0.05
SNB (°)	71.03 ± 4.33	77.94 ± 4.28	< 0.05
ANB (°)	7.18 ± 1.99	4.06 ± 2.08	< 0.05
MPH (mm)	22.88 ± 4.5	16.53 ± 3.86	< 0.05
PAS (mm)	12.56 ± 2.82	16.87 ± 3.78	< 0.05

SNA, SNB, PAS in the group OSAS (+) lower than in the group OSAS (−), $p < 0.05$
ANB, MPH in the group OSAS (+) higher than in the group OSAS (−), $p < 0.05$

Table 6 Lateral cephalometric X-ray characteristic and AHI

Parameters	AHI (event/h)			*p*
	5–15 (n = 8)	16–30 (n = 16)	> 30 (n = 9)	
SNA (°)	78.75 ± 3.85	78.19 ± 3.66	77 ± 3.28	> 0.05
SNB (°)	73.25 ± 3.62	71.38 ± 4.47	68.44 ± 3.64	> 0.05
ANB (°)	5.5 ± 2.07	7.19 ± 1.68	8.67 ± 1.23	< 0.05
MPH (mm)	20.35 ± 3.35	23.06 ± 4.2	24.84 ± 5.23	> 0.05
PAS (mm)	13.43 ± 2.62	13.25 ± 3.17	10.57 ± 0.83	< 0.05

ANB and PAS had a significant difference between the AHI classification $p < 0.05$

Table 7 The correlation between AHI, Epworth scale, BMI and SNA, SNB, ANB, MPH, PAS in the group OSAS (+)

		AHI	Epworth	BMI	SNA	SNB	ANB	MPH	PAS
AHI	r	1.00	0.650	– 0.06	– 0.23	– 0.452	0.575	0.392	– 0.459
	p		*0.001*	0.75	0.19	*0.01*	*0.001*	*0.02*	*0.01*
Epworth	r	0.650	1.00	– 0.04	– 0.15	– 0.30	0.32	0.21	– 0.491
	p	0.00		0.80	0.41	0.09	0.07	0.24	*0.001*
BMI	r	– 0.06	– 0.04	1.00	– 0.26	– 0.11	– 0.15	0.12	0.10
	p	0.75	0.80		0.14	0.54	0.40	0.52	0.57
SNA	r	– 0.23	– 0.15	– 0.26	1.00	0.894	– 0.15	– 0.06	– 0.21
	p	0.19	0.41	0.14		*0.001*	0.40	0.76	0.23
SNB	r	– 0.452	– 0.30	– 0.11	0.894	1.00	– 0.527	– 0.19	– 0.10
	p	*0.01*	0.09	0.54	*0.001*		*0.001*	0.29	0.58
ANB	r	0.575	0.32	– 0.15	– 0.15	– 0.527	1.00	0.397	– 0.28
	p	*0.001*	0.07	0.40	0.40	*0.001*		*0.02*	0.12
MPH	r	0.392	0.21	0.12	– 0.06	– 0.19	0.397	1.00	– 0.545
	p	*0.02*	0.24	0.52	0.76	0.29	*0.02*		*0.001*
PAS	r	– 0.459	– 0.491	0.10	– 0.21	– 0.10	– 0.28	– 0.545	1.00
	p	*0.01*	*0.001*	0.57	0.23	0.58	0.12	*0.001*	

There was a correlation between AHI and Epworth, SNB, ANB, MPH, PAS $p < 0.05$

4 Discussion

4.1 Characteristics of Age, Gender, Neck Circumference

There was no statistically significant difference in age between the two groups OSAS (+) and OSAS (–). The proportion of patients > 60 years in the OSAS group (+) accounts for the majority with 87.9%.

This result was consistent with the study of Prasanthi Ratnakumar [11], the frequency of OSAS increases with age, the proportion of OSAS (+) in patients over 65 years was 3 times higher than patients 30–64 years of age. This was the result of age-related adipose tissue deposits, especially the oropharyngeal adipose tissue, along with structural weakness and elasticity of pharyngeal muscle tissue over time [11].

The incidence of obstructive sleep apnea syndrome was higher in men than in women. This might be due to differences in anatomical characteristics, upper respiratory tract function, hyperventilation response during sleep, and a high concentration of fat in the oropharynx area. In our research results, men predominated in both patient groups (Table 1).

4.2 *Relationship Between OSAS and Lateral Cephalometric X-Ray Characteristics*

Our research results show that there was a statistically significant difference: SNA, SNB, PAS values in OSAS group (+) were smaller than OSAS group (−), ANB and MPH values in OSAS group (+) were greater than in the OSAS group (−) with $p < 0.05$. The average value in OSAS (+) group of SNA, SNB, ANB, MPH, PAS were $78 \pm 3.55°$, $71.03 \pm 4.33°$, $7.18 \pm 1.99°$, 22.88 ± 4.5 mm, 12.56 ± 2.82 mm.

Neelapu et al. [12] concluded that the SNB angle measure in the OSAS group (+) was 1.45° lower than the control group. The study also showed that there was a lower downward movement of the basal bone compared to the anterior layer and a decrease in the post-pharyngeal air space in OSAS (+) patients compared to the control group [12].

The study by Ahsan et al. [13] concluded that the measurement of SNA angle, SNB angle of Japanese patients was smaller than European patients at the same severity level of OSAS with statistical significance with $p < 0.01$. This result suggests that there was a backward reduction in the lower jaw bone in Asians compared to Europeans in facial bone anatomy [13].

4.3 *Correlation Between AHI and SNB and ANB Values*

In our study, there was a negative correlation between AHI and SNB values in OSAS patients (+), $r = -0.45$, $p < 0.05$. The smaller the SNB angle, the worse the severity of the disease, suggesting that the patients with lower jaw bone backward relative to the skull base, the higher the severity of the disease. This result was different from that of Paulo et al. [2] in 102 patients diagnosed with OSAS determination, no correlation between SNB value and AHI ($p > 0.05$) [2].

Our study results on the correlation between AHI and ANB values showed a relatively positive correlation between AHI and ANB values in OSAS (+) patients with r = 0.58, $p < 0.05$. The larger the ANB angle, the greater the AHI, which suggests that the larger the ANB angle, the more severe the degree of pathology. Therefore, patients who are still young, are in the age of growth, if the lower jaw bone back, it is advisable to treat early, stimulate the lower jaw to develop first to reduce the risk of disease correlation between AHI and MPH and PAS values.

In our study, there was a moderate positive correlation between AHI and MPH values in OSAS (+) patients. The regression equation $y = 0.89\ x + 2.42$, $r = 0.39$ with $p < 0.05$. This result is quite similar to that of Paulo et al. (2014), there is a positive correlation between AHI and MPH values with $r = 0.251, p = 0.011$ [2].

Our results show a moderately inverse correlation between AHI and PAS values among OSAS (+) patients. The regression equation $y = -\ 1.66x + 43.61$, $r = -0.46$ with $p < 0.05$. This result is different from that of Paulo et al. (2014), there is no correlation between AHI and PAS value with $p > 0.05$ [2].

5 Conclusion

The lateral head X-ray provides SNB, ANB, MPH, PAS values correlated with AHI in obstructive sleep apnea syndrome patients. Lateral cephalometric by X-ray (SNA, SNB, ANB, MPH, PAS) can be used to evaluate the severity of obstructive sleep apnea syndrome.

Conflicts of Interest The authors have no conflict of interest to declare.

References

1. Valipour A (2012) Gender-related differences in the obstructive sleep apnea syndrome. Pneumologie 66(10):584–588. https://doi.org/10.1055/s-0032-1325664
2. Paulo de Tarso MB, Edson SFF, de Telma MEA, Jose MMN, de Nubia ESB, Baltasar MN, Viriato C, Jorge RP, Li ML (2013) Correlation of cephalometric and anthropometric measures with obstructive sleep apnea severity. Int Arch Otorhinolaryngol 17(3):321–328. https://doi.org/10.7162/S1809-97772013000300013
3. DeFronzo RA, Ferrannini E, Zimmet P, Alberti G (2015) International textbook of diabetes mellitus. Wiley. ISBN: 978-0-470-65861-1
4. Arita A, Sasanabe R, Hasegawa R, Nomura A, Hori R, Mano M, Shiomi T (2015) Risk factors for automobile accidents caused by falling asleep while driving in obstructive sleep apnea syndrome. Sleep Breath Schlaf Atmung 19(4):1229–1234. https://doi.org/10.1007/s11325-015-1145-7
5. Duong QS, Dang TMK, Tran VN, Nguyen XBH, Hua HT, Chalumeau F, Dinh XTA, Soyez F, Martin F (2018) Study about the prevalence of the obstructive sleep apnoea syndrome in Vietnam. Rev Mal Respir 35(1):14–24. https://doi.org/10.1016/j.rmr.2017.10.006

6. Penha D, Rosado E, Cabral P, Joao P (2013) Obstructive sleep apnea syndrome—what every radiologist should know. Eur Cong Radiol 33:10–31. https://doi.org/10.1594/ecr2013/C-1643
7. Jung SH, Koo SK, Choi JW, Moon JS, Lee SH (2017) Upper airway structural changes induced by CPAP in OSAS patients: a study using drug-induced sleep endoscopy. Eur Arch Otorhinolaryngol 274(1):247–252. https://doi.org/10.1007/s00405-016-4233-9
8. McNicholas WT, Bassetti CL, Ferini-Strambi L, Pépin JL, Pevernagie D, Verbraecken J, Randerath W (2018) Challenges in obstructive sleep apnoea. Lancet Respir Med 6(3):170–172. https://doi.org/10.1016/S2213-2600(18)30059-6
9. Vishesh KK, Dennis HA, Susmita C, David CK, Reena M, Kannan R, Christopher GH (2017) Clinical practice guideline for diagnostic testing for adult obstructive sleep apnea. An American academy of sleep medicine clinical practice guideline. J Clin Sleep Med 13(3). https://doi.org/10.5664/jcsm.6506
10. Burkhard JPM, Dietrich AD, Jacobsen C, Roos M, Lübbers H-T, Obwegeser JA (2014) Cephalometric and three-dimensional assessment of the posterior airway space and imaging software reliability analysis before and after orthognathic surgery. J Craniomaxillofac Surg 42(7):1428–1436. https://doi.org/10.1016/j.jcms.2014.04.005
11. Prasanthi Ratnakumar AM, Manuel A (2016) Management of obstructive sleep apnoea syndrome. Prescriber 27(8):23–27. https://doi.org/10.1002/psb.1488
12. Neelapu BC, Kharbanda OP, Sardana HK, Balachandran R, Sardana V, Kapoor P, Vasamsetti S (2017) Craniofacial and upper airway morphology in adult obstructive sleep apnea patients: a systematic review and meta-analysis of cephalometric studies. Sleep Med Rev 31:79–90. https://doi.org/10.1016/j.smrv.2016.01.007
13. Ahsan A, Yamaki M, Hossain Z (2013) Craniofacial cephalometric analysis of Bangladeshi and Japanese adults with normal occlusion and balanced faces: a comparative study. J Orthodont Sci 2:7–15. https://doi.org/10.4103/2278-0203.110327.Wine Guardian: Proper Wine Storage Temperature and Humidity. Available: https://wineguardian.com/proper-wine-storage-temperature-and-humidity, last accessed 2019/12/27

Sleep Apnea Syndrome in Patients with Atrial Fibrillation

Hoang Anh Tien, Mai Tran Phuoc Loc, and Tran Viet An

Abstract Sleep apnea syndrome is a common sleep disorder, which has a negative impact on health and quality of life. Atrial fibrillation (AF) is the most common cardiac arrhythmia and is associated with significant morbidity and mortality. 52 patients with atrial fibrillation in the Department of Internal Cardiology at Hue University of Medicine and Pharmacy Hospital had loud snoring and/or evidence of sleep apnea. A case–control study of 52 patients (case group: 38, control group: 14) with symptoms of loud snoring and/or evidence of suspected obstructive sleep apnea syndrome (OSAS) admitted to the Hue University of Medicine and Pharmacy's Hospital, from May 2018 to May 2019. Sleep apnea syndrome was measured with Embletta Gold, USA. Data analysis used SPSS 20.0 software. Common age is above 60 years old (69.7%); men account for the majority (63.46%). Loud snoring symptoms during sleep, daytime sleepiness and difficulty concentrating at work are the majority in patients with sleep apnea syndrome (71.80%, 82.60% and 92.40% respectively). There is a positive correlation between the AHI index and the severity of atrial fibrillation by EHRA (r = 0.485; $p < 0.05$), left atrial diameter (r = 0.020; $p > 0.05$) and cardiovascular factors (r = 0.354; $p < 0.05$). Atrial fibrillation patients have a higher incidence of sleep apnea syndrome due to the general population and a correlation between the severity of atrial fibrillation and the AHI index, with cardiovascular risk factors.

Keywords Sleep apnea syndrome · Atrial fibrillation

H. A. Tien (✉) · M. T. P. Loc
Hue University Of Medicine and Pharmacy, Hue University, Hue, Vietnam
e-mail: hatien@hueuni.edu.vn; hatien@huemed-univ.edu.vn

T. V. An
Can Tho University of Medicine and Pharmacy, Can Tho, Vietnam

V. Van Toi et al. (eds.), *8th International Conference on the Development of Biomedical Engineering in Vietnam*, IFMBE Proceedings 85,
https://doi.org/10.1007/978-3-030-75506-5_84

1 Background

Sleep apnea syndrome is a common sleep disorder [1] that has a negative impact on health and quality of life [2], which is the cause of drowsiness and fatigue during the day, leading to neurophysiological changes such as memory loss and concentration ability [3], and increase the risk of occupational accidents and traffic accidents [4]. This disease is common in both developed and developing countries. In Asia, the prevalence is about 8.5% [5], similar to that in Europeans and Americans [4].

Sleep apnea syndrome causing many dangers to the cardiovascular, nervous and respiratory systems. However, the diagnosis is often missed, and until it is discovered, the sleep apnea syndrome has left serious complications [6]. Many studies in the world have shown a close association between obstructive sleep apnea syndrome and conditions such as obesity, hypertension, coronary artery disease, cardiac arrhythmia, systolic heart failure, stroke, and type 2 diabetes [7, 8].

In Vietnam, there was no study about this issue. So, we conducted the study "Sleep apnea syndrome in patients with atrial fibrillation" with two objectives:

1. *Evaluate the parameters of portable polysomnography in patients with atrial fibrillation.*
2. *Assess the correlation between the AHI index and the severity of atrial fibrillation by EHRA score, the cardiovascular risk factors.*

2 Subjects and Research Methods

2.1 Subjects

Patients admitted to the cardiology department, Hue University of medicine and pharmacy Hospital, diagnosed atrial fibrillation with loud snoring symptoms and/or evidence of suspected sleep apnea syndrome (SAS).

Disease group. 38 patients with SAS (SAS (+))
Control group. 14 patients without SAS (SAS (−))
Exclusion criteria. Patients in critical condition; chronic lung disease.

2.2 Methods

Case–control study. The patient was diagnosed with atrial fibrillation, according to the European Society of Cardiology (ESC 2016) [9]. Polysomnography was performed by Embletta Gold (USA). Data analysis by Statistical Package for the Social Sciences (SPSS) software 20.0.

3 Results

3.1 General Characteristics

Characteristics of age, gender and body measurements

See Table 1.

Cardiovascular risk factors

See Table 2.

3.2 Clinical and Subclinical Characteristics

The symptoms of sleep apnea syndrome

See Table 3.

EHRA score

See Table 4.

Results of polysomnography

See Tables 5 and 6.

Echocardiography characteristics

See Table 7.

3.3 Correlations Between AHI and Left Atrial Diameter, The Severity of Atrial Fibrillation According to EHRA Score and Cardiovascular Risk Factors

Correlations between AHI and left atrial diameter

See Fig. 1.

The area under the ROC curve was 0.806 with $p < 0.001$. Sensitivity (Se): 78.95%; Specificity (Sp): 78.57%. Positive predictive value (PPV): 90.90%; Negative predictive value (NPV): 57.89%.

Correlation between AHI and severity of atrial fibrillation according to the EHRA score, the number of cardiovascular risk factors

There was a positive correlation between the AHI index and the severity of atrial fibrillation according to EHRA score with $r = 0.485$, $p < 0.05$; the correlation between

Table 1 Distribution by age group, gender, anthropometric

General characteristics		SAS (+) 4n = 38		SAS (−) n = 14		Total n = 52		p
		n	%	n	%	n	%	
Age	≤ 45	1	100	0	0	1	100	> 0.05
	46–60	14	77.8	4	22.2	18	100	
	> 60	23	69.7	10	30.3	33	100	
	Total	38	73.1	14	26.9	52	100	
Gender	Male	23	69.7	10	30.3	33	100	> 0.05
	Female	15	78.9	4	21.1	19	100	
	Total	38	73.1	14	26.9	52	100	
Anthropometric	Neck circumference	37.61 ± 1.44		36.29 ± 0.46		37.25 ± 1.38		< 0.05
	Waist circumference	90.25 ± 3.26		83.14 ± 2.83		88.85 ± 4.69		< 0.05
	BMI	23.26 ± 3.68		22.43 ± 2.63		23.04 ± 3.42		> 0.05

The mean circumference of the neck and waist in the SAS (+) was higher than the group of patients SAS (−), $p < 0.05$

Table 2 Cardiovascular risk factors

Cardiovascular risk factors	SAS (+) n = 38		SAS (−) n = 14		Total n = 52		*p*
	n	%	n	%	n	%	
Smoke	22	88.0	3	12.0	25	100	< 0.05
Dyslipidemia	6	100	0	0	6	100	> 0.05
Diabetes	11	100	0	0	11	100	< 0.05
Hypertension	15	75.0	5	25.0	20	100	> 0.05
Coronary artery disease	9	75.0	3	25.0	12	100	> 0.05
Heart failure	13	68.4	6	31.6	19	100	> 0.05
Arrhythmia	16	69.6	7	30.4	23	100	> 0.05
Stroke	4	57.1	3	42.9	7	100	> 0.05

Two factors smoking and diabetes were higher in SAS (+) than SAS (−), $p < 0.05$

Table 3 The functional symptoms of SAS

Symptoms	SAS (+) (n = 38)		SAS (−) (n = 14)		Total (n = 52)		*p*
	n	%	n	%	N	%	
Snoring	28	71.8	11	28.2	39	100	> 0.05
Sleep apnea	12	66.7	6	33.3	18	100	> 0.05
Wake up	10	71.4	4	28.6	14	100	> 0.05
Nocturia	4	80,0	1	20.0	5	100	> 0.05
Morning headache	20	66.7	10	33.3	30	100	> 0.05
Daytime sleepiness	19	82.6	4	17.4	23	100	< 0.05
Difficulty concentrating	17	94.4	1	5.6	18	100	< 0.05

Symptoms of "daytime sleepiness" and "difficulty concentrating" accounted for a higher frequency in the SAS group (+) than in the SAS group (−) $p < 0.05$

Table 4 EHRA score in group SAS (+) and group SAS (−)

EHRA score	SAS (+) n = 38		SAS (−) n = 14		Total n = 52		*p*
	n	%	n	%	n	%	
1	1	33.3	2	66.7	3	100	< 0.05
2a	2	40.0	3	60.0	5	100	
2b	4	50.0	4	50.0	8	100	
3	21	84.0	4	16.0	25	100	
4	10	90.9	1	9.1	11	100	

The greater the severity of atrial fibrillation according to the EHRA score, the rate of having SAS (+) increased gradually $p < 0.05$

Table 5 Parameters on polysomnographic device

Parameters	SAS (+) (n = 38)	SAS (−) (n = 14)	Total (n = 52)	p
AHI (event/h)	63.13 ± 78.85	1.71 ± 1.93	46.60 ± 72.89	< 0.05
Time (h)	8.12 ± 0.51	8.04 ± 0.13	8.09 ± 0.44	> 0.05
Apnea index	67.74 ± 10.49	1.64 ± 1.98	30.53 ± 60.37	< 0.05
Hypopnea index	21.95 ± 30.11	0.07 ± 0.26	16.06 ± 27.45	< 0.05
Time supine (min)	197.81 ± 120.61	191.70 ± 111.24	196.16 ± 117.11	> 0.05
Time recumbent (min)	277.14 ± 113.07	282.52 ± 113.98	278.59 ± 112.21	> 0.05
SpO_2 (%)	93.98 ± 1.94	95.78 ± 1.07	94.47 ± 1.91	< 0.05
SpO_2 < 90% (min)	22.42 ± 60.44	0.28 ± 0.36	16.46 ± 52.42	> 0.05
Epworth scale	12.84 ± 1.89	8.43 ± 1.56	11.65 ± 2.66	< 0.05

The mean of AHI, apnea index and hypopnea index, SpO_2, Epworth score in SAS (+) was significantly differenced from SAS (−) $p < 0.05$

Table 6 Classification of sleep apnea syndrome

AHI (event/h)	n	%
Obstructive	40	76.9
Central	12	23.1
Mixed	11	21.2
Hypopnea	35	67.3
Total	52	100

The majority of patients had obstructive sleep apnea syndrome (76.9%)

Table 7 Echocardiography characteristics

Characteristics echocardiography	SAS (+) (n = 38)	SAS (−) (n = 14)	Total (n = 52)	p
EF (%)	60.84 ± 10.48	70.57 ± 4.79	63.46 ± 10.22	< 0.05
FS (%)	33.16 ± 7.28	35.93 ± 5.48	33.90 ± 6.90	< 0.05
Left atrial diameter (mm)	42.21 ± 10.98	30.29 ± 10.43	39.00 ± 11.99	< 0.05

The mean of EF, FS in SAS (+) was lower than in SAS (−), $p < 0.05$
The mean of left atrial diameter in SAS (+) was higher than in SAS (−), $p < 0.05$

the AHI and the number of cardiovascular risk factors with $r = 0.354$, $p < 0.05$ (Table 8).

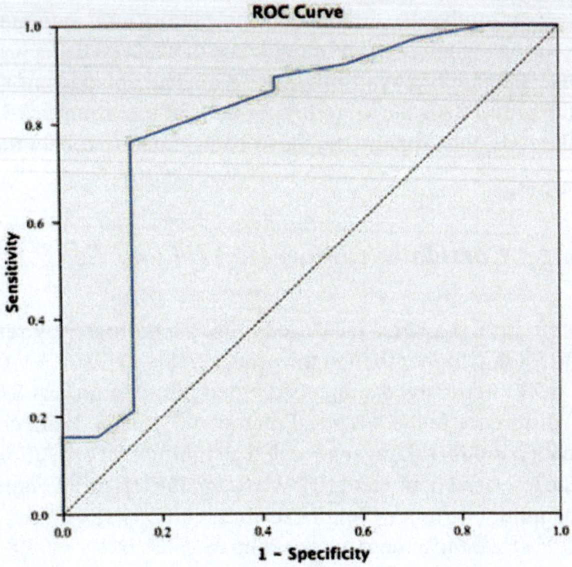

Fig. 1 ROC curve in diagnostic SAS based on the left atrial diameter

Table 8 The correlation between AHI and EHRA score, number of risk factors

The correlation	AHI	EHRA score	Number of risk factors
AHI	1	0.485	0.354
		0.002	0.0029
EHRA score	0.485	1	0.141
	0.002		0.399
Number of risk factors	0.354	0.141	1
	0.0029	0.399	

4 Discussion

4.1 Parameters on the Portable Polysomnography

The AHI was calculated by the total number of episodes of apnea and hypopnea during 1 h of sleep; these events must last at least 10 s. AHI was used to diagnose sleep apnea syndrome and assess its severity. In our study, AHI was the most important indicator for evaluating the severity of SAS and was the main objective of this study. The results showed that the average AHI was 46.60 ± 72.89 events/hour and AHI in SAS (+) (63.13 ± 78.85 events/h) was higher than the control group (1.71 ± 1.93 events/h) with $p < 0.05$.

Our study also found that the majority of patients (atrial fibrillation with loud snoring symptoms and/or evidence of suspected sleep apnea syndrome) had obstructive sleep apnea syndrome (76.9%). The episodes of apnea, hypopnea and SpO_2 in the group SAS (+) were higher than in the group SAS (−), $p < 0.05$. In addition to the AHI, these parameters also contributed to reflect the severity of the SAS patient.

4.2 Correlation Between AHI and Left Atrial Diameter

Left atrial diameter measured on echocardiography results in the SAS (+) 42.21 ± 10.98 mm higher than in the group SAS (−) 30.29 ± 10.43, $p < 0.05$. Drager studied on 73 middle-aged subjects divided into two groups SAS (+), and SAS (−) showed a difference in the left atrial diameter 37.6 ± 3.3 mm compared with 34.7 ± 3.2 mm with $p < 0.001$. Drager revealed a significant correlation between the AHI index and the left atrial diameter [10]. We found the left atrial diameter cut-off point of 31.5 mm had the sensitivity (Se): 78.95%; specificity (Sp): 78.57%; positive predictive value (PPV): 90.90%; negative predictive value (NPV): 57.89%.

4.3 Correlation Between AHI and EHRA Score of Atrial Fibrillation

Patients with Sleep Apnea Syndrome (SAS) may be susceptible to cardiac arrhythmias due to altered sympathetic and parasympathetic nervous system activity that occurs with SAS-related hypoxia, acidosis, apnea and stimulate vasoconstriction, increased catecholamine as tachycardia, stress O_2 chemical and metabolic disorders leading to a risk of hypertension, congestive heart failure, arrhythmias, ischemic myocardial disease. At that time, the increase in AHI was strongly related to cardiovascular disease.

In our study, the proportion of patients with SAS tends to increase with the severity of atrial fibrillation, the highest in EHRA III has 55.3% of patients with SAS (+). Szymanski showed that sleep apnea syndrome was found in 33.3% of EHRA I group, 43.9% of EHRA II group, 58.1% of EHRA III group and 61.5% EHRA IV group ($p < 0.01$) in a continuous 177 patient survey [11]. In the group SAS (+), there was a positive correlation between the AHI index and the severity of atrial fibrillation according to EHRA with the correlation coefficient $r = 0.485$, $p < 0.05$.

4.4 *Correlation Between Sleep Apnea Syndrome and Cardiovascular Risk Factors*

OSA is associated with a significant increase in sympathetic activity during sleep, which in turn influences heart rate and blood pressure [12]. Increased sympathetic activity appears to be induced through a variety of different mechanisms, including chemoreflex stimulation by hypoxia and hypercapnia, baroreflexes, pulmonary afferents, impairment in venous return to the heart, alterations in cardiac output, and possibly the arousal response. Endothelial dysfunction may also play a role [13].

In our study, smoking factors (88.0%), arrhythmia (69.6%) and hypertension (75.0%) accounted for a high proportion in both groups SAS(+) and SAS(−). Smoking and diabetes were higher in group SAS (+) than in group SAS (−), $p < 0.05$. The number of cardiovascular risk factors in group SAS (+) is higher than that of group SAS (−). This difference is statistically significant with $p < 0.05$. In the group SAS (+), there was a positive correlation between the AHI and the number of cardiovascular risk factors with $r = 0.354$, $p < 0.05$. The study of Nguyen Huu Duc in Vietnam showed the risk factors suggested severe obstructive sleep apnea syndrome, including hypertension, neck circumference, waist circumference, obese ($p < 0.05$) [14].

5 Conclusion

In atrial fibrillation, the group with sleep apnea syndrome have higher EHRA score, more cardiovascular risk factors and lower cardiac function.

Conflicts of Interest The authors have no conflict of interest to declare.

References

1. Valipour A (2012) Gender-related differences in the obstructive sleep apnea syndrome. Pneumologie 66(10):584–588. https://doi.org/10.1055/s-0032-1325664
2. Paulo de Tarso MB, Edson SFF, de Telma MEA, Jose MMN, de Nubia ESB, Baltasar MN, Viriato C, Jorge RP, Li ML (2013) Correlation of cephalometric and anthropometric measures with obstructive sleep apnea severity. Int Arch Otorhinolaryngol 17(3):321–328. https://doi.org/10.7162/S1809-97772013000300013
3. DeFronzo RA, Ferrannini E, Zimmet P, Alberti G (2015) International textbook of diabetes mellitus. Wiley. ISBN: 978-0-470-65861-1
4. Arita A, Sasanabe R, Hasegawa R, Nomura A, Hori R, Mano M, Shiomi T (2015) Risk factors for automobile accidents caused by falling asleep while driving in obstructive sleep apnea syndrome. Sleep Breath Schlaf Atmung 19(4):1229–1234. https://doi.org/10.1007/s11325-015-1145-7

5. Duong QS, Dang TMK, Tran VN, Nguyen XBH, Hua HT, Chalumeau F, Dinh XTA, Soyez F, Martin F (2018) Study about the prevalence of the obstructive sleep apnoea syndrome in Vietnam. Rev Mal Respir 35(1):14–24. https://doi.org/10.1016/j.rmr.2017.10.006
6. Heinzer R, Vat S, Marques-Vidal P et al (2015) Prevalence of sleep-disordered breathing in the general population: the HypnoLaus study. Lancet Respir Med 3(4):310–318. https://doi.org/10.1016/S2213-2600(15)00043-0
7. Ashaari ZA, Rahman JA, Mohamed AH et al (2017) Association between severity of obstructive sleep apnea and number and sites of upper airway operations with surgery complications. JAMA Otolaryngol Head Neck Surg 143(3):239–246. https://doi.org/10.1001/jamaoto.2016.3268
8. Jeremy RT, David C, Kezirian EJ, De Marco T, Mirzayan A, Sadroonri B, Goldberg AN, Long C, Gerstenfeld EP, Yeghiazarians Y (2018) Obstructive sleep apnea in cardiovascular disease: a review of the literature and proposed multidisciplinary clinical management strategy. J Am Heart Assoc 8:e010440. https://doi.org/10.1161/JAHA.118.010440
9. Paulus Kirchhof SB, Kotecha D, Ahlsson A, Atar D, Casadei B, Castella M, Diener H-C, Heidbuchel H, Hendriks J (2016) ESC Guidelines for the management of atrial fibrillation developed in collaboration with EACTS. Eur Heart J 37(38):2893–2962. https://doi.org/10.1093/eurheartj/ehw210
10. Drager LF, Bortolotto LA Pedrosa RP, Krieger EM, Lorenzi-Filho G (2010) Left atrial diameter is independently associated with arterial stiffness in patients with obstructive sleep apnea: potential implications for atrial fibrillation. Int J Cardiol 144(2):257–259. https://doi.org/10.1016/j.ijcard.2009.01.018
11. Szymański FM, Platek A, Karpiński G, Koźluk E, Puchalski B, Filipiak KJ (2014) Obstructive sleep apnoea in patients with atrial fibrillation: prevalence, determinants and clinical characteristics of patients in Polish population. Kardiol Pol 72(8):716–724. https://doi.org/10.5603/KP.a2014.0070
12. Friedman O, Logan A (2009) The price of obstructive sleep apnea-hypopnea: hypertension and other ill effects. Am J Hypertens 22. https://doi.org/10.1038/ajh.2009.43
13. Seif F, Patel S, Walia H et al (2013) Association between obstructive sleep apnea severity and endothelial dysfunction in an increased background of cardiovascular burden. J Sleep Res 22:443. https://doi.org/10.1111/jsr.12026
14. Duc NH, Thuc TTD, Toan VD, Tien HA (2012) Study on sleep apnea syndrome in the patients with cardiovascular risk factors. J Med Pharm 2(1):52–59. https://trichdan.link/bai-bao/HAT2012763